T0324168

Personalized Psychiatry

Edited by

Bernhard T. Baune

ACADEMIC PRESS

An imprint of Elsevier

Academic Press is an imprint of Elsevier
125 London Wall, London EC2Y 5AS, United Kingdom
525 B Street, Suite 1650, San Diego, CA 92101, United States
50 Hampshire Street, 5th Floor, Cambridge, MA 02139, United States
The Boulevard, Langford Lane, Kidlington, Oxford OX5 1GB, United Kingdom

Notices
Knowledge and best practice in this field are constantly changing. As new research and experience broaden our understanding, changes in research methods, professional practices, or medical treatment may become necessary.

Practitioners and researchers must always rely on their own experience and knowledge in evaluating and using any information, methods, compounds, or experiments described herein. In using such information or methods they should be mindful of their own safety and the safety of others, including parties for whom they have a professional responsibility.

To the fullest extent of the law, neither the Publisher nor the authors, contributors, or editors, assume any liability for any injury and/or damage to persons or property as a matter of products liability, negligence or otherwise, or from any use or operation of any methods, products, instructions, or ideas contained in the material herein.

Library of Congress Cataloging-in-Publication Data
A catalog record for this book is available from the Library of Congress

British Library Cataloguing-in-Publication Data
A catalogue record for this book is available from the British Library

ISBN 978-0-12-813176-3

For information on all Academic Press publications
visit our website at https://www.elsevier.com/books-and-journals

Publisher: Nikki Levi
Acquisition Editor: Joslyn T. Chaiprasert-Paguio
Editorial Project Manager: Timothy Bennett
Production Project Manager: Omer Mukthar
Cover Designer: Christian J. Bilbow

Typeset by SPi Global, India

Contents

17. Personalized mental health: Artificial intelligence technologies for treatment response prediction in anxiety disorders

Ulrike Lueken and Tim Hahn

18. The genetic architecture of bipolar disorder: Entering the road of discoveries

Olav B. Smeland, Andreas J. Forstner, Alexander Charney, Eli A. Stahl and Ole A. Andreassen

19. Genomics of borderline personality disorder

Fabian Streit, Lucía Colodro-Conde, Alisha S.M. Hall and Stephanie H. Witt

25. Proteomics for diagnostic and therapeutic blood biomarker discovery in schizophrenia and other psychotic disorders

David R. Cotter, Sophie Sabherwal and Klaus Oliver Schubert

26. Molecular biomarkers in depression: Toward personalized psychiatric treatment

Anand Gururajan, John F Cryan and Timothy G Dinan

27. Neuroimaging biomarkers of late-life major depressive disorder pathophysiology, pathogenesis, and treatment response

Helmet T. Karim, Charles F. Reynolds, III and Stephen F. Smagula

33. Pharmacogenomics of treatment response in major depressive disorder

Joanna M. Biernacka, Ahmed T. Ahmed, Balwinder Singh and Mark A. Frye

34. Genomic treatment response prediction in schizophrenia

Sophie E. Legge, Antonio F. Pardiñas and James T.R. Walters

35. Personalized treatment in bipolar disorder

Estela Salagre, Eduard Vieta and Iria Grande

36. Genetic testing in psychiatry: State of the evidence

Chad A. Bousman, Lisa C. Brown, Ajeet B. Singh, Harris A. Eyre and Daniel J. Müller

Contributors

Numbers in paraentheses indicate the pages on which the authors' contributions begin.

Ryan Abbott (549), School of Law, University of Surrey, Guildford, United Kingdom; David Geffen School of Medicine at UCLA, Los Angeles, CA, United States

Ahmed T. Ahmed (403), Department of Psychiatry and Psychology, Mayo Clinic, Rochester, MN, United States

Martin Alda (393), Department of Psychiatry, Dalhousie University, Halifax, NS, Canada

Ananda B. Amstadter (285), Department of Psychiatry, Virginia Institute of Psychiatric and Behavioral Genetics; Department of Human and Molecular Genetics, Virginia Commonwealth University, Richmond, VA, United States

Ole A. Andreassen (215, 275), NORMENT, KG Jebsen Centre for Psychosis Research; Institute of Clinical Medicine, University of Oslo, Oslo, Norway

Victoria K. Arnet (537), Discipline of Psychiatry, University of Adelaide, Adelaide, SA, Australia

Cedric Bardy (127), South Australian Health and Medical Research Institute (SAHMRI); The College of Medicine and Public Health, Flinders University, Adelaide, SA, Australia

Csaba Barta (239), Institute of Medical Chemistry, Molecular Biology and Pathobiochemistry, Semmelweis University, Budapest, Hungary

Bernhard T. Baune (1, 3, 13, 385, 521, 537, 557), Department of Psychiatry, University of Münster, Münster, Germany; Department of Psychiatry, Melbourne Medical School, The University of Melbourne, Melbourne; The Florey Institute of Neuroscience and Mental Health, The University of Melbourne, Parkville, VIC, Australia; Department of Psychiatry and Psychotherapy, University of Münster, Münster, Germany; Department of Psychiatry, Melbourne Medical School, The University of Melbourne, Melbourne, Australia; Discipline of Psychiatry, School of Medicine, The University of Adelaide, Adelaide, SA, Australia; Department of Psychiatry and Psychotherapy, University of Münster, Münster, Germany

Elisabeth R.B. Becker (63), Department of Health Promotion and Behavioral Sciences, University of Texas Health Science Center at Houston, School of Public Health, Houston, TX, United States

Bettina H. Bewernick (83), Department of Psychiatry and Psychotherapy; Department of Geriatric Psychiatry, University Hospital Bonn, Bonn, Germany

Joanna M. Biernacka (403), Department of Psychiatry and Psychology; Department of Health Sciences Research, Mayo Clinic, Rochester, MN, United States

Stefan Bleich (375), Department of Psychiatry, Social Psychiatry and Psychotherapy, Hannover Medical School (MHH), Hannover, Germany

Marissa S. Blumenthal (63), Munson Medical Center, Traverse City, MI, United States

Cristian Bonvicini (253), IRCCS Istituto Centro San Giovanni di Dio Fatebenefratelli, Brescia, Italy

Chad A. Bousman (63, 437, 449), Department of Medical Genetics; Department of Psychiatry; Department of Physiology & Pharmacology, University of Calgary, Calgary, AB, Canada; Alberta Children's Hospital Research Institute; Hotchkiss Brain Institute, Calgary, AB, Canada; Innovation Institute, Texas Medical Center, Houston, TX, United States

Lisa C. Brown (449), Myriad Neuroscience, Mason, OH, United States

Christie L. Burton (239), Neurosciences & Mental Health, Hospital for Sick Children, Toronto, ON, Canada

Han Cao (117), Department of Psychiatry and Psychotherapy, Central Institute of Mental Health, Medical Faculty Mannheim, University of Heidelberg, Mannheim, Germany

Joanne S. Carpenter (39), Youth Mental Health Team, Brain and Mind Centre, University of Sydney, Sydney, NSW, Australia

Danielle Cath (239), Department of Psychiatry, Groningen University and University Medical Center Groningen, Groningen; Department of Specialist Training, Drenthe Mental Health Institution, Assen, The Netherlands

Micah Cearns (385, 521), Discipline of Psychiatry, School of Medicine, The University of Adelaide; Discipline of Psychiatry, University of Adelaide, Adelaide, SA, Australia

Donald D. Chang (549), The University of Queensland School of Medicine, Ochsner Clinical School, New Orleans, LA, United States

Alexander Charney (215), Division of Psychiatric Genomics, Department of Genetics and Genomic Sciences, Institute for Genomics and Multiscale Biology, Icahn School of Medicine at Mount Sinai, New York, NY, United States

Christopher R.K. Ching (483), Imaging Genetics Center, Mark & Mary Stevens Neuroimaging and Informatics Institute, Keck School of Medicine of USC, Los Angeles, CA, United States

Kate M. Chitty (39), Youth Mental Health Team, Brain and Mind Centre, University of Sydney, Sydney, NSW, Australia

Liliana G. Ciobanu (385), Discipline of Psychiatry, School of Medicine, The University of Adelaide, Adelaide, SA, Australia

Scott R. Clark (3, 13, 521), Discipline of Psychiatry, University of Adelaide, Adelaide, SA, Australia

A.M. Cole (173), Department of Psychiatry and Neuropsychiatric Genetics Research Group, Trinity College Dublin, Dublin, Ireland

Jane L. Collinson (13), Discipline of Psychiatry, University of Adelaide, Adelaide, SA, Australia

Lucía Colodro-Conde (227), Genetics & Computational Biology Department, QIMR Berghofer Medical Research, Brisbane, QLD, Australia

A. Corvin (173), Department of Psychiatry and Neuropsychiatric Genetics Research Group, Trinity College Dublin, Dublin, Ireland

David R. Cotter (307), Psychiatry, RCSI, Dublin, Ireland

Shane P.M. Cross (39), Youth Mental Health Team, Brain and Mind Centre, University of Sydney, Sydney, NSW, Australia

Jacob J. Crouse (39), Youth Mental Health Team, Brain and Mind Centre, University of Sydney, Sydney, NSW, Australia

John F Cryan (319), Department of Anatomy & Neuroscience, University College Cork; APC Microbiome Ireland, Cork, Ireland

Darina Czamara (103), Department of Translational Psychiatry, Max Planck Institute of Psychiatry, Munich, Germany

Shareefa Dalvie (297), Department of Psychiatry and Mental Health, University of Cape Town; South African Medical Research Council (SAMRC) Unit on Risk & Resilience in Mental Disorders, Cape Town, South Africa

Franziska Degenhardt (357), Institute of Human Genetics, University of Bonn, Bonn, Germany

Douglas L. Delahanty (285), Department of Psychological Sciences, Kent State University, Kent; Department of Psychiatry, Northeastern Ohio Medical University, Rootstown, OH, United States

Timothy G Dinan (319), APC Microbiome Ireland; Department of Psychiatry & Neurobehavioural Science, University College Cork, Cork, Ireland

Valsamma Eapen (239), Department of Psychiatry, University of New South Wales; Academic Unit of Child Psychiatry South West Sydney, Ingham Institute and Liverpool Hospital, Sydney, NSW, Australia

Harris A. Eyre (63, 437, 549), Innovation Institute, Texas Medical Center, Houston, TX, United States; Discipline of Psychiatry, University of Adelaide, Adelaide, SA, Australia; IMPACT SRC, Deakin University, Geelong, VIC, Australia; Department of Psychiatry, University of Melbourne, Melbourne, VIC, Australia; Brainstorm Lab for Mental Health Innovation, Stanford Medical School, Stanford University, Palo Alto, CA, United States; The University of Queensland School of Medicine, Ochsner Clinical School, New Orleans, LA, United States; Texas Medical Center, Innovation Institute, Yarraville, VIC, Australia; Department of Psychiatry, University of Melbourne, Parkville, VIC, Australia; Disciplines of Psychiatry, University of Adelaide, Adelaide, SA, Australia; IMPACT SRC, School of Medicine, Deakin University, Geelong, VIC, Australia; Brainstorm Lab for Mental Health Innovation, Department of Psychiatry and Behavioral Sciences, Stanford University, Palo Alto, CA, United States

Andreas J. Forstner (215), Institute of Human Genetics, University of Bonn, School of Medicine & University Hospital Bonn; Department of Genomics, Life & Brain Center, University of Bonn, Bonn, Germany; Human Genomics Research Group, Department of Biomedicine; Department of Psychiatry (UPK), University of Basel, Basel, Switzerland

Helge Frieling (375), Department of Psychiatry, Social Psychiatry and Psychotherapy, Hannover Medical School (MHH), Hannover, Germany

Mark A. Frye (403), Department of Psychiatry and Psychology, Mayo Clinic, Rochester, MN, United States

Lillian J. Gehue (39), Youth Mental Health Team, Brain and Mind Centre, University of Sydney, Sydney, NSW, Australia

Daniel Geller (239), Department of Psychiatry, Massachusetts General Hospital and Harvard Medical School, Boston, MA, United States

Nicholas Glozier (39), Youth Mental Health Team, Brain and Mind Centre, University of Sydney, Sydney, NSW, Australia

Beata R. Godlewska (471), Oxford Health NHS Foundation Trust, Warneford Hospital, Oxford, United Kingdom

Andrea N. Goldstein-Piekarski (499), Department of Psychiatry and Behavioral Sciences, Stanford University, Stanford; Sierra-Pacific Mental Illness Research, Education, and Clinical Center (MIRECC), Veterans Affairs Palo Alto Health Care System, Palo Alto, CA, United States

Ilona Gorbovskaya (449), Pharmacogenetics Research Clinic, Centre for Addiction and Mental Health, Toronto, Canada

Hans Jörgen Grabe (363), Department of Psychiatry and Psychotherapy, University Medicine of Greifswald, Greifswald, Germany

Iria Grande (423), Barcelona Bipolar and Depressive Disorders Unit, Institute of Neurosciences, Hospital Clinic, University of Barcelona, IDIBAPS, CIBERSAM, Barcelona, Spain

Paul Grant (161), Section on Behavioral Pediatrics, National Institute of Mental Health, NIH, Bethesda, MD, United States

Zarina Greenberg (127), South Australian Health and Medical Research Institute (SAHMRI), Adelaide, SA, Australia

Edna Grünblatt (239), Department of Child and Adolescent Psychiatry and Psychotherapy, Psychiatric Hospital, University of Zurich; Neuroscience Center Zurich, University of Zurich and ETH Zurich; Zurich Center for Integrative Human Physiology, University of Zurich, Zurich, Switzerland

Adam J. Guastella (39), Youth Mental Health Team, Brain and Mind Centre, University of Sydney, Sydney, NSW, Australia

Anand Gururajan (319), Department of Anatomy & Neuroscience, University College Cork, Cork, Ireland

Tim Hahn (201), Artificial Intelligence and Machine Learning Group, Translational Psychiatry Lab, Department of Psychiatry, University of Münster, Münster, Germany

Alisha S.M. Hall (227), Department of Genetic Epidemiology in Psychiatry, Central Institute of Mental Health, Medical Faculty Mannheim, University of Heidelberg, Mannheim, Germany

Blake A. Hamilton (39), Youth Mental Health Team, Brain and Mind Centre, University of Sydney, Sydney, NSW, Australia

Alfons O. Hamm (69), Department of Psychology, University of Greifswald, Greifswald, Germany

Catherine J. Harmer (471), Oxford Health NHS Foundation Trust, Warneford Hospital, Oxford, United Kingdom

Jessica Hartmann (27), Orygen, The National Centre of Excellence in Youth Mental Health; Centre for Youth Mental Health, The University of Melbourne, Parkville, VIC, Australia

Michael A. Hauser (285), Department of Medicine, Duke Molecular Physiology Institute, Duke University Medical Center, Durham, NC, United States

Daniel F. Hermens (39), Youth Mental Health Team, Brain and Mind Centre, University of Sydney, Sydney, NSW, Australia

Ian B. Hickie (39), Youth Mental Health Team, Brain and Mind Centre, University of Sydney, Sydney, NSW, Australia

Frank Iorfino (39), Youth Mental Health Team, Brain and Mind Centre, University of Sydney, Sydney, NSW, Australia

Neda Jahanshad (483), Imaging Genetics Center, Mark & Mary Stevens Neuroimaging and Informatics Institute, Keck School of Medicine of USC; Department of Neurology, University of Southern California; Department of Biomedical Engineering, University of Southern California, Los Angeles, CA, United States

Iris E. Jansen (275), Department of Complex Trait Genetics, Center for Neurogenomics and Cognitive Research, Amsterdam Neuroscience, VU University, Amsterdam, The Netherlands; Alzheimer Center and Department of Neurology, Amsterdam Neuroscience, VU University, Amsterdam UMC, Amsterdam, The Netherlands

Angela G. Junglen (285), Department of Psychological Sciences, Kent State University, Kent, OH, United States

Helmet T. Karim (339), Department of Psychiatry, University of Pittsburgh School of Medicine, Pittsburgh, PA, United States

Manreena Kaur (39), Youth Mental Health Team, Brain and Mind Centre, University of Sydney, Sydney, NSW, Australia

Nastassja Koen (297), Department of Psychiatry and Mental Health, University of Cape Town; South African Medical Research Council (SAMRC) Unit on Risk & Resilience in Mental Disorders, Cape Town, South Africa

Dagmar Koethe (39), Youth Mental Health Team, Brain and Mind Centre, University of Sydney, Sydney, NSW, Australia

Jim Lagopoulos (39), Youth Mental Health Team, Brain and Mind Centre, University of Sydney, Sydney, NSW, Australia

Rico S.Z. Lee (39), Youth Mental Health Team, Brain and Mind Centre, University of Sydney, Sydney, NSW, Australia

Sophie E. Legge (413), MRC Centre for Neuropsychiatric Genetics and Genomics, Division of Psychological Medicine and Clinical Neurosciences, Cardiff University, Cardiff, United Kingdom

Douglas F. Levinson (187), Department of Psychiatry and Behavioral Sciences, Program on the Genetics of Brain Function, Stanford University, Palo Alto, CA, United States

F. Markus Leweke (39), Youth Mental Health Team, Brain and Mind Centre, University of Sydney, Sydney, NSW, Australia

Julio Licinio (127), Psychiatry, Neuroscience & Physiology, Pharmacology, and Medicine; College of Medicine, SUNY Upstate Medical University, Syracuse, NY, United States

David E.J. Linden (465), Faculty of Health, Medicine and Life Sciences, School of Mental Health and Neuroscience, Maastricht University, Maastricht, The Netherlands; Division of Psychological Medicine and Clinical Neurosciences, School of Medicine, Cardiff University, Cardiff, United Kingdom

Jonathan C.W. Liu (449), Pharmacogenetics Research Clinic, Centre for Addiction and Mental Health; Department of Pharmacology & Toxicology, University of Toronto, Toronto, ON, Canada

Anita Lo (537), Discipline of Psychiatry, University of Adelaide, Adelaide, SA, Australia

Falk W. Lohoff (147), Section on Clinical Genomics and Experimental Therapeutics, National Institute on Alcohol Abuse and Alcoholism, National Institutes of Health, Bethesda, MD, United States

Adriana Lori (285), Department of Psychiatry and Behavioral Sciences, Emory University School of Medicine, Atlanta, GA, United States

Ulrike Lueken (201), Department of Psychology, Humboldt-Universität zu Berlin, Berlin, Germany

Carlo Maj (253), IRCCS Istituto Centro San Giovanni di Dio Fatebenefratelli, Brescia, Italy; Institute of Genomic Statistics and Bioinformatics, University of Bonn, Bonn, Germany

Patrick D. McGorry (27), Orygen, The National Centre of Excellence in Youth Mental Health, Parkville, VIC, Australia; Centre for Youth Mental Health, The University of Melbourne, Parkville, VIC, Australia

Roger S. McIntyre (459), Mood Disorders Psychopharmacology Unit, University Health Network; Department of Psychiatry; Department of Pharmacology; Institute of Medical Science, University of Toronto, Toronto, ON, Canada

Agnes B. McMahon (483), Imaging Genetics Center, Mark & Mary Stevens Neuroimaging and Informatics Institute, Keck School of Medicine of USC, Los Angeles, CA, United States

Divya Mehta (103), School of Psychology and Counselling, Faculty of Health, Institute of Health and Biomedical Innovation, Queensland University of Technology, Kelvin Grove, QLD, Australia

Cristina Mei (27), Orygen, The National Centre of Excellence in Youth Mental Health; Centre for Youth Mental Health, The University of Melbourne, Parkville, VIC, Australia

Rajendra A. Morey (285), National Center for PTSD, Behavioral Science Division, VA Boston Healthcare System; Department of Psychiatry and Biomedical Genetics, Boston University School of Medicine, Boston, MA; Mid-Atlantic Mental Illness Research, Education and Clinical Center, Durham VA Medical Center; Duke-UNC Brain Imaging and Analysis Center, Duke University, Durham, NC, United States

Daniel J. Müller (437, 449), Centre for Addiction and Mental Health, Campbell Family Research Institute; Department of Psychiatry, University of Toronto; Pharmacogenetics Research Clinic, Centre for Addiction and Mental Health; Department of Pharmacology & Toxicology, University of Toronto, Toronto, ON, Canada

Sharon L. Naismith (39), Youth Mental Health Team, Brain and Mind Centre, University of Sydney, Sydney, NSW, Australia

Barnaby Nelson (27), Orygen, The National Centre of Excellence in Youth Mental Health; Centre for Youth Mental Health, The University of Melbourne, Parkville, VIC, Australia

Alexandra Neyazi (375), Department of Psychiatry, Social Psychiatry and Psychotherapy, Hannover Medical School (MHH), Hannover, Germany

Alissa Nichles (39), Youth Mental Health Team, Brain and Mind Centre, University of Sydney, Sydney, NSW, Australia

Caroline M. Nievergelt (285), Department of Psychiatry, University of California, San Diego; VA San Diego Healthcare System, VA Center of Excellence for Stress and Mental Health (CESAMH), La Jolla, CA, United States

Nicole R. Nugent (285), Bradley Hasbro Children's Research Center of Rhode Island Hospital; Department of Psychiatry and Human Behavior, Alpert Medical School of Brown University, Providence, RI, United States

C. Ormond (173), Department of Psychiatry and Neuropsychiatric Genetics Research Group, Trinity College Dublin, Dublin, Ireland

Antonio F. Pardiñas (413), MRC Centre for Neuropsychiatric Genetics and Genomics, Division of Psychological Medicine and Clinical Neurosciences, Cardiff University, Cardiff, United Kingdom

Seth W. Perry (127), Psychiatry, and Neuroscience & Physiology; College of Medicine, SUNY Upstate Medical University, Syracuse, NY, United States

Claudia Pisanu (393), Department of Biomedical Sciences, Section of Neuroscience and Clinical Pharmacology, University of Cagliari, Cagliari, Italy

Danielle Posthuma (275), Department of Complex Trait Genetics, Center for Neurogenomics and Cognitive Research, Amsterdam Neuroscience, VU University; Department of Clinical Genetics, VU University, Amsterdam UMC, Amsterdam, The Netherlands

Renee-Marie Ragguett (459), Mood Disorders Psychopharmacology Unit, University Health Network, Toronto, ON, Canada

Cyrus Raji (63), Department of Radiology and Biomedical Imaging, Division of Neuroradiology, University of California, San Francisco, San Francisco, CA, United States

Margarita Raygada (161), Division of Intramural Research, National Institute of Child Health and Human Development, National Institutes of Health (NIH); Pediatric Oncology Branch, National Cancer Institute, NIH, Bethesda, MD; Department of Pediatrics, Georgetown University Medical School, Washington, DC, United States

Owen M. Rennert (161), Division of Intramural Research, National Institute of Child Health and Human Development, National Institutes of Health (NIH), Bethesda, MD; Department of Pediatrics, Georgetown University Medical School, Washington, DC, United States

Charles F. Reynolds, III (339), Department of Psychiatry, University of Pittsburgh School of Medicine, Pittsburgh, PA, United States; Department of Behavioral and Community Health Sciences, Graduate School of Public Health, University of Pittsburgh, Pittsburgh, PA, United States

Sophie Sabherwal (307), Psychiatry, RCSI, Dublin, Ireland

Estela Salagre (423), Barcelona Bipolar and Depressive Disorders Unit, Institute of Neurosciences, Hospital Clinic, University of Barcelona, IDIBAPS, CIBERSAM, Barcelona, Spain

Pamela H. Saunders (537), SABR Manager, SAHMRI, Adelaide, SA, Australia

Catia Scassellati (253), IRCCS Istituto Centro San Giovanni di Dio Fatebenefratelli, Brescia, Italy

Thomas E. Schlaepfer (83), Department of Interventional Biological Psychiatry; Department of Psychiatry and Psychotherapy, Medical Center—University of Freiburg, Faculty of Medicine, University of Freiburg, Freiburg, Germany; Departments of Psychiatry and Mental Health, The Johns Hopkins University, Baltimore, MD, United States

Lianne Schmaal (483), Orygen, The National Centre of Excellence in Youth Mental Health; Centre for Youth Mental Health, The University of Melbourne, Parkville, VIC, Australia

Klaus Oliver Schubert (521), Discipline of Psychiatry, University of Adelaide; Northern Adelaide Mental Health Services, Northern Adelaide Local Health Network, Lyell McEwin Hospital, Elizabeth Vale, SA, Australia; Mental Health Service, Adelaide, SA, Australia

Emanuel Schwarz (117), Department of Psychiatry and Psychotherapy, Central Institute of Mental Health, Medical Faculty Mannheim, University of Heidelberg, Mannheim, Germany

Elizabeth M. Scott (39), Youth Mental Health Team, Brain and Mind Centre, University of Sydney, Sydney, NSW, Australia

Jan Scott (39), Youth Mental Health Team, Brain and Mind Centre, University of Sydney, Sydney, NSW, Australia

Jonathan Sebat (285), Department of Psychiatry; Beyster Center for Genomics of Psychiatric Diseases; Department of Cellular and Molecular Medicine, University of California, San Diego, La Jolla, CA, United States

Giovanni Severino (393), Department of Biomedical Sciences, Section of Neuroscience and Clinical Pharmacology, University of Cagliari, Cagliari, Italy

Christina Sheerin (285), Department of Psychiatry, Virginia Institute of Psychiatric and Behavioral Genetics, Virginia Commonwealth University, Richmond, VA, United States

Ajeet B. Singh (63, 437), IMPACT SRC, Deakin University, Geelong, VIC, Australia

Balwinder Singh (403), Department of Psychiatry and Psychology, Mayo Clinic, Rochester, MN, United States

Stephen F. Smagula (339), Department of Psychiatry, University of Pittsburgh School of Medicine; Department of Epidemiology, Graduate School of Public Health, University of Pittsburgh, Pittsburgh, PA, United States

Olav B. Smeland (215), NORMENT, KG Jebsen Centre for Psychosis Research; Institute of Clinical Medicine, University of Oslo, Oslo, Norway

Alicia K. Smith (285), Department of Psychiatry and Behavioral Sciences, Emory University School of Medicine, Atlanta, GA, United States

Jill L. Sorcher (147), Section on Clinical Genomics and Experimental Therapeutics, National Institute on Alcohol Abuse and Alcoholism, National Institutes of Health, Bethesda, MD, United States

Rachael Spooner (27), Orygen, The National Centre of Excellence in Youth Mental Health; Centre for Youth Mental Health, The University of Melbourne, Parkville, VIC, Australia

Alessio Squassina (393), Department of Biomedical Sciences, Section of Neuroscience and Clinical Pharmacology, University of Cagliari, Cagliari, Italy; Department of Psychiatry, Dalhousie University, Halifax, NS, Canada

Eli A. Stahl (215), Division of Psychiatric Genomics, Department of Genetics and Genomic Sciences, Institute for Genomics and Multiscale Biology, Icahn School of Medicine at Mount Sinai, New York, NY, United States

Dan J. Stein (297), Department of Psychiatry and Mental Health, University of Cape Town; South African Medical Research Council (SAMRC) Unit on Risk & Resilience in Mental Disorders, Cape Town, South Africa

Fabian Streit (227), Department of Genetic Epidemiology in Psychiatry, Central Institute of Mental Health, Medical Faculty Mannheim, University of Heidelberg, Mannheim, Germany

Patrick F. Sullivan (91), Department of Psychiatry; Department of Genetics, The University of North Carolina at Chapel Hill, Chapel Hill, NC, United States; Department of Medical Epidemiology and Biostatistics, Karolinska Institutet, Stockholm, Sweden

Jennifer A. Sumner (285), Center for Cardiovascular Behavioral Health, Columbia University Medical Center, New York, NY, United States

Paul M. Thompson (483), Imaging Genetics Center, Mark & Mary Stevens Neuroimaging and Informatics Institute, Keck School of Medicine of USC; Department of Neurology; Department of Psychiatry; Department of Radiology; Department of Pediatrics; Department of Biomedical Engineering, University of Southern California, Los Angeles, CA, United States

Ashleigh M. Tickell (39), Youth Mental Health Team, Brain and Mind Centre, University of Sydney, Sydney, NSW, Australia

Catherine Toben (537), Discipline of Psychiatry, University of Adelaide, Adelaide, SA, Australia

Monica Uddin (285), Carl R Woese Institute for Genomic Biology, University of Illinois Urbana-Champaign, Urbana; Department of Psychology, University of Illinois Urbana-Champaign, Champaign, IL, United States

Arshya Vahabzadeh (63), Psychiatry Academy, Massachusetts General Hospital, Boston; Brain Power, Cambridge, MA, United States

Linh K. Van (13), Discipline of Psychiatry, University of Adelaide, Adelaide, SA, Australia

Margot P. van de Weijer (275), Department of Biological Psychology, VU University; Amsterdam Public Health Research Institute, Amsterdam University Medical Centre, Amsterdam, The Netherlands

Odile A. van den Heuvel (239), Amsterdam University Medical Centers, Department of Psychiatry and Department of Anatomy & Neuroscience, Vrije Universiteit Amsterdam, Amsterdam Neuroscience, Amsterdam, The Netherlands

Sandra Van der Auwera (363), Department of Psychiatry and Psychotherapy, University Medicine of Greifswald, Greifswald, Germany

Anouk H.A. Verboven (275), Department of Human Genetics, Radboudumc, Donders Institute for Brain, Cognition, and Behaviour, Nijmegen, The Netherlands

Eduard Vieta (423), Barcelona Bipolar and Depressive Disorders Unit, Institute of Neurosciences, Hospital Clinic, University of Barcelona, IDIBAPS, CIBERSAM, Barcelona, Spain

Zoe Wainer (63), Department of Surgery, St Vincent's Hospital, University of Melbourne, Melbourne, VIC, Australia

James T.R. Walters (413), MRC Centre for Neuropsychiatric Genetics and Genomics, Division of Psychological

Medicine and Clinical Neurosciences, Cardiff University, Cardiff, United Kingdom

Hunna J. Watson (91), Department of Psychiatry, The University of North Carolina at Chapel Hill, Chapel Hill, NC, United States; School of Psychology, Curtin University; Division of Paediatrics, School of Medicine, The University of Western Australia, Perth, WA, Australia

Django White (39), Youth Mental Health Team, Brain and Mind Centre, University of Sydney, Sydney, NSW, Australia

Leanne M. Williams (499), Department of Psychiatry and Behavioral Sciences, Stanford University, Stanford; Sierra-Pacific Mental Illness Research, Education, and Clinical Center (MIRECC), Veterans Affairs Palo Alto Health Care System, Palo Alto, CA, United States

Stephanie H. Witt (227), Department of Genetic Epidemiology in Psychiatry, Central Institute of Mental Health,

Medical Faculty Mannheim, University of Heidelberg, Mannheim, Germany

Yin Yao (239), Lab of Statistical Genomics and Data Analysis, Intramural Program, National Institute of Mental Health, National Institutes of Health, Bethesda, MD, United States

Zeynep Yilmaz (91), Department of Psychiatry; Department of Genetics, The University of North Carolina at Chapel Hill, Chapel Hill, NC, United States

Gwyneth Zai (239), Neurogenetics Section, Molecular Brain Science Department, Campbell Family Mental Health Research Institute, Centre for Addiction and Mental Health; Department of Psychiatry, University of Toronto, Toronto, ON, Canada

Natalia Zmicerevska (39), Youth Mental Health Team, Brain and Mind Centre, University of Sydney, Sydney, NSW, Australia

Preface

Personalized psychiatry, referring to the selection of a treatment best suited for an individual patient, involves the integration and translation into clinical practice of several new technologies from a range of fields such as neuroscience, genomics, deep clinical phenotyping, measurement-based assessments, and bioinformatics. The scope is much broader than indicated by the terms "genomic" or "precision psychiatry," as many nongenomic factors are taken into consideration in developing personalized psychiatry. In addition, despite considerable advances that have taken place in molecular biology and biotechnology, fields that form the basis for personalized psychiatry, personalized psychiatry still requires much development to become a viable clinical option.

The concept of personalized psychiatry as presented in this book brings the four pillars of personalized medicine to the field of Psychiatry: *Prevention, Prediction, Personalization,* and *Participation*. With this approach, the fundamental questions and requirements for personalized psychiatry will be addressed from a research as well as a clinical perspective. The concept includes knowledge of and extensive developments in basic science methodologies (e.g., genomics, epigenomics, transcriptomics), systems biology, bioinformatics, and prediction modeling as well as innovative clinical trials and phenotypic underpinnings in regard to clinical disease and response to treatment relevant for personalized psychiatry.

Specifically, the book aims to achieve key goals in the novel field of personalized psychiatry: (a) it conceptualizes personalized psychiatry; (b) it provides state-of-the-art knowledge on the biological, neuroscience, and bioinformatics methodologies and integrated clinical phenomenology relevant to personalized psychiatry; (c) it presents the most recent findings on clinical genetics of a broad range of psychiatric disorders; and (d) the book discusses important principles of prevention, participation, health care, genetic testing, and ethical implications.

By bringing together experts from a variety of disciplines of various scientific and clinical backgrounds from around the world, this comprehensive collection of up-to-date knowledge is directed to a broad audience of basic and clinical investigators, neuroscientists, psychiatrists, psychologists, residents, and medical and graduate students.

Bernhard T. Baune
Editor

Chapter 1

What is personalized psychiatry and why is it necessary?

Bernhard T. Baune

Department of Psychiatry, University of Münster, Münster, Germany; Department of Psychiatry, Melbourne Medical School, The University of Melbourne, Melbourne, VIC, Australia; The Florey Institute of Neuroscience and Mental Health, The University of Melbourne, Parkville, VIC, Australia

Personalized medicine, often interchangeably referred to as precision or individualized medicine, is an emerging approach for disease classification and treatment stratification as well as for the prevention of illness that takes into account individual variability in genes, environment, and lifestyle. Following the personalized medicine approach, patients are separated or stratified into different groups—with medical decisions, practices, interventions, and/or products being tailored to the individual patient based on their predicted response to treatment or risk of disease.

Most of the current treatments are approved and developed on the basis of their performance in a large population of people and each treatment is prescribed to all patients with a certain diagnosis. However, psychiatry is now developing personalized solutions for a particular patient's needs that will become more readily available. In case of complex psychiatric disorders, the conventional "one-drug-fits-all" approach involves trial and error before appropriate treatment is found. Clinical trial data for new treatments merely show the average response of a study group. Given the considerable individual clinical and biological variation, some patients show no response whereas others show dramatic response. Although approximately 99.9% of our DNA sequence is identical, the 0.1% difference between any two individuals (except identical twins) is medically significant. Within this small percentage of difference lie the clues to hereditary susceptibility to psychiatric disorders. At the DNA level, this 0.1% difference translates into 3 million sites of genomic variation.

It is obvious that the concept of "one medicine fits all patients with the same disease" does not hold true and a more individualized approach is needed. The aim of personalized psychiatry is to match the right treatment to the right patient at the right time, and in some cases even to design the treatment for a patient according to genotype and other individual biological, psychological, or environmental characteristics.

Personalized medicine in the context of psychiatry seeks to identify environmental, biological, and clinical factors that contribute to disease vulnerability, the onset and course of mental disorders, and the predicted response to pharmacological and nonpharmacological interventions. In psychiatry, this task is complicated by the syndromic nature of diagnoses. These diagnostic categories, as represented in the Diagnostic and Statistical Manual of Mental Disorders (DSM), have been developed without a detailed understanding of underlying biological mechanisms. While they have provided the basis for the development of the science of psychiatry, they are not sufficient or reliable descriptions of an individual's clinical presentation, response to treatment, illness, or functional trajectories, resulting in uncertainty in psychiatric diagnosis and prediction of treatment outcomes. Clinically, this may lead to a "wait and watch" approach where multiple trials of different medications with unnecessary side effects or nonefficient nonpharmacological interventions become an additional burden due to poor efficacy.

New research and clinical approaches in psychiatry have become necessary. These should aim to combine advances from clinical phenotyping, biology, neuroscience, and bioinformatic methodologies into a single but complex approach to seek better understanding of complex mental illnesses, better translate research findings into clinical practice, and stimulate reverse translation from clinical to basic science research.

By applying a personalized psychiatry approach, it is not proposed that outcomes are completely predetermined at disease onset, but that structured assessment and modeling of multivariate cross-sectional and longitudinal predictors can be used to describe the risk of specific outcome trajectories and to predict response to specific treatments. As a result, such a predictive approach could aid decision-making processes in clinical psychiatry. For this purpose, predictor variables are grouped into key psychiatric systems that can be differentiated into (a) individual clinical characteristics of a patient, including risk factors such as maternal pregnancy complications, early neurodevelopmental history, and socioeconomic status; (b) their neurocognitive, affective, and functional profile; (c) brain structure and neural function; (d) molecular profile; and (e) modulating prognostic factors such as personality, insight, and resilience. In a complex modeling approach, data from these systems are modeled to derive descriptions of illness trajectories and outcomes. In a translational next step, structured assessments feed into computerized prognostic models that are used to determine the best current treatment based on the most likely response and illness trajectory. The basic idea of these exciting developments in the field of psychiatry is conceptualized in this book.

The concept of personalized psychiatry that is presented here brings the four pillars of personalized medicine to the field of psychiatry: *Prevention, Prediction, Personalization* and *Participation*. With this approach, the fundamental questions and requirements for personalized psychiatry from a research as well as a clinical perspective will be addressed. The concept includes knowledge and extensive developments in basic science methodologies (e.g., genomics, epigenomics, transcriptomics) and neurobiological and clinical conceptual underpinnings of clinical disease relevant for personalized psychiatry. Importantly, the key principles of translation into clinical practice for the individual patient are embedded into personalized clinical trials and in complex prediction modeling. A personalized approach to mental health requires extensive developments in statistical modeling to decipher the complex interacting biological and clinical information as well as the development of a digitalized measurement-based assessment of diagnosis and treatment progression, response to treatment, and trajectories of illness. Taken together, these could transform clinical practice in psychiatry.

Chapter 2

The modeling of trajectories in psychotic illness

Scott R. Clark[a], Klaus Oliver Schubert[a,b] and Bernhard T. Baune[c,d,e]

[a]*Discipline of Psychiatry, University of Adelaide, Adelaide, SA, Australia,* [b]*Northern Adelaide Local Health Network, Mental Health Service, Adelaide, SA, Australia,* [c]*Department of Psychiatry and Psychotherapy, University of Münster, Münster, Germany,* [d]*Department of Psychiatry, Melbourne Medical School, The University of Melbourne, Melbourne, Australia,* [e]*The Florey Institute of Neuroscience and Mental Health, The University of Melbourne, Parkville, VIC, Australia*

1   Introduction

The natural history of psychotic illness is one of fluctuating symptoms and function. Its course may take a number of paths through prodromal illness, from first episode, to recovery; or though episodic relapse, or chronic illness (Clark, Schubert, & Baune, 2015; Schubert, Clark, & Baune, 2015). For some cases, these trajectories may be clinically intuitive; for example, a poor prognosis for the patient with first presentation psychosis, prominent negative symptoms, and cognitive impairment. However, recent studies have shown that complex multiple trajectory patterns layered from physiological to phenomenological, and overlapping across time, may underlie higher order phenotypic presentations. For example, brain indices of gray and white matter volume and connectivity evolve in different patterns across childhood and youth in patients that develop schizophrenia or bipolar disorder in comparison with siblings and controls (Baker, Holmes, Masters, et al., 2014; Katagiri, Pantelis, Nemoto, et al., 2015; Liberg, Rahm, Panayiotou, et al., 2016; Ordonez, Luscher, & Gogtay, 2016; Schmidt, Crossley, Harrisberger, et al., 2016). Consequently, the identification of key illness and functional trajectory signatures from longitudinal data may be critical to improve the prediction of long-term illness prognosis, facilitating personalized treatment of psychosis (Clark et al., 2015; Schubert et al., 2015).

A trajectory can be simply defined as the pattern of change in a measure of interest over time. Trajectories can be easily visualized by plotting mean values or raw data by time, and these plots can be used to grossly determine between subject variability and trend in the measure of interest (Wu, Selig, & Tood, 2013; Xiu, 2015). Simple calculations can provide values for the mean change between baseline and follow up, or the mean slope (change per unit time) where there are two or more time points. These crude analyses are suitable for exploration, but fail to capture extremes and patterns of fluctuation that are likely to be important in the prediction of long-term outcomes in mental illness (Adamis, 2009). For example, Fig. 1A reproduced from Clark et al. (2015) illustrates four possible trajectories of psychotic symptoms for those at clinical high risk (CHR) of psychosis. Patients may oscillate between CHR and normal status, may remain at CHR risk without transition, or may take a standard path from CHR to first episode psychosis. In a large clinical trial involving people at CHR of psychosis, Polari, Lavoie, Yuen, et al. (2018) noted 17 different trajectories in a 12-month period, which were classified as: recovery (35.7%), remission (7.5%), any recurrence (20%), no remission (17.3%), relapse (4.0%), and transition to psychosis (15.8%). In a hypothetical study designed to determine the outcome of these patients, the number and frequency of assessments will determine the accuracy of trajectory analysis. Predicting diagnosis and outcome given fluctuations in mood and psychotic symptoms in bipolar disorder is similarly difficult (Hennen, 2003). With only a few data points, it is easy to miss the underlying trajectory. At least three data points are required to identify linear versus curvilinear change (Curran, Obeidat, & Losardo, 2010).

The statistical analysis of longitudinal data has evolved considerably over the past 30 years, providing better estimates of trends and the determinants of growth and change at the level of the individual (Wu et al., 2013). In comparison with cross sectional analysis, longitudinal methods are more statistically efficient, requiring fewer subjects to achieve a similar level of statistical power. Each subject can serve as their own control, and intra-subject variability is substantially less than inter-subject variability (Gibbons, Hedeker, & DuToit, 2010). Prospective longitudinal analysis allows assessment of event contiguity, leading to more robust conclusions regarding causality when appropriate confounding and covariates are

Personalized Psychiatry. **https://doi.org/10.1016/B978-0-12-813176-3.00002-X**

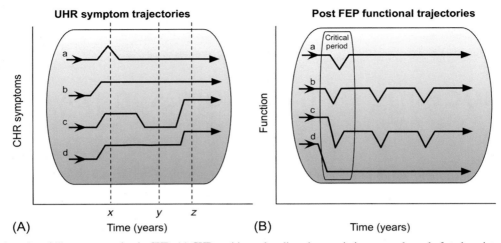

FIG. 1 (A) Trajectories of illness progression in CHR. (a) CHR positive at baseline, then remission at *y* and *z* end of study points. (b) CHR positive without transition at all points. (c) CHR positive at baseline, then remission at end point *y*, but transition at end point *z*. (d) CHR positive at baseline and end point *y*, transition at end point *z*. (B) Trajectories of function from first psychotic episode. (a) Single episode; (b) multiple episode, full inter-episode function; (c) multiple episode, partial return of inter-episode function; (d) treatment resistance with no functional improvement.

considered (Adamis, 2009). However, the logistics and cost of longitudinal studies can be prohibitive. There are also barriers to follow up of patients with mental illness, due to fluctuating symptoms, disengagement, lack of insight, and stigma that result in drop out and missing data (Mazumdar, Tang, Houck, et al., 2007; Woodall, Morgan, Sloan, et al., 2010).

Increasing complexity of techniques has contributed to the slow uptake of more advanced longitudinal analyses (Locascio & Atri, 2011). Modeling software has proliferated, leading to the risk of inappropriate use by those with limited statistical experience (Gibbons et al., 2010). The purpose of this chapter then is twofold: (1) to investigate the use of trajectory modeling techniques in psychosis research, and (2) to increase awareness of the advantages and disadvantages of these methods. We hope that readers of this chapter will be encouraged to seek expert statistical support when selecting and implementing the more complex techniques discussed in this chapter.

2 Sample trajectories (between-subject analyses)

Traditional statistical approaches to longitudinal analysis, such as the paired *t*-test, analysis of variance (ANOVA), analysis of covariance (ANCOVA), multivariate analysis of variance (MANOVA), and multivariate analysis of covariance (MANCOVA) and multiple regression all conform to the general linear model (GLM) (Abdi, 2010). The equation for the common basic form of the GLM can be written as $Y = B_0 + B_1 + e$ for each subject, where Y = dependent variable; B_0 = the intercept or baseline value; B_1 = slope or change in dependent variable associated each unit on the independent variable (factor); e = error or residual variance Y unexplained by intercept and slope (see Fig. 2) (Poline & Brett, 2012). A line of best fit to the residuals is defied using the ordinary least squares method (Abdi, 2010; Poline & Brett, 2012). Each independent variable adds a further B term to the equation. Time can be represented categorically (e.g., ANOVA, MANOVA) or continuously (e.g., Regression). Factors in the GLM can vary between subjects around the mean, but not within subjects over time, hence these are termed "fixed effects" models (Gelman, 2005; Gueorguieva & Krystal, 2004). GLMs are therefore useful for identifying sample, but not individual, trajectories of psychiatric symptoms or outcomes. GLMs require balanced numbers at all measurement points, and are unable to handle missing data or variable time points (Locascio & Atri, 2011). Traditional methods of managing missing data, such as imputation or using the last observation carried forward, can introduce errors and bias into the analysis (Gueorguieva & Krystal, 2004).

Missing data is particularly problematic if cases are missing systematically due to factors that are not measured during the study (missing not at random); for example, new onset medication side effects just prior to study drop out (Mazumdar et al., 2007). GLMs assume normally distributed and homogenous residuals across time points, and do not correct for within-subject correlation between repeated assessments, which is usually higher when measures are closer together in time (e.g., psychotic symptoms across a hospital treatment episode) (Locascio & Atri, 2011; Nayak Savla, Moore, Roesch, et al., 2006; Twisk, 2004; Xiu, 2015).

Each GLM has some strengths. The *t*-test is simple, but is limited to comparisons between sample means at two time points or between two samples. It can be used sequentially to compare multiple time points within a longitudinal data set.

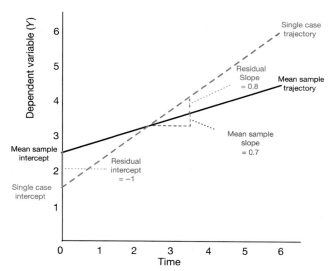

FIG. 2 Components of linear model mean and case trajectories.

However, the use of multiple *t*-tests leads to type 1 statistical error, and increases the risk of false positive findings (Gueorguieva & Krystal, 2004). In contrast, ANOVA can be used to determine the association of one or more independent variables (factors) with a change in a continuous dependent variable over occasions of measurement (Locascio & Atri, 2011). The related ANCOVA procedure also allows control for covariates that may confound the relationship between the independent variables and the outcome of interest (Locascio & Atri, 2011). An example is the use of baseline cognitive scores and the total number of years of education as covariates for the assessment of change in cognitive function following first episode psychosis (Kenney, Anderson-Schmidt, Scanlon, et al., 2015). MANOVA and MANCOVA model the relationship among multiple independent and dependent variables; the MANCOVA also includes covariates (Xiu, 2015). The inclusion of multiple dependent variables in a single test reduces the total number of tests required to analyze a dataset, and therefore reduces type 1 error. For example, Flahault, Schaer, Ottet, et al. (2012) found a relationship between psychosis risk, chromosome 22q11 deletion syndrome, and a reduction in hippocampal body size over 3 years. They used a MANCOVA to compare variation in hippocampal sub-regional volumes (6 sites, bilaterally) between patients and controls, and adjusted for total gray matter volume covaried for age, gender, and full-scale IQ.

Other GLM-based methods for nonlinear data types such as count and binomial data use the generalized linear model. This generalized model enables the analysis of non-normally distributed residuals by including a link function for exponential distributions (e.g., logit, Poisson), for example, in logistic regression or parametric survival analysis using proportional hazard modeling (Adamis, 2009; Agresti, 2015). Agnew-Blais et al. (Agnew-Blais, Buka, Fitzmaurice, et al., 2015; MacCabe, Wicks, Lofving, et al., 2013; Welham, Scott, Williams, et al., 2010) used logistic regression to assess the relationship between IQ at ages 4 and 7, and the subsequent risk of psychosis. Lower childhood IQ, a small increase in IQ between age 4 and 7, and male gender were all found to be predictive of transition to non-affective psychosis.

In contrast, generalized estimating equations (GEE) extend the GLM by using residuals to estimate the correlation structure between repeated measures (Burton, Gurrin, & Sly, 1998; Zeger & Liang, 1986). For example, Devylder, Ben-David, Schobel, et al. (2013) found that impaired stress tolerance, but not life events, promoted psychotic symptoms, depression, anxiety, and poor function in a CHR sample. Goghari, Harrow, Grossman, et al. (2013) found that the 20-year course of hallucinations varied with diagnosis, which was worse with schizophrenia over schizoaffective or bipolar disorder, and least for major depression. Early hallucinations and increased frequency were associated with poor recovery and poor occupational outcomes. The procedure, however, is best limited to estimate sample mean trajectories (Adamis, 2009).

Survival analysis is used to calculate trajectories of event timing, for example, transition to first psychotic episode from CHR (Nelson, Yuen, Wood, et al., 2013; Ruhrmann, Schultze-Lutter, Salokangas, et al., 2010), or the risk of adverse events associated with antipsychotic medications (Stroup, Gerhard, Crystal, et al., 2016). These analyses can investigate four parameters of event timing: (1) survival, i.e., the probability of an individual surviving, or an event not occurring in a particular time period; (2) Hazard function, i.e., the risk that an individual will experience the event in the next interval of time; (3) probability density function, i.e., the unconditional probability of the event occurring at given time; (4) the mean time to the event of interest from any point within the study (Klein & Moeschberger, 2003). Methods include nonparametric (Kaplan-Meier curves), semiparametric (Cox proportional hazards), and fully parametric (survival hazard function) analyses (Adamis, 2009; Hennen, 2003). Semi- and nonparametric models make fewer assumptions about the shape of underlying data distributions, while semiparametric and parametric analyses allow the inclusion of covariates (Klein & Moeschberger, 2003). For example, Nelson et al. (2013) calculated a Kaplan-Meier curve to estimate transition rates and then used step-wise cox regression, finding that year of assessment, delay to assessment from symptom onset, baseline prodromal and psychotic symptoms, quality of life and function predicted transition to first episode psychosis from UHR. The Cox regression equation has also been used to develop a risk stratification model for transition from UHR to FEP (Ruhrmann et al., 2010). Curves can be compared for different groups (e.g., treatment versus control, presence or absence of different risk factors) (Klein & Moeschberger, 2003). Advanced techniques can model fluctuations in outcome states, such as remission and relapse of psychotic symptoms. Survival analyses can adjust for missing data or "censored" data (Adamis, 2009; Hennen, 2003). Cases are right censored if the event of interest does not occur in the time frame of the study, and left censored if the event occurs before the start of the study. Interval censoring occurs when the event happens between study sample points and the exact timing is not available (Klein & Moeschberger, 2003).

2.1 Simple Bayesian trajectories

Recently, our group has used the odds ratio form of Bayes' rule to calculate individual patient risk trajectories based on the sequential investigation results (see Fig. 5) (Clark et al., 2015; Clark, Baune, Schubert, et al., 2016; Schmidt, Cappucciati, Radua, et al., 2016; Schubert et al., 2015). The model follows a simple formula: baseline odds of outcome multiplied by the likelihood ratio of each predictor variable to calculate post-test odds. Post-test odds can then be converted to probability of outcome (McGee, 2002). The model allows stepwise calculations of the probability of binary outcomes, such as transition to psychosis from UHR, or high or low function following FEP. Individual trajectories can be displayed as plots of probability of outcome over time. These calculations are also referred to as "naïve" Bayes models, and are linear in nature, similar to GLM analyses. The model assumes that predictors have independent effects on probability of outcome, but classification accuracy is not affected by violations of this assumption (Rish, 2001). The tradeoff is poor calibration leading to over-confidence of calculated probabilities (Monti & Cooper, 1998). The strength of odds ratio form of Bayes rule models lies in the simplicity of calculations and flexibility of modeling (Clark et al., 2015; Clark et al., 2016; Schubert et al., 2015), characteristics that make this approach attractive for use in day to day clinical practice. Missing data is handled implicitly as the conditional probability at any point in time for any given individual case is related only to the data available.

Fig. 3 shows individual trajectories of probability of transition to FEP from CHR derived using an odd ratio form of Bayes' rule model. Probability is calculated following individual assessments of history, symptoms (clinical assessment), and blood fatty acid analysis (total omega-3 and nervonic acid). Fig. 3A shows cases that remained as CHR, and 3b those that transitioned at 1 year. The addition of new types of information sequentially changes the individual probability of outcome. Cases at intermediate risk require further investigation to improve the certainty of predicted outcome. The combined model predicted outcome with AUROC = 0.919, Sensitivity = 72.73%, Specificity = 96.43%.

In summary, simple GLM, GEE, and survival analyses are useful for representing average sample or subgroup trajectories. Survival analysis measures rates, risk, and timing of events in a sample or subsamples defined by covariates, while GLM and GEE analyses are used to explore changes in dependent variables over time. These techniques focus on between-subject differences, rather than patterns of change within individual subjects over time. In contrast, simple Bayesian models can be used to calculate individual patient risk trajectories based on sample conditional probabilities (e.g., the probability of psychosis given prodromal symptoms at baseline). These models generate predictive trajectories rather than describing change over time, and are useful for analysis of binary investigation and treatment decisions. None of these approaches model within-subjects patterns of variance in measured variables over time. Methods that account for within-subjects variation on key predictor measures (e.g., brain volumes, connectivity, and psychotic symptoms) may provide a better understanding of the predictors of individual outcomes in comparison with these analyses of population parameters.

3 Subgroup and individual trajectories (between- and within-subject analyses)

To describe subgroup and individual trajectories, models must account for both between- and within-subject variation over time. There are two broad approaches to this problem: mixed modeling (MM) and structural equation modeling (SEM). The field is complex, with a number of similar techniques and confusing redundant terminology that has evolved in different disciplines (Gibbons et al., 2010; Wu et al., 2013). For example, multilevel modeling, random coefficient modeling, and hierarchical linear modeling (HLM) are all forms of MM (Huta, 2014). On the other hand, growth models, latent growth

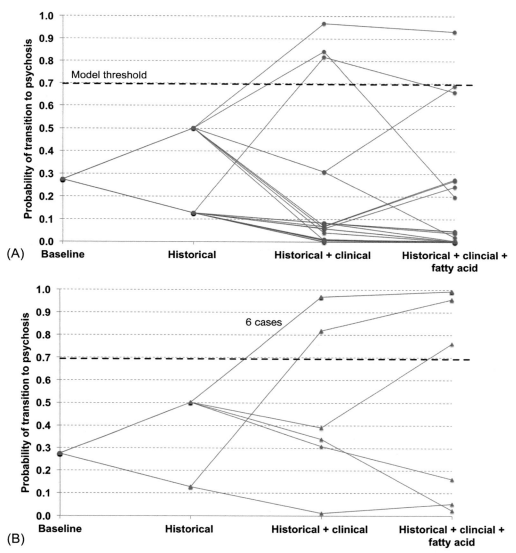

FIG. 3 Plots of probability of transition to FEP given historical, clinical, and biomarker information. (A) Stepwise probability of transition for individual cases not transitioned to psychosis at 1 year. (B) Stepwise probability of transition for individual cases transitioned to psychosis at 1 year.

models, latent trajectory models, and latent curve analysis are all forms of SEM (Curran & Hussong, 2003). MM and SEM models themselves are complex, and lack a single solution; rather, versions of each model are compared on indices of fit to identify the most appropriate form (Penga & Lu, 2012; Preacher & Merkle, 2012). Missing data can be handled using maximum likelihood estimation methods, but is assumed to be missing at random (Xiu, 2015).

3.1 Mixed modeling

Mixed models combine a fixed effect analysis of sample means for intercept and slope with a random effects model of the variance around the mean of these values for each case, thus accounting for within-subject variation over time. Each individual's intercept and slope is composed of a mean value for the sample, and a residual component unique to each case and time point (see Fig. 2) (Wu et al., 2013). This residual variation is considered random, as the individual is a random selection from the analysis sample; hence the term "random effects." Correlation of these residuals between time points is used to reflect the association between repeated measures over time. In HLM, multiple levels of data can be nested using a similar approach. For example, there is a clear hierarchical structure to data in multisite clinical trials or observational studies (between study sites, between patients, within patients over time) (Morris, Bloom, & Kang, 2007). Mixed models can manage unbalanced numbers at each time point of assessment, enabling the use of all data regardless of timing, missing assessments, or dropout. Polynomial, exponential, and many other trajectory shapes may be modeled by adding components to the equation (Long & Ryoo, 2010; Wu et al., 2013). MM, however, assumes that all individuals come from a population with a common estimate of growth parameters, and will undergo similar growth trajectories under a given set of conditions (Jung & Wickrama, 2008).

In psychosis literature, MM analyses have been used to identify trajectories of brain changes leading up to FEP, and in cognitive function post-FEP. Studies of adolescents at high genetic risk of psychosis based on familial incidence show reduced frontoparietal gyral surface area and cortical thickness at baseline, and greater reductions in whole brain, left and right prefrontal, and temporal lobe volume longitudinally to 8 years in comparison with controls (McIntosh, Owens, Moorhead, et al., 2011; Prasad, Goradia, Eack, et al., 2010). Steeper prefrontal reductions proportional to the severity of symptoms may be indicative of neurotoxic processes closer to the onset of psychosis (McIntosh et al., 2011). In comparison, healthy siblings of patients with childhood onset schizophrenia show trajectories of reduction in left parietal white matter and left prefrontal, bilateral temporal, right prefrontal, and inferior parietal cortical gray matter and in cerebellar structures that normalize with age (Gogtay, Greenstein, Lenane, et al., 2007; Gogtay, Hua, Stidd, et al., 2012; Greenstein, Lenroot, Clausen, et al., 2011). MM analysis shows differing cognitive trajectories post-FEP with poor outcomes for baseline low and deteriorated groups, and higher cognitive function associated with stable symptom resolution 1-year post-FEP (Leeson, Sharma, Harrison, et al., 2011; Rund, Barder, Evensen, et al., 2016).

3.2 Structural equation modeling

In contrast to MM, SEM uses latent variables to account for residual variance in the intercept and slope of dependent measures over time. The key processes within SEM include pathway analysis to investigate the relationships between observed variables, and factor analysis to identify underlying latent variables that account for variance in the pathway analysis model (Rahman, Shah, & Rasli, 2015; Xiu, 2015). Complex relationships among multiple independent, latent, and dependent variables can be modeled. Model structure is developed a priori, and formalized in diagrams that include observed and latent variables, directionality of relationships, and covariance (Rahman et al., 2015; Xiu, 2015). In path diagrams, boxes are used to indicate observed variables, while circles or ovals designate the latent factors, and causal relationships are denoted by arrows (see Fig. 3).

In a latent growth model (LGM), the intercept and slope of change are modeled as latent variables. The LGM curve produced is the average of all cases over time, and accounts for within-subject correlation among repeated measures. Covariates can alter the shape of the curve, and polynomial curves can be modeled to suit the underlying data (Wu et al., 2013; Xiu, 2015). Growth mixture modeling (GMM) expands on LGM to allow the identification of latent subclasses of individuals with similar trajectories within the population trajectory. Each class may have a different intercept, shape, and slope. Covariates in these analyses affect each latent trajectory differently (Jung & Wickrama, 2008; Willke, Zheng, Subedi, et al., 2012). Once identified, further analyses can determine specific risk factors for membership of each trajectory group (Leiby, 2012). Particular clinical trajectories may be associated with important exposures and outcomes, such as the impact of substance use on psychotic symptoms, treatment response, and general function post-FEP. For example, Mackie, O'Leary-Barrett, Al-Khudhairy, et al. (2013) used GMM to identify 3 trajectories of psychotic experiences in a large sample of adolescents: elevated, increasing, and low. Cannabis use was associated with increasing psychotic symptom group membership, and with increased psychotic symptoms in the low cannabis use group. Psychotic symptoms at baseline did not predict subsequent cannabis use. Lecomte, Mueser, MacEwan, et al. (2013) found two trajectories of psychotic symptoms in methamphetamine users over 6 months: persistent psychotic symptoms (30%) and low psychotic symptoms. Those with persistent psychosis were significantly older, had more severe psychotic symptoms, misused methamphetamine for more years, had more antisocial personality traits, and had more sustained depressive symptoms.

Studies using GMM identify three to four trajectories of change in psychosis symptoms across clinical trials of antipsychotic treatment (see Fig. 4) (Case, Stauffer, Ascher-Svanum, et al., 2011; Chen, Johnston, Kinon, et al., 2013; Levine, Rabinowitz, Faries, et al., 2012). Consistent subgroups include: rapid significant response, slower moderate response, and nonsustained improvement or nonresponse. Olanzapine treatment was associated with the best response trajectory in one study (Levine et al., 2012), while depressive symptoms were associated with treatment resistance in another (Case et al., 2011). One GMM study showed that assertive community treatment (ACT) produced 2 trajectories of response over 2 years: low stable function (80%), and higher baseline with improvement over time (20%). Low stable function was predicted by a higher number of psychiatric comorbidities (Wilk, Vingilis, Bishop, et al., 2013). The stabilizing function of work was apparent in another GMM study that identified two trajectories, steady and nonworkers, in patients with psychotic

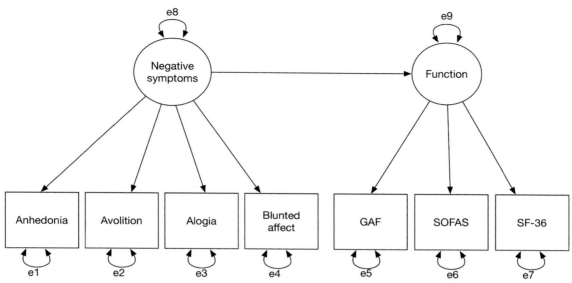

FIG. 4 Simple hypothetical structural equation modeling diagram for the relationship of negative symptoms and function in psychosis. Rectangles = Measured variables or scales; Circles = latent variables; Single headed arrows = variance in outcome; Double headed arrows (e) = error (variance or residuals).

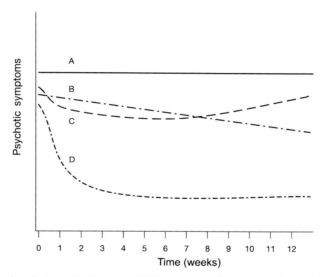

FIG. 5 Trajectories of response to antipsychotic medication from GMM studies: (A) No response; (B) marginal response; (C) response then relapse; (D) rapid and sustained response.

illness and comorbid substance abuse (McHugo, Drake, Xie, et al., 2012). Steady workers achieved independent housing and remission of substance use much earlier.

4 Discussion

The key advantage of trajectory over cross-sectional analysis is the ability to understand important time-dependent patterns in the evolution of mental illness. Longitudinal studies of brain imaging, symptoms, cognitive and general function in CHR and FEP samples highlight within- and between-subject variation with a potential impact on treatment outcomes. Trajectory analysis techniques allow the extraction of higher order features of this variation that may not be obvious to an individual clinician at cross-sectional, or even longitudinal review. We have identified two broad groups of trajectory modeling techniques: those that use between-subjects analysis to derive mean sample trajectories, and those that derive subgroup trajectories using both between- and within-subject analyses. Sample level and subgroup trajectories may contribute to the understanding of the evolution of mental illness in different ways. GLM and GEE may be of use when there is limited heterogeneity in illness course, and an understanding of the average change process or outcome rate is adequate. Survival analyses and logistic regression are useful for predicting discrete events and MANOVA for defining relationships among multiple dependent variables at a sample level. Simple Bayesian sequential probability models are useful for understanding the evolution of risk over time with sequential assessments.

MM and SEM techniques model both within- and between-subject variation in longitudinal measures. MM is useful for testing hypotheses about the trajectory of subgroups defined a priori by criteria of interest. Only GMM is able to extract multiple trajectory groups with different intercept, shape, and slope. Importantly, GMM techniques have been used to identify subgroups of patients with different antipsychotic response profiles, a finding with potential for early stratification of patients, using trajectory predictors, to personalize care.

In terms of limitations, the implementation of GMM and MM require specialized software and analysis skills. In contrast, GLM-based models can be much simpler, and are useful for data exploration before undertaking more complex analyses. GLM methods may also be used to determine predictors of trajectory membership derived using MM or GMM. Regardless of the modeling technique, studies with more data points are likely to better fit the underlying trajectory of the variable in question. There is a trade off in terms of the increasing costs of data collection and the increased risk of missing data with more sequential assessments. An understanding of underlying fluctuations and trend in the variable of interest through exploratory analysis is important for the selection of frequency and number of data points. Missing data and variation in assessment timing are both common in longitudinal studies of mental illness. Advanced trajectory methods such as MM and SEM can include cases with missing data points or variable timing of assessments, thus increasing the sample size available for analysis, and therefore the power of the study.

4.1 Future designs in trajectory research

More complex trajectory analysis methods, such as GMM and SEM, are still underutilized in longitudinal studies. These methods have the potential to identify groups of patients with specific risks for transition to psychosis, treatment resistance, and of poor functional outcomes. Future research into the clinical and biological basis of these subgroups may lead to new treatments. Early identification of specific trajectories can then be used to personalize care. Ultimately, clinical trials are required to assess the utility of trajectory prediction for psychosis care.

New developments in wearable technology, from smart phones, to EEG, and implantable measuring devices has greatly increased the feasibility of the collection of both event-based and continuous longitudinal data that will become a rich source for mental illness trajectory modeling. Modeling techniques are also evolving rapidly. Fulcher et al. (Fulcher, Little, & Jones, 2013) recently analyzed more than 35,000 real-world and model-generated time series data sets, and more than 9000 time-series analysis algorithms to develop a tool that can automatically categorize features of longitudinal data and suggest particular analysis techniques based on the output from an extensive range of models. The difficulty of selecting the optimal longitudinal modeling method for a given problem may be greatly simplified by using such tools.

4.2 Conclusions

Trajectory analysis of people with psychotic symptoms has revealed multiple patterns of longitudinal variation in biological, clinical, and functional features that may be useful to better personalize care. Trajectory modeling methods have evolved significantly in recent times, but complex methods and redundant terminology make the techniques difficult to access. Newer techniques, such as MM and GMM, can model both between- and within-subjects variance, and are efficient at handling missing data, and flexible regarding the timing of assessments. GMM can be used to extract multiple trajectory groups that may be of clinical relevance. Simpler GLM, survival analysis, and naïve Bayesian techniques are useful for sample level exploration and event analyses. Rapidly evolving technology for the continuous and event-related collection of longitudinal data will provide increasing opportunity to collect multimodal data that can be used for trajectory modeling. Such complex multimodal trajectories may be key to the personalization of psychosis treatment.

References

Abdi, H. (2010). Encyclopedia of research design. In N. J. Salkind (Ed.), *Encyclopedia of research design chapter title: "General linear model"* (pp. 539–541). Thousand Oaks: SAGE Publications, Inc.

Adamis, D. (2009). Statistical methods for analysing longitudinal data in delirium studies. *International Review of Psychiatry, 21*, 74–85.

Agnew-Blais, J. C., Buka, S. L., Fitzmaurice, G. M., et al. (2015). Early childhood IQ trajectories in individuals later developing schizophrenia and affective psychoses in the New England family studies. *Schizophrenia Bulletin, 41*, 817–823.

Agresti, A. (2015). *Foundations of linear and generalized linear models.* New Jersy: John Wiley & Sons Inc.

Baker, J. T., Holmes, A. J., Masters, G. A., et al. (2014). Disruption of cortical association networks in schizophrenia and psychotic bipolar disorder. *JAMA Psychiatry, 71*, 109–118.

Burton, P., Gurrin, L., & Sly, P. (1998). Extending the simple linear regression model to account for correlated responses: An introduction to generalized estimating equations and multi-level mixed modelling. *Statistics in Medicine, 17*, 1261–1291.

Case, M., Stauffer, V. L., Ascher-Svanum, H., et al. (2011). The heterogeneity of antipsychotic response in the treatment of schizophrenia. *Psychological Medicine, 41*, 1291–1300.

Chen, L., Johnston, J. A., Kinon, B. J., et al. (2013). The longitudinal interplay between negative and positive symptom trajectories in patients under antipsychotic treatment: A post hoc analysis of data from a randomized, 1-year pragmatic trial. *BMC Psychiatry, 13*, 320.

Clark, S. R., Baune, B. T., Schubert, K. O., et al. (2016). Prediction of transition from ultra-high risk to first-episode psychosis using a probabilistic model combining history, clinical assessment and fatty-acid biomarkers. *Translational Psychiatry, 6*, e897.

Clark, S. R., Schubert, K. O., & Baune, B. T. (2015). Towards indicated prevention of psychosis: Using probabilistic assessments of transition risk in psychosis prodrome. *Journal of Neural Transmission (Vienna), 122*, 155–169.

Curran, P. J., & Hussong, A. M. (2003). The use of latent trajectory models in psychopathology research. *Journal of Abnormal Psychology, 112*, 526–544.

Curran, P. J., Obeidat, K., & Losardo, D. (2010). Twelve frequently asked questions about growth curve modeling. *Journal of Cognition and Development, 11*, 121–136.

Devylder, J. E., Ben-David, S., Schobel, S. A., et al. (2013). Temporal association of stress sensitivity and symptoms in individuals at clinical high risk for psychosis. *Psychological Medicine, 43*, 259–268.

Flahault, A., Schaer, M., Ottet, M. C., et al. (2012). Hippocampal volume reduction in chromosome 22q11.2 deletion syndrome (22q11.2DS): A longitudinal study of morphometry and symptomatology. *Psychiatry Research, 203*, 1–5.

Fulcher, B. D., Little, M. A., & Jones, N. S. (2013). Highly comparative time-series analysis: The empirical structure of time series and their methods. *Journal of the Royal Society, Interface, 10*.

Gelman, A. (2005). Analysis of variance—Why it is more important than ever. *The Annals of Statistics, 33*, 1–53.

Gibbons, R. D., Hedeker, D., & DuToit, S. (2010). Advances in analysis of longitudinal data. *Annual Review of Clinical Psychology, 6*, 79–107.

Goghari, V. M., Harrow, M., Grossman, L. S., et al. (2013). A 20-year multi-follow-up of hallucinations in schizophrenia, other psychotic, and mood disorders. *Psychological Medicine, 43*, 1151–1160.

Gogtay, N., Greenstein, D., Lenane, M., et al. (2007). Cortical brain development in nonpsychotic siblings of patients with childhood-onset schizophrenia. *Archives of General Psychiatry, 64*, 772–780.

Gogtay, N., Hua, X., Stidd, R., et al. (2012). Delayed white matter growth trajectory in young nonpsychotic siblings of patients with childhood-onset schizophrenia. *Archives of General Psychiatry, 69*, 875–884.

Greenstein, D., Lenroot, R., Clausen, L., et al. (2011). Cerebellar development in childhood onset schizophrenia and non-psychotic siblings. *Psychiatry Research, 193*, 131–137.

Gueorguieva, R., & Krystal, J. H. (2004). Move over ANOVA: Progress in analyzing repeated-measures data and its reflection in papers published in the archives of general psychiatry. *Archives of General Psychiatry, 61*, 310–317.

Hennen, J. (2003). Statistical methods for longitudinal research on bipolar disorders. *Bipolar Disorders, 5*, 156–168.

Huta, V. (2014). When to use hierarchical linear modeling. *The Quantitative Methods for Psychology, 10*, 13–28.

Jung, T., & Wickrama, K. (2008). An introduction to latent class growth analysis and growth mixture modeling. *Social and Personality Psychology Compass, 2*, 302–317.

Katagiri, N., Pantelis, C., Nemoto, T., et al. (2015). A longitudinal study investigating sub-threshold symptoms and white matter changes in individuals with an 'at risk mental state' (ARMS). *Schizophrenia Research, 162*, 7–13.

Kenney, J., Anderson-Schmidt, H., Scanlon, C., et al. (2015). Cognitive course in first-episode psychosis and clinical correlates: A 4 year longitudinal study using the MATRICS Consensus Cognitive Battery. *Schizophrenia Research, 169*, 101–108.

Klein, J. P., & Moeschberger, M. L. (2003). *Survival analysis techniques for censored and truncated data.* New York: Springer.

Lecomte, T., Mueser, K. T., MacEwan, W., et al. (2013). Predictors of persistent psychotic symptoms in persons with methamphetamine abuse receiving psychiatric treatment. *The Journal of Nervous and Mental Disease, 201*, 1085–1089.

Leeson, V. C., Sharma, P., Harrison, M., et al. (2011). IQ trajectory, cognitive reserve, and clinical outcome following a first episode of psychosis: A 3-year longitudinal study. *Schizophrenia Bulletin, 37*, 768–777.

Leiby, B. E. (2012). Growth curve mixture models. *Shanghai Archives of Psychiatry, 24*, 355–358.

Levine, S. Z., Rabinowitz, J., Faries, D., et al. (2012). Treatment response trajectories and antipsychotic medications: Examination of up to 18 months of treatment in the CATIE chronic schizophrenia trial. *Schizophrenia Research, 137*, 141–146.

Liberg, B., Rahm, C., Panayiotou, A., et al. (2016). Brain change trajectories that differentiate the major psychoses. *European Journal of Clinical Investigation, 46*, 658–674.

Locascio, J. J., & Atri, A. (2011). An overview of longitudinal data analysis methods for neurological research. *Dementia and Geriatric Cognitive Disorders Extra, 1*, 330–357.

Long, J., & Ryoo, J. (2010). Using fractional polynomials to model non-linear trends in longitudinal data. *The British Journal of Mathematical and Statistical Psychology, 63*, 177–203.

MacCabe, J. H., Wicks, S., Lofving, S., et al. (2013). Decline in cognitive performance between ages 13 and 18 years and the risk for psychosis in adulthood: A Swedish longitudinal cohort study in males. *JAMA Psychiatry, 70*, 261–270.

Mackie, C. J., O'Leary-Barrett, M., Al-Khudhairy, N., et al. (2013). Adolescent bullying, cannabis use and emerging psychotic experiences: A longitudinal general population study. *Psychological Medicine, 43*, 1033–1044.

Mazumdar, S., Tang, G., Houck, P. R., et al. (2007). Statistical analysis of longitudinal psychiatric data with dropouts. *Journal of Psychiatric Research, 41*, 1032–1041.

McGee, S. (2002). Simplifying likelihood ratios. *Journal of General Internal Medicine, 17*, 646–649.

McHugo, G. J., Drake, R. E., Xie, H., et al. (2012). A 10-year study of steady employment and non-vocational outcomes among people with serious mental illness and co-occurring substance use disorders. *Schizophrenia Research, 138*, 233–239.

McIntosh, A. M., Owens, D. C., Moorhead, W. J., et al. (2011). Longitudinal volume reductions in people at high genetic risk of schizophrenia as they develop psychosis. *Biological Psychiatry, 69*, 953–958.

Monti, S., & Cooper, G. F. (1998). The impact of modeling the dependencies among patient findings on classification accuracy and calibration. In: *Proceedings of the AMIA Symposium*, pp. 592–596.

Morris, A., Bloom, J. R., & Kang, S. (2007). Organizational and individual factors affecting consumer outcomes of care in mental health services. *Administration and Policy in Mental Health, 34*, 243–253.

Nayak Savla, G., Moore, D. J., Roesch, S. C., et al. (2006). An evaluation of longitudinal neurocognitive performance among middle-aged and older schizophrenia patients: Use of mixed-model analyses. *Schizophrenia Research, 83*, 215–223.

Nelson, B., Yuen, H. P., Wood, S. J., et al. (2013). Long-term follow-up of a group at ultra high risk ("prodromal") for psychosis: The PACE 400 study. *JAMA Psychiatry, 70*, 793–802.

Ordonez, A. E., Luscher, Z. I., & Gogtay, N. (2016). Neuroimaging findings from childhood onset schizophrenia patients and their non-psychotic siblings. *Schizophrenia Research, 173*, 124–131.

Penga, H., & Lu, Y. (2012). Model selection in linear mixed effect models. *Journal of Multivariate Analysis, 109*, 109–129.

Polari, A., Lavoie, S., Yuen, H. P., et al. (2018). Clinical trajectories in the ultra-high risk for psychosis population. *Schizophrenia Research, 197*, 550–556.

Poline, J. B., & Brett, M. (2012). The general linear model and fMRI: Does love last forever? *NeuroImage, 62*, 871–880.

Prasad, K. M., Goradia, D., Eack, S., et al. (2010). Cortical surface characteristics among offspring of schizophrenia subjects. *Schizophrenia Research, 116*, 143–151.

Preacher, K. J., & Merkle, E. C. (2012). The problem of model selection uncertainty in structural equation modeling. *Psychological Methods, 17*, 1–14.

Rahman, W., Shah, A. F., & Rasli, A. (2015). Use of structural equation modeling in social science research. *Asian Social Science, 11*, 371–377.

Rish, I. (2001). An empirical study of the naive Bayes classifier. *IJCAI 2001 Workshop on Empirical Methods in Artificial Intelligence, 3*, 41–46.

Ruhrmann, S., Schultze-Lutter, F., Salokangas, R. K., et al. (2010). Prediction of psychosis in adolescents and young adults at high risk: Results from the prospective European prediction of psychosis study. *Archives of General Psychiatry, 67*, 241–251.

Rund, B. R., Barder, H. E., Evensen, J., et al. (2016). Neurocognition and duration of psychosis: A 10-year follow-up of first-episode patients. *Schizophrenia Bulletin, 42*, 87–95.

Schmidt, A., Cappucciati, M., Radua, J., et al. (2016). Improving prognostic accuracy in subjects at clinical high risk for psychosis: Systematic review of predictive models and meta-analytical sequential testing simulation. *Schizophrenia Bulletin, 43*, 375–388.

Schmidt, A., Crossley, N. A., Harrisberger, F., et al. (2016). Structural network disorganization in subjects at clinical high risk for psychosis. *Schizophrenia Bulletin, 43*, 583–591.

Schubert, K. O., Clark, S. R., & Baune, B. T. (2015). The use of clinical and biological characteristics to predict outcome following first episode psychosis. *The Australian and New Zealand Journal of Psychiatry, 49*, 24–35.

Stroup, T. S., Gerhard, T., Crystal, S., et al. (2016). Comparative effectiveness of clozapine and standard antipsychotic treatment in adults with schizophrenia. *The American Journal of Psychiatry, 173*, 166–173.

Twisk, J. W. (2004). Longitudinal data analysis. A comparison between generalized estimating equations and random coefficient analysis. *European Journal of Epidemiology, 19*, 769–776.

Welham, J., Scott, J., Williams, G. M., et al. (2010). The antecedents of non-affective psychosis in a birth-cohort, with a focus on measures related to cognitive ability, attentional dysfunction and speech problems. *Acta Psychiatrica Scandinavica, 121*, 273–279.

Wilk, P., Vingilis, E., Bishop, J. E., et al. (2013). Distinctive trajectory groups of mental health functioning among assertive community treatment clients: An application of growth mixture modelling analysis. *Canadian Journal of Psychiatry, 58*, 670–678.

Willke, R. J., Zheng, Z., Subedi, P., et al. (2012). From concepts, theory, and evidence of heterogeneity of treatment effects to methodological approaches: A primer. *BMC Medical Research Methodology, 12*, 185.

Woodall, A., Morgan, C., Sloan, C., et al. (2010). Barriers to participation in mental health research: Are there specific gender, ethnicity and age related barriers? *BMC Psychiatry, 10*, 103.

Wu, W., Selig, J., & Tood, D. L. (2013). Longitudinal data analysis. In D. L. Tood (Ed.), *Statistical analysis: Vol. 2. The Oxford handbook of quantitative methods in psychology* (pp. 378–410). Oxford: Oxford University Press.

Xiu, L. (2015). *Methods and applications of longitudinal data analysis*. London: Academic Press.

Zeger, S. L., & Liang, K. Y. (1986). Longitudinal data analysis for discrete and continuous outcomes. *Biometrics, 42*, 121–130.

Chapter 3

Mood trajectories as a basis for personalized psychiatry in young people

Klaus Oliver Schubert[a,b], Scott R. Clark[a], Linh K. Van[a], Jane L. Collinson[a] and Bernhard T. Baune[c,d,e]

[a]Discipline of Psychiatry, University of Adelaide, Adelaide, SA, Australia; [b]Northern Adelaide Local Health Network, Mental Health Service, Adelaide, SA, Australia; [c]Department of Psychiatry and Psychotherapy, University of Münster, Münster, Germany; [d]Department of Psychiatry, Melbourne Medical School, The University of Melbourne, Melbourne, VIC, Australia; [e]The Florey Institute of Neuroscience and Mental Health, The University of Melbourne, Parkville, VIC, Australia

1 Introduction

Depressive disorders, including major depressive disorder (MDD) and dysthymic disorder (DD), are a leading cause of the global burden of disease (Ferrari, Charlson, Norman, et al., 2013). While MDD is a relatively rare diagnosis in childhood, the incidence of the disorder rises rapidly during adolescence and early adulthood (Hankin, Abramson, Moffitt, et al., 1998; Lewinsohn, Hops, Roberts, et al., 1993; Lewinsohn, Rohde, & Seeley, 1998). Forty percent of MDD patients have suffered their first depressive episode by the age of 20 (Eaton, Shao, Nestadt, et al., 2008), and 50% of patients experiencing MDD before the age of 18 will have further mood episodes in adulthood (Kessler, Avenevoli, & Ries Merikangas, 2001). These numbers have provided an argument for early intervention strategies in MDD that can target symptoms early and prevent progression to a chronic illness stage (Allen, Hetrick, Simmons, et al., 2007). However, intermittent subthreshold depressive symptoms affect 10%–20% of all young people (for review see Wesselhoeft, Sorensen, Heiervang, et al., 2013), and may represent a maturational process occurring in a developmental window similar to memory and fear development (King, Pattwell, Sun, et al., 2013). Targeting evidence-based interventions to those most in need, therefore, remains a major challenge for clinicians and policy-makers (Purcell, Jorm, Hickie, et al., 2015), and there are currently no clear clinical, demographic, or biological markers that could stratify young patients with mood symptoms into prognostic subgroups. Accurate prediction tools that indicate the likelihood of progression to full-blown depressive syndrome or to chronic depressive states would represent a major advance toward personalized interventions in young people.

Longitudinal studies tracking the fluctuations of depressive symptoms in young people over time provide an avenue toward a better understanding of the biological, psychological, and social correlates of unfolding mood psychopathology. Generally referred to as "developmental trajectory research," these investigations may also help untangle the considerable heterogeneity in the course of depressive symptoms among young people (Rhebergen, Lamers, Spijker, et al., 2012). Compared with any other age, mood symptoms are most dynamic in late adolescence and early adulthood, and surge to a peak between age 15 and 17 (Kessler et al., 2001).

In this chapter, we first describe the types of analyses carried out in symptom trajectory research. Second, we describe the findings in the field of depression research in young people, for each study type, in the form of a narrative review. Third, we summarize these findings and propose implications for future research and clinical practice.

2 Three types of studies investigate depressive symptom trajectories in young people

A developmental trajectory describes the course of a behavior or state over age or time. In mental health research, examples include symptom trajectories, organic trajectories (e.g., brain development), or functional trajectories (e.g., time spent in employment).

Thematically, approaches to developmental trajectory research can be grouped as follows: (1) studies describing the number and shape of depressive symptom trajectories, and describing risk factors for specific mood symptom trajectories (Fig. 1A); (2) studies examining the contribution of specific and predefined risk factors to depressive symptom trajectories (Fig. 1B); (3) studies investigating trajectories concurrent with depressive symptoms (Fig. 1C).

Personalized Psychiatry. https://doi.org/10.1016/B978-0-12-813176-3.00003-1

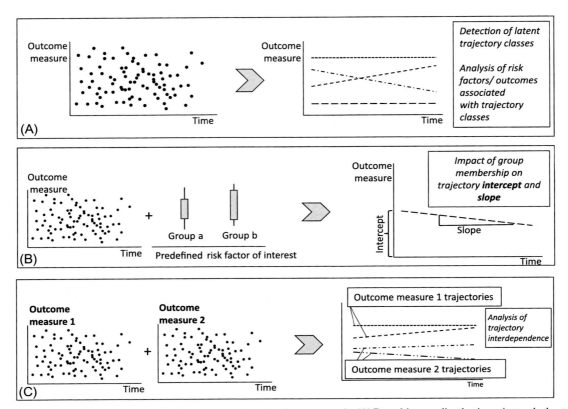

FIG. 1 Three types of studies investigate depressive symptom trajectories in young people. (A) Data-driven studies that investigate whether trajectories of the outcome variable (e.g., a measurement of mood) can be statistically grouped. In a second step, studies often aim to identify predictor variables that put people at increased risk of belonging to a particular trajectory class. (B) Hypothesis-driven studies that define experimental groups a priori, according to characteristics of interest, and depending on the original research question (e.g., female vs male). The impact of these dichotomies on the outcome variable of interest (e.g., depressive symptoms) at baseline ("intercept") and over time ("slope") is assessed. (C) Studies investigating the co-occurrence and interdependence of two outcome variables of interest over time (e.g., mood symptoms and substance use). *(Adapted with permission from Schubert, K. O., Clark, S. R., Van, L. K., et al. (2017). Depressive symptom trajectories in late adolescence and early adulthood: A systematic review. The Australian and New Zealand Journal of Psychiatry, 51, 477–499.)*

The first group of studies is data-driven and investigates whether trajectories of the outcome variable (here: a measurement of mood state) can be statistically grouped according to a "best fit" model (Fig. 1A). These trajectory groups are then assigned descriptive labels, such as an "increasing" group, a "decreasing group" or a "no change" group (Nagin, 1999). Often, predictor variables (e.g., age, gender, ethnicity) that put people at increased risk of following a particular trajectory are identified in these studies.

The second group of studies is hypothesis-driven and defines experimental groups a priori, according to risk factors of interest, and depending on the original research question (Fig. 1B). For example, predefined experimental groups may include "male" vs "female," or "Caucasian" vs "Asian." Then, the impact of these dichotomies on the outcome variable of interest (e.g., depressive symptoms) at baseline ("intercept") and over time ("slope") is assessed.

The third group of trajectory analyses investigates the co-occurrence and relationship of two outcome variables of interest over time (Fig. 1C). For example, such a study may track the joint trajectories of depressive symptoms and anxiety symptoms, and come to conclusions about their interdependence (Nagin, 2005).

In research practice, there is some overlap between the areas, because many studies investigating concurrent trajectories also provide separate information on depressive symptom trajectories alone. The following sections describe the main findings identified by these three study types.

3 The shape and the number of depressive symptom trajectories in young people

In the first group of studies (Fig. 1A), growth-modeling statistics are applied to longitudinal depressive symptoms data, for the identification of distinct trajectories or "classes" among participants over time. If many studies arrive at a statistically valid model describing distinct classes, the idea that such a differentiation could represent a true phenomenon within the examined populations becomes plausible.

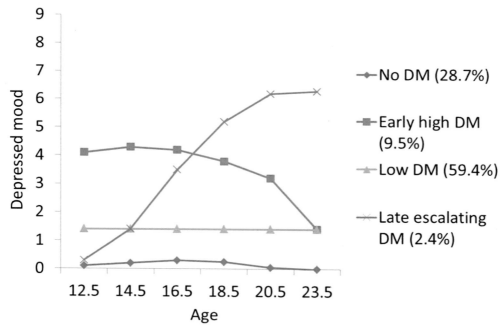

FIG. 2 Example of depressive symptom trajectories calculated by a semiparametric group-based method. The graph shows four trajectory classes of depressed mood (DM) in a sample of 11,559 American youths aged 12–25, measured by a three-item questionnaire abbreviated from CES-D. Of note is a transient increase of depressive symptoms in late adolescence, even in the "no depressive mood" class, peaking about age 16.5. Depressed mood was measured with three items indicating how often in the past week adolescents had felt sad, depressed, or could not shake off the blues (0 = never or rarely, 3 = most of the time or all of the time). Responses were summed to create a composite score for each adolescent that ranged from 0 to 9. *(Reprinted with permission from Costello, D. M., Swendsen, J., Rose, J. S., et al. (2008). Risk and protective factors associated with trajectories of depressed mood from adolescence to early adulthood. Journal of Consulting and Clinical Psychology, 76, 173–183.)*

3.1 The shape of depressive symptom trajectories in young people

Studies conducted in this field to date confirm that mood and depressive symptoms are dynamic during emerging adulthood, and that symptoms reach their peak around age 15–17 (e.g., Fig. 2; Costello et al., 2008). This is consistent with previous literature, describing that depressive symptoms increase during early to mid-adolescence, and then continuously decline in the transition from late adolescence to young adulthood (Ge, Natsuaki, & Conger, 2006; Rawana & Morgan, 2014) and further into the third decade of life (Pettit, Lewinsohn, Seeley, et al., 2010). There is evidence that females reach the depressive symptom peak slightly earlier than males (Edwards, Joinson, Dick, et al., 2014). Studies also concur that the peak occurs independent of trajectory group membership, indicating that even youths with overall low levels of depression experience an increase of symptoms at this age.

Studies differ considerably in their descriptions of the shape of identified trajectories. Whereas latent class analyses tend to find cross-diagonal (i.e., distinctive upward or downward) curves, more recent analyses seem to support a model of increasing then decreasing depression across all groups in parallel trajectories. Therefore, a risk group for a young person is likely to be determined relatively early in adolescence, whereas the course of depressive symptoms throughout adolescence is primarily a function of normal youth development (Ferro, Gorter, & Boyle, 2015b). Interestingly, studies that included a large number of young people from disadvantaged social backgrounds (e.g., ethnic minorities or socioeconomic disadvantage) were more likely to report cross-diagonal trajectory groups than studies that were more representative of the general population. One explanation for these cross-diagonal curves is that if the presence or absence of severe and enduring psychosocial stressors could "shift" people from a low-risk trajectory, they would normally follow into a higher risk group, or vice versa.

3.2 The number of depressive symptom trajectories in young people

The number of individual trajectories or "classes" identified by studies to date ranges between three and six. The majority of studies report either three or four distinct trajectories. Notably, studies that recruited younger patients from the age of 10–12 years, and followed them into late adolescence (Chaiton, Contreras, Brunet, et al., 2013; Diamantopoulou, Verhulst, &

van der Ende, 2011; Ferro et al., 2015b; Sabiston, O'Loughlin, Brunet, et al., 2013; Yaroslavsky, Pettit, Lewinsohn, et al., 2013), tend to report fewer trajectories than studies that covered the course of depressive symptoms into middle adulthood (Colman, Ploubidis, Wadsworth, et al., 2007; Costello et al., 2008; Olino, Klein, Lewinsohn, et al., 2010; Stoolmiller, Kim, & Capaldi, 2005).

Two types of depressive symptom trajectories feature in virtually all of these studies. First, there is evidence for a class of individuals that experience only minimal depression throughout adolescence and early adulthood. Authors label this class variably as representing "no depressed mood" (Costello et al., 2008), "absence of symptoms" (Colman et al., 2007), or "stable low [symptoms]" (Mezulis, Salk, Hyde, et al., 2014). In most studies, a large proportion of assessed subjects fall into this minimal depression class, typically between 40% and 55%.

Second, most studies identified a group of patients suffering from consistently high depressive symptoms throughout adolescence and early adulthood. This class is labeled as "persistent depression" (Olino et al., 2010), "high stable" (Yaroslavsky et al., 2013), "high persistent" (Stoolmiller et al., 2005), or "chronically high" (Wickrama, Wickrama, & Lott, 2009) class. Compared with the low depression class, there is more heterogeneity among studies with regard to the frequency of high persistent symptoms in the study populations, ranging between 1.7% in a sample spanning ages 13–53 years (Colman et al., 2007) and 32% in a study covering a narrower age range from 14 to 18 years (Yaroslavsky et al., 2013), with most studies reporting that about 5%–15% of participants fall into this group. It is also notable that this high persistent symptom group tends to show little variability over time in most studies (Stoolmiller et al., 2005).

Moreover, most studies come to the conclusion that there is at least one "intermediate" group of young people with less severe depressive symptoms over time. These intermediate trajectories tend to be less stable than the high and low trajectory groups, and often show considerable positive or negative slope (variation) throughout the study period. Given the considerable heterogeneity between studies in describing the trajectory of this group (e.g., stable intermediate vs increasing vs decreasing), it is possible that external environmental factors play a more significant role for patients following these intermediate trajectories.

4 Risk factors for specific mood symptom trajectory membership

Most studies reporting distinct depressive symptom trajectories over time also analyzed risk factors for falling into these groups. In the following section, the most commonly identified risk factors are discussed.

4.1 Gender

Several studies identified female gender as a risk factor for following high depression trajectories (Costello et al., 2008; Ferro et al., 2015b; Yaroslavsky et al., 2013). However, some studies have come to more complex conclusions. Mezulis et al. reported that female gender was a risk factor for a trajectory of increasing depressive symptoms between the ages of 12 and 18, but not for a trajectory of early high and then decreasing symptoms. Olino et al. found that female participants were more likely to belong to classes characterized by fluctuations in the course of depressive and anxiety disorders; however, sex differences were not observed in classes characterized by persistent depressive and anxiety disorders (Olino et al., 2010). Briere et al. reported that male gender was the most informative prognostic factor, resulting in increased odds of membership in a high-persistent depression trajectory, relative to other trajectories (Briere, Rohde, Stice, et al., 2015).

Interestingly, one study found that although more girls than boys were likely to follow high-level trajectories of depression, the adult outcome of adolescents on high-level trajectories was poorer for boys than for girls (Diamantopoulou et al., 2011). Similarly, a study by Chaiton et al. reported that following a high depression trajectory was a statistically significant independent predictor of depression, stress, and self-rated mental health in young adulthood in boys and girls. However, only boys, but not girls, in the high trajectory group had a statistically significant increase in the likelihood of seeking psychiatric care (Chaiton et al., 2013).

Therefore, while most studies concurred that females are at higher risk of following high depressive symptom trajectories in adolescence and early adulthood, it is possible that the longer-term consequences of such a trajectory are more functionally damaging and disruptive in affected males.

4.2 Physical developmental factors

In a large 1946 birth cohort with mental health data spanning early adolescence to middle adulthood, Colman et al. found that heavier babies had a lower likelihood of elevated depressive and anxious symptom trajectories. Delays in first standing and walking were associated with subsequent higher likelihood of adverse depressive and anxious symptoms (Colman et al., 2007).

4.3 Parental history of mental disorders and other parental factors

One study reported that parent's depressive symptoms significantly discriminated a "high chronic" symptom class from three lower symptom classes (Stoolmiller et al., 2005). Another study pointed at an interesting differentiation between mood and anxiety disorders, as far as trajectory risk for the offspring was concerned: children of parents with depression were more likely to follow a high depression trajectory, whereas offspring of parents with anxiety disorders tended to have a course characterized by anxiety disorders (Olino et al., 2010).

In contrast, a two-parent family structure and feelings of connectedness toward parents were identified as protective factors against depressive symptoms, increasing the likelihood of following a low-depression trajectory (Costello et al., 2008).

4.4 Interpersonal relationship factors

Two studies identified poor interpersonal relationships or poor interpersonal functioning as a risk factor for consistently elevated symptom trajectories (Ferro et al., 2015b; Yaroslavsky et al., 2013).

4.5 Physical health factors

Two studies reported that chronic health conditions (Ferro et al., 2015b) and high volatility of physical problems over time (Wickrama et al., 2009) increased the likelihood of following high depressive symptom trajectories, highlighting the potential interactions between depression trajectories and physical health outcomes in adolescents and young adults.

4.6 Socioeconomic factors

Low socioeconomic status was associated with depressed mood trajectory groups in three studies (Costello et al., 2008; Ferro et al., 2015b; Wickrama et al., 2009).

4.7 Psychosocial stressors in childhood

Multiple parental transitions, negative life events, and poor childhood academic achievements were identified as risk factors for a high chronic symptom trajectory by one study (Stoolmiller et al., 2005).

4.8 Ethnic factors

One study identified that African-American, Hispanic, Pacific Islander, and Asian-American ethnicities increased the likelihood of following high depression trajectories, compared with Caucasian Americans (Costello et al., 2008).

4.9 Psychological factors

One study identified high infant negative affectivity as a risk factor for both genders for increasing depressive symptom class throughout adolescence, whereas it was a risk factor for an early high class only for boys. For girls, a high level of rumination was an additional risk factor for the early high symptom class (Mezulis et al., 2014). Lower self-concept was a risk factor for the high symptom classes in another study (Ferro et al., 2015b). Further, negative cognitive style and a higher motivation to reduce symptoms were associated with elevated symptom trajectories (Briere et al., 2015).

4.10 Lifestyle factors

"Risky" lifestyle factors between ages 12 and 19, such as having multiple sex partners, having been arrested/having committed crime, being an excessive drinker, being a smoker, and being unmarried were all identified as risk factors for chronically high, increasing, and decreasing depressive symptoms groups, compared with a consistently low trajectory group. In the same cohort, using alcohol, tobacco, or other drugs on a weekly basis were risk factors for high depressive trajectory groups (Costello et al., 2008).

5 The contribution of specific and predefined risk factors to depressive symptom trajectories

A considerable body of research has investigated the effects of specific, predefined risk factors on the longitudinal course of depressive symptoms in adolescents and young adults, using statistical methods including growth curve modeling, growth mixture modeling, or structural equation modeling (Fig. 1B) (Nagin, 2005). These studies do not directly identify "high" or "low" symptom trajectories and their risk factors. Rather, this type of analysis generates knowledge of the effects of known risk factors of risk status (i.e., cross-sectional diagnosis) on longitudinal disease activity. The most commonly investigated factors influencing mood trajectory slope and intercept are discussed in the following section.

5.1 Gender

Several studies have explored the impact of gender on depressive symptom trajectories in young people, based on the well-documented sex differences in the incidence and prevalence of depressive disorders.

These investigations overall seem to confirm that young women and men differ considerably with regard to depressive symptom trajectories in adolescence and early adulthood. Studies suggest that adolescent girls have overall higher symptom levels (Adkins, Wang, & Elder, 2009; Ge et al., 2006) and a more pronounced increase of symptoms than boys from about age 12, peaking around age 17 (Burstein, Ginsburg, Petras, et al., 2010; Edwards et al., 2014; Garber, Keiley, & Martin, 2002; Hankin, 2009). In early adulthood, the gap between genders seems to narrow again, and there are some reports that at age 25, the overall difference may no longer be statistically significant (Galambos, Barker, & Krahn, 2006). Interestingly, females show a sharper reduction of depressive symptoms between the ages of 17 and 25 (Chen, Haas, Gillmore, et al., 2011; Galambos et al., 2006; Stapinski, Montgomery, Heron, et al., 2013). What also emerges is that well-established risk factors for depressive disorders, such as socioeconomic disadvantage and negative life events, have more profound effects on young females than on males (Adkins et al., 2009; St Clair, Goodyer, Dunn, et al., 2012), particularly when they are amplified by personal psychological traits such as negative cognitive style and a tendency to ruminate (Hankin, 2009). On the individual symptom level, females may be more vulnerable to increases in "sad mood," "sleep disturbance," and "low self-esteem/guilt," whereas males were found more susceptible to reductions over time in "concentration/decision-making" (Kouros & Garber, 2014).

One study investigated the consequences of increasing trajectories of depressive symptoms between the ages of 12 and 17 on alcohol use at age 18 (Edwards et al., 2014). For females, increasing depressive symptom trajectories were associated with increased alcohol use; in contrast, no such relationship was identified in males.

5.2 Ethnicity and minority group membership

Minority group membership in childhood and adolescence is regarded as a risk factor for many adverse mental health outcomes in adulthood, including psychosis and MDD. Several studies reviewed for this chapter set out to explore the impact of these social factors on depressive symptoms over time.

In the United States, several studies found that adolescents of African-American, Hispanic, and Asian minorities experience higher levels of depressive symptoms across early life compared with Caucasians (Adkins et al., 2009; Brown, Meadows, & Elder, 2007; Chen et al., 2011). Childhood SES and stressful life events explained much of the disparity between minority groups and Caucasians, suggesting that minorities show greater sensitivity to the effects of low childhood socioeconomic status (Adkins et al., 2009; Brown et al., 2007). Regardless of ethnic background, adolescents reporting high levels of ethnic/racial discrimination combined with poor sleep reported a corresponding increase in depressive symptoms and lower levels of self-esteem over time, whereas other "combinations" of sleep quality and ethnic/racial discrimination reported more positive adjustment over time (Yip, 2015). It is therefore possible that sleep disturbance, and perhaps other indicators of general well-being, are an important mediator of the ethnicity-depression relationship. Notably, one study reported that higher levels of *maternal support* were related to lower levels of depressive symptoms among all race-ethnic groups, suggesting that maternal support may act as a "race-ethnicity equalizer" (Brown et al., 2007).

Two studies investigated the impact of *minority sexual orientation* on depressive symptom trajectories, and reported that the rates of depressive symptoms and suicidality in early adolescence were higher among sexual minority youth than among heterosexual youth. These disparities persisted over time as participants transitioned into young adulthood, but did not increase at older ages (Needham, 2012). The observed longitudinal disparities were largest for females and for bisexually identified youth (Marshal, Dermody, Cheong, et al., 2013).

5.3 Physical health

One study found that youth with chronic illness had significantly less favorable trajectories and significantly higher proportions of clinically relevant depressive symptoms over time, compared with their peers without chronic illness (Ferro, Gorter, & Boyle, 2015a).

5.4 Genetic factors

Three studies to date have explored the impact of genetic variation on depressive symptom trajectories in young people. Taken together, these studies indicate that genetic variance could play a substantial role in determining the course of depressive symptoms in adolescence and young adulthood.

In young people aged 13–25, robust associations of genetic dopamine receptor D2 (DRD2) and dopamine receptor D4 (DRD4) variants with high depressive symptom trajectories among male adolescents from the age of 13, and among young adults of both genders between ages 24–26 were identified (Guo & Tillman, 2009). These genetic associations remained significant after controlling for a wide range of psychosocial parameters. The authors reported that the DRD2*304/178 and the DRD4*379/379 genotypes raised mean depressive symptoms by 3%–5% and 17%, respectively.

In another sample of US high school students, an association between the relatively uncommon R5 allele of DRD4 and depressive symptom trajectories was described, for both males and females (Adkins, Daw, McClay, et al., 2012). Individuals with any 5R alleles, representing 2.87% of the full sample, appeared to follow a unique trajectory with relatively low symptom levels through late adolescence, before experiencing rapid increases in early adulthood. Thus, carriers of the DRD4 5R appear to navigate their high school years with relative psychological ease compared with others, but begin to experience elevated distress as they transition into adult roles. In contrast, in males only, the 3.5R allele of the Monoamine Oxidase A Variable Number Tandem Repeat promoter was associated with higher symptom peaks in late adolescence, and sharper declines of symptom intensity in early adulthood. Therefore, males with the 3.5 genotype appeared to have a particularly stressful time during high school and the subsequent transition to adulthood, but converged with their peers in early adulthood.

A heritability analysis of adolescents between the ages of 14 and 18 found that the trajectory intercept factor (i.e., baseline stability) of depressive and anxiety symptoms is substantially more heritable than cross-sectional scores observed at any age, ranging between 72% (childhood) and 83% (adolescence) for males and 64% (childhood) and 84% (adolescence) for females (Lubke, Miller, Verhulst, et al., 2015). In comparison, cross-sectionally assessed age-specific heritabilities ranged from 43% to 54% between the ages of 7 and 18. The authors concluded that their findings support the existence of a latent genetic "intercept factor" of depressive and anxiety symptoms that is less liable to measurement error and may therefore represent a reliable phenotype for further genetic studies. Further, they concluded that three time points of measuring symptoms are sufficient to extract this highly heritable phenotype.

5.5 Psychological factors

Investigating psychosocial mechanisms that may account for sex differences in internalizing symptoms of depression and anxiety in adolescents aged 14–18, one study reported that initial levels of depressive symptoms were mediated by an interaction of rumination and stressors, as well as by an interaction of negative cognitive style and stressors (Hankin, 2009). The authors concluded that some of the well-documented sex differences in depressive symptom trajectories can be accounted for by particular cognitive vulnerabilities (particularly negative cognitive style and rumination) and stressors as risk mechanisms.

Another study investigated whether within-individual relations between depression vulnerability factors (childhood trauma, dysfunctional attitudes, maladaptive coping) and depressive symptom trajectories varied as a function of the number of prior major depressive episodes experienced in the lifetime of young people aged 18–31. The authors reported a significant interaction of "coping," and the number of previous major depressive episodes and time (Morris, Kouros, Fox, et al., 2014). Results indicated that among individuals with less adaptive coping (i.e., lower primary or lower secondary control coping scores), depressive symptoms increased significantly in relation to the number of prior depressive episodes.

6 Concurrent trajectories of depressive symptoms and other longitudinally measured factors

The third type of trajectory analysis tracks the interdependence over time of depressive symptoms with other clinical or behavioral outcome measures, such as delinquency, anxiety, or substance use (Fig. 1C). These investigations address the problem that conventional reports of associations between two problem behaviors are typically represented by summary statistics (e.g., a correlation coefficient or an odds ratio is computed at each wave of assessment), which makes little use of information about group and individual developmental courses of the two constructs over time (Nagin & Tremblay, 2001). Moreover, summary statistics are unable to distinguish subgroups within the examined cohort, and cannot reveal possible subgroup differences in the degree or developmental patterns of co-occurring problems. For example, associations may be very weak for some subgroups, but much stronger for others. As a consequence, studies on co-occurring trajectory studies provide information on the associations between levels, concurrent change, and episodic expressions of both phenomena, as well as the predictive relationships between level and growth (Fleming, Mason, Mazza, et al., 2008). Studies may yield important information for prevention programs tackling problem behaviors (e.g., crime prevention), enabling them to develop tailored strategies for identified subgroups that take co-occurring problems into account. The most commonly investigated concurrent trajectories are described in the following section.

6.1 Concurrent trajectories of mood symptoms and trauma/life events/stress

One study investigated the question of whether exposure to major life events (e.g., illness, loss, trauma) and minor hassles (e.g., money problems, social difficulties, work or study pressure) coincides with an increase of depressive symptoms as people advance from adolescence (aged 14–18) into adulthood (age 30) (Pettit et al., 2010). The authors reported strong evidence for a bidirectional relationship between trait-like trajectories of stress (hassles), and the development or maintenance of depressive symptoms over time. Conversely, there were only weak to modest bidirectional relations between major life events and depressive symptoms, when latent trajectories were taken into account. The authors therefore suggested that strategies for the prevention of depression in young adults might be most effective if they focus on coping and stress inoculation strategies for "hassle-like" events.

6.2 Concurrent trajectories of mood and delinquency/oppositional defiance

Three studies investigated the relationship between depressive symptoms and delinquent or oppositional defiant (OD) behaviors.

Two studies found that adolescents following a low-level trajectory of depression were most likely to also follow a low-level trajectory on delinquency, and vice versa; adolescents following a low-level trajectory on delinquency were most likely to also follow a low-level trajectory on depression (Diamantopoulou et al., 2011; Wiesner & Kim, 2006). Findings appear to confirm the notion that the co-occurrence of depression and delinquency does not occur for the whole population, but mainly for those who display high levels of these problem behaviors. Studies differed in their conclusions of gender impact on trajectory interplay. One study indicated that delinquent behavior was more predictive of depressive symptoms than vice versa for boys, whereas both problem behaviors were mutually predictive of each other for girls (Wiesner & Kim, 2006). In contrast, another study concluded that while high depression levels appeared to always be related with high delinquency levels in both genders, high delinquency levels were more likely to be related to depression in girls than in boys (Diamantopoulou et al., 2011).

Another study tracked co-occurring trajectories of depressive symptoms, anxiety, and OD behavior in a Canadian cohort aged 12–18 (Leadbeater, Thompson, & Gruppuso, 2012). The authors reported that expressions of anxiety, depression, and oppositional symptoms are distinctive within domains and invariant across time. Moderate correlations among adolescent levels for all symptom combinations were found, such that adolescents who started high in one domain were also consistently high in the others at each assessment point, suggesting that there is considerable consolidation of psychopathology, and that this persists across the transition to young adulthood.

6.3 Concurrent trajectories of depressive symptoms and body mass index

Two studies to date investigated the concurrent trajectories of mood body mass index (BMI). In the 1946 British Birth Cohort, Gaysina et al. reported that women with adolescent-onset elevated depressive symptoms had a lower mean BMI at age 15 years, faster rates of increase across adulthood, and higher BMI at age 53 years than those with no symptoms.

Men with adolescent-onset symptoms had a lower BMI than their low symptom peers at all ages from 15 to 53 years. The BMI trajectories of men and women with adult-onset symptoms did not differ from those with absence of symptoms at all ages (Gaysina, Hotopf, Richards, et al., 2011).

Another study of young people aged 15–27 found five profiles that characterize concurring depression and weight trajectories (Mumford, Liu, Hair, et al., 2013). African-American or Hispanic respondents had consistently greater odds of being in any class other than the normative reference class ("stable normal weight, stable good mental health"—82% of the cohort). Males had higher odds of being in the "consistently obese, stable good mental health" class (6.8% of the cohort), and lower odds of being in the "overweight becoming obese, declining mental health" class (5.6%). Respondents reporting a baseline income of less than 100% of the poverty level had higher odds of being in classes of increasing obesity risks, with variations in mental health. Overall, the findings suggest that there is a detectable subgroup of young people who are particularly vulnerable to comorbid depression and obesity, supporting the idea that a "metabolic" phenotype could exist among mood disorder patients (Mansur, Brietzke, & McIntyre, 2015).

6.4 Concurrent trajectories of depressive symptoms and substance use

Five recent studies described co-joint trajectories of depressive symptoms and substance use, and aspects of their relationship in adolescents and young adults. Overall, this literature paints a picture of a complex, rather than simplistic, relationship between both phenomena. All studies reviewed here identified subgroups of 10%–15% of youths in which consistently high levels of co-morbidity are seen (Brook, Brook, & Zhang, 2014; Fleming et al., 2008; Needham, 2007; Rodriguez, Moss, & Audrain-McGovern, 2005; Willoughby & Fortner, 2015). For these "high-risk" youths, the co-morbidity is either causally linked or co-mediated through common confounders. On the other hand, the evidence reviewed here indicates that, for the majority of affected young people, depressive symptoms and substance use develop independently of each other.

Several studies identified a surprising impact of depression scores on substance use over time, indicating that while high initial depression scores in teenagers were associated with higher initial levels of substance use, they were also associated with decreased acceleration of substance use over time (Needham, 2007; Rodriguez et al., 2005).

Complementing these findings, another study reported that levels of substance use at age 14 did not predict a change in depressive symptoms until age 18, nor did higher levels of depressive symptoms at age 14 predict greater increases in substance use. However, evidence was found for a positive association between episodic expressions of depressive symptoms and alcohol use that fell outside developmental trajectories (Fleming et al., 2008).

It is likely that the co-occurrence of depressive symptoms and substance use is influenced by other factors, such as psychological traits. In a recent study of 14–18 year-olds, high ratings on the "delay of gratification" trait increased participants' odds of belonging to a depressive symptoms and alcohol "low co-occurrence" group as compared with higher risk groups, indicating that difficulties in delaying gratification is a risk factor for both depression and substance abuse in this age group (Willoughby & Fortner, 2015). In contrast, high "novelty seeking" emerged as a risk factor of "high alcohol use only" in boys, whereas low "novelty seeking" in girls was associated with increased odds of being in the "depressive symptoms only" group. The authors concluded that programs for adolescents that focus on improving delay of gratification and self-regulation, in general, may be particularly helpful for reducing co-occurring depressive symptoms and alcohol use, but also for reducing at-risk depressive symptoms and alcohol use that occur independently of each other.

7 Summary of findings on depressive symptom trajectories in young people

This chapter described the current literature on mood symptom trajectories in young people between the ages of about 15–25. The main findings are summarized in Fig. 3.

Studies found strong evidence that depressive symptoms are not uniformly distributed across adolescent and young adult populations. Rather, young people follow distinct subgroups that are characterized by differential courses of depressive symptoms over time. The majority of young people (60%–80%) report consistently low levels of depressive symptoms, with slight normative increases from early to late adolescence, peaking at around age 17. In contrast, studies consistently point to a group representing 5%–12% of the population, who report consistently high levels of symptoms. Between these extremes, intermediate trajectory groups are described that are more variable across studies in terms of their shape and number, possibly depending on the specific characteristics of the population studied. For example, smaller studies involving minority groups have tended to identify higher numbers of distinct trajectories than larger studies more representative of the entire population (Fig. 3A).

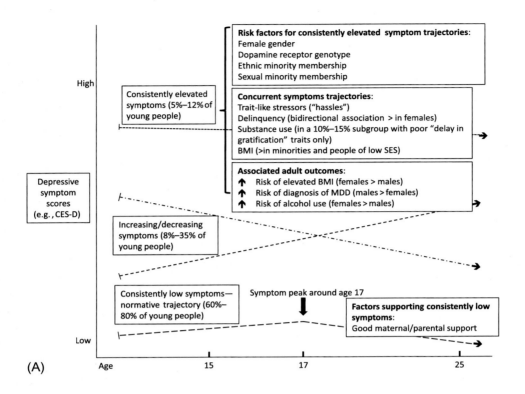

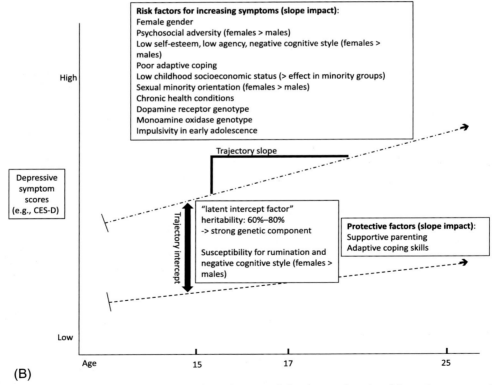

FIG. 3   Summary of mood trajectory research in young people. (A) Studies report distinct latent trajectories of depressive symptoms in young people. Frequently identified latent groups include "consistently elevated symptoms," "consistently low symptoms," and "increasing/decreasing symptoms." The total number of latent classes varies among studies. Frequently identified risk factors for the high and low symptom classes are summarized. (B) Frequently identified factors that impact on baseline symptom levels (intercept) and direction of the trajectory curve (slope) are summarized. Overall, findings support a stress-vulnerability model for depressive symptom trajectories in young people. *(Adapted with permission from Schubert, K. O., Clark, S. R., Van, L. K., et al. (2017). Depressive symptom trajectories in late adolescence and early adulthood: A systematic review. The Australian and New Zealand Journal of Psychiatry, 51, 477–499.)*

Etiologically, the literature reviewed here supports a "stress-vulnerability" model for depressive symptom trajectories in young people. Analyses of genetic heritability and specific genetic variants suggest a genetically mediated biological basis to experiencing higher than normal levels of depressive emotions over time, expressed as a "latent intercept factor" (Adkins et al., 2012; Guo & Tillman, 2009; Lubke et al., 2015) (Fig. 3B). When individuals with genetically determined vulnerability are exposed to certain internal or external environmental risk factors, their odds of experiencing chronically elevated levels of depressive affect are further enhanced. Internal risk factors include sex hormones (testosterone may have a protective effect) and certain cognitive-psychological traits (e.g., ruminations or negative cognitive style). External factors seem invariably associated with some form of chronic psychosocial stress, as encountered by people of low socio-economic status, racial or sexual minority groups, or by those reporting trait-like chronic "hassles."

A subgroup of adolescents following high depressive symptom trajectories is additionally at increased risk for problem behaviors such as substance use or delinquency (Fig. 3B). The interactions between depressive affect and these behaviors are complex, and almost certainly involve mediating psychological traits such as novelty seeking and capacity for delayed gratification. Interestingly, the bidirectional links between high depressive symptoms and problem behaviors are more pronounced in females than in males, again suggesting mediating roles of sex hormones, and possibly societal norms and expectations.

Some limited conclusions can be drawn from the reviewed literature about factors that are protective and that characterize people who follow the normative low symptom trajectory. Adequate parental support, intact family structures, advantageous socioeconomic status, and psychological traits such as adaptive coping have all been associated with the low symptom group. However, the literature reveals a bias toward exploration and description of high-risk outcomes and associated risks. A thorough examination of resilience factors is urgently required, particularly in those young people with a biological/genetic predisposition to high symptom trajectories.

8   Implications for clinical practice and research

Considering the strong evidence for distinct latent classes of depressive symptom trajectories in adolescence and young adulthood, and the equally strong evidence for a range of adverse health and behavioral outcomes for young people following consistently elevated depression trajectories, one might ask whether a shift in psychiatric nosology toward a "trajectory-centric" understanding of mood disorders might lead toward more personalized clinical approaches to this group. Current DSM criteria for MDD are based on cross-sectional assessments by clinicians combined with patients' subjective recall of mood symptoms over the recent weeks. It is clear that such an approach has limited capacity to delineate the considerable heterogeneity in severity and outcome of depressive disorders, particularly in young people, where symptoms are highly dynamic. Additionally, attempts to dissect MDD heterogeneity by symptom characteristics alone have been only moderately successful (Flint & Kendler, 2014). In this situation, a more systematic consideration of depressive symptom trajectories may offer a rewarding new avenue for clinical practice and research.

A truly "longitudinal" approach to assessment and diagnosis of mood symptoms may ultimately support a novel classification of MDD that may be better aligned with the underlying architecture of population heterogeneity, thereby representing a more valid target for causal enquiry into the roles of genes, environmental factors, and their interaction (Colman et al., 2007). Additionally, reliable assignment of patients to prognostic groups might result in markedly different treatment approaches. A patient likely to enter a trajectory of "high and persistent" symptoms might, for example, benefit from more aggressive and early pharmacological treatments, in combination with assertive psychosocial interventions. In contrast, patients on a "declining" symptom trajectory might do best if "left alone," or with more generic and nonpharmacological interventions, such as exercise or self-guided online therapy programs. Such personalized recommendations based on prognosis stand in contrast to the current paradigm in young people, where generic interventions are offered first and to all (e.g., at the school or primary care level), while escalation of treatment intensity only occurs after a "watch and wait" period, and failure to respond to low-intensity approaches (Cheung, Zuckerbrot, Jensen, et al., 2018). The latter approach inherently leads to longer periods of poorly treated depression, which in turn is associated with poorer long-term outcomes (Ghio, Gotelli, Cervetti, et al., 2015; Ghio, Gotelli, Marcenaro, et al., 2014).

Assessment of symptom severity over time on a population level seems increasingly feasible with the advent and proliferation of personal health tracking devices. The suggestion that three measurements over time could be sufficient to detect a latent heritable "intercept factor" is particularly encouraging in this context (Lubke et al., 2015). Future large-scale longitudinal studies should aim to sample a range of biological specimens and seek "deeper" phenotypical assessment of participants, to allow for a more comprehensive biopsychosocial characterization of "high" and "low" depressive symptom trajectory groups.

Further, as described by our group before, Bayesian statistics offer relatively simple paradigms that can account for biological, social, and psychological risk factors of individuals, and that can produce improved predictions of an illness trajectory a patient is likely to follow (Clark, Baune, Schubert, et al., 2016; Clark, Schubert, & Baune, 2015; Schubert,

Clark, & Baune, 2015). If these tools can be successfully developed in the area of youth mental health, they would aid services in the provision of indicated prevention programs to high-risk individuals. At the same time, better reassurance could be provided to low-risk patients and their families to avoid the medicalization of normal developmental processes.

9   Limitations

The majority of studies reviewed for this chapter have relied on self-report of depressive symptoms. This approach may be problematic in young people who may be easily influenced in their answers, or prone to under-or over-reporting. The use of clinician-rated questionnaires may have provided different results. However, given the high numbers of subjects enrolled in these studies, clinician interviews for each individual are hardly feasible.

Because these studies have investigated continuous symptom scores on mood questionnaires, it is difficult for clinicians to discern whether "high symptom trajectories" were actually associated with the presence of a categorical diagnosis of MDD according to DSM criteria.

We found only one study that makes reference to this problem. Here, 29.2% of young men following a high persistent symptom trajectory and 20% following a high decreasing trajectory had been diagnosed with MDD at age 26, compared with only 5.7% and 0% in moderate-decreasing and very low symptom classes (Stoolmiller et al., 2005). These numbers indicate that "high symptom trajectories" may increase the risk for a DSM diagnosis of MDD, but also show that the relationship between "trajectory" and "categorical" concepts of depression is complex.

Another potential caveat of the studies reviewed here is that they investigated depressive symptoms without prior regard to specific symptom clusters that could underlie subtypes of depressive disorders. The stratification of study populations into these proposed subtypes, identified in adult populations by using latent class analysis (Lamers, de Jonge, Nolen, et al., 2010; Li, Aggen, Shi, et al., 2014) or genetic linkage could therefore substantially alter the findings reported in these articles.

10   Conclusions

A considerable body of large-scale investigations exists that describes the development of depressive symptoms over time in adolescent and young adult populations. Longitudinal assessment of mood symptoms in large samples may help untangle the substantial clinical and biological heterogeneity that characterizes mood disorders. The findings of these studies have the potential to serve as valuable guidance and starting points for clinicians, researchers, and public health administrators in improving future strategies toward early intervention, indicated prevention, and personalized treatment in mood disorders in young people.

References

Adkins, D. E., Daw, J. K., McClay, J. L., et al. (2012). The influence of five monoamine genes on trajectories of depressive symptoms across adolescence and young adulthood. *Development and Psychopathology, 24*, 267–285.

Adkins, D. E., Wang, V., & Elder, G. H., Jr. (2009). Structure and stress: Trajectories of depressive symptoms across adolescence and young adulthood. *Social Forces, 88*, 31.

Allen, N. B., Hetrick, S. E., Simmons, J. G., et al. (2007). Early intervention for depressive disorders in young people: The opportunity and the (lack of) evidence. *The Medical Journal of Australia, 187*, S15–S17.

Briere, F. N., Rohde, P., Stice, E., et al. (2015). Group-based symptom trajectories in indicated prevention of adolescent depression. *Depression and Anxiety, 33*, 444–451.

Brook, D. W., Brook, J. S., & Zhang, C. (2014). Joint trajectories of smoking and depressive mood: Associations with later low perceived self-control and low well-being. *Journal of Addictive Diseases, 33*, 53–64.

Brown, J. S., Meadows, S. O., & Elder, G. H., Jr. (2007). Race-ethnic inequality and psychological distress: Depressive symptoms from adolescence to young adulthood. *Developmental Psychology, 43*, 1295–1311.

Burstein, M., Ginsburg, G. S., Petras, H., et al. (2010). Parent psychopathology and youth internalizing symptoms in an urban community: A latent growth model analysis. *Child Psychiatry and Human Development, 41*, 61–87.

Chaiton, M., Contreras, G., Brunet, J., et al. (2013). Heterogeneity of depressive symptom trajectories through adolescence: Predicting outcomes in young adulthood. *Journal of Canadian Academy of Child and Adolescent Psychiatry, 22*, 96–105.

Chen, A. C., Haas, S., Gillmore, M. R., et al. (2011). Trajectories of depressive symptoms from adolescence to young adulthood: Chinese Americans versus non-Hispanic whites. *Research in Nursing & Health, 34*, 176–191.

Cheung, A. H., Zuckerbrot, R. A., Jensen, P. S., et al. (2018). Guidelines for adolescent depression in primary care (GLAD-PC): Part II. Treatment and ongoing management. *Pediatrics, 141*(3), e20174082.

Clark, S. R., Baune, B. T., Schubert, K. O., et al. (2016). Prediction of transition from ultra-high risk to first-episode psychosis using a probabilistic model combining history, clinical assessment and fatty-acid biomarkers. *Translational Psychiatry, 6*, e897.

Clark, S. R., Schubert, K. O., & Baune, B. T. (2015). Towards indicated prevention of psychosis: Using probabilistic assessments of transition risk in psychosis prodrome. *Journal of Neural Transmission (Vienna), 122*, 155–169.

Colman, I., Ploubidis, G. B., Wadsworth, M. E., et al. (2007). A longitudinal typology of symptoms of depression and anxiety over the life course. *Biological Psychiatry, 62*, 1265–1271.

Costello, D. M., Swendsen, J., Rose, J. S., et al. (2008). Risk and protective factors associated with trajectories of depressed mood from adolescence to early adulthood. *Journal of Consulting and Clinical Psychology, 76*, 173–183.

Diamantopoulou, S., Verhulst, F. C., & van der Ende, J. (2011). Gender differences in the development and adult outcome of co-occurring depression and delinquency in adolescence. *Journal of Abnormal Psychology, 120*, 644–655.

Eaton, W. W., Shao, H., Nestadt, G., et al. (2008). Population-based study of first onset and chronicity in major depressive disorder. *Archives of General Psychiatry, 65*, 513–520.

Edwards, A. C., Joinson, C., Dick, D. M., et al. (2014). The association between depressive symptoms from early to late adolescence and later use and harmful use of alcohol. *European Child & Adolescent Psychiatry, 23*, 1219–1230.

Ferrari, A. J., Charlson, F. J., Norman, R. E., et al. (2013). Burden of depressive disorders by country, sex, age, and year: Findings from the global burden of disease study 2010. *PLoS Medicine, 10*.

Ferro, M. A., Gorter, J. W., & Boyle, M. H. (2015a). Trajectories of depressive symptoms during the transition to young adulthood: The role of chronic illness. *Journal of Affective Disorders, 174*, 594–601.

Ferro, M. A., Gorter, J. W., & Boyle, M. H. (2015b). Trajectories of depressive symptoms in Canadian emerging adults. *American Journal of Public Health, 105*, 2322–2327.

Fleming, C. B., Mason, W. A., Mazza, J. J., et al. (2008). Latent growth modeling of the relationship between depressive symptoms and substance use during adolescence. *Psychology of Addictive Behaviors, 22*, 186–197.

Flint, J., & Kendler, K. S. (2014). The genetics of major depression. *Neuron, 81*, 484–503.

Galambos, N. L., Barker, E. T., & Krahn, H. J. (2006). Depression, self-esteem, and anger in emerging adulthood: Seven-year trajectories. *Developmental Psychology, 42*, 350–365.

Garber, J., Keiley, M. K., & Martin, C. (2002). Developmental trajectories of adolescents' depressive symptoms: Predictors of change. *Journal of Consulting and Clinical Psychology, 70*, 79–95.

Gaysina, D., Hotopf, M., Richards, M., et al. (2011). Symptoms of depression and anxiety, and change in body mass index from adolescence to adulthood: Results from a British birth cohort. *Psychological Medicine, 41*, 175–184.

Ge, X., Natsuaki, M. N., & Conger, R. D. (2006). Trajectories of depressive symptoms and stressful life events among male and female adolescents in divorced and nondivorced families. *Development and Psychopathology, 18*, 253–273.

Ghio, L., Gotelli, S., Cervetti, A., et al. (2015). Duration of untreated depression influences clinical outcomes and disability. *Journal of Affective Disorders, 175*, 224–228.

Ghio, L., Gotelli, S., Marcenaro, M., et al. (2014). Duration of untreated illness and outcomes in unipolar depression: A systematic review and meta-analysis. *Journal of Affective Disorders, 152-154*, 45–51.

Guo, G., & Tillman, K. H. (2009). Trajectories of depressive symptoms, dopamine D2 and D4 receptors, family socioeconomic status and social support in adolescence and young adulthood. *Psychiatric Genetics, 19*, 14–26.

Hankin, B. L. (2009). Development of sex differences in depressive and co-occurring anxious symptoms during adolescence: Descriptive trajectories and potential explanations in a multiwave prospective study. *Journal of Clinical Child and Adolescent Psychology, 38*, 460–472.

Hankin, B. L., Abramson, L. Y., Moffitt, T. E., et al. (1998). Development of depression from preadolescence to young adulthood: Emerging gender differences in a 10-year longitudinal study. *Journal of Abnormal Psychology, 107*, 128–140.

Kessler, R. C., Avenevoli, S., & Ries Merikangas, K. (2001). Mood disorders in children and adolescents: An epidemiologic perspective. *Biological Psychiatry, 49*, 1002–1014.

King, E. C., Pattwell, S. S., Sun, A., et al. (2013). Nonlinear developmental trajectory of fear learning and memory. *Annals of the New York Academy of Sciences, 1304*, 62–69.

Kouros, C. D., & Garber, J. (2014). Trajectories of individual depressive symptoms in adolescents: Gender and family relationships as predictors. *Developmental Psychology, 50*, 2633–2643.

Lamers, F., de Jonge, P., Nolen, W. A., et al. (2010). Identifying depressive subtypes in a large cohort study: Results from the Netherlands Study of Depression and Anxiety (NESDA). *The Journal of Clinical Psychiatry, 71*, 1582–1589.

Leadbeater, B., Thompson, K., & Grupuso, V. (2012). Co-occurring trajectories of symptoms of anxiety, depression, and oppositional defiance from adolescence to young adulthood. *Journal of Clinical Child and Adolescent Psychology, 41*, 719–730.

Lewinsohn, P. M., Hops, H., Roberts, R. E., et al. (1993). Adolescent psychopathology: I. Prevalence and incidence of depression and other DSM-III-R disorders in high school students. *Journal of Abnormal Psychology, 102*, 133–144.

Lewinsohn, P. M., Rohde, P., & Seeley, J. R. (1998). Major depressive disorder in older adolescents: Prevalence, risk factors, and clinical implications. *Clinical Psychology Review, 18*, 765–794.

Li, Y., Aggen, S., Shi, S., et al. (2014). Subtypes of major depression: Latent class analysis in depressed Han Chinese women. *Psychological Medicine, 44*, 3275–3288.

Lubke, G. H., Miller, P. J., Verhulst, B., et al. (2015). A powerful phenotype for gene-finding studies derived from trajectory analyses of symptoms of anxiety and depression between age seven and 18. *American Journal of Medical Genetics Part B, Neuropsychiatric Genetics, 171*, 948–957.

Mansur, R. B., Brietzke, E., & McIntyre, R. S. (2015). Is there a "metabolic-mood syndrome"? A review of the relationship between obesity and mood disorders. *Neuroscience and Biobehavioral Reviews, 52*, 89–104.

Marshal, M. P., Dermody, S. S., Cheong, J., et al. (2013). Trajectories of depressive symptoms and suicidality among heterosexual and sexual minority youth. *Journal of Youth and Adolescence, 42*, 1243–1256.

Mezulis, A., Salk, R. H., Hyde, J. S., et al. (2014). Affective, biological, and cognitive predictors of depressive symptom trajectories in adolescence. *Journal of Abnormal Child Psychology, 42*, 539–550.

Morris, M. C., Kouros, C. D., Fox, K. R., et al. (2014). Interactive models of depression vulnerability: The role of childhood trauma, dysfunctional attitudes, and coping. *The British Journal of Clinical Psychology, 53*, 245–263.

Mumford, E. A., Liu, W., Hair, E. C., et al. (2013). Concurrent trajectories of BMI and mental health patterns in emerging adulthood. *Social Science & Medicine, 98*, 1–7.

Nagin, D. S. (1999). Analyzing developmental trajectories: A semiparametric, group-based approach. *Psychological Methods, 4*, 139–157.

Nagin, D. S. (2005). *Group-based modelling of development*. Cambridge: Harvard University Press.

Nagin, D. S., & Tremblay, R. E. (2001). Analyzing developmental trajectories of distinct but related behaviors: A group-based method. *Psychological Methods, 6*, 18–34.

Needham, B. L. (2007). Gender differences in trajectories of depressive symptomatology and substance use during the transition from adolescence to young adulthood. *Social Science & Medicine, 65*, 1166–1179.

Needham, B. L. (2012). Sexual attraction and trajectories of mental health and substance use during the transition from adolescence to adulthood. *Journal of Youth and Adolescence, 41*, 179–190.

Olino, T. M., Klein, D. N., Lewinsohn, P. M., et al. (2010). Latent trajectory classes of depressive and anxiety disorders from adolescence to adulthood: Descriptions of classes and associations with risk factors. *Comprehensive Psychiatry, 51*, 224–235.

Pettit, J. W., Lewinsohn, P. M., Seeley, J. R., et al. (2010). Developmental relations between depressive symptoms, minor hassles, and major events from adolescence through age 30 years. *Journal of Abnormal Psychology, 119*, 811–824.

Purcell, R., Jorm, A. F., Hickie, I. B., et al. (2015). Demographic and clinical characteristics of young people seeking help at youth mental health services: Baseline findings of the Transitions Study. *Early Intervention in Psychiatry, 9*, 487–497.

Rawana, J. S., & Morgan, A. S. (2014). Trajectories of depressive symptoms from adolescence to young adulthood: The role of self-esteem and body-related predictors. *Journal of Youth and Adolescence, 43*, 597–611.

Rhebergen, D., Lamers, F., Spijker, J., et al. (2012). Course trajectories of unipolar depressive disorders identified by latent class growth analysis. *Psychological Medicine, 42*, 1383–1396.

Rodriguez, D., Moss, H. B., & Audrain-McGovern, J. (2005). Developmental heterogeneity in adolescent depressive symptoms: Associations with smoking behavior. *Psychosomatic Medicine, 67*, 200–210.

Sabiston, C. M., O'Loughlin, E., Brunet, J., et al. (2013). Linking depression symptom trajectories in adolescence to physical activity and team sports participation in young adults. *Preventive Medicine, 56*, 95–98.

Schubert, K. O., Clark, S. R., & Baune, B. T. (2015). The use of clinical and biological characteristics to predict outcome following First Episode Psychosis. *The Australian and New Zealand Journal of Psychiatry, 49*, 24–35.

St Clair, M. C., Goodyer, I. M., Dunn, V., et al. (2012). Depressive symptoms during adolescence: Comparison between epidemiological and high risk sampling. *Social Psychiatry and Psychiatric Epidemiology, 47*, 1333–1341.

Stapinski, L. A., Montgomery, A. A., Heron, J., et al. (2013). Depression symptom trajectories and associated risk factors among adolescents in Chile. *PLoS One, 8*.

Stoolmiller, M., Kim, H. K., & Capaldi, D. M. (2005). The course of depressive symptoms in men from early adolescence to young adulthood: Identifying latent trajectories and early predictors. *Journal of Abnormal Psychology, 114*, 331–345.

Wesselhoeft, R., Sorensen, M. J., Heiervang, E. R., et al. (2013). Subthreshold depression in children and adolescents—A systematic review. *Journal of Affective Disorders, 151*, 7–22.

Wickrama, K. A., Wickrama, T., & Lott, R. (2009). Heterogeneity in youth depressive symptom trajectories: Social stratification and implications for young adult physical health. *The Journal of Adolescent Health, 45*, 335–343.

Wiesner, M., & Kim, H. K. (2006). Co-occurring delinquency and depressive symptoms of adolescent boys and girls: A dual trajectory modeling approach. *Developmental Psychology, 42*, 1220–1235.

Willoughby, T., & Fortner, A. (2015). At-risk depressive symptoms and alcohol use trajectories in adolescence: A person-centred analysis of co-occurrence. *Journal of Youth and Adolescence, 44*, 793–805.

Yaroslavsky, I., Pettit, J. W., Lewinsohn, P. M., et al. (2013). Heterogeneous trajectories of depressive symptoms: Adolescent predictors and adult outcomes. *Journal of Affective Disorders, 148*, 391–399.

Yip, T. (2015). The effects of ethnic/racial discrimination and sleep quality on depressive symptoms and self-esteem trajectories among diverse adolescents. *Journal of Youth and Adolescence, 44*, 419–430.

Further reading

Schubert, K. O., Clark, S. R., Van, L. K., et al. (2017). Depressive symptom trajectories in late adolescence and early adulthood: A systematic review. *The Australian and New Zealand Journal of Psychiatry, 51*, 477–499.

Chapter 4

Transdiagnostic early intervention, prevention, and prediction in psychiatry

Cristina Mei[a,b], Barnaby Nelson[a,b], Jessica Hartmann[a,b], Rachael Spooner[a,b] and Patrick D. McGorry[a,b]

[a]Orygen, The National Centre of Excellence in Youth Mental Health, Parkville, VIC, Australia, [b]Centre for Youth Mental Health, The University of Melbourne, Parkville, VIC, Australia

1 Introduction

Mental disorders represent a major global health concern and challenge, reflecting their potentially severe, persistent, and recurrent nature, which can lead to substantial disability and burden. These impacts, including lost productivity at an individual and broader societal level, are often long-lasting (Gibb, Fergusson, & Horwood, 2010). This is largely due to the emergence of most mental disorders (75%) between childhood and young adulthood (Kessler et al., 2005), a period that underpins adult achievement across a range of life domains, including social, occupation, and education. This critical life stage for mental health also represents a key opportunity to predict and possibly alter the potentially deleterious trajectory of mental disorders (McGorry, 2011). Traditional models for prediction and care have been built upon the conceptualization of mental disorders as distinct diagnostic categories that follow a predictable and fixed pathway from subthreshold symptoms to a first episode of clear illness. This narrow approach is now understood to misrepresent the epidemiology, natural history, and dimensionality of major mental disorders, which are characterized by shared etiological mechanisms, and overlapping and undifferentiated early symptomatology that can evolve in a number of different patterns with substantial comorbidity (Anttila et al., 2018; Kim-Cohen et al., 2003; Merikangas et al., 2010). This has meant that a new transdiagnostic approach for both clinical care and research is needed that acknowledges the fluidity and heterogeneity of emerging mental disorders (McGorry & Nelson, 2016). In this chapter, transdiagnostic approaches for early intervention, prevention, and prediction of mental disorders are discussed with reference to clinical staging, dynamic modeling, and broad youth service delivery models such as *headspace*.

2 Emergence and progression of mental disorders

The rationale for a transdiagnostic approach in psychiatry is best exemplified through the emergence and evolution of mental illness. Symptoms of major psychiatric disorders begin to emerge prior to formal diagnostic criteria being met (e.g., Diagnostic and Statistical Manual of Mental Disorders or International Classification of Diseases). While these symptoms may be transient in some cases, they can represent an increased risk for a potentially serious mental disorder and can, in some cases, be seen as a disorder in their own right (Fusar-Poli et al., 2015). This stage, seen as subthreshold in relation to the major traditional syndromes, is characterized by a fluctuating mixture of clinical features that can include depression, anxiety, and other nonspecific symptoms, such as sleep disturbance, withdrawal, and apathy (Hartmann, Nelson, Ratheesh, Treen, & McGorry, 2019; Harvey, Murray, Chandler, & Soehner, 2011). These symptoms can either recede or evolve in several, sometimes unpredictable, directions (Kim-Cohen et al., 2003; Shankman et al., 2009), often overlapping traditional diagnostic boundaries, before meeting clinical criteria for a specific (although often comorbid) psychiatric diagnosis. This fluidity and heterogeneity results in a complex and unclear progression to a full-threshold disorder (van Os, 2013), underscoring the need for a transdiagnostic approach that transcends traditional diagnostic silos, and can appropriately inform early intervention and prevention strategies (Hartmann, Nelson, Ratheesh, et al., 2019; McGorry & Nelson, 2016).

Despite not meeting the intensity or severity for a full-threshold disorder of the traditional type, early symptoms of mental illness are often associated with significant functional impairment, risk of self-harm, suicidal thoughts and behaviors, and substance abuse (Fusar-Poli et al., 2015; Rickwood et al., 2015; Rickwood, Telford, Parker, Tanti, &

Personalized Psychiatry. https://doi.org/10.1016/B978-0-12-813176-3.00004-3

McGorry, 2014; Scott, Hermens, Glozier, et al., 2012; Scott, Hermens, Naismith, et al., 2012). The distress and functional impairment commonly experienced at this early phase, which can persist over time (Addington et al., 2011), warrants a genuine need for care (Hamilton, Naismith, Scott, Purcell, & Hickie, 2011; Rickwood et al., 2014). However, current mental health service structures marginalize young people, often excluding their access to care, largely due to the adoption of service entry criteria that favor established mental disorders, as opposed to emerging and transdiagnostic phenotypes that are typically seen in young individuals (see Section 5 of this chapter for further discussion).

A key factor perpetuating this state of affairs is that current psychiatric diagnostic systems fail to capture the fluidity of symptom sets and functional impairment associated with common mental disorders, and the tendency for these to recur, persist, or progress to more severe profiles. The traditional categorical and static approach to diagnosis lacks utility (Kendell & Jablensky, 2003), and fails to differentiate early vs late stage clinical symptomatology. These traditional systems are grounded on cross-sectional symptoms (instead of longitudinal syndromal trajectories) that are typically reported by individuals with well-established and longstanding mental disorders (McGorry et al., 2007). As a result, they conflate early and late stage clinical features, and offer limited value for early intervention, prevention, and prediction efforts, particularly in view of the pluripotent trajectories of mental disorders (i.e., the potential of early clinical symptoms to evolve into a range of different syndromes). This is especially relevant to the life stage of adolescence and young adulthood, in which the onset of mental illness peaks (Kessler et al., 2005).

3 Transdiagnostic approaches to diagnosis and classification

A new diagnostic approach is required to improve the utility and validity of psychiatric diagnosis, as well as appropriately guide intervention, prevention, and prediction approaches. To address the limitations of discrete diagnostic categories, dimensional approaches have been proposed, including clinical staging (McGorry, Hickie, Yung, Pantelis, & Jackson, 2006), the Research Domain Criteria (RDoC) initiative (Cuthbert & Insel, 2013), and the Hierarchical Taxonomy of Psychopathology (HiTOP) project (Kotov et al., 2017). RDoC attempts to provide a neuroscience-based research framework for psychiatric classification that currently de-emphasizes the role of clinical-based research in guiding diagnosis. HiTOP proposes a hierarchical model that is driven by a quantitative approach to nosology. Although this approach is supported by the dimensional and hierarchical structure of psychopathology (Krueger et al., 2018), it has been contended that categories are intrinsically linked to clinical practice (Kendler, 2018). In line with this view, the transferability of RDoC and HiTOP to clinical practice has been questioned, given their limited capacity to guide clinical decision-making, which is inherently binary or categorical based (McGorry, Hartmann, Spooner, & Nelson, 2018). In comparison with these approaches, staging offers a clinically focused framework that couples each illness stage to intervention and preemptive strategies. This clinical utility is a key advantage of clinical staging, and reflects its dimensional approach while retaining a categorical structure that can guide treatment decisions (McGorry, 2013). In view of this, this chapter focuses on clinical staging.

3.1 Clinical staging

The clinical staging model in psychiatry, based on those successfully applied within the medical field (e.g., oncology), conceptualizes the transdiagnostic trajectories of mental disorders (McGorry et al., 2006). Within the clinical staging model, individuals are placed on a continuum of illness ranging from asymptomatic (stage 0), nonspecific (stage 1a), and attenuated mental syndromes (stage 1b), full-threshold disorder (stage 2), recurrence and persistence (stage 3), and severe, chronic mental illness (stage 4) (see Fig. 1). Compared with traditional diagnostic systems, the model differentiates early and mild clinical profiles (stages 1a and 1b) from those that are more severe and established (stages 2+), thus improving scope to reorient early intervention. This differentiation acknowledges the need for care prior to a first episode, and enables the selection of treatments that are more effective, simpler, and safer for earlier stages (Cross, Hermens, & Hickie, 2016). Interventions of greater intensity are typically implemented during later stages, where the phenotype is characterized by increasing symptom specificity, severity, persistence, and disability (Hamilton et al., 2011; Purcell et al., 2015; Scott, Hermens, Glozier, et al., 2012). However, the model does not assume an inevitable illness progression (Hickie et al., 2013); recovery is attainable at any stage, although the likelihood of complete recovery reduces with advancing illness (McGorry, Keshavan, et al., 2014). A key advantage of the model is that it moves beyond traditional fixed categories to a stage-based continuum that can appropriately guide treatment planning and prognosis. Notably, it provides a prevention-orientated framework to assist the implementation of interventions that aim to target the progression of presenting psychopathology to more advanced stages.

There has been ongoing debate regarding the utility of a universal clinical staging model (as proposed here) vs disorder-specific frameworks (i.e., separate models for bipolar disorder, schizophrenia, etc.) (Hartmann, Nelson,

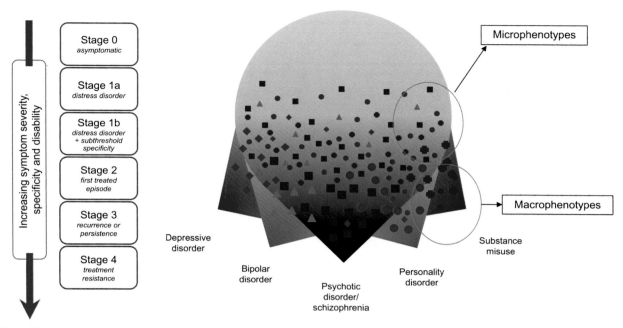

FIG. 1 Clinical staging model depicting the progression of mental illness. Early clinical phenotypes are broad and nonspecific, with clearer syndromes emerging in more advanced stages, coinciding with increased disability. *(Reproduced from McGorry, P. D., Hartmann, J. A., Spooner, R., & Nelson, B. (2018). Beyond the "at risk mental state" concept: Transitioning to transdiagnostic psychiatry. World Psychiatry, 17(2), 133–142 with permission from John Wiley and Sons.)*

Ratheesh, et al., 2019). A transdiagnostic clinical staging model advocates a "lumping" approach largely because it is most useful for providing health care to young people early in the course of their disorders when undifferentiated clinical phenotypes are the typical presentations. As illness advances over time, and as typically seen in mid-life, symptoms tend to coalesce into disorders that can progressively be split from each other, subtyped or stratified into more conventional diagnostic categories (although often presenting comorbidly). Some authors have argued that not all common mood or psychotic disorders are amenable to a broad staging approach, given their heterogeneous pattern of progression (Duffy, Malhi, & Grof, 2017), with these proponents preferring to promote more conventional single-disorder staging models (Cosci & Fava, 2013; Duffy, 2014). However, these single-disorder models will likely fail to include early presentations that are expressed by mixed and fluid symptomatology, overlap discrete syndromal boundaries, or do not meet criteria for an established, clinically diagnosable disorder. Further, inclusion of earlier broader phenotypes is more likely to unravel the complex pathophysiology (i.e., the shared and unique pathways) of major mental disorders, thus increasing capacity to prevent a range of disorders (McGorry, 2010).

4 Prediction of mental disorder onset and progression

A crucial requirement of prevention efforts is being able to accurately predict the trajectory of mental disorder, particularly, which cases will develop into more severe stages of mental disorder. Despite extensive attempts, silo-based diagnostic approaches have not substantially facilitated research into the mechanisms, or objective markers, underpinning the onset and longitudinal course of common mental disorders. Rather, this research has emphasized the shared neurobiological, genetic, and cognitive markers of major psychiatric disorders (Anttila et al., 2018; Caspi & Moffitt, 2018; Goodkind et al., 2015; Henriksen, Nordgaard, & Jansson, 2017; Lee et al., 2015; McTeague et al., 2017). Over recent years, considerable efforts have been made to predict the onset of mental disorders in high-risk populations (Hartmann, Nelson, Ratheesh, et al., 2019). This originated within the field of psychosis through attempts to predict the emergence of a psychotic disorder based on ultra-high risk (UHR) criteria (Fusar-Poli et al., 2013; Yung et al., 1996, 2003). Outcomes from this approach have, however, identified the need to refine prediction strategies. First, transition rates to psychosis have been declining (Nelson et al., 2016; Yung et al., 2007), reducing the statistical power needed to predict low incidence disorders (Cuijpers, 2003). Second, those at UHR for psychosis often transition to or have persistent nonpsychotic disorders (Lim et al., 2015; Lin et al., 2015; Wigman et al., 2012), and conversely, frank psychosis may evolve via other pathways, such as nonpsychotic at-risk states (Lee, Lee, Kim, Choe, & Kwon, 2018; Shah et al., 2017). Together, these outcomes emphasize the need for a broad-based prediction approach, allowing for a wider range of subthreshold symptoms and full-threshold disorders, as opposed to at-risk criteria and outcomes being limited to a single disorder (Cuijpers, 2003; McGorry et al., 2018). At present, transdiagnostic criteria, based on the UHR model and consistent with the pluripotent pathways to mental disorders, are undergoing validation to identify help-seeking young people who are at risk of developing a range of mental disorders (Hartmann, Nelson, Spooner, et al., 2019). This broad approach poses considerable advantages, including increased statistical power to accurately predict outcomes and improved capacity for prevention that targets populations who present with a broad range of subthreshold symptoms.

This pluripotent risk identification approach also aligns with the clinical staging model, which provides a transdiagnostic framework for both prediction of mental disorder onset and progression to a range of established syndromes. The model is guided by the notion that a range of risk factors, including biomarkers, underlie mental disorder onset, persistence, and progression, particularly in the context of youth mental health (McGorry, Keshavan, et al., 2014). These risk factors, especially those that are modifiable, can then become primary targets to prevent or delay the onset and progression of mental disorders. A number of clinical variables have been identified in transdiagnostic samples that are associated with progression from attenuated syndromes (stage 1b) to a first episode (stage 2), including not being in education, employment, or training (NEET) status, and increased negative symptoms at baseline (Cross, Scott, & Hickie, 2017). Additionally, a range of neurobiological markers have been linked to stage of illness, including neuroimaging, neurocognitive, and circadian changes (Hermens et al., 2013; Lagopoulos, Hermens, Naismith, Scott, & Hickie, 2012; Naismith et al., 2012). Overall, findings from these studies are consistent with illness progression proposed by the clinical staging model in that discrete and established disorders were associated with more severe neurobiological and neurocognitive markers (e.g., decreased gray matter volume, delayed sleep phase, and reduced neuropsychological performance) compared with early attenuated syndromes (Hermens et al., 2013; Lagopoulos et al., 2012; Naismith et al., 2012). These findings support the relevance of clinical staging to prognosis and early intervention.

4.1 Dynamic prediction techniques

Dynamic prediction approaches, based on cross-disciplinary methods, have been proposed that are consistent with the fluid, fluctuating, and transdiagnostic emergence of psychiatric symptomatology (Nelson, McGorry, Wichers, Wigman, & Hartmann, 2017). These approaches, including network theory and dynamical systems theory, as well as the statistical technique of joint modeling, offer promising methods to enhance prediction of mental disorder onset, given their suitability for capturing the time-varying and reactive quality of psychiatric symptoms (McGorry et al., 2018; Nelson et al., 2017).

4.1.1 Network theory

Network theory attempts to move beyond approaches that are based on identifying a common cause underlying symptoms of mental disorder (Borsboom, 2017), which have been largely unsuccessful (Kendler, Zachar, & Craver, 2011). Instead, it adopts a more dynamic perspective by positing that psychiatric symptoms are the result of causal interactions among each other (Borsboom & Cramer, 2013). From this perspective, symptoms do not arise from a common latent cause or mental disorder (e.g., schizophrenia), but rather through the dynamic and reciprocal influence of symptoms on each other (e.g., auditory hallucinations may evoke paranoia, which may, in turn, exacerbate the auditory hallucinations, rather than both symptoms being caused by an underlying disorder, such as schizophrenia). If symptoms engage in patterns of mutual reinforcement and feedback loops, the system as a whole may become trapped or locked in a state of extended symptom activation, a point at which a mental disorder may be diagnosed. Across the evolution of mental illness, the interactions between symptoms ("nodes") in networks may graduate from one that is less active to one that is strongly connected and entrenched (Borsboom, 2017). This has implications for prevention, because targeting early symptoms that are central to networks with a particular outcome (e.g., depression) may avert progression to a full episode (McNally, 2016). This approach has shown potential, with preliminary evidence indicating that symptom networks of healthy people show early warning signals prior to shifting to a disordered state (Fried et al., 2017). Patients whose symptom networks display these early warning signals can be identified and provided with targeted preventative treatment. This has implications across diagnostic categories, with emerging evidence indicating that network theory is apt to model psychopathology transdiagnostically (Wigman, de Vos, Wichers, van Os, & Bartels-Velthuis, 2017). Additionally, its integration with clinical staging may further refine the emergence and progression of symptoms (Wigman et al., 2013), enhancing opportunities for personalized interventions.

4.1.2 Dynamical systems theory

A related approach that has recently been applied in the field of psychiatry is dynamical systems theory (Nelson et al., 2017; Scheffer, 2009; Scheffer et al., 2012). Within this approach, mental health/disorder is conceptualized as a system in which changes or transitions over time can occur based on the interaction between the structure of a system and the perturbations ("stressors") to that system (Scheffer et al., 2012). In systems where constituent elements are heterogeneous and loosely connected, change may occur gradually (Fig. 2A). However, in systems with highly interconnected elements, change may initially be resisted until a tipping point or critical threshold for change is reached (e.g., a major life stressor), resulting in abrupt change (e.g., onset of a mental disorder) (Scheffer et al., 2012) (Fig. 2B). In some instances, early warning signs may foreshadow this tipping point (Fig. 2C). An example of this is the phenomenon of "critical slowing down." This refers to a system taking an increasingly long period of time to recover or return to a state of equilibrium in response to stressors the closer it is to a critical threshold (Scheffer et al., 2009). There is emerging proof of concept evidence for dynamic systems theory being applicable to mental disorder. For example, the phenomenon of critical slowing down as an early warning sign of illness onset/relapse has been observed in mood disorders (Bayani, Hadaeghi, Jafari, & Murray, 2017; van de Leemput et al., 2014; Wichers & Groot, 2016). This suggests the potential for this approach to be used to anticipate clinically relevant shifts in mental states to guide the implementation of preventive interventions. There is a need to now apply this approach in a variety of diagnostic samples, and in the pluripotent early stages, prior to a first episode of illness.

4.1.3 Joint modeling

Another promising prediction approach is joint modeling, a statistical technique that examines the association between variables assessed repeatedly over time and time-dependent outcomes (Yuen, Mackinnon, & Nelson, 2018). This approach has been found to be of utility in other areas of medicine. For example, Brankovic et al. (2018) showed that evolution of renal markers dynamically predicted clinical outcomes in patients with chronic heart failure. Cao et al. (2018) showed that the longitudinal values of a cancer antigen could credibly predict overall survival of ovarian cancer patients. Andrinopoulou et al. (2015) found that an increasing trend of the biomarker brain natriuretic peptide was a significant predictor of death in patients suffering from severe aortic stenosis, and showed that it is possible to construct individualized dynamic prediction models as evidence-based tools for use in medical practice. Our group has investigated the performance of joint modeling and dynamic prediction in the context of transition to psychosis in UHR individuals (Yuen & Mackinnon, 2016; Yuen, Mackinnon, Hartmann, et al., 2018; Yuen, Nelson, et al., 2018). We have shown that compared with traditional prediction models that rely exclusively on baseline data, joint modeling offers a superior approach in predicting psychosis in UHR individuals (Yuen, Mackinnon, Hartmann, et al., 2018).

With their emphasis on the evolving and potentially fluctuating pathways to disorder, the preceding models provide encouraging avenues for strengthening the prediction of mental disorder onset, particularly within a transdiagnostic perspective, while also still being applicable to a single-disorder approach (McGorry et al., 2018; Nelson et al., 2017). Other dynamic prediction approaches not outlined herein, including chaos and catastrophe theories, can also be applied to study the transdiagnostic emergence of symptoms (Nelson et al., 2017). A key method of the models described here is the repeated measurement of key variables longitudinally ("time-dependent predictors" or time series data), which may aid the identification of critical periods preceding transitions in mental state (Nelson, Hartmann, & Parnas, 2018). Thus, these models can potentially overcome the shortfalls of traditional static approaches, and enable preventive interventions to be adapted and personalized to individual clinical profiles (Wichers & Groot, 2016; Yuen, Mackinnon, Hartmann, et al., 2018). At a prediction level, these models can inform the development of a dynamic risk calculator that can be continuously updated based on repeated time-to-event data (Nelson et al., 2017), as opposed to the current method, that relies on static or single baseline data (Cannon et al., 2016; Fusar-Poli et al., 2017).

5 Transdiagnostic early intervention and clinical services

The pluripotent nature of mental disorders also raises possibilities for reforming early intervention, particularly in terms of implementing broad-based, transdiagnostic services with a preventive focus. Long-standing mental health service structures, where entry criteria for specialized care rely on meeting specific diagnostic or severity criteria, often deny access to individuals with early subthreshold symptoms that are characteristically mild, undifferentiated, and frequently changing. This, essentially, has had the greatest impact on young people, particularly those who are transitioning to an adult mental health service, or are attempting to access care for the first time, as they face potential exclusion from adult services due to their unstable and evolving symptoms. The paradox of this is that treatment outcomes are differential for individuals at early and late stages of mental disorders, with more positive outcomes attainable for the former (Berk et al., 2011; Hegelstad et al., 2012; Scott et al., 2006). In addition to the weaknesses of diagnostic and severity-based criteria, specialist services

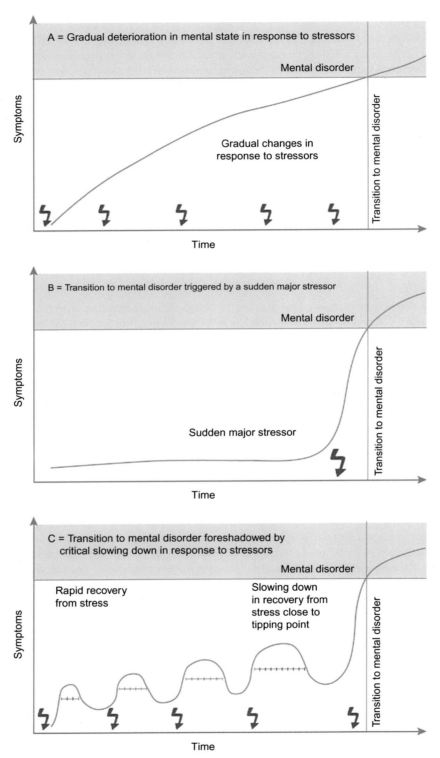

FIG. 2 Symptom progression in the onset of mental disorders as reflected by dynamical systems theory. Stressors are represented by the *red arrows*. *(Reproduced from Nelson, B., McGorry, P. D., Wichers, M., Wigman, J. T. W., & Hartmann, J. A. (2017). Moving from static to dynamic models of the onset of mental disorder a review. JAMA Psychiatry, 74(5), 528–534 with permission from the American Medical Association.)*

that target specific high-risk populations may also carry their own limitations. For example, UHR for psychosis clinics may not capture all of the varied pathways to psychosis, such as those emerging from nonpsychotic states (Lee et al., 2018).

A potential solution is the embedding of transdiagnostic models within stigma-free, youth-friendly services, such as *headspace*. Here, the focus would be on early intervention and prevention that is inclusive of the broad range of emerging mental health symptoms that can diverge into a range of different syndromes. *Headspace*, an enhanced primary care model for youth mental health, targets a range of sub- and full-threshold mental disorders between the ages of 12–25 years across four core areas: mental health, physical health, vocational and educational support, and drug and alcohol services (Rickwood et al., 2014). *Headspace* provides a broad-based entry point to care for young people that is backed by specialist services for those with more severe or complex presentations (McGorry, Goldstone, Parker, Rickwood, & Hickie, 2014). The services provided at *headspace* are integrated and underpinned by youth participation, ensuring a holistic approach to care that is delivered by a multidisciplinary team of clinical and nonclinical professionals. This approach creates a highly accessible and engaging service for young people that results in improved mental health outcomes (Hetrick et al., 2017; Rickwood et al., 2015).

A core component of the *headspace* model is ensuring that the delivery of evidence-based treatment is preemptive and is matched to the patient's need for care, rather than arbitrary diagnostic thresholds. This is best provided through a stepped care approach that is guided by risk-benefit considerations and shared decision-making, with a key target being improved functional outcomes (McGorry, Goldstone, et al., 2014). Within this approach, care should be targeted toward those at greatest risk of mental disorders (i.e., help-seeking young people), and should be suitable to the earliest stages of mental illness, where nonspecific symptoms are common, thus preventing further illness progression, and the onset of a full-threshold disorder.

5.1 Transdiagnostic, stage-based care

Such a stepwise approach to treatment is represented through a transdiagnostic clinical staging model, which recognizes the need for care well before a distinct diagnosis emerges. Staging is sensitive to risk-benefit considerations, and facilitates the selection of interventions, with a preemptive and preventive focus to treatment (Cross et al., 2016; McGorry et al., 2007). Specifically, earlier stages are initially treated via simple and benign approaches (e.g., brief psychosocial approaches, education, and support), with more intensive and drug-based interventions employed for later stages, or if first-line treatments are ineffective. This stage-based model is distinct to stepped care approaches based on severity (Australian Government Department of Health, 2015). The latter approach offers a more reactive strategy to symptom reduction, in which treatment failure at each step permits progression to the following step that involves a higher intensity of intervention. Staged-based stepped care, guided by a transdiagnostic clinical staging model, has been proposed as a more suitable model due to its efficiency, appropriate allocation of resources, and preemptive focus (Cross & Hickie, 2017). That is, the treatment provided is proportional to the presenting clinical need, as well as the risk of progression to later stages that are associated with greater severity and impairment. Clinical staging has the potential to evolve into an evidence-based clinicopathological framework, whereby each stage is matched to comprehensive psychological, neurobiological, and clinical phenotyping. This would form the basis for highly personalized and stage-specific intervention selection, secondary prevention, and prediction of illness trajectory (Hickie et al., 2013; McGorry et al., 2006).

5.2 Applying a transdiagnostic clinical staging model within youth mental health services

The clinical staging model has been implemented within various *headspace* services in Australia (Cross, Scott, Hermens, & Hickie, 2018; Hickie et al., 2013; Purcell et al., 2015), with key components of this primary care stage-based model including comprehensive baseline assessment, treatment planning using shared decision-making, stage-matched interventions, routine outcome assessment, and integration with specialist mental health services (Cross et al., 2014). Preliminary work indicates that between 64% and 88% of young people attending *headspace* present with subthreshold (stage 1a or 1b) symptoms (Hickie et al., 2013; Purcell et al., 2015), providing key opportunities at a clinical level to delay or prevent a full-threshold disorder, and at a research level to identify markers of illness onset and progression. Of this group with subthreshold symptoms, approximately 19% progress to a major mental disorder (Cross et al., 2017; Hickie et al., 2013), with this critical transition most likely to occur within 12 months of initial service contact (Hickie et al., 2013). This has implications on early intervention where a 12-month service program may be more appropriate and effective for those at stage 1b (Cross et al., 2016). With regard to treatment outcomes, young people entering care at earlier stages (stage 1a) show more favorable symptomatic and functional outcomes following treatment that is more benign and less intensive, compared with those with attenuated or discrete mental disorders (stage 1b+) (Cross et al., 2016). This is in line with the model's underlying assumption that delivering effective evidence-based interventions at the right time (i.e., early during the course of mental illness) can result in improved outcomes (McGorry et al., 2006).

6 Conclusion

Traditional categorical approaches to mental health have neither substantially advanced prediction nor facilitated early intervention of mental disorders to prevent their onset, progression, and associated disability. New transdiagnostic approaches congruent with the pluripotent trajectories of mental disorders have the potential to address these theoretical and service gaps. Clinical staging provides a relevant framework for implementing transdiagnostic early intervention with a preventive focus to young people due to its recognition of the wide range of early clinical symptoms that are often excluded by traditional service models, where the need for care is based on arbitrary diagnostic thresholds. While clinical staging in itself has the capacity to facilitate prediction research, its integration with cross-disciplinary dynamic prediction models that capture the fluidity of early phenotypes may further advance prospects for prevention. Together, these approaches may improve the targeting and personalization of interventions, as well as aid the development of powerful predictive tools.

References

Addington, J., Cornblatt, B. A., Cadenhead, K. S., Cannon, T. D., McGlashan, T. H., Perkins, D. O., … Heinssen, R. (2011). At clinical high risk for psychosis: Outcome for nonconverters. *The American Journal of Psychiatry, 168*(8), 800–805.

Andrinopoulou, E. R., Rizopoulos, D., Geleijnse, M. L., Lesaffre, E., Bogers, A. J. J. C., & Takkenberg, J. J. M. (2015). Dynamic prediction of outcome for patients with severe aortic stenosis: Application of joint models for longitudinal and time-to-event data. *BMC Cardiovascular Disorders, 15*(1), 28.

Anttila, V., Bulik-Sullivan, B., Finucane, H. K., Walters, R. K., Bras, J., Duncan, L., … Neale, B. M. (2018). Analysis of shared heritability in common disorders of the brain. *Science, 360*(6395).

Australian Government Department of Health. (2015). *Australian government response to contributing lives, thriving communities—Review of mental health programmes and services.* Canberra: Department of Health.

Bayani, A., Hadaeghi, F., Jafari, S., & Murray, G. (2017). Critical slowing down as an early warning of transitions in episodes of bipolar disorder: A simulation study based on a computational model of circadian activity rhythms. *Chronobiology International, 34*(2), 235–245.

Berk, M., Brnabic, A., Dodd, S., Kelin, K., Tohen, M., Malhi, G. S., … McGorry, P. D. (2011). Does stage of illness impact treatment response in bipolar disorder? Empirical treatment data and their implication for the staging model and early intervention. *Bipolar Disorders, 13*(1), 87–98.

Borsboom, D. (2017). A network theory of mental disorders. *World Psychiatry, 16*(1), 5–13.

Borsboom, D., & Cramer, A. O. J. (2013). Network analysis: An integrative approach to the structure of psychopathology. *Annual Review of Clinical Psychology, 9*, 91–121.

Brankovic, M., Akkerhuis, K. M., van Boven, N., Anroedh, S., Constantinescu, A., Caliskan, K., … Kardys, I. (2018). Patient-specific evolution of renal function in chronic heart failure patients dynamically predicts clinical outcome in the Bio-SHiFT study. *Kidney International, 93*(4), 952–960.

Cannon, T. D., Yu, C., Addington, J., Bearden, C. E., Cadenhead, K. S., Cornblatt, B. A., … Kattan, M. W. (2016). An individualized risk calculator for research in prodromal psychosis. *The American Journal of Psychiatry, 173*(10), 980–988.

Cao, Y., Jiang, Y., Lin, X., Liu, J., Lu, T., Cheng, W., & Yan, F. (2018). Dynamic prediction of outcome for patients with ovarian cancer: Application of a joint model for longitudinal cancer antigen 125 values. *International Journal of Gynecological Cancer, 28*(1), 85–91.

Caspi, A., & Moffitt, T. E. (2018). All for one and one for all: Mental disorders in one dimension. *The American Journal of Psychiatry, 175*(9), 831–844.

Cosci, F., & Fava, G. A. (2013). Staging of mental disorders: Systematic review. *Psychotherapy and Psychosomatics, 82*(1), 20–34.

Cross, S. P. M., Hermens, D. F., & Hickie, I. B. (2016). Treatment patterns and short-term outcomes in an early intervention youth mental health service. *Early Intervention in Psychiatry, 10*(1), 88–97.

Cross, S. P. M., Hermens, D. F., Scott, E. M., Ottavio, A., McGorry, P. D., & Hickie, I. B. (2014). A clinical staging model for early intervention youth mental health services. *Psychiatric Services, 65*(7), 939–943.

Cross, S. P. M., & Hickie, I. (2017). Transdiagnostic stepped care in mental health. *Public Health Research and Practice, 27*(2).

Cross, S. P., Scott, J. L., Hermens, D. F., & Hickie, I. B. (2018). Variability in clinical outcomes for youths treated for subthreshold severe mental disorders at an early intervention service. *Psychiatric Services, 69*(5), 555–561.

Cross, S. P. M., Scott, J., & Hickie, I. B. (2017). Predicting early transition from sub-syndromal presentations to major mental disorders. *BJPsych Open, 3*(5), 223–227.

Cuijpers, P. (2003). Examining the effects of prevention programs on the incidence of new cases of mental disorders: The lack of statistical power. *The American Journal of Psychiatry, 160*(8), 1385–1391.

Cuthbert, B. N., & Insel, T. R. (2013). Toward the future of psychiatric diagnosis: The seven pillars of RDoC. *BMC Medicine, 11*(1), 126.

Duffy, A. (2014). Toward a comprehensive clinical staging model for bipolar disorder: Integrating the evidence. *Canadian Journal of Psychiatry, 59*(12), 659–666.

Duffy, A., Malhi, G. S., & Grof, P. (2017). Do the trajectories of bipolar disorder and schizophrenia follow a universal staging model? *Canadian Journal of Psychiatry, 62*(2), 115–122.

Fried, E. I., van Borkulo, C. D., Cramer, A. O. J., Boschloo, L., Schoevers, R. A., & Borsboom, D. (2017). Mental disorders as networks of problems: A review of recent insights. *Social Psychiatry and Psychiatric Epidemiology, 52*(1), 1–10.

Fusar-Poli, P., Borgwardt, S., Bechdolf, A., Addington, J., Riecher-Rössler, A., Schultze-Lutter, F., … Yung, A. (2013). The psychosis high-risk state: A comprehensive state-of-the-art review. *Archives of General Psychiatry, 70*(1), 107–120.

Fusar-Poli, P., Rocchetti, M., Sardella, A., Avila, A., Brandizzi, M., Caverzasi, E., … McGuire, P. (2015). Disorder, not just state of risk: Meta-analysis of functioning and quality of life in people at high risk of psychosis. *The British Journal of Psychiatry*, 207(3), 198–206.

Fusar-Poli, P., Rutigliano, G., Stahl, D., Davies, C., Bonoldi, I., Reilly, T., & McGuire, P. (2017). Development and validation of a clinically based risk calculator for the transdiagnostic prediction of psychosis. *JAMA Psychiatry*, 74(5), 493–500.

Gibb, S. J., Fergusson, D. M., & Horwood, L. J. (2010). Burden of psychiatric disorder in young adulthood and life outcomes at age 30. *The British Journal of Psychiatry*, 197(2), 122–127.

Goodkind, M., Eickhoff, S. B., Oathes, D. J., Jiang, Y., Chang, A., Jones-Hagata, L. B., … Etkin, A. (2015). Identification of a common neurobiological substrate for mental Illness. *JAMA Psychiatry*, 72(4), 305–315.

Hamilton, B. A., Naismith, S. L., Scott, E. M., Purcell, S., & Hickie, I. B. (2011). Disability is already pronounced in young people with early stages of affective disorders: Data from an early intervention service. *Journal of Affective Disorders*, 131(1–3), 84–91.

Hartmann, J. A., Nelson, B., Ratheesh, A., Treen, D., & McGorry, P. D. (2019). At-risk studies and clinical antecedents of psychosis, bipolar disorder and depression: A scoping review in the context of clinical staging. *Psychological Medicine*, 49, 177–189.

Hartmann, J. A., Nelson, B., Spooner, R., Amminger, G. P., Chanen, A., Davey, C. G., … McGorry, P. D. (2019). Broad clinical high-risk mental state (CHARMS): Methodology of a cohort study validating criteria for pluripotent risk. *Early Intervention in Psychiatry*, 13, 379–386.

Harvey, A. G., Murray, G., Chandler, R. A., & Soehner, A. (2011). Sleep disturbance as transdiagnostic: Consideration of neurobiological mechanisms. *Clinical Psychology Review*, 31(2), 225–235.

Hegelstad, W. T. V., Larsen, T. K., Auestad, B., Evensen, J., Haahr, U., Joa, I., … McGlashan, T. (2012). Long-term follow-up of the TIPS early detection in psychosis study: Effects on 10-year outcome. *The American Journal of Psychiatry*, 169(4), 374–380.

Henriksen, M. G., Nordgaard, J., & Jansson, L. B. (2017). Genetics of schizophrenia: Overview of methods, findings and limitations. *Frontiers in Human Neuroscience*, 11, 322.

Hermens, D. F., Naismith, S. L., Lagopoulos, J., Lee, R. S. C., Guastella, A. J., Scott, E. M., & Hickie, I. B. (2013). Neuropsychological profile according to the clinical stage of young persons presenting for mental health care. *BMC Psychology*, 1(1), 8.

Hetrick, S. E., Bailey, A. P., Smith, K. E., Malla, A., Mathias, S., Singh, S. P., … McGorry, P. D. (2017). Integrated (one-stop shop) youth health care: Best available evidence and future directions. *The Medical Journal of Australia*, 207(10), S5–S18.

Hickie, I. B., Scott, E. M., Hermens, D. F., Naismith, S. L., Guastella, A. J., Kaur, M., … McGorry, P. D. (2013). Applying clinical staging to young people who present for mental health care. *Early Intervention in Psychiatry*, 7(1), 31–43.

Kendell, R., & Jablensky, A. (2003). Distinguishing between the validity and utility of psychiatric diagnoses. *The American Journal of Psychiatry*, 160(1), 4–12.

Kendler, K. S. (2018). Classification of psychopathology: Conceptual and historical background. *World Psychiatry*, 17(3), 241–242.

Kendler, K. S., Zachar, P., & Craver, C. (2011). What kinds of things are psychiatric disorders? *Psychological Medicine*, 41(6), 1143–1150.

Kessler, R. C., Berglund, P., Demler, O., Jin, R., Merikangas, K. R., & Walters, E. E. (2005). Lifetime prevalence and age-of-onset distributions of DSM-IV disorders in the national comorbidity survey replication. *Archives of General Psychiatry*, 62(6), 593–602.

Kim-Cohen, J., Caspi, A., Moffitt, T. E., Harrington, H., Milne, B. J., & Poulton, R. (2003). Prior juvenile diagnoses in adults with mental disorder: Developmental follow-back of a prospective-longitudinal cohort. *Archives of General Psychiatry*, 60(7), 709–717.

Kotov, R., Waszczuk, M. A., Krueger, R. F., Forbes, M. K., Watson, D., Clark, L. A., … Zimmerman, M. (2017). The hierarchical taxonomy of psychopathology (HiTOP): A dimensional alternative to traditional nosologies. *Journal of Abnormal Psychology*, 126(4), 454–477.

Krueger, R. F., Kotov, R., Watson, D., Forbes, M. K., Eaton, N. R., Ruggero, C. J., … Zimmermann, J. (2018). Progress in achieving quantitative classification of psychopathology. *World Psychiatry*, 17(3), 282–293.

Lagopoulos, J., Hermens, D. F., Naismith, S. L., Scott, E. M., & Hickie, I. B. (2012). Frontal lobe changes occur early in the course of affective disorders in young people. *BMC Psychiatry*, 12, 4.

Lee, R. S. C., Hermens, D. F., Naismith, S. L., Lagopoulos, J., Jones, A., Scott, J., … Hickie, I. B. (2015). Neuropsychological and functional outcomes in recent-onset major depression, bipolar disorder and schizophrenia-spectrum disorders: A longitudinal cohort study. *Translational Psychiatry*, 5(4).

Lee, T. Y., Lee, J., Kim, M., Choe, E., & Kwon, J. S. (2018). Can we predict psychosis outside the clinical high-risk state? A systematic review of non-psychotic risk syndromes for mental disorders. *Schizophrenia Bulletin*, 44(2), 276–285.

Lim, J., Rekhi, G., Rapisarda, A., Lam, M., Kraus, M., Keefe, R. S., & Lee, J. (2015). Impact of psychiatric comorbidity in individuals at ultra high risk of psychosis—Findings from the Longitudinal Youth at Risk Study (LYRIKS). *Schizophrenia Research*, 164(1–3), 8–14.

Lin, A., Wood, S. J., Nelson, B., Beavan, A., McGorry, P., & Yung, A. R. (2015). Outcomes of nontransitioned cases in a sample at ultra-high risk for psychosis. *The American Journal of Psychiatry*, 172(3), 249–258.

McGorry, P. D. (2010). Staging in neuropsychiatry: A heuristic model for understanding, prevention and treatment. *Neurotoxicity Research*, 18(3–4), 244–255.

McGorry, P. (2011). Transition to adulthood: The critical period for pre-emptive, disease-modifying care for schizophrenia and related disorders. *Schizophrenia Bulletin*, 37(3), 524–530.

McGorry, P. D. (2013). The next stage for diagnosis: Validity through utility. *World Psychiatry*, 12(3), 213–215.

McGorry, P. D., Goldstone, S. D., Parker, A. G., Rickwood, D. J., & Hickie, I. B. (2014). Cultures for mental health care of young people: An Australian blueprint for reform. *Lancet Psychiatry*, 1(7), 559–568.

McGorry, P. D., Hartmann, J. A., Spooner, R., & Nelson, B. (2018). Beyond the "at risk mental state" concept: Transitioning to transdiagnostic psychiatry. *World Psychiatry*, 17(2), 133–142.

McGorry, P. D., Hickie, I. B., Yung, A. R., Pantelis, C., & Jackson, H. J. (2006). Clinical staging of psychiatric disorders: A heuristic framework for choosing earlier, safer and more effective interventions. *The Australian and New Zealand Journal of Psychiatry*, 40(8), 616–622.

McGorry, P., Keshavan, M., Goldstone, S., Amminger, P., Allott, K., Berk, M., … Hickie, I. (2014). Biomarkers and clinical staging in psychiatry. *World Psychiatry*, *13*(3), 211–223.

McGorry, P., & Nelson, B. (2016). Why we need a transdiagnostic staging approach to emerging psychopathology, early diagnosis, and treatment. *JAMA Psychiatry*, *73*(3), 191–192.

McGorry, P. D., Purcell, R., Hickie, I. B., Yung, A. R., Pantelis, C., & Jackson, H. J. (2007). Clinical staging: A heuristic model for psychiatry and youth mental health. *The Medical Journal of Australia*, *187*(7), S40–S42.

McNally, R. J. (2016). Can network analysis transform psychopathology? *Behaviour Research and Therapy*, *86*, 95–104.

McTeague, L. M., Huemer, J., Carreon, D. M., Jiang, Y., Eickhoff, S. B., & Etkin, A. (2017). Identification of common neural circuit disruptions in cognitive control across psychiatric disorders. *The American Journal of Psychiatry*, *174*(7), 676–685.

Merikangas, K. R., He, J.-P., Burstein, M., Swanson, S. A., Avenevoli, S., Cui, L., … Swendsen, J. (2010). Lifetime prevalence of mental disorders in US adolescents: Results from the national comorbidity study-adolescent supplement (NCS-A). *Journal of the American Academy of Child and Adolescent Psychiatry*, *49*(10), 980–989.

Naismith, S. L., Hermens, D. F., Ip, T. K. C., Bolitho, S., Scott, E., Rogers, N. L., & Hickie, I. B. (2012). Circadian profiles in young people during the early stages of affective disorder. *Translational Psychiatry*, *2*.

Nelson, B., Hartmann, J. A., & Parnas, J. (2018). Detail, dynamics and depth: Useful correctives for some current research trends. *The British Journal of Psychiatry*, *212*(5), 262–264.

Nelson, B., McGorry, P. D., Wichers, M., Wigman, J. T. W., & Hartmann, J. A. (2017). Moving from static to dynamic models of the onset of mental disorder: A review. *JAMA Psychiatry*, *74*(5), 528–534.

Nelson, B., Yuen, H. P., Lin, A., Wood, S. J., McGorry, P. D., Hartmann, J. A., & Yung, A. R. (2016). Further examination of the reducing transition rate in ultra high risk for psychosis samples: The possible role of earlier intervention. *Schizophrenia Research*, *174*(1–3), 43–49.

Purcell, R., Jorm, A. F., Hickie, I. B., Yung, A. R., Pantelis, C., Amminger, G. P., … McGorry, P. D. (2015). Demographic and clinical characteristics of young people seeking help at youth mental health services: Baseline findings of the transitions study. *Early Intervention in Psychiatry*, *9*(6), 487–497.

Rickwood, D. J., Mazzer, K. R., Telford, N. R., Parker, A. G., Tanti, C. J., & Mc Gorry, P. D. (2015). Changes in psychological distress and psychosocial functioning in young people accessing headspace centres for mental health problems. *The Medical Journal of Australia*, *202*(10), 537–543.

Rickwood, D. J., Telford, N. R., Parker, A. G., Tanti, C. J., & McGorry, P. D. (2014). Headspace—Australia's innovation in youth mental health: Who are the clients and why are they presenting? *The Medical Journal of Australia*, *200*(2), 108–111.

Scheffer, M. (2009). *Critical transitions in nature and society*. Princeton: Princeton University Press.

Scheffer, M., Bascompte, J., Brock, W. A., Brovkin, V., Carpenter, S. R., Dakos, V., … Sugihara, G. (2009). Early-warning signals for critical transitions. *Nature*, *461*(7260), 53–59.

Scheffer, M., Carpenter, S. R., Lenton, T. M., Bascompte, J., Brock, W., Dakos, V., … Vandermeer, J. (2012). Anticipating critical transitions. *Science*, *338*(6105), 344–348.

Scott, E. M., Hermens, D. F., Glozier, N., Naismith, S. L., Guastella, A. J., & Hickie, I. B. (2012). Targeted primary care-based mental health services for young Australians. *The Medical Journal of Australia*, *196*(2), 136–140.

Scott, E. M., Hermens, D. F., Naismith, S. L., White, D., Whitwell, B., Guastella, A. J., … Hickie, I. B. (2012). Thoughts of death or suicidal ideation are common in young people aged 12 to 30 years presenting for mental health care. *BMC Psychiatry*, *12*, 234.

Scott, J., Paykel, E., Morriss, R., Bentall, R., Kinderman, P., Johnson, T., & Hayhurst, R. A. H. (2006). Cognitive-behavioural therapy for severe and recurrent bipolar disorders: Randomised controlled trial. *The British Journal of Psychiatry*, *188*, 313–320.

Shah, J. L., Crawford, A., Mustafa, S. S., Iyer, S. N., Joober, R., & Malla, A. K. (2017). Is the clinical high-risk state a valid concept? Retrospective examination in a first-episode psychosis sample. *Psychiatric Services*, *68*(10), 1046–1052.

Shankman, S. A., Lewinsohn, P. M., Klein, D. N., Small, J. W., Seeley, J. R., & Altman, S. E. (2009). Subthreshold conditions as precursors for full syndrome disorders: A 15-year longitudinal study of multiple diagnostic classes. *Journal of Child Psychology and Psychiatry*, *50*(12), 1485–1494.

van de Leemput, I. A., Wichers, M., Cramer, A. O. J., Borsboom, D., Tuerlinckx, F., Kuppens, P., … Scheffer, M. (2014). Critical slowing down as early warning for the onset and termination of depression. *Proceedings of the National Academy of Sciences of the United States of America*, *111*(1), 87–92.

van Os, J. (2013). The dynamics of subthreshold psychopathology: Implications for diagnosis and treatment. *The American Journal of Psychiatry*, *170*(7), 695–698.

Wichers, M., & Groot, P. C. (2016). Critical slowing down as a personalized early warning signal for depression. *Psychotherapy and Psychosomatics*, *85*(2), 114–116.

Wigman, J. T., de Vos, S., Wichers, M., van Os, J., & Bartels-Velthuis, A. A. (2017). A transdiagnostic network approach to psychosis. *Schizophrenia Bulletin*, *43*(1), 122–132.

Wigman, J. T., van Nierop, M., Vollebergh, W. A., Lieb, R., Beesdo-Baum, K., Wittchen, H. U., & van Os, J. (2012). Evidence that psychotic symptoms are prevalent in disorders of anxiety and depression, impacting on illness onset, risk, and severity—Implications for diagnosis and ultra-high risk research. *Schizophrenia Bulletin*, *38*(2), 247–257.

Wigman, J. T., van Os, J., Thiery, E., Derom, C., Collip, D., Jacobs, N., & Wichers, M. (2013). Psychiatric diagnosis revisited: Towards a system of staging and profiling combining nomothetic and idiographic parameters of momentary mental states. *PLoS One*, *8*(3).

Yuen, H. P., & Mackinnon, A. (2016). Performance of joint modelling of time-to-event data with time-dependent predictors: An assessment based on transition to psychosis data. *PeerJ*, *4*.

Yuen, H. P., Mackinnon, A., Hartmann, J., Amminger, G. P., Markulev, C., Lavoie, S., … Nelson, B. (2018). Dynamic prediction of transition to psychosis using joint modelling. *Schizophrenia Research*, *202*, 333–340.

Yuen, H. P., Mackinnon, A., & Nelson, B. (2018). A new method for analysing transition to psychosis: Joint modelling of time-to-event outcome with time-dependent predictors. *International Journal of Methods in Psychiatric Research*, *27*(1).

Yung, A. R., McGorry, P. D., McFarlane, C. A., Jackson, H. J., Patton, G. C., & Rakkar, A. (1996). Monitoring and care of young people at incipient risk of psychosis. *Schizophrenia Bulletin*, *22*(2), 283–303.

Yung, A. R., Phillips, L. J., Yuen, H. P., Francey, S. M., McFarlane, C. A., Hallgren, M., & McGorry, P. D. (2003). Psychosis prediction: 12-month follow up of a high-risk ("prodromal") group. *Schizophrenia Research*, *60*(1), 21–32.

Yung, A. R., Yuen, H. P., Berger, G., Francey, S., Hung, T.-C., Nelson, B., ... McGorry, P. (2007). Declining transition rate in ultra high risk (prodromal) services: Dilution or reduction of risk? *Schizophrenia Bulletin*, *33*(3), 673–681.

Chapter 5

Early intervention, prevention, and prediction in mood disorders: Tracking multidimensional outcomes in young people presenting for mental health care

Elizabeth M. Scott, Joanne S. Carpenter, Frank Iorfino, Shane P.M. Cross, Daniel F. Hermens, Django White, Rico S.Z. Lee, Sharon L. Naismith, Adam J. Guastella, Nicholas Glozier, F. Markus Leweke, Dagmar Koethe, Jim Lagopoulos, Jan Scott, Blake A. Hamilton, Jacob J. Crouse, Ashleigh M. Tickell, Alissa Nichles, Natalia Zmicerevska, Lillian J. Gehue, Manreena Kaur, Kate M. Chitty and Ian B. Hickie
Youth Mental Health Team, Brain and Mind Centre, University of Sydney, Sydney, NSW, Australia

The premature death and disability attributable to major mental disorders, and most notably the anxiety, mood, and psychotic disorders, derives from their early age of onset, population prevalence, chronicity, risk to deliberate self-harm (DSH), suicidal thoughts and behavior (STB), and comorbidity with physical illness and alcohol and other substance misuse (Erskine et al., 2015; Whiteford, Ferrari, Degenhardt, Feigin, & Vos, 2015). Consequently, a key population-health priority is to reduce that burden. From an individual clinical and health services planning perspective, this includes identification and enhanced long-term care of those who are in the early phases of life threatening or chronic-disabling mood disorders (Hickie et al., 2013; Insel, 2007; Insel, 2009; McGorry, Goldstone, Parker, Rickwood, & Hickie, 2014; Scott et al., 2012).

1 Developing an early intervention mood disorders service

Over the past decade, we have recruited young people from targeted early intervention services that prioritize young people presenting with various mood disorders and DSH and STB, comorbid medical or mental disorders, or alcohol or substance misuse. These young people have participated in various cross-sectional, longitudinal, or interventional studies and, often concurrently, relevant neuropsychological, brain imaging, circadian, and metabolic assessments. Here, we report the baseline demographic and clinical characteristics of this unique cohort, and the ways in which the use of a multidimensional outcomes framework can be applied to facilitate more personalized health and social care, early intervention, and secondary prevention (see Tables 1–4).

As this cohort has unique characteristics in terms of its: (i) size; (ii) emphasis on early clinical phases of common mood disorders; (iii) recruitment based on presentation for care rather than specific diagnoses; (iv) patterns of comorbidity; and (v) concurrent collection of relevant neurobiological data, we have been able to use it to test various approaches to early detection, secondary prevention, and prediction of the course (at an individual and sub-group level) of those who experience common mood disorders.

Our emphasis on common pathways to recruitment rather than arbitrary inclusion or exclusion by specific diagnostic criteria (for anxiety, mood, or psychotic symptoms, i.e., a "transdiagnostic" approach—see Hickie et al., 2013; Hickie, Hermens, Naismith, Guastella, et al., 2013; Hickie, Scott, Hermens, Naismith, et al., 2013; Scott et al., 2013b; Scott, Scott, et al., 2014) is consistent with the National Institute of Mental Health (NIMH)-developed "Research Domain Criteria" (RDoC) (Cuthbert & Insel, 2013; Insel et al., 2010; Kozak & Cuthbert, 2016). RDoC was designed to establish new ways of classifying mental disorders based on correlations with independent neurobiological measures. Our emphasis on non-selective recruitment from early intervention services is also consistent with other NIMH recommendations (see Bruce Cuthbert in Casey et al. (2013)) that advanced clinical research should recruit large cohorts from common service settings. These cohorts have the advantage that they have the capacity to demonstrate appropriate variance along relevant dimensions of interest (e.g., neuropsychological function, cortical, or subcortical brain structure), cross-sectionally and longitudinally.

Personalized Psychiatry. https://doi.org/10.1016/B978-0-12-813176-3.00005-5

TABLE 1 Demographics, symptoms, and social and occupational function (SaOF) data in females and males across the Brain and Mind Centre youth cohort.

	Female (n = 4256)	Male (n = 3140)	Statistics
Age, years (n=7397)	18.5±3.8	18.6±4.0	$t_{(7395)}=1.38, p=.17$
K-10 (n=2844)	30.3±8.7	26.4±8.9	$t_{(2842)}=11.87, p<.001$
SOFAS (n=2727)	61.6±12.6	60.7±13.1	$t_{(2725)}=1.77, p=.07$
WSAS (n=2103)	20.5±8.3	18.3±8.5	$t_{(2101)}=5.94, p<.001$
BDQ "days out of role" (n=1850)	8.7±9.1	7.3±9.3	$t_{(1848)}=3.47, p=.001$

Statistics are independent samples t-tests. *SOFAS*, social and occupational functioning assessment scale (Goldman, Skodol, & Lave, 1992); *K-10*, Kessler psychological distress scale (Kessler et al., 2002); *WSAS*, work and social adjustment scale (Mundt, Marks, Shear, & Greist, 2002); *BDQ*, brief disability questionnaire (Von Korff, Ustun, Ormel, Kaplan, & Simon, 1996); *Days out of role*, in the past 30 days, for how many days were you totally unable to carry out your usual activities or work because of any health condition?

TABLE 2 Comparison of demographics, symptoms, and social and occupational function (SaOF) across primary diagnostic subgroups.

Primary diagnosis	Depression n = 1197	Bipolar n = 211	Anxiety n = 633	Psychosis n = 205	Other[a] n = 718	Statistics	Pairwise comparisons
Age, years	18.1±3.5 n=1197	20.3±3.7 n=211	18.0±3.8 n=633	20.5±3.9 n=205	17.1±3.7 n=718	$F_{(4,2959)}=54.5, p<.001$	O<A, D<B, P
Sex (% female)	62.4% n=747	62.6% n=132	58.1% n=368	34.6% n=71	41.9% n=301	$X^2=117.5, p<.001$	
K-10	31.4±7.9 n=978	29.8±8.8 n=144	27.8±8.3 n=499	27.4±10.2 n=133	24.9±9.2 n=556	$F_{(4,2305)}=55.9 p<.001$	O<P, A, B, D; P, A<D
SOFAS	61.0±11.9 n=1024	59.6±13.2 n=165	61.9±12.4 n=533	53.9±16.3 n=127	62.3±13.4 n=609	$F_{(4,2453)}=12.7 p<.001$	P<B, D, A, O;
WSAS	21.6±7.9 n=688	22.3±8.6 n=104	18.8±7.9 n=352	20.1±8.9 n=89	16.6±8.2 n=378	$F_{(4,1606)}=27.3, p<.001$	O<A, P, D, B; A<D, B
BDQ "days out of role"	9.3±9.0 n=602	10.4±10.5 n=94	7.5±9.1 n=322	10.4±10.9 n=75	6.1±8.8 n=326	$F_{(4,1414)}=9.4, p<.001$	O, A<D, B, P[a] note AvB not sig

Statistics are ANOVA or Chi-squared tests. Pairwise comparisons are Bonferroni corrected. *SOFAS*, social and occupational functioning assessment scale; *K-10*, Kessler psychological distress scale; *WSAS*, work and social adjustment scale; *BDQ*, brief disability questionnaire; *Days out of role*, In the past 30 days, for how many days were you totally unable to carry out your usual activities or work because of any health condition?
[a] *"Other" diagnosis includes personality disorders, eating disorders, neurodevelopmental problems (attention-deficit hyperactivity disorder, autism spectrum disorders), etc.*

Our clinical and research approach also overtly places emphasis on:

(i) recognition of developmental trajectories (see Fig. 1). Here, key transitions from anxiety, attentional, or neurodevelopmental states in childhood to post-pubertal depressive, bipolar, or psychotic states (and other comorbid conditions) are critical;

(ii) recognition of ongoing complex brain developmental processes from childhood to early puberty, and on to mid- and later-adolescence and early adulthood. These are often recognized phenotypically (or by objective testing) in differing patterns of neurocognitive, circadian, hormonal, immune, metabolic, or brain structural development; and

(iii) the impacts of active bidirectional interaction with the environment at critical stages of the transitions from childhood to post-pubertal states and later transitions to adulthood. Key factors such as the changes in the social environment (e.g., development of intimate relationships outside of close family; use of alcohol and other substances; new education and employment-based experiences; or exposure to new infectious agents) need to be recognized.

TABLE 3 Multidimensional outcome frameworks for the Brain and Mind Centre youth cohort.

Outcome construct	Contributory factors	Measurement tools
1. Social and occupational function	Employment status	NEET status and receipt of government benefits or income support (Cross, Scott, & Hickie, 2017; Hamilton, Naismith, Scott, Purcell, & Hickie, 2011; Lee et al., 2017; O'Dea et al., 2014, 2016; Scott et al., 2014)
	Participation in age-appropriate education or training	
	Engagement with family, peers, and social network	Work and social adjustment scale (WSAS) (Hamilton et al., 2011; Scott et al., 2013b)
		Brief disability questionnaire (BDQ)—days out of role (Hamilton et al., 2011; Scott, Scott, et al., 2014)
		Clinician-rated social and occupational functional assessment scale (SOFAS) (Cross, Hermens, & Hickie, 2016; Cross, Hermens, Scott, Salvador-Carulla, & Hickie, 2017; Cross, Scott, Hermens, & Hickie, 2018; Hamilton et al., 2011; Iorfino et al., 2018a; Kaur et al., 2013; Lee et al., 2013; O'Dea et al., 2016; Scott, Scott, et al., 2014)
		WHO disability assessment schedule 2.0 (WHODAS) (Lee et al., 2015; O'Dea et al., 2016)
		WHO quality of life (WHOQoL-BREF) Scale (Lee, Hermens, et al., 2015)
2. Deliberate self-harm, accidents and injury, and suicidal thoughts and behaviors	Suicidal thoughts and behaviors	Suicidal ideation attributes scale (SIDAS) (Iorfino et al., 2017)
	Self-harming behaviors	Hamilton depression rating scale (HDRS)—suicide item (Scott et al., 2012)
	Serious accidents and injury, typically requiring medical care	Clinician ratings of suicide ideation and behaviors and self-harm
3. Concurrent alcohol or other substance misuse	Alcohol use	Alcohol use disorders identification test (AUDIT)—risky drinking status (Chitty, Kaur, Lagopoulos, Hickie, & Hermens, 2011; Chitty, Kaur, Lagopoulos, Hickie, & Hermens, 2014; Chitty, Lagopoulos, Hickie, & Hermens, 2013; Chitty, Lagopoulos, Hickie, & Hermens, 2015; Hermens et al., 2013; Hermens et al., 2013; Hermens et al., 2015; Lee et al., 2015; Pesa et al., 2012; van Zwieten et al., 2013)
	Cannabis use	Drug use disorders identification test (DUDIT)
	Stimulant use	WHO alcohol, smoking, and substance involvement screening test (WHO-ASSIST) (Bogaty, Crouse, Hickie, & Hermens, 2019; Crouse et al., 2018; Crouse, Moustafa, Bogaty, Hickie, & Hermens, 2018; Hermens, Scott, et al., 2013; Pesa et al., 2012)
	Other substance use	Severity of dependence scale (SDS)
4. Physical health	Tobacco status	Tobacco smoking (Hermens, Scott, et al., 2013; Scott et al., 2015)
	Metabolic health	Body mass index (BMI) (Banihashemi et al., 2016)
	Cardiovascular risk	Waist circumference
	Immune status	Fasting glucose, insulin (HOMA-IR calculations), and lipids (Scott et al., 2019; Scott et al., 2015)
	Medical conditions	Immune activation screening
		Medical history
5. Illness type, clinical stage, and lifetime trajectory	*See* Table 2	*See* Table 2

NEET, not in education, employment, or training; *WHO*, World Health Organization; *HOMA-IR*, homeostasis model assessment of insulin resistance.

TABLE 4 Construct five: Illness type, clinical stage, and lifetime trajectory.

Construct	Concepts	Measurement tools
Illness type	"Tripartite" model, differentiating "anxious-depression," "circadian (mania-fatigue)," and "developmental-psychotic" subtypes	Clinical history: Allocating cases on the basis of life-time phenotypes first to developmental-psychotic, then mania-fatigue and the residual non-specific depressive cases to anxious (arousal)-depression (Hickie et al., 2013; Scott et al., 2015)
Clinical stage	Differentiated early Stages 1a and 1b from 2+	Clinical ratings (Hickie, Scott, Hermens, Naismith, et al., 2013; McGorry, Hickie, Yung, Pantelis, & Jackson, 2006)
		Application of algorithms to self-rating scales
Lifetime trajectory	Objective measures of neurodevelopment, cognitive function, sleep-wake and circadian rhythms, age-specific biological changes, and other developmental factors or influences (see Tables 6A and 6B)	Neuropsychological function—emphasizing both classical neurocognitive features, but also social cognition
		Levels of educational achievement
		Structural and functional brain imaging (e.g., MRI, MRS)
		Sleep-wake cycle monitoring through 2-week recordings of motor activity (see Table 7 for references utilizing these objective measurement tools)

MRI, magnetic resonance imaging; *MRS*, magnetic resonance spectroscopy.

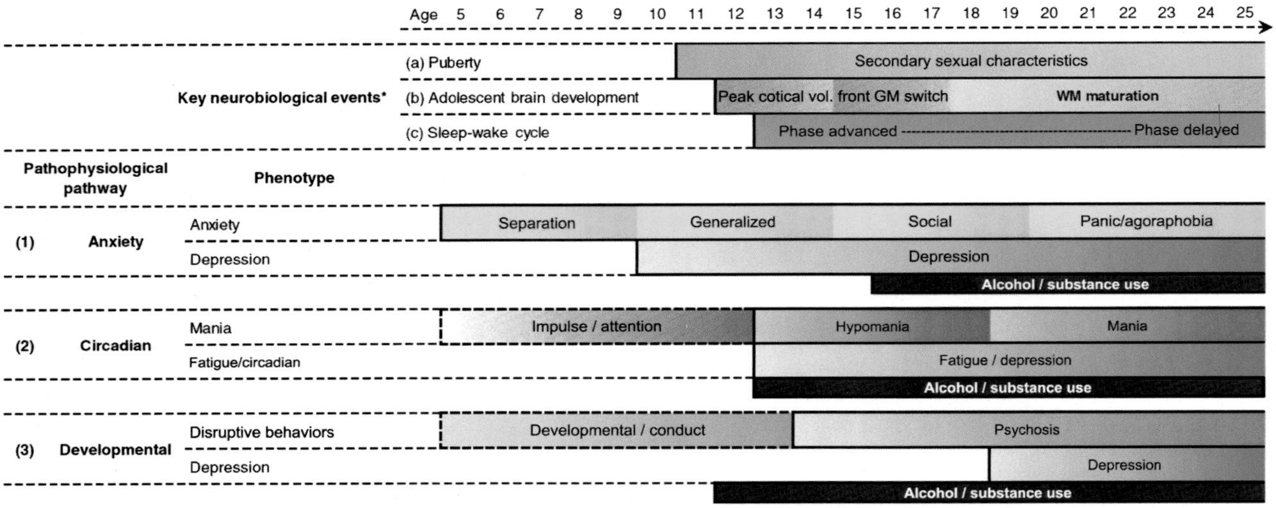

FIG. 1 Proposed pathophysiological pathways to early-onset depressive disorders. There are at least three common trajectories that lead to depression in the teenage and early-adult years. These are characterized by (1) "anxiety-central nervous system reactivity," (2) "circadian and 24- hour sleep-wake cycle dysfunction," and (3) "developmental brain abnormalities." The differing phenotypic patterns have distinct ages of onset and variable comorbidities with alcohol and other substance misuse. From age 8 to 10 years onward, these processes are accompanied by key neurobiological phenomena: (A) puberty, (B) adolescent brain development, and (C) sleep-wake cycle. *(Reproduced from McGorry, P. D., Tanti, C., Stokes, R., Hickie, I. B., Carnell, K., Littlefield, L. K., et al. (2007). Headspace: Australia's National Youth Mental Health Foundation—Where young minds come first. The Medical Journal of Australia, 187, S68–S70.)*

2 Characteristics of the Brain and Mind Center youth cohort

Our cohort consists of individuals who presented to the Brain and Mind Centre's youth mental health clinics in the Sydney suburbs of Camperdown and Campbelltown. These clinics consist of an integrated mix of primary-level services branded as headspace (McGorry, Tanti, et al., 2007), as well as more specialized services. These clinics attract young people with a range of mental health problems, but primarily focus on those with sub-threshold and full threshold anxiety and depressive disorders. They have been self-referred, referred via a family member or friend, or via the community, including external general practitioners, schools, or university (Scott, Hermens, Glozier, et al., 2012). The young people in this cohort were recruited to a research register between January 2005 and December 2017. All young people received clinician-based case management and relevant psychological, social, and/or medical interventions over the duration of their time in care, which may also include referral to/from higher-tier mental health services or hospitalization for those whose needs exceed the capacity of the primary care services. Individuals were included in this cohort if they were aged between 12 and 30 years at the time of initial assessment. Exclusion criteria for all potential participants were: lack of capacity to give informed consent (as determined by a psychiatrist), and/or clinically evident intellectual disability, and/or insufficient English to participate in the research protocol. The cohort register (and subsequent sub-studies) was approved by the University of Sydney Human Research Ethics Committee.

3   Data collection process

Clinical and administrative staff collected the relevant demographic, symptom, and impairment data at entry to the clinical services, making use of standardized administrative data sets, standard self-report instruments, and a clinical proforma. The clinical proforma captures key clinical information about the current episode and specific illness course characteristics, and has been reported in previous publications (Hickie, Scott, Hermens, Naismith, et al., 2013; Scott, Hermens, Glozier, et al., 2012). The proforma collects information about; (i) basic demographics (age, gender, receipt of government benefits); (ii) mental health diagnoses (based on DSM-IV or V criteria); (iii) clinical course information (hospitalizations, childhood diagnoses); (iv) comorbidities (physical health diagnoses, such as autoimmune, endocrine, metabolic, etc., and suicidal thoughts and behaviors); and (v) functioning (assessed using the Social Occupational Functional Assessment Scale [SOFAS, Goldman et al., 1992] and engagement in part-time or full-time education, employment, or training, used to determine not in education, employment, or training [NEET] status). The SOFAS is a clinician-rated measure that assesses functioning on a 0–100 scale, with lower scores suggesting more severe impairment. The instructions emphasize that the rater should aim to avoid confounding the rating with clinical symptoms. Clinical stages are assigned to all people who present for clinical care. In extended versions of this model, we use "stage 0" to describe population groups of interest (e.g., first-degree relatives of young probands) who are at increased risk of major anxiety, mood, or psychotic disorders; but are not presenting for clinical care. Additionally, we outline stages 3 and 4 for recurrent, persistent, and chronic illness courses (McGorry et al., 2006).

4   Demographic, symptom, and social and occupational functioning of the brain and mind youth mood disorders cohort

The available demographic, symptom, and social and occupational function (SaOF) data for the whole cohort ($n = 7397$, 58% female) are shown in Table 1. A subgroup of these participants ($n = 3087$; 59% female, mean age $= 18.52 \pm 3.8$) also have outcome data at more than 3 months from the initial assessment. With regard to specific primary clinical diagnosis at presentation, the proportions were approximately 40% depressive disorders, 21% anxiety disorders, 7% bipolar disorders, 7% psychotic disorders, and 25% other (with a strong focus on prior developmental or attentional disorders). The demographic and clinical characteristics of the cohort by primary diagnosis are shown in Table 2.

5   Developing a multidimensional outcome framework for young people with emerging mood disorders

To facilitate the health care and social needs of young people with emerging mood disorders, we have developed a specific multidimensional outcome framework to underpin our baseline assessments, selection of interventions, and tracking of long-term illness course. Five key constructs are included in this multidimensional framework. The central characteristics (summarized in Tables 3 and 4 with associated metric options) include:

1. Social and occupational function (SaOF)
2. Deliberate self-harm (DSH), accidents and injury, and suicidal thoughts and behaviors (STB)
3. Concurrent alcohol or other substance misuse
4. Physical health
5. Illness type, clinical stage, and lifetime trajectory

Elsewhere we have reported neuropsychological and neurobiological correlates of these five constructs (Iorfino, Hickie, Lee, Lagopoulos, & Hermens, 2016). While much research in youth mental health, and notably, pathways to early psychosis, has focused on construct 5: Illness type, clinical stage and lifetime trajectory, we have placed far greater emphasis on constructs 1: social and occupational function (SaOF); and 2: deliberate self-harm (DSH), accidents and injury and suicidal thoughts and behaviors (STB). This has been influenced by the importance placed on these constructs by young people and their families, as well as population-based national mental health policies (Australian Government National Mental Health Commission, 2014; Smith-Merry et al., 2018; World Health Organization, 2013–2020).

Much of our work has placed the greatest emphasis on delineating the longitudinal course (and clinical, neuropsychological, and neurobiological predictors) of functional impairment. As we recruit broadly, we are then able to examine the relative predictive power of specific diagnostic features (or comorbidity) as compared with other demographic, social, clinical, neurobiological, or treatment-dependent factors (see Figs. 2 and 3, for an example related to neuropsychological function). That is, by contrast with most traditional modeling approaches, we do not simply assume that the course of a diagnosed mental illness (construct 5: Illness type, clinical stage and lifetime trajectory) is the major driver of the other four outcomes. For example, we specifically test the extent to which baseline SaOF, or alcohol or other substance misuse, predict STB, independent of illness course (see Fig. 8).

6 Illness type, clinical stage, and lifetime trajectory

Diagnostic symptom sets for "single" disorders (e.g., schizophrenia, bipolar disorder, or unipolar depression) prioritize phenomena such as delusions, hallucinations, depressed or elevated mood, anhedonia, motor activation or retardation for schizophrenia, bipolar disorder, or severe depression. Data from community studies that assess subjects longitudinally from childhood or adolescence emphasize the extent to which many clinical phenomena (most notably anxiety, depressed mood, cognitive impairment, psychotic-like experiences, and brief periods of activation) are shared across diagnoses, both at onset and across the whole illness trajectory (Copeland et al., 2013; Kelleher et al., 2012; Merikangas et al., 2008, 2010; Merikangas et al., 2012; Murray & Jones, 2012; Ormel et al., 2015). Additionally, earlier childhood risk states such as anxiety, conduct, and developmental disorders that are evident before age 12 years predict the full spectrum of later depressive, bipolar, and psychotic disorders (Kim-Cohen et al., 2003). Our great clinical challenge is to derive new dynamic diagnostic systems for the common mood disorders that are not only consistent with developmental epidemiology (and mediating neurobiology), but are also useful when applied in everyday clinical practice.

This multidimensional framework promotes the utility of using a dimensional, rather than categorical, approach to diagnostic systems. Within this framework (see Fig. 4), we place a primary emphasis on the dimension of severity of anxious-depressed mood (which is unidirectional). Across all models explored, depression severity (from among the symptom domains) in these early cohorts is the most robust symptom-based predictor of impairment (Kohn, Glozier, Hickie, Durrant-Whyte, & Cripps, 2018). The *depressive dimension* is most easily differentiated from less severe or impairing anxiety-based temperamental or symptom factors, and is strongly associated with help-seeking behavior. In that sense, it is both the common gateway to care, and the major shared phenomenon in these cohorts seeking mental health care in designated early intervention services.

FIG. 2 A conceptual model of the interrelationships between symptom and neuropsychological profiles and functioning using structural equation modeling. The model tests whether the relationship between neuropsychological profile at baseline and functioning at follow up (path C) is mediated by neuropsychological functioning at follow up (path A/B) and/or symptoms (path D/E). Circles are unmeasured latent variables, rectangles are measured observed variables. *T1*, timepoint 1; *T2*, timepoint 2; *NΨ*, neuropsychological profile; *Fx*, functioning; *Sx*, symptoms; *ATT*, attention; *MEM*, memory; *EF*, executive functioning; *SOFAS*, social and occupational functioning assessment scale; *WHO-DAS*, World Health Organization Disability Assessment Schedule; *SFS*, social functioning scale; *D/S/T*, duration, severity, and typology of psychopathology; *STG*, stage of illness; *SUI*, suicidal ideation and self-harm.

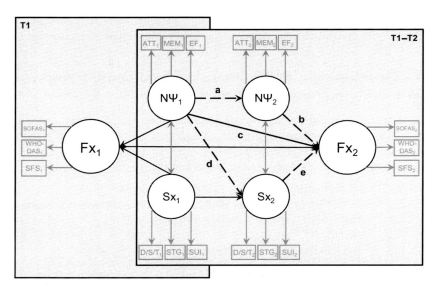

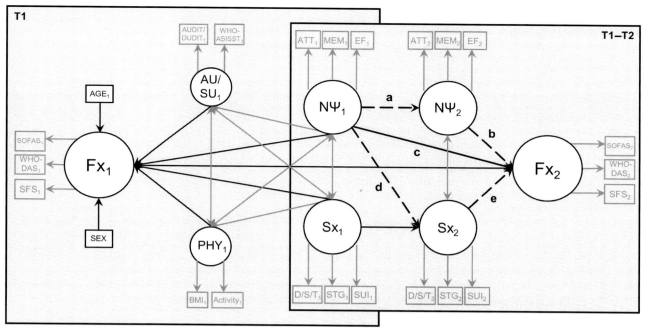

FIG. 3   Proposed working model to test the relationships between neuropsychological and symptom profiles, functioning, and potential demographic and health related-moderators. In addition to the paths outlined in Fig. 2, interactions with alcohol and substance use and physical health are also included. Circles are unmeasured latent variables, rectangles are measured observed variables. *T1*, timepoint 1; *T2*, timepoint 2; *AU/SU*, alcohol use/substance use; *NΨ*, neuropsychological profile; *Fx*, functioning; *PHY*, physical health; *Sx*, symptoms; *AUDIT/DUDIT*, alcohol use disorders identification test, drug use disorders identification test; *WHO-ASSIST*, alcohol, smoking and substance involvement screening test; *ATT*, attention; *MEM*, memory; *EF*, executive functioning; *SOFAS*, social and occupational functioning assessment scale; *WHO-DAS*, World Health Organization Disability Assessment Schedule; *SFS*, social functioning scale; *BMI*, body mass index; *Med-Con*, medical conditions; *D/S/T*, duration, severity, and typology of psychopathology; *STG*, stage of illness; *SUI*, suicidal ideation and self-harm.

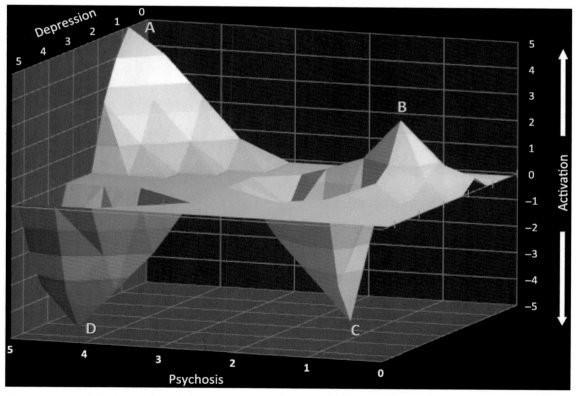

FIG. 4   Using new family studies to model the bipolar disorder concept. This model depicts the potential relationships between the three independent dimensions (mania-activation, psychosis, and depression) evident in new family studies of common mood disorders and observed clinical syndromes. The shared "floor" of this model is no activation, no depression, and no psychotic symptoms. Clinically, Peak A represents the mania-hypomania (high activation, moderate psychosis, and low depression) syndrome; Peak B represents the "mixed states" (moderate activation, low psychosis, and moderate depression); Peak C represents "atypical" depression (low activation/retardation, low psychosis, and moderate depression); Peak D represents psychotic depression (very low activation/retardation, moderate psychosis, and high depression). In this model, each clinical syndrome can occur independently. Over the life course, an individual may experience one or more of these specific clinical syndromes, but does not necessarily experience both low and high activation states. *(Reproduced from Scott, J. (2011). Bipolar disorder: From early identification to personalized treatment. Early Intervention in Psychiatry, 5(2), 89–90.)*

We do place a secondary, but clinically important, emphasis on two other (potentially orthogonal) dimensions of *psychotic symptoms* (which is unidirectional) and *motor activation* (which is bidirectional). This framework informs both our "tripartite" diagnostic modeling (Hickie, Hermens, Naismith, Guastella, et al., 2013), as well as our independent neurobiological, neuropsychological, and targeted intervention studies. The key relevance of motor activation to identifying true bipolar disorders has been emphasized in relevant clinical and family studies (Duffy, Vandeleur, Heffer, & Preisig, 2017; Scott et al., 2017; Shou et al., 2017).

6.1 Illness type

Consequently, (see Table 4) we have utilized a simple "top-down" and "tripartite" model for common mood disorders that proposes three key pathophysiological pathways (anxious-arousal, circadian dysregulation, and neurodevelopmental) that may underpin the three common syndromal constructs—namely anxious-depression, bipolar-type, and psychotic-like disorders (see Fig. 1). We have explored the symptom sets, developmental trajectories, cognitive characteristics, and neurobiological and longitudinal characteristics of these proposed subtypes (detailed in Tables 6A, 6B, and 7). We now also utilize this approach as a provisional basis for optimal sequencing of relevant treatments (see Table 5).

6.2 Clinical stage

To advance the matching of best and safest interventions to early phases of illness, we have supported the application of the general medical concept of clinical staging in mental health practice. That is, as in other areas of medicine, it is commonly accepted that it is totally inadequate to choose treatments, or plan health care, for persons who suffer from conditions that are likely to progress or recur simply on the basis of currently meeting criteria for a very broad illness category (e.g., breast cancer or cardiovascular disease). We suggest that it is equally meaningless in mental health to expect to plan early intervention or secondary preventive strategies, or select specific treatments, on the basis of recognition of broad diagnostic categories such as schizophrenia, bipolar disorder, or major depression. There is indeed a wealth of evidence indicating that subjects at different points along the illness continuum of all of these conditions show quite different patterns of response to psychological or pharmacological interventions (McGorry et al., 2006, 2007; Scott, 2011; Scott et al., 2006).

Consequently, we have proposed a general framework for clinical staging that can be applied to the more severe mood or psychotic disorders (Hickie, Scott, Hermens, Naismith, et al., 2013; McGorry et al., 2006). The clinical staging framework, when applied to young people (ages 12–30), proposes that earlier stages are characterized by lower rates of impairment, and are at lower risk of progression to later, more severe disorders (that are also characterized by greater comorbidity and ongoing disability). We apply this framework to young people who present for health care, and clearly differentiate those in early phases (stages 1a "seeking help" or 1b "attenuated syndromes") from those who have reached a "full-threshold" specific or non-specific, and potentially progressive or recurring, disorder (stage 2 and above). Within the earlier stage 1 disorders, we differentiated the stage 1b "attenuated syndromes" (which often meet DSM-IV or V or ICD-10 criteria for

TABLE 5 Putative stepped-care therapies for relevant depressive subtypes.

Depression type	First line therapy: Psychological/behavioral	Second line therapy: Pharmacological	Experimental therapies
1. Anxious-depression	CBT, IPT, problem-solving, e-health-based anxiety management, exposure therapy, CBCM, MCT	SSRIs, SNRIs	Fish oils, DCS, oxytocin, ketamine
2. Circadian-fatigue/depression	Behavioral regulation, physical activity, sleep-wake cycle/circadian-CBT, rumination-focused CBT, DBT, CBCM, MCT	Melatonin, melatonin analogues, lithium, pregabalin, lamotrigine	Sleep deprivation suvorexant, stimulants, modafinil, TMS, tDCS, ketamine, fish oils
3. Developmental/psychosis	Problem-solving, social skills training, cognitive training, social recovery therapy, CBCM, MCT, IPS	Atypical antipsychotics	Ketamine, cannabidiol, oxytocin, novel neuropeptides, hormonal therapies, fish oils

CBT, cognitive behavior therapy; *IPT*, interpersonal therapy; *CBCM*, cognitive behavioral case management; *MCT*, meta-cognitive therapy; *SSRI*, selective serotonin reuptake inhibitor; *SNRI*, selective serotonin and norepinephrine reuptake inhibitors; *DCS*, d-cycloserine; *DBT*, dialectic behavior therapy; *TMS*, transcranial magnetic stimulation; *tDCS*, transcranial direct current stimulation; *IPS*, individual placement and support.

specific anxiety or mood disorders) from the stage 1a more non-specific, and less severe or impairing anxiety and depressive syndromes. Most important, the domain that has dominated most research work in the early psychosis field, namely the transition from "at-risk" states to "full-threshold" psychotic syndromes (i.e., mental illness trajectory), constitutes only one of our concerns here. In fact, within a transdiagnostic perspective, transition from earlier (stage 1b) to later (2+) stages is the equivalent concept. In our work, the key "clinical transition," therefore, is for young people who largely present with admixtures of anxious and depressive symptomatology from either stages 1a or 1b to stage 2.

6.3 Lifetime trajectory

For the major anxiety, mood, or psychotic disorders, a progressive illness trajectory typically has its onset in late childhood or early puberty, and then recurs or continues progressively into adult life (Hafner, der Heiden, & Maurer, 2008; Merikangas et al., 2010; Paus et al., 2008). Although 75% of major mental disorders begin before the age of 25 years (Gore et al., 2011), current diagnostic and research criteria used to identify these subjects are derived largely from the experiences reported by middle-aged persons with recurring or chronic illness. These mid-life phenotypes often map poorly onto earlier, and often less specific phases of the illness experience (Hickie, Scott, Hermens, Naismith, et al., 2013; McGorry, 2007, 2010; McGorry, Yung, Bechdolf, & Amminger, 2008). Most current research systems also assume that multiple parallel pathways or pathophysiologies underpin each "independent" or "clinical" category—an assumption that is not readily supported by modern family, genetic, and neurobiological risk factor studies (Buckholtz & Meyer-Lindenberg, 2012; Lichtenstein et al., 2009; Sullivan, Daly, & O'Donovan, 2012; Waszczuk, Zavos, Gregory, & Eley, 2014). Consequently, developing criteria for separate early recognition syndromes for anxiety, depression, bipolar, or psychotic disorders makes the demonstrably false assumption that an individual can simply be assigned initially only to one of these trajectories, and that they will stay on that trajectory throughout their entire illness course. Additionally, our own longitudinal work mapping the course of somatic and psychological symptoms associated with common mood disorders in young adolescent twins indicates the extent to which the patterns of symptoms and individual trajectories are highly variable, particularly early in the course of illness (Hansell et al., 2012; Scott et al., 2017).

7 Testing the robustness and clinical utility of a multidimensional outcomes framework

We have now established a long-term longitudinal and interventional framework for testing the robustness and clinical utility of our multidimensional outcomes framework. The conclusions that can be drawn, to date, from our cross-sectional, longitudinal, and predictive work are summarized in Tables 6A, 6B, and 7. Our current understanding with regard to each of the key constructs are now presented:

(i) *Social and occupational function (SaOF).* The most important findings in the testing of our multidimensional framework relate to the course of functional impairment (see Figs. 5 and 6). Contrary to expectations, the majority of those who present early in the course of common anxiety and mood disorders already have established impairment in major areas of educational, employment, or social function (O'Dea et al., 2014, 2016; Purcell et al., 2015). While such findings have been common in the at-risk to psychosis and first-episode psychosis literature, they have not been as extensively delineated in these early mood disorder populations. Most important, there is a significant sub-population (that we characterize as clinical stage 1b—about 30% of those who present) who have not yet developed a persistent or "full-threshold" disorder, but are already impaired *and* often remain largely impaired, or deteriorate further, despite the provision of standard clinical care (Cross et al., 2016). In our view, this population specifically should be the target of more detailed investigation, more intensive and prolonged intervention, and be provided with additional educational or employment support (that is highly individualized) to achieve better outcomes.

Developing new therapeutic approaches to promote SaOF

Beyond non-specific psychological (e.g., CBT) or pharmacological (e.g., SSRIs) treatments for depressive syndromes, we have studied the specific interventions that target SaOF: (i) "DIRECTLY" (e.g., vocational intervention, social recovery, or behavioral activation) or (ii) "INDIRECTLY" via a focus on one or more of the predictive factors (e.g., cognitive training, social skills training, exposure, circadian behavioral management, sleep-wake cycle, activity exercises, and clinic-level and individual-level strategies to reduce STB). Consequently, we work with a model that recognizes the potential value of using a range of cognitive, behavioral, and social interventions that, when deployed (in a timely sequence), have a high chance, directly or indirectly, of improving SaOF (see Fig. 7). Medications are also deployed to target those specific symptom profiles (e.g., anergia, positive symptoms, cognitive changes, sleep-wake

TABLE 6A Robustness of a multidimensional outcomes framework: Constructs 1–4.

	Cross-sectional data	Longitudinal data	Predictive or interventional data
Social and occupational function	– Impairment is common at presentation for care, even in services designed to promote early intervention (Purcell et al., 2015) – Typically a quarter of young people are already disconnected from education or employment and in receipt of financial assistance (O'Dea et al., 2014, 2016). This is at least twice as high as age-matched subjects from the general population in Australia – Baseline impairment is predicted by neurocognitive features, severity of mood, or psychotic symptoms and concurrent substance (notably cannabis) misuse (O'Dea et al., 2014)	– While the longitudinal course of impairment may vary considerably for individuals, the overall pattern is largely one of stable trajectories, whereby the degree of impairment at entry to care is the main predictor of long-term course (Iorfino et al., 2018a)	– Longer-term impairment is best predicted by baseline impairment (Lee et al., 2017) – Severe impairment early in the course of disorders is predictive of both worsening course of illness and very poor long-term function (Scott et al., 2019; WHO, 2011) – Current short-term or less intensive treatment programs (focused on symptom reduction) do not have major impacts on longer-term impairment (Cross, Scott, & Hickie, 2017) – Some interventional studies suggest that a greater impact on functional recovery may be achieved by more intensive or better supported care (Cross, Scott, & Hickie, 2017)
Deliberate self-harm, accidents and injury, and suicidal thoughts and behaviors	– DSH and STB are common at presentation for care, with up to a third of young people are already experiencing STB (Scott, Hermens, Naismith, et al., 2012) – These compare with Australian data of 7.5% of teenagers expressing suicidal ideation and 2.4% attempting suicide in the previous 12 months (Zubrick et al., 2016)	– Participation in care is associated with a marked reduction in STB in those engaged with care (Iorfino et al., 2018b) – New experiences of STB emerge during care, particularly among those whose illness course and trajectory is worsening, and in association with alcohol and other substance misuse (Iorfino et al., 2018b)	– STB at entry to care does predict further risk of later STB (Iorfino et al., 2018b) – STB at entry is also predictive of the emergence of bipolar-type (mood instability) syndromes (Iorfino et al., 2018b) – STB at entry is predictive of worse social and economic function at longer-term follow-up (Iorfino et al., 2018b)
Alcohol or other substance misuse	– Alcohol or other substance misuse is common at service entry (Hermens, Scott, et al., 2013) – Up to a third of young people have established alcohol or substance misuse at entry) (Hermens, Scott, et al., 2013) – Prevalence of at least weekly use of alcohol across three age bands are 12% (12–17 years), 39% (18–19 years), and 45% (20–30 years), and for at least weekly cannabis are 7% (12–17 years), 14% (18–19 years), and 18% (20–30 year olds)(Hermens, Scott, et al., 2013) – More frequent substance use is predicted by older age, male gender, and having psychotic or bipolar disorders (Hermens, Scott, et al., 2013)	– Early first alcohol use (≤age 14) is associated with lower clinician-rated sociooccupational functioning at service entry, while early cannabis (first use ≤ age 15) or amphetamine-type stimulant (first use ≤ age 16) use are not (independent of early alcohol status) (Crouse et al., 2019). Sociooccupational differences between early and non-early drinkers do not appear sustained in the long-term (Crouse et al., 2019)	– Early-onset alcohol use is predictive of manic-like experiences, suicidal ideation, and risky drinking status (Crouse et al., n.d.) – Some research suggests an associative model of poor outcome and early alcohol use, whereby common factors increase risk for both early drinking and poor outcomes (Crouse et al., 2019)

Physical health			
	– Physical health problems and risk factors to later poor physical health are common in young people presenting to care (Scott et al., 2015)	– Youth cohorts tend to gain weight and develop insulin resistance in association with engagement with care (Goldstein, Blanco, He, & Merikangas, 2016)	– Depressive syndromes are predictive of early death due to cardiovascular disease (Goldstein et al., 2015)
	– A third of young people smoke tobacco daily (Scott et al., 2015). This is up to three times higher than the age-matched Australian population	– The emergence of comorbid physical health problems is notable in those who continue to engage with care, particularly among those who are treatment-resistant or develop significant (neurological or metabolic) side-effects from medications (Scott et al., 2015)	– Interventional studies are underway in this cohort to attempt to reduce risks to later cardiovascular disease (Gehue, Scott, Hermens, Scott, & Hickie, 2015)
	– 20% of young people are overweight and 10% are obese (Scott et al., 2015)		– More proactive approaches (e.g., metformin therapy) may need to be tested in those identified as having or being at-risk of developing insulin resistance (Rosenblat & McIntyre, 2017)
	– The patterns of overweight and obesity are the same as age-matched subjects from the general population in Australia		
	– Up to 10% of young women are underweight (Scott et al., 2015). This rate is higher than that expected in the age-matched population		
	– Increasing BMI is associated with evidence of emerging insulin resistance (Scott et al., 2019)		

BMI, body mass index.

TABLE 6B Robustness of a multidimensional outcomes framework, construct 5: Illness type, clinical stage and lifetime trajectory.

	Cross-sectional data	Longitudinal data	Predictive or interventional data
Illness type	– Young people with unipolar mood disorders are as impaired as those with bipolar disorders (Scott, Scott, et al., 2014) and have similar initial clinical presentations (Scott et al., 2013a) – The validity of three illness sub-groups (anxious-depression, circadian, developmental psychotic) is supported by demographic, family history, and neuropsychological data (Hickie, Hermens, Naismith, Guastella, et al., 2013; Scott et al., 2015) – Bipolar-type disorders are associated with objective evidence of sleep phase delay an intrinsic shifts in underlying circadian rhythms (Robillard et al., 2013; Robillard et al., 2013)	– Progression to more severe disorders (i.e., later clinical stages) is typically characterized by the development of comorbidity and further functional impairment, rather than the progression to a specific diagnostic (e.g., bipolar, schizophrenia) category (Iorfino et al., 2019; Hickie, Scott, Hermens, Naismith, et al., 2013). – The emergence of bipolar-type disorders is common in this cohort, but is characterized largely by mood instability or brief periods of activation as distinct from clear manic episodes (Grierson et al., 2016; Scott et al., 2013a) – Severe depressive disorders are often associated with psychotic-like experiences without clear progression to schizophrenia or manic episodes. The delineation of affective vs nonaffective first-episode psychosis is not clear-cut in these young people (Iorfino et al., 2019; Hickie, Scott, Hermens, Naismith, et al., 2013)	– Some illness-type specificity is suggested by the extent to which reduction in depressive symptoms in those with circadian dysfunction is correlated with correction of underlying circadian rhythm disturbance (Robillard et al., 2018; Grierson et al., n.d.)
Clinical stage	– The differentiation of early stages (i.e., stages 1a, 1b and 2+) of anxiety, mood, and psychotic disorder is supported by independent functional impairment (Hamilton et al., 2011; Purcell et al., 2015; Scott, Hermens, Glozier, et al., 2012), neuropsychological (Hermens et al., 2013; Tickell, RSC, Hickie, & Hermens, 2017), structural brain imaging (Lagopoulos et al., 2013; Lagopoulos, Hermens, Naismith, Scott, & Hickie, 2012), and circadian biology (Naismith et al., 2012; Scott et al., 2014) data	– Differential rates of transition across clinical stage over longitudinal tracking support the validity of the model (Iorfino et al., 2019; Hickie, Scott, Hermens, Naismith, et al., 2013) – Over 2 years, less than 5% of stage 1a progress to stage 2+, compared with approximately 20% of stage 1b persons (Iorfino et al., 2019) – Up to half of all transitions to later stages occur within 12 months of presentation to care, with the majority in the first 3 months after presentation (Iorfino et al., 2019; Cross, Scott, & Hickie, 2017)	– Progression across clinical stage is predicted by greater functional impairment at baseline, and to a lesser degree, by manic-like or psychotic-like symptoms at baseline (Cross, Scott, & Hickie, 2017)
Lifetime trajectory	– Young people are entering care at highly variable stages of cognitive, social, educational, physical, and neurobiological development – Current age is a relatively poor predictor of many key aspects of individual development (Paus, Keshavan, & Giedd, 2008)	– Individual clinical course is highly variable in terms of symptoms, functioning, and changes in neuropsychological or neurobiological function (Cross et al., 2018)	– It is unclear the extent to which highly-personalized interventions have a positive impact on key developmental constructs associated with mood disorders

TABLE 7 Robustness of a multidimensional outcomes framework: Mechanistic data from sleep-wake and circadian, neuropsychological, and neuroimaging studies.

	Cross-sectional data	Longitudinal data	Predictive or interventional data
Sleep-wake and circadian function	– Delayed sleep is prominent in young people presenting for mental health care (Robillard et al., 2014; Robillard, Naismith, Rogers, Ip, et al., 2013), is more pronounced across illness stages (Scott, Robillard, et al., 2014), and also shows associations with alcohol and substance use (Glozier et al., 2014) – Disrupted and more variable sleep is common, but not specific to any diagnostic category or stage of illness (Robillard et al., 2015; Scott, Robillard, et al., 2014) – Abnormally long sleep may indicate a group at high-risk for neuropsychological deficits, potentially related to psychotropic medication use (Carpenter et al., 2015) – Biological markers of underlying circadian rhythms are also delayed in a subgroup (Carpenter et al., 2017) and at later clinical stages (Naismith et al., 2012) – Specific sleep disturbances and delayed circadian phase appear to be more indicative of a bipolar-type phenotype (Robillard, Naismith, Rogers, Scott, et al., 2013; Scott et al., 2016)	– Poorer baseline sleep efficiency predicts worsening in manic symptoms at follow up (Robillard et al., 2016) – Both shorter sleep and poorer circadian rhythmicity are predictive of worsening in verbal memory (Robillard et al., 2016)	– Correction of circadian abnormalities appears to be associated with depressive symptom improvement in response to chronobiological interventions (Robillard et al., 2018; Grierson et al., n.d.)
Neuropsychological function	– Impairments in neuropsychological performance are progressive across clinical stages (Hermens, Naismith, et al., 2013) and are heterogeneous, with evidence for distinct patterns of dysfunction that are independent of clinical symptoms (Hermens et al., 2011) – Impairments in both lower order (e.g., emotion recognition) and higher order (e.g., theory of mind, attributional biases) are impaired in those disorders characterized by social impairment (autism and psychosis (Couture et al., 2010)). Such social cognition impairments may also provide a broad cross-diagnostic phenotype across psychiatric populations (Cotter et al., 2018) that predict social skill and social disability	– Improvement in neuropsychological function is associated with clinical and functional improvement (Tickell et al., 2017), with baseline cognitive performance independently predictive of functional outcomes (Lee, Hermens, et al., 2013; Lee, Hermens, et al., 2015) – While social cognition is influenced by other cognitive variables (executive function (Couture et al., 2010)), it remains unique with underpinning circuitry that typically predates psychiatric symptoms (Addington, Penn, Woods, Addington, & Perkins, 2008; Chung, Kang, Shin, Yoo, & Kwon, 2008; Constantino et al., 2017) and separately predicts social disability (Constantino et al., 2017; Nuske, Vivanti, & Dissanayake, 2013)	– Evidence suggests that cognitive training can improve neuropsychological and psychosocial functioning (Lee et al., 2013). Targeting such interventions to those with significant impairment has the potential to make a significant impact on outcomes – Interventions that improve social cognition (Guastella et al., 2010, 2015) also improve social reciprocity, social skill, behavior, and functioning (Inchausti et al., 2017; Yatawara, Einfeld, Hickie, Davenport, & Guastella, 2016). These include oxytocin studies (Yatawara et al., 2016) and social skills training (Inchausti et al., 2017). These studies suggest promise, particularly for disorders of social impairment, including those with psychosis and autism

Continued

TABLE 7 Robustness of a multidimensional outcomes framework: Mechanistic data from sleep-wake and circadian, neuropsychological, and neuroimaging studies.—cont'd

	Cross-sectional data	Longitudinal data	Predictive or interventional data
Neuroimaging	– Frontal gray matter volume is decreased across clinical stages (Eggins, Hatton, Hermens, Hickie, & Lagopoulos, 2018; Lagopoulos et al., 2012), and reduced gray matter volume (anterior insula) is associated with poorer neuropsychological function, as well as clinical illness characteristics (Hatton et al., 2012) – White matter changes are present across all illness stages, indicating that such changes likely occur early in the course of disorder (Hatton et al., 2014a; Lagopoulos et al., 2013; Lagopoulos, Hermens, Hatton, Battisti, et al., 2013) – Multiple neurometabolic profiles are present in this population, independent of clinical, demographic, and functional characteristics (Hermens et al., 2015; Hermens, Lagopoulos, Naismith, Tobias-Webb, & Hickie, 2012) – Neurometabolic abnormalities are associated with alcohol and substance use (Chitty et al., 2013; Chitty, Lagopoulos, Hickie, & Hermens, 2014; Hermens, Chitty, et al., 2015), as well as delayed sleep and melatonin rhythms (Naismith et al., 2014; Robillard et al., 2017) – Structural neuroimaging indicates that cortical thinning and some white matter changes are associated with neuropsychological deficits, as well as psychiatric symptoms (Hatton et al., 2013; Hatton et al., 2014b; Hermens et al., 2017)	– Longitudinally, changes in alcohol and tobacco use in bipolar disorder are associated with neurometabolic changes indicative of oxidative stress (Chitty et al., 2015)	

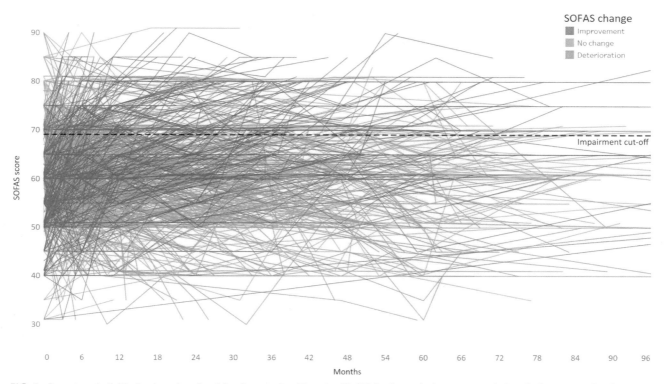

FIG. 5 Long-term individual trajectories of social and occupational function (SaOF) for those who improve, remain largely the same, or deteriorate over the course of engagement with clinical services. *SOFAS*, clinician-rated social and occupational functional assessment scale; Scores <70 are indicative of significant impairment in daily function.

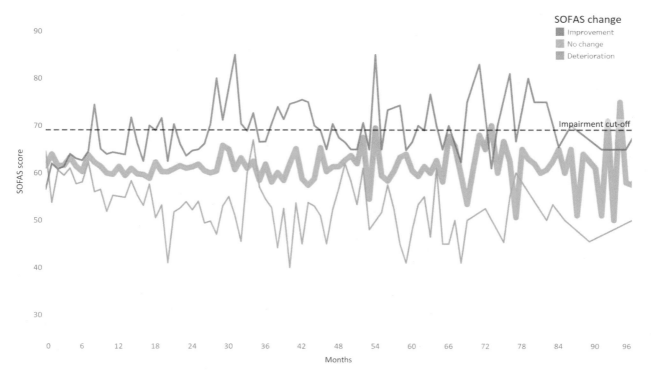

FIG. 6 Long-term group-based variations in social and occupational function (SaOF) for those who improve, remain largely the same, or deteriorate over the course of engagement with clinical services. *SOFAS*, clinician-rated social and occupational functional assessment scale; Scores <70 are indicative of significant impairment in daily function.

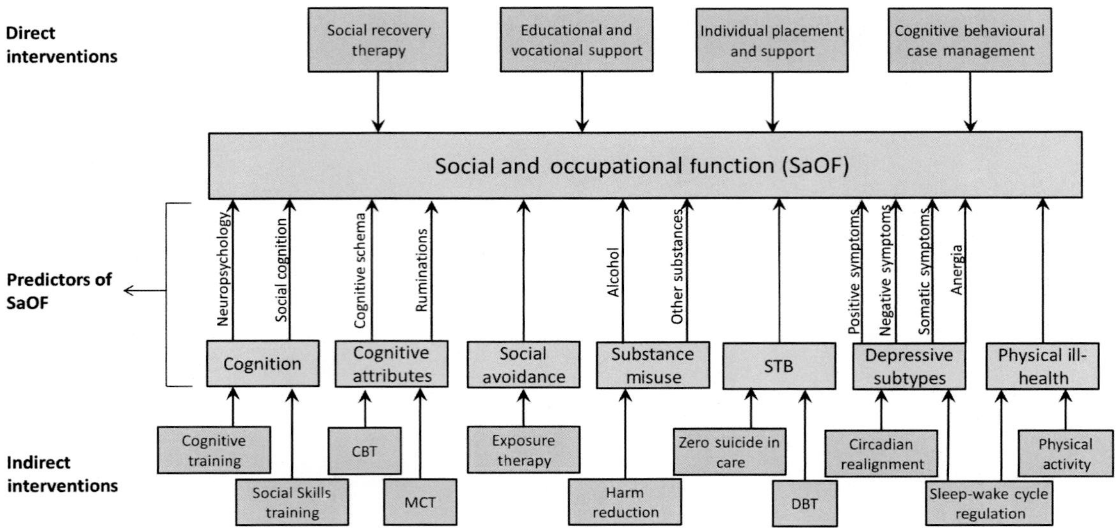

FIG. 7 Evidence-based direct and indirect interventions to improve SaOF. *STB*, suicidal thoughts and behavior; *CBT*, cognitive behavioral therapy; *MCT*, metacognitive therapy; *DBT*, dialectical behavior therapy.

cycle disturbances), or depressive subtypes (anxious-depression, circadian-depression) of greatest relevance to promoting SaOF. This may include specifically considering the ways in which adverse effects of some medications can have major impacts on SaOF through mechanisms such as drowsiness, cognitive slowing, motor impairments, or reduced physical activity.

(ii) *Deliberate self-harm, accidents and injury, and suicidal thoughts and behaviors:* The importance of a specific focus on DSH and STB is emphasized. While some epidemiologically-based reports emphasize that such behaviors in the wide community may be self-limiting, in this population with early disorders and help-seeking behavior, they are not only a determinant of immediate distress and additional impairment, but also a predictor of later onset of more severe (bipolar-type or mood unstable) illness, and greater functional impairment (Iorfino et al., 2018b) (see Fig. 8). Importantly, at entry to services, STB are not restricted to those with more severe illness types (Scott, Hermens, Naismith, et al., 2012). These findings have important implications for an enhanced focus on reducing DSH and STB throughout the course of clinical care for *all* of those who present to such services—not just those who are rated as "higher-risk."

(iii) *Alcohol and other substance misuse.* Concurrent alcohol and other substance misuse are often recognized as important co-morbid conditions for anxiety and depressive disorders, but are rarely subject to systematic evaluation, tracking, or intervention within such mental health care services. These results emphasize the need to assess adequately the contributions at entry of substance misuse to current impairment, DSH, and STB, and the potential relationships with physical health, illness type (particularly bipolar disorders), and illness course. Additionally, these data emphasize the potential for secondary prevention of alcohol and other substance misuse by active management of common anxiety and depressive syndromes *before* the age of likely exposure to alcohol and other substances. Again, such broad-based secondary prevention strategies should be a core feature of service delivery—particularly to the younger persons (i.e., under the age of 18 years) who present for care. Active management of comorbid disorders early in the course of illness has the potential to not only reduce longer term substance misuse-determined morbidity and mortality, but significantly improve functional outcomes more broadly, and reduce the morbidity due to DSH and STB.

(iv) *Physical health.* While the adverse physical health consequences of persistent mental illness in middle-aged persons is now well-recognized, the opportunity to prevent this morbidity and premature mortality in those who present with early mood disorders has not yet been realized. These data highlight key opportunities, notably related to cessation of smoking tobacco, preventing weight gain, and providing early active interventions for metabolic disturbance, hormone dysfunction, or concurrent autoimmune and other inflammatory conditions. It is likely that we will need to develop and evaluate new screening and active intervention protocols to achieve better outcomes in this very "at-risk" population. While a current focus on dealing with these risks in "early psychosis" cohorts is laudable, it is clear that we need to adopt a much more proactive and evidence-informed approach to the clinical care of those with early anxiety-depressive syndromes.

(v) *Illness type, clinical stage, and lifetime trajectory.* Within this cohort, we have extensively evaluated the utility of clinical staging as an adjunct to clinical care. While we have demonstrated its fundamental predictive validity (i.e.,

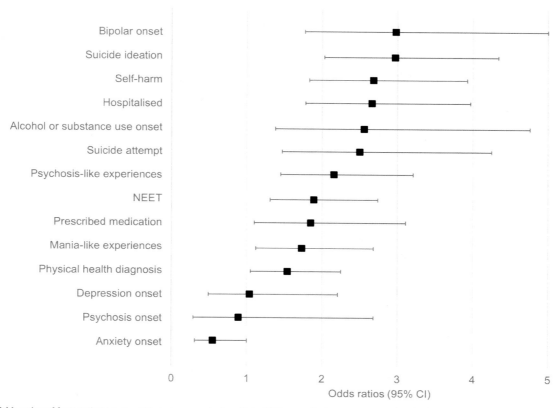

FIG. 8 Odds ratios of factors that are more likely to occur during clinical follow-up in those young people who had attempted suicide before or at the time that they presented for mental health care. *NEET*, not in education employment or training. Odds ratios and 95% CIs for each follow up outcome presented in descending order (increased odds to reduced odds). The values between 0 and 1 on the *x*-axis are logarithmic.

early stages have a differential capacity to predict transitions to "full-threshold" disorders), and detailed supporting neuropsychological, structural brain imaging, and circadian measures, we have not yet demonstrated:

(a) which clinical factors or other objective markers—at an individual level—can predict, robustly, transition to a later stage. Here, the search for differentiating psychosocial or neurobiological "markers" *within* subjects who are at the same clinical stage at baseline is an urgent priority. The matching of such markers to the provision of targeted or novel interventions would then be justified; and

(b) that active clinical care, or type of care, has the capacity to prevent transitions to later stages. This requires the development of specific health service intervention trials evaluating the impact of more intensive, longer-duration, more personalized or better-targeted care packages. From our perspective, the most obvious cohort to focus on is that we have classed as stage 1b (about 30% of our cohort) with "attenuated" clinical syndromes, established impairment, a high risk of transition to stage 2 (approx. 20% over 2 years), and poor functional outcomes under current care conditions.

8 Designing personalized treatments based on illness type and clinical stage

Internationally, there is an increasing recognition of the need for assertive management of adolescents and young adults who present for care in the early phases of major mood or psychotic disorders. However, the symptom complexes presented are often an admixture of anxiety, depressive, hypomanic, psychotic, or substance misuse-related related symptoms (Hickie, Scott, Hermens, Naismith, et al., 2013; Hickie, Scott, Hermens, Scott, et al., 2013) and, typically, these syndromes do not meet the diagnostic thresholds employed for more specific disorders. The evidence-base for providing specific treatments for many of these sub-threshold or first episode-type disorders is sparse (Scott, Hickie, & McGorry, 2012), and there is also an increasing desire to link interventions more closely to underlying developmental or specific pathophysiological pathways (Hickie, Scott, Hermens, Scott, et al., 2013).

Complementary to our clinical staging and "tripartite models," we have commenced the development and evaluation of a treatment-selection model (see Table 3), demonstrating the capacity to, first, prioritize psychological, social, and behavioral approaches, and, second, only later choose pharmacological approaches that may be most relevant to the underlying pathophysiological pathway (inferred from the observed phenotype or concurrent neurobiological testing). For example, 24-h sleep-wake cycle behavioral interventions or melatonin-based antidepressants may be preferred for some depressive disorders in those who have phenotypic, actigraphic, or laboratory-based evidence of underlying circadian disturbance. This approach is the subject of ongoing clinical testing and refinement. We do have preliminary evidence from an open trial that resolution of depressive symptoms in those with "circadian-type" disorders is strongly correlated with correction of abnormal underlying circadian (laboratory-determined dim-light melatonin onset times) parameters (Robillard et al., 2018).

9   Designing health services to respond to early mood disorders: Opportunities for early intervention and secondary prevention

In addition to our development of individual-level approaches (i.e., multidimensional outcomes framework, clinical staging, tripartite diagnostic system) to early intervention and secondary prevention for young people with emerging mood disorders, we have placed considerable emphasis on taking these interventions to scale through both traditionally place-based (e.g., headspace network (Cross et al., 2014; McGorry, Tanti, et al., 2007)) and new online assessment and delivery (Project Synergy (Burns, Davenport, Durkin, Luscombe, & Hickie, 2010; Christensen & Hickie, 2010; Iorfino et al., 2017)) systems. Concurrently, therefore, we are testing whether our new clinical insights can be used by relevant enhanced primary care services (Cross et al., 2014, 2016), as well as secondary care services and hospital-based service for young people (Tickell et al., n.d., 2019) to deliver more effective care options to young people presenting early in the course of major mood disorders. Inevitably, questions then arise as to the appropriate allocation of clinical resources (i.e., duration, intensity, and characteristics of interventions) relative to demand for care. While placing considerable emphasis on risk management, and active approaches to reducing DSH and STB for the whole cohort, our current focus is whether prior identification of those classes as being at stage 1b "attenuated syndromes" combined with 12-months of more intensive and personalized care options may result in improved longer-term social and economic function, reduced DSH and STB, improved course of depressive symptoms, and lower transition rates to clinical stage 2 "full-threshold" depressive, bipolar, psychotic, or comorbid disorders.

10   Testing novel treatments

There is clearly an urgent need to develop and test new treatments for the major mood disorders that affect young people. We have argued that any approach to clinical trials in this cohort will need to address substantive issues related to clinical stage, baseline impairment, patterns of comorbidity, and potential pathophysiological mechanisms, and stratify or select participants accordingly. A range of therapeutic options now exist, and are deserving of much more systematic evaluation (see Fig. 9). We have supported the development of specific cognitively informed models of care that may be particularly relevant to those with ruminations (Grierson et al., n.d.; Grierson, Hickie, Naismith, & Scott, 2016), which themselves have a considerable impact on sleep-wake cycle patterns (Grierson, Hickie, Naismith, Hermens, et al., 2016; Scott et al., 2016). We also continue to evaluate whether other novel treatment approaches, such as oxytocin-assisted or d-cycloserine (DCS)-enhanced behavioral or group therapies are particularly relevant to those with impaired social cognition or severe social anxiety. Novel immune or hormonal therapies may also be relevant, not only to those with objective markers or clinical indications for such approaches, but also to broader categories, such as "circadian-type" or "atypical" depressive syndromes. Novel cannabinoid-based strategies may be particularly relevant to those with psychotic phenomena, or those with significant comorbid anxiety.

11   Conclusions

Over the past decade, we have accumulated substantial cross-sectional and longitudinal evidence detailing how early intervention and secondary prevention services can be developed, structured, and evaluated for young people with emerging mood disorders. The key findings support the utility of clinical staging as an adjunct to traditional diagnostic frameworks, the recognition of at least three pathophysiological underlying paths (resulting in different age-dependent phenotypes in childhood and adolescence), and a strong focus on implementing a genuinely multidimensional assessment and longitudinal outcomes framework. A clear conclusion is that impairment and concurrent morbidity is well-established in these

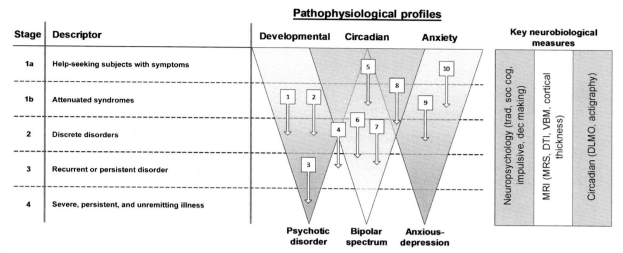

FIG. 9 Stepped-care or phase II interventional studies related to clinical stage and depressive subtype. Specific intervention studies are conducted at various stages of illness, but also with respect to one of three proposed pathophysiological paths. (1) cannbidiol; (2) social cognitive training augmented by oxytocin; (3) adjunctive estrogen therapy; (4) ketamine; (5) rumination focused CBT; (6) transcranial magnetic stimulation; (7) melatonin-enhanced vs suvorexant-enhanced behavioral regulation therapies; (8) fish oils; (9) oxytocin- or d-cycloserine-enhanced group-based social anxiety therapy; (10) e-health-based anxiety management. *Trad*, traditional; *soc. cog*, social cognition; *dec*, decision; *MRI*, magnetic resonance imaging; *MRS*, magnetic resonance spectroscopy; *DTI*, diffusion tensor imaging; *VBM*, voxel-based morphometry; *DLMO*, dim light melatonin onset.

young people (mean age 18 years) by the time they present for mental health-focused care. A clear caveat, however, is that early identification followed by standard care does not automatically result in improved outcomes. Much work is yet to be done in the development and evaluation of specific individual-level and service-level care packages to reduce the morbidity and mortality due to early-onset major mood disorders.

REFERENCES

Addington, J., Penn, D., Woods, S. W., Addington, D., & Perkins, D. O. (2008). Facial affect recognition in individuals at clinical high risk for psychosis. *The British Journal of Psychiatry*, *192*(1), 67–68.

Australian Government National Mental Health Commission. (2014). *Contributing Lives, Thriving Communities*. Report of the national review of mental health programmes and services.

Banihashemi, N., Robillard, R., Yang, J., Carpenter, J. S., Hermens, D. F., Naismith, S. L., et al. (2016). Quantifying the effect of body mass index, age, and depression severity on 24-h activity patterns in persons with a lifetime history of affective disorders. *BMC Psychiatry*, *16*(1), 317.

Bogaty, S. E. R., Crouse, J. J., Hickie, I. B., & Hermens, D. F. (2019). The neuropsychological profiles of young psychosis patients with and without current cannabis use. *Cognitive neuropsychiatry*, *24*, 40–53. Under Review.

Buckholtz, J. W., & Meyer-Lindenberg, A. (2012). Psychopathology and the human connectome: Toward a transdiagnostic model of risk for mental illness. *Neuron*, *74*(6), 990–1004.

Burns, J. M., Davenport, T. A., Durkin, L. A., Luscombe, G. M., & Hickie, I. B. (2010). The internet as a setting for mental health service utilisation by young people. *The Medical Journal of Australia*, *192*, S22–S26.

Carpenter, J. S., Robillard, R., Hermens, D. F., Naismith, S. L., Gordon, C., Scott, E. M., et al. (2017). Sleep-wake profiles and circadian rhythms of core temperature and melatonin in young people with affective disorders. *Journal of Psychiatric Research*, *94*, 131–138.

Carpenter, J. S., Robillard, R., Lee, R. S., Hermens, D. F., Naismith, S. L., White, D., et al. (2015). The relationship between sleep-wake cycle and cognitive functioning in young people with affective disorders. *PLoS One*, *10*(4).

Casey, B. J., Craddock, N., Cuthbert, B. N., Hyman, S. E., Lee, F. S., & Ressler, K. J. (2013). DSM-5 and RDoC: Progress in psychiatry research? *Nature Reviews Neuroscience*, *14*(11), 810–814.

Chitty, K. M., Kaur, M., Lagopoulos, J., Hickie, I. B., & Hermens, D. F. (2011). Alcohol use and mismatch negativity in young patients with psychotic disorder. *Neuroreport*, *22*(17), 918–922.

Chitty, K. M., Kaur, M., Lagopoulos, J., Hickie, I. B., & Hermens, D. F. (2014). Risky alcohol use predicts temporal mismatch negativity impairments in young people with bipolar disorder. *Biological Psychology*, *99*, 60–68.

Chitty, K. M., Lagopoulos, J., Hickie, I. B., & Hermens, D. F. (2013). Risky alcohol use in young persons with emerging bipolar disorder is associated with increased oxidative stress. *Journal of Affective Disorders*, *150*(3), 1238–1241.

Chitty, K. M., Lagopoulos, J., Hickie, I. B., & Hermens, D. F. (2014). The impact of alcohol and tobacco use on in vivo glutathione in youth with bipolar disorder: An exploratory study. *Journal of Psychiatric Research*, *55*, 59–67.

Chitty, K. M., Lagopoulos, J., Hickie, I. B., & Hermens, D. F. (2015). A longitudinal proton magnetic resonance spectroscopy study investigating oxidative stress as a result of alcohol and tobacco use in youth with bipolar disorder. *Journal of Affective Disorders, 175*, 481–487.

Christensen, H., & Hickie, I. B. (2010). Using e-health applications to deliver new mental health services. *The Medical Journal of Australia, 192*, S53–S56.

Chung, Y. S., Kang, D. H., Shin, N. Y., Yoo, S. Y., & Kwon, J. S. (2008). Deficit of theory of mind in individuals at ultra-high-risk for schizophrenia. *Schizophrenia Research, 99*(1–3), 111–118.

Constantino, J. N., Kennon-McGill, S., Weichselbaum, C., Marrus, N., Haider, A., Glowinski, A. L., et al. (2017). Infant viewing of social scenes is under genetic control and is atypical in autism. *Nature, 547*(7663), 340–344.

Copeland, W. E., Adair, C. E., Smetanin, P., Stiff, D., Briante, C., Colman, I., et al. (2013). Diagnostic transitions from childhood to adolescence to early adulthood. *Journal of Child Psychology and Psychiatry, 54*(7), 791–799.

Cotter, J., Granger, K., Backx, R., Hobbs, M., Looi, C. Y., & Barnett, J. H. (2018). Social cognitive dysfunction as a clinical marker: A systematic review of meta-analyses across 30 clinical conditions. *Neuroscience and Biobehavioral Reviews, 84*, 92–99.

Couture, S. M., Penn, D. L., Losh, M., Adolphs, R., Hurley, R., & Piven, J. (2010). Comparison of social cognitive functioning in schizophrenia and high functioning autism: More convergence than divergence. *Psychological Medicine, 40*(4), 569–579.

Cross, S. P., Hermens, D. F., & Hickie, I. B. (2016). Treatment patterns and short-term outcomes in an early intervention youth mental health service. *Early Intervention in Psychiatry, 10*(1), 88–97.

Cross, S. P. M., Hermens, D. F., Scott, E. M., Ottavio, A., McGorry, P. D., & Hickie, I. B. (2014). A clinical staging model for early intervention youth mental health services. *Psychiatric Services, 65*(7), 939–943.

Cross, S. P. M., Hermens, D. F., Scott, J., Salvador-Carulla, L., & Hickie, I. B. (2017). Differential impact of current diagnosis and clinical stage on attendance at a youth mental health service. *Early Intervention in Psychiatry, 11*(3), 255–262.

Cross, S. P., Scott, J. L., Hermens, D. F., & Hickie, I. B. (2018). Variability in clinical outcomes for youths treated for subthreshold severe mental disorders at an early intervention service. *Psychiatric Services, 69*, 555–561.

Cross, S. P. M., Scott, J., & Hickie, I. B. (2017). Predicting early transition from sub-syndromal presentations to major mental disorders. *British Journal of Psychiatry Open, 3*(5), 223–227.

Crouse, J. J., Chitty, K. M., Iorfino, F., White, D., Nichles, A., Zmicerevska, N., et al. (2019). Exploring associations between early substance use and longitudinal socio-occupational functioning in young people engaged in a mental health service. *PLoS ONE, 14*(1), e0210877.

Crouse, J.J., Chitty, K. M., Iorfino, F., White, D., Nichles, A., Zmicerevska, N., et al. (n.d.). Longitudinal associations between early alcohol initiation, future risky use, and clinical course in young people with mental illness. In preparation.

Crouse, J. J., Lee, R. S. C., White, D., Moustafa, A. A., Hickie, I. B., & Hermens, D. F. (2018). Distress and sleep quality in young amphetamine-type stimulant users with an affective or psychotic illness. *Psychiatry Research, 262*, 254–261.

Crouse, J. J., Moustafa, A. A., Bogaty, S. E. R., Hickie, I. B., & Hermens, D. F. (2018). Parcellating cognitive heterogeneity in early psychosis-spectrum illnesses: A cluster analysis. *Schizophrenia Research, 202*, 91–98. Under Review.

Cuthbert, B. N., & Insel, T. R. (2013). Toward the future of psychiatric diagnosis: The seven pillars of RDoC. *BMC Medicine, 11*, 126.

Duffy, A., Vandeleur, C., Heffer, N., & Preisig, M. (2017). The clinical trajectory of emerging bipolar disorder among the high-risk offspring of bipolar parents: Current understanding and future considerations. *International Journal of Bipolar Disorders, 5*(1), 37.

Eggins, P. S., Hatton, S. N., Hermens, D. F., Hickie, I. B., & Lagopoulos, J. (2018). Subcortical volumetric differences between clinical stages of young people with affective and psychotic disorders. *Psychiatry Research, 271*, 8–16.

Erskine, H. E., Moffitt, T. E., Copeland, W. E., Costello, E. J., Ferrari, A. J., Patton, G., et al. (2015). A heavy burden on young minds: The global burden of mental and substance use disorders in children and youth. *Psychological Medicine, 45*(7), 1511–1563.

Gehue, L. J., Scott, E., Hermens, D. F., Scott, J., & Hickie, I. (2015). Youth early-intervention study (YES)—Group interventions targeting social participation and physical well-being as an adjunct to treatment as usual: Study protocol for a randomized controlled trial. *Trials, 16*, 333.

Glozier, N., O'Dea, B., McGorry, P. D., Pantelis, C., Amminger, G. P., Hermens, D. F., et al. (2014). Delayed sleep onset in depressed young people. *BMC Psychiatry, 14*, 33.

Goldman, H. H., Skodol, A. E., & Lave, T. R. (1992). Revising axis V for DSM-IV: A review of measures of social functioning. *The American Journal of Psychiatry, 149*, 1148–1156.

Goldstein, B. I., Blanco, C., He, J. P., & Merikangas, K. (2016). Correlates of overweight and obesity among adolescents with bipolar disorder in the National Comorbidity Survey-Adolescent Supplement (NCS-A). *Journal of the American Academy of Child and Adolescent Psychiatry, 55*(12), 1020–1026.

Goldstein, B. I., Carnethon, M. R., Matthews, K. A., McIntyre, R. S., Miller, G. E., Raghuveer, G., et al. (2015). Major depressive disorder and bipolar disorder predispose youth to accelerated atherosclerosis and early cardiovascular disease: A scientific statement from the American Heart Association. *Circulation, 132*(10), 965–986.

Gore, F. M., Bloem, P. J. N., Patton, G. C., Ferguson, J., Joseph, V., Coffey, C., et al. (2011). Global burden of disease in young people aged 10–24 years: A systematic analysis. *Lancet, 377*(9783), 2093–2102.

Grierson, A. B., Hickie, I. B., Naismith, S. L., Hermens, D. F., Scott, E. M., & Scott, J. (2016). Circadian rhythmicity in emerging mood disorders: State or trait marker? *International Journal of Bipolar Disorders, 4*(1), 3.

Grierson, A. B., Hickie, I. B., Naismith, S. L., & Scott, J. (2016). The role of rumination in illness trajectories in youth: Linking trans-diagnostic processes with clinical staging models. *Psychological Medicine, 46*(12), 2467–2484.

Grierson AB, Scott J, Carpenter JS, White D, Naismith SL, Scott EM, et al. A pilot study of a brief intervention program (RECHARGE) to reduce sleep-wake and circadian rhythm disturbances in youth being treated for depression n.d. In preparation.

Guastella, A. J., Einfeld, S. L., Gray, K. M., Rinehart, N. J., Tonge, B. J., Lambert, T. J., et al. (2010). Intranasal oxytocin improves emotion recognition for youth with autism spectrum disorders. *Biological Psychiatry, 67*(7), 692–694.

Guastella, A. J., Ward, P. B., Hickie, I. B., Shahrestani, S., Hodge, M. A., Scott, E. M., et al. (2015). A single dose of oxytocin nasal spray improves higher-order social cognition in schizophrenia. *Schizophrenia Research, 168*(3), 628–633.

Hafner H, an der Heiden W, Maurer K. Evidence for separate diseases?: Stages of one disease or different combinations of symptom dimensions? European Archives of Psychiatry and Clinical Neuroscience. 2008;258 Suppl. 2:85–96.

Hamilton, B. A., Naismith, S. L., Scott, E. M., Purcell, S., & Hickie, I. B. (2011). Disability is already pronounced in young people with early stages of affective disorders: Data from an early intervention service. *Journal of Affective Disorders, 131*(1–3), 84–91.

Hansell, N. K., Wright, M. J., Medland, S. E., Davenport, T. A., Wray, N. R., Martin, N. G., et al. (2012). Genetic co-morbidity between neuroticism, anxiety/depression and somatic distress in a population sample of adolescent and young adult twins. *Psychological Medicine, 42*(6), 1249–1260.

Hatton, S. N., Lagopoulos, J., Hermens, D. F., Hickie, I. B., Scott, E., & Bennett, M. R. (2014a). Short association fibres of the insula-temporoparietal junction in early psychosis: A diffusion tensor imaging study. *PLoS One, 9*(11).

Hatton, S. N., Lagopoulos, J., Hermens, D. F., Hickie, I. B., Scott, E., & Bennett, M. R. (2014b). White matter tractography in early psychosis: Clinical and neurocognitive associations. *Journal of Psychiatry & Neuroscience, 39*(6), 417–427.

Hatton, S. N., Lagopoulos, J., Hermens, D. F., Naismith, S. L., Bennett, M. R., & Hickie, I. B. (2012). Correlating anterior insula gray matter volume changes in young people with clinical and neurocognitive outcomes: An MRI study. *BMC Psychiatry, 12*, 45.

Hatton, S. N., Lagopoulos, J., Hermens, D. F., Scott, E., Hickie, I. B., & Bennett, M. R. (2013). Cortical thinning in young psychosis and bipolar patients correlate with common neurocognitive deficits. *International Journal of Bipolar Disorders, 1*, 3.

Hermens, D. F., Chitty, K. M., Lee, R. S., Tickell, A., Haber, P. S., Naismith, S. L., et al. (2015). Hippocampal glutamate is increased and associated with risky drinking in young adults with major depression. *Journal of Affective Disorders, 186*, 95–98.

Hermens, D. F., Hatton, S. N., Lee, R. S., Naismith, S. L., Duffy, S. L., Paul Amminger, G., et al. (2017). In vivo imaging of oxidative stress and fronto-limbic white matter integrity in young adults with mood disorders. *European Archives of Psychiatry and Clinical Neuroscience, 268*, 145–156.

Hermens, D. F., Lagopoulos, J., Naismith, S. L., Tobias-Webb, J., & Hickie, I. B. (2012). Distinct neurometabolic profiles are evident in the anterior cingulate of young people with major psychiatric disorders. *Translational Psychiatry, 2*, e110.

Hermens, D. F., Lee, R. S., De Regt, T., Lagopoulos, J., Naismith, S. L., Scott, E. M., et al. (2013). Neuropsychological functioning is compromised in binge drinking young adults with depression. *Psychiatry Research, 210*(1), 256–262.

Hermens, D. F., Naismith, S. L., Chitty, K. M., Lee, R. S., Tickell, A., Duffy, S. L., et al. (2015). Cluster analysis reveals abnormal hippocampal neurometabolic profiles in young people with mood disorders. *European Neuropsychopharmacology, 25*(6), 836–845.

Hermens, D. F., Naismith, S. L., Lagopoulos, J., RSC, L., Guastella, A. J., Scott, E. M., et al. (2013). Neuropsychological profile according to the clinical stage of young persons presenting for mental health care. *BMC Psychology, 1*(8).

Hermens, D. F., Redoblado Hodge, M. A., Naismith, S. L., Kaur, M., Scott, E., & Hickie, I. B. (2011). Neuropsychological clustering highlights cognitive differences in young people presenting with depressive symptoms. *Journal of the International Neuropsychological Society, 17*(2), 267–276.

Hermens, D. F., Scott, E. M., White, D., Lynch, M., Lagopoulos, J., Whitwell, B. G., et al. (2013). Frequent alcohol, nicotine or cannabis use is common in young persons presenting for mental healthcare: A cross-sectional study. *BMJ Open, 3*(2), e002229.

Hickie, I. B., Hermens, D. F., Naismith, S. L., Guastella, A. J., Glozier, N., Scott, J., et al. (2013). Evaluating differential developmental trajectories to adolescent-onset mood and psychotic disorders. *BMC Psychiatry, 13*, 303.

Hickie, I. B., Scott, E. M., Hermens, D. F., Naismith, S. L., Guastella, A. J., Kaur, M., et al. (2013). Applying clinical staging to young people who present for mental health care. *Early Intervention in Psychiatry, 7*(1), 31–43.

Hickie, I. B., Scott, J., Hermens, D. F., Scott, E. M., Naismith, S. L., Guastella, A. J., et al. (2013). Clinical classification in mental health at the cross-roads: Which direction next? *BMC Medicine, 11*, 125.

Inchausti, F., Garcia-Poveda, N. V., Ballesteros-Prados, A., Ortuno-Sierra, J., Sanchez-Reales, S., Prado-Abril, J., et al. (2017). The effects of metacognition-oriented social skills training on psychosocial outcome in schizophrenia-spectrum disorders: A randomized controlled trial. *Schizophrenia Bulletin, 44*, 1235–1244.

Insel, T. R. (2007). The arrival of preemptive psychiatry. *Early Intervention in Psychiatry, 1*(1), 5–6.

Insel, T. R. (2009). Translating scientific opportunity into public health impact: A strategic plan for research on mental illness. *Archives of General Psychiatry, 66*(2), 128–133.

Insel, T., Cuthbert, B., Garvey, M., Heinssen, R., Pine, D. S., Quinn, K., et al. (2010). Research domain criteria (RDoC): Toward a new classification framework for research on mental disorders. *The American Journal of Psychiatry, 167*(7), 748–751.

Iorfino, F., Davenport, T. A., Ospina-Pinillos, L., Hermens, D. F., Cross, S., Burns, J., et al. (2017). Using new and emerging technologies to identify and respond to suicidality among help-seeking young people: A cross-sectional study. *Journal of Medical Internet Research, 19*(7).

Iorfino, F., Hermens, D., Cross, S. P. M., Zmicerevska, N., Nichles, A., Badcock, C., et al. (2018a). Delineating the trajectories of social and occupational functioning of young people attending early intervention mental health services in Australia: A longitudinal study. *BMJ Open, 8*(3), e020678.

Iorfino, F., Hermens, D. F., Cross, S. P. M., Zmicerevska, N., Nichles, A., Groot, J., et al. (2018b). Longer-term functional outcomes among young people who attempt suicide and attend early intervention mental health services. *Journal of Affective Disorders, 238*, 563–569.

Iorfino, F., Hickie, I. B., Lee, R. S., Lagopoulos, J., & Hermens, D. F. (2016). The underlying neurobiology of key functional domains in young people with mood and anxiety disorders: A systematic review. *BMC Psychiatry, 16*, 156.

Iorfino, F., Scott, E. M., Carpenter, J. S., Cross, S. P., Hermens, D. F., Killedar, M., et al. (2019). Clinical stage transitions in persons aged 12 to 25 years presenting to early intervention mental health services with anxiety, mood and psychotic disorders. *JAMA Psychiatry.* https://doi.org/10.1001/jamapsychiatry.2019.2360.

Kaur, M., Lagopoulos, J., Lee, R. S., Ward, P. B., Naismith, S. L., Hickie, I. B., et al. (2013). Longitudinal associations between mismatch negativity and disability in early schizophrenia- and affective-spectrum disorders. *Progress in Neuro-Psychopharmacology & Biological Psychiatry*, *46*, 161–169.

Kelleher, I., Keeley, H., Corcoran, P., Lynch, F., Fitzpatrick, C., Devlin, N., et al. (2012). Clinicopathological significance of psychotic experiences in non-psychotic young people: Evidence from four population-based studies. *The British Journal of Psychiatry*, *201*(1), 26–32.

Kessler, R. C., Andrews, G., Colpe, L. J., Hiripi, E., Mroczek, D. K., Normand, S. L. T., et al. (2002). Short screening scales to monitor population prevalences and trends in non-specific psychological distress. *Psychological Medicine*, *32*(6), 959–976.

Kim-Cohen, J., Caspi, A., Moffitt, T. E., Harrington, H., Milne, B. J., & Poulton, R. (2003). Prior juvenile diagnoses in adults with mental disorder: Developmental follow-back of a prospective-longitudinal cohort. *Archives of General Psychiatry*, *60*(7), 709–717.

Kohn, D., Glozier, N., Hickie, I. B., Durrant-Whyte, H., & Cripps, S. (2018). *Irreproducibility: Nothing is more predictable*. arXiv preprint. arXiv:1803.04481.

Kozak, M. J., & Cuthbert, B. N. (2016). The NIMH research domain criteria initiative: Background, issues, and pragmatics. *Psychophysiology*, *53*(3), 286–297.

Lagopoulos, J., Hermens, D. F., Hatton, S. N., Battisti, R. A., Tobias-Webb, J., White, D., et al. (2013). Microstructural white matter changes are correlated with the stage of psychiatric illness. *Translational Psychiatry*, *3*, e248.

Lagopoulos, J., Hermens, D. F., Hatton, S. N., Tobias-Webb, J., Griffiths, K., Naismith, S. L., et al. (2013). Microstructural white matter changes in the corpus callosum of young people with bipolar disorder: A diffusion tensor imaging study. *PLoS One*, *8*(3).

Lagopoulos, J., Hermens, D. F., Naismith, S. L., Scott, E. M., & Hickie, I. B. (2012). Frontal lobe changes occur early in the course of affective disorders in young people. *BMC Psychiatry*, *12*, 4.

Lee, R. S., Dore, G., Juckes, L., De Regt, T., Naismith, S. L., Lagopoulos, J., et al. (2015). Cognitive dysfunction and functional disability in alcohol-dependent adults with or without a comorbid affective disorder. *Cognitive Neuropsychiatry*, *20*(3), 222–231.

Lee, R. S., Hermens, D. F., Naismith, S. L., Lagopoulos, J., Jones, A., Scott, J., et al. (2015). Neuropsychological and functional outcomes in recent-onset major depression, bipolar disorder and schizophrenia-spectrum disorders: A longitudinal cohort study. *Translational Psychiatry*, *5*, e555.

Lee, R. S., Hermens, D. F., Redoblado-Hodge, M. A., Naismith, S. L., Porter, M. A., Kaur, M., et al. (2013). Neuropsychological and socio-occupational functioning in young psychiatric outpatients: A longitudinal investigation. *PLoS One*, *8*(3).

Lee, R. S. C., Hermens, D. F., Scott, J., O'Dea, B., Glozier, N., Scott, E. M., et al. (2017). A transdiagnostic study of education, employment, and training outcomes in young people with mental illness. *Psychological Medicine*, *47*(12), 2061–2070.

Lee, R. S., Redoblado-Hodge, M. A., Naismith, S. L., Hermens, D. F., Porter, M. A., & Hickie, I. B. (2013). Cognitive remediation improves memory and psychosocial functioning in first-episode psychiatric out-patients. *Psychological Medicine*, *43*(6), 1161–1173.

Lichtenstein, P., Yip, B. H., Björk, C., Pawitan, Y., Cannon, T. D., Sullivan, P. F., et al. (2009). Common genetic determinants of schizophrenia and bipolar disorder in Swedish families: A population-based study. *Lancet*, *373*, 234–239.

McGorry, P. (2007). Issues for DSM-V: Clinical staging: A heuristic pathway to valid nosology and safer, more effective treatment in psychiatry. *The American Journal of Psychiatry*, *164*(6), 859–860.

McGorry, P. D. (2010). Risk syndromes, clinical staging and DSM V: New diagnostic infrastructure for early intervention in psychiatry. *Schizophrenia Research*, *120*(1–3), 49–53.

McGorry, P. D., Goldstone, S. D., Parker, A. G., Rickwood, D. J., & Hickie, I. B. (2014). Cultures for mental health care of young people: An Australian blueprint for reform. *Lancet Psychiatry*, *1*(7), 559–568.

McGorry, P. D., Hickie, I. B., Yung, A. R., Pantelis, C., & Jackson, H. J. (2006). Clinical staging of psychiatric disorders: A heuristic framework for choosing earlier, safer and more effective interventions. *The Australian and New Zealand Journal of Psychiatry*, *40*, 616–622.

McGorry, P. D., Purcell, R., Hickie, I. B., Yung, A. R., Pantelis, C., & Jackson, H. J. (2007). Clinical staging: A heuristic model for psychiatry and youth mental health. *The Medical Journal of Australia*, *187*, S40–S42.

McGorry, P. D., Tanti, C., Stokes, R., Hickie, I. B., Carnell, K., Littlefield, L. K., et al. (2007). Headspace: Australia's National Youth Mental Health Foundation—Where young minds come first. *The Medical Journal of Australia*, *187*, S68–S70.

McGorry, P. D., Yung, A. R., Bechdolf, A., & Amminger, P. (2008). Back to the future: Predicting and reshaping the course of psychotic disorder. *Archives of General Psychiatry*, *65*(1), 25–26.

Merikangas, K. R., Cui, L., Kattan, G., Carlson, G. A., Youngstrom, E. A., & Angst, J. (2012). Mania with and without depression in a community sample of US adolescents. *Archives of General Psychiatry*, *69*(9), 943–951.

Merikangas, K. R., He, J.-P., Burstein, M., Swanson, S. A., Avenevoli, S., Cui, L., et al. (2010). Lifetime prevalence of mental disorders in U.S. adolescents: Results from the national comorbidity survey replication—Adolescent Supplement (NCS-A). *Journal of the American Academy of Child and Adolescent Psychiatry*, *49*(10), 980–989.

Merikangas, K. R., Herrell, R., Swendsen, J., Rossler, W., Ajdacic-Gross, V., & Angst, J. (2008). Specificity of bipolar spectrum conditions in the comorbidity of mood and substance use disorders: Results from the Zurich cohort study. *Archives of General Psychiatry*, *65*(1), 47–52.

Mundt, J. C., Marks, I. M., Shear, M. K., & Greist, J. M. (2002). The work and social adjustment scale: A simple measure of impairment in functioning. *The British Journal of Psychiatry*, *5*, 461–464.

Murray, G. K., & Jones, P. B. (2012). Psychotic symptoms in young people without psychotic illness: Mechanisms and meaning. *The British Journal of Psychiatry*, *201*(1), 4–6.

Naismith, S. L., Hermens, D. F., Ip, T. K., Bolitho, S., Scott, E., Rogers, N. L., et al. (2012). Circadian profiles in young people during the early stages of affective disorder. *Translational Psychiatry*, *2*(5).

Naismith, S. L., Lagopoulos, J., Hermens, D. F., White, D., Duffy, S. L., Robillard, R., et al. (2014). Delayed circadian phase is linked to glutamatergic functions in young people with affective disorders: A proton magnetic resonance spectroscopy study. *BMC Psychiatry, 14*, 345.

Nuske, H. J., Vivanti, G., & Dissanayake, C. (2013). Are emotion impairments unique to, universal, or specific in autism spectrum disorder? A comprehensive review. *Cognition & Emotion, 27*(6), 1042–1061.

O'Dea, B., Glozier, N., Purcell, R., McGorry, P. D., Scott, J., Feilds, K. L., et al. (2014). A cross-sectional exploration of the clinical characteristics of disengaged (NEET) young people in primary mental healthcare. *BMJ Open, 4*(12).

O'Dea, B., Lee, R. S., McGorry, P. D., Hickie, I. B., Scott, J., Hermens, D. F., et al. (2016). A prospective cohort study of depression course, functional disability, and NEET status in help-seeking young adults. *Social Psychiatry and Psychiatric Epidemiology, 51*(10), 1395–1404.

Ormel, J., Raven, D., van Oort, F., Hartman, C. A., Reijneveld, S. A., Veenstra, R., et al. (2015). Mental health in Dutch adolescents: A TRAILS report on prevalence, severity, age of onset, continuity and co-morbidity of DSM disorders. *Psychological Medicine, 45*(2), 345–360.

Paus, T., Keshavan, M., & Giedd, J. N. (2008). Why do many psychiatric disorders emerge during adolescence? *Nature Reviews Neuroscience, 9*(12), 947–957.

Pesa, N., Hermens, D. F., Battisti, R. A., Kaur, M., Hickie, I. B., & Solowij, N. (2012). Delayed preattentional functioning in early psychosis patients with cannabis use. *Psychopharmacology, 222*(3), 507–518.

Purcell, R., Jorm, A. F., Hickie, I. B., Yung, A. R., Pantelis, C., Amminger, G. P., et al. (2015). Demographic and clinical characteristics of young people seeking help at youth mental health services: Baseline findings of the transitions study. *Early Intervention in Psychiatry, 9*(6), 487–497.

Robillard, R., Carpenter, J. S., Feilds, K. L., Hermens, D. F., White, D., Naismith, S. L., et al. (2018). Parallel changes in mood and melatonin rhythm following an adjunctive multimodal chronobiological intervention with agomelatine in people with depression: A proof of concept open label study. *Frontiers in Psychiatry, 9*, 624.

Robillard, R., Hermens, D. F., Lee, R. S., Jones, A., Carpenter, J. S., White, D., et al. (2016). Sleep-wake profiles predict longitudinal changes in manic symptoms and memory in young people with mood disorders. *Journal of Sleep Research, 25*(5), 549–555.

Robillard, R., Hermens, D. F., Naismith, S. L., White, D., Rogers, N. L., Ip, T. K., et al. (2015). Ambulatory sleep-wake patterns and variability in young people with emerging mental disorders. *Journal of Psychiatry & Neuroscience, 40*(1), 28–37.

Robillard, R., Lagopoulos, J., Hermens, D. F., Naismith, S. L., Rogers, N. L., White, D., et al. (2017). Lower in vivo myo-inositol in the anterior cingulate cortex correlates with delayed melatonin rhythms in young persons with depression. *Frontiers in Neuroscience, 11*.

Robillard, R., Naismith, S. L., Rogers, N. L., Ip, T. K., Hermens, D. F., Scott, E. M., et al. (2013). Delayed sleep phase in young people with unipolar or bipolar affective disorders. *Journal of Affective Disorders, 145*(2), 260–263.

Robillard, R., Naismith, S. L., Rogers, N. L., Scott, E. M., Ip, T. K., Hermens, D. F., et al. (2013). Sleep-wake cycle and melatonin rhythms in adolescents and young adults with mood disorders: Comparison of unipolar and bipolar phenotypes. *European Psychiatry, 28*(7), 412–416.

Robillard, R., Naismith, S. L., Smith, K. L., Rogers, N. L., White, D., Terpening, Z., et al. (2014). Sleep-wake cycle in young and older persons with a lifetime history of mood disorders. *PLoS One, 9*(2).

Rosenblat, J. D., & McIntyre, R. S. (2017). Pharmacological approaches to minimizing cardiometabolic side effects of mood stabilizing medications. *Current Treatment Options in Psychiatry, 4*(4), 319–332.

Scott, J. (2011). Bipolar disorder: From early identification to personalized treatment. *Early Intervention in Psychiatry, 5*(2), 89–90.

Scott, E. M., Carpenter, J. S., Iorfino, F., Cross, S. P. M., Hermens, D. F., Gehue, J., et al. (2019). What is the prevalence, and what are the clinical correlates, of insulin resistance in young people presenting for mental health care? A cross-sectional study. *BMJ Open, 9*, e025674.

Scott, J., Davenport, T. A., Parker, R., Hermens, D. F., Lind, P. A., Medland, S. E., et al. (2017). Pathways to depression by age 16 years: Examining trajectories for self-reported psychological and somatic phenotypes across adolescence. *Journal of Affective Disorders, 230*, 1–6.

Scott, E. M., Hermens, D. F., Glozier, N., Naismith, S. L., Guastella, A. J., & Hickie, I. B. (2012). Targeted primary care-based mental health services for young Australians. *The Medical Journal of Australia, 196*(2), 136–140.

Scott, E. M., Hermens, D. F., Naismith, S. L., Guastella, A. J., De Regt, T., White, D., et al. (2013a). Distinguishing young people with emerging bipolar disorders from those with unipolar depression. *Journal of Affective Disorders, 144*(3), 208–215.

Scott, E. M., Hermens, D. F., Naismith, S. L., Guastella, A. J., White, D., Whitwell, B. G., et al. (2013b). Distress and disability in young adults presenting to clinical services with mood disorders. *International Journal of Bipolar Disorders, 1*, 23.

Scott, E. M., Hermens, D. F., Naismith, S. L., White, D., Whitwell, B., Guastella, A. J., et al. (2012). Thoughts of death or suicidal ideation are common in young people aged 12 to 30 years presenting for mental health care. *BMC Psychiatry, 12*, 234.

Scott, E. M., Hermens, D. F., White, D., Naismith, S. L., GeHue, J., Whitwell, B. G., et al. (2015). Body mass, cardiovascular risk and metabolic characteristics of young persons presenting for mental healthcare in Sydney, Australia. *BMJ Open, 5*(3), e007066.

Scott, J., Hickie, I. B., & McGorry, P. (2012). Pre-emptive psychiatric treatments: Pipe dream or a realistic outcome of clinical staging models? *Neuropsychiatry, 2*(4), 263–266.

Scott, J., Murray, G., Henry, C., Morken, G., Scott, E., Angst, J., et al. (2017). Activation in bipolar disorders: A systematic review. *JAMA Psychiatry, 74*(2), 189–196.

Scott, J., Naismith, S., Grierson, A., Carpenter, J., Hermens, D., Scott, E., et al. (2016). Sleep-wake cycle phenotypes in young people with familial and non-familial mood disorders. *Bipolar Disorders, 18*(8), 642–649.

Scott, J., Paykel, E., Morriss, R., Bentall, R., Kinderman, P., Johnson, T., et al. (2006). Cognitive-behavioural therapy for severe and recurrent bipolar disorders: Randomised controlled trial. *The British Journal of Psychiatry, 188*, 313–320.

Scott, E. M., Robillard, R., Hermens, D. F., Naismith, S. L., Rogers, N. L., Ip, T. K., et al. (2014). Dysregulated sleep-wake cycles in young people are associated with emerging stages of major mental disorders. *Early Intervention in Psychiatry, 10*, 63–70.

Scott, J., Scott, E. M., Hermens, D. F., Naismith, S. L., Guastella, A. J., White, D., et al. (2014). Functional impairment in adolescents and young adults with emerging mood disorders. *The British Journal of Psychiatry*, *205*(5), 362–368.

Shou, H., Cui, L., Hickie, I., Lameira, D., Lamers, F., Zhang, J., et al. (2017). Dysregulation of objectively assessed 24-hour motor activity patterns as a potential marker for bipolar I disorder: Results of a community-based family study. *Translational Psychiatry*, *7*(8).

Smith-Merry, J., Hancock, N., Bresnan, A., Yen, I., Gilroy, J., & Llewellyn, G. (2018). *Mind the gap: The national disability insurance scheme and psychosocial disability*. Final report: Stakeholder identified gaps and solutions. Lidcombe: University of Sydney.

Sullivan, P. F., Daly, M. J., & O'Donovan, M. (2012). Genetic architectures of psychiatric disorders: The emerging picture and its implications. *Nature Reviews Genetics*, *13*(8), 537–551.

Tickell A, Hermens D, Davenport T, Ospina-Pinillos L, Scott E, Iorfino F, et al. Neurocognitive assessment with personalized feedback: A pilot study of young people with affective disorders in an inpatient facility Journal of Affective Disorders 242, 80–86. Under Review.

Tickell, A. M., RSC, L., Hickie, I. B., & Hermens, D. F. (2017). The course of neuropsychological functioning in young people with attenuated vs discrete mental disorders. *Early Intervention in Psychiatry*, *13*, 425–433.

Tickell, A. M., Scott, E. M., Davenport, T., Iorfino, F., Ospina-Pinillos, L., White, D., et al. (2019). Developing neurocognitive standard clinical care: A study of young adult inpatients. *Psychiatry Research*, *276*, 232–238. Under Review.

van Zwieten, A., Meyer, J., Hermens, D. F., Hickie, I. B., Hawes, D. J., Glozier, N., et al. (2013). Social cognition deficits and psychopathic traits in young people seeking mental health treatment. *PLoS One*, *8*(7).

Von Korff, M., Ustun, T. B., Ormel, J., Kaplan, I., & Simon, G. E. (1996). Self-report disability in an international primary care study of psychological illness. *Journal of Clinical Epidemiology*, *49*(3), 297–303.

Waszczuk, M. A., Zavos, H. M., Gregory, A. M., & Eley, T. C. (2014). The phenotypic and genetic structure of depression and anxiety disorder symptoms in childhood, adolescence, and young adulthood. *JAMA Psychiatry*, *71*(8), 905–916.

Whiteford, H. A., Ferrari, A. J., Degenhardt, L., Feigin, V., & Vos, T. (2015). The global burden of mental, neurological and substance use disorders: An analysis from the global burden of disease study 2010. *PLoS One*, *10*(2).

WHO. (2011). Use of glycated haemoglobin (HbA1c) in the diagnosis of diabetes mellitus. *Diabetes Research and Clinical Practice*, *93*(3), 299–309.

World Health Organization. (2013–2020). *Comprehensive mental health action plan* (p. 2013). Geneva: World Health Organization.

Yatawara, C. J., Einfeld, S. L., Hickie, I. B., Davenport, T. A., & Guastella, A. J. (2016). The effect of oxytocin nasal spray on social interaction deficits observed in young children with autism: A randomized clinical crossover trial. *Molecular Psychiatry*, *21*(9), 1225–1231.

Zubrick, S. R., Hafekost, J., Johnson, S. E., Lawrence, D., Saw, S., Sawyer, M., et al. (2016). Suicidal behaviours: Prevalence estimates from the second Australian child and adolescent survey of mental health and wellbeing. *The Australian and New Zealand Journal of Psychiatry*, *50*(9), 899–910.

Chapter 6

Consumer participation in personalized psychiatry

Harris A. Eyre[a,b,c,d,e], Elisabeth R.B. Becker[f], Marissa S. Blumenthal[g], Ajeet B. Singh[c], Cyrus Raji[h], Arshya Vahabzadeh[i,j], Zoe Wainer[k] and Chad Bousman[d,l,m]

[a]Innovation Institute, Texas Medical Center, Houston, TX, United States, [b]Discipline of Psychiatry, University of Adelaide, Adelaide, SA, Australia, [c]IMPACT SRC, Deakin University, Geelong, VIC, Australia, [d]Department of Psychiatry, University of Melbourne, Melbourne, VIC, Australia, [e]Brainstorm Lab for Mental Health Innovation, Stanford Medical School, Stanford University, Palo Alto, CA, United States, [f]Department of Health Promotion and Behavioral Sciences, University of Texas Health Science Center at Houston, School of Public Health, Houston, TX, United States, [g]Munson Medical Center, Traverse City, MI, United States, [h]Department of Radiology and Biomedical Imaging, Division of Neuroradiology, University of California, San Francisco, San Francisco, CA, United States, [i]Psychiatry Academy, Massachusetts General Hospital, Boston, MA, United States, [j]Brain Power, Cambridge, MA, United States, [k]Department of Surgery, St Vincent's Hospital, University of Melbourne, Melbourne, VIC, Australia, [l]Departments of Medical Genetics, Psychiatry, and Physiology and Pharmacology, University of Calgary, Calgary, AB, Canada, [m]Alberta Children's Hospital Research Institute, Calgary, AB, Canada

1 Introduction

The field of personalized psychiatry focuses on an individual's unique attributes to help improve risk assessment, diagnosis, treatment, and prevention of psychiatric disorders. It is a broad, interdisciplinary field that focuses on a patient's biology, lifestyle, and environment. It seeks to both enhance patient engagement with their health care, and improve an individual's health outcomes. The successful development and implementation of personalized psychiatry into practice will require consumer participation in research, novel solution development, clinical care, and policy.

Key examples of the move toward personalized psychiatry are seen in research and clinical care. The Research Domain Criteria project supported by the National Institute of Mental Health is a global effort to develop a new diagnostic and treatment approach that combines biological, behavioral, and social factors. Digital phenotyping is gaining prominence in research analysis, and in some early clinical applications. It involves collecting movement and activity, as well as voice and speech data from smartphones to measure behavior, cognition, and mood. Pharmacogenetic-guided clinical decision support tools provision support for both selecting and dosing of medications; these are increasingly being used clinically (Bousman et al., 2017).

A novel approach to medicine, stemming from systems biology, outlines the importance of four "P"s—predictive, preventive, personalized, and participatory (Sagner et al., 2017). The focus on consumer participation is seen as key to bringing transformative personalized medicine into healthcare. Authors of this model note trends in healthcare consumers increasingly wanting to manage their own health, leverage the power of personal smart devices, use the internet to gather information, and self-organize using social network tools (Sagner et al., 2017). Consumer participation is relevant to a number of facets of personalized psychiatry including research, novel solution development, clinical care, online patient communities, and policy. This chapter outlines the rationale for consumer participation in the aforementioned areas, as well as models and mechanisms that facilitate this participation.

2 Consumer participation in research

Research is important to further the evidence base of a field by spurring discovery, and by informing the utility of novel solutions. In this section we outline a number of methodologies and initiatives shaping the role of consumers in personalized psychiatry research. Consumer engagement in personalized psychiatry research requires educating consumers about the availability to participate in various cutting-edge research topics. This section is kept brief as the topics are covered further in other chapters.

Personalized Psychiatry. https://doi.org/10.1016/B978-0-12-813176-3.00006-7

2.1 Sociodemographic considerations

Sociodemographic factors, such as sex, gender, ethnicity, and life-stage, are important considerations in psychiatry research. Sex and gender differences in psychiatry are increasingly recognized as an area for further exploration. Evidence suggests differences in prevalence, symptomatology, risk factors, influencing factors, course of illness, pathophysiology, and optimal treatment based on sex and/or gender. More research is needed on gender differences in illness behavior, coping, help-seeking, and compliance, as well as on sex-specific aspects of psychopharmacology, hormonal therapies, or gender-sensitive psychotherapy (Riecher-Rossler, 2017).

Differences in patient ethnicity are demonstrated by treatment outcomes of mood stabilizers. With pharmacogenetic-guided clinical decision support tools, HLA genotyping in patients of Asian background who may be treated with the mood stabilizer and anticonvulsant carbamazepine offer a good example of the importance of ethnicity in personalized psychiatry. Carbamazepine has an up to 75-fold risk of causing the potentially fatal skin reaction Stevens-Johnson syndrome and its related disease, toxic epidermal necrolysis, in patients of Asian ethnicity, compared with Caucasians.

2.2 Citizen science as a model

Citizen science engages consumers in a number of core ways into the research process (Follett & Strezov, 2015). The methods of engagement vary across the span of a study. There is contributory engagement, where consumers contribute to data collection, and help analyze the data and disseminate results. Collaborative engagement is a process whereby consumers contribute, but also help in designing the study, interpreting the data, and drawing conclusions. Co-created engagement is a process whereby consumers participate at all stages of the project, including answering new questions after study completion.

2.3 Patient-Centered Outcomes Research Institute (PCORI)

PCORI is an independent nonprofit, nongovernmental organization in the United States that provides a model for enhanced consumer participation in psychiatric research (PCORI, 2017). The PCORI mandate is to "*improve the quality and relevance of evidence available to help patients, caregivers, clinicians, employers, insurers, and policy makers make better-informed health decisions.*" To do this, PCORI works with healthcare stakeholders to identify critical research questions and answer them through comparative clinical effectiveness research focused on real-world outcomes important to patients. Throughout the research process, there is ongoing consumer participation in recruitment, advertising, and dissemination of results.

2.4 Precision Medicine Initiative

A large precision medicine initiative was recently announced by the US government (NIH, 2017). This project is described as "*a bold new research effort to revolutionize how we improve health and treat disease.*" Precision medicine is this context is described as an innovative approach that takes into account individual differences in people's genes, environments, and lifestyles. One proposed project for this initiative includes linking data from electronic health records, biological samples, and behavioral data among a longitudinal cohort of 1 million Americans. Personalized psychiatry is seen as synonymous with precision psychiatry, the psychiatric derivative of precision medicine. The Precision Medicine Initiative is expected to yield data that supports the objectives of 4P medicine for psychiatry—predictive, preventive, personalized, and participatory.

2.5 Online patient communities as clinical and research tools

Online patient communities allow consumers to identify other patient peers, find support, and share information about illness and treatment options (Chiauzzi & Lowe, 2016). Social media use and online engagement is high among psychiatric patients. A recent study found that almost half of psychiatric patients use social media (Kalckreuth, Trefflich, & Rummel-Kluge, 2014). Social media activity is focused on searching for information about mental disorders, services, and medications. However, much of the information available on websites, related to mental health, is nonspecific and not personalized.

2.5.1 The example of PatientsLikeMe

PatientsLikeMe has established itself as a prominent online patient community and platform for collecting patient-generated health data (PatientsLikeMe, 2017). The *PatientsLikeMe* platform was designed as *"an open registry, which allows patients and caregivers to share medical data with each other, and facilitates connecting with and learning from others who have similar experiences."* This platform has more than 500,000 members, and is the largest online patient-powered research and clinical network. The platform began as an amyotrophic lateral sclerosis specific community; however, patients are now sharing their data and experiences with more than 2500 different conditions, including mental disorders. As of 2017, there are more than 67,000 patients sharing data about mental health conditions.

By crowdsourcing patient experiences across a variety of chronic diseases, *PatientsLikeMe* helps patients answer questions that they often ask about their conditions: *"What is this illness I have? What will this diagnosis or treatment do to me? Am I alone? What might help me get better? What might help me live with it? How do I deal with problems caused by my illness?"*

3 Consumer participation in novel solution development

Consumer participation in the solution development process is key to ensuring novel solutions are adding value to patients. Novel solutions include both tools and devices for diagnosis, treatment, and prevention. Design thinking is design methodology that provides a solution-based approach to solving problems. It is a creative, human-centered, problem-solving approach that leverages empathy, collective idea generation, rapid prototyping, and continuous testing to tackle complex challenges (Bhui, 2017). Design thinkers undertake a major effort to understand patients and their experiences, as well as all users of a given technology; this thorough understanding is what guides the rest of the development process. Design thinking involves continuously testing and refining ideas, and feedback is sought early and often from multiple stakeholders. Design thinking teams are ideally multidisciplinary, and use research methods such as ethnography, focus groups, and surveys to better understand user experiences. Rapid prototyping is also a key component, with critical feedback being sought along the way. Furthermore, the approach allows challenging of preconceived solutions that may be held by those who are experts in the field, and hence drives lateral thinking and important innovation. To ensure the creation of novel preventive, diagnostic, and treatment solutions that maximize value at the lowest cost, all aspects of the healthcare system need to align with patient needs and preferences.

4 Consumer participation in clinical care

There are significant opportunities for consumer participation in the clinical deployment of personalized psychiatry. Key areas in which clinical tools can be deployed in personalized psychiatry include the identification of "at-risk" individuals (predictive), assisting with diagnosis, and guiding optimal treatment. This is covered in greater depth in other chapters.

4.1 Shared decision making (SDM)

SDM is a patient-centered approach to care in which the clinician and patient (including family members or caregivers) share a balanced role in treatment decisions based on evidence-based practices aligned with the patient's values, goals, experiences, prioritized health outcomes, and preferences (Charles, Gafni, & Whelan, 1999; Raue et al., 2010; Slade, 2017). This occurs through an open discussion of advantages, disadvantages, and uncertainties of various treatment options (Charles et al., 1999; Raue et al., 2010; Slade, 2017). SDM acknowledges the patient as a valued partner and contributor to the treatment team, and rests on the assumption that greater patient involvement in treatment decisions may lead to increased involvement and engagement with care, increased treatment adherence, greater satisfaction with care, and reduced mental health stigma—all of which contribute to better outcomes (Slade, 2017). This approach is of particular importance for minority populations.

Given the challenges of antidepressant management, patient engagement by way of SDM may be beneficial in alleviating depressive symptoms such as helplessness and hopelessness, and enhancing autonomy, self-efficacy, and empowerment (LeBlanc et al., 2015). This is particularly germane to a disorder such as depression, characterized by disempowerment and a poor sense of self-efficacy, whereby SDM may reduce a cycle of demoralization and helplessness exacerbating distress (Tecuta, Tomba, Grandi, & Fava, 2015). There is a suggestion that individuals who are given autonomy to indicate a preference for either antidepressants or psychotherapy are more likely to respond favorably to their chosen modality of their preference, as opposed to if they are not given the choice (Dunlop et al., 2012). It is essential for treatment needs to be concordant with the patient's desires and worldview. Additionally, the Treatment Initiation and Participation program, a protocolized and pragmatic psychosocial intervention to understand and target psychological barriers to depression care, such as beliefs about medications, has shown to improve both adherence and depression outcomes with antidepressants in primary care (Sirey et al., 2017; Sirey, Bruce, & Kales, 2010).

5 Consumer participation in policy

Appropriate public policies are vital to promoting the development of personalized psychiatry innovations. Further active consumer participation and advocacy is a critical component for successful policy initiatives around reimbursement, regulatory affairs, and health information system upgrades. Local, state, and national advocacy agencies are a potential vehicle to harness consumer participation. A few key policy topics are considered here; however, this topic is covered in more detail in other chapters.

5.1 Reimbursement

Patient access to, and further development of, personalized psychiatry will largely depend on payer reimbursement policies. Adequate reimbursement will ensure private sector investments into research and development, and ensure the translation of technologies into global markets.

Transparent reimbursement criteria are vital to development efforts. Developers of novel solutions need to understand the evidentiary burden required to qualify for reimbursement. Given the rapidity of personalized psychiatry innovations, it has been difficult for payers to keep up with robust and clear criteria for reimbursement. A key component of reimbursement requires cost-effectiveness data demonstrating the value of the innovative therapeutic approach, and in turn justifying the economic impacts of novel solutions. This is in line with the trend toward value-based healthcare.

6 Discussion

Consumer participation involves an important and multifaceted agenda for enhancing the field of personalized psychiatry in the 21st century. Fig. 1 summarizes the number of ways consumers can participate in the development of the field of personalized psychiatry.

Moving the field of personalized psychiatry forward involves engaging consumers in a number of ways.

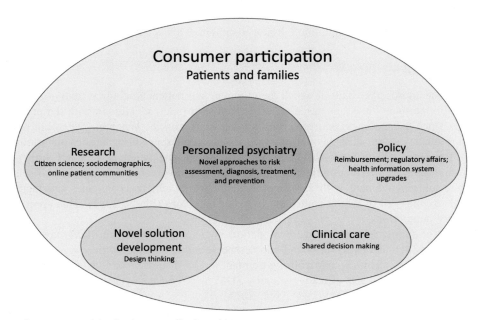

FIG. 1   Key elements of consumer participation in personalized psychiatry.

6.1 Future roles of online patient communities as clinical and research tools

- Patient education: These platforms can allow patients to journal, track, and graph their progression and symptom profile. They also allow for expert-led or consumer-led patient education information. This type of consumer engagement is important, given the current drawbacks of intermittent, retrospective, and low fidelity symptom tracking that occur in typical clinical practice.
- Medication management and tracking: After starting a new treatment, patients can track their experiences regarding side effects and effectiveness. These data can be shared with their clinician, allowing for a richer dataset, based on more robust patient-specific information and parameters, to adjust clinical decisions. This is useful in psychiatric settings where it is helpful to enhance adherence, as well as understand the impact of medications on various symptom domains (e.g., various aspects of mood and cognition).
- Mutual support: Platforms can allow patients to provide mutual peer support. They are able to find consumers who fit similar profiles, disease journeys, or treatment experiences. There is increasing evidence to support the effectiveness of peer-driven problem solving and counseling.
- Designing pragmatic clinical trials: While clinical trials should contribute to informed decision-making, consumers frequently find participating in clinical trials meaningless or disempowering. The lack of patient centeredness in clinical trials may be partially addressed through trial design, and patient perspectives that can be quickly cultivated via these platforms.
- Optimizing recruitment and retention for clinical trials: Many consumers are eager and willing to enroll in a clinical trial, yet very few are invited, or know what resources are available related to clinical trial enrollment. Even so, trial participation is often burdensome and not optimized for the patient and their caretakers. Currently, there is dialogue and efforts toward patient-involved design of clinical trials to help overcome recruitment retention and experience issues. These online platforms synergize with the increasing attention to multisite and multinational trials.

6.2 Future roles of design thinking in novel solution development and shared decision making in clinical care

Consumer engagement in design thinking, including patients, families, and clinicians, is important to ensure novel technologies are appropriate for clinical use. An example of the importance of design thinking comes from the field of clinical perceptual/immersive reality technologies. There has been an increasing number of consumer-grade immersive technologies, such as virtual reality (VR) and augmented reality (AR) apps running on modern head-mounted devices. The modern version of these technologies have greater usability, are lighter, and are less expensive than their older counterparts. VR/AR technologies have already demonstrated promise as tools to aid with psychiatric and behavioral conditions (Freeman et al., 2017), including phobias (Parsons & Rizzo, 2008), schizophrenia (Freeman et al., 2016, 2017), autism (Keshav, Salisbury, Vahabzadeh, & Sahin, 2017; Liu, Salisbury, Vahabzadeh, & Sahin, 2017), and posttraumatic stress disorder (Reger et al., 2011). Consumer feedback during development and postrelease is critical to ensuring that VR/AR apps are usable, effective, and safe. Ensuring the user experience is optimized and that consumer engagement remains high are important considerations in the use and development of these perceptual/immersive technologies.

Pharmacogenetic decision support tools may represent an important facet of SDM by moving forward a more scientifically-driven prescribing strategy, and enriching two-way conversations between doctors and patients.

7 Conclusion

Consumers, clinicians, health executives, entrepreneurs, and policy experts need to be aware of the multitude of ways consumers can be engaged in the processes of personalized psychiatry research, novel solution development, clinical care, and policy. Consumer engagement is a vital driver of growing the field, and involved parties should work to support and further facilitate their engagement. Personalized psychiatry can reach its full potential only with in-depth consumer participation.

References

Bhui, K. C. (2017). Quality improvement and psychiatric research: Can design thinking bridge the gap? *The British Journal of Psychiatry*, *210*(5), 377–378. https://doi.org/10.1192/bjp.210.5.377.

Bousman, C. A., Forbes, M., Jayaram, M., Eyre, H., Reynolds, C. F., Berk, M., ... Ng, C. (2017). Antidepressant prescribing in the precision medicine era: A prescriber's primer on pharmacogenetic tools. *BMC Psychiatry*, *17*(1), 60. https://doi.org/10.1186/s12888-017-1230-5.

Charles, C., Gafni, A., & Whelan, T. (1999). Decision-making in the physician-patient encounter: Revisiting the shared treatment decision-making model. *Social Science & Medicine*, *49*(5), 651–661.

Chiauzzi, E., & Lowe, M. (2016). *PatientsLikeMe: Crowdsourced patient health data as a clinical tool in psychiatry*. Retrieved from http://www.psychiatrictimes.com/telepsychiatry/patientslikeme-crowdsourced-patient-health-data-clinical-tool-psychiatry/page/0/3.

Dunlop, B. W., Kelley, M. E., Mletzko, T. C., Velasquez, C. M., Craighead, W. E., & Mayberg, H. S. (2012). Depression beliefs, treatment preference, and outcomes in a randomized trial for major depressive disorder. *Journal of Psychiatric Research*, *46*(3), 375–381. https://doi.org/10.1016/j.jpsychires.2011.11.003.

Follett, R., & Strezov, V. (2015). An analysis of citizen science based research: Usage and publication patterns. *PLoS One*, *10*(11). https://doi.org/10.1371/journal.pone.0143687.

Freeman, D., Bradley, J., Antley, A., Bourke, E., DeWeever, N., Evans, N., ... Dunn, G. (2016). Virtual reality in the treatment of persecutory delusions: Randomised controlled experimental study testing how to reduce delusional conviction. *The British Journal of Psychiatry*. https://doi.org/10.1192/bjp.bp.115.176438.

Freeman, D., Reeve, S., Robinson, A., Ehlers, A., Clark, D., Spanlang, B., & Slater, M. (2017). Virtual reality in the assessment, understanding, and treatment of mental health disorders. *Psychological Medicine*, *47*(14), 2393–2400. https://doi.org/10.1017/S003329171700040X.

Kalckreuth, S., Trefflich, F., & Rummel-Kluge, C. (2014). Mental health related internet use among psychiatric patients: A cross-sectional analysis. *BMC Psychiatry*, *14*, 368. https://doi.org/10.1186/s12888-014-0368-7.

Keshav, N. U., Salisbury, J. P., Vahabzadeh, A., & Sahin, N. T. (2017). Social communication coaching smartglasses: Well tolerated in a diverse sample of children and adults with autism. *JMIR mHealth and uHealth*, *5*(9).

LeBlanc, A., Herrin, J., Williams, M. D., Inselman, J. W., Branda, M. E., Shah, N. D., ... Montori, V. M. (2015). Shared decision making for antidepressants in primary care: A cluster randomized trial. *JAMA Internal Medicine*, *175*(11), 1761–1770. https://doi.org/10.1001/jamainternmed.2015.5214.

Liu, R., Salisbury, J. P., Vahabzadeh, A., & Sahin, N. T. (2017). Feasibility of an autism-focused augmented reality smartglasses system for social communication and behavioral coaching. *Frontiers in Pediatrics*, *5*, 145.

NIH. (2017). *Precision Medicine Initiative*. Retrieved from https://allofus.nih.gov/.

Parsons, T. D., & Rizzo, A. A. (2008). Affective outcomes of virtual reality exposure therapy for anxiety and specific phobias: A meta-analysis. *Journal of Behavior Therapy and Experimental Psychiatry*, *39*(3), 250–261.

PatientsLikeMe. (2017). *PatientsLikeMe*. Retrieved from https://www.patientslikeme.com/.

PCORI. (2017). *Patient-Centered Outcomes Research Institute*. Retrieved from https://www.pcori.org/.

Raue, P. J., Schulberg, H. C., Lewis-Fernandez, R., Boutin-Foster, C., Hoffman, A. S., & Bruce, M. L. (2010). Shared decision-making in the primary care treatment of late-life major depression: A needed new intervention? *International Journal of Geriatric Psychiatry*, *25*(11), 1101–1111. https://doi.org/10.1002/gps.2444.

Reger, G. M., Holloway, K. M., Candy, C., Rothbaum, B. O., Difede, J., Rizzo, A. A., & Gahm, G. A. (2011). Effectiveness of virtual reality exposure therapy for active duty soldiers in a military mental health clinic. *Journal of Traumatic Stress*, *24*(1), 93–96. https://doi.org/10.1002/jts.20574.

Riecher-Rossler, A. (2017). Sex and gender differences in mental disorders. *Lancet Psychiatry*, *4*(1), 8–9. https://doi.org/10.1016/S2215-0366(16)30348-0.

Sagner, M., McNeil, A., Puska, P., Auffray, C., Price, N. D., Hood, L., ... Arena, R. (2017). The P4 health spectrum—A predictive, preventive, personalized and participatory continuum for promoting healthspan. *Progress in Cardiovascular Diseases*, *59*(5), 506–521. https://doi.org/10.1016/j.pcad.2016.08.002.

Sirey, J. A., Banerjee, S., Marino, P., Bruce, M. L., Halkett, A., Turnwald, M., ... Kales, H. C. (2017). Adherence to depression treatment in primary care: A randomized clinical trial. *JAMA Psychiatry*. https://doi.org/10.1001/jamapsychiatry.2017.3047.

Sirey, J., Bruce, M., & Kales, H. (2010). Improving antidepressant adherence and depression outcomes in primary care: The treatment initiation and participation program. *American Journal of Geriatric Psychiatry*, *18*(6), 554–562.

Slade, M. (2017). Implementing shared decision making in routine mental health care. *World Psychiatry*, *16*(2), 146–153. https://doi.org/10.1002/wps.20412.

Tecuta, L., Tomba, E., Grandi, S., & Fava, G. A. (2015). Demoralization: A systematic review on its clinical characterization. *Psychological Medicine*, *45*(4), 673–691. https://doi.org/10.1017/s0033291714001597.

Chapter 7

Experimental validation of psychopathology in personalized psychiatry

Alfons O. Hamm

Department of Psychology, University of Greifswald, Greifswald, Germany

1 Introduction

Although substantial progress has been made in investigating how the brain produces adaptive behavior, and how normal functioning becomes disrupted in various forms of mental disorders, the translation of this basic psychobiological research to clinical issues is still in its earliest stages. One problem for this translation is that the classification of mental disorders is still almost exclusively *descriptive*, based on patients' phenomenological symptom reports. To put it in the words from Tom Insel, the former Director of the National Institute of Mental Health (NIMH), "While the DSM (Diagnostic and Statistical Manual of Mental Disorders) has been described as a 'Bible' for the field, it is, at best, a dictionary, creating a set of labels and defining each. The strength of each of the editions of the DSM has been reliability" (Insel, 2013). However, due to the effort to maximize reliability, that is, interdiagnostician agreement, validity has been unintentionally sacrificed. To put it again in words from Insel, "The weakness is the lack of validity. Unlike our definitions of ischemic heart disease, lymphoma, or AIDS, the DSM diagnoses are based on consensus about clusters of clinical symptoms, not any objective laboratory measures. In the rest of medicine, this would be equivalent to creating diagnostic systems based on the nature of chest pain or the quality of fever" (Insel, 2013).

Accordingly, instead of starting with an illness definition and then seeking it's neurobiological underpinnings (e.g., looking at brain responses and genetic interaction in well categorized patient groups), this chapter starts with psychobiological understandings of brain-behavior relationships based on experimental studies, both with animals and humans, and links them to clinical problems, cutting across disorder categories. Thus, the focus of the chapter will be primarily on research in experimental psychopathology that is relevant for a better understanding of mental disorders, with an emphasis on anxiety disorders. At the end of the chapter, however, I will also discuss some possible implications of this research for conceptualizing personalized psychiatry. Particularly, it will be described how defensive behavior assessed in experimental tasks might be used to guide decisions for personalized therapy. Moreover, some evidence will be presented on how such defensive behaviors are modulated by genetic measures, thus providing potential target areas for stratified pharmacological treatment. Of course, only a selection of findings in experimental psychopathology can be reported here due to space limitations. Thus, the current chapter will focus on a personalized approach for conceptualizing anxiety disorders.

2 Psychopathology of anxiety disorders

From a clinical and health care perspective, anxiety disorders are among the most frequent and most cost-intensive disorders of all mental disorders (Bandelow & Michaelis, 2015). According to DSM 5 (APA, 2013), all mental disorders included in the section of anxiety disorders "*share features of excessive fear and anxiety and related behavioral disturbances.*" While fear is defined as an emotional response to real or perceived imminent threat, anxiety refers to the anticipation of future threat. Accordingly, the pattern of symptoms associated with both feeling states differ. While fear is "*more often associated with surges of autonomic arousal, necessary for fight or flight, thoughts of immediate danger, and escape behaviors, anxiety is more often associated with muscle tension and vigilance in preparation for future danger and cautious avoidance behavior*" (DSM 5, p. 189). However, although both states differ in their profile of symptoms, it is also stated that both emotional states can obviously overlap. Thus, phobic disorders are defined as marked fear *or* anxiety elicited by *specific situations.* Accordingly, phobic disorders are then further categorized based on the *type of the feared situations* into *specific phobia* (fear or anxiety about specific objects or situations such as animals, heights, etc.), *social phobia* (fear or anxiety about one or more social situations in which the individual is exposed to possible scrutiny by others, such as social interactions, being observed or performing in front of others) and *agoraphobia*. Here, fear and anxiety is more generalized, concerning two or more of five situations that can be grouped into two clusters (a) situations of entrapment (e.g., public transportation, enclosed places, being in a crowd, or standing in line), or (b) situations in which, in case of potential emergencies (e.g., heart attack, stroke, etc.), no immediate help is available (e.g., being outside of the home alone, being in open spaces). The reason for this rather generalized fear and avoidance behavior might be explained by the fact that the source of threat is not in the feared situation itself (agoraphobic patients are not afraid of the shopping mall itself), but rather that the feared situation provides a context in which a potential threat arising from inside the body might be more fatal. Accordingly, agoraphobia is often associated with unexpected (i.e., the individual does not perceive any obvious cue or trigger) panic pathology (Kessler et al., 2006; Pané-Farré et al., 2014). The majority of panic attacks, clinically defined as an abrupt surge of intense fear that reaches a peak within minutes, during which time 4 (or more) of a list of 13 cardiorespiratory, autonomic-somatic, and cognitive symptoms occur, are experienced during the day (85%) and outside of the home (73%), thus occurring in the contexts that are typical for agoraphobic situations (Johnson, Ferici, & Shekhar, 2014; Pané-Farré et al., 2014). If such panic attacks are followed by persistent concerns and worries about additional panic attacks or their consequences (e.g., losing control, having a heart attack, going crazy), a panic disorder is diagnosed. If such worries are more generalized concerning a number of events (health, finances, job performance, etc.), persistent, difficult to control, and associated with symptoms such as irritability, restlessness, muscle tension, and so forth, a generalized anxiety disorder is diagnosed.

As stated herein, such classifications of anxiety disorders have been progressively refined across different versions of the nosological systems to increase reliability of clinical diagnoses, basically by providing more specific descriptive criteria based on patient's phenomenological symptom reports. However, to provide a more mechanism-based approach for understanding fear, anxiety, and its pathological manifestations, one should go beyond the level of symptom reports.

2.1 Experimental validation of fear and anxiety

From the perspective of behavioral science, emotions such as fear and anxiety can be functionally conceived as action dispositions evoked by particular patterns of threat-related stimuli that in turn, cause patterns of adaptive behaviors to cope with the threat in order to survive (Adolphs, 2013; Lang & Bradley, 2010). Thus, to understand anxiety disorders from a perspective of experimental psychopathology, one has to describe how threat information is encoded, and how defensive behaviors are organized. Starting from this behavioral perspective, one can then investigate the neural networks that are involved in the detection and integration of threat information, and to explore how these networks communicate with those circuits organizing defensive action, including the accompanying adjustments of the autonomic nervous system. Such an approach follows the strategic aim of the Research Domain Criteria (RDoC) program (Cuthbert & Insel, 2013) proposed by the NIMH and the European Roadmap for Mental Health Research (ROAMER; Wittchen et al., 2014). The starting point of this research strategy is models of dynamic defense network activation, and its corresponding behavioral adjustments derived from basic research with animals. The translation of these models to humans allows a new approach that should finally lead to a better understanding of adaptive, but also *maladaptive,* human defensive behaviors.

2.1.1 A transdiagnostic dimensional model of defensive behaviors

Animal data suggest that defensive behaviors are organized along a dimension depending upon the distance, proximity, or imminence of the threat (see Blanchard & Blanchard, 1990; Fanselow, 1994). When the organism is in an environment or context in which a threat has been encountered before (or—in the case of humans—that the individual has been informed that a threat might possibly occur), but has not yet been detected (*preencounter defense*; Fanselow, 1994), a class of adaptive defensive behaviors is engaged. These include emotional-motivational and cognitive processes that preempt hostile encounters, for example, inhibition of appetitive behaviors, threat-nonspecific hypervigilance, and increased autonomic arousal. Humans, in such situations, that is, if they encounter such uncertain, novel, and ambiguous environments or contexts in which potential threats might occur, report feelings of anxiety, apprehension, or worry (see Davis, Walker, Miles, & Grillon, 2010). Pharmacological and lesion studies with rodents measuring defensive behavior during open field tests or contextual conditioning (i.e., entering a context where unpredictable aversive events might occur) demonstrated that the bed nucleus of the stria terminalis (BNST)—a major target area of projections from the basolateral and central amygdala—is critically involved in the regulation of anxiety-like preemptive safety or avoidance behavior (e.g., staying cautiously at the wall of the open field; Tovote, Fadok, & Lüthi, 2015).

As soon as the threatening stimulus is detected (*postencounter defense*), but is still distant, a defensive response is activated that is characterized by an increase in selective attention to the threatening cue, accompanied by heart rate deceleration ("fear bradycardia," Campbell, Wood, & McBride, 1997), potentiation of the startle reflex gated through the central nucleus of the amygdala (for review see Hamm, 2015), and motor "freezing," depending on projections from the amygdala to the ventral periaqueductal gray (vPAG) in the midbrain (Fanselow, 1994; Maren, 2001; Morgan & Carrive, 2001; Tovote et al., 2015). In humans, these defensive responses are often labeled as a feeling of fear, and individuals mostly are able to report specific fear triggers, for example, a snake or a dog. If the reported fear also causes significant distress or impairment, or is out of proportion of the actual danger (according to the cultural context), these defensive responses are diagnosed as symptoms of Specific Phobia, as specified and categorized based on the threat cues.

With the increasing imminence of the threat (*circa-strike defense*), defensive response mobilization changes into active action which—depending on the behavioral options at hand—can result in a flight/fight (if possible) response, or during a traumatic event, where fight/flight is no option, in tonic immobility (fright; Schauer & Elbert, 2010). Humans often report feelings of panic or terror during circa-strike defense. Animal data suggest that the initiation of such defensive actions is mediated by the dorsal periaqueductal gray (dlPAG) directing the expression of escape behavior (Kim et al., 2013; LeDoux, 2012). Circa-strike defense mobilization is also accompanied by a general discharge of the sympathetic nervous system (Cannon, 1932) that also stimulates the adrenal medulla to secrete the two catecholamines, epinephrine (adrenaline) and norepinephrine (noradrenaline) into the bloodstream. In addition to direct physiological effects (e.g., increased heart rate, sweating, etc.), adrenergic system activity modulates cognitive processes, especially the formation and consolidation of emotional memories (McGaugh, 2004; Weymar et al., 2009), thus resulting in a better memory of emotionally arousing events—but unfortunately, also traumatic events. Again, these data show the close connection between emotional and cognitive processes involved in the development of psychopathology.

Such dynamic organization of defensive behaviors has been repeatedly advocated by many researchers, and has been formalized as the threat imminence model (Blanchard & Blanchard, 1990; Fanselow, 1994), the defense cascade model (Lang, Bradley, & Cuthbert, 1997), or the action-action tendency model (Schauer & Elbert, 2010; for reviews see Adolphs, 2013; Lowe & Ziemke, 2011; McNaughton, 2011) (Fig. 1).

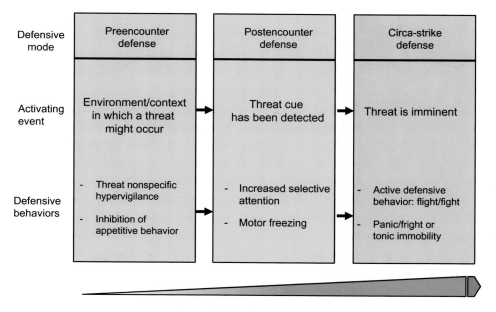

FIG. 1 Transdiagnostic dimensional model of defensive behaviors suggesting that defensive reactivity is dynamically organized along a continuum depending on the imminence or proximity of the threat (Threat-Imminence Model, Fanselow, 1994; Defense Cascade Model, Blanchard & Blanchard, 1990; Lang et al., 1997). (*Adapted from Hamm, A. O., Richter, J., Pané-Farré, C. M., Westphal, D., Wittchen, H.-U., & Deckert, J. (2016). Panic disorder with agoraphobia from a behavioral neuroscience perspective: Applying the research principles formulated by the research domain criteria (RDoC) initiative. Psychophysiology, 53, 312–322.*)

2.2   Empirical evaluation of the model in humans

2.2.1   Preencounter defense: Generalized hypervigilance

The predictions of the theoretical model outlined in the previous section were tested in our laboratory by measuring event related potentials during contexts in which a potential threat might occur. Informing individuals with high spider fear that pictures of spiders might be presented in the following experiment resulted in greater P1 amplitudes over occipital areas, not only for fear-relevant spider pictures, but also for fear-irrelevant neutral stimuli (Michalowski et al., 2009), supporting the hypothesis that preencounter defense is characterized by a generalized hypervigilance to all cues in the potentially dangerous environment. Following up on this research, Michalowski, Pané-Farré, Löw, and Hamm (2015) tested whether *context change* would be explicitly associated with changes in P1 amplitudes. In this study, 32 neutral pictures were presented in a safe context to individuals with elevated spider fear, and controls and P1-amplitudes were measured. A cover story was prepared (all participants were instructed that they had been selected because of their good selective attention performance) to ensure that animal-fearful individuals did not establish any relation between their spider fear and the study. Fig. 2A shows the P1 amplitudes and the distribution of the scalp potential difference map (spider phobia individuals minus controls) for this condition. There were no group differences between individuals with spider phobia and controls in the P1 amplitude in this condition. Then participants were given an informed consent about viewing of spider pictures, and they were asked whether they wanted to continue with the experiment. After this instruction, again, a block of 32 neutral stimuli was presented, followed by the presentation of spider pictures. The upper part of Fig. 2B shows the P1-amplitudes to neutral pictures after the instruction, and the lower part of the P1-amplitudes to the block of spider pictures presented after the neutral block. High spider-fearful individuals showed significantly increased P1-amplitudes to neutral pictures compared

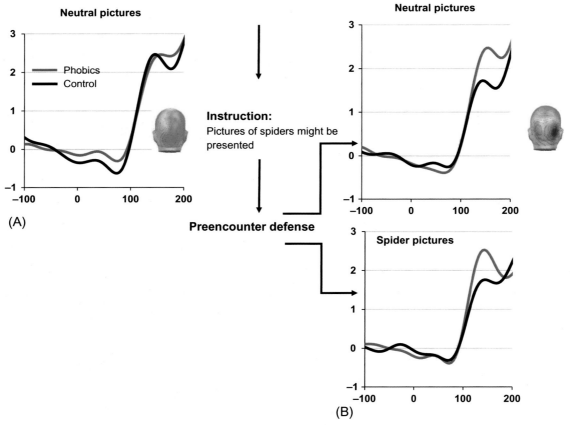

FIG. 2   (A) P1 amplitudes of the evoked brain potentials to neutral pictures in individuals with spider phobia (*red*) and healthy controls (*black*) prior to the instruction that pictures of spiders might be presented. The distribution of the scalp potential difference map (spider phobia individuals minus controls) for this condition is depicted in the inlay. (B) P1 amplitudes to the same neutral pictures (*upper panel*) for both groups after the instruction that phobia-relevant pictures might be presented. The *lower panel* shows P1 amplitudes of both groups to phobia-relevant pictures. Again, distribution of the scalp difference maps is presented in the inlay of panel (B).

with controls over the occipital cortex (see inlay) in a context in which pictures of spiders could occur. This increase in amplitude was comparable to that observed for the P1-to spider pictures (see lower panel of Fig. 2B). These data clearly support the hypothesis that the change from a safe context to an environment where potential threat stimuli might occur resulted in a generalized hypervigilance in those individuals for whom the context was threatening.

Evidence that these changes in vigilance might be implemented by an increase of the afferent sensory gain came from an earlier study by Weymar, Bradely, Hamm, and Lang (2013) and Weymar, Keil, and Hamm (2013). In this study, relative to controls high spider-fearful individuals showed enhanced C1 amplitudes—the earliest cortical component in visual processing in the V1—to fearful, but also nonfearful targets in a visual search task. That is, whenever there is a possibility that a spider stimulus might occur in the visual field (preencounter defense), high fearful individuals showed increased vigilance to all visual targets. The fact that sensory gain might generally be increased during this phase of preencounter defense is also supported by a more recent study from Cornwell, Garrido, Overstreet, Pine, and Grillon (2017). These investigators used a threat of unpredictable shock paradigm to create a potentially dangerous context. Participants had to complete an auditory oddball task either during the potentially dangerous context, and during a safe context, while whole-head MEG was recorded. Threat of shock increased auditory cortical responses to deviant tones, which was an effect that could be reduced by the administration of the anxiolytic alprazolam. Using dynamic causal modeling of the MEG data revealed that threat of shock increased the gain of the primary auditory cortex, thus gating feed-forward connectivity. Interestingly, using the same threat of shock paradigm, Bublatzky and Schupp (2012) also found increased P1-amplitudes to pictures—irrespective of their emotional content—presented in a context in which an aversive electric shock might be delivered relative to the P1-amplitudes to visual stimuli presented in a safe context. The same unspecific enhancement of the visual evoked potentials can also be observed when words are encoded under threat of shock (Weymar, Bradely, et al., 2013; Weymar, Keil, et al., 2013).

These data suggest that the generalized hypervigilance (or broadening attention; Eysenck, 1992) during preencounter defense is realized by an increase in the gain of the primary sensory processing areas in the brain. This hypothesis is also supported by a study from Straube, Mentzel, and Miltner (2007) who measured brain activation using functional resonance imaging (fMRI) while high spider-fearful individuals were waiting for pictures of spiders. During this stage of preencounter defense, the authors found enhanced activity in the extrastriate visual cortex and in the thalamus. Moreover, increased activation during this stage was also found in the BNST, supporting pharmacological and lesion studies with rodents (Davis et al., 2010; Tovote et al., 2015). Direct projections from the BNST to cholinergic neurons in the midbrain may mediate increases in synaptic transmission in thalamic sensory relay neurons. This cholinergic activation, along with increases in thalamic transmission, which is accompanied by activation of the locus coeruleus, may thus lead to generalized hypervigilance during preencounter defense (see Davis & Whalen, 2001; Rosen & Schulkin, 1998). Of course, preencounter defense can be evoked in all contexts in which potential threats might occur, irrespective of whether they come from outside or inside the body, or from inanimate or from conspecific sources. I will come back to potential sources of threat in the next section.

2.2.2 Postencounter defense: Defensive freezing

As soon as the threat cue is detected, but is still distant (e.g., a spider is sitting in the upper corner of the room), increased selective attention is allocated to the threat cue, accompanied by augmented sympathetic arousal, heart deceleration, and potentiation of protective reflexes, a response pattern that can be described as defensive freezing, or attentive immobility (Galdwin, Hashemi, van Ast, & Roelofs, 2016), in contrast to passive tonic immobility ("playing dead"), which often occurs during sexual or physical assault (Schauer & Elbert, 2010). Accordingly, individuals who report excessive spider fear show enhanced late positive potentials during viewing of spider pictures compared with nonfearful individuals (Kolassa, Musial, Mohr, Trippe, & Miltner, 2005; Michalowski et al., 2009, 2015; see Fig. 3), supporting the view that more attentional resources are allocated to phobia-relevant stimuli. However, the LPP was similarly increased for other unpleasant pictures (in this case, in both fearful and nonfearful individuals; see Fig. 3), demonstrating that selective attention capture is not specific for phobia-relevant cues, but is a more general phenomenon for all motivationally relevant stimuli, including pleasant ones (Schupp et al., 2000, 2004).

Moreover, high fearful individuals consistently show larger potentiation of their startle responses and increased skin conductance responses during viewing of phobia-relevant pictures, relative to controls (Hamm, Cuthbert, Globisch, & Vaitl, 1997). In this study, startle potentiation and increase in sympathetic arousal excelled blink magnitudes and skin conductance evoked by other unpleasant pictures. The same defensive response pattern is observed during fear conditioning, that is, during viewing a cue that predicts the occurrence of an exteroceptive threat—a moderately painful stimulus (Hamm, Greenwald, Bradley, & Lang, 1993; Hamm & Vaitl, 1996; Lipp, Sheridan, & Siddle, 1994). Interestingly, while

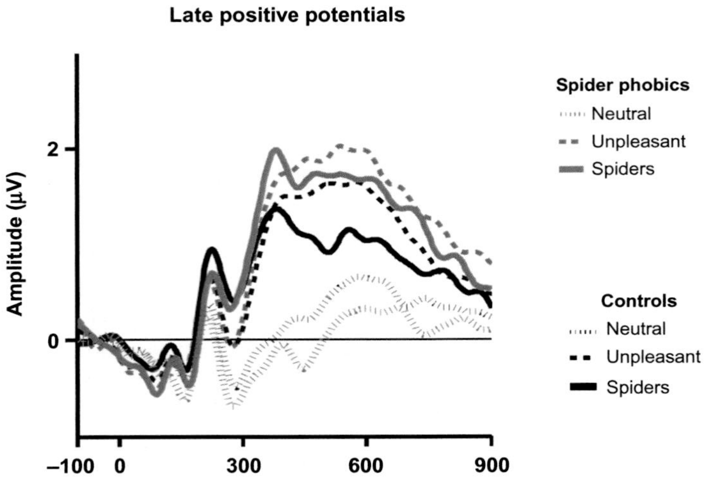

Late positive potentials

FIG. 3 Late positive potentials (LPPs) occurring in the time window between 300 and 550 ms over parietal brain regions in the evoked potentials to phobic (spider pictures), unpleasant, and neutral pictures in individuals with elevated spider fear (*red*) and control subjects (*black*). LPPs evoked by spider pictures were specifically increased in the phobia group, but did not exceed the LPPs evoked by other unpleasant pictures. LPPs to emotionally relevant contents were larger than those evoked by neutral pictures. *(Adapted from Michalowski, J. M., Pané-Farré, C. M., Löw, A., & Hamm, A. O. (2015). Brain dynamics of visual attention during anticipation and encoding of threat- and safe-cues in spider phobic individuals. Social Cognitive and Affective Neuroscience, 10, 1177–1186.)*

sweat gland activity increases with sympathetic activation, heart rate—which is mainly under parasympathetic control—decreases, at least for some individuals, during the conditioned stimulus (Hamm & Vaitl, 1996; Hodes, Cook, & Lang, 1985)—resembling fear bradycardia that is observed in prey animals confronting a predator at a distance (Campbell et al., 1997). On the other hand, some individuals show an acceleration of their heart rate during viewing a cue predicting the occurrence of an aversive electrical stimulus (Hamm & Vaitl, 1996; Hodes et al., 1985; Moratti & Keil, 2005). Moreover, high spider-fearful individuals also showed heart rate acceleration during viewing of phobia-relevant pictures (Hamm et al., 1997). These data suggest that during defensive freezing, both branches of the autonomic nervous system become activated, and that the pattern of defensive responses varies depending on which system is dominant at a certain point in time. Which autonomic pattern is dominant during defensive response activation may be shaped by trait factors, as well as environmental ones (for review see Roelofs, 2017). Animal data suggest that the distance from the predator, but also the presence of escape routes, play an important role in this respect (Blanchard, Griebel, Robbe, & Blanchard, 2011). Recent findings from our laboratory support these animal data. Löw, Weymar, and Hamm (2015) traced the dynamics of defensive response activation assessing autonomic arousal, brain stem reflex activity, and evoked brain potentials depending upon threat proximity and the presence of an escape route using an instructed-fear paradigm. In this experiment, the color of a frame signaled whether there was a behavioral option to escape from the threat (a moderately painful stimulus) or not (active vs passive). In addition, a symbol presented after 500 ms in the frame signaled whether or not the threat would occur (safe vs threat). This cue increased in size in five stages as though it was approaching the participant, signaling the decreasing distance of the threat. In the safe condition, not aversive stimulus was delivered. Fig. 4 shows the modulation of the startle response (upper panel) when there was not an escape route available (passive condition; left), and when the individual had the option to actively avoid the electrical stimulus by quickly pressing a response button (active condition, right). The lower part of Fig. 4 shows the changes in heart rate during these conditions.

When there was no possibility of actively avoiding the aversive stimulus, humans—as rodents do—showed defensive freezing, characterized by a linear increase in startle potentiation with increasing proximity of the inevitable threat. In addition, heart rate decreased substantially starting at stage five, that is, immediately prior to the delivery of the aversive stimulus. Interestingly, skin conductance level as a measure of sympathetic activation linearly increased throughout the entire defense cascade, supporting the view that both counteracting branches of the autonomic nervous system (the sympathetic and the parasympathetic nervous system) are activated during defensive freezing. Indeed, individual difference factors, such as vagally mediated heart rate variability, might shape the physiological response pattern that is elicited during defensive freezing. Recent data from our laboratory suggest that individuals with high heart variability seem to be more flexible to adapting to environmental demands, that is, showing more conditioned inhibition in an extinction paradigm (Wendt, Neubert, Koenig, Thayer, & Hamm, 2015).

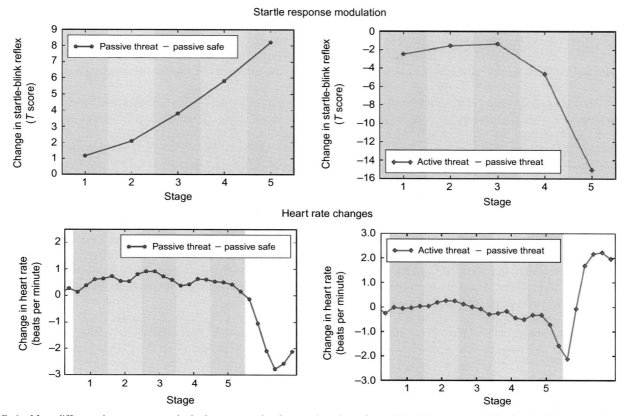

FIG. 4 Mean difference in response magnitudes between passive threat and passive safe condition (there was not a behavioral option to escape) across five different stages of increasing threat imminence for fear potential startle (*upper left*) and hear rate (*lower left*). Mean difference in response magnitudes between active (escape was possible) and passive (no escape was possible) threat conditions for fear potentiated startle (*upper right*) and heart rate (*lower right*). (*Adapted from Löw, A., Weymar, M., & Hamm, A. O. (2015). When threat is near, get out of here: Dynamics of defensive behavior during freezing and active avoidance. Psychological Science, 26, 1706–1716.*)

2.2.3 Circa-strike defense: Initiation of action

A different pattern of defensive reactivity was observed if an escape route was present, and the individuals had the option to actively avoid the painful stimulation. A sharp increase in skin conductance, combined with a strong heart rate acceleration occurred (see lower right panel of Fig. 4). This autonomic pattern was associated with the preparation of motor responses that were systematically sped up in the behavioral context of the circa-strike defense. Notably, a drastic reversal of the fear potentiated startle became evident—an abrupt inhibition of the startle reflex just prior to the preparation for active action (see upper right panel of Fig. 4). Moreover, the P3-component of the evoked potentials elicited by the invariant acoustic startle probes was also significantly reduced, suggesting that more attentional resources were allocated to the action-relevant foreground stimuli blocking the processing of irrelevant stimuli in the context of action. Interestingly, animals show hypoalgesia during activation of a circa-strike defense by stimulation of the dorsolateral periaqueductal gray (PAG) (Fanselow, 1994).

3 Neural network activation

Detecting and responding effectively to various threats is a fundamental requirement of life. Thus, the neural networks organizing defensive action are rooted in the survival circuits of the brain (LeDoux, 2012). As discussed herein, these defense circuits feed back to the sensory systems, increasing the ability to detect potentially threatening and harmful stimuli. Increased activation during this stage or preencounter defense was found in the primary sensory areas, as well as in the BNST promoting hypervigilance to external stimuli (see Cornwell et al., 2017; Rosen & Schulkin, 1998). The neural circuits underlying postencounter defense having been extensively explored in animal research using Pavlovian fear conditioning as a model. The conditioned stimulus that predicts the occurrence of threat activates neurons in the lateral amygdala, which propagates neural activity to the central nucleus of the amygdala (CeA). Efferents from the CeA to the ventrolateral part of the PAG then interrupt ongoing behavior, and promote attentive freezing and potentiation of the startle reflex (Gross & Canteras, 2012; for reviews see Hamm, 2015; Tovote et al., 2015), as well as bradycardia (Bandler, Keasy, Floyd, & Price, 2000). Human fear conditioning studies—using functional magnetic resonance imaging techniques—found transient activation of the amygdala during early phases of learning (Sehlmeyer et al., 2009). More recent meta-analyses of fMRI-data obtained from human fear conditioning studies suggest that conditioned stimuli activate a larger network, including increased activation of the anterior cingulate and of the anterior insula—it is a hallmark finding in pain research that anticipation of pain is associated with the activation of the anterior insula (Wiech et al., 2010)—accompanied by a decreased activation of the ventromedial prefrontal cortex (Fullana et al., 2015; Lindner et al., 2015; Milad & Quirk, 2012). The change from defensive freezing to action during a circa strike (e.g., flight, fight, active avoidance) is mediated by the dorsolateral section of the PAG (Fanselow, 1994; Kim et al., 2013). Lesions of the dorsolateral PAG increase freezing in rodents (DeOca, DeCola, Maren, & Fanselow, 1998), while damage of the ventrolateral PAG disrupts freezing (Fanselow & Poulos, 2005), suggesting that attentive freezing might obstruct active avoidance and vice versa (Benaroch, 2012). Guided by these animal findings, Mobbs and coworkers investigated brain activation in humans as a function threat proximity. In their fMRI study, they used an active avoidance video game setting in which participants could escape from a virtual "predator" (in this case red circle) that chased the individual (symbolized by a triangle) by moving the triangle in a two-dimensional space. The findings of this study supported the animal data in showing that brain activity shifted from prefrontal brain areas to the amygdala and the PAG (Mobbs et al., 2009, 2007). In a recent study, we followed up on this research in our laboratory, explicitly varying the proximity of threat, but also whether individuals had the option to actively avoid the threat or not, using the threat imminence paradigm developed by Löw et al. (2015). The physiological data replicated the findings observed by Löw and coworkers (see Wendt, Löw, Weymar, Lotze, & Hamm, 2017). Activity in the ventromedial prefrontal cortex increased at the beginning of the defense cascade, that is, when the anticipated threat was still distant. This prefrontal activation decreased with increasing proximity of the upcoming threat. In contrast, activation of the PAG increased with increasing proximity of the threat, particularly when individuals had the option to actively avoid the threat, thus supporting previous data by Mobbs and coworkers. The main findings of this study are summarized in the left panel of Fig. 5, while the switch from prefrontal to midbrain activation is illustrated in a schematic model in the right panel of Fig. 5.

Taken together, these recent imaging data support the view that not only physiological changes during fear and anxiety are dynamically organized according to the proximity of a potential threat and the behavioral options at hand, but that neural

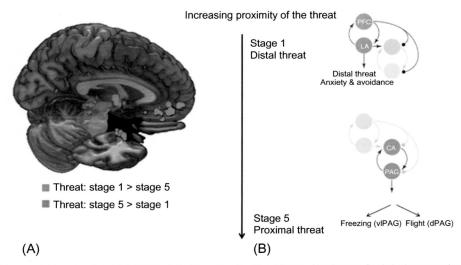

FIG. 5   (A) Prefrontal, amygdala, and periaqueductal gray activations as a function of threat imminence. Statistical parametric maps (SPMs) illustrate brain areas whose activity *decreased* from stage 1 to stage 5 in threat conditions (*green*) and brain areas whose activity *increased* from stage 1 to stage 5 with increasing threat imminence (*purple*). (B) Schematic depiction of brain circuits activated during distal threat including prefrontal cortex (PFC) and lateral amygdala (LA) and those circuits activated during proximal threat including the central amygdala (CA) and the periaqueductal gray. *(Panel (A) From Wendt, J., Löw, A., Weymar, M., Lotze, M., & Hamm, A. O. (2017). Active avoidance and attentive freezing in the face of approaching threat. Neuro-Image, 158, 196–204. Panel (B) Adapted from Maren, S. (2007). The threatened brain. Science, 317, 1043–1044.)*

network activation is also rather distributed, and changes from prefrontal activation during distal threat to more midbrain activation when the threat is getting more proximal. Such dynamic conceptualization of defensive adaptions would render debates on whether cognitive or emotional processes are more important for the understanding of psychopathology unnecessary. In fact, the dynamic model outlined herein would rather suggest that both processes work hand in hand, and that the imminence of the threat would determine which process is more dominant.

4 Avoidance behavior as an index of functional impairment

The transdiagnostic dimensional model outlined herein can also be used to advance our understanding of the mechanisms involved in maladaptive avoidance behaviors. According to the model, *passive avoidance* occurs very early in the defense cascade, even prior to preencounter defense, that is, individuals avoid the *context* in which a potential threat might occur. Clinically, avoidance is related to functional impairment as the main criterion for defining mental disorders, other than the distress criterion (APA, 2013; Barlow, 2000). There is evidence that such maladaptive passive avoidance behavior is motivated rather by the effort to avoid uncertainty or to maintain safety (Talkovsky & Norton, 2016), than by elevated fear or anxiety (see also early work by Solomon, Kamin, & Wynne, 1953). Supporting this view, Craske, Rapee, and Barlow (1988) demonstrated that not the rated fear level, but rather the perceived probability of panicking in a specific situation was the strongest predictor ($r = 0.49$) for active and passive avoidance of a variety of behavioral avoidance tests. Indeed, persistent avoidance prevents the disconfirmation of central concerns about the increased risk to encounter a potential threat. A recent study from our laboratory investigating the development of stable passive avoidance across time supports this view (see Benke, Krause, Hamm, & Pané-Farré, 2018). In this study, individuals breathed through a tightly fitting mask. The inspiratory port was connected to a flow resistor that increased the inspiratory resistance, producing a feeling of dyspnea in three steps from slight, strong, and maximally tolerable unpleasant feelings of dyspnea, each lasting for 60 s. The sequence was terminated by a postexpiratory breathing occlusion of 15 s. A subsample of the participants ($N = 24$) repeatedly terminated the sequence of exposure dyspnea. In accordance with animal data (see Solomon et al., 1953), individuals ceased the exposure to increasing interoceptive threat at increasingly lower threat levels and at shorter time periods of exposure. Moreover, relative to a matched control group who did not avoid the exposure to threat, initial avoidance behavior was accompanied by a linear increase in autonomic arousal (heart rate and skin conductance) prior to termination. This autonomic activation was completely absent during the last termination of the exposure, showing that autonomic arousal did not trigger repetitive passive avoidance. Reported anxiety also linearly increased prior to initial termination of the exposure. In contrast to physiological data, reported anxiety increased from the first to the last termination, particularly at a longer time period before the actual time of termination. These data support the view that expectations and concerns about the occurrence of a potential threat are more important triggers of passive avoidance than autonomic arousal. Thus, changing persistent avoidance by exposure therapy should probably focus more on experiences that explicitly induce expectancy violations to increase prediction errors promoting extinction learning, rather than to focus on fear augmentation to increase autonomic arousal to induce habituation processes (see Hamm et al., 2016).

5 Clinical implications: Anxiety disorders

In this chapter we have started to describe the psychopathology of anxiety disorders using a transdiagnostic model of defensive behaviors derived from animal research. I have provided empirical evidence that human defensive behaviors are also dynamically organized depending upon the proximity of the relevant threat. It was also explained that this dynamic organization of encoding threat-relevant information and organizing the behavioral adaptation, including the accompanying physiological adjustments, is regulated by distributed neural networks in the brain, including prefrontal and midbrain circuits. In the last chapter I want to provide some data to show how such neuroscience-based models might help to better understand the psychopathology of anxiety disorders. We tested defensive reactivity in a subgroup of 345 (259 females) patients from a total sample of 369 patients, all of whom were diagnosed with a principal diagnosis of panic disorder with agoraphobia (PD/AG) and enrolled for a multicenter, randomized controlled clinical trial study (for details see Gloster et al., 2009, 2011). Because marked fear of entrapment and avoidance of being in enclosed places is a prominent symptom in patients with PD/AG (Arrindell, Cox, Van der Ende, & Kwee, 1995; Kwon, Evans, & Oei, 1990; Rodriguez, Pagano, & Keller, 2007), we measured defensive reactivity during anticipation (10 min) and exposure (maximum duration of 10 min) of a behavioral avoidance test (i.e., sitting in a small [75 cm wide, 120 cm long, and 190 cm high] dark, closed test chamber with the door locked from the outside by the experimenter; for details see Richter et al., 2012). Defensive reactivity differed substantially, although the same principal diagnoses was assigned to all patients based on their symptom reports. Thirty-nine patients (11.3%) refused to enter the chamber after anticipation, thus showing strong passive avoidance, 72 patients

(20.9%) entered the test chamber, but terminated the exposure prematurely (active avoidance), and 125 patients (36.2%) remained in the chamber, despite medium to high levels of reported anxiety. The analysis of defensive reactivity of patients who showed active avoidance revealed that heart and skin conductance increased linearly during the last 60 s just prior to escape (see upper panel of Fig. 6) similar to the autonomic response profile observed in the defense cascade study by Löw et al. (2015) when there was a behavioral option to actively avoid the threat stimulus. Moreover, also supporting the data from Löw et al. (2015), startle response magnitudes were elevated when patients entered the chamber, but were actively inhibited a few seconds prior to escape again, supporting the idea that reduced attention is allocated to stimuli that are irrelevant for the behavioral option. This pattern of defensive behavior resembled the response pattern that was observed during acute panic attacks. Twenty-six patients reported 34 panic attacks during exposure by pressing the "panic button." The lower panel of Fig. 6 depicts startle magnitudes and heart rate changes averaged in 5-s bins starting 30 s prior to button press and 60 s during and after button press for those 11 panic attacks for which reliable data from both physiological variables were available (for more detailed analysis see Richter et al., 2012).

Heart data increased prior to the button press suggesting that the increase in autonomic arousal indeed triggered the experience of a panic attack. As during active avoidance, startle magnitudes were inhibited during panic attacks, suggesting that acute panic attacks are instances of circa-strike defense, possibly triggered by the perception of acute threat signals

FIG. 6 (A) Mean heart rate and startle response magnitudes of 55 PD/AG patients during the initial 60 s of the exposure period and the dark room, and the 60 s prior to escape from the dark room. (B) Heart rate changes and startle responses 30 s prior and 60 s after the button press signaling that the individuals were experiencing a panic attack. The figure shows the averaged data of 11 panic attacks reported by 11 PD/AG patients for which both variables could be reliably scored. *(From Hamm, A. O., Richter, J., Pané-Farré, C. M., Westphal, D., Wittchen, H.-U., & Deckert, J. (2016). Panic disorder with agoraphobia from a behavioral neuroscience perspective: Applying the research principles formulated by the research domain criteria (RDoC) initiative. Psychophysiology, 53, 312–322.)*

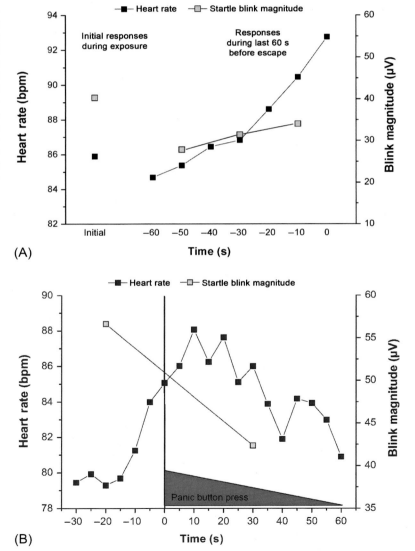

arising from inside the body. Interestingly, there are early reports by Nashold, Wilson, and Slaughter (1974) that electrical stimulations of the PAG evoked acute panic attacks in humans (see Schenberg et al., 2014), supporting the view that panic attacks might be instances of circa-strike defense modulated by the dlPAG. In contrast, processing of cues that predict the occurrence of mild body symptoms (postencounter defense) is associated with stronger activation of prefrontal areas of the brain (Holtz, Melzig, Wendt, Lotze, & Hamm, 2012).

5.1   Toward personalized treatment: Genetic and behavioral data

Finally, we tested whether defensive behaviors might be useful in predicting clinical outcomes in two multicenter, randomized controlled trials (Gloster et al., 2011; Hamm et al., 2016). A group of 369 patients with the principal diagnosis of panic disorder with agoraphobia were randomized to two manual-based variants of CBT, or wait list control (WL; $N = 68$). CBT variants were identical in content, structure, and length, except for implementation of exposure in situ. In one variant (T+; $N = 163$), exposure exercises were accompanied by the therapist (e.g., taking a bus ride together); in the other variant (T−; $N = 138$), the exposure exercises were planned and discussed with the patient, but exercises outside the therapy room were not accompanied by the therapist. A second multicenter trial including 124 patients with panic disorder with agoraphobia followed up on this study. In this trial, one CBT variant was comparable to the T+ variant of the first clinical trial ($N = 61$), in the other variant ($N = 63$), exposure sessions were also accompanied by the therapist, but in addition, patients were instructed to provoke body symptoms to increase their anxiety during exposure. We collected behavioral data from the standardized behavioral avoidance test for 397 patients allocated to the active treatment conditions in both clinical trials. We found larger attrition rated in 116 PD/AG patients who showed active and passive avoidance behavior (i.e., either did not enter the chamber or escaped prematurely) in the standardized BAT prior to therapy, with nearly twice as many drop outs ($N = 35$; 30.2%) as compared with the 281 patients who did not show any avoidance behavior ($N = 50$; 17.8%). In a recent study, we analyzed cardiac vagal tone of a subgroup of these patients prior to therapy (Wendt et al., 2018), and found that patients who dropped out of treatment had overall a lower cardiac vagal tone. Interestingly, patients showing active or passive avoidance in this task also respond with larger heart rate acceleration during this task. Such increased defensive cardiovascular responding is particularly observed in those PD/AG patients who carry two T-alleles of the noradrenalin transporter gene (SLC6A2) (Hommers et al., 2018) and the risk variant in the promoter region of the gene encoding monoamine oxidase A. Interestingly, all but one patient reporting a panic attack in the chamber (33 out of 34) were high-risk genotype carriers (Reif et al., 2014).

On the other hand, if these high-risk patients completed the CBT, they showed larger improvement, but only if the therapist accompanied the patients during exposure exercise outside the therapy room (see Hamm et al., 2016). These data suggest a multilevel analysis of defensive responding including behavioral, physiological, and genetic information might be useful for selecting personalized treatment within a group of patients who were all diagnosed with the same categorical diagnosis.

References

Adolphs, R. (2013). The biology of fear. *Current Biology, 23*, R79–R93.

American Psychiatric Association, (2013). *Diagnostic and statistical manual of mental disorders* (5th ed.). Washington, DC: American Psychiatry Association.

Arrindell, W. A., Cox, B. J., Van der Ende, J., & Kwee, M. G. (1995). Phobic dimensions—II. Cross national confirmation of the multidimensional structure underlying the Mobility Inventory (MI). *Behavior Research and Therapy, 33*, 711–724.

Bandelow, B., & Michaelis, S. (2015). Epidemiology of anxiety disorders in the 21st century. *Dialogues in Clinical Neuroscience, 17*, 327–335.

Bandler, R., Keasy, K. A., Floyd, N., & Price, J. (2000). Central circuits mediating patterned autonomic activity during active vs. passive emotional coping. *Brain Research Bulletin, 53*, 95–104.

Barlow, D. H. (2000). *Anxiety and its disorders* (2nd ed.). New York: Guilford Press.

Benaroch, E. E. (2012). Periaqueductal gray: An interface for behavioral control. *Neurology, 78*, 210–217.

Benke, C., Krause, E., Hamm, A. O., & Pané-Farré, C. A. (2018). Dynamics of defensive response mobilization during repeated terminations of exposure to increasing interoceptive threat. *International Journal of Psychophysiology, 131*, 44–56.

Blanchard, R. J., & Blanchard, D. C. (1990). An ethoexperimental analysis of defense, fear and anxiety. In N. McNaughton & G. Andrews (Eds.), *Anxiety* (pp. 124–133). Dunedin: Otago University Press.

Blanchard, D. C., Griebel, G., Robbe, R., & Blanchard, R. J. (2011). Risk assessment as an evolved threat detection and analysis process. *Neuroscience and Biobehavioral Reviews, 35*, 991–998.

Bublatzky, F., & Schupp, H. T. (2012). Pictures cueing threat: Brain dynamics in viewing explicitly instructed danger cues. *Social Cognitive and Affective Neuroscience, 7*, 611–622.

Campbell, B. A., Wood, G., & McBride, T. (1997). Origins of orienting and defensive responses: An evolutionary perspective. In P. J. Lang, R. F. Simons, & M. T. Balaban (Eds.), *Attention and orienting: Sensory and motivational processes* (pp. 41–67). Hillsdale, NJ: Erlbaum.

Cannon, W. (1932). *The wisdom of the body.* W.W. Norton & Company.

Cornwell, B. R., Garrido, M. I., Overstreet, C., Pine, D. S., & Grillon, C. (2017). The unpredictive brain under threat: A neurocomputational account of anxious hypervigilance. *Biological Psychiatry, 82,* 447–454.

Craske, M. G., Rapee, R. M., & Barlow, D. H. (1988). The significance of panic-expectancy for individual patterns of avoidance. *Behavior Therapy, 19,* 577–592.

Cuthbert, B. N., & Insel, T. (2013). Toward the future of psychiatric diagnosis: The seven pillars of RDoC. *BMC Medicine, 11,* 126. https://doi.org/10.1186/1741-7015-11-126.

Davis, M., Walker, D. L., Miles, L., & Grillon, C. (2010). Phasic vs. sustained fear in rats and humans: Role of the extended amygdala in fear vs anxiety. *Neuropsychopharmacology, 35,* 105–135.

Davis, M., & Whalen, P. J. (2001). The amygdala: Vigilance and emotion. *Molecular Psychiatry, 6,* 13–34.

DeOca, B. M., DeCola, J. P., Maren, S., & Fanselow, M. S. (1998). Distinct regions of the periaqueductal gray are involved in the acquisition and expression of defensive responses. *Journal of Neuroscience, 18,* 3426–3432.

Eysenck, M. W. (1992). *Anxiety: The cognitive perspective.* Hillsdale, NJ: Erlbaum.

Fanselow, M. S. (1994). Neural organization of the defensive behavior system responsible for fear. *Psychonomic Bulletin & Review, 1,* 429–439.

Fanselow, M. S., & Poulos, A. M. (2005). The neuroscience of mammalian associative learning. *Annual Review of Psychology, 56,* 207–234.

Fullana, M. A., Harrison, B. J., Soriano-Mas, C., Vervliet, B., Cardoner, N., Avila-Parcet, A., & Radua, J. (2015). Neural signatures of human fear conditioning: An updated and extended meta-analysis of fMRI studies. *Molecular Psychiatry, 21,* 500–508.

Galdwin, T. E., Hashemi, M. M., van Ast, V., & Roelofs, K. (2016). Ready and waiting: Freezing as active action preparation under threat. *Neuroscience Letters, 619,* 182–188.

Gloster, A. T., Wittchen, H.-U., Einsle, F., Höfler, M., Lang, T., Helbig-Lang, S., … Arolt, V. (2009). Mechanisms of action in CBT (MAC): Methods of a multi-center randomized controlled trial in 369 patients with panic disorder and agoraphobia. *European Archives of Psychiatry and Clinical Neuroscience, 259*(Suppl. 2), 155–166.

Gloster, A. T., Wittchen, H. U., Einsle, F., Lang, T., Helbig-Lang, S., Fydrich, T., … Arolt, V. (2011). Psychological treatment for panic disorder with agoraphobia: A randomized controlled trial to examine the role of therapist-guided exposure in-situ in CBT. *Journal of Consulting and Clinical Psychology, 79,* 406–420.

Gross, C. T., & Canteras, N. S. (2012). The many paths to fear. *Nature Review Neuroscience, 13,* 651–658.

Hamm A.O. (2015). Fear-potentiated startle, In: *J. D. Wright (Ed. in Chief), International encyclopedia of social & behavioral sciences,* 2nd ed., Vol. 8, Elsevier, Oxford, 860–867.

Hamm, A. O., Cuthbert, B. N., Globisch, J., & Vaitl, D. (1997). Fear and the startle reflex: Blink modulation and autonomic response patterns in animal and mutilation fearful subjects. *Psychophysiology, 34,* 97–107.

Hamm, A. O., Greenwald, M. K., Bradley, M. M., & Lang, P. J. (1993). Emotional learning, hedonic change, and the startle probe. *Journal of Abnormal Psychology, 102,* 453–465.

Hamm, A. O., Richter, J., Pané-Farré, C. M., Westphal, D., Wittchen, H.-U., & Deckert, J. (2016). Panic disorder with agoraphobia from a behavioral neuroscience perspective: Applying the research principles formulated by the Research Domain Criteria (RDoC) initiative. *Psychophysiology, 53,* 312–322.

Hamm, A. O., & Vaitl, D. (1996). Affective learning: Awareness and aversion. *Psychophysiology, 33,* 698–710.

Hodes, R. L., Cook, E. W., III, & Lang, P. J. (1985). Individual differences in autonomic response: Conditioned association or conditioned fear? *Psychophysiology, 22,* 545–560.

Holtz, K., Melzig, C. A., Wendt, J., Lotze, M., & Hamm, A. O. (2012). Brain activation during interoceptive threat. *NeuroImage, 61,* 857–863.

Hommers, L. G., Richter, J., Yang, Y., Raab, A., Baumann, C., Lang, K., … Deckert, J. (2018). A functional genetic variation of SLC6A2 repressor hsa-miR-579-3p upregulates sympathetic noradrenergic process of fear and anxiety. *Translational Psychiatry, 8,* 226. https://doi.org/10.1038/s41398-018-0278-4.

Insel, T. (2013). *Transforming diagnosis. Retrieved from (2013). https://www.nimh.nih.gov/about/directors/thomas-insel/blog/2013/transforming-diagnosis.shtml.*

Johnson, P. L., Ferici, L. M., & Shekhar, A. (2014). Etiology, triggers and neurochemical circuits associated with unexpected, expected, and laboratory-induced panic attacks. *Neuroscience & Biobehavioral Reviews, 46,* 429–454.

Kessler, R. C., Chiu, W. T., Jin, R., Ruscio, A. M., Shear, K., & Walters, E. E. (2006). The epidemiology of panic attacks, panic disorder, and agoraphobia in the National Comorbidity Survey Replication. *Archives of General Psychiatry, 63,* 415–424.

Kim, E. J., Horovitz, O., Pellman, B. A., Tan, L. M., Li, Q., Richter-Levin, G., & Kim, J. J. (2013). Dorsal periaqueductal gray–amygdala pathway conveys both innate and learned fear responses in rats. *Proceedings of the National Academy of Sciences of the United States of America, 110,* 14795–14800.

Kolassa, I. T., Musial, F., Mohr, A., Trippe, R. H., & Miltner, W. H. R. (2005). Electrophysiological correlates of threat processing in spider phobics. *Psychophysiology, 42,* 520–530.

Kwon, S. M., Evans, L., & Oei, T. P. (1990). Factor structure of the Mobility Inventory for agoraphobia: A validational study with Australian samples of agoraphobic patients. *Journal of Psychopathology and Behavioral Assessment, 12,* 365–374.

Lang, P. J., & Bradley, M. M. (2010). Emotion and the motivational brain. *Biological Psychology, 184,* 437–450.

Lang, P. J., Bradley, M. M., & Cuthbert, B. N. (1997). Motivated attention: Affect, activation and action. In P. J. Lang, R. F. Simons, & M. T. Balaban (Eds.), *Attention and orienting: Sensory and motivational processes* (pp. 97–135). Hillsdale, NJ: Erlbaum.

LeDoux, J. (2012). Rethinking the emotional brain. *Neuron, 73,* 653–676.

Lindner, K., Neubert, J., Pfannmöller, J., Lotze, M., Hamm, A. O., & Wendt, J. (2015). Fear potentiated startle processing in humans: Parallel fMRI and orbicularis EMG assessment during cue conditioning and extinction. *International Journal of Psychophysiology, 98*, 535–545.

Lipp, O. V., Sheridan, J., & Siddle, D. A. (1994). Human blink startle during aversive and nonaversive Pavlovian conditioning. *Journal of Experimental Psychology: Animal Behavior Processes, 20*, 380–389.

Löw, A., Weymar, M., & Hamm, A. O. (2015). When threat is near, get out of here: Dynamics of defensive behavior during freezing and active avoidance. *Psychological Science, 26*, 1706–1716.

Lowe, R., & Ziemke, T. (2011). The feeling of action tendencies: On the emotional regulation of goal-directed behavior. *Frontiers in Psychology, 2*, 346.

Maren, S. (2001). Neurobiology of Pavlovian fear conditioning. *Annual Review of Neuroscience, 24*, 897–931.

McGaugh, J. L. (2004). The amygdala modulates the consolidation of memories of emotional arousing experiences. *Annual Review of Neuroscience, 27*, 1–28.

McNaughton, N. (2011). Fear, anxiety and their disorders: Past, present and future neural theories. *Psychology & Neuroscience, 4*, 173–181.

Michalowski, J. M., Melzig, C. M., Weike, A. I., Stockburger, J., Schupp, H. T., & Hamm, A. O. (2009). Brain dynamics in spider phobic individuals exposed to phobia-relevant and other emotional stimuli. *Emotion, 9*, 306–315.

Michalowski, J. M., Pané-Farré, C. M., Löw, A., & Hamm, A. O. (2015). Brain dynamics of visual attention during anticipation and encoding of threat- and safe-cues in spider phobic individuals. *Social Cognitive and Affective Neuroscience, 10*, 1177–1186.

Milad, M. R., & Quirk, G. J. (2012). Fear extinction as a model for translational neuroscience: Ten years of progress. *Annual Review of Psychology, 63*, 129–151.

Mobbs, D., Marchant, J. L., Hassabis, D., Seymour, B., Tan, F., Gray, M., & Frith, C. D. (2009). From threat to fear: The neural organization of defensive fear systems in humans. *Journal of Neuroscience, 29*, 12236–12243.

Mobbs, D., Petrovic, P., Marchant, J. L., Hassabis, D., Weiskopf, N., Seymour, B., ... Frith, C. D. (2007). When fear is near: Threat imminence elicits prefrontal-periaqueductal gray shifts in humans. *Science, 317*, 1079–1083.

Moratti, S., & Keil, A. (2005). Cortical activation during Pavlovian fear conditioning depends on heart rate response patterns: An MEG study. Brain Research. *Cognitive Brain Research, 25*, 459–471.

Morgan, M. M., & Carrive, P. (2001). Activation of the ventrolateral periaqueductal gray reduces locomotion but not mean arterial pressure in awake, freely moving rats. *Neuroscience, 102*, 905–910.

Nashold, B. S., Jr., Wilson, W. P., & Slaughter, G. S. (1974). The midbrain and pain. In: J. J. Bonica (Ed.), *Vol. 4. Advances in neurology* (pp. 191–196). New York: Raven Press.

Pané-Farré, C. A., Stender, J. P., Fenske, K., Deckert, J., Reif, A., John, U., ... Hamm, A. O. (2014). The phenomenology of the first panic attack in clinical and community-based samples. *Journal of Anxiety Disorders, 28*, 522–529.

Reif, A., Richter, J., Straube, B., Höfler, M., Lueken, U., Gloster, A. T., ... Deckert, J. (2014). MAOA and mechanisms of panic disorder revisited: From bench to molecular psychotherapy. *Molecular Psychiatry, 19*, 122–128.

Richter, J., Hamm, A. O., Pané-Farré, C. A., Gerlach, A., Gloster, A. T., Wittchen, H.-U., ... Arolt, V. (2012). Dynamics of defensive reactivity in patients with panic disorder and agoraphobia: Implications for the etiology of panic disorder. *Biological Psychiatry, 72*, 512–520.

Rodriguez, B. F., Pagano, M. E., & Keller, M. B. (2007). Psychometric characteristics of the Mobility Inventory in a longitudinal study of anxiety disorders: Replicating and exploring a three-component solution. *Journal of Anxiety Disorders, 21*, 752–761.

Roelofs, K. (2017). Freeze for action: Neurobiological mechanisms in animal and human freezing. *Philosophical Transactions of the Royal Society B, 372*, 20160206.

Rosen, J. B., & Schulkin, J. (1998). From normal fear to pathological anxiety. *Psychological Review, 105*, 325–350.

Schauer, M., & Elbert, T. (2010). Dissociation following traumatic stress: Etiology and treatment. *Journal of Psychology, 218*, 109–127.

Schenberg, L. C., Schimitel, F. G., de Souza Armini, R., Bernabé, C. S., Rosa, C. A., Tufik, S., ... Quintano-dos-Santos, J. W. (2014). Translational approach to studying panic disorder in rats: Hits and misses. *Neuroscience & Biobehavioral Reviews, 46*, 472–496.

Schupp, H. T., Cuthbert, B. N., Bradley, M. M., Cacioppo, J. T., Ito, T., & Lang, P. J. (2000). Affective picture processing: The late positive potential is modulated by motivational relevance. *Psychophysiology, 37*, 257–261.

Schupp, H. T., Cuthbert, B. N., Bradley, M. M., Hillman, C. H., Hamm, A. O., & Lang, P. J. (2004). Brain processes in emotional perception: Motivated attention. *Cognition and Emotion, 18*, 593–611.

Sehlmeyer, C., Schöning, S., Zwisterlood, P., Pfleiderer, B., Kircher, T., Arolt, V., & Konrad, C. (2009). Human fear conditioning and extinction in neuroimaging: A systematic review. *PLoS One, 4*.

Solomon, R. L., Kamin, L. J., & Wynne, L. C. (1953). Traumatic avoidance learning. The outcome of several extinction procedures with dogs. *Journal of Abnormal and Social Psychology, 48*, 291–302.

Straube, T., Mentzel, H.-J., & Miltner, W. H. R. (2007). Waiting for spiders: Brain activation during anticipatory anxiety in spider phobics. *NeuroImage, 37*, 1427–1436.

Talkovsky, A. M., & Norton, P. J. (2016). Intolerance of uncertainty and transdiagnostic group cognitive behavioral therapy for anxiety. *Journal of Anxiety Disorders, 41*, 108–114.

Tovote, P., Fadok, J. P., & Lüthi, A. (2015). Neuronal circuits for fear and anxiety. *Nature Reviews Neuroscience, 16*, 317–331.

Wendt, J., Hamm, A. O., Pané-Farré, C. M., Thayer, J. F., Gerlach, A., Gloster, A., ... Richter, J. (2018). Pretreatment cardiac vagal tone predicts dropout from and residual symptoms after exposure therapy in patients with panic disorder and agoraphobia. *Psychotherapy and Psychosomatics*. https://doi.org/10.1159/000487599.

Wendt, J., Löw, A., Weymar, M., Lotze, M., & Hamm, A. O. (2017). Active avoidance and attentive freezing in the face of approaching threat. *NeuroImage, 158*, 196–204.

Wendt, J., Neubert, J., Koenig, J., Thayer, J. F., & Hamm, A. O. (2015). Resting heart rate variability is associated with inhibition of conditioned fear. *Psychophysiology, 52*, 1161–1166.

Weymar, M., Bradely, M. M., Hamm, A. O., & Lang, P. J. (2013). When fear forms memories: Threat of shock and brain potentials during encoding and recognition. *Cortex, 49*, 819–826.

Weymar, M., Keil, A., & Hamm, A. O. (2013). Timing the fearful brain: Unspecific hypervigilance and spatial attention in early visual perception. *Social Cognitive and Affective Neuroscience, 9*, 723–729.

Weymar, M., Löw, A., Modess, C., Engel, G., Gründling, M., Petersmann, A., ... Hamm, A. O. (2009). Propranolol selectively blocks the enhanced parietal old/new effect during long-term recollection of unpleasant pictures: A high density ERP study. *NeuroImage, 49*, 2800–2806.

Wiech, K., Lin, C. S., Brodersen, K. H., Bingel, U., Ploner, M., & Tracey, I. (2010). Anterior insula integrates information about salience into perceptual decisions about pain. *Journal of Neuroscience, 30*, 16324–16331.

Wittchen, H.-U., Knappe, S., Anderson, G., Araya, R., Rivera, R. M. B., Barkman, M., ... Schumann, G. (2014). The need for a behavioural science focus in research on mental health and mental disorders. *International Journal of Methods in Psychiatric Research, 23*, 38–40.

Further reading

Cacioppo, J. T., Tassinary, L. G., & Berntson, G. G. (2000). Psychophysiological science. In J. T. Cacioppo, L. G. Tassinary, & G. G. Bernstein (Eds.), *Handbook of psychophysiology* (2nd ed., pp. 3–23). Cambridge: Cambridge University Press.

Dawson, M. E. (1990). Psychophysiology at the interface of clinical science, cognitive science, and neuroscience. *Psychophysiology, 27*, 243–255.

Kozak, M. J., & Cuthbert, B. N. (2016). The NIMH Research Domain Criteria Initiative: Background, issues, and pragmatics. *Psychophysiology, 53*, 286–297.

Maren, S. (2007). The threatened brain. *Science, 317*, 1043–1044.

Chapter 8

Deep brain stimulation for major depression: A prototype of a personalized treatment in psychiatry

Thomas E. Schlaepfer[a,b,c] and Bettina H. Bewernick[d,e]

[a]Department of Interventional Biological Psychiatry, Medical Center—University of Freiburg, Faculty of Medicine, University of Freiburg, Freiburg, Germany, [b]Department of Psychiatry and Psychotherapy, Medical Center—University of Freiburg, Faculty of Medicine, University of Freiburg, Freiburg, Germany, [c]Departments of Psychiatry and Mental Health, The Johns Hopkins University, Baltimore, MD, United States, [d]Department of Psychiatry and Psychotherapy, University Hospital Bonn, Bonn, Germany, [e]Department of Geriatric Psychiatry, University Hospital Bonn, Bonn, Germany

1 The need for personalized treatment in psychiatry

The need for personalized treatment in psychiatry is obvious: regarding major depression, a third of patients are still not treated effectively, even after several treatment attempts with conventional antidepressant treatments such as pharmacotherapy, psychotherapy, or electroconvulsive therapy (Rush, Trivedi, Wisniewski, et al., 2006). A large variety of classes of antidepressants, novel pharmacological approaches such as ketamine, and special psychotherapeutic treatments, for example, cognitive behavioral analysis system of psychotherapy are available for the treatment of depression. In addition, several approved brain stimulation techniques, for example, electroconvulsive therapy (ECT), transcranial magnetic stimulation (TMS), and vagus nerve stimulation (VNS) are often used later in the treatment course. Finally, after treatment resistance to conventional approaches, deep brain stimulation (DBS) is currently under research in clinical studies as an experimental and invasive option for some patients.

It is well known that the probability of response decreases after each failed treatment attempt, with the time being in a depressive episode and with residual symptoms after recovery (Hung, Liu, Yang, et al., 2017; Judd, Schettler, Rush, et al., 2016; Rush et al., 2006). The higher the degree of treatment resistance, the more experimental and riskier the interventions are that are chosen. Thus, a systematic assignment of a patient to a personalized, probably efficient treatment, would significantly impact psychic health.

Unfortunately, there are no systematically directed assignments of patients to a specific treatment currently. Reasons are the lack of (a) predictors of response to a specific treatment, as well as the lack of (b) risk factors for side effects; even in ECT as a historical "old" treatment, there is little agreement on predictors of response or side effects (Pinna, Manchia, Oppo, et al., 2018). The identification of a patient being at high-risk for treatment resistance would lead to a more intense treatment course (e.g., combination of psychotherapy and pharmacotherapy, or ECT earlier in the treatment course).

Conflicting results in studies on biological markers of depression (Strawbridge, Young, & Cleare, 2017) suggest that major depression is a very heterogeneous disease with different clinical subtypes (e.g., melancholic vs. atypical features), and combinations of different etiological biological and psychological factors (e.g., stress, deprivations during childhood, disbalance of neurotransmitters, dysfunction of neuronal networks). This makes the search for predictors of response, and of predictors of side effects challenging.

DBS will be presented as a treatment option for severe treatment-resistant patients suffering from depression. DBS can be seen as prototypal of personalized treatment in major depression. Very likely, a combination of different biomarkers and clinical features will be useful to assign patients to a treatment regimen (Strawbridge et al., 2017).

2 Deep brain stimulation

DBS refers to the stereotaxic placement of unilateral or bilateral electrodes connected to a permanently implanted, battery-powered neurostimulator, usually placed subcutaneously in the chest area. One or two leads connect the actual stimulating electrodes in the brain with the impulse generator (Lozano et al., 2019).

Personalized Psychiatry. https://doi.org/10.1016/B978-0-12-813176-3.00008-0

The technique of DBS has been developed in the past century, and its clinical application began in the late 1980s with patients suffering from Parkinson's disease and tremors. Today, thousands of neurological patients can benefit from DBS. This led to the evaluation of DBS in psychiatry (Schlaepfer & Lieb, 2005).

DBS is currently under research for chronic, treatment-resistant psychiatric disorders. Most data are available on major depression and obsessive-compulsive disorder (OCD); for the latter, DBS has been approved in several countries (Youngerman, Chan, Mikell, et al., 2016).

Many small pilot trials and two randomized-controlled trials (Dougherty, Carpenter, Bhati, et al., 2012; Holtzheimer, Husain, Lisanby, et al., 2017) have been published assessing depression and bipolar disorder, but have not led to an approval for this indication yet. Other indications such as addiction, Alzheimer's disease, schizophrenia, and anorexia are still in a very early stage of clinical research (Naesstrom, Blomstedt, & Bodlund, 2016).

2.1 Targets for depression

Different targets for DBS have been suggested as a consequence of our knowledge about brain functioning. The subgenual cingulate cortex (Cg25) is hyperactive during a depressive episode, likely has dysfunctional connections to the dorsal and ventral compartments of the emotion regulation circuit in depression (Mayberg, 1997), and plays a critical role in remission (Mayberg, Lozano, Voon, et al., 2005). The anterior limb of the capsula interna (ALIC) has been proposed because of historic lesion studies that put forward the hypothesis that the inactivation of larger brain areas inhibits dysfunctional connections through this region (Greenberg & Malone, 2007).

The nucleus accumbens (NAcc) was stimulated due to its prominent role in the reward system (Schlaepfer, Cohen, Frick, et al., 2008). Recently, the superolateral medial forebrain bundle (slMFB) has been proposed (Coenen, Schlaepfer, Maedler, et al., 2011), because it is neuroanatomical and functional connection with the other DBS targets in depression. This has been demonstrated by fiber tracking, and emulation of the electrical field of existing targets (Coenen, Schlaepfer, Allert, et al., 2012). The medial forebrain bundle is an important structure of the reward system, connecting several brain areas (e.g., ventral tegmental area, NAcc) involved in reward seeking and reward processing (Coenen, Panksepp, Hurwitz, et al., 2012). In addition, the slMFB is connected with currently used targets for DBS in depression (ALIC, NAcc, Cg25). Stimulating the slMFB was assumed to produce antidysphoric effects by inducing the motivation to engage in rewarding situations (Coenen, Panksepp, et al., 2012).

2.2 Antidepressant efficacy

Open-label pilot studies have documented promising antidepressant short- and long-term effects of DBS to the subgenual cingulate gyrus (Cg25) (Lozano, Mayberg, Giacobbe, et al., 2008; Mayberg et al., 2005; McNeely, Mayberg, Lozano, et al., 2008), the ALIC (Malone, 2011; Malone, Dougherty, Rezai, et al., 2009), and the NAcc (Bewernick, Hurlemann, Matusch, et al., 2010; Schlaepfer et al., 2008) in about 50% of patients (Bewernick, Kayser, Sturm, et al., 2012; Holtzheimer, Kelley, Gross, et al., 2012; Kennedy, Giacobbe, Rizvi, et al., 2011; Lozano, Giacobbe, Hamani, et al., 2012; Malone et al., 2009; Puigdemont, Perez-Egea, Portella, et al., 2012), and in about 70% of patients stimulating the slMFB (Bewernick et al., 2012; Fenoy, Schulz, Selvaraj, et al., 2016; Schlaepfer, Bewernick, Kayser, et al., 2013). This led to strong enthusiasm regarding this method, because these patients were treatment-resistant to all conventional treatments.

Two sham controlled clinical trials stimulating ALIC (Dougherty et al., 2012) or Cg25 (Holtzheimer et al., 2017; Morishita, Fayad, Higuchi, et al., 2014) have failed to show superiority of DBS to sham stimulation in short-time trials. This stands in contrast to the long-term efficacy of DBS at several targets in intraindividual comparisons (Bewernick et al., 2010; Lozano et al., 2012; Malone et al., 2009; Schlaepfer et al., 2013), and resulted in a discussion about the adequate study design for these chronic, severely affected patients (Schlaepfer, 2015). Thus, further studies with larger sample sizes are needed to decide about antidepressant efficacy and predictors of response.

2.3 DBS for treatment-resistant depression—State of the art

Currently, DBS for psychiatric indications is not broadly available. Although DBS for OCD has been approved in Europe and in the United States (Youngerman et al., 2016), availability is limited because of the small number of specialized centers (specialized neurosurgeons, psychiatrists, and psychologists), and because of the reimbursement process of the respective health systems. More data is needed to evaluate predictors of response, long-term clinical effects, and safety.

For depression and bipolar disorder, DBS has not been approved by the national authorities, and is therefore only available in clinical studies. Current research is challenged by the long approval processes for clinical studies.

Due to large costs of DBS and lack of governmental funding, only small numbers of patients have been treated so far. This limits the analysis of predictors of response. In addition, because of the invasive nature of the intervention and the risk of possible severe side effects (e.g., intracranial bleeding, infections of the IPG), studies currently select patients very restrictively. Extremely treatment-resistant, severely affected patients lacking substantial clinical comorbidity are currently included in clinical trials. Typical exclusion criteria are psychiatric comorbidity (e.g., substance abuse, PTSD, comorbid disorder if equally severe as the main diagnosis), personality disorder, acute suicidal ideation, somatic risk factors related to the surgery, and neurological disorder (Schlaepfer et al., 2008).

Due to small sample sizes in DBS studies, predictors of response are difficult to determine. We have analyzed possible predictors of response known from other antidepressant treatments in 24 patients stimulated at the slMFB. We have analyzed treatment resistance, response to prior brain stimulation, subtypes of depression, and clinical symptoms, as well as psychosocial factors and personality.

We could demonstrate that a combination of predictors is more predictive than single factors. In our study, response to other brain stimulation treatments, interpersonal experiences during childhood, and social support during treatment were relevant for the prediction of a response to slMFB-DBS (Bewernick et al., in preparation). Individual symptoms of depression were not related to the response to DBS at our target. Personality was differentially related to response to slMFB-DBS (Bewernick, Kilian, Schmidt, et al., 2018). Currently, there is not enough data to exclude patients from DBS in relation to predictors, and unipolar as well as bipolar patients seem to benefit equally as well from the treatments. The contribution of further biomarkers to response, such as functional connectivity, inflammation, or epigenetics needs to be analyzed in larger samples.

2.4 Stimulating with high spatial precision

DBS allows precise stimulation of the brain at a selected target region. Thus, DBS is the antidepressant treatment with the highest localization (Coenen, Schlaepfer, et al., 2012). In addition, due to its invasive character, brain regions in the midbrain or brainstem can be reached.

This high spatial precision is in contrast to other brain stimulation techniques: ECT elicits general seizures, VNS stimulates the vagus nerve, which is interconnected with many brain regions, and TMS or tDCS (Jog, Smith, Jann, et al., 2016) stimulates a larger area of the brain.

2.4.1 How important is the high spatial precision?

Progress in stereotactic surgery, especially DTI-guided targeting (Coenen, Schlaepfer, et al., 2012), enables us to precisely operate the electrodes at the patient's individual fiber tract, which is what we want to manipulate. In our study, we were able to precisely reach each patient's individual superolateral branch of the medial forebrain bundle. Possibly, the high number of responders is also related to surgical precision. In line with this assumption, it has been demonstrated by colleagues stimulating at the subgenual cingulate region that the antidepressant response was related to the precision with which these fibers were reached (Riva-Posse, Choi, Holtzheimer, et al., 2014, 2018).

2.5 From stimulating targets to modulating networks

DBS allows the precise targeting of small areas in the brain. Early theories assumed the inhibition of nuclear structures to be the underlying mode of action in the DBS (Gradinaru, Mogri, Thompson, et al., 2009; Hamani & Nobrega, 2010). Also, changes in neurotransmitter release (glutamate, dopamine) have been reported (Hilker, Voges, Thiel, et al., 2002; Stefani, Fedele, Galati, et al., 2006).

Neurophysiologic recordings during stimulation have demonstrated that the oscillatory activity between brain structures is modulated by DBS (Kringelbach, Jenkinson, Owen, et al., 2007). Functional neuroimaging data have demonstrated that DBS changes the activity of brain areas far beyond the targeted region. Thus, in addition to local changes of neuronal activity, complex neural networks are probably modulated (Kringelbach et al., 2007). For example, a normalization of brain metabolism in frontostriatal networks was observed shortly after surgery (1 week) (Schlaepfer et al., 2008), and after 1 year (Bewernick et al., 2010) as a result of stimulation of the NAcc in patients suffering from depression.

Thus, actual theories on depression assume the dysfunction of neuronal networks as a biological basis for depression (Gong & He, 2015), and not a pathology in a single brain area or cell type.

In the following, several network approaches will be described. In most approaches, depression is mediated by altered functioning across an integrated corticolimbic circuit in the forebrain (Akil, Gordon, Hen, et al., 2018). Important network nodes include regions of the prefrontal cortex (PFC) connected with numerous subcortical structures, including the amygdala and NAcc (Akil et al., 2018).

2.6 Relating clinical symptom clusters to dysfunctional brain networks

The idea of associating a subset of clinical symptoms with dysfunctions in brain circuits has stimulated research. Several models of depression have been proposed.

Data from human imaging studies, lesional studies, animal studies, and clinical symptom classification have led to a distinction of attentional-cognitive symptoms (e.g., apathy, psychomotor slowing, impairments in attention) localized in dorsal brain regions ("dorsal compartment," meaning the dorsolateral PFC, anterior cingulate cortex, and striatum) from vegetative and somatic symptoms (e.g., sleep disturbance, loss of appetite) localized in paralimbic, subcortical, and brainstem regions ("ventral compartment") (Mayberg, 1997). In depression, mood dysregulation is conceptualized as a dysregulation of both compartments (Mayberg, 1997). A prominent role has been attributed to the anterior cingulate cortex, which was thus selected as a target for DBS in depression.

In a recent approach including parallel findings from resting-state functional connectivity, arterial spin labeling, and diffusion imaging studies, two dysfunctional networks have been proposed in depression (Phillips, Chase, Sheline, et al., 2015). First, an implicit emotion regulation circuitry including the amygdala and regions of the medial PFC, which is primarily serotonergic. Second, a reward circuitry, including the ventral striatum and areas of the medial PFC, which is primarily dopaminergic. It is suggested that abnormalities in these parallel circuits might be associated with different symptom dimensions of depression. Depending on the circuit, different treatments could be selected, for example, abnormalities in implicit emotion regulation circuitry may be related to low mood and anxiety, while dysfunctions in the reward circuitry could be associated with apathy and anhedonia (Phillips et al., 2015).

Recently, it has been demonstrated with resting-state functional connectivity, that patients with depression can be subdivided into four neurophysiological subtypes (so-called "biotypes") by distinct patterns of dysfunctional connectivity in limbic and frontostriatal networks (Drysdale, Grosenick, Downar, et al., 2017). Interestingly, these subtypes could not be differentiated on the basis of clinical features alone, but an association with specific symptom profiles has been found. In this study, the biotype predicted the response to TMS (Drysdale et al., 2017).

The aim of all these approaches is to identify critical relationships between symptoms and brain functioning, which then may be cured, even across conventional diagnostic categories of psychiatric diseases.

2.7 Is there a target—Specific effect for subtypes of depression?

The challenges in DBS in psychiatry are (a) to find the network related to a specific symptom (cluster), and then (b) to select the optimum stimulation site for the modification of the network activity.

Historically, several subtypes of depression have been classified according to clinical features such as melancholic or atypical depression, or metabolic syndrome. It is not clear yet how these subtypes can be related to dysfunctional brain circuits, but first attempts have been made, and might lead to new, biomarker-driven categories.

To date, sample sizes in DBS in depression are too small to decide if DBS at a specific target (e.g., Cg25, slMFB, or ALIC) is associated with a better antidepressant response of a subgroup of depressed patients. Maybe, due to the highly interconnected structure of the brain, different targets are only different entrances to the same networks, and do not lead to dissociable clinical effects.

A first attempt has been made in a study with pooled data analyzing predictors of response to DBS in OCD. It describes older age at OCD onset and sexual or religions obsessions as general predictors of response to DBS (Alonso, Cuadras, Gabriels, et al., 2015). Due to the small numbers of patients, these predictors have to be considered with caution, especially as they were analyzed over all targets and not specifically for each target.

3 Current research strategies for the individualized treatment of depression

The future of personalized antidepressant therapy includes the careful analysis of clinical symptoms and biological factors. Many biomarkers were related to depression: reduced gray matter volume in the hippocampus, in the PFC, and in the basal ganglia; hypercortisolism, thyroid dysfunction, reduced dopamine, or noradrenaline, increased glutamate, increased proinflammatory cytokines, alterations to tryptophan, as well as genetic polymorphism or changes in epigenetics (Strawbridge et al., 2017). Probably, a combination of different biomarkers and clinical features, as well as sociodemographic factors, will be necessary to assign patients to a treatment.

If a *neurotransmitter disturbance* is diagnosed, pharmacotherapy would be selected. First attempts have been made to individualize therapy in pharmacotherapy. In a study, antidepressants in general were more effective for core emotional symptoms than for sleep or atypical symptoms (Chekroud, Zotti, Shehzad, et al., 2016).

Pharmacogenomics assesses how a person's genetic makeup at the nucleotide level influences their reaction to medications (Amare, Schubert, & Baune, 2017). It is established that mood disorders are polygenic. The interaction of multiple genetic variants with environmental factors contribute to depression (Amare et al., 2017).

Increased levels of various proinflammatory markers are associated with about 30% of depressed patients, and could constitute another subset of patient (Strawbridge et al., 2017). The inflammatory system is complex, and there are numerous biomarkers representing different aspects of this system that can be associated differently with current depressive episodes and response to a treatment. Up to now, clinical symptoms (e.g., atypical depression) have not been agreed upon in relation to elevated inflammation, although some authors propose an "inflammation subtype" of depression (Strawbridge et al., 2017). Possibly, these patients could profit from add-on antiinflammatory treatment.

Psychotherapy is an effective treatment option in addition, or instead of, biological approaches, and has been demonstrated to also alter brain metabolism (Kennedy & Giacobbe, 2007). First meta-analyses emerge on the relevance of biological markers, for example, cortisol levels, as predictors of the antidepressant response to psychotherapy (Fischer, Strawbridge, Vives, et al., 2017).

Disturbed network functioning will probably be treated with brain stimulation methods. DBS will then be a treatment option for a subgroup of patients that have a distinct disturbance in neuronal networks. Depending on the degree and localization of the network disturbance, a specific treatment (tDCS, TMS or DBS) could be selected. Being able to select individual patients for modulating, at a high local precision, a node of a disturbed network will be a large step toward personalized medicine if we compare the broad approaches in pharmacotherapy or ECT.

Four types of depression have been proposed relating aberrant biomarkers to clinical symptoms (Kunugi, Hori, & Ogawa, 2015): (1) hypercortisolism related to melancholic depression, (2) hypocortisolism related to atypical depression, (3) dopamine-dysfunction associated with anhedonia, and (4) the inflammatory subtype with elevated inflammation that has not related to clinical correlates yet (Kunugi et al., 2015).

Machine-learning algorithms might be an option for response prediction because a variety of variables can be included (LIT).

Technical developments in brain stimulation, such as closed-loop approaches, are under research (Deeb, Giordano, Rossi, et al., 2016). Because biomarkers as feedback parameters are lacking so far in depression, certain behaviors might have to be used as markers for the electrophysiological manipulation of brain activity in closed-loop stimulation.

Being able to stimulate, with a high spatial resolution, targets that are deep in the brain without the need for invasive neurosurgery is another arm of research. Focused ultrasound is currently under development as a new noninvasive method for brain stimulation (Fini & Tyler, 2017).

4 Summary and outlook

Pharmacological treatments, brain stimulation, and psychotherapy are options for the treatment of depression. Individualizing the assignment of patients to a specific treatment is difficult due to the heterogeneity of the disease and lack specific biomarkers. Although, DBS allows the stimulation of a specific brain target with a high spatial resolution, criteria selection of patients and the selection of the brain target to be stimulated are lacking so far. Many biomarkers and network models are currently under research in relation to clinical symptoms, for treatment efficacy and better resistance to side effects. In the future, progress in these research areas will guide individualized treatments and allow patients to be treated more successfully.

REFERENCES

Akil, H., Gordon, J., Hen, R., et al. (2018). Treatment resistant depression: A multi-scale, systems biology approach. *Neuroscience and Biobehavioral Reviews, 84*, 272–288.

Alonso, P., Cuadras, D., Gabriels, L., et al. (2015). Deep brain stimulation for obsessive-compulsive disorder: A meta-analysis of treatment outcome and predictors of response. *PLoS One, 10*(7).

Amare, A. T., Schubert, K. O., & Baune, B. T. (2017). Pharmacogenomics in the treatment of mood disorders: Strategies and opportunities for personalized psychiatry. *The EPMA Journal, 8*(3), 211–227.

Bewernick, B. H., Hurlemann, R., Matusch, A., et al. (2010). Nucleus accumbens deep brain stimulation decreases ratings of depression and anxiety in treatment-resistant depression. *Biological Psychiatry, 67*(2), 110–116.

Bewernick, B., Kayser, S., Sturm, V., et al. (2012). Long-term effects of nucleus accumbens deep brain stimulation in treatment-resistant depression: Evidence for sustained efficacy. *Neuropsychopharmacology*, *37*(9), 1975–1985. https://doi.org/10.1038/npp.2012.44.

Bewernick, B. H., Kilian, H. M., Schmidt, K., et al. (2018). Deep brain stimulation of the supero-lateral branch of the medial forebrain bundle does not lead to changes in personality in patients suffering from severe depression. *Psychological Medicine*, *48*, 2684–2692.

Bewernick, B.H., et al., Personalizing treatment of depression with slMFB deep brain stimulation in severely depressed patients by predictors of response (in preparation).

Chekroud, A. M., Zotti, R. J., Shehzad, Z., et al. (2016). Cross-trial prediction of treatment outcome in depression: A machine learning approach. *Lancet Psychiatry*, *3*(3), 243–250.

Coenen, V. A., Panksepp, J., Hurwitz, T. A., et al. (2012). Human medial forebrain bundle (MFB) and anterior thalamic radiation (ATR): Imaging of two major subcortical pathways and the dynamic balance of opposite affects in understanding depression. *The Journal of Neuropsychiatry and Clinical Neurosciences*, *24*(2), 223–236.

Coenen, V. A., Schlaepfer, T. E., Allert, N., et al. (2012). Diffusion tensor imaging and neuromodulation: DTI as key technology for deep brain stimulation. *International Review of Neurobiology*, *107*, 207–234.

Coenen, V. A., Schlaepfer, T. E., Maedler, B., et al. (2011). Cross-species affective functions of the medial forebrain bundle—Implications for the treatment of affective pain and depression in humans. *Neuroscience and Biobehavioral Reviews*, *35*, 1971–1981.

Deeb, W., Giordano, J. J., Rossi, P. J., et al. (2016). Proceedings of the fourth annual deep brain stimulation think tank: A review of emerging issues and technologies. *Frontiers in Integrative Neuroscience*, *10*, 38.

Dougherty, D., Carpenter, L., Bhati, M., et al. (2012). 720-A randomized sham-controlled trial of DBS of the VC/VS for treatment-resistant depression. *Biological Psychiatry*, *71*(8), 230S.

Drysdale, A. T., Grosenick, L., Downar, J., et al. (2017). Resting-state connectivity biomarkers define neurophysiological subtypes of depression. *Nature Medicine*, *23*(1), 28–38.

Fenoy, A. J., Schulz, P., Selvaraj, S., et al. (2016). Deep brain stimulation of the medial forebrain bundle: Distinctive responses in resistant depression. *Journal of Affective Disorders*, *203*, 143–151.

Fini, M., & Tyler, W. J. (2017). Transcranial focused ultrasound: A new tool for non-invasive neuromodulation. *International Review of Psychiatry*, *29*(2), 168–177.

Fischer, S., Strawbridge, R., Vives, A. H., et al. (2017). Cortisol as a predictor of psychological therapy response in depressive disorders: Systematic review and meta-analysis. *The British Journal of Psychiatry*, *210*(2), 105–109.

Gong, Q., & He, Y. (2015). Depression, neuroimaging and connectomics: A selective overview. *Biological Psychiatry*, *77*(3), 223–235.

Gradinaru, V., Mogri, M., Thompson, K. R., et al. (2009). Optical deconstruction of parkinsonian neural circuitry. *Science*, *324*(5925), 354–359.

Greenberg, B., & Malone, D. (2007). *Preliminary results from DBS multicenter study in depression*. Providence, Rhode Island, Personal Communication.

Hamani, C., & Nobrega, J. N. (2010). Deep brain stimulation in clinical trials and animal models of depression. *The European Journal of Neuroscience*, *32*(7), 1109–1117.

Hilker, R., Voges, J., Thiel, A., et al. (2002). Deep brain stimulation of the subthalamic nucleus versus levodopa challenge in Parkinson's disease: Measuring the on- and off-conditions with FDG-PET. *Journal of Neural Transmission (Vienna)*, *109*(10), 1257–1264.

Holtzheimer, P. E., Husain, M. M., Lisanby, S. H., et al. (2017). Subcallosal cingulate deep brain stimulation for treatment-resistant depression: A multisite, randomised, sham-controlled trial. *Lancet Psychiatry*, *4*, 839–849.

Holtzheimer, P. E., Kelley, M. E., Gross, R. E., et al. (2012). Subcallosal cingulate deep brain stimulation for treatment-resistant unipolar and bipolar depression. *Archives of General Psychiatry*, *69*(2), 150–158.

Hung, C. I., Liu, C. Y., & Yang, C. H. (2017). Untreated duration predicted the severity of depression at the two-year follow-up point. *PLoS One*, *12*(9).

Jog, M. V., Smith, R. X., Jann, K., et al. (2016). In-vivo imaging of magnetic fields induced by transcranial direct current stimulation (tDCS) in human brain using MRI. *Scientific Reports*, *6*, 34385.

Judd, L. L., Schettler, P. J., Rush, A. J., et al. (2016). A new empirical definition of major depressive episode recovery and its positive impact on future course of illness. *The Journal of Clinical Psychiatry*, *77*(8), 1065–1073.

Kennedy, S. H., & Giacobbe, P. (2007). Treatment resistant depression—Advances in somatic therapies. *Annals of Clinical Psychiatry*, *19*(4), 279–287.

Kennedy, S. H., Giacobbe, P., Rizvi, S. J., et al. (2011). Deep brain stimulation for treatment-resistant depression: Follow-up after 3 to 6 years. *The American Journal of Psychiatry*, *168*(5), 502–510.

Kringelbach, M. L., Jenkinson, N., Owen, S. L., et al. (2007). Translational principles of deep brain stimulation. *Nature Reviews. Neuroscience*, *8*(8), 623–635.

Kunugi, H., Hori, H., & Ogawa, S. (2015). Biochemical markers subtyping major depressive disorder. *Psychiatry and Clinical Neurosciences*, *69*(10), 597–608.

Lozano, A. M., Giacobbe, P., Hamani, C., et al. (2012). A multicenter pilot study of subcallosal cingulate area deep brain stimulation for treatment-resistant depression. *Journal of Neurosurgery*, *116*(2), 315–322.

Lozano, A. M., Lipsman, N., Bergman, H., et al. (2019). Deep brain stimulation: Current challenges and future directions. *Nature Reviews Neurology*, *15*(3), 148–160.

Lozano, A. M., Mayberg, H. S., Giacobbe, P., et al. (2008). Subcallosal cingulate gyrus deep brain stimulation for treatment-resistant depression. *Biological Psychiatry*, *64*(6), 461–467.

Malone, D. A. (2011). Use of deep brain stimulation in treatment-resistant depression. *Cleveland Clinic Journal of Medicine*, *77*, 77.

Malone, D. A., Jr., Dougherty, D. D., Rezai, A. R., et al. (2009). Deep brain stimulation of the ventral capsule/ventral striatum for treatment-resistant depression. *Biological Psychiatry*, *65*(4), 267–275.

Mayberg, H. S. (1997). Limbic-cortical dysregulation: A proposed model of depression. *The Journal of Neuropsychiatry and Clinical Neurosciences, 9*(3), 471–481.

Mayberg, H. S., Lozano, A. M., Voon, V., et al. (2005). Deep brain stimulation for treatment-resistant depression. *Neuron, 45*(5), 651–660.

McNeely, H. E., Mayberg, H. S., Lozano, A. M., et al. (2008). Neuropsychological impact of Cg25 deep brain stimulation for treatment-resistant depression: Preliminary results over 12 months. *The Journal of Nervous and Mental Disease, 196*(5), 405–410.

Morishita, T., Fayad, S. M., Higuchi, M. A., et al. (2014). Deep brain stimulation for treatment-resistant depression: Systematic review of clinical outcomes. *Neurotherapeutics, 11*(3), 475–484.

Naesstrom, M., Blomstedt, P., & Bodlund, O. (2016). A systematic review of psychiatric indications for deep brain stimulation, with focus on major depressive and obsessive-compulsive disorder. *Nordic Journal of Psychiatry, 70*(7), 483–491.

Phillips, M. L., Chase, H. W., Sheline, Y. I., et al. (2015). Identifying predictors, moderators, and mediators of antidepressant response in major depressive disorder: Neuroimaging approaches. *The American Journal of Psychiatry, 172*(2), 124–138.

Pinna, M., Manchia, M., Oppo, R., et al. (2018). Clinical and biological predictors of response to electroconvulsive therapy (ECT): A review. *Neuroscience Letters, 669*, 32–42.

Puigdemont, D., Perez-Egea, R., Portella, M. J., et al. (2012). Deep brain stimulation of the subcallosal cingulate gyrus: Further evidence in treatment-resistant major depression. *The International Journal of Neuropsychopharmacology, 15*(1), 121–133.

Riva-Posse, P., Choi, K. S., Holtzheimer, P. E., et al. (2014). Defining critical white matter pathways mediating successful subcallosal cingulate deep brain stimulation for treatment-resistant depression. *Biological Psychiatry, 76*(12), 963–969.

Riva-Posse, P., Choi, K. S., Holtzheimer, P. E., et al. (2018). A connectomic approach for subcallosal cingulate deep brain stimulation surgery: Prospective targeting in treatment-resistant depression. *Molecular Psychiatry, 23*, 843–849.

Rush, A. J., Trivedi, M. H., Wisniewski, S. R., et al. (2006). Acute and longer-term outcomes in depressed outpatients requiring one or several treatment steps: A STAR*D report. *The American Journal of Psychiatry, 163*(11), 1905–1917.

Schlaepfer, T. E. (2015). Deep brain stimulation for major depression-steps on a long and winding road. *Biological Psychiatry, 78*(4), 218–219.

Schlaepfer, T., Bewernick, B., Kayser, S., et al. (2013). Rapid effects of deep brain stimulation for treatment resistant major depression. *Biological Psychiatry, 73*, 1204–1212.

Schlaepfer, T. E., Cohen, M. X., Frick, C., et al. (2008). Deep brain stimulation to reward circuitry alleviates anhedonia in refractory major depression. *Neuropsychopharmacology, 33*, 368–377.

Schlaepfer, T. E., & Lieb, K. (2005). Deep brain stimulation for treatment of refractory depression. *Lancet, 366*(9495), 1420–1422.

Stefani, A., Fedele, E., Galati, S., et al. (2006). Deep brain stimulation in Parkinson's disease patients: Biochemical evidence. *Journal of Neural Transmission. Supplementum, 70*, 401–408.

Strawbridge, R., Young, A. H., & Cleare, A. J. (2017). Biomarkers for depression: Recent insights, current challenges and future prospects. *Neuropsychiatric Disease and Treatment, 13*, 1245–1262.

Youngerman, B. E., Chan, A. K., Mikell, C. B., et al. (2016). A decade of emerging indications: Deep brain stimulation in the United States. *Journal of Neurosurgery, 125*(2), 461–471.

Chapter 9

The Psychiatric Genomics Consortium: History, development, and the future

Hunna J. Watson[a,b,c], Zeynep Yilmaz[a,d] and Patrick F. Sullivan[a,d,e]

[a]*Department of Psychiatry, The University of North Carolina at Chapel Hill, Chapel Hill, NC, United States,* [b]*School of Psychology, Curtin University, Perth, WA, Australia,* [c]*Division of Paediatrics, School of Medicine, The University of Western Australia, Perth, WA, Australia,* [d]*Department of Genetics, The University of North Carolina at Chapel Hill, Chapel Hill, NC, United States,* [e]*Department of Medical Epidemiology and Biostatistics, Karolinska Institutet, Stockholm, Sweden*

The World Health Organization estimates that upward of 450 million people in the world currently, and one in four people in their lifetime, will experience a psychiatric disorder (World Health Organization, 2001a, 2001b). Psychiatric disorders account for ~32% of years lived with a disability, and carry substantial morbidity and mortality-related risks (Vigo, Thornicroft, & Atun, 2016). Clearly, this picture presents a compelling and urgent need for new and more effective treatment options. The Psychiatric Genomics Consortium (PGC) is a large collaborative scientific effort, and its goal is to generate high-confidence "maps" of the genetic architecture of psychiatric diseases to yield major breakthroughs in treatment and clinical care. This chapter discusses the work of the consortium, summarizes key contributions of the research program, and highlights future directions.

1 Background

1.1 Heritability in psychiatric disorders

Before the modern genomics era, genetic epidemiological studies had illuminated the substantial heritability of psychiatric disorders and vastly remolded our understanding of psychiatric disorders from merely "mental" illnesses to biologically-based (Shih, Belmonte, & Zandi, 2004). Family and twin studies showed substantial heritability for all major psychiatric disorders, including major depression, schizophrenia, bipolar disorder, anorexia nervosa, attention-deficit/hyperactivity disorder (ADHD), autism spectrum disorders, and anxiety disorders. Heritability estimates from twin studies are: major depression 33%–45%, anxiety disorders 30%–50%, anorexia nervosa 48%–74%, autism spectrum 64%–91%, ADHD 75%, bipolar disorder 80%, and schizophrenia 80% (Baker, Schaumberg, & Munn-Chernoff, 2017; Polderman et al., 2015; Schachar, 2014; Shih et al., 2004; Smoller, Cerrato, & Weatherall, 2015; Tick, Bolton, Happé, Rutter, & Rijsdijk, 2016). This evidence undeniably sparked a major shift in the conceptualization and science of psychiatric disorders, but conferred no insight into biology.

1.2 Linkage and candidate gene studies

There needed to be a deeper understanding at a molecular and physiological level of the genetic basis of psychiatric illness. Linkage approaches and candidate gene studies attempted to reconcile this gap from the mid-1980s to the mid-2000s. Linkage studies attempt to find correlations between psychiatric diagnosis and the inheritance of certain alleles within families with affected individuals. At the time, these studies were expected to yield "the gene for" particular psychiatric disorders (Kendler, 2015). Later, they were shown to be poor tools for this purpose, because linkage studies are far better suited to Mendelian diseases and traits, but have had limited success with traits that do not follow Mendelian inheritance patterns, that is, "complex traits," such as virtually all psychiatric disorders. Candidate gene association studies test the hypothesis that a particular allele of a given polymorphism is associated with a trait. The hypotheses are essentially "best guesses" at genetic loci. In spite of early excitement and promise, it turns out that for complex traits, candidate gene studies produced a succession of nonreplicable findings, and have proven to be no better than chance at identifying susceptibility variants (Farrell et al., 2015; Johnson et al., 2017; Sullivan, 2017). In fact, the US NIMH recently indicated that it would no longer fund candidate gene studies due to their abysmal track record in identifying secure associations for psychiatric disorders. The common problem underlying linkage and candidate gene study approaches was that it was assumed that the genetic loci involved would be few, and carry large genetic risks—both of these assumptions were incorrect (Sullivan et al., 2018).

1.3   Polygenicity

We now know that at inception of the foray into molecular discovery, the polygenicity of psychiatric disorders was vastly underestimated. The true landscape is that for common variation, the biological basis consists of thousands of single-nucleotide polymorphisms (SNPs) that each contribute a small effect to disease risk (Kendler, 2015; Sullivan, Daly, & O'Donovan, 2012). In fact, for many major psychiatric disorders (schizophrenia, major depression, bipolar disorder, anorexia nervosa, autism spectrum) where sufficiently large sample sizes have been accumulated, there is virtually 100% power to find common genetic variants with an odds ratio >1.35, so, if there existed "the gene for" any particular psychiatric disorder, it would already have been discovered (Farrell et al., 2015; Yao et al., 2016). Influential early genome-wide association studies (GWASs) (e.g., the discovery of complement factor H (CFH) for age-related macular degeneration) (Klein et al., 2005) encouraged the expectation that clinically meaningful discovery through GWAS might be straightforward, and possible at small sample sizes, but these were exceptional, and this was not to be the case (Kendler, 2015).

1.4   High-throughput assays

Advances in technology and reduced costs associated with genomic analyses have made it possible for the use of higher coverage methods in the examination of the genome for assessing genetic risk (Psychiatric GWAS Consortium Coordinating Committee, 2010). As a result, genome-wide data obtained from unrelated individuals have been successfully utilized to study the heritability of psychiatric disorders, especially through GWAS designs (Bulik-Sullivan, Finucane, et al., 2015; Yang, Lee, Goddard, & Visscher, 2011). These studies commonly compare cases (i.e., individuals affected with a condition) to controls on millions of common SNPs (population frequency >1%) across the entire genome, to identify susceptibility and protective alleles.

One of the first GWAS arrays had 317K SNPs, and cost well over US$1,000 (circa 2007). The current generation of GWAS arrays assess around 700K SNPs at a cost of $35.

2   History, development, and contributions of the Psychiatric Genomics Consortium

2.1   History and aims

In 2007, the PGC was formed to expedite large-scale genetic analyses for five major psychiatric disorders: ADHD, autism, bipolar disorder, major depressive disorder, and schizophrenia (Psychiatric GWAS Consortium Coordinating Committee, 2010; Psychiatric GWAS Consortium Steering Committee, 2009; Sullivan, 2010). A consortium model was adopted, and around 800+ scientists from >40 countries are participating as of this writing. DNA collection and phenotype data originate from clinical and case populations recruited through hospitals, population-based studies/sources, health maintenance organizations, and national medical registers.

The first major area of work of the PGC, and still a major focus, are GWASs. The first ever GWAS in the world was on age-related macular degeneration in 2005 (Klein et al., 2005). Since then, nearly 2,000 GWAS papers have been published (MacArthur et al., 2016). Many GWAS findings, broadly, have been unexpected and demonstrative, providing new clues about the genetic basis of diseases and health. A key ingredient to success for discovery in GWAS of complex traits is large sample sizes, in the order of thousands of individuals. These numbers are usually beyond the reach of any single research group, and rallying scientists from around the world to speed up the process of discovery though large-scale collaboration has been important for advancing GWAS research, and was the impetus for creating the PGC.

At its inception, PGC focused on bringing together clinical and research scientists and GWAS data from patients from all over the world as a part of one massive research resource, with the initial goal of conducting GWAS mega-analyses to understand the role of common variation in the etiology of ADHD, autism spectrum disorder, bipolar disorder, major depression, and schizophrenia (Psychiatric GWAS Consortium Steering Committee, 2009). Over time, and in light of success in gene discovery for the core five disorders, the number of disorders and study aims has expanded. Working groups for eating disorders, obsessive-compulsive disorder/Tourette's, posttraumatic stress disorder, substance use disorders, anxiety disorders, and Alzheimer's disease have been added.

The scientific scope has expanded the aims to include the following—currently underway—*aim 1*: even larger GWAS mega-analyses, cross-disorder GWASs, and pathway analyses to clarify the biological mechanisms and gene sets pertinent to GWAS findings; *aim 2*: developing genetic risk scores (GRS) and evaluating the developmental effects of GRS, evaluating genotype (GRS) × environmental interactions; *aim 3*: estimating genetic correlations between psychiatric disorders and other brain diseases and quantitative traits to shed light on pleiotropy, comorbidity, and nosology; and *aims 4, 5, and 6*: investigating the role of rare variation through copy number variants (CNVs), targeted sequencing of exons in genes known to contain rare variation, and in gene sets implicated in schizophrenia, and studies of unusual pedigrees with a high incidence of psychiatric disorders (for more about each aim, see Sullivan et al., 2018).

In the past decade, psychiatric genetics has made incredible progress, in contrast to the decades prior to the advent of GWAS and high-throughput sequencing studies that were marked by few replicable and robust findings (Psychiatric GWAS Consortium Coordinating Committee, 2010).

2.2 Contributions

2.2.1 General contributions

PGC has assembled many of the largest GWAS to date in psychiatry and identified many loci associated with psychiatric disorders (see Fig. 1) (Demontis et al., 2019; Duncan et al., 2017; Hinney et al., 2011; Major Depressive Disorder Working Group of the Psychiatric GWAS Consortium, 2013; Psychiatric GWAS Consortium Bipolar Disorder Working Group, 2011; Ripke et al., 2014, 2013; Schizophrenia Psychiatric Genome-Wide Association Study Consortium, 2011; Stahl et al., 2019; Watson et al., 2019). Refinement of these findings will allow genes and biological mechanisms to be identified. For schizophrenia and major depression, enrichment of particular gene sets and pathways is a major discovery yielded from GWAS results (de Jong, Vidler, Mokrab, Collier, & Breen, 2016; Gaspar & Breen, 2016; Wray et al., 2018). For decades, drug development in psychiatry has been stalled because of a dearth of credible, novel drug targets. These findings could hold the key to the discovery of novel drug compounds, and long-awaited therapeutic advances.

PGC studies have brought to light the clear association between psychiatric disorders and common variation, and have provided groundbreaking evidence of the polygenicity (Sullivan et al., 2012). Fig. 2 shows SNP-based heritability estimates for ADHD, autism spectrum disorder, bipolar disorder, major depression, and schizophrenia, and coheritability estimates. The odds ratios for risk-increasing common variants has been modest, typically in the vicinity of 1.10–1.20 (Demontis et al., 2019; Ripke et al., 2014; Wray et al., 2018). Nevertheless, common variants that explain <1% of variance can lead to biological and therapeutic intervention breakthroughs (Hirschhorn, 2009).

2.2.2 Schizophrenia

In the early days of the PGC, initial GWAS efforts in schizophrenia were significantly underpowered to detect genome-wide significant risk variants (International Schizophrenia Consortium et al., 2009), and the relationship between the increase in sample size and the number of loci identified was not linear until an inflection point was reached at ∼9,000 cases (Schizophrenia Psychiatric Genome-Wide Association Study Consortium, 2011). In a 2014 PGC GWAS of schizophrenia (36,989 cases and 113,075 controls), 108 independent genomic loci were significantly associated with schizophrenia risk (see Fig. 3) (Ripke et al., 2014). A novel finding from this study is that associations were enriched not only in the brain, but in regions linked to immunity. Previously established pathophysiology, specifically the antipsychotic target dopamine D2 receptor, also had a signal. In the more recent GWAS by Pardiñas et al. (2018) that combined CLOZUK + PGC samples (40,675 cases and 64,643 controls), the results implicated 145 independent loci and six gene sets concerning loss-of-function intolerant genes, voltage-gated calcium channel complexes, the $5HT_{2c}$ receptor complex, and three gene sets related to neurophysiological and behavioral learning processes. Complement component 4 protein has also been newly documented to have a synaptic pruning function in brain tissues, which may also be a molecular explanation for increased risk for schizophrenia (Sekar et al., 2016).

2.2.3 Bipolar disorder

PGCs earliest mega-analysis of bipolar disorder (11,974 cases and 51,792 controls) illustrated genome-wide significant associations with a calcium channel gene (*CACNA1C*) (Psychiatric GWAS Consortium Bipolar Disorder Working Group, 2011). More recently, 30 independent loci have been detected (20,352 cases and 31,358 controls), and these loci contain genes encoding ion channels and neurotransmitter transporters, synaptic processes, immune and energy functions, and multiple potential drug targets for mood stabilizers (Stahl et al., 2019).

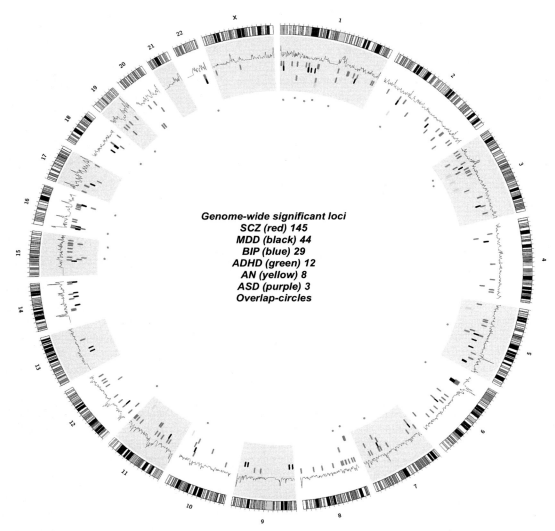

FIG. 1   GWAS "hits" to date. Shown are genome-wide significant "clumps" as a Circos plot. The genome is depicted as a *circle*, from the start of chromosome 1 at the *top*, clockwise through chromosome 22 and the end of chromosome X. Eight tracks are shown. The outermost track is a standard ideogram. The next track inward shows GENCODE gene density per Mb. The next six tracks show the genome-wide significant loci from publications in the following order: (a) SCZ, 145 loci (Pardiñas et al., 2018); (b) PGC MDD2, 44 loci (Wray et al., 2018); (c) PGC BIP2, 29 loci (Stahl et al., 2019); (d) PGC/iPSYCH ADHD2, 12 loci (Demontis et al., 2019); (e) PGC AN2, 8 loci (Watson et al., 2019); and (f) PGC ASD2, 3 loci, (Grove et al., 2019). Manhattan plot of SNPs with genome-wide *P* values of $<5 \times 10^{-8}$ for ADHD, anorexia nervosa, autism spectrum disorder, bipolar disorder, major depression, and schizophrenia in mega-analyses. *ADHD*, attention-deficit/hyperactivity disorder; *AN*, anorexia nervosa; *ASD*, autism spectrum disorder; *BIP*, bipolar disorder; *MDD*, major depressive disorder; *SCZ*, schizophrenia.

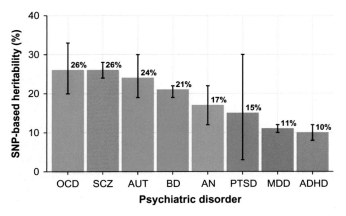

FIG. 2   Heritability explained by common variation. SNP-based heritability estimates illustrate the role of common variation in the genomic and biological architecture of psychiatric disorders. The mean and 95% confidence interval are shown. Estimates were taken from Anttila et al. (2018). *ADHD*, attention-deficit/hyperactivity disorder; *AN*, anorexia nervosa; *AUT*, autism spectrum disorder; *BD*, bipolar disorder; *MDD*, major depressive disorder; *OCD*, obsessive-compulsive disorder; *PTSD*, posttraumatic stress disorder; *SCZ*, schizophrenia.

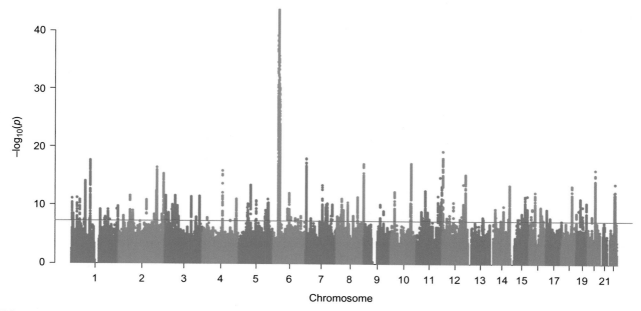

FIG. 3 Manhattan plot of the schizophrenia GWAS meta-analysis that identified 145 independent loci. The data include CLOZUK and PGC datasets (40,675 cases and 64,643 controls) (Pardiñas et al., 2018). Chromosomes depicted range from 1 to 22. The red bar indicates the P threshold for genome-wide significance ($P < 5 \times 10^{-8}$ which is 7.3 on $-\log 10$ scale). *Green* and *blue* differentiate adjacent chromosomes.

2.2.4 Autism

The first PGC GWAS of autism identified one genome-wide significant association (7,387 cases and 8,567 controls) (Autism Spectrum Disorders Working Group of The Psychiatric Genomics Consortium et al., 2017). As the sample size reached 18,381 cases and 27,969 controls, the number of genome-wide significant hits increased to five, and seven additional loci shared with other traits were also identified with the leveraging of GWAS results from schizophrenia, major depression, and educational attainment—all of which have notable genetic overlap with autism spectrum disorder (Autism Spectrum Disorders Working Group of The Psychiatric Genomics Consortium et al., 2017). Small sample size and limited power are constraining discovery, and enlarged samples will prove helpful, as in schizophrenia (see Fig. 4).

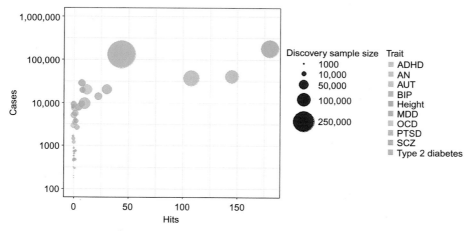

FIG. 4 In GWASs of complex traits, the number of genome-wide significant hits increases as the number of cases and the discovery sample size increase. Many GWAS in psychiatric genomics are not well-powered to detect associated genetic loci. Over time, data collections have been increasing so that associated genetic loci can be identified.

2.2.5 Attention-deficit/hyperactivity disorder

The most recent GWAS mega-analysis (20,183 cases and 35,191 controls) yielded the first significant risk loci for ADHD (Demontis et al., 2019). In total, 12 significant loci were found, and none of these corresponded with candidate genes proposed for ADHD preceding the era of GWAS. Signals in regions involved with synapse formation, language development, and neural processes were noted. Functional studies are needed to uncover the mechanisms involved. An increased burden of rare CNVs compared with controls was shown, though none individually were genome-wide significant (Yang et al., 2013).

2.2.6 Major depressive disorder

Unlike the case of schizophrenia, for which ~9,000 cases signified an inflection point for gene discovery, initial GWAS efforts in major depression with >18,000 cases proved to be underpowered for the identification of risk loci, therefore underscoring potentially important differences in the genetic architecture of major depression (Major Depressive Disorder Working Group of the Psychiatric GWAS Consortium, 2013). Since sample sizes reached 135,458 cases and 344,901 controls, 44 independent risk loci have become significant genome-wide for major depression risk (Wray et al., 2018). The two strongest signals are in or near genes mediating neurite outgrowth and synaptic plasticity in the brain. GWAS findings also implicated genes that are current targets of antidepressant medications, and genes that have been implicated previously in schizophrenia and autism spectrum disorder, involved in gene splicing, and in creating isoforms.

2.2.7 Anorexia nervosa

GWAS for anorexia nervosa conducted under the PGC umbrella (16,992 cases and 55,525 controls) have identified eight genome-wide significant loci (Watson et al., 2019). Correlations suggested that genetic risk for anorexia nervosa was positively associated with genetic risk for beneficial metabolic traits and other psychiatric illnesses, and negatively associated with genetic risk for body mass index and negative metabolic traits. The implication is that anorexia nervosa may need potential reconceptualization as a disorder of psychiatric and metabolic origins.

2.2.8 Other disorders

The PGCs workgroups for other disorders (posttraumatic stress disorder, substance use disorders, obsessive-compulsive disorder/Tourette's syndrome, anxiety disorders, Alzheimer's disease) initially had smaller sample sizes that did not result in genome-wide significant hits. Manuscripts from all of these groups incorporating larger sample sizes are in preparation or have been published (Walters et al., 2018), and nearly all groups have genome-wide significant associations.

2.2.9 Rare variants

The PGC is also contributing evidence for associations between rare genomic variants and psychiatric disorders (Sullivan et al., 2018). For example, copy number variation burden was implicated in schizophrenia in a study of 21,094 cases and 20,227 controls (Marshall et al., 2017). Rare variation in complex traits is somewhat more challenging to study than common variation because of larger sample requirements.

2.2.10 Cross-disorder GWAS

Genetic studies that examine the shared genetic risk among different psychiatric disorders is another key contribution. The rationale of cross-disorder analyses is that it is widely thought that the separable diagnostic categories in DSM and ICD insufficiently "carve nature at its joints" (Sullivan et al., 2012). Over the years, twin studies have provided strong evidence for the coaggregation of psychiatric disorders in families, demonstrating that the presence of one psychiatric disorder may also increase risk for another psychiatric disorder (Cederlöf et al., 2015; Lichtenstein et al., 2009; Mataix-Cols et al., 2013; Mathews & Grados, 2011; Meier et al., 2015; Ronald, Simonoff, Kuntsi, Asherson, & Plomin, 2008). Research findings clearly show substantial genetic overlap among the psychiatric disorders (Cross-Disorder Group of the Psychiatric Genomics Consortium, 2013; Cross-Disorder Group of the Psychiatric Genomics Consortium et al., 2013). This holds for childhood- and adult-onset disorders. Additionally, similar gene sets and pathways (e.g., calcium channel signaling) have been implicated in multiple psychiatric disorders, suggesting strong pleiotropic effects (Cross-Disorder Group of the Psychiatric Genomics Consortium et al., 2013). One notable implication of cross-disorder GWAS efforts is that the findings may lead to refined diagnostic nomenclature.

2.2.11 Genetic risk scores

In 2009, the International Schizophrenia Consortium (which was subsequently collapsed into the PGC) introduced GRS—a weighted sum of common risk variants per individual as a measure of genetic liability. As each genome-wide significant locus accounts for <5% of variance (OR ≤ 1.2), GRS aims to provide insight into the genetic architecture of psychiatric disorders using evidence for association from large numbers of genetic variants below the stringent threshold for genome-wide significance, allowing for the utilization of more of the genetic information obtained through GWAS (International Schizophrenia Consortium et al., 2009). Briefly, risk alleles and their effect sizes from GWAS results in a discovery sample are used to generate a risk score for each individual in an independent "target" sample, the validity of which is then evaluated through regression of the target sample phenotype (which does not have to be the same as the discovery sample phenotype) on the GRS (Wray et al., 2014). Since its initial conceptualization in 2007 (Wray, Goddard, & Visscher, 2007) and its first application in schizophrenia (International Schizophrenia Consortium et al., 2009), validity of GRS has been consistently demonstrated across studies and different psychiatric phenotypes (Cross-Disorder Group of the Psychiatric Genomics Consortium et al., 2013; Psychiatric GWAS Consortium Bipolar Disorder Working Group, 2011; Ripke et al., 2014, 2013). In fact, it is so robust that it is used for quality control.

The predictive validity of GRS for illness onset, disease severity, course, treatment response, and application to personalized precision medicine is under investigation. As examples of findings, GRS for alcohol dependence significantly predicted the trajectory of heavy episodic drinking in males in adolescence (Li et al., 2017), and individuals with treatment-resistant schizophrenia had a higher schizophrenia GRS score than individuals with nontreatment-resistant schizophrenia (Frank et al., 2015).

As sample sizes continue to increase, GRS has the potential to become much more powerful for the prediction of illness onset, severity, course, treatment response, and for application to personalized precision medicine.

2.2.12 Genetic correlations

A complementary method to cross-disorder association is the study of genetic correlations between two phenotypes to examine their shared genetic architecture (Bulik-Sullivan, Finucane, et al., 2015; Bulik-Sullivan, Loh, et al., 2015; Cross-Disorder Group of the Psychiatric Genomics Consortium et al., 2013). A positive genetic correlation means that the genetic risk variants have the same directional effect on both phenotypes, while a negative genetic correlation means that the cumulative effect of risk variants for one phenotype is in the opposite direction for the other phenotype. The magnitude of the genetic correlation reflects the degree to which the genetic basis is shared between disorders.

Psychiatric disorders show patterns of genetic overlap with each other (and with nonpsychiatric phenotypes). SNP-based coheritability estimates for five of the psychiatric disorders studied as a part of the PGC are shown in Fig. 2. Additionally, genetic correlations have revealed that (1) anorexia nervosa may have a metabolic component to its etiology (Watson et al., 2019), and (2) bipolar disorder I is more closely associated with schizophrenia genetically than is bipolar disorder II ($r_g = 0.71$ vs $r_g = 0.51$) and, bipolar disorder II is more closely associated with major depression genetically than is bipolar disorder I ($r_g = 0.69$ v $r_g = 0.30$) (Stahl et al., 2019).

3 What does the future hold?

Many of the future considerations for PGC apply broadly to psychiatric genetics. Going forward in psychiatric genetics, we need to increase the statistical power of our GWASs. Some of the ways this can be done are through increased sample size, selection of notably severe cases, improved SNP array coverage, and improved imputation reference sets.

Beyond association signals, the next challenge underway is identifying what variants are causally related to disease, and through which mechanisms do the genetic signals act. In schizophrenia, Sekar et al. (2016) have implicated complement component 4 genes, which intersects with a longstanding theory of synaptic pruning proposed in the 1980s (Feinberg, 1982). Integration of functional genomic data, as has been done by the CommonMind and PsychENCODE consortia, has proven highly important (Fromer et al., 2016). Giusti-Rodríguez et al. (2018) used brain chromatin interactions to connect schizophrenia GWAS results to 80 specific genes. Fine-mapping, employing large and dense array datasets, and functional studies of gene expression in mouse and human models will help us understand the biology behind the risk variants.

How do we understand hundreds or thousands of genetic signals per disorder, such as hundreds of genome-wide significant SNPs per disorder, or genome-wide significant SNPs that tag gene-dense regions of the genome (for example, the major histocompatibility complex region in schizophrenia, which contains hundreds of genes) (Ripke et al., 2014)? Analyses of molecular pathways and analysis of gene sets—tests of associations with genetic variants aggregated into functionally related gene sets—are two approaches implemented in our GWASs (Cross-Disorder Group of the Psychiatric

Genomics Consortium et al., 2013; Duncan et al., 2017; O'Dushlaine et al., 2015; Ripke et al., 2014; Watson et al., 2019). Intriguingly, recent studies have found that GWAS results for psychiatric illnesses connect to brain cell types (i.e., pyramidal cells in cortex and hippocampus and medium spiny neurons) (Skene et al., 2018; Watson et al., 2019). Increased statistical power will lead to greater success in pinpointing the molecular mechanisms that are truly involved and not simply statistically associated because of linkage disequilibrium or physical proximity to the causal variant.

How do we understand the many significant hits in noncoding regions of the genome, such as for schizophrenia? One example of the strategies PGC is using is to study chromatin interactions from Hi-C experiments, which is based on the idea that SNPs tag genes in physically distal parts of the genome because of the way DNA is tightly folded and packed inside the cell (Giusti-Rodríguez et al., 2018; Watson et al., 2019). The broad answer is that this challenge requires integration of many methods, including transcriptomic data, bioinformatics resources, and in vitro and in vivo experimental approaches (for a review of strategies, see Xiao, Chang, & Li, 2017).

Rare variation has a role in psychiatric disorders and needs ongoing consideration (Sullivan et al., 2018, 2012). Most convincingly demonstrated thus far for schizophrenia and autism spectrum disorder is a greater burden of rare variation (CNVs) in cases compared with controls (Marshall et al., 2017; Weiner et al., 2017). Larger study samples are needed for multigenerational studies of densely affected pedigrees. Using this method, Khan et al. (2018) reported a greater burden of rare CNVs in exonic regions in schizophrenia, and Rao, Yourshaw, Christensen, Nelson, and Kerner (2017) found an increased burden of rare deleterious mutations in bipolar disorder. Replication efforts and experimental validation need to test the robustness of rare variation findings and their causal relationships to disease.

The monumental shift in the field toward greater data sharing and collaboration has been climactic to scientific discovery in psychiatric genetics. Further progress will be aided by enlarged GWAS meta- and mega-analyses, attention to phenotype harmonization, deeper phenotyping (i.e., beyond diagnosis), functional studies, and studies of gene-environment (G × E) interactions (Sullivan et al., 2018). There have been single-team efforts to study G × E in psychiatric genetics, but the development of consortia is essential for even minimally adequate statistical power (sample size requirements to detect statistical interactions are much larger than those needed to detect the main effects of the same magnitude) and accelerate harmonization to combine environmental data across studies. Past G × E studies examined candidate genetic loci, which bears the inherent limitations that candidate gene studies did not yield loci with replicable effects (Assary, Vincent, Keers, & Pluess, 2017; Duncan & Keller, 2011). Now that it is known that many common variants of low effect inform the genetic architecture, polygenic predictor variables (i.e., G [genetic risk score] × E) may usher in a new plethora of findings.

Conventional psychiatric drug discovery has been at an impasse for half a century, and early successful discoveries around the 1950s were based largely on serendipity (Griebel & Holmes, 2013). As the genetic risk map for psychiatric disorders is built, the opportunity to identify therapeutic drugs is ripe. PGC is implementing drug pathway analyses that test for enrichment of variants identified by GWAS in druggable gene targets (Gaspar & Breen, 2017). Schizophrenia, for example, shows enrichment for antipsychotic drugs, selective calcium channel blockers, and antiepileptics (Gaspar & Breen, 2017), and similar results for MDD GWAS and antidepressant targets (Wray et al., 2018). Partnerships with the pharmaceutical industry might help to identify novel compounds, or existing drugs that can be repurposed for illnesses they were not initially intended to treat (Sanseau et al., 2012). To further this end, the PGC holds annual meetings concerning the potential implications of the emerging findings (Gaspar et al., 2018).

4   Conclusions

The PGC has been a driving force of progress and discovery in psychiatry in the past decade. Molecular genetics research has complemented family and twin studies in confirming that psychiatric disorders are moderately heritable. Psychiatric disorders are unequivocally highly polygenic—with GWASs showing that a substantial portion of genetic risk is conferred by many common variants of small effect, plus a role for rare genetic variation. Science is poised to pinpoint the molecular pathways and mechanisms involved in the genetic architecture, with several promising findings already.

The PGC is the largest effort in the history of psychiatric genomics. The PGCs research program has made immeasurable contributions to our understanding of the etiology of psychiatric disorders. In line with its founding principles of open science and international collaboration, PGC will continue to spearhead transparency and data sharing to encourage the rapid progress of science. The potential for making significant discoveries pertaining to the genetic architecture of severe psychiatric illness has never been greater. These discoveries carry significant promise for leading to new and more effective treatments that will improve the lives of patients. The translational benefit of understanding and predicting illness onset, course, and response to treatment, to implement personalized care, is undeniable. Reducing morbidity, disability, burden of disease, and personal suffering of millions of people globally is what is at stake.

References

Anttila, V., Bulik-Sullivan, B., Finucane, H. K., Walters, R. K., Bras, J., Duncan, L., & Murray, R. (2018). Analysis of shared heritability in common disorders of the brain. *Science, 360*(6395).

Assary, E., Vincent, J. P., Keers, R., & Pluess, M. (2017). Gene-environment interaction and psychiatric disorders: Review and future directions. *Seminars in Cell and Developmental Biology, 77*, 133–143.

Autism Spectrum Disorders Working Group of The Psychiatric Genomics Consortium, Anney, R. J., Ripke, S., Anttila, V., Grove, J., Holmans, P., & Daly, M. J. (2017). Meta-analysis of GWAS of over 16,000 individuals with autism spectrum disorder highlights a novel locus at 10q24.32 and a significant overlap with schizophrenia. *Molecular Autism, 8*, 1–17.

Baker, J. H., Schaumberg, K., & Munn-Chernoff, M. A. (2017). Genetics of anorexia nervosa. *Current Psychiatry Reports, 19*(11), 84.

Bulik-Sullivan, B. K., Finucane, H. K., Anttila, V., Gusev, A., Day, F. R., Loh, P. R., & Neale, B. M. (2015). An atlas of genetic correlations across human diseases and traits. *Nature Genetics, 47*(11), 1236–1241.

Bulik-Sullivan, B. K., Loh, P. R., Finucane, H. K., Ripke, S., Yang, J., Patterson, N., & Neale, B. M. (2015). LD score regression distinguishes confounding from polygenicity in genome-wide association studies. *Nature Genetics, 47*(3), 291–295.

Cederlöf, M., Thornton, L. M., Baker, J., Lichtenstein, P., Larsson, H., Ruck, C., & Mataix-Cols, D. (2015). Etiological overlap between obsessive-compulsive disorder and anorexia nervosa: A longitudinal cohort, multigenerational family and twin study. *World Psychiatry, 14*(3), 333–338.

Cross-Disorder Group of the Psychiatric Genomics Consortium. (2013). Genetic relationship between five psychiatric disorders estimated from genome-wide SNPs. *Nature Genetics, 45*(9), 984–994.

Cross-Disorder Group of the Psychiatric Genomics Consortium, Smoller, J. W., Craddock, N., Kendler, K., Lee, P. H., Neale, B. M., & Sullivan, P. F. (2013). Identification of risk loci with shared effects on five major psychiatric disorders: A genome-wide analysis. *Lancet, 381*(9875), 1371–1379.

de Jong, S., Vidler, L. R., Mokrab, Y., Collier, D. A., & Breen, G. (2016). Gene-set analysis based on the pharmacological profiles of drugs to identify repurposing opportunities in schizophrenia. *Journal of Psychopharmacology, 30*(8), 826–830.

Demontis, D., Walters, R. K., Martin, J., Mattheisen, M., Als, T. D., Agerbo, E., & Neale, B. M. (2019). Discovery of the first genome-wide significant risk loci for attention deficit/hyperactivity disorder. *Nature Genetics, 51*(1), 63–75.

Duncan, L. E., & Keller, M. C. (2011). A critical review of the first 10 years of candidate gene-by-environment interaction research in psychiatry. *American Journal of Psychiatry, 168*(10), 1041–1049.

Duncan, L. E., Yilmaz, Z., Gaspar, H., Walters, R., Goldstein, J., Anttila, V., & Bulik, C. M. (2017). Significant locus and metabolic genetic correlations revealed in genome-wide association study of anorexia nervosa. *American Journal of Psychiatry, 174*(9), 850–858.

Farrell, M., Werge, T., Sklar, P., Owen, M. J., Ophoff, R., O'Donovan, M. C., & Sullivan, P. F. (2015). Evaluating historical candidate genes for schizophrenia. *Molecular Psychiatry, 20*(5), 555–562.

Feinberg, I. (1982). Schizophrenia: Caused by a fault in programmed synaptic elimination during adolescence? *Journal of Psychiatric Research, 17*(4), 319–334.

Frank, J., Lang, M., Witt, S. H., Strohmaier, J., Rujescu, D., Cichon, S., & Rietschel, M. (2015). Identification of increased genetic risk scores for schizophrenia in treatment-resistant patients. *Molecular Psychiatry, 20*(2), 150–151.

Fromer, M., Roussos, P., Sieberts, S. K., Johnson, J. S., Kavanagh, D. H., Perumal, T. M., & Sklar, P. (2016). Gene expression elucidates functional impact of polygenic risk for schizophrenia. *Nature Neuroscience, 19*(11), 1442–1453.

Gaspar, H., & Breen, G. (2016). Pathways analyses of schizophrenia GWAS focusing on known and novel drug targets. *bioRxiv*.

Gaspar, H., & Breen, G. (2017). Drug enrichment and discovery from schizophrenia genome-wide association results: An analysis and visualisation approach. *Scientific Reports, 7*(1), 12460.

Gaspar, H., Collier, D. A., Geschwind, D. H., Lewis, C. M., Li, Q., Roth, B. L., & Breen, G. (2018). Conference report: Psychiatric Genomics Consortium meeting: Pathways to drugs, London, March 2017. *Biological Psychiatry, 84*(6), e49–e50.

Giusti-Rodríguez, P., Lu, L., Yang, Y., Crowley, C., Liu, X., Juric, I., & Sullivan, P. (2018). Schizophrenia and a high-resolution map of the three-dimensional chromatin interactome of adult and fetal cortex. *bioRxiv*, 406330.

Griebel, G., & Holmes, A. (2013). 50 years of hurdles and hope in anxiolytic drug discovery. *Nature Reviews Drug Discovery, 12*(9), 667–687.

Grove, J., Ripke, S., Als, T. D., Mattheisen, M., Walters, R., Won, H., & Børglum, A. D. (2019). Identification of common genetic risk variants for autism spectrum disorder. *Nature Genetics, 51*, 431–444.

Hinney, A., Scherag, A., Jarick, I., Albayrak, Ö., Pütter, C., Pechlivanis, S., & Psychiatric GWAS Consortium: ADHD Subgroup. (2011). Genome-wide association study in German patients with attention deficit/hyperactivity disorder. *American Journal of Medical Genetics Part B: Neuropsychiatric Genetics, 156*(8), 888–897.

Hirschhorn, J. N. (2009). Genomewide association studies: Illuminating biologic pathways. *New England Journal of Medicine, 360*(17), 1699–1701.

International Schizophrenia Consortium, Purcell, S. M., Wray, N. R., Stone, J. L., Visscher, P. M., O'Donovan, M. C., & Sklar, P. (2009). Common polygenic variation contributes to risk of schizophrenia and bipolar disorder. *Nature, 460*(7256), 748–752.

Johnson, E. C., Border, R., Melroy-Greif, W. E., de Leeuw, C. A., Ehringer, M. A., & Keller, M. C. (2017). No evidence that schizophrenia candidate genes are more associated with schizophrenia than noncandidate genes. *Biological Psychiatry, 82*(10), 702–708.

Kendler, K. (2015). A joint history of the nature of genetic variation and the nature of schizophrenia. *Molecular Psychiatry, 20*(1), 77–83.

Khan, F. F., Melton, P. E., McCarthy, N. S., Morar, B., Blangero, J., Moses, E. K., & Jablensky, A. (2018). Whole genome sequencing of 91 multiplex schizophrenia families reveals increased burden of rare, exonic copy number variation in schizophrenia probands and genetic heterogeneity. *Schizophrenia Research, 197*(18), 337–345.

Klein, R. J., Zeiss, C., Chew, E. Y., Tsai, J.-Y., Sackler, R. S., Haynes, C., & Hoh, J. (2005). Complement factor H polymorphism in age-related macular degeneration. *Science, 308*(5720), 385–389.

Li, J. J., Cho, S. B., Salvatore, J. E., Edenberg, H. J., Agrawal, A., Chorlian, D. B., & Dick, D. M. (2017). The impact of peer substance use and polygenic risk on trajectories of heavy episodic drinking across adolescence and emerging adulthood. *Alcoholism: Clinical and Experimental Research, 41*(1), 65–75.

Lichtenstein, P., Yip, B. H., Björk, C., Pawitan, Y., Cannon, T. D., Sullivan, P. F., & Hultman, C. M. (2009). Common genetic determinants of schizophrenia and bipolar disorder in Swedish families: A population-based study. *Lancet, 373*(9659), 234–239.

MacArthur, J., Bowler, E., Cerezo, M., Gil, L., Hall, P., Hastings, E., & Parkinson, H. (2016). The new NHGRI-EBI catalog of published genome-wide association studies (GWAS catalog). *Nucleic Acids Research, 45*(D1), D896–D901.

Major Depressive Disorder Working Group of the Psychiatric GWAS Consortium. (2013). A mega-analysis of genome-wide association studies for major depressive disorder. *Molecular Psychiatry, 18*(4), 497–511.

Mataix-Cols, D., Boman, M., Monzani, B., Rück, C., Serlachius, E., Långström, N., & Lichtenstein, P. (2013). Population-based, multigenerational family clustering study of obsessive-compulsive disorder. *JAMA Psychiatry, 70*(7), 709–717.

Marshall, C. R., Howrigan, D. P., Merico, D., Thiruvahindrapuram, B., Wu, W., & Greer, D. S. (2017). CNV and Schizophrenia Working Groups of the Psychiatric Genomics Consortium. Contribution of copy number variants to schizophrenia from a genome-wide study of 41,321 subjects. *Nature Genetics, 49*(1), 27–35.

Mathews, C. A., & Grados, M. A. (2011). Familiality of Tourette syndrome, obsessive-compulsive disorder, and attention-deficit/hyperactivity disorder: Heritability analysis in a large sib-pair sample. *Journal of American Academy of Child and Adolescent Psychiatry, 50*(1), 46–54.

Meier, S. M., Bulik, C. M., Thornton, L. M., Mattheisen, M., Mortensen, P. B., & Petersen, L. (2015). Diagnosed anxiety disorders and the risk of subsequent anorexia nervosa: A Danish population register study. *European Eating Disorders Review, 23*(6), 524–530.

O'Dushlaine, C., Rossin, L., Lee, P. H., Duncan, L. E., Parikshak, N. N., Newhouse, S., & Posthuma, D. (2015). Psychiatric genome-wide association study analyses implicate neuronal, immune and histone pathways. *Nature Neuroscience, 18*(2), 199–209.

Pardiñas, A. F., Holmans, P., Pocklington, A. J., Escott-Price, V., Ripke, S., Carrera, N., & Walters, J. T. R. (2018). Common schizophrenia alleles are enriched in mutation-intolerant genes and in regions under strong background selection. *Nature Genetics, 50*(3), 381–389.

Polderman, T. J., Benyamin, B., de Leeuw, C. A., Sullivan, P. F., van Bochoven, A., Visscher, P. M., & Posthuma, D. (2015). Meta-analysis of the heritability of human traits based on fifty years of twin studies. *Nature Genetics, 47*(7), 702–709.

Psychiatric GWAS Consortium Bipolar Disorder Working Group. (2011). Large-scale genome-wide association analysis of bipolar disorder identifies a new susceptibility locus near ODZ4. *Nature Genetics, 43*(10), 977–983.

Psychiatric GWAS Consortium Coordinating Committee. (2010). Genomewide association studies: History, rationale, and prospects for psychiatric disorders. *Focus, 8*(3), 417–434.

Psychiatric GWAS Consortium Steering Committee. (2009). A framework for interpreting genome-wide association studies of psychiatric disorders. *Molecular Psychiatry, 14*(1), 10–17.

Rao, A. R., Yourshaw, M., Christensen, B., Nelson, S. F., & Kerner, B. (2017). Rare deleterious mutations are associated with disease in bipolar disorder families. *Molecular Psychiatry, 22*(7), 1009–1014.

Ripke, S., Neale, B. M., Corvin, A., Walters, J. T., Farh, K.-H., Holmans, P. A., & Huang, H. (2014). Biological insights from 108 schizophrenia-associated genetic loci. *Nature, 511*(7510), 421–427.

Ripke, S., O'Dushlaine, C., Chambert, K., Moran, J. L., Kähler, A. K., Akterin, S., & O'Donovan, M. C. (2013). Genome-wide association analysis identifies 13 new risk loci for schizophrenia. *Nature Genetics, 45*(10), 1150–1159.

Ronald, A., Simonoff, E., Kuntsi, J., Asherson, P., & Plomin, R. (2008). Evidence for overlapping genetic influences on autistic and ADHD behaviours in a community twin sample. *Journal of Child Psychology and Psychiatry, 49*(5), 535–542.

Sanseau, P., Agarwal, P., Barnes, M. R., Pastinen, T., Richards, J. B., Cardon, L. R., & Mooser, V. (2012). Use of genome-wide association studies for drug repositioning. *Nature Biotechnology, 30*(4), 317–320.

Schachar, R. (2014). Genetics of attention deficit hyperactivity disorder (ADHD): Recent updates and future prospects. *Current Developmental Disorders Reports, 1*(1), 41–49.

Schizophrenia Psychiatric Genome-Wide Association Study Consortium. (2011). Genome-wide association study identifies five new schizophrenia loci. *Nature Genetics, 43*(10), 969–976.

Sekar, A., Bialas, A. R., De Rivera, H., Davis, A., Hammond, T. R., Kamitaki, N., & McCarroll, S. A. (2016). Schizophrenia risk from complex variation of complement component 4. *Nature, 530*(7589), 177–183.

Shih, R. A., Belmonte, P. L., & Zandi, P. P. (2004). A review of the evidence from family, twin and adoption studies for a genetic contribution to adult psychiatric disorders. *International Review of Psychiatry, 16*(4), 260–283.

Skene, N. G., Bryois, J., Bakken, T. E., Breen, G., Crowley, J. J., Gaspar, H. A., & Hjerling-Leffler, J. (2018). Genetic identification of brain cell types underlying schizophrenia. *Nature Genetics, 50*(6), 825–833.

Smoller, J., Cerrato, F., & Weatherall, S. (2015). The genetics of anxiety disorders. In *Anxiety disorders: Translational perspectives on diagnosis and treatment* (pp. 47–61). Oxford: Oxford University Press.

Stahl, E. A., Breen, G., Forstner, A. J., McQuillian, A., Ripke, S., Trubetskoy, V., & Bipolar Disorder Working Group of the Psychiatric Genomics Consortium. (2019). Genome-wide association study identifies 30 loci associated with bipolar disorder. *Nature Genetics, 51*(5), 793–803.

Sullivan, P. F. (2010). The psychiatric GWAS consortium: Big science comes to psychiatry. *Neuron, 68*(2), 182–186.

Sullivan, P. F. (2017). How good were candidate gene guesses in schizophrenia genetics? *Biological Psychiatry, 82*(10), 696–697.

FIG. 1 A Manhattan plot depicting genome-wide association resulting *P*-values, ordered according to their position in the genome.

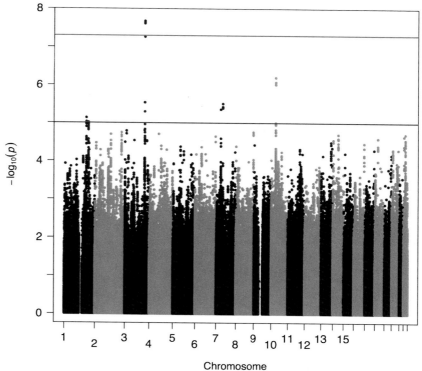

Several genome-wide significant susceptibility loci were identified for a variety of complex traits. In meta-analysis including over 130,000 MDD cases, for example, and over 330,000 controls, 14 genome-wide hits were found (Wray et al., 2017).

1.4 CNV analysis

Another important part of the human genome are copy number variations (CNVs). They occur if chromosomal regions present multiple times, the number of copies differing between individuals (Iafrate et al., 2004; Sebat et al., 2004). Although about three times more SNPs than CNVs are present in the human genome, the relative contribution to genomic variation (as measured in nucleotides) is comparable (Malhotra & Sebat, 2012). Roughly 12% of our genome is affected by copy number change (Redon et al., 2006).

While CNVs can also be analyzed by themselves, one major approach is to calculate the so-called burden of rare variants (i.e., the number of CNVs carried by an individual).

In the Autism Genome Project, for example, Pinto et al. (2010) found that cases had a higher burden of rare CNVs as compared with healthy controls.

1.5 Family-based association studies

While family-based studies are mainly used for linkage analysis, association analysis can also be performed in families. One of the most applied approaches is the transmission disequilibrium test (TDT; Spielman, McGinnis, & Ewens, 1993). This method is used in trio data, where genotypes of affected offspring and their parents are investigated. The TDT measures the overtransmission of an allele from heterozygous parents to affected offspring. If the marker is not associated to the disease, the transmission rate of a specific allele should vary around 0.5. The TDT assesses if the actual transmission rate differs significantly from this value, and hence, if an association between disease and marker is present. Several studies looking at familial association with psychiatric diseases using TDTs have been published (e.g., Curran et al., 2006; Wei & Hemmings, 2000). One advantage is that each family serves as its own control, hence the TDT is robust to population stratification that arises if allele frequencies differ among ethnicities (see also Section 2). A disadvantage of this approach is that at least one parent has to be heterozygous for the marker. Furthermore, family-based association studies generally present with lower power, as compared with case-control studies (Risch & Merikangas, 1996).

Several extensions of the TDT, including FBAT (family-based association test, see, for example, Laird & Lange, 2008), allowing, for example, for the inclusion of unaffected siblings, have been proposed.

1.6 Twin studies

Monozygotic (MZ) twins are genetically identical. Therefore, any differences between MZ twins hint at environmental differences (Plomin, DeFries, McClearn, & Rutter, 1997). Dizygotic (DZ) twins share, on average, 50% of the genome. Hence, any differences between DZ twins might be due to genetics and/or environment. Studying twins is based on the comparison of correlation within MZ twins and DZ twins. If the correlation in MZ twins is higher as compared with DZ twins, a certain genetic component is implied.

2   Genetic architecture

2.1   Population stratification

Population stratification occurs due to the fact that allele frequencies differ between ethnicities. Combining different ethnicities in an association study without any correction might give false-positive results. In the most extreme scenario, cases have a different ethnic background than controls. In this situation, we cannot disentangle if significant associations are due to differences in allele frequencies between cases and controls, or between ethnicities in general. However, in most studies, there is a certain ethnic diversity in cases as well as in controls. Different methods can be used to correct for population stratification in this scenario. Devlin and Roeder (1999) proposed the genomic control procedure. They define the genomic inflation factor λ as median of the association test statistics divided by the theoretical median under the null hypothesis of no association. Resulting values larger than 1 are indicators for population stratification or other confounders such as cryptic relatedness (Price, Zaitlen, Reich, & Patterson, 2010). Dividing the test statistics with λ and deriving P-values based on these statistics can then be performed to correct for population stratification. One disadvantage of this approach, however, is that, as all P-values are corrected with the same value, a constant inflation across the whole genome is assumed, which does not necessarily need to be true.

A different approach performs principal-component analysis or multidimensional scaling on the genome-wide SNPs. Doing so, main axes of variations can be identified and used as covariates in subsequent analyses. However, these axes do not always reflect population heterogeneity, they could also depict long-range LD (Tian et al., 2008) or familial relatedness (Patterson, Price, & Reich, 2006), for example. Therefore, linear mixed models, which can model stratification as well as cryptic relatedness and family structure have been proposed (Price et al., 2010).

2.2   Imputation of missing genotypes

High-throughput technology made large-scale whole-genome genotyping feasible and cost-effective, however, not all SNPs are covered on genotyping arrays. We can make use of the fact that the human genome is structured in linkage-disequilibrium (LD)-blocks, which means that not all SNPs are totally independent of each other. Nowadays, large samples from different ethnicities that have been sequenced are available (Genomes Project Consortium et al., 2010; International HapMap Consortium, 2003). We can assess the LD-pattern in these samples and use these to impute missing genotypes in our own samples (Marchini, Howie, Myers, McVean, & Donnelly, 2007). The imputation technique is a standard method, and it has been shown that the prediction works quite accurately (Howie, Donnelly, & Marchini, 2009).

2.3   Heritability, genetic correlation, and polygenic risk scores

Genome-wide genotype data has enabled direct estimation of additive heritability attributable to common genetic variation ("SNP heritability" or h^2_{SNP}) (Lee, Wray, Goddard, & Visscher, 2011; Yang et al., 2010), and aggregation of these variants into a single empirical polygenic risk score (Ruderfer et al., 2014). The polygenic risk score method takes statistically independent genetic variants from a GWAS (discovery or training set), ranks the results by the P-value of significance, and applies the weights from these to an independent target sample (Wray et al., 2014). The statistical power for such analyses depends on the training set, and this power can be harnessed to test into smaller target samples.

Polygenic risk score $= \sum x_i \times \log(\mathrm{OR}_i)$, where OR_i is the allelic odds ratio in the discovery dataset and x_i is the number of risk alleles present at a single genetic locus in each individual (Iyegbe, Campbell, Butler, Ajnakina, & Sham, 2014). The allelic log (odds ratio) for the association of SNP_i vs trait is multiplied by the number of risk alleles. This process is repeated for each SNP within a specified P-value range. The aggregate count of weighted risk alleles creates a polygenic score per individual. The distribution of score values is normalized by fitting to a standard normal distribution curve, which helps with the interpretation of the score in downstream analyses. The polygenic method has been widely used, both within and across psychiatric disorders analyses (Cross-Disorder Group of the Psychiatric Genomics Consortium, 2013; International Schizophrenia Consortium et al., 2009). The large number of null GWAS effects encouraged common practice to present the change in the association statistic at different inclusion thresholds, as provided by programs such as PRSice (Euesden, Lewis, & O'Reilly, 2015) and PLINK (Chang et al., 2015; Purcell et al., 2007).

Another method called genome-wide complex trait analysis (GCTA) utilizes the genetic variant information from GWAS to assess the degree of genetic relatedness between individuals, assuming that cases are genetically more similar to each other than to controls (Yang, Lee, Goddard, & Visscher, 2011). The genetic relationship matrix is then correlated with dichotomous or quantitative phenotypes. Most of the preceding methods are based on genetic restricted maximum likelihood analysis, and are implemented in software packages such as GCTA and LDAK (Lee et al., 2011; Speed, Hemani, Johnson, & Balding, 2012; Yang et al., 2010, 2011). An extension of these methods estimates genetic correlation (rg_{SNP}) explained by GWAS SNPs between two disorders (Lee, Yang, Goddard, Visscher, & Wray, 2012), with a positive correlation indicating that the cases of one disorder show higher genetic similarity to the cases of the other disorder than to their own controls. When the covariance term between the traits is similar in magnitude to the variance terms, a high correlation (rg_{SNP}) is obtained. This method reports SNP-based coheritability across pairs of disorders. A disadvantage of these methods is that they are computationally intensive and require individual-level genotype data.

Cross-trait LD score regression allows us to estimate genetic correlation using only summary statistics from GWAS (Bulik-Sullivan, Loh, et al., 2015). This computationally fast method does not require individual genotypes, genome-wide significant SNPs or LD-pruning, that is, no independent genetic variants. The approach harnesses the LD information under the assumption that if a trait is genetically influenced, then variants in high LD would tag causal variants, and have higher test statistics than variants with low LD. This method is flexible and can be adapted to estimate SNP heritability (Bulik-Sullivan, Loh, et al., 2015), partition SNP heritability by functional categories (Finucane et al., 2015), and estimate genetic correlation between different complex traits (Bulik-Sullivan, Finucane, et al., 2015). Methods to calculate LD score regression include the publicly available Python code (https://github.com/bulik/ldsc/wiki/Heritability-and-Genetic-Correlation) and LD Hub web interface and centralized database of summary-level GWAS results for 173 diseases/traits (http://ldsc.broadinstitute.org/). The LD Hub allows for calculation of heritability and genetic correlation via LD score regression analysis of user GWAS data against these traits (Zheng et al., 2017).

These methods will pave the way to gain more insights into the genetic architecture of psychiatric disorders.

3 Gene-environment interactions

Gene-by-environment interactions ($G \times E$) studies assess environmental effects on a phenotype that differ depending on the genetic background. A $G \times E$ is identified when the risk for the disorder, if exposed to both the risk gene (G) and risk environment (E) differs from the sum (additive model) or the product (multiplicative model) of the risks, compared with exposure to only the G or E (Sharma, Powers, Bradley, & Ressler, 2016).

Methods modeling interactions range from conventional approaches to exploratory novel methods. Analysis of single gene $G \times E$ is straightforward, results are presented in 2×2 or 3×2 tables of relative risks, of the environmental exposure in the risk and nonrisk genotype groups (Little et al., 2002). Modeling both main effects and $G \times E$ interactions via a two degree of freedom joint test is preferable over the traditional approach of main effect testing followed by testing for an interaction conditional on the main effects (Kraft, Yen, Stram, Morrison, & Gauderman, 2007). An overview of common methods used to detect $G \times E$ is presented in Table 1, these methods can also be implemented for gene-gene ($G \times G$) interactions (Chen, Liu, Zhang, & Zhang, 2007; Chen et al., 2008; Cook et al., 2004; Culverhouse et al., 2004; Hahn et al., 2003; Millstein et al., 2006; Moore & Hahn, 2002; Moore et al., 2004; Nelson et al., 2001; Park & Hastie, 2008; Ritchie et al., 2003; Strobl et al., 2009; Tomita et al., 2004; Zhang & Liu, 2007). Methods include conventional parametric logistic regression driven approaches or Cox regression models where multiple confounders can be adjusted for (Thomas, 2010). Haplotypes incorporating multi-SNP genetic risk can be tested in interactions with environmental exposure (haplotype $\times$ E), as an extension of the $G \times E$ analysis.

While most $G \times E$ studies have been conducted on hypothesis-driven candidate genes, recently focus has shifted to using the polygenic risk score for a disorder as the G in the $G \times E$ (Arloth et al., 2015; Mullins et al., 2016; Peyrot et al., 2014).

TABLE 1 An overview of methods used to detect G × E.

Method type	Name	Reference
Tree-based methods	Multivariate adaptive regression (MARS)	Cook, Zee, and Ridker (2004)
	Random forests	Lunetta, Hayward, Segal, and Van Eerdewegh (2004)
	Classification and regression trees (CART)	Strobl, Malley, and Tutz (2009)
Data reduction	Multifactor dimensionality reduction (MDR)	Hahn, Ritchie, and Moore (2003)
	Focused interaction testing framework	Millstein, Conti, Gilliland, and Gauderman (2006)
	Combinatorial partitioning method	Nelson, Kardia, Ferrell, and Sing (2001)
	Restricted partitioning method	Culverhouse, Klein, and Shannon (2004)
Pattern recognition and data mining	Support vector machines (SVMs)	Chen et al. (2008)
	Penalized regression	Park and Hastie (2008)
	Bayesian methods	Zhang and Liu (2007)
	Parameter decreasing method (PDM)	Tomita et al. (2004)
	Genetic programming optimized neural network (GPNN)	Ritchie, White, Parker, Hahn, and Moore (2003)
	Genetic algorithm strategies	Moore, Hahn, Ritchie, Thornton, and White (2004)
	Cellular automata (CA) approach	Moore and Hahn (2002)

Resampling-based tests derived from the Bayesian models (Wakefield, Haneuse, Dobra, & Teeple, 2011; Yu et al., 2012) allow testing of complicated gene-environment interactions, where genetic variants in multiple loci within the region interact with the environmental risk factor(s). Tree-based methods such as classification and regression trees (Breiman, Friedman, Olshen, & Stone, 1984) and random forests (Lunetta et al., 2004) can also deal with many parameters, but cannot account for available prior information, and hence have limited value. Unbiased approaches include genome-wide gene by environment interaction studies (GWEIs) (Dunn et al., 2016), although GWEIs require larger samples, and have several statistical complications (Almli et al., 2014). Given the "multidimensionality curse" of assessing high-dimensional interactions, dimension-reduction methods have recently been favored. Multidimensional interactions can be reliably explored via tools such as the Multifactor Dimension Reduction (Edwards, Lewis, Velez, Dudek, & Ritchie, 2009), which scans across the multiway contingency table to derive the optimal classifier of disease risk based on multiple training sets, and tests their predictions on the remaining data. Similarly, the Focused Interaction Testing Framework (Millstein et al., 2006) builds through sequences of main effects and higher order interactions. Such methods allow more high-dimensional modeling for diseases involving multiple genes, multiple environmental risk factors, and epistatic (G × G) interactions.

G × E studies face many methodological challenges (Duncan & Keller, 2011; Karg & Sen, 2012; Mehta & Binder, 2012). As the number of factors per model increase, there is an exponential increase in the computational time and number of hypotheses tested. Failure to control for all covariate interactions, including gene-by-covariate and environment-by-covariate interaction terms, can result in spurious interactions (Keller, 2014). Hybrid methods avoid the potential genetic-environment correlation bias (Dai et al., 2012) while combining the estimates derived from the case-control and case-only designs (Mukherjee & Chatterjee, 2008).

Finally, the power required to detect G × E associations is very high, with an interaction requiring at least a fourfold larger sample size than a main effect of comparable magnitude (Smith & Day, 1984) to detect even moderate effect sizes (Murcray, Lewinger, Conti, Thomas, & Gauderman, 2011). Case-only studies that assume G-E independence among controls (Weinberg & Umbach, 2000) and two-stage design tests using use a case-only comparison for screening, followed by a family-based comparison that does not require G-E independence (Chen, Lin, & Liu, 2009) are complementary approaches aimed at gaining power to detect interactions.

Future G × E studies in larger, deep-phenotyped, longitudinal cohorts, assessing for multiple genetic and environmental risk factors across different developmental stages are warranted.

4 Gene expression and DNA methylation analysis

Statistical analysis of gene expression and DNA methylation can be performed at individual gene/locus or at the genome-wide level. Candidate gene analysis is generally performed using quantitative real-time PCR (qPCR) for gene expression and pyrosequencing, or Sequenom Epityper for DNA methylation. The data output can be analyzed using simple analysis tools such as Excel and SPSS. Complex high-throughput genome-wide analysis is performed via microarrays (e.g., Illumina, Agilent, and Affymetrix), or deep sequencing using next-generation applications such as RNA-seq or Bisulfite-seq. Open-source, flexible R, and Bioconductor environments provide free statistical packages with prebuilt functions, workflows, and documentation that can be customized (Huber et al., 2015).

The main steps for analyzing DNA methylation and gene expression microarray data are similar (Fig. 2), and have been previously described (Dedeurwaerder et al., 2014; Michels et al., 2013; Wright et al., 2016).

Following are the basic steps for analysis of RNA sequencing data, similar approaches are available for analysis of Bisulfite sequencing data (Akman, Haaf, Gravina, Vijg, & Tresch, 2014; Rackham et al., 2017; Wreczycka et al., 2017):

(a) Assembly: Aligning the genomic reads to a reference genome and analysis of the number of reads that map to specific regions/windows to determine the abundance levels of genes, transcripts, enhancers, or other sequence intervals. Common programs include BWA (Li & Durbin, 2009), Bowtie (Langmead, Trapnell, Pop, & Salzberg, 2009), and SOAP (Li, Li, Kristiansen, & Wang, 2008). Bioconductor packages include RSamtools and GenomicAlignments.

(b) Filtering: Removal of regions with low coverage using a cutoff is performed to ensure good read depth for downstream analysis.

(c) Normalization: Reads per kilobase per million reads is a simple read coverage adjustment that considers gene counts standardized by the gene length and the total number of reads in each library as expression values (Mortazavi, Williams, McCue, Schaeffer, & Wold, 2008). Similarly, FPKM, or fragments per kilobase per million (Trapnell et al., 2010), adjusts for the total number of reads, or fragments mapped per kilobase, per million mapped reads.

(d) Differential expression/methylation analysis: Differential gene expression or DNA methylation is performed via packages such as edgeR (Nikolayeva & Robinson, 2014; Robinson, McCarthy, & Smyth, 2010), limma (Diboun, Wernisch, Orengo, & Koltzenburg, 2006; Ritchie et al., 2015), and DESeq2 (Love, Huber, & Anders, 2014; Varet, Brillet-Gueguen, Coppee, & Dillies, 2016). These methods have been reviewed in detail (Law, Alhamdoosh, Su, Smyth, & Ritchie, 2016).

Integrated pipelines are also available that perform all the data analysis steps for RNA-seq (Gao et al., 2015; Lim, Lee, & Kim, 2017; Torres-Garcia et al., 2014) and Bisulfite-seq (Jiang et al., 2014; LoVerso & Cui, 2015) data.

Technical constraints in data analysis exist because many samples are assessed and are generally processed in batches, which might cause systematic errors due to nonbiological effects. Several methods correct for these (Chen et al., 2011; Kupfer et al., 2012), including Combat, which is an empirical Bayes method that adjusts for batch effects (Johnson, Li, & Rabinovic, 2007), and surrogate variable analysis (SVA) (Leek & Storey, 2007), which uses a linear model analysis

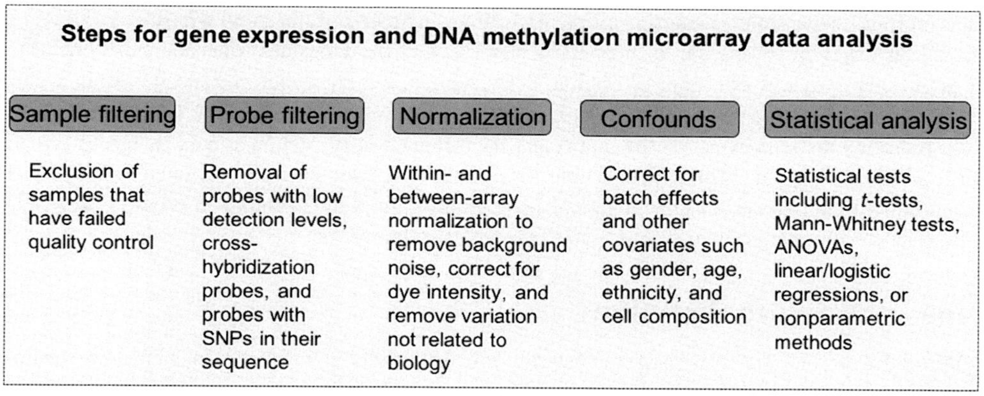

Steps for gene expression and DNA methylation microarray data analysis

Sample filtering	Probe filtering	Normalization	Confounds	Statistical analysis
Exclusion of samples that have failed quality control	Removal of probes with low detection levels, cross-hybridization probes, and probes with SNPs in their sequence	Within- and between-array normalization to remove background noise, correct for dye intensity, and remove variation not related to biology	Correct for batch effects and other covariates such as gender, age, ethnicity, and cell composition	Statistical tests including t-tests, Mann-Whitney tests, ANOVAs, linear/logistic regressions, or nonparametric methods

FIG. 2 Workflow outlining the main steps for analyzing DNA methylation and gene expression microarray data.

to estimate eigenvalues from a residual expression matrix from which biological variation has already been removed. Combat corrects for known batch effects, while SVA corrects for known and unknown confounds, making the method far superior, but highly conservative. Other technical artifacts including skewedness and heteroscedasticity can be overcome by log transformation of the data.

Selected scientific journals have made it mandatory for authors to deposit their gene expression and DNA methylation data on websites such as Gene Expression Omnibus or ArrayExpress. Moreover, larger international consortia such as the Psychiatric Genomics Consortium also have gene expression and epigenomics working groups, which allow integration of these data within and across different cohorts.

5 Gene enrichment and network analysis

High-throughput technologies generate large numbers of genes with differences in genetic sequence or gene activity between groups at a genome-wide scale. The list of genes obtained is often difficult to interpret, hence pathway and network-based methods are used to aggregate these genetic variants and/or genes based on shared function and biology. Computational methods have been developed to predict and prioritize genetic variants and genes based on functional annotations, and these methods have been explicitly reviewed (Kao, Leung, Chan, Yip, & Yap, 2017).

At the genetic sequence level, gene set analysis allows us to prioritize loci where multiple SNPs show evidence of association with a trait. Gene set analysis methods bypass stringent multiple-testing corrections needed for analysis of individual signals, and can account for gene size, SNP density, and LD information across the genome, thereby increasing the power of the study. Pathway-based genome-wide association analysis tools include INRICH (INterval enRICHment analysis) that test for enriched association signals of predefined gene sets across independent genomic intervals (Lee, O'Dushlaine, Thomas, & Purcell, 2012), and MAGENTA (Segre et al., 2010), a computational tool that tests for enrichment of genetic associations in predefined biological processes or sets of functionally related genes. Outputs are gene set enrichment analysis P-values and a false discovery rate that corrects for each tested gene set or pathway.

For biological measures of gene activity, such as gene expression and DNA methylation, public databases with preavailable gene sets can be used for pathway analyses and functional annotation. These include the Pathway Commons database (Cerami et al., 2011), the Kyoto Encyclopedia of Genes and Genomes (Kanehisa & Goto, 2000), Gene Ontology (Ashburner et al., 2000), DAVID (da Huang, Sherman, & Lempicki, 2009; Huang et al., 2007), and Molecular Signatures Database (Liberzon et al., 2011). Protein-protein interaction (PPI) networks can also be used to define gene sets by selecting genes that interact at the protein level by plotting direct PPI neighbors and the interactions between these proteins to visualize large protein networks. Software to build PPIs include STRING (Szklarczyk et al., 2015) and PINA (Cowley et al., 2012).

Weighted gene coexpression network analysis (WGCNA) focuses on exploring correlation between probes in gene expression data and comparing the gene networks to clinical data (Langfelder & Horvath, 2008). WGCNA reflects the notion that genes within the same biological pathway will have similar patterns of expression, and has been successfully used to unravel orchestrated networks of genes in psychiatric disorders (Breen et al., 2017; de Jong et al., 2016; Kim, Hwang, Webster, & Lee, 2016). First, genes are clustered based on the dissimilarity measure based on average linkage hierarchical clustering. Next, gene coexpression modules are detected by applying a branch cutting method. Eigen-values for coexpression networks can finally be integrated with external information, such as phenotypes and covariates, to identify networks associated with specific traits.

Pathway-based methods pose several shortcomings, including intrinsic redundancy, whereby gene sets might only have a partial overlap, but are tagged by the same biological processes at the annotation levels. Furthermore, genes within a defined gene set do not always have coherent biological activity, hence there is heterogeneity within a gene set that might not be reflected by combining these together within the curated sets.

Network-based methods also allow a systems approach by integration of results across different types of data, in conjunction with clinical and biological information, to identify genes associated with psychiatric disorders. This in turn will allow elucidation of coordinated gene activity of sets of biologically annotated genes at a broader network level, and offer deeper insight into disease mechanisms and biology of psychiatric disorders.

6 Conclusions and future directions

Psychiatry has entered the era of "Big Data" where vast amounts of clinical, genomic, transcriptomic, brain imaging, endocrine, environmental, and other types of data are routinely assessed. Analyses of these data and integrating them across different strata is a challenging task, therefore the establishment and implementation of appropriate statistical methods are key to the interpretation of the data.

In this chapter, we have introduced and described some of the current statistical methods that are widely used in genomics studies of complex traits. Some of these methods can be further tuned to psychiatric disorders, by incorporating environmental risk factors with genetic risk factors, thereby providing an avenue for identification of individual biological and genetic risk markers for disease.

The "one-size-fits-all" notion has been shown to be ineffective for psychiatric disorders, as evidenced by increased rates of relapses, treatment-resistant disorders, and adverse reaction to psychiatric drugs; hence it is clear that psychiatry will benefit enormously from tailored treatments. Utilization of the statistical methods outlined in this chapter will allow us to integrate knowledge from different types of biological and clinical information, and usage of this combined information will facilitate improved diagnosis classification and personalized psychiatric interventions.

Future longitudinal studies aimed at collection and analysis of large, well-characterized patient information and incorporating clinical data with observed biological measures will help uncover disease processes and trajectories. By understanding how a disease progresses, predictive statistical models can be built to identify high-risk individuals and monitor them closely, subsequently allowing for timely intervention.

In conclusion, by leveraging simple and advanced statistical methods to interrogate patient data, researchers and clinicians will be able to get a deeper insight into the etiology of the psychiatric disorder, an important step toward the goal of personalized psychiatry.

References

Akman, K., Haaf, T., Gravina, S., Vijg, J., & Tresch, A. (2014). Genome-wide quantitative analysis of DNA methylation from bisulfite sequencing data. *Bioinformatics, 30*(13), 1933–1934. https://doi.org/10.1093/bioinformatics/btu142.

Almli, L. M., Duncan, R., Feng, H., Ghosh, D., Binder, E. B., Bradley, B., ... Epstein, M. P. (2014). Correcting systematic inflation in genetic association tests that consider interaction effects: Application to a genome-wide association study of posttraumatic stress disorder. *JAMA Psychiatry, 71*(12), 1392–1399. https://doi.org/10.1001/jamapsychiatry.2014.1339.

Arloth, J., Bogdan, R., Weber, P., Frishman, G., Menke, A., Wagner, K. V., ... Major Depressive Disorder Working Group of the Psychiatric Genomics Consortium PGC. (2015). Genetic differences in the immediate transcriptome response to stress predict risk-related brain function and psychiatric disorders. *Neuron, 86*(5), 1189–1202. https://doi.org/10.1016/j.neuron.2015.05.034.

Ashburner, M., Ball, C. A., Blake, J. A., Botstein, D., Butler, H., Cherry, J. M., ... Sherlock, G. (2000). Gene ontology: Tool for the unification of biology. The Gene Ontology Consortium. *Nature Genetics, 25*(1), 25–29. https://doi.org/10.1038/75556.

Breen, M. S., Tylee, D. S., Maihofer, A. X., Neylan, T. C., Mehta, D., Binder, E., ... Glatt, S. J. (2017). PTSD blood transcriptome mega-analysis: Shared inflammatory pathways across biological sex and modes of trauma. *Neuropsychopharmacology.* https://doi.org/10.1038/npp.2017.220.

Breiman, L., Friedman, J. H., Olshen, R. A., & Stone, C. J. (1984). *Classification and regression trees.* Belmont, CA: Wadsworth International.

Bulik-Sullivan, B., Finucane, H. K., Anttila, V., Gusev, A., Day, F. R., Loh, P. R., ... Neale, B. M. (2015). An atlas of genetic correlations across human diseases and traits. *Nature Genetics, 47*(11), 1236–1241. https://doi.org/10.1038/ng.3406.

Bulik-Sullivan, B. K., Loh, P. R., Finucane, H. K., Ripke, S., Yang, J., Schizophrenia Working Group of the Psychiatric Genomics Consortium, ... Neale, B. M. (2015). LD score regression distinguishes confounding from polygenicity in genome-wide association studies. *Nature Genetics, 47*(3), 291–295. https://doi.org/10.1038/ng.3211.

Cerami, E. G., Gross, B. E., Demir, E., Rodchenkov, I., Babur, O., Anwar, N., ... Sander, C. (2011). Pathway Commons, a web resource for biological pathway data. *Nucleic Acids Research, 39*(Database issue), D685–D690. https://doi.org/10.1093/nar/gkq1039.

Chang, C. C., Chow, C. C., Tellier, L. C., Vattikuti, S., Purcell, S. M., & Lee, J. J. (2015). Second-generation PLINK: Rising to the challenge of larger and richer datasets. *Gigascience, 4,* 7. https://doi.org/10.1186/s13742-015-0047-8.

Chen, C., Grennan, K., Badner, J., Zhang, D., Gershon, E., Jin, L., & Liu, C. (2011). Removing batch effects in analysis of expression microarray data: An evaluation of six batch adjustment methods. *PLoS One, 6*(2). https://doi.org/10.1371/journal.pone.0017238.

Chen, Y. H., Lin, H. W., & Liu, H. (2009). Two-stage analysis for gene-environment interaction utilizing both case-only and family-based analysis. *Genetic Epidemiology, 33*(2), 95–104. https://doi.org/10.1002/gepi.20357.

Chen, X., Liu, C. T., Zhang, M., & Zhang, H. (2007). A forest-based approach to identifying gene and gene gene interactions. *Proceedings of the National Academy of Sciences of the United States of America, 104*(49), 19199–19203. https://doi.org/10.1073/pnas.0709868104.

Chen, S. H., Sun, J., Dimitrov, L., Turner, A. R., Adams, T. S., Meyers, D. A., ... Hsu, F. C. (2008). A support vector machine approach for detecting gene-gene interaction. *Genetic Epidemiology, 32*(2), 152–167. https://doi.org/10.1002/gepi.20272.

Cook, N. R., Zee, R. Y., & Ridker, P. M. (2004). Tree and spline based association analysis of gene-gene interaction models for ischemic stroke. *Statistics in Medicine, 23*(9), 1439–1453. https://doi.org/10.1002/sim.1749.

Cowley, M. J., Pinese, M., Kassahn, K. S., Waddell, N., Pearson, J. V., Grimmond, S. M., ... Wu, J. (2012). PINA v2.0: Mining interactome modules. *Nucleic Acids Research, 40*(Database issue), D862–D865. https://doi.org/10.1093/nar/gkr967.

Cross-Disorder Group of the Psychiatric Genomics Consortium. (2013). Identification of risk loci with shared effects on five major psychiatric disorders: A genome-wide analysis. *Lancet, 381*(9875), 1371–1379. https://doi.org/10.1016/S0140-6736(12)62129-1.

Culverhouse, R., Klein, T., & Shannon, W. (2004). Detecting epistatic interactions contributing to quantitative traits. *Genetic Epidemiology, 27*(2), 141–152. https://doi.org/10.1002/gepi.20006.

Curran, S., Powell, J., Neale, B. M., Dworzynski, K., Li, T., Murphy, D., & Bolton, P. F. (2006). An association analysis of candidate genes on chromosome 15 q11-13 and autism spectrum disorder. *Molecular Psychiatry, 11*(8), 709–713. https://doi.org/10.1038/sj.mp.4001839.

da Huang, W., Sherman, B. T., & Lempicki, R. A. (2009). Systematic and integrative analysis of large gene lists using DAVID bioinformatics resources. *Nature Protocols, 4*(1), 44–57. https://doi.org/10.1038/nprot.2008.211.

Dai, J. Y., Logsdon, B. A., Huang, Y., Hsu, L., Reiner, A. P., Prentice, R. L., & Kooperberg, C. (2012). Simultaneously testing for marginal genetic association and gene-environment interaction. *American Journal of Epidemiology, 176*(2), 164–173. https://doi.org/10.1093/aje/kwr521.

Dedeurwaerder, S., Defrance, M., Bizet, M., Calonne, E., Bontempi, G., & Fuks, F. (2014). A comprehensive overview of Infinium HumanMethylation450 data processing. *Briefings in Bioinformatics, 15*(6), 929–941. https://doi.org/10.1093/bib/bbt054.

de Jong, S., Newhouse, S. J., Patel, H., Lee, S., Dempster, D., Curtis, C., … Breen, G. (2016). Immune signatures and disorder-specific patterns in a cross-disorder gene expression analysis. *The British Journal of Psychiatry, 209*(3), 202–208. https://doi.org/10.1192/bjp.bp.115.175471.

Devlin, B., & Roeder, K. (1999). Genomic control for association studies. *Biometrics, 55*(4), 997–1004.

Diboun, I., Wernisch, L., Orengo, C. A., & Koltzenburg, M. (2006). Microarray analysis after RNA amplification can detect pronounced differences in gene expression using limma. *BMC Genomics, 7*, 252. https://doi.org/10.1186/1471-2164-7-252.

Duncan, L. E., & Keller, M. C. (2011). A critical review of the first 10 years of candidate gene-by-environment interaction research in psychiatry. *The American Journal of Psychiatry, 168*(10), 1041–1049. https://doi.org/10.1176/appi.ajp.2011.11020191.

Dunn, E. C., Wiste, A., Radmanesh, F., Almli, L. M., Gogarten, S. M., Sofer, T., … Smoller, J. W. (2016). Genome-wide association study (GWAS) and genome-wide by environment interaction study (GWEIS) of depressive symptoms in African American and Hispanic/Latina women. *Depression and Anxiety, 33*(4), 265–280. https://doi.org/10.1002/da.22484.

Easton, D. F., Pooley, K. A., Dunning, A. M., Pharoah, P. D., Thompson, D., Ballinger, D. G., … Ponder, B. A. (2007). Genome-wide association study identifies novel breast cancer susceptibility loci. *Nature, 447*(7148), 1087–1093. https://doi.org/10.1038/nature05887.

Edwards, T. L., Lewis, K., Velez, D. R., Dudek, S., & Ritchie, M. D. (2009). Exploring the performance of Multifactor Dimensionality Reduction in large scale SNP studies and in the presence of genetic heterogeneity among epistatic disease models. *Human Heredity, 67*(3), 183–192. https://doi.org/10.1159/000181157.

Euesden, J., Lewis, C. M., & O'Reilly, P. F. (2015). PRSice: Polygenic Risk Score software. *Bioinformatics, 31*(9), 1466–1468. https://doi.org/10.1093/bioinformatics/btu848.

Finucane, H. K., Bulik-Sullivan, B., Gusev, A., Trynka, G., Reshef, Y., Loh, P. R., … Price, A. L. (2015). Partitioning heritability by functional annotation using genome-wide association summary statistics. *Nature Genetics, 47*(11), 1228–1235. https://doi.org/10.1038/ng.3404.

Gao, S., Zou, D., Mao, L., Zhou, Q., Jia, W., Huang, Y., … Sorensen, K. D. (2015). SMAP: A streamlined methylation analysis pipeline for bisulfite sequencing. *Gigascience, 4*, 29. https://doi.org/10.1186/s13742-015-0070-9.

Genomes Project Consortium, Abecasis, G. R., Altshuler, D., Auton, A., Brooks, L. D., Durbin, R. M., … McVean, G. A. (2010). A map of human genome variation from population-scale sequencing. *Nature, 467*(7319), 1061–1073. https://doi.org/10.1038/nature09534.

Graham, D. S. C., & Vyse, T. J. (2005). The common disease common variant concept. In *Encyclopedia of genetics, genomics, proteomics and bioinformatics*. Wiley. https://doi.org/10.1002/047001153X.g105205.

Hahn, L. W., Ritchie, M. D., & Moore, J. H. (2003). Multifactor dimensionality reduction software for detecting gene-gene and gene-environment interactions. *Bioinformatics, 19*(3), 376–382.

Howie, B. N., Donnelly, P., & Marchini, J. (2009). A flexible and accurate genotype imputation method for the next generation of genome-wide association studies. *PLoS Genetics, 5*(6). https://doi.org/10.1371/journal.pgen.1000529.

Huang, D. W., Sherman, B. T., Tan, Q., Kir, J., Liu, D., Bryant, D., … Lempicki, R. A. (2007). DAVID Bioinformatics Resources: Expanded annotation database and novel algorithms to better extract biology from large gene lists. *Nucleic Acids Research, 35*, W169–W175. https://doi.org/10.1093/nar/gkm415 (Web Server issue).

Huber, W., Carey, V. J., Gentleman, R., Anders, S., Carlson, M., Carvalho, B. S., … Morgan, M. (2015). Orchestrating high-throughput genomic analysis with Bioconductor. *Nature Methods, 12*(2), 115–121. https://doi.org/10.1038/nmeth.3252.

Iafrate, A. J., Feuk, L., Rivera, M. N., Listewnik, M. L., Donahoe, P. K., Qi, Y., … Lee, C. (2004). Detection of large-scale variation in the human genome. *Nature Genetics, 36*(9), 949–951. https://doi.org/10.1038/ng1416.

International HapMap Consortium. (2003). The International HapMap Project. *Nature, 426*(6968), 789–796. https://doi.org/10.1038/nature02168.

International Schizophrenia Consortium, Purcell, S. M., Wray, N. R., Stone, J. L., Visscher, P. M., O'Donovan, M. C., … Sklar, P. (2009). Common polygenic variation contributes to risk of schizophrenia and bipolar disorder. *Nature, 460*(7256), 748–752. https://doi.org/10.1038/nature08185.

Iyegbe, C., Campbell, D., Butler, A., Ajnakina, O., & Sham, P. (2014). The emerging molecular architecture of schizophrenia, polygenic risk scores and the clinical implications for GxE research. *Social Psychiatry and Psychiatric Epidemiology, 49*(2), 169–182. https://doi.org/10.1007/s00127-014-0823-2.

Jiang, P., Sun, K., Lun, F. M., Guo, A. M., Wang, H., Chan, K. C., … Sun, H. (2014). Methy-Pipe: An integrated bioinformatics pipeline for whole genome bisulfite sequencing data analysis. *PLoS One, 9*(6). https://doi.org/10.1371/journal.pone.0100360.

Johnson, W. E., Li, C., & Rabinovic, A. (2007). Adjusting batch effects in microarray expression data using empirical Bayes methods. *Biostatistics, 8*(1), 118–127. https://doi.org/10.1093/biostatistics/kxj037.

Kanehisa, M., & Goto, S. (2000). KEGG: Kyoto encyclopedia of genes and genomes. *Nucleic Acids Research, 28*(1), 27–30.

Kao, P. Y., Leung, K. H., Chan, L. W., Yip, S. P., & Yap, M. K. (2017). Pathway analysis of complex diseases for GWAS, extending to consider rare variants, multi-omics and interactions. *Biochimica et Biophysica Acta, 1861*(2), 335–353. https://doi.org/10.1016/j.bbagen.2016.11.030.

Karg, K., & Sen, S. (2012). Gene x environment interaction models in psychiatric genetics. *Current Topics in Behavioral Neurosciences, 12*, 441–462. https://doi.org/10.1007/7854_2011_184.

Keller, M. C. (2014). Gene x environment interaction studies have not properly controlled for potential confounders: The problem and the (simple) solution. *Biological Psychiatry*, 75(1), 18–24. https://doi.org/10.1016/j.biopsych.2013.09.006.

Kim, S., Hwang, Y., Webster, M. J., & Lee, D. (2016). Differential activation of immune/inflammatory response-related co-expression modules in the hippocampus across the major psychiatric disorders. *Molecular Psychiatry*, 21(3), 376–385. https://doi.org/10.1038/mp.2015.79.

Kraft, P., Yen, Y. C., Stram, D. O., Morrison, J., & Gauderman, W. J. (2007). Exploiting gene-environment interaction to detect genetic associations. *Human Heredity*, 63(2), 111–119. https://doi.org/10.1159/000099183.

Kupfer, P., Guthke, R., Pohlers, D., Huber, R., Koczan, D., & Kinne, R. W. (2012). Batch correction of microarray data substantially improves the identification of genes differentially expressed in rheumatoid arthritis and osteoarthritis. *BMC Medical Genomics*, 5, 23. https://doi.org/10.1186/1755-8794-5-23.

Laird, N. M., & Lange, C. (2008). Family-based methods for linkage and association analysis. *Advances in Genetics*, 60, 219–252. https://doi.org/10.1016/S0065-2660(07)00410-5.

Lander, E., & Kruglyak, L. (1995). Genetic dissection of complex traits: Guidelines for interpreting and reporting linkage results. *Nature Genetics*, 11(3), 241–247. https://doi.org/10.1038/ng1195-241.

Lander, E. S., Linton, L. M., Birren, B., Nusbaum, C., Zody, M. C., Baldwin, J., ... International Human Genome Sequencing Consortium. (2001). Initial sequencing and analysis of the human genome. *Nature*, 409(6822), 860–921. https://doi.org/10.1038/35057062.

Langfelder, P., & Horvath, S. (2008). WGCNA: An R package for weighted correlation network analysis. *BMC Bioinformatics*, 9, 559. https://doi.org/10.1186/1471-2105-9-559.

Langmead, B., Trapnell, C., Pop, M., & Salzberg, S. L. (2009). Ultrafast and memory-efficient alignment of short DNA sequences to the human genome. *Genome Biology*, 10(3), R25. https://doi.org/10.1186/gb-2009-10-3-r25.

Law, C. W., Alhamdoosh, M., Su, S., Smyth, G. K., & Ritchie, M. E. (2016). RNA-seq analysis is easy as 1-2-3 with limma, Glimma and edgeR. *F1000Res*, 5, 1408. https://doi.org/10.12688/f1000research.9005.2.

Lee, P. H., O'Dushlaine, C., Thomas, B., & Purcell, S. M. (2012). INRICH: Interval-based enrichment analysis for genome-wide association studies. *Bioinformatics*, 28(13), 1797–1799. https://doi.org/10.1093/bioinformatics/bts191.

Lee, S. H., Wray, N. R., Goddard, M. E., & Visscher, P. M. (2011). Estimating missing heritability for disease from genome-wide association studies. *American Journal of Human Genetics*, 88(3), 294–305. https://doi.org/10.1016/j.ajhg.2011.02.002.

Lee, S. H., Yang, J., Goddard, M. E., Visscher, P. M., & Wray, N. R. (2012). Estimation of pleiotropy between complex diseases using single-nucleotide polymorphism-derived genomic relationships and restricted maximum likelihood. *Bioinformatics*, 28(19), 2540–2542. https://doi.org/10.1093/bioinformatics/bts474.

Leek, J. T., & Storey, J. D. (2007). Capturing heterogeneity in gene expression studies by surrogate variable analysis. *PLoS Genetics*, 3(9), 1724–1735. https://doi.org/10.1371/journal.pgen.0030161.

Levinson, D. F. (2006). The genetics of depression: A review. *Biological Psychiatry*, 60(2), 84–92. https://doi.org/10.1016/j.biopsych.2005.08.024.

Li, H., & Durbin, R. (2009). Fast and accurate short read alignment with Burrows-Wheeler transform. *Bioinformatics*, 25(14), 1754–1760. https://doi.org/10.1093/bioinformatics/btp324.

Li, R., Li, Y., Kristiansen, K., & Wang, J. (2008). SOAP: Short oligonucleotide alignment program. *Bioinformatics*, 24(5), 713–714. https://doi.org/10.1093/bioinformatics/btn025.

Liberzon, A., Subramanian, A., Pinchback, R., Thorvaldsdottir, H., Tamayo, P., & Mesirov, J. P. (2011). Molecular Signatures Database (MSigDB) 3.0. *Bioinformatics*, 27(12), 1739–1740. https://doi.org/10.1093/bioinformatics/btr260.

Lim, J. H., Lee, S. Y., & Kim, J. H. (2017). TRAPR: R package for statistical analysis and visualization of RNA-Seq data. *Genomics Inform*, 15(1), 51–53. https://doi.org/10.5808/GI.2017.15.1.51.

Little, J., Bradley, L., Bray, M. S., Clyne, M., Dorman, J., Ellsworth, D. L., ... Weinberg, C. (2002). Reporting, appraising, and integrating data on genotype prevalence and gene-disease associations. *American Journal of Epidemiology*, 156(4), 300–310.

Love, M. I., Huber, W., & Anders, S. (2014). Moderated estimation of fold change and dispersion for RNA-seq data with DESeq2. *Genome Biology*, 15(12), 550. https://doi.org/10.1186/s13059-014-0550-8.

LoVerso, P. R., & Cui, F. (2015). A computational pipeline for cross-species analysis of RNA-seq data using R and Bioconductor. *Bioinformatics and Biology Insights*, 9, 165–174. https://doi.org/10.4137/BBI.S30884.

Lunetta, K. L., Hayward, L. B., Segal, J., & Van Eerdewegh, P. (2004). Screening large-scale association study data: Exploiting interactions using random forests. *BMC Genetics*, 5, 32. https://doi.org/10.1186/1471-2156-5-32.

Malhotra, D., & Sebat, J. (2012). CNVs: Harbingers of a rare variant revolution in psychiatric genetics. *Cell*, 148(6), 1223–1241. https://doi.org/10.1016/j.cell.2012.02.039.

Marchini, J., Howie, B., Myers, S., McVean, G., & Donnelly, P. (2007). A new multipoint method for genome-wide association studies by imputation of genotypes. *Nature Genetics*, 39(7), 906–913. https://doi.org/10.1038/ng2088.

McClellan, J., & King, M. C. (2010). Genetic heterogeneity in human disease. *Cell*, 141(2), 210–217. https://doi.org/10.1016/j.cell.2010.03.032.

McQueen, M. B., Devlin, B., Faraone, S. V., Nimgaonkar, V. L., Sklar, P., Smoller, J. W., ... Laird, N. M. (2005). Combined analysis from eleven linkage studies of bipolar disorder provides strong evidence of susceptibility loci on chromosomes 6q and 8q. *American Journal of Human Genetics*, 77(4), 582–595. https://doi.org/10.1086/491603.

Mehta, D., & Binder, E. B. (2012). Gene x environment vulnerability factors for PTSD: The HPA-axis. *Neuropharmacology*, 62(2), 654–662. https://doi.org/10.1016/j.neuropharm.2011.03.009.

Michels, K. B., Binder, A. M., Dedeurwaerder, S., Epstein, C. B., Greally, J. M., Gut, I., ... Irizarry, R. A. (2013). Recommendations for the design and analysis of epigenome-wide association studies. *Nature Methods*, *10*(10), 949–955. https://doi.org/10.1038/nmeth.2632.

Millstein, J., Conti, D. V., Gilliland, F. D., & Gauderman, W. J. (2006). A testing framework for identifying susceptibility genes in the presence of epistasis. *American Journal of Human Genetics*, *78*(1), 15–27. https://doi.org/10.1086/498850.

Moore, J. H., & Hahn, L. W. (2002). A cellular automata approach to detecting interactions among single-nucleotide polymorphisms in complex multifactorial diseases. *Pacific Symposium on Biocomputing*, 53–64.

Moore, J. H., Hahn, L. W., Ritchie, M. D., Thornton, T. A., & White, B. C. (2004). Routine discovery of complex genetic models using genetic algorithms. *Applied Soft Computing*, *4*(1), 79–86. https://doi.org/10.1016/j.asoc.2003.08.003.

Mortazavi, A., Williams, B. A., McCue, K., Schaeffer, L., & Wold, B. (2008). Mapping and quantifying mammalian transcriptomes by RNA-Seq. *Nature Methods*, *5*(7), 621–628. https://doi.org/10.1038/nmeth.1226.

Morton, N. E. (1955). Sequential tests for the detection of linkage. *American Journal of Human Genetics*, *7*(3), 277–318.

Mukherjee, B., & Chatterjee, N. (2008). Exploiting gene-environment independence for analysis of case-control studies: An empirical Bayes-type shrinkage estimator to trade-off between bias and efficiency. *Biometrics*, *64*(3), 685–694. https://doi.org/10.1111/j.1541-0420.2007.00953.x.

Mullins, N., Power, R. A., Fisher, H. L., Hanscombe, K. B., Euesden, J., Iniesta, R., ... Lewis, C. M. (2016). Polygenic interactions with environmental adversity in the aetiology of major depressive disorder. *Psychological Medicine*, *46*(4), 759–770. https://doi.org/10.1017/S0033291715002172.

Murcray, C. E., Lewinger, J. P., Conti, D. V., Thomas, D. C., & Gauderman, W. J. (2011). Sample size requirements to detect gene-environment interactions in genome-wide association studies. *Genetic Epidemiology*, *35*(3), 201–210. https://doi.org/10.1002/gepi.20569.

Nelson, M. R., Kardia, S. L., Ferrell, R. E., & Sing, C. F. (2001). A combinatorial partitioning method to identify multilocus genotypic partitions that predict quantitative trait variation. *Genome Research*, *11*(3), 458–470. https://doi.org/10.1101/gr.172901.

Ng, M. Y., Levinson, D. F., Faraone, S. V., Suarez, B. K., DeLisi, L. E., Arinami, T., ... Lewis, C. M. (2009). Meta-analysis of 32 genome-wide linkage studies of schizophrenia. *Molecular Psychiatry*, *14*(8), 774–785. https://doi.org/10.1038/mp.2008.135.

Nikolayeva, O., & Robinson, M. D. (2014). edgeR for differential RNA-seq and ChIP-seq analysis: An application to stem cell biology. *Methods in Molecular Biology*, *1150*, 45–79. https://doi.org/10.1007/978-1-4939-0512-6_3.

Ott, J., & Terwiliger, J. D. (1994). *Handbook for human genetic linkage*. Baltimore, MD: Johns Hopkins University Press.

Park, M. Y., & Hastie, T. (2008). Penalized logistic regression for detecting gene interactions. *Biostatistics*, *9*(1), 30–50. https://doi.org/10.1093/biostatistics/kxm010.

Patterson, N., Price, A. L., & Reich, D. (2006). Population structure and eigenanalysis. *PLoS Genetics*, *2*(12). https://doi.org/10.1371/journal.pgen.0020190.

Pe'er, I., Yelensky, R., Altshuler, D., & Daly, M. J. (2008). Estimation of the multiple testing burden for genomewide association studies of nearly all common variants. *Genetic Epidemiology*, *32*(4), 381–385. https://doi.org/10.1002/gepi.20303.

Peyrot, W. J., Milaneschi, Y., Abdellaoui, A., Sullivan, P. F., Hottenga, J. J., Boomsma, D. I., & Penninx, B. W. (2014). Effect of polygenic risk scores on depression in childhood trauma. *The British Journal of Psychiatry*, *205*(2), 113–119. https://doi.org/10.1192/bjp.bp.113.143081.

Pinto, D., Pagnamenta, A. T., Klei, L., Anney, R., Merico, D., Regan, R., ... Betancur, C. (2010). Functional impact of global rare copy number variation in autism spectrum disorders. *Nature*, *466*(7304), 368–372. https://doi.org/10.1038/nature09146.

Plomin, R., DeFries, J. C., McClearn, G., & Rutter, M. (1997). *Behavioural genetics*. W.H. Freeman.

Price, A. L., Zaitlen, N. A., Reich, D., & Patterson, N. (2010). New approaches to population stratification in genome-wide association studies. *Nature Reviews. Genetics*, *11*(7), 459–463. https://doi.org/10.1038/nrg2813.

Pulst, S. M. (1999). Genetic linkage analysis. *Archives of Neurology*, *56*(6), 667–672.

Purcell, S., Neale, B., Todd-Brown, K., Thomas, L., Ferreira, M. A., Bender, D., ... Sham, P. C. (2007). PLINK: A tool set for whole-genome association and population-based linkage analyses. *American Journal of Human Genetics*, *81*(3), 559–575. https://doi.org/10.1086/519795.

Rackham, O. J., Langley, S. R., Oates, T., Vradi, E., Harmston, N., Srivastava, P. K., ... Petretto, E. (2017). A Bayesian approach for analysis of whole-genome bisulfite sequencing data identifies disease-associated changes in DNA methylation. *Genetics*, *205*(4), 1443–1458. https://doi.org/10.1534/genetics.116.195008.

Redon, R., Ishikawa, S., Fitch, K. R., Feuk, L., Perry, G. H., Andrews, T. D., ... Hurles, M. E. (2006). Global variation in copy number in the human genome. *Nature*, *444*(7118), 444–454. https://doi.org/10.1038/nature05329.

Reich, D. E., & Lander, E. S. (2001). On the allelic spectrum of human disease. *Trends in Genetics*, *17*(9), 502–510.

Risch, N., & Merikangas, K. (1996). The future of genetic studies of complex human diseases. *Science*, *273*(5281), 1516–1517.

Ritchie, M. E., Phipson, B., Wu, D., Hu, Y., Law, C. W., Shi, W., & Smyth, G. K. (2015). Limma powers differential expression analyses for RNA-sequencing and microarray studies. *Nucleic Acids Research*, *43*(7), e47. https://doi.org/10.1093/nar/gkv007.

Ritchie, M. D., White, B. C., Parker, J. S., Hahn, L. W., & Moore, J. H. (2003). Optimization of neural network architecture using genetic programming improves detection and modeling of gene-gene interactions in studies of human diseases. *BMC Bioinformatics*, *4*, 28. https://doi.org/10.1186/1471-2105-4-28.

Robinson, M. D., McCarthy, D. J., & Smyth, G. K. (2010). edgeR: A Bioconductor package for differential expression analysis of digital gene expression data. *Bioinformatics*, *26*(1), 139–140. https://doi.org/10.1093/bioinformatics/btp616.

Ruderfer, D. M., Fanous, A. H., Ripke, S., McQuillin, A., Amdur, R. L., Schizophrenia Working Group of the Psychiatric Genomics Consortium, ... Kendler, K. S. (2014). Polygenic dissection of diagnosis and clinical dimensions of bipolar disorder and schizophrenia. *Molecular Psychiatry*, *19*(9), 1017–1024. https://doi.org/10.1038/mp.2013.138.

Sebat, J., Lakshmi, B., Troge, J., Alexander, J., Young, J., Lundin, P., ... Wigler, M. (2004). Large-scale copy number polymorphism in the human genome. *Science*, *305*(5683), 525–528. https://doi.org/10.1126/science.1098918.

Segre, A. V., DIAGRAM Consortium, MAGIC Investigators, Groop, L., Mootha, V. K., Daly, M. J., & Altshuler, D. (2010). Common inherited variation in mitochondrial genes is not enriched for associations with type 2 diabetes or related glycemic traits. *PLoS Genetics, 6*(8). https://doi.org/10.1371/journal.pgen.1001058.

Sharma, S., Powers, A., Bradley, B., & Ressler, K. J. (2016). Gene x environment determinants of stress- and anxiety-related disorders. *Annual Review of Psychology, 67*, 239–261. https://doi.org/10.1146/annurev-psych-122414-033408.

Sladek, R., Rocheleau, G., Rung, J., Dina, C., Shen, L., Serre, D., … Froguel, P. (2007). A genome-wide association study identifies novel risk loci for type 2 diabetes. *Nature, 445*(7130), 881–885. https://doi.org/10.1038/nature05616.

Smith, P. G., & Day, N. E. (1984). The design of case-control studies: The influence of confounding and interaction effects. *International Journal of Epidemiology, 13*(3), 356–365.

Smoller, J. W. (2014). Psychiatric genetics and the future of personalized treatment. *Depression and Anxiety, 31*(11), 893–898. https://doi.org/10.1002/da.22322.

Speed, D., Hemani, G., Johnson, M. R., & Balding, D. J. (2012). Improved heritability estimation from genome-wide SNPs. *American Journal of Human Genetics, 91*(6), 1011–1021. https://doi.org/10.1016/j.ajhg.2012.10.010.

Spielman, R. S., McGinnis, R. E., & Ewens, W. J. (1993). Transmission test for linkage disequilibrium: The insulin gene region and insulin-dependent diabetes mellitus (IDDM). *American Journal of Human Genetics, 52*(3), 506–516.

Strobl, C., Malley, J., & Tutz, G. (2009). An introduction to recursive partitioning: Rationale, application, and characteristics of classification and regression trees, bagging, and random forests. *Psychological Methods, 14*(4), 323–348. https://doi.org/10.1037/a0016973.

Sullivan, P. F. (2010). The psychiatric GWAS consortium: Big science comes to psychiatry. *Neuron, 68*(2), 182–186. https://doi.org/10.1016/j.neuron.2010.10.003.

Szklarczyk, D., Franceschini, A., Wyder, S., Forslund, K., Heller, D., Huerta-Cepas, J., … von Mering, C. (2015). STRING v10: Protein-protein interaction networks, integrated over the tree of life. *Nucleic Acids Research, 43*(Database issue), D447–D452. https://doi.org/10.1093/nar/gku1003.

Tamiya, G., Shinya, M., Imanishi, T., Ikuta, T., Makino, S., Okamoto, K., … Inoko, H. (2005). Whole genome association study of rheumatoid arthritis using 27 039 microsatellites. *Human Molecular Genetics, 14*(16), 2305–2321. https://doi.org/10.1093/hmg/ddi234.

Thomas, D. (2010). Methods for investigating gene-environment interactions in candidate pathway and genome-wide association studies. *Annual Review of Public Health, 31*, 21–36. https://doi.org/10.1146/annurev.publhealth.012809.103619.

Tian, C., Plenge, R. M., Ransom, M., Lee, A., Villoslada, P., Selmi, C., … Seldin, M. F. (2008). Analysis and application of European genetic substructure using 300 K SNP information. *PLoS Genetics, 4*(1). https://doi.org/10.1371/journal.pgen.0040004.

Tomita, Y., Tomida, S., Hasegawa, Y., Suzuki, Y., Shirakawa, T., Kobayashi, T., & Honda, H. (2004). Artificial neural network approach for selection of susceptible single nucleotide polymorphisms and construction of prediction model on childhood allergic asthma. *BMC Bioinformatics, 5*, 120. https://doi.org/10.1186/1471-2105-5-120.

Torres-Garcia, W., Zheng, S., Sivachenko, A., Vegesna, R., Wang, Q., Yao, R., … Verhaak, R. G. (2014). PRADA: Pipeline for RNA sequencing data analysis. *Bioinformatics, 30*(15), 2224–2226. https://doi.org/10.1093/bioinformatics/btu169.

Trapnell, C., Williams, B. A., Pertea, G., Mortazavi, A., Kwan, G., van Baren, M. J., … Pachter, L. (2010). Transcript assembly and quantification by RNA-Seq reveals unannotated transcripts and isoform switching during cell differentiation. *Nature Biotechnology, 28*(5), 511–515. https://doi.org/10.1038/nbt.1621.

Uhr, M., Tontsch, A., Namendorf, C., Ripke, S., Lucae, S., Ising, M., … Holsboer, F. (2008). Polymorphisms in the drug transporter gene ABCB1 predict antidepressant treatment response in depression. *Neuron, 57*(2), 203–209. https://doi.org/10.1016/j.neuron.2007.11.017.

Varet, H., Brillet-Gueguen, L., Coppee, J. Y., & Dillies, M. A. (2016). SARTools: A DESeq2- and EdgeR-based R pipeline for comprehensive differential analysis of RNA-Seq data. *PLoS One, 11*(6). https://doi.org/10.1371/journal.pone.0157022.

Visscher, P. M., Brown, M. A., McCarthy, M. I., & Yang, J. (2012). Five years of GWAS discovery. *American Journal of Human Genetics, 90*(1), 7–24. https://doi.org/10.1016/j.ajhg.2011.11.029.

Wakefield, J., Haneuse, S., Dobra, A., & Teeple, E. (2011). Bayes computation for ecological inference. *Statistics in Medicine, 30*(12), 1381–1396. https://doi.org/10.1002/sim.4214.

Wei, J., & Hemmings, G. P. (2000). The NOTCH4 locus is associated with susceptibility to schizophrenia. *Nature Genetics, 25*(4), 376–377. https://doi.org/10.1038/78044.

Weinberg, C. R., & Umbach, D. M. (2000). Choosing a retrospective design to assess joint genetic and environmental contributions to risk. *American Journal of Epidemiology, 152*(3), 197–203.

Wellcome Trust Case Control Consortium. (2007). Genome-wide association study of 14,000 cases of seven common diseases and 3,000 shared controls. *Nature, 447*(7145), 661–678. https://doi.org/10.1038/nature05911.

Wray, N. R., Lee, S. H., Mehta, D., Vinkhuyzen, A. A., Dudbridge, F., & Middeldorp, C. M. (2014). Research review: Polygenic methods and their application to psychiatric traits. *Journal of Child Psychology and Psychiatry, 55*(10), 1068–1087. https://doi.org/10.1111/jcpp.12295.

Wray, N. R., Sullivan, P. F., & Major Depressive Disorder Working Group of the PGC. (2017). Genome-wide association analyses identify 44 risk variants and refine the genetic architecture of major depressive disorder. *bioRxiv*. https://doi.org/10.1101/167577.

Wreczycka, K., Gosdschan, A., Yusuf, D., Gruning, B., Assenov, Y., & Akalin, A. (2017). Strategies for analyzing bisulfite sequencing data. *Journal of Biotechnology, 261*, 105–115. https://doi.org/10.1016/j.jbiotec.2017.08.007.

Wright, M. L., Dozmorov, M. G., Wolen, A. R., Jackson-Cook, C., Starkweather, A. R., Lyon, D. E., & York, T. P. (2016). Establishing an analytic pipeline for genome-wide DNA methylation. *Clinical Epigenetics, 8*, 45. https://doi.org/10.1186/s13148-016-0212-7.

Yang, J., Benyamin, B., McEvoy, B. P., Gordon, S., Henders, A. K., Nyholt, D. R., … Visscher, P. M. (2010). Common SNPs explain a large proportion of the heritability for human height. *Nature Genetics, 42*(7), 565–569. https://doi.org/10.1038/ng.608.

Yang, J., Lee, S. H., Goddard, M. E., & Visscher, P. M. (2011). GCTA: A tool for genome-wide complex trait analysis. *American Journal of Human Genetics, 88*(1), 76–82. https://doi.org/10.1016/j.ajhg.2010.11.011.

Yu, K., Wacholder, S., Wheeler, W., Wang, Z., Caporaso, N., Landi, M. T., & Liang, F. (2012). A flexible Bayesian model for studying gene-environment interaction. *PLoS Genetics, 8*(1). https://doi.org/10.1371/journal.pgen.1002482.

Zhang, Y., & Liu, J. S. (2007). Bayesian inference of epistatic interactions in case-control studies. *Nature Genetics, 39*(9), 1167–1173. https://doi.org/10.1038/ng2110.

Zheng, J., Erzurumluoglu, A. M., Elsworth, B. L., Kemp, J. P., Howe, L., Haycock, P. C., ... Neale, B. M. (2017). LD Hub: A centralized database and web interface to perform LD score regression that maximizes the potential of summary level GWAS data for SNP heritability and genetic correlation analysis. *Bioinformatics, 33*(2), 272–279. https://doi.org/10.1093/bioinformatics/btw613.

Further reading

Keverne, J., Czamara, D., Cubells, J. F., & Binder, E. B. (2016). Genetics and genomics. In A. Schatzberg, & C. B. Nemeroff (Eds.), *Textbook of psychopharmacology*. (5th ed.). The American Psychiatric Association Publishing.

Chapter 11

Opportunities and challenges of machine learning approaches for biomarker signature identification in psychiatry

Han Cao and Emanuel Schwarz

Department of Psychiatry and Psychotherapy, Central Institute of Mental Health, Medical Faculty Mannheim, University of Heidelberg, Mannheim, Germany

1 Introduction

Machine learning (ML) is a field of artificial intelligence that automatically learns a functional mapping $Y = f(X)$ from data X and outcome (i.e., diagnosis) Y. With the recent explosion of available data, machine learning has been applied widely in every data-intensive area, including speech and natural language processing (Sebastiani, 2002), image processing and computer vision (Goodfellow et al., 2016), biomedical informatics (Libbrecht & Noble, 2015), and web-based applications (Das et al., 2015). The specific challenges faced in different application areas have led to variants of ML algorithms with different focuses. In addition to predictive ability, biomedical applications, in particular, require models to be biologically interpretable, to explore which underlying mechanisms contribute to accurate prediction (Lin et al., 2013). In this context, multivariate predictive patterns may elucidate biological mechanisms that are not easily derived from univariate alterations.

ML approaches are of fundamental importance for modern biomarker discovery, which aims to improve the clinical management of illnesses through biological differentiation, down to the individual patient level and toward a personalized approach to medicine. To achieve this goal, ML tools have been applied to diverse, clinically relevant questions, ranging from differentiation of patients from healthy controls, and stratification of patients into subgroups, to prediction of clinical outcomes based on biological data (Majumder et al., 2015; Singal et al., 2013; Yu et al., 2016). These efforts have highlighted challenges tied to the complexity of the disorders, the high dimensionality and variability of the data, and the low effect sizes of individual markers. It is becoming increasingly clear that the biological complexity of mental disorders necessitates the use of high-dimensional, multimodal data, which creates substantial challenges for machine learning analysis. In this chapter, we will discuss the opportunities and challenges surrounding the application of machine learning for such biomarker discovery objectives, with a particular focus on psychiatric disorders.

Diagnosis of psychiatric disorders is based on the fifth edition of the Diagnostic and Statistical Manual of Mental Disorder (DSM5 (American Psychiatric Association, 2013)) and the broadly similar ICD10, which define the disorders based on the presence of certain symptom constellations. However, due to the lack of accurate biological hallmarks, there is uncertainty about whether the diagnostic boundaries delineate etiologically different disorders. It appears plausible that the current diagnostic systems instead differentiate conditions that are biologically heterogeneous, causing a reduction in effect sizes of marker candidates, and creating substantial challenges for investigation of etiological factors. Genome-wide association studies have highlighted the genetic complexity of highly heritable disorders, such as schizophrenia or bipolar disorder, and the necessity of very large sample numbers to identify significant illness associations. Gene-environment interaction that includes obstetric complications (Byrne et al., 2007), urban births (Haddad et al., 2015), or seasonal effects (Mortensen et al., 1999) further contribute to the etiological complexity and the challenges faced by machine learning approaches applied on biological data. Beyond such factors with biological impact, there are numerous challenges arising from the nature of the acquired data as such, particularly their dimensionality, as well as missing values.

A central challenge for machine learning analysis of high-dimensional data is the "curse of dimensionality" (Shultz et al., 2011). The efficient identification of good classifiers requires the available data to appropriately "sample" the data space. However, in high dimensions, the volume of the data space grows exponentially, causing the available data to sparsely populate the space, limiting the ability to identify good classifiers. In a scenario where the number of biological

Personalized Psychiatry. https://doi.org/10.1016/B978-0-12-813176-3.00011-0

predictors by far exceeds the number of subjects, more complex algorithms are typically affected by overfitting, leading to poor generalizability in independent data. In high dimensions, there is often a large (or infinite) number of classifiers with similar performance observed in the training data. ML algorithms employ different strategies to select among these solutions (such as the classifier with the maximal marginal distance for SVM (Joachims, 1998)). Cross-validation is typically used during model selection, such that a given classifier is trained and tested separately on nonoverlapping data subsets, to prevent overfitting and control model complexity. Despite this, it is difficult to guarantee that the selected classifier is optimal as feature dimensions increase (Chávez et al., 2001). For example, a naïve table look-up ML technique requires the number of training data points to exponentially increase with the feature dimensions to preserve the same expectation of error rate (Jain, Duin, & Jianchang, 2000). A multiomics integration study (Phan et al., 2012) has indicated that the statistical confidence of parameter estimation is limited by curse of dimensionality because of low data density in the feature space. Another study (Clarke et al., 2008) has reviewed the negative implications of high data dimensionality in genetic studies. For example, the pairwise distance between subjects tends to be equivalent in high dimensions, limiting the performance of distance-based clustering methods. Spurious correlations exist among genetic features in high-dimensional space, restricting the accuracy of feature selection and dimension reduction approaches. Furthermore, the low data density in high dimensions leads to a "flat" data distribution, making missing value imputation challenging.

The inference of missing data is generally a challenge for biomarker discovery approaches. It is assumed that if more than 15% of data are missing, model interpretability is potentially biased (Acuna & Rodriguez, 2004). The sources of missing values in genomic data are diverse. For example, in microarray data, missing values may be due to insufficient resolution, image corruption, or artifacts on the microarray (Moorthy, Saberi Mohamad, & Deris, 2014; Troyanskaya et al., 2001). In several studies, missing values have been demonstrated to hinder (Aittokallio, 2010) and negatively (Liew, Law, & Yan, 2011) affect the downstream analysis, such as hierarchical clustering and discriminant analysis using SVM (Liew et al., 2011). Moreover, numerous computational methods cannot be easily applied on data with missing values, such as principle component analysis or singular value decomposition.

These challenges introduce substantial uncertainty during analysis, and hinder the identification of reproducible and biologically interpretable biomarkers. To overcome this, one potential solution is to incorporate prior (biological) knowledge into machine learning algorithms (Cun & Frohlich, 2012). The additional knowledge can be specific for a given application, such as spatial information of Single Nucleotide Polymorphisms (SNPs), ontological information, biological networks (i.e., coexpression networks), or integration derived from other data modalities. In this chapter, we summarize different strategies to incorporate such knowledge into the machine learning framework to improve biomarker identification. A further focus of this chapter is the discussion of the missing data problem, including mechanisms that lead to missing values, and current countermeasures.

2   Opportunities

The ML community has developed numerous flexible learning frameworks to incorporate biological prior knowledge via the embedded biomarker selection process. For example, a coexpression network can help to select as predictors functionally relevant sets of genes that are linked to useful predictors as part of the network (Cun & Frohlich, 2012). In this section, we will discuss knowledge incorporation strategies via regularization, Bayesian, and kernel methods.

2.1   Knowledge incorporation via regularization

Regularization is a common strategy in ML to incorporate prior knowledge. By adding a so-called penalty term ($\Omega(w)$) to the loss function (data fitting term), the classifier is biased toward the prior knowledge. A penalty is a simple function of variables that constrains the structure of the variables. For example, the $l2$-penalty ($\Omega(w) = \|w\|_2^2$) (Hoerl & Kennard, 1970) constrains the value scale of each variable, and has been repeatedly used to prevent model overfitting and to stabilize the numerical solution (Friedman, Hastie, & Tibshirani, 2001). The $l1$-penalty (Lasso) ($\Omega(w) = \|w\|_1$) (Tibshirani, 2011), another popular regularization method, is used as a feature selection process embedded in ML algorithms because the coefficients of irreverent features are automatically penalized to 0 (creating a "sparse" pattern of important predictors), and thus the prediction is only dependent on the selected biomarkers. The two penalty types are often combined, resulting, for example, in the elastic net penalty $\Omega(w) = \alpha\|w\|_1 + (1 - \alpha)\|w\|_2^2$ (Zou & Hastie, 2005), which linearly combines the effect of $l1$-penalty and $l2$-penalty.

To incorporate prior knowledge via regularization for biomarker selection, it is important to utilize a meaningful penalty to constrain the structure of the features. These methods can be categorized into two classes. In one, the feature values are penalized, while in the other, the difference between feature values is penalized.

2.1.1 Penalization of feature value

In this category, the Lasso (Tibshirani, 2011) has been extended to consider the group structure of features, where the groups are predefined according to additional biological information. For example, genes can be mapped to function through ontological information (Ashburner et al., 2000). Functionally related genes may, as a cluster, improve the robustness of the biomarker selection process (He & Yu, 2010). For this strategy, a popular algorithm is the group Lasso (Kim & Xing, 2012; Ma, Song, & Huang, 2007; Meier, Van De Geer, & Bühlmann, 2008; Silver et al., 2012) ($\Omega(w) = \sum_{g=1}^{G} \|w_{I_g}\|_2$, where G is a predefined group structure over the feature space). One study (Ma et al., 2007) applied the group Lasso for cancer biomarker identification and survival analysis using expression data. The authors assumed cluster structures existing among coexpressed genes, as such genes are likely to share biological functions. To select such important gene clusters, K-means clustering was applied, and the optimal number of clusters was determined by the Gap method. The second step was to remove irrelevant genes in a given cluster using the conventional lasso method. Finally, a group lasso was used to remove irrelevant clusters. As a result, 20–60 genes were finally identified among 500 candidates. The conventional group lasso method makes the naïve assumption that no sparse pattern exists in a given cluster, which typically does not hold in practice. Therefore, the authors applied an extra step to create an in-cluster sparse pattern. Such an in-cluster sparse pattern has also been used in another study by introducing a joint penalty, called the sparse group lasso ($\Omega(w) = \alpha\|w\|_1 + (1-\alpha)\sum_{g=1}^{G}\|w_{I_g}\|_2$) (Simon et al., 2013). This algorithm simultaneously searches gene-wise and cluster-wise for biomarkers in one learning procedure. Compared with the separated steps, which remove irrelevant in-cluster genes independently, this algorithm considers the impact of such removals to other clusters, and may lead to a better solution. This algorithm has been used in a genetic association study (Zhou et al., 2010) and for genomic data integration (Lin et al., 2013). In the latter study, canonical correlation analysis (CCA) with the sparse group lasso method was used to identify correlations between gene expression and genetic variation. The authors compared the proposed method with conventional sparse CCA on simulated and real data, and found CCA with the group lasso penalty to be capable of more accurately identifying correlated features and controlling a low level of discordance.

Another extension of the group lasso method is to introduce hierarchical structure over groups of predictors, such as the tree lasso (Kim & Xing, 2012). This study aimed to identify genetic biomarkers with pleiotropic effects based on eQTL analysis (Kim & Xing, 2012). A typical strategy is to incorporate gene-gene interactions after the eQTL analysis. However, the authors argued that such a strategy cannot exploit the entire statistical power from the incorporated knowledge. Therefore, the eQTL analysis was reformulated as a multitask learning problem, where each gene represents a task. A tree structure was predefined based on hierarchical clustering of expression data. Such a structure can capture the gene-gene associations at multiple granularities. For example, genes are tightly correlated in a leaf cluster, but loosely correlated in the root cluster. Multitask regression with a tree penalty was then performed on all genes. During eQTL analysis, such an algorithm could identify significant SNP sets for a given gene, and these SNP sets are linked to gene-gene interactions in the expression domain.

2.1.2 Penalization of the difference between features

The methods in this category attempt to penalize the difference between features and encourage structural smoothness. Such a strategy is suitable to incorporate different kinds of biological networks (i.e., coexpression networks, PPI networks) and spatial/temporal information. The primary approach is the fused lasso, which assumes the features to be ordered, and the difference between adjacent features to be small. Consequently, the classifier is biased by penalizing the difference between adjacent features ($\Omega(w) = \lambda_1\|w\|_1 + \lambda_2\sum_{i=1}^{p}|w_i - w_{i=1}|$). A variant of fused lasso has been applied to genetic association data to incorporate information in linkage disequilibrium (LD) (Liu et al., 2013). In GWAS studies, it is common to test the association of each SNP with a given phenotype, and then to combine the results. However, such a procedure ignores the structure of SNPs that is, those related to LD patterns. A previous study (Liu et al., 2013) assumed that SNPs are ordered on the chromosome according to physical position, thus the adjacent SNPs are strongly affected by LD, and expected to have similar model coefficients. The method was tested on the rheumatoid arthritis genetic data. Due to the computational complexity, only SNPs of chromosome 6 (~2000 subjects, 31,670 SNPs) were included in the analysis. Compared with original single-SNP linear regression or Lasso, the proposed method demonstrated an improved ability to select illness-associated SNPs.

Another important extension of the fused lasso method is the network regularized lasso ($\Omega(w) = \lambda_1\|w\|_1 + \lambda_2 w^T L w$, where L is the graph Laplacian matrix calculated from the given network). The strategy penalizes the difference between two features directly connected in the network, and thus encourages feature smoothness over the network. For biomarker identification, such a technique has been applied to incorporate biological network information. In one study (Li & Li, 2008), researchers have explored gene expression biomarkers of cancer by incorporating the pathway structure defined by KEGG (Kanehisa et al., 2017), and aimed to identify illness-relevant genes and subnetworks. Specifically, the network was constructed based on 33 regulatory pathways, where each node represented a gene or gene product. To highlight the importance of hub genes, the node degrees were considered as weights. Combined with gene expression data, 1533 genes remained for further analysis. Finally, the authors demonstrated the subnetwork identified by the method to be more plausible than that derived from the conventional Lasso method.

A weakness of the fused and network regularized lasso is that negatively coregulated or correlated biomarkers are incorrectly penalized by the model. For example, assume the coefficient of a given gene to be 1, and the coefficient of a negatively correlated gene to be -1. Such values would be distorted through strong penalization. One can filter the negative correlations from the network, but then the information contained in the biological network cannot be fully utilized. An interesting study (Yang et al., 2016) analyzed such a scenario and proposed an improved version of the fused lasso, called absolute fused lasso. Such a penalty ($\Omega(w) = \lambda_1 \|w\|_1 + \lambda_2 \sum_{i=1}^{p} \||w_i| - |w_{i=1}|\|$) penalizes the difference between the absolute value of features, and thus encourages the smoothness of both positively and negatively coregulated biomarkers. Solving such a penalty is challenging, because of its nonconvex nature, and therefore, the authors proposed to identify a locally optimal solution by utilizing the difference of convex function (DC) programming.

2.2 Knowledge incorporation via Bayesian methods

2.2.1 Bayesian statistical modeling

Bayesian methods have been used in biomarker discovery for a long time (Bhattacharjee, Botting, & Sillanpaa, 2008; Dridi et al., 2017; Hernández, Pennington, & Parnell, 2015; Yeung, Bumgarner, & Raftery, 2005). The flexibility of Bayesian methods allows researchers to embed additional information via priors. Such prior information could be SNP-SNP distances (Bhattacharjee et al., 2008), protein-protein correlations (Dridi et al., 2017), or robustness to irrelevant features (Yang & Song, 2010). A previous study (Dridi et al., 2017) considered a novel statistical model for biomarker identification using proteomics data. Instead of univariate testing of proteins, the authors aimed to identify an illness-associated partition of the protein set. In such a partition, case and control distributions were assumed to be different. The data of illness-irrelevant proteins were "collected" in an additional partition and explained by another distribution. The entire model has several advantages. First, the model considers multipredictor effects. For complex illnesses such as schizophrenia, associated features are not independently related to the disorder. Therefore, it is advantageous to test features as a group. Second, the colinearity between features is considered intrinsically. The nondiscriminant proteins are collected and explained in the second partition, while all correlated and discriminant proteins are selected by the first partition.

2.2.2 Bayesian supervised learning

Another interesting approach to Bayesian ML is based on the Bayesian Lasso (Park & Casella, 2008). Similar to Lasso, it assumes numerous disease-irrelevant biomarkers to exist in the data, and the algorithms aim to filter out such irrelevant features by introducing a sparseness penalty. In a Bayesian framework, the objective of Lasso can be seen as maximizing the posterior probability of the parameters, while the prior is the Laplace prior. However, Bayesian Lasso can output the full posterior distribution of the parameters instead of a point estimation. Therefore, the prediction can be performed considering all possible values of parameters with the corresponding probabilities, and thus the estimates of credible intervals are provided. There are many variant algorithms based on Bayesian Lasso (Cawley & Talbot, 2006; Li et al., 2011; Yang & Song, 2010). For example, one study (Yang & Song, 2010) considered a probit model with the g-prior for the coefficients' distribution using gene expression data. Then Markov Chain Monte Carlo was used to calculate the posterior distribution of gene coefficients. The method was tested on colon cancer and leukemia data, and the performance was found to be comparable to that of other popular methods, such as SVM. The method selected tens of genes among a total of 2000 genes, and the selection was shown to be robust against the initial starting point of the method (i.e., an initialized risk gene indicator). Another study (Cawley & Talbot, 2006) using a Bayesian approach to Lasso considered the combination of logistic regression with the Lasso penalty, and treated the sparsity controlling parameter as a random variable that was thus integrated out. This led to efficient model selection (2–3 orders faster than the original methods). The authors tested the method on a colon cancer dataset, and obtained a substantial improvement of efficiency and comparable predictive performance.

2.2.3 Bayesian networks

Another interesting set of Bayesian methods for identification of biomarker candidates are Bayesian networks (Assawamakin et al., 2013; Sherif, Zayed, & Fakhr, 2015). Such methods treat biological predictors and diagnostic status as a set of random variables and attempt to train a probabilistic model with a directional acyclic graph structure to represent the conditional independence between variables. Similar to other machine learning tools, the model can output the most useful predictors and make inferences in new data. One study (Sherif et al., 2015) started from all SNPs of the top 10 Alzheimer-related genes identified by Bertram et al. (2007), and then tested four Bayesian networks with different structure constraints on case/control datasets: Naïve Bayes structure (NB), Tree augmented naïve Bayes (TANB), Markov Blanket (MB), and Minimal Augmented Markov Blanket (MAMB). The four structures embedded two different types of prior knowledge. Specifically, NB assumed no SNP-SNP interactions, and all SNPs participated in the model during estimation. TANB started from the NB structure and added the SNP-SNP interactions sequentially. MB and MAMB only considered a small set of SNPs who blocked the node of diagnostic status from the remaining SNPs. These SNPs can be seen as the intermediate "phenotypes" between diagnostic status and the remaining SNPs. On the basis of MB, MAMB additionally considered the SNP-SNP interactions. The models were tested using 10-fold cross-validation. The predictive accuracy was quite similar for the four methods, and was in the range from 60% to 65%. MB and MAMB achieved higher sensitivity (~80%) and lower specificity (~20%), whereas NB and TANB yielded balanced results (~60%). According to the different priors, NB and TANB captured more illness-associated predictors (~400) compared with the other methods (~10).

2.3 Knowledge incorporation via Kernel methods

2.3.1 Customized kernels to incorporate biological network information

A kernel is a mathematical mapping function in the form of $k(x, y)$, which can be seen as the similarity measure between any two subjects x and y (Schölkopf & Smola, 2002; Shawe-Taylor & Cristianini, 2004). To incorporate biological knowledge, a kernel matrix of all subjects is precalculated to encode such knowledge, and thus the classification in the feature space is transferred to the knowledge-related space. For example, several studies have incorporated biological network information into the kernel matrix, such as coexpression (Gao et al., 2009; Rapaport et al., 2007) and PPI networks (Nitsch et al., 2009). In such frameworks, the kernel is represented as $k(x, y) = x^T Q y$, $Q = p \times p$, where p is the number of the network's nodes (i.e., genes). Q is the kernel matrix encoding the network information. For example, when $Q = I$, no coexpression information is encoded, thus the classification is performed based on subject-subject similarity in the original feature space. Once Q is calculated from the coexpression data, subject-subject similarity considers coexpression information. Two subjects are similar if their high/low expressed genes are closely connected in the network. Regarding the choice of Q, there are two popular options: the diffusion kernel (Kondor & Lafferty, 2002) and the Laplacian-based kernel (Gao et al., 2009). The diffusion kernel measures the similarity between two nodes in a network by averaging all connected paths from one node to another. In another study (Rapaport et al., 2007), authors aimed to identify a set of expression profiles from irradiated and nonirradiated yeast strains incorporating metabolic network information. Here, Q was initialized to the Laplacian matrix of the metabolic network. By assuming biological information is concentrated in the low-frequency component, the large eigenvalues of Q were attenuated. The kernel has been compared with the linear kernel using SVM, and was found to achieve the same predictive accuracy, but the proposed kernel led to better interpretability. The proposed model indicated more biologically meaningful gene models, given the metabolic network.

2.3.2 Feature selection using nonlinear kernels

Another popular property of a kernel is nonlinear functional mapping. This property has the potential to improve the ability of the model to represent complex data. For biomarker identification, the most common choice is linear mapping, because of the associated ease of interpretability and the embedded feature selection. In nonlinear kernels, classification does not occur in the original feature space, but in a higher dimensional space. Thus, it is difficult to control the model's interpretability by tuning model complexity. Moreover, because the data of subjects x and y is replaced with $k(x, y)$, the subject vector works as a solid unit, and thus the general feature selection process does not work. However, it can be achieved by separating the learning procedure and feature selection. For example, one study (Cho, Lin, & Wang, 2014) achieved the highest rank in the psoriasis subchallenge of the IMPROVER (industrial methodology for process verification of research) challenge, launched by IBM and Philip Morris International. The authors used an interesting combination of techniques: kernel Fisher discriminant (KFD) (Mika et al., 1999) classifier, scaled alignment selection (SAS) (Ramona, Richard, & David, 2012), and recursive feature elimination (RFE). KFD, kernelized linear discriminant analysis, introduced nonlinear mapping to the classifier. SAS is a feature selection process specialized for the nonlinear kernel method. This method calculates features' importance by maximizing the consistency between the kernel matrix and yy^T, where y is the diagnosis vector. Such alignment indicates how suitable the kernel-defined similarity measure is to match to the ideal diagnostic similarity. At last, RFE was used to remove unimportant features. The authors used radial basis function as the kernel function, and achieved high predictive performance. However, such results depend on several strong assumptions. For example, the authors did not input the full set of genes (~12,000) to the classifier, but preselected 747 genes according to differential expression analysis, which decreased the difficulty for the classifier to locate the real hyperplane. All subjects were measured on the same microarray platform, so cross-platform batch effects did not have to be considered. In addition, the number of cases and controls were well balanced, and the subject size was sufficiently large (238).

3 Challenges

Data missingness is a challenge in all data analysis areas. According to the mechanism of missingness, we can roughly classify the problem into three categories: missing data at random, missing data conditional on observed predictors, and missing data conditional on unobserved predictors. It is increasingly difficult to impute the missingness along this sequence. Missing data at random: in this category, every missing value is independent, and not affected by other factors. Such random missingness widely occurs in data derived from high-throughput technologies. For example, in microarray experiments, dust and scratches on the slide contaminate the data randomly, and lead to suspicious values, which are usually declared as missing (Lee, 2004). Missing data conditional on observed predictors: in this class, data is not randomly missing, but can be "imputed" based on available data. Examples of this effect include missingness in genetic association data (Li et al., 2009). The SNPs with missing values can be imputed based on an external reference, and adjacent SNPs with known values due to LD effects (Marchini & Howie, 2010). Missing data conditional on unobserved predictors: this problem is the most difficult to address, as data missingness is caused by unknown factors, such as unobserved batch effects.

Data missingness is also a severe challenge in biomarker discovery because, during biological data acquisition and pre-processing, missingness is typically not generated by individual mechanisms, but by a combination of different mechanisms. For example, microarray experiments suffer from data missingness due to image noise, batch effects, and hybridization failures (Troyanskaya et al., 2001). Due to this complexity, suitable imputation of missing values depends on experience, and a combination of procedures, as well as trial and error. However, many strategies for data imputation have been developed in the last decade (Liew et al., 2011; Oba et al., 2003; Troyanskaya et al., 2001), and have a substantial benefit for downstream machine learning analysis. Because missingness is typically generated by a complex mechanism, it is more meaningful to categorize imputation methods by the information type required for resolving missingness, than the mechanism that gave rise to the missingness. Thus, imputation approaches can be categorized as methods based on correlation structure or on additional knowledge.

3.1 Imputation approaches based on correlation structure

Correlation structure in the data is an important piece of information for estimation of missing values. Assuming there is no extra data providing missingness-related information, estimation can only depend on similar entries with known values. Furthermore, according to the type of correlation structure used, three kinds of methods can be differentiated: global, local, and hybrid approaches.

3.1.1 Global approach

The global approach assumes a global "trend" existing in an entire data set that constrains all data values. Thus, imputation is performed by recognizing and utilizing this trend. Such an assumption is denominated differently in different areas, that is, low rank (Chen & Suter, 2004) in computer vision, and low entropy (Moorthy et al., 2014) in bioinformatics. This approach is suitable for imputing missing data generated randomly, because no local structures exists, and all data values are equally useful for imputation.

Most global approaches are based on orthogonal matrix decomposition, that is, the singular value decomposition impute algorithm (SVDimpute) (Troyanskaya et al., 2001) and Bayesian principle component analysis (BPCA) (Oba et al., 2003). SVDimpute is a classical approach, and has been applied successfully on gene expression data. The method is performed iteratively, and imputed values are updated until convergence. The initial value of missing entries is the row average of known values. In each iteration, the core step is to calculate the so-called eigengenes (Alter, Brown, & Botstein, 2000), and then to use these as predictors for estimating missing values. Eigengenes are genes associated with the top eigenvalues of the expression data matrix. In theory, any gene can be seen as the linear combination of eigengenes, thus missing value imputation is formulated as a linear regression problem.

3.1.2 Local approach

The local approach assumes that a few highly correlated subjects or features can provide significant missingness-related information, while the remaining data is not useful for this purpose. Thus, this approach is useful to impute missingness conditional on observed predictors, because only local correlation structures are considered as imputation references.

Most local approaches are based on distance-based methods, such as k-nearest neighbor imputation (KNNimpute) (Troyanskaya et al., 2001). Here, the distance can be the correlation coefficient, or the Euclidian distance between subject or feature pairs. The idea is to select k-nearest subjects/features around the target subject/feature with missing values, and then perform a weighted imputation according to the pairwise distance between the target subject/feature and its neighbor. Another widely used local approach is based on least square regression (LSimpute) (Bo, Dysvik, & Jonassen, 2004). For the target gene, the top k correlated genes are selected, and then univariate regression is performed using the target gene as responses, and the correlated genes as predictors. After that, k predicted missing values are linearly combined to obtain the final result.

3.1.3   Hybrid approach

For complicated imputation problems, such as microarray imputation, it is difficult to decide whether missingness is conditional on any factors. Therefore, a hybrid approach (Jornsten et al., 2005) may be helpful for analysis. This approach combines numerous popular local and global methods in a convex manner, such as SVDimpute (Troyanskaya et al., 2001), BPCA (Oba et al., 2003), and KNNimpute (Troyanskaya et al., 2001), and then learns an optimal weight allocation for each method, and imputes missingness using the combined model. The advantage of the hybrid approach is the automatic learning of missingness-related correlation structure in data, and thus provides a precise estimation according to the structure.

3.2   Imputation approach based on additional knowledge

All approaches described herein are data-driven, using data-intrinsic information for imputation, and are usually limited for several reasons. For example, the performance is limited when the percentage of missing values is large, the number of subjects is small, the data space is high-dimensional, or the data is substantially affected by noise. Moreover, when the missingness is conditional on unobserved predictors, data-driven approaches do not work well. To overcome these issues, one can incorporate additional biological knowledge during imputation. Thus, imputation may remove variance in the data, and estimate values more accurately. Such a strategy is used in several methods that differ depending on the specific knowledge type, such as gene ontology information (Tuikkala et al., 2006) or regulatory mechanism (Xiang et al., 2008).

For the former method (Tuikkala et al., 2006), a novel combined distance measure of gene pairs was used, and the measure was embedded into the KNNimpute framework. The proposed distance measure was a weighted product of data-driven distance and functional dissimilarity, with the latter being derived from GeneOntology information. For each gene, the associated biological process and molecular function were extracted from the GeneOntology, and then used for calculating functional dissimilarity. The data-driven distance was estimated as described herein. Using the knowledge of GeneOntology, the missing values can be estimated, even if the number of subjects is small, or the rate of missingness is large. The authors suggested that the proposed method outperformed others when the subject size was smaller than 10, or the missing rate larger than 10%. The latter method (Xiang et al., 2008) was based on the assumption that coexpressed genes have similar histone acetylation values, and this was incorporated into the LSimpute framework. Specifically, clustering was performed over genes using histone acetylation features. Then the mean expression value of each cluster was used as a predictor for univariate regression and, therefore, in the same cluster, missingness of target genes could be estimated.

4   Conclusion

The integration of multiple types of biological knowledge has evolved as a crucial step during biomarker discovery of complex illnesses. Here, we have discussed ML frameworks to incorporate such biological knowledge, and highlighted benefits regarding selection of useful predictors, model interpretability, and the possibility of addressing challenges arising from missing values. It should be noted that ML analysis of complex data may profit from a combination of the approaches presented here. For example, Bayesian supervised learning with regularization (Bayesian Lasso (Park & Casella, 2008)) provides an estimation of posterior distributions and structural constraints of features. Multitask multiple kernel learning (Jawanpuria & Nath, 2011; Widmer et al., 2015; Williams et al., 2004) combines regularization and kernel methods useful for multiomics analysis. Regularization has been used to exchange information across data modalities, and kernel methods have been applied due to their nonlinear mapping abilities. Regularized multitask learning has been applied to link genetic association to expression data for biomarker identification (Lin et al., 2014), and to combine expression, CNV, and mutation data for drug sensitivity prediction (Yuan et al., 2016). Furthermore, as an extension of the kernel method, multiple kernel learning has been applied to integrate genetic, demographic, and multiple types of neuroimaging data for biomarker identification in Alzheimer's disease (Ye et al., 2011). These examples demonstrate that the repertoire of machine learning methods is starting to be applied for complex medical questions, moving from single data sources to integrative analyses of multiple modalities. These efforts can profit from integration of external biological knowledge, and we anticipate ML approaches to make increasing use of the enormous, already available data resources.

Acknowledgment

The work contributing to this chapter was supported by the Deutsche Forschungsgemeinschaft (DFG), SCHW 1768/1-1.

References

Acuna, E., & Rodriguez, C. (2004). The treatment of missing values and its effect on classifier accuracy. In D. Banks, F. R. McMorris, P. Arabie, & W. Gaul (Eds.), *Classification, clustering, and data mining applications* (pp. 639–647). Berlin: Springer.

Aittokallio, T. (2010). Dealing with missing values in large-scale studies: Microarray data imputation and beyond. *Briefings in Bioinformatics, 11*(2), 253–264.

Alter, O., Brown, P. O., & Botstein, D. (2000). Singular value decomposition for genome-wide expression data processing and modeling. *Proceedings of the National Academy of Sciences of the United States of America, 97*(18), 10101–10106.

American Psychiatric Association. (2013). *Diagnostic and statistical manual of mental disorders (DSM-5®)*. American Psychiatric Pub.

Ashburner, M., et al. (2000). Gene ontology: Tool for the unification of biology. The Gene Ontology Consortium. *Nature Genetics, 25*(1), 25–29.

Assawamakin, A., et al. (2013). Biomarker selection and classification of "-omics" data using a two-step Bayes classification framework. *BioMed Research International, 2013.*

Bertram, L., et al. (2007). Systematic meta-analyses of Alzheimer disease genetic association studies: The AlzGene database. *Nature Genetics, 39*(1), 17–23.

Bhattacharjee, M., Botting, C. H., & Sillanpaa, M. J. (2008). Bayesian biomarker identification based on marker-expression proteomics data. *Genomics, 92*(6), 384–392.

Bo, T. H., Dysvik, B., & Jonassen, I. (2004). LSimpute: Accurate estimation of missing values in microarray data with least squares methods. *Nucleic Acids Research, 32*(3).

Byrne, M., et al. (2007). Obstetric conditions and risk of first admission with schizophrenia: A Danish national register based study. *Schizophrenia Research, 97*(1–3), 51–59.

Cawley, G. C., & Talbot, N. L. (2006). Gene selection in cancer classification using sparse logistic regression with Bayesian regularization. *Bioinformatics, 22*(19), 2348–2355.

Chávez, E., et al. (2001). Searching in metric spaces. *ACM Computing Surveys, 33*(3), 273–321.

Chen, P., & Suter, D. (2004). Recovering the missing components in a large noisy low-rank matrix: Application to SFM. *IEEE Transactions on Pattern Analysis and Machine Intelligence, 26*(8), 1051–1063.

Cho, J.-H., Lin, A., & Wang, K. (2014). Kernel-based method for feature selection and disease diagnosis using transcriptomics data. *Systems Biomedicine, 1*(4), 254–260.

Clarke, R., et al. (2008). The properties of high-dimensional data spaces: Implications for exploring gene and protein expression data. *Nature Reviews. Cancer, 8*(1), 37–49.

Cun, Y., & Frohlich, H. (2012). Biomarker gene signature discovery integrating network knowledge. *Biology (Basel), 1*(1), 5–17.

Das, S., et al. (2015). Applications of artificial intelligence in machine learning: Review and prospect. *International Journal of Computer Applications, 115*(9), 31–41.

Dridi, N., et al. (2017). Bayesian inference for biomarker discovery in proteomics: An analytic solution. *EURASIP Journal on Bioinformatics and Systems Biology, 2017*(1), 9.

Friedman, J., Hastie, T., & Tibshirani, R. (2001). The elements of statistical learning. *Springer series in statistics: Vol. 1.* New York: Springer-Verlag New York.

Gao, C., et al. (2009). Graph ranking for exploratory gene data analysis. *BMC Bioinformatics, 10*(Suppl. 11), S19.

Goodfellow, I., et al. (2016). *Deep learning. Vol. 1.* Cambridge: MIT Press.

Haddad, L., et al. (2015). Brain structure correlates of urban upbringing, an environmental risk factor for schizophrenia. *Schizophrenia Bulletin, 41*(1), 115–122.

He, Z., & Yu, W. (2010). Stable feature selection for biomarker discovery. *Computational Biology and Chemistry, 34*(4), 215–225.

Hernández, B., Pennington, S. R., & Parnell, A. C. (2015). Bayesian methods for proteomic biomarker development. *EuPA Open Proteomics, 9*, 54–64.

Hoerl, A. E., & Kennard, R. W. (1970). Ridge regression: Biased estimation for nonorthogonal problems. *Technometrics, 12*(1), 55.

Jain, A. K., Duin, P. W., & Jianchang, M. (2000). Statistical pattern recognition: A review. *IEEE Transactions on Pattern Analysis and Machine Intelligence, 22*(1), 4–37.

Jawanpuria, P., & Nath, J. S. (2011). Multi-task multiple kernel learning. In *SIAM international conference on data mining* (pp. 828–838).

Joachims, T. (1998). Making large-scale SVM learning practical. In *Komplexitätsreduktion in Multivariaten Datenstrukturen*: Universität Dortmund. Technical report, SFB 475.

Jornsten, R., et al. (2005). DNA microarray data imputation and significance analysis of differential expression. *Bioinformatics, 21*(22), 4155–4161.

Kanehisa, M., et al. (2017). KEGG: New perspectives on genomes, pathways, diseases and drugs. *Nucleic Acids Research, 45*(D1), D353–D361.

Kim, S., & Xing, E. P. (2012). Tree-guided group lasso for multi-response regression with structured sparsity, with an application to eQTL mapping. *The Annals of Applied Statistics, 6*(3), 1095–1117.

Kondor, R. I., & Lafferty, J. (2002). Diffusion kernels on graphs and other discrete input spaces. In *ICML proceedings of the nineteenth international conference on machine learning* (pp. 315–322). San Francisco, CA: Morgan Kaufmann Publishers Inc.

Lee, M. L. T. (2004). Missing values in microarray data. In Analysis of microarray gene expression data. Boston, MA: Springer.

Li, C., & Li, H. (2008). Network-constrained regularization and variable selection for analysis of genomic data. *Bioinformatics, 24*(9), 1175–1182.

Li, Y., et al. (2009). Genotype imputation. *Annual Review of Genomics and Human Genetics, 10,* 387–406.

Li, J., et al. (2011). The Bayesian lasso for genome-wide association studies. *Bioinformatics, 27*(4), 516–523.

Libbrecht, M. W., & Noble, W. S. (2015). Machine learning applications in genetics and genomics. *Nature Reviews. Genetics, 16*(6), 321–332.

Liew, A. W., Law, N. F., & Yan, H. (2011). Missing value imputation for gene expression data: Computational techniques to recover missing data from available information. *Briefings in Bioinformatics, 12*(5), 498–513.

Lin, D., et al. (2013). Group sparse canonical correlation analysis for genomic data integration. *BMC Bioinformatics, 14,* 245.

Lin, D., et al. (2014). Integrative analysis of multiple diverse omics datasets by sparse group multitask regression. *Frontiers in Cell and Development Biology, 2,* 62.

Liu, J., et al. (2013). Accounting for linkage disequilibrium in genome-wide association studies: A penalized regression method. *Statistics and Its Interface, 6*(1), 99–115.

Ma, S., Song, X., & Huang, J. (2007). Supervised group Lasso with applications to microarray data analysis. *BMC Bioinformatics, 8,* 60.

Majumder, B., et al. (2015). Predicting clinical response to anticancer drugs using an ex vivo platform that captures tumour heterogeneity. *Nature Communications, 6,* 6169.

Marchini, J., & Howie, B. (2010). Genotype imputation for genome-wide association studies. *Nature Reviews. Genetics, 11*(7), 499–511.

Meier, L., Van De Geer, S., & Bühlmann, P. (2008). The group lasso for logistic regression. *Journal of the Royal Statistical Society: Series B, 70*(1), 53–71.

Mika, S., et al. (1999). Fisher discriminant analysis with kernels. In *Proceedings of the 1999 IEEE signal processing society workshop* (pp. 41–48). Madison, WI: IEEE.

Moorthy, K., Saberi Mohamad, M., & Deris, S. (2014). A review on missing value imputation algorithms for microarray gene expression data. *Current Bioinformatics, 9*(1), 18–22.

Mortensen, P. B., et al. (1999). Effects of family history and place and season of birth on the risk of schizophrenia. *The New England Journal of Medicine, 340*(8), 603–608.

Nitsch, D., et al. (2009). Network analysis of differential expression for the identification of disease-causing genes. *PLoS One, 4*(5).

Oba, S., et al. (2003). A Bayesian missing value estimation method for gene expression profile data. *Bioinformatics, 19*(16), 2088–2096.

Park, T., & Casella, G. (2008). The Bayesian Lasso. *Journal of the American Statistical Association, 103*(482), 681–686.

Phan, J. H., et al. (2012). Multiscale integration of -omic, imaging, and clinical data in biomedical informatics. *IEEE Reviews in Biomedical Engineering, 5,* 74–87.

Ramona, M., Richard, G., & David, B. (2012). Multiclass feature selection with kernel Gram-matrix-based criteria. *IEEE Transactions on Neural Networks and Learning Systems, 23*(10), 1611–1623.

Rapaport, F., et al. (2007). Classification of microarray data using gene networks. *BMC Bioinformatics, 8,* 35.

Schölkopf, B., & Smola, A. J. (2002). *Learning with kernels: Support vector machines, regularization, optimization, and beyond.* MIT Press.

Sebastiani, F. (2002). Machine learning in automated text categorization. *ACM Computing Surveys, 34*(1), 1–47.

Shawe-Taylor, J., & Cristianini, N. (2004). *Kernel methods for pattern analysis.* Cambridge University Press.

Sherif, F. F., Zayed, N., & Fakhr, M. (2015). Discovering Alzheimer genetic biomarkers using Bayesian networks. *Advances in Bioinformatics, 2015,* 639367.

Shultz, T. R., et al. (2011). *Curse of dimensionality* (pp. 257–258). Cornell University.

Silver, M., Montana, G., & Alzheimer's Disease Neuroimaging Initiative. (2012). Fast identification of biological pathways associated with a quantitative trait using group lasso with overlaps. *Statistical Applications in Genetics and Molecular Biology, 11*(1), 7.

Simon, N., et al. (2013). A sparse-group lasso. *Journal of Computational and Graphical Statistics, 22*(2), 231–245.

Singal, A. G., et al. (2013). Machine learning algorithms outperform conventional regression models in predicting development of hepatocellular carcinoma. *The American Journal of Gastroenterology, 108*(11), 1723–1730.

Tibshirani, R. (2011). Regression shrinkage and selection via the lasso: A retrospective. *Journal of the Royal Statistical Society: Series B, 73*(3), 273–282.

Troyanskaya, O., et al. (2001). Missing value estimation methods for DNA microarrays. *Bioinformatics, 17*(6), 520–525.

Tuikkala, J., et al. (2006). Improving missing value estimation in microarray data with gene ontology. *Bioinformatics, 22*(5), 566–572.

Widmer, C., et al. (2015). *Framework for multi-task multiple kernel learning and applications in genome analysis.* arXiv preprint arXiv:1506.09153 Cornell University.

Williams, N. M., et al. (2004). Support for RGS4 as a susceptibility gene for schizophrenia. *Biological Psychiatry, 55*(2), 192–195.

Xiang, Q., et al. (2008). Missing value imputation for microarray gene expression data using histone acetylation information. *BMC Bioinformatics, 9,* 252.

Yang, A. J., & Song, X. Y. (2010). Bayesian variable selection for disease classification using gene expression data. *Bioinformatics, 26*(2), 215–222.

Yang, T., et al. (2016). Absolute fused lasso and its application to genome-wide association studies. In *KDD 2016—Proceedings of the 22nd ACM SIGKDD international conference on knowledge discovery and data mining* (pp. 1955–1964).

Ye, J., et al. (2011). Machine learning approaches for the neuroimaging study of Alzheimer's disease. *Computer, 44*(4), 99–101.

Yeung, K. Y., Bumgarner, R. E., & Raftery, A. E. (2005). Bayesian model averaging: Development of an improved multi-class, gene selection and classification tool for microarray data. *Bioinformatics, 21*(10), 2394–2402.

Yu, J. S., et al. (2016). A support vector machine model provides an accurate transcript-level-based diagnostic for major depressive disorder. *Translational Psychiatry*, *6*(10), e931.

Yuan, H., et al. (2016). Multitask learning improves prediction of cancer drug sensitivity. *Scientific Reports*, *6*, 31619.

Zhou, H., et al. (2010). Association screening of common and rare genetic variants by penalized regression. *Bioinformatics*, *26*(19), 2375–2382.

Zou, H., & Hastie, T. (2005). Regularization and variable selection via the elastic net. *Journal of the Royal Statistical Society: Series B*, *67*(2), 301–320.

Further reading

Hastie, T., Tibshirani, R., & Friedman, J. (2009a). *The elements of statistical learning* (2nd ed.). Springer.

Hastie, T., Tibshirani, R., & Friedman, J. (2009b). *The elements of statistical learning*. Springer Science & Business Media.

Chapter 12

Personalized psychiatry with human iPSCs and neuronal reprogramming

Cedric Bardy[a,b], Zarina Greenberg[a], Seth W. Perry[c] and Julio Licinio[d]

[a]South Australian Health and Medical Research Institute (SAHMRI), Adelaide, SA, Australia, [b]The College of Medicine and Public Health, Flinders University, Adelaide, SA, Australia, [c]Psychiatry, and Neuroscience & Physiology; College of Medicine, SUNY Upstate Medical University, Syracuse, NY, United States, [d]Psychiatry, Neuroscience & Physiology, Pharmacology, and Medicine; College of Medicine, SUNY Upstate Medical University, Syracuse, NY, United States

1 Introduction

Human induced pluripotent stem cells (iPSCs) are, in a nutshell, a type of pluripotent stem cell that have been artificially derived (i.e., induced or "reprogrammed") from a mature (i.e., post-differentiated, non-pluripotent) human cell. Their major conceptual and practical advance over previously available stem cell technologies is their capacity to be derived from non-embryonic human tissue—that is, iPSCs can be derived from many types of cells that are easily obtained from adult human volunteer donors—thus unexpectedly demonstrating that fully mature adult cells can be "reversed" or "reprogrammed" to behave like embryonic stem cells capable of developing into all cell types of the body, and also significantly ameliorating ethical concerns of embryonic stem cell technologies in the process. This discovery has revolutionized medical research, and earned the 2012 Nobel Prize in Physiology or Medicine for its inventor Dr. Shinya Yamanaka, and his predecessor Dr. John B. Gurdon, who paved the way in 1962 by successfully cloning a frog using the nucleus from an adult frog cell inserted into a frog egg cell, thus showing that mature adult cells contained all the information necessary for creating any cells of the organism (Gurdon, 1962; Takahashi et al., 2007; Takahashi & Yamanaka, 2006). That all of the work to be described here has occurred since Takahashi and Yamanaka's (2006) pioneering and Nobel prize winning discovery of iPSC techniques is a testament to just how rapidly this field has grown, and how important it is to personalized psychiatry and medicine in general (Papapetrou, 2016).

This chapter will begin with a historical and contemporary overview of the importance of iPSC methods and technologies to personalized psychiatry, followed by a discussion of some of the challenges and opportunities for their application to personalized psychiatry and medicine. Next we will provide a focused exploration of how iPSC technologies have thus far been used to model and better understand psychiatric diseases, and finally, we will close by addressing what we consider to be the key needs and opportunities for maximizing the impact of iPSC research and technology on personalized psychiatry.

1.1 Psychiatry needs precision medicine

All disease therapies benefit from advances in precision medicine, but psychiatry is one of the medical fields that may need it the most. The etiology of psychiatric disorders lies in a dynamic combination of genetic predispositions, epigenetic factors, and brain plasticity. Despite decades of efforts to categorize psychiatric disorders to improve and standardize treatments (e.g., diagnostic and statistical manual of mental disorders, DSM-V), each psychiatric disorder is most accurately described as a spectrum. The severity and diversity of psychiatric symptoms are unique to individuals, and require personalized treatments.

Individual genetic predispositions result in vast phenotypical changes that require specific therapeutics. These findings have led to the pursuit of precision medicine and pharmacogenomics. The concept of personalized medicine is rooted in the idea that individuals possess unique genetic, epigenetic, physiological, and molecular predispositions, and therefore patient-tailored interventions are required for the optimal care of patients, and also to accurately evaluate novel treatment outcomes in clinical trials. With rapidly increasing accessibility to DNA sequencing and proteomics analysis, precision

Personalized Psychiatry. https://doi.org/10.1016/B978-0-12-813176-3.00012-2

medicine is already becoming routine clinical practice in several areas of medicine, including oncology (Gil, Laczmanska, Pesz, & Sasiadek, 2018; Schilsky, 2010). Currently, the majority of personalized medicine revolves heavily around symptomatic and DNA profiling. Patient-centric neural cell reprogramming approaches provide new opportunities to implement precision medicine, and may prove particularly useful in brain disorders. This method can take into consideration patients' genetic variations, helping us to elucidate the dynamic links between genotype, clinical phenotypes, and neuronal phenotypes in psychiatric disorders. Herein we examine the use and future of cell reprogramming technologies, specifically iPSCs, for personalized psychiatry.

Moreover, a difficult challenge in treating psychiatric disorders today is the high degree of variability in drug efficacy among patients. Despite the availability of a broad range of pharmacological interventions, the heterogeneous nature of these disorders, along with individual variabilities, result in inconsistent patient outcomes. Trial and error strategies for identifying effective drugs for patients are standard practice, and it can take months or years to figure out appropriate treatment regimes. Given the rates of non-compliance and the high-risk nature of some patients with psychiatric disorders, an extended unmedicated (or suboptimally medicated) time frame may have hazardous consequences (Bockting et al., 2008; Weich, Nazareth, Morgan, & King, 2007). Inappropriate treatment for an individual may trigger adverse side effects without significant improvement in symptoms (Kane, Kishimoto, & Correll, 2013; Lim et al., 2012; Liu-Seifert, Adams, & Kinon, 2005). Ineffective therapeutics also exacerbate economic and societal burdens. The advent of high-throughput genomic analysis has identified a large amount of inter-individual variation, and highlighted the importance of considering a patient's genetics to guide the choice in treatment.

1.2 Limitations of existing (non-iPSC) models

Our current understanding of psychiatric disorders relies heavily on research performed in post-mortem patient brains and animal models, which, despite clear demonstrated values, have significant shortcomings. Human post-mortem brain tissue only offers a glimpse into the late stages of a disease, and reveals little of the dynamics or pathogenesis. Animal models provide valuable insights into the disease pathogenesis, and have been useful in ascertaining the function of particular causal genes and cellular signaling pathways; however, there is a clear gap between the neurobiology of rodents and humans, particularly at the psychiatric level (Kaiser & Feng, 2015). Psychiatric symptoms arise from aberrant high order functioning, or dysfunction in limbic processing, and few symptoms can be outwardly observed and quantified, which makes accurate diagnosis challenging in humans, and arduous to transpose to rodents. Furthermore, psychiatric disorders rarely originate from a monogenic cause. Instead, they typically arise from several genetic mutations that synergistically drive or predispose the underlying pathophysiology (Gratten, Wray, Keller, & Visscher, 2014). Multigenic diseases are hard to recapitulate in animal models. Because of the shortcomings in past experimental models, slow progress has been made in translating basic neurobiological findings to improve patient outcomes in psychiatry. New assays of personalized neuronal tissue, generated from patient-derived iPSCs, provide unprecedented opportunities to overcome the limitations of animal and postmortem models.

2 iPSC-derived neurons for personalized psychiatry: Opportunities and challenges

2.1 Genetics, epigenetics, and iPSCs in psychiatric disease

Our fundamental understanding of the biological mechanisms underlying psychiatric disorders remains tenuous. To this date, the field of psychiatry still endures slow translational advances with minimal effective treatments available. Today's most widely used antipsychotic drugs, such as chlorpromazine and haloperidol, were the result of serendipitous findings more than 60 years ago, rather than a detailed understanding of the pathophysiology of psychosis (Baumeister, 2013; Preskorn, 2007; Watmuff et al., 2016). However, progress in human genomics provides new insights into the biological roots of psychiatric diseases, and has established links with multi-genic polymorphisms (Glatt & Lee, 2016; Licinio & Wong, 2011; Reynolds, McGowan, & Dalton, 2014; Wong, Dong, Andreev, Arcos-Burgos, & Licinio, 2012; Wong, Dong, Maestre-Mesa, & Licinio, 2008; Xia et al., 2018). The genetic burden associated with the risk factors for developing a psychiatric disorder may vary for individuals, but genetic predispositions, coupled with epigenetic influences, are generally well accepted to constitute the initial causes of most psychiatric disorders, including major depression, bipolar disorder, autism, and schizophrenia (Gratten et al., 2014; McGuffin et al., 2003; Plomin, Owen, & McGuffin, 1994; Sullivan, Kendler, & Neale, 2003). An in-depth understanding of genomic studies is very important, and strongly tied with the use of iPSCs, because the value of iPSC models relies on the presence of genomic predispositions captured in the cells that contain the full genetic background of the donor. Using monogenic mutant cell lines, or animal models with isogenic

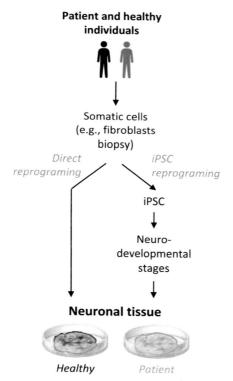

FIG. 1 Generating live human brain tissue with cell reprogramming technologies. Individual-specific neuronal cells can be generated through iPSC reprogramming or via direct conversion. Somatic cells taken from patients and healthy controls can be converted to iPSCs via exposure to specific transcription factors (such as Oct4, Sox2, Klf4, and c-Myc) (right-hand pathway). From this stage, iPSCs can be converted to a variety of neuronal and/or glial subtypes based on exposure to different culture conditions. Alternatively, somatic cells taken from patient biopsies can bypass pluripotency and be directly converted into neuronal tissue (left-hand pathway). Both pathways allow for direct comparison of neurons derived from both patients and healthy controls in vitro in a variety of applications.

controls, are valuable to study a variety of diseases. However, psychiatric disorders are rarely caused by a single highly penetrant mutation. For this reason, patient-derived iPSCs are particularly attractive for experimentally harnessing the complex genomic background of brain disorders.

However, because epigenetic factors are thought to play a significant role in psychiatric diseases, one potential disadvantage of iPSCs is that the epigenetic, i.e., "memory" of the cells may be "erased" during the reprogramming process (Lee, Hore, & Reik, 2014; Nashun, Hill, & Hajkova, 2015), to varying degrees dependent upon donor cell type (Kim et al., 2011) and methodology (Kim et al., 2010). iPSC sister technologies that allow "direct" reprogramming of somatic cells to a transdifferentiated state without need for a pluripotent cell intermediate (Fig. 1)—e.g., the generation of functional-induced neurons (iNs) directly from fibroblasts (Vierbuchen et al., 2010)—have been shown to maintain much of the cells' original epigenetic "memory" (Yang, Ng, Pang, Sudhof, & Wernig, 2011), and for these reasons may be a useful complementary method for some personalized psychiatry applications. However, iNs are not without their own limitations and trade-offs (discussed in Kalman, Hathy, & Rethelyi, 2016; Soliman, Aboharb, Zeltner, & Studer, 2017), as is any model system, and to our knowledge they have not yet been widely applied to psychiatric research. Nonetheless, this similar technology may have future applications in personalized psychiatry, and therefore we will touch upon it briefly herein.

2.1.1 iPSC technology, step 1: Reprogramming of patient-derived somatic cells to pluripotent states (iPSCs)

Embryonic stem cells (ESCs) have the potential to become any somatic cell type (Chen & Lai, 2015; Thomson et al., 1998). However, once ESCs differentiate, they lose their pluripotent capacity. Yamanaka, Takahashi, and colleagues demonstrated in a landmark study that adult fibroblasts could be reprogrammed into a pluripotent cellular state, which they called induced pluripotent stem cells (iPSCs) (Takahashi et al., 2007; Takahashi & Yamanaka, 2006). Like ESCs, iPSCs possess the ability to self-renew and differentiate into any somatic cell type. Pluripotency can be induced in biopsied cells, such as human fibroblasts, by upregulating four key transcription factors known to have essential roles during early development:

Oct4, Sox2, Klf4, and c-Myc. The discovery of iPSC technology has generated tremendous hope in the field of medical research for at least two reasons: (1) iPSCs bypass some of the bioethical and practical limitations surrounding the use of embryonic tissues and (2) unlike ESCs, iPSCs can be generated from adult individuals, such as psychiatric patients and matched healthy subjects (Fig. 1).

Since their inception, iPSC reprogramming technologies have undergone continuous revisions and improvements, including increases in efficacy and quality (Beers et al., 2015; Malik & Rao, 2013). Initially, retrovirus transduction was used to upregulate the necessary factors for the cells to revert to a pluripotent state. Because retrovirus results in viral integration into the host genome, several papers have since expanded on different methods of transduction to reduce the genomic footprint of the reprogramming process, while maintaining or improving efficiency and reliability (Beers et al., 2015; Carey et al., 2009; Sommer et al., 2009). For example, integration-free methods, such as the Sendai virus, plasmids, or small molecules, are now more frequently used (Chen et al., 2013; Fusaki, Ban, Nishiyama, Saeki, & Hasegawa, 2009; Nishimura et al., 2017). Rigorous cell quality control with the additional safeguards and optimizations may increase costs and the length of the reprogramming process, but are essential for translational applications. Skin cell punch biopsy is most commonly used to obtain the initial patient-derived somatic cell sample, however, other (non-skin) somatic cell types may also be used. The choice of primary cell types might be guided by the importance of minimally invasive techniques, especially when dealing with donors suffering from paranoia and psychosis. Also, because certain cell types used for generating iPSCs can introduce epigenetic changes that may not be reflective of patient neurons, a tradeoff needs to be evaluated prior to each study (González, Boué, & Belmonte, 2011; Mertens, Marchetto, Bardy, & Gage, 2016).

2.1.2 iPSC technology, step 2: Generating neurons relevant for psychiatric disorders from iPSCs

Neuronal tissue can be generated from iPSCs with several protocols. (Herein, for simplicity, we will refer to neurons so derived as "iPSC-neurons.") A popular strategy consists of mimicking, in vitro, the sequence of human neural development that occurs in the embryo and fetus, from the blastocyst (iPSC embryonic bodies), neural tube (neural rosette-like progenitors), and radial glia (neural glia progenitors), to the brain (neurons and astrocytes). Another widely used strategy is the direct reprogramming of fibroblasts into neurons, which bypasses pluripotency and most neuronal developmental stages. This latter approach (Fig. 1) significantly reduces the amount of time required to generate neurons, with a tradeoff on other aspects that we have discussed elsewhere (Mertens et al., 2016) (see iNs in Section 2.1). Both iPSC differentiation and direct conversion have been used to generate a variety of neuronal subtypes, including GABAergic, glutamatergic, dopaminergic, and serotonergic neurons, and glial cells, usually by manipulation of a few key transcription factors and signaling pathways (Table 1). For example, we developed a protocol to differentiate serotoninergic neurons from human fibroblasts (Vadodaria et al., 2016), because serotonergic circuits are implicated in major depression, and many other brain disorders, but had previously been very difficult to recapitulate in vitro. Moreover, advances in stem cell protocols have allowed for more sophisticated modeling by recapitulating the molecular signatures of various brain regions affected in disease (Arber et al., 2015; Miskinyte et al., 2017; Sarkar et al., 2018; Shi et al., 2012; Victor et al., 2014; Yu et al., 2014). Cortical neurons were one of the first regional models generated (Shi et al., 2012), with the cortex, particularly the prefrontal and orbitofrontal cortices, being heavily implicated in psychiatric disorders.

Since then, several other specialized brain regions important to psychiatric disease pathology have been modeled with iPSCs. Yu and colleagues used iPSCs to model the hippocampal dentate gyrus, which is responsible for learning and

TABLE 1 Signalling factors used for differentiation into specific neuronal cell types.

Neural cell type	iPSC methods	iN methods
Dopaminergic	FGF2, FGF8, SHH (Ma, Liu, & Zhang, 2011)	MASH1, NURR1, LMX1A (Caiazzo et al., 2011)
GABAergic	ASCL1, DLX2, NKX2.1, LHX6 (Sun et al., 2016)	ASCL1, DLX2 (Yang et al., 2017)
Glutamatergic	RA, SMAD inhibition (Shi, Kirwan, Smith, Robinson, & Livesey, 2012)	BRN2, MYT1L, FEZF2 (Miskinyte et al., 2017)
Serotonergic	SHH, FGF4, WNT (Lu et al., 2016)	NKX2.2, FEV, GATA2, LMX1B, ASCL1, NGN2 (Vadodaria et al., 2016; Xu et al., 2016)
Astrocytes	FGFs, RA, CNTF (Krencik & Zhang, 2011)	NFIA, NFIB, SOX9 (Caiazzo et al., 2015)
Oligodendrocytes	RA, PDGF, NT3 (Gorris et al., 2015)	SOX10, OLIG2, ZFP536 (Yang et al., 2013)

3   Using iPSCs for psychiatric disease modeling

Psychiatric disease modeling using iPSCs generally falls into one of two broad categories: (1) Using iPSCs derived from healthy donors to study the impact of established drugs or therapies on molecular mechanisms known or believed to be pertinent to the disease under investigation, or (2) Comparing iPSCs between healthy-diseased or diseased-diseased donors to uncover differences in molecular mechanisms, physiology, phenotypes, and/or response to therapies that can inform the quest for more targeted or effective treatments for psychiatric disease. It is this latter application, depicted in Fig. 2, that may yield the most direct or immediate benefits to personalized psychiatry, by uncovering disease-relevant phenotypic or mechanistic differences between healthy donors and those suffering from psychiatric illnesses (Fig. 2A); identifying drug responders vs nonresponders, and the pathways that predict drug response, in those with a particular disease (Fig. 2B); and/or by identifying subtypes or individually unique phenotypes or biology among patients with the disease (Fig. 2C). This, in turn, will help guide which existing drugs or other therapies may be most effective on an individual or population basis, as well as identify novel molecular targets and therapies with improved efficacy and tolerability.

Thus far, a significant portion of the work on iPSCs for modeling psychiatric disease has focused on schizophrenia, bipolar disorder, and autism spectrum disorder (ASD), and yielded a wealth of information that has significantly advanced our understanding of these disorders (Table 2). Fewer studies have utilized iPSC technologies to directly explore depression and anxiety disorders, and these areas are ripe with opportunity and need. We will very briefly cover major advances in each of these areas, and refer the reader to excellent detailed reviews for further discussions.

3.1   iPSCs in bipolar disorder

Bipolar disorder (BD) is a psychiatric illness characterized by emotional highs (mania) and lows (depression) and is the sixth leading cause of disability worldwide, afflicting about 60 million people (Hoffmann, Sportelli, Ziller, & Spengler, 2018, and references therein), with an estimated prevalence and lifetime prevalence of about 1.5% and 2.4%, respectively (Yatham et al., 2018). It carries high rates of morbidity and mortality, accounting for 3.4%–14% of suicide deaths, and roughly 25% of BD patients attempt suicide (Hoffmann et al., 2018; Schaffer et al., 2015). There are few new or specific therapies for BD, and often around half of BD patients may not be effectively treated by the available medications (Nierenberg, 2010), making development of more effective BD therapies a critical need.

However, before there were, to our knowledge, any studies comparing iPSCs from healthy vs bipolar disorder (BD) subjects, as in the kinds of paradigms depicted in Fig. 2, the Haggarty group used iPSCs reprogrammed from a normal human fibroblast cell line to derive neural progenitor cells (NPCs) in high volume, with the potential to then be subsequently differentiated into large numbers of neuronal or glial cells (Zhao et al., 2012). Using this model, they reported demonstrating for the first time that human iPSC-derived NPCs had a functionally active Wnt/β-catenin pathway with expected responsivity to the well-established BD therapeutic lithium, and could be used in large numbers in a high throughput screening (HTS) assay to identify multiple novel and known small molecule compounds that modulated the Wnt/β-catenin pathway in this model system (Zhao et al., 2012). Lithium inhibits glycogen synthase kinase-three beta (GSK-3β) to activate Wnt/β-catenin signaling (Hedgepeth et al., 1997; Jope, 2003), and as such, these intertwined pathways are increasingly thought to be integral to BD pathophysiology (Valvezan & Klein, 2012). Hence, this work was an elegant early example of how some of the challenges of population heterogeneity and low cell numbers were overcome to provide a robust and scalable method for producing self-renewing and genomically stable NPC precursors, with the potential to be subsequently differentiated into large numbers of functional post-mitotic neurons (Zhao et al., 2012), and used for exploring questions relevant to psychiatric disease, particularly BD.

Two years later, the first reports using iPSCs derived from individuals with BD began to emerge. Several studies have identified changes in gene expression of numerous proteins related to neurodevelopment and synaptic function in iPSC-derived neurons from BD compared with healthy subjects (Bavamian et al., 2015; Chen et al., 2014; Kim et al., 2015; Madison et al., 2015; Wang et al., 2014), and a subset of these studies investigated and found that these differences seen in the BD subjects were partially or largely prevented by lithium treatment (Chen et al., 2014; Madison et al., 2015; Wang et al., 2014). These findings validate both the utility of iPSCs for modeling BD, and the importance of lithium-responsive pathways (e.g., Wnt, GSK-3β) in its molecular pathophysiology, and help identify additional novel therapeutic targets. Other elegant studies have found distinct differences in synaptic function between lithium responsive (LR) and lithium non-responsive (NR) BD subjects. Mertens et al. found hyperactive synaptic activity in BD, but not healthy iPSC-derived hippocampal neurons, that could be selectively normalized by lithium (with greater accompanying gene expression changes) in only LR subjects, but not in NR subjects (Mertens et al., 2015). Moreover, in a follow-up study, this group

TABLE 2 Psychiatric diseases modeled with iPSC technologies, organized by neuronal subtypes and brain regions studied.

Disease	Brain region affected	Neuronal types affected	Brain region modeled with iPSCs	Neuronal types modeled with iPSCs
Schizophrenia	Hippocampus (Lodge & Grace, 2005; Weiss et al., 2003) Cortex (Abi-Dargham, 2004; Pierri, Volk, Auh, Sampson, & Lewis, 2003) Prefrontal cortex (Farzan et al., 2010; Volk, Austin, Pierri, Sampson, & Lewis, 2000) Ventricles, striatum (Kegeles, Abi-Dargham, Frankle, et al., 2010; Kessler et al., 2009)	GABAergic (Beasley & Reynolds, 1997; Farzan et al., 2010; Perry et al., 1979; Volk et al., 2000; Yoon et al., 2010) Dopaminergic (Abi-Dargham, 2004; Abi-Dargham et al., 2000; Bogerts, Hantsch, & Herzer, 1983; Kegeles et al., 2010; Laruelle, Abi-Dargham, Gil, Kegeles, & Innis, 1999; Lodge & Grace, 2005) Glutamatergic (Javitt, 1987; Olney & Farber, 1995) Serotonergic (Stahl, 2018; Xia et al., 2018)	Hippocampus-CA3 (Sarkar et al., 2018) Hippocampus dentate gyrus PROX1 (Yu et al., 2014) Forebrain (Liu et al., 2013)	GABAergic (Brennand et al., 2011a; Liu et al., 2013; Sun et al., 2016) Dopaminergic (Brennand et al., 2011a; Robicsek et al., 2013) Glutamatergic (Robicsek et al., 2013)
Bipolar disorder	Amygdala (DelBello, Zimmerman, Mills, Getz, & Strakowski, 2004) Prefrontal cortex (Lyoo et al., 2006; Rajkowska, 2000) Hippocampus (Bertolino et al., 2003; Rajkowska, 2000) Basal ganglia (Caligiuri et al., 2003) Thalamus (Caligiuri et al., 2003; DelBello et al., 2004)	Pyramidal (Rajkowska, 2000; Rajkowska, Halaris, & Selemon, 2001) Glia (Rajkowska, 2000; Rajkowska et al., 2001) Serotonergic (Young, Warsh, Kish, Shannak, & Hornykeiwicz, 1994)	Hippocampus dentate gyrus PROX1 (Mertens et al., 2015; Stern, Santos, et al., 2018)	Hippocampal dentate gyrus PROX1 (Mertens et al., 2015; Stern, Santos, et al., 2018)
William syndrome	Hippocampal (Meyer-Lindenberg et al., 2005) Visual cortex (Mobbs et al., 2007) Orbitofrontal cortex (Meyer-Lindenberg et al., 2005) Amygdala (Capitão et al., 2011) Intraparietal sulcus (Meyer-Lindenberg et al., 2005)	Pyramidal (Lew, Brown, Bellugi, & Semendeferi, 2016)	Forebrain (Chailangkarn et al., 2016; Khattak et al., 2015)	Cortical layer V/VI (Chailangkarn et al., 2016; Khattak et al., 2015)
Autism	Prefrontal cortex (Goldberg et al., 2011) Amygdala (Schumann et al., 2004) Primary motor cortex (Nebel et al., 2014; Cerebellum (Ritvo et al., 1986; Yip, Soghomonian, & Blatt, 2007)	Purkinje cells (Ritvo et al., 1986; Yip et al., 2007) GABAergic (Yip et al., 2007)	Forebrain (Karina Griesi-Oliveira et al., 2015)	Forebrain (Karina Griesi-Oliveira et al., 2015)
Major depression	Anterior cingulate (Cotter, Mackay, Landau, Kerwin, & Everall, 2001) Hippocampus (Klumpers et al., 2010)	Glia (Cotter et al., 2001) Serotonergic (Åsberg et al., 1984; Coccaro, Siever, Klar, et al., 1989; Träskman, Åsberg, Bertilsson, & Sjüstrand, 1981) GABAergic (Klumpers et al., 2010) Pyramidal (Cotter et al., 2001)	Midbrain (Vadodaria et al., 2016)	Serotonergic (Lu et al., 2016; Vadodaria et al., 2016; Xu et al., 2016)
Anxiety	Anterior cingulate (Etkin, Prater, Hoeft, Menon, & Schatzberg, 2010; Thomaes et al., 2013) Hippocampus (Hettema et al., 2012) Thalamus (Giménez et al., 2012) Amygdala (Etkin et al., 2010; Schienle, Ebner, & Schäfer, 2011)	Serotonergic (Murphy, 1990) GABAergic (Klumpers et al., 2010)	Midbrain (Vadodaria et al., 2016)	Serotonergic (Lu et al., 2016; Vadodaria et al., 2016; Xu et al., 2016)

from Fred Gage's laboratory found that these hyper-excitable BD neurons had distinctly different electrophysiological properties in the LR vs NR groups, which enabled highly accurate prediction of a patient's responsiveness to lithium based on the electrophysiological "signature" of the neurons derived from their iPSCs (Stern, Santos, et al., 2018). This exciting and seminal finding is an excellent example of how iPSCs can be used to predict drug responsiveness for personalized psychiatry. Another seminal study using iPSC-neurons from LR, NR, control, and unrelated psychiatric disease donors identified collapsin response mediator protein-2 (CRMP2) as perhaps the most critical major molecular player underlying LR BD pathogenesis (Tobe et al., 2017). We refer the reader to several excellent reviews for very detailed discussions of these and other studies that have used iPSCs to model BD (Hoffmann et al., 2018; Liu, Lu, & Yao, 2017; Watmuff et al., 2016).

Studies such as these and others (Oedegaard et al., 2016) utilizing iPSCs to uncover the genetics and functional neurobiology of lithium responsivity in BD have provided exceptional insights that should lead to improved therapies in the coming years, and have reinforced the notion that differing responses to lithium or other mood stabilizers reflect multiple different disease mechanisms that contribute to BD pathology (Leckband, McCarthy, & Kelsoe, 2012). On the basis of their findings (Tobe et al., 2017), some even speculate "that cases of bipolar disorder that do not respond to the drug [lithium] are actually a different disease altogether" (Leckband et al., 2012). Whether this viewpoint will gain traction as more research unfolds remains to be seen.

3.2 iPSCs in depression and anxiety disorders

Depression and anxiety disorders, respectively, afflicted about 268 and 275 million people globally in 2016, compared with 40 and 21 million people for BD and schizophrenia, respectively (Ritchie & Roser, 2018). That's about nine times the number of people affected by depression and anxiety disorders combined, vs BD and schizophrenia combined. Depressive disorders are also the leading cause of disability worldwide, and a key trigger for suicide, which causes one million deaths per year, and rising. BD and schizophrenia are indeed devastating and serious disorders, also linked to high suicide rates, and by no means is it our intent to diminish their significance and toll on public health. At the same time, given these discrepancies in prevalence, it is perhaps surprising that, to our knowledge, no studies have yet sought to use iPSCs derived from depressed subjects to model depressive disorders, as has been done for BD and schizophrenia. For example, given the vast variability in patients' therapeutic responses to antidepressants, it would be worthwhile to investigate whether depression-derived iPSC-neurons can be used to model or predict antidepressant response, similar to what has been done for lithium using BD-derived iPSC-neurons.

This is not to say, however, that no progress has been made using iPSCs to model depression and anxiety disorders. Several studies have utilized iPSCs derived from healthy subjects to model mechanisms or drug responses relevant to depression. Using iPSCs from healthy donors differentiated into midbrain dopamine neurons, Collo et al. found that ropinirole and pramipexole—two dopamine D3 receptor agonists used to treat Parkinson's disease and as adjunctive therapeutics for treatment resistant depression—dose-dependently increased dendritic arborization and soma size, likely through brain-derived neurotrophic factor (BDNF) dependent pathways (Collo et al., 2018). BDNF has been consistently implicated in depression, particularly treatment resistant depression (TRD), and increased BDNF signaling is believed to be central to ketamine's therapeutic effect on TRD (Allen et al., 2015; Haile et al., 2014; Lepack, Fuchikami, Dwyer, Banasr, & Duman, 2014). This work suggests ropinirole and pramipexole may act through similar mechanisms to benefit treatment resistant depression, and in fact, this same group, using the same human iPSC-dopaminergic neuron system, demonstrated similar neuroplasticity effects and mechanisms with ketamine (Cavalleri et al., 2018). Another group used human iPSC models (again, not from depressed subjects) to explore the potential mechanistic role of fibroblast growth factor 2 (FGF2) and its upstream and downstream mediators, and effects in major depressive disorder (MDD) (Gupta et al., 2018), and at least two other groups have used iPSC model systems to explore the toxicology of select antidepressants (Huang et al., 2017; Pei et al., 2016).

However, perhaps the most significant accomplishments to date involving the use human iPSCs in depression research have been methodological developments (Licinio & Wong, 2016). The serotonin system is integrally involved in depression and antidepressant actions, yet much like dopamine neurons, primary central nervous system serotonergic neurons have historically been difficult to reliably culture in large numbers, even from animals, let alone human sources. Surmounting these obstacles, dopamine neurons have now been cultivated from human iPSCs, and also recently, two groups independently reported methods for generating functionally active serotonergic (5HT) neurons from primary human fibroblasts (Vadodaria et al., 2016; Xu et al., 2016). Both teams used direct conversion methods (i.e., iN; Fig. 1, Fig. 3B), with Vadodaria et al. demonstrating that primary human dermal fibroblasts obtained from skin biopsies could be transdifferentiated to 5HT neurons by overexpression of four transcription factors (NKX2.2, FEV, GATA2, and LMX1B), along with ASCL1 and NGN2 (Vadodaria et al., 2016). Xu et al. showed that human lung fibroblasts (obtained from three different human lung fibroblast cell lines) could be induced to serotonergic neurons with just four transcription factors—ASCL1, FOXA2, LMX1B, and FEV (AFLV) (Xu et al., 2016)—three of which were shared with the Vadodaria method, and one (FOXA2) different. These recently developed methods will, for example, facilitate personalized prediction of drug response for serotonergic antidepressants, as has been done for lithium in bipolar (Stern, Santos, et al., 2018), and open a realm of exciting opportunities for understanding and treating depression (Vadodaria, Stern, Marchetto, & Gage, 2018).

3.3 iPSCs in schizophrenia

Both BD and schizophrenia (SCZ) are understood to have complex genetic, epigenetic, and neurodevelopmental origins, and accordingly, use of iPSCs to model SCZ has largely mirrored the work done with iPSCs in BD. As such, we will not discuss these studies in great detail here, as they have been touched upon in the preceding sections and tables herein, and have been thoroughly reviewed elsewhere. However, we will highlight a few general concepts. Several groups have identified abnormalities in gene expression, neurodevelopment, mitochondrial function, oxidative stress, neuronal connectivity (typically reduced), neuronal activity, and other elements of synaptic function in iPSC-neurons derived from individuals with SCZ compared with healthy controls (reviewed in Ahmad, Sportelli, Ziller, Spengler, & Hoffmann, 2018; Liu, Lu, & Yao, 2017; Watmuff et al., 2016). In these models, a few studies have had success blocking some of their observed effects with drugs sometimes prescribed for SCZ such as loxapine (Brennand et al., 2011b) or valproate (Paulsen Bda et al., 2012), but perhaps less reliably so than has been demonstrated for lithium in the BD iPSCs experiments described in the previous section. For example, of five antipsychotics tested—clozapine, loxapine, olanzapine, risperidone, and thioridazine—only loxapine ameliorated the reduced neuronal connectivity and altered gene expression seen in the SCZ-derived vs healthy-derived iPSC-neurons (Brennand et al., 2011b). Intriguingly, this suggests that loxapine's therapeutic effect in this model system may be mediated by yet-unknown off-target effects, which in turn may suggest an even greater etiogenic and mechanistic complexity for SCZ than previously imagined. Both of these possibilities require further exploration in future experiments such as these using human iPSCs to uncover the neurobiology of SCZ.

3.4 iPSCs in autism spectrum disorder

As defined in the DSM-5, autism spectrum disorder (ASD) is a neurodevelopmental disorder characterized by persistent abnormalities or deficits in social communication and interaction, and restricted and/or repetitive patterns of behavior, interests, or activities. Intellectual or language disabilities may or may not be present (American Psychiatric Association, 2013). It is, in effect, a disorder characterized or defined by a pattern of related behaviors or phenotypes, but that may have multiple causes or etiologies of known (traditionally referred to as syndromic ASD) or unknown (traditionally referred to as non-syndromic or idiopathic ASD) origin. As with BD and SCZ, most cases of ASD are believed to arise from an interaction of multiple genetic, environmental, and epigenetic factors; that is, they are nonsyndromic or idiopathic ASD. Some cases of ASD arise from known genetic mutations that also cause other identified syndromes, such as Rett syndrome, Fragile X syndrome, and Timothy syndrome; that is, they are syndromic ASD. However, this "syndromic" and "nonsyndromic" nomenclature system for ASD can quickly become both complex and blurred, leading some to propose alternative classification systems (Fernandez & Scherer, 2017). For the purposes of our discussions on iPSCs and personalized psychiatry herein, a more useful division is simply whether the ASD's origin is monogenic (i.e., resulting from a single known gene mutation) or polygenic (resulting from multiple known or unknown genetic influences) in nature.

Both monogenic and polygenic ASD have been modeled by human iPSCs, yielding useful insights for understanding ASD disease mechanisms and personalized psychiatry. Using iPSC-neurons derived from patients with monogenic ASDs such as Rett, Fragile X, Timothy, and Phelan-McDermid syndromes, a number of studies have demonstrated deficits in iPSC-neuron synaptic architecture and function that often mimic clinical disease phenotypes, and have identified numerous potential genetic and molecular therapeutic targets that, in some cases, could rescue these aberrant ASD phenotypes (reviewed in Adegbola, Bury, Fu, Zhang, & Wynshaw-Boris, 2017; Brennand, Simone, Tran, & Gage, 2012; Falk et al., 2016; Kim, Kim, Oh, Lee, & Kim, 2016; Lim et al., 2015; Marchetto, Brennand, Boyer, & Gage, 2011; Shen, Yeung, & Lai, 2018; Vitrac & Cloez-Tayarani, 2018). In some respects, studies modeling monogenic ASD may have the potential to offer the most immediate benefits to personalized psychiatry and medicine, because the iPSCs derived from these monogenic ASD subjects will, by definition, carry a single identifiable genetic mutation (Yoo, 2015), which may therefore offer the most direct insights into disease mechanisms and targets that are explicitly responsible for the ASD phenotype. On the other hand, iPSC studies of polygenic ASD have also identified potential molecular targets for novel therapeutic intervention (Bury & Wynshaw-Boris, 2018; Liu et al., 2017). One such study has identified TRPC6 (transient receptor potential cation channel, subfamily C, member 6) as an attractive potential therapeutic target for non-syndromic (polygenic) ASD (Griesi-Oliveira et al., 2015). This and other research modeling ASD with human iPSCs has been covered in depth elsewhere (Ben-Reuven & Reiner, 2016; Brito, Russo, Muotri, & Beltrao-Braga, 2018). The ongoing challenge for ASD personalized psychiatry will be to uncover iPSC-neuron phenotypes or characteristics that predict drug response and/ or identify druggable targets that will provide novel treatments for ASD on either a population or individualized basis, as has thus far perhaps been done most successfully for BD.

4 Conclusions and future directions

Continuing to advance iPSC technology and methods for improved understanding of psychiatric diseases will benefit personalized medicine in at least three ways: (1) Identifying novel and clarifying known or suspected molecular disease mechanisms for developing more targeted and effective therapies, (2) better *individualized* prediction of response to drug therapies for psychiatric disease (Stern, Linker, Vadodaria, Marchetto, & Gage, 2018), and (3) conceivably, iPSCs could ultimately serve as stem cell therapies for severe or refractory psychiatric disease (Chen, Song, & Ming, 2018; Samoilova et al., 2018). With the exception of Section 2.3.4, most of our discussions herein have focused on the first two areas, while the latter area has thus far received relatively little attention in the literature, as applicable to psychiatric disease, and would benefit from further research. The comparative lack of attention to this third area thus far may reflect a perception that neurorestorative stem cell therapies for psychiatric illnesses are highly invasive and risky treatments of last resort. Yet, when we consider that large numbers of individuals worldwide suffer from severe and/or refractory forms of psychiatric diseases such as schizophrenia, bipolar disorder, obsessive-compulsive disorder (OCD), or treatment resistant depression, for which they may have found no effective treatment or may be considering ablative surgical options, then neurorestorative stem cell therapies certainly appear more attractive. And with continued progress in iPSC technologies, they will also become more feasible as viable treatment options with reasonable risk-reward. Here, experimentation in chimera models (i.e., transfer of human iPSCs to animal models) will likely prove both necessary and informative for advancing iPSC-based stem cell therapies for psychiatric diseases. In these endeavors, further progress in three dimensional organoid- and tissue-based iPSC approaches that better recapitulate native neural architectures (Quadrato, Brown, & Arlotta, 2016), and including understanding and harnessing the integral role of glia in modulating neuronal development and behavior in iPSC models (Gonzalez, Gregory, & Brennand, 2017), will be of utmost importance. We expect that these kinds of ongoing research into the use of iPSCs for modeling and treating psychiatric disease will yield exciting new personalized treatment opportunities, improved outcomes, and reduction in disease burdens, thus improving individual and global health.

References

Abi-Dargham, A. (2004). Do we still believe in the dopamine hypothesis? New data bring new evidence. *International Journal of Neuropsychopharmacology, 7*(Supplement_1), S1–S5.

Abi-Dargham, A., Rodenhiser, J., Printz, D., Zea-Ponce, Y., Gil, R., Kegeles, L. S., ... Laruelle, M. (2000). Increased baseline occupancy of D(2) receptors by dopamine in schizophrenia. *Proceedings of the National Academy of Sciences of the United States of America, 97*(14), 8104–8109.

Adegbola, A., Bury, L. A., Fu, C., Zhang, M., & Wynshaw-Boris, A. (2017). Concise review: Induced pluripotent stem cell models for neuropsychiatric diseases. *Stem Cells Translational Medicine, 6*(12), 2062–2070.

Ahmad, R., Sportelli, V., Ziller, M., Spengler, D., & Hoffmann, A. (2018). Tracing early neurodevelopment in schizophrenia with induced pluripotent stem cells. *Cell, 7*(9).

Allen, A. P., Naughton, M., Dowling, J., Walsh, A., Ismail, F., Shorten, G., ... Clarke, G. (2015). Serum BDNF as a peripheral biomarker of treatment-resistant depression and the rapid antidepressant response: A comparison of ketamine and ECT. *Journal of Affective Disorders, 186*, 306–311.

American Psychiatric Association (2013). *Diagnostic and statistical manual of mental disorders* (5th ed.). Washington, DC: American Psychiatric Association.

Arber, C., Precious, S. V., Cambray, S., Risner-Janiczek, J. R., Kelly, C., Noakes, Z., ... Li, M. (2015). Activin A directs striatal projection neuron differentiation of human pluripotent stem cells. *Development (Cambridge, England), 142*(7), 1375–1386.

Åsberg, M., Bertilsson, L., Mårtensson, B., Scalia-Tomba, G. P., Thorén, P., & Träskman-Bendz, L. (1984). CSF monoamine metabolites in melancholia. *Acta Psychiatrica Scandinavica, 69*(3), 201–219.

Avior, Y., Sagi, I., & Benvenisty, N. (2016). Pluripotent stem cells in disease modelling and drug discovery. *Nature Reviews Molecular Cell Biology, 17*(3), 170–182.

Bardy, C., Van den Hurk, M., Eames, T., Marchand, C., Hernandez, R. V., Kellogg, M., ... Gage, F. H. (2015). Neuronal medium that supports basic synaptic functions and activity of human neurons in vitro. *Proceedings of the National Academy of Sciences, 112*(20), E2725–E2734.

Baumeister, A. A. (2013). The chlorpromazine enigma. *Journal of the History of the Neurosciences, 22*(1), 14–29.

Bavamian, S., Mellios, N., Lalonde, J., Fass, D. M., Wang, J., Sheridan, S. D., ... Haggarty, S. J. (2015). Dysregulation of miR-34a links neuronal development to genetic risk factors for bipolar disorder. *Molecular Psychiatry, 20*(5), 573–584.

Beasley, C. L., & Reynolds, G. P. (1997). Parvalbumin-immunoreactive neurons are reduced in the prefrontal cortex of schizophrenics. *Schizophrenia Research, 24*(3), 349–355.

Beers, J., Linask, K. L., Chen, J. A., Siniscalchi, L. I., Lin, Y., Zheng, W., … Chen, G. (2015). A cost-effective and efficient reprogramming platform for large-scale production of integration-free human induced pluripotent stem cells in chemically defined culture. *Scientific Reports, 5,* 11319.

Ben-David, U., & Benvenisty, N. (2011). The tumorigenicity of human embryonic and induced pluripotent stem cells. *Nature Reviews Cancer, 11,* 268.

Bener, A., Dafeeah, E. E., & Salem, M. O. (2013). A study of reasons of non-compliance of psychiatric treatment and patients' attitudes towards illness and treatment in Qatar. *Issues in Mental Health Nursing, 34*(4), 273–280.

Ben-Reuven, L., & Reiner, O. (2016). Modeling the autistic cell: iPSCs recapitulate developmental principles of syndromic and nonsyndromic ASD. *Development, Growth & Differentiation, 58*(5), 481–491.

Bertolino, A., Frye, M., Callicott, J. H., Mattay, V. S., Rakow, R., Shelton-Repella, J., … Weinberger, D. R. (2003). Neuronal pathology in the hippocampal area of patients with bipolar disorder: A study with proton magnetic resonance spectroscopic imaging. *Biological Psychiatry, 53*(10), 906–913.

Bockting, C. L. H., ten Doesschate, M. C., Spijker, J., Spinhoven, P., Koeter, M. W. J., & Schene, A. H. (2008). Continuation and maintenance use of antidepressants in recurrent depression. *Psychotherapy and Psychosomatics, 77*(1), 17–26.

Bogerts, B., Hantsch, J., & Herzer, M. (1983). A morphometric study of the dopamine-containing cell groups in the mesencephalon of normals, Parkinson patients, and schizophrenics. *Biology Psychiartry, 18*(9), 951–969.

Brederlau, A., Correia, A. S., Anisimov, S. V., Elmi, M., Paul, G., Roybon, L., … Li, J. Y. (2006). Transplantation of human embryonic stem cell-derived cells to a rat model of Parkinson's disease: Effect of in vitro differentiation on graft survival and teratoma formation. *Stem Cells, 24*(6), 1433–1440.

Brennand, K., Simone, A., Jou, J., Gelboin-Burkhart, C., Tran, N., Sangar, S., … Gage, F. H. (2011a). Modeling schizophrenia using hiPSC neurons. *Nature, 473*(7346), 221–225.

Brennand, K. J., Simone, A., Jou, J., Gelboin-Burkhart, C., Tran, N., Sangar, S., … Gage, F. H. (2011b). Modelling schizophrenia using human induced pluripotent stem cells. *Nature, 473*(7346), 221–225.

Brennand, K. J., Simone, A., Tran, N., & Gage, F. H. (2012). Modeling psychiatric disorders at the cellular and network levels. *Molecular Psychiatry, 17*(12), 1239–1253.

Brickman, J. M., & Burdon, T. G. (2002). Pluripotency and tumorigenicity. *Nature Genetics, 32,* 557.

Brito, A., Russo, F. B., Muotri, A. R., & Beltrao-Braga, P. C. B. (2018). Autism spectrum disorders and disease modeling using stem cells. *Cell and Tissue Research, 371*(1), 153–160.

Bury, L. A., & Wynshaw-Boris, A. (2018). Modeling non-syndromic autism with human-induced pluripotent stem cells. *Neuropsychopharmacology, 43*(1), 219–220.

Caiazzo, M., Dell'Anno, M. T., Dvoretskova, E., Lazarevic, D., Taverna, S., Leo, D., … Broccoli, V. (2011). Direct generation of functional dopaminergic neurons from mouse and human fibroblasts. *Nature, 476,* 224.

Caiazzo, M., Giannelli, S., Valente, P., Lignani, G., Carissimo, A., Sessa, A., … Broccoli, V. (2015). Direct conversion of fibroblasts into functional astrocytes by defined transcription factors. *Stem Cell Reports, 4*(1), 25–36.

Caligiuri, M. P., Brown, G. G., Meloy, M. J., Eberson, S. C., Kindermann, S. S., Frank, L. R., … Lohr, J. B. (2003). An fMRI study of affective state and medication on cortical and subcortical brain regions during motor performance in bipolar disorder. *Psychiatry Research: Neuroimaging, 123*(3), 171–182.

Capitão, L., Sampaio, A., Sampaio, C., Vasconcelos, C., Férnandez, M., Garayzábal, E., … Gonçalves, Ó. F. (2011). MRI amygdala volume in Williams syndrome. *Research in Developmental Disabilities, 32*(6), 2767–2772.

Carey, B. W., Markoulaki, S., Hanna, J., Saha, K., Gao, Q., Mitalipova, M., & Jaenisch, R. (2009). Reprogramming of murine and human somatic cells using a single polycistronic vector. *Proceedings of the National Academy of Sciences of the United States of America, 106*(1), 157–162.

Cavalleri, L., Merlo Pich, E., Millan, M. J., Chiamulera, C., Kunath, T., Spano, P. F., & Collo, G. (2018). Ketamine enhances structural plasticity in mouse mesencephalic and human iPSC-derived dopaminergic neurons via AMPAR-driven BDNF and mTOR signaling. *Molecular Psychiatry, 23*(4), 812–823.

Chailangkarn, T., Trujillo, C. A., Freitas, B. C., Hrvoj-Mihic, B., Herai, R. H., Yu, D. X., … Muotri, A. R. (2016). A human neurodevelopmental model for Williams syndrome. *Nature, 536*(7616), 338–343.

Chan, M. K., Gottschalk, M. G., Haenisch, F., Tomasik, J., Ruland, T., Rahmoune, H., … Bahn, S. (2014). Applications of blood-based protein biomarker strategies in the study of psychiatric disorders. *Progress in Neurobiology, 122,* 45–72.

Chen, H. M., DeLong, C. J., Bame, M., Rajapakse, I., Herron, T. J., McInnis, M. G., & O'Shea, K. S. (2014). Transcripts involved in calcium signaling and telencephalic neuronal fate are altered in induced pluripotent stem cells from bipolar disorder patients. *Translational Psychiatry, 4.*

Chen, I. P., Fukuda, K., Fusaki, N., Iida, A., Hasegawa, M., Lichtler, A., & Reichenberger, E. J. (2013). Induced pluripotent stem cell reprogramming by integration-free Sendai virus vectors from peripheral blood of patients with craniometaphyseal dysplasia. *Cellular Reprogramming, 15*(6), 503–513.

Chen, Y., & Lai, D. (2015). Pluripotent states of human embryonic stem cells. *Cellular Reprogramming, 17*(1), 1–6.

Chen, H. I., Song, H., & Ming, G. L. (2018). Applications of human brain organoids to clinical problems. *Developmental Dynamics, 248*(1), 53–64.

Clevers, H. (2016). Modeling development and disease with organoids. *Cell, 165*(7), 1586–1597.

Coccaro, E. F., Siever, L. J., Klar, H. M., et al. (1989). Serotonergic studies in patients with affective and personality disorders: Correlates with suicidal and impulsive aggressive behavior. *Archives of General Psychiatry, 46*(7), 587–599.

Collo, G., Cavalleri, L., Bono, F., Mora, C., Fedele, S., Invernizzi, R. W., … Spano, P. (2018). Ropinirole and Pramipexole promote structural plasticity in human iPSC-derived dopaminergic neurons via BDNF and mTOR signaling. *Neural Plasticity, 2018,* 4196961.

Cotter, D., Mackay, D., Landau, S., Kerwin, R., & Everall, I. (2001). Reduced glial cell density and neuronal size in the anterior cingulate cortex in major depressive disorder. *Archives of General Psychiatry, 58*(6), 545–553.

DelBello, M. P., Zimmerman, M. E., Mills, N. P., Getz, G. E., & Strakowski, S. M. (2004). Magnetic resonance imaging analysis of amygdala and other subcortical brain regions in adolescents with bipolar disorder. *Bipolar Disorders, 6*(1), 43–52.

Di Lullo, E., & Kriegstein, A. R. (2017). The use of brain organoids to investigate neural development and disease. *Nature Reviews. Neuroscience, 18*(10), 573–584.

Donegan, J. J., Tyson, J. A., Branch, S. Y., Beckstead, M. J., Anderson, S. A., & Lodge, D. J. (2017). Stem cell-derived interneuron transplants as a treatment for schizophrenia: Preclinical validation in a rodent model. *Molecular Psychiatry, 22*(10), 1492–1501.

Dyer, M. A. (2016). Stem cells expand insights into human brain evolution. *Cell Stem Cell, 18*(4), 425–426.

Etkin, A., Prater, K. E., Hoeft, F., Menon, V., & Schatzberg, A. F. (2010). Failure of anterior cingulate activation and connectivity with the amygdala during implicit regulation of emotional processing in generalized anxiety disorder. *The American Journal of Psychiatry, 167*(5), 545–554.

Falk, A., Heine, V. M., Harwood, A. J., Sullivan, P. F., Peitz, M., Brustle, O., … Djurovic, S. (2016). Modeling psychiatric disorders: From genomic findings to cellular phenotypes. *Molecular Psychiatry, 21*(9), 1167–1179.

Farzan, F., Barr, M. S., Levinson, A. J., Chen, R., Wong, W., Fitzgerald, P. B., & Daskalakis, Z. J. (2010). Evidence for gamma inhibition deficits in the dorsolateral prefrontal cortex of patients with schizophrenia. *Brain, 133*(5), 1505–1514.

Fava, M. (2000). New approaches to the treatment of refractory depression. *The Journal of Clinical Psychiatry, 61*, 26–32.

Fernandez, B. A., & Scherer, S. W. (2017). Syndromic autism spectrum disorders: Moving from a clinically defined to a molecularly defined approach. *Dialogues in Clinical Neuroscience, 19*(4), 353–371.

Fusaki, N., Ban, H., Nishiyama, A., Saeki, K., & Hasegawa, M. (2009). Efficient induction of transgene-free human pluripotent stem cells using a vector based on Sendai virus, an RNA virus that does not integrate into the host genome. *Proceedings of the Japan Academy. Series B, Physical and Biological Sciences, 85*(8), 348–362.

Gil, J., Laczmanska, I., Pesz, K. A., & Sasiadek, M. M. (2018). Personalized medicine in oncology. New perspectives in management of gliomas. *Contemporary Oncology, 22*(1A), 1–2.

Giménez, M., Pujol, J., Ortiz, H., Soriano-Mas, C., López-Solà, M., Farré, M., … Martín-Santos, R. (2012). Altered brain functional connectivity in relation to perception of scrutiny in social anxiety disorder. *Psychiatry Research: Neuroimaging, 202*(3), 214–223.

Glatt, C. E., & Lee, F. S. (2016). Common polymorphisms in the age of research domain criteria (RDoC): Integration and translation. *Biological Psychiatry, 79*(1), 25–31.

Goldberg, M. C., Spinelli, S., Joel, S., Pekar, J. J., Denckla, M. B., & Mostofsky, S. H. (2011). Children with high functioning autism show increased prefrontal and temporal cortex activity during error monitoring. *Developmental Cognitive Neuroscience, 1*(1), 47–56.

González, F., Boué, S., & Belmonte, J. C. I. (2011). Methods for making induced pluripotent stem cells: Reprogramming à la carte. *Nature Reviews Genetics, 12*(4), 231–242.

Gonzalez, D. M., Gregory, J., & Brennand, K. J. (2017). The importance of non-neuronal cell types in hiPSC-based disease modeling and drug screening. *Frontiers in Cell and Development Biology, 5*, 117.

Gorris, R., Fischer, J., Erwes, K. L., Kesavan, J., Peterson, D. A., Alexander, M., … Brüstle, O. (2015). Pluripotent stem cell-derived radial glia-like cells as stable intermediate for efficient generation of human oligodendrocytes. *Glia, 63*(12), 2152–2167.

Gratten, J., Wray, N. R., Keller, M. C., & Visscher, P. M. (2014). Large-scale genomics unveils the genetic architecture of psychiatric disorders. *Nature Neuroscience, 17*, 782.

Griesi-Oliveira, K., Acab, A., Gupta, A. R., Sunaga, D. Y., Chailangkarn, T., Nicol, X., … Muotri, A. R. (2015). Modeling non-syndromic autism and the impact of TRPC6 disruption in human neurons. *Molecular Psychiatry, 20*(11), 1350–1365.

Gupta, S., M-Redmond, T., Meng, F., Tidball, A., Akil, H., Watson, S., … Uhler, M. (2018). Fibroblast growth factor 2 regulates activity and gene expression of human post-mitotic excitatory neurons. *Journal of Neurochemistry, 145*(3), 188–203.

Gurdon, J. B. (1962). The developmental capacity of nuclei taken from intestinal epithelium cells of feeding tadpoles. *Journal of Embryology and Experimental Morphology, 10*(4), 622.

Haile, C. N., Murrough, J. W., Iosifescu, D. V., Chang, L. C., Al Jurdi, R. K., Foulkes, A., … Mathew, S. J. (2014). Plasma brain derived neurotrophic factor (BDNF) and response to ketamine in treatment-resistant depression. *The International Journal of Neuropsychopharmacology, 17*(2), 331–336.

Heckers, S., Stone, D., Walsh, J., Shick, J., Koul, P., & Benes, F. M. (2002). Differential hippocampal expression of glutamic acid decarboxylase 65 and 67 messenger rna in bipolar disorder and schizophrenia. *Archives of General Psychiatry, 59*(6), 521–529.

Hedgepeth, C. M., Conrad, L. J., Zhang, J., Huang, H. C., Lee, V. M., & Klein, P. S. (1997). Activation of the Wnt signaling pathway: A molecular mechanism for lithium action. *Developmental Biology, 185*(1), 82–91.

Hettema, J. M., Kettenmann, B., Ahluwalia, V., McCarthy, C., Kates, W. R., Schmitt, J. E., … Fatouros, P. (2012). A pilot multimodal twin imaging study of generalized anxiety disorder. *Depression and Anxiety, 29*(3), 202–209.

Hoffmann, A., Sportelli, V., Ziller, M., & Spengler, D. (2018). From the Psychiatrist's couch to induced pluripotent stem cells: Bipolar disease in a dish. *International Journal of Molecular Sciences, 19*(3), E770.

Huang, J., Liu, F., Tang, H., Wu, H., Li, L., Wu, R., … Chen, J. (2017). Tranylcypromine causes neurotoxicity and represses BHC110/LSD1 in human-induced pluripotent stem cell-derived cerebral organoids model. *Frontiers in Neurology, 8*, 626.

Javitt, D. (1987). Negative schizophrenic symptomatology and the PCP (phencyclidine) model of schizophrenia. *The Hillside Journal of Clinical Psychiatry, 9*(1), 12–35.

Jope, R. S. (2003). Lithium and GSK-3: One inhibitor, two inhibitory actions, multiple outcomes. *Trends in Pharmacological Sciences, 24*(9), 441–443.

Kaiser, T., & Feng, G. (2015). Modeling psychiatric disorders for developing effective treatments. *Nature Medicine, 21*(9), 979–988.

Kalia, M., & Costa, E. S. J. (2015). Biomarkers of psychiatric diseases: Current status and future prospects. *Metabolism, 64*(3 Suppl 1), S11–S15.

Kalman, S., Hathy, E., & Rethelyi, J. M. (2016). A dishful of a troubled mind: Induced pluripotent stem cells in psychiatric research. *Stem Cells International, 2016*, 7909176.

Kane, J. M., Kishimoto, T., & Correll, C. U. (2013). Non-adherence to medication in patients with psychotic disorders: Epidemiology, contributing factors and management strategies. *World Psychiatry, 12*(3), 216–226.

Kegeles, L. S., Abi-Dargham, A., Frankle, W., et al. (2010). Increased synaptic dopamine function in associative regions of the striatum in schizophrenia. *Archives of General Psychiatry, 67*(3), 231–239.

Kessler, R. M., Woodward, N. D., Riccardi, P., Li, R., Ansari, M. S., Anderson, S., … Meltzer, H. Y. (2009). Dopamine D2 receptor levels in striatum, thalamus, Substantia Nigra, limbic regions, and cortex in schizophrenic subjects. *Biological Psychiatry, 65*(12), 1024–1031.

Khattak, S., Brimble, E., Zhang, W., Zaslavsky, K., Strong, E., Ross, P. J., … Ellis, J. (2015). Human induced pluripotent stem cell derived neurons as a model for Williams-Beuren syndrome. *Molecular Brain, 8*(1), 77.

Kida, S., & Kato, T. (2015). Microendophenotypes of psychiatric disorders: Phenotypes of psychiatric disorders at the level of molecular dynamics, synapses, neurons, and neural circuits. *Current Molecular Medicine, 15*(2), 111–118.

Kim, K., Doi, A., Wen, B., Ng, K., Zhao, R., Cahan, P., … Daley, G. Q. (2010). Epigenetic memory in induced pluripotent stem cells. *Nature, 467*(7313), 285–290.

Kim, S., Kim, M. K., Oh, D., Lee, S. H., & Kim, B. (2016). Induced pluripotent stem cells as a novel tool in psychiatric research. *Psychiatry Investigation, 13*(1), 8–17.

Kim, K. H., Liu, J., Sells Galvin, R. J., Dage, J. L., Egeland, J. A., Smith, R. C., … Paul, S. M. (2015). Transcriptomic analysis of induced pluripotent stem cells derived from patients with bipolar disorder from an old order Amish pedigree. *PLoS One, 10*(11).

Kim, K., Zhao, R., Doi, A., Ng, K., Unternaehrer, J., Cahan, P., … Daley, G. Q. (2011). Donor cell type can influence the epigenome and differentiation potential of human induced pluripotent stem cells. *Nature Biotechnology, 29*(12), 1117–1119.

Klumpers, U. M. H., Veltman, D. J., Drent, M. L., Boellaard, R., Comans, E. F. I., Meynen, G., … Hoogendijk, W. J. G. (2010). Reduced parahippocampal and lateral temporal GABAA-[11C]flumazenil binding in major depression: Preliminary results. *European Journal of Nuclear Medicine and Molecular Imaging, 37*(3), 565–574.

Krencik, R., & Zhang, S.-C. (2011). Directed differentiation of functional Astroglial subtypes from human pluripotent stem cells. *Nature Protocols, 6*(11), 1710–1717.

Lancaster, M. A., & Knoblich, J. A. (2014). Organogenesis in a dish: Modeling development and disease using organoid technologies. *Science, 345*(6194).

Laruelle, M., Abi-Dargham, A., Gil, R., Kegeles, L., & Innis, R. (1999). Increased dopamine transmission in schizophrenia: Relationship to illness phases. *Biological Psychiatry, 46*(1), 56–72.

Leckband, S. G., McCarthy, M., & Kelsoe, J. R. (2012). The pharmacogenomics of mood stabilizer response in bipolar disorder. *Mental Health Clinician, 1*(9), 217–221.

Lee, H. J., Hore, T. A., & Reik, W. (2014). Reprogramming the methylome: Erasing memory and creating diversity. *Cell Stem Cell, 14*(6), 710–719.

Lepack, A. E., Fuchikami, M., Dwyer, J. M., Banasr, M., & Duman, R. S. (2014). BDNF release is required for the behavioral actions of ketamine. *The International Journal of Neuropsychopharmacology, 18*(1).

Lew, C. H., Brown, C., Bellugi, U., & Semendeferi, K. (2016). Neuron density is decreased in the prefrontal cortex in Williams syndrome. *Autism Research, 10*(1), 99–112.

Licinio, J., & Wong, M. L. (2011). Pharmacogenomics of antidepressant treatment effects. *Dialogues in Clinical Neuroscience, 13*(1), 63–71.

Licinio, J., & Wong, M. L. (2016). Serotonergic neurons derived from induced pluripotent stem cells (iPSCs): A new pathway for research on the biology and pharmacology of major depression. *Molecular Psychiatry, 21*(1), 1–2.

Lim, S.-W., Kwon, Y.-S., Ha, J., Yoon, H.-G., Bae, S.-M., Shin, D.-W., … Oh, K.-S. (2012). Comparison of treatment adherence between selective serotonin reuptake inhibitors and moclobemide in patients with social anxiety disorder. *Psychiatry Investigation, 9*(1), 73–79.

Lim, C. S., Yang, J. E., Lee, Y. K., Lee, K., Lee, J. A., & Kaang, B. K. (2015). Understanding the molecular basis of autism in a dish using hiPSCs-derived neurons from ASD patients. *Molecular Brain, 8*(1), 57.

Liu, X., Campanac, E., Cheung, H. H., Ziats, M. N., Canterel-Thouennon, L., Raygada, M., … Rennert, O. M. (2017). Idiopathic autism: Cellular and molecular phenotypes in pluripotent stem cell-derived neurons. *Molecular Neurobiology, 54*(6), 4507–4523.

Liu, Y., Liu, H., Sauvey, C., Yao, L., Zarnowska, E. D., & Zhang, S.-C. (2013). Directed differentiation of forebrain GABA interneurons from human pluripotent stem cells. *Nature Protocols, 8*(9), 1670–1679.

Liu, Y. N., Lu, S. Y., & Yao, J. (2017). Application of induced pluripotent stem cells to understand neurobiological basis of bipolar disorder and schizophrenia. *Psychiatry and Clinical Neurosciences, 71*(9), 579–599.

Liu-Seifert, H., Adams, D. H., & Kinon, B. J. (2005). Discontinuation of treatment of schizophrenic patients is driven by poor symptom response: A pooled post-hoc analysis of four atypical antipsychotic drugs. *BMC Medicine, 3*(1), 21.

Lodge, D. J., & Grace, A. A. (2005). The hippocampus modulates dopamine neuron responsivity by regulating the intensity of phasic neuron activation. *Neuropsychopharmacology, 31*, 1356.

Lu, J., Zhong, X., Liu, H., Hao, L., Huang, C. T.-L., Sherafat, M. A., … Zhang, S.-C. (2016). Generation of serotonin neurons from human pluripotent stem cells. *Nature Biotechnology, 34*(1), 89–94.

Lyoo, I. K., Sung, Y. H., Dager, S. R., Friedman, S. D., Lee, J. Y., Kim, S. J., … Renshaw, P. F. (2006). Regional cerebral cortical thinning in bipolar disorder. *Bipolar Disorders, 8*(1), 65–74.

Ma, L., Liu, Y., & Zhang, S.-C. (2011). Directed differentiation of dopamine neurons from human pluripotent stem cells. In P. H. Schwartz, & R. L. Wesselschmidt (Eds.), *Human pluripotent stem cells: Methods and protocols* (pp. 411–418). Totowa, NJ: Humana Press.

Machado-Vieira, R., Baumann, J., Wheeler-Castillo, C., Latov, D., Henter, I. D., Salvadore, G., & Zarate, C. A. (2010). The timing of antidepressant effects: A comparison of diverse pharmacological and somatic treatments. *Pharmaceuticals (Basel, Switzerland), 3*(1), 19–41.

Madison, J. M., Zhou, F., Nigam, A., Hussain, A., Barker, D. D., Nehme, R., … Haggarty, S. J. (2015). Characterization of bipolar disorder patient-specific induced pluripotent stem cells from a family reveals neurodevelopmental and mRNA expression abnormalities. *Molecular Psychiatry, 20*(6), 703–717.

Malik, N., & Rao, M. S. (2013). A review of the methods for human iPSC derivation. *Methods in molecular biology (Clifton, NJ), 997*, 23–33.

Marchetto, M. C., Brennand, K. J., Boyer, L. F., & Gage, F. H. (2011). Induced pluripotent stem cells (iPSCs) and neurological disease modeling: Progress and promises. *Human Molecular Genetics, 20*(R2), R109–R115.

McGuffin, P., Rijsdijk, F., Andrew, M., Sham, P., Katz, R., & Cardno, A. (2003). The heritability of bipolar affective disorder and the genetic relationship to unipolar depression. *Archives of General Psychiatry, 60*(5), 497–502.

Mertens, J., Marchetto, M. C., Bardy, C., & Gage, F. H. (2016). Evaluating cell reprogramming, differentiation and conversion technologies in neuroscience. *Nature Reviews Neuroscience, 17*, 424.

Mertens, J., Wang, Q. W., Kim, Y., Yu, D. X., Pham, S., Yang, B., … Yao, J. (2015). Differential responses to lithium in hyperexcitable neurons from patients with bipolar disorder. *Nature, 527*(7576), 95–99.

Meyer-Lindenberg, A., Mervis, C. B., Sarpal, D., Koch, P., Steele, S., Kohn, P., … Berman, K. F. (2005). Functional, structural, and metabolic abnormalities of the hippocampal formation in Williams syndrome. *Journal of Clinical Investigation, 115*(7), 1888–1895.

Miskinyte, G., Devaraju, K., Grønning Hansen, M., Monni, E., Tornero, D., Woods, N. B., … Kokaia, Z. (2017). Direct conversion of human fibroblasts to functional excitatory cortical neurons integrating into human neural networks. *Stem Cell Research & Therapy, 8*, 207.

Mobbs, D., Eckert, M. A., Menon, V., Mills, D., Korenberg, J., Galaburda, A. M., … Reiss, A. L. (2007). Reduced parietal and visual cortical activation during global processing in Williams syndrome. *Developmental Medicine and Child Neurology, 49*(6), 433–438.

Murphy, D. L. (1990). Neuropsychiatric disorders and the multiple human brain serotonin receptor subtypes and subsystems. *Neuropsychopharmacology, 3*(5–6), 457–471.

Nashun, B., Hill, P. W., & Hajkova, P. (2015). Reprogramming of cell fate: Epigenetic memory and the erasure of memories past. *The EMBO Journal, 34*(10), 1296–1308.

Nebel, M. B., Joel, S. E., Muschelli, J., Barber, A. D., Caffo, B. S., Pekar, J. J., & Mostofsky, S. H. (2014). Disruption of functional organization within the primary motor cortex in children with autism. *Human Brain Mapping, 35*(2), 567–580.

Nierenberg, A. A. (2010). A critical appraisal of treatments for bipolar disorder. *Prim Care Companion Journal of Clinical Psychiatry, 12*(Suppl 1), 23–29.

Nishimura, K., Ohtaka, M., Takada, H., Kurisaki, A., Tran, N. V. K., Tran, Y. T. H., … Nakanishi, M. (2017). Simple and effective generation of transgene-free induced pluripotent stem cells using an auto-erasable Sendai virus vector responding to microRNA-302. *Stem Cell Research, 23*, 13–19.

Oedegaard, K. J., Alda, M., Anand, A., Andreassen, O. A., Balaraman, Y., Berrettini, W. H., … Kelsoe, J. R. (2016). The Pharmacogenomics of Bipolar Disorder study (PGBD): Identification of genes for lithium response in a prospective sample. *BMC Psychiatry, 16*, 129.

Okita, K., Ichisaka, T., & Yamanaka, S. (2007). Generation of germline-competent induced pluripotent stem cells. *Nature, 448*, 313.

Olney, J. W., & Farber, N. B. (1995). Glutamate receptor dysfunction and schizophrenia. *Archives of General Psychiatry, 52*(12), 998–1007.

Papapetrou, E. P. (2016). Induced pluripotent stem cells, past and future. *Science, 353*(6303), 991–992.

Paulsen Bda, S., de Moraes Maciel, R., Galina, A., Souza da Silveira, M., dos Santos Souza, C., Drummond, H., … Rehen, S. K. (2012). Altered oxygen metabolism associated to neurogenesis of induced pluripotent stem cells derived from a schizophrenic patient. *Cell Transplantation, 21*(7), 1547–1559.

Pei, S., Liu, L., Zhong, Z., Wang, H., Lin, S., & Shang, J. (2016). Risk of prenatal depression and stress treatment: Alteration on serotonin system of offspring through exposure to Fluoxetine. *Scientific Reports, 6*, 33822.

Perry, T., Kish, S., Buchanan, J., & Hansen, S. (1979). Gamma-aminobutyric-acid deficiency in brain of schizophrenic patients. *Lancet, 1*(8110), 237–239.

Perry, S. W., Norman, J. P., Litzburg, A., & Gelbard, H. A. (2004). Antioxidants are required during the early critical period, but not later, for neuronal survival. *Journal of Neuroscience Research, 78*(4), 485–492.

Pierri, J. N., Volk, C. L. E., Auh, S., Sampson, A., & Lewis, D. A. (2003). Somal size of prefrontal cortical pyramidal neurons in schizophrenia. *Biological Psychiatry, 54*(2), 111–120.

Plomin, R., Owen, M. J., & McGuffin, P. (1994). The genetic basis of complex human behaviors. *Science, 264*(5166), 1733.

Preskorn, S. H. (2007). The evolution of antipsychotic drug therapy: Reserpine, chlorpromazine, and haloperidol. *Journal of Psychiatric Practice, 13*(4), 253–257.

Qiu, M., Zhang, H., Mellor, D., Shi, J., Wu, C., Huang, Y., … Peng, D. (2018). Aberrant neural activity in patients with bipolar depressive disorder distinguishing to the unipolar depressive disorder: A resting-state functional magnetic resonance imaging study. *Frontiers in Psychiatry, 9*, 238.

Quadrato, G., Brown, J., & Arlotta, P. (2016). The promises and challenges of human brain organoids as models of neuropsychiatric disease. *Nature Medicine, 22*(11), 1220–1228.

Rajkowska, G. (2000). Postmortem studies in mood disorders indicate altered numbers of neurons and glial cells. *Biological Psychiatry, 48*(8), 766–777.

Rajkowska, G., Halaris, A., & Selemon, L. D. (2001). Reductions in neuronal and glial density characterize the dorsolateral prefrontal cortex in bipolar disorder. *Biological Psychiatry, 49*(9), 741–752.

Razafsha, M., Khaku, A., Azari, H., Alawieh, A., Behforuzi, H., Fadlallah, B., … Gold, M. S. (2015). Biomarker identification in psychiatric disorders: From neuroscience to clinical practice. *Journal of Psychiatric Practice, 21*(1), 37–48.

Reynolds, G. P., McGowan, O. O., & Dalton, C. F. (2014). Pharmacogenomics in psychiatry: The relevance of receptor and transporter polymorphisms. *British Journal of Clinical Pharmacology, 77*(4), 654–672.

Riggs, J. W., Barrilleaux, B. L., Varlakhanova, N., Bush, K. M., Chan, V., & Knoepfler, P. S. (2013). Induced pluripotency and oncogenic transformation are related processes. *Stem Cells and Development, 22*(1), 37–50.

Ritchie, H., & Roser, M. (2018). *Mental health.* Published online at OurWorldInData.org. Retrieved from: https://ourworldindata.org/mental-health. (Accessed 22 October 2018).

Ritvo, E. R., Freeman, B. J., Scheibel, A. B., Duong, T., Robinson, H., & Guthrie, D. (1986). Lower Purkinje cell counts in the cerebella of four autistic subjects: Initial findings of the UCLA-NSAC autopsy research report. *American Journal of Psychiatry, 143*(7), 862–866.

Robicsek, O., Karry, R., Petit, I., Salman-Kesner, N., Müller, F. J., Klein, E., … Ben-Shachar, D. (2013). Abnormal neuronal differentiation and mitochondrial dysfunction in hair follicle-derived induced pluripotent stem cells of schizophrenia patients. *Molecular Psychiatry, 18*, 1067.

Roffman, J. L. (2011). Biomarkers and personalized psychiatry. *Harvard Review of Psychiatry, 19*(3), 99–101.

Samoilova, E. M., Kalsin, V. A., Kushnir, N. M., Chistyakov, D. A., Troitskiy, A. V., & Baklaushev, V. P. (2018). Adult neural stem cells: Basic research and production strategies for neurorestorative therapy. *Stem Cells International, 2018*, 4835491.

Sarkar, A., Mei, A., Paquola, A. C. M., Stern, S., Bardy, C., Klug, J. R., … Gage, F. H. (2018). Efficient generation of CA3 neurons from human pluripotent stem cells enables modeling of hippocampal connectivity in vitro. *Cell Stem Cell, 22*(5). 684–697.e689.

Scarr, E., Millan, M. J., Bahn, S., Bertolino, A., Turck, C. W., Kapur, S., … Dean, B. (2015). Biomarkers for psychiatry: The journey from fantasy to fact, a report of the 2013 CINP think tank. *The International Journal of Neuropsychopharmacology, 18*(10).

Schaffer, A., Isometsa, E. T., Tondo, L., Moreno, H. D., Turecki, G., Reis, C., … Yatham, L. N. (2015). International society for bipolar disorders task force on suicide: Meta-analyses and meta-regression of correlates of suicide attempts and suicide deaths in bipolar disorder. *Bipolar Disorders, 17*(1), 1–16.

Schienle, A., Ebner, F., & Schäfer, A. (2011). Localized gray matter volume abnormalities in generalized anxiety disorder. *European Archives of Psychiatry and Clinical Neuroscience, 261*(4), 303–307.

Schilsky, R. L. (2010). Personalized medicine in oncology: The future is now. *Nature Reviews Drug Discovery, 9*, 363.

Schumann, C. M., Hamstra, J., Goodlin-Jones, B. L., Lotspeich, L. J., Kwon, H., Buonocore, M. H., … Amaral, D. G. (2004). The amygdala is enlarged in children but not adolescents with Autism: The hippocampus is enlarged at all ages. *The Journal of Neuroscience, 24*(28), 6392.

Shen, X., Yeung, H. T., & Lai, K. O. (2018). Application of human-induced pluripotent stem cells (hiPSCs) to study synaptopathy of neurodevelopmental disorders. *Developmental Neurobiology, 79*(1), 20–35.

Shi, Y., Kirwan, P., Smith, J., Robinson, H. P. C., & Livesey, F. J. (2012). Human cerebral cortex development from pluripotent stem cells to functional excitatory synapses. *Nature Neuroscience. 15*(3). https://doi.org/10.1038/nn.3041.

Singh, V. K., Kalsan, M., Kumar, N., Saini, A., & Chandra, R. (2015). Induced pluripotent stem cells: Applications in regenerative medicine, disease modeling, and drug discovery. *Frontiers in Cell and Development Biology, 3*, 2.

Sokolowska, I., Ngounou Wetie, A. G., Wormwood, K., Thome, J., Darie, C. C., & Woods, A. G. (2015). The potential of biomarkers in psychiatry: Focus on proteomics. *Journal of Neural Transmission (Vienna), 122*(Suppl 1), S9–18.

Soldner, F., & Jaenisch, R. (2018). Stem cells, genome editing, and the path to translational medicine. *Cell, 175*(3), 615–632.

Soliman, M. A., Aboharb, F., Zeltner, N., & Studer, L. (2017). Pluripotent stem cells in neuropsychiatric disorders. *Molecular Psychiatry, 22*(9), 1241–1249.

Sommer, C. A., Stadtfeld, M., Murphy, G. J., Hochedlinger, K., Kotton, D. N., & Mostoslavsky, G. (2009). iPS cell generation using a single lentiviral stem cell cassette. *Stem Cells (Dayton, Ohio), 27*(3), 543–549.

Southwell, D. G., Nicholas, C. R., Basbaum, A. I., Stryker, M. P., Kriegstein, A. R., Rubenstein, J. L., & Alvarez-Buylla, A. (2014). Interneurons from embryonic development to cell-based therapy. *Science (New York, NY), 344*(6180), 1240622.

Spatazza, J., Mancia Leon, W. R., & Alvarez-Buylla, A. (2017). Chapter 3—Transplantation of GABAergic interneurons for cell-based therapy. In S. B. Dunnett, & A. Björklund (Eds.), *Progress in brain research* (pp. 57–85): Elsevier.

Stahl, S. M. (2018). Beyond the dopamine hypothesis of schizophrenia to three neural networks of psychosis: Dopamine, serotonin, and glutamate. *CNS Spectrums, 23*(3), 187–191.

Stern, S., Linker, S., Vadodaria, K. C., Marchetto, M. C., & Gage, F. H. (2018). Prediction of response to drug therapy in psychiatric disorders. *Open Biology, 8*(5), 180031.

Stern, S., Santos, R., Marchetto, M. C., Mendes, A. P. D., Rouleau, G. A., Biesmans, S., … Gage, F. H. (2018). Neurons derived from patients with bipolar disorder divide into intrinsically different sub-populations of neurons, predicting the patients' responsiveness to lithium. *Molecular Psychiatry, 23*(6), 1453–1465.

Sullivan, P. F., Kendler, K. S., & Neale, M. C. (2003). Schizophrenia as a complex trait: Evidence from a meta-analysis of twin studies. *Archives of General Psychiatry, 60*(12), 1187–1192.

Sun, A. X., Yuan, Q., Tan, S., Xiao, Y., Wang, D., Khoo, A. T., … Je, H. S. (2016). Direct induction and functional maturation of forebrain GABAergic neurons from human pluripotent stem cells. *Cell Reports, 16*(7), 1942–1953.

Takahashi, K., Tanabe, K., Ohnuki, M., Narita, M., Ichisaka, T., Tomoda, K., & Yamanaka, S. (2007). Induction of pluripotent stem cells from adult human fibroblasts by defined factors. *Cell, 131*(5), 861–872.

Takahashi, K., & Yamanaka, S. (2006). Induction of pluripotent stem cells from mouse embryonic and adult fibroblast cultures by defined factors. *Cell, 126*(4), 663–676.

Tang, Y., Stryker, M. P., Alvarez-Buylla, A., & Espinosa, J. S. (2014). Cortical plasticity induced by transplantation of embryonic somatostatin or parvalbumin interneurons. *Proceedings of the National Academy of Sciences of the United States of America, 111*(51), 18339–18344.

Tapia, N., & Scholer, H. R. (2016). Molecular obstacles to clinical translation of iPSCs. *Cell Stem Cell, 19*(3), 298–309.

Thomaes, K., Dorrepaal, E., Draijer, N., de Ruiter, M. B., Elzinga, B. M., Sjoerds, Z., … Veltman, D. J. (2013). Increased anterior cingulate cortex and hippocampus activation in complex PTSD during encoding of negative words. *Social Cognitive and Affective Neuroscience, 8*(2), 190–200.

Thomson, J. A., Itskovitz-Eldor, J., Shapiro, S. S., Waknitz, M. A., Swiergiel, J. J., Marshall, V. S., & Jones, J. M. (1998). Embryonic stem cell lines derived from human blastocysts. *Science, 282*(5391), 1145.

Tobe, B. T. D., Crain, A. M., Winquist, A. M., Calabrese, B., Makihara, H., Zhao, W. N., … Snyder, E. Y. (2017). Probing the lithium-response pathway in hiPSCs implicates the phosphoregulatory set-point for a cytoskeletal modulator in bipolar pathogenesis. *Proceedings of the National Academy of Sciences of the United States of America, 114*(22), E4462–E4471.

Träskman, L., Åsberg, M., Bertilsson, L., & Sjüstrand, L. (1981). Monoamine metabolites in csf and suicidal behavior. *Archives of General Psychiatry, 38*(6), 631–636.

Trivedi, M. H., Fava, M., Wisniewski, S. R., Thase, M. E., Quitkin, F., Warden, D., … Rush, A. J. (2006). Medication augmentation after the failure of SSRIs for depression. *New England Journal of Medicine, 354*(12), 1243–1252.

Vadodaria, K. C., Mertens, J., Paquola, A., Bardy, C., Li, X., Jappelli, R., … Gage, F. H. (2016). Generation of functional human serotonergic neurons from fibroblasts. *Molecular Psychiatry, 21*(1), 49–61.

Vadodaria, K. C., Stern, S., Marchetto, M. C., & Gage, F. H. (2018). Serotonin in psychiatry: in vitro disease modeling using patient-derived neurons. *Cell and Tissue Research, 371*(1), 161–170.

Valvezan, A. J., & Klein, P. S. (2012). GSK-3 and Wnt signaling in neurogenesis and bipolar disorder. *Frontiers in Molecular Neuroscience, 5*, 1.

Victor, M. B., Richner, M., Hermanstyne, T. O., Ransdell, J. L., Sobieski, C., Deng, P.-Y., … Yoo, A. S. (2014). Generation of human striatal neurons by MicroRNA-dependent direct conversion of fibroblasts. *Neuron, 84*(2), 311–323.

Vierbuchen, T., Ostermeier, A., Pang, Z. P., Kokubu, Y., Sudhof, T. C., & Wernig, M. (2010). Direct conversion of fibroblasts to functional neurons by defined factors. *Nature, 463*(7284), 1035–1041.

Vitrac, A., & Cloez-Tayarani, I. (2018). Induced pluripotent stem cells as a tool to study brain circuits in autism-related disorders. *Stem Cell Research & Therapy, 9*(1), 226.

Volk, D. W., Austin, M. C., Pierri, J. N., Sampson, A. R., & Lewis, D. A. (2000). Decreased glutamic acid decarboxylase67 messenger rna expression in a subset of prefrontal cortical γ-aminobutyric acid neurons in subjects with schizophrenia. *Archives of General Psychiatry, 57*(3), 237–245.

Wang, J. L., Shamah, S. M., Sun, A. X., Waldman, I. D., Haggarty, S. J., & Perlis, R. H. (2014). Label-free, live optical imaging of reprogrammed bipolar disorder patient-derived cells reveals a functional correlate of lithium responsiveness. *Translational Psychiatry, 4*.

Wang, C., Ward, M. E., Chen, R., Liu, K., Tracy, T. E., Chen, X., … Gan, L. (2017). Scalable production of iPSC-derived human neurons to identify tau-lowering compounds by high-content screening. *Stem Cell Reports, 9*(4), 1221–1233.

Watmuff, B., Berkovitch, S. S., Huang, J. H., Iaconelli, J., Toffel, S., & Karmacharya, R. (2016). Disease signatures for schizophrenia and bipolar disorder using patient-derived induced pluripotent stem cells. *Molecular and Cellular Neurosciences, 73*, 96–103.

Weich, S., Nazareth, I., Morgan, L., & King, M. (2007). Treatment of depression in primary care: Socio-economic status, clinical need and receipt of treatment. *British Journal of Psychiatry, 191*(2), 164–169.

Weiss, A. P., Schacter, D. L., Goff, D. C., Rauch, S. L., Alpert, N. M., Fischman, A. J., & Heckers, S. (2003). Impaired hippocampal recruitment during normal modulation of memory performance in schizophrenia. *Biological Psychiatry, 53*(1), 48–55.

Wium-Andersen, I. K., Vinberg, M., Kessing, L. V., & McIntyre, R. S. (2017). Personalized medicine in psychiatry. *Nordic Journal of Psychiatry, 71*(1), 12–19.

Wong, M. L., Dong, C., Andreev, V., Arcos-Burgos, M., & Licinio, J. (2012). Prediction of susceptibility to major depression by a model of interactions of multiple functional genetic variants and environmental factors. *Molecular Psychiatry, 17*(6), 624–633.

Wong, M. L., Dong, C., Maestre-Mesa, J., & Licinio, J. (2008). Polymorphisms in inflammation-related genes are associated with susceptibility to major depression and antidepressant response. *Molecular Psychiatry, 13*(8), 800–812.

Xia, X., Ding, M., Xuan, J. F., Xing, J. X., Pang, H., Wang, B. J., & Yao, J. (2018). Polymorphisms in the human serotonin receptor 1B (HTR1B) gene are associated with schizophrenia: A case control study. *BMC Psychiatry, 18*(1), 303.

Xu, Z., Jiang, H., Zhong, P., Yan, Z., Chen, S., & Feng, J. (2016). Direct conversion of human fibroblasts to induced serotonergic neurons. *Molecular Psychiatry, 21*(1), 62–70.

Yang, N., Chanda, S., Marro, S., Ng, Y.-H., Janas, J. A., Haag, D., … Wernig, M. (2017). Generation of pure GABAergic neurons by transcription factor programming. *Nature Methods, 14*(6), 621–628.

Yang, N., Ng, Y. H., Pang, Z. P., Sudhof, T. C., & Wernig, M. (2011). Induced neuronal cells: How to make and define a neuron. *Cell Stem Cell, 9*(6), 517–525.

Yang, N., Zuchero, J. B., Ahlenius, H., Marro, S., Ng, Y. H., Vierbuchen, T., … Wernig, M. (2013). Generation of oligodendroglial cells by direct lineage conversion. *Nature Biotechnology, 31*(5), 434–439.

Yatham, L. N., Kennedy, S. H., Parikh, S. V., Schaffer, A., Bond, D. J., Frey, B. N., … Berk, M. (2018). Canadian network for mood and anxiety treatments (CANMAT) and international society for bipolar disorders (ISBD) 2018 guidelines for the management of patients with bipolar disorder. *Bipolar Disorders, 20*(2), 97–170.

Yip, J., Soghomonian, J.-J., & Blatt, G. J. (2007). Decreased GAD67 mRNA levels in cerebellar Purkinje cells in autism: Pathophysiological implications. *Acta Neuropathologica, 113*(5), 559–568.

Yoo, H. (2015). Genetics of autism spectrum disorder: Current status and possible clinical applications. *Experimental Neurobiology, 24*(4), 257–272.

Yoon, J. H., Maddock, R. J., Rokem, A., Silver, M. A., Minzenberg, M. J., Ragland, J. D., & Carter, C. S. (2010). GABA concentration is reduced in visual cortex in schizophrenia and correlates with orientation-specific surround suppression. *The Journal of Neuroscience, 30*(10), 3777–3781.

Young, L. T., Warsh, J. J., Kish, S. J., Shannak, K., & Hornykeiwicz, O. (1994). Reduced brain 5-HT and elevated NE turnover and metabolites in bipolar affective disorder. *Biological Psychiatry, 35*(2), 121–127.

Yu, D. X., Giorgio, D., Francesco, P., Yao, J., Marchetto, M. C., Brennand, K., … Gage, F. H. (2014). Modeling hippocampal neurogenesis using human pluripotent stem cells. *Stem Cell Reports, 2*(3), 295–310.

Zhao, W. N., Cheng, C., Theriault, K. M., Sheridan, S. D., Tsai, L. H., & Haggarty, S. J. (2012). A high-throughput screen for Wnt/beta-catenin signaling pathway modulators in human iPSC-derived neural progenitors. *Journal of Biomolecular Screening, 17*(9), 1252–1263.

Further reading

Singh, I., & Rose, N. (2009). Biomarkers in psychiatry. *Nature, 460*, 202.

Chapter 13

Genetics of alcohol use disorder

Jill L. Sorcher and Falk W. Lohoff
Section on Clinical Genomics and Experimental Therapeutics, National Institute on Alcohol Abuse and Alcoholism, National Institutes of Health, Bethesda, MD, United States

1 Introduction

AUD is a relapsing disorder characterized by excessive and chronic use of alcohol, leading to negative physiological, psychological, and societal outcomes. AUD is a common disorder with high morbidity and mortality, contributing to approximately 88,000 deaths annually in the United States (Stahre, Roeber, Kanny, Brewer, & Zhang, 2014), and 4% of deaths worldwide (Association, 2013; Rehm et al., 2014, 2009; Stahre et al., 2014). To diagnose individuals with AUD, the Diagnostic and Statistical Manual of Mental Disorders, Fifth Edition (Mizokawa et al., 2013) utilizes 11 criteria pertaining to excessive alcohol use, alcohol abuse, and alcohol dependence. These 11 criteria incorporate an array of symptoms, including drinking more or for longer than intended, or continuing to drink despite psychological or health problems. The range of symptoms encompassed by these criteria demonstrates the heterogeneous clinical presentation of the disorder.

The prevalence of AUD in the Unites States has increased in the past decade in individuals 18 years of age or older (Reilly, Noronha, Goldman, & Koob, 2017). The National Epidemiologic Survey on Alcohol and Related Conditions (NESARC) revealed that lifetime DSM IV AUD diagnosis increased from 30.3% in 2001–2002 to 43.6% in 2012–2013 (Grant et al., 2016). AUD was found to be more common in men (42%) than women (19.5%) (Grant et al., 2016), and frequently co-occurs with other psychiatric disorders, including mood and anxiety disorders (Regier et al., 1990), post-traumatic stress disorder (Sampson et al., 2015), and other substance use disorders (Kessler et al., 1997). The Epidemiologic Catchment Area Study reported that roughly half of all individuals with AUD have a "dual diagnosis," or comorbid psychiatric disorder (Regier et al., 1990). Furthermore, the National Comorbidity Study found that three-fourths of men and women with AUD met lifetime criteria for the diagnosis of a psychiatric illness (Kessler et al., 1997). These psychiatric disorders often have strong genetic heritability estimates. Therefore, this data not only highlights the phenotypic variability of AUD, but also indicates how this heterogeneity complicates the genetics of AUD.

Various pathways to the development of AUD include an interaction of environmental and genetic risk factors. This interaction between genetics and environmental factors influences the phenotypic variability associated with AUD. While the pathophysiology of AUD is not yet fully understood, there is strong evidence suggesting a genetic component (Tawa, Hall, & Lohoff, 2016). Approximately 50% of the risk for developing AUD is due to genetics, while the remaining percent may be due to either environmental factors, or gene-environment interactions (Tawa et al., 2016). Therefore, a genetic predisposition to the addictive effects of alcohol, combined with other environmental risk factors, may cause harmful lifetime drinking patterns and AUD development. Lifetime drinking history is an important behavioral risk factor, which includes the age of first alcohol use, average number of drinks per day, and number of years of heavy drinking (Grant & Dawson, 1998). Environmental factors, such early life stressors (including physical or sexual abuse) increase the risk for alcohol use disorders later in life (Enoch, 2011). High trait anxiety and other psychological factors are associated with an increased risk for developing AUD (Poikolainen, 2000). Physiological factors, such as alcohol withdrawal, also influence the risk of developing AUD (Becker & Mulholland, 2014). These environmental risk factors for AUD have been previously predicted to have an underlying genetic cause, further complicating the ability to define a specific universal set of genetic and environmental factors that influence risk of AUD.

This chapter examines the genetic underpinnings of alcohol use disorder (AUD), with an emphasis on recent genome-wide association studies (GWAS) (Table 1). Because AUD is a common psychiatric disorder characterized by a complex array of traits, GWAS may provide an effective experimental approach to identifying associated genetic variants. The heterogeneity observed in (Goodwin, 1979) AUD may be best treated by personalized medicine, which has started to replace

TABLE 1 Positive findings from GWAS on alcohol use disorder and alcohol-related phenotypes.

Author	Phenotype	Gene/SNP	*P* value	Sample	Ethnicity
Bae et al. (2012)	Alcohol dependence	Chr20 (rs61195302–61,195,978)	$p=5.02e-05$	1138	Korean population
Baik, Cho, Kim, Han, and Shin (2011)	Alcohol consumption	*C12ORF24* (rs2074356)	$p=9.49e-59$	1721	Korean men
Biernacka, Geske, Schneekloth, et al. (2013) and Biernacka, Geske, Jenkins, et al. (2013)	Alcohol dependence	KEGG pathway ID 72-synthesis and degradation of keton bodies	$p=.003$	SAGE (2544)	European- and African-American
Bierut et al. (2010)	Alcohol dependence	*GABRA2*	$p<0.05$	SAGE (3829)	European- and African American
		PBX/knotted 1 homeobox 2, *PKNOX2*	$p=1.93e-07$		
Blednov et al. (2015)	Alcohol dependence	*PPARGC1A*	$p<0.001$	COGA (2322)	European-American
Chen et al. (2012)	Alcohol drinking	Ankyrin repeat domain 7, *ANKRD 7*, and cytokine-like 1, *CYTL1*, (rs6466686-rs4295599-rs12531086) (halotype)	$p=6.51e-8$	904	Caucasian
Edenberg et al. (2010)	Alcohol dependence	Bobby sox homolog, *BBX*	$p=3.4e-06$	COGA (2604)	European-American
		Solute carrier family 22 member 18, *SLC22A18*, cluster on chromo 11	$p=3.4e-07$		
		Pleckstrin homology-like domain family A member 2, *PHLDA2*, cluster on chromo 11	$p=3.4e-08$		
		Nucleosome assembly protein 1-like 4, *NAP1L4*, cluster on chromo 11	$p=3.4e-09$		
		Small nucleolar RNA, H/ACA box 54, *SNORA54*, cluster on chromo 11	$p=3.4e-10$		
		Cysteine-tRNA ligase, *CARS*, cluster on chromo 11	$p=3.4e-11$		
		Oxysterol-binding protein-related protein 5, *OSBPL5*, cluster on chromo 11	$p=3.4e-12$		
Frank et al. (2012)	Alcohol dependence	*ALDH2* (rs671)	$p=1.27e-08$	3501	German descent
		ADH1 between *ADH1B* and *ADH1C* (rs1789891)	$p=1.27e-08$		

TABLE 1 Positive findings from GWAS on alcohol use disorder and alcohol-related phenotypes—cont'd

Author	Phenotype	Gene/SNP	*P* value	Sample	Ethnicity
Gelernter et al. (2014)	Alcohol dependence	*ADH1B* (rs1229984)	$p=1.17e-31$	379 European Americans	European- and African American
		ADH1B (rs1789882)	$p=6.33e-17$	3318 African Americans (total=16,087)	
		ADH1C (Thr151Thr)	$p=4.94e-10$		
		Between *MTIF2* and *CCDC88A* on chromosome 2 (rs1437396)	$p=1.17e-10$		
Heath et al. (2011)	Drinking severity	Ankyrin repeat and sterile alpha motif domain containing 1A, *ANKS1A* (rs2140418)	$p=3.4e-04$	8754	Australian
Johnson et al. (2006)	Alcohol dependence	51 gene loci, including *CDH11*, *CDH13*	$p=0.00034$	COGA (280)	European-American
Kendler et al. (2011)	Alcohol dependence	Calcium-activated potassium channel subunit alpha-1, *KCNMA1*	$p=2.17e-05$	3169	Caucasians and African American
		Carcinoembryonic antigen-related cell adhesion molecule 6 precursor, *CEACAM6*	$p=1.05e-05$		
Lind et al. (2010)	Alcohol and nicotine co-dependence	Near MAP/microtubule affinity-regulating kinase 1, *MARK1*	$p=1.90e-09$	1087	Australian
		Near DEAD (Asp-Glu-Ala-Asp) box helicase 6, *DDX6*	$p=2.6e-09$		
		KIAA1409	$p=4.86e-08$		
Lind et al. (2010)	Alcohol and nicotine co-dependence	Near semaphorin 3E, *SEMA3E*	$p=6.23e-06$	OZALC (2386)	Australian and Dutch
Pan et al. (2013)	Maximum number of drinks	Near *SGOL1* (rs11128951)	$p=4.27e-8$	COGA (1059), SAGE (1628)	European- and African-American
Park et al. (2013)	Alcohol dependence	*ALDH2* (rs671)	$p=8.42e-08$	396	Korean
		ADH1B (rs1229984)	$p=2.63e-21$		
		ADH7 (rs1442492)	$p=6.28e-8$		
Pei et al. (2012)	Alcohol drinking	Near FYVE, PhoGEF, and PH domain containing protein 4, *FGD4*	$p=3.05e-03$	2286	Caucasian
		Near latent-transforming growth factor beta-binding protein 1, *LTBP1*	$p=1.91e-03$		
Quillen et al. (2014)	Alcohol dependence	*ALDH2* (rs671)	$p=4.55e-08$	595	Chinese
Schumann et al. (2011)	Alcohol consumption	*AUTS2* (rs6943555)	$p=4e-08$	12 population-based samples (26,316)	European

Continued

TABLE 1 Positive findings from GWAS on alcohol use disorder and alcohol-related phenotypes—cont'd

Author	Phenotype	Gene/SNP	P value	Sample	Ethnicity
Takeuchi et al. (2011)	Alcohol consumption	ALDH2 (rs 671)	$p = 3.6e-211$	2974 drinkers, 1521 occasional drinkers, 1351 non-drinkers	Japanese
		ADH1B (rs1229984)	$p = 3.6e-4$		
Treutlein et al. (2009)	Alcohol dependence	Near peroxisomal trans-2-enoyl-CoA reductase, PECR (rs7590720 and rs1344694)	$p = 9.72e-09$	1845	German
		CDH13 (rs11640875)	$p = 1.84e-5$	2020	German
		ADH1C (rs1614972)	$p = 1.41e-4$		
		GATA4 (rs13273672)	$p = 4.75e-4$		
Wang et al. (2011)	Alcohol dependence	DSCMAL1	$p < 10e-08$	COGA, OZALC 272 nuclear families	European-American and Australian
Wang et al. (2011)	Alcohol dependence	Near endothelin receptor type B, EDNRB	$p = 8.51e-06$	COGA, OZALC 272 nuclear families	European-American, African-American, and Australian
		TPARP, CYFIP2, THEMIS, PSG11	$p = 2.31e-5$		
Wang et al. (2011)	Alcohol dependence	KIAA0040, THSD7B, NRD1	$p = 1.86e-07$	COGA (1594), SAGE (1669), OZALC (3334)	European-American, African-American, and Australian
Wang et al. (2013)	Alcohol dependence symptom count	3 SNPs in C15ORF53 gene	$p = 4.5e-8$	COGA (2322)	European-American
Zuo et al. (2011)	Alcohol dependence	PHD finger protein 3, PHF3, − Protein tyrosine phosphatase type IVA 1, PTP4A1, locus	$p < 10e-4$	COGA, SAGE (4116)	European- and African American
Zuo et al. (2012)	Alcohol dependence	KIAA0040	$p = 2.8e-07$	COGA, SAGE (4116)	European- and African-American
Zuo et al. (2012)	Alcohol and nicotine co-dependence	SH3 domain binding protein 5, SH3BP5	$p = 6.9e-6$	SAGE (3143)	European- and African American
		Nuclear receptor subfamily 2, group C, member 2, NR2C2	$p = 5.3e-4$		
		Plasminogen-like B2, PLGLB2	$p = 3.1e-08$		
Zuo et al. (2013)	Alcohol Dependence	NKAIN1-SERINC2	$p = 1.7e-07$	COGA, SAGE (2927)	European- and African-American
Zuo et al. (2013)	Alcohol and nicotine co-dependence	IPO11-HTR1A region on chromosome 5q	$p = 6.2e-9$	COGA, SAGE (2214)	European- and African-American

the unsuccessful one-size-fits-all treatment approach (Reilly et al., 2017). Further exploration of SNPs and other genetic variants associated with AUD could be used to advise at-risk individuals and develop more effective pharmacological interventions.

2 Twin studies

The assumed role of genetics in AUD was founded on observations that alcohol problems are common in families (Cotton, 1979). AUD has been documented in families by almost every civilization dating back to the ancient Greeks (Goodwin, 1979). However, this observational evidence of familial history of AUD was not sufficient to provide conclusive proof that familial association of AUD is mitigated by genetics.

Research using family and adoption studies clarified the role of genetics in AUD. Early studies reported that relatives of individuals with AUD exhibited a three to fourfold increase in risk for AUD (Goodwin, 1979). Further research sought to clarify whether this risk was based on environment, genetics, or a combination of the two. Studies found that adverse alcohol outcomes in offspring was more closely related to AUD in a biological parent than AUD in foster families (Schuckit, Goodwin, & Winokur, 1972). This was confirmed by studies demonstrating a threefold or greater increased risks for AUD in sons of individuals with AUD, even when they were adopted as infants (Goodwin et al., 1974; Goodwin, Schulsinger, Hermansen, Guze, & Winokur, 1973; Goodwin, Schulsinger, Knop, Mednick, & Guze, 1977).

Further evidence of genetics in AUD was demonstrated through research utilizing twin studies. The Australian twin study of AUD (OZALC) found a greater concordance of alcohol dependence in monozygotic (56% for males) compared with dizygotic twins (33% for males), and a heritability estimate of 64% (Heath et al., 1997). More recent twin studies have established that AUD heritability ranges from 40% to 70% (Agrawal & Lynskey, 2008; Enoch & Goldman, 2001; Kendler et al., 2012), with similar heritability estimates in both males and females (Heath et al., 1997; Prescott, Aggen, & Kendler, 1999). Another study using twins from the Vietnam Era Twin Registry reported heritability estimates for 23 symptoms of alcohol dependence, further highlighting the heterogeneity of AUD (Slutske et al., 1999). The heritability estimates of these twin studies strongly suggest that genetic components contribute to AUD. Given that concordance rates are lower than 50%, other factors, such as rare somatic de novo mutations, environmental influences, and gene-environment interactions might contribute to the remaining heritability.

Although family and twin studies provide compelling evidence for the role of genetics in AUD, these studies fail to indicate specific genes involved in AUD heritability (Reilly et al., 2017).

3 Linkage studies

Because of the strong epidemiological evidence for the role of genetic factors in AUD, the field has hoped for a straightforward identification of AUD risk alleles. The first comprehensive investigation into the genetics of AUD used linkage studies. Linkage studies are useful for identifying broad regions of the genome associated with a large increase in risk for a disorder. The term *linkage* refers to the observation that two genetic markers on the same chromosome are often inherited together. This approach uses families with multiple affected members to determine chromosomal regions with genetic risk variants. Previous linkage studies have been most successful in rare autosomal dominant diseases with high penetrance, such as cystic fibrosis.

One of the first linkage studies on alcohol dependence from the Collaborative Study on the Genetics of Alcoholism (COGA) (Reich et al., 1998) was a family based study formed to map and characterize the genetic variants associated with alcoholism (Bierut et al., 2002). The results from COGA and a sib-pair study from a Southwest American Indian tribe (Long et al., 1998) reported a broad risk locus on chromosome 4q. This region contains genes that encode isoforms of alcohol dehydrogenase. Other linkage studies have implicated regions encoding GABA-A A (Wang et al., 2004) and CHRM2 (Wang et al., 2004), among other variants (Edenberg & Foroud, 2014). These studies were unable to identify specific risk variants associated with AUD, and instead found chromosomal regions with multiple linkage peaks (Tawa et al., 2016), which highlight the complex, polygenetic biology and heterogeneic heritability of AUD.

3.1 Candidate gene association studies

Case/control association studies can be used to test candidate genes within chromosomal regions identified through linkage studies or based on neurological plausibility. Candidate gene association studies identify alleles of a gene that have different frequency in a population with a trait or disease vs those without (Reilly et al., 2017), and can identify small increases in genetic risk variants associated with a disorder. Despite the promise of these studies, many candidate gene studies on AUD have yielded results with small effect sizes, and are often difficult to replicate.

Based on previous linkage studies, the strongest associations have been identified in the alcohol metabolism genes, alcohol dehydrogenase (*ADH*) and aldehyde dehydrogenase (*ALDH*). Alcohol metabolism is a two-step process in which ethanol is first oxidized to acetaldehyde by ADH. Acetaldehyde is then oxidized by ALDH to form acetate. Disruptions in this process can lead to accumulation of acetaldehyde, a toxic intermediate that can cause adverse physiological symptoms, including flushing syndrome, tachycardia, and nausea. Genetic variations in ADH and ALDH influence the rate at which acetaldehyde is produced and converted to acetate. Evidence indicates that individuals with isoforms of ADH that oxidize ethanol at a faster rate and/or isoforms of ALDH that oxidize acetaldehyde at a slower rate are protected against AUD due to the unpleasant effects that result from acetaldehyde accumulation after alcohol consumption (Tawa et al., 2016).

There are several isoforms of ADH that metabolize alcohol in the liver (ADH1A, ADH1B, ADH1C, and ADH4–7). The majority of alcohol oxidation is performed by the isoform ADH1B. Research demonstrates that individuals with the ADH1B*2 single-nucleotide polymorphism (SNP) rs1229984 (Arg48His) metabolize alcohol at a much faster rate than those with the ADH1B*1 variant (Tawa et al., 2016). ADH1B*2 is more common in East Asians than other populations, and has been found to have a protective effect on the risk for developing AUD (Edenberg, 2007; Li, Zhao, & Gelernter, 2012; Luczak, Glatt, & Wall, 2006). The underlying mechanism of the hypothesis is attributed to the aversive reaction associated with toxic acetaldehyde accumulation. ADH1B has been shown to reduce the risk for AUD in Asians, Native Americans, Europeans, and African Americans (Reilly et al., 2017). ADH1A and ADH1C have also been shown to contribute to alcohol metabolism at lower ethanol concentrations, and several studies have found a protective effect of the ADH1C*2 SNP rs698 (Ile350Val) (Biernacka, Geske, Schneekloth, et al., 2013). Variations in isoforms of ALDH have also been indicated to affect the risk of AUD. A majority of the oxidation of acetaldehyde to acetate is performed by the acetaldehyde dehydrogenase isoform ALDH2. Individuals with the ALDH2*2 SNP rs671 (Glu504Lys) metabolize acetaldehyde at a decreased rate. Slow metabolism of acetaldehyde causes alcohol flushing syndrome, which is prevalent in East Asian populations (Tawa et al., 2016). Partially due to this negative symptom, an ALDH2*2 variant has been associated with a decreased risk of AUD (Edenberg & Foroud, 2006; Hurley & Edenberg, 2012; Li et al., 2012). Both the ADH and ALDH2 variants have been associated with protective effects against the risk of developing AUD.

Several candidate gene studies conducted by COGA have found associations with more than 20 genes affecting AUD and endophenotypes associated with alcoholism (Edenberg & Foroud, 2006). Other studies conducted in various countries and populations have identified genetic variants contributing to vulnerability to alcoholism (Goldman, Oroszi, O'Malley, & Anton, 2005; Kalsi, Prescott, Kendler, & Riley, 2009). Although alcohol metabolizing enzyme polymorphisms are the most consistent candidate genes, genetic variants in several neurotransmitter systems have been associated with AUD (Reilly et al., 2017; Tawa et al., 2016). These include the glutamate receptor (*GIRK1*), gamma-aminobutyric acid receptor (*GABA-A*), D2 dopamine receptor (*DRD2*), dopamine transporter (*SLC6A3*), serotonin transporter (*SLC6A4* and *5-HTTLPR*), tryptophan hydroxylase 1 (*TPH1*), catechol-*O*-methyltransferase (*COMT*), cholinergic muscarinic receptor (*CHRM2*), and u-opioid receptor (*OPRM1*, among others) (Chen et al., 2012; Du, Nie, Li, & Wan, 2011; Edenberg et al., 2004; Kranzler et al., 2009; McHugh, Hofmann, Asnaani, Sawyer, & Otto, 2010; Munafo, Matheson, & Flint, 2007; Sahni et al., 2018; Tammimaki & Mannisto, 2010; Wang et al., 2004; Xu & Lin, 2011). However, many of these genetic risk variants are dependent on gene-environment interactions, and replication of findings has been difficult due to the generally small expected effect size, lack of power, and clinical and genetic heterogeneity.

3.2 Genome-wide association studies

Unlike candidate gene studies, which require prior knowledge about the neurobiology underlying potential genetic risk variants, genome-wide association studies (GWAS) provide high resolution analysis of the entire genome without an *a priori* hypothesis. GWAS are used to analyze several hundred thousand or millions of SNPs across the genome to identify differences in genotype frequencies among case-controlled individuals, without selecting for only a few specific genes. The GWAS approach is theoretically capable of identifying all the common genetic variants associated with a disease or trait, and is particularly useful in identifying genetic risk variants in complex disorders where small effect sizes are expected. However, the heritability explained by SNP associations that reach statistical association in GWAS studies is significantly less than heritability derived from family studies (Reilly et al., 2017), which is referred to as the "missing heritability" problem (Manolio et al., 2009). Nonetheless, GWAS studies have been successful in identifying SNPs associated with disease processes, such as rheumatoid arthritis (Suzuki, Kochi, Okada, & Yamamoto, 2011). Emerging evidence suggests that GWAS may be successful in identifying genetic underpinnings of mental illnesses, including schizophrenia and bipolar disorder (Moskvina et al., 2009; Schizophrenia Working Group of the Psychiatric Genomics, 2014). Therefore, GWAS may be useful in elucidating the genetic variants associated with AUD. The studies presented here assess AUD based on the DSM-IV-TR criteria for alcohol dependence.

A study analyzing 104, 268 SNPs from 120 patients and 160 controls in a European-American population from the COGA sample identified 51 chromosomal regions in the genome that harbor at least three genetic risk markers, including genes of the cell adhesion molecules cadherin 11 (*CDH11*) and cadherin 13 (*CDH13*) (Johnson et al., 2006). However, the results of this study were only nominally significant. The first positive GWAS of alcohol dependence found genome-wide significance for two intergenic loci on chromosome 2q35 (rs7590720 and rs1344694) close to the gene for peroxisomal trans-2-enoyl-CoA reductase (*PECR*) (Treutlein et al., 2009). This study examined 487 alcohol dependent patients and 1358 controls from a German population (Treutlein et al., 2009). A follow-up study replicated 15 significant SNPs, including rs11640875 in *CDH13*, rs1614972 in *ADH1C*, and rs13273672 in the GATA binding protein 4 (*GATA4*) (Treutlein et al., 2009). Previous studies have indicated an association among GATA4 and alcohol dependence, relapse risk (Kiefer et al., 2011), and limbic gray matter volume, which was predictive of an increased risk for alcohol relapse (Zois et al., 2016). Although this evidence indicates an association between the *GATA4* gene and a diagnosis of AUD, a recent replication study failed to find confirm this association in an independent sample (Mauro et al., 2018). Reanalysis of the GWAS data focusing on peroxisome proliferator-activated receptors (PPARRs), which have been associated with neuroinflammatory processes associated with alcohol (Varga, Czimmerer, & Nagy, 2011), found an association of SNPs in PPARA and PPARG with alcohol withdrawal, and PPARGC1A with alcohol dependence (Blednov et al., 2015).

In contrast to early GWAS studies, recent efforts have attempted to increase sample size by analyzing genome-wide SNPs from the COGA, OZALC, and the Study of Addiction: Genetics and Environment (SAGE) samples (Tawa et al., 2016). Although the first complete GWAS of the COGA failed to find single SNPs with genome-wide significance, it provided converging evidence for a chromosome 11 gene cluster (Edenberg et al., 2010). A separate study analyzing 11,120 SNPs from the COGA and OZALC samples found genome-wide significance at $p < 10^{-8}$ for DSCMAL1 (Wang et al., 2011). Despite this promising result, further research studies analyzing the COGA, OZALC, and SAGE samples have failed to find genome-wide significant markers (Biernacka, Geske, Jenkins, et al., 2013; Bierut et al., 2010; Lind et al., 2010; Wang et al., 2011; Zuo et al., 2011, 2012, 2013).

Despite the lack of positive results from the COGA, OZALC, and SAGE samples, a group of GWAS of AUD published in the last several years found significant SNPs in genes encoding alcohol metabolizing enzymes to be among variants with the largest effects on AUD risk (Reilly et al., 2017). These GWAS finding have confirmed previous findings from candidate gene and linkage studies indicating association for genetic risk variants in *ADH* and *ALDH* genes. For example, a GWAS of alcohol risk variants in a German population found genome-wide significance for rs1789891 in the *ADH1* gene cluster (Frank et al., 2012). The SNP was found to be in linkage disequilibrium with the functional variant ADH1C (Arg272Gln). A GWAS in an East Asian sample confirmed the association with the ADH gene cluster by finding multiple nominally significant SNPs in the ADH gene cluster on chromosome 4q22-q23, as well as genome-wide significance for rs1442492 and rs10516441 in *ADH7* and rs671 in *ALDH2* (Park et al., 2013). Additionally, other GWAS have shown that the *ALDH2*2* variant rs671 (Glu504Lys) is associated with a decreased risk of AUD in East Asian populations (Frank et al., 2012; Quillen et al., 2014; Takeuchi et al., 2011). GWAS studies have found that the ADH1B*2 SNP rs1229984 in particular is protective against AUD development (Park et al., 2013; Takeuchi et al., 2011). This relationship was supported by a study that reported that the SNP rs1229984 decreased the risk of AUD in a European-American population and theADH1C variant (Thr151Thr), and the ADH1B SNP rs1789882 (Arg369Cyc) decreased the risk of AUD in an African-American population (Gelernter et al., 2014).

GWAS analyzing other alcohol phenotypes other than AUD and alcohol dependence, such as alcohol and nicotine codependence, and alcohol consumption, have found genome-wide significant SNPs. For example, genome-wide significance for rs7530302 near MAP/microtubule affinity-regulating kinase 1 (*MARK1*), rs1784300 near DEAD box helicase 6 (*DDX6*), and rs12882384 in *KIAA1409* in a study population analyzing individuals with alcohol and nicotine comorbidity (Lind et al., 2010). Another study found SNPs in SH3 domain binding protein 5 (*SH3BP5*), nuclear receptor subfamily 2 group C member 2 (*NR2C2*), Plasminogen-like B2 (*PLGLB2*), and rs7445832 in IPO11-HRT1A region on chromosome 5q associated with alcohol and nicotine dependence codependence (Zuo et al., 2012, 2013). Positive associations between genetic risk variants and alcohol consumption have also been indicated. A study of 26,316 individuals from 12 population-based samples identified genome-wide significance of alcohol consumption for *AUT2* rs6943555 (Schumann et al., 2011). Another study analyzing genetic variants associated with alcohol consumption found significance of SNPs on chromosome 12q24, including *C12ORF51* rs2074356, which is LD with *ALDH2*, *CCDC63*, and *MYL2* (Baik et al., 2011). Additionally, genome-wide significance for SNP clusters in Ankyrin repeat domain 7 (*ANKRD7*) Cytokine-like 1 (*CYTL1*) were associated with alcohol consumption (Chen et al., 2012). Alcohol dependent symptom count have also been used as criteria in GWAS, which indicated three SNPs in the *C15ORF53* as significant (Wang et al., 2013).

The recent use of GWAS to identify genetic risk factors of AUD has been promising, but must be interpreted with caution. There are several limitations of GWAS that must be considered. First, GWAS genotypes hundreds of thousands to 2 million markers simultaneously in cases and controls. This "hypothesis-free" design generates large amounts of data and complicates the ability to account for multiple comparisons. The current statistical correction for GWAS is a p value of 10^{-8} (Tawa et al., 2016). Relative to this stringent correction factor, early GWAS in psychiatric phenotypes failed to yield significant results (Craddock & Sklar, 2013; Sklar et al., 2008). Despite sample sizes in the range of 1000–2000, these early GWAS were underpowered to detect risk variants of the small effect mandated by the statistical correction factor. Current power and sample size estimates for GWAS with effect sizes of 1.05–1.2 range from 30,000 to 120,000 (Owen, Craddock, & O'Donovan, 2010; Schizophrenia Working Group of the Psychiatric Genomics, 2014). The use of a strict p-value avoids false positive finding, but may also result in false negatives. To prevent missing "true" variants, recent attempts have used pathway analysis and polygenic risk score approaches (Gelernter et al., 2014). However, these approaches have not been applied to AUD genetic analyses on a global scale.

AUD is a complex disorder, and thus, it is likely that hundreds, if not thousands, of genes contribute to its pathogenesis and varied phenotype. Due to this heterogeneity, it is unlikely that GWAS will detect genes of large effect. Furthermore, the current chip-based GWAS methodology misses rare de novo mutations or insertion/deletion variants that could contribute to the development of AUD (Clarke & Cooper, 2010; Stefansson et al., 2008; Walsh et al., 2008). Despite these methodologic shortcomings, it is expected that GWAS will continue to be the standard of investigation in the quest to understand the genetic underpinnings of AUD. In order to advance the current understanding of the genetic mechanisms of AUD, GWAS will need to be expanded to include large meta-analyses. Although large meta-analyses have been successful for GWAS in other psychiatric phenotypes (Psychiatric, 2011, 2009), no AUD GWAS meta-analysis currently exists literature (Dick & Agrawal, 2008).

The changing psychiatric clinical diagnoses criteria regarding AUD may pose a challenge to sample continuity, and thus complicate identification of genetic variants using GWAS. The GWAS studies presented in this chapter employ the DSM-IV-TR criteria for alcohol dependence. However, the field of psychiatry has recently transitioned to the DSM-5 criteria for AUD. This shift may cause changes in the functional variants identified by GWAS. As a result of this changing definition, future GWAS should focus on the endophenotypes of AUD in order to better understand the genetic connections to specific behavioral symptoms. Clinically, AUD often presents as a comorbidity associated with other substance use and psychiatric disorders. As the field progresses, it will be important to understand and differentiate genetic variants involved in multiple disorders and addiction relative to those solely implicated in AUD development.

3.3 GCTA/GREML methods

The genome-wide complex trait analysis (GCTA)/genomic restricted maximum likelihood (GREML) has been adopted by researchers to address the "missing heritability" associated with GWAS studies (Yang, Lee, Goddard, & Visscher, 2011). GCTA/GREML, or GCTA, is a statistical method that estimates variance in genetics by quantifying the chance genetic similarity of individuals and comparing their similarity in trait measurements. If the GCTA estimate of SNP heritability is consistent with the total genetic heritability, it is implicated that those genetics variants have a causal effect on the observed phenotype (Yang et al., 2011). GCTA can corroborate GWAS results by relating SNP to trait heritability; thus addressing the "missing heritability" component of GWAS studies. With regard to AUD, GCTA could be used to assess heritability of SNPs that were correlated with alcohol-dependent phenotypes and reached genome-wide significance. GCTA estimates of heritability could be useful for diagnostic purposes, and elucidate which genetic variants may contribute to an individual's risk of developing AUD.

Previous research has employed GCTA to investigate the heritability of various substance disorders, including tobacco, cannabis, and illicit drugs. This evidence indicated that common SNPs contribute to at least 20% of the variance in substance dependence heritability risk (Palmer et al., 2015). Additional GCTA research has examined the heritability of behavioral disinhibition, which has been previously linked to substance use disorder development. Behavioral inhibition was highly genetically correlated with all substance use traits, including nicotine use/dependence, alcohol consumption, alcohol dependence, and drug use (Vrieze, McGue, Miller, Hicks, & Iacono, 2013). In particular, heritability was as high as 56%, and the aggregate additive SNP effects estimated by GCTA accounted for 16% of the variance associated with alcohol dependence (Vrieze et al., 2013). Overall, these studies indicate the utility of GCTA methods in identifying heritability of substance use disorders via aggregate effects of genetic variants, and address an important gap in literature created by GWAS studies. Because GWAS findings on substance dependence have been limited by the "missing heritability" problem, these GCTA studies contribute to a deeper understanding of the genetic mechanisms underlying substance use disorders.

3.4 Whole genome sequencing

With the advent of new technology, the cost of whole genome sequencing has decreased, making it a more feasible research method. Although more expensive than GWAS, whole genome sequencing is more likely to identify rare mutations, particularly recessive mutations, in exonic regions of the genome. Exome analysis has been successful in identifying genetic mutations associated with autism spectrum disorders, as well as mood and psychiatric disorders (Kato, 2015). Currently, exome/whole genome sequencing for substance use disorders does not exist in the literature. Given the value of genetic information obtained from whole genome sequencing, the financial burden could be outweighed by the value of genetic information that could obtained.

4 Conclusions

Previous research supports an interaction between environmental and genetic factors, as in the development of AUD. Despite strong evidence indicating the heritability of AUD, identification of a single risk genetic risk variant of AUD has been elusive. As the field of genomics has advanced and new technology has developed, GWAS have identified new variants associated with AUD. The results from GWAS studies have been the most successful in advancing the field of genetics and substance use disorders. In particular, the most significant SNPs were in the alcohol metabolism enzyme genes, *ADH* and *ALDH*. The prevalence of the various isoforms of ADH and ALDH differs among ethnicities and populations. Lower alcohol use in certain population may be due to gene-environment interactions mitigated by the protective effect of alcohol metabolism SNPs.

Knowledge of the neurobiology and genetics of AUD could greatly enhance prevention and treatment of the disorder. Currently, family history can be used to identify at-risk individuals. However, understanding the complex genetic architecture of AUD could provide a better understanding of at-risk individuals' biology, and facilitate the development of new and more targeted medications. Despite the strong evidence supporting the heritability of AUD, identifying underlying genetic risk factors has proved challenging due to small effect sizes, heterogeneity, and complex modes of inheritance (Tawa et al., 2016). However, the field of epigenetics in AUD is just developing (Robison & Nestler, 2011), and new advances are facilitating epigenome-wide association studies of disease phenotypes using DNA methylation. Given the known interplay of genetic and environmental risk factors in AUD pathogenesis, exploring epigenetics underlying AUD could be a promising direction for the field. A recent cross-tissue and cross-phenotypic epigenetic analysis of genome-wide methylomic variation in AUD identified proprotein convertase subtilisin/kexin 9 (*PCSK9*) (Lohoff et al., 2018) and phosphatidylinositol-4-phosphate 5-kinase (*PIP5K1C*) (Lee et al., 2018) as top targets associated with AUD. Understanding the epigenetic and genetic architecture of AUD will improve both the understanding of at-risk individuals' biology and help facilitate the development of new pharmacological interventions.

REFERENCES

Agrawal, A., & Lynskey, M. T. (2008). Are there genetic influences on addiction: Evidence from family, adoption and twin studies. *Addiction, 103*(7), 1069–1081. https://doi.org/10.1111/j.1360-0443.2008.02213.x.

Association, A (2013). *Diagnotic and statistical manual of mental disorders.* (Vol. 5). Washington, DC: American Psychiatric Association.

Bae, J. S., Jung, M. H., Lee, B. C., Cheong, H. S., Park, B. L., Kim, L. H., ... Choi, I. G. (2012). The genetic effect of copy number variations on the risk of alcoholism in a Korean population. *Alcoholism: Clinical and Experimental Research, 36*(1), 35–42. https://doi.org/10.1111/j.1530-0277.2011.01578.x.

Baik, I., Cho, N. H., Kim, S. H., Han, B. G., & Shin, C. (2011). Genome-wide association studies identify genetic loci related to alcohol consumption in Korean men. *The American Journal of Clinical Nutrition, 93*(4), 809–816. https://doi.org/10.3945/ajcn.110.001776.

Becker, H. C., & Mulholland, P. J. (2014). Neurochemical mechanisms of alcohol withdrawal. *Handbook of Clinical Neurology, 125*, 133–156. https://doi.org/10.1016/B978-0-444-62619-6.00009-4.

Biernacka, J. M., Geske, J., Jenkins, G. D., Colby, C., Rider, D. N., Karpyak, V. M., ... Fridley, B. L. (2013). Genome-wide gene-set analysis for identification of pathways associated with alcohol dependence. *The International Journal of Neuropsychopharmacology, 16*(2), 271–278. https://doi.org/10.1017/S1461145712000375.

Biernacka, J. M., Geske, J. R., Schneekloth, T. D., Frye, M. A., Cunningham, J. M., Choi, D. S., … Karpyak, V. M. (2013). Replication of genome wide association studies of alcohol dependence: Support for association with variation in ADH1C. *PLoS One*, *8*(3). https://doi.org/10.1371/journal.pone.0058798.

Bierut, L. J., Agrawal, A., Bucholz, K. K., Doheny, K. F., Laurie, C., Pugh, E., … Gene, E. A. S. C. (2010). A genome-wide association study of alcohol dependence. *Proceedings of the National Academy of Sciences of the United States of America*, *107*(11), 5082–5087. https://doi.org/10.1073/pnas.0911109107.

Bierut, L. J., Saccone, N. L., Rice, J. P., Goate, A., Foroud, T., Edenberg, H., … Reich, T. (2002). Defining alcohol-related phenotypes in humans. The Collaborative Study on the Genetics of Alcoholism. *Alcohol Research & Health*, *26*(3), 208–213.

Blednov, Y. A., Benavidez, J. M., Black, M., Ferguson, L. B., Schoenhard, G. L., Goate, A. M., … Harris, R. A. (2015). Peroxisome proliferator-activated receptors alpha and gamma are linked with alcohol consumption in mice and withdrawal and dependence in humans. *Alcoholism, Clinical and Experimental Research*, *39*(1), 136–145. https://doi.org/10.1111/acer.12610.

Chen, D., Liu, F., Yang, C., Liang, X., Shang, Q., He, W., & Wang, Z. (2012). Association between the TPH1 A218C polymorphism and risk of mood disorders and alcohol dependence: Evidence from the current studies. *Journal of Affective Disorders*, *138*(1–2), 27–33. https://doi.org/10.1016/j.jad.2011.04.018.

Clarke, A. J., & Cooper, D. N. (2010). GWAS: Heritability missing in action? *European Journal of Human Genetics*, *18*(8), 859–861. https://doi.org/10.1038/ejhg.2010.35.

Cotton, N. S. (1979). The familial incidence of alcoholism: A review. *Journal of Studies on Alcohol*, *40*(1), 89–116.

Craddock, N., & Sklar, P. (2013). Genetics of bipolar disorder. *Lancet*, *381*(9878), 1654–1662. https://doi.org/10.1016/S0140-6736(13)60855-7.

Dick, D. M., & Agrawal, A. (2008). The genetics of alcohol and other drug dependence. *Alcohol Research & Health*, *31*(2), 111–118.

Du, Y., Nie, Y., Li, Y., & Wan, Y. J. (2011). The association between the SLC6A3 VNTR 9-repeat allele and alcoholism-a meta-analysis. *Alcoholism, Clinical and Experimental Research*, *35*(9), 1625–1634. https://doi.org/10.1111/j.1530-0277.2011.01509.x.

Edenberg, H. J. (2007). The genetics of alcohol metabolism: Role of alcohol dehydrogenase and aldehyde dehydrogenase variants. *Alcohol Research & Health*, *30*(1), 5–13.

Edenberg, H. J., Dick, D. M., Xuei, X., Tian, H., Almasy, L., Bauer, L. O., … Begleiter, H. (2004). Variations in GABRA2, encoding the alpha 2 subunit of the GABA(A) receptor, are associated with alcohol dependence and with brain oscillations. *American Journal of Human Genetics*, *74*(4), 705–714. https://doi.org/10.1086/383283.

Edenberg, H. J., & Foroud, T. (2006). The genetics of alcoholism: Identifying specific genes through family studies. *Addiction Biology*, *11*(3–4), 386–396. https://doi.org/10.1111/j.1369-1600.2006.00035.x.

Edenberg, H. J., & Foroud, T. (2014). Genetics of alcoholism. *Handbook of Clinical Neurology*, *125*, 561–571. https://doi.org/10.1016/B978-0-444-62619-6.00032-X.

Edenberg, H. J., Koller, D. L., Xuei, X., Wetherill, L., McClintick, J. N., Almasy, L., … Foroud, T. (2010). Genome-wide association study of alcohol dependence implicates a region on chromosome 11. *Alcoholism, Clinical and Experimental Research*, *34*(5), 840–852. https://doi.org/10.1111/j.1530-0277.2010.01156.x.

Enoch, M. A. (2011). The role of early life stress as a predictor for alcohol and drug dependence. *Psychopharmacology*, *214*(1), 17–31. https://doi.org/10.1007/s00213-010-1916-6.

Enoch, M. A., & Goldman, D. (2001). The genetics of alcoholism and alcohol abuse. *Current Psychiatry Reports*, *3*(2), 144–151.

Frank, J., Cichon, S., Treutlein, J., Ridinger, M., Mattheisen, M., Hoffmann, P., … Rietschel, M. (2012). Genome-wide significant association between alcohol dependence and a variant in the ADH gene cluster. *Addiction Biology*, *17*(1), 171–180. https://doi.org/10.1111/j.1369-1600.2011.00395.x.

Gelernter, J., Kranzler, H. R., Sherva, R., Almasy, L., Koesterer, R., Smith, A. H., … Farrer, L. A. (2014). Genome-wide association study of alcohol dependence: Significant findings in African- and European-Americans including novel risk loci. *Molecular Psychiatry*, *19*(1), 41–49. https://doi.org/10.1038/mp.2013.145.

Goldman, D., Oroszi, G., O'Malley, S., & Anton, R. (2005). COMBINE genetics study: The pharmacogenetics of alcoholism treatment response: genes and mechanisms. *Journal of Studies on Alcohol Supplement*, *15*, 56–64 (discussion 33).

Goodwin, D. W. (1979). The cause of alcoholism and why it runs in families. *The British Journal of Addiction to Alcohol and Other Drugs*, *74*(2), 161–164.

Goodwin, D. W., Schulsinger, F., Hermansen, L., Guze, S. B., & Winokur, G. (1973). Alcohol problems in adoptees raised apart from alcoholic biological parents. *Archives of General Psychiatry*, *28*(2), 238–243.

Goodwin, D. W., Schulsinger, F., Knop, J., Mednick, S., & Guze, S. B. (1977). Psychopathology in adopted and nonadopted daughters of alcoholics. *Archives of General Psychiatry*, *34*(9), 1005–1009.

Goodwin, D. W., Schulsinger, F., Moller, N., Hermansen, L., Winokur, G., & Guze, S. B. (1974). Drinking problems in adopted and nonadopted sons of alcoholics. *Archives of General Psychiatry*, *31*(2), 164–169.

Grant, B. F., & Dawson, D. A. (1998). Age of onset of drug use and its association with DSM-IV drug abuse and dependence: Results from the national longitudinal alcohol epidemiologic survey. *Journal of Substance Abuse*, *10*(2), 163–173.

Grant, B. F., Saha, T. D., Ruan, W. J., Goldstein, R. B., Chou, S. P., Jung, J., … Hasin, D. S. (2016). Epidemiology of DSM-5 drug use disorder: Results from the national epidemiologic survey on alcohol and related conditions-III. *JAMA Psychiatry*, *73*(1), 39–47. https://doi.org/10.1001/jamapsychiatry.2015.2132.

Heath, A. C., Bucholz, K. K., Madden, P. A., Dinwiddie, S. H., Slutske, W. S., Bierut, L. J., … Martin, N. G. (1997). Genetic and environmental contributions to alcohol dependence risk in a national twin sample: Consistency of findings in women and men. *Psychological Medicine*, *27*(6), 1381–1396.

Heath, A. C., Whitfield, J. B., Martin, N. G., Pergadia, M. L., Goate, A. M., Lind, P. A., … Montgomery, G. W. (2011). A quantitative-trait genome-wide association study of alcoholism risk in the community: Findings and implications. *Biological Psychiatry*, *70*(6), 513–518. https://doi.org/10.1016/j.biopsych.2011.02.028.

Hurley, T. D., & Edenberg, H. J. (2012). Genes encoding enzymes involved in ethanol metabolism. *Alcohol Research: Current Reviews*, *34*(3), 339–344.

Johnson, C., Drgon, T., Liu, Q. R., Walther, D., Edenberg, H., Rice, J., … Uhl, G. R. (2006). Pooled association genome scanning for alcohol dependence using 104,268 SNPs: Validation and use to identify alcoholism vulnerability loci in unrelated individuals from the collaborative study on the genetics of alcoholism. *American Journal of Medical Genetics. Part B, Neuropsychiatric Genetics*, *141B*(8), 844–853. https://doi.org/10.1002/ajmg.b.30346.

Kalsi, G., Prescott, C. A., Kendler, K. S., & Riley, B. P. (2009). Unraveling the molecular mechanisms of alcohol dependence. *Trends in Genetics*, *25*(1), 49–55. https://doi.org/10.1016/j.tig.2008.10.005.

Kato, T. (2015). Whole genome/exome sequencing in mood and psychotic disorders. *Psychiatry and Clinical Neurosciences*, *69*(2), 65–76. https://doi.org/10.1111/pcn.12247.

Kendler, K. S., Chen, X., Dick, D., Maes, H., Gillespie, N., Neale, M. C., & Riley, B. (2012). Recent advances in the genetic epidemiology and molecular genetics of substance use disorders. *Nature Neuroscience*, *15*(2), 181–189. https://doi.org/10.1038/nn.3018.

Kendler, K. S., Kalsi, G., Holmans, P. A., Sanders, A. R., Aggen, S. H., Dick, D. M., … Gejman, P. V. (2011). Genomewide association analysis of symptoms of alcohol dependence in the molecular genetics of schizophrenia (MGS2) control sample. *Alcoholism: Clinical and Experimental Research*, *35*(5), 963–975. https://doi.org/10.1111/j.1530-0277.2010.01427.x.

Kessler, R. C., Crum, R. M., Warner, L. A., Nelson, C. B., Schulenberg, J., & Anthony, J. C. (1997). Lifetime co-occurrence of DSM-III-R alcohol abuse and dependence with other psychiatric disorders in the National Comorbidity Survey. *Archives of General Psychiatry*, *54*(4), 313–321.

Kiefer, F., Witt, S. H., Frank, J., Richter, A., Treutlein, J., Lemenager, T., … Mann, K. (2011). Involvement of the atrial natriuretic peptide transcription factor GATA4 in alcohol dependence, relapse risk and treatment response to acamprosate. *The Pharmacogenomics Journal*, *11*(5), 368–374. https://doi.org/10.1038/tpj.2010.51.

Kranzler, H. R., Gelernter, J., Anton, R. F., Arias, A. J., Herman, A., Zhao, H., … Covault, J. (2009). Association of markers in the 3′ region of the GluR5 kainate receptor subunit gene to alcohol dependence. *Alcoholism, Clinical and Experimental Research*, *33*(5), 925–930. https://doi.org/10.1111/j.1530-0277.2009.00913.x.

Lee, J. S., Sorcher, J. L., Rosen, A. D., Damadzic, R., Sun, H., Schwandt, M., … Lohoff, F. W. (2018). Genetic association and expression analyses of the phosphatidylinositol-4-phosphate 5-kinase (PIP5K1C) gene in alcohol use disorder—Relevance for pain signaling and alcohol use. *Alcoholism, Clinical and Experimental Research*. https://doi.org/10.1111/acer.13751.

Li, D., Zhao, H., & Gelernter, J. (2012). Strong protective effect of the aldehyde dehydrogenase gene (ALDH2) 504lys (*2) allele against alcoholism and alcohol-induced medical diseases in Asians. *Human Genetics*, *131*(5), 725–737. https://doi.org/10.1007/s00439-011-1116-4.

Lind, P. A., Macgregor, S., Vink, J. M., Pergadia, M. L., Hansell, N. K., de Moor, M. H., … Madden, P. A. (2010). A genomewide association study of nicotine and alcohol dependence in Australian and Dutch populations. *Twin Research and Human Genetics*, *13*(1), 10–29. https://doi.org/10.1375/twin.13.1.10.

Lohoff, F. W., Sorcher, J. L., Rosen, A. D., Mauro, K. L., Fanelli, R. R., Momenan, R., … Kaminsky, Z. A. (2018). Methylomic profiling and replication implicates deregulation of PCSK9 in alcohol use disorder. *Molecular Psychiatry*, *23*(9), 1900–1910. https://doi.org/10.1038/mp.2017.168.

Long, J. C., Knowler, W. C., Hanson, R. L., Robin, R. W., Urbanek, M., Moore, E., … Goldman, D. (1998). Evidence for genetic linkage to alcohol dependence on chromosomes 4 and 11 from an autosome-wide scan in an American Indian population. *American Journal of Medical Genetics*, *81*(3), 216–221.

Luczak, S. E., Glatt, S. J., & Wall, T. L. (2006). Meta-analyses of ALDH2 and ADH1B with alcohol dependence in Asians. *Psychological Bulletin*, *132*(4), 607–621. https://doi.org/10.1037/0033-2909.132.4.607.

Manolio, T. A., Collins, F. S., Cox, N. J., Goldstein, D. B., Hindorff, L. A., Hunter, D. J., … Visscher, P. M. (2009). Finding the missing heritability of complex diseases. *Nature*, *461*(7265), 747–753. https://doi.org/10.1038/nature08494.

Mauro, K. L., Helton, S. G., Rosoff, D. B., Luo, A., Schwandt, M., Jung, J., … Lohoff, F. W. (2018). Association analysis between genetic variation in GATA binding protein 4 (GATA4) and alcohol use disorder. *Alcohol and Alcoholism*. https://doi.org/10.1093/alcalc/agx120.

McHugh, R. K., Hofmann, S. G., Asnaani, A., Sawyer, A. T., & Otto, M. W. (2010). The serotonin transporter gene and risk for alcohol dependence: A meta-analytic review. *Drug and Alcohol Dependence*, *108*(1–2), 1–6. https://doi.org/10.1016/j.drugalcdep.2009.11.017.

Mizokawa, T., Wakisaka, Y., Sudayama, T., Iwai, C., Miyoshi, K., Takeuchi, J., … Sawatzky, G. A. (2013). Role of oxygen holes in Li(x)CoO(2) revealed by soft X-ray spectroscopy. *Physical Review Letters*. *111*(5). https://doi.org/10.1103/PhysRevLett.111.056404.

Moskvina, V., Craddock, N., Holmans, P., Nikolov, I., Pahwa, J. S., Green, E., … O'Donovan, M. C. (2009). Gene-wide analyses of genome-wide association data sets: Evidence for multiple common risk alleles for schizophrenia and bipolar disorder and for overlap in genetic risk. *Molecular Psychiatry*, *14*(3), 252–260. https://doi.org/10.1038/mp.2008.133.

Munafo, M. R., Matheson, I. J., & Flint, J. (2007). Association of the DRD2 gene Taq1A polymorphism and alcoholism: A meta-analysis of case-control studies and evidence of publication bias. *Molecular Psychiatry*, *12*(5), 454–461. https://doi.org/10.1038/sj.mp.4001938.

Owen, M. J., Craddock, N., & O'Donovan, M. C. (2010). Suggestion of roles for both common and rare risk variants in genome-wide studies of schizophrenia. *Archives of General Psychiatry*, *67*(7), 667–673. https://doi.org/10.1001/archgenpsychiatry.2010.69.

Palmer, R. H., Brick, L., Nugent, N. R., Bidwell, L. C., McGeary, J. E., Knopik, V. S., & Keller, M. C. (2015). Examining the role of common genetic variants on alcohol, tobacco, cannabis and illicit drug dependence: Genetics of vulnerability to drug dependence. *Addiction*, *110*(3), 530–537. https://doi.org/10.1111/add.12815.

Pan, Y., Luo, X., Liu, X., Wu, L. Y., Zhang, Q., Wang, L., ... Wang, K. S. (2013). Genome-wide association studies of maximum number of drinks. *Journal of Psychiatric Research, 47*(11), 1717–1724. https://doi.org/10.1016/j.jpsychires.2013.07.013.

Park, B. L., Kim, J. W., Cheong, H. S., Kim, L. H., Lee, B. C., Seo, C. H., ... Choi, I. G. (2013). Extended genetic effects of ADH cluster genes on the risk of alcohol dependence: From GWAS to replication. *Human Genetics, 132*(6), 657–668. https://doi.org/10.1007/s00439-013-1281-8.

Pei, Y. F., Zhang, L., Yang, T. L., Han, Y., Hai, R., Ran, S., ... Deng, H. W. (2012). Genome-wide association study of copy number variants suggests LTBP1 and FGD4 are important for alcohol drinking. *PLoS ONE, 7*(1), e30860. https://doi.org/10.1371/journal.pone.0030860.

Poikolainen, K. (2000). Risk factors for alcohol dependence: A case-control study. *Alcohol and Alcoholism, 35*(2), 190–196.

Prescott, C. A., Aggen, S. H., & Kendler, K. S. (1999). Sex differences in the sources of genetic liability to alcohol abuse and dependence in a population-based sample of U.S. twins. *Alcoholism, Clinical and Experimental Research, 23*(7), 1136–1144.

Psychiatric, G. C. S. C. (2009). A framework for interpreting genome-wide association studies of psychiatric disorders. *Molecular Psychiatry, 14*(1), 10–17. https://doi.org/10.1038/mp.2008.126.

Psychiatric, G. C. B. D. W. G. (2011). Large-scale genome-wide association analysis of bipolar disorder identifies a new susceptibility locus near ODZ4. *Nature Genetics, 43*(10), 977–983. https://doi.org/10.1038/ng.943.

Quillen, E. E., Chen, X. D., Almasy, L., Yang, F., He, H., Li, X., ... Gelernter, J. (2014). ALDH2 is associated to alcohol dependence and is the major genetic determinant of "daily maximum drinks" in a GWAS study of an isolated rural Chinese sample. *American Journal of Medical Genetics Part B, Neuropsychiatric Genetics, 165B*(2), 103–110. https://doi.org/10.1002/ajmg.b.32213.

Regier, D. A., Farmer, M. E., Rae, D. S., Locke, B. Z., Keith, S. J., Judd, L. L., & Goodwin, F. K. (1990). Comorbidity of mental disorders with alcohol and other drug abuse. Results from the Epidemiologic Catchment Area (ECA) Study. *JAMA, 264*(19), 2511–2518.

Rehm, J., Dawson, D., Frick, U., Gmel, G., Roerecke, M., Shield, K. D., & Grant, B. (2014). Burden of disease associated with alcohol use disorders in the United States. *Alcoholism, Clinical and Experimental Research, 38*(4), 1068–1077. https://doi.org/10.1111/acer.12331.

Rehm, J., Mathers, C., Popova, S., Thavorncharoensap, M., Teerawattananon, Y., & Patra, J. (2009). Alcohol and Global Health 1 Global burden of disease and injury and economic cost attributable to alcohol use and alcohol-use disorders. *Lancet, 373*(9682), 2223–2233.

Reich, T., Edenberg, H. J., Goate, A., Williams, J. T., Rice, J. P., Van Eerdewegh, P., ... Begleiter, H. (1998). Genome-wide search for genes affecting the risk for alcohol dependence. *American Journal of Medical Genetics, 81*(3), 207–215.

Reilly, M. T., Noronha, A., Goldman, D., & Koob, G. F. (2017). Genetic studies of alcohol dependence in the context of the addiction cycle. *Neuropharmacology, 122*, 3–21. https://doi.org/10.1016/j.neuropharm.2017.01.017.

Robison, A. J., & Nestler, E. J. (2011). Transcriptional and epigenetic mechanisms of addiction. *Nature Reviews. Neuroscience, 12*(11), 623–637. https://doi.org/10.1038/nrn3111.

Sahni, S., Tickoo, M., Gupta, R., Vaswani, M., Ambekar, A., Grover, T., & Sharma, A. (2018). Association of serotonin and GABA pathway gene polymorphisms with alcohol dependence: A preliminary study. *Asian Journal of Psychiatry.* https://doi.org/10.1016/j.ajp.2018.04.023.

Sampson, L., Cohen, G. H., Calabrese, J. R., Fink, D. S., Tamburrino, M., Liberzon, I., ... Galea, S. (2015). Mental health over time in a military sample: The impact of alcohol use disorder on trajectories of psychopathology after deployment. *Journal of Traumatic Stress, 28*(6), 547–555. https://doi.org/10.1002/jts.22055.

Schizophrenia Working Group of the Psychiatric Genomics, C (2014). Biological insights from 108 schizophrenia-associated genetic loci. *Nature, 511* (7510), 421–427. https://doi.org/10.1038/nature13595.

Schuckit, M. A., Goodwin, D. A., & Winokur, G. (1972). A study of alcoholism in half siblings. *The American Journal of Psychiatry, 128*(9), 1132–1136. https://doi.org/10.1176/ajp.128.9.1132.

Schumann, G., Coin, L. J., Lourdusamy, A., Charoen, P., Berger, K. H., Stacey, D., ... Elliott, P. (2011). Genome-wide association and genetic functional studies identify autism susceptibility candidate 2 gene (AUTS2) in the regulation of alcohol consumption. *Proceedings of the National Academy of Sciences of the United States of America, 108*(17), 7119–7124. https://doi.org/10.1073/pnas.1017288108.

Sklar, P., Smoller, J. W., Fan, J., Ferreira, M. A., Perlis, R. H., Chambert, K., ... Purcell, S. M. (2008). Whole-genome association study of bipolar disorder. *Molecular Psychiatry, 13*(6), 558–569. https://doi.org/10.1038/sj.mp.4002151.

Slutske, W. S., True, W. R., Scherrer, J. F., Heath, A. C., Bucholz, K. K., Eisen, S. A., ... Tsuang, M. T. (1999). The heritability of alcoholism symptoms: "Indicators of genetic and environmental influence in alcohol-dependent individuals" revisited. *Alcoholism, Clinical and Experimental Research, 23* (5), 759–769.

Stahre, M., Roeber, J., Kanny, D., Brewer, R. D., & Zhang, X. (2014). Contribution of excessive alcohol consumption to deaths and years of potential life lost in the United States. *Preventing Chronic Disease, 11.* https://doi.org/10.5888/pcd11.130293.

Stefansson, H., Rujescu, D., Cichon, S., Pietilainen, O. P., Ingason, A., Steinberg, S., ... Stefansson, K. (2008). Large recurrent microdeletions associated with schizophrenia. *Nature, 455*(7210), 232–236. https://doi.org/10.1038/nature07229.

Suzuki, A., Kochi, Y., Okada, Y., & Yamamoto, K. (2011). Insight from genome-wide association studies in rheumatoid arthritis and multiple sclerosis. *FEBS Letters, 585*(23), 3627–3632. https://doi.org/10.1016/j.febslet.2011.05.025.

Takeuchi, F., Isono, M., Nabika, T., Katsuya, T., Sugiyama, T., Yamaguchi, S., ... Kato, N. (2011). Confirmation of ALDH2 as a major locus of drinking behavior and of its variants regulating multiple metabolic phenotypes in a Japanese population. *Circulation Journal, 75*(4), 911–918.

Tammimaki, A. E., & Mannisto, P. T. (2010). Are genetic variants of COMT associated with addiction? *Pharmacogenetics and Genomics, 20*(12), 717–741. https://doi.org/10.1097/FPC.0b013e328340bdf2.

Tawa, E. A., Hall, S. D., & Lohoff, F. W. (2016). Overview of the genetics of alcohol use disorder. *Alcohol and Alcoholism, 51*(5), 507–514. https://doi.org/10.1093/alcalc/agw046.

Treutlein, J., Cichon, S., Ridinger, M., Wodarz, N., Soyka, M., Zill, P., ... Rietschel, M. (2009). Genome-wide association study of alcohol dependence. *Archives of General Psychiatry, 66*(7), 773–784. https://doi.org/10.1001/archgenpsychiatry.2009.83.

2.2 Nonsyndromic

What defines "nonsyndromic autism" (Sztainberg & Zoghbi, 2016) is simply that it identifies those patients in whom the physician fails to make a diagnosis of an entity with a *definable, recognizable disease*. This classification varies over time as numbers of cases reach a threshold that allows the clinical features of patients to be clustered so that criteria are established that lead to recognition of its etiology as being associated with a recognizable genetic mutation (or etiology).

The identification of more than 800 gene mutations and 2000 CNVs in patients with social communication/language deficits and repetitive/restricted behavior has provided the basis for future genotype-phenotype correlations (Chaste, Roeder, & Devlin, 2017; Finucane & Myers, 2016). It provides a rationale for an additional approach, specifically to accumulate many affected individuals' DNA variants, and to subsequently characterize these individuals in order to define the phenotype and its variable expression. The study of the CHD8 mutation in ASD patients (Bernier et al., 2014) confirms the validity of such endeavors, as do subsequent studies of patients with 22q13.3 deletion (Phelan-McDermid syndrome), which led to the recognition of the extreme variability of the phenotype (Zwanberg, Ruiter, van den Heuvel, Flapper, & Van Ravenswaaij-Arts, 2016).

Numerous studies have taken this approach, and thousands of ASD patients and nonrelated controls have been analyzed. These studies have demonstrated that ASD patients are 10–20 times more likely to have CNVs (Jaquemont et al., 2006; Sebat et al., 2007), and CNVs are more prevalent in the population of cognitively impaired individuals (Autism Genome Project Consortium, 2007; Christian et al., 2008; Gilman et al., 2011; Glessner et al., 2009; Itsara et al., 2010; Marshall et al., 2008; Pinto et al., 2010; Sanders et al., 2011). Each of these "associations" occurs in approximately 1% of patients (Finucane & Myers, 2016). These facts highlight the importance of focusing research efforts into the characterization of CNVs. The challenge is now to understand how so many variations across the genome lead to one phenotype. Several attempts have been made to identify common molecular and functional pathways that unify the numerous CNVs involved in ASD. These studies have defined three common cellular networks; specifically, neurotransmission, synapse formation, and ubiquitination (Ben-David & Shifman, 2012, 2013; Parikshak et al., 2013; Ziats & Rennert, 2016) (Fig. 1).

Similar research efforts have focused on whole exome sequencing (WES) and whole genome sequencing (WGS). These studies have proven to be equally informative in identifying specific single gene defects associated with ASD. So far, approximately 800 such defects have been described in the literature (Jiang et al., 2013). These mutations occur throughout the genome, in both coding and noncoding genes (Wu, Parikshak, Belgard, & Geschwind, 2016; Ziats & Rennert, 2014). Thus, again, the challenge remains to understand how such diverse and numerous defects are responsible for one common phenotype. As with CNVs, attempts have been made to identify common denominators that may shed light on this matter. These investigations (genome sequencing plus CNV) have identified three broad underlying mechanisms (Fig. 1): synaptic and neuronal adhesion components, RNA processing, and transcriptional regulation (Bernier et al., 2014; Codina-Solà et al., 2015; de la torre-Ubieta, Won, Stein, & Geschwind, 2016; Liu et al., 2013).

Can one define distinct anatomical or functional abnormalities of the brain in ASD? A literature survey yields inconclusive data on anatomic, regional, or embryological deficits. However, recently Ecker, Schmeisser, Loth, and Murphy (2017) and Varghese et al. (2017) proposed that growth patterns characteristic of ASD may be identified in patients, as well as in certain animal models. General studies in patients with ASD suggest that 20% of these children exhibit early brain overgrowth (2–4 years of age), as evidenced both by imaging studies of brain volume, as well as by fronto-occipital

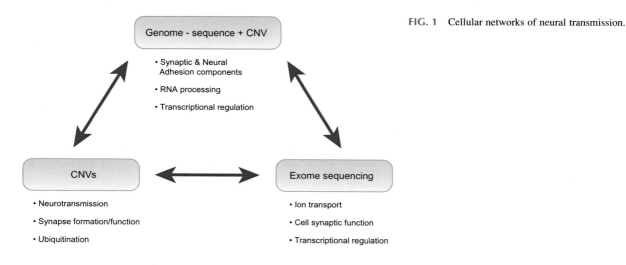

FIG. 1   Cellular networks of neural transmission.

circumference measurement (Carper & Courchesne, 2005; Lainhart et al., 2006; Redcay & Courchesne, 2005). Between 5 and 6 years of age, growth velocity declines or plateaus in these children, and beyond 6 years of age, no increase is noted. This phenomenon may be indicative of an "atypical brain maturation trajectory." These data suggest that aberrant brain developmental patterns may be most prominent during the period from mid-gestation through the second postnatal years of life. However, it should be stressed that other investigations identify that 80% of children with ASD do not display this feature. Additionally, the finding is highly variable, relatively small populations have been studied (predominantly retrospective studies), and correlation with somatic "overgrowth" (particularly in males) would suggest this finding may not be specific for brain development only. Assuming the brain maturation trajectory is deranged in autism, it would suggest altered brain connectivity is a pathophysiological characteristic of this disorder, and that this phenomenon occurs during the period between mid-gestation and 2–3 years of life.

3 Diagnostic work-up for nonsyndromic autism spectrum disorders

Nevertheless, based on the preceding considerations, there are no established guidelines for conducting a diagnostic work up for nonsyndromic ASD. The set of evaluations and tests to initiate assessment should be tailored to each individual based on medical and family history. However, there are some tests that are used as a part of the standard work-up that have been shown to have good diagnostic yields for the population of ASD patients. These include the following (Fig. 2 depicts the algorithm for the diagnostic work-up for nonsydromic ASD):

1. *DNA-based chromosome microarrays (CMAs):* This technique has a 14%–18% positive yield for detecting CNVs in the ASD patient population. Some studies report an increase in positive findings as high as 30% (Schaefer et al., 2010). CMAs are now established as first-tier tests for patients with developmental delays, ASD, and other neurobehavioral disorders. The advent of CMA use in the ASD population has flooded the field with more than 2000 CNVs that cover the entire genome, and contain genes involved in diverse pathways. These data have confirmed the profound diversity at the genotypic level of the already recognized heterogeneity of the clinical phenotype characteristic of the autism spectrum

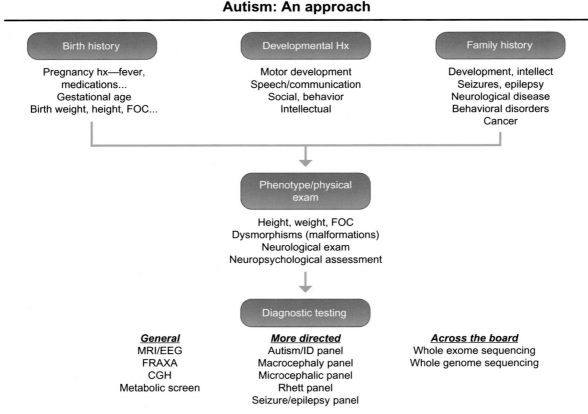

FIG. 2 Diagnostic algorithm for autism.

TABLE 1 Newer CNVs associated with ASD.

• 1q21.1	del
• 5q35	del
• 7q11.23	del
• 15q11–13	del (Maternal—Angelman/Paternal—PWS)
• 16p13.11	del
• 16p11.2	del
• 17p11.2	del
• 17q12	
• 22q11.2	del/dup
• 22q13.3	del (PMS)

Reference for preceding *CNV*, Leppa; V. M., Kravitz, S. N., Martin, C, L., Andrieux, J., Le Caignec, C., Martin-Coignard, D., et al. (2016). Rare inherited and de novo CNVs reveal complex contributions to ASD risk in multiplex families. *American Journal of Human Genetics, 99*(3), 540–554.

disorders. At the same time, it has resulted in the identification of new CNV syndromes, elucidated the function of new genes important in brain development, and has allowed the recognition of at-risk individuals. (Table 1 is a list of newly identified CNV syndromes in ASD.)

2. *Fragile X testing:* The phenotype of Fragile X syndrome has evolved significantly in the past several years. There is no longer a specific constellation of physical findings that define a Fragile X phenotype. Due to the dynamic nature of these mutations, there is extreme variability in its phenotypes. For these reason, a Fragile X (FRAXA) test is recommended as part of the standard diagnostic work-up of ASD patients.

3. *Next generation sequencing (NGS) panels:* The vast number of gene mutations and CNVs associated with ASD (800+ genes, 2100+ CNVs) have increased the complexity of defining molecular events that give rise to autism. Indeed, at times, it seems that the individual mutations appear to be unique for each family. Additionally, associated features of intellectual disability, seizures, and other co-morbid features lead to difficulties in assessing prognosis and potential therapeutic interventions. Advances in the technology available for assessing individual patients and their families have been greatly enhanced by the development of NGS panels that specifically identify genes that have been identified based on the concurrence of clinical features, such as intellectual disability and "autism," or seizures and "autism." This approach has allowed the recognition of genetic etiologies in a significantly greater number of patients (5%–7%) (Poultney et al., 2013) at less cost than whole exome/genome sequencing, and it more readily allows identification of significant mutations that are, as a rule, missed by CMAs. There are numerous examples of relevant NGS panels that highlight the coexistence of autism with easily recognizable clinical features such seizures, microcephaly, intellectual disability, and motor impairments. These targeted panels, because of their increased specificity, result in a higher yield in diagnosis than would occur with WES. The use of these panels has identified both de novo and heritable variants that are now well defined as part of the autism spectrum. The genes included in these panels are involved in diverse products and functions, including post-synaptic adhesion proteins, chromatin modifiers, ubiquitin proteins, cellular proliferation, and methylation.

4. *Brain imaging and neuroanatomical studies:* Regions thought to be associated with autism include the amygdala, hypothalamus, the anterior cingulate cortex, the prefrontal cortex, the cerebellum, the cingulate, and the limbic system. Individuals with autism lack central coherence, defined as the cognitive ability to bind together a jumble of separate features into a single coherent object or concept. Reduced information transfer is thought to be a consequence of local "over-connectivity" or *long range* "under-connectivity." Over-connectivity occurs when sensory input yields abnormally larger deviations with a reduction in selectivity of regions activated. Therefore, brain regions subserving integrative functions should manifest reductions in activation, and functional correlation with sensory regions (Belmonte et al., 2004). Changes in connectivity have been noted as early as 6 months of age in ASD patients and high-risk infants (Lewis et al., 2017). In addition, the corpus callosum facilitates long-distance integration within large brains, and it is postulated that brain size reflects the brain's microstructure and the relationship between structural versus functional connectivity (Kennedy, Paul, & Adolphs, 2015). Thus, the use of brain imaging in the diagnostic work-up for

nonsyndromic ASD patients should be individually tailored to each patient, based on clinical and family history. Overall, brain MRIs are the technique most commonly used due to the noninvasive nature of this diagnostic modality. The potential for a dramatic application of diagnostic imaging to study the pathophysiology of autism is a consequence of employing this modality to study brain connectivity during fetal development. Earlier work established the methodology (Seshamani et al., 2016). And the study published recently (Thomason et al., 2017) documented its applicability, presenting data on 32 pregnancies studied during weeks 22–36 of pregnancy. These latter investigators provided data suggesting that altered brain connectivity in the "pre-language region" was demonstrable prior to birth, and that it related to prematurity and an altered prenatal milieu. These observations establish the potential both to study and identify the underlying prenatal pathophysiology of autism, as well as opening new vistas for intervention.

5. *Additional tests* may be useful to define the particular clinical phenotype in each patient. Such tests could include metabolic panels, organic acids, amino acids, and mitochondrial studies.

"We haven't been able to determine, in terms of genes, what makes a human being a human, and not another mammal" (to quote Walter Gilbert, PhD, Nobel prize in chemistry). "Consider the *human brain*, a collection of 86×10^9 neurons, with trillions of connections—yet it orchestrates everything from understanding, and memory, to movement and sleep."

4 Pathway analysis and molecular networks

The typical psychological features of autism (dysfunctions in social perception/cognition/communication, and the restricted, repetitive patterns of behavior, interest and activities) appear to be independent from the "causes" of autism, yet they seem to be part of a common pathophysiological pathway, that is, a specific altered global network or altered connectivity. Potential genetic mechanisms possibly accounting for this phenomenon during brain neurodevelopment include:

○ Gene haploinsufficiency
○ Gene mutation leading to truncation and loss of alternative transcripts
○ Misregulation of activity-dependent splicing networks
○ Loss of transcription binding sites
○ Altered regulation by DNA methylation
○ Imbalanced transcriptional regulation involving mRNAs
○ LncRNAs' roles in epigenetic regulation, gene expression regulation
○ Dysregulation of development secondary to nutritional, toxic stimuli.

The majority of the data supports the hypothesis that genes involved in synaptic elimination are enriched in de novo autism mutations, a potential mechanism for autism. Examples include SHANK, PARK2, CASP6, MAPK3, NLGN, NRXN, CTNNB1, NGFR, and PTEN (Provenzano, Chelini, & Bozzi, 2017). In addition, transcriptome analysis by several investigators (Ram Venkataraman, O'Connell, Egawa, Katherine-haghighi, & Wall, 2016) have documented altered expression of gene networks involved in synapse development, neuronal activity, and immune functions, while others have documented changes in glutamatergic and GABAergic brain receptors in ASD (Ram Venkataraman et al., 2016). Likewise, the role of chromatin modifiers, postsynaptic proteins, epilepsy, dysregulation of protein synthesis, and protein degradation have all been associated with the pathogenesis of ASD. The following is a summary of the most common subcategories identified to date (Chahrour et al., 2016; O'Roak et al., 2014):

○ High confidence ASD genes: chromatin modifiers, including CHD8, CHD2, ARID1B
○ Embryonic expression genes: TBR1, DYRK1A, PTEN
○ Postsynaptic proteins: GRIN2B, GABRB3, SHANK3
○ Recessive loss of function genes: CNTNAP2, SLC9A9, NHE9
○ Epilepsy, ASD, intellectual disability associated genes: BCKDA, CC2D1A, hypomorphic missense variants: AMT, PEX7, VPS13B, gene clusters, dysregulation of protein synth.: FMR1, TSC1/2, PTEN, protein degradation (ubiquitination): UBE3A, HUWE1, UBE3C, USP7)

Some of the pathways incriminated in such synaptic events include Akt/mTOR (rapamycin) signaling, the downstream effects of this network on molecular nodes including FMR1, PTEN, TSC1, and TSC2, as well as the Wnt signaling pathway. The consequences of these effects impact cellular growth, proliferation, and cytokine production. mTOR, decreased glycogen synthase kinase 3alpha activity, decreased tuberin, and increased activity of the Akt/mTOR pathway have all been reported (Mellios et al., 2017; Onore, Yang, van dear Water, & Ashwood, 2017).

Additional studies have identified changes in various networks, as well as changes in miRNA, lncRNA, and differences in the cellular composition of various brain regions (Edmonson, Ziats, and Rennert, 2014; Mahfouz, Ziats, Rennert, Lelieveldt, & Reinders, 2015; Ziats & Rennert, 2011; Ziats & Rennert, 2013; Ziats & Rennert, 2014). Newer approaches utilized neurons differentiated from induced pluripotent stem (iPS) cells to study differences in neural electrophysiology and transcription in ASD (Kathuria, Sala, Verpelli, & Price, 2017; Lin et al., 2016; Liu et al., 2017; Nagy et al., 2017; Tamburini & Li, 2017; Yazawa & Dolmetsch, 2013).

Prenatal and postnatal studies have elucidated the role of essential genes (EGs) in ASD. Ji, Kember, Brown, and Bucan (2016) postulated that genes essential for completion of prenatal and postnatal development are enriched, and when mutated, give rise to human disease. Data reveal that approximately 30% of protein coding genes ($\sim$6000) are essential for pre- and postnatal survival. One hundred of these essential genes (EGs) are associated with autism; their function is central to the processes of transcriptional regulation, chromatin modeling, and synaptic functions. Transcriptional regulation genes predominantly act as chromatin modifiers during brain development, whereas synapse protein/genes tend to code for post-synaptic adhesion molecules. These play a role in protein-protein interaction networks. Their impact is reflected in the consequence of their haploinsufficiency. Last, haploinsufficiency of the neuronal splice regulator nsR100/SRRM4 gene (microexon splicing program misregulated) has been identified in 30% of analyzed individuals with ASD (Irinia et al., 2016).

Additional pathways continue to be proposed as possibly functional in the pathophysiology of ASD. For example, Konopka and Roberts (Konopka & Roberts, 2016) have identified central genes in networks that are fundamental in the processes of speech and social communication, such as FOXP2,1, CNTNAP2, GNPTG, NAGPA, GNPTAB, and TSC1 involving the basal ganglia, cortex, and cerebellar pathways (Dempsey & Sawtell, 2016). Another example includes the mutations in human accelerated regions (HARs) that disrupt cognition and social behavior (Doan et al., 2016); these contain conserved genomic loci and display elevated divergence in humans. Mutations in these genes are likely to impact cognitive and social disorders. Doan et al. (2016) identified a significant excess in individuals with ASD in cases whose parents share common ancestry, compared with familial controls, suggesting a contribution of consanguinity in 5% of ASD cases. Another evolving pathway is MIR137, which has also been implicated in the pathophysiology of ASD (Doan et al., 2016; Quesnet-Vallieres et al., 2016). The correlation of MIR137 SNPs (brain) and synapse function has been documented; it targets the expression of CPLX1, NSF, SYN3, and SYT1, SYN3 (22q12-13), cohesin neuronal phosphoprotein in Rett syndrome. More recently, Mellios and Sur (2012) demonstrated that miRNAs regulate early neurogenesis by modulating the ERK/AKT signaling pathways.

These investigations, if replicated, have important future applications both to define the pathophysiology of these disorders, and to suggest new approaches for diagnosis, as well as potential models to develop "precision" therapy. The current classification system of neurodevelopmental disorders is based on clinical criteria; however, this approach fails to incorporate what is known about genomic similarities and differences among closely related clinical neurodevelopmental disorders. Ziats, Rennert, and Ziats (2019) proposed an alternative clinical molecular classification that contrasts (ASD), and syndromes with autistic features, based upon molecular pathways that define fundamental pathophysiologic mechanisms. This schema (Fig. 1) of *known genetic mutations* identifies molecular pathways that characterize autism syndromes into three clinical molecular nodes: *ion transport disorders, cellular synaptic function disorders, and transcriptional regulation disorders*. It further identifies salient clinical characteristics that may guide diagnosis, prognosis, and treatment. Approaches such as these may lead to a shift from the current clinical classification toward one that identifies molecularly-driven pathways. This may result in more precisely guided clinical decision making, and may herald more informative future clinical trials and drug development.

5 Genetic counseling of autism spectrum disorders

The phenotype of the autism spectrum has evolved from a rare condition into a complex heterogeneous neurobehavioral disorder (Chaste et al., 2017) with a heritability of approximately 60%–95% (Finucane & Myers, 2016), and an incidence of 1 in 68. Assistance of these families requires a comprehensive plan of interventions and consults that are best managed by genetic counselors (GCs). The diagnosis of ASD sets forth a chain of reactions that begins the long process of resetting goals and expectations. GCs play an important role in helping families during this process by educating the family, and managing the patient's care. There are two main areas of focus for the GC.

5.1 Navigating the diagnostic work-up

Performing diagnostic tests (genetic and nongenetic) in patients newly diagnosed with ASD is a standard of care (see the preceding section describing the diagnostic work-up). The current guidelines for genetic testing of ASD patients include deletion/duplication analysis, next generation sequencing panels, and/or whole exome sequencing (WES). Usually, these tests are done in tiers to minimize cost. This, however, increases the waiting time, and prolongs the uncertainty. There are three possible outcomes in this battery of tests. The first one is that a causative link is established, be it a mutation or a deletion, or a duplication that has been seen before in patients who are on the spectrum. The second one is that there are no genetic defects identified. And the last one is that a variant of unknown significance (VUS) has been identified, or a deletion/duplication not previously identified has been found. The pre-test counseling session should include assessment of the family's understanding of the purpose of the tests, explanation of these three possible outcomes and implications of results, a review of possible inheritance patterns of ASD, and a discussion of the psychosocial impact of genetic testing.

The majority of ASD patients will fall in the second category where no genetic defects are found. Thus, the uncertainty about the cause of the disease and lack of clinical framework remains with the family. In this case scenario, the parents need to be reassured that all available tools for finding a diagnosis have been used, and that yearly follow-ups will address re-testing with newer technology, if available. The group of patients that fall in the VUS category will be left with a degree of uncertainty. The genetic counselor should work with the testing laboratory to request additional testing of family members that may help shed light into the VUS in question. In addition, most laboratories that perform WES and/or NGS offer free re-analysis after one year. This will include the newest data on previously unclassified variants, and may prove to be informative. Those in the third category who are found to have a genetic defect that has been previously associated with ASD will have a clearer understanding of the etiology of ASD in their family. The GC needs to make sure that genetic and phenotypic heterogeneity are discussed in detail when counseling this group.

5.2 Managing the patient and helping the family cope

Every family has the expectation of having a "normal" child, and when the child is diagnosed with ASD, the process of resetting goals and expectations begins. The GC assists the family in this process by helping them manage the symptoms, and by resetting specific goals. Examples of this process would be from having a child that speaks to having a child that communicates, from having a normal child to having a happy child, reducing spasticity or hypotonia, and so forth. A patient with ASD requires a comprehensive plan of interventions that includes many specialists and multiple evaluations and resources. Parents often feel overwhelmed, and they frequently doubt themselves in their role as caretakers of a child with ASD. It is thus important to reassure them and assist them in this process.

1. *Discussing recurrence risk:* One can characterize recurrence risk in autism into three broad categories, specifically, sporadic, low risk families, and those in whom there is a relatively high risk of recurrence. In sporadic autism (low-risk), spontaneous mutation with high penetrance in males and relatively poor penetrance in females is generally the case. In high-risk families, the mother often carries a new causative mutation, but she is unaffected, and transmits the mutation in dominant fashion to her offspring (Lin et al., 2016). Most cases of syndromic autism have an established pattern of inheritance (autosomal dominant, autosomal recessive, or X linked), with an associated recurrence risk. Thus, discussion of recurrence in future offspring is simple. In contrast, the recurrence risk in cases of idiopathic ASD is complex, and may involve genetic, epigenetic, and environmental factors. Thus, the GC will rely on empirical recurrence risk data, and available information from previously reported cases. Again, genetic and phenotypic heterogeneity needs to be included in this discussion.
2. *Discussion of prognosis:* The clinical phenotype of ASD is highly variable, even in patients who share the same genetic defects. Thus, predicting clinical outcomes in this patient population is filled with uncertainty. ASD patients share core clinical findings that define them as autistic (some degree of social and communication impairment, and repetitive/restricted behaviors). However, the spectrum of clinical presentations ranges from mild to severe in every aspect of the disease. When a child is newly diagnosed with autism, parents are haunted by fears of what the future holds. It is thus difficult to engage the family in focusing on immediate goals, and on celebrating small victories. With time, most families learn to appreciate the progress the child is making, and to avoid comparing their child with typical children.

6 Summary

Autism spectrum disorders are a group of syndromes that share the behavior pattern identified by Kanner (1943) as "inborn autistic disturbances of affective contact." The application of molecular diagnostics/technology has resulted in the identification of hundreds of gene variants and mutations that give rise to its genetic heterogeneity—the challenge for the clinician/investigator is to define the molecular networks (pathways) that reveal the pathogenesis of the social communication deficit, and its overlap with the pathways that give rise to the syndromic phenotype. The development and use of therapeutics to treat patients is the objective of precision medicine/psychiatry. Such an approach will lead to therapeutics that will alter disease manifestations based on the molecular homologies that define the syndrome. To establish "precision psychiatry" for a complex array of diseases known as ASD, future research is vital. "Big data," the establishment of databases that contain carefully acquired phenotypic information (including historical, medical/clinical evaluations, neuropsychological characterization), and genetic assessment, including whole exome/genome sequencing, will be essential to define the molecular basis of the expression of ASD's behavioral and phenotypic elements. Biological computation delineating elements of neural molecular pathways and their correlation to behavioral traits highlighting the nodes and edges of these networks will be required. A shift in our traditional mode of analysis of diseases is requisite—we now focus on disease manifestations and correlate them to identified molecular (genomic and proteomic) pathways. Because "autism" is an exaggerated expression of behavioral traits, it is essential to explore the genetic correlates by recognizing genes, and their polymorphisms, to understand the neural networks and their intersections (nodes) with pathways operative in diseases such as FRAXA, NF1, tuberous sclerosis, Rett syndrome, and others. Targeted therapeutics can be developed for those molecular nodes. Similarly, research efforts to better define psychiatry syndromes on the basis of their molecular pathways (computational analysis) will lead to development of more effective and new therapies. Exciting new horizons will result from the increasing use of transgenic animal models of ASD, as well as the increased development of iPS-derived neuronal cultures and organoid cultures for the study of the pathophysiology of ASD, and as model systems for development and assessment of pharmaceuticals for therapy.

References

Autism Genome Project Consortium. (2007). Mapping autism risk loci using genetic linkage and chromosome rearrangement. *Nature Genetics, 39*, 319–328.

Belmonte, M. K., Allen, G., Beckel-Mitchener, A., Boulanger, L. M., Carper, R. A., & Webb, S. J. (2004). Autism and abnormal development of brain connectivity. *The Journal of Neuroscience, 24*, 9228–9231.

Ben-David, E., & Shifman, S. (2012). Networks of neuronal genes affected by common and rare variants in autism spectrum disorders. *PLoS Genetics, 8*(3).

Ben-David, E., & Shifman, S. (2013). Combined analysis of exome sequencing points toward a major role for transcription regulation during brain development in autism. *Molecular Psychiatry, 18*(10), 1054–1056.

Bernier, R., Golzio, C., Xiong, B., Stessman, H. A., Coe, B. P., Penn, O., et al. (2014). Disruptive CHD8 mutations define a subtype of autism early in development. *Cell, 158*(2), 263–276.

Buescher, A. V., Cidav, Z., Knapp, M., & Mandell, D. S. (2014). Costs of autism spectrum disorders in the United Kingdom and the United States. *JAMA Pediatrics, 168*(8), 721–728.

Carper, R. A., & Courchesne, E. (2005). Localized enlargement of the frontal cortex in early autism. *Biological Psychiatry, 57*(2), 126–133.

Chahrour, M., O'Roak, B. J., Santini, E., Samaco, R. C., Kleinman, R. J., & Mancini, M. C. (2016). Current perspectives in ASD: From genes to therapy. *The Journal of Neuroscience, 9*, 142–11410.

Chaste, P., Roeder, B., & Devlin, B. (2017). The yin and the yang of Autism GENETICS: How rare de novo and common variants affect liability. *Annual Review of Genomics and Human Genetics, 18*, 9.1–9.21.

Christensen, D. L., Baio, J., Van Naarden Braun, K., Bilder, D., Charles, J., et al. (2016). Prevalence of autism spectrum disorders. *MMWR, 65*(3), 1–23.

Christian, S. L., Brune, C. W., Sudi, J., Kumar, R. A., Liu, S., Karamohamed, S., et al. (2008). Novel submicroscopic chromosomal abnormalities in ASD. *Biological Psychiatry, 63*, 1111–1117.

Codina-Solà, M., Rodríguez-Santiago, B., Homs, A., Santoyo, J., Rigau, M., Aznar-Laín, G., et al. (2015). Integrated analysis of whole-exome sequencing and transcriptome profiling in males with autism spectrum disorders. *Molecular Autism, 6*, 21.

de la torre-Ubieta, L., Won, H., Stein, J. L., & Geschwind, D. (2016). Advancing the understanding of autism. *Nature Medicine, 22*(4), 345–346.

Dempsey, C., & Sawtell, N. B. (2016). Cerebellar learning. *Neuron, 92*, 931–933.

Doan, R. N., Bae, B. I., Cubelos, B., Chang, C., Hossain, A. A., Al-Saad, S., et al. (2016). Mutations in human accelerated regions disrupt cognition and social behavior. *Cell, 167*, 341–3554.

Ecker, C., Schmeisser, M. J., Loth, E., & Murphy, D. G. (2017). Neuroanatomy and neuropathology of autism spectrum disorder in humans. *Advances in Anatomy, Embryology, and Cell Biology, 224*, 27–48.

Edmonson, C., Ziats, M. N., & Rennert, O. M. (2014). Altered glial marker expression in autistic post-mortem prefrontal cortex and cerebellum. *Molecular Autism, 5*, 3–10.

Finucane, B., & Myers, S. M. (2016). Genetic counseling for autism spectrum disorders in an evolving theoretical landscape. *Current Genetic Medicine Reports, 4*, 147–153.

Gilman, S. R., Iossifov, I., Levy, D., Ronemus, M., Wigler, M., Vitkup, D., et al. (2011). Rare de novo variants associated with ASD implicate large functional network of genes involved in formation and function of synapses. *Neuron, 70*, 898–907.

Glessner, J. T., Wang, K., Cai, G., Korvatska, O., Kim, C. E., Wood, S., et al. (2009). Autism genome-wide CNV reveals ubiquitin and neuronal genes. *Nature*, *459*, 569–573.

Irinia, M., Weatheritt, R. J., Ellis, J. D., Parikshak, N. N., Gonatopoulos-Pournatzis, T., Babor, M., et al. (2016). Highly conserved program of neuronal microexons misregulated inautistic brains. *Cell*, *159*, 1511–1523.

Itsara, A., Wu, H., Smith, J. D., Nickerson, D. A., Romieu, I., London, S. J., et al. (2010). De novo rates and selection of large copy number variation. *Genome Research*, *20*(11), 1469–1481.

Jaquemont, M. L., Sanlaville, D., Redon, R., Raoul, O., Cormier-Daire, V., Lyonnet, S., et al. (2006). Array-based comparative genomic hybridisation identifies high frequency of cryptic chromosomal rearrangements in patients with syndromic autism spectrum disorders. *Journal of Medical Genetics*, *43*, 843.

Ji, X., Kember, R. L., Brown, C. D., & Bucan, M. (2016). Increased burden of deleterious variants in essential genes in autism spectrum disorders. *Proceedings of the National Academy of Sciences of the United States of America*, *113*, 15054–15059.

Jiang, Y. H., Yuen, R. K., Jin, X., Wang, M., Chen, N., Wu, X., et al. (2013). Detection of clinically relevant genetic variants in autism spectrum disorder by whole-genome sequencing. *American Journal of Human Genetics*, *93*(2), 249–263.

Kanner, L. (1943). Autistic disturbances of affective contact. *The Nervous Child*, *2*, 217–250.

Kathuria, A., Sala, C., Verpelli, C., & Price, J. (2017). Modelling autistic neurons with induced pluripotent stem cells. *Advances in Anatomy, Embryology, and Cell Biology*, *224*, 49–64.

Kennedy, D. P., Paul, L. K., & Adolphs, R. (2015). Brain connectivity in autism: The Significance of null findings. *Biological Psychiatry*, *78*(2), 81–82.

Konopka, G., & Roberts, T. F. (2016). Insights into the neural and genetic basis of vocal communication. *Cell*, *164*, 1269–1276.

Lainhart, J. E., Bigler, E. D., Bocian, M., Coon, H., Dinh, E., Dawson, G., et al. (2006). Head circumference and height in autism: A study by the collaborative program of excellence in Autism. *American Journal of Medical Genetics. Part A*, *140*, 2257–2274.

Lewis, J. D., Evans, A. C., Pruett, J. R., Botteron, K. N., McKinstry, R. C., Zwaigenbaum, L., et al. (2017). The emergence of network inefficiencies in infants with autism spectrum disorder. *Biological Psychiatry*, *82*, 176–185.

Lin, M., Pedrosa, E., Hrabovsky, A., Chen, J., Puliafito, B. R., Gilbert, S. R., et al. (2016). Integrative transcriptome network analysis of iPSC-derived neurons from schizophrenia and schizoaffective disorder patients with 22q11.2 deletion. *BMC Systems Biology*, *10*(1), 105.

Liu, X., Campanac, E., Cheung, H. H., Ziats, M. N., Canterel-Thouennon, L., et al. (2017). Idiopathic autism: Cellular and molecular phenotypes in pluripotent stem cell-derived neurons. *Molecular Neurobiology*, *54*, 4507–4523.

Liu, L., Jouan, L., Rochefort, D., Dobrzeniecka, S., Lachapelle, K., Dion, P. A., et al. (2013). Analysis of rare exomic variation among patients with ASD. *PLoS Genetics*, *9*(4), e1003443.

Mahfouz, A., Ziats, M. N., Rennert, O. M., Lelieveldt, B. P., & Reinders, M. J. (2015). Shared pathways among autism candidate genes determined by co-expression network analysis of the developing human brain transcriptome. *Journal of Molecular Neuroscience*, *57*(4), 580–594.

Marshall, C. R., Noor, A., Vincent, J. B., Lionel, A. C., Feuk, L., Skaug, J., et al. (2008). Structural variation of chromosomes in autism spectrum disorders. *American Journal of Human Genetics*, *82*, 477–488.

Mellios, N., Feldman, D. A., Sheridan, S. D., JPK, I., Kwok, S., Amoah, S. K., et al. (2017). MeCP2-regulated miRNAs control early human neurogenesis through differential effects on ERK and AKT signaling. *Molecular Psychiatry*.

Mellios, N., & Sur, M. (2012). The emerging role of microRNAs in schizophrenia and autism spectrum disorders. *Frontiers in Psychiatry*, *3*, 39–43.

Nagy, J., Kobolák, J., Berzsenyi, S., Ábrahám, Z., Avci, H. X., Bock, I., et al. (2017). Altered neurite morphology and cholinergic function of induced pluripotent stem cell-derived neurons from a patient with Kleefstra syndrome and autism. *Translational Psychiatry*, *7*(7).

O'Roak, B. J., Stessman, H. A., Boyle, E. A., Witherspoon, K. T., Martin, B., Lee, C., et al. (2014). Recurrent de novo mutations implicate novel genes underlying simplex autism risk. *Nature Communications*, *24*, 5595.

Onore, C., Yang, H., van dear Water, J., & Ashwood, P. (2017). Dynamic Akt/mTOR signaling in children with ASD. *Frontiers in Pediatrics*, *5*, 43–57.

Parikshak, N. N., Luo, R., Zhang, A., Won, H., Lowe, J. K., Chandran, V., et al. (2013). Integrative functional genomic analyses implicate specific molecular pathways and circuits in autism. *Cell*, *155*(5), 1008–1021.

Pinto, D., Pagnamenta, A. T., Klei, L., Anney, R., Merico, D., Regan, R., et al. (2010). Functional impact of global rare CNVs in ASD. *Nature*, *466*, 368–372.

Poultney, C. S., Goldberg, A. P., Drapeau, E., Kou, Y., Harony-Nicolas, H., Kajiwara, Y., et al. (2013). Identification of small exonic CNV from whole-exome sequence data and application to autism spectrum disorder. *American Journal of Human Genetics*, *93*(4), 607–619.

Provenzano, G., Chelini, G., & Bozzi, Y. (2017). Genetic control of social behavior: Lessons from mutant mice. *Behavior Brain Research*, *325*, 237–250.

Quesnet-Vallieres, M., Dargaei, Z., Irimia, M., Gonatopoulos-Pournatzis, T., Ip, J. Y., Wu, M., et al. (2016). Misregulation of an activity-dependent splicing network as a common mechanism underlying ASD. *Molecular Cell*, *64*, 1023–1034.

Ram Venkataraman, G., O'Connell, C., Egawa, F., Katherine-haghighi, D., & Wall, D. P. (2016). De novo mutations in autism implicate the synaptic elimination network. *Pacific Symposium on Biocomputing*, *22*, 521–532.

Redcay, E., & Courchesne, E. (2005). When is the brain enlarged in autism? A meta-analysis of all brain size reports. *Biological Psychiatry*, *58*, 1–9.

Sanders, S. J., Ercan-Sencicek, A. G., Hus, V., Luo, R., Murtha, M. T., Moreno-De-Luca, D., et al. (2011). Multiple recurrent de novo CNVs include duplication of 7q11.23 Williams syndrome are strongly associated with autism. *Neuron*, *70*, 863–885.

Schaefer, G. B., Starr, L., Pickering, D., Skar, G., Dehaai, K., & Sanger, W. G. (2010). Array comparative genomic hybridization findings in a cohort referred for an autism evaluation. *Journal of Child Neurology*, *25*(12), 1498–1503.

Sebat, J., Lakshmi, B., Malhotra, D., Troge, J., Lese-Martin, C., Walsh, T., et al. (2007). Strong association of de novo copy number mutations with autism. *Science*, *316*, 445–449.

Seshamani, S., Blazejewska, A. I., Mckown, S., Caucutt, J., Dighe, M., Gatenby, C., et al. (2016). Detecting default mode networks in utero by integrated 4D fMRI reconstruction and analysis. *Human Brain Mapping, 37*, 4158–4178.

Spence, S., Grant, P., Thurm, A., & Swedo, S. (2009). In D. S. Carney & E. J. Nestler (Eds.), *Chapter 65 in neurobiology of mental disorders. Neurobiology of autism and the pervasive developmental disorders*: Oxford Press.

Sztainberg, Y., & Zoghbi, H. Y. (2016). Lessons learned from studying syndromic autism spectrum disorders. *Nature Neuroscience, 19*, 1408–1417.

Tamburini, C., & Li, M. (2017). Understanding neurodevelopmental disorders using human pluripotent stem cell-derived neurons. *Brain Pathology, 27*(4), 508–517.

Thomason, M. E., Scheinost, D., Manning, J. H., Grove, L. E., Hect, J., Marshall, N., et al. (2017). Weak functional connectivity in the human fetal brain prior to preterm birth. *Scientific Reports, 7*, 39286.

Varghese, M., Keshav, N., Jacot-Descombes, S., Warda, T., Wicinski, B., Dickstein, D. L., et al. (2017). Autism spectrum disorder: Neuropathology and animal models. *Acta Neuropathologica, 134*, 1–30.

Wu, Y. E., Parikshak, N. N., Belgard, T. G., & Geschwind, D. H. (2016). Genome-wide, integrative analysis implicates microRNA dysregulation in autism spectrum disorder. *Nature Neuroscience, 19*, 1463–1476.

Yazawa, M., & Dolmetsch, R. E. (2013). Modeling timothy syndrome with iPS cells. *Cardiovascular Translational Research, 6*(1), 1–9.

Ziats, M. N., & Rennert, O. M. (2011). Expression profiling of autism candidate genes during human brain development implicates central immune signaling pathways. *PLoS One, 6*, e24691.

Ziats, M. N., & Rennert, O. M. (2013). Aberrant expression of long noncoding RNAs in autistic brain. *Journal of Molecular Neuroscience, 49*, 589–593.

Ziats, M. N., & Rennert, O. M. (2014). Identification of differentially expressed microRNAs across the developing human brain. *Molecular Psychiatry, 19*, 848–852.

Ziats, M. N., & Rennert, O. M. (2016). The evolving diagnostic and genetic landscapes of autism spectrum disorder. *Frontiers in Genetics, 7*, 65–72.

Ziats, C. A., Rennert, O. M., & Ziats, M. N. (2019). Toward a pathway driven clinical-molecular framework for classifying autism spectrum disorders. *Pediatric Neurology, 98*, 46–52.

Zwanberg, R. J., Ruiter, S. A., van den Heuvel, E. R., Flapper, B. C., & Van Ravenswaaij-Arts, C. M. (2016). Developmental phenotype in Phelan-McDermid (22q13.3 deletion) syndrome: A systematic and prospective study in 34 children. *Journal of Neurodevelopmental Disorders, 8*, 16.

Further reading

Han, J., Sarkar, A., & Gage, F. H. (2015). MIR137: Big impacts from small changes. *Nature Neuroscience, 18*, 931–932.

Leppa, V. M., Kravitz, S. N., Martin, C. L., Andrieux, J., Le Caignec, C., Martin-Coignard, D., et al. (2016). Rare inherited and de novo CNVs reveal complex contributions to ASD risk in multiplex families. *American Journal of Human Genetics, 99*(3), 540–554.

Chapter 15

Genomics of schizophrenia

A. Corvin, C. Ormond and A.M. Cole

Department of Psychiatry and Neuropsychiatric Genetics Research Group, Trinity College Dublin, Dublin, Ireland

1 Introduction

Schizophrenia is an enigmatic syndrome that challenges people to confront their perception and understanding of the world around them. The core symptoms affect perception of reality, thinking, behavior, and motivation. Schizophrenia typically emerges in early adulthood, and for many sufferers, presages a long-term course characterized by episodic, or continuous illness, with significant functional impairment (Hegarty et al., 1994).

The disorder is heritable, but social factors, including childhood adversity and migration, are also involved (Cantor-Graae & Selten, 2005; Matheson et al., 2013). Decades of research indicate a neurodevelopmental contribution to the etiology (reviewed in Birnbaum & Weinberger, 2017). We have learned that schizophrenia is associated with structural, functional, and neurochemical brain changes, particularly involving the dopaminergic system (Tandon, Keshavan, & Nasrallah, 2008). Progress has been made in treatment: antipsychotic medications targeting the dopamine D2 receptor effectively treat psychotic symptoms and reducing the risk of relapse (Miyamoto et al., 2012). However, this has little impact on debilitating behavioral and cognitive deficits and overall illness outcomes (Leucht et al., 2013).

Schizophrenia affects almost 1% of adults, causing significant morbidity and a reduction in average life expectancy measured in decades (Tiihonen et al., 2009). The diagnosis, although reliable, is clinical, and may capture a heterogeneous population of patients with a range of symptoms and outcomes, and may reflect different underlying disease processes. This has hampered efforts to develop new treatments or early intervention strategies, as it is difficult to predict who will respond to particular therapies, or even which individuals will go on to develop the full syndrome among those at risk (Fischer & Carpenter Jr., 2009). Understanding the genetic etiology of a disease can be a powerful enabler of drug development (Kamb, Harper, & Stefansson, 2013; Nioi et al., 2016). Schizophrenia is a substantially heritable disorder, and the last decade has seen significant progress in illuminating the genetic architecture involved. In this chapter, we review the evidence for heritability, and the evolution of methods that have enabled detailed exploration of the genome to investigate molecular etiology. We also consider how these findings may translate into better patient care.

2 Schizophrenia heritability

The current conceptualization of schizophrenia (described in the Diagnostic Statistical Manual of Mental Disorders [DSM-5] and International Classification of Diseases [ICD-10]) is deeply rooted in the 19th century work of Emil Kraepelin. He postulated that schizophrenia (then termed *dementia praecox*) was a disease with a unifying pathophysiology, and he made a distinction between this condition and manic-depression. Kraepelin recognized that schizophrenia clustered in families, and identified a general hereditary predisposition to mental disorders in about 70% of his cases. Reporting his data in 1907, Heron reached a similar conclusion. Their work confirmed familial clustering of schizophrenia, but did not address whether this was explained by a contribution from shared environmental or genetic factors, or some combination of both.

Pivotal twin and adoption studies beginning in the 1960s confirmed the role of risk genes (reviewed in Cardno & Gottesman, 2000; Ingraham & Kety, 2000). Furthermore, heritability could be estimated and proved remarkably consistent across individual twin studies and meta-analyses of the data ($h^2 = 80\%–85\%$) (Sullivan, Kendler, & Neale, 2003). With heritability confirmed, speculation moved on to how many genomic loci contribute to risk; their population frequencies; their effect sizes, and their interactions with each other and with environmental risk. Two major theories emerged. The first, from Rubin and others, proposed a single major locus, or a number of rare, highly penetrant risk loci. The second, based on model fitting of risk data from different classes of relatives of affected probands, proposed that a large proportion of cases were polygenic, involving many common variants of small effect (reviewed in Plomin, Owen, & McGuffin, 1994).

Personalized Psychiatry. https://doi.org/10.1016/B978-0-12-813176-3.00015-8

3 An emerging genetic architecture of schizophrenia

As molecular methods evolved and sample sizes increased, linkage studies of family pedigrees became sufficiently powered to identify large genetic effects, and to localize them to genomic regions (Ng et al., 2009). Such studies resolved that schizophrenia was not a single gene disorder, but provided a small number of signals, suggesting a role for moderately large effects in a subset of cases (Ng et al., 2009). Cytogenetic studies added support for such effects. Deletions involving the chr22q11.2 locus (velocardiofacial syndrome [VCFS]) (reviewed in Murphy, Jones, & Owen, 1999) and a reciprocal chromosomal t(1:11) translocation (St Clair et al., 1990) disrupting the *DISC1* locus were the most prominent examples. Both loci represented rare events; the former occurring at a rate of 1 in 4000 live births, and the latter being identified in a single, large Scottish pedigree. Over the next decade, hundreds of potentiation functional or positional candidate genes were investigated for evidence of association with common risk alleles. These studies were generally small, and underpowered to definitively identify the common, small genetic effects likely to be involved (Sullivan, Daly, & O'Donovan, 2012). However, this wave of molecular research indicated that a spectrum of genetic variation, from the common to the very rare, was likely to contribute to schizophrenia risk across the population.

Significantly, the pedigrees assessed in the cytogenetic studies included individuals with a range of psychiatric phenotypes, raising a further question. What were the genetic boundaries of the schizophrenia syndrome? In fact, this question had been recognized early in the history of the field. Kraepelin acknowledged cases of "latent schizophrenia" within family members of affected probands, and Rudin identified an increased prevalence of what we now call bipolar disorder, and what he termed "eccentric personalities" among the relatives of affected probands (Zerbin-Rudin & Kendler, 1996). Investigating families more systematically, Kendler and colleagues identified a familial predisposition to a spectrum of clinical syndromes, including schizoaffective disorder, other nonaffective psychosis, schizotypal personality disorder, and psychotic affective illness in relatives of schizophrenia probands (Kendler et al., 1993a, 1993b). There was also emerging evidence of overlap between schizophrenia and affective disorder in families (Maier et al., 1993). In 2009, Swedish investigators reported a register-based analysis of more than 2 million nuclear families, finding that first-degree relatives of probands with either schizophrenia ($n = 35,985$) or bipolar disorder ($n = 40,487$) were at substantially increased risk of both disorders. They estimated that 63% of co-morbidity between disorders was due to additive genetic effects common to both disorders (Lichtenstein et al., 2009).

4 Genome-wide association studies

The development of array-based methods in the past two decades offered a new vista for genetics research, where it became possible to test hundreds of thousands of genetic variants in a single experiment. Genome-wide association studies (GWAS) could comprehensively assay common single nucleotide polymorphisms (SNPs) that occur predictably, at a given minor allele frequency (MAF) (e.g., <5%) across the human genome. By systematically testing for differences in SNP frequencies between cases and controls, risk or protective alleles could be identified (Welter et al., 2014). Notable early successes for other common disorders (e.g., diabetes, inflammatory bowel disease) spurred efforts to generate the large sample sizes required to power schizophrenia GWAS (Corvin, Craddock, & Sullivan, 2010). In 2009, three independent GWAS reported robust, replicable association signals at a 5.5 megabase (Mb) region around the major histocompatibility complex (MHC) on chromosome 6p (Purcell et al., 2009; Shi et al., 2009; Stefansson et al., 2009). This locus, one of the most complex and genetically diverse genomic regions, encodes ~250 genes, including the classical and transplantation HLA alleles, but also many immune and nonimmune genes. Subsequent GWAS have provided further association evidence, while highlighting the challenge posed in mapping this locus (reviewed in Corvin & Morris, 2014).

Most profoundly, these studies showed genomics could be applied fruitfully in psychiatry. This provided impetus for the Psychiatric GWAS Consortium (PGC) to coordinate a much larger effort to combine all available GWAS data for five disorders, with the largest sample size being for schizophrenia (Rees et al., 2014a; Schizophrenia Working Group of the Psychiatric Genomics, C., 2014). The Schizophrenia Working Group of the PGC has seen a trajectory of discovery similar to that of other common disorders, with seven significant loci being identified by PGC1 in a sample of 9394 cases published in Vacic et al. (2011); this increased to 108 loci in an analysis of 36,989 cases and 113,075 controls published in 2014 (Rees et al., 2014a); and approximately 150 by early 2018 (Fig. 1). The identified association signals localized the search to genes that are current targets for treatment (*DRD2*); genes more widely involved in glutamatergic neurotransmission (*GRIN2A*, *SRR*, *CLCN3*, and *GRIA1*); and unexpected mechanisms involving neuronal calcium signaling (e.g., *CACNA1C*, *CACNA1l*, *CACNB2*, *RIMS1*) or broader synaptic function (*KCTD13*, *CNTN4*, *PAK6*) (Schizophrenia Working Group of the Psychiatric Genomics, C., 2014). Among these loci, 83 were newly implicated

Manhattan plot showing schizophrenia associations

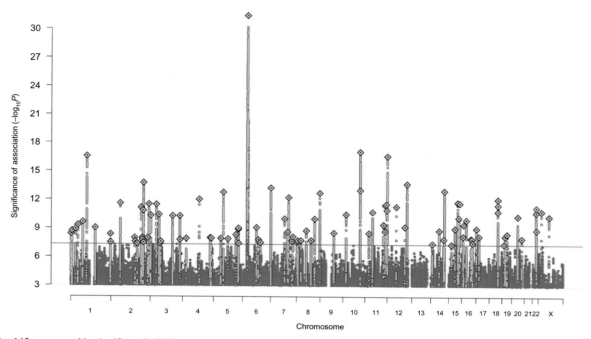

FIG. 1 145 genome-wide significant loci discovered by the Schizophrenia Working Group of the PGC. The genome-wide significance threshold ($p = 5 \times 10^{-8}$) is shown in green. *From Ripke, S., et al. (2014). Biological insights from 108 schizophrenia-associated genetic loci. Nature, 511, 421–427.*

in schizophrenia, highlighting the discovery power of GWAS. By adding a dataset of 11,260 cases and 24,542 controls, Pardinas and colleagues (Pardinas et al., 2018) identified 50 novel associated loci and, with systems genomics methods, identified candidate causal genes at 33 of these loci.

5 Rare structural risk variants

Array-based studies have shown that rare chromosomal rearrangements involving deletion, duplication, inversion, or translocation of DNA (termed copy number variants (CNVs)) make a sizable contribution to human genetic variation (Iafrate et al., 2004; Sebat et al., 2004). Typically defined as being greater than 1 kb in size, in truth, they follow a continuous distribution from very large or complex events to simple insertions/deletions (Malhotra & Sebat, 2012). These submicroscopic events make a sizable and clinically significant contribution to neurodevelopmental disorders.

In 2008, Walsh and colleagues reported enrichment of deletions and duplications (>100 kb) in schizophrenia patients (Walsh et al., 2008). Larger consortia studies, and the recent Psychiatric Genomics Consortium (PGC), centralized analysis of 21,094 cases, and 22,227 controls confirmed this finding, and have shown that the CNVs that drive this effect overlap genes (International Schizophrenia Consortium, 2008; Marshall et al., 2017; Stefansson et al., 2008). The implicated loci overlap CNV regions known to be involved in intellectual disability (ID) (Rees et al., 2016), and genes involved in *N*-methyl-D-aspartate receptor (NMDAR) and neuronal activity-regulated cytoskeleton-associated (ARC) postsynaptic signaling complexes (Kirov et al., 2012). These studies have also identified specific loci that were significantly associated with schizophrenia. Confirming the known association with chr22q11.2 deletions, the early consortia studies also reported novel associations with deletions at 1q21.1, 15q11.2, and 15q13.3 (International Schizophrenia Consortium, 2008; Stefansson et al., 2008) (Fig. 2). Subsequent studies implicated at least 10 other regions with either risk or protective loci (reviewed in Kirov, 2015). The PGC analysis provided genome-wide significant evidence for eight of these (1q21.1, 2p16.3 (*NRXN1*), 3q29, 7q11.2, 15q13.3, distal 16p11.2, proximal 16p11.2 and 22q11.2), and suggestive support for eight more (Marshall et al., 2017).

Regions with excess large deletions in cases

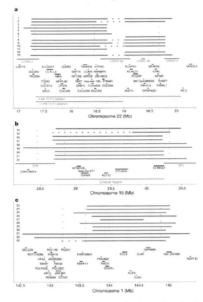

FIG. 2 Three large deletions associated with schizophrenia. *From International Schizophrenia Consortium. (2008). Rare chromosomal deletions and duplications increase risk of schizophrenia. Nature, 455 (7210), 237–241.*

Collectively, this list of known/probable risk loci are carried by <2.5% of patients (Kirov, 2015), and are of moderate penetrance for schizophrenia (OR = 2–30) (Douglas et al., 2014; Kirov et al., 2014; Malhotra & Sebat, 2012; Rees et al., 2014b; Rujescu et al., 2009; Tansey et al., 2016). Most represent new events in the people affected, likely explaining the higher rate of *de novo* CNVs reported in schizophrenia, compared with controls (Kirov et al., 2012). Structurally, most of the implicated CNVs recur in the population at the same genomic locations mediated by nonallelic homologous recombination (NAHR), where repeat sequences flanking the critical region make it prone to rearrangement in the population. But this is only one of a number of mechanisms for genome rearrangements (reviewed in Liu et al., 2012), and deletions at the genes *NRXN1* and *VIPR2* were identified because of different, disruptive CNV events with different breakpoints in individual carriers, rather than a more universal genomic mechanism (Rujescu et al., 2009; Vacic et al., 2011).

6 Genome-sequencing studies

Next-generation sequencing (NGS) technology allows comprehensive investigation of sequence variation across the genome (MacArthur et al., 2012). NGS studies have established that gene-disrupting and protein-damaging rare, or ultra-rare variants (dURVs) play a substantial role in neurodevelopmental disorders (e.g., ID (de Ligt et al., 2012; Rauch et al., 2012), autism (De Rubeis et al., 2014; O'Roak et al., 2012), and epilepsy (Allen et al., 2013)). As with schizophrenia, many of these conditions significantly compromise fecundity (Power et al., 2013). Because affected individuals have fewer offspring, high-risk alleles are likely to be selected against in the populations. If such mutations are observed, they are likely to be new, or at most, a few generations old. Therefore, these mutations are very rare in the general population, and this has significant power implications for the case-control approaches that proved so successful for GWAS of other diseases and disorders (Zuk et al., 2014).

So what has been learned so far? Gene-disruptive and putatively protein-damaging dURVs, but not synonymous URVs, are more abundant in people with schizophrenia. This excess is several-fold larger than the increased rate of *de novo* mutations (DNMs) seen in schizophrenia patients, suggesting that most dURVs are inherited, even if only a few generations old (Genovese et al., 2016). dURVs are enriched in genes' overlapping regions near common variants associated with schizophrenia, indicating at least some convergent genetic architecture. Mutations are also concentrated in genes implicated in X-linked intellectual disability and other developmental disorders. Such genes are known to be risk factors for syndromal forms of autism, and the same may be true for schizophrenia.

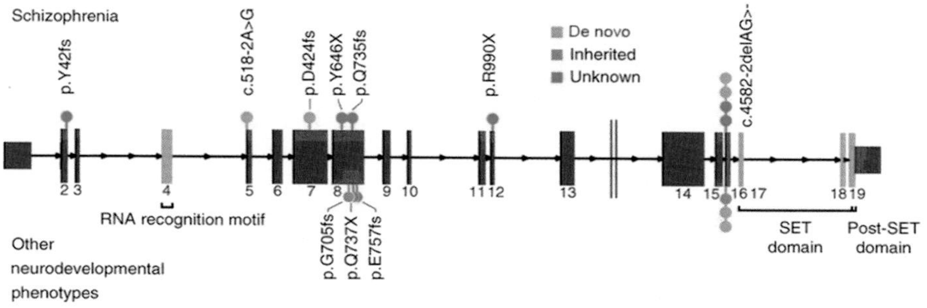

Variants discovered in patients with schizophrenia are plotted above the gene and those discovered in individuals with other neurodevelopmental disorders (from DDD and SISu) are plotted below. Each variant is colored according to its mode of inheritance. All LoF variants appeared before the conserved SET domain, which is responsible for catalyzing methylation. Seven LoF variants occurred at the same two-base deletion at the exon 16 splice acceptor (c.4582-2delAG>-).

FIG. 3 16 loss of function variants in *SETD1A*, colored by mode of inheritance. Variants from schizophrenia cases are plotted above the gene, and those from other neurodevelopmental disorders are plotted. *From Singh, T., et al. (2016). Rare loss-of-function variants in SETD1A are associated with schizophrenia and developmental disorders Nature Neuroscience, 19 (4), 571–577.*

A large combined analysis of case-control ($n = 6669$ cases) and trio ($n = 1077$) exome data identified an association between rare variants in *SETD1A* and risk of schizophrenia (Singh et al., 2016). In more than 20,000 exomes, the investigators observed 7 loss-of-function (LoF) variants, and three DNMs in cases, and none in controls (Fig. 3). *SETD1A* encodes a methyltransferase that catalyzes the methylation of lysine residues in histone H3 and is a regulator of gene transcription. The same *SETD1A* LoF mutation was also identified in an independent exome sequencing study of 231 schizophrenia trios (Takata et al., 2014), and recurrent DNMs were also identified in the genes *LAMA2, DPYD, TRRAP*, and *VPS* in the same cohort (Xu et al., 2012).

As an alternative to case-control studies, family-based studies may be a powerful way to identify the rare variation that may be contributing to schizophrenia risk. Steinberg and colleagues recently reported an NGS study of an Icelandic family with ten individuals with psychosis. They found exactly one variant, a nonsense mutation in the *RNA-binding-motif protein 12 (RBM12)* gene, that was carried by all 10 individuals showing significant association with psychosis ($P = 2.2 \times 10^{-4}$) (Steinberg et al., 2017). They were also able to identify a risk variant in this gene in a Finnish pedigree.

Other investigators have looked at large, multiplex families affected by schizophrenia. Homann, Misura (Homann et al., 2016) performed WGS on nine multiplex families with psychoses, where individuals diagnosed with schizophrenia of schizoaffective were classified as "affected." In one family, seven siblings with schizophrenia spectrum disorders each carried a novel private missense variant within *SHANK2*, while another family with four affected siblings and their unaffected mother carried a novel private missense variant in the *SMARCA1* on the X chromosome. Both variants represent candidates that may be causal for psychotic disorders, given their known disease biology (Homann et al., 2016). Timms, Dorschner (Timms et al., 2013), investigated the genetic architecture of 12 patients from five large pedigrees using exome sequencing. One pedigree shared a missense and frameshift substitution of *GRM5*. Another pedigree transmitted a missense substitution in *PPEF2*. Three pedigrees demonstrated different missense substitutions within *LRP1B*, encoding a low-density lipoprotein receptor-related protein tied to both the NMDA receptor and located in a chromosome 2q22 region previously linked to schizophrenia. Stoll and colleagues also reported enrichment of a rare chromosome 22q11.22 deletion in individuals with schizophrenia, and ID in a Northern Finnish isolate. They detected association with a gene TOP3B that encoded a topoisomerase that is a component of cytosolic messenger ribonucleoproteins.

Monozygotic twins (MZ) discordant for schizophrenia have been another focus for many investigators. One study investigated eight pairs of MZ twins, and identified four CNVs implicated with schizophrenia in four families (Tang et al., 2017). Another study specifically set out to identify somatic variants in two pairs of discordant twins. They identified five genes with variants in both affected twins included: *GUSBP3* (glucuronidase, β-pseudogene 3), *SMA4* (glucuronidase, β-pseudogene), *CMIP* (C-Mafinducing protein), *CYP2A7* (cytochrome P450 family 2 subfamily A member 7), and *MIR663AHG* (MIR663A host gene). Of particular interest, the same variant was identified in both affected twins in *CMIP*, which is involved in the T-cell signaling pathway, and supports the role of an altered immune system in schizophrenia (Reble et al., 2017).

7 From gene discovery to etiology

At this juncture, the 150 common loci and smaller number of rare risk variants that have been identified collectively explain <10% of variance in schizophrenia susceptibility. We know that SNP heritability across the genome is largely uniform, and it is likely that more than 71% of 1 Mb genomic regions contain at least one risk variant (Loh et al., 2015). For most affected individuals, schizophrenia has a polygenic architecture in which hundreds, or even thousands of variants collectively contribute to risk (Gratten et al., 2014). Additional loci are likely to be confirmed by ongoing GWAS efforts, but as effect sizes and allele frequencies become smaller, study power will diminish. More is likely to be learned about the contribution of rare, more penetrant risk variants. These findings will continue to inform our understanding of: (i) the genetic architecture of schizophrenia; (ii) the molecular mechanisms by which these genes contribute; and (iii) how these effects are collectively orchestrated to impact brain function and disease.

8 Genetic architecture

In their GWAS analysis, the International Schizophrenia Consortium (ISC) demonstrated for the first time that SNP data could be used to empirically test the role of polygenic inheritance in schizophrenia susceptibility (Purcell et al., 2009). By summing variation across a large number of nominally associated loci into quantitative scores, they asked whether these polygenic risk scores (PRS) could predict disease state in independent samples. Their results, supported by a subsequent analysis, using a different approach (Genome-wide Complex Trait Analysis (GCTA)), provided molecular evidence that polygenic inheritance contributes to schizophrenia susceptibility (Lee et al., 2012). With larger sample sizes, it has become possible to estimate that common alleles explain one-third to half of the genetic variance in schizophrenia risk (Purcell et al., 2014).

This ISC paper provided a second important observation. By extending PRS analysis to other cohorts, they showed that common risk variants were shared with bipolar disorder, but not with six common nonpsychiatric diseases (Fig. 4). This was molecular validation of the epidemiological evidence of overlap reported from the Swedish register study. With the PGC

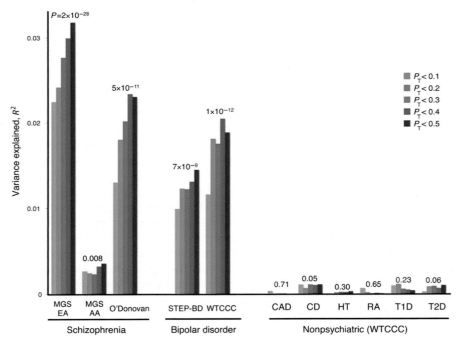

Replication of the ISC-derived polygenic component in independent SCZ and BPD samples

Variance explained in the target samples based on scores derived in the entire ISC for five significance thresholds (p_T < 0.1, 0.2, 0.3, 0.4 and 0.5, plotted left to right in each study). The y-axis indicates Nagelkerke's pseudo-R^2; the number above each set of bars is the P-value for the p_T < 0.5 target sample analysis. Numbers for cases/controls: MGS-EA 2687 / 2656; MGS-AA 1287 / 973; O'Donovan 479 / 2938; STEP-BD 955 / 1498; WTCCC 1829 / 2935; CAD 1926 / 2935; CD 1748 / 2935; HT 1952 / 2935; RA 1860 / 2935 ; T1D 1963 / 2935 ; T2D 1924 / 2935.

FIG. 4 Variance explained by common risk variants in the target samples on the basis of scores derived from schizophrenia and bipolar disorder cases, for five difference significance thresholds. *CAD*, coronary artery disease; *CD*, Crohn's disease; *HT*, hypertension; *RA*, rheumatoid arthritis; *T1D*, type I diabetes; *T2D*, type II diabetes.

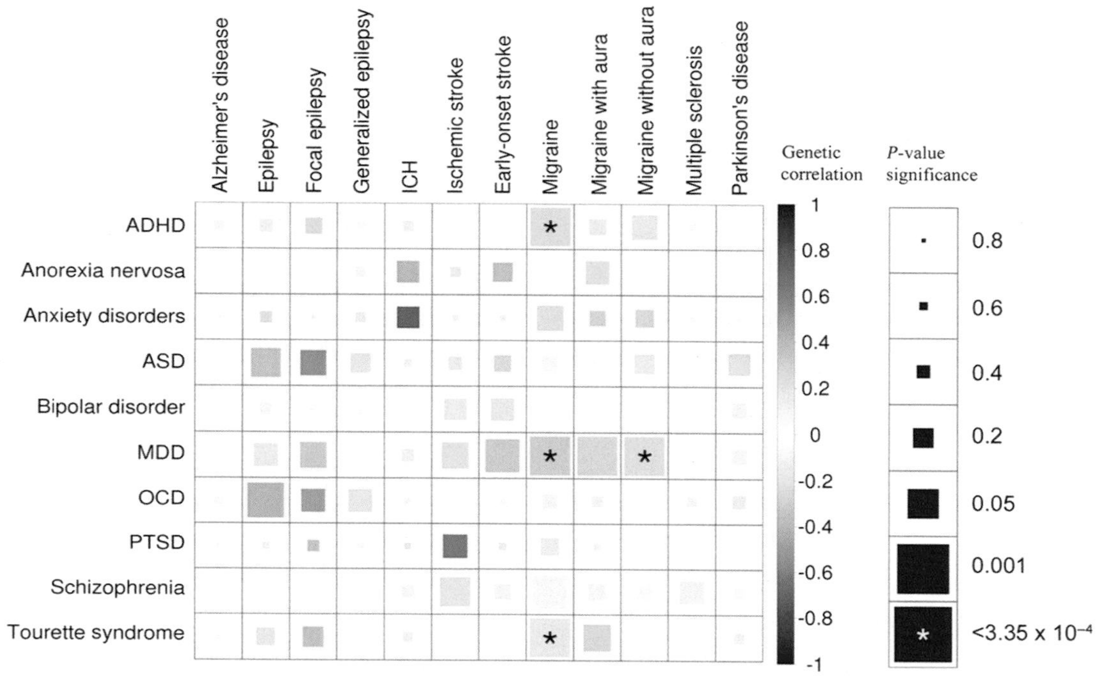

FIG. 5 Genetic correlation matrix between neurological and psychiatric disorders. *ADHD*, attention deficit hyperactivity disorder; *ASD*, autism spectrum disorder; *ICH*, intracerebral hemorrhage; *MDD*, major depressive disorder; *OCD*, obsessive-compulsive disorder.

dataset, it was possible to test whether genetic risk variation is unique to individual disorders, or is shared across schizophrenia, bipolar disorder, major depressive disorder (MDD), autistic spectrum disorder (ASD), and attention-deficit hyperactivity disorder (ADHD). This confirmed moderate levels of covariance between schizophrenia and MDD, with modest overlap between schizophrenia and childhood onset disorders such as ASD (Lee et al., 2013). So it is likely that distinctions made between these psychiatric disorders mask overlapping genetic architecture, and presumably shared molecular etiology. The Brainstorm Consortium has recently completed an ambitious project to test the extent of shared genetic contributions across 23 brain disorders and 15 traits of interest. They found substantial sharing of common risk variants across almost all psychiatric disorders, but little sharing between psychiatric and neurological disorders (https://www.biorxiv.org/content/early/2016/04/16/048991) (Fig. 5). This raises the interesting possibility that common genetic mechanisms contribute to the development of psychiatric morbidity across disorders. Further research is required to understand these mechanisms and whether they can be targeted for future treatment strategies.

9   Diagnostics

Most cases of schizophrenia are likely to involve polygenic risk spread across many variants. In the PGC3 paper, the authors looked at how the transition from carrying relatively small numbers of risk variants (first decile) to almost all known risk variants (10th decile) impacted risk to the individual. They confirmed that odds ratios increased almost linearly across this transition in all participating populations. However, in each subsample, there were individuals in the control population who were unaffected, despite harboring many common risk variants. The area under the receiver operating curve (AUC) ranged from 0.65 to 0.8 across samples; significantly below the threshold for clinical diagnostics utility (Schizophrenia Working Group of the Psychiatric Genomics, C., 2014). As more loci are identified, these values will likely improve, but are unlikely to provide the basis for a simple diagnostic test.

For some people, the situation will be more nuanced. Chromosomal microarray (CMA) testing for identification of clinically significant CNVs is recommended as a first tier genetic test for patients with autism spectrum disorder, developmental delay, and intellectual disability (O'Byrne et al., 2016; Schaefer & Mendelsohn, 2013). Rates of CNVs are lower in schizophrenia, but carrying such a mutation will have implications, even if much is still to be done to understand the pathogenicity and penetrance of individual events. Eight rare CNV loci that surpass genome-wide significance explain <1% of schizophrenia variance in the population, and are carried by <2% of patients (Kirov, 2015; Marshall et al.,

2017). These loci are of moderate penetrance for schizophrenia (OR = 2–30) (Douglas et al., 2014; Kirov et al., 2014; Malhotra & Sebat, 2012; Rees et al., 2014b; Rujescu et al., 2009; Tansey et al., 2016), but almost all have pleiotrophic effects conferring risk for other developmental phenotypes, including ASD, ID, epilepsy, and congenital anomalies. In fact, the risk of developing any early developmental disorder (e.g., ID, ASD, developmental delay) is significantly higher (penetrance 10%–88%) than the risk of developing schizophrenia (penetrance 2%–18%) (Kirov et al., 2014). Even in the absence of a psychiatric diagnosis, carriers of CNVs associated with schizophrenia have significant, but variable cognitive deficits (Kendall et al., 2016; Stefansson et al., 2014). This is potentially important information for individuals, their families, and treating clinicians.

Pick up rates for CNVs in the general schizophrenia population are likely to be low. In autism and intellectual disability (ID), "syndromal" cases where additional phenotypes and dysmorphic features are also present are more likely to have a genomic etiology (Miles et al., 2005). Our Trinity College Dublin team has started to explore whether syndromal features, including co-morbid ID, autism, epilepsy, developmental delay, or learning difficulties identify a schizophrenia population at greater risk. In a pilot study of 1215 schizophrenia cases, we found that having a co-morbid neurodevelopmental disorder, developmental delay, or learning difficulties increased risk of being a CNV carrier eight-fold. Similar findings have been reported from cohorts with ID and co-morbid psychosis (Thygesen et al., 2018). As more data becomes available from sequencing studies, it is likely that we will be able to identify patient subgroups for whom genetic testing will be of diagnostic significance.

10 Localizing risk genes

Most common risk variants are likely to have subtle, regulatory effects on the function, rather than the structure of proteins (Maurano et al., 2012; Schork et al., 2013). At present, defining which genes or alleles explain association at a locus can be a formidable task (see Fig. 5 for illustration). Improving annotation, and understanding of the genome is helping. Common risk variants identified by polygene score methods in the PGC are significant enrichment among genic elements, particularly the 5'UTR, exon, and 3'UTR compared with intergenic SNPs, suggesting that associated loci are likely to impact genes. Furthermore, this pool of variants is enriched for expression of quantitative trait loci (eQTL), specifically loci that affect gene expression in the human brain (Richards et al., 2012). As with other neurodevelopmental disorders (e.g., ASD, ID, and developmental delay), it appears that most of the common and rare variant associations are enriched at the subset of genes that are most intolerant of loss-of-function mutations (Pardinas et al., 2018). From a recent study by Genovese and colleagues, potentially synaptic genes appear to explain more than 70% of the exome enrichment in dURVs (Genovese et al., 2016). More extensive investigation of the protein complexes and structures at the synapse is likely to be critical in parsing out the molecular etiology, or etiologies, involved.

11 Molecular mechanisms and pathways

(i) *Immune function:* Perhaps one of the more surprising discoveries from the GWAS was identification of enrichment of genetic variants in immunological pathways (Network and Pathway Analysis Subgroup of Psychiatric Genomics Consortium, 2015). The top schizophrenia GWAS signals come from the major histocompatibility complex (MHC) locus spanning 4 megabases of chromosome 6, but other immune genes are also implicated (Corvin & Morris, 2014; Irish Schizophrenia Genomics Consortium and the Wellcome Trust Case Control Consortium 2, 2012; Psychiatric GWAS Consortium Bipolar Disorder Working Group, 2011; Schizophrenia Working Group of the Psychiatric Genomics, C., 2014; Shi et al., 2009). The MHC is one of the most genetically diverse regions of the genome, and in its extended form, encodes more than 400 genes critical to immune function, but also involved in many other functions. The challenges of mapping this locus have been discussed elsewhere, but progress is being made (Corvin & Morris, 2014; Sekar et al., 2016). Steve McCarroll and his team at the Broad Institute identified three genome-wide significant association signals contributing to schizophrenia risk at the MHC locus. The most strongly associated of these was near a complex multi-allelic form of structural variation that affects the C4 gene encoding complement component 4 (Sekar et al., 2016). The team identified many common, structurally distinct C4 alleles that affect expression of C4A and C4B in the brain, and established that each allele was associated with schizophrenia risk in proportion to its effect on C4A expression. They showed that C4 is expressed in neurons, and in mice, C4 promoted synaptic elimination during development, opening up a new avenue for molecular research.

(ii) *Glutamatergic neurotransmission and synaptic plasticity:* Among the loci identified in the PGC-schizophrenia GWAS, a number contained genes implicated in glutamatergic neurotransmission and synaptic function (Rees et al., 2014a).

These data suggest that disruption of postsynaptic signaling is important in the genetic etiology of schizophrenia. Calcium influx at the *N*-methyl-d-aspartate receptor (NMDAR) activates pathways that converge on extracellular signal-regulated kinases, which have a critical role in synaptic plasticity through the transcription of activity-regulated cytoskeleton-associated protein (ARC) (reviewed in Hall et al., 2015). Kirov and colleagues identified significant enrichment at CNV loci for NMDAR and ARC genes, and this has been supported by exome sequencing studies.

The NMDA receptor plays a key role in the control of synaptic plasticity at the post-synaptic density (PSD). NMDA activation triggers a series of secondary messenger systems which, among other things, alter regulation of gene expression through transcription factors (reviewed in Blanke & VanDongen, 2009). This leads to rapid transcription of the Arc gene, a regulator of synaptic plasticity. Blocking Arc activity impairs major forms of synaptic plasticity, including long-term potentiation and long-term depression. Arc is regulated by translational inhibition by the FMRP-CYFIP1 protein complex. Mutations that have been identified at ARC and NMDA genes in schizophrenia patients are now an avenue of intensive research to identify the molecular mechanisms at play (Fromer et al., 2014; Pocklington et al., 2015).

(iii) *Calcium channel genes:* In addition to the GWAS and CNV findings, rare disruptive mutations within calcium ion channels are enriched in schizophrenia patients. Calcium influx into cells triggers the NMDA activation and cascade critically involved in synaptic plasticity. Most of the work published to date has focused on the most robust genetic finding, the association with *CACNA1C* of the gene encoding the aplha1c subunit of the Cav1.2 L-type voltage-gated calcium channel (LTCC). This protein regulates signaling cascades related to the brain derived neurotrophic factor (BDNF) and cAMP response element-binding protein (CREB). LTCC antagonists reduce induction of long-term potentiation in the CA1 of the rat hippocampus. Cav 1.2. knockdowns have reduced CREB transcription and hippocampal LTP. Findings from these animal models suggest that the risk mutations are involved in regulation of cognitive function and learning in a molecular pathway that involves calcium channels, NMDA receptor function, ARC, and FMRP complexes.

12 Drug discovery

Knowing the molecular etiological mechanism of a disease is critical to drug discovery. We are in an era where the tools available to dissect these mechanisms are rapidly advancing. We are beginning to consider the functional organization of the genome and how the regulation of transcription and spatial configuration of DNA may be important in determining the translation of genes (Rajarajan et al., 2016). The development of CRISPR screening is making it easier to identify novel therapeutic targets, and even to identify synergistic drugs to target more than one molecular pathway at a time (Han et al., 2017). This may be important, given the extensive list of target genes emerging from gene discovery.

Functional assays in model systems, including iPS cells (Lee et al., 2015; Wen et al., 2014), will be important in trying to understand whether genetic findings converge at the level of molecular pathways. This, in turn, will inform the critical question for personalized medicine as to whether schizophrenia represents a single, or multiple disease processes. The tools are also becoming available to evolve from cellular studies to more complex systems investigating neural circuitry in vivo (Kim, Adhikari, & Deisseroth, 2017). There is also an emerging roadmap for the required steps in bringing together genomics, other "omics" data, and environmental data to inform translational research (e.g., see Akil et al., 2010; Corvin & Sullivan, 2016).

The GWAS era and CNV studies have provided a framework identifying post-synaptic signaling complexes (ARC, NMDAR), FMRP targets, and voltage-gated calcium channels as a starting point for discovery. Common variants have been the focus of most scrutiny, but as with GWAS from other complex traits, resolving the mechanism of risk at each locus is a significant challenge. Rare variants directly impacting gene function may be more obvious targets for drug development. Although sequencing studies have been a relatively recent addition, the identification of loss-of-function effects involving *SETD1A* indicates that histone H3 methylation may be an important avenue for research. In previous sections, we have noted the evidence of molecular overlap between schizophrenia and other neurodevelopmental disorders. Large-scale sequencing efforts in ID and autism may identify novel therapeutic pathways relevant to schizophrenia. For example, an exonic mutation of *CACNA1C* causes Timothy syndrome, a condition characterized by autistic features and long QT syndrome. The genetic mechanism involved is understood, and the subject of intense molecular research in model systems, including iPSC cells from patients (Krey et al., 2013). Common and rare variants at the gene are implicated in schizophrenia, and this body of research may inform understanding of its role in schizophrenia if the molecular mechanism involved can be elucidated.

13 Personalized treatment

The extent to which gene discovery will inform personalized treatment in schizophrenia is unclear. We are beginning to understand a fraction of the molecular etiology involved. Empowered by the information we have, we can begin to ask important questions about the onset of the disorder, severity, and possible treatment response. There is a significant history of pharmacogenetics research in the field investigating the genetic contribution to variance in drug response, genetic risk of developing side-effects, and aiming to improve treatment efficacy. This literature has been the subject of substantial reviews elsewhere (e.g., Zhang & Malhotra, 2011, 2013), and has identified that a genetic variant at the *DRD2* gene identifies patients with 54% greater likelihood of antipsychotic treatment response (Zhang & Malhotra, 2011). This body of work highlights the point that the genetic architectures underlying illness severity, treatment response, or side-effects are likely to be complex and require similar genomics, with statistically rigorous methods, to those being applied to understand pathophysiology.

The PRS method is being applied to look at common genetic risk in aggregate. A recent study has shown that PRS-SCZ predicts distinct types of prepubertal developmental impairments at different stages of childhood; higher risk score is associated with lower performance IQ, poorer speech intelligibility and fluency, and more "headstrong" behavior at age from 7 to 9 years, or with social difficulties and behavior problems as early as 4 years (Riglin et al., 2017). Furthermore, PRS-SCZ has a greater power to explain the heritability of childhood-onset schizophrenia, a severe form of psychotic symptoms beginning before age 13, compared with that of late-onset schizophrenia (Ahn et al., 2016). The PRS-SCZ may help to detect individuals who potentially would benefit from attending high-risk clinics, or for preventative strategies. A recent study has made a comparison of PRS-SCZ between carriers of a known pathogenic CNV from cases and controls, as well as between the cases carrying the CNV and controls. The CNV carrier cases show higher PRS-SCZ than controls, suggesting that common variants contribute to the susceptibility in the presence of a CNV. Even within the CNV carriers, PRS-SCZ is able to statistically differentiate between cases and controls. Because carriers of a pathogenic CNV are prone to have other developmental disorders or cognitive deficits, the PRS-SCZ calculation could be useful for the detection of CNV carriers at increased risk for schizophrenia (Tansey et al., 2016).

The second question is if we can use PRS-SCZ to predict a patient's response to drugs or treatment. Around 30% of schizophrenia patients are known to be treatment-resistant (Leucht et al., 2013; Meltzer, 1997). The atypical antipsychotic medication clozapine is associated with severe adverse events, and hence only prescribed to the treatment-resistant patients (Nielsen, Nielsen, & Correll, 2012). Because a delay in providing proper treatments and a lack of efficacy for the initial treatment lead to a poorer prognosis, there is an urgent need for new approaches to the identification of patients who will require treatment with clozapine. Zhang et al. have reviewed pharmacogenetic findings for two important antipsychotic adverse effects: antipsychotic weight gain and clozapine induced agranulocytosis (CIA) (Zhang & Malhotra, 2013). Variations in the human leukocyte antigens (HLA) genomic region have been associated with clozapine induced agranulocytosis (CIA) (Lieberman et al., 1990; Yunis et al., 1995). Association of one variant of HLA-DQB1 with CIA (OR 16.9, $P < 0.001$) was identified by Athanasiou and colleagues, with high specificity, but low sensitivity (Athanasiou et al., 2011). The low sensitivity indicates that other genetic variants likely contribute to risk of CIA and further studies with large sample sizes will be required to identify these other genetic risks (Zhang & Malhotra, 2013). The PRS-SCZ is being used to explore whether genetic information can help identify this patient group early, as treatment-resistant patients tend to have higher risk scores than those who can respond to standard medications (Frank et al., 2015). Follow-up studies with a large set of samples are necessary to enable genetic testing for the early intervention with clozapine.

Although we discussed case studies that illustrate potential applications of genomic findings for translational medicine, the practical use of PRS-SCZ has several caveats (Chatterjee, Shi, & Garcia-Closas, 2016). First, because the number of susceptible SNPs that can be detected from GWAS is limited, a choice of SNPs to be included in the calculation has a critical effect on the estimate. Second, the use of only SNPs above genome-wide significance doesn't always show the best performance, suggesting that a threshold for the inclusion/exclusion of SNPs has to be optimized in empirical ways, such as using independent sets of samples or cross-validation. Third, given the highly polygenic architecture of schizophrenia, the precision of the PRS-SCZ model strongly depends on the sample size used for GWAS; a huge sample size is required to significantly increase the accuracy of PRS-SCZ.

14 Conclusion

Despite substantial progress, much of the heritability of schizophrenia is yet to be explained. The advent of genome sequencing is likely to lead to greater understanding. A substantial contribution of common risk variants across all populations, even controls, is an important message in destigmatizing this disorder, but also suggests limitations for molecular diagnostics. Whether a subgroup of patients with more highly penetrant variants, often with other neurodevelopmental pathologies, will benefit from molecular diagnostics is a focus of active translational research.

As yet it is unclear whether the schizophrenia syndrome captures one or more disease states, but the evidence is converging on a number of molecular pathways at a time when neuroscience is developing more sophisticated tools to understand how neural circuits function in vivo. Already there is a significant body of work developing on investigation of the impact of genetic variation. For example, on variants that affect calcium channel function (e.g., CACNA1C) and resultant behavioral phenotypes for cognition and anxiety in rodent model systems (reviewed in Moon et al., 2018). The ability to target specific cell populations in animal models is allowing more focused research on specific brain regions and functions (Daigle et al., 2018). Coupled with optogenetics methods, this is facilitating increasingly sophisticated, focused investigation of brain circuitry in the living animal (Deisseroth, 2015). And the genetics data is providing evidence as to which are the most important cell types to investigate. Using single cell RNA sequencing, Skene and colleagues (Skene et al., 2018) have recently demonstrated that the common variant genomic results consistently map to pyramidal cells, medium spiny neurons (MSNs), and certain interneurons. This is a time when substantial progress is being made in developing greater cellular specificity using induced pluripotent stem cell (iPSC) technology to generate patient-derived neurons to model schizophrenia and other psychiatric disorders (Vadodaria et al., 2018). Such convergence is likely to represent the start of an exciting new era of opportunity for drug discovery in this devastating disorder.

References

Ahn, K., et al. (2016). Common polygenic variation and risk for childhood-onset schizophrenia. *Molecular Psychiatry, 21*(1), 94–96.

Akil, H., et al. (2010). Medicine. The future of psychiatric research: Genomes and neural circuits. *Science, 327*(5973), 1580–1581.

Allen, A. S., et al. (2013). De novo mutations in epileptic encephalopathies. *Nature, 501*(7466), 217–221.

Athanasiou, M. C., et al. (2011). Candidate gene analysis identifies a polymorphism in HLA-DQB1 associated with clozapine-induced agranulocytosis. *The Journal of Clinical Psychiatry, 72*(4), 458–463.

Birnbaum, R., & Weinberger, D. R. (2017). Genetic insights into the neurodevelopmental origins of schizophrenia. *Nature Reviews. Neuroscience, 18*(12), 727–740.

Blanke, M.L. and A.M.J. VanDongen, Chapter 13 activation mechanisms of the NMDA receptor, Biology of the NMDA Receptor, V.D. Am, Editor. 2009, Boca Raton (FL): CRC Press/Taylor & Francis: Duke University Medical Center, North Carolina.

Cantor-Graae, E., & Selten, J.-P. (2005). Schizophrenia and migration: A meta-analysis and review. *American Journal of Psychiatry, 162*(1), 12–24.

Cardno, A. G., & Gottesman, I. I. (2000). Twin studies of schizophrenia: From bow-and-arrow concordances to star wars Mx and functional genomics. *American Journal of Medical Genetics, 97*(1), 12–17.

Chatterjee, N., Shi, J., & Garcia-Closas, M. (2016). Developing and evaluating polygenic risk prediction models for stratified disease prevention. *Nature Reviews. Genetics, 17*(7), 392–406.

Corvin, A., Craddock, N., & Sullivan, P. F. (2010). Genome-wide association studies: A primer. *Psychological Medicine, 40*(7), 1063–1077.

Corvin, A., & Morris, D. W. (2014). Genome-wide association studies: Findings at the major histocompatibility complex locus in psychosis. *Biological Psychiatry, 75*(4), 276–283.

Corvin, A., & Sullivan, P. F. (2016). What next in schizophrenia genetics for the psychiatric genomics consortium? *Schizophrenia Bulletin, 42*(3), 538–541.

Daigle, T. L., et al. (2018). A suite of transgenic driver and reporter mouse lines with enhanced brain-cell-type targeting and functionality. *Cell, 174*(2), 465–480. e22.

de Ligt, J., et al. (2012). Diagnostic exome sequencing in persons with severe intellectual disability. *The New England Journal of Medicine, 367*(20), 1921–1929.

De Rubeis, S., et al. (2014). Synaptic, transcriptional and chromatin genes disrupted in autism. *Nature, 515*(7526), 209–215.

Deisseroth, K. (2015). Optogenetics: 10 years of microbial opsins in neuroscience. *Nature Neuroscience, 18*(9), 1213–1225.

Douglas, A. P., et al. (2014). Next-generation sequencing of the mitochondrial genome and association with IgA nephropathy in a renal transplant population. *Scientific Reports, 4*, 7379.

Fischer, B. A., & Carpenter, W. T., Jr. (2009). Will the Kraepelinian dichotomy survive DSM-V? *Neuropsychopharmacology, 34*(9), 2081–2087.

Frank, J., et al. (2015). Identification of increased genetic risk scores for schizophrenia in treatment-resistant patients. *Molecular Psychiatry, 20*(2), 150–151.

Fromer, M., et al. (2014). De novo mutations in schizophrenia implicate synaptic networks. *Nature, 506*(7487), 179–184.

Genovese, G., et al. (2016). Increased burden of ultra-rare protein-altering variants among 4,877 individuals with schizophrenia. *Nature Neuroscience, 19*(11), 1433–1441.

Gratten, J., et al. (2014). Large-scale genomics unveils the genetic architecture of psychiatric disorders. *Nature Neuroscience, 17*(6), 782–790.

Hall, J., et al. (2015). Genetic risk for schizophrenia: Convergence on synaptic pathways involved in plasticity. *Biological Psychiatry, 77*(1), 52–58.

Han, K., et al. (2017). Synergistic drug combinations for cancer identified in a CRISPR screen for pairwise genetic interactions. *Nature Biotechnology, 35*(5), 463–474.

Hegarty, J. D., et al. (1994). One hundred years of schizophrenia: A meta-analysis of the outcome literature. *The American Journal of Psychiatry, 151*(10), 1409–1416.

Homann, O. R., et al. (2016). Whole-genome sequencing in multiplex families with psychoses reveals mutations in the SHANK2 and SMARCA1 genes segregating with illness. *Molecular Psychiatry, 21*(12), 1690–1695.

Iafrate, A. J., et al. (2004). Detection of large-scale variation in the human genome. *Nature Genetics, 36*(9), 949–951.

Ingraham, L. J., & Kety, S. S. (2000). Adoption studies of schizophrenia. *American Journal of Medical Genetics, 97*(1), 18–22.

International Schizophrenia Consortium. (2008). Rare chromosomal deletions and duplications increase risk of schizophrenia. *Nature, 455*(7210), 237–241.

Irish Schizophrenia Genomics Consortium and the Wellcome Trust Case Control Consortium 2. (2012). Genome-wide association study implicates HLA-C*01:02 as a risk factor at the major histocompatibility complex locus in schizophrenia. *Biological Psychiatry, 72*(8), 620–628.

Kamb, A., Harper, S., & Stefansson, K. (2013). Human genetics as a foundation for innovative drug development. *Nature Biotechnology, 31*(11), 975–978.

Kendall, K. M., et al. (2016). Cognitive performance among carriers of pathogenic copy number variants: Analysis of 152,000 UK biobank subjects. *Biological Psychiatry, 82*(2), 103–110.

Kendler, K. S., et al. (1993). The Roscommon Family Study. III. Schizophrenia-related personality disorders in relatives. *Archives of General Psychiatry, 50*(10), 781–788.

Kendler, K. S., et al. (1993b). The Roscommon Family Study. I. Methods, diagnosis of probands, and risk of schizophrenia in relatives. *Archives of General Psychiatry, 50*(7), 527–540.

Kim, C. K., Adhikari, A., & Deisseroth, K. (2017). Integration of optogenetics with complementary methodologies in systems neuroscience. *Nature Reviews Neuroscience, 18*(4), 222–235.

Kirov, G. (2015). CNVs in neuropsychiatric disorders. *Human Molecular Genetics, 24*(R1), R45–R49.

Kirov, G., et al. (2012). De novo CNV analysis implicates specific abnormalities of postsynaptic signalling complexes in the pathogenesis of schizophrenia. *Molecular Psychiatry, 17*(2), 142–153.

Kirov, G., et al. (2014). The penetrance of copy number variations for schizophrenia and developmental delay. *Biological Psychiatry, 75*(5), 378–385.

Krey, J. F., et al. (2013). Timothy syndrome is associated with activity-dependent dendritic retraction in rodent and human neurons. *Nature Neuroscience, 16*(2), 201–209.

Lee, S. H., et al. (2012). Estimating the proportion of variation in susceptibility to schizophrenia captured by common SNPs. *Nature Genetics, 44*(3), 247–250.

Lee, S. H., et al. (2013). Genetic relationship between five psychiatric disorders estimated from genome-wide SNPs. *Nature Genetics, 45*(9), 984–994.

Lee, I. S., et al. (2015). Characterization of molecular and cellular phenotypes associated with a heterozygous CNTNAP2 deletion using patient-derived hiPSC neural cells. *NPJ Schizophrenia, 1*, 15019.

Leucht, S., et al. (2013). Comparative efficacy and tolerability of 15 antipsychotic drugs in schizophrenia: A multiple-treatments meta-analysis. *Lancet, 382*(9896), 951–962.

Lichtenstein, P., et al. (2009). Common genetic determinants of schizophrenia and bipolar disorder in Swedish families: A population-based study. *Lancet, 373*(9659), 234–239.

Lieberman, J. A., et al. (1990). HLA-B38, DR4, DQw3 and clozapine-induced agranulocytosis in Jewish patients with schizophrenia. *Archives of General Psychiatry, 47*(10), 945–948.

Liu, P., et al. (2012). Mechanisms for recurrent and complex human genomic rearrangements. *Current Opinion in Genetics & Development, 22*(3), 211–220.

Loh, P. R., et al. (2015). Contrasting genetic architectures of schizophrenia and other complex diseases using fast variance-components analysis. *Nature Genetics, 47*(12), 1385–1392.

MacArthur, D. G., et al. (2012). A systematic survey of loss-of-function variants in human protein-coding genes. *Science, 335*(6070), 823–828.

Maier, W., et al. (1993). Continuity and discontinuity of affective disorders and schizophrenia. Results of a controlled family study. *Archives of General Psychiatry, 50*(11), 871–883.

Malhotra, D., & Sebat, J. (2012). CNVs: Harbingers of a rare variant revolution in psychiatric genetics. *Cell, 148*(6), 1223–1241.

Marshall, C. R., et al. (2017). Contribution of copy number variants to schizophrenia from a genome-wide study of 41,321 subjects. *Nature Genetics, 49*(1), 27–35.

Matheson, S., et al. (2013). Childhood adversity in schizophrenia: A systematic meta-analysis. *Psychological Medicine, 43*(2), 225.

Maurano, M. T., et al. (2012). Systematic localization of common disease-associated variation in regulatory DNA. *Science, 337*(6099), 1190–1195.

Meltzer, H. Y. (1997). Treatment-resistant schizophrenia—The role of clozapine. *Current Medical Research and Opinion, 14*(1), 1–20.

Miles, J. H., et al. (2005). Essential versus complex autism: Definition of fundamental prognostic subtypes. *American Journal of Medical Genetics. Part A, 135*(2), 171–180.

Miyamoto, S., et al. (2012). Pharmacological treatment of schizophrenia: A critical review of the pharmacology and clinical effects of current and future therapeutic agents. *Molecular Psychiatry, 17*(12), 1206–1227.

Moon, A. L., et al. (2018). CACNA1C: Association with psychiatric disorders, behavior, and neurogenesis. *Schizophrenia Bulletin, 44*(5), 958–965.

Murphy, K. C., Jones, L. A., & Owen, M. J. (1999). High rates of schizophrenia in adults with velo-cardio-facial syndrome. *Archives of General Psychiatry, 56*(10), 940–945.

Network and Pathway Analysis Subgroup of Psychiatric Genomics Consortium. (2015). Psychiatric genome-wide association study analyses implicate neuronal, immune and histone pathways. *Nature Neuroscience, 18*(2), 199–209.

Ng, M. Y., et al. (2009). Meta-analysis of 32 genome-wide linkage studies of schizophrenia. *Molecular Psychiatry*, *14*(8), 774–785.

Nielsen, J., Nielsen, R. E., & Correll, C. U. (2012). Predictors of clozapine response in patients with treatment-refractory schizophrenia: Results from a Danish Register Study. *Journal of Clinical Psychopharmacology*, *32*(5), 678–683.

Nioi, P., et al. (2016). Variant ASGR1 associated with a reduced risk of coronary artery disease. *The New England Journal of Medicine*, *374*(22), 2131–2141.

O'Byrne, J. J., et al. (2016). Unexplained developmental delay/learning disability: Guidelines for best practice protocol for first line assessment and genetic/metabolic/radiological investigations. *Irish Journal of Medical Science*, *185*(1), 241–248.

O'Roak, B. J., et al. (2012). Multiplex targeted sequencing identifies recurrently mutated genes in autism spectrum disorders. *Science*, *338*(6114), 1619–1622.

Pardinas, A. F., et al. (2018). Common schizophrenia alleles are enriched in mutation-intolerant genes and in regions under strong background selection. *Nature Genetics*, *50*(3), 381–389.

Plomin, R., Owen, M. J., & McGuffin, P. (1994). The genetic basis of complex human behaviors. *Science*, *264*(5166), 1733–1739.

Pocklington, A. J., et al. (2015). Novel findings from CNVs implicate inhibitory and excitatory signaling complexes in schizophrenia. *Neuron*, *86*(5), 1203–1214.

Power, R. A., et al. (2013). Fecundity of patients with schizophrenia, autism, bipolar disorder, depression, anorexia nervosa, or substance abuse vs their unaffected siblings. *JAMA Psychiatry*, *70*(1), 22–30.

Psychiatric GWAS Consortium Bipolar Disorder Working Group. (2011). Large-scale genome-wide association analysis of bipolar disorder identifies a new susceptibility locus near ODZ4. *Nature Genetics*, *43*(10), 977–983.

Purcell, S. M., et al. (2009). Common polygenic variation contributes to risk of schizophrenia and bipolar disorder. *Nature*, *460*(7256), 748–752.

Purcell, S. M., et al. (2014). A polygenic burden of rare disruptive mutations in schizophrenia. *Nature*, *506*(7487), 185–190.

Rajarajan, P., et al. (2016). Spatial genome organization and cognition. *Nature Reviews. Neuroscience*, *17*(11), 681–691.

Rauch, A., et al. (2012). Range of genetic mutations associated with severe non-syndromic sporadic intellectual disability: An exome sequencing study. *Lancet*, *380*(9854), 1674–1682.

Reble, E., et al. (2017). VarScan2 analysis of de novo variants in monozygotic twins discordant for schizophrenia. *Psychiatric Genetics*, *27*(2), 62–70.

Rees, E., et al. (2014). Evidence that duplications of 22q11.2 protect against schizophrenia. *Molecular Psychiatry*, *19*(1), 37–40.

Rees, E., et al. (2014b). CNV analysis in a large schizophrenia sample implicates deletions at 16p12.1 and SLC1A1 and duplications at 1p36.33 and CGNL1. *Human Molecular Genetics*, *23*(6), 1669–1676.

Rees, E., et al. (2016). Analysis of intellectual disability copy number variants for association with schizophrenia. *JAMA Psychiatry*, *73*(9), 963–969.

Richards, A. L., et al. (2012). Schizophrenia susceptibility alleles are enriched for alleles that affect gene expression in adult human brain. *Molecular Psychiatry*, *17*(2), 193–201.

Riglin, L., et al. (2017). Schizophrenia risk alleles and neurodevelopmental outcomes in childhood: A population-based cohort study. *Lancet Psychiatry*, *4*(1), 57–62.

Rujescu, D., et al. (2009). Disruption of the neurexin 1 gene is associated with schizophrenia. *Human Molecular Genetics*, *18*(5), 988–996.

Schaefer, G. B., & Mendelsohn, N. J. (2013). Clinical genetics evaluation in identifying the etiology of autism spectrum disorders: 2013 guideline revisions. *Genetics in Medicine*, *15*(5), 399–407.

Schizophrenia Working Group of the Psychiatric Genomics, C. (2014). Biological insights from 108 schizophrenia-associated genetic loci. *Nature*, *511*(7510), 421–427.

Schork, A. J., et al. (2013). All SNPs are not created equal: Genome-wide association studies reveal a consistent pattern of enrichment among functionally annotated SNPs. *PLoS Genetics*, *9*(4).

Sebat, J., et al. (2004). Large-scale copy number polymorphism in the human genome. *Science*, *305*(5683), 525–528.

Sekar, A., et al. (2016). Schizophrenia risk from complex variation of complement component 4. *Nature*, *530*(7589), 177–183.

Shi, J., et al. (2009). Common variants on chromosome 6p22.1 are associated with schizophrenia. *Nature*, *460*(7256), 753–757.

Singh, T., et al. (2016). Rare loss-of-function variants in SETD1A are associated with schizophrenia and developmental disorders. *Nature Neuroscience*, *19*(4), 571–577.

Skene, N. G., et al. (2018). Genetic identification of brain cell types underlying schizophrenia. *Nature Genetics*, *50*(6), 825–833.

St Clair, D., et al. (1990). Association within a family of a balanced autosomal translocation with major mental illness. *Lancet*, *336*(8706), 13–16.

Stefansson, H., et al. (2008). Large recurrent microdeletions associated with schizophrenia. *Nature*, *455*(7210), 232–236.

Stefansson, H., et al. (2009). Common variants conferring risk of schizophrenia. *Nature*, *460*(7256), 744–747.

Stefansson, H., et al. (2014). CNVs conferring risk of autism or schizophrenia affect cognition in controls. *Nature*, *505*(7483), 361–366.

Steinberg, S., et al. (2017). Truncating mutations in RBM12 are associated with psychosis. *Nature Genetics*, *49*(8), 1251–1254.

Sullivan, P. F., Daly, M. J., & O'Donovan, M. (2012). Genetic architectures of psychiatric disorders: The emerging picture and its implications. *Nature Reviews. Genetics*, *13*(8), 537–551.

Sullivan, P. F., Kendler, K. S., & Neale, M. C. (2003). Schizophrenia as a complex trait: Evidence from a meta-analysis of twin studies. *Archives of General Psychiatry*, *60*(12), 1187–1192.

Takata, A., et al. (2014). Loss-of-function variants in schizophrenia risk and SETD1A as a candidate susceptibility gene. *Neuron*, *82*(4), 773–780.

Tandon, R., Keshavan, M. S., & Nasrallah, H. A. (2008). Schizophrenia, "Just the Facts": What we know in 2008 part 1: Overview. *Schizophrenia Research*, *100*(1–3), 4–19.

Tang, J., et al. (2017). Whole-genome sequencing of monozygotic twins discordant for schizophrenia indicates multiple genetic risk factors for schizophrenia. *Journal of Genetics and Genomics, 44*(6), 295–306.

Tansey, K. E., et al. (2016). Common alleles contribute to schizophrenia in CNV carriers. *Molecular Psychiatry, 21*(8), 1085–1089.

Thygesen, J. H., et al. (2018). Neurodevelopmental risk copy number variants in adults with intellectual disabilities and comorbid psychiatric disorders. *The British Journal of Psychiatry, 212*(5), 287–294.

Tiihonen, J., et al. (2009). 11-year follow-up of mortality in patients with schizophrenia: A population-based cohort study (FIN11 study). *Lancet, 374*(9690), 620–627.

Timms, A. E., et al. (2013). Support for the N-methyl-D-aspartate receptor hypofunction hypothesis of schizophrenia from exome sequencing in multiplex families. *JAMA Psychiatry, 70*(6), 582–590.

Vacic, V., et al. (2011). Duplications of the neuropeptide receptor gene VIPR2 confer significant risk for schizophrenia. *Nature, 471*(7339), 499–503.

Vadodaria, K. C., et al. (2018). Modeling psychiatric disorders using patient stem cell-derived neurons: A way forward. *Genome Medicine, 10*(1), 1.

Walsh, T., et al. (2008). Rare structural variants disrupt multiple genes in neurodevelopmental pathways in schizophrenia. *Science, 320*(5875), 539–543.

Welter, D., et al. (2014). The NHGRI GWAS catalog, a curated resource of SNP-trait associations. *Nucleic Acids Research, 42*(Database issue), 6.

Wen, Z., et al. (2014). Synaptic dysregulation in a human iPS cell model of mental disorders. *Nature, 515*(7527), 414–418.

Xu, B., et al. (2012). De novo gene mutations highlight patterns of genetic and neural complexity in schizophrenia. *Nature Genetics, 44*(12), 1365–1369.

Yunis, J. J., et al. (1995). HLA associations in clozapine-induced agranulocytosis. *Blood, 86*(3), 1177–1183.

Zerbin-Rudin, E., & Kendler, K. S. (1996). Ernst Rudin (1874-1952) and his genealogic-demographic department in Munich (1917-1986): An introduction to their family studies of schizophrenia. *American Journal of Medical Genetics, 67*(4), 332–337.

Zhang, J.-P., & Malhotra, A. K. (2011). Pharmacogenetics and antipsychotics: Therapeutic efficacy and side effects prediction. *Expert Opinion on Drug Metabolism & Toxicology, 7*(1), 9–37.

Zhang, J.-P., & Malhotra, A. K. (2013). Pharmacogenetics of antipsychotics: Recent progress and methodological issues. *Expert Opinion on Drug Metabolism & Toxicology, 9*(2), 183–191.

Zuk, O., et al. (2014). Searching for missing heritability: Designing rare variant association studies. *Proceedings of the National Academy of Sciences of the United States of America, 111*(4), E455–E464.

Chapter 16

Genomics of major depressive disorder

Douglas F. Levinson

Department of Psychiatry and Behavioral Sciences, Program on the Genetics of Brain Function, Stanford University, Palo Alto, CA, United States

Major depressive disorder (MDD) is the most common major psychiatric disorder, and is one of the leading causes of disability worldwide (http://www.who.int/healthinfo/global_burden_disease/estimates/en/).

Current medications and psychotherapies can be dramatically effective, yet many patients experience little benefit. Current antidepressant medications were discovered serendipitously. The discovery of genetically-based biological risk factors could provide better targets for development of therapeutic drugs, improve diagnostic frameworks, and improve biological inferences about etiology. In the future, genetic sequencing data might be useful in tailoring strategies for prevention, diagnosis, and treatment.

Results of genome-wide association studies (GWAS) are currently driving this field forward. It has taken some years to achieve the very large sample sizes needed for MDD—as predicted based on the high prevalence and moderate heritability of MDD (Levinson et al., 2014; Major Depressive Disorder Working Group of the Psychiatric Genomics Consortium et al., 2013). The results of the most recent GWAS analyses will be featured in this chapter.

1 What is the phenotype?

Most genetic studies have focused on the MDD category (American Psychiatric Association, 2013; Feighner et al., 1972; World Health Organisation, 1992). A major depressive episode is defined as two or more weeks with four or more of eight additional criteria (disturbances of sleep, appetite, energy level, concentration, self-esteem/guilt, capacity for pleasure, psychomotor agitation/retardation, and suicidal thoughts/acts), in the absence of other diagnoses in which depression can occur (e.g., bipolar disorder, certain psychotic disorders). MDD can be diagnosed with high inter-rater reliability. It significantly predicts course of illness, familial risks, and treatment response, although a purely categorical approach is limited by substantial overlap with other categories (anxiety, bipolar, and psychotic disorders). Lifetime prevalence is high (12%–20%) (Hasin, Goodwin, Stinson, & Grant, 2005), with recurrence in around two-thirds of cases. There is no "genetically distinct" depression subtype.

Depression can also be measured with self-rating questionnaires about symptoms during the past week or month, or about longstanding tendencies to experience unpleasant moods and anxiety ("Neuroticism") (Kendler, Gardner, & Prescott, 2002). Depression and neuroticism ratings have been collected in many large epidemiological studies, often along with GWAS data. The relationship between these phenotypes has been addressed by recent GWAS analyses (see the following).

2 Genetic epidemiology

There are no known Mendelian (dominant or recessive) forms of MDD. It is a *complex trait*: risk is influenced by many DNA sequence variants, throughout the genome, and psychological and physiological environmental factors.

Heritability is the proportion of risk attributable to genetic differences, on a scale of 0–1, usually estimated from concordance rates for identical vs nonidentical twin pairs (monozygotic [MZ] vs dizygotic [DZ]). MDD twin heritability averages ~37%, perhaps higher (>50%) in severe clinical cases (Sullivan, Neale, & Kendler, 2000). Heritability is similar in men and women, with a 75% correlation in genetic risk factors (Fernandez-Pujals et al., 2015). Thus, there should be some sex-specific factors.

Relative risk (RR) is an odds ratio estimate of risk to a class of relatives (e.g., siblings) vs the population risk (e.g., from relatives of controls). The RR of MDD in cases' first-degree relatives is ~3 (Sullivan et al., 2000). The comparable value for schizophrenia or bipolar disorder is ~10.

Personalized Psychiatry. https://doi.org/10.1016/B978-0-12-813176-3.00016-X

3 Do proband characteristics predict familial risk?

Family studies have reported higher RR of MDD to first-degree relatives when the proband has an earlier age at onset (AAO) and/or recurrent episodes (Levinson et al., 2003). A larger, population-based twin study reported that AAO, recurrence, and severity (number of criteria) only modestly predicted familial risk (Kendler, Gatz, Gardner, & Pedersen, 2007). A recent large family study confirmed increased heritability for recurrent MDD probands (42%, vs 28% for a single episode), but not for early AAO, with high familial correlations between early- vs late-onset and recurrent vs single-episode MDD (Fernandez-Pujals et al., 2015). These features do not mark genetically distinct disorders, but when very large samples of recurrent, severe cases can be achieved, they could be particularly useful for genetic studies (Flint, Chen, Shi, Kendler, & CONVERGE Consortium, 2012).

4 Relationship of MDD to bipolar disorder and schizophrenia: Epidemiology

Major depressive episodes are also common in individuals with other primary diagnoses, including bipolar and psychotic disorders. Early family studies reported similar risks of MDD in relatives of probands with bipolar disorder (14.1%) and MDD (17.9%) (vs 5.2% for relatives of controls); but risk of bipolar I or II was low in relatives of MDD (2.2%) vs bipolar probands (8.7%) (0.7% in controls) (Smoller & Finn, 2003). Twin studies showed that MDD and bipolar disorders share some, but not most of their genetic risk factors (Craddock & Forty, 2006).

The clinical impression that *increased* sleep and appetite are more common in "bipolar depression," is not supported empirically (Blacker et al., 1996; Zimmermann et al., 2009). In one study, transition to mania was predicted by "subthreshold" manic symptoms (seen in 40% of MDD cases) or bipolar family history (7.2% transitioned vs 1.7% without these features) (Zimmermann et al., 2009). Psychomotor retardation and difficulty thinking may also be more common in MDD relatives of bipolar patients (Mitchell et al., 2011).

MDD risk is also elevated in relatives of schizophrenia probands (Maier et al., 1993), while schizophrenia spectrum psychosis was elevated in relatives of probands with MDD with psychotic features (Kendler et al., 1993; Maier et al., 1993). Genetic relationships among classical diagnostic categories are now a major focus of molecular genetic studies.

5 Childhood trauma and other stressful events

Stressful life events increase the onset of MDD (Nanni, Uher, & Danese, 2012) and of other adult psychiatric disorders (Nelson et al., 2002). Adult MDD risk is predicted by early parental loss (Kendler et al., 2002), and childhood abuse (which also predicts poorer treatment response) (Nanni et al., 2012), and particularly severe sexual abuse (e.g., with penetration) (Kendler, Kuhn, & Prescott, 2004). There also appears to be an interaction between psychological adversity and MDD family history in increasing risk (Zimmermann et al., 2008).

6 Molecular genetic methods

Four molecular genetic strategies have been employed in studies of MDD:

(1) Linkage analysis detects the approximate genomic location of disease mutations with strong (dominant or recessive) effects on risk, using as few as several hundred "marker" sequences to study families with multiple cases. Replicable evidence for linkage in MDD was never discovered in cohorts or in single families, as reviewed previously (Levinson, 2013).

(2) Many studies of psychiatric disorders have focused on putatively functional variants in "candidate genes" encoding receptor proteins and metabolic enzymes for neurotransmitters impacted by psychotropic drugs (see the following discussion).

(3) *GWAS* has been successful for MDD and other common disorders whose familial patterns suggest *polygenic* effects (Visscher, Brown, McCarthy, & Yang, 2012). The Common Disease/Common Variant hypothesis (Lander, 1996; Reich & Lander, 2001) posited that multiple common sequence variants underlie genetic risks for common diseases. These variants are ancient mutations whose high frequency (>5% of chromosomes in a population) tells us that they do not have large effects on mortality or reproduction. The HapMap project (Altshuler et al., 2005) identified millions of common single nucleotide polymorphisms (SNPs) in different populations, and promoted the commercial development of "SNP chip" assays of 200,000–5,000,000 SNPs. Only a small proportion of common SNPs have to be assayed to interrogate almost all common SNPs (and other types of common variants). This is because of *linkage disequilibrium* (LD), for example, a parent's two copies of chromosome 1 wind around each other during *meiosis*,

resulting in a merger of large segments (millions of nucleic acids) of each copy into a single new chromosome for the sperm or egg cell. Over many generations, these segments are broken into smaller pieces of the "ancestral" chromosome containing a set of the ancestral SNP alleles. Typically, there are several common ancient "haplotypes" in each LD "block," so that genotypes of SNPs within each block are highly correlated, and assaying one or two of them will "tag" the entire haplotype. Genotypes for most other common SNPs can be "imputed" statistically based on knowledge of LD block haplotypes in a reference cohort (McCarthy et al., 2016).

The online GWAS Catalogue (5/6/2018) lists 61,173 unique SNP-trait associations (http://www.ebi.ac.uk/gwas/), including many neuropsychiatric findings. Because common variants have individually small effects on harmful traits, large samples are needed for GWAS. Multiple cohorts of cases and controls with similar geographical ancestry are usually genotyped and meta-analyzed, using extensive quality control procedures prior to association analysis (Purcell et al., 2014) (and see https://gigascience.biomedcentral.com/articles/10.1186/s13742-015-0047-8; https://www.cog-genomics.org/plink2).

GWAS data are typically analyzed in three ways:

(1) *Association of individual SNPs or genes.* Binary logistic regression is generally used to test case-control difference (0, 1, or 2 of a specified allele, for each subject) for each SNP (or common insertion-deletion), correcting for covariates reflecting ancestry differences (principal components or similar scores from a set of genome-wide SNP genotypes). Based on estimates of the number of "independent" tests among common SNPs based on LD patterns, $P < 5 \times 10^{-8}$ is considered "genome-wide significant" for each test (Dudbridge & Gusnanto, 2008). Sometimes "gene-wise" tests are also used to test whether SNPs in or near each gene are more significant than expected by chance (de Leeuw, Mooij, Heskes, & Posthuma, 2015).

For a given trait, underpowered cohorts produce no or few significant results, but above some N (the "inflection point"), N linearly predicts "hits," because adequate power has been achieved. The strongest associations for MDD have ORs $\sim$1.04 (see the following), vs $\sim$1.10 for schizophrenia (Schizophrenia Working Group of the Psychiatric Genomics Consortium, 2014), thus much larger samples are required.

(2) *Polygenic analyses.* These methods utilize genome-wide common-SNP information to learn more about the disease and its relationships with other diseases/traits (Wray et al., 2014):

(a) *Polygenic risk scores* (PRS; or genetic risk scores, GRS) (Purcell et al., 2009; Wray et al., 2014) require association statistics to be used as weights (e.g., log(OR) from a large *discovery sample*). Then in a *target sample*, if, for example, SNP allele "A" has weight W, and subject 1 carries 2 A alleles, 2*W is added to that subject's score, and the score is summed across a set of SNPs with little SNP-SNP LD. PRS can evaluate consistency of genetic affects across cohorts ("leave-one-out" analyses) or across cohorts with different diagnoses or traits. Larger discovery samples predict more variance in PRS, permitting the study of individual outcomes.

(b) Total common-SNP heritability of one trait, or shared heritability across two traits or subgroups, can be estimated using genomic-relationship matrix restricted maximum likelihood (*GREML*) (Wray et al., 2014), or the less computationally-intensive *LD score regression* (Bulik-Sullivan et al., 2015).

(c) *Pathway analyses* determine whether lower *P*-values than expected by chance are observed within sets of genes with related biological functions (de Leeuw et al., 2015).

(3) *Gene expression* can be measured genome-wide using microarrays with oligonucleotide probes, or by sequencing messenger RNA, in the blood, brain, or other tissues. Expression reflects both inherited DNA sequence variation and current physiological state, with many potentially confounding variables. Large-scale studies of white blood cells in MDD will be discussed briefly in the following section.

7 GWAS of depression

7.1 Large-scale GWAS analyses

The failure of GWAS analyses of smaller cohorts (thousands of cases) led to efforts to generate large single datasets, or combinations of much larger datasets. Table 1 summarizes the six such studies published to date. The Psychiatric Genomics Consortium reported no significant associations in >9200 cases (plus controls) from 7 cohorts, plus data from an additional 6783 cases for the best 819 SNPs (Major Depressive Disorder Working Group of the Psychiatric Genomics Consortium et al., 2013). It was quite unusual for a GWAS of this size to have no findings, but the authors noted that given the high prevalence and moderate heritability of MDD, if the number of associated SNPs was similar to that for schizophrenia, the

TABLE 1 Genome-wide significant GWAS findings for MDD and depressive symptoms.

Study	Depression datasets	Significant loci	Comment
PGC-MDD1[a]	9 cohorts (9240 cases, 9519 controls); replication (top SNPs) in 7 cohorts (6783 cases, 50,695 controls). *Total* 16,023 cases, 60,214 controls	0	Clinical and population-based cohorts; individually interviewed for diagnosis
CHARGE[b]	17 population-based cohorts ($N=34,549$); replication 5 cohorts ($N=16,709$)	1	Discovery: age > 40, mean 66.5, CES-D scale; replication: other scales, older adults
CONVERGE[c]	5303 cases, 5337 controls	2	Chinese women, clinical cases, recurrent MDD, individually interviewed. 2 additional findings in subjects without severe adversity[d]
SSGAC-1[e]	UKB self-reported depressive symptoms ($N=105,739$); PGC-MDD1 cohort; GERA (7231 cases, 49,316 controls) *Total* 16,471 cases, 58,835 controls, plus UKB	2	UKB (depression score from self-reported depressed mood and anhedonia items); PGC-MDD1; GERA (medical records diagnoses; also included in SSGAC-2 and PGC-MDD2). Also studied self-reported Subjective Well-Being and Neuroticism
23andMe[f]	75,607 cases, 231,747 controls. Replication: 45,773 cases, 106,354 controls, plus PGC-MDD1 *Total* 130,620 cases, 347,630 controls	15	Cases reported diagnosis of clinical depression on 23andMe customer research questionnaires; controls denied such a diagnosis. Overlap with PGC-MDD1
SSGAC-2[g]	23andMe discovery cohort; GERA cohort from SSGAC-1; UKB cohort from SSGAC-1 with updated depression score. *Total* 82,838 cases and 281,063 controls plus 105,739 UKB subjects	64	UKB depression score from 5 items (vs 2 in SSGAC-1). 32 significant loci for depression, increasing to 64 using the MTAG approach to add information from correlated traits (Subjective Well-Being and Neuroticism)—see text
PGC-MDD2[h]	PGC29 (16,823 cases, 25,632 controls) and 6 additional cohorts (118,635 cases, 319,269 controls). *Total* 135,458 cases, 344,901 controls	44	PGC29 (all directly-interviewed cases) included PGC-MDD1; additional cohorts included GERA, 23andMe discovery cohort, and cases/controls drawn from UKB based on diagnoses

[a]*Major Depressive Disorder Working Group of the Psychiatric Genomics Consortium et al. (2013).*
[b]*Hek et al. (2013).*
[c]*CONVERGE Consortium (2015).*
[d]*Peterson et al. (2018).*
[e]*Okbay et al. (2016).*
[f]*Hyde et al. (2016).*
[g]*Turley et al. (2018).*
[h]*Wray et al. (2018).*
Shown are the characteristics and numbers of genome-wide significant findings (independent loci) detected by each of the six published large-scale GWAS analyses of MDD and/or depressive symptoms. CES-D, Center for Epidemiological Studies-Depression scale; GERA, Resource for Genetic Epidemiology Research on Aging; MTAG, Multitrait analysis of GWAS; PGC, Psychiatric Genomics Consortium; SSGAC, Social Science Genetic Association Consortium; UKB, UK Biobank.

individual effect sizes must be much smaller, so that the "inflection point should be 3–5 times that for schizophrenia." The latter value is ~15,000–20,000 cases, so one should have expected to need 45,000–100,000 cases.

Several significant findings were soon reported. The CHARGE consortium carried out a GWAS of recent depressive symptom scores from the Center for Epidemiological Studies-Depression scale (CES-D), in 17 population-based cohorts of older adults (Hek et al., 2013). After meta-analysis with five additional cohorts using similar self-rating scales, there was one significant finding that has been replicated in MDD (see the following). The CONVERGE consortium detected two associations in 5303 Chinese women with recurrent, clinically-severe MDD (CONVERGE Consortium, 2015), and two additional findings in subjects without severe life stress or childhood sexual abuse before MDD onset (Peterson et al., 2018). These associations have not yet been replicated in Europeans. The investigators noted that their sample could have been more powerful because of the severe phenotype (which is supported by SNP heritability estimates, discussed in the following section); they are currently recruiting a much larger Chinese sample. Finally, the Social Science Genetic Association Consortium (SSGAC-1 in Table 1) analyzed publicly-available GWAS data for three genetically-related traits (depression, neuroticism, and subjective well-being) (Okbay et al., 2016). For depression, they combined two under-powered MDD datasets (PGC-MDD1 and the GERA dataset (Kvale et al., 2015) of cases diagnosed from electronic medical records) with a large UK Biobank cohort with very brief questionnaire data on depressed mood and/or loss of enjoyment. There were two significant findings.

The larger 23andMe analysis (Hyde et al., 2016) clearly surpassed the "inflection point" for depression, with 15 significant findings. 23andMe is a direct-to-consumer genetics company that offers customers information about ancestry and disease risks. Most customers also give consent for research using responses to optional online surveys on aspects of medical history. Cases had responded "yes" to any of several questions such as, "Have you ever been diagnosed with clinical depression?" Remarkably, this type of brief self-report has worked well in 23andMe GWAS analyses of disease phenotypes, perhaps because their generally well-educated clientele has good access to medical care. The significant findings were detected by a meta-analysis of 75,607 23andMe cases ("discovery sample"), the PGC-MDD1 results, and (for top findings) another 45,773 23andMe cases (total of 130,620 cases). Note that 23andMe provides GWAS discovery cohort results to other scientists who agree to specific privacy restrictions, but not for the second cohort.

Two recent publications used the 23andMe discovery data to achieve very large sample sizes and much larger numbers of findings. The second SSGAC analysis dropped the PGC-MDD1 cohort, added the 23andMe discovery cohort, retained GERA, and, for UK Biobank, added items for the depression score (Turley et al., 2018). They labeled the phenotype as "depressive symptoms," but it is probably more an approximation of lifetime depressive episodes. Depression alone yielded 32 significant findings. They then introduced "multi-trait analysis of GWAS" (MTAG). Because depression, neuroticism, and subjective well-being are genetically correlated, they argue that association information can be drawn from each trait (weighted according to the correlation) to enhance the power of each of the other two analyses. This analysis yielded 64 significant findings.

Finally, the PGC-MDD2 analysis combined 29 cohorts (using direct diagnostic interviews or, in one cohort, extensive medical records information) with 6 additional cohorts, including the 23andMe discovery cohort, totaling 135,458 cases plus controls (Wray et al., 2018). The phenotype was termed "major depression" because each study attempted to select cases likely to have had an episode of MDD, whether by interview, records information, or self-report (23andMe and most of UK Biobank). The analysis detected 44 independent associations.

7.2   What can we learn from the results so far?

These are the first known *etiological* risk factors for depression. But common-SNP associations are difficult to interpret. The largest ORs are ~1.04 (i.e., each SNP explaining a 4% increase in risk). Larger cohorts will detect even smaller ORs. Some of the implicated genes have many known functions, yet our understanding of gene functions and interactions ("annotation") is incomplete. Further, a proposed "omnigenic" hypothesis suggests that variants in virtually *all* genes influence each trait, depending on context (tissue, time of life, internal and external environment, gene-gene interactions, etc.) (Boyle, Li, & Pritchard, 2017). Thus, identifying specific risk remains a difficult task.

The problem of "missing heritability" (Maher, 2008) also remains: total common-SNP heritabilities are smaller than twin estimates, thus, full understanding of risk mechanisms might require assaying types of genomic variation (e.g., rare SNPs, complex structural variants) in very large cohorts using whole-genome sequencing. But "missing" heritability might actually be "phantom heritability," because SNP-SNP interactions inflate twin heritability estimates (the computations assume no epistasis) (Zuk, Hechter, Sunyaev, & Lander, 2012). GWAS findings might therefore be most critical.

Beyond individual associations, the findings provide additional information: estimates of total SNP heritability (h^2_{SNP}); correlations in common-SNP heritability among depression samples and between depression and other psychiatric and non-psychiatric phenotypes; and identification of biological pathways underlying risk.

7.3   Common-SNP heritability and genetic correlations among datasets

For the full PGC-MDD2 meta-analysis, h^2_{SNP} was 0.08 (assuming population prevalence of 0.15); it was 0.14 for MDD29 (direct interviews or extensive medical records); 0.18 for the iPsych Danish registry cohort (systematic clinical records); and 0.08 for 23andMe (self-reported clinical diagnosis); with other cohorts in between (Wray et al., 2018). (Generation Scotland was an outlier at 0.26 with small N [~1000 cases] and large standard error [0.14], thus the estimate is unreliable.) By comparison, h^2_{SNP} was 0.04 for CHARGE (self-reported depressive symptoms in older adults), and much higher for CONVERGE (Chinese women with severe, clinically-treated, recurrent MDD): 0.20 (SE = 0.09) using LDSC (assuming prevalence of 0.036), in the PGC-MDD2 paper; 0.30 by GCTA (Peterson et al., 2018). Thus, SNP heritability was indeed largest for cohorts with more "clinical," severe MDD, and lowest for more diverse cohorts (CHARGE, brief symptom ratings; 23andMe, self-reported diagnosis).

Yet, cross-cohort estimates of common-SNP genetic correlation (r_g) are high, which simply indicates that SNPs tend to have *relatively* similar effects on disease risk in two cohorts, even if they are stronger in one cohort. In PGC-MDD2, for example, r_g for 23andMe was 0.94 with GERA, 0.80 with deCODE or UK Biobank, 0.78 with iPsych, 0.67 with MDD29 and 0.40 with Generation Scotland. The r_g was 0.91 between all of PGC-MDD2 and CHARGE. Thus, there is high common-SNP genetic relatedness between cohorts with quite different characteristics.

7.4 Genetic pathway findings

PGC-MDD2 produced significant evidence for overrepresentation of low (more significant) *P*-values in functionally-related "gene sets" ("pathways"), suggesting their involvement in risk for depression. Significant pathways, shown in detail in their Supplementary Table 11, include multiple sets related to neuronal and synaptic development and function; targets of three regulators of alternative mRNA splicing with widespread roles in neuronal development and function (RBFOX1 and RBFOX3; FMRP [Fragile X Mental Retardation Protein, encoded by FMR1]; and CELF); regulation of cytokine production in immune response; and voltage-gated calcium channel activity. Examples are discussed in the following sections. This provides a "macrolevel" overview of common-variant determinants of risk, with most details to be filled in by future research.

7.5 Single-locus associations

The three largest analyses of European-ancestry cohorts (23andMe, SSGAC-2, PGC-MDD2) have substantial overlap in both in their significant findings and in the datasets that were included: the 23andMe discovery sample was in all three, PGC-MDD1 was in the first and third, and UK Biobank was in the latter two. Of 44 significant loci in PGC-MDD2, 19 were among the 32 findings in the SSGAC-2 analysis of depression alone; 21 were among 64 significant SSGAC-2 multi-trait findings (24 of the latter, because 3 were nonindependent in PGC-MDD2). Twelve of 15 significant 23andMe findings were significant in PGC-MDD2, plus the significant CHARGE finding. Experience with GWAS suggests that many of the "non-replicated" loci will eventually be confirmed in much larger analyses.

The complexities of interpretation are illustrated by considering the 44 findings in PGC-MDD2 (summarized in Table 2)—the largest GWAS, to date, of major depression, with substantial contributions from cohorts with more in-depth clinical information. The top SNP was within a gene in 26 of the loci, although there were other genes in LD with the top SNP, and often there were quantitative trait loci for gene expression (eQTLs) for more than one gene. Only eight findings were within a single gene with no other gene within 200 kb, simplifying interpretation; seven were in intergenic regions (>200 kb from any gene), without a clear indication of gene effects (including the CHARGE locus, chr5:103.672–104.092). We will discuss a few individual findings to illustrate their promise and complexity. The following summaries are based on the reviews that this author helped to prepare for Supplementary Table 6 in the PGC-MDD2 paper (https://www.nature.com/articles/s41588-018-0090-3#Sec24).

The two top SNPs in PGC-MDD2 were linked to genes that were also highly significant in 23andMe and SSGAC-2: OLFM4 and NEGR1. Both genes are associated with obesity (Bradfield et al., 2012; Willer et al., 2009). OLFM4 (olfactomedin 4) is one of the olfactomedin proteins that have diverse roles in neural differentiation and in immunity (Anholt, 2014). Although the known functional roles of OLFM4 are primarily in immunity and cancer (Liu & Rodgers, 2016), it has a likely role in inhibiting neurite outgrowth (Saunders et al., 2014). NEGR1 (neuronal growth regulator 1) is involved in axon extension and synaptic plasticity (Sanz, Ferraro, & Fournier, 2015), including modulation of synapse formation in hippocampal neurons (Hashimoto, Maekawa, & Miyata, 2009) via regulation of neurite outgrowth (Pischedda & Piccoli, 2015). The nutritional state modified its expression in brain areas related to feeding behavior (Boender, van Rozen, & Adan, 2012).

Two independent loci implicated RBFOX1 (RNA binding protein fox-1 homolog 1), which regulates alternative splicing of many genes in neuronal nuclei, including genes that are associated with autism and/or synaptic activity (Lee et al., 2016). There is intriguing evidence that RBFOX1 terminates the release of CRH (corticotropic releasing hormone) after stress (Amir-Zilberstein et al., 2012). Impaired activity of RBFOX1 could therefore contribute to the excessive corticosteroid levels that are observed in MDD (Pariante & Lightman, 2008). But pathway analysis (described herein) found a genome-wide signal for the many RBFOX1 and RBFOX3 target genes, thus it is premature to link this association to its effects on CRH.

TABLE 2 Significant GWAS findings in the PGC-MDD2 cohort.

Chr	SNP/indel	Location	P	OR	Gene context
1	rs159963	8,504,421	3.2E−08	0.97	[RERE]
1	rs1432639	72,813,218	4.6E−15	1.04	NEGR1 (−64,941)
1	rs12129573	73,768,366	4.0E−12	1.04	LINC01360 (−3486)
1	rs9427672	197,754,741	3.1E−08	0.97	DENND1B (−10,118)
2	rs1226412	157,111,313	2.4E−08	1.03	[LINC01876]
3	chr3_44287760_I	44,287,760	4.6E−08	1.03	[TOPAZ1]
3	rs7430565	158,107,180	2.9E−09	0.97	[RSRC1]
4	rs34215985	42,047,778	3.1E−09	0.96	[SLC30A9]
5	chr5_87992715_I	87,992,715	7.9E−11	0.97	LINC00461 (−12,095); MEF2C (21,342)
5	rs116755193	124,251,883	7.0E−09	0.97	LOC101927421 (−120,640)
5	rs4869056	166,992,078	6.8E−09	0.97	[TENM2]
6	rs115507122	30,737,591	3.3E−11	0.96	Major histocompatibility complex
7	rs10950398	12,264,871	2.6E−08	1.03	[TMEM106B]
9	rs7856424	119,733,595	8.5E−09	0.97	[ASTN2]
9	rs7029033	126,682,068	2.7E−08	1.05	[DENND1A]
10	rs61867293	106,563,924	7.0E−10	0.96	[SORCS3]
11	rs1806153	31,850,105	1.2E−09	1.04	[DKFZp686K1684]; [PAUPAR]
12	rs4074723	23,947,737	3.1E−08	0.97	[SOX5]
13	rs4143229	44,327,799	2.5E−08	0.95	[ENOX1]
13	rs12552	53,625,781	6.1E−19	1.04	[OLFM4]
14	rs4904738	42,179,732	2.6E−09	0.97	[LRFN5]
14	rs915057	64,686,207	7.6E−10	0.97	[SYNE2]
14	chr14_75356855_I	75,356,855	3.8E−09	1.03	[DLST]
14	rs10149470	104,017,953	3.1E−09	0.97	BAG5 (4927); APOPT1 (−11,340)
16	rs8063603	6,310,645	6.9E−09	0.97	[RBFOX1]
16	rs7198928	7,666,402	1.0E−08	1.03	[RBFOX1]
16	rs7200826	13,066,833	2.4E−08	1.03	[SHISA9]
16	rs11643192	72,214,276	3.4E−08	1.03	PMFBP1 (−7927)
17	rs17727765	27,576,962	8.5E−09	0.95	[CRYBA1]
18	rs62099069	36,883,737	1.3E−08	0.97	[MIR924HG]
18	rs11663393	50,614,732	1.6E−08	1.03	[DCC]
18	rs1833288	52,517,906	2.6E−08	1.03	[RAB27B]
18	rs12958048	53,101,598	3.6E−11	1.03	[TCF4]
22	rs5758265	41,617,897	7.6E−09	1.03	[L3MBTL2]

Shown are 34 findings from PGC-MDD2 (Wray et al., 2018) that were in or close to genes. The listed SNP had the most significant *P*-value in the LD region. Names of genes are in brackets if the SNP is within the gene; otherwise, the distance of the gene from the SNP is shown in parentheses as a positive number if the gene is downstream (3′), or as a negative number if the gene is upstream (5′) from the SNP. The other 10 findings, with similar *P*-values and ORs, were not within 100,000 bp of any gene (chr1:80,799,329, chr1:90,796,053, chr2:57,987,593, chr5:103,942,055, chr5:164,523,472 (replicating the CHARGE finding), chr6:99,566,521, chr7:109,105,611, chr9:2,983,774, chr9:11,544,964, chr15:37,648,402). *Chr*, chromosome. OR-odds ratio of the allele tested in the analysis. Locations are in base pairs, HG19.

There were independent associations in two connecden family genes, DENND1A and 1B (DENN domain containing 1A, 1B). DENND1A may be involved in endocytosis of synaptic vesicles (Allaire et al., 2006); DENND1B has known immune system functions, and has been associated with autoimmune diseases (Yang et al., 2016). (The functions of these genes have probably not been fully characterized.) Thus the connecdens, like the olfactomedins discussed herein, and other gene families, such as the immunoglobulin superfamily (Zinn & Ozkan, 2017), are involved in both immunity and neuronal development. These two organ systems require proteins that permit many combinatorial possibilities (for responses to diverse antigens; for diverse neuronal connectivity). Another example is LRFN5 (Leucine Rich Repeat And Fibronectin Type III Domain Containing 5, one of five "synaptic adhesion-like molecules," or SALMs), which participate in synapse formation (Nam, Mah, & Kim, 2011), particularly presynaptic differentiation in contacting axons (Mah et al., 2010). LRFN5 also limits T-cell-induced neuroinflammation (CNS "immune privilege") by binding to the herpes virus entry mediator (HVEM, TNFRSF14), an effect that is rescued by the LRFN5-specific monoclonal antibody (Zhu et al., 2016). There is a great deal of literature on immune system dysfunction and neuroinflammation in depression (Zunszain, Hepgul, & Pariante, 2013), but in the PGC-MDD2 cohort, the strength of association signals is primarily predicted by gene expression in neurons rather than other tissues. Thus, it is not yet known which genes with overlapping functions influence depression risk via neuronal or immune functions, or both.

7.6 Genetic correlation with psychiatric and nonpsychiatric traits

The landmark PGC "cross-disorder" study reported significant genetic correlations among schizophrenia, bipolar disorder, and MDD (Cross-Disorder Group of the Psychiatric Genomics Consortium et al., 2013). In PGC-MDD2, r_g was 0.34 for MDD-schizophrenia and 0.32 for MDD-bipolar disorder. Mendelian randomization analysis showed a bidirectional relationship, suggesting shared biology. Further, more significant P-values were enriched in significantly schizophrenia-associated loci (Schizophrenia Working Group of the Psychiatric Genomics Consortium, 2014), including 7 of the 44 significant PGC-MDD2 loci.

There are also significant common-variant genetic correlations between MDD and anxiety disorders (0.80), neuroticism (0.70), ADHD (0.42), autism spectrum disorders (0.44), anorexia nervosa (0.13), and smoking (ever vs never) (0.29); and inversely with subjective well-being (−0.65), years of education (−0.13), and college completion (−0.17). (These correlations are between sets of GWAS association statistics, not with phenotypes.)

The interpretation of these correlations remains unclear. They could be evidence of a genetic basis for some general vulnerability to psychopathology (which has long been hypothesized in psychology) (Caspi et al., 2014). It is unclear whether the development of new treatment strategies will be facilitated by understanding these common factors, or by identifying factors that are more specific to some version of what we now call diagnostic categories. Or, there could be more disorder-specific genetic factors that are not captured by common-variant GWAS. GWAS is proving to be a powerful method for raising such questions and forcing us to address them.

The theme of genetic correlation between depression and obesity (r_g = 0.15 for body fat, 0.09 for BMI, 0.20 for obesity class 3) is illustrated by the associations to OLFM4 and NEGR1 (discussed herein). "Mendelian randomization" analysis of the directionality of this correlation showed that genetic risk factors for BMI are significant predictors of depression (but not the reverse). Coleman et al. (2018) reported intriguing relevant analyses: in the UK Biobank cohort, MDD PRS were *higher* in individuals exposed to childhood trauma, although genetic correlation was high between exposed and nonexposed individuals. But the genetic correlations between depression and obesity traits were seen only in the trauma-exposed group. There are many possible interpretations, including a complex interaction between genetic risk of obesity, the combined effects of childhood trauma and of phenotypic obesity on risk of depression, and interaction of the latter effects with genetic risk of depression. An analysis of severe premorbid life stress and/or childhood sexual abuse in the CONVERGE cohort also suggested an interaction of psychological adversity and common-variant effects on depression, possibly accounting for >10% of the etiology of depression (Peterson et al., 2018).

8 Large-scale studies of gene expression

There have been many small, underpowered studies of mRNA expression levels of various candidate genes in MDD, but only two large genome-wide expression studies of peripheral white blood cells in MDD patients vs controls. The first studied 463 individuals with recurrent MDD and 459 never-depressed controls using deep RNA sequencing (Mostafavi et al., 2014). Pathway analysis detected significant elevation in the interferon alpha/beta signaling pathway, although without clinical correlations with that profile. These genes are activated by interferons in response to infection or cancer. MDD is common in patients treated with interferon drugs (Raison et al., 2009).

The second study measured expression levels with microarrays in 882 current MDD cases, 635 remitted cases, and 331 controls. Expression was increased in interleukin-6 signaling genes, and decreased in natural killer cell pathways in currently-depressed individuals vs controls (Jansen et al., 2016). DVL3 (Disheveled Segment Polarity Protein 3) was also upregulated with current depression. These findings partially overlap with those of a recent, smaller microarray study ($N \sim 200$ cases) (Leday et al., 2018), but the latter study offers little information about methods.

9 Studies of candidate genes and polymorphisms

This chapter will not review candidate gene studies in detail. Before GWAS, most candidate genes were neurotransmitters and receptors relevant to antidepressant drug effects. These studies have been problematic. All were enormously underpowered, thus the a priori probability of false positive results was overwhelming (Sullivan, Eaves, Kendler, & Neale, 2001), but proper correction for multiple testing was rare. Also, in 325 candidate gene studies, error rates (inferred from deviation from expected genotype distributions—Hardy-Weinberg equilibrium) were significantly correlated with N, suggesting high genotyping error rates (particularly for 5-HTTLPR, see the following section) (Sen & Burmeister, 2008). Studies focused on one or two polymorphisms per gene, rather than systematically studying sequence variation. These studies failed to produce a robust, replicable MDD association finding.

9.1 5-HTTLPR genotype and the risk of depressive episodes following stress

The majority of candidate genes studies focused on one polymorphism, 5-HTTLPR. Many antidepressants block the presynaptic serotonin reuptake transporter, encoded by SLC6A4 (Solute Carrier Family 6 Member 4). Serotonin is 5-HT (5-hydroxytryptamine), and the transporter is called 5-HTT, so an insertion-deletion polymorphism in the gene promoter (a region that regulates gene expression) is termed 5-HTTLPR ("5-HTT-linked promoter region") (Heils et al., 1996). Associations were reported between "s" (short, deleted) alleles, and either MDD, or onset of MDD after stress.

A meta-analysis suggested a small effect of s alleles on MDD diagnosis ($OR = 1.76$, $P = 0.001$) in 46 studies of ~7800 cases and 16,000 controls (Clarke, Flint, Attwood, & Munafo, 2010). The authors raised the *caveat* that there was no effect in five studies that also genotyped a third allele ("long" with a SNP that made it function like "short"). They also observed the genotyping problems noted herein. Also, because 5-HTTLPR is in LD with surrounding SNPs (Wray et al., 2009), an association should be—but is not—observed in large-scale GWAS.

The more influential finding was that MDD onset was predicted by an interaction between the number of s alleles and stress (recent stressful life events or childhood maltreatment) in 847 individuals from a longitudinal study (Caspi et al., 2003). Dozens of subsequent studies tested this appealing hypothesis for different stressors and clinical contexts. A controversy emerged when two sets of meta-analyses, using methods to select studies and analyze results, arrived at different conclusions (Table 3). Some meta-analyses selected studies that were as comparable as possible, and requested individual-level data for uniform quality control and analysis. The other approach was to meta-analyze the P-values or ORs of published studies.

The first two "selective" meta-analyses considered only study designs similar to Caspi et al. The first one identified 14 studies, but could obtain individual-level data for only 5 ($N = 6699$) (Munafo, Durrant, Lewis, & Flint, 2009). The second one (Risch et al., 2009) obtained data for 14 "comparable" studies ($N = 12,520$). The third study published a protocol inviting investigators to join (Culverhouse et al., 2013), which attracted inappropriately ad hominem criticism (Moffitt & Caspi, 2014), with some investigators declining to participate, eventually reporting on 31 datasets ($N = 38,802$) (Culverhouse, Saccone, Horton, et al., 2018). All three papers found that MDD onset was significantly predicted by stress, but not by 5-HTTLPR genotype alone, or in interaction with stress. The Culverhouse paper conducted separate analyses of child maltreatment and of "broad stress," subdivided by outcomes of lifetime or current (recent) depression.

The meta-analyses of published statistics included a P-value meta-analysis of 54 studies that detected a significant interaction effect in all studies combined, and in studies of childhood trauma and medical disease stress (but not studies of recent stressful life events); and an OR meta-analysis of 51 studies (48 effect sizes) (Bleys, Luyten, Soenens, & Claes, 2018), which reported a significant interaction.

The present author agrees with Culverhouse and colleagues (with whom this author does not collaborate) who wrote: "(T)he best evidence to date indicates that there is either no interaction or only a modest interaction effect that is not robust and not easily generalizable. Further questions regarding the potential interplay between genetic vulnerability and stress will likely be answered definitively using the new, large data sets that are becoming available, and include hundreds of thousands participants: the UK Biobank, the Million Veterans Program, the National Institutes of Health All of Us Research

TABLE 3 Meta-analyses of 5-HTTLPR genotype×stress interaction and depression risk.

First author, year	N studies	N subjects	Effects		
			Stress	Genotype	Interaction
Meta-analyses of individual-level data from selected comparable studies					
Munafo, 2009[a]	5	6699	$OR = 2.08$ (1.77–2.44) $P < 0.001$	$OR=1.07$ (0.9–1.27) $P=0.42$	$OR=1.16$ (0.89–1.49) $P=0.42$
Risch, 2009[b]	14	12,520	$OR=1.41$ (1.25–1.57)	$OR=1.05$ (0.98–1.13)	$OR=1.01$ (0.94–1.10)
Culverhouse, 2018[c]	31	38,802	Significant in 6 subanalyses	n.s. in all analyses	n.s. in all analyses
Meta-analyses of statistical results from published studies					
Karg, 2011[d]	54	40,749	NA	NA	$P= 0.00002$
Bleys, 2018[e]	51	51,449			$OR= 1.18$ (1.09–1.28)

[a]*Munafo et al. (2009).*
[b]*Risch et al. (2009).*
[c]*Culverhouse, Saccone, and Bierut (2018).*
[d]*Karg, Burmeister, Shedden, and Sen (2011).*
[e]*Bleys et al. (2018).*

Program, and others. We urge the field to move forward to analyze these large modern datasets and move past the meta-analysis of published results of previous small studies" (Culverhouse, Saccone, & Bierut, 2018).

10 Pharmacogenomic prediction of antidepressant medication response

The preceding studies have attempted to find genetic factors influencing who develops MDD. Separate literature exists on genetic predictors of antidepressant drug response and selection of optimal drug therapy, focused on candidate genes selected for potential pharmacokinetic (drug metabolism) or pharmacodynamic (drug effect mechanism) relevance. A full review is beyond the scope of this chapter, but the reader is referred to recent reviews (Fabbri, Crisafulli, Calabro, Spina, & Serretti, 2016; Peterson et al., 2017; Rosenblat, Lee, & McIntyre, 2017).

A current trend is toward commercialization of tests that claim to predict optimal drug selection based on multiple genetic polymorphisms. All of these "kits" include polymorphisms that influence the activity of cytochrome p450 enzymes that metabolize drugs (always CYP2D6 and CYP2C19 and sometimes others), and some tests include genes such as ABCB1 and ABCC1, which influence the entry of some antidepressants into the CNS (Singh, 2015). Overall, there is some evidence to support the utility of these tests—primarily due to pharmacokinetic, rather than pharmacodynamic effects, but most papers do not separate these effects. But most available data are from industry-sponsored studies, many with open designs. There are only a few randomized control trials, with mostly nonsignificant results. The prediction methods are typically proprietary. Greden et al. (2019) reported on a double-blind industry-sponsored study of 8 weeks of antidepressant treatment in 1167 outpatients who had responded inadequately to at least one previous drug, comparing treatment "guided" by the GENESIGHT pharmacogenomic test vs. unguided treatment. The primary outcome (symptom improvement) showed no significant advantage for guided treatment ($P = 0.107$), although two secondary outcomes were modestly significant (meeting defined criteria for response [26.0% vs. 19.9%, $P = 0.013$] and remission [15.3% vs. 10.1%, $P = 0.007$]). This author's view is that marketing for this test over-exaggerates the importance of a study whose primary outcome was not supported. There are no significant GWAS findings for antidepressant response, perhaps they are still underpowered (Biernacka et al., 2015; Fabbri et al., 2018; Li, Tian, Seabrook, Drevets, & Narayan, 2016).

Genetic testing strategies could ultimately improve short-term antidepressant treatment outcome by avoiding drugs that are predictably under- or overmetabolized or poorly taken up into the brain. It is unclear whether pharmacodynamic findings will play a significant role. From the geneticist's perspective, it seems unfortunate to rely on small numbers of polymorphisms, rather than complete sequencing of genes (Wendt, Sajantila, Moura-Neto, Woerner, & Budowle, 2017). From a clinician's perspective, it seems unlikely that treatment-resistant depression can be solved by genetic testing, because many TRD patients try virtually all options. In the long run, entirely new antidepressants are needed.

11 Conclusions and future directions

GWAS methods are in the process of revolutionizing our understanding of the role of genetics in risk of depression, as well as the relationships among depression and other phenotypes. Depression studies have required very large sample sizes because of the small effect sizes of individual contributing alleles. Fortunately, investigators have developed clinical, genetic, and computational technologies to recruit and evaluate larger cohorts than were ever thought possible, and to carry out new types of genome-wide analyses. Researchers in the field believe that cohorts of at least 1,000,000 cases will be feasible in the foreseeable future, which will dramatically increase the rate of discovery. Deeper knowledge of the full range of common-variant risk factors should produce some initial understanding of biological mechanisms that underlie the genetic risk for depression, which (along with large sample sizes) will also facilitate research into environmental risk factors and their interactions with genes.

It is important to keep in mind that so far, progress is related to common genetic variants (SNPs, common insertion-deletions, and common structural variations with which they are in LD). We can detect a 34% genetic correlation in common-variant effects on schizophrenia and MDD, but common variants do not predict all of the heritability of MDD (although it remains possible, as proposed by Zuk et al., that genetic interactions have led to overestimation of the heritability). And even if the true genetic correlation proved to be around 34%, we do not yet know how to distinguish the "correlated" 34% from the "uncorrelated" 66%, or how to determine which component is more critical in producing the familiar clinical syndrome of depression, and in determining which biological targets would lead to the most useful treatments for those with the most severe depression. Whereas researchers in the candidate gene era made the mistake of confusing "the gene" (such as the serotonin transporter or the dopamine receptor) with one or two known sequence variants in that gene, we should not confuse the correlated component of genetic risk with all genetic risk.

Similarly, this writer would urge caution about rushing to abandon longstanding diagnostic categories such as MDD, as imperfect as they may be. GWAS—the most successful research method in the history of psychiatry—has relied almost entirely on these categories, which confirms the assumptions on which they were originally developed: if a measurement is able to predict family history, course of illness, and treatment response, it is highly likely to bear some relationship with underlying causation. The existing categories were developed primarily on the basis of what were then considered large-scale epidemiological-family studies, coupled with observations of treatment responses to available medications that significantly corresponded to the evolving diagnostic categories (lithium in bipolar disorder; antidepressants as sole treatment in MDD; antipsychotics as sole treatment in schizophrenia spectrum psychoses). To fully harness new genetic findings, a new era of both epidemiological and treatment research will be needed to refine both categorical and dimensional constructs that fit the data.

Disclosure Statement

No conflicts. The author receives financial support from NIH grants U19MH104172, U01MH109501 and R01MH106595, from the Walter E. Nichols, M.D., Professorship at the Stanford School of Medicine, and from the Stanford Schizophrenia Genetics Research fund from an anonymous donor.

References

Allaire, P. D., Ritter, B., Thomas, S., Burman, J. L., Denisov, A. Y., Legendre-Guillemin, V., ... McPherson, P. S. (2006). Connecdenn, a novel DENN domain-containing protein of neuronal clathrin-coated vesicles functioning in synaptic vesicle endocytosis. *The Journal of Neuroscience, 26*(51), 13202–13212.

Altshuler, D., Brooks, L. D., Chakravarti, A., Collins, F. S., Daly, M. J., & Donnelly, P. (2005). A haplotype map of the human genome. *Nature, 437*(7063), 1299–1320.

American Psychiatric Association (2013). *Diagnostic and statistical manual of mental disorders (5th ed.: Dsm-5)*. Washington, DC: American Psychiatric Press.

Amir-Zilberstein, L., Blechman, J., Sztainberg, Y., Norton, W. H., Reuveny, A., Borodovsky, N., ... Levkowitz, G. (2012). Homeodomain protein otp and activity-dependent splicing modulate neuronal adaptation to stress. *Neuron, 73*(2), 279–291.

Anholt, R. R. (2014). Olfactomedin proteins: Central players in development and disease. *Frontiers in Cell and Development Biology, 2*, 6.

Biernacka, J. M., Sangkuhl, K., Jenkins, G., Whaley, R. M., Barman, P., Batzler, A., ... Weinshilboum, R. (2015). The International SSRI Pharmacogenomics Consortium (ISPC): A genome-wide association study of antidepressant treatment response. *Translational Psychiatry, 5*.

Blacker, D., Faraone, S. V., Rosen, A. E., Guroff, J. J., Adams, P., Weissman, M. M., & Gershon, E. S. (1996). Unipolar relatives in bipolar pedigrees: A search for elusive indicators of underlying bipolarity. *American Journal of Medical Genetics, 67*(5), 445–454.

Bleys, D., Luyten, P., Soenens, B., & Claes, S. (2018). Gene-environment interactions between stress and 5-HTTLPR in depression: A meta-analytic update. *Journal of Affective Disorders, 226*, 339–345.

Boender, A. J., van Rozen, A. J., & Adan, R. A. (2012). Nutritional state affects the expression of the obesity-associated genes Etv5, Faim2, Fto, and Negr1. *Obesity, 20*(12), 2420–2425.

Boyle, E. A., Li, Y. I., & Pritchard, J. K. (2017). An expanded view of complex traits: From polygenic to omnigenic. *Cell, 169*(7), 1177–1186.

Bradfield, J. P., Taal, H. R., Timpson, N. J., Scherag, A., Lecoeur, C., Warrington, N. M., … Grant, S. F. (2012). A genome-wide association meta-analysis identifies new childhood obesity loci. *Nature Genetics, 44*(5), 526–531.

Bulik-Sullivan, B. K., Loh, P. R., Finucane, H. K., Ripke, S., Yang, J., Patterson, N., … Neale, B. M. (2015). LD Score regression distinguishes confounding from polygenicity in genome-wide association studies. *Nature Genetics, 47*(3), 291–295.

Caspi, A., Houts, R. M., Belsky, D. W., Goldman-Mellor, S. J., Harrington, H., Israel, S., … Moffitt, T. E. (2014). The p factor: One general psychopathology factor in the structure of psychiatric disorders? *Clinical Psychological Science, 2*(2), 119–137.

Caspi, A., Sugden, K., Moffitt, T., Taylor, A., Craig, I., Harrington, H., … Poulton, R. (2003). Influence of life stress on depression: Moderation by a polymorphism in the 5-htt gene. *Science, 301*, 386–389.

Clarke, H., Flint, J., Attwood, A. S., & Munafo, M. R. (2010). Association of the 5-HTTLPR genotype and unipolar depression: A meta-analysis. *Psychological Medicine, 40*(11), 1767–1778.

Coleman, J. R. I., Purves, K. L., Davis, K. A. S., Rayner, C., Choi, S. W., Hübel, C., … Breen, G. (2018). Genome-wide gene-environment analyses of depression and reported lifetime traumatic experiences in UK Biobank. https://www.biorxiv.org/content/10.1101/247353v2, https://doi.org/10.1101/247353.

CONVERGE Consortium (2015). Sparse whole-genome sequencing identifies two loci for major depressive disorder. *Nature, 523*(7562), 588–591.

Craddock, N., & Forty, L. (2006). Genetics of affective (mood) disorders. *European Journal of Human Genetics, 14*(6), 660–668.

Cross-Disorder Group of the Psychiatric Genomics Consortium, Lee, S. H., Ripke, S., Neale, B. M., Faraone, S. V., Purcell, S. M., … Wray, N. R. (2013). Genetic relationship between five psychiatric disorders estimated from genome-wide SNPs. *Nature Genetics, 45*(9), 984–994.

Culverhouse, R. C., Bowes, L., Breslau, N., Nurnberger, J. I., Jr., Burmeister, M., Fergusson, D. M., … Bierut, L. J. (2013). Protocol for a collaborative meta-analysis of 5-HTTLPR, stress, and depression. *BMC Psychiatry, 13*, 304.

Culverhouse, R. C., Saccone, N. L., & Bierut, L. J. (2018). The state of knowledge about the relationship between 5-HTTLPR, stress, and depression. *Journal of Affective Disorders, 228*, 205–206.

Culverhouse, R. C., Saccone, N. L., Horton, A. C., Ma, Y., Anstey, K. J., Banaschewski, T., … Bierut, L. J. (2018). Collaborative meta-analysis finds no evidence of a strong interaction between stress and 5-HTTLPR genotype contributing to the development of depression. *Molecular Psychiatry, 23*(1), 133–142.

de Leeuw, C. A., Mooij, J. M., Heskes, T., & Posthuma, D. (2015). MAGMA: Generalized gene-set analysis of GWAS data. *PLoS Computational Biology, 11*(4).

Dudbridge, F., & Gusnanto, A. (2008). Estimation of significance thresholds for genomewide association scans. *Genetic Epidemiology, 32*(3), 227–234.

Fabbri, C., Crisafulli, C., Calabro, M., Spina, E., & Serretti, A. (2016). Progress and prospects in pharmacogenetics of antidepressant drugs. *Expert Opinion on Drug Metabolism & Toxicology, 12*(10), 1157–1168.

Fabbri, C., Tansey, K. E., Perlis, R. H., Hauser, J., Henigsberg, N., Maier, W., … Lewis, C. M. (2018). New insights into the pharmacogenomics of antidepressant response from the GENDEP and STAR*D studies: Rare variant analysis and high-density imputation. *The Pharmacogenomics Journal, 18*(3), 413–421.

Feighner, J. P., Robins, E., Guze, S. B., Woodruff, R. A., Jr., Winokur, G., & Munoz, R. (1972). Diagnostic criteria for use in psychiatric research. *Archives of General Psychiatry, 26*(1), 57–63.

Fernandez-Pujals, A. M., Adams, M. J., Thomson, P., McKechanie, A. G., Blackwood, D. H., Smith, B. H., … McIntosh, A. M. (2015). Epidemiology and heritability of major depressive disorder, stratified by age of onset, sex, and illness course in Generation Scotland: Scottish Family Health Study (GS:SFHS). *PLoS One, 10*(11).

Flint, J., Chen, Y., Shi, S., Kendler, K. S., & CONVERGE Consortium. (2012). Epilogue: Lessons from the CONVERGE study of major depressive disorder in China. *Journal of Affective Disorders, 140*(1), 1–5.

Greden, J. F., Parikh, S. V., Rothschild, A. J., Thase, M. E., Dunlop, B. W., Debattista, C., … Dechairo, B. (2019). Impact of pharmacogenomics on clinical outcomes in major depressive disorder in the guided trial: A large, patient- and rater-blinded, randomized, controlled study. *Journal of Psychiatric Research, 111*, 59–67.

Hashimoto, T., Maekawa, S., & Miyata, S. (2009). IgLON cell adhesion molecules regulate synaptogenesis in hippocampal neurons. *Cell Biochemistry and Function, 27*(7), 496–498.

Hasin, D. S., Goodwin, R. D., Stinson, F. S., & Grant, B. F. (2005). Epidemiology of major depressive disorder: Results from the National Epidemiologic Survey on alcoholism and related conditions. *Archives of General Psychiatry, 62*(10), 1097–1106.

Heils, A., Teufel, A., Petri, S., Stober, G., Riederer, P., Bengel, D., & Lesch, K. P. (1996). Allelic variation of human serotonin transporter gene expression. *Journal of Neurochemistry, 66*(6), 2621–2624.

Hek, K., Demirkan, A., Lahti, J., Terracciano, A., Teumer, A., Cornelis, M. C., … Murabito, J. (2013). A genome-wide association study of depressive symptoms. *Biological Psychiatry, 73*(7), 667–678.

Hyde, C. L., Nagle, M. W., Tian, C., Chen, X., Paciga, S. A., Wendland, J. R., … Winslow, A. R. (2016). Identification of 15 genetic loci associated with risk of major depression in individuals of European descent. *Nature Genetics, 48*(9), 1031–1036.

Jansen, R., Penninx, B. W., Madar, V., Xia, K., Milaneschi, Y., Hottenga, J. J., … Sullivan, P. F. (2016). Gene expression in major depressive disorder. *Molecular Psychiatry, 21*(3), 339–347.

Karg, K., Burmeister, M., Shedden, K., & Sen, S. (2011). The serotonin transporter promoter variant (5-HTTLPR), stress, and depression meta-analysis revisited: Evidence of genetic moderation. *Archives of General Psychiatry, 68*(5), 444–454.

Kendler, K. S., Gardner, C. O., & Prescott, C. A. (2002). Toward a comprehensive developmental model for major depression in women. *The American Journal of Psychiatry, 159*(7), 1133–1145.

Kendler, K. S., Gatz, M., Gardner, C. O., & Pedersen, N. L. (2007). Clinical indices of familial depression in the Swedish Twin Registry. *Acta Psychiatrica Scandinavica, 115*(3), 214–220.

Kendler, K. S., Kuhn, J. W., & Prescott, C. A. (2004). Childhood sexual abuse, stressful life events and risk for major depression in women. *Psychological Medicine, 34*(8), 1475–1482.

Kendler, K. S., McGuire, M., Gruenberg, A. M., O'Hare, A., Spellman, M., & Walsh, D. (1993). The Roscommon Family Study. IV. Affective illness, anxiety disorders, and alcoholism in relatives. *Archives of General Psychiatry, 50*(12), 952–960.

Kvale, M. N., Hesselson, S., Hoffmann, T. J., Cao, Y., Chan, D., Connell, S., … Risch, N. (2015). Genotyping informatics and quality control for 100,000 subjects in the Genetic Epidemiology Research on Adult Health and Aging (GERA) cohort. *Genetics, 200*(4), 1051–1060.

Lander, E. S. (1996). The new genomics: Global views of biology. *Science, 274*(5287), 536–539.

Leday, G. G. R., Vertes, P. E., Richardson, S., Greene, J. R., Regan, T., Khan, S., … Bullmore, E. T. (2018). Replicable and coupled changes in innate and adaptive immune gene expression in two case-control studies of blood microarrays in major depressive disorder. *Biological Psychiatry, 83*(1), 70–80.

Lee, J. A., Damianov, A., Lin, C. H., Fontes, M., Parikshak, N. N., Anderson, E. S., … Martin, K. C. (2016). Cytoplasmic Rbfox1 regulates the expression of synaptic and autism-related genes. *Neuron, 89*(1), 113–128.

Levinson, D. F. (2013). Genetics of depression. In Charney, D. S., Sklar, P., Buxbaum, J. D., & Nestler, E. J. (Eds.), *Neurobiology of mental illness.* (4th ed.). New York: Oxford University Press.

Levinson, D. F., Mostafavi, S., Milaneschi, Y., Rivera, M., Ripke, S., Wray, N. R., & Sullivan, P. F. (2014). Genetic studies of major depressive disorder: Why are there no genome-wide association study findings and what can we do about it? *Biological Psychiatry, 76*(7), 510–512.

Levinson, D. F., Zubenko, G. S., Crowe, R. R., DePaulo, R. J., Scheftner, W. S., Weissman, M. M., … Chellis, J. (2003). Genetics of recurrent early-onset depression (GenRED): Design and preliminary clinical characteristics of a repository sample for genetic linkage studies. *American Journal of Medical Genetics. Part B, Neuropsychiatric Genetics, 119B*(1), 118–130.

Li, Q. S., Tian, C., Seabrook, G. R., Drevets, W. C., & Narayan, V. A. (2016). Analysis of 23andMe antidepressant efficacy survey data: Implication of circadian rhythm and neuroplasticity in bupropion response. *Translational Psychiatry, 6*(9).

Liu, W., & Rodgers, G. P. (2016). Olfactomedin 4 expression and functions in innate immunity, inflammation, and cancer. *Cancer Metastasis Reviews, 35*(2), 201–212.

Mah, W., Ko, J., Nam, J., Han, K., Chung, W. S., & Kim, E. (2010). Selected SALM (synaptic adhesion-like molecule) family proteins regulate synapse formation. *The Journal of Neuroscience, 30*(16), 5559–5568.

Maher, B. (2008). Personal genomes: The case of the missing heritability. *Nature, 456*(7218), 18–21.

Maier, W., Lichtermann, D., Minges, J., Hallmayer, J., Heun, R., Benkert, O., & Levinson, D. F. (1993). Continuity and discontinuity of affective disorders and schizophrenia. Results of a controlled family study. *Archives of General Psychiatry, 50*(11), 871–883.

Major Depressive Disorder Working Group of the Psychiatric Genomics Consortium, Ripke, S., Wray, N. R., Lewis, C. M., Hamilton, S. P., Weissman, M. M., … Sullivan, P. F. (2013). A mega-analysis of genome-wide association studies for major depressive disorder. *Molecular Psychiatry, 18*(4), 497–511.

McCarthy, S., Das, S., Kretzschmar, W., Delaneau, O., Wood, A. R., Teumer, A., … Durbin, R. (2016). A reference panel of 64,976 haplotypes for genotype imputation. *Nature Genetics, 48*(10), 1279–1283.

Mitchell, P. B., Frankland, A., Hadzi-Pavlovic, D., Roberts, G., Corry, J., Wright, A., … Breakspear, M. (2011). Comparison of depressive episodes in bipolar disorder and in major depressive disorder within bipolar disorder pedigrees. *The British Journal of Psychiatry, 199*(4), 303–309.

Moffitt, T. E., & Caspi, A. (2014). Bias in a protocol for a meta-analysis of 5-HTTLPR, stress, and depression. *BMC Psychiatry, 14*, 179.

Mostafavi, S., Battle, A., Zhu, X., Potash, J. B., Weissman, M. M., Shi, J., … Levinson, D. F. (2014). Type I interferon signaling genes in recurrent major depression: Increased expression detected by whole-blood RNA sequencing. *Molecular Psychiatry, 19*(12), 1267–1274.

Munafo, M. R., Durrant, C., Lewis, G., & Flint, J. (2009). Gene X environment interactions at the serotonin transporter locus. *Biological Psychiatry, 65*(3), 211–219.

Nam, J., Mah, W., & Kim, E. (2011). The SALM/Lrfn family of leucine-rich repeat-containing cell adhesion molecules. *Seminars in Cell & Developmental Biology, 22*(5), 492–498.

Nanni, V., Uher, R., & Danese, A. (2012). Childhood maltreatment predicts unfavorable course of illness and treatment outcome in depression: A meta-analysis. *The American Journal of Psychiatry, 169*(2), 141–151.

Nelson, E. C., Heath, A. C., Madden, P. A., Cooper, M. L., Dinwiddie, S. H., Bucholz, K. K., … Martin, N. G. (2002). Association between self-reported childhood sexual abuse and adverse psychosocial outcomes: Results from a twin study. *Archives of General Psychiatry, 59*(2), 139–145.

Okbay, A., Baselmans, B. M., De Neve, J. E., Turley, P., Nivard, M. G., Fontana, M. A., … Cesarini, D. (2016). Genetic variants associated with subjective well-being, depressive symptoms, and neuroticism identified through genome-wide analyses. *Nature Genetics, 48*(6), 624–633.

Pariante, C. M., & Lightman, S. L. (2008). The HPA axis in major depression: Classical theories and new developments. *Trends in Neurosciences, 31*(9), 464–468.

Peterson, R. E., Cai, N., Dahl, A. W., Bigdeli, T. B., Edwards, A. C., Webb, B. T., … Kendler, K. S. (2018). Molecular genetic analysis subdivided by adversity exposure suggests etiologic heterogeneity in major depression. *The American Journal of Psychiatry, 175*, 545–554.

Peterson, K., Dieperink, E., Anderson, J., Boundy, E., Ferguson, L., & Helfand, M. (2017). Rapid evidence review of the comparative effectiveness, harms, and cost-effectiveness of pharmacogenomics-guided antidepressant treatment versus usual care for major depressive disorder. *Psychopharmacology, 234*(11), 1649–1661.

Pischedda, F., & Piccoli, G. (2015). The IgLON family member Negr1 promotes neuronal arborization acting as soluble factor via FGFR2. *Frontiers in Molecular Neuroscience, 8*, 89.

Purcell, S. M., Moran, J. L., Fromer, M., Ruderfer, D., Solovieff, N., Roussos, P., ... Sklar, P. (2014). A polygenic burden of rare disruptive mutations in schizophrenia. *Nature, 506*(7487), 185–190.

Purcell, S. M., Wray, N. R., Stone, J. L., Visscher, P. M., O'Donovan, M. C., Sullivan, P. F., ... Sklar, P. (2009). Common polygenic variation contributes to risk of schizophrenia and bipolar disorder. *Nature, 460*(7256), 748–752.

Raison, C. L., Borisov, A. S., Majer, M., Drake, D. F., Pagnoni, G., Woolwine, B. J., ... Miller, A. H. (2009). Activation of central nervous system inflammatory pathways by interferon-alpha: Relationship to monoamines and depression. *Biological Psychiatry, 65*(4), 296–303.

Reich, D. E., & Lander, E. S. (2001). On the allelic spectrum of human disease. *Trends in Genetics, 17*(9), 502–510.

Risch, N., Herrell, R., Lehner, T., Liang, K. Y., Eaves, L., Hoh, J., ... Merikangas, K. R. (2009). Interaction between the serotonin transporter gene (5-HTTLPR), stressful life events, and risk of depression: A meta-analysis. *JAMA, 301*(23), 2462–2471.

Rosenblat, J. D., Lee, Y., & McIntyre, R. S. (2017). Does pharmacogenomic testing improve clinical outcomes for major depressive disorder? A systematic review of clinical trials and cost-effectiveness studies. *The Journal of Clinical Psychiatry, 78*(6), 720–729.

Sanz, R., Ferraro, G. B., & Fournier, A. E. (2015). IgLON cell adhesion molecules are shed from the cell surface of cortical neurons to promote neuronal growth. *The Journal of Biological Chemistry, 290*(7), 4330–4342.

Saunders, N. R., Noor, N. M., Dziegielewska, K. M., Wheaton, B. J., Liddelow, S. A., Steer, D. L., ... Dietrich, W. D. (2014). Age-dependent transcriptome and proteome following transection of neonatal spinal cord of Monodelphis domestica (South American grey short-tailed opossum). *PLoS One, 9*(6).

Schizophrenia Working Group of the Psychiatric Genomics Consortium. (2014). Biological insights from 108 schizophrenia-associated genetic loci. *Nature, 511*(7510), 421–427.

Sen, S., & Burmeister, M. (2008). Hardy-Weinberg analysis of a large set of published association studies reveals genotyping error and a deficit of heterozygotes across multiple loci. *Human Genomics, 3*(1), 36–52.

Singh, A. B. (2015). Improved antidepressant remission in major depression via a pharmacokinetic pathway polygene pharmacogenetic report. *Clinical Psychopharmacology and Neuroscience, 13*(2), 150–156.

Smoller, J. W., & Finn, C. T. (2003). Family, twin, and adoption studies of bipolar disorder. *American Journal of Medical Genetics. Part C, Seminars in Medical Genetics, 123C*(1), 48–58.

Sullivan, P. F., Eaves, L. J., Kendler, K. S., & Neale, M. C. (2001). Genetic case-control association studies in neuropsychiatry. *Archives of General Psychiatry, 58*(11), 1015–1024.

Sullivan, P. F., Neale, M. C., & Kendler, K. S. (2000). Genetic epidemiology of major depression: Review and meta-analysis. *The American Journal of Psychiatry, 157*(10), 1552–1562.

Turley, P., Walters, R. K., Maghzian, O., Okbay, A., Lee, J. J., Fontana, M. A., ... Benjamin, D. J. (2018). Multi-trait analysis of genome-wide association summary statistics using MTAG. *Nature Genetics, 50*(2), 229–237.

Visscher, P. M., Brown, M. A., McCarthy, M. I., & Yang, J. (2012). Five years of GWAS discovery. *American Journal of Human Genetics, 90*(1), 7–24.

Wendt, F. R., Sajantila, A., Moura-Neto, R. S., Woerner, A. E., & Budowle, B. (2017). Full-gene haplotypes refine CYP2D6 metabolizer phenotype inferences. *International Journal of Legal Medicine, 132*, 1007–1024.

Willer, C. J., Speliotes, E. K., Loos, R. J., Li, S., Lindgren, C. M., Heid, I. M., ... Hirschhorn, J. N. (2009). Six new loci associated with body mass index highlight a neuronal influence on body weight regulation. *Nature Genetics, 41*(1), 25–34.

World Health Organisation (1992). *ICD-10 classifications of mental and behavioural disorder: Clinical descriptions and diagnostic guidelines.* Geneva: World Health Organisation.

Wray, N. R., James, M. R., Gordon, S. D., Dumenil, T., Ryan, L., Coventry, W. L., ... Martin, N. G. (2009). Accurate, large-scale genotyping of 5HTTLPR and flanking single nucleotide polymorphisms in an association study of depression, anxiety, and personality measures. *Biological Psychiatry, 66*(5), 468–476.

Wray, N. R., Lee, S. H., Mehta, D., Vinkhuyzen, A. A., Dudbridge, F., & Middeldorp, C. M. (2014). Research review: Polygenic methods and their application to psychiatric traits. *Journal of Child Psychology and Psychiatry, and Allied Disciplines, 55*(10), 1068–1087.

Wray, N. R., Ripke, S., Mattheisen, M., Trzaskowski, M., Byrne, E. M., Abdellaoui, A., ... Sullivan, P. F. (2018). Genome-wide association analyses identify 44 risk variants and refine the genetic architecture of major depression. *Nature Genetics, 50*(5), 668–681.

Yang, C. W., Hojer, C. D., Zhou, M., Wu, X., Wuster, A., Lee, W. P., ... Chan, A. C. (2016). Regulation of T cell receptor signaling by DENND1B in TH2 cells and allergic disease. *Cell, 164*(1–2), 141–155.

Zhu, Y., Yao, S., Augustine, M. M., Xu, H., Wang, J., Sun, J., ... Chen, L. (2016). Neuron-specific SALM5 limits inflammation in the CNS via its interaction with HVEM. *Science Advances, 2*(4).

Zimmermann, P., Bruckl, T., Lieb, R., Nocon, A., Ising, M., Beesdo, K., & Wittchen, H. U. (2008). The interplay of familial depression liability and adverse events in predicting the first onset of depression during a 10-year follow-up. *Biological Psychiatry, 63*(4), 406–414.

Zimmermann, P., Bruckl, T., Nocon, A., Pfister, H., Lieb, R., Wittchen, H. U., ... Angst, J. (2009). Heterogeneity of DSM-IV major depressive disorder as a consequence of subthreshold bipolarity. *Archives of General Psychiatry, 66*(12), 1341–1352.

Zinn, K., & Ozkan, E. (2017). Neural immunoglobulin superfamily interaction networks. *Current Opinion in Neurobiology, 45*, 99–105.

Zuk, O., Hechter, E., Sunyaev, S. R., & Lander, E. S. (2012). The mystery of missing heritability: Genetic interactions create phantom heritability. *Proceedings of the National Academy of Sciences of the United States of America, 109*(4), 1193–1198.

Zunszain, P. A., Hepgul, N., & Pariante, C. M. (2013). Inflammation and depression. *Current Topics in Behavioral Neurosciences, 14*, 135–151.

Chapter 17

Personalized mental health: Artificial intelligence technologies for treatment response prediction in anxiety disorders

Ulrike Lueken[a] and Tim Hahn[b]

[a]*Department of Psychology, Humboldt-Universität zu Berlin, Berlin, Germany;* [b]*Artificial Intelligence and Machine Learning Group, Translational Psychiatry Lab, Department of Psychiatry, University of Münster, Münster, Germany*

1 Introduction

1.1 Interindividual variability in normal to pathological forms of fear and anxiety

Fear and anxiety are primary emotions of high evolutional relevance. Serving as our very basic "alarm system," they ensure survival of the organism in a potentially harmful and dynamically changing environment. The organism's defensive system is both shaped by (epi)genetic factors and environmental influences, including individual learning experiences (Gottschalk & Domschke, 2016; Schiele & Domschke, 2017). Therefore, a considerably high rate of interindividual variation in normal to pathological forms of fear and anxiety can be observed (Lonsdorf & Merz, 2017) that may explain why individuals experience more or less fear—and some patients benefit more from anxiety treatments than others.

Anxiety disorders constitute the largest group of mental disorders, with an estimated 12-month prevalence of 14% in Europe (Wittchen et al., 2011). Anxiety disorders are characterized by an early onset during childhood and adolescence (Beesdo-Baum & Knappe, 2012), with a chronic and relapsing course, and represent a well-known risk factor and precursor for the development of comorbid conditions such as depression (Meier et al., 2015). As a consequence, they are a leading cause of disability, with an exceptionally high individual and societal burden (Gustavsson et al., 2011): In 2010, more than 60 million individuals in Europe suffered from anxiety disorders, resulting in health care costs of more than 74 billion Euros (Gustavsson et al., 2011; Wittchen et al., 2011). Worldwide, anxiety disorders explain a substantial proportion of the global burden of disease (14.6% of disability-adjusted life years [DALYs]) attributable to mental and substance use disorders (Whiteford et al., 2013). Targeting anxiety disorders by improving their treatment options, thus, is a key priority for researchers, clinicians, mental health care systems, and societies.

1.2 What is a predictive biomarker?

According to definitions from the National Institutes of Health (De Gruttola et al., 2001) and the US Food and Drug Administration (2014), a predictive biomarker is a biological characteristic that is objectively measured prior to treatment, and categorizes patients according to their likelihood to respond to a particular treatment. A predictive biomarker should be characterized by its reliability, sensitivity, and specificity. Biomarker properties should be replicated and confirmed in independent samples and—in order to be suitable to inform clinical decision making—it should allow for predictions on the single-patient level, with quantified prediction accuracies (Savitz, Rauch, & Drevets, 2013). Finally, it is crucial to assess the generalizability of a given biomarker's properties across different samples, research contexts, and populations (Woo, Chang, Lindquist, & Wager, 2017). Once different biomarkers with proven validity and reliability are available, safety and efficiency considerations have to be made, for example, including health economic as well as ethical aspects (under which clinical circumstances do expensive biomarkers "pay off"? Are other biomarkers or proxies thereof with a better safety or cost-ratio profile available? How "expensive"—both in terms of patient suffering and economic cost—is a misclassified nonresponder in terms of direct and indirect costs and a misclassified responder in terms of allocated resources?).

Personalized Psychiatry. https://doi.org/10.1016/B978-0-12-813176-3.00017-1

1.3 Targeting nonrespondent patients by personalizing treatments?

Exposure-based cognitive-behavioral therapy (CBT) represents a first-line treatment for all anxiety disorders (Bandelow et al., 2014; National Institute of Health and Care Excellence, 2011). The high level of evidence that CBT works for the average patient (Hofmann, Asnaani, Vonk, Sawyer, & Fang, 2012) is, however, counteracted by the fact that a significant proportion of patients drop out early, do not respond in a clinically sufficient way, or relapse in the long term (Fernandez, Salem, Swift, & Ramtahal, 2015; Taylor, Abramowitz, & McKay, 2012). It is estimated that more than one-third of patients may be unresponsive to CBT (Taylor et al., 2012), with severe consequences for patients and increasing costs for societies. Individual patient characteristics may act as one major source of variability underlying these differences in treatment response. Knowledge about markers of treatment (non)response could thus open up new avenues, allowing clinicians to custom-tailor treatments and clinical processes: knowing a priori if a patient is likely to respond before a particular treatment is initiated could help in sparing ineffective treatments, associated side effects on patient compliance, disease chronification or aggravation, as well as direct and indirect financial costs (Richter, Pittig, Hollandt, & Lueken, 2017). The lack of validated clinical predictors (Dunlop, Kelley, et al., 2017) serves as an impetus to search for biologically derived markers of treatment (non)response that may more closely match the underlying neural mechanisms of change following CBT (Lueken & Hahn, 2016).

The gap between the average and the individual patient is a phenomenon that is now increasingly acknowledged in clinical research and practice. Originally inspired by the domain of genetics and pharmacology (Hamburg & Collins, 2010), a paradigm shift toward more individualized or personalized medicine can be observed for mental health also (Fig. 1).

This development also impacts the major research lines in this field where the logic of clinical trials designed to test the effectiveness of new treatments for the average patient is supplemented by a stronger focus on individual patient characteristics that moderate outcomes, thus bearing potential to custom-tailor treatments. These avenues toward novel research strategies do, however, demand the development and application of innovative methods particularly suitable for capturing individual variability on the single-case level. While group-based studies (for example, on the effects that CBT exerts on the brain) inform us about the (neural) mechanisms of a given treatment; these are more recently complemented by research approaches employing predictive analytics for the individual patient (Hahn, Nierenberg, & Whitfield-Gabrieli, 2017). As outlined in Lueken and Hahn (2016), both research strategies aim to overcome the pressing problem of better understanding treatment (non)response. However, they differ in their approach. While the former aims to improve our understanding of the general mechanisms of (non)response, the latter seeks to equip clinicians with a system providing valid and reliable information about the very patient sitting in front of them. Based on an improved understanding of which neural circuits are modulated by CBT, mechanistic studies are invaluable for the development of novel and innovative CBT treatment strategies. For example, testing a new drug employing classic group-level statistics, as is done in clinical trials today— while providing a good estimate of the average effect of treatment in the population recruited—does not provide sufficient information regarding its effectiveness of the treatment in a new, unseen patient. Perhaps even more harmful in the current state of treatment research in psychiatry, the approach may obscure effects present only in a small subset of individuals. If,

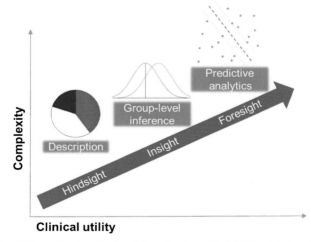

FIG. 1 Predictive analytics in mental health is moving from the description of patients (hindsight) and the investigation of statistical group differences or associations (insight) toward models capable of predicting current or future characteristics for individual patients (foresight), thereby allowing for a direct assessment of a model's clinical utility. *(From Hahn, T., Nierenberg, A. A., & Whitfield-Gabrieli, S. (2017). Predictive analytics in mental health: Applications, guidelines, challenges and perspectives.* Molecular Psychiatry, 22(1), 37–43.)

for example, an intervention was highly effective in the 10% of patients who would otherwise be treatment resistant, but did not work, or was even harmful to the other 90%, it would never be discovered using the current approach. If we could, however, find a marker to safely select the 10% of beneficiaries, any clinician would instantly see the great utility of the new intervention. Augmentation or add-on interventions, for example, by neurofeedback (Zilverstand, Sorger, Sarkheil, & Goebel, 2015) or noninvasive brain stimulation (Zwanzger, Herrmann, & Baeken, 2016), suffer from similar issues. Testing augmentations on the average patient group may greatly underestimate the true utility of a treatment for a subgroup of patients due to ceiling effects (a substantial proportion of patients already respond to CBT alone). As a consequence, treatment add-ons should be tested specifically on the vulnerable group of nonrespondent patients—which can be identified a priori by predictive models. Utilizing the wealth of information often available today for an individual patient—from his genetic makeup over individual neural dynamics to personal patterns of behavior—biomarker research not only aims to help plan individual treatments, but is invaluable as a tool to stratify responders and nonresponders such that novel interventions and add-ons can be developed specifically for the most problematic patients today. More generally, both mechanistic and predictive research strategies can thus be seen as complementary to improve existing treatments and address the problem of nonresponse (Fig. 2).

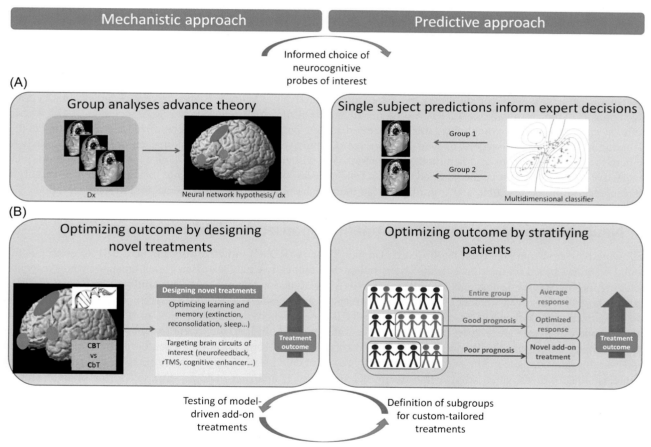

FIG. 2 Illustration of the mechanistic vs predictive approach. (A) Research strategies pursued by mechanistic studies use group analyses to advance etiopathophysiological models of a given disorder or process. In contrast, multivariate pattern recognition embedded within a machine learning framework uses neural information to predict disease status or treatment response for the individual patient. Input from mechanism-oriented studies is essential for predictive approaches, as this knowledge enables an informed choice regarding which neurofunctional probe(s) of interest should be used for prediction. (B) Based on an improved understanding of how a particular treatment works at the level of the brain, existing treatments may be supplemented by mechanism-based add-ons, or novel treatments that may be designed to directly target the proposed underlying brain circuits. Complementing this perspective, the predictive approach aims to stratify patients a priori based on a valid and reliable marker suitable for single-patient predictions. Patients with a poor prognosis could be identified prior to undergoing a probably unsuccessful treatment, thus reducing nonresponse and chronification, as well as lost time and expenses for both the patient and health care system. Input of model-driven add-on treatments is essential to inform clinicians about which add-on treatments could represent a valuable alternative for otherwise nonresponsive patients. In summary, both perspectives complement each other and, together, are crucial in optimizing treatments and response rates as a cornerstone goal for clinical research. *(From Lueken, U., & Hahn, T. (2016). Functional neuroimaging of psychotherapeutic processes in anxiety and depression: From mechanisms to predictions. Current Opinion in Psychiatry, 29, 25–31.)*

1.4 Novel methods for "mental health predictomics"

In the field of clinical neuroimaging, multivariate pattern recognition and machine learning is gaining increased attention, as these methods allow a better representation of neural alterations as complex system or network dysfunctions (which matches the assumed nature of mental disorders much better than traditional univariate brain-mapping approaches do). The flexibility with regard to modeling, as well as the multivariate nature of these approaches, also leads to often increased test sensitivity compared with conventional mass-univariate analyses. Most important, however, is the fact that such approaches are capable of deriving predictions that can be applied to the single patient (Doyle, Mehta, & Brammer, 2015; Orru, Pettersson-Yeo, Marquand, Sartori, & Mechelli, 2012; Woo et al., 2017). Woo et al. (2017) recently showed that the number of neuroimaging publications employing these methods has sharply increased within the past few years; however, the field is still dominated by investigations on purely neurological disorders, such as Alzheimer's disease, while anxiety disorders are represented in only 2.5% of these publications. While the vast majority of research utilizes cross-sectional classification approaches (e.g., prediction of patients vs controls), longitudinal investigations on disease prognosis or treatment response are investigated only by a minority of studies (approx. 14.4%) across all disorders of the brain, possibly owing to the fact that longitudinal studies are much more cost-intensive than cross-sectional investigations.

1.5 Aims and contents of this chapter

This chapter serves to provide the reader with a summary of the current body of evidence regarding neurobiological markers of treatment response crosscutting pharmacological, as well as CBT treatments of anxiety disorders. In line with previous work from our groups (Lueken & Hahn, 2016, 2017; Lueken et al., 2016), we will present evidence from both group-based mechanistic studies, as well as preliminary evidence from predictive modeling accounts. In correspondence with current developments in reconceptualizing anxiety disorders within the DSM-5 (American Psychiatric Association, 2013), we here focus on specific phobia, agoraphobia, social anxiety disorder, panic disorder, and generalized anxiety disorder, thus excluding trauma-related disorders, as well as obsessive-compulsive disorders. Crosscutting multiple levels of analysis, we here focus on (epi)genetic and neuroimaging markers. We will close this chapter with an executive outlook on future perspectives on how to improve the new and dynamic field of predictive analytics in order to foster clinical applicability of neuroscience-informed research, and to further bridge the translational gap between clinical research and practice.

2 Neurobiological markers of treatment response: Available evidence

Following new developments from the Research Domain Criteria (RDoC) initiative (Insel et al., 2010) to conceptualize mental disorders based on their core dysfunctions rather than clinically established categories, treatment (non)response, including its biobehavioral substrates, can also be seen as a (dys)function that may cut across different diagnostic entities. It also is largely unresolved whether the nature of treatment (non)response is a general, overarching feature shared by different treatment approaches, such as pharmacotherapy, CBT, or psychodynamic treatments, or if those processes underlying treatment response are, rather, treatment- and disorder-specific. In line, the extent to which pretreatment markers predetermining a patient's likelihood to respond could be generalized across different disorders is still elusive. In order to answer this question, and in line with the RDoC idea, studies should ideally focus on treatment response as a key process of interest that is investigated from a transdiagnostic perspective, and that includes different forms of treatments—encompassing multiple levels of analysis, such as molecular-genetic, neural systems, psychological processes and environmental factors.

2.1 (Epi)genetic markers

In a recent review by our group, we identified 27 studies based on 12 independent samples that investigated therapygenetic effects for both pharmacological, as well as psychological therapies for anxiety disorders (see Lueken et al., 2016 for a complete list of references). Among 18 studies, the *5-HTT*LPR/rs25531 variant was the most often studied genetic marker. Although the highest number of studies with a positive effect was found for this variant, evidence for the predictive value of the serotonin transporter gene remained mixed, with 8 studies showing a positive effect, while 10 studies reported a null finding. Since then, another replication study on the putative therapygenetic effects of this polymorphism also failed to show a positive association (Lester et al., 2016). It is noteworthy that a recent meta-analysis (Culverhouse et al., 2017) testing the hypothesis about genome-environment (G × E) interactions of the serotonin transporter gene in the etiopathogenesis of affective disorders, a hypothesis originally posed by Caspi et al. (2003), was not able to find evidence for a strong G × E interaction contributing to the development of depression.

Based on these candidate gene studies on the serotonin transporter genotype, we may draw as a preliminary conclusion that the putative therapygenetic effects of singular candidate genes are either small-scale, such that individual studies so far were underpowered to find an association, or that other factors, such as gene-gene interactions, epigenetics, or "noise" due to heterogeneous study quality may contribute to the missing therapygenetic effect of the *5-HTT*LPR polymorphism so far. In line with Culverhouse et al. (2017), sufficiently powered meta-analytic approaches could be highly valuable in shedding further light on the genetic properties of this candidate gene relevant to treatment response. While a direct therapygenetic effect has yet failed to show clear evidence, imaging genetics studies, however, reported that the *5-HTT*LPR polymorphism acts upon brain structure and function in those circuits associated with emotion regulation (e.g., anterior cingulate cortex (ACC)-amygdala connectivity) in healthy and clinical populations (Heinz et al., 2005; Lueken et al., 2015; Pezawas et al., 2005), thus serving as promising intermediate phenotypes on the level of neural structure and functioning.

Other serotonin-related gene variants, such as *TPH2* G703T, *TPH* A218C, *HTR1A*-1019C/G, *HTR2A* (several variants), and *MAOA*-uVNTR, as identified in the previous review (Lueken et al., 2016), were investigated in eight studies, with six studies showing a positive effect. Regarding the latter, allelic variation in the *MAO-A* gene has been associated with treatment response in a sample of PD/AG patients (Reif et al., 2014). Providing translational evidence about the putative mechanisms underlying this therapygenetic effect using neurofunctional, autonomic, and behavioral readouts, it was shown that *MAO-A* risk allele carriers showed stronger elevated heart rates and increased fear during an standardized anxiety-provoking situation (behavioral avoidance task), did not habituate to the situation when repeatedly exposed, and showed less discriminatory learning during fear conditioning on the level of ACC activation, a signature that has been previously associated with treatment nonresponse in this sample (Lueken et al., 2013).

Beyond serotonergic neurotransmission, only singular studies are available, making it difficult to draw conclusions on their potential as a predictive biomarker (see Lueken et al., 2016 for details). Two recent studies additionally showed either no association of genes regulating HPA axis functioning with treatment outcome, or some limited evidence for genetic variation in the endocannabinoid system to explain individual differences in response to CBT (Lester et al., 2017; Roberts et al., 2015). Replicated evidence must be obtained before further conclusions can be drawn.

In contrast to these candidate gene studies that are prone to an a priori selection bias, a first hypotheses-free genome-wide analysis study (GWAS) has recently been published; it suggests, however, no common variants of very high effect underlying response to CBT (Coleman et al., 2016). As a limiting factor, the smaller sample sizes of treatment studies, typically averaging at some hundreds of patients, is not comparable to sample sizes usually achieved in GWASs, thus possibly resulting in underpowered analyses. Again, meta-analytic approaches transcending the limitations of singular studies are highly warranted.

Epigenetic processes are increasingly investigated as a possible mechanism explaining the "hidden heritability" (Manolio et al., 2009) that describes the gap between expected heritability rates based on family and twin studies (which is estimated between 30%–67%; see Domschke & Deckert, 2012) and the limited amount of variance that is actually explained by identified genetic risk variants. Epigenetic mechanisms such as DNA methylation, histone modifications, or noncoding RNAs modulate gene expression without changing the DNA sequence itself. DNA methylation as the most prominent candidate under investigation regulates gene expression by recruiting proteins involved in gene repression, or by inhibiting the binding of transcription factor(s) to DNA (Moore, Le, & Fan, 2013). As outlined by Schiele and Domschke (2017), epigenetics are a prime candidate to confer G × E interactions in anxiety disorders by acting as a possible "hinge" between them, as they are temporally highly dynamic and sensitive to environmental influences. In this sense, positive, as well as aversive life events may alter epigenetic signatures, rendering vulnerable subjects prone to the development of an anxiety disorder. Furthermore, treatments can act as important environmental influences with the potential to reverse altered methylation patterns. First evidence is available that anxiety disordered patients show a hypomethylation of gene variants that are perceived as "risk genes" for a given disorder (e.g., MAO-A hypomethylation in PD; see Domschke et al., 2012; Ziegler et al., 2016). Also, these epigenetic patterns seem to be modifiable by CBT with differential changes in responders vs nonresponders (Roberts et al., 2015, 2014; Ziegler et al., 2016). However, studies on epigenetic signatures serving as pretreatment biomarkers for treatment outcome in anxiety disorders are missing thus far, although first evidence in depression is available that hypomethylation of the serotonin transporter and *MAO-A* gene predicts response to pharmacotherapeutic interventions (Domschke et al., 2014, 2015).

2.2 Neuroimaging markers

Based on our recent review, we identified 17 studies investigating neuroimaging-based markers of treatment response (see Lueken et al., 2016 for a full list of references). The evaluation of these markers is limited by the heterogeneity of assessment (from task-based regional to structural or intrinsic connectivity measures) and analysis methods. Regarding the latter, hypothesis-driven region-of-interest (ROI) approaches are difficult to compare to more exploratory whole-brain approaches, because they may induce an a priori selection bias regarding particular brain regions (e.g., amygdala). Taking the analysis approach into account, the most compelling evidence was found for the ACC function to serve as a predictive biomarker, with eight studies (47.1%) reporting a positive effect, and three studies (17.7%) showing a null finding, while only two studies (11.8%) did not explicitly investigate ACC function (ROI analyses excluding the ACC were counted as not having studied this brain region). Furthermore, predictive information could be obtained from different regions subsumed within the temporal lobe (seven [41.2%] studies with a positive finding, vs two [11.8%] studies with a null finding, vs four [23.5%] studies [including ROI approaches] not investigating temporal lobe structures). For the hippocampus and frontal lobe, an equal number of studies favored a positive finding vs reporting a null effect, while for the insula, amygdala, medial prefrontal cortex (mPFC)/orbitofrontal cortex (OFC), dorsolateral (dl) PFC, parietal lobe, and occipital lobe, the number of studies with a null finding was higher than those reporting a positive effect (see Fig. 3 for a summary). Meanwhile, two

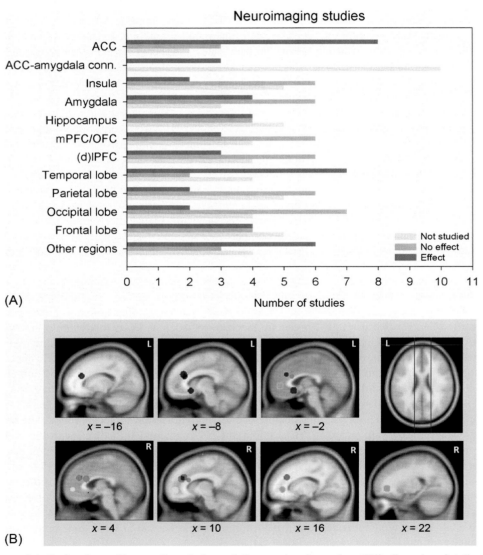

FIG. 3   (A) Frequency of studies favoring positive vs null results for predictive neuroimaging markers. "Effect": number of studies that reported a relationship of this marker with treatment response; "no effect": number of studies that did not find a relationship with treatment response; "not studied": number of studies that did not analyze this marker. Please note: if a region was not analyzed in region-of-interest studies, it was coded as "not studied"; if a region was not detected in a whole-brain analysis, it was coded as "no effect." (B) Mapping of anterior cingulate cortex coordinates related to treatment response. If available, coordinates were derived from the respective studies and an 8 mm sphere was drawn. *ACC*, anterior cingulate cortex. *(Modified from Lueken, U., Zierhut, K. C., Hahn, T., Straube, B., Kircher, T., Reif, A., et al. (2016). Neurobiological markers predicting treatment response in anxiety disorders: A systematic review and implications for clinical application.* Neuroscience and Biobehavioral Reviews, 66, *143–162.)*

further studies support ACC function as a predictor of treatment outcome in youth anxiety disorders for both pharmacological, as well as CBT treatments (Burkhouse et al., 2017), and in SAD as a specific predictor for CBT, while response toward acceptance and commitment therapy was predicted by posterior insula activation (Burklund, Torre, Lieberman, Taylor, & Craske, 2017).

Results so far indicate that brain regions most commonly associated with threat processing and the pathophysiology of anxiety disorders (e.g., amygdala, insula, [dorsal] ACC and PFC/OFC cortex) (Etkin & Wager, 2007; Shin & Liberzon, 2010) are only partly overlapping with those areas that bear predictive potential for treatment outcome. For example, the amygdala and the insular cortex (although investigated by virtually all studies) did not show pronounced evidence as a useful predictive marker of symptom reduction. On the contrary, repeated evidence points toward the ACC (depending on study, in the dorsal (d) ACC vs rostral (r) or subgenual (sg) parts, i.e., BA 25), followed by the temporal lobe.

Related studies from the field of depression corroborate the relevance of the ACC as an overarching marker of treatment response (Dunlop, Rajendra, et al., 2017; Pizzagalli, 2011; Siegle et al., 2012). Being involved in neural circuits and networks that are associated with salience processing and negative affectivity (Williams, 2016), the ACC may represent a critical hub within the default mode network and associated self-referential processes, possibly rebalancing different resting-state (rs) networks (Pizzagalli, 2011). Supporting previous reports on sgACC activation change as a function of treatment choice (Kennedy et al., 2007), recent findings from the PReDICT study indicate that sgACC connectivity may serve as a prescriptive marker for treatment response toward CBT vs antidepressant medication, thus offering potential to support decisions about the first-line treatment choice for major depression (Dunlop, Rajendra, et al., 2017). The direction of effects seems, however, to depend on the neurocognitive probe of interest (e.g., task effects), type of treatment, and location within the ACC.

In conclusion, preliminary evidence indicates that function of the ACC and its connectivity to limbic structures such as the amygdala may represent a critical hub moderating a common mechanism of treatment response crosscutting different internalizing disorders (Dunlop, Rajendra, et al., 2017; Lueken et al., 2016; Pizzagalli, 2011). However, most studies to date have limited power due to small sample sizes and, due to their predominant post hoc nature, show a limited overlap of employed tasks, assessment protocols, and associated findings. As this compromises any direct comparisons, conclusions regarding the common and distinct signatures of (non)response for different disorders that are treated with CBT are limited. Better standardization of neuroimaging tasks and methods, as well as a transdiagnostic focus, should be a prime priority for future studies in order to improve comparability and clinical usability.

2.3 Single-case predictions

Thus far, the vast majority of studies on neurobiological markers characterizing treatment response has been conducted solely on a group inference level. While these studies are informative about potential mechanisms underlying treatment (non)response, they are inherently limited regarding their translational value for predictions on the single-case level (see Figs. 1 and 2). From the perspective of personalized medicine, however, research on indictors of treatment response necessarily have to bridge the translational gap from group-based research to the individual patient level. As outlined earlier, machine learning represents a valuable and promising new tool for transcending the group interference level to single-case predictions. The multivariate nature of neurobiological measures, such as neuroimaging, seems to be particularly well suited to serve as input data for multidimensional algorithms. There is yet preliminary, but increasing, evidence from neuroimaging studies that treatment response in anxiety disorders can be predicted with sufficient accuracy between 79% and 92% using different paradigms, measurements, and multivariate algorithms (Ball, Stein, Ramsawh, Campbell-Sills, & Paulus, 2014; Doehrmann et al., 2013; Hahn et al., 2015; Mansson et al., 2015; Whitfield-Gabrieli et al., 2015). Of note, neuroimaging-based prediction accuracies appear to outperform predictions based on clinical data only (Ball et al., 2014; Doehrmann et al., 2013; Whitfield-Gabrieli et al., 2015), although it has to be noted that the multivariate nature of neuroimaging vs clinical data sets may significantly differ, thus still limiting direct comparisons about the predictive power of these different domains from a methodological point of view. While these initial results crucially call for independent replications, it has to be noted that single-case prediction approaches do not automatically yield promising results.

As the use of machine learning algorithms in the field of mental health "predictomics" is still in its infancy, the following paragraph will give an executive summary of the current status quo regarding its methodological development and future challenges awaiting this field of research.

3 Toward personalized clinical decision support systems: Current status quo and future challenges

As evidenced by a plethora of research articles, reviews, and opinion pieces in leading psychiatric journals in recent years, researchers are starting to shift focus away from solely group-based, classic statistical inference toward analysis strategies capable of single-subject prediction, realizing that—while the two approaches are by no means opposites, but regularly supplement each other (cf. Fig. 2)—direct clinical utility of any research finding will spring, first and foremost, from predictive technology.

3.1 Model construction and validation: The relevance of real-life, heterogeneous patient samples

To better understand this emerging field, including its opportunities as well as its risks, we first need to understand the notion of model construction and validation in predictive modeling. While in classic statistical inference, model parameters are estimated with the goal of determining their probability under a given hypothesis and making specific assumptions regarding the process that generated the data, predictive approaches estimate model parameters aiming to minimize a given error (i.e., a loss function), and test the thus trained model on new, previously unseen data. Concretely, while we would use classic statistical approaches to tell us how likely a model or its parameters are given a hypothesis (thereby enabling us to "test" said hypothesis), we would use a predictive approach to estimate model parameters for the purpose of predicting new data. Thus, we could use the very same method—for example, a simple linear regression model—to perform classic statistical inference (for example, determining whether a predictor in the model explains an amount of variance exceeding what we would expect by random chance alone), as well as predictive modeling (i.e., predicting new data using the regression line determined using our training data). For all clinical purposes, the crucial difference, thus, does not necessarily lie in the models themselves, but in the way they are employed. Briefly, what sets apart predictive approaches is their ability to train (i.e., estimate their parameters) on a dataset and generalize to novel data. This is mirrored by the fact that, for any predictive project, we will always need at least two datasets: one to train on, and one with which to test our model. This has far reaching implications for the way in which we draw patient samples in clinical studies: While we traditionally aim to gather data from well-defined, homogeneous patient groups (decreasing noise and other sources of undesired variance to allow us to detect statistical effects), building a practically useful predictive model requires us to draw real-life, heterogeneous patient samples. It is easy to understand that we cannot build a working model predicting patient characteristics if we consider only a subgroup of patients we see in our daily practice. However, current recruiting practice—more often than not—does exactly that. While decreasing noise variance and increasing statistical power in classic statistical inference, this practice crucially limits the scope and utility of a predictive model. In fact, only a heterogeneous, real-life sample of patients can give us a true estimate of the model's performance in day-to-day clinical work. To avoid bias in our model performance estimates, we need to start recruiting a sample of all patients under all conditions and all circumstances (Sundermann et al., 2017).

3.2 Novel approaches and technologies provided by machine learning and artificial intelligence

As mentioned herein, any statistical model could be used for predictive purposes. In practice, however, the sheer number of variables, as well as the complexity of the predictive rule we aim to infer often compel us to move toward more sophisticated methodology. Given that in psychiatry, we cannot expect a single variable—be it a gene, a voxel in the brain, or a psychometric score—to provide a meaningful amount of information with regard to any given clinically relevant question, we additionally need to consider the multivariate patterns present in our data. To this end, the fields of machine learning and artificial intelligence provide us with a wealth of approaches and technologies suitable for virtually all types of data and data structures (Goodfellow, Bengio, & Courville, 2016; Hastie, Tibshirani, & Friedman, 2001). The often challenging mathematics and the at least rudimentary programming skills needed to perform state-of-the-art analyses have led to an unfortunate focus, not only in psychiatry, on extremely few machine learning algorithms. For example, a recent review of neuroimaging studies employing predictive technology revealed that of the 173 publications listed, 126 used the same approach—namely the well-known Support Vector Machine (Arbabshirani, Plis, Sui, & Calhoun, 2017). This is particularly disturbing as the No-Free-Lunch-Theorem unambiguously states that if an algorithm performs well on a certain class of problems, then it necessarily pays for that with degraded performance on the set of all remaining problems (Wolpert, 2001). With freely available software such as Tensorflow (tensorflow.org) and scikit learn (scikit-learn.org), as well as with high-level packages such as Keras (keras.io) and PHOTON (photon-ai.com), however, the use of even the most sophisticated machine learning approaches is within reach already today. Thus, we are confident that the quality of machine learning analysis pipelines and strategies will substantially increase in the near future due to novel tools and a heightened interest in the field. Nonetheless, numerous data analytic challenges in psychiatry still await a proper solution (for a more in-depth view on the field, see Hahn et al., 2017). In short, while techniques for domain-knowledge based feature engineering might be successful for many applications in the short-term, automatic feature construction using Deep Learning approaches appear most promising in the mid-term—especially considering the growing amount of high-quality data from multicenter projects in all areas of psychiatry (see Heinig et al., 2017 for the German research initiative PROTECT-AD).

3.3 Sharing pretrained models rather than original data

While hugely important, advances in software platforms and growing data sharing initiatives alone will not suffice. Especially when successful models have been discovered, the need for rigorous replication will require that the models nearing clinical application be available for all researchers and clinicians to evaluate independently. Thus, sharing pretrained models is crucial. Only when model access is sufficiently simple so that everyone in the research community can upload data and receive results to evaluate on their own patients, will we be able to create models with the generality and robustness necessary for clinical everyday use. First efforts to establish model-sharing platforms such as the PHOTON Repository (photon-ai.com/repo) have been initiated.

3.4 Legal issues, data privacy, security, and ownership associated with predictive analytics

With the wide availability of pretrained and, most importantly, validated models, numerous legal issues will arise: For example, it is currently unclear who could be held responsible for potential errors a model might make. Whereas the programmer or the company employing him could commit to certain procedures for quality assurance, it is at the heart of a machine learning model that it is not programmed, but trained on data. Thus, there simply is no classic software development process that could be evaluated and controlled by the proper authorities. Moreover, even perfectly suited model architectures will be outright useless if trained on insufficient data (cf. the problem of limited model scope arising from homogeneous patient samples discussed herein). In this regard, clear regulation and certification procedures for medical artificial intelligence products are needed; preferably based on international consensus. Taking the perspective of the individual patient, questions of data privacy, security, and ownership come into focus. While privacy is a much debated topic, and security is already on every clinic's agenda, the importance of clarifying data ownership is fatally underestimated today. Given that our data is currently flowing to companies more or less freely via our smartphones, browsers, or online payment systems to name but a few, regulations in science are much more stringent. No matter what the scenario, the individual either explicitly allows others to use his or her data, or is more or less swiftly coerced by convenience and opaque terms and conditions. While this is—for the most part—perfectly legal, individuals who contributed the data do not participate in the profits at all. This seems particularly problematic, as artificial intelligence models, as outlined herein, are not programmed by a software developer, but learn directly from individual data. With large-scale patient data generating much, or even most of the added value of the products we will see in medicine and elsewhere in the future, there is currently no system that could compensate individuals in any way. We believe that a broad debate in societies around the world is necessary to ensure a balanced solution benefiting developing companies, as well as individuals providing the data. Technological solutions for individual compensation have recently started to emerge based on smart contracts in the block chain using, for example, IOTA (iota.org), but are still in their infancy.

3.5 Turning robust, validated biomarker models into clinical decision support systems

Even with validated, large-scale models and all technological, legal, and ethical issues resolved, we will then face the final challenge of implementing the new tools into clinical practice. Generally, this step of turning robust, already working biomarker models into clinical decision support systems seamlessly integrated with the workflow of health professionals will be crucial when it comes to utility in daily practice, change management during the transition process toward artificial intelligence driven medicine, and acceptance of the new technology. On one hand, clinics, as well as funding agencies and governing bodies, will need to foster Big Data infrastructure projects, which fully digitalize patient data and make it directly available to clinicians and for model training. To this end, cloud-based systems with fast response times seem most appropriate to ensure the highest data security standards, as well as scalable infrastructure. On the other hand, any Big Data effort requires at least an equal amount of technological development and prior planning to ensure that the wealth of data gathered is indeed helpful to the clinician in the field. To take this step of turning data into valuable information able to guide and improve clinical decision making, it is crucial to design digital assistance systems on top of the biomarker models described. In the simplest case, such a system might provide a short summary of the patient data in textual form or graphs displaying trends for symptoms or parameters at a glance. More sophisticated systems will not only summarize data, but also predict clinical outcomes under different circumstances and scenarios, providing quantitative information (e.g., in the form of response probabilities for a certain intervention) to the clinician. Furthermore, digital assistant systems may not only learn from patient data, but also be equipped to learn behavioral and cognitive patterns of health professionals, pointing out, for example, a consistent bias in diagnoses assigned over the past 6 months, or directing attention toward information inconsistent with the current decision. Finally, interaction with such assistants might happen via classic graphical user interfaces, or using more sophisticated, for example, immersive three-dimensional rendering technology. Given today's clinical workflows, assistants able to engage in natural language conversation with whom patients could be discussed in much the same way as we would consult a colleague today seem most intriguing.

4 Conclusions and outlook

Anxiety disorders pose an enormous burden with devastating effects for affected individuals and societies (Gustavsson et al., 2011; Whiteford et al., 2013). The good news is, as evidenced by international guidelines and expert consensus (Smits & Hofmann, 2009), effective treatments based on psycho- and pharmacotherapy are available; access is, however, limited, and patients—if treated at all—usually receive delayed treatments years after disorder onset, when conditions become chronic and complicated by psychosocial consequences, and comorbid disorders such as depression or substance use. While the dissemination of evidence-based treatments is targeted by other domains such as public health, clinical research within the past few years has been heavily influenced by the upcoming paradigm of personalized medicine: finding out which treatments or combinations thereof fit best to the individual (e.g., its psychosocial, symptomatic, and neurobiological characteristics).

Generally, valid predictive models would be instrumental, both for minimizing patient suffering, and for maximizing the efficient allocation of resources. Specifically, the prediction of therapeutic response can support the selection of optimized interventions through comparative effectiveness research, thereby improving the trial-and-error-based approach common in psychiatry. This maximizes adherence and minimizes undesired side effects. Importantly, it also allows clinicians to focus resources on patients who will most likely not benefit from the planned treatment. Of greatest scientific importance is that fact that identifying treatment-resistant individuals with high accuracy would also simplify the development and evaluation of novel drugs and interventions, as research efforts could be custom-tailored to homogeneous subpopulations. Put simply, an intervention tested using today's group-statistical approach will be deemed effective if, and only if, it has an effect on most participants and/or exerts very large effects on a minority. Identifying nonresponders to first-line treatment, however, would enable the development of interventions for this subgroup, making potential beneficial effects statistically detectable for the first time.

While the implementation of artificial intelligence-based clinical decision support systems might, in some cases, require, for example, neuroimaging and/or genome-wide data (eliciting costs of ~500€/patient today), such methods would already pay off if an inpatient stay could be shortened by just 3 days. Thus, even predictive models with moderate performance—while not substantially improving the situation of the individual patient—could be hugely beneficial from a health economic point of view.

Neurobiological research is now making increasing efforts in identifying the neural pathways and mechanisms by which treatment response may be conferred in order to identify novel and innovative treatments that can be offered to those patients not responding to first-line treatments. In addition, the rapid development, but also challenges posed by new technologies and "Big Data" that can be applied to the mental health sector will, without doubt, impact clinical research strategies and foster clinical applicability of neuroscience-informed research to further bridge the translational gap between clinical research and practice. Above and beyond the further professionalization of large-scale artificial intelligence approaches to mental health issues, numerous novel challenges arise with regard to the dissemination and regulation of the new technologies. First, validated predictive models need to be made available to all. While seemingly simple, this will require a massive data infrastructure and computational capabilities to serve potentially thousands of hospitals and millions of patients every day. Second, we need to ensure the safety and security of these models. For example, we need to ensure the model used by a clinician has neither been compromised—whether by technical failure or by intent—nor altered in a way that could violate patients' rights, for example, regarding data confidentiality. While current efforts rooted in cryptography and block chain technology certainly have the potential to solve these issues, scalable, industry-grade implementations of these technologies seem to be several years in the future. Third, predictive and machine learning technology is largely unregulated at this point. Moreover, even a consistent strategy toward nationally or internationally harmonized standards is not in sight. To this end, technological innovators, legislators, and patient representatives need to initiate an open discussion regarding such regulations. Finally, while it is becoming clear that artificial intelligence technology will change daily practice in all areas of medicine more profoundly than even the most enthusiastic futurologists would have predicted 10 years ago, researchers must communicate the fact that the next generation of artificial intelligence models will provide tools, not solutions. Thus, such tools are not meant to substitute for clinicians, but will become part of integrated decision support systems augmenting health professionals helping them help their patients.

References

American Psychiatric Association. (2013). *Diagnostic and statistical manual of mental disorders (DSM-5®)* (4th ed.). Arlington, VA: American Psychiatric Publishing.

Arbabshirani, M. R., Plis, S., Sui, J., & Calhoun, V. D. (2017). Single subject prediction of brain disorders in neuroimaging: Promises and pitfalls. *NeuroImage, 145*(Pt. B), 137–165.

Ball, T. M., Stein, M. B., Ramsawh, H. J., Campbell-Sills, L., & Paulus, M. P. (2014). Single-subject anxiety treatment outcome prediction using functional neuroimaging. *Neuropsychopharmacology, 39*(5), 1254–1261.

Bandelow, B., Wiltink, J., Alpers, G. W., Benecke, C., Deckert, J., Eckhardt-Henn, A., … Beutel, M. E. (2014). *Deutsche S3-Leitlinie Behandlung von Angststörungen*. Retrieved from www.awmf.org/leitlinien.html.

Beesdo-Baum, K., & Knappe, S. (2012). Developmental epidemiology of anxiety disorders. *Child and Adolescent Psychiatric Clinics of North America, 21*(3), 457–478.

Burkhouse, K. L., Kujawa, A., Klumpp, H., Fitzgerald, K. D., Monk, C. S., & Phan, K. L. (2017). Neural correlates of explicit and implicit emotion processing in relation to treatment response in pediatric anxiety. *Journal of Child Psychology and Psychiatry, 58*(5), 546–554.

Burklund, L. J., Torre, J. B., Lieberman, M. D., Taylor, S. E., & Craske, M. G. (2017). Neural responses to social threat and predictors of cognitive behavioral therapy and acceptance and commitment therapy in social anxiety disorder. *Psychiatry Research, 261*, 52–64.

Caspi, A., Sugden, K., Moffitt, T. E., Taylor, A., Craig, I. W., Harrington, H., … Poulton, R. (2003). Influence of life stress on depression: Moderation by a polymorphism in the 5-HTT gene. *Science, 301*(5631), 386–389.

Coleman, J. R., Lester, K. J., Keers, R., Roberts, S., Curtis, C., Arendt, K., … Eley, T. C. (2016). Genome-wide association study of response to cognitive-behavioural therapy in children with anxiety disorders. *The British Journal of Psychiatry, 209*(3), 236–243.

Culverhouse, R. C., Saccone, N. L., Horton, A. C., Ma, Y., Anstey, K. J., Banaschewski, T., … Bierut, L. J. (2017). Collaborative meta-analysis finds no evidence of a strong interaction between stress and 5-HTTLPR genotype contributing to the development of depression. *Molecular Psychiatry, 23*(1), 133–142.

De Gruttola, V. G., Clax, P., DeMets, D. L., Downing, G. J., Ellenberg, S. S., Friedman, L., … Zeger, S. L. (2001). Considerations in the evaluation of surrogate endpoints in clinical trials. Summary of a National Institutes of Health workshop. *Controlled Clinical Trials, 22*(5), 485–502.

Doehrmann, O., Ghosh, S. S., Polli, F. E., Reynolds, G. O., Horn, F., Keshavan, A., … Gabrieli, J. D. (2013). Predicting treatment response in social anxiety disorder from functional magnetic resonance imaging. *JAMA Psychiatry, 70*(1), 87–97.

Domschke, K., & Deckert, J. (2012). Genetics of anxiety disorders—Status quo and quo vadis. *Current Pharmaceutical Design, 18*(35), 5691–5698.

Domschke, K., Tidow, N., Kuithan, H., Schwarte, K., Klauke, B., Ambree, O., … Deckert, J. (2012). Monoamine oxidase A gene DNA hypomethylation—A risk factor for panic disorder? *The International Journal of Neuropsychopharmacology, 15*(9), 1217–1228.

Domschke, K., Tidow, N., Schwarte, K., Deckert, J., Lesch, K.-P., Arolt, V., … Baune, B. T. (2014). Serotonin transporter gene hypomethylation predicts impaired antidepressant treatment response. *The International Journal of Neuropsychopharmacology, 17*, 1167–1176.

Domschke, K., Tidow, N., Schwarte, K., Ziegler, C., Lesch, K. P., Deckert, J., … Baune, B. T. (2015). Pharmacoepigenetics of depression: No major influence of MAO-A DNA methylation on treatment response. *Journal of Neural Transmission, 122*(1), 99–108.

Doyle, O. M., Mehta, M. A., & Brammer, M. J. (2015). The role of machine learning in neuroimaging for drug discovery and development. *Psychopharmacology, 232*, 4179–4189.

Dunlop, B. W., Kelley, M. E., Aponte-Rivera, V., Mletzko-Crowe, T., Kinkead, B., Ritchie, J. C., … Mayberg, H. S. (2017). Effects of patient preferences on outcomes in the Predictors of Remission in Depression to Individual and Combined Treatments (PReDICT) study. *The American Journal of Psychiatry, 174*(6), 546–556.

Dunlop, B. W., Rajendra, J. K., Craighead, W. E., Kelley, M. E., McGrath, C. L., Choi, K. S., … Mayberg, H. S. (2017). Functional connectivity of the subcallosal cingulate cortex and differential outcomes to treatment with cognitive-behavioral therapy or antidepressant medication for major depressive disorder. *The American Journal of Psychiatry, 174*(6), 533–545.

Etkin, A., & Wager, T. D. (2007). Functional neuroimaging of anxiety: A meta-analysis of emotional processing in PTSD, social anxiety disorder, and specific phobia. *The American Journal of Psychiatry, 164*(10), 1476–1488.

Fernandez, E., Salem, D., Swift, J. K., & Ramtahal, N. (2015). Meta-analysis of dropout from cognitive behavioral therapy: Magnitude, timing, and moderators. *Journal of Consulting and Clinical Psychology, 83*(6), 1108–1122.

Goodfellow, I., Bengio, Y., & Courville, A. (2016). *Deep learning.* Cambridge, MA: MIT Press.

Gottschalk, M. G., & Domschke, K. (2016). Novel developments in genetic and epigenetic mechanisms of anxiety. *Current Opinion in Psychiatry, 29*(1), 32–38.

Gustavsson, A., Svensson, M., Jacobi, F., Allgulander, C., Alonso, J., Beghi, E., … Olesen, J. (2011). Cost of disorders of the brain in Europe 2010. *European Neuropsychopharmacology, 21*(10), 718–779.

Hahn, T., Kircher, T., Straube, B., Wittchen, H. U., Konrad, C., Strohle, A., ... Lueken, U. (2015). Predicting treatment response to cognitive behavioral therapy in panic disorder with agoraphobia by integrating local neural information. *JAMA Psychiatry, 72*(1), 68–74.

Hahn, T., Nierenberg, A. A., & Whitfield-Gabrieli, S. (2017). Predictive analytics in mental health: Applications, guidelines, challenges and perspectives. *Molecular Psychiatry, 22*(1), 37–43.

Hamburg, M. A., & Collins, F. S. (2010). The path to personalized medicine. *The New England Journal of Medicine, 363*(4), 301–304.

Hastie, T., Tibshirani, R., & Friedman, J. (2001). *The elements of statistical learning: Data mining, inference and prediction.* New York, NY: Springer-Verlag.

Heinig, I., Pittig, A., Richter, J., Hummel, K., Alt, I., Dickhover, K., ... Wittchen, H. U. (2017). Optimizing exposure-based CBT for anxiety disorders via enhanced extinction: Design and methods of a multicentre randomized clinical trial. *International Journal of Methods in Psychiatric Research, 26*(2) https://doi.org/10.1002/mpr.1560.

Heinz, A., Braus, D. F., Smolka, M. N., Wrase, J., Puls, I., Hermann, D., ... Buchel, C. (2005). Amygdala-prefrontal coupling depends on a genetic variation of the serotonin transporter. *Nature Neuroscience, 8*(1), 20–21.

Hofmann, S. G., Asnaani, A., Vonk, I. J. J., Sawyer, A. T., & Fang, A. (2012). The efficacy of cognitive behavioral therapy: A review of meta-analyses. *Cognitive Therapy and Research, 36*(5), 427–440.

Insel, T. R., Cuthbert, B. N., Garvey, M., Heinssen, R., Pine, D. S., Quinn, K., ... Wang, P. (2010). Research domain criteria (RDoC): Toward a new classification framework for research on mental disorders. *The American Journal of Psychiatry, 167*(7), 748–751.

Kennedy, S. H., Konarski, J. Z., Segal, Z. V., Lau, M. A., Bieling, P. J., McIntyre, R. S., & Mayberg, H. S. (2007). Differences in brain glucose metabolism between responders to CBT and venlafaxine in a 16-week randomized controlled trial. *The American Journal of Psychiatry, 164*(5), 778–788.

Lester, K. J., Coleman, J. R., Roberts, S., Keers, R., Breen, G., Bogels, S., ... Eley, T. C. (2017). Genetic variation in the endocannabinoid system and response to Cognitive Behavior Therapy for child anxiety disorders. *American Journal of Medical Genetics Part B, Neuropsychiatric Genetics, 174*(2), 144–155.

Lester, K. J., Roberts, S., Keers, R., Coleman, J. R., Breen, G., Wong, C. C., ... Eley, T. C. (2016). Non-replication of the association between 5HTTLPR and response to psychological therapy for child anxiety disorders. *The British Journal of Psychiatry, 208*(2), 182–188.

Lonsdorf, T. B., & Merz, C. J. (2017). More than just noise: Inter-individual differences in fear acquisition, extinction and return of fear in humans—Biological, experiential, temperamental factors, and methodological pitfalls. *Neuroscience and Biobehavioral Reviews, 80*, 703–728.

Lueken, U., & Hahn, T. (2016). Functional neuroimaging of psychotherapeutic processes in anxiety and depression: From mechanisms to predictions. *Current Opinion in Psychiatry, 29*, 25–31.

Lueken, U., & Hahn, T. (2017). Prädiktive Analytik: Neue Ansätze zur Entwicklung klinischer Vorhersage-Modelle am Beispiel der Angststörungen. *Neurotransmitter, 28*(9), 27–33.

Lueken, U., Straube, B., Konrad, C., Wittchen, H. U., Strohle, A., Wittmann, A., ... Kircher, T. (2013). Neural substrates of treatment response to cognitive-behavioral therapy in panic disorder with agoraphobia. *The American Journal of Psychiatry, 170*(11), 1345–1355.

Lueken, U., Straube, B., Wittchen, H. U., Konrad, C., Strohle, A., Wittmann, A., ... Reif, A. (2015). Therapygenetics: Anterior cingulate cortex-amygdala coupling is associated with *5-HTT*LPR and treatment response in panic disorder with agoraphobia. *Journal of Neural Transmission, 122*(1), 135–144.

Lueken, U., Zierhut, K. C., Hahn, T., Straube, B., Kircher, T., Reif, A., ... Domschke, K. (2016). Neurobiological markers predicting treatment response in anxiety disorders: A systematic review and implications for clinical application. *Neuroscience and Biobehavioral Reviews, 66*, 143–162.

Manolio, T. A., Collins, F. S., Cox, N. J., Goldstein, D. B., Hindorff, L. A., Hunter, D. J., ... Visscher, P. M. (2009). Finding the missing heritability of complex diseases. *Nature, 461*(7265), 747–753.

Mansson, K. N. T., Frick, A., Boraxbekk, C. J., Marquand, A. F., Williams, S. C. R., Carlbring, P., ... Furmark, T. (2015). Predicting long-term outcome of Internet-delivered cognitive behavior therapy for social anxiety disorder using fMRI and support vector machine learning. *Translational Psychiatry, 5*.

Meier, S. M., Petersen, L., Mattheisen, M., Mors, O., Mortensen, P. B., & Laursen, T. M. (2015). Secondary depression in severe anxiety disorders: A population-based cohort study in Denmark. *Lancet Psychiatry, 2*(6), 515–523.

Moore, L. D., Le, T., & Fan, G. (2013). DNA methylation and its basic function. *Neuropsychopharmacology, 38*(1), 23–38.

National Institute of Health and Care Excellence. (2011). *Generalised anxiety disorder and panic disorder in adults: Management [CG113].* Retrieved from. www.nice.org.uk/guidance/cg113.

Orru, G., Pettersson-Yeo, W., Marquand, A. F., Sartori, G., & Mechelli, A. (2012). Using Support Vector Machine to identify imaging biomarkers of neurological and psychiatric disease: A critical review. *Neuroscience and Biobehavioral Reviews, 36*(4), 1140–1152.

Pezawas, L., Meyer-Lindenberg, A., Drabant, E. M., Verchinski, B. A., Munoz, K. E., Kolachana, B. S., ... Weinberger, D. R. (2005). 5-HTTLPR polymorphism impacts human cingulate-amygdala interactions: A genetic susceptibility mechanism for depression. *Nature Neuroscience, 8*(6), 828–834.

Pizzagalli, D. A. (2011). Frontocingulate dysfunction in depression: Toward biomarkers of treatment response. *Neuropsychopharmacology, 36*(1), 183–206.

Reif, A., Richter, J., Straube, B., Hofler, M., Lueken, U., Gloster, A. T., ... Deckert, J. (2014). MAOA and mechanisms of panic disorder revisited: From bench to molecular psychotherapy. *Molecular Psychiatry, 19*(1), 122–128.

Richter, J., Pittig, A., Hollandt, M., & Lueken, U. (2017). Bridging the gaps between basic science and cognitive-behavioral treatments for anxiety disorders in routine care. *Zeitschrift für Psychologie, 225*(3), 252–267.

Roberts, S., Keers, R., Lester, K. J., Coleman, J. R., Breen, G., Arendt, K., ... Wong, C. C. (2015). HPA axis related genes and response to psychological therapies: Genetics and epigenetics. *Depression and Anxiety, 32*(12), 861–870.

Roberts, S., Lester, K. J., Hudson, J. L., Rapee, R. M., Creswell, C., Cooper, P. J., ... Eley, T. C. (2014). Serotonin transporter methylation and response to cognitive behaviour therapy in children with anxiety disorders. *Translational Psychiatry, 4*.

Savitz, J., Rauch, S., & Drevets, W. (2013). Clinical application of brain imaging for the diagnosis of mood disorders: The current state of play. *Molecular Psychiatry, 18*(5), 528–539.

Schiele, M. A., & Domschke, K. (2017). Epigenetics at the crossroads between genes, environment and resilience in anxiety disorders. *Genes, Brain, and Behavior, 17.* https://doi.org/10.1111/gbb.12423.

Shin, L. M., & Liberzon, I. (2010). The neurocircuitry of fear, stress, and anxiety disorders. *Neuropsychopharmacology, 35*(1), 169–191.

Siegle, G. J., Thompson, W. K., Collier, A., Berman, S. R., Feldmiller, J., Thase, M. E., & Friedman, E. S. (2012). Toward clinically useful neuroimaging in depression treatment: Prognostic utility of subgenual cingulate activity for determining depression outcome in cognitive therapy across studies, scanners, and patient characteristics. *Archives of General Psychiatry, 69*(9), 913–924.

Smits, J. A. J., & Hofmann, S. G. (2009). A meta-analytic review of the effects of psychotherapy control conditions for anxiety disorders. *Psychological Medicine, 39*(2), 229–239.

Sundermann, B., Bode, J., Lueken, U., Westphal, D., Gerlach, A. L., Straube, B., … Pfleiderer, B. (2017). Support vector machine analysis of functional magnetic resonance imaging of interoception does not reliably predict individual outcomes of cognitive behavioral therapy in panic disorder with agoraphobia. *Frontiers in Psychiatry, 8*, 99. https://doi.org/10.3389/fpsyt.2017.00099.

Taylor, S., Abramowitz, J. S., & McKay, D. (2012). Non-adherence and non-response in the treatment of anxiety disorders. *Journal of Anxiety Disorders, 26*(5), 583–589.

US Food and Drug Administration. (2014). *Qualification process for drug development tools.* Retrieved from http://www.fda.gov/downloads/drugs/guidancecomplianceregulatoryinformation/guidances/ucm230597.pdf.

Whiteford, H. A., Degenhardt, L., Rehm, J., Baxter, A. J., Ferrari, A. J., Erskine, H. E., … Vos, T. (2013). Global burden of disease attributable to mental and substance use disorders: Findings from the Global Burden of Disease Study 2010. *Lancet, 382*(9904), 1575–1586.

Whitfield-Gabrieli, S., Ghosh, S. S., Nieto-Castanon, A., Saygin, Z., Doehrmann, O., Chai, X. J., … Gabrieli, J. D. (2015). Brain connectomics predict response to treatment in social anxiety disorder. *Molecular Psychiatry, 21*(5), 680–685.

Williams, L. M. (2016). Precision psychiatry: A neural circuit taxonomy for depression and anxiety. *Lancet Psychiatry, 3*(5), 472–480.

Wittchen, H. U., Jacobi, F., Rehm, J., Gustavsson, A., Svensson, M., Jonsson, B., … Steinhausen, H. C. (2011). The size and burden of mental disorders and other disorders of the brain in Europe 2010. *European Neuropsychopharmacology, 21*(9), 655–679.

Wolpert, D. H. (2001). The supervised learning no-free-lunch theorems. In *Proceedings of the 6th online world conference on soft computing in industrial applications.*

Woo, C. W., Chang, L. J., Lindquist, M. A., & Wager, T. D. (2017). Building better biomarkers: Brain models in translational neuroimaging. *Nature Neuroscience, 20*(3), 365–377.

Ziegler, C., Richter, J., Mahr, M., Gajewska, A., Schiele, M. A., Gehrmann, A., … Domschke, K. (2016). MAO-A hypomethylation in panic disorder—Reversibility of an epigenetic risk pattern by psychotherapy. *Translational Psychiatry, 6.* https://doi.org/10.1038/tp.2016.41.

Zilverstand, A., Sorger, B., Sarkheil, P., & Goebel, R. (2015). fMRI neurofeedback facilitates anxiety regulation in females with spider phobia. *Frontiers in Behavioral Neuroscience, 9*, 148.

Zwanzger, P., Herrmann, M. J., & Baeken, C. (2016). "Torpedo" for the brain: Perspectives in neurostimulation. *Journal of Neural Transmission, 123*(10), 1119–1120.

The genetic architecture of bipolar disorder: Entering the road of discoveries

Olav B. Smeland[a,b], Andreas J. Forstner[c,d,e,f], Alexander Charney[g], Eli A. Stahl[g] and Ole A. Andreassen[a,b]

[a]NORMENT, KG Jebsen Centre for Psychosis Research, University of Oslo, Oslo, Norway; [b]Institute of Clinical Medicine, University of Oslo, Oslo, Norway; [c]Institute of Human Genetics, University of Bonn, School of Medicine & University Hospital Bonn, Bonn, Germany; [d]Department of Genomics, Life & Brain Center, University of Bonn, Bonn, Germany; [e]Human Genomics Research Group, Department of Biomedicine, University of Basel, Basel, Switzerland; [f]Department of Psychiatry (UPK), University of Basel, Basel, Switzerland; [g]Division of Psychiatric Genomics, Department of Genetics and Genomic Sciences, Institute for Genomics and Multiscale Biology, Icahn School of Medicine at Mount Sinai, New York, NY, United States

1 Introduction

Bipolar disorder (BD) is a mental disorder characterized by recurrent episodes of pathological mood alternating between mania, hypomania, and depression, and the core symptoms also include abnormal activity, thought, perception, and social behavior (Grande et al., 2016). BD was previously known as manic depressive illness, and it affects people irrespective of nationality, ethnicity, or socioeconomic status. With a lifetime prevalence of around 1%–2%, elevated morbidity and mortality, typical onset in young adulthood, and frequently a chronic course (Goodwin et al., 2016), BD is a major public health problem and a leading cause of the global burden of disease (Ferrari et al., 2016). It often leads to recurrent suicidality, impaired functional level and quality of life, with a high degree of comorbidity (Grande et al., 2016). The classification of BD includes two main clinical subtypes, bipolar I disorder (BD1) and bipolar II disorder (BD2). They are defined based on the presence of specific types of lifetime mood episodes. In BD1, manic episodes typically alternate with depressive episodes, while a diagnosis of BD2 is based on the occurrence of at least one depressive and one hypomanic episode during the lifetime, but no manic episode.

The treatment depends on the current state, and differs across hypomania, mania, depression, and normal mood (euthymia), which defines whether it requires acute treatment or prevention of new episodes. Mood stabilizers, including lithium and antipsychotics, are the main pharmacological treatment alternatives, while psychosocial interventions are also effective (Grande et al., 2016).

Several lines of evidence strongly suggest that BD is a multifactorial disorder (Lichtenstein et al., 2009), but the disease mechanisms remain mainly unknown, despite the great progress we have seen recent years in neurobiology, including stem cells (Soliman et al., 2017) and neuroimaging (Hibar et al., 2016, 2017). However, perhaps the most illuminating discoveries have been the genetic findings, which will be reviewed here.

2 Family studies

A large number of studies since 1960 have shown that there is an accumulation of BD in families (Smoller & Finn, 2003), and the risk of the disease is about 10 times higher in BD patients' first-degree relatives (siblings, children/parents, Table 1). Family members also have an increased risk of unipolar depression, about three times higher than relatives of healthy controls. There have been some studies of clinical subgroups, and there are some indications that BD1 and BD2 have somewhat different genetic heredity (Heun & Maier, 1993), but there are also findings that support overlapping genetic mechanisms (Smoller & Finn, 2003). Familiar risk of subgroups of BD or special disease characteristics has also been investigated, and there are indications of an increased family risk of early onset, psychosis, lithium response, frequency of episodes of BD, and suicidality (Barnett & Smoller, 2009; Craddock & Sklar, 2009). But accumulation of a disease in families can also be due to social conditions, and to distinguish this, twin and adoption studies must be performed.

Personalized Psychiatry. https://doi.org/10.1016/B978-0-12-813176-3.00018-3

TABLE 1 Genetic epidemiology.

Population risk	1–2%
Risk first-degree relatives	10%
Monozygotic twin concordance	40%–45%
Heritability	60%–80%
Risk unipolar depression relative	3 × vs healthy

3 Twin and adoption studies

The combination of many twin studies in BD has given concordance rates of 4.5%–5.6% in dizygotic and 39%–43% in monozygotic twins (Barnett & Smoller, 2009; Craddock & Sklar, 2009; Edvardsen et al., 2008; McGuffin et al., 2010). On average, this gives an estimated heritability, that is, the proportion of cause of disease due to genetic factors, of approximately 70%–80% (Table 1). Family-based heritability of BD calculated from Swedish registry data suggests lower heritability estimates, closer to 59% (Lichtenstein et al., 2009). Heritability indicates only the possibility of developing disease, and not the disease itself. Therefore, the concordance of single twins may be lower than estimated heritability. In addition, heritability estimates are based on constant environmental factors, so that it can only be transferred from one population to another where there are reasonably comparable environmental conditions. The significance of genetic disposition has also been demonstrated in adoption studies investigating patients with BD that were adopted in childhood, and found a higher degree of concordance with disease in the original biological family (inheritance) than in the adoptive family (environment) (Craddock & Sklar, 2009).

4 Molecular genetics

Family and twin studies have shown that heritability plays a major role in the etiology of BD, but it has been difficult to identify the specific disease genes that are the cause of increased vulnerability. However, this has changed dramatically in the past 10 years, driven by the foundation of international consortia, such as the Psychiatric Genomics Consortium (PGC), and the rapid development in genotyping technology, which has allowed the systematic investigation of >1,000,000 single-nucleotide polymorphisms (SNPs) in large cohorts of BD patients and controls.

4.1 Linkage studies

In short, linkage studies are investigating whether one or more specific loci of the genome are inherited together with a family's disease. For this purpose, genetic markers are used, which are specific DNA sequences at known chromosomal loci where genetic variation in the population exists, so that a unique genetic profile for each individual can be obtained. A number of linkage studies have been conducted in BD. The main conclusion of these studies is that there is no single gene with great effect that causes BD. Although there are several reports of significant linkage at loci 6q, 8q, 13q, and 22q (Barnett & Smoller, 2009), there is reasonable agreement that linkage studies have given little in terms of identifying the specific genetic mechanisms that cause the high heritability of BD. Therefore, although very recently, new family studies have used sequencing (see the following section), the focus in recent years has been on association studies.

4.2 Association studies

In association studies, different genetic variants are investigated to determine whether a specific genetic variant is associated with group-level disease. The most common approach is to study the frequency of variants in a group of patients compared with a control group, the so-called case-control design. Both groups must be unrelated. Association studies are effective in identifying susceptibility variants with low effect, and thus fit well for BD. Initially, the candidate gene approach was most popular, where the choice of genes to be investigated was based on hypotheses of disease cause. Typically, the number of cases and controls was rarely more than a few hundred, and studies focused on functional gene variants. Most commonly, candidate genes were chosen from predominant pathophysiological models of BD, especially the neurotransmitter systems dopamine, norepinephrine, and serotonin, as well as known mechanisms of action for mood stabilizing drugs. Several genes have been found to be associated with BD by this approach, such as monoamine oxidase A (Preisig et al., 2000), catechol *O*-methyltransferase (Funke et al., 2005), serotonin transporter (*SLC6A4*) (Anguelova, Benkelfat, & Turecki, 2003), and various "clock genes" (Nievergelt et al., 2006) involved in daytime rhythm regulation (Barnett & Smoller, 2009). However, it has been difficult to replicate these findings, and currently there is little support for any of these genes being associated with BD.

5   Genome-wide association studies

New advancements in genotyping technology have led to the development of high-capacity methods that make it possible to perform genome-wide association studies (GWAS). By such methods, hundreds of thousands of genetic variants (DNA polymorphisms) scattered throughout the genome can be genotyped at a reasonable cost using so-called "chip" technologies (Hirschhorn & Daly, 2005). The method provides a screening of the genetic variation of the individual, and a large proportion of a person's overall genetic variation is mapped. This makes it possible to hunt for variants located in or near all known genes, to investigate association with disease in a hypothesis-free manner. GWAS have great potential for examining genetic causes of complex disorders, such as BD, where it can be assumed that a large number of common variants with small effect contribute to disease etiology. Such GWAS have enabled major breakthroughs in biomedical knowledge, especially in the case of complex diseases (Visscher et al., 2017), and in June 2018, 3395 publications and 62,156 unique SNP-trait associations from GWAS have been reported (http://www.ebi.ac.uk/gwas). Because GWAS include a very large number of variables (recently several million genetic variants), it must be corrected for multiple tests in the statistical analyses, with the consequence that a large number of individuals are required in GWAS to gain sufficient statistical power. It is now common to define genetic associations as "genome-wide significant" only if they have a *P*-value of $<5 \times 10^{-8}$.

6   GWAS in BD

Recent GWAS in BD have identified a number of significant associations between disease status and common genetic variants. The decisive factor for these successes has been the large sample sizes achieved by international consortia such as the PGC (launched in the fall of 2010), in which GWAS datasets were merged and jointly analyzed with standardized methods (Budde et al., 2017). The first robust findings came from a meta-analysis that included 4387 BD cases and 6209 controls (Ferreira et al., 2008). In this study, Ferreira and colleagues reported associations between BD and common variants at the genes ankyrin 3 (*ANK3*) and calcium voltage-gated channel subunit alpha1 C (*CACNA1C*) (Ferreira et al., 2008). These findings have been replicated in subsequent studies (e.g., Anguelova et al., 2003).

The first large, collaborative BD GWAS by the PGC BD Working Group was comprised of 7481 BD patients and 9250 controls, and identified four genome-wide significant loci (Sklar et al., 2011). Three subsequent studies performed meta-analyses that included the PGC BD data (Baum et al., 2008a; Chen et al., 2013; Muhleisen et al., 2014), identifying five additional associated loci. Up to 2017, BD GWAS have reported a total of 20 genome-wide significant loci (Baum et al., 2008b; Charney et al., 2017; Chen et al., 2013; Cichon et al., 2011; Ferreira et al., 2008; Green et al., 2012, 2013; Hou et al., 2016a; Muhleisen et al., 2014; Schulze et al., 2009; Scott et al., 2009; Sklar et al., 2008, 2011; Smith et al., 2009; Wellcome Trust Case Control Consortium, 2007). Notably, the identification of each individual genetic variant promises a great benefit for the search of disease causes, because it points to a genomic site that contributes to the biology of BD (Budde et al., 2017). Subsequent analyses provided evidence that the association signals seem to converge into specific biological pathways. The implicated gene sets include calcium and glutamate signaling, microRNAs, and histone pathways, among others (Forstner et al., 2015; Sklar et al., 2011).

7   Genetic architecture of BD

Estimates of the proportion of variance attributable to common genome-wide SNPs (SNP heritability) indicate that approximately 1/4 of the heritability for BD is due to common genetic variants (Cross-Disorder Group of the Psychiatric Genomics Consortium et al., 2013). To date, only a small fraction of these variants has been identified. Based upon the experience from other human complex traits, it is expected that by increasing the sample sizes, many more loci contributing to the development of BD will be detected. GWAS results thus will continue to provide valuable insights into the biology of BD, and can form the basis for future exploration of the molecular pathomechanisms and potential drug targets.

7.1 Recent GWAS of BD

In the most recent GWAS, the PGC Bipolar Disorder Working Group reports results from their second GWAS of BD, including 20,352 cases and 31,358 controls of European descent in a single, systematic analysis (Fig. 1), with a follow-up sample of 9412 cases and 137,760 controls (Stahl et al., 2019). In total, SNPs in 30 loci achieved genome-wide significance ($P < 5 \times 10^{-8}$) in this combined analysis, including 18 loci that were not previously genome-wide significant (Table 2). The combined analysis identified three additional previously known loci as genome-wide significant, and a total of 20 novel BD risk loci. Previous BD GWAS have reported a total of 20 loci (Baum et al., 2008b; Charney et al., 2017; Chen et al., 2013; Cichon et al., 2011; Ferreira et al., 2008; Green et al., 2012, 2013; Hou et al., 2016a; Muhleisen et al., 2014; Schulze et al., 2009; Scott et al., 2009; Sklar et al., 2008, 2011; Smith et al., 2009; Wellcome Trust Case Control Consortium, 2007); 9 of which were not genome-wide significant in the most recent GWAS combined analysis.

7.2 Genetic overlap

Investigating genetic overlap between complex traits and disorders may offer insights into shared etiology, which can inform disease nosology, diagnostic practices, and drug development (Visscher et al., 2017). BD shows substantial clinical and genetic overlap with other psychiatric disorders (Cross-Disorder Group of the Psychiatric Genomics Consortium et al., 2013; Ruderfer et al., 2014). Using genome-wide genotype data, the PGC cross-disorders group systematically investigated the genetic correlation between BD and four other psychiatric disorders (schizophrenia [SCZ], major depressive disorder [MDD], autism spectrum disorders, and attention-deficit/hyperactivity disorder) (Hou et al., 2016a). Notably, BD showed the strongest genetic correlation with SCZ (0.68). In addition, a moderate correlation between BD and MDD (0.47) was found in this analysis, similar to that observed for SCZ and MDD (0.43).

SCZ and BD are two distinct diagnoses that share clinical characteristics and treatment options, and have been suggested to be part of the psychosis continuum, with prototypical SCZ on one end and BD on the other (Craddock & Owen, 2010). Several studies have investigated the genetic factors overlapping between the two disorders, enabled by the polygenic risk scores (International Schizophrenia Consortium et al., 2009) derived from GWAS. These findings provide further evidence for overlapping genetic risk (Hirschhorn & Daly, 2005), also across the psychosis continuum (Tesli et al., 2014). Interestingly, the shared SCZ-BD-associated SNPs seem to accumulate in particular pathways, such as calmodulin binding or calcium- and glutamate signaling (Forstner et al., 2017).

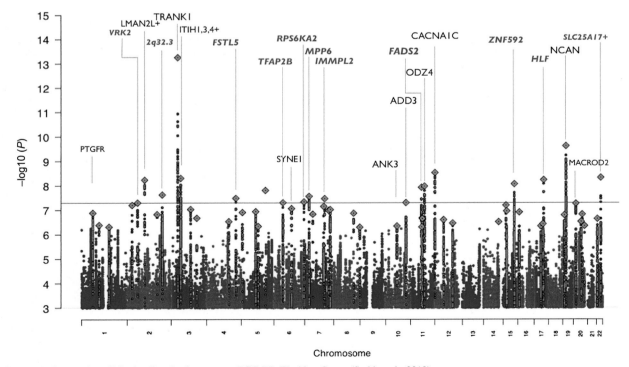

FIG. 1 Manhattan plot of bipolar disorder from recent PGC BD Working Group (Stahl et al., 2019).

TABLE 2 Genetic loci associated with bipolar disorder.

Locus	Lead SNP	Chr	OR	*P*-value	Nearest gene
1	rs7544145	1	1.085	4.8E−08	*PLEKHO1*
2	chr2_97376407_I	2	0.93	3.8E−09	*LMAN2L*
3	rs17183814	2	0.88	2.0E−09	*SCN2A*
4	chr2_194465711_D	2	0.93	7.9E−10	*PCGEM1*
5	rs9834970	3	0.93	5.7E−12	*TRANK1*
6	rs2302417	3	0.93	6.6E−11	*ITIH1*
7	rs3804640	3	1.065	2.0E−08	*CD47*
8	rs11724116	4	0.92	2.4E−08	*FSTL5*
9	chr5_7587236_D	5	0.92	1.5E−08	*ADCY2*
10	rs10035291	5	1.070	2.7E−08	*SSBP2*
11	chr6_72519394_D	6	1.064	3.5E−08	*RIMS1*
12	rs2388334	6	0.94	4.0E−09	*POU3F2*
13	rs10455979	6	0.94	4.3E−08	*RPS6KA2*
14	rs113779084	7	1.076	2.5E−09	*THSD7A*
15	rs73188321	7	0.92	1.1E−09	*SRPK2*
16	chr7_140700006_I	7	0.92	6.2E−10	*MRPS33*
17	rs10994318	10	1.145	6.8E−09	*ANK3*
18	chr10_111745562_I	10	1.090	1.2E−08	*ADD3*
19	rs12226877	11	1.085	9.9E−10	*FADS2*
20	rs10896090	11	1.084	1.9E−08	*PACS1*
21	rs7122539	11	0.94	3.8E−08	*PC*
22	rs12575685	11	1.073	7.7E−09	*SHANK2*
23	rs10744560	12	1.076	3.6E−10	*CACNA1C*
24	rs4447398	15	1.099	9.4E−09	*STARD9*
25	chr15_85357857_I	15	0.93	2.7E−08	*ALPK3*
26	rs11647445	16	0.93	1.1E−10	*GRIN2A*
27	rs112114764	17	0.93	2.5E−08	*HDAC5*
28	rs11557713	18	1.069	3.6E−08	*ZCCHC2*
29	rs111444407	19	1.097	1.3E−09	*NCAN*
30	chr20_43682549_I	20	0.929	1.1E−08	*STK4*

Genome-wide significant loci associated with bipolar disorder. Loci are ordered by genomic position. *Chr*, chromosome; *OR*, odds ratio; *SNP*, single-nucleotide polymorphism.
(From PGC BD Working group: Stahl, E., et al. (2019). Genome-wide association study identifies 30 loci associated with bipolar disorder. *Nat. Genet. 51(5), 793–803.*)

Many genome-wide significant BD loci also harbor SCZ associations, and conditional analyses in the most recent BD GWAS (Stahl et al., 2019) confirmed that the BD associations are most often not independent of SCZ associations, with exceptions including the NCAN and TRANK1 loci. This is in line with the high positive genetic correlation between the two diagnostic groups. Genetic overlap between mood disorders has also been a focus of great interest, given the clear overlap in clinical characteristics between unipolar depression and BD. The recent large GWAS (Wray et al., 2018) on 130,664 MDD cases and 330,470 controls provided better power to investigate genetic overlap, and revealed a significant genetic correlation of 0.32 with BD. Yet, there were no overlapping genome-wide significant loci between the BD and MDD GWAS (Wray et al., 2018), indicating that the overlapping genetic effects between BD and MDD may be too weak to be identified at the current GWAS sample size using standard tools.

Polygenic risk score prediction based on the latest BD GWAS reaches $R^2 \sim 8\%$, allowing the use of the polygenic risk scores in a variety of study designs. Polygenic scoring confirmed differential sharing of genetic risk between the common subtypes of BD with SCZ and MDD (Stahl et al., 2019). In particular, BD1 was genetically more closely related to SCZ, while BD2 was more closely related to MDD. This clarifies clinical hypotheses in a much-debated area. Further studies may be necessary to examine the genetic relationship between severe recurrent MDD, for which we did not have GWAS data available, and BD2.

Recently, a comprehensive analysis was performed to identify shared and disorder-specific symptoms based on genetic data consisting of 53,555 cases (20,129 BD, 33,426 SCZ) and 54,065 controls. Here, Ruderfer et al. identified 114 genome-wide significant loci implicating synaptic and neuronal pathways shared between the disorders, and discovered specific loci that distinguish between BD and SCZ, pointing to the utility of genetics to inform symptomology, and potentially, treatment (Bipolar Disorder and Schizophrenia Working Group of the Psychiatric Genomics Consortium, 2018). This study also identified several significant correlations of diagnostic PRS with subphenotypes in either diagnostic group, including SCZ PRS with the presence of psychosis, and age of onset in BD cases (Bipolar Disorder and Schizophrenia Working Group of the Psychiatric Genomics Consortium, 2018).

Across a broader range of brain disorders, strikingly, neurological disorders show much clearer genetic boundaries than observed between psychiatric disorders (Anttila et al., 2018). Moreover, there is little genetic overlap between neurological and psychiatric disorders (Anttila et al., 2018). Hence, the genetic data seem to align with the large degree of clinical overlap between psychiatric disorders (Smoller et al., 2018), supporting the notion that the current psychiatric diagnostic classification may not reflect distinct disease entities. Moreover, converging genetic evidence suggests that susceptibility to psychiatric disorders may be dimensional by nature, rather than categorically different from normality (Insel, 2014). For example, a recent study quantified significant positive genetic correlations between BD and the personality traits openness and extraversion, while there was a negative genetic correlation with conscientiousness (Lo et al., 2017). These findings further emphasize that the etiology of BD may be continuous with variation in normal neurobiological and behavior dimensions (Lo et al., 2017). It remains to be seen whether the emerging genetic data may inform the development or evolution of new diagnostic classification systems in psychiatry (Smoller et al., 2018).

7.3 Novel polygenic analytical approaches

It is not just heritability and sample size that govern the power to detect novel GWAS loci. Additionally, the effect size distribution across the causal loci substantially influences the *discoverability* of the loci (Park et al., 2010). The term *discoverability* was recently introduced to represent the number of genetic variants that can be consistently and reproducibly detected by GWAS for a given phenotype (Fan et al., 2018). Specifically, when sample size is fixed, GWAS power to detect novel loci is determined by the effect size distribution of true causal loci (Park et al., 2010). The larger effect size per causal locus, the higher the *discoverability*. At higher levels of polygenicity for a given heritability, more true signals will diffuse into the background, and make SNP effects harder to detect (Yang et al., 2011a). Additionally, the correlation among SNPs, that is, linkage-disequilibrium (LD), mingles the diffuse background with true signals from causal SNPs, making the signals even harder to differentiate, while increasing power for causal SNPs with high LD (Yang et al., 2011a). Novel statistical frameworks using mixture modeling can estimate this effect size distribution (Holland et al., 2017), and better characterize the power needed to detect novel loci for a given phenotype (Holland et al., 2017). Recent power analyses indicate that the genomic architectures of SCZ and BD are similarly polygenic with similar discoverability, and most of the SNP effects are weak (Holland et al., 2017). This indicates that GWAS power and discovery of novel loci for BD is expected to follow that of SCZ, as the GWAS sample size continues to grow larger (Holland et al., 2017).

The flexibility of mixture modeling provides opportunities to increase SNP discovery in BD. Because GWAS are approximated by mixtures of effect size distributions, enrichment priors can be incorporated into the mixture components to improve the accuracy of effect size estimation (Gianola, 2013). By better estimating the true effect size of each SNP, we can improve *discoverability* without relying on significance thresholds defined by assuming all SNPs are equal (Thompson et al., 2016; Wang et al., 2016; Zablocki et al., 2014). Enrichment methodologies have been extensively explored in the quantitative genetics field, and polygenic enrichment methodologies hold promise for human complex disease genetics (Schork et al., 2016). One of the key insights motivating the use of enrichment methods is that genetic variants have attributes that predict their effect sizes across many complex traits (Finucane et al., 2015; Schork et al., 2013; Wang et al., 2016). To name a few, GWAS associations are systematically overrepresented in coding and regulatory regions, and depleted in intergenic regions (Gusev et al., 2014; Hindorff et al., 2009; Maurano et al., 2012; Schork et al., 2013; Yang et al., 2011b); SNPs with a higher level of LD with nearby regions tend to have larger effect sizes in GWAS (Bulik-Sullivan et al., 2015; Schork et al., 2013), and SNPs found to be associated with one phenotype tend to have larger effect sizes in another GWAS of a closely related phenotype (Liu et al., 2013; Schork et al., 2013). Leveraging these attributes, borrowing strength from across sources of prior information has led to better estimation of effect sizes for GWAS, and resulted in increased probability to detect and replicate causal loci (Giambartolomei et al., 2014; Kichaev et al., 2014; Pickrell, 2014; Wang et al., 2016; Zablocki et al., 2014). For example, by using SNP effect sizes obtained from a large-scale GWAS of intelligence (Savage et al., 2018) as a SNP attribute (Smeland et al., 2019), a recent analysis discovered 31 loci associated with BD that were not identified in the recent BD GWAS, 22 of which were also associated with SCZ at $P < 0.05$. Like many psychiatric disorders (Kahn & Keefe, 2013), BD is associated with cognitive dysfunction, which may influence treatment and functional outcome (Green, 2006). However, it has been difficult to detect shared genetic influences between BD and cognitive function, and there appears to be no genetic correlation between the phenotypes using standard tools (Savage et al., 2018; Stahl et al., 2019). Using the mixture modeling framework, the combined GWAS analysis of BD (Stahl et al., 2019) and intelligence (Savage et al., 2018) identified 12 shared loci between BD and intelligence, where 9 BD risk alleles were associated with higher cognitive performance (Smeland et al., 2019). This contrasts the findings for SCZ where the majority of SCZ risk alleles are associated with lower cognitive performance at shared loci (Savage et al., 2018; Smeland et al., 2017). The converging results suggest that BD and SCZ differ in their genetic relationship with cognitive function, potentially pointing to unique genetic effects underlying these disorders.

8 From genetic loci to disease mechanisms

Some recent genetic findings reinforce specific hypotheses regarding BD neurobiology; however, the majority of the findings provide new biological insights. The most robust disease genes in BD today seem to be *ANK3* and *CACNA1C* (Table 2). *ANK3* belongs to a class of proteins that connect membrane proteins to the underlying cytoskeleton (Spectrin-Actin), and play an important role in cell motility, proliferation, contact with other cells, and maintenance of specific membrane regions (Bennett & Healy, 2009). *ANK3* is also known to modulate sodium channels. *CACNA1C* is a gene encoding part of the calcium channel molecule, with important roles in a series of neurophysiological functions (Nanou & Catterall, 2018). Both genes are involved in ion channel function, which is an important basis for neuronal excitability, and thus neurotransmission. It has therefore been proposed that ion channel dysfunction may be a cause of BD characterized by periodic mood problems (Cipriani et al., 2016). This resembles epilepsy, where there are indications that ion channel pathology is an important part of the disease mechanism. This has given researchers hope with regard to the development of new and more targeted drugs for BD (Cipriani et al., 2016).

Another strong association signal in the PGC GWAS was located at the previously reported *TRANK1* locus, but little is known about its function. Pathway analyses provided further support for the role of calcium and other ion channels, and newly implicate endocannabinoid signaling (Stahl et al., 2019). Intriguingly, many implicated genes are also associated with immune system functions (*TLR9*, *CD47*, *SART1*, *CD248*) and energy metabolism (*PC*, *G6PC3*, *ACOT12*). BD loci also include neurotransmitter receptors (*GRIN2A*), additional ion channels and transporters (*CACNA1C*, *SCN2A*, *SLC4A1*), and synaptic components (*RIMS1*, *ANK3*). This is of particular interest in light of the recent findings of hyperexcitable iPSC-derived neurons from BD patients, where lithium, a classic mood stabilizing drug, differentially affects neuronal excitability (Mertens et al., 2015). Similarly, the confirmed association with calcium voltage-gated channels supports the rekindled interest in calcium channel antagonists as potential treatments for BD.

Increasingly, studies of gene expression in the brain are being performed in order to understand pathogenesis in BD. The largest such study used postmortem brain RNA sequencing data from approximately 100 BD cases and found that BD-associated loci from GWAS are enriched in modules of genes that are coexpressed together in BD brain specimens (Gandal et al., 2018). Interestingly, this study also found that overlaps between psychiatric disorder expression signatures (including BD and SCZ) are correlated with the genetic overlaps discussed herein. Human brain and disease-specific data from larger cohorts will soon be available that are comprised of transcriptome and other "omics" data from hundreds of BD cases and controls, promising to shed further light on the disease mechanisms that proceed from genetic risk.

9 Contribution of rare variants

BD is a complex genetic disease, to which both common and rare genetic variants likely contribute. Rare variants can be particularly relevant for understanding disease development if they show a higher penetrance that makes them particularly suitable for subsequent functional analyses (Budde et al., 2017). Rare variants (minor allele frequency <1%) can be divided into different classes according to their genomic size: copy number variants (CNVs, >1 kb), smaller-sized insertions or deletions (InDels), and single-nucleotide variants (SNVs). While the contribution of CNVs to psychiatric disorders such as SCZ is well documented (e.g., Marshall et al., 2017), findings on CNVs in BD are less clear. In a meta-analysis, Green and colleagues showed that duplications at the chromosomal locus 16p11.2 were significantly associated with BD after correction for multiple testing (Green et al., 2016). Interestingly, the duplications at 16p11.2 were also found in healthy individuals at a much lower frequency (Goes, 2016), so that reduced penetrance can be assumed (Budde et al., 2017). The first next-generation sequencing studies of BD patients (multiplex families, case-control studies) have identified rare SNVs in a number of candidate genes (Kato, 2015). Interestingly, the identified rare variants seem to accumulate in specific pathways, including calcium signaling, potassium channels, and G protein-coupled receptors (Cruceanu et al., 2017; Kato, 2015). However, the overlap of the identified candidate genes between the individual studies was limited. In addition, no rare sequence variants with a genome-wide association with BD have been identified so far (Budde et al., 2017). Therefore, the investigation of further BD patients and controls is necessary in order to better understand the contribution of rare sequence variants to BD.

10 Clinical implications of genetic knowledge

The main consequence of the new discoveries in BD genetics will probably be a better understanding of neurobiological disease mechanisms, as described herein. Drug development within the psychiatric field has so far been mainly based on trial and error, and knowledge of disease mechanisms can lead to rational development of new drugs and other treatment methods. New genetic knowledge may also be of significance for the diagnosis and classification of BD. By investigating whether certain genetic variants are associated with symptom profiles and specific subphenotypes, one can further develop better diagnostic criteria, reduce heterogeneity, and acquire better knowledge about comorbidity. This can help improve the diagnosis system, by identifying genetic variants associated with clinical subgroups, as described herein for family studies. A good example is disease genes that overlap between SCZ and BD, such as *ZNF804A* (O'Donovan et al., 2008), which suggests that there may be a specific genetic vulnerability to psychosis development across our classical diagnostic categories. Another example is age of onset, where several studies have identified possible genetic variants that are specifically associated with early onset of BD (Barnett & Smoller, 2009; Craddock & Sklar, 2013). However, despite recent developments in psychiatric genetics, the field remains far from using genetic knowledge to help diagnose the individual patient. The main reason for this is that the different variants identified so far all have low effect sizes (odds ratio between 1.1 and 1.3), meaning that a large proportion of healthy controls also carry disease variants. Thus, today's knowledge of vulnerability genes has no practical significance in clinical diagnostics. But low effect genes are certainly of great value to detect and understand disease mechanisms.

Another possible translation of genetic findings into clinical practice is the field of pharmacogenomics. As an example, in a recent large GWAS of lithium response in BD, the international Consortium on Lithium Genetics (ConLiGen) investigated 2563 BD patients and identified a genome-wide associated locus on chromosome 21 containing two genes for long, noncoding RNAs (Hou et al., 2016b). The results of this study suggest that GWAS are a powerful tool to understand how genetic variants affect treatment response. However, further studies in larger samples are needed to establish the biological context and the potential clinical utility of these findings (Hou et al., 2016b).

The ethical implications of genetic discoveries must be evaluated continuously. New knowledge that emerges in psychiatric genetics must be managed in a proper manner so that the individual patient's concern is taken into consideration and research breakthroughs are used to the benefit of this patient group. Importantly, genetic knowledge of BD should contribute to a proper emphasis on the role of stress vulnerability in clinical activities. As indicated by the continuous genetic load, all individuals are susceptible to developing a psychiatric disorder, some more than others. It is not bad parents or inadequate upbringing that leads to BD, but mainly the inherited genetic vulnerability. In this way, genetic counseling may reduce the guilt and stigma that many patients and parents are struggling with. Moreover, genetic variation does not explain all risk to psychiatric disorders, and it is important to emphasize and educate about the interaction between individual risk for disease and stress factors, that is, the environment. This can promote better treatment adherence and therapeutic alliance, and by reducing stress factors such as drowsiness, irregular daily rhythm, and workload, decrease the risk of relapse and severe progression.

11 Conclusion

BD has a high degree of heritability, and there is a clear accumulation of the disease in families. Many disease genes have been identified, especially in recent years through the new technique of genome-wide genotyping. However, the genetic variants identified so far explain too little variance to be used for clinical diagnostics or for individual prediction of disease risk, and the specific molecular genetic mechanisms underlying susceptibility to BD are still poorly understood. Yet, new genetic knowledge is providing a better understanding of disease mechanisms, which in the long term, can provide a basis for developing better drugs and inform new diagnostic tools. It is likely that BD patients inherit a number of genes that increase vulnerability to disease, and are exposed to stress conditions to trigger the disease and cause relapses. Therefore, it is important to communicate the stress-vulnerability model to patients and relatives, which can promote better acceptance and involvement from patients and the community, leading to better treatment adherence and outcome.

Acknowledgments

The authors thank the Psychiatric Genomics Consortium Bipolar Disorder Working Group for their contribution. The work was supported by grants from NIH, Research Council of Norway, and Stiftelsen KG Jebsen.

Competing Financial Interests

None.

References

Anguelova, M., Benkelfat, C., & Turecki, G. (2003). A systematic review of association studies investigating genes coding for serotonin receptors and the serotonin transporter: II. Suicidal behavior. *Molecular Psychiatry, 8*(7), 646–653.

Anttila, V., et al. (2018). Analysis of shared heritability in common disorders of the brain. *Science, 360*(6395) pii: eaap8757.

Barnett, J. H., & Smoller, J. W. (2009). The genetics of bipolar disorder. *Neuroscience, 164*(1), 331–343.

Baum, A. E., et al. (2008a). Meta-analysis of two genome-wide association studies of bipolar disorder reveals important points of agreement. *Molecular Psychiatry, 13*(5), 466–467.

Baum, A. E., et al. (2008b). A genome-wide association study implicates diacylglycerol kinase eta (DGKH) and several other genes in the etiology of bipolar disorder. *Molecular Psychiatry, 13*(2), 197–207.

Bennett, V., & Healy, J. (2009). Membrane domains based on ankyrin and spectrin associated with cell-cell interactions. *Cold Spring Harbor Perspectives in Biology, 1*(6), a003012.

Bipolar Disorder and Schizophrenia Working Group of the Psychiatric Genomics Consortium (2018). Genomic dissection of bipolar disorder and schizophrenia, including 28 subphenotypes. *Cell, 173*(7) 1705–1715.e16.

Budde, M., et al. (2017). Genetics of bipolar disorder. *Nervenarzt, 88*(7), 755–759.

Bulik-Sullivan, B. K., et al. (2015). LD Score regression distinguishes confounding from polygenicity in genome-wide association studies. *Nature Genetics, 47*, 291.

Charney, A. W., et al. (2017). Evidence for genetic heterogeneity between clinical subtypes of bipolar disorder. *Translational Psychiatry, 7*(1).

Chen, D. T., et al. (2013). Genome-wide association study meta-analysis of European and Asian-ancestry samples identifies three novel loci associated with bipolar disorder. *Molecular Psychiatry, 18*(2), 195–205.

Cichon, S., et al. (2011). Genome-wide association study identifies genetic variation in neurocan as a susceptibility factor for bipolar disorder. *American Journal of Human Genetics, 88*(3), 372–381.

Cipriani, A., et al. (2016). A systematic review of calcium channel antagonists in bipolar disorder and some considerations for their future development. *Molecular Psychiatry, 21*(10), 1324–1332.

Craddock, N., & Owen, M. J. (2010). The Kraepelinian dichotomy—Going, going.. but still not gone. *The British Journal of Psychiatry, 196*(2), 92–95.

Craddock, N., & Sklar, P. (2009). Genetics of bipolar disorder: Successful start to a long journey. *Trends in Genetics, 25*(2), 99–105.

Craddock, N., & Sklar, P. (2013). Genetics of bipolar disorder. *Lancet, 381*(9878), 1654–1662.

Cross-Disorder Group of the Psychiatric Genomics Consortium, et al. (2013). Genetic relationship between five psychiatric disorders estimated from genome-wide SNPs. *Nature Genetics, 45*(9), 984–994.

Cruceanu, C., et al. (2017). Rare susceptibility variants for bipolar disorder suggest a role for G protein-coupled receptors. *Molecular Psychiatry, 23*, 2050–2056.

Edvardsen, J., et al. (2008). Heritability of bipolar spectrum disorders. Unity or heterogeneity? *Journal of Affective Disorders, 106*(3), 229–240.

Fan, C. C., et al. (2018). Beyond heritability: Improving discoverability in imaging genetics. *Human Molecular Genetics, 27*(R1), R22–R28.

Ferrari, A. J., et al. (2016). The prevalence and burden of bipolar disorder: Findings from the Global Burden of Disease Study 2013. *Bipolar Disorders, 18*(5), 440–450.

Ferreira, M. A., et al. (2008). Collaborative genome-wide association analysis supports a role for ANK3 and CACNA1C in bipolar disorder. *Nature Genetics, 40*(9), 1056–1058.

Finucane, H. K., et al. (2015). Partitioning heritability by functional annotation using genome-wide association summary statistics. *Nature Genetics, 47*(11), 1228–1235.

Forstner, A. J., et al. (2015). Genome-wide analysis implicates microRNAs and their target genes in the development of bipolar disorder. *Translational Psychiatry, 5*.

Forstner, A. J., et al. (2017). Identification of shared risk loci and pathways for bipolar disorder and schizophrenia. *PLoS One, 12*(2).

Funke, B., et al. (2005). COMT genetic variation confers risk for psychotic and affective disorders: A case control study. *Behavioral and Brain Functions, 1*, 19.

Gandal, M. J., et al. (2018). Shared molecular neuropathology across major psychiatric disorders parallels polygenic overlap. *Science, 359*(6376), 693–697.

Giambartolomei, C., et al. (2014). Bayesian test for colocalisation between pairs of genetic association studies using summary statistics. *PLoS Genetics, 10*(5).

Gianola, D. (2013). Priors in whole-genome regression: The Bayesian alphabet returns. *Genetics, 194*(3), 573.

Goes, F. S. (2016). Genetics of bipolar disorder: Recent update and future directions. *The Psychiatric Clinics of North America, 39*(1), 139–155.

Goodwin, G. M., et al. (2016). Evidence-based guidelines for treating bipolar disorder: Revised third edition recommendations from the British Association for Psychopharmacology. *Journal of Psychopharmacology, 30*(6), 495–553.

Grande, I., et al. (2016). Bipolar disorder. *Lancet, 387*(10027), 1561–1572.

Green, M. F. (2006). Cognitive impairment and functional outcome in schizophrenia and bipolar disorder. *The Journal of Clinical Psychiatry, 67*(Suppl. 9), 3–8 (discussion 36–42).

Green, E. K., et al. (2012). Association at SYNE1 in both bipolar disorder and recurrent major depression. *Molecular Psychiatry, 18*, 614–617.

Green, E. K., et al. (2013). Replication of bipolar disorder susceptibility alleles and identification of two novel genome-wide significant associations in a new bipolar disorder case-control sample. *Molecular Psychiatry, 18*(12), 1302–1307.

Green, E. K., et al. (2016). Copy number variation in bipolar disorder. *Molecular Psychiatry, 21*(1), 89–93.

Gusev, A., et al. (2014). Partitioning heritability of regulatory and cell-type-specific variants across 11 common diseases. *American Journal of Human Genetics, 95*(5), 535–552.

Heun, R., & Maier, W. (1993). The distinction of bipolar II disorder from bipolar I and recurrent unipolar depression: Results of a controlled family study. *Acta Psychiatrica Scandinavica, 87*(4), 279–284.

Hibar, D. P., et al. (2016). Subcortical volumetric abnormalities in bipolar disorder. *Molecular Psychiatry, 21*(12), 1710–1716.

Hibar, D. P., et al. (2017). Cortical abnormalities in bipolar disorder: An MRI analysis of 6503 individuals from the ENIGMA Bipolar Disorder Working Group. *Molecular Psychiatry, 23*, 932–942.

Hindorff, L. A., et al. (2009). Potential etiologic and functional implications of genome-wide association loci for human diseases and traits. *Proceedings of the National Academy of Sciences of the United States of America, 106*(23), 9362–9367.

Hirschhorn, J. N., & Daly, M. J. (2005). Genome-wide association studies for common diseases and complex traits. *Nature Reviews Genetics, 6*(2), 95–108.

Holland, D., et al. (2017). Estimating phenotypic polygenicity and causal effect size variance from GWAS summary statistics while accounting for inflation due to cryptic relatedness. *bioRxiv*.

Hou, L., et al. (2016a). Genome-wide association study of 40,000 individuals identifies two novel loci associated with bipolar disorder. *Human Molecular Genetics, 25*(15), 3383–3394.

Hou, L., et al. (2016b). Genetic variants associated with response to lithium treatment in bipolar disorder: A genome-wide association study. *Lancet, 387* (10023), 1085–1093.

Insel, T. R. (2014). The NIMH Research Domain Criteria (RDoC) Project: Precision medicine for psychiatry. *The American Journal of Psychiatry, 171*(4), 395–397.

International Schizophrenia Consortium,et al. (2009). Common polygenic variation contributes to risk of schizophrenia and bipolar disorder. *Nature, 460* (7256), 748–752.

Kahn, R. S., & Keefe, R. S. (2013). Schizophrenia is a cognitive illness: Time for a change in focus. *JAMA Psychiatry, 70*(10), 1107–1112.

Kato, T. (2015). Whole genome/exome sequencing in mood and psychotic disorders. *Psychiatry and Clinical Neurosciences, 69*(2), 65–76.

Kichaev, G., et al. (2014). Integrating functional data to prioritize causal variants in statistical fine-mapping studies. *PLoS Genetics, 10*(10).

Lichtenstein, P., et al. (2009). Common genetic determinants of schizophrenia and bipolar disorder in Swedish families: A population-based study. *Lancet, 373*, 234–239.

Liu, J. Z., et al. (2013). Dense genotyping of immune-related disease regions identifies nine new risk loci for primary sclerosing cholangitis. *Nature Genetics, 45*(6), 670–675.

Lo, M. T., et al. (2017). Genome-wide analyses for personality traits identify six genomic loci and show correlations with psychiatric disorders. *Nature Genetics, 49*(1), 152–156.

Marshall, C. R., et al. (2017). Contribution of copy number variants to schizophrenia from a genome-wide study of 41,321 subjects. *Nature Genetics, 49*(1), 27–35.

Maurano, M. T., et al. (2012). Systematic localization of common disease-associated variation in regulatory DNA. *Science, 337*(6099), 1190–1195.

McGuffin, P., et al. (2010). The genetics of affective disorder and suicide. *European Psychiatry, 25*(5), 275–277.

Mertens, J., et al. (2015). Differential responses to lithium in hyperexcitable neurons from patients with bipolar disorder. *Nature, 527*(7576), 95–99.

Muhleisen, T. W., et al. (2014). Genome-wide association study reveals two new risk loci for bipolar disorder. *Nature Communications, 5*, 3339.

Nanou, E., & Catterall, W. A. (2018). Calcium channels, synaptic plasticity, and neuropsychiatric disease. *Neuron, 98*(3), 466–481.

Nievergelt, C. M., et al. (2006). Suggestive evidence for association of the circadian genes PERIOD3 and ARNTL with bipolar disorder. *American Journal of Medical Genetics. Part B, Neuropsychiatric Genetics, 141B*(3), 234–241.

O'Donovan, M. C., et al. (2008). Identification of loci associated with schizophrenia by genome-wide association and follow-up. *Nature Genetics, 40*(9), 1053–1055.

Park, J.-H., et al. (2010). Estimation of effect size distribution from genome-wide association studies and implications for future discoveries. *Nature Genetics, 42*, 570.

Pickrell, J. K. (2014). Joint analysis of functional genomic data and genome-wide association studies of 18 human traits. *The American Journal of Human Genetics, 94*(4), 559–573.

Preisig, M., et al. (2000). Association between bipolar disorder and monoamine oxidase A gene polymorphisms: Results of a multicenter study. *The American Journal of Psychiatry, 157*(6), 948–955.

Ruderfer, D. M., et al. (2014). Polygenic dissection of diagnosis and clinical dimensions of bipolar disorder and schizophrenia. *Molecular Psychiatry, 19*(9), 1017–1024.

Savage, J. E., et al. (2018). Genome-wide association meta-analysis in 269,867 individuals identifies new genetic and functional links to intelligence. *Nature Genetics, 50*, 912–919.

Schork, A. J., et al. (2013). All SNPs are not created equal: Genome-wide association studies reveal a consistent pattern of enrichment among functionally annotated SNPs. *PLoS Genetics, 9*(4).

Schork, A. J., et al. (2016). New statistical approaches exploit the polygenic architecture of schizophrenia—Implications for the underlying neurobiology. *Current Opinion in Neurobiology, 36*, 89–98.

Schulze, T. G., et al. (2009). Two variants in Ankyrin 3 (ANK3) are independent genetic risk factors for bipolar disorder. *Molecular Psychiatry, 14*(5), 487–491.

Scott, L. J., et al. (2009). Genome-wide association and meta-analysis of bipolar disorder in individuals of European ancestry. *Proceedings of the National Academy of Sciences of the United States of America, 106*(18), 7501–7506.

Sklar, P., et al. (2008). Whole-genome association study of bipolar disorder. *Molecular Psychiatry, 13*(6), 558–569.

Sklar, P., et al. (2011). Large-scale genome-wide association analysis of bipolar disorder identifies a new susceptibility locus near ODZ4. *Nature Genetics, 43*(10), 977–983.

Smeland, O. B., et al. (2017). Identification of genetic loci jointly influencing schizophrenia risk and the cognitive traits of verbal-numerical reasoning, reaction time, and general cognitive function. *JAMA Psychiatry, 74*(10), 1065–1075.

Smeland, O. B., et al. (2019). Genome-wide analysis reveals extensive genetic overlap between schizophrenia, bipolar disorder, and intelligence. *Molecular Psychiatry. 2019*, https://doi.org/10.1038/s41380-018-0332-x [Epub ahead of print].

Smith, E. N., et al. (2009). Genome-wide association study of bipolar disorder in European American and African American individuals. *Molecular Psychiatry, 14*(8), 755–763.

Smoller, J. W., & Finn, C. T. (2003). Family, twin, and adoption studies of bipolar disorder. *American Journal of Medical Genetics Part C, Seminars in Medical Genetics, 123C*(1), 48–58.

Smoller, J. W., et al. (2018). Psychiatric genetics and the structure of psychopathology. *Molecular Psychiatry, 24*, 409–420.

Soliman, M. A., et al. (2017). Pluripotent stem cells in neuropsychiatric disorders. *Molecular Psychiatry, 22*(9), 1241–1249.

Stahl, E., et al. (2019). Genome-wide association study identifies 30 loci associated with bipolar disorder. *Nat. Genet., 51*(5), 793–803.

Tesli, M., et al. (2014). Polygenic risk score and the psychosis continuum model. *Acta Psychiatrica Scandinavica, 130*(4), 311–317.

Thompson, W. K., et al. (2016). An empirical Bayes mixture model for effect size distributions in genome-wide association studies. *PLoS Genetics, 11*(12).

Visscher, P. M., et al. (2017). 10 Years of GWAS discovery: Biology, function, and translation. *American Journal of Human Genetics, 101*(1), 5–22.

Wang, Y., et al. (2016). Leveraging genomic annotations and pleiotropic enrichment for improved replication rates in schizophrenia GWAS. *PLoS Genetics, 12*(1).

Wellcome Trust Case Control Consortium. (2007). Genome-wide association study of 14,000 cases of seven common diseases and 3,000 shared controls. *Nature, 447*(7145), 661–678.

Wray, N. R., et al. (2018). Genome-wide association analyses identify 44 risk variants and refine the genetic architecture of major depression. *Nature Genetics, 50*(5), 668–681.

Yang, J., et al. (2011a). Genomic inflation factors under polygenic inheritance. *European Journal of Human Genetics, 19*, 807.

Yang, J., et al. (2011b). Genome partitioning of genetic variation for complex traits using common SNPs. *Nature Genetics, 43*(6), 519–525.

Zablocki, R. W., et al. (2014). Covariate-modulated local false discovery rate for genome-wide association studies. *Bioinformatics, 30*(15), 2098–2104.

Chapter 19

Genomics of borderline personality disorder

Fabian Streit[a], Lucía Colodro-Conde[b], Alisha S.M. Hall[a] and Stephanie H. Witt[a]

[a]Department of Genetic Epidemiology in Psychiatry, Central Institute of Mental Health, Medical Faculty Mannheim, University of Heidelberg, Mannheim, Germany, [b]Genetics & Computational Biology Department, QIMR Berghofer Medical Research, Brisbane, QLD, Australia

1 Clinical aspects of borderline personality disorder

Borderline personality disorder (BPD) is a mental illness marked by an ongoing pattern of varying self-image, moods, and behavior. The median prevalence of BPD is around 3% (Tomko, Trull, Wood, & Sher, 2014), and BPD patients constitute a relatively large portion of the population receiving inpatient (20%; American Psychiatric Association, 2013) and outpatient (10%; Zimmerman, Rothschild, & Chelminski, 2005) mental health care. Approximately 75% of patients diagnosed with BPD are women, and the prevalence of BPD may decrease with age (American Psychiatric Association, 2013). However, different presentation patterns and help-seeking behavior, as well as history of abuse and gender bias, may explain why women are diagnosed with BPD more often (Sansone & Sansone, 2011).

According to the Diagnostic and Statistical Manual of Mental Disorders (DSM-5), BPD is characterized by marked impulsivity, as well as unstable interpersonal relationships, self-image, and affect (American Psychiatric Association, 2013). Instability of affect can be manifested in feelings of rage or problems controlling anger, often followed by feelings of guilt and shame. In addition, BPD patients display impulsivity in activities that are potentially self-damaging. A tendency to think about and exhibit nonsuicidal self-injurious and suicidal behavior is often the reason that these individuals seek help: around 46% attempt suicide (Soloff & Chiappetta, 2018), and the rate of completed suicide attempts among BPD patients is around 10% (Paris & Zweig-Frank, 2001).

Dimensional as well as categorical diagnostic approaches have been suggested to assess BPD. Heterogeneity of the BPD patient population is very high, and patients often display distinctly different sets of symptoms. Due to this, categorical diagnosis suffers from limitations related to, for example, high diagnostic comorbidity and considerable within-disorder heterogeneity, which results in low reliability of clinicians' diagnoses and makes it difficult to design and plan interventions (highly specific vs general) (Clark, 2007; Herpertz et al., 2017; Hopwood et al., 2018; Morey, Benson, Busch, & Skodol, 2015; Trull & Durrett, 2005). Although still predominantly based on categorical diagnosis, the DSM-5 includes a dimensional alternative to the categorical classification, proposing impairments in personality functioning (self-functioning, interpersonal functioning) and seven pathological personality traits from the domains negative affectivity, disinhibition, and antagonism as diagnostic criteria for BPD (American Psychiatric Association, 2013). Also, since the introduction of the Research Domain Criteria (RDoC) framework by the American National Institutes of Health in 2008, there has been a greater focus on integrating different levels of information (observable behavior and neurobiological measures) in order to explore the basic dimensions of functioning that span the range of human behavior (from normal to abnormal), thus creating new ways of classifying mental illnesses (Insel et al., 2010). The investigation of neurobiological risk mechanisms underlying BPD—distinct or overlapping with other disorders—may inform researchers about specific effects of genetic risk variants and further our understanding of pathomechanisms. Reconstructing the categorical diagnosis into sets of symptoms (e.g., features or subphenotypes) will help better characterize the individual patient and enable more accurate prediction of course of illness (that is, the development of the disease in a patient) and treatment response, ultimately paving the way for precision medicine.

Borderline personality features (BPF) typically include affective instability, identity problems, negative relationships, and self-harm and show phenotypic (or observed) as well as genetic associations with the full BPD diagnosis (Rojas et al., 2014; Stein, Pinsker-Aspen, & Hilsenroth, 2007). Because BPF can be measured as continuous traits at the population level and thus leverage the whole range of manifestations (including subclinical levels), they represent a valuable approach to investigating the underlying mechanisms of BPD. The self-report Personality Assessment Inventory Borderline (PAI-BOR) questionnaire, a clinical scale of the Personality Assessment Inventory (Morey, 2007, 2011), is by far the most commonly used continuous measure of BPF. It includes the subscales affective instability, identity problems, negative relationships, and self-harm.

A subphenotype is a subset of a phenotype or an observable trait that is characteristic of a subset of a population. Some proposed subphenotypes of BPD are emotional dysregulation (with additional components impulsivity and interpersonal sensitivity) and aggression (Goodman, New, Triebwasser, Collins, & Siever, 2010; Siever, 2005). Negative emotion dysregulation, defined as the neurobiologically-based inability to flexibly respond to and manage negative emotions, involves several processes, including emotion sensitivity, heightened and labile negative affect, a deficit of appropriate regulation strategies, and a surplus of maladaptive regulation strategies (Carpenter & Trull, 2013), and has been studied extensively. Subphenotypes may also help in deconstructing BPD, which could result in more successful genetic analyses, redefinition of diagnosis, better study of the course of illness, biomarkers, and ultimately precision medicine (Traylor, Markus, & Lewis, 2015).

BPD patients often display comorbidities with other psychiatric disorders such as panic disorder (up to 48%; Zanarini et al., 1998) and posttraumatic stress disorder (up to 56%; Zanarini et al., 1998), but especially with mood disorders such as depression (up to ~80%; Zanarini et al., 1998) and bipolar disorder (up to ~20%; Agius, Lee, Gardner, & Wotherspoon, 2012; Bassett, 2012; Fornaro et al., 2016). Comorbid alcohol and substance use and dependency, as well as eating disorders, are also highly common (up to 56%; McGlashan et al., 2000; Zanarini et al., 1998). Moreover, there appears to be a substantial overlap with attention deficit hyperactivity disorder (ADHD; Matthies & Philipsen, 2014). It has been suggested that some etiological risk factors, such as neurobiological features (e.g., alterations in structure and function of the limbic system), as well as adverse early environment, are shared at least between BPD and bipolar disorder (Agius et al., 2012).

The aim of this chapter is to summarize findings on the genetic as well as epigenetic underpinnings of BPD and BPF. After presenting findings from formal genetic studies investigating the (co)heritability of BPD and BPF, we will discuss molecular genetic approaches, aiming to identify implicated genetic variants as well as epigenetic alterations as paths to aid precision medicine.

2 Genetic studies

Like all common diseases, BPD is a complex disorder involving both genetic and environmental risk factors in its development. Additionally, genetics and environment commonly interact in complex traits, and environmental factors may affect disease development via epigenetic factors, such as DNA methylation. However, unlike other common diseases such as cancer, diabetes, or hypertension, there are currently no objective biological markers that support the diagnosis or prognosis of the disease (Sullivan, Daly, & O'Donovan, 2012). Thus, so far, diagnosis relies exclusively on clinically observable symptoms. A diagnosis is traditionally made when a defined minimum number of specific symptoms from a catalog (like the DSM-5) have occurred over a defined period of time. As a result, patients with the same diagnosis can have different symptoms and courses of illness. In addition to within-disorder heterogeneity, there is a high degree of comorbidity among mental disorders. In particular, affective disorders, which exhibit high comorbidities with BPD, also show high comorbidity with schizophrenic disorders and alcohol dependence. It can be assumed that this clinical heterogeneity is also associated with genetic heterogeneity. Due to their design, genetic studies can not only find risk factors for BPD, but can also help disentangle the heterogeneity of the disorder. The following sections summarize the main findings from formal and molecular genetic studies to date.

2.1 Formal genetic studies

Twin and family studies have been the main method to study genetic risk factors in behavior for decades. They are based on the fact that family members share different degrees of genetic relatedness. These designs allow the calculation of heritability, which is defined as the proportion of variation in a trait that is due to heritable differences between individuals in a population (van Dongen, Slagboom, Draisma, Martin, & Boomsma, 2012), and can be distinguished from the effects of shared and unique environment.

In a family study, diagnosis of BPD, as well as single symptoms, were significantly more frequent in the first-degree relatives of individuals with BPD than those without BPD, which is compatible with a familial, potentially genetic risk component for BPD (Gunderson et al., 2011; Zanarini et al., 2004). A twin study by Torgersen (2000) investigated the heritability of different personality disorders in 221 twin pairs. BPD presented a strong heritability estimate of ~69%, while the remaining variance was attributed to unique environmental experiences and measurement error. Studies investigating

BPF show heritabilities of $\sim$35%–50%, a number that is similar across different countries (Distel, Carlier, et al., 2011; Distel et al., 2008; Torgersen et al., 2012).

In addition to investigating the heritability of disorders and traits, formal genetic studies can be used to investigate whether the underlying genetic and environmental factors are shared between disorders and traits, and by that inform researchers about shared or distinct etiological pathways. There is a partially genetically-driven overlap between BPF, personality traits, and other psychiatric symptom dimensions or disorders, which indicates the existence of shared etiological pathways (Rojas et al., 2014; Stein et al., 2007).

Twin studies show that genetic variation in BPF is shared with the genetic variation underlying the dimensions of personality traits, supporting the dimensional proposal for the study of personality disorders and the use of personality questionnaires for their assessment (Distel et al., 2009; Kendler, Myers, & Reichborn-Kjennerud, 2011). A similar overlap in the contribution of genetic and environmental factors has been observed for the phenotypic association between BPF and trait anger (Distel et al., 2012).

Furthermore, twin studies indicate that the genetic and environmental risk for the DSM-IV symptoms of BPD and other personality disorders is partially shared (Ørstavik et al., 2012). Dimensional symptoms of BPD are not only phenotypically, but also genetically associated with major depressive disorder (MDD; Reichborn-Kjennerud et al., 2010) and anxiety disorders (Welander-Vatn et al., 2016). Similarly, there is a moderate correlation between BPF and ADHD symptoms (Distel, Carlier, et al., 2011).

Recent twin studies indicate that among DSM-IV criteria for personality disorders, those for BPD and antisocial personality disorder (APD) were the strongest predictors of phenotypic and genotypic liability to alcohol use and alcohol use disorder (Long et al., 2017) and of cocaine use (Gillespie et al., 2018). Results from a previous twin study indicate that the—in part genetic—association between BPF and substance use disorders is attributable to normal variation in personality traits, especially neuroticism (Few et al., 2014). Interestingly, available data suggests that the contribution of genetic factors might depend on developmental phases. For example, a study of adolescent female twins showed that the association of BPF with substance abuse at the age of 14 was accounted for by environmental factors, while at the age of 18, genetic factors accounted for the association (Bornovalova, Hicks, Iacono, & McGue, 2013).

The interaction between genetic risk and environmental risk factors (known as $G \times E$) plays an especially important role in the etiology of complex traits and specifically, of BPD. Aversive experiences, such as abuse, might have especially harmful consequences in subjects with a genetic vulnerability. Distel, Middeldorp, et al. (2011) showed in a sample of twins and nontwin siblings that the genetic predisposition for BPF interacted with the exposure to specific life events.

In summary, twin and family studies indicate that the risk for BPD and increased BPF has a significant genetic component that is, to some degree, shared with personality traits and other (comorbid) psychiatric disorders such as MDD, ADHD, and substance abuse. Identifying the underlying genetic variants and corresponding genes will enable us to link specific biological mechanisms to the risk unique for BPD and that shared with other psychiatric disorders. Molecular genetic studies are designed to identify such variants.

2.2 Molecular genetic studies

The identification of genetic factors at the molecular level is expected to yield completely new insights into biological disease processes. Here, it is examined whether BPD patients differ in the occurrence of genetic variants compared with unaffected persons, or whether the occurrence of genetic variants predicts higher BPF.

Genetic variants consist of changes in the deoxyribonucleic acid (DNA) in a sequence of 3.2×10^9 base pairs (A, T, C, G). The greatest source of genetic variability arises when single base pairs are replaced in individual variants, so-called single-nucleotide polymorphisms (SNPs). It is estimated that the human genome contains 11 million such SNPs, of which about 7 million have a frequency of >5% (and the remainder a frequency of 1%–5%) in the normal population. The study of these genetic variants and their association with a phenotype of interest and follow-up analyses allows a better understanding of the genetic basis of specific phenotypes—in our case, borderline personality.

2.2.1 Candidate genes

Molecular genetic studies focus on common variants with small effects on the trait of interest, and they usually use patients and controls or community samples. In the past, when it was not yet possible to study the entire genome, these studies focused on so-called candidate genes, which are specific genes that were thought to play a role in the disease based on prior knowledge, for example, from animal studies or pharmacological observations.

However, as the candidate gene approach makes use of already existing hypotheses, the gain in knowledge about the other molecular genetic causes of disorders is low.

In candidate gene studies, genetic variants in, or in proximity to, genes coding for proteins hypothesized to contribute to the disorder are tested for association with a trait or case-control status. Candidate genes that have been investigated in regard to BPD include genes related to monoamine metabolism, such as the serotonin transporter (*5-HTTLPR*) or receptor genes, genes involved in the biological stress reaction such as the gene coding for the glucocorticoid receptor (*NR3C1*) or FK506 binding protein 5 (*FKBP5*), neuropeptide signaling genes such as the oxytocin receptor gene (*OXTR*), and genes previously shown to be associated in other psychiatric disorders such as the gene encoding calcium voltage-gated channel subunit alpha1 C (*CACNA1C*). As many reports used fairly small sample sizes—which makes it difficult to reliably interpret the results—we exemplarily describe large studies or meta-analyses in the following paragraphs.

Several studies have implicated monoamine metabolism genes in BPD (Dammann et al., 2011; Ni, Chan, Chan, McMain, & Kennedy, 2009; Ni et al., 2007; Yang et al., 2014). However, meta-analyses could not identify stable association signals (Amad, Ramoz, Thomas, Jardri, & Gorwood, 2014; Calati, Gressier, Balestri, & Serretti, 2013; Liu et al., 2017). A recent study investigating 10 genetic variants of the dopa-decarboxylase gene (*DDC*) found a significant association of one variant, rs12718541, with BPD (Mobascher et al., 2014). Another recent study investigated 47 polymorphisms in 10 genes involved in the regulation of the hypothalamus-pituitary-adrenal (HPA) axis, a core component of the stress response, and found an association of single variants as well as specific allele combinations (i.e., haplotypes) in the *FKBP5* and corticotropin-releasing hormone receptor 1 gene (*CRHR1*) with BPD (Martin-Blanco et al., 2016).

Moreover, subphenotypes of borderline personality have been associated with genetic risk factors. For example, decision-making in a neuropsychological test (Iowa Gambling Task, ITG) showed an association with a *TPH1* haplotype in BPD patients (Maurex et al., 2009). Similarly, a small study using functional magnetic resonance imaging (fMRI) indicated that, in BPD patients, brain processing was modulated by the Val158Met polymorphism in the *COMT* gene (Schmahl et al., 2012).

Other studies have followed up on the genetic correlations and comorbidities of BPD, investigating genes implicated in other psychiatric disorders for their association with BPD. In a study testing >1000 variants in >70 candidate genes for heroin dependence for their association with BPD symptoms in a sample of heroin dependency cases and neighborhood controls, the best signal was observed for the neurexin-3-alpha gene (*NRXN3*) (Panagopoulos et al., 2013). In another study, five SNPs that previously showed genome-wide significance for bipolar disorder were investigated for their association with BPD. Here, rs1006737 in *CACNA1C* showed a nominally significant association with BPD, a signal driven by the female subsample (Witt et al., 2014). This indicates that some of the genetic risk for other disorders also increases the risk for BPD.

However, it must be noted that the candidate gene studies investigating the genetic underpinnings of BPD carried out so far have investigated genetic variants in only a handful of around 20,000 genes in the human genome. The selection of classical candidate genes is based on preexisting knowledge of the neurobiological mechanisms underlying psychiatric disorders. By testing what is already known or suggested, it is difficult to generate new hypotheses and identify still-unknown risk mechanisms. Moreover, assuming the genetic architecture of BPD is comparable to other psychiatric disorders, most of the candidate studies for BPD are statistically underpowered. Therefore, to discover the molecular genetic mechanisms underlying BPD, large-scale studies that systematically investigate the whole genome are needed. The next section will describe genome-wide approaches.

2.2.2 Genome-wide approaches

As a result of biotechnological advances, it became possible in 2005 to carry out genome-wide association studies (GWAS) in samples of several hundred to thousands of individuals (Hirschhorn & Daly, 2005). In GWAS, genetic variants, primarily SNPs, are examined without hypothesizing about the role of any particular gene or biological function (Psychiatric GWAS Consortium Coordinating Committee et al., 2009). This means that a large number of SNPs, typically 1–2 million today, are being investigated to identify differences in the frequency of genetic variants between groups (The International HapMap Consortium, 2007). In recent years, systematic screening of the whole genome in GWAS has been established as a successful method to identify genetic risk variants for traits and disorders. An increasing number of variants, genes, pathways, and previously unknown biological processes have been identified in the past decade for a large number of complex diseases.

However, substantial information is contained in the results of GWAS beyond the identified genome-wide hits. The information from findings below the genome-wide significance threshold is increasingly finding application in newly developed statistical evaluation methods. These include gene and gene set-based evaluation methods, as well as the application of polygenic risk scores (PRS) or LD-score regression (see the following paragraphs).

In the first BPD case-control GWAS, the genome-wide data of 998 BPD patients and 1545 controls was analyzed (Witt et al., 2017). While the single variant analysis did not reveal a genome-wide association signal, gene-based analyses taking into account all variants in each gene indicated a significant enrichment of associated variants in the genes Plakophilin-4 (*PKP4*) and dihydropyrimidine dehydrogenase (*DPYD*). *DPYD* has shown a pleiotropic effect, that is, it has been associated with other psychiatric disorders, such as schizophrenia and bipolar disorder (Duan et al., 2014; Schizophrenia Working Group of the Psychiatric Genomics Consortium, 2014). The pleiotropic effects of genetic variants might predispose an individual to an increased risk for a range of disorders. Additionally, gene set-based analyses implicated the gene set "Exocytosis" to be involved in the pathophysiology of BPD (Witt et al., 2017).

Apart from identifying genetic risk factors for the investigated phenotypes in order to understand the underlying pathophysiology, this study also used the obtained results to investigate the degree to which BPD and other psychiatric disorders share genetic influences, that is, to assess the shared etiology between disorders using polygenic methods. The first of them was the estimation of PRS, which uses GWAS results to predict the genetic load for a trait of an individual. PRS are estimated as the sum of risk alleles weighted by their respective independently estimated effect sizes (Wray et al., 2014) and thus provide a quantitative measure of the cumulative genetic risk or vulnerability that an individual possesses for a disorder or a trait (Fig. 1). Witt et al. estimated the PRS for bipolar disorder, schizophrenia, and MDD in their sample of BPD cases and controls and observed that the genetic load for these disorders was increased in the case group compared with the controls. As comorbidity with MDD was observed in around 70% of the BPD cases (bipolar disorder and schizophrenia were exclusion criteria), they tested whether this affected the PRS enrichment. There was no evidence for differences in the genetic risk for MDD between the BPD cases with and without MDD, and the genetic risk was increased in both subgroups compared with the controls (Witt et al., 2017).

To further explore the genetic relatedness of BPD with other psychiatric disorders, Witt et al. also applied LD-score regression (LDSC; Bulik-Sullivan et al., 2015). This method uses the fact that SNPs can be associated to other SNPs (they are in "linkage disequilibrium" or LD) (Slatkin, 2008). The GWAS effect-size estimate for a given SNP thus incorporates the effects of other SNPs in LD. This information can be used to estimate the heritability explained by SNPs (SNP-based heritability) and the genetic correlations between phenotypes, that is, to what extent their genetic risk factors are shared. In line with the PRS analysis, they found a genetic overlap between BPD and bipolar disorder ($r_g = 0.28$), schizophrenia ($r_g = 0.34$), and MDD ($r_g = 0.57$).

The first GWAS of BPF investigated 7125 subjects from the general population assessed with the PAI-BOR (Lubke et al., 2014). The SNP-based heritability of BPF was 23%, primarily based on the affect instability subscale ($h^2 = 42.7\%$), while no significant heritability was detected for self-harm, negative relations, or identity problems. The genome-wide analysis indicated genetic variation in the gene *SERINC5*, which encodes a protein involved in myelination, to be associated with BPF. While this signal failed to reach genome-wide significance, it was replicated in an independent cohort. In a large population-based GWAS in a sample of 108,038 subjects, PRS derived from this GWAS were significantly correlated with neuroticism (Gale et al., 2016), in line with results from twin studies (Distel et al., 2009; Kendler et al., 2011). The Borderline Personality Consortium is currently working on identifying the genetic variants associated with BPF (Colodro-Conde et al., 2017) by pooling these and other results of multiple GWAS conducted by different research teams to perform a meta-analysis. Preliminary results show a highly significant positive genetic correlation with the results published by Witt et al. (2017), indicating that BPF and BPD are likely to share a large part of their genetic architecture and therefore partially cover the same construct.

In agreement with the results from formal genetic studies, the molecular genetic studies indicate that a genetically influenced shared etiology exists between BPD, BPF, and other psychiatric disorders. However, compared with genetic studies

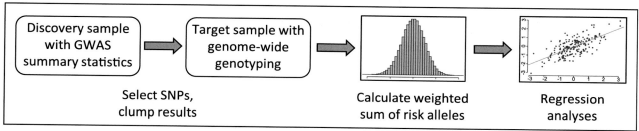

FIG. 1 Stages involved in calculating PRS. *(Adapted from Wray, N. R., Lee, S. H., Mehta, D., Vinkhuyzen, A. A., Dudbridge, F., & Middeldorp, C. M. (2014). Research review: Polygenic methods and their application to psychiatric traits.* Journal of Child Psychology and Psychiatry, 55(10), 1068–1087. *doi:10.1111/jcpp.12295.)*

of other psychiatric disorders such as MDD, where some sample sizes have reached >130,000 cases (Major Depressive Disorder Working Group of the Psychiatric Genomics Consortium, 2017), the samples for BPD and BPF are still small—especially for reliable gene discovery and predictive use of derived PRS, for which substantially larger samples are needed.

While studies of genetic variants can be used to target biological mechanisms underlying the inherited genetic risk for BPD, molecular genetic methods can also be applied to investigate the mechanisms by which environmental risk factors impact BPD etiology.

2.3 Epigenetic factors

Environmental influences, especially early life trauma, maltreatment, and physical and/or sexual abuse (American Psychiatric Association, 2013) have been shown to be an important risk factor for BPD. One potential mechanism that has been suggested to mediate the effect of the environment on disease risk later in life is epigenetic modification of DNA—that is, modification that does not alter the genetic code, but instead regulates gene expression in a time- and cell-type-dependent fashion. One such modification is DNA methylation, in which a methyl group ($-CH_3$) is covalently bound to the DNA, mostly at so-called cytosine-phosphate-guanine (CpG) sites, which can result in altered gene expression.

Animal studies have demonstrated that epigenetic processes are an important mediator of the effect of environmental risk factors on biological systems, behavior, and mental health. For example, in a rat model, reduced maternal care behavior resulted in altered stress reactivity and HPA axis regulation in offspring. It was shown that this was mediated via epigenetic programming, in particular by methylation changes occurring in exons 1–7 of the glucocorticoid receptor gene (*NR3C1*) (Weaver et al., 2004).

In humans, differential methylation at the corresponding site can be linked to adverse environments and subsequent alterations of the endocrine stress response (Oberlander et al., 2008). Epigenetic mechanisms can mediate the effects of early life experiences on adult behavior and susceptibility to psychiatric disorders (Dammann et al., 2011; Labonte et al., 2012; McGowan et al., 2009; Uher, 2011) and influence an individual's response to therapy (Uher, 2011). Methylation levels, despite being relatively stable, can change over time (McGowan, Meaney, & Szyf, 2008). There is first evidence that psychotherapy as well as antidepressants may change the methylation of specific genes (Fries, Walss-Bass, Soares, & Quevedo, 2016; Lopez et al., 2013; Ludwig & Dwivedi, 2016; Menke & Binder, 2014; Perroud et al., 2013; Wilkinson & Goodyer, 2011). In summary, epigenetic signatures may be related to risk and resilience and prognosis of treatment.

In regard to the high incidence of traumatic experiences in BPD subjects and their strong implication in its etiology, the role of methylation processes in BPD has been subject to research. Several studies on methylation patterns of candidate genes in BPD have been carried out: comparing methylation levels of *NR3C1* in subjects with BPD and MDD, BPD patients with higher levels of childhood sexual abuse showed increased methylation levels (Perroud et al., 2011). In line with this observation, a study comparing the methylation levels of sites in 14 candidate genes in BPD patients and controls found an increased methylation level in BPD patients for *NR3C1*, but also for *HTR2A*, *MAOA*, *MAOB*, and *S-COMT*, showing that epigenetic modulations of more than one gene might mediate the influence of environmental factors (Dammann et al., 2011). A study by Radtke et al. (2015) showed that methylation of *NR3C1* was also associated with childhood maltreatment in subjects from the general population and contributed to BPD related symptoms. In another study, the difference in *BDNF* methylation between BPD patients and controls was investigated longitudinally (Perroud et al., 2013). At baseline, methylation levels were increased in BPD patients, and childhood maltreatment was associated with higher methylation. Interestingly, increased methylation levels in BPD cases only decreased in patients who successfully responded to psychotherapy. This indicates that methylation changes might mediate both the effects of adverse early experience and psychotherapeutic treatment on BPD.

However, similar to genetic association studies, methylation studies investigating known candidate genes cannot unveil previously unknown biological pathways, and it must be postulated that a multitude of genes are involved. For this, studies systematically investigating genome-wide methylation patterns need to be carried out.

So far, two studies have been published that investigated genome-wide methylation changes associated with BPD. The first study, assessing 27,578 CpG sites, showed increased methylation in six genes in 24 female BPD patients, compared with 11 healthy controls (Dammann et al., 2011). Using a denser chip covering 450,000 methylation sites, Prados et al. (2015) investigated methylation in 96 BPD cases with high levels, and 93 MDD cases with low levels of childhood maltreatment. Methylation differences between BPD and MDD, as well as in association with the severity of childhood maltreatment, were found in or in proximity to 14 genes. For example, miR124-3, a microRNA, regulates genes including *NR3C1*, and therefore might mediate altered stress regulation in subjects with BPD.

3   Summary

BPF and the risk for BPD are heritable, complex traits. The genetic factors influencing BPF and BPD appear to be partially shared with the ones for disorders such as bipolar, schizophrenia, MDD, and ADHD and personality traits such as neuroticism. In combination with environmental risk factors (early and current), which can be mediated by epigenetic mechanisms, they modulate disease mechanisms, leading to an increase in BPF and BPD (see Fig. 2).

First findings stem from candidate gene association and methylation studies linking specific genes to the risk for BPD. The genome-wide studies carried out in recent years used sample sizes that were too small to have the power to reliably detect contributing common genetic risk variants—each accounting for only a small amount of variance. It has been hypothesized, and subsequently empirically shown, that several 10,000–100,000 subjects are needed to reliably identify these genetic effects (Levinson et al., 2014; Sullivan et al., 2018). Therefore, while the current genome-wide studies give first insight into the disease etiology and the genetic basis for comorbidity with other disorders, it is clear that they only represent a first step. At this point, no relevant predictions or paths for treatment can be based on their results.

Similarly, early GWAS of other psychiatric disorders in hundreds to thousands of subjects lacked the necessary statistical power, and did not reliably identify replicable results. A systematic increase of sample size, primarily due to the pursuit of meta-analytic approaches in large international consortiums, such as the Psychiatric Genomics Consortium (PGC), have resulted in the reliable identification of genetic variants associated with schizophrenia (Schizophrenia Working Group of the Psychiatric Genomics Consortium, 2014), bipolar disorder (Stahl et al., 2018), and major depressive disorder (Major Depressive Disorder Working Group of the Psychiatric Genomics Consortium, 2017). Efforts to pursue the same approach for the study of borderline personality are currently being implemented.

4   Future directions

The risk for BPD is considered to be determined by a broad range of interacting mechanisms (see Fig. 2), and patients vary with respect to affected mechanisms beyond the categorical diagnosis of BPD, displaying heterogeneity in symptom categories or subphenotypes. These include changes in functioning or structure of the brain, stress processing and regulation, synaptic transmission, and patterns of behavior and cognition.

Advances in psychiatric genetics in respect to BPD are very exciting. However, if molecular genetic studies are to make a substantial contribution to precision medicine for BPD, drastically larger genome- and epigenome-wide studies must be performed. While this is a huge task for the field, it is not an impossible challenge. In recent years, collaboration efforts of the international psychiatric genetics community, for example, in the PGC, have led to a vast increase of sample sizes for various disorders, such as schizophrenia (>35,000 cases) or MDD (>130,000 cases; Major Depressive Disorder Working Group of the Psychiatric Genomics Consortium, 2017). In these efforts, the collection of severely affected individuals by clinicians can complement studies of BPF and data sets from newly established large population-based cohorts. One such

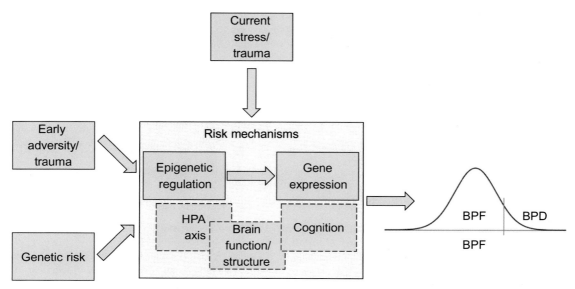

FIG. 2   Schematic depiction of the interaction of genetic and environmental risk factors and risk mechanisms.

cohort is the UK Biobank, a long-term biobank study in the United Kingdom investigating the respective contributions of genetic predisposition and environmental exposure to the development of disease, which includes genetic data, self-reports, interviews, and health repository data. Additionally, new strategies, such as the participation of patients and healthy subjects in web-based surveys, including the mailing of saliva sampling kits, need to be explored. Next, the use of the wealth of data collected by companies offering direct-to-consumer genetic testing, such as 23andMe, have the potential to enlarge sample sizes. In this context, the option to study BPF in the general population with standardized questionnaires broadens the opportunities to assess very large samples.

Future molecular genetic research should focus on the dimensional nature and possible subphenotypes of the complex disorder in affected individuals and general population. As the development of BPD symptomatology over time is influenced by genetic factors (Bornovalova et al., 2013; Reichborn-Kjennerud et al., 2015), changes over time should also be assessed in longitudinal studies.

A detailed assessment of the symptom presentation (symptomics) is of crucial importance for these approaches, as this enables the dissection of which molecular or neurobiological mechanisms underlie which symptoms or symptom clusters.

As described, the options to investigate the underlying mechanisms of BPD using molecular genetic approaches are many and various. Methods such as PRS analyses or LD-score regression offer the opportunity to leverage the results of large GWAS of subphenotypes carried out in independent samples for the subsequent analysis of BPD target samples. For example, large GWAS meta-analyses are performed for structural and functional MRI subphenotypes within the ENIGMA Network—a consortium collecting brain imaging and genetics data—and GWAS of personality measures are accessible through large data sets, such as the UK Biobank.

At the time of writing, due to the constraints mentioned herein, molecular genetic methods cannot yet contribute to precision medicine to treat BPD. However, this has the potential to change in the next few years, with an increased understanding of risk mechanisms, an increased predictive value of genetic measures, and the application of sophisticated biostatistical models to model the complex associations. Integrating different levels of data, including genetic, epigenetic, phenotypic, and clinical data, will be needed to derive algorithms and risk profiles predicting, for example, disease risk, course of disease, and treatment response.

Integrating molecular genetic data, specifically for subdimensions or mechanisms of BPD, into algorithms to assess risk profiles could be used to predict treatment response and therefore inform clinicians of the optimal assignment of affected individuals to specific combinations of psychotherapeutic or pharmacological treatment choices. Thus, in the future, genetic analyses may contribute to prediction algorithms for patient subgroups with life-threatening BPD, as well as individuals with different loads of BPF, and accordingly enable personalization of treatment.

References

Agius, M., Lee, J., Gardner, J., & Wotherspoon, D. (2012). Bipolar II disorder and borderline personality disorder—Co-morbidity or spectrum? *Psychiatria Danubina*, 24(Suppl. 1), 197–201.

Amad, A., Ramoz, N., Thomas, P., Jardri, R., & Gorwood, P. (2014). Genetics of borderline personality disorder: Systematic review and proposal of an integrative model. *Neuroscience and Biobehavioral Reviews*, 40, 6–19. https://doi.org/10.1016/j.neubiorev.2014.01.003.

American Psychiatric Association (2013). *Diagnostic and statistical manual of mental disorders (DSM-5®)*. American Psychiatric Pub.

Bassett, D. (2012). Borderline personality disorder and bipolar affective disorder. Spectra or spectre? A review. *The Australian and New Zealand Journal of Psychiatry*, 46(4), 327–339. https://doi.org/10.1177/0004867411435289.

Bornovalova, M. A., Hicks, B. M., Iacono, W. G., & McGue, M. (2013). Longitudinal twin study of borderline personality disorder traits and substance use in adolescence: Developmental change, reciprocal effects, and genetic and environmental influences. *Personality Disorders*, 4(1), 23–32. https://doi.org/10.1037/a0027178.

Bulik-Sullivan, B. K., Loh, P. R., Finucane, H. K., Ripke, S., Yang, J., Schizophrenia Working Group of the Psychiatric Genomics Consortium, ... Neale, B. M. (2015). LD Score regression distinguishes confounding from polygenicity in genome-wide association studies. *Nature Genetics*, 47(3), 291–295. https://doi.org/10.1038/ng.3211.

Calati, R., Gressier, F., Balestri, M., & Serretti, A. (2013). Genetic modulation of borderline personality disorder: Systematic review and meta-analysis. *Journal of Psychiatric Research*, 47(10), 1275–1287. https://doi.org/10.1016/j.jpsychires.2013.06.002.

Carpenter, R. W., & Trull, T. J. (2013). Components of emotion dysregulation in borderline personality disorder: A review. *Current Psychiatry Reports*, 15(1), 335. https://doi.org/10.1007/s11920-012-0335-2.

Clark, L. A. (2007). Assessment and diagnosis of personality disorder: Perennial issues and an emerging reconceptualization. *Annual Review of Psychology*, 58, 227–257. https://doi.org/10.1146/annurev.psych.57.102904.190200.

Colodro-Conde, L. C., Amin, N., Hottenga, J.-J., Gizer, I., Trull, T., van Duijn, C., ... Medland, S. (2017). The first genome-wide association meta-analysis of borderline personality disorder features. *European Neuropsychopharmacology*, 27, S504–S505. https://doi.org/10.1016/j.euroneuro.2016.09.609.

Dammann, G., Teschler, S., Haag, T., Altmuller, F., Tuczek, F., & Dammann, R. H. (2011). Increased DNA methylation of neuropsychiatric genes occurs in borderline personality disorder. *Epigenetics*, 6(12), 1454–1462. https://doi.org/10.4161/epi.6.12.18363.

Distel, M. A., Carlier, A., Middeldorp, C. M., Derom, C. A., Lubke, G. H., & Boomsma, D. I. (2011). Borderline personality traits and adult attention-deficit hyperactivity disorder symptoms: A genetic analysis of comorbidity. *American Journal of Medical Genetics, Part B, Neuropsychiatric Genetics, 156B*(7), 817–825. https://doi.org/10.1002/ajmg.b.31226.

Distel, M. A., Middeldorp, C. M., Trull, T. J., Derom, C. A., Willemsen, G., & Boomsma, D. I. (2011). Life events and borderline personality features: The influence of gene-environment interaction and gene-environment correlation. *Psychological Medicine, 41*(4), 849–860. https://doi.org/10.1017/S0033291710001297.

Distel, M. A., Roeling, M. P., Tielbeek, J. J., van Toor, D., Derom, C. A., Trull, T. J., & Boomsma, D. I. (2012). The covariation of trait anger and borderline personality: A bivariate twin-siblings study. *Journal of Abnormal Psychology, 121*(2), 458–466. https://doi.org/10.1037/a0026393.

Distel, M. A., Trull, T. J., Derom, C. A., Thiery, E. W., Grimmer, M. A., Martin, N. G., … Boomsma, D. I. (2008). Heritability of borderline personality disorder features is similar across three countries. *Psychological Medicine, 38*(9), 1219–1229. https://doi.org/10.1017/S0033291707002024.

Distel, M. A., Trull, T. J., Willemsen, G., Vink, J. M., Derom, C. A., Lynskey, M., … Boomsma, D. I. (2009). The five-factor model of personality and borderline personality disorder: A genetic analysis of comorbidity. *Biological Psychiatry, 66*(12), 1131–1138. https://doi.org/10.1016/j.biopsych.2009.07.017.

Duan, J., Shi, J., Fiorentino, A., Leites, C., Chen, X., Moy, W., … Gejman, P. V. (2014). A rare functional noncoding variant at the GWAS-implicated MIR137/MIR2682 locus might confer risk to schizophrenia and bipolar disorder. *American Journal of Human Genetics, 95*(6), 744–753. https://doi.org/10.1016/j.ajhg.2014.11.001.

Few, L. R., Grant, J. D., Trull, T. J., Statham, D. J., Martin, N. G., Lynskey, M. T., & Agrawal, A. (2014). Genetic variation in personality traits explains genetic overlap between borderline personality features and substance use disorders. *Addiction, 109*(12), 2118–2127. https://doi.org/10.1111/add.12690.

Fornaro, M., Orsolini, L., Marini, S., De Berardis, D., Perna, G., Valchera, A., … Stubbs, B. (2016). The prevalence and predictors of bipolar and borderline personality disorders comorbidity: Systematic review and meta-analysis. *Journal of Affective Disorders, 195*, 105–118. https://doi.org/10.1016/j.jad.2016.01.040.

Fries, G. R., Walss-Bass, C., Soares, J. C., & Quevedo, J. (2016). Non-genetic transgenerational transmission of bipolar disorder: Targeting DNA methyltransferases. *Molecular Psychiatry, 21*(12), 1653–1654. https://doi.org/10.1038/mp.2016.172.

Gale, C. R., Hagenaars, S. P., Davies, G., Hill, W. D., Liewald, D. C., Cullen, B., … Harris, S. E. (2016). Pleiotropy between neuroticism and physical and mental health: Findings from 108 038 men and women in UK Biobank. *Translational Psychiatry. 6*, https://doi.org/10.1038/tp.2016.56.

Gillespie, N. A., Aggen, S. H., Gentry, A. E., Neale, M. C., Knudsen, G. P., Krueger, R. F., … Kendler, K. S. (2018). Testing genetic and environmental associations between personality disorders and cocaine use: A population-based twin study. *Twin Research and Human Genetics, 21*(1), 24–32. https://doi.org/10.1017/thg.2017.73.

Goodman, M., New, A. S., Triebwasser, J., Collins, K. A., & Siever, L. (2010). Phenotype, endophenotype, and genotype comparisons between borderline personality disorder and major depressive disorder. *Journal of Personality Disorders, 24*(1), 38–59. https://doi.org/10.1521/pedi.2010.24.1.38.

Gunderson, J. G., Zanarini, M. C., Choi-Kain, L. W., Mitchell, K. S., Jang, K. L., & Hudson, J. I. (2011). Family study of borderline personality disorder and its sectors of psychopathology. *Archives of General Psychiatry, 68*(7), 753–762. https://doi.org/10.1001/archgenpsychiatry.2011.65.

Herpertz, S. C., Huprich, S. K., Bohus, M., Chanen, A., Goodman, M., Mehlum, L., … Sharp, C. (2017). The challenge of transforming the diagnostic system of personality disorders. *Journal of Personality Disorders, 31*(5), 577–589. https://doi.org/10.1521/pedi_2017_31_338.

Hirschhorn, J. N., & Daly, M. J. (2005). Genome-wide association studies for common diseases and complex traits. *Nature Reviews Genetics, 6*(2), 95–108. https://doi.org/10.1038/nrg1521.

Hopwood, C. J., Kotov, R., Krueger, R. F., Watson, D., Widiger, T. A., Althoff, R. R., … Zimmermann, J. (2018). The time has come for dimensional personality disorder diagnosis. *Personality and Mental Health, 12*(1), 82–86. https://doi.org/10.1002/pmh.1408.

Insel, T., Cuthbert, B., Garvey, M., Heinssen, R., Pine, D. S., Quinn, K., … Wang, P. (2010). Research domain criteria (RDoC): Toward a new classification framework for research on mental disorders. *The American Journal of Psychiatry, 167*(7), 748–751. https://doi.org/10.1176/appi.ajp.2010.09091379.

Kendler, K. S., Myers, J., & Reichborn-Kjennerud, T. (2011). Borderline personality disorder traits and their relationship with dimensions of normative personality: A web-based cohort and twin study. *Acta Psychiatrica Scandinavica, 123*(5), 349–359. https://doi.org/10.1111/j.1600-0447.2010.01653.x.

Labonte, B., Suderman, M., Maussion, G., Navaro, L., Yerko, V., Mahar, I., … Turecki, G. (2012). Genome-wide epigenetic regulation by early-life trauma. *Archives of General Psychiatry, 69*(7), 722–731. https://doi.org/10.1001/archgenpsychiatry.2011.2287.

Levinson, D. F., Mostafavi, S., Milaneschi, Y., Rivera, M., Ripke, S., Wray, N. R., & Sullivan, P. F. (2014). Genetic studies of major depressive disorder: Why are there no genome-wide association study findings and what can we do about it? *Biological Psychiatry, 76*(7), 510–512. https://doi.org/10.1016/j.biopsych.2014.07.029.

Liu, J., Nie, G., Guo, W., Gong, J., Xiao, B., Cui, X., … Wu, J. (2017). Association between the COMT gene val158met polymorphism and borderline personality disorder: A meta-analysis. *Psychiatry Research, 258*, 614–615. https://doi.org/10.1016/j.psychres.2017.03.039.

Long, E. C., Aggen, S. H., Neale, M. C., Knudsen, G. P., Krueger, R. F., South, S. C., … Reichborn-Kjennerud, T. (2017). The association between personality disorders with alcohol use and misuse: A population-based twin study. *Drug and Alcohol Dependence, 174*, 171–180. https://doi.org/10.1016/j.drugalcdep.2017.01.022.

Lopez, J. P., Mamdani, F., Labonte, B., Beaulieu, M. M., Yang, J. P., Berlim, M. T., … Turecki, G. (2013). Epigenetic regulation of BDNF expression according to antidepressant response. *Molecular Psychiatry, 18*(4), 398–399. https://doi.org/10.1038/mp.2012.38.

Lubke, G. H., Laurin, C., Amin, N., Hottenga, J. J., Willemsen, G., van Grootheest, G., … Boomsma, D. I. (2014). Genome-wide analyses of borderline personality features. *Molecular Psychiatry, 19*(8), 923–929. https://doi.org/10.1038/mp.2013.109.

Ludwig, B., & Dwivedi, Y. (2016). Dissecting bipolar disorder complexity through epigenomic approach. *Molecular Psychiatry, 21*(11), 1490–1498. https://doi.org/10.1038/mp.2016.123.

Major Depressive Disorder Working Group of the Psychiatric Genomics Consortium. (2017). Genome-wide association analyses identify 44 risk variants and refine the genetic architecture of major depression. *bioRxiv.* https://doi.org/10.1101/167577.

Martin-Blanco, A., Ferrer, M., Soler, J., Arranz, M. J., Vega, D., Calvo, N., … Pascual, J. C. (2016). The role of hypothalamus-pituitary-adrenal genes and childhood trauma in borderline personality disorder. *European Archives of Psychiatry and Clinical Neuroscience, 266*(4), 307–316. https://doi.org/10.1007/s00406-015-0612-2.

Matthies, S. D., & Philipsen, A. (2014). Common ground in attention deficit hyperactivity disorder (ADHD) and borderline personality disorder (BPD)-review of recent findings. *Borderline Personality Disorder and Emotion Dysregulation, 1,* 3. https://doi.org/10.1186/2051-6673-1-3.

Maurex, L., Zaboli, G., Wiens, S., Asberg, M., Leopardi, R., & Ohman, A. (2009). Emotionally controlled decision-making and a gene variant related to serotonin synthesis in women with borderline personality disorder. *Scandinavian Journal of Psychology, 50*(1), 5–10. https://doi.org/10.1111/j.1467-9450.2008.00689.x.

McGlashan, T. H., Grilo, C. M., Skodol, A. E., Gunderson, J. G., Shea, M. T., Morey, L. C., … Stout, R. L. (2000). The Collaborative Longitudinal Personality Disorders Study: Baseline Axis I/II and II/II diagnostic co-occurrence. *Acta Psychiatrica Scandinavica, 102*(4), 256–264.

McGowan, P. O., Meaney, M. J., & Szyf, M. (2008). Diet and the epigenetic (re)programming of phenotypic differences in behavior. *Brain Research, 1237,* 12–24. https://doi.org/10.1016/j.brainres.2008.07.074.

McGowan, P. O., Sasaki, A., D'Alessio, A. C., Dymov, S., Labonte, B., Szyf, M., … Meaney, M. J. (2009). Epigenetic regulation of the glucocorticoid receptor in human brain associates with childhood abuse. *Nature Neuroscience, 12*(3), 342–348. https://doi.org/10.1038/nn.2270.

Menke, A., & Binder, E. B. (2014). Epigenetic alterations in depression and antidepressant treatment. *Dialogues in Clinical Neuroscience, 16*(3), 395–404.

Mobascher, A., Bohus, M., Dahmen, N., Dietl, L., Giegling, I., Jungkunz, M., … Lieb, K. (2014). Association between dopa decarboxylase gene variants and borderline personality disorder. *Psychiatry Research, 219*(3), 693–695. https://doi.org/10.1016/j.psychres.2014.06.031.

Morey, L. C. (2007). *Personality Assessment Inventory (PAI): Professional manual.*

Morey, L. C. (2011). *Personality Assessment Inventory.* PsycTESTS Dataset. https://doi.org/10.1037/t03903-000.

Morey, L. C., Benson, K. T., Busch, A. J., & Skodol, A. E. (2015). Personality disorders in DSM-5: Emerging research on the alternative model. *Current Psychiatry Reports, 17*(4), 558. https://doi.org/10.1007/s11920-015-0558-0.

Ni, X., Chan, D., Chan, K., McMain, S., & Kennedy, J. L. (2009). Serotonin genes and gene-gene interactions in borderline personality disorder in a matched case-control study. *Progress in Neuro-Psychopharmacology & Biological Psychiatry, 33*(1), 128–133. https://doi.org/10.1016/j.pnpbp.2008.10.022.

Ni, X., Sicard, T., Bulgin, N., Bismil, R., Chan, K., McMain, S., & Kennedy, J. L. (2007). Monoamine oxidase a gene is associated with borderline personality disorder. *Psychiatric Genetics, 17*(3), 153–157. https://doi.org/10.1097/YPG.0b013e328016831c.

Oberlander, T. F., Weinberg, J., Papsdorf, M., Grunau, R., Misri, S., & Devlin, A. M. (2008). Prenatal exposure to maternal depression, neonatal methylation of human glucocorticoid receptor gene (NR3C1) and infant cortisol stress responses. *Epigenetics, 3*(2), 97–106.

Ørstavik, R. E., Kendler, K. S., Roysamb, E., Czajkowski, N., Tambs, K., & Reichborn-Kjennerud, T. (2012). Genetic and environmental contributions to the co-occurrence of depressive personality disorder and DSM-IV personality disorders. *Journal of Personality Disorders, 26*(3), 435–451. https://doi.org/10.1521/pedi.2012.26.3.435.

Panagopoulos, V. N., Trull, T. J., Glowinski, A. L., Lynskey, M. T., Heath, A. C., Agrawal, A., … Nelson, E. C. (2013). Examining the association of NRXN3 SNPs with borderline personality disorder phenotypes in heroin dependent cases and socio-economically disadvantaged controls. *Drug and Alcohol Dependence, 128*(3), 187–193. https://doi.org/10.1016/j.drugalcdep.2012.11.011.

Paris, J., & Zweig-Frank, H. (2001). A 27-year follow-up of patients with borderline personality disorder. *Comprehensive Psychiatry, 42*(6), 482–487. https://doi.org/10.1053/comp.2001.26271.

Perroud, N., Paoloni-Giacobino, A., Prada, P., Olie, E., Salzmann, A., Nicastro, R., … Malafosse, A. (2011). Increased methylation of glucocorticoid receptor gene (NR3C1) in adults with a history of childhood maltreatment: A link with the severity and type of trauma. *Translational Psychiatry. 1,* https://doi.org/10.1038/tp.2011.60.

Perroud, N., Salzmann, A., Prada, P., Nicastro, R., Hoeppli, M. E., Furrer, S., … Malafosse, A. (2013). Response to psychotherapy in borderline personality disorder and methylation status of the BDNF gene. *Translational Psychiatry. 3,* https://doi.org/10.1038/tp.2012.140.

Prados, J., Stenz, L., Courtet, P., Prada, P., Nicastro, R., Adouan, W., … Perroud, N. (2015). Borderline personality disorder and childhood maltreatment: A genome-wide methylation analysis. *Genes, Brain, and Behavior, 14*(2), 177–188. https://doi.org/10.1111/gbb.12197.

Psychiatric GWAS Consortium Coordinating Committee, Cichon, S., Craddock, N., Daly, M., Faraone, S. V., Gejman, P. V., … Sullivan, P. F. (2009). Genomewide association studies: History, rationale, and prospects for psychiatric disorders. *The American Journal of Psychiatry, 166*(5), 540–556. https://doi.org/10.1176/appi.ajp.2008.08091354.

Radtke, K. M., Schauer, M., Gunter, H. M., Ruf-Leuschner, M., Sill, J., Meyer, A., & Elbert, T. (2015). Epigenetic modifications of the glucocorticoid receptor gene are associated with the vulnerability to psychopathology in childhood maltreatment. *Translational Psychiatry. 5.* https://doi.org/10.1038/tp.2015.63.

Reichborn-Kjennerud, T., Czajkowski, N., Roysamb, E., Orstavik, R. E., Neale, M. C., Torgersen, S., & Kendler, K. S. (2010). Major depression and dimensional representations of DSM-IV personality disorders: A population-based twin study. *Psychological Medicine, 40*(9), 1475–1484. https://doi.org/10.1017/S0033291709991954.

Reichborn-Kjennerud, T., Czajkowski, N., Ystrøm, E., Ørstavik, R., Aggen, S. H., Tambs, K., … Kendler, K. S. (2015). A longitudinal twin study of borderline and antisocial personality disorder traits in early to middle adulthood. *Psychological Medicine, 45*(14), 3121–3131. https://doi.org/10.1017/S0033291715001117.

Rojas, E. C., Cummings, J. R., Bornovalova, M. A., Hopwood, C. J., Racine, S. E., Keel, P. K., … Klump, K. L. (2014). A further validation of the Minnesota borderline personality disorder scale. *Personality Disorders*, *5*(2), 146–153. https://doi.org/10.1037/per0000036.

Sansone, R. A., & Sansone, L. A. (2011). Gender patterns in borderline personality disorder. *Innovations in Clinical Neuroscience*, *8*(5), 16–20.

Schizophrenia Working Group of the Psychiatric Genomics Consortium. (2014). Biological insights from 108 schizophrenia-associated genetic loci. *Nature*, *511*(7510), 421–427. https://doi.org/10.1038/nature13595.

Schmahl, C., Ludascher, P., Greffrath, W., Kraus, A., Valerius, G., Schulze, T. G., … Bohus, M. (2012). COMT val158met polymorphism and neural pain processing. *PLoS One*. *7*(1). https://doi.org/10.1371/journal.pone.0023658.

Siever, L. J. (2005). Endophenotypes in the personality disorders. *Dialogues in Clinical Neuroscience*, *7*(2), 139–151.

Slatkin, M. (2008). Linkage disequilibrium—Understanding the evolutionary past and mapping the medical future. *Nature Reviews. Genetics*, *9*(6), 477–485. https://doi.org/10.1038/nrg2361.

Soloff, P. H., & Chiappetta, L. (2018). 10-Year outcome of suicidal behavior in borderline personality disorder. *Journal of Personality Disorders*, 1–19. https://doi.org/10.1521/pedi_2018_32_332.

Stahl, E., Breen, G., Forstner, A., McQuillin, A., Ripke, S., Cichon, S., … Sklar, P. (2018). Genomewide association study identifies 30 loci associated with bipolar disorder. *bioRxiv*. https://doi.org/10.1101/173062.

Stein, M. B., Pinsker-Aspen, J. H., & Hilsenroth, M. J. (2007). Borderline pathology and the Personality Assessment Inventory (PAI): An evaluation of criterion and concurrent validity. *Journal of Personality Assessment*, *88*(1), 81–89. https://doi.org/10.1080/00223890709336838.

Sullivan, P. F., Agrawal, A., Bulik, C. M., Andreassen, O. A., Borglum, A. D., Breen, G., … Psychiatric Genomics, C. (2018). Psychiatric genomics: An update and an agenda. *The American Journal of Psychiatry*, *175*(1), 15–27. https://doi.org/10.1176/appi.ajp.2017.17030283.

Sullivan, P. F., Daly, M. J., & O'Donovan, M. (2012). Genetic architectures of psychiatric disorders: The emerging picture and its implications. *Nature Reviews. Genetics*, *13*(8), 537–551. https://doi.org/10.1038/nrg3240.

The International HapMap Consortium. (2007). A second generation human haplotype map of over 3.1 million SNPs. *Nature*, *449*, 851. https://doi.org/10.1038/nature06258.

Tomko, R. L., Trull, T. J., Wood, P. K., & Sher, K. J. (2014). Characteristics of borderline personality disorder in a community sample: Comorbidity, treatment utilization, and general functioning. *Journal of Personality Disorders*, *28*(5), 734–750. https://doi.org/10.1521/pedi_2012_26_093.

Torgersen, S. (2000). Genetics of patients with borderline personality disorder. *The Psychiatric Clinics of North America*, *23*(1), 1–9. https://doi.org/10.1016/S0193-953X(05)70139-8.

Torgersen, S., Myers, J., Reichborn-Kjennerud, T., Røysamb, E., Kubarych, T. S., & Kendler, K. S. (2012). The heritability of cluster B personality disorders assessed both by personal interview and questionnaire. *Journal of Personality Disorders*, *26*(6), 848–866. https://doi.org/10.1521/pedi.2012.26.6.848.

Traylor, M., Markus, H., & Lewis, C. M. (2015). Homogeneous case subgroups increase power in genetic association studies. *European Journal of Human Genetics*, *23*(6), 863–869. https://doi.org/10.1038/ejhg.2014.194.

Trull, T. J., & Durrett, C. A. (2005). Categorical and dimensional models of personality disorder. *Annual Review of Clinical Psychology*, *1*, 355–380. https://doi.org/10.1146/annurev.clinpsy.1.102803.144009.

Uher, R. (2011). Genes, environment, and individual differences in responding to treatment for depression. *Harvard Review of Psychiatry*, *19*(3), 109–124. https://doi.org/10.3109/10673229.2011.586551.

van Dongen, J., Slagboom, P. E., Draisma, H. H., Martin, N. G., & Boomsma, D. I. (2012). The continuing value of twin studies in the omics era. *Nature Reviews. Genetics*, *13*(9), 640–653. https://doi.org/10.1038/nrg3243.

Weaver, I. C., Cervoni, N., Champagne, F. A., D'Alessio, A. C., Sharma, S., Seckl, J. R., … Meaney, M. J. (2004). Epigenetic programming by maternal behavior. *Nature Neuroscience*, *7*(8), 847–854. https://doi.org/10.1038/nn1276.

Welander-Vatn, A., Ystrom, E., Tambs, K., Neale, M. C., Kendler, K. S., Reichborn-Kjennerud, T., & Knudsen, G. P. (2016). The relationship between anxiety disorders and dimensional representations of DSM-IV personality disorders: A co-twin control study. *Journal of Affective Disorders*, *190*, 349–356. https://doi.org/10.1016/j.jad.2015.09.038.

Wilkinson, P. O., & Goodyer, I. M. (2011). Childhood adversity and allostatic overload of the hypothalamic-pituitary-adrenal axis: A vulnerability model for depressive disorders. *Development and Psychopathology*, *23*(4), 1017–1037. https://doi.org/10.1017/S0954579411000472.

Witt, S. H., Kleindienst, N., Frank, J., Treutlein, J., Muhleisen, T., Degenhardt, F., … Bohus, M. (2014). Analysis of genome-wide significant bipolar disorder genes in borderline personality disorder. *Psychiatric Genetics*, *24*(6), 262–265. https://doi.org/10.1097/YPG.0000000000000060.

Witt, S. H., Streit, F., Jungkunz, M., Frank, J., Awasthi, S., Reinbold, C. S., … Rietschel, M. (2017). Genome-wide association study of borderline personality disorder reveals genetic overlap with bipolar disorder, major depression and schizophrenia. *Translational Psychiatry*. *7*(6)https://doi.org/10.1038/tp.2017.115.

Wray, N. R., Lee, S. H., Mehta, D., Vinkhuyzen, A. A., Dudbridge, F., & Middeldorp, C. M. (2014). Research review: Polygenic methods and their application to psychiatric traits. *Journal of Child Psychology and Psychiatry*, *55*(10), 1068–1087. https://doi.org/10.1111/jcpp.12295.

Yang, M., Mamy, J., Wang, Q., Liao, Y. H., Seewoobudul, V., Xiao, S. Y., & Hao, W. (2014). The association of 5-HTR2A-1438A/G, COMTVal158Met, MAOA-LPR, DATVNTR and 5-HTTVNTR gene polymorphisms and borderline personality disorder in female heroin-dependent Chinese subjects. *Progress in Neuro-Psychopharmacology & Biological Psychiatry*, *50*, 74–82. https://doi.org/10.1016/j.pnpbp.2013.12.005.

Zanarini, M. C., Frankenburg, F. R., Dubo, E. D., Sickel, A. E., Trikha, A., Levin, A., & Reynolds, V. (1998). Axis I comorbidity of borderline personality disorder. *The American Journal of Psychiatry*, *155*(12), 1733–1739. https://doi.org/10.1176/ajp.155.12.1733.

Zanarini, M. C., Frankenburg, F. R., Yong, L., Raviola, G., Bradford Reich, D., Hennen, J., … Gunderson, J. G. (2004). Borderline psychopathology in the first-degree relatives of borderline and Axis II comparison probands. *Journal of Personality Disorders*, *18*(5), 439–447. https://doi.org/10.1521/pedi.18.5.439.51327.

Zimmerman, M., Rothschild, L., & Chelminski, I. (2005). The prevalence of DSM-IV personality disorders in psychiatric outpatients. *The American Journal of Psychiatry*, *162*(10), 1911–1918. https://doi.org/10.1176/appi.ajp.162.10.1911.

Chapter 20

Genetics of obsessive-compulsive disorder and Tourette disorder

Christie L. Burton[a], Csaba Barta[b], Danielle Cath[c,d], Daniel Geller[e], Odile A. van den Heuvel[f], Yin Yao[g], (Obsessive Compulsive Disorder and Tourette Syndrome Working Group of the Psychiatric Genomics Consortium), Valsamma Eapen[h,i,*], Edna Grünblatt[j,k,l,*] and Gwyneth Zai[m,n,*]

[a]Neurosciences & Mental Health, Hospital for Sick Children, Toronto, ON, Canada, [b]Institute of Medical Chemistry, Molecular Biology and Pathobiochemistry, Semmelweis University, Budapest, Hungary, [c]Department of Psychiatry, Groningen University and University Medical Center Groningen, Groningen, The Netherlands, [d]Department of Specialist Training, Drenthe Mental Health Institution, Assen, The Netherlands, [e]Department of Psychiatry, Massachusetts General Hospital and Harvard Medical School, Boston, MA, United States, [f]Amsterdam University Medical Centers, Department of Psychiatry and Department of Anatomy & Neuroscience, Vrije Universiteit Amsterdam, Amsterdam Neuroscience, Amsterdam, The Netherlands, [g]Lab of Statistical Genomics and Data Analysis, Intramural Program, National Institute of Mental Health, National Institutes of Health, Bethesda, MD, United States, [h]Department of Psychiatry, University of New South Wales, Sydney, NSW, Australia, [i]Academic Unit of Child Psychiatry South West Sydney, Ingham Institute and Liverpool Hospital, Sydney, NSW, Australia, [j]Department of Child and Adolescent Psychiatry and Psychotherapy, Psychiatric Hospital, University of Zurich, Zurich, Switzerland, [k]Neuroscience Center Zurich, University of Zurich and ETH Zurich, Zurich, Switzerland, [l]Zurich Center for Integrative Human Physiology, University of Zurich, Zurich, Switzerland, [m]Neurogenetics Section, Molecular Brain Science Department, Campbell Family Mental Health Research Institute, Centre for Addiction and Mental Health, Toronto, ON, Canada, [n]Department of Psychiatry, University of Toronto, Toronto, ON, Canada

1 Introduction

Obsessive-compulsive disorder (OCD) and Tourette disorder (TD) are chronic and debilitating neuropsychiatric disorders, affecting 1%–3% (Ruscio, Stein, Chiu, & Kessler, 2010) and 0.5%–1% (Robertson, 2015; Scharf et al., 2015) of the general population, respectively. OCD is characterized by obsessions (irrational, unwanted, and intrusive thoughts, images, or urges that generate distress) and/or compulsions (repetitive rituals to alleviate distress caused by the obsessions; American Psychiatric Association, 2013). Individuals with TD have one or more vocal or motor tics that began before age 18, and were present for more than 1 year (American Psychiatric Association, 2013; Bertelsen et al., 2014). A higher rate of comorbid OCD and TD has previously been well-documented (Hirschtritt et al., 2015; Peterson, Pine, Cohen, & Brook, 2001) and these disorders also share genetic susceptibility (Davis et al., 2013; Grados et al., 2008; Mathews & Grados, 2011; McGrath et al., 2014). This chapter discusses the genetics of OCD and TD.

1.1 Heritability and familiality of OCD and TD

Family and twin studies clearly demonstrate a genetic component for both OCD and TD. First-degree relatives of individuals affected by TD are 10–100 times more likely to have TD than the general population (O'Rourke, Scharf, Yu, & Pauls, 2009). The largest clinical twin study for TD included 30 monozygotic and 13 dizygotic twins (Price, Kidd, Cohen, Pauls, & Leckman, 1985). The concordance of monozygotic twins with TD or chronic tics was 77%, but the concordance of dizygotic twins for TD or chronic tics was only 23%, suggesting a dominant genetic component in TD (O'Rourke et al., 2009). Two recent twin studies in tics and TD found heritability rates between 0.25 and 0.77, depending on phenotypic definitions (Mataix-Cols et al., 2015; Zilhao et al., 2017). Heritability estimates for OCD from twin studies are somewhat lower (.30–.65), depending on age and sex of the sample (van Grootheest, Cath, Beekman, & Boomsma, 2005; Van Grootheest, Cath, Beekman, & Boomsma, 2007). TD and OCD are also highly comorbid. Fifty percent of individuals with TD exhibit obsessive-compulsive behaviors (Albin & Mink, 2006), and conversely, about 20% of individuals

---

* These senior authors contributed equally.

Personalized Psychiatry. https://doi.org/10.1016/B978-0-12-813176-3.00020-1

with OCD show tics (de Vries et al., 2016). Additionally, a family study reported that obsessive-compulsive symptoms (OCS) were significantly correlated in sibling pairs concordant for TD (Leckman et al., 2003). Twin studies indicate that tics and OCD may share some genetic liabilities, with genetic correlations between 0.37 and 0.58 (Guo et al., 2012; Zilhao, Smit, Boomsma, & Cath, 2016). Similarly, genome-wide association studies (GWAS) data in large clinical samples of TD and OCD indicate some shared but also distinct genetic architectures for OCD and TD, revealing a genetic correlation of 0.41, and a heritability point estimate of 0.37 and 0.58, respectively, for OCD and TD (Davis et al., 2013). A recent study from the Brainstorm consortium reported a higher genetic correlation between OCD and TD (0.7) using LD score regression (Antilla et al., 2016). Although highly correlated, each disorder had distinct correlations with other psychiatric disorders. For example, OCD was highly genetically correlated with anorexia nervosa, bipolar disorder, and schizophrenia, while TD was highly correlated with attention-deficit/hyperactivity disorder (ADHD).

1.2 Candidate gene studies in OCD and TD

In the past two decades, many hypothesis-based association studies have been conducted, searching for an association between the serotonergic, glutamatergic, dopaminergic, and other systems with OCD and TD (see a selection in Table 1). The serotonergic system has been the most widely studied system in OCD because selective serotonin reuptake inhibitors (SSRIs) are the primary pharmacological OCD treatment (Bandelow et al., 2016; Davis et al., 2013; Murphy et al., 2013). Furthermore, glutamate abnormalities have been implicated in the etiology of OCD, specifically in the cortico-striato-thalamo-cortical (CSTC) circuits (Pauls, Abramovitch, Rauch, & Geller, 2014). Last, the role of the dopaminergic system in OCD has come from the effects of antipsychotics as adjunct agents to antidepressants (Pauls et al., 2014). Animal, neuroimaging, and neurochemical studies also implicate dopaminergic dysfunction in the pathophysiology of OCD (Koo, Kim, Roh, & Kim, 2010). Similarly, these systems were investigated in TD, as the glutamate receptor, ionotropic N-methyl D-aspartate (NMDA) 2B gene (*GRIN2B*), known to be expressed in the hippocampus, basal ganglia, and cerebral cortex, codes for subunit 2 of the NMDA receptor as well as acts as a binding site for glutamate, thus it is involved in excitatory neurotransmission (Schito et al., 1997).

1.3 Genome-wide association studies in OCD and TD

With four studies to date, OCD GWAS is still in its early days. The first GWAS was conducted by the International Obsessive-Compulsive Disorder Foundation (IOCDF) Genetics Collaborative (Stewart, Yu, et al., 2013), including 1,465 OCD patients with mixed age of onset, 5,557 controls, and 400 trios. In the trio analysis, a SNP near the BTB domain containing 3 (*BTBD3*) reached genome-wide significance ($P = 3.84 \times 10^{-8}$), while intronic SNPs within the DLG associated protein 1 (*DLGAP1*), a gene previously associated with OCD, approached genome-wide significance in the case-control analysis. No genome-wide SNPs were identified in the combined trio-case/control analysis, but the top SNPs were enriched for frontal lobe and methylation expression quantitative trait loci (eQTLs).

The second GWAS was conducted by the OCD Collaborative Genetics Association Study (OCGAS) consortium on a sample of 5,061 individuals, including 1,065 families (with 1,406 childhood-onset OCD patients) and unrelated population-based controls (Mattheisen et al., 2015). A locus near the protein tyrosine phosphatase receptor type D (*PTPRD*) approached significance, while markers with next-smallest P-value were near cadherin genes, *CDH9* and *CDH10*.

TABLE 1 Summary of some candidate gene studies in OCD and TD

System	Gene (SNPS/ polymorphism)	Risk allele	
		OCD	**TD**
Serotonergic	*SLC6A4* (5-HTTLPR, rs25531)	L$_A$ allele (Brem, Grunblatt, Drechsler, Riederer, & Walitza, 2014; Taylor, 2013, 2016; Walitza et al., 2014)	L$_A$ allele (Moya et al., 2013)
	HTR2A (rs6311; rs6313)	A allele (Taylor, 2016; Walitza et al., 2012)	—
	MAOA	Trend, mostly in males (Brem et al., 2014; Di Nocera, Colazingari, Trabalza, Mamazza, & Bevilacqua, 2014; Liu, Yin, Wang, Zhang, & Ma,	Trend (Diaz-Anzaldua et al., 2004; Gade et al., 1998)

TABLE 1 Summary of some candidate gene studies in OCD and TD—cont'd

System	Gene (SNPS/ polymorphism)	Risk allele	
		OCD	TD
		2013; Mas et al., 2014; Sampaio et al., 2015; Taylor, 2013; Walitza et al., 2004)	
	TPH1; TPH2	Trend (Brem et al., 2014; Di Nocera et al., 2014; Liu et al., 2013; Mas et al., 2014; Sampaio et al., 2015; Taylor, 2013; Walitza et al., 2004)	Trend (Comings et al., 1996; Dehning et al., 2010; Mossner, Muller-Vahl, Doring, & Stuhrmann, 2007)
	HTR1B; HTR2C	Trend (Brem et al., 2014; Di Nocera et al., 2014; Liu et al., 2013; Mas et al., 2014; Sampaio et al., 2015; Taylor, 2013; Walitza et al., 2004)	Trend (Comings et al., 1996; Dehning et al., 2010; Mossner et al., 2007)
	TDO2	—	Trend (Comings et al., 1996; Dehning et al., 2010; Mossner et al., 2007)
Glutamatergic	SLC1A1 (rs301443, rs12682897, rs378041)	Trend, particularly gender specific (Dickel et al., 2006; Stewart, Fagerness, et al., 2007; Stewart, Mayerfeld, et al., 2013; Taylor, 2013)	—
	GRIN2B	Trend (Qin et al., 2016)	Trend (Che et al., 2015)
	ADORA1; ADORA2A	—	Trend (Ciruela et al., 2006; Hettinger, Lee, Linden, & Rosin, 2001; Janik, Berdynski, Safranow, & Zekanowski, 2015)
Dopaminergic	COMT (rs4680)	Trend, particularly gender specific (Melo-Felippe et al., 2016; Taylor, 2013, 2016)	—
	DRD2, ANKK1 (rs1800497)	Symmetry associated (Lochner et al., 2016), while no association with OCD (Taylor, 2013)	Risk allele (Yuan et al., 2015)
	DRD4	No association with OCD (Taylor, 2013)	Trend (Asghari et al., 1995; Diaz-Anzaldua et al., 2004; Grice et al., 1996; Liu et al., 2014)
	SLC6A3 (DAT1)	No association with OCD (Taylor, 2013)	Trend (Comings et al., 1996; Diaz-Anzaldua et al., 2004; Tarnok et al., 2007; Yoon et al., 2007)
Others	BDNF	Inconsistent (Taylor, 2013; Zai et al., 2015)	—
	MOG, OLIG2	Trend (Stewart, Platko, et al., 2007; Zai et al., 2004)	—
	SLITRK1-5	Trend (Ozomaro et al., 2013; Song et al., 2017)	Trend (Abelson et al., 2005)
	HDC	—	Trend (Ercan-Sencicek et al., 2010; Karagiannidis et al., 2013; Lei et al., 2012)
	NRXN1	Trend (Noh et al., 2017)	Trend (Sundaram, Huq, Wilson, & Chugani, 2010)
	BTBD9	—	Trend (Guo et al., 2012; Riviere et al., 2009)

Abbreviation: ADORA1, ADORA2A, adenosine receptor A1 or A2A; ANKK1, ankyrin repeat and kinase domain containing 1; BDNF, brain-derived neurotrophic factor; COMT, catechol-O-methyltransferase; DRD2, dopamine D2 receptor; DRD4, dopamine D4 receptor; GRIN2B, ionotropic N-methyl D-aspartate (NMDA) 2B receptor; HDC, histidine decarboxylase; HTR1B, serotonin 1D-beta receptor; HTR2A, serotonin 2A receptor; HTR2C, serotonin 2C receptor; MAOA, monoamine oxidase A; MOG, myelin-oligodendrocyte glycoprotein; NRXN1, neurexin 1; OLIG2, oligodendrocyte lineage transcription factor 2; SLC1A1, glutamate transporter; SLC6A3, DAT1, dopamine transporter; SLC6A4, serotonin transporter; SLITRK1, SLIT and NTRK like family member 1; TDO2, tryptophan 2,3-dioxygenase; TPH1 & 2, tryptophan hydroxylase.

Recently, the two previous GWAS were meta-analyzed with a total sample of 2,688 Caucasian OCD patients and 7,037 genomically matched controls (International Obsessive Compulsive Disorder Foundation Genetics Collaborative (IOCDF-GC) and OCD Collaborative Genetics Association Studies (OCGAS), 2017). Although no markers were genome-wide significant, the SNPs with the lowest *P*-values tagged haplotype blocks close to, or in, cancer susceptibility 8 (*CASC8/CASC11*), glutamate ionotropic receptor delta type subunit 2 (*GRID2*), and KIT proto-oncogene receptor tyrosine kinase (*KIT*). The top markers were also near or within genes identified in previous genome-wide studies: Ankyrin repeat and SOCS box containing 13 (*ASB13*), *GRIK2*, *CHD20*, *DLGAP1*, fas apoptotic inhibitory molecule 2 (*FAIM2*), *PTPRD*, and R-spondin 4 (*RSPO4*). Larger samples should lead to genome-wide hits, as this sample was underpowered.

Finally, GWAS of OCS and traits in community, rather than clinic, samples have identified genome-wide variants. A GWAS of OCS with 6,931 participants from the Netherlands Twin Registry (NTR) study identified a genome-wide significant marker (rs8100480) in the BLOC-1 related complex subunit 8 (*BORCS8* or *MEF2BNB*) gene and four significant genes in the myocyte enhancer factor 2B (*MEF2B*) family (den Braber et al., 2016). Polygenic risk based on the IOCDF GWAS significantly predicted OCS, suggesting that OCS may be a useful subclinical phenotype in gene discovery for OCD. A GWAS of obsessive-compulsive traits in a sample of Caucasian children and adolescents from the community ($n = 4,945$) identified genome-wide significant markers in *NPAS2* and *PTPRD*, a gene identified a previous OCD GWAS study. A hypothesis-driven GWAS demonstrated that SNPs linked to central nervous system (CNS) development, but not glutamate, were associated with obsessive-compulsive traits (Burton et al., 2015).

In TD, there are no genome-wide significant loci to date from the two published GWAS studies, likely because of lack of power. The first in 2012 by the Tourette Syndrome Association International Consortium for Genetics (TSAICG) included 1,285 cases and 4,964 ancestry-matched controls (Scharf et al., 2013). The top signal was an SNP located in the collagen type XXVII alpha 1 chain (*COL27A1*) gene ($P = 1.85 \times 10^{-6}$).

A replication study with the top 42 SNPs of the original GWAS was performed with a sample of 609 cases and 610 controls. The top signal in the meta-analysis was at rs2060546 ($P = 5.8 \times 10^{-7}$), in proximity of the netrin 4 (*NTN4*) gene on chromosome 12q22, which codes for netrin 4, an axon guidance protein expressed in the developing striatum. Many of the previous findings (26 out of 42) showed a similar trend underlining the reliability of the GWAS hits as true risk factors for TD (Paschou et al., 2014).

1.4 Copy number variation in OCD and TD

To date, few copy number variation (CNV) studies have focused on OCD associations. One recent study of 307 unrelated OCD patients (including 174 cases from complete trios) and 3,681 population controls (Gazzellone et al., 2016) reported the rate of de novo CNVs in OCD was lower than other neurodevelopmental disorders (2.3%). OCD patients had CNVs in genes previously associated with OCD and TD in candidate and GWAS studies, such as *PTPRD* and *BTBD9* (Guo et al., 2012; Mattheisen et al., 2015; Riviere et al., 2009). CNVs identified in OCD patients were also enriched in several brain relevant gene-sets, including targets of fragile X mental retardation protein, neuronal migration, and synapse formation (Gazzellone et al., 2016). A recent publication in 121 early-onset pediatric OCD patients and 124 controls, including another 820 in-house healthy controls and 1,038 Affymetrix controls screened for rare small CNVs, reported a significantly higher frequency of rare small CNVs affecting brain-related genes in the OCD patients (Grunblatt et al., 2017). Enrichment analysis of CNVs gene content confirmed the involvement of genes in synaptic and brain-related functional pathways in OCD patients that was not observed in controls. Two patients demonstrated de novo CNVs encompassing genes previously associated with neurodevelopmental disorders (*NRXN1*, *ANKS1B*, *UHRF1BP1*).

In a pilot genome-wide study screening for CNVs in 16 adults with early-onset OCD and 12 controls, a rare small deletion encompassing *FMN1* gene (Chr. 15q13.3), which was paternally inherited, was detected in a male OCD patient (Cappi et al., 2014). The *FMN1* gene is known to be involved in the glutamatergic pathway, which supports the hypothesis of the involvement of this system in OCD.

Another study performed a cross-disorder genome-wide CNV analysis in OCD and TD using a case-control design (2,699 cases; $n = 1,613$ with OCD and $n = 1,086$ with TD, $n = 1,789$ controls) (McGrath et al., 2014). A 3.3-fold increase of large deletions was observed among OCD/TD cases compared with controls. Most deletions were located in the 16p13.11 locus, which has been linked to other neurodevelopmental disorders. Evidence was weaker in TD than in OCD cases (McGrath et al., 2014). In 188 TD cases, intronic deletions in the inner mitochondrial membrane peptidase subunit 2 (*IMMP2L*) gene were identified significantly more frequently than in the study and reference controls (Affymetrix) cohort (Bertelsen et al., 2014). SNP-based heritability indicates that rare variants including CNVs (minor allele frequency of < 1%) account for more variance, and thus play a bigger role, in TD than OCD (Davis et al., 2013).

In a case-control study of 460 individuals with TD, including 148 parent-child trios and 1,131 controls, TD cases were not enriched for de novo or transmitted rare CNVs. Pathway analysis showed enrichment of genes within histamine receptor (subtypes 1 and 2) signaling pathways, and several brain-related pathways (e.g., nervous system development, and synaptic structure and function). Several CNVs overlapped with genes previously identified in autism spectrum disorders. TD cases also had three de novo CNVs that were likely to be pathogenic, one of which disrupted multiple GABA receptor genes (Fernandez et al., 2012).

A GWAS of CNVs in TD using a Latin American sample (210 cases and 285 controls) showed that large CNVs were increased among cases compared with controls (Nag et al., 2013). Of the 24 large CNVs in the cases, four duplications and two deletions were located in the collagen type VII alpha 1 chain (*COL8A1*) and *NRXN1* gene regions—both genes have been implicated in other neurodevelopmental disorders, including autism. Two out of three *NRXN1* deletions in the TD cases were de novo mutations.

A recent study of rare CNVs used SNP microarray data from the TSAICG GWAS sample (Scharf et al., 2013), with a total of 2,434 TD cases and 4,093 ancestry-matched controls, reported an enrichment of global CNV burden that was prominent for large (>1 Mb), singleton events, and known, pathogenic CNVs (Huang et al., 2017). Two individual, genome-wide significant loci were also identified, each conferring a substantial increase in TD risk (*NRXN1* deletions, OR = 20.3; contactin 6 [*CNTN6*] duplications, OR = 10.1).

1.5 Whole exome sequencing in OCD and TD

Only one study has examined 20 sporadic OCD cases and their unaffected parents using whole exome sequencing (WES), and it detected a high rate of de novo single-nucleotide variants (SNVs; Cappi et al., 2016). Additionally, nearly all de novo SNVs were in genes expressed in the human brain, and were enriched in immunological, CNS functioning and development using a Degree-Aware Disease Gene Prioritization to rank the protein-protein interaction network genes.

A recent study using WES was completed in 325 TD trios from the TIC Genetics cohort and a replication sample of 186 trios from the TSA International Consortium on Genetics ($n = 511$ total; Willsey et al., 2017). Robust evidence indicated the contribution of de novo likely gene-disrupting variants to TD. Additionally, de novo damaging variants were overrepresented in probands (RR 1.37, $P = 0.003$) with these variants in approximately 400 genes contributing risk in 12% of clinical cases.

1.6 Pharmacogenetics of antidepressants in OCD and TD

Pharmacogenetics has become increasingly important in the treatment of psychiatric disorders. Interindividual genetic variation may partly determine drug response and tolerability.

(i) *Pharmacokinetic Factors*

Pharmacokinetics (PK) refers to the body's handling of medication, including gastrointestinal absorption, relative and absolute extracellular water volume (distribution), liver metabolism, renal clearance, protein binding, fat solubility, and active transport into the brain. PK parameters such as drug half-life ($T_{1/2}$), time to maximum concentration (T_{max}), as well as peak serum concentrations (C_{max}) of medication are highly variable and individualized. Both PK and pharmacodynamic parameters change with age, which may lead to differences in clinical response, and adverse effect profiles in children vs adults (Geller, 1991; Murry, Crom, Reddick, Bhargava, & Evans, 1995; Vitiello & Jensen, 1995). Our understanding of genetic markers regulating oxidative hepatic pathways for drug metabolism has increased considerably in the past decade (Mrazek, 2010).

Cytochromes P450 (CYP450) are a large family of proteins extensively involved in drug metabolism. Many drugs are not only substrates for these enzymes, but may also inhibit or induce enzyme activity. Polymorphisms in genes coding for CYP450 can lead to altered metabolism, with consequent influences on serum PK parameters, clinical response, and adverse events (Elliott et al., 2017). In psychiatric pharmacogenomics, genes coding for CYP450 enzymes that metabolize SSRIs are the most salient. CYP1A2, 2B6, 2C19, 2C9, 3A4, and 2D6 may all be important in drug metabolism in OCD treatment (see Table 2 for details). Only CYP2D6 and CYP2C19 have been studied in OCD using small sample sizes (Brandl et al., 2014; Muller et al., 2012; Van Nieuwerburgh, Denys, Westenberg, & Deforce, 2009). For venlafaxine, CYP2D6 nonextensive metabolism was associated with higher number of antidepressant trials (48% vs 22% with ≥4 trials; $P = 0.007$), and with greater side effects ($P = 0.022$; Brandl et al., 2014).

TABLE 2 CYP-450 enzymes responsible for serotonergic medication metabolism.

	CYP 1A2	CYP 2B6	CYP 2C19	CYP 2C9	CYP 3A4	CYP 2D6
Citalopram						
Clomipramine	+				+	+
Desvenlafaxine			+		+	
Duloxetine	+					
Escitalopram			+		+	
Fluoxetine			+	+	+	+
Fluvoxamine	+					+
Levomilnacipran			+		+	+
Mirtazapine	+					+
Paroxetine					+	+
Sertraline		+	+	+	+	+
Venlafaxine				+		+
Vilazodone			+		+	+
Vortioxetine		+	+	+	+	+

Some drugs have more than one metabolic enzymatic pathway, and inhibition of one enzyme can lead to unpredictable PK parameters utilizing alternate pathways. Many medications also inhibit CYP-450 enzymes. Autoinhibition produces nonlinear pharmacokinetics, as the drug is both an inhibitor and substrate for an enzyme.

(ii) *Pharmacodynamic Factors*

Psychotropic medications have various pharmacodynamic mechanisms of action in which neurotransmitter systems, including serotonin, glutamate, and dopamine are involved. These interacting systems have been implicated in both anti-obsessive drug response and adverse effects in OCD (Zai, Brandl, Muller, Richter, & Kennedy, 2014). Serotonergic candidate genes previously examined in OCD include *SLC6A4* as well as its promoter (*HTTLPR*), *HTR2A*, *HTR2C*, *HTR1B*, and *TPH*. Only one study (Corregiari, Bernik, Cordeiro, & Vallada, 2012) reported a significant finding between *HTR2A* rs6305 and nonresponders. Real, Gratacos, and Alonso (2010) and Zhang et al. (2015) both examined the glutamatergic gene *SLC1A1* in relation to prospective SSRI response. They reported a significant association with rs301434 and SSRI nonresponse, and between rs301430 and fluoxetine response. Five studies investigating the dopaminergic *DRD2*, *DRD4*, *COMT*, and *MAOA*, have been conducted, but no associations were detected (Miguita, Cordeiro, Shavitt, Miguel, & Vallada, 2011; Umehara et al., 2015; Viswanath et al., 2013; Vulink et al., 2012; L. Zhang et al., 2004), except for *COMT* rs4680 Met/Met genotype with citalopram response (Vulink et al., 2012). Limited pharmacogenetic candidate studies examining other genes have revealed inconsistent findings (Zai et al., 2014). The first pharmacogenetic GWAS of OCD (Qin et al., 2016) in 804 OCD cases detected a significant association between antidepressant response and *DISP1* rs17162912 ($P = 1.76 \times 10^{-8}$). Further research is needed to clarify these genes' potential roles in OCD pharmacogenetics. No study to date has investigated the pharmacogenetics of TD.

1.7 Imaging genetics of OCD and TD

Imaging genetics aims to provide insight into genetic influences on brain structure and function. Only very recently, large multinational initiatives started to investigate imaging genetics with large sample sizes (e.g., ENIGMA, IMAGEN, ADNI, CHARGE) (Bearden & Thompson, 2017; Bogdan et al., 2017; Medland, Jahanshad, Neale, & Thompson, 2014; Thompson et al., 2014). No imaging genetics studies exist for TD, while few imaging genetics publications exist in OCD, mostly using candidate genes (Arnold, Macmaster, Hanna, et al., 2009; Arnold, Macmaster, Richter, et al., 2009; Atmaca et al., 2010, 2011; Gasso et al., 2015; Hesse et al., 2011; Honda et al., 2017; Mas et al., 2016; Ortiz et al., 2016; Scherk et al., 2009; Wolf et al., 2014; Wu et al., 2013). The main genetic pathways investigated in OCD are the serotonergic, glutamatergic, and dopaminergic systems (Grunblatt, Hauser, & Walitza, 2014). For the serotonin system, the orbitofrontal cortex (OFC) and the raphe nuclei are associated with the *5-HTTLPR* gene variant (Grunblatt et al., 2014). Further, gray matter volumes in the right frontal pole showed a trend of being reduced in *5-HTTLPR* gene L_A allele carriers with OCD compared with controls (Honda et al., 2017). For the glutamatergic system, the CSTC loops, the OFC, the thalamus, and the ACC were associated with several gene variations in glutamatergic genes (see Honda et al., 2017 for review). Increased concentrations of choline measured by proton magnetic resonance spectroscopy (^{1}H MRS) in ACC of OCD patients were associated with SNPs in the glutamate receptor AMPA1 (*GRIA1*) (Ortiz et al., 2016). Mean diffusivity (MD) in the right anterior and posterior cerebellar lobes was associated with *SLC1A1* rs3087879 in OCD patients (Gasso et al., 2015). The few studies on dopaminergic markers and imaging correlates reported a positive association between the dopamine transporter gene (*SLC6A3*) and the metabolism of *N*-acetylaspartate in the putamen (Grunblatt et al., 2014), as well as MD of the white matter in right anterior and posterior cerebellar lobes measured by MRS (Gasso et al., 2015).

The complex, heterogeneous OCD phenotype is likely to be influenced by a plethora of common SNPs and multiple genetic biological pathways (Guo et al., 2012; Yu et al., 2015). Therefore, current efforts are focused on the combination of polygenic risk scores with neuroimaging, particularly in large consortia, such as the Enhancing NeuroImaging and Genetics through Meta-Analysis (ENIGMA) consortium, to overcome power issues (Bearden & Thompson, 2017). Recently, this group reported volume asymmetry between pediatric 501 OCD cases and 439 controls in the thalamus and pallidum (Kong et al., 2019). No group differences were observed in the 1,777 adults with OCD and 1,654 controls. Furthermore, pathway enrichment analysis (Mattingsdal et al., 2013), multivariate parallel independent component analysis (Meda et al., 2012), clustering analysis, weight voxel coactivation network analysis, principal component analysis (Nymberg, Jia, Ruggeri, & Schumann, 2013), machine learning (Mas et al., 2016) and other techniques (Bearden & Thompson, 2017; Bogdan et al., 2017) are being developed for diagnostic and course prediction purposes. These techniques hold great promise for the future investigation of genetic and neuronal factors underlying OCD and TD.

1.8 Epigenetics of OCD and TD

The few studies conducted to date suggest some role of epigenetic alterations in the development of OCD. In a genome-wide DNA methylation analysis of blood cells from 65 OCD patients and 96 healthy controls, several differentially methylated genes were identified, including previously implicated candidate genes for OCD (e.g., *BCYRN1*, *BCOR*, *FGF13*, *HLA-DRB1*, *ARX*, etc.) (Yue et al., 2016). These results were not replicated in a smaller study that analyzed only 14 candidate genes (Nissen et al., 2016). OCD patients also showed more DNA methylation than controls in exons of the oxytocin receptor (*OXTR*) using peripheral blood leukocytes in one study (Cappi et al., 2016). In a study combining methylation alterations of an amplicon at the beginning of the first intron of SLC6A4 gene concomitantly with mRNA levels in peripheral samples, pediatric OCD patients had hypermethylated levels of the amplicon compared with age-matched controls, and to adults with OCD (Grunblatt et al., 2018); however, no changes in transcription were observed. Further, in a study examining the effects of allelic variation on mRNA expression in OCD, Jaffe et al. (2014) explored genes that were differentially expressed in OCD patients using eQTLs analysis of postmortem human brain tissue involving the dorsolateral prefrontal cortex. While this study identified significant effects of genetic variation on gene expression and differentially expressed genes linked to the broad diagnosis of OCD, no clinically significant SNP-expression pairs were found. Future studies involving larger samples are indicated to elucidate the molecular mechanisms involved.

The first of three association studies on DNA methylation in TD found differences in methylation in TD patients and controls in a region on chromosome 8, recently identified by genome-wide screens and mapping mutations in single families (Sanchez Delgado et al., 2014). This region includes potassium two-pore domain channel subfamily K member 9 (*KCNK9*) as well as trafficking protein particle complex 9 (*TRAPPC9*) genes. The Epigenome-Wide Association Study on tic phenotype in 1,678 controls and tic-positive individuals from the NTR (Zilhao et al., 2015) found no genome-wide significant methylation (top hits, e.g., *GABBRI*, *BLM*, and *ADAM10*). In the last study, TD patients demonstrated hypermethylation in the promoter and first exon of *DRD2*, which increased with tic severity.

Few studies have assessed the role of microRNA-mediated regulation of gene expression in TD. The first identified a nucleotide variant (var321) in the *SLITRK1* 3′ UTR, putatively leading to its stronger binding, thus with more effective repression by miR-189 (Abelson et al., 2005). However, the role of this variant in TD pathogenesis is questionable due to its very low frequency in patients (Keen-Kim & Freimer, 2006). A recent pilot study profiled the expression of 754 miRNAs in the sera of six TD patients and three unaffected controls (Rizzo et al., 2015). miR-429, involved in midbrain and hindbrain differentiation as well as synaptic transmission, was significantly underexpressed in TD patients. If this finding is replicated, measurement of circulating miR-429 could be a useful molecular biomarker to augment TD diagnosis.

1.9 Gene × environment of OCD and TD

Heritability estimates of OCD and TD indicate that specific environmental factors that increase risk and interact with genetic vulnerability are at least as important as genes to the etiology of these disorders. A population-based study of environmental risk factors for TD (Brander et al., 2017) and OCD (Brander, Perez-Vigil, Larsson, & Mataix-Cols, 2016) found that impaired fetal growth, preterm birth, breech presentation, cesarean section, and maternal smoking during pregnancy were associated with the development of both TD and OCD (Brander, Rydell, et al., 2016, 2017). The study also identified a dose-response relationship between the environmental exposures and the development of both disorders. Maternal smoking and low birth weight had previously been implicated in the development of TD (Chao, Hu, & Pringsheim, 2014). Group A streptococcal infections as a risk-modifier for TD and OCD have not been conclusively demonstrated to date (Brander, Perez-Vigil, et al., 2016; Hoekstra, Dietrich, Edwards, Elamin, & Martino, 2013). A recent systematic review acknowledged the possibility that environmental factors may only increase risk in genetically susceptible individuals (Brander, Perez-Vigil, et al., 2016).

2 Conclusion and future perspective

OCD and TD are polygenic disorders that are clinically and genetically heterogeneous, with common (in OCD) and rare (in TD) inherited or de novo risk variants, in addition to nongenetic factors, playing a substantial role in the etiology of both of these disorders (Pauls et al., 2014; Robertson et al., 2017). Elucidating the genetic underpinnings of these complex conditions has been a major challenge, and will require the combination of clinical research, genomics, gene-by-environment, and epigenomics. To date, no clinically relevant genetic markers can explain the genetic etiology of OCD and TD. Candidate gene studies have become less of a focus in psychiatric genetics, given the advancement in genomic technology and biostatistical modeling techniques to handle large dataset including GWAS, sequencing, neuroimaging, and epigenomic studies. Therefore, future directions should focus on expanding sample size and the collection of clinically meaningful data with whole genome analysis that may improve the current perspectives of these genetically complex disorders. Once we are able to identify multiple regions of interest within the whole genome analysis, fine mapping these regions with the combination of clinical, neuroimaging, epigenomics, transcriptomics, and proteomics data will ultimately refine our understanding of the functionality and role of these genetic variations in the underlying etiology of OCD and TD. To develop targeted treatments for OCD and TD, we will need to 1) dissect the genetic, epigenetic, and environmental factors on neuronal development and circuitry formation, as well as 2) understand the shared and unique pathogenesis of these two conditions.

Acknowledgments

The authors would like to sincerely thank Drs. Paul D. Arnold, James A. Knowles, Carol A. Mathews, Gerald Nestadt, and Jeremiah M. Scharf for their contribution in the final editing of this chapter. We would also like to thank the Obsessive Compulsive Disorder and Tourette Syndrome Working Group of the Psychiatric Genomics Consortium for their ongoing support.

References

Abelson, J. F., Kwan, K. Y., O'Roak, B. J., Baek, D. Y., Stillman, A. A., Morgan, T. M., … State, M. W. (2005). Sequence variants in SLITRK1 are associated with Tourette's syndrome. *Science*, *310*(5746), 317–320. https://doi.org/10.1126/science.1116502.

Albin, R. L., & Mink, J. W. (2006). Recent advances in Tourette syndrome research. *Trends in Neurosciences*, *29*(3), 175–182. https://doi.org/10.1016/j.tins.2006.01.001.

American Psychiatric Association. (2013). *Diagnostic and statistical manual of mental disorders* (5th ed.). Washington, DC: American Psychiatric Association.

Antilla, V., Bulik-Sullivan, B., Finucane, H., Bras, J., Duncan, L., & Escott-Price, V. (2016). Analysis of shared heritability in common disorders of the brain. *BioRxiv*. https://doi.org/10.1101/048991.

Arnold, P. D., Macmaster, F. P., Hanna, G. L., Richter, M. A., Sicard, T., Burroughs, E., … Rosenberg, D. R. (2009). Glutamate system genes associated with ventral prefrontal and thalamic volume in pediatric obsessive-compulsive disorder. *Brain Imaging and Behavior*, *3*(1), 64–76. https://doi.org/10.1007/s11682-008-9050-3.

Arnold, P. D., Macmaster, F. P., Richter, M. A., Hanna, G. L., Sicard, T., Burroughs, E., … Rosenberg, D. R. (2009). Glutamate receptor gene (GRIN2B) associated with reduced anterior cingulate glutamatergic concentration in pediatric obsessive-compulsive disorder. *Psychiatry Research*, *172*(2), 136–139. https://doi.org/10.1016/j.pscychresns.2009.02.005.

Asghari, V., Sanyal, S., Buchwaldt, S., Paterson, A., Jovanovic, V., & Van Tol, H. H. (1995). Modulation of intracellular cyclic AMP levels by different human dopamine D4 receptor variants. *Journal of Neurochemistry*, *65*(3), 1157–1165.

Atmaca, M., Onalan, E., Yildirim, H., Yuce, H., Koc, M., & Korkmaz, S. (2010). The association of myelin oligodendrocyte glycoprotein gene and white matter volume in obsessive-compulsive disorder. *Journal of Affective Disorders, 124*(3), 309–313. https://doi.org/10.1016/j.jad.2010.03.027.

Atmaca, M., Onalan, E., Yildirim, H., Yuce, H., Koc, M., Korkmaz, S., & Mermi, O. (2011). Serotonin transporter gene polymorphism implicates reduced orbito-frontal cortex in obsessive-compulsive disorder. *Journal of Anxiety Disorders, 25*(5), 680–685. https://doi.org/10.1016/j.janxdis.2011.03.002.

Bandelow, B., Baldwin, D., Abelli, M., Altamura, C., Dell'Osso, B., Domschke, K., … Riederer, P. (2016). Biological markers for anxiety disorders, OCD and PTSD—A consensus statement. Part I: Neuroimaging and genetics. *The World Journal of Biological Psychiatry, 17*(5), 321–365. https://doi.org/10.1080/15622975.2016.1181783.

Bearden, C. E., & Thompson, P. M. (2017). Emerging global initiatives in neurogenetics: The enhancing neuroimaging genetics through meta-analysis (ENIGMA) consortium. *Neuron, 94*(2), 232–236. https://doi.org/10.1016/j.neuron.2017.03.033.

Bertelsen, B., Melchior, L., Jensen, L. R., Groth, C., Glenthoj, B., Rizzo, R., … Tumer, Z. (2014). Intragenic deletions affecting two alternative transcripts of the IMMP2L gene in patients with Tourette syndrome. *European Journal of Human Genetics, 22*(11), 1283–1289. https://doi.org/10.1038/ejhg.2014.24.

Bogdan, R., Salmeron, B. J., Carey, C. E., Agrawal, A., Calhoun, V. D., Garavan, H., … Goldman, D. (2017). Imaging genetics and genomics in psychiatry: A critical review of progress and potential. *Biological Psychiatry, 82*(3), 165–175. https://doi.org/10.1016/j.biopsych.2016.12.030.

Brander, G., Perez-Vigil, A., Larsson, H., & Mataix-Cols, D. (2016). Systematic review of environmental risk factors for obsessive-compulsive disorder: A proposed roadmap from association to causation. *Neuroscience and Biobehavioral Reviews, 65,* 36–62. https://doi.org/10.1016/j.neubiorev.2016.03.011.

Brander, G., Rydell, M., Kuja-Halkola, R., Fernandez de la Cruz, L., Lichtenstein, P., Serlachius, E., … Mataix-Cols, D. (2016). Association of perinatal risk factors with obsessive-compulsive disorder: A population-based birth cohort, sibling control study. *JAMA Psychiatry, 73*(11), 1135–1144. https://doi.org/10.1001/jamapsychiatry.2016.2095.

Brander, G., Rydell, M., Kuja-Halkola, R., Fernandez de la Cruz, L., Lichtenstein, P., Serlachius, E., … Mataix-Cols, D. (2017). Perinatal risk factors in Tourette's and chronic tic disorders: A total population sibling comparison study. *Molecular Psychiatry.* https://doi.org/10.1038/mp.2017.31.

Brandl, E. J., Tiwari, A. K., Zhou, X., Deluce, J., Kennedy, J. L., Muller, D. J., & Richter, M. A. (2014). Influence of CYP2D6 and CYP2C19 gene variants on antidepressant response in obsessive-compulsive disorder. *The Pharmacogenomics Journal, 14*(2), 176–181. https://doi.org/10.1038/tpj.2013.12.

Brem, S., Grunblatt, E., Drechsler, R., Riederer, P., & Walitza, S. (2014). The neurobiological link between OCD and ADHD. *Attention Deficit Hyperactivity Disorders, 6*(3), 175–202. https://doi.org/10.1007/s12402-014-0146-x.

Burton, C. L., Crosbie, J., Erdman, L., Dupuis, A., Park, L. S., Sinopoli, V., … Arnold, P. D. (2015). A hypothesis-driven genome-wide association study of an obsessive-compulsive quantitative trait in a community-based sample of children and adolescents. In *Paper presented at the World Congress of Psychiatric Genetics, Toronto.*

Cappi, C., Diniz, J. B., Requena, G. L., Lourenco, T., Lisboa, B. C., Batistuzzo, M. C., … Brentani, H. (2016). Epigenetic evidence for involvement of the oxytocin receptor gene in obsessive-compulsive disorder. *BMC Neuroscience, 17*(1), 79. https://doi.org/10.1186/s12868-016-0313-4.

Cappi, C., Hounie, A. G., Mariani, D. B., Diniz, J. B., Silva, A. R., Reis, V. N., … Brentani, H. (2014). An inherited small microdeletion at 15q13.3 in a patient with early-onset obsessive-compulsive disorder. *PLoS One, 9*(10). https://doi.org/10.1371/journal.pone.0110198.

Chao, T. K., Hu, J., & Pringsheim, T. (2014). Prenatal risk factors for Tourette syndrome: A systematic review. *BMC Pregnancy and Childbirth, 14,* 53. https://doi.org/10.1186/1471-2393-14-53.

Che, F., Zhang, Y., Wang, G., Heng, X., Liu, S., & Du, Y. (2015). The role of GRIN2B in Tourette syndrome: Results from a transmission disequilibrium study. *Journal of Affective Disorders, 187,* 62–65. https://doi.org/10.1016/j.jad.2015.07.036.

Ciruela, F., Casado, V., Rodrigues, R. J., Lujan, R., Burgueno, J., Canals, M., … Franco, R. (2006). Presynaptic control of striatal glutamatergic neurotransmission by adenosine A1-A2A receptor heteromers. *The Journal of Neuroscience, 26*(7), 2080–2087. https://doi.org/10.1523/JNEUROSCI.3574-05.2006.

Comings, D. E., Gade, R., Muhleman, D., Chiu, C., Wu, S., To, M., … MacMurray, J. P. (1996). Exon and intron variants in the human tryptophan 2,3-dioxygenase gene: Potential association with Tourette syndrome, substance abuse and other disorders. *Pharmacogenetics, 6*(4), 307–318.

Corregiari, F. M., Bernik, M., Cordeiro, Q., & Vallada, H. (2012). Endophenotypes and serotonergic polymorphisms associated with treatment response in obsessive-compulsive disorder. *Clinics (São Paulo, Brazil), 67*(4), 335–340.

Davis, L. K., Yu, D., Keenan, C. L., Gamazon, E. R., Konkashbaev, A. I., Derks, E. M., … Scharf, J. M. (2013). Partitioning the heritability of Tourette syndrome and obsessive compulsive disorder reveals differences in genetic architecture. *PLoS Genetics, 9*(10). https://doi.org/10.1371/journal.pgen.1003864.

Dehning, S., Muller, N., Matz, J., Bender, A., Kerle, I., Benninghoff, J., … Zill, P. (2010). A genetic variant of HTR2C may play a role in the manifestation of Tourette syndrome. *Psychiatric Genetics, 20*(1), 35–38. https://doi.org/10.1097/YPG.0b013e32833511ce.

den Braber, A., Zilhao, N. R., Fedko, I. O., Hottenga, J. J., Pool, R., Smit, D. J., … Boomsma, D. I. (2016). Obsessive-compulsive symptoms in a large population-based twin-family sample are predicted by clinically based polygenic scores and by genome-wide SNPs. *Translational Psychiatry. 6,* https://doi.org/10.1038/tp.2015.223.

de Vries, F. E., Cath, D. C., Hoogendoorn, A. W., van Oppen, P., Glas, G., Veltman, D. J., … van Balkom, A. J. (2016). Tic-related versus tic-free obsessive-compulsive disorder: Clinical picture and 2-year natural course. *The Journal of Clinical Psychiatry, 77*(10), e1240–e1247. https://doi.org/10.4088/JCP.14m09736.

Diaz-Anzaldua, A., Joober, R., Riviere, J. B., Dion, Y., Lesperance, P., Richer, F., … Montreal Tourette Syndrome Study Group (2004). Tourette syndrome and dopaminergic genes: A family-based association study in the French Canadian founder population. *Molecular Psychiatry, 9*(3), 272–277. https://doi.org/10.1038/sj.mp.4001411.

Dickel, D. E., Veenstra-VanderWeele, J., Cox, N. J., Wu, X., Fischer, D. J., Van Etten-Lee, M., ... Hanna, G. L. (2006). Association testing of the positional and functional candidate gene SLC1A1/EAAC1 in early-onset obsessive-compulsive disorder. *Archives of General Psychiatry, 63*(7), 778–785. https://doi.org/10.1001/archpsyc.63.7.778.

Di Nocera, F., Colazingari, S., Trabalza, A., Mamazza, L., & Bevilacqua, A. (2014). Association of TPH2 and dopamine receptor gene polymorphisms with obsessive-compulsive symptoms and perfectionism in healthy subjects. *Psychiatry Research, 220*(3), 1172–1173. https://doi.org/10.1016/j.psychres.2014.09.015.

Elliott, L. S., Henderson, J. C., Neradilek, M. B., Moyer, N. A., Ashcraft, K. C., & Thirumaran, R. K. (2017). Clinical impact of pharmacogenetic profiling with a clinical decision support tool in polypharmacy home health patients: A prospective pilot randomized controlled trial. *PLoS One, 12*(2). https://doi.org/10.1371/journal.pone.0170905.

Ercan-Sencicek, A. G., Stillman, A. A., Ghosh, A. K., Bilguvar, K., O'Roak, B. J., Mason, C. E., ... State, M. W. (2010). L-histidine decarboxylase and Tourette's syndrome. *The New England Journal of Medicine, 362*(20), 1901–1908. https://doi.org/10.1056/NEJMoa0907006.

Fernandez, T. V., Sanders, S. J., Yurkiewicz, I. R., Ercan-Sencicek, A. G., Kim, Y. S., Fishman, D. O., ... State, M. W. (2012). Rare copy number variants in Tourette syndrome disrupt genes in histaminergic pathways and overlap with autism. *Biological Psychiatry, 71*(5), 392–402. https://doi.org/10.1016/j.biopsych.2011.09.034.

Gade, R., Muhleman, D., Blake, H., MacMurray, J., Johnson, P., Verde, R., ... Comings, D. E. (1998). Correlation of length of VNTR alleles at the X-linked MAOA gene and phenotypic effect in Tourette syndrome and drug abuse. *Molecular Psychiatry, 3*(1), 50–60.

Gasso, P., Ortiz, A. E., Mas, S., Morer, A., Calvo, A., Bargallo, N., ... Lazaro, L. (2015). Association between genetic variants related to glutamatergic, dopaminergic and neurodevelopment pathways and white matter microstructure in child and adolescent patients with obsessive-compulsive disorder. *Journal of Affective Disorders, 186*, 284–292. https://doi.org/10.1016/j.jad.2015.07.035.

Gazzellone, M. J., Zarrei, M., Burton, C. L., Walker, S., Uddin, M., Shaheen, S. M., ... Scherer, S. W. (2016). Uncovering obsessive-compulsive disorder risk genes in a pediatric cohort by high-resolution analysis of copy number variation. *Journal of Neurodevelopmental Disorders, 8*, 36. https://doi.org/10.1186/s11689-016-9170-9.

Geller, B. (1991). Psychopharmacology of children and adolescents: Pharmacokinetics and relationships of plasma/serum levels to response. *Psychopharmacology Bulletin, 27*(4), 401–409.

Grados, M. A., Mathews, C. A., & Tourette Syndrome Association International Consortium for Genetics. (2008). Latent class analysis of Gilles de la Tourette syndrome using comorbidities: Clinical and genetic implications. *Biological Psychiatry, 64*(3), 219–225. https://doi.org/10.1016/j.biopsych.2008.01.019.

Grice, D. E., Leckman, J. F., Pauls, D. L., Kurlan, R., Kidd, K. K., Pakstis, A. J., ... Gelernter, J. (1996). Linkage disequilibrium between an allele at the dopamine D4 receptor locus and Tourette syndrome, by the transmission-disequilibrium test. *American Journal of Human Genetics, 59*(3), 644–652.

Grunblatt, E., Hauser, T. U., & Walitza, S. (2014). Imaging genetics in obsessive-compulsive disorder: Linking genetic variations to alterations in neuroimaging. *Progress in Neurobiology, 121*, 114–124. https://doi.org/10.1016/j.pneurobio.2014.07.003.

Grunblatt, E., Marinova, Z., Roth, A., Gardini, E., Ball, J., Geissler, J., ... Walitza, S. (2018). Combining genetic and epigenetic parameters of the serotonin transporter gene in obsessive-compulsive disorder. *Journal of Psychiatric Research, 96*, 209–217. https://doi.org/10.1016/j.jpsychires.2017.10.010.

Grunblatt, E., Oneda, B., Ekici, A. B., Ball, J., Geissler, J., Uebe, S., ... Walitza, S. (2017). High resolution chromosomal microarray analysis in paediatric obsessive-compulsive disorder. *BMC Medical Genomics, 10*(1), 68. https://doi.org/10.1186/s12920-017-0299-5.

Guo, Y., Su, L., Zhang, J., Lei, J., Deng, X., Xu, H., ... Deng, H. (2012). Analysis of the BTBD9 and HTR2C variants in Chinese Han patients with Tourette syndrome. *Psychiatric Genetics, 22*(6), 300–303. https://doi.org/10.1097/YPG.0b013e32835862b1.

Hesse, S., Stengler, K., Regenthal, R., Patt, M., Becker, G. A., Franke, A., ... Sabri, O. (2011). The serotonin transporter availability in untreated early-onset and late-onset patients with obsessive-compulsive disorder. *The International Journal of Neuropsychopharmacology, 14*(5), 606–617. https://doi.org/10.1017/S1461145710001604.

Hettinger, B. D., Lee, A., Linden, J., & Rosin, D. L. (2001). Ultrastructural localization of adenosine A2A receptors suggests multiple cellular sites for modulation of GABAergic neurons in rat striatum. *The Journal of Comparative Neurology, 431*(3), 331–346.

Hirschtritt, M. E., Lee, P. C., Pauls, D. L., Dion, Y., Grados, M. A., Illmann, C., ... Tourette Syndrome Association International Consortium for Genetics. (2015). Lifetime prevalence, age of risk, and genetic relationships of comorbid psychiatric disorders in Tourette syndrome. *JAMA Psychiatry, 72*(4), 325–333. https://doi.org/10.1001/jamapsychiatry.2014.2650.

Hoekstra, P. J., Dietrich, A., Edwards, M. J., Elamin, I., & Martino, D. (2013). Environmental factors in Tourette syndrome. *Neuroscience and Biobehavioral Reviews, 37*(6), 1040–1049. https://doi.org/10.1016/j.neubiorev.2012.10.010.

Honda, S., Nakao, T., Mitsuyasu, H., Okada, K., Gotoh, L., Tomita, M., ... Kanba, S. (2017). A pilot study exploring the association of morphological changes with 5-HTTLPR polymorphism in OCD patients. *Annals of General Psychiatry, 16*, 2. https://doi.org/10.1186/s12991-017-0126-6.

Huang, A. Y., Yu, D., Davis, L. K., Sul, J. H., Tsetsos, F., Ramensky, V., ... Gilles de la Tourette Syndrome GWAS Replication Initiative (GGRI). (2017). Rare copy number variants in NRXN1 and CNTN6 increase risk for Tourette syndrome. *Neuron, 94*(6), 1101–1111. e1107. https://doi.org/10.1016/j.neuron.2017.06.010.

International Obsessive Compulsive Disorder Foundation Genetics Collaborative (IOCDF-GC) and OCD Collaborative Genetics Association Studies (OCGAS). (2017). Revealing the complex genetic architecture of obsessive-compulsive disorder using meta-analysis. *Molecular Psychiatry*, https://doi.org/10.1038/mp.2017.154.

Jaffe, A. E., Deep-Soboslay, A., Tao, R., Hauptman, D. T., Kaye, W. H., Arango, V., ... Kleinman, J. E. (2014). Genetic neuropathology of obsessive psychiatric syndromes. *Translational Psychiatry, 4*, e432. https://doi.org/10.1038/tp.2014.68.

Janik, P., Berdynski, M., Safranow, K., & Zekanowski, C. (2015). Association of ADORA1 rs2228079 and ADORA2A rs5751876 polymorphisms with Gilles de la Tourette syndrome in the Polish population. *PLoS One, 10*(8). https://doi.org/10.1371/journal.pone.0136754.

Karagiannidis, I., Dehning, S., Sandor, P., Tarnok, Z., Rizzo, R., Wolanczyk, T., ... Paschou, P. (2013). Support of the histaminergic hypothesis in Tourette syndrome: Association of the histamine decarboxylase gene in a large sample of families. *Journal of Medical Genetics, 50*(11), 760–764. https://doi.org/10.1136/jmedgenet-2013-101637.

Keen-Kim, D., & Freimer, N. B. (2006). Genetics and epidemiology of Tourette syndrome. *Journal of Child Neurology, 21*(8), 665–671. https://doi.org/10.1177/08830738060210081101.

Kong, X. Z., Boedhoe, P. S. W., Abe, Y., Alonso, P., Ameis, S. H., Arnold, P. D., ... Francks, C. (2019). Mapping cortical and subcortical asymmetry in obsessive-compulsive disorder: Findings from the ENIGMA consortium. *Biological Psychiatry,* https://doi.org/10.1016/j.biopsych.2019.04.022 [Epub ahead of print].

Koo, M. S., Kim, E. J., Roh, D., & Kim, C. H. (2010). Role of dopamine in the pathophysiology and treatment of obsessive-compulsive disorder. *Expert Review of Neurotherapeutics, 10*(2), 275–290. https://doi.org/10.1586/ern.09.148.

Leckman, J. F., Pauls, D. L., Zhang, H., Rosario-Campos, M. C., Katsovich, L., Kidd, K. K., ... Tourette Syndrome Association International Consortium for Genetics. (2003). Obsessive-compulsive symptom dimensions in affected sibling pairs diagnosed with Gilles de la Tourette syndrome. *American Journal of Medical Genetics. Part B, Neuropsychiatric Genetics, 116B*(1), 60–68. https://doi.org/10.1002/ajmg.b.10001.

Lei, J., Deng, X., Zhang, J., Su, L., Xu, H., Liang, H., ... Deng, H. (2012). Mutation screening of the HDC gene in Chinese Han patients with Tourette syndrome. *American Journal of Medical Genetics. Part B, Neuropsychiatric Genetics, 159B*(1), 72–76. https://doi.org/10.1002/ajmg.b.32003.

Liu, S., Cui, J., Zhang, X., Wu, W., Niu, H., Ma, X., ... Yi, M. (2014). Variable number tandem repeats in dopamine receptor D4 in Tourette's syndrome. *Movement Disorders, 29*(13), 1687–1691. https://doi.org/10.1002/mds.26027.

Liu, S., Yin, Y., Wang, Z., Zhang, X., & Ma, X. (2013). Association study between MAO-A gene promoter VNTR polymorphisms and obsessive-compulsive disorder. *Journal of Anxiety Disorders, 27*(4), 435–437. https://doi.org/10.1016/j.janxdis.2013.04.005.

Lochner, C., McGregor, N., Hemmings, S., Harvey, B. H., Breet, E., Swanevelder, S., & Stein, D. J. (2016). Symmetry symptoms in obsessive-compulsive disorder: Clinical and genetic correlates. *Revista Brasileira de Psiquiatria, 38*(1), 17–23. https://doi.org/10.1590/1516-4446-2014-1619.

Mas, S., Gasso, P., Morer, A., Calvo, A., Bargallo, N., Lafuente, A., & Lazaro, L. (2016). Integrating genetic, neuropsychological and neuroimaging data to model early-onset obsessive compulsive disorder severity. *PLoS One, 11*(4). https://doi.org/10.1371/journal.pone.0153846.

Mas, S., Pagerols, M., Gasso, P., Ortiz, A., Rodriguez, N., Morer, A., ... Lazaro, L. (2014). Role of GAD2 and HTR1B genes in early-onset obsessive-compulsive disorder: Results from transmission disequilibrium study. *Genes, Brain, and Behavior, 13*(4), 409–417. https://doi.org/10.1111/gbb.12128.

Mataix-Cols, D., Isomura, K., Perez-Vigil, A., Chang, Z., Ruck, C., Larsson, K. J., ... Lichtenstein, P. (2015). Familial risks of Tourette syndrome and chronic tic disorders. A population-based cohort study. *JAMA Psychiatry, 72*(8), 787–793. https://doi.org/10.1001/jamapsychiatry.2015.0627.

Mathews, C. A., & Grados, M. A. (2011). Familiality of Tourette syndrome, obsessive-compulsive disorder, and attention-deficit/hyperactivity disorder: Heritability analysis in a large sib-pair sample. *Journal of the American Academy of Child and Adolescent Psychiatry, 50*(1), 46–54. https://doi.org/10.1016/j.jaac.2010.10.004.

Mattheisen, M., Samuels, J. F., Wang, Y., Greenberg, B. D., Fyer, A. J., McCracken, J. T., ... Nestadt, G. (2015). Genome-wide association study in obsessive-compulsive disorder: Results from the OCGAS. *Molecular Psychiatry, 20*(3), 337–344. https://doi.org/10.1038/mp.2014.43.

Mattingsdal, M., Brown, A. A., Djurovic, S., Sonderby, I. E., Server, A., Melle, I., ... Andreassen, O. A. (2013). Pathway analysis of genetic markers associated with a functional MRI faces paradigm implicates polymorphisms in calcium responsive pathways. *NeuroImage, 70*, 143–149. https://doi.org/10.1016/j.neuroimage.2012.12.035.

McGrath, L. M., Yu, D., Marshall, C., Davis, L. K., Thiruvahindrapuram, B., Li, B., ... Scharf, J. M. (2014). Copy number variation in obsessive-compulsive disorder and Tourette syndrome: A cross-disorder study. *Journal of the American Academy of Child and Adolescent Psychiatry, 53*(8), 910–919. https://doi.org/10.1016/j.jaac.2014.04.022.

Meda, S. A., Narayanan, B., Liu, J., Perrone-Bizzozero, N. I., Stevens, M. C., Calhoun, V. D., ... Pearlson, G. D. (2012). A large scale multivariate parallel ICA method reveals novel imaging-genetic relationships for Alzheimer's disease in the ADNI cohort. *NeuroImage, 60*(3), 1608–1621. https://doi.org/10.1016/j.neuroimage.2011.12.076.

Medland, S. E., Jahanshad, N., Neale, B. M., & Thompson, P. M. (2014). Whole-genome analyses of whole-brain data: Working within an expanded search space. *Nature Neuroscience, 17*(6), 791–800. https://doi.org/10.1038/nn.3718.

Melo-Felippe, F. B., de Salles Andrade, J. B., Giori, I. G., Vieira-Fonseca, T., Fontenelle, L. F., & Kohlrausch, F. B. (2016). Catechol-O-methyltransferase gene polymorphisms in specific obsessive-compulsive disorder patients' subgroups. *Journal of Molecular Neuroscience, 58*(1), 129–136. https://doi.org/10.1007/s12031-015-0697-0.

Miguita, K., Cordeiro, Q., Shavitt, R. G., Miguel, E. C., & Vallada, H. (2011). Association study between genetic monoaminergic polymorphisms and OCD response to clomipramine treatment. *Arquivos de Neuro-Psiquiatria, 69*(2B), 283–287.

Mossner, R., Muller-Vahl, K. R., Doring, N., & Stuhrmann, M. (2007). Role of the novel tryptophan hydroxylase-2 gene in Tourette syndrome. *Molecular Psychiatry, 12*(7), 617–619. https://doi.org/10.1038/sj.mp.4002004.

Moya, P. R., Wendland, J. R., Rubenstein, L. M., Timpano, K. R., Heiman, G. A., Tischfield, J. A., ... Murphy, D. L. (2013). Common and rare alleles of the serotonin transporter gene, SLC6A4, associated with Tourette's disorder. *Movement Disorders, 28*(9), 1263–1270. https://doi.org/10.1002/mds.25460.

Mrazek, D. A. (2010). Psychiatric pharmacogenomic testing in clinical practice. *Dialogues in Clinical Neuroscience, 12*(1), 69–76.

Muller, D. J., Brandl, E. J., Hwang, R., Tiwari, A. K., Sturgess, J. E., Zai, C. C., ... Richter, M. A. (2012). The AmpliChip(R) CYP450 test and response to treatment in schizophrenia and obsessive compulsive disorder: A pilot study and focus on cases with abnormal CYP2D6 drug metabolism. *Genetic Testing and Molecular Biomarkers, 16*(8), 897–903. https://doi.org/10.1089/gtmb.2011.0327.

Murphy, D. L., Moya, P. R., Fox, M. A., Rubenstein, L. M., Wendland, J. R., & Timpano, K. R. (2013). Anxiety and affective disorder comorbidity related to serotonin and other neurotransmitter systems: Obsessive-compulsive disorder as an example of overlapping clinical and genetic heterogeneity. *Philosophical Transactions of the Royal Society of London Series B, Biological Sciences, 368*(1615), 20120435. https://doi.org/10.1098/rstb.2012.0435.

Murry, D. J., Crom, W. R., Reddick, W. E., Bhargava, R., & Evans, W. E. (1995). Liver volume as a determinant of drug clearance in children and adolescents. *Drug Metabolism and Disposition, 23*(10), 1110–1116.

Nag, A., Bochukova, E. G., Kremeyer, B., Campbell, D. D., Muller, H., Valencia-Duarte, A. V., ... Ruiz-Linares, A. (2013). CNV analysis in Tourette syndrome implicates large genomic rearrangements in COL8A1 and NRXN1. *PLoS One, 8*(3). https://doi.org/10.1371/journal.pone.0059061.

Nissen, J. B., Hansen, C. S., Starnawska, A., Mattheisen, M., Borglum, A. D., Buttenschon, H. N., & Hollegaard, M. (2016). DNA methylation at the neonatal state and at the time of diagnosis: Preliminary support for an association with the estrogen receptor 1, gamma-aminobutyric acid B receptor 1, and myelin oligodendrocyte glycoprotein in female adolescent patients with OCD. *Frontiers in Psychiatry, 7*, 35. https://doi.org/10.3389/fpsyt.2016.00035.

Noh, H. J., Tang, R., Flannick, J., O'Dushlaine, C., Swofford, R., Howrigan, D., ... Lindblad-Toh, K. (2017). Integrating evolutionary and regulatory information with a multispecies approach implicates genes and pathways in obsessive-compulsive disorder. *Nature Communications, 8*(1), 774. https://doi.org/10.1038/s41467-017-00831-x.

Nymberg, C., Jia, T., Ruggeri, B., & Schumann, G. (2013). Analytical strategies for large imaging genetic datasets: Experiences from the IMAGEN study. *Annals of the New York Academy of Sciences, 1282*, 92–106. https://doi.org/10.1111/nyas.12088.

O'Rourke, J. A., Scharf, J. M., Yu, D., & Pauls, D. L. (2009). The genetics of Tourette syndrome: A review. *Journal of Psychosomatic Research, 67*(6), 533–545. https://doi.org/10.1016/j.jpsychores.2009.06.006.

Ortiz, A. E., Gasso, P., Mas, S., Falcon, C., Bargallo, N., Lafuente, A., & Lazaro, L. (2016). Association between genetic variants of serotonergic and glutamatergic pathways and the concentration of neurometabolites of the anterior cingulate cortex in paediatric patients with obsessive-compulsive disorder. *The World Journal of Biological Psychiatry, 17*(5), 394–404. https://doi.org/10.3109/15622975.2015.1111524.

Ozomaro, U., Cai, G., Kajiwara, Y., Yoon, S., Makarov, V., Delorme, R., ... Grice, D. E. (2013). Characterization of SLITRK1 variation in obsessive-compulsive disorder. *PLoS One, 8*(8), e70376. https://doi.org/10.1371/journal.pone.0070376.

Paschou, P., Yu, D., Gerber, G., Evans, P., Tsetsos, F., Davis, L. K., ... Scharf, J. M. (2014). Genetic association signal near NTN4 in Tourette syndrome. *Annals of Neurology, 76*(2), 310–315. https://doi.org/10.1002/ana.24215.

Pauls, D. L., Abramovitch, A., Rauch, S. L., & Geller, D. A. (2014). Obsessive-compulsive disorder: An integrative genetic and neurobiological perspective. *Nature Reviews Neuroscience, 15*(6), 410–424. https://doi.org/10.1038/nrn3746.

Peterson, B. S., Pine, D. S., Cohen, P., & Brook, J. S. (2001). Prospective, longitudinal study of tic, obsessive-compulsive, and attention-deficit/hyperactivity disorders in an epidemiological sample. *Journal of the American Academy of Child and Adolescent Psychiatry, 40*(6), 685–695. https://doi.org/10.1097/00004583-200106000-00014.

Price, R. A., Kidd, K. K., Cohen, D. J., Pauls, D. L., & Leckman, J. F. (1985). A twin study of Tourette syndrome. *Archives of General Psychiatry, 42*(8), 815–820.

Qin, H., Samuels, J. F., Wang, Y., Zhu, Y., Grados, M. A., Riddle, M. A., ... Shugart, Y. Y. (2016). Whole-genome association analysis of treatment response in obsessive-compulsive disorder. *Molecular Psychiatry, 21*(2), 270–276. https://doi.org/10.1038/mp.2015.32.

Real, E., Gratacos, M., Alonso, P., et al. (2010). Pharmacological resistance level in OCD patients without a stressful life event at onset of the disorder is associated with a polymorphism of the glutamate transporter gene (SLC1A1). In *Paper presented at the 18th world congress of psychiatric genetics*. Athens, Greece.

Riviere, J. B., Xiong, L., Levchenko, A., St-Onge, J., Gaspar, C., Dion, Y., ... Montreal Tourette Study Group. (2009). Association of intronic variants of the BTBD9 gene with Tourette syndrome. *Archives of Neurology, 66*(10), 1267–1272. https://doi.org/10.1001/archneurol.2009.213.

Rizzo, R., Ragusa, M., Barbagallo, C., Sammito, M., Gulisano, M., Cali, P. V., ... Purrello, M. (2015). Circulating miRNAs profiles in Tourette syndrome: Molecular data and clinical implications. *Molecular Brain, 8*, 44. https://doi.org/10.1186/s13041-015-0133-y.

Robertson, M. M. (2015). Corrections. A personal 35 year perspective on Gilles de la Tourette syndrome: Assessment, investigations, and management. *Lancet Psychiatry, 2*(4), 291. https://doi.org/10.1016/S2215-0366(15)00131-5.

Robertson, M. M., Eapen, V., Singer, H. S., Martino, D., Scharf, J. M., Paschou, P., ... Leckman, J. F. (2017). Gilles de la Tourette syndrome. Review-*Nature Reviews. Disease Primers, 3*, 16097. https://doi.org/10.1038/nrdp.2016.97.

Ruscio, A. M., Stein, D. J., Chiu, W. T., & Kessler, R. C. (2010). The epidemiology of obsessive-compulsive disorder in the National Comorbidity Survey Replication. *Molecular Psychiatry, 15*(1), 53–63. https://doi.org/10.1038/mp.2008.94.

Sampaio, A. S., Hounie, A. G., Petribu, K., Cappi, C., Morais, I., Vallada, H., ... Miguel, E. C. (2015). COMT and MAO-A polymorphisms and obsessive-compulsive disorder: A family-based association study. *PLoS One, 10*(3). https://doi.org/10.1371/journal.pone.0119592.

Sanchez Delgado, M., Camprubi, C., Tumer, Z., Martinez, F., Mila, M., & Monk, D. (2014). Screening individuals with intellectual disability, autism and Tourette's syndrome for KCNK9 mutations and aberrant DNA methylation within the 8q24 imprinted cluster. *American Journal of Medical Genetics. Part B, Neuropsychiatric Genetics, 165B*(6), 472–478. https://doi.org/10.1002/ajmg.b.32250.

Scharf, J. M., Miller, L. L., Gauvin, C. A., Alabiso, J., Mathews, C. A., & Ben-Shlomo, Y. (2015). Population prevalence of Tourette syndrome: A systematic review and meta-analysis. *Movement Disorders, 30*(2), 221–228. https://doi.org/10.1002/mds.26089.

Scharf, J. M., Yu, D., Mathews, C. A., Neale, B. M., Stewart, S. E., Fagerness, J. A., … Pauls, D. L. (2013). Genome-wide association study of Tourette's syndrome. *Molecular Psychiatry, 18*(6), 721–728. https://doi.org/10.1038/mp.2012.69.

Scherk, H., Backens, M., Schneider-Axmann, T., Kraft, S., Kemmer, C., Usher, J., … Gruber, O. (2009). Dopamine transporter genotype influences N-acetyl-aspartate in the left putamen. *The World Journal of Biological Psychiatry, 10*(4 Pt. 2), 524–530. https://doi.org/10.1080/15622970701586349.

Schito, A. M., Pizzuti, A., Di Maria, E., Schenone, A., Ratti, A., Defferrari, R., … Mandich, P. (1997). mRNA distribution in adult human brain of GRIN2B, a N-methyl-D-aspartate (NMDA) receptor subunit. *Neuroscience Letters, 239*(1), 49–53.

Song, M., Mathews, C. A., Stewart, S. E., Shmelkov, S. V., Mezey, J. G., Rodriguez-Flores, J. L., … Glatt, C. E. (2017). Rare synaptogenesis-impairing mutations in SLITRK5 are associated with obsessive compulsive disorder. *PLoS One, 12*(1). https://doi.org/10.1371/journal.pone.0169994.

Stewart, S. E., Fagerness, J. A., Platko, J., Smoller, J. W., Scharf, J. M., Illmann, C., … Pauls, D. L. (2007). Association of the SLC1A1 glutamate transporter gene and obsessive-compulsive disorder. *American Journal of Medical Genetics. Part B, Neuropsychiatric Genetics, 144B*(8), 1027–1033. https://doi.org/10.1002/ajmg.b.30533.

Stewart, S. E., Mayerfeld, C., Arnold, P. D., Crane, J. R., O'Dushlaine, C., Fagerness, J. A., … Mathews, C. A. (2013). Meta-analysis of association between obsessive-compulsive disorder and the 3' region of neuronal glutamate transporter gene SLC1A1. *American Journal of Medical Genetics. Part B, Neuropsychiatric Genetics, 162B*(4), 367–379. https://doi.org/10.1002/ajmg.b.32137.

Stewart, S. E., Platko, J., Fagerness, J., Birns, J., Jenike, E., Smoller, J. W., … Pauls, D. L. (2007). A genetic family-based association study of OLIG2 in obsessive-compulsive disorder. *Archives of General Psychiatry, 64*(2), 209–214. https://doi.org/10.1001/archpsyc.64.2.209.

Stewart, S. E., Yu, D., Scharf, J. M., Neale, B. M., Fagerness, J. A., Mathews, C. A., … Pauls, D. L. (2013). Genome-wide association study of obsessive-compulsive disorder. *Molecular Psychiatry, 18*(7), 788–798. https://doi.org/10.1038/mp.2012.85.

Sundaram, S. K., Huq, A. M., Wilson, B. J., & Chugani, H. T. (2010). Tourette syndrome is associated with recurrent exonic copy number variants. *Neurology, 74*(20), 1583–1590. https://doi.org/10.1212/WNL.0b013e3181e0f147.

Tarnok, Z., Ronai, Z., Gervai, J., Kereszturi, E., Gadoros, J., Sasvari-Szekely, M., & Nemoda, Z. (2007). Dopaminergic candidate genes in Tourette syndrome: Association between tic severity and 3' UTR polymorphism of the dopamine transporter gene. *American Journal of Medical Genetics. Part B, Neuropsychiatric Genetics, 144B*(7), 900–905. https://doi.org/10.1002/ajmg.b.30517.

Taylor, S. (2013). Molecular genetics of obsessive-compulsive disorder: A comprehensive meta-analysis of genetic association studies. *Molecular Psychiatry, 18*(7), 799–805. https://doi.org/10.1038/mp.2012.76.

Taylor, S. (2016). Disorder-specific genetic factors in obsessive-compulsive disorder: A comprehensive meta-analysis. *American Journal of Medical Genetics. Part B, Neuropsychiatric Genetics, 171B*(3), 325–332. https://doi.org/10.1002/ajmg.b.32407.

Thompson, P. M., Stein, J. L., Medland, S. E., Hibar, D. P., Vasquez, A. A., Renteria, M. E., … Alzheimer's Disease Neuroimaging Initiative (2014). The ENIGMA Consortium: Large-scale collaborative analyses of neuroimaging and genetic data. *Brain Imaging and Behavior, 8*(2), 153–182. https://doi.org/10.1007/s11682-013-9269-5.

Umehara, H., Numata, S., Tajima, A., Kinoshita, M., Nakaaki, S., Imoto, I., … Ohmori, T. (2015). No association between the COMT Val158Met polymorphism and the long-term clinical response in obsessive-compulsive disorder in the Japanese population. *Human Psychopharmacology, 30*(5), 372–376. https://doi.org/10.1002/hup.2485.

van Grootheest, D. S., Cath, D. C., Beekman, A. T., & Boomsma, D. I. (2005). Twin studies on obsessive-compulsive disorder: A review. *Twin Research and Human Genetics, 8*(5), 450–458. https://doi.org/10.1375/183242705774310060.

Van Grootheest, D. S., Cath, D. C., Beekman, A. T., & Boomsma, D. I. (2007). Genetic and environmental influences on obsessive-compulsive symptoms in adults: A population-based twin-family study. *Psychological Medicine, 37*(11), 1635–1644. https://doi.org/10.1017/S0033291707000980.

Van Nieuwerburgh, F. C., Denys, D. A., Westenberg, H. G., & Deforce, D. L. (2009). Response to serotonin reuptake inhibitors in OCD is not influenced by common CYP2D6 polymorphisms. *International Journal of Psychiatry in Clinical Practice, 13*(1), 345–348. https://doi.org/10.3109/13651500902903016.

Viswanath, B., Taj, M. J. R., Purushottam, M., Kandavel, T., Shetty, P. H., Reddy, Y. C., & Jain, S. (2013). No association between DRD4 gene and SRI treatment response in obsessive compulsive disorder: Need for a novel approach. *Asian Journal of Psychiatry, 6*(4), 347–348. https://doi.org/10.1016/j.ajp.2013.03.003.

Vitiello, B., & Jensen, P. S. (1995). Developmental perspectives in pediatric psychopharmacology. *Psychopharmacology Bulletin, 31*(1), 75–81.

Vulink, N. C., Westenberg, H. G., van Nieuwerburgh, F., Deforce, D., Fluitman, S. B., Meinardi, J. S., & Denys, D. (2012). Catechol-O-methyltransferase gene expression is associated with response to citalopram in obsessive-compulsive disorder. *International Journal of Psychiatry in Clinical Practice, 16*(4), 277–283. https://doi.org/10.3109/13651501.2011.653375.

Walitza, S., Bove, D. S., Romanos, M., Renner, T., Held, L., Simons, M., … Grunblatt, E. (2012). Pilot study on HTR2A promoter polymorphism, -1438G/A (rs6311) and a nearby copy number variation showed association with onset and severity in early onset obsessive-compulsive disorder. *Journal of Neural Transmission (Vienna), 119*(4), 507–515. https://doi.org/10.1007/s00702-011-0699-1.

Walitza, S., Marinova, Z., Grunblatt, E., Lazic, S. E., Remschmidt, H., Vloet, T. D., & Wendland, J. R. (2014). Trio study and meta-analysis support the association of genetic variation at the serotonin transporter with early-onset obsessive-compulsive disorder. *Neuroscience Letters, 580*, 100–103. https://doi.org/10.1016/j.neulet.2014.07.038.

Walitza, S., Wewetzer, C., Gerlach, M., Klampfl, K., Geller, F., Barth, N., … Hinney, A. (2004). Transmission disequilibrium studies in children and adolescents with obsessive-compulsive disorders pertaining to polymorphisms of genes of the serotonergic pathway. *Journal of Neural Transmission (Vienna), 111*(7), 817–825. https://doi.org/10.1007/s00702-004-0134-y.

Willsey, A. J., Fernandez, T. V., Yu, D., King, R. A., Dietrich, A., Xing, J., … Heiman, G. A. (2017). De novo coding variants are strongly associated with Tourette disorder. *Neuron, 94*(3), 486–499. e489. https://doi.org/10.1016/j.neuron.2017.04.024.

Wolf, C., Mohr, H., Schneider-Axmann, T., Reif, A., Wobrock, T., Scherk, H., ... Gruber, O. (2014). CACNA1C genotype explains interindividual differences in amygdala volume among patients with schizophrenia. *European Archives of Psychiatry and Clinical Neuroscience, 264*(2), 93–102. https://doi.org/10.1007/s00406-013-0427-y.

Wu, K., Hanna, G. L., Easter, P., Kennedy, J. L., Rosenberg, D. R., & Arnold, P. D. (2013). Glutamate system genes and brain volume alterations in pediatric obsessive-compulsive disorder: A preliminary study. *Psychiatry Research, 211*(3), 214–220. https://doi.org/10.1016/j.pscychresns.2012.07.003.

Yoon, D. Y., Rippel, C. A., Kobets, A. J., Morris, C. M., Lee, J. E., Williams, P. N., ... Singer, H. S. (2007). Dopaminergic polymorphisms in Tourette syndrome: Association with the DAT gene (SLC6A3). *American Journal of Medical Genetics. Part B, Neuropsychiatric Genetics, 144B*(5), 605–610. https://doi.org/10.1002/ajmg.b.30466.

Yu, D., Mathews, C. A., Scharf, J. M., Neale, B. M., Davis, L. K., Gamazon, E. R., ... Pauls, D. L. (2015). Cross-disorder genome-wide analyses suggest a complex genetic relationship between Tourette's syndrome and OCD. *The American Journal of Psychiatry, 172*(1), 82–93. https://doi.org/10.1176/appi.ajp.2014.13101306.

Yuan, A., Su, L., Yu, S., Li, C., Yu, T., & Sun, J. (2015). Association between DRD2/ANKK1 TaqIA polymorphism and susceptibility with Tourette syndrome: A meta-analysis. *PLoS One, 10*(6). https://doi.org/10.1371/journal.pone.0131060.

Yue, W., Cheng, W., Liu, Z., Tang, Y., Lu, T., Zhang, D., ... Huang, Y. (2016). Genome-wide DNA methylation analysis in obsessive-compulsive disorder patients. *Scientific Reports, 6*, 31333. https://doi.org/10.1038/srep31333.

Zai, G., Bezchlibnyk, Y. B., Richter, M. A., Arnold, P., Burroughs, E., Barr, C. L., & Kennedy, J. L. (2004). Myelin oligodendrocyte glycoprotein (MOG) gene is associated with obsessive-compulsive disorder. *American Journal of Medical Genetics. Part B, Neuropsychiatric Genetics, 129B*(1), 64–68. https://doi.org/10.1002/ajmg.b.30077.

Zai, G., Brandl, E. J., Muller, D. J., Richter, M. A., & Kennedy, J. L. (2014). Pharmacogenetics of antidepressant treatment in obsessive-compulsive disorder: An update and implications for clinicians. *Pharmacogenomics, 15*(8), 1147–1157. https://doi.org/10.2217/pgs.14.83.

Zai, G., Zai, C. C., Arnold, P. D., Freeman, N., Burroughs, E., Kennedy, J. L., & Richter, M. A. (2015). Meta-analysis and association of brain-derived neurotrophic factor (BDNF) gene with obsessive-compulsive disorder. *Psychiatric Genetics, 25*(2), 95–96. https://doi.org/10.1097/YPG.0000000000000077.

Zhang, K., Cao, L., Zhu, W., Wang, G., Wang, Q., Hu, H., & Zhao, M. (2015). Association between the efficacy of fluoxetine treatment in obsessive-compulsive disorder patients and SLC1A1 in a Han Chinese population. *Psychiatry Research, 229*(1–2), 631–632. https://doi.org/10.1016/j.psychres.2015.06.039.

Zhang, L., Liu, X., Li, T., Yang, Y., Hu, X., & Collier, D. (2004). Molecular pharmacogenetic studies of drug responses to obsessive-compulsive disorder and six functional genes. *Zhonghua Yi Xue Yi Chuan Xue Za Zhi, 21*(5), 479–481.

Zilhao, N. R., Olthof, M. C., Smit, D. J., Cath, D. C., Ligthart, L., Mathews, C. A., ... Dolan, C. V. (2017). Heritability of tic disorders: A twin-family study. *Psychological Medicine, 47*(6), 1085–1096. https://doi.org/10.1017/S0033291716002981.

Zilhao, N. R., Padmanabhuni, S. S., Pagliaroli, L., Barta, C., BIOS Consortium, Smit, D. J., ... Boomsma, D. I. (2015). Epigenome-wide association study of Tic disorders. *Twin Research and Human Genetics, 18*(6), 699–709. https://doi.org/10.1017/thg.2015.72.

Zilhao, N. R., Smit, D. J., Boomsma, D. I., & Cath, D. C. (2016). Cross-disorder genetic analysis of Tic disorders, obsessive-compulsive, and hoarding symptoms. *Frontiers in Psychiatry, 7*, 120. https://doi.org/10.3389/fpsyt.2016.00120.

Chapter 21

Genetics and pharmacogenetics of attention deficit hyperactivity disorder in childhood and adulthood

Cristian Bonvicini[a], Carlo Maj[a,b] and Catia Scassellati[a]

[a]IRCCS Istituto Centro San Giovanni di Dio Fatebenefratelli, Brescia, Italy, [b]Institute of Genomic Statistics and Bioinformatics, University of Bonn, Bonn, Germany

1 Introduction

Attention deficit hyperactivity disorder (ADHD) has been a diagnosis of some controversy over the past century in psychiatric and psychological circles. Over the past 30 years, however, a consensus has come about regarding both the existence and clarity of the symptoms that make up the ADHD phenomenon, and more recently, how these manifest as the condition persists into adulthood. As of 2013, with the introduction of the Diagnostic and Statistical Manual of Mental Disorders, Fifth Edition (DSM-5), it is no longer classified as a childhood disorder, but as a chronic, lifelong disorder.

In the 1960s, the modern history of ADHD began with the publication of the DSM-II, in which clinical observations of school-aged children who were overtly overactive, distractible, and restless led to increasing recognition of what appeared to be a common syndrome. "Hyperkinetic reaction of childhood" was typically characterized by impaired, unruly behavior at home and school, and professionals were instructed that the disorder typically diminished by adolescence. In the 1970s and 1980s, a handful of longitudinal studies followed children with ADHD into adolescence and adulthood (rev Klein & Mannuzza, 1991). These studies laid to rest claims that symptoms of ADHD fully diminished by the end of childhood. By adulthood, many individuals with a childhood diagnosis of ADHD still experienced symptoms resembling the attention and self-control deficits they experienced as children. Some even still met full criteria for the disorder. Even when ADHD symptoms appeared to relent by adulthood, major life outcomes for children with ADHD were substantially impaired, compared with typically developing peers. These early longitudinal studies uncovered clinically meaningful risks for criminality, substance abuse, and emergent psychiatric disorders among adults who had been diagnosed with ADHD in childhood. During the 1990s, the field accepted the notion that ADHD is a chronic disorder affecting individuals through the lifespan (Barkley, 1998). With increased public attention on the disorder, it also became increasingly common for individuals to present clinically with complaints of suspected ADHD for the first time in adulthood (Faraone, Biederman, & Mick, 2006). In 2013, a new classification system was approved by NIMH, in which the Research Domain Criteria (RDoC) project proposed a set of assumptions, by integrating genetics, imaging, and cognitive information (Insel, 2013).

Although there is substantial variability between countries, worldwide, an estimated 5%–7% of school-aged children are diagnosed with ADHD, which makes ADHD the most often used psychiatric classification assigned to children (Polanczyk, Willcutt, Salum, Kieling, & Rohde, 2014). In adulthood, the prevalence was estimated to be 2.5%–4.9% (Ramos-Quiroga, Nasillo, Fernández-Aranda, & Casas, 2014).

ADHD is characterized by age-inappropriate symptoms of inattention or hyperactivity and impulsivity, with difficulties of sustaining attention, organization, planning, and restlessness, and with a heterogeneous clinical phenotype (American Psychiatric Association, 2013). Severity level and presentation of ADHD can change during a person's lifetime, with adult patients displaying less obvious symptoms of hyperactivity and impulsivity (Haavik, Halmøy, Lundervold, & Fasmer, 2010). The phenotypic heterogeneity of the disorder is apparent from a large diversity of psychiatric comorbidities, frequently seen both in children and in adults.

Neuroimaging studies indicate that ADHD is a result of abnormal anatomical functioning and connectivity throughout fronto-striatal, fronto-temporal, fronto-parietal, and/or fronto-striato-parieto-cerebellar circuitry. In addition to circuitry, specific structures and areas of the brain have also received attention, including the prefrontal cortex, anterior cingulate

Personalized Psychiatry. https://doi.org/10.1016/B978-0-12-813176-3.00021-3

cortex, caudate, globus pallidus, parietal regions, temporal regions, corpus callosum, splenium, cerebellar vermis, and cerebellum. Some structural brain findings differentiate ADHD in children and adults, suggesting potential differential etiological pathways for both subjects (Shaw et al., 2013).

Pharmacotherapy has an essential role in the treatment of ADHD (Kornfield et al., 2013). Many studies have documented the efficacy of stimulants (methylphenidate—MPH, amphetamines) and nonstimulants (atomoxetine) in reducing ADHD symptoms, as well as improving neuropsychological performance on the measure of executive functions. MPH is clinically effective also in adults, with meta-analyses showing improvements over placebos with medium-to-large effect sizes. The alpha-2 agonists clonidine and guanfacine, which were developed and initially utilized as antihypertensive agents, also showed improvement in ADHD symptoms (Sallee, Connor, & Newcorn, 2013).

In the era of precision (personalized) medicine, the biomarker approach to diagnosis and treatment, as supported by the RDoC, represents a more valid way to classify complex mental disorders such as ADHD, and offer the opportunity to standardize and improve diagnostic assessment, provide insights into etiological mechanisms, and contribute to developing individualized therapies. To date, although biomarkers are successfully used in predicting diseases such as cancer, there is no laboratory test that is used clinically for diagnosis of ADHD. Moreover, although there are several treatments for ADHD, the mechanisms of action of these agents are still unclear, and no biological predictors of treatment response are available that can be used in treatment planning.

1.1 Genetics: State of the art

ADHD is a complex and heterogeneous disorder, and its etiology is not yet completely understood. Despite evidence that environmental factors (i.e., maternal smoking, low birth weight, prematurity) play a significant role in its etiology, classical genetics studies support a strong genetic contribution for ADHD. The risk of ADHD among parents of children with ADHD is increased by two- to eightfold compared with the population rate (Faraone et al., 2005). A meta-analysis of 20 pooled twin studies estimated an average heritability of 76%, suggesting that ADHD is one of the disorders with the strongest genetic component in psychiatry (Faraone et al., 2005). Data about the heritability of ADHD in children and adults had been inconsistent, with some suggestion that heritability is higher in children (75%–90%) (Faraone & Mick, 2010) compared with adults (30%–50%) (Boomsma et al., 2010; Kan et al., 2013; Larsson et al., 2013), and other studies supporting greater heritability in adults (Biederman et al., 1996; Faraone et al., 2000). A more recent review suggests that the heritability of ADHD in adults and in childhood is better explained by rater effects (Brikell, Kuja-Halkola, & Larsson, 2015).

Despite these high heritability estimates, identification of genes that confer susceptibility to ADHD has been a slow and difficult process. Current findings suggest that ADHD is a polygenic disorder determined not only by common variants, but also by rare variants, including deletions or duplications known as copy number variants (CNVs) (Martin, O'Donovan, Thapar, Langley, & Williams, 2015). A useful database on the genomic studies in ADHD is available (http://adhd. psych.ac.cn/index.do).

1.1.1 Genomics approaches and results

Genetic linkage studies provided a first possibility to perform hypothesis-free genetic studies in the early 2000s. However, gene identification through linkage analysis has been limited (Banaschewski, Becker, Scherag, Franke, & Coghill, 2010). A robust association was found for 5p13, locus for *SLC6A3* (dopamine transporter) gene (Ogdie et al., 2006), and 16q23.3 (*CDH13*, T-cadherin 3) was found in ADHD children (Zhou et al., 2008) and in adults (Lesch et al., 2008). Other associations regard three SNPs in the *LPHN3* (latrophin-3) locus, and 4q13.1, 9q22, 9q33 in ADHD children (Li, Chang, Zhang, Gao, & Wang, 2014).

More recently, most studies based on hypothesis-driven *associated candidate genes (CGA), meta-analyses,* and hypothesis-free *genome-wide association studies (GWAS)* illustrate an involvement of common variants in dopaminergic (among which is the *SLC6A3* gene), serotonergic, and glutamatergic signaling, neuronal crosstalk, neuronal migration, and cell adhesion pathways in childhood and adulthood ADHD etiology (revs Bonvicini, Faraone, & Scassellati, 2016; Hawi et al., 2015).

Studies of rare variants were also performed for ADHD. To date, 10 (rev Hawi et al., 2015) and 2 (Akutagava-Martins et al., 2014; Ramos-Quiroga, Sánchez-Mora, et al., 2014) CNVs-GWAS have been published for ADHD children and adults, respectively. The glutamatergic system was affected by CNVs in both ADHD children and adults (rev Bonvicini, Faraone, & Scassellati, 2017). Other genes involved in specific deletions were *BCHE, CDH13, CPLX2, CTNNA2, ASTN2* (neurodevelopmental system), *HTR1B* (serotoninergic receptor), *COMT* (Catechol-*O*-Methyltransferase), and *DISC1* (Disrupted In Schizophrenia 1) genes in ADHD children and in adults, respectively (rev Bonvicini et al., 2017). CNVs associated to 15q13 and 16p13.11 were detected (rev Bonvicini et al., 2017).

Recently, alternative and bioinformatics approaches have been performed (rev Akutagava-Martins, Rohde, & Hutz, 2016). By clustering ADHD-related genes within functional *networks* or *pathways*, the main biological processes resulted in neurite outgrowth, neuron projection morphogenesis, oxogenesis, cell-cell communication, glutamatergic synapse/receptor signaling (Hawi et al., 2015), neuron projections, and synaptic components (Yang et al., 2013), all consistent with a neurodevelopmental pathophysiology of ADHD. These findings are strengthened by a recent study (Mooney et al., 2016), that by a cross-method convergent approach, revealed a number of brain-relevant pathways. Other pathways, such as immune/inflammatory systems, needed more investigation (rev Akutagava-Martins et al., 2016).

The *polygenic risk score analyses*, unlike pathways analyses, consists of building a risk score based on common genetic variants, being not restricted to a set of functionally related genes or pathways. The genetic risk score was conceptualized based on the number of risk alleles at each loci, and weighted by the odds ratio log from a discovery sample. The scores are then applied to a target sample to verify association with the phenotype. The variants to be analyzed may be selected from the discovery sample at several different significance thresholds. Hamshere et al. (2013) demonstrated that ADHD cases present higher polygenic risk scores, and Stergiakouli et al. (2015) showed that ADHD symptoms and ADHD traits in the general population arise from shared genetic factors.

Several studies have proposed that genetic investigation may best proceed with the use of *endophenotypes* or *intermediate phenotypes*. Indeed, the literature suggests that combining endophenotypes with relevant genotypes may lead to a clearer understanding of the association between these candidate genes and ADHD. Dopaminergic pathway genes, among which *SLC6A3* has been found to be associated with specific neuropsychological (rev Kebir & Joober, 2011) and neuroimaging (rev Vilor-Tejedor, Cáceres, Pujol, Sunyer, & González, 2017) endophenotypes.

Another interesting scenario helping to disclose the specific factors that determine the development of a specific phenotype from a pleiotropic genotype is "gene × gene" and "gene × environment" interaction, but also epigenetic mechanisms. Methylation is the best characterized epigenetic mechanism. Through the addition of methyl groups to the DNA, gene transcription is regulated in response to environmental stimuli, without changes in DNA sequence. A study demonstrated that DNA methylation levels of dopaminergic/serotoninergic (*DRD4*, *5HTT*) genes at birth were negatively associated with ADHD symptoms assessed at 6 years of age (rev Akutagava-Martins et al., 2016). More recently, *next generation sequencing* also provides an excellent opportunity to identify rare, new, and de novo variants associated with the complex diseases. Some studies performed exome sequencing in ADHD, both in children (Hawi et al., 2017; Kim et al., 2017) and adults (Demontis et al., 2016; Zayats et al., 2016).

Several lines of evidence demonstrate the usefulness of *expression studies* in the identification of molecular alterations in psychiatric disorders. In this regard, an overlap has been observed between gene expression profiles in blood cells and in postmortem brains, supporting the hypothesis that studies in peripheral tissues may provide new information on the disease pathogenesis and innovative biomarkers for the diagnostic assessment, and for personalization of therapies. A recent review (Bonvicini et al., 2017) summarized the studies where altered expression levels have been observed in *SLC6A3*, but also in other dopaminergic genes and circadian rhythms in children/adult patients and controls. Even alterations in specific sets of microRNA, with small RNA species (*22 nucleotides) playing a key role in posttranscriptional regulation of gene expression, have been observed in ADHD.

1.1.2 Metabolomics

In line with expression studies, the identification of peripheral biochemical markers (metabolomics), measurable in vivo with noninvasive methods, might facilitate the differential diagnosis of ADHD and the personalization of therapies.

As reported in a recent review (Bonvicini et al., 2017), different levels in plasma/serum of several proteins such as the neurotrophins BDNF and NGF, cortisol and oxidative stress proteins were observed between children/adult patients and controls. In relation to polyunsaturated fatty acids, negative results were observed for serum arachidonic acid, docosahexaenoic acid, and eicosapentaenoic acid levels in children/adult patients compared with controls. Contrasting results for kynurenine pathways and two positive studies for adiponectin were available in children/adults with ADHD.

1.2 Pharmacogenetics: State of the art

The term "pharmacogenetics/omics" has been used to encompass studies investigating gene variations and their relationship to medication response, and to emphasize these insights to discover new therapeutic targets and optimize drug therapy (Evans & Johnson, 2001). The great potential of ADHD pharmacogenetics for clinical application lies in predicting better medication choices, avoiding adverse effects, maximizing individual treatment outcomes, and determining the most appropriate drug dosage.

It is known that a considerable proportion (35%) of ADHD patients do not respond to treatment, or present with adverse effects. Thus, the identification of accurate predictors of response to medication should be beneficial for clinical practice.

A recent review of the literature on ADHD pharmacogenetics by a *CGA studies* approach in childhood supported a significant effect of neurodevelopmental and noradrenergic systems in the MPH response, whereas no or contrasting results were obtained for pathways belonging to dopamine (including the *SLC6A3* gene) and serotonin, synaptosomal-associated protein 25 (*SNAP25*), and metabolic enzymes (rev Bruxel et al., 2014). In adults, a recent meta-analysis (Bonvicini et al., 2016) showed no association of *SLC6A3* gene and MPH response, and single studies supported the lack of association with dopaminergic (*DRD4*), serotoninergic (*HTR1B*, *SLC6A4*), adrenergic (*SLC6A2*, *ADRA2A*) systems, metabolic enzymes (*TPH2*, *DBH*, *COMT*), and *SNAP25* genes.

As far as using a *GWAS* approach, only a few studies have been performed (rev Bruxel et al., 2014). These, in general, have failed to identify genes at the stringent genome-wide significance level. However, among the top findings, the *GRM7* (glutamate metabotropic receptor) and *NET1* (noradrenergic transporter) genes showed potential involvement in MPH response.

There is an increasing number of studies assessing "gene × gene" and "gene × environment" interactions. As reviewed in Bruxel et al. (2014), some studies demonstrated a positive interaction of *DRD4* (dopamine receptor D4) and *ADRA2A* (adrenergic receptor 2A) and/or *NET1* genes on the MPH response. In addition, some genes such as *LPHN3* and *NET1* have been found to be involved in MPH response, taking into account maternal smoking and stress during pregnancy. In addition to the study of response to medication itself, adverse events have attracted great interest lately. For example, some studies reported the involvement of genes such as *SLC6A3* in emotionality, *CES1* (carboxylesterase 1) in appetite reduction, and *NET1* and *ADRA2A* in cardiovascular adverse effect (revs Bruxel et al., 2014; Joensen, Meyer, & Aagaard, 2017).

Applying a convergent functional genomics approach, in our previous study (Faraone, Bonvicini, & Scassellati, 2014), we selected studies investigating the association among susceptibility, specific endophenotypes (neuropsychological/neurophysiological tasks, structural brain regions and functioning), metabolomics, gene expression alterations, and drug treatment response with candidate genes for ADHD. In this work, we built a hypothetical signature of a set of genetic and biochemical markers useful for ADHD diagnosis and treatment. On the top of this pyramid, we localize *SLC6A3* as a potential genetic biomarker associated with diagnosis and treatment, for its prevalently positive association with susceptibility, neuropsychology, neuroimaging, altered expression, and treatment response.

In the following paragraphs, through a meta-analytic approach, we aim: (a) to confirm the involvement of the *SLC6A3* gene in the ADHD etiopathogenesis and diagnosis in childhood and in adulthood; (b) to confirm the role of this gene as a potential mediator/predictor of the MPH drug response in childhood and in adulthood.

2 Role of *SLC6A3* gene as a potential genetic biomarker in ADHD diagnosis in children and adults

2.1 Methods

2.1.1 Search strategy and selection

To identify eligible studies for the meta-analysis, we searched the electronic databases PubMed and the "ADHDgene Database," from 1995 (first study, Cook et al., 1995) until September 2017. We searched for all available original studies regarding *SLC6A3* in ADHD children. The following keywords were employed: children, child, attention deficit/hyperactivity disorder or ADHD combined with: *SLC6A3*/dopamine transporter/DAT1, polymorphism, VNTR, variable tandem repeats, association, TDT, Transmission Disequilibrium Test, family-based.

The initial search produced 220 records on PubMed. Because we investigated eight polymorphisms, our Bonferroni corrected significance level was 0.006.

2.1.2 Inclusion and exclusion criteria

We selected articles that met the following inclusion criteria: ADHD diagnosis according to the DSM-IV or the International Classification of Diseases 10th Revision (ICD-10); having a case-control or family-based study design; all languages. We excluded studies: (a) performed in different years by the same authors when using the same samples; (b) with not enough data reported; (c) lacking statistical data; (d) using comparisons with a familiar control (healthy siblings).

2.1.3 Data extraction for meta-analyses and statistical analyses

The literature search was performed independently by three individuals (C.S., C.B., C.M.). We extracted the following data from the original included publications: first author, year of publication, populations studied, sample size, location, and polymorphisms (Table 1).

TABLE 1 Studies investigating the association of *SLC6A3* gene polymorphisms and ADHD in children, included in the meta-analyses

| | Population | Sample size | | | 5'-UTR rs2652511 | 5'-UTR rs2975226 | Intron 8 30 bp VNTR | Exon 9 rs6347 | Intron 9 rs8179029 | Intron 14 rs40184 | 3'-UTR rs27072 | 3'-UTR 40 bp VNTR |
Author (year)		Case (n)	Control (n)	TDT (2n)								
Cook et al. (1995)	84% Caucasian, 15% African American, 1% Hispanic			84								TDT
Gill, Daly, Heron, Hawi, and Fitzgerald (1997)	Irish			62								TDT
Waldman et al. (1998)	68% Caucasian, 12% African American, 4% Hispanic, 16% Mixed			122								TDT
Palmer et al. (1999)	79.9% Caucasian			209								TDT
Daly, Hawi, Fitzgerald, and Gill (1999)	Irish			187								TDT
Holmes et al. (2000)	UK	133	295	86								CC/ TDT
Swanson et al. (2000)	USA			80								TDT
Roman et al. (2001)	European-Brazilian	66	112	135								CC/ TDT
Todd et al. (2001)	USA			119								TDT
Barr et al. (2001)	UK, Ireland, Germany, French, Italian, Polish, Dutch			103				TDT	TDT			TDT

Continued

TABLE 1 Studies investigating the association of *SLC6A3* gene polymorphisms and ADHD in children, included in the meta-analyses—cont'd

Author (year)	Population	Sample size			5'-UTR rs2652511	5'-UTR rs2975226	Intron 8 30bp VNTR	Exon 9 rs6347	Intron 9 rs8179029	Intron 14 rs40184	3'-UTR rs27072	3'-UTR 40bp VNTR
		Case (n)	Control (n)	TDT (2n)								
Maher, Marazita, Ferrell, and Vanyukov (2002)	Caucasian			33								TDT
Kirley et al. (2002)	Irish			118								TDT
Smith et al. (2003)	Caucasian	105	68									CC
Hawi et al. (2003)	Irish			187							TDT	TDT
Chen et al. (2003)	Taiwanese			37								TDT
Kustanovich et al. (2004)	79% Caucasian, 2% Black, 4% Hispanic, 2% Asian, 13% Mixed			293								TDT
Qian et al. (2007)	Chinese	332	216	376								CC/TDT
Langley et al. (2005)	UK	263	287	522								CC/TDT
Bobb et al. (2005)	75% Caucasian, 12% African American, 10% Hispanic, 2% Asian, 1% others	163	129	110				CC/TDT				CC/TDT
Kim et al. (2005)	Korean			33								TDT

Study	Ancestry								
Hawi et al. (2005)	Irish			104					TDT
Feng et al. (2005)	UK, Ireland, Germany, French, Italian, Polish, Dutch			156			TDT	TDT	TDT
Bakker et al. (2005)	Dutch			212	TDT	TDT			TDT
Simsek, Al-Sharbati, Al-Adawi, and Lawatia (2006)	Oman	92	110		TDT				CC
Kim, Kim, and Cho (2006)	Korean	85	100						CC
Cheuk, Li, and Wong (2006)	Chinese	64	64	64					CC/TDT
Lim et al. (2006)	Korean			21					TDT
Brookes et al. (2006)	UK and Taiwan			37/97					TDT
Shaw et al. (2007)	78% Caucasian, 12% African American, 7% Hispanic, 2% Asian, 1% others	105	103						CC
Laurin et al. (2007)	European descent			266					TDT
Das and Mukhopadhyay (2007)	Indian			37					TDT
Asherson et al. (2007)	Caucasian			791	TDT	TDT			TDT
Friedel et al. (2007)	German			329			TDT	TDT	

Continued

TABLE 1 Studies investigating the association of SLC6A3 gene polymorphisms and ADHD in children, included in the meta-analyses—cont'd

Author (year)	Population	Case (n)	Control (n)	TDT (2n)	5'-UTR rs2652511	5'-UTR rs2975226	Intron 8 30bp VNTR	Exon 9 rs6347	Intron 9 rs8179029	Intron 14 rs40184	3'-UTR rs27072	3'-UTR 40bp VNTR
Genro et al. (2007)	Caucasian			454	TDT							TDT
Wang et al. (2008)	Chinese	54	66	108								CC/TDT
Kopecková et al. (2008)	Czechs	100	100									CC
Banoei et al. (2008)	Iranian	100	130									CC
Wohl et al. (2008)	Caucasian			146								TDT
Brookes et al. (2008)	Caucasian			776	TDT		TDT			TDT		TDT
Genro et al. (2008)	European descent			252		TDT	TDT	TDT	TDT	TDT	TDT	
Xu et al. (2009)	UK and Taiwan			197/212	TDT	TDT						
Doyle et al. (2009)	UK and Irish			183/267						TDT		
Gharaibeh, Batayneh, Khabour, and Daoud (2010)	Jordanian	50	50									CC
Hawi et al. (2010)	Irish 1, Irish 2, UK, Caucasian			103/36/41/703			TDT					TDT
Das et al. (2011)	Indian	126	96	123			TDT					
El-Tarras et al. (2012)	Saudi	120	160									CC/TDT
Šerý et al. (2015)	Czech	218	253									CC

Study	Population									
Agudelo et al. (2015)	Colombian	73	54							CC
Yu et al. (2016)	Taiwan	97	110			CC				
Stanley, Chavda, Subramanian, Prabhu, and Ashavaid (2017)	Indian	44	44		CC					CC
Das Bhowmik et al. (2017)	Indo-Caucasian	152	96		CC					

CC, case-control studies; *TDT*, Transmission Disequilibrium Test.

Review Manager was used to analyze the data (RevMan Version 5.1.6; Copenhagen, The Nordic Cochrane Centre, The Cochrane Collaboration, 2008). We used the fixed-effects model to generate a pooled effect size and 95% confidence interval (CI) from individual study effect sizes (the odd ratios for genetics studies using the Mantel-Haenszel (M-H)). The significance of the pooled effect sizes was determined by the z-test. Between-study heterogeneity was assessed using a $\chi2$ test of goodness of fit test and the I^2 statistic. We used a P-value of 0.05 to assert statistical significance. Where the results showed a significant effect in the presence of significant between-study heterogeneity, a random-effects model was used, with effect sizes pooled using the DerSimonian and Laird method.

Publication bias was estimated by the method of Egger, Smith, and Phillips (1997), which uses a linear regression approach to measure funnel plot asymmetry on the natural logarithm scale of the effect size. The significance of the intercept (a) was determined by the t-test (Egger et al., 1997). The rank correlation method and regression method tests were conducted by MIX version 1.7 (http://www.mix-for-meta-analysis.info).

2.2   Results

During the screening of 220 records, we selected 51 studies meeting our eligibility criteria (Table 1). The most studied *SLC6A3* variant is a VNTR of 40 base pairs located at the 3′-untranslated region (3′-UTR) of the gene. The ten repeat (10R) and nine repeat (9R) alleles are the most common. In Table 1, we reported 35 family-based studies (TDT) and 19 studies case-control (CC) in ADHD children. Preferential transmission of the 10R was detected ($Z = 3.94$, $P < 0.0001$) in TDT studies, which was significant even after the Bonferroni correction. Heterogeneity in effect sizes among the studies was observed ($P < 0.00001$, $I^2 = 91\%$) (Fig. 1).

No association was observed in CC ($Z = 0.77$; $P = 0.44$) (Fig. 2).

In adulthood, our previous meta-analysis (Bonvicini et al., 2016) indicated that, if we did not apply the Bonferroni correction, the carriers of 9R alleles were more present in patients.

Another VNTR in intron 8 of the gene was investigated by several studies. This VNTR is a 30-bp repeat sequence with two common alleles of five (5R) and six (6R) repeats. For this polymorphism, six family-based studies and two case-controls were available. Preferential transmission of the 6R allele was detected ($Z = 2.96$, $P = 0.003$) in TDT studies, which was significant even after Bonferroni correction. Heterogeneity was observed ($P < 0.00001$, $I^2 = 79\%$) (Fig. 1).

No association was detected in CC ($Z = 0.81$, $P = 0.42$) (Fig. 2).

In adulthood, our previous meta-analysis (Bonvicini et al., 2016) did not find an association after Bonferroni correction.

Other SNPs (single-nucleotide polymorphisms) were investigated. We meta-analyzed rs2652511 (3 TDT), rs2975226 (2 TDT), rs6347 (6 TDT, 2 CC), rs8179029 (4 TDT), rs40184 (3 TDT), and rs27072 (4 TDT). Allele A of rs6347 was overtransmitted in ADHD children ($Z = 2.70$, $P = 0.007$), indicating a trend after Bonferroni correction. The presence of heterogeneity was observed ($P = 0.001$, $I^2 = 75\%$) (Fig. 3). CC studies were not enough and thus were excluded.

If we consider all polymorphisms, the respective risk alleles of each polymorphism go in the same direction, showing overtransmission in probands ($Z = 3.87$, $P = 0.0001$). This is even after Bonferroni correction. In adulthood, the SNPs analyzed were rs6347, rs40184, and rs2652511, for which not enough studies were available to perform meta-analysis (Bonvicini et al., 2016). In general, they were not found to be associated with ADHD.

According to Egger's test, the results indicated no publication bias for the meta-analyses: Fig. 4: $P = 0.90$ for 40 bp VNTR 3′-UTR, $P = 0.59$ for 30 bp VNTR intron 8; Fig. 5: $P = 0.51$ for 40 bp VNTR 3′-UTR; Fig. 6: $P = 0.37$ for rs26525511; $P = 0.49$ for rs2975226; $P = 0.45$ for rs6347; $P = 0.88$ for rs8179029; $P = 0.05$ for rs40184; $P = 0.52$ for rs27072.

3   Role of *SLC6A3* gene as potential mediator/predictor of the MPH drug response in children and in adults

3.1   Methods

3.1.1   Search strategy and selection

To identify eligible studies for the meta-analysis, we searched PubMed and the "ADHDgene Database" from inception until September 2017, for all available studies on *SLC6A3* and ADHD MPH treatments in childhood. "ADHD" and "Attention Deficit Hyperactivity Disorder" were used to conduct searches with *SLC6A3*/dopamine transporter/DAT1, methylphenidate, MPH, drugs, treatments, clinical trials, polymorphism, and VNTR. The initial search produced 125 records on PubMed.

3.1.2   Inclusion and exclusion criteria

We included studies that provided the features needed for performing meta-analyses, thus the studies with not enough or lacking statistical data were excluded.

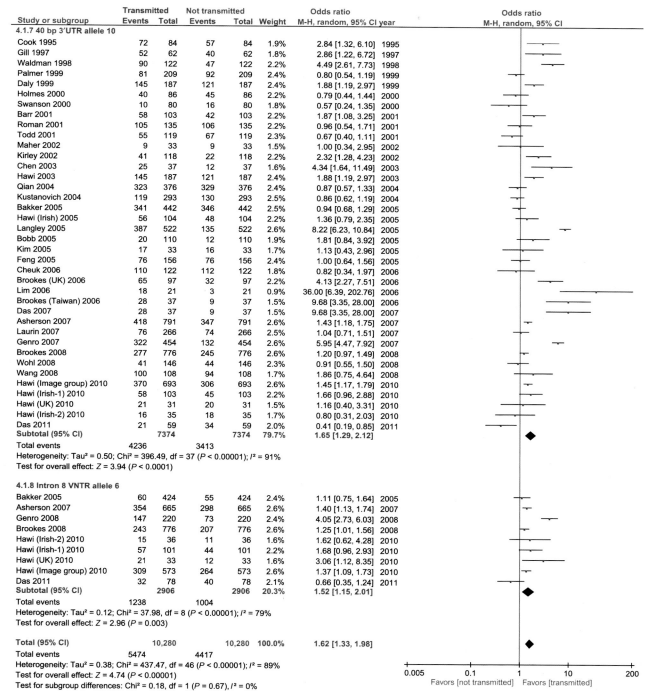

FIG. 1 Random forest plot for odds ratio from meta-analyses of TDT (Transmission Disequilibrium Test) studies of 40 bp VNTR 3′-UTR and 30 bp intron 8 VNTR in *SLC6A3* gene. *Chi2*, χ2 test of goodness of fit; *Tau2*, estimate of the between-study variance in a random-effects meta-analysis.

3.1.3 Data extraction for meta-analysis

The literature search was performed independently by three individuals (C.S., C.B., C.M.). For all studies suitable for meta-analysis, we extracted the following data from the original publications: first author and year of publication, ethnicity, the number of participants, study designs, dosage, and treatment time duration (Table 2).

3.1.4 Statistical analyses

Review Manager was used. Publication bias was estimated by the method of Egger et al. (1997).

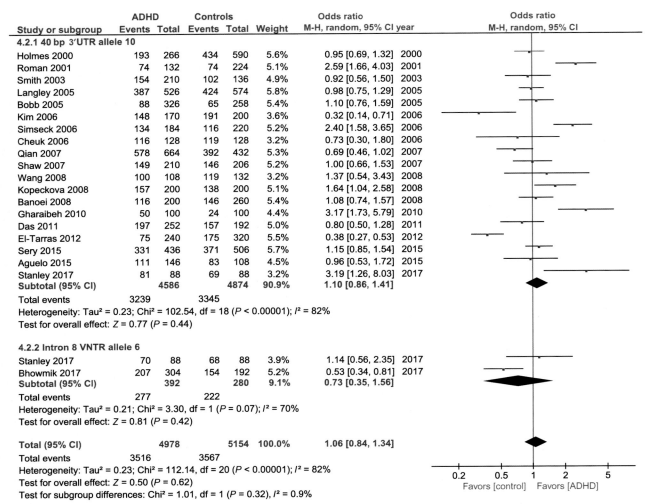

FIG. 2   Random forest plot for odds ratio from meta-analyses of case-control studies of 40 bp VNTR 3′-UTR and 30 bp intron 8 VNTR in *SLC6A3* gene. *Chi2*, χ2 test of goodness of fit; *Tau2*, estimate of the between-study variance in a random-effects meta-analysis.

3.2   Results

After the screening of papers according to the inclusion/exclusion criteria, 10 studies on MPH response and *SLC6A3* 40 bp VNTR polymorphism were available (Table 2). No association was observed after MPH treatment ($Z = 1.86$, $P = 0.06$). Significant heterogeneity was observed ($P = 0.001$, $I^2 = 67\%$) (Fig. 7).

Although this negative result occurred, an effect of the 10R/10R on the poor response can be observed.

In adulthood, this polymorphism was not found to be associated with MPH response according to our previous meta-analysis (Bonvicini et al., 2016). According to Egger's test, the results indicated no publication bias (Fig. 8, $P = 0.08$).

4   Discussion and future research

After an overview of the genomics, metabolomics, and pharmacogenetics in ADHD, from which results suggest a strong involvement of the *SLC6A3* gene, this chapter aims: (a) to confirm the role of *SLC6A3* as potential genetic biomarker associated with ADHD diagnosis in children vs adults; and (b) to determine the role of *SLC6A3* as a potential mediator/predictor of MPH response in children vs adults.

SLC6A3 is a well-known gene with a crucial role in the pathophysiology of different psychiatric illnesses, including ADHD. Dopamine is an important neuromodulator that is released prominently in frontal and striatal areas, known to have an important role in cognitive function. Moreover, it has a key role in regulating striatal dopamine levels, and the amount of DAT1 protein expression in striatum has, in previous meta-analyses, been shown to be associated with the 40 bp VNTR

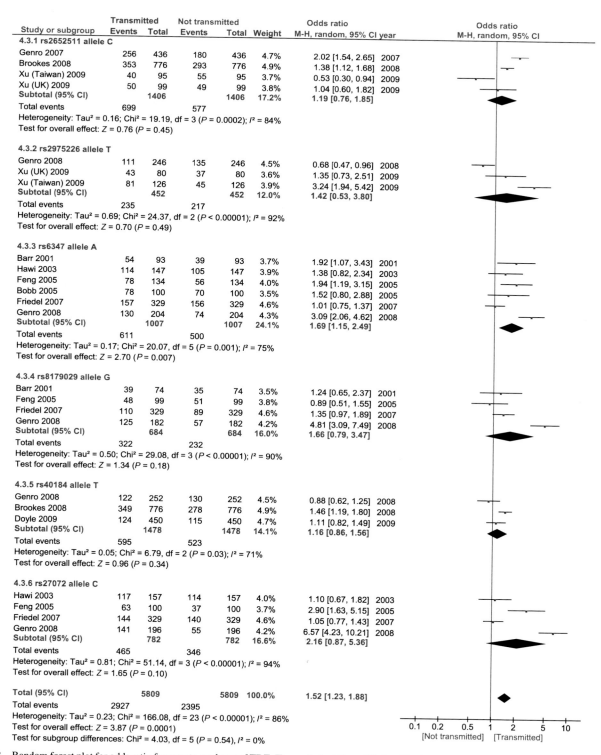

FIG. 3 Random forest plot for odds ratio from meta-analyses of TDT (Transmission Disequilibrium Test) studies of SNPs in *SLC6A3* gene. *Chi2*, χ2 test of goodness of fit; *Tau2*, estimate of the between-study variance in a random-effects meta-analysis.

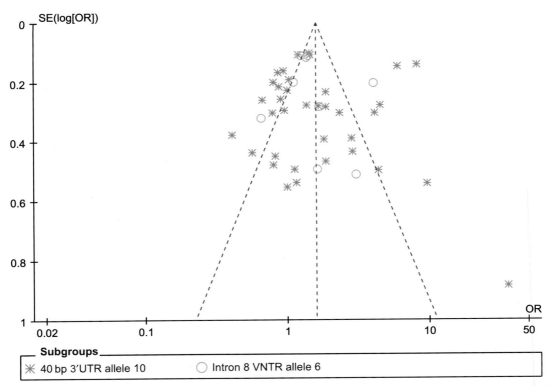

FIG. 4 Funnel plot of studies included in the meta-analysis of 40 bp VNTR 3′-UTR and 30 bp intron 8 VNTR in *SLC6A3* gene in TDT (Transmission Disequilibrium Test) studies.

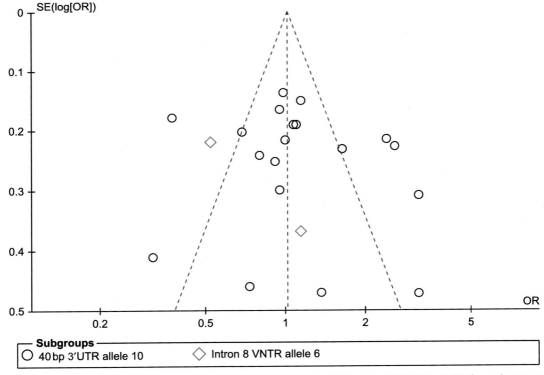

FIG. 5 Funnel plot of studies included in the meta-analysis of 40 bp VNTR 3′-UTR and 30 bp intron 8 VNTR in *SLC6A3* gene in case-control studies.

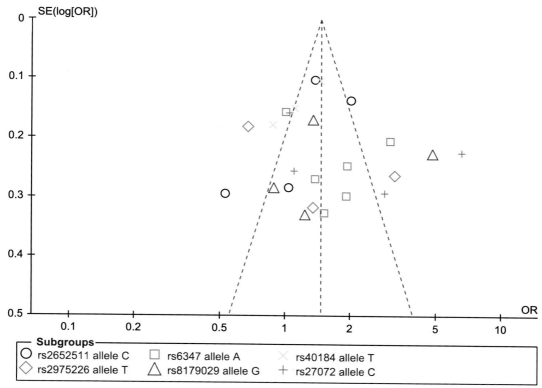

FIG. 6 Funnel plot of studies included in the meta-analysis of SNPs in *SLC6A3* gene in TDT (Transmission Disequilibrium Test) studies.

TABLE 2 Studies investigating the association of VNTR 40 bp 3′-UTR polymorphism in *SLC6A3* gene and methylphenidate response in children, included in the meta-analyses

Study (year)	Ethnicity	n	Response measure Scale	MPH treatment Design	Duration	Dose
Winsberg and Comings (1999)	African American	30	ABRS	Naturalistic	NA	40–60 mg day^{-1}; <0.7 mg kg^{-1} day^{-1}
Roman et al. (2002)	Brazilian	50	ABRS, CGAS	Naturalistic	~30 days	0.3–0.7 mg kg^{-1} day^{-1}
Kirley et al. (2003)	Irish	95	CPRS	Naturalistic	NA	NA
Bellgrove et al. (2005)	Irish	46		Naturalistic	NA	NA
Cheon, Ryu, Kim, and Cho (2005)	Korean	11	ARS	Naturalistic	8 weeks	0.3 mg kg^{-1} day^{-1}
Langley et al. (2005)	British	168	CGI	Naturalistic	≥3 months	NA
Stein et al. (2005)	Mixed	47	CGI-S	Clinical trial	4 weeks	Placebo or 18–54 mg day^{-1}
Kereszturi et al. (2008)	Hungarian	121	ARS	Naturalistic	4 weeks	0.22–0.95 mg kg^{-1} day^{-1}
Purper-Ouakil et al. (2008)	NA	141	ARS, CGI	Naturalistic	55.5 days	Mean of 31.19 mg day^{-1}
Tharoor, Lobos, Todd, and Reiersen (2008)	American	156	MAGIC	Naturalistic	NA	NA

ABRS, Conners' Abbreviated Rating Scale; *ARS*, ADHD Rating Scale; *CGAS*, Clinical Global Assessment Scale; *CGI*, Conner's Global Index; *CGI-S*, Clinical Global Impression-Severity of Impairment; *CPRS*, Conners' Parents Rating Scale Revised; *MAGIC*, Missouri Assessment for Genetics Interview for children; *MPH*, methylphenidate; *NA*, not available.

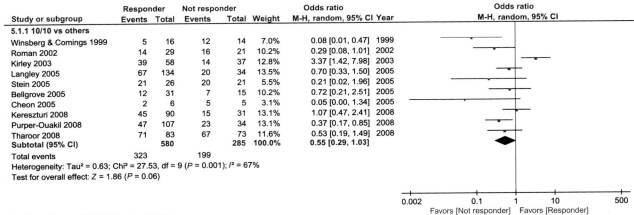

FIG. 7   Random forest plot for odds ratio from meta-analyses of studies of 40 bp VNTR 3′-UTR in *SLC6A3* gene in relation to the response to methylphenidate. *Chi2*, χ2 test of goodness of fit; *Tau2*, estimate of the between-study variance in a random-effects meta-analysis.

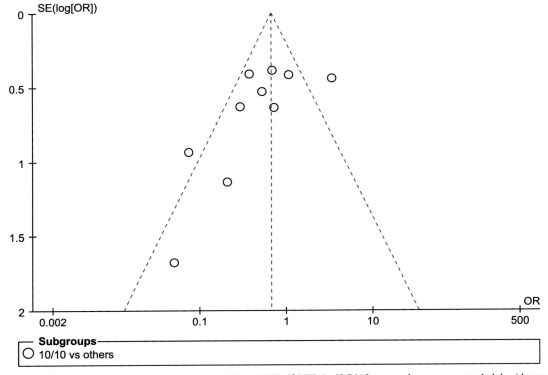

FIG. 8   Funnel plot of studies included in the meta-analysis of 40 bp VNTR 3′-UTR in *SLC6A3* gene and response to methylphenidate treatment.

polymorphism (Faraone, Spencer, Madras, Zhang-James, & Biederman, 2014). Knockout mice for *SLC6A3* show hyperactivity and deficits in inhibitory behavior, and the gene has been mapped near to a susceptibility locus for ADHD, 5p13. Interestingly, the DAT1 protein is the main target for two ADHD medications: MPH and amphetamine (AMP). MPH blocks the transporter, elevating the extracellular dopamine level in the mammalian brain.

From a convergent functional genomics approach (Faraone, Bonvicini, et al., 2014), allele 10R, a susceptibility risk allele (Gizer, Ficks, & Waldman, 2009) associated with higher DA concentrations and higher DAT1 density in basal ganglia, has been suggested to be linked to different and specific neuropsychological tasks, generate more activation in specific brain areas, and be associated with reduced brain structure, even with contrasting results, in MPH response. Moreover, 10R allele conferred increased risk for ADHD when exposed to prenatal smoke (Becker, El-Faddagh, Schmidt, Esser, & Laucht, 2008; Kahn, Khoury, Nichols, & Lanphear, 2003), psychosocial adversity (Laucht et al.,

2007), low parental expressed emotion (Sonuga-Barke et al., 2009), institutionalized deprivation (Stevens et al., 2009), and maltreatment in girls (Li & Lee, 2012). Also, 6R allele of the intron 8 VNTR has been indicated as a risk factor (Gizer et al., 2009), and the haplotype formed with 40 bp VNTR (10/6) was associated with deficits in dopaminergic reward-processing in ADHD (Paloyelis, Mehta, Faraone, Asherson, & Kuntsi, 2012).

We extended the last meta-analysis performed between this gene and childhood ADHD susceptibility (Gizer et al., 2009) by adding new studies and dividing family-based from case-control studies. We confirm that the 10R and 6R are risk alleles in TDT studies. In adulthood, our previous meta-analysis (Bonvicini et al., 2016) showed an association of 9R as risk allele. Interestingly, ADHD adults carrying the 9R allele had 5.9% larger striatum volume, as compared with subjects not carrying this allele (Onnink et al., 2016), probably for the increased activity/expression of the *SLC6A3* gene in this region, due to the presence of the 9R allele. Thus, these findings suggest an allelic heterogeneity where *SLC6A3* could modulate the ADHD phenotype across the lifespan, with differential associations depending on age.

Contrary to Gizer et al. (2009), rs6347 was found to be associated with ADHD, and interestingly, the significant effect of the risk alleles of rs2652511 (C), rs2975226 (T), rs8179029 (G), rs40184 (T), rs27072 (C) on ADHD strengthens the involvement of the *SLC6A3* gene as potential biomarker associated with diagnosis.

The pronounced action of MPH at the DAT1, reducing the symptoms of ADHD, has led to numerous pharmacogenetics studies involving the gene. In particular, a number of studies have investigated the VNTR as a possible source of variation in clinical response to this drug in patients with ADHD. The last meta-analysis performed in 2014 (Kambeitz, Romanos, & Ettinger, 2014) reported no significant summary effect for the *SLC6A3* VNTR on the response to MPH treatment. This work also included three studies performed on ADHD in adults (Contini et al., 2010; Kooij et al., 2008; Mick, Biederman, Spencer, Faraone, & Sklar, 2006). Given the importance of possible differences between ADHD in children vs adults, we performed a new meta-analysis by updating the studies in children and excluding the studies on adults. The new analysis did not change the results in children, and similarly, in adults, the same 40 bp VNTR was not found to be associated with MPH response (Bonvicini et al., 2016). Thus, these data suggest that the 40 bp VNTR is not related to variability in clinical response to MPH treatment in childhood and in adulthood. However, although the results are negative, an effect of 10R/10R on poor response can be observed. Because of the correlation between increased DAT1 density and the presence of homozygosity for the 10R/10R, this result suggests the possibility that more MPH is needed, and that 10R could be utilized as a factor predicting treatment response to MPH. Further investigations are mandatory.

Limitations. Although no significant Egger's test for funnel plot asymmetry was observed, which indicates no publication bias, this work has some limitations. We found significant heterogeneity in effect size across the studies due to several issues: (a) differences in sample and methodological characteristics; (b) not all studies reported quality control analyses, other than tests of Hardy-Weinberg equilibrium and analyses of the quality of the genotyping conducted (call rates, similarity of obtained allele frequencies to those in public databases, repeated genotyping consistency); (c) no complete homogeneity in the study designs: most studies have a naturalistic design, others are clinical trials; (d) different concentrations of drugs used in the several studies; (e) different durations of treatment; (f) varying methods to evaluate the treatment response; (g) very small sample size, mainly in the pharmacogenetics studies; (h) given the limited number of studies available, the prediction of treatment response is likely to involve not only genetic factors, but also clinical, environmental, and sociodemographic factors; (i) low significant effect sizes.

Future research should be focused on the replication of these findings, to assess their specificity for ADHD, and to quantify the degree to which they are sufficiently precise to be useful in clinical settings. Thus, more work is needed to determine whether the statistical significance of our findings translate into diagnostic utility. Moreover, future studies will have to take into account the deep integration of "omics" sciences such as the "pharmacogenomics," "phenomics," "epigenomic," "proteomics," "transcriptomics," and "metabolomics" in a convergent functional genomics approach. In fact, a better understanding of the interaction network of genes, proteins, and biochemical processes in relation to more accurate clinical profiles and definition of therapeutic response, by using new high-throughput computational methods and multicenter collaborations, will allow us to identify a list of biomarkers both for the optimization of the diagnostic assessment, as well as for the personalization of therapies.

5 Conclusion

Currently, no biomarkers for ADHD have achieved the status of clinical utility as a diagnostic tool, as well as a predictor of drug response. This chapter focuses on the *SLC6A3* gene and demonstrated, through a meta-analytic approach for eight polymorphisms, an age-specific allelic heterogeneity implicating 10R as risk allele, possibly associated with methylphenidate nonresponse in childhood, and 9R in adulthood. Along with the evidence of associations between other polymorphisms and ADHD in childhood and not in adulthood, we speculate that the *SLC6A3* gene could be the best potential candidate biomarker for children with ADHD.

References

Agudelo, J. A., Gálvez, J. M., Fonseca, D. J., Mateus, H. E., Talero-Gutiérrez, C., & Velez-Van-Meerbeke, A. (2015). Evidence of an association between 10/10 genotype of DAT1 and endophenotypes of attention deficit/hyperactivity disorder. *Neurología*, *30*(3), 137–143.

Akutagava-Martins, G. C., Rohde, L. A., & Hutz, M. H. (2016). Genetics of attention-deficit/hyperactivity disorder: An update. *Expert Review of Neurotherapeutics*, *16*(2), 145–156.

Akutagava-Martins, G. C., Salatino-Oliveira, A., Genro, J. P., Contini, V., Polanczyk, G., Zeni, C., ... Hutz, M. H. (2014). Glutamatergic copy number variants and their role in attention-deficit/hyperactivity disorder. *American Journal of Medical Genetics, Part B: Neuropsychiatric Genetics*, *165B*(6), 502–509.

American Psychiatric Association. (2013). *Diagnostic and statistical manual of mental disorders* (5th ed.). Washington, DC: American Psychiatric Association Press.

Asherson, P., Brookes, K., Franke, B., Chen, W., Gill, M., Ebstein, R. P., ... Faraone, S. V. (2007). Confirmation that a specific haplotype of the dopamine transporter gene is associated with combined-type ADHD. *American Journal of Psychiatry*, *164*(4), 674–677.

Bakker, S. C., van der Meulen, E. M., Oteman, N., Schelleman, H., Pearson, P. L., Buitelaar, J. K., & Sinke, R. J. (2005). DAT1, DRD4, and DRD5 polymorphisms are not associated with ADHD in Dutch families. *American Journal of Medical Genetics, Part B: Neuropsychiatric Genetics*, *132B*(1), 50–52.

Banaschewski, T., Becker, K., Scherag, S., Franke, B., & Coghill, D. (2010). Molecular genetics of attention-deficit/hyperactivity disorder: An overview. *European Child and Adolescent Psychiatry*, *19*(3), 237–257.

Banoei, M. M., Majidizadeh, T., Shirazi, E., Moghimi, N., Ghadiri, M., Najmabadi, H., & Ohadi, M. (2008). No association between the DAT1 10-repeat allele and ADHD in the Iranian population. *American Journal of Medical Genetics, Part B: Neuropsychiatric Genetics*, *147B*(1), 110–111.

Barkley, R. A. (1998). How should attention deficit disorder be described? *The Harvard Mental Health Letter/From Harvard Medical School*, *14*(8), 8.

Barr, C. L., Xu, C., Kroft, J., Feng, Y., Wigg, K., Zai, G., ... Kennedy, J. L. (2001). Haplotype study of three polymorphisms at the dopamine transporter locus confirm linkage to attention-deficit/hyperactivity disorder. *Biological Psychiatry*, *49*(4), 333–339.

Becker, K., El-Faddagh, M., Schmidt, M. H., Esser, G., & Laucht, M. (2008). Interaction of dopamine transporter genotype with prenatal smoke exposure on ADHD symptoms. *The Journal of Pediatrics*, *152*(2), 263–269.

Bellgrove, M. A., Hawi, Z., Kirley, A., Fitzgerald, M., Gill, M., & Robertson, I. H. (2005). Association between dopamine transporter (DAT1) genotype, left-sided inattention, and an enhanced response to methylphenidate in attention-deficit hyperactivity disorder. *Neuropsychopharmacology*, *30*(12), 2290–2297.

Biederman, J., Faraone, S., Milberger, S., Curtis, S., Chen, L., Marrs, A., ... Spencer, T. (1996). Predictors of persistence and remission of ADHD into adolescence: Results from a four-year prospective follow-up study. *Journal of the American Academy of Child and Adolescent Psychiatry*, *35*(3), 343–351.

Bobb, A. J., Addington, A. M., Sidransky, E., Gornick, M. C., Lerch, J. P., Greenstein, D. K., ... Rapoport, J. L. (2005). Support for association between ADHD and two candidate genes: NET1 and DRD1. *American Journal of Medical Genetics, Part B: Neuropsychiatric Genetics*, *134B*(1), 67–72.

Bonvicini, C., Faraone, S. V., & Scassellati, C. (2016). Attention-deficit hyperactivity disorder in adults: A systematic review and meta-analysis of genetic, pharmacogenetic and biochemical studies. *Molecular Psychiatry*, *21*(7), 872–884.

Bonvicini, C., Faraone, S. V., & Scassellati, C. (2017). Common and specific genes and peripheral biomarkers in children and adults with attention-deficit/hyperactivity disorder. *The World Journal of Biological Psychiatry*, *24*, 1–21.

Boomsma, D. I., Saviouk, V., Hottenga, J. J., Distel, M. A., de Moor, M. H., Vink, J. M., ... Willemsen, G. (2010). Genetic epidemiology of attention deficit hyperactivity disorder (ADHD index) in adults. *PLoS One*, *5*(5).

Brikell, I., Kuja-Halkola, R., & Larsson, H. (2015). Heritability of attention-deficit hyperactivity disorder in adults. *American Journal of Medical Genetics, Part B: Neuropsychiatric Genetics*, *168*(6), 406–413.

Brookes, K. J., Mill, J., Guindalini, C., Curran, S., Xu, X., Knight, J., ... Asherson, P. (2006). A common haplotype of the dopamine transporter gene associated with attention-deficit/hyperactivity disorder and interacting with maternal use of alcohol during pregnancy. *Archives of General Psychiatry*, *63*(1), 74–81.

Brookes, K. J., Xu, X., Anney, R., Franke, B., Zhou, K., Chen, W., ... Asherson, P. (2008). Association of ADHD with genetic variants in the 5′-region of the dopamine transporter gene: Evidence for allelic heterogeneity. *American Journal of Medical Genetics, Part B: Neuropsychiatric Genetics*, *147B*(8), 1519–1523.

Bruxel, E. M., Akutagava-Martins, G. C., Salatino-Oliveira, A., Contini, V., Kieling, C., Hutz, M. H., & Rohde, L. A. (2014). ADHD pharmacogenetics across the life cycle: New findings and perspectives. *American Journal of Medical Genetics, Part B: Neuropsychiatric Genetics*, *165B*(4), 263–282.

Chen, C. K., Chen, S. L., Mill, J., Huang, Y. S., Lin, S. K., Curran, S., ... Asherson, P. (2003). The dopamine transporter gene is associated with attention deficit hyperactivity disorder in a Taiwanese sample. *Molecular Psychiatry*, *8*(4), 393–396.

Cheon, K. A., Ryu, Y. H., Kim, J. W., & Cho, D. Y. (2005). The homozygosity for 10-repeat allele at dopamine transporter gene and dopamine transporter density in Korean children with attention deficit hyperactivity disorder: Relating to treatment response to methylphenidate. *European Neuropsychopharmacology*, *15*(1), 95–101.

Cheuk, D. K., Li, S. Y., & Wong, V. (2006). No association between VNTR polymorphisms of dopamine transporter gene and attention deficit hyperactivity disorder in Chinese children. *American Journal of Medical Genetics, Part B: Neuropsychiatric Genetics*, *141B*(2), 123–125.

Contini, V., Victor, M. M., Marques, F. Z., Bertuzzi, G. P., Salgado, C. A., Silva, K. L., ... Bau, C. H. (2010). Response to methylphenidate is not influenced by DAT1 polymorphisms in a sample of Brazilian adult patients with ADHD. *Journal of Neural Transmission*, *117*(2), 269–276.

Cook, E. H., Jr., Stein, M. A., Krasowski, M. D., Cox, N. J., Olkon, D. M., Kieffer, J. E., & Leventhal, B. L. (1995). Association of attention-deficit disorder and the dopamine transporter gene. *American Journal of Human Genetics, 56*(4), 993–998.

Daly, G., Hawi, Z., Fitzgerald, M., & Gill, M. (1999). Mapping susceptibility loci in attention deficit hyperactivity disorder: Preferential transmission of parental alleles at DAT1, DBH and DRD5 to affected children. *Molecular Psychiatry, 4*(2), 192–196.

Das, M., Das Bhowmik, A., Bhaduri, N., Sarkar, K., Ghosh, P., Sinha, S., … Mukhopadhyay, K. (2011). Role of gene-gene/gene-environment interaction in the etiology of eastern Indian ADHD probands. *Progress in Neuro-Psychopharmacology and Biological Psychiatry, 35*(2), 577–587.

Das, M., & Mukhopadhyay, K. (2007). DAT1 3'-UTR 9R allele: Preferential transmission in Indian children with attention deficit hyperactivity disorder. *American Journal of Medical Genetics, Part B: Neuropsychiatric Genetics, 144B*(6), 826–829.

Das Bhowmik, A., Sarkar, K., Ghosh, P., Das, M., Bhaduri, N., Sarkar, K., … Mukhopadhyay, K. (2017). Significance of dopaminergic gene variants in the male biasness of ADHD. *Journal of Attention Disorders, 21*(3), 200–208.

Demontis, D., Lescai, F., Børglum, A., Glerup, S., Østergaard, S. D., Mors, O., … Franke, B. (2016). Whole-exome sequencing reveals increased burden of rare functional and disruptive variants in candidate risk genes in individuals with persistent attention-deficit/hyperactivity disorder. *Journal of the American Academy of Child and Adolescent Psychiatry, 55*(6), 521–523.

Doyle, C., Brookes, K., Simpson, J., Park, J., Scott, S., Coghill, D. R., … Kent, L. (2009). Replication of an association of a promoter polymorphism of the dopamine transporter gene and attention deficit hyperactivity disorder. *Neuroscience Letters, 462*(2), 179–181.

Egger, M., Smith, G. D., & Phillips, A. N. (1997). Meta-analysis: Principles and procedures. *British Medical Journal, 315*(7121), 1533–1537.

El-Tarras, A. E., Alsulaimani, A. A., Awad, N. S., Mitwaly, N., Said, M. M., & Sabry, A. M. (2012). Association study between the dopamine-related candidate gene polymorphisms and ADHD among Saudi Arabia population via PCR technique. *Molecular Biology Reports, 39*(12), 11081–11086.

Evans, W. E., & Johnson, J. A. (2001). Pharmacogenomics: The inherited basis for interindividual differences in drug response. *Annual Review of Genomics and Human Genetics, 2*, 9–39.

Faraone, S., Biederman, J., & Mick, E. (2006). The age-dependent decline of attention deficit hyperactivity disorder: A meta-analysis of follow-up studies. *Psychological Medicine, 36*(2), 159–165.

Faraone, S. V., Biederman, J., Spencer, T., Wilens, T., Seidman, L. J., Mick, E., & Doyle, A. E. (2000). Attention-deficit/hyperactivity disorder in adults: An overview. *Biological Psychiatry, 48*(1), 9–20.

Faraone, S. V., Bonvicini, C., & Scassellati, C. (2014). Biomarkers in the diagnosis of ADHD—Promising directions. *Current Psychiatry Reports, 16*(11), 497.

Faraone, S. V., & Mick, E. (2010). Molecular genetics of attention deficit hyperactivity disorder. *The Psychiatric Clinics of North America, 33*(1), 159–180.

Faraone, S. V., Perlis, R. H., Doyle, A. E., Smoller, J. W., Goralnick, J. J., Holmgren, M. A., & Sklar, P. (2005). Molecular genetics of attention-deficit/hyperactivity disorder. *Biological Psychiatry, 57*(11), 1313–1323.

Faraone, S. V., Spencer, T. J., Madras, B. K., Zhang-James, Y., & Biederman, J. (2014). Functional effects of dopamine transporter gene genotypes on in vivo dopamine transporter functioning: A meta-analysis. *Molecular Psychiatry, 19*(8), 880–889.

Feng, Y., Wigg, K. G., Makkar, R., Ickowicz, A., Pathare, T., Tannock, R., … Barr, C. L. (2005). Sequence variation in the 3'-untranslated region of the dopamine transporter gene and attention-deficit hyperactivity disorder (ADHD). *American Journal of Medical Genetics, Part B: Neuropsychiatric Genetics, 139B*(1), 1–6.

Friedel, S., Saar, K., Sauer, S., Dempfle, A., Walitza, S., Renner, T., … Hebebrand, J. (2007). Association and linkage of allelic variants of the dopamine transporter gene in ADHD. *Molecular Psychiatry, 12*(10), 923–933.

Genro, J. P., Polanczyk, G. V., Zeni, C., Oliveira, A. S., Roman, T., Rohde, L. A., & Hutz, M. H. (2008). A common haplotype at the dopamine transporter gene 5' region is associated with attention-deficit/hyperactivity disorder. *American Journal of Medical Genetics, Part B: Neuropsychiatric Genetics, 147B*(8), 1568–1575.

Genro, J. P., Zeni, C., Polanczyk, G. V., Roman, T., Rohde, L. A., & Hutz, M. H. (2007). A promoter polymorphism (-839 C > T) at the dopamine transporter gene is associated with attention deficit/hyperactivity disorder in Brazilian children. *American Journal of Medical Genetics, Part B: Neuropsychiatric Genetics, 144B*(2), 215–219.

Gharaibeh, M. Y., Batayneh, S., Khabour, O. F., & Daoud, A. (2010). Association between polymorphisms of the DBH and DAT1 genes and attention deficit hyperactivity disorder in children from Jordan. *Experimental and Therapeutic Medicine, 1*(4), 701–705.

Gill, M., Daly, G., Heron, S., Hawi, Z., & Fitzgerald, M. (1997). Confirmation of association between attention deficit hyperactivity disorder and a dopamine transporter polymorphism. *Molecular Psychiatry, 2*(4), 311–313.

Gizer, I. R., Ficks, C., & Waldman, I. D. (2009). Candidate gene studies of ADHD: A meta-analytic review. *Human Genetics, 126*(1), 51–90.

Haavik, J., Halmøy, A., Lundervold, A. J., & Fasmer, O. B. (2010). Clinical assessment and diagnosis of adults with attention-deficit/hyperactivity disorder. *Expert Review of Neurotherapeutics, 10*(10), 1569–1580.

Hamshere, M. L., Langley, K., Martin, J., Agha, S. S., Stergiakouli, E., Anney, R. J., … Thapar, A. (2013). High loading of polygenic risk for ADHD in children with comorbid aggression. *American Journal of Psychiatry, 170*(8), 909–916.

Hawi, Z., Cummins, T. D., Tong, J., Arcos-Burgos, M., Zhao, Q., Matthews, N., … Bellgrove, M. A. (2017). Rare DNA variants in the brain-derived neurotrophic factor gene increase risk for attention-deficit hyperactivity disorder: A next-generation sequencing study. *Molecular Psychiatry, 22*(4), 580–584.

Hawi, Z., Cummins, T. D., Tong, J., Johnson, B., Lau, R., Samarrai, W., & Bellgrove, M. A. (2015). The molecular genetic architecture of attention deficit hyperactivity disorder. *Molecular Psychiatry, 20*(3), 289–297.

Hawi, Z., Kent, L., Hill, M., Anney, R. J., Brookes, K. J., Barry, E., … Gill, M. (2010). ADHD and DAT1: Further evidence of paternal over-transmission of risk alleles and haplotype. *American Journal of Medical Genetics, Part B: Neuropsychiatric Genetics, 153B*(1), 97–102.

Hawi, Z., Lowe, N., Kirley, A., Gruenhage, F., Nöthen, M., Greenwood, T., … Gill, M. (2003). Linkage disequilibrium mapping at DAT1, DRD5 and DBH narrows the search for ADHD susceptibility alleles at these loci. *Molecular Psychiatry*, *8*(3), 299–308.

Hawi, Z., Segurado, R., Conroy, J., Sheehan, K., Lowe, N., Kirley, A., … Gill, M. (2005). Preferential transmission of paternal alleles at risk genes in attention-deficit/hyperactivity disorder. *American Journal of Human Genetics*, *77*(6), 958–965.

Holmes, J., Payton, A., Barrett, J. H., Hever, T., Fitzpatrick, H., Trumper, A. L., … Thapar, A. (2000). A family-based and case-control association study of the dopamine D4 receptor gene and dopamine transporter gene in attention deficit hyperactivity disorder. *Molecular Psychiatry*, *5*(5), 523–530.

Insel, T. (2013). *Director's blog: Transforming diagnosis*. http://www.nimh.nih.gov/about/director/2013/transforming-diagnosis.shtml.

Joensen, B., Meyer, M., & Aagaard, L. (2017). Specific genes associated with adverse events of methylphenidate use in the pediatric population: A systematic literature review. *Journal of Research in Pharmacy Practice*, *6*(2), 65–72.

Kahn, R. S., Khoury, J., Nichols, W. C., & Lanphear, B. P. (2003). Role of dopamine transporter genotype and maternal prenatal smoking in childhood hyperactive-impulsive, inattentive, and oppositional behaviors. *The Journal of Pediatrics*, *143*(1), 104–110.

Kambeitz, J., Romanos, M., & Ettinger, U. (2014). Meta-analysis of the association between dopamine transporter genotype and response to methylphenidate treatment in ADHD. *The Pharmacogenomics Journal*, *14*(1), 77–84.

Kan, K. J., Dolan, C. V., Nivard, M. G., Middeldorp, C. M., van Beijsterveldt, C. E., Willemsen, G., & Boomsma, D. I. (2013). Genetic and environmental stability in attention problems across the lifespan: Evidence from the Netherlands twin register. *Journal of the American Academy of Child and Adolescent Psychiatry*, *52*(1), 12–25.

Kebir, O., & Joober, R. (2011). Neuropsychological endophenotypes in attention-deficit/hyperactivity disorder: A review of genetic association studies. *European Archives of Psychiatry and Clinical Neuroscience*, *261*(8), 583–594.

Kereszturi, E., Tarnok, Z., Bognar, E., Lakatos, K., Farkas, L., Gadoros, J., … Nemoda, Z. (2008). Catechol-O-methyltransferase Val158Met polymorphism is associated with methylphenidate response in ADHD children. *American Journal of Medical Genetics, Part B: Neuropsychiatric Genetics*, *147B*(8), 1431–1435.

Kim, D. S., Burt, A. A., Ranchalis, J. E., Wilmot, B., Smith, J. D., Patterson, K. E., … University of Washington Center for Mendelian Genomics. (2017). Sequencing of sporadic attention-deficit hyperactivity disorder (ADHD) identifies novel and potentially pathogenic de novo variants and excludes overlap with genes associated with autism spectrum disorder. *American Journal of Medical Genetics, Part B: Neuropsychiatric Genetics*, *174*(4), 381–389.

Kim, J. W., Kim, B. N., & Cho, S. C. (2006). The dopamine transporter gene and the impulsivity phenotype in attention deficit hyperactivity disorder: A case-control association study in a Korean sample. *Journal of Psychiatric Research*, *40*(8), 730–737.

Kim, Y. S., Leventhal, B. L., Kim, S. J., Kim, B. N., Cheon, K. A., Yoo, H. J., … Cook, E. H. (2005). Family-based association study of DAT1 and DRD4 polymorphism in Korean children with ADHD. *Neuroscience Letters*, *390*(3), 176–181.

Kirley, A., Hawi, Z., Daly, G., McCarron, M., Mullins, C., Millar, N., … Gill, M. (2002). Dopaminergic system genes in ADHD: Toward a biological hypothesis. *Neuropsychopharmacology*, *27*(4), 607–619.

Kirley, A., Lowe, N., Hawi, Z., Mullins, C., Daly, G., Waldman, I., … Gill, M. (2003). Association of the 480 bp DAT1 allele with methylphenidate response in a sample of Irish children with ADHD. *American Journal of Medical Genetics, Part B: Neuropsychiatric Genetics*, *121B*(1), 50–54.

Klein, R. G., & Mannuzza, S. (1991). Long-term outcome of hyperactive children: A review. *Journal of the American Academy of Child and Adolescent Psychiatry*, *30*(3), 383–387.

Kooij, J. S., Boonstra, A. M., Vermeulen, S. H., Heister, A. G., Burger, H., Buitelaar, J. K., & Franke, B. (2008). Response to methylphenidate in adults with ADHD is associated with a polymorphism in SLC6A3 (DAT1). *American Journal of Medical Genetics, Part B: Neuropsychiatric Genetics*, *147B*(2), 201–208.

Kopecková, M., Paclt, I., Petrásek, J., Pacltová, D., Malíková, M., & Zagatová, V. (2008). Some ADHD polymorphisms (in genes DAT1, DRD2, DRD3, DBH, 5-HTT) in case-control study of 100 subjects 6-10 age. *Neuroendocrinology Letters*, *29*(2), 246–251.

Kornfield, R., Watson, S., Higashi, A. S., Conti, R. M., Dusetzina, S. B., Garfield, C. F., … Alexander, G. C. (2013). Effects of FDA advisories on the pharmacologic treatment of ADHD, 2004-2008. *Psychiatric Services*, *64*(4), 339–346.

Kustanovich, V., Ishii, J., Crawford, L., Yang, M., McGough, J. J., McCracken, J. T., … Nelson, S. F. (2004). Transmission disequilibrium testing of dopamine-related candidate gene polymorphisms in ADHD: Confirmation of association of ADHD with DRD4 and DRD5. *Molecular Psychiatry*, *9*(7), 711–717.

Langley, K., Turic, D., Peirce, T. R., Mills, S., Van Den Bree, M. B., Owen, M. J., … Thapar, A. (2005). No support for association between the dopamine transporter (DAT1) gene and ADHD. *American Journal of Medical Genetics, Part B: Neuropsychiatric Genetics*, *139B*(1), 7–10.

Larsson, H., Asherson, P., Chang, Z., Ljung, T., Friedrichs, B., Larsson, J. O., & Lichtenstein, P. (2013). Genetic and environmental influences on adult attention deficit hyperactivity disorder symptoms: A large Swedish population-based study of twins. *Psychological Medicine*, *43*(1), 197–207.

Laucht, M., Skowronek, M. H., Becker, K., Schmidt, M. H., Esser, G., Schulze, T. G., & Rietschel, M. (2007). Interacting effects of the dopamine transporter gene and psychosocial adversity on attention-deficit/hyperactivity disorder symptoms among 15-year-olds from a high-risk community sample. *Archives of General Psychiatry*, *64*(5), 585–590.

Laurin, N., Feng, Y., Ickowicz, A., Pathare, T., Malone, M., Tannock, R., … Barr, C. L. (2007). No preferential transmission of paternal alleles at risk genes in attention-deficit/hyperactivity disorder. *Molecular Psychiatry*, *12*(3), 226–229.

Lesch, K. P., Timmesfeld, N., Renner, T. J., Halperin, R., Röser, C., Nguyen, T. T., … Jacob, C. (2008). Molecular genetics of adult ADHD: Converging evidence from genome-wide association and extended pedigree linkage studies. *Journal of Neural Transmission*, *115*(11), 1573–1585.

Li, Z., Chang, S. H., Zhang, L. Y., Gao, L., & Wang, J. (2014). Molecular genetic studies of ADHD and its candidate genes: A review. *Psychiatry Research*, *219*(1), 10–24.

Li, J. J., & Lee, S. S. (2012). Interaction of dopamine transporter (DAT1) genotype and maltreatment for ADHD: A latent class analysis. *Journal of Child Psychology and Psychiatry, and Allied Disciplines, 53*(9), 997–1005.

Lim, M. H., Kim, H. W., Paik, K. C., Cho, S. C., Yoon, D. Y., & Lee, H. J. (2006). Association of the DAT1 polymorphism with attention deficit hyperactivity disorder (ADHD): A family-based approach. *American Journal of Medical Genetics, Part B: Neuropsychiatric Genetics, 141B*(3), 309–311.

Maher, B. S., Marazita, M. L., Ferrell, R. E., & Vanyukov, M. M. (2002). Dopamine system genes and attention deficit hyperactivity disorder: A meta-analysis. *Psychiatric Genetics, 12*(4), 207–215.

Martin, J., O'Donovan, M. C., Thapar, A., Langley, K., & Williams, N. (2015). The relative contribution of common and rare genetic variants to ADHD. *Translational Psychiatry, 10*(5), e506.

Mick, E., Biederman, J., Spencer, T., Faraone, S. V., & Sklar, P. (2006). Absence of association with DAT1 polymorphism and response to methylphenidate in a sample of adults with ADHD. *American Journal of Medical Genetics, Part B: Neuropsychiatric Genetics, 141B*(8), 890–894.

Mooney, M. A., McWeeney, S. K., Faraone, S. V., Hinney, A., Hebebrand, J., IMAGE2 Consortium, German ADHD GWAS Group, … Wilmot, B. (2016). Pathway analysis in attention deficit hyperactivity disorder: An ensemble approach. *American Journal of Medical Genetics, Part B: Neuropsychiatric Genetics, 171*(6), 815–826.

Ogdie, M. N., Bakker, S. C., Fisher, S. E., Francks, C., Yang, M. H., Cantor, R. M., … Smalley, S. L. (2006). Pooled genome-wide linkage data on 424 ADHD ASPs suggests genetic heterogeneity and a common risk locus at 5p13. *Molecular Psychiatry, 11*(1), 5–8.

Onnink, A. M., Franke, B., van Hulzen, K., Zwiers, M. P., Mostert, J. C., Schene, A. H., … Hoogman, M. (2016). Enlarged striatal volume in adults with ADHD carrying the 9-6 haplotype of the dopamine transporter gene DAT1. *Journal of Neural Transmission, 123*(8), 905–915.

Palmer, C. G., Bailey, J. N., Ramsey, C., Cantwell, D., Sinsheimer, J. S., Del'Homme, M., … Smalley, S. L. (1999). No evidence of linkage or linkage disequilibrium between DAT1 and attention deficit hyperactivity disorder in a large sample. *Psychiatric Genetics, 9*(3), 157–160.

Paloyelis, Y., Mehta, M. A., Faraone, S. V., Asherson, P., & Kuntsi, J. (2012). Striatal sensitivity during reward processing in attention-deficit/hyperactivity disorder. *Journal of the American Academy of Child and Adolescent Psychiatry, 51*(7), 722–732.

Polanczyk, G. V., Willcutt, E. G., Salum, G. A., Kieling, C., & Rohde, L. A. (2014). ADHD prevalence estimates across three decades: An updated systematic review and meta-regression analysis. *International Journal of Epidemiology, 43*(2), 434–442.

Purper-Ouakil, D., Wohl, M., Orejarena, S., Cortese, S., Boni, C., Asch, M., … Gorwood, P. (2008). Pharmacogenetics of methylphenidate response in attention deficit/hyperactivity disorder: Association with the dopamine transporter gene (SLC6A3). *American Journal of Medical Genetics, Part B: Neuropsychiatric Genetics, 147B*(8), 1425–1430.

Qian, Q., Wang, Y., Li, J., Yang, L., Wang, B., Zhou, R., … Faraone, S. V. (2007). Evaluation of potential gene-gene interactions for attention deficit hyperactivity disorder in the Han Chinese population. *American Journal of Medical Genetics, Part B: Neuropsychiatric Genetics, 144B*(2), 200–206.

Ramos-Quiroga, J. A., Nasillo, V., Fernández-Aranda, F., & Casas, M. (2014). Addressing the lack of studies in attention-deficit/hyperactivity disorder in adults. *Expert Review of Neurotherapeutics, 14*(5), 553–567.

Ramos-Quiroga, J. A., Sánchez-Mora, C., Casas, M., Garcia-Martínez, I., Bosch, R., Nogueira, M., … Ribasés, M. (2014). Genome-wide copy number variation analysis in adult attention-deficit and hyperactivity disorder. *Journal of Psychiatric Research, 49*, 60–67.

Roman, T., Schmitz, M., Polanczyk, G., Eizirik, M., Rohde, L. A., & Hutz, M. H. (2001). Attention-deficit hyperactivity disorder: A study of association with both the dopamine transporter gene and the dopamine D4 receptor gene. *American Journal of Medical Genetics, 105*(5), 471–478.

Roman, T., Szobot, C., Martins, S., Biederman, J., Rohde, L. A., & Hutz, M. H. (2002). Dopamine transporter gene and response to methylphenidate in attention-deficit/hyperactivity disorder. *Pharmacogenetics, 12*(6), 497–499.

Sallee, F., Connor, D. F., & Newcorn, J. H. (2013). A review of the rationale and clinical utilization of α2-adrenoceptor agonists for the treatment of attention-deficit/hyperactivity and related disorders. *Journal of Child and Adolescent Psychopharmacology, 23*(5), 308–319.

Šerý, O., Paclt, I., Drtílková, I., Theiner, P., Kopečková, M., Zvolský, P., & Balcar, V. J. (2015). A 40-bp VNTR polymorphism in the 3′-untranslated region of DAT1/SLC6A3 is associated with ADHD but not with alcoholism. *Behavioral and Brain Functions, 11*, 21.

Shaw, P., Gornick, M., Lerch, J., Addington, A., Seal, J., Greenstein, D., … Rapoport, J. L. (2007). Polymorphisms of the dopamine D4 receptor, clinical outcome, and cortical structure in attention-deficit/hyperactivity disorder. *Archives of General Psychiatry, 64*(8), 921–931.

Shaw, P., Malek, M., Watson, B., Greenstein, D., de Rossi, P., & Sharp, W. (2013). Trajectories of cerebral cortical development in childhood and adolescence and adult attention-deficit/hyperactivity disorder. *Biological Psychiatry, 74*(8), 599–606.

Simsek, M., Al-Sharbati, M., Al-Adawi, S., & Lawatia, K. (2006). The VNTR polymorphism in the human dopamine transporter gene: Improved detection and absence of association of VNTR alleles with attention-deficit hyperactivity disorder. *Genetic Testing, 10*(1), 31–34.

Smith, K. M., Daly, M., Fischer, M., Yiannoutsos, C. T., Bauer, L., Barkley, R., & Navia, B. A. (2003). Association of the dopamine beta hydroxylase gene with attention deficit hyperactivity disorder: Genetic analysis of the Milwaukee longitudinal study. *American Journal of Medical Genetics, Part B: Neuropsychiatric Genetics, 119B*(1), 77–85.

Sonuga-Barke, E. J. S., Oades, R. D., Psychogiou, L., Chen, W., Franke, B., Buitelaar, J., … Faraone, S. V. (2009). Dopamine and serotonin transporter genotypes moderate sensitivity to maternal expressed emotion: The case of conduct and emotional problems in attention deficit/hyperactivity disorder. *Journal of Child Psychology and Psychiatry, and Allied Disciplines, 50*(9), 1052–1063.

Stanley, A., Chavda, K., Subramanian, A., Prabhu, S. V., & Ashavaid, T. F. (2017). DRD4 and DAT1 VNTR genotyping in children with attention deficit hyperactivity disorder. *Indian Journal of Clinical Biochemistry, 32*(2), 239–242.

Stein, M. A., Waldman, I. D., Sarampote, C. S., Seymour, K. E., Robb, A. S., Conlon, C., … Cook, E. H. (2005). Dopamine transporter genotype and methylphenidate dose response in children with ADHD. *Neuropsychopharmacology, 30*(7), 1374–1382.

Stergiakouli, E., Martin, J., Hamshere, M. L., Langley, K., Evans, D. M., St Pourcain, B., … Davey Smith, G. (2015). Shared genetic influences between attention-deficit/hyperactivity disorder (ADHD) traits in children and clinical ADHD. *Journal of the American Academy of Child and Adolescent Psychiatry, 54*(4), 322–327.

Stevens, S. E., Kumsta, R., Kreppner, J. M., Brookes, K. J., Rutter, M., & Sonuga-Barke, E. J. S. (2009). Dopamine transporter gene polymorphism moderates the effects of severe deprivation on ADHD symptoms: Developmental continuities in gene-environment interplay. *American Journal of Medical Genetics, Part B: Neuropsychiatric Genetics, 150B*(6), 753–761.

Swanson, J. M., Flodman, P., Kennedy, J., Spence, M. A., Moyzis, R., Schuck, S., … Posner, M. (2000). Dopamine genes and ADHD. *Neuroscience and Biobehavioral Reviews, 24*(1), 21–25.

Tharoor, H., Lobos, E. A., Todd, R. D., & Reiersen, A. M. (2008). Association of dopamine, serotonin, and nicotinic gene polymorphisms with methylphenidate response in ADHD. *American Journal of Medical Genetics, Part B: Neuropsychiatric Genetics, 147B*(4), 527–530.

Todd, R. D., Jong, Y. J., Lobos, E. A., Reich, W., Heath, A. C., & Neuman, R. J. (2001). No association of the dopamine transporter gene 3' VNTR polymorphism with ADHD subtypes in a population sample of twins. *American Journal of Medical Genetics, 105*(8), 745–748.

Vilor-Tejedor, N., Cáceres, A., Pujol, J., Sunyer, J., & González, J. R. (2017). Imaging genetics in attention-deficit/hyperactivity disorder and related neurodevelopmental domains: State of the art. *Brain imaging and behavior, 11*(6), 1922–1931.

Waldman, I. D., Rowe, D. C., Abramowitz, A., Kozel, S. T., Mohr, J. H., Sherman, S. L., … Stever, C. (1998). Association and linkage of the dopamine transporter gene and attention-deficit hyperactivity disorder in children: Heterogeneity owing to diagnostic subtype and severity. *American Journal of Human Genetics, 63*(6), 1767–1776.

Wang, Y., Wang, Z., Yao, K., Tanaka, K., Yang, Y., Shirakawa, O., & Maeda, K. (2008). Lack of association between the dopamine transporter gene 3' VNTR polymorphism and attention deficit hyperactivity disorder in Chinese Han children: Case-control and family-based studies. *The Kobe Journal of Medical Sciences, 53*(6), 327–333.

Winsberg, B. G., & Comings, D. E. (1999). Association of the dopamine transporter gene (DAT1) with poor methylphenidate response. *Journal of the American Academy of Child and Adolescent Psychiatry, 38*(12), 1474–1477.

Wohl, M., Boni, C., Asch, M., Cortese, S., Orejarena, S., Mouren, M. C., … Purper-Ouakil, D. (2008). Lack of association of the dopamine transporter gene in a French ADHD sample. *American Journal of Medical Genetics, Part B: Neuropsychiatric Genetics, 147B*(8), 1509–1510.

Xu, X., Mill, J., Sun, B., Chen, C. K., Huang, Y. S., Wu, Y. Y., & Asherson, P. (2009). Association study of promoter polymorphisms at the dopamine transporter gene in attention deficit hyperactivity disorder. *BMC Psychiatry, 9*, 3.

Yang, L., Neale, B. M., Liu, L., Lee, S. H., Wray, N. R., Ji, N., … Psychiatric GWAS Consortium: ADHD Subgroup. (2013). Polygenic transmission and complex neuro developmental network for attention deficit hyperactivity disorder: Genome-wide association study of both common and rare variants. *American Journal of Medical Genetics, Part B: Neuropsychiatric Genetics, 162B*(5), 419–430.

Yu, C. J., Du, J. C., Chiou, H. C., Chung, M. Y., Yang, W., Chen, Y. S., … Chen, M. L. (2016). Increased risk of attention-deficit/hyperactivity disorder associated with exposure to organophosphate pesticide in Taiwanese children. *Andrology, 4*(4), 695–705.

Zayats, T., Jacobsen, K. K., Kleppe, R., Jacob, C. P., Kittel-Schneider, S., Ribasés, M., … Johansson, S. (2016). Exome chip analyses in adult attention deficit hyperactivity disorder. *Translational Psychiatry, 6*(10), e923.

Zhou, K., Dempfle, A., Arcos-Burgos, M., Bakker, S. C., Banaschewski, T., Biederman, J., … Asherson, P. (2008). Meta-analysis of genome-wide linkage scans of attention deficit hyperactivity disorder. *American Journal of Medical Genetics, Part B: Neuropsychiatric Genetics, 147B*(8), 1392–1398.

Chapter 22

Genomics of Alzheimer's disease

Margot P. van de Weijer[a,b], Iris E. Jansen[c,d], Anouk H.A. Verboven[e], Ole A. Andreassen[f,g] and Danielle Posthuma[c,h]

[a]Department of Biological Psychology, VU University, Amsterdam, The Netherlands; [b]Amsterdam Public Health Research Institute, Amsterdam University Medical Centre, Amsterdam, The Netherlands; [c]Department of Complex Trait Genetics, Center for Neurogenomics and Cognitive Research, Amsterdam Neuroscience, VU University, Amsterdam, The Netherlands; [d]Alzheimer Center and Department of Neurology, Amsterdam Neuroscience, VU University, Amsterdam UMC, Amsterdam, The Netherlands; [e]Department of Human Genetics, Radboudumc, Donders Institute for Brain, Cognition, and Behaviour, Nijmegen, The Netherlands; [f]NORMENT, KG Jebsen Centre for Psychosis Research, University of Oslo, Oslo, Norway; [g]Institute of Clinical Medicine, University of Oslo, Oslo, Norway; [h]Department of Clinical Genetics, VU University, Amsterdam UMC, Amsterdam, The Netherlands

1 Introduction

With about 1 in 10 people affected above the age of 65, Alzheimer's disease (AD) is currently the most prevalent neuro-degenerative disease (Alzheimer's Association, 2017). AD is the most common cause of dementia, characterized by progressive decline in memory, language, and other cognitive skills. In addition, basic bodily functions, such as walking, might become impaired in later stages of the disorder (Alzheimer's Association, 2017). Given its high prevalence and the aging of the general population, AD is one of the most investigated neurodegenerative disorders: a simple look-up of the term "Alzheimer" in the PubMed search engine (March 2018) results in an output of 96,874 publications. However, while there are several thousands of papers published on AD each year, there is still no consensus on the biological mechanisms contributing to the disease etiology. In addition, no effective treatment or prevention method is currently available.

Brains of AD patients are characterized by specific neuropathological hallmarks: the extracellular aggregation of amyloid-beta (Aβ) protein into amyloid plaques, and formation of intraneuronal neurofibrillary tangles (NFT) that consist of hyperphosphorylated tau protein (Tanzi & Bertram, 2005). Cellular consequences include synaptic loss, selective neuronal death, loss of neurotransmitters, and neuro-inflammation (Spires-Jones & Hyman, 2014). These neuropathological hallmarks can only be assessed in postmortem brain tissue. Nevertheless, clinical diagnoses, often aided by several biomarkers such as medial temporal brain atrophy identified with structural magnetic resonance imaging (MRI) (Salvatore, Battista, & Castiglioni, 2016), or reduced Aβ and elevated tau protein quantified from cerebrospinal fluid (CSF) (Galasko & Shaw, 2017), are already 78%–87% accurate (Beach, Monsell, Phillips, & Kukull, 2012).

Multiple theories exist on the biological mechanisms underlying AD. The first proposed hypothesis was the *amyloid cascade hypothesis*, stating that the deposition of Aβ in the brain is the causal mechanism that initiates a cascade of events leading to AD (Sommer, 2002). While Aβ depositions are certainly an important factor for AD pathogenesis, there is increasing debate on whether it is the sole causative mechanism: the association between Aβ load and AD is more complicated and diverse than initially expected, and there has been no successful clinical trial based on this hypothesis (Herrup, 2015). Another theory is that AD is immune-related and that neuroinflammation is the driving force behind AD (Heppner, Ransohoff, & Becher, 2015). There is evidence from mouse models suggesting that neuroinflammation might be a driving factor for AD (Krstic et al., 2012; Lim et al., 2000), and, more recently, genetic studies using cohorts with Alzheimer patients are supporting this theory due to the identification of genetic factors of which the biological involvement seems to converge on immune-related mechanisms and body tissues (Jansen et al., 2019; Sims et al., 2017). Last, an important role has also been attributed to phospholipids: brains of AD patients show biochemical alterations in lipid composition, and high serum cholesterol levels have been associated with increased AD risk (Di Paolo & Kim, 2011). Genetic research can help elucidate the relative importance of, and the interaction between, these biological mechanisms by examining the genetic pathways contributing to AD. Moreover, by taking a data-driven approach instead of the hypothesis-driven approach that dominated AD research in the pregenomic era, new, previously overlooked mechanisms might be identified.

Based on the age of onset, two forms of AD can be distinguished: early-onset AD (EOAD, age of onset < 65) and late-onset AD (LOAD, age of onset ≥ 65). EOAD occurs much less frequently than LOAD, and is characterized by a more aggressive disease course, most often the presence of a family history, and shorter survival time (Seltzer & Sherwin,

Personalized Psychiatry. https://doi.org/10.1016/B978-0-12-813176-3.00022-5

1983). Moreover, if an AD case has an obvious familial history of AD, it is categorized as familial AD (FAD), while cases with no familial history of AD are categorized as sporadic AD (SAD). FAD is mostly observed in individuals with EOAD, while sporadic cases are more frequently diagnosed with LOAD.

Based on twin research (see Box 1), the heritability of AD is estimated at around 60%–80% (Gatz et al., 2006). Given the high estimated contribution of genetic factors to AD, geneticists have attempted to pin down the genetic architecture for AD for many decades, and broadly two types of genetic inheritance have been observed. First, EOAD is mostly inherited as an autosomal dominant disease, meaning that if one inherits a causative variant from a parent, the odds of developing AD are extremely high. Highly penetrant causal mutations with severe associated risk have been observed in at least three different genes (*APP*, *PSEN1*, and *PSEN2*) that are sufficient for causing EOAD (Tang & Gershon, 2003). These mutations seem to be in line with the amyloid cascade hypothesis: mutations in the amyloid precursor protein (*APP*) lead to the generation of the $A\beta_{1-42}$ peptide (as opposed to the $A\beta_{1-40}$ peptide), which is more aggregation-prone, and found in high amounts in amyloid plaques. Presenilin-1 (*PSEN1*) and Presenilin-2 (*PSEN2*) are components of γ-secretase, and when mutated, influence APP cleavage in favor of $A\beta_{1-42}$. This Mendelian inheritance pattern is observed in only about 5%–10% of EOAD cases (Selkoe, 2001). Second, for the majority of the cases, and especially in LOAD, the genetic etiology of AD is vastly more complex: LOAD is influenced by the combination of many genetic risk variants, with each risk variant influencing susceptibility for the disease only slightly. The strongest genetic risk factor for LOAD is the ε4 allele of the apolipoprotein E (*APOE*) gene: the frequency of this gene is increased to ~40% in AD patients (Liu, Kanekiyo, Xu, Bu, & Bu, 2013). Compared with individuals with an ε3/ε3 genotype, people carrying one copy the ε4 allele and one copy of the ε2 allele or ε3 allele have an increased risk of AD (OR = 2.6, OR = 3.2, respectively). People carrying two ε4 alleles have the highest relative risk (OR = 14.9). The presence of the allele also increases risk for EOAD. Identifying the genetic risk factors influencing LOAD helps unravel the complex biological mechanisms underlying the disorder, which will eventually help

BOX 1 Terms and definitions in complex trait genetics.

Term	Definition
Twin heritability	To partition the total variance of a trait into genetic (additive/dominance) and (unique and shared) environmental components, twin studies compare the concordance in monozygotic (MZ) twin pairs with dizygotic (DZ) twin pairs. When MZ twins have higher concordance rates for AD than DZ twins, there is evidence for the contribution of genetic factors.
SNP heritability	The SNP heritability reflects the variance explained by all SNPs tested by a population-based genome-wide association study.
Linkage analysis	Tracks the transmission of genetic markers in families and links these markers, and their corresponding regions, to the disorder if markers are present in cases, and absent in unaffected family members.
Candidate gene association studies	Identification of risk variants associated with a trait or disorder by using prior knowledge on the biological functions/mechanisms of genes to select particular genes. An association test studies the difference in allele frequency of variants in preselected genes between cases and controls.
Genome-wide association study	Identification of genetic markers along the entire genome associated with complex traits or disorders. Millions of SNPs are genotyped for large groups of individuals. Each SNP is regressed against a phenotype to examine potential association.
Next-generation sequencing	Massively parallel or deep sequencing of preselected regions or the total genome, which allows for fast DNA sequence assessment.

Term	Definition
Transcriptome-wide association study	Correlates gene expression to a trait of interest to identify significant expression-trait associations, by comparison of gene-expression profiles between cases and controls.
Epigenome-wide association study	Correlates epigenetic marks (e.g., methylation at CpG sites) with a trait of interest to identify significant associations between epigenetic changes in the genome and a trait of interest.
Functional annotation	Follow-up analyses of genetic studies to assign biological explanation/information to identified genomic elements. This term is generally used to describe in vivo/vitro experiments, but could also indicate in silico analysis.
Positional gene-mapping	Using the physical location of a genetic variant to relate it to a gene.
eQTL gene-mapping	Identification of genes with an expression that is influenced by a genetic variant, either on the same chromosome (cis-eQTL), or on another chromosome (trans-eQTL).
Chromatin interaction mapping	Mapping genetic variants to genes based on three-dimensional DNA-DNA interactions between the variant's genomic region and nearby or distant genes.
Gene-set analysis	Tests for the aggregated effect of multiple variants within genes belonging to the same predefined gene set, which could be based on biological mechanisms or gene-expression profiles (e.g., body-tissue specific).
Cross-trait genetic analysis	Examines the genetic overlap (e.g., genetic correlation) based on GWAS-association statistics between two or more different traits to gain functional insights into the genetic etiology and overlap of the different traits.

design and improve therapies. Because EOAD is often inherited as an autosomal dominant disorder, and is genetically different from LOAD, we only review the genetic factors implicated in LOAD in subsequent parts of this chapter.

The goal of this chapter is to highlight what the past decade of genomics research has revealed about the genetic architecture of LOAD. Moreover, we provide a brief overview of current findings using other -omics approaches, and the merits of multiomics approaches to uncovering disease mechanisms. We end with a brief discussion on the questions that are still left unanswered, and how future research can best address these questions.

2 The genetics of AD

More than a decade ago, genetic research of AD was mostly based on linkage studies and candidate gene studies (see Box 1). While family-based linkage analysis was fruitful for identifying rare genetic variants with large effects, this method is underpowered for examining the contribution of common SNPs with small effect sizes. Candidate gene studies in larger sample sizes of unrelated samples enabled the identification of variants with lower impact, though limited by the prior selection of interesting genes based on previous biological knowledge. Genome-wide association (GWA) studies, on the other hand, can be used to examine complex disorders that are influenced by many genetic variants with small effects using a hypothesis-free approach. These studies typically measure millions of single-nucleotide polymorphisms (SNPs) of many thousands of individuals, and try to associate these SNPs to trait levels or disease status (Box 1). If any genetic variant is significantly associated with a phenotype, this association is often explored in more detail by examining its genomic region, and the potential biological mechanism by which the variant might influence the phenotype.

To have enough statistical power in GWA studies to identify associated variants, large sample sizes are needed. Therefore, summary statistics from several cohorts are often combined into meta-analyses. The first large GWA meta-analysis for LOAD ($N = 74,046$) combined case-control data from four consortia to identify SNPs potentially associated with LOAD, which were further evaluated for association in independent cohorts including individuals originating from 11 countries (Lambert et al., 2013). This first large-scale investigation resulted in the identification of 19 associated loci (see Table 1), in addition to the *APOE* locus. Eight of these loci had already been identified in previous GWA studies (*ABCA7*,

TABLE 1 Genetic variants identified for AD in the studies discussed in this chapter

Nearest gene	Location	Study	SNP	Odds ratio	Beta	P-value
ABCA7	19p13.3	Lambert et al. (2013)	rs4147929	1.15	—	1.1×10^{-15}
		Steinberg et al. (2015)	[a]	[a]	—	[a]
		Jansen et al. (2019)	rs111278892	—	0.019	7.9×10^{-11}
ABI3	17q21.32	Sims et al. (2017)	rs616338	1.43	—	4.6×10^{-10}
ADAM10	15q21.3	Jansen et al. (2019)	rs442495	0.98	−0.013	1.3×10^{-9}
ADAMTS4	1q23.3	Jansen et al. (2019)	rs4575098	1.02	0.016	2.1×10^{-10}
APH1B	15q22.2	Jansen et al. (2019)	rs117618017	1.02	0.018	3.4×10^{-8}
APOE	19q13.32	Several				
BIN1	2q14.3	Lambert et al. (2013)	rs6733839	1.22	—	6.9×10^{-44}
		Jansen et al. (2019)	rs4663105	1.06	0.030	3.4×10^{-44}
BZRAP1-AS1	17q22	Jun et al. (2017)	rs2632516	0.92	—	4.4×10^{-8}
CASS4	20q13.31	Lambert et al. (2013)	rs7274581	0.88	—	2.5×10^{-8}
		Jansen et al. (2019)	rs6014724	0.95	−0.022	6.56×10^{-10}
CELF1	11p11.2	Lambert et al. (2013)	rs10838725	1.08	—	1.1×10^{-8}
CD2AP	6p12.3	Lambert et al. (2013)	rs10948363	1.10	—	5.2×10^{-11}
CD33	19q13.41	Jansen et al. (2019)	rs3865444	0.97	−0.013	6.3×10^{-9}
CLU	8p21.1	Lambert et al. (2013)	rs9331896	0.86	—	2.8×10^{-25}
		Jansen et al. (2019)	rs4236673	0.95	−0.020	2.6×10^{-19}

Continued

TABLE 1 Genetic variants identified for AD in the studies discussed in this chapter—cont'd

Nearest gene	Location	Study	SNP	Odds ratio	Beta	*P*-value
CNTNAP2	7q35-q36.1	Jansen et al. (2019)	rs114360492	–	0.174	2.1×10^{-9}
CR1	1q32.2	Lambert et al. (2013) Jansen et al. (2019)	rs6656401 rs2093760	1.18 –	– 0.024	5.7×10^{-24} 1.1×10^{-18}
ECHDC3	10p14	Desikan et al. (2015) Liu, Erlich, and Pickrell (2017)	rs7920721 rs7920721	1.07 1.07	– –	3.4×10^{-8} 4.3×10^{-8}
ENSG00000237452	19q13.32	Jansen et al. (2019)	rs76320948	1.07	0.034	4.6×10^{-8}
ENSG00000271046	10p14	Jansen et al. (2019)	rs11257238	–	0.013	1.3×10^{-8}
ENSG00000249334	4p16.1	Jansen et al. (2019)	rs6448453	1.01	0.014	1.9×10^{-9}
ENSG00000262039	17q21.33	Jansen et al. (2019)	rs28394864	1.02	0.012	1.8×10^{-8}
ENSG00000266330	6p21.31	Jansen et al. (2019)	rs9381563	1.03	0.014	2.5×10^{-10}
EPHA1	7q34-q35	Lambert et al. (2013) Jansen et al. (2019)	rs11771145 rs7810606	0.90 –	– −0.014	1.1×10^{-13} 3.6×10^{-11}
FERMT2	14q22.1	Lambert et al. (2013)	rs17125944	1.14	–	7.9×10^{-9}
HBEGF	5q31.3	Jun et al. (2017) Liu et al. (2017)	rs11168036 rs2074612	1.08 1.08	– –	7.1×10^{-9} 8.0×10^{-9}
HESX1	3p14.3	Jansen et al. (2019)	rs184384746	–	0.195	1.2×10^{-8}
HLA-DRB5–HLA-DRB1	6p21.32	Lambert et al. (2013)	rs9271192	1.11	–	2.9×10^{-12}
HLA-DQA1	6p21.32	Jansen et al. (2019)	rs6931277	–	−0.019	8.4×10^{-11}
HS3ST1	4p15.33	Desikan et al. (2015)	rs13113697	1.07	–	2.9×10^{-8}
INPP5D	2q37.1	Lambert et al. (2013) Jansen et al. (2019)	rs35349669 rs10933431	1.08 0.97	– −0.015	3.2×10^{-8} 8.9×10^{-10}
KAT8	16p11.2	Jansen et al. (2019)	rs59735493	–	−0.013	4.0×10^{-8}
MEF2C	5q14.3	Lambert et al. (2013)	rs190982	0.93	–	3.2×10^{-8}
MS4A6A	11q12.2	Lambert et al. (2013) Jansen et al. (2019)	rs983392 rs2081545	0.90 –	– −0.017	6.1×10^{-16} 1.6×10^{-15}
NME8	7p14.1	Lambert et al. (2013)	rs2718058	0.93	–	4.8×10^{-9}
PICALM	11q14.2	Lambert et al. (2013) Jansen et al. (2019)	rs10792832 rs867611	0.87 –	– −0.020	9.3×10^{-26} 2.2×10^{-18}
PILRA	7q22.1	Jansen et al. (2019)	rs1859788	0.97	−0.018	2.2×10^{-15}
PLCG2	16q23.3	Sims et al. (2017)	rs72824905	0.68	–	5.38×10^{-10}
PTK2B	8p21.2	Lambert et al. (2013)	rs28834970	1.10	–	7.4×10^{-14}
PVRL2	19q13.32	Jansen et al. (2019)	rs41289512	1.52	0.200	5.8×10^{-276}
RP11-333E1.1	17p13.2	Jansen et al. (2019)	rs113260531	1.04	0.019	9.2×10^{-10}
SCIMP	17p13.2	Liu et al. (2017)	rs77493189	1.11	–	9.6×10^{-10}
SLC24A4/RIN3	14q32.12	Lambert et al. (2013) Jansen et al. (2019)	rs10498633 rs12590654	0.91 0.97	– −0.014	5.5×10^{-9} 1.7×10^{-10}
SORL1	11q24.1	Lambert et al. (2013) Jansen et al. (2019)	rs11218343 rs11218343	0.77 0.92	– −0.035	9.7×10^{-15} 1.1×10^{-11}
SPPL2A	15q21.2	Liu et al. (2017)	rs59685680	0.92	–	7.3×10^{-9}

TABLE 1 Genetic variants identified for AD in the studies discussed in this chapter—cont'd

Nearest gene	Location	Study	SNP	Odds ratio	Beta	*P*-value
TREM2	6p21.1	Jonsson et al. (2013) Sims et al. (2017) Guerreiro et al. (2013)	rs75932628 rs75932628 rs143332484[a]	2.92 2.46 1.67[a]	– – –	3.4×10^{-10} 5.4×10^{-24} 1.6×10^{-14}[a]
UNC5CL	6p21.1	Jansen et al. (2019)	rs187370608	–	0.226	1.5×10^{-16}
USP6NL	10p14	Jun et al. (2017)	rs7920721	1.08	–	3.0×10^{-8}
ZCWPW1	7q22.1	Lambert et al. (2013)	rs1476679	0.91	–	5.6×10^{-10}

Note. The genes in this table are the genes that were closest (in physical distance) to the associated SNPs, and thus do not necessarily reflect the causal genes. In addition, we only report the genes reported in the GWA studies discussed in this chapter. The odds ratio and beta are given whenever provided in the original study.
[a]*Several rare variants.*

BIN1, CLU, CR1, CD2AP, EPHA1, MS4A6A, and *PICALM*), and 11 of these loci were novel findings (*HLA-DRB5–HLA-DRB1, SORL1, PTK2B, SLC24A4, ZCWPW1, CELF1, NME8, FERMT2, CASS4, INPP5D,* and *MEF2C*). While a detailed discussion of the mechanisms and pathways involved with these genes is beyond the scope of this chapter, there are some interesting highlights: the newly implicated loci were involved in some relevant pathways previously suggested for AD, such as APP and tau pathology, inflammation, and lipid transport and endocytosis. Moreover, some new potential pathways were suggested, such as hippocampal synaptic function and regulation of gene expression. However, the identification of causal genes was solely based on proximity to the associated SNPs, but the closest gene is not necessarily the causal one. Since this study, more GWA meta-analyses have been performed for LOAD, and they are becoming increasingly successful at identifying the relevant genetic variants and the most plausible genes these variants operate on.

Genetic effects are not necessarily fixed across different ethnic groups, demonstrating the need for a comparison of allelic effects across different ethnicities. In addition, combining data from multiple ancestries can help improve the power to detect effects by increasing the number of included subjects. For LOAD, one such effort has recently been taken in a trans-ethnic GWAS using a sample including Caucasians of European ancestry, African-Americans, Japanese, and Israeli-Arabs (Jun et al., 2017). This led to the identification of four novel SNPs (in close proximity to *PFDN1/HBEGF, USP6NL/ECHDC3, BZRAP1-AS1,* and *NFIC*) (Table 1) and one novel gene (*TPBG*). Interestingly, these associations were not genome-wide significant in previous GWA studies that used only the individual discovery cohorts (Alzheimer's Disease Genetics Consortium datasets with participants from European Ancestry, African-Americans, Japanese, and Israeli-Arabs), revealing that a multiethnic approach can indeed help boost power in analyses.

The search for common variants contributing to AD risk is driven by the *common disease common variant hypothesis*, stating that common variants contribute significantly to the genetic risk for common disorders, such as LOAD (Lander et al., 1999). Alternatively, researchers have studied the contribution of rare variants to common disorders (also known as the *common disease rare variant hypothesis*). The advance of high-throughput sequencing technology has allowed for whole-exome/genome studies, and has thereby powered the search for these rare variants. Two distinct studies simultaneously and independently associated rare variants in the same gene to AD susceptibility. A genome-wide association study that employed whole-genome sequencing to include variants that are likely to affect protein function (Jonsson et al., 2013) showed that a risk variant in *TREM2* increased the risk for AD with an effect size similar to the ε4 allele of the *APOE* gene (but less frequently than the ε4 allele). The association between LOAD and *TREM2* was furthermore observed by Guerreiro and colleagues, who selected the *TREM2* locus to investigate its relation to LOAD (Guerreiro et al., 2013) based on its previous association with other forms of dementia. The study revealed that heterozygous rare variants in *TREM2* contribute to AD risk. This gene is involved in the immune system (Paloneva et al., 2002), and altered function of the gene likely influences the removal of Aβ in LOAD.

Steinberg and colleagues undertook a sequencing effort by examining possible rare variant associations near genes located in regions harboring common variants that were previously associated with LOAD (Steinberg et al., 2015), resulting in the identification of a second causal gene with rare variants influencing AD susceptibility. The researchers focused on two classes of rare variants, missense (causing an amino acid change in the resulting protein) and loss of function (resulting in a shorter transcript), because these classes likely have a large influence on gene function. The study found strong loss-of-function associations in *ABCA7*, a gene expressed primarily in brain microglia (the immune defense cells of the brain), involved in transporting lipids across membranes. Moreover, evidence from cell cultures indicates that this gene promotes the efflux of phospholipids and cholesterol to apoA-I and apoE, while evidence from mouse models suggests a role in phagocytosis.

In a similar fashion, a study performed by Sims and colleagues examined the contribution of rare variants in a well-powered, three-stage exome genotyping array design (Sims et al., 2017). This led to the identification of rare coding variants in three genes (*PLCG2*, *ABI3*, and *TREM2*), also highly expressed in brain microglia, implicated in AD. These three variants turned out to be part of a common innate immune response.

Linking these findings back to the different theories on the biology of LOAD mentioned in the introduction of this chapter, this suggests that immune- and phospholipid-dysregulation might interact to drive the biology of AD. Moreover, these findings further demonstrate the complexity of LOAD: not only common, but also rare variants contribute to the biology of LOAD. Instead of focusing only on one phenotype, one can also take a multivariate approach and use GWA summary statistics to study the genetic overlap between different phenotypes. Given the results mentioned herein, a useful application of such an approach is to study the genetic overlap between AD and immune- and phospholipid-related phenotypes. Using data from several large GWA studies, Desikan and colleagues examined the genetic overlap between AD, C-reactive protein (CRP) levels, and plasma lipid levels (Desikan et al., 2015). In addition to identifying two novel loci associated with AD (near *HS3ST1* and near *ECHDC3*) (Table 1), they found that SNPs associated with CRP and plasma lipids are also associated with increased risk for AD. The relationship between variants associated with AD and CRP levels had a consistent direction of effect, in line with the hypothesis that inflammation influences AD pathogenesis.

The most prominent limiting factor of gene finding for complex disorders is the ascertainment of cases. Using a case-control design, researchers need: (1) a priori information on disease status for many individuals, (2) an a priori decision on which phenotype they want to examine, and (3) sufficient resources to reach this select group of individuals. While (1) and (2) are less problematic in cohort studies (where data is obtained from a general, unbiased sample of the population), this type of design often contains too few cases because a disorder occurs infrequently or only in specific subpopulations. A recent development in the field of neuropsychiatric genetics is the use of *proxy-cases*: instead of identifying "true" cases, researchers for example identify the relatives of affective individuals (Liu et al., 2017). In this context, individuals who have one or two affected parents can be classified as proxy-cases. The controls in these types of studies are individuals without any known family history. This type of approach is useful with the availability of data from large cohorts such as the UK Biobank (Sudlow et al., 2015), where many individuals are not diagnosed with AD themselves because they have not reached the critical age yet, but do indicate that one or both parents are affected. Genome-wide association study by proxy (GWAX) was recently validated by Liu and colleagues using simulations and data from UK Biobank (Liu et al., 2017) (Table 1). Following this validation, a large GWAX where AD-by-proxy data from UK Biobank was meta-analyzed with case-control data with clinical AD cases, led to the identification of nine novel loci for AD (Jansen et al., 2019) (Table 1), revealing that this method holds great promise not only for AD, but as a methodology for complex disorders in neuropsychiatric genetics in general.

In the study by Jansen and colleagues, GWAX analyses were followed by a series of in silico functional follow-up analyses, including functional annotation, gene mapping (positional/eQTL/chromatin interaction), gene-set analyses, and cross-trait genetic influences (Box 1). In contrast to earlier GWA studies (e.g., Lambert et al., 2013), causative genes are mapped on more criteria than mainly physical proximity, such as DNA interactions, often leading to more accurate mapping of genes. Enhanced functional follow-up methods lead to a better understanding of the influence of genetic factors on disorders, because they provide insights into the biological pathways and mechanisms leading to disease. In Jansen et al. (2019), two main biological mechanisms (protein lipid complex and regulation of APP catabolic process) were associated with AD risk. In addition, in accordance with previous studies, the associated genes are strongly expressed in immune-related tissues and cell types. Future investigations into these pathways and their interaction might lead to better insights into the causative mechanism for AD.

3 Other omics approaches

The studies we have discussed so far all fall under the broad category "genomics," a field of study that focuses on identifying genetic variants associated with disease, response to treatment, or future patient prognosis (Hasin, Seldin, & Lusis, 2017). However, there are several other "-omics" data types that can be used to study the biology of complex disorders.

Studying multiple -omics levels simultaneously has the potential to provide a more detailed understanding of the genetic architecture of complex disorders. The advent of large, publicly available datasets (GTEx, NIH Roadmap Epigenomics) has fueled different types of -omics investigations, enabling researchers to access different data types.

A way to examine the functional genes involved in AD, in addition to the GWA approach, is to look at gene expression profiling (GEP) by quantifying the transcriptome of certain cell types or tissues (Cooper-Knock et al., 2012). A review examining GEP studies from 2005 to 2012 (Cooper-Knock et al., 2012) lists a few interesting observations from CNS mixed-cell studies: aberrant gene expression in AD was most likely to occur (1) in the temporal lobe-hippocampus and frontal prefrontal area, (2) during progression from mild to moderate dementia, and (3) in intracellular signaling pathways (especially calcium signaling) and neuroinflammation. Evidence from studies in peripheral cells also supports the involvement of neuroinflammation.

Another approach is to examine the epigenomics of AD. Research into epigenetic changes in AD was for a long period mostly constrained to animal studies because of practical limitations related to the accessibility of human brain tissue. However, newer methods, such as epigenome-wide (or methylome-wide) association studies (EWAS), have made it possible to get more insights into the epigenomics of complex disorders. Relatively large methylation studies have been performed for AD. One such study examined 708 autopsied brains to assess the methylation state of the brain as a response to AD (De Jager et al., 2014). In this study, the level of methylation at 71 CpG sites was associated with AD pathology. Interestingly, some of these CpG sites were in known AD susceptibility loci (e.g., *ABCA7* and *BIN1*), showing that these loci operate through multiple genomic mechanisms to affect AD. These findings were replicated in a study where CpG sites in 11 AD susceptibility loci were associated with a quantitative measure of neuritic plaques (Chibnik et al., 2015). In this study, six regions reached significance (independent of genetic variation): *BIN1, CLU, MS4A6A, ABCA7, CD2AP,* and *APOE*.

Other epigenetic processes are microRNA (miRNA) or histone modifications. With respect to miRNAs, it has been hypothesized that miRNA might lead to upregulation of AD-associated genes, such as *APP, PSEN1,* and *PSEN2*. In a systematic literature review investigating the value of miRNAs in converting AD or mild cognitive impairment (MCI) from cognitively healthy individuals in 18 studies (Wu et al., 2015), 20 upregulating and 32 downregulating miRNAs were found in AD patients compared with controls. However, the effects were not always consistent over different studies. Concerning the role of histone modifications in AD, most evidence comes from studies in rodents. These studies showed that histone acetylation plays a role in memory consolidation, and that inhibiting histone acetylation could lead to improvements in memory consolidation (Fischer, 2014). Moreover, the administration of histone-deacetylases (HDAC) rescued some of the memory deficits and tau formation in mouse models for amyloid deposition (Fischer, 2014). While there has not been much research on histone modifications in AD outside of rodent studies, there have also been reports of increased acetyl histone levels in postmortem AD brains (Narayan, Lill, Faull, Curtis, & Dragunow, 2015) and in a cellular model of AD (Lu et al., 2014).

Proteomics is another -omics area that holds large promise for the future of AD research. Proteomics analyses typically organize disease-related proteins into meaningful modules or networks of co-expressed proteins. This adds additional information over gene expression, as the gene-protein expression correlation is low to moderate. A recent study examining the AD-proteome identified 487 differentially expressed proteins in AD patients ($N = 8$) versus controls ($N = 8$), further demonstrating the complexity of the disorder (Zhang et al., 2018). Once the relevant proteins are identified, a next step is to examine how the different proteins interact in protein networks. In an attempt to create such a network, Hallock and Thomas (2012) reviewed the primary literature and web sources related to AD. The initial network was expanded by adding proteins that interact with core proteins, and gene expression studies were also mapped to this network. Two cellular pathways were identified that had clusters of differentially regulated genes whose protein products interacted in the network: the MAPK/ERK pathway and clathrin-mediated receptor endocytosis, involved in long-term potentiation and synaptic plasticity, and the internalization of APP, respectively.

As cellular mechanisms are dynamic processes where distinct levels of "omics" units are constantly interacting, the most optimal and informative strategy for assessing AD is to take a *multiomics* approach where there is an integration of different omics data types. However, there are several factors that one should take into account when examining a complex disorder such as AD. Most importantly, different combinations of different factors (SNPs, CpG sites, environmental factors, etc.) could combine to phenotypically similar states, and the reverse is also true: the same factors influence different people in different ways. To be able to predict and understand AD on an individual level, we need much more extensive insights into the interaction of different genetic, biological, and environmental factors at different time points in different people. Moreover, there are multiple strategies for combining different data types, and it is up to the researchers to determine which strategy is optimal (Hasin et al., 2017).

4 Conclusion

The past few decades of research have certainly shed new light on the genetics and biology of AD. Most important, as shown in Table 1, there is a rapid increase in the number of identified genetic variants from before 2013 to now. Newer methods and approaches, such as GWAX and trans-ethnic investigations, will continue to power the search for more variants. Yet, there are many questions still left unanswered. Now that we have increased knowledge on the genetic etiology of AD, a next step is to get more insight into the mechanisms and functions under which these variants operate. This will require in-depth functional follow-up analyses both in silico (like in Jansen et al., 2019) and in vitro/vivo (Shulman et al., 2014), but also the integration of multiple -omics approaches and data types. In the future, these developments could lead to classifications of different types of patients based on their biological profile, and the development of personalized treatments that affect certain biological mechanisms.

References

Alzheimer's Association. (2017). 2017 Alzheimer's disease facts and figures. *Alzheimer's & Dementia, 13*, 325–373.

Beach, T. G., Monsell, S. E., Phillips, L. E., & Kukull, W. (2012). Accuracy of the clinical diagnosis of Alzheimer disease at national institute on aging Alzheimer disease centers, 2005–2010. *Journal of Neuropathology & Experimental Neurology, 71*(4), 266–273.

Chibnik, L. B., Yu, L., Eaton, M. L., Srivastava, G., Schneider, J. A., Kellis, M., ... De Jager, P. L. (2015). Alzheimer's loci: Epigenetic associations and interaction with genetic factors. *Annals of Clinical and Translational Neurology, 2*(6), 636–647.

Cooper-Knock, J., Kirby, J., Ferraiuolo, L., Heath, P. R., Rattray, M., & Shaw, P. J. (2012). Gene expression profiling in human neurodegenerative disease. *Nature Reviews Neurology, 8*(9), 518–530.

De Jager, P. L., Srivastava, G., Lunnon, K., Burgess, J., Schalkwyk, L. C., Yu, L., ... Bennett, D. A. (2014). Alzheimer's disease: Early alterations in brain DNA methylation at ANK1, BIN1, RHBDF2 and other loci. *Nature Neuroscience, 17*(9), 1156–1163.

Desikan, R. S., Schork, A. J., Wang, Y., Thompson, W. K., Dehghan, A., Ridker, P. M., ... Dale, A. M. (2015). Polygenic overlap between C-reactive protein, plasma lipids, and Alzheimer disease. *Circulation, 131*(23), 2061–2069.

Di Paolo, G., & Kim, T.-W. (2011). Linking lipids to Alzheimer's disease: Cholesterol and beyond. *Nature Reviews Neuroscience, 12*(5), 284–296.

Fischer, A. (2014). Targeting histone-modifications in Alzheimer's disease. What is the evidence that this is a promising therapeutic avenue? *Neuropharmacology, 80*, 95–102.

Galasko, D. R., & Shaw, L. M. (2017). CSF biomarkers for Alzheimer disease—Approaching consensus. *Nature Reviews Neurology, 13*(3), 131–132.

Gatz, M., Reynolds, C. A., Fratiglioni, L., Johansson, B., Mortimer, J. A., Berg, S., ... Pedersen, N. L. (2006). Role of genes and environments for explaining Alzheimer disease. *Archives of General Psychiatry, 63*(2), 168.

Guerreiro, R., Wojtas, A., Bras, J., Carrasquillo, M., Rogaeva, E., Majounie, E., ... Hardy, J. (2013). TREM2 variants in Alzheimer's disease. *The New England Journal of Medicine, 368*(2), 117–127.

Hallock, P., & Thomas, M. A. (2012). Integrating the Alzheimer's disease proteome and transcriptome: A comprehensive network model of a complex disease. *Omics: A Journal of Integrative Biology, 16*(1–2), 37–49.

Hasin, Y., Seldin, M., & Lusis, A. (2017). Multi-omics approaches to disease. *Genome Biology, 18*(1), 83.

Heppner, F. L., Ransohoff, R. M., & Becher, B. (2015). Immune attack: The role of inflammation in Alzheimer disease. *Nature Reviews Neuroscience, 16*(6), 358–372.

Herrup, K. (2015). The case for rejecting the amyloid cascade hypothesis. *Nature Neuroscience, 18*(6), 794–799.

Jansen, I., Savage, J., Watanabe, K., Bryois, J., Williams, D. M., Steinberg, S., ... Posthuma, D. (2019). Genetic meta-analysis identifies 9 novel loci and functional pathways for Alzheimers disease risk. *Nature Genetics, 51*(3), 404–413.

Jonsson, T., Stefansson, H., Steinberg, S., Jonsdottir, I., Jonsson, P. V., Snaedal, J., ... Stefansson, K. (2013). Variant of TREM2 associated with the risk of Alzheimer's disease. *New England Journal of Medicine, 368*(2), 107–116.

Jun, G. R., Chung, J., Mez, J., Barber, R., Beecham, G. W., Bennett, D. A., ... Farrer, L. A. (2017). Transethnic genome-wide scan identifies novel Alzheimer's disease loci. *Alzheimer's & Dementia: The Journal of the Alzheimer's Association, 13*(7), 727–738.

Krstic, D., Madhusudan, A., Doehner, J., Vogel, P., Notter, T., Imhof, C., ... Knuesel, I. (2012). Systemic immune challenges trigger and drive Alzheimer-like neuropathology in mice. *Journal of Neuroinflammation, 9*(1), 699.

Lambert, J.-C., Ibrahim-Verbaas, C. A., Harold, D., Naj, A. C., Sims, R., Bellenguez, C., ... Amouyel, P. (2013). Meta-analysis of 74,046 individuals identifies 11 new susceptibility loci for Alzheimer's disease. *Nature Genetics, 45*(12), 1452–U206.

Lander, E. S., Cargill, M., Altshuler, D., Ireland, J., Sklar, P., Ardlie, K., ... Daley, G. Q. (1999). Characterization of single-nucleotide polymorphisms in coding regions of human genes. *Nature Genetics, 22*(3), 231–238.

Lim, G. P., Yang, F., Chu, T., Chen, P., Beech, W., Teter, B., ... Cole, G. M. (2000). Ibuprofen suppresses plaque pathology and inflammation in a mouse model for Alzheimer's disease. *The Journal of Neuroscience: The Official Journal of the Society for Neuroscience, 20*(15), 5709–5714.

Liu, J. Z., Erlich, Y., & Pickrell, J. K. (2017). Case–control association mapping by proxy using family history of disease. *Nature Genetics, 49*(3), 325–331.

Liu, C.-C., Kanekiyo, T., Xu, H., Bu, G., & Bu, G. (2013). Apolipoprotein E and Alzheimer disease: Risk, mechanisms and therapy. *Nature Reviews Neurology, 9*(2), 106–118.

Lu, X., Deng, Y., Yu, D., Cao, H., Wang, L., Liu, L., ... Yu, G. (2014). Histone acetyltransferase p300 mediates histone acetylation of PS1 and BACE1 in a cellular model of Alzheimer's disease. *PLoS One, 9*(7), e103067.

Narayan, P. J., Lill, C., Faull, R., Curtis, M. A., & Dragunow, M. (2015). Increased acetyl and total histone levels in post-mortem Alzheimer's disease brain. *Neurobiology of Disease*, *74*, 281–294.

Paloneva, J., Manninen, T., Christman, G., Hovanes, K., Mandelin, J., Adolfsson, R., … Peltonen, L. (2002). Mutations in two genes encoding different subunits of a receptor signaling complex result in an identical disease phenotype. *The American Journal of Human Genetics*, *71*(3), 656–662.

Salvatore, C., Battista, P., & Castiglioni, I. (2016). Frontiers for the early diagnosis of AD by means of MRI brain imaging and support vector machines. *Current Alzheimer Research*, *13*(5), 509–533.

Selkoe, D. J. (2001). Alzheimer's disease: Genes, proteins, and therapy. *Physiological Reviews*, *81*(2), 741–766.

Seltzer, B., & Sherwin, I. (1983). A comparison of clinical features in early- and late-onset primary degenerative dementia. *Archives of Neurology*, *40*(3), 143.

Shulman, J. M., Imboywa, S., Giagtzoglou, N., Powers, M. P., Hu, Y., Devenport, D., … Feany, M. B. (2014). Functional screening in Drosophila identifies Alzheimer's disease susceptibility genes and implicates Tau-mediated mechanisms. *Human Molecular Genetics*, *23*(4), 870–877.

Sims, R., van der Lee, S. J., Naj, A. C., Bellenguez, C., Badarinarayan, N., Jakobsdottir, J., … Schellenberg, G. D. (2017). Rare coding variants in PLCG2, ABI3, and TREM2 implicate microglial-mediated innate immunity in Alzheimer's disease. *Nature Genetics*, *49*(9), 1373–1384.

Sommer, B. (2002). Alzheimer's disease and the amyloid cascade hypothesis: Ten years on. *Current Opinion in Pharmacology*, *2*(1), 87–92.

Spires-Jones, T. L., & Hyman, B. T. (2014). The intersection of amyloid beta and tau at synapses in Alzheimer's disease. *Neuron*, *82*(4), 756–771.

Steinberg, S., Stefansson, H., Jonsson, T., Johannsdottir, H., Ingason, A., Helgason, H., … Stefansson, K. (2015). Loss-of-function variants in ABCA7 confer risk of Alzheimer's disease. *Nature Genetics*, *47*(5), 445–447.

Sudlow, C., Gallacher, J., Allen, N., Beral, V., Burton, P., Danesh, J., … Collins, R. (2015). UK Biobank: An open access resource for identifying the causes of a wide range of complex diseases of middle and old age. *PLoS Medicine*, *12*(3), e1001779.

Tang, Y.-P., & Gershon, E. S. (2003). Genetic studies in Alzheimer's disease. *Dialogues in Clinical Neuroscience*, *5*(1), 17–26.

Tanzi, R. E., & Bertram, L. (2005). Twenty years of the Alzheimer's disease amyloid hypothesis: A genetic perspective. *Cell*, *120*(4), 545–555.

Wu, H. Z. Y., Ong, K. L., Seeher, K., Armstrong, N. J., Thalamuthu, A., Brodaty, H., … Mather, K. (2015). Circulating microRNAs as biomarkers of Alzheimer's disease: A systematic review. *Journal of Alzheimer's Disease*, *49*(3), 755–766.

Zhang, Q., Ma, C., Gearing, M., Wang, P. G., Chin, L.-S., & Li, L. (2018). Integrated proteomics and network analysis identifies protein hubs and network alterations in Alzheimer's disease. *Acta Neuropathologica Communications*, *6*(1), 19.

Chapter 23

Current progress and future direction in the genetics of PTSD: Focus on the development and contributions of the PGC-PTSD working group

Angela G. Junglen[a], Christina Sheerin[b], Douglas L. Delahanty[a,c], Michael A. Hauser[d], Adriana Lori[e], Rajendra A. Morey[f,g,h,i], Caroline M. Nievergelt[j,k], Nicole R. Nugent[l,m], Jonathan Sebat[j,n,o], Alicia K. Smith[e], Jennifer A. Sumner[p], Monica Uddin[q,r] and Ananda B. Amstadter[b,s]

[a]Department of Psychological Sciences, Kent State University, Kent, OH, United States, [b]Department of Psychiatry, Virginia Institute of Psychiatric and Behavioral Genetics, Virginia Commonwealth University, Richmond, VA, United States, [c]Department of Psychiatry, Northeastern Ohio Medical University, Rootstown, OH, United States, [d]Department of Medicine, Duke Molecular Physiology Institute, Duke University Medical Center, Durham, NC, United States, [e]Department of Psychiatry and Behavioral Sciences, Emory University School of Medicine, Atlanta, GA, United States, [f]National Center for PTSD, Behavioral Science Division, VA Boston Healthcare System, Boston, MA, United States, [g]Department of Psychiatry and Biomedical Genetics, Boston University School of Medicine, Boston, MA, United States, [h]Mid-Atlantic Mental Illness Research, Education and Clinical Center, Durham VA Medical Center, Durham, NC, United States, [i]Duke-UNC Brain Imaging and Analysis Center, Duke University, Durham, NC, United States, [j]Department of Psychiatry, University of California, San Diego, La Jolla, CA, United States, [k]VA San Diego Healthcare System, VA Center of Excellence for Stress and Mental Health (CESAMH), La Jolla, CA, United States, [l]Bradley Hasbro Children's Research Center of Rhode Island Hospital, Providence, RI, United States, [m]Department of Psychiatry and Human Behavior, Alpert Medical School of Brown University, Providence, RI, United States, [n]Beyster Center for Genomics of Psychiatric Diseases, University of California, San Diego, La Jolla, CA, United States, [o]Department of Cellular and Molecular Medicine, University of California, San Diego, La Jolla, CA, United States, [p]Center for Cardiovascular Behavioral Health, Columbia University Medical Center, New York, NY, United States, [q]Carl R Woese Institute for Genomic Biology, University of Illinois Urbana-Champaign, Urbana, IL, United States, [r]Department of Psychology, University of Illinois Urbana-Champaign, Champaign, IL, United States, [s]Department of Human and Molecular Genetics, Virginia Commonwealth University, Richmond, VA, United States

1 Introduction

Posttraumatic stress disorder (PTSD) is unique among psychiatric disorders in that it includes an environmental exposure (i.e., a traumatic experience, such as assault, disaster, etc.) as a required diagnostic criterion. A diagnosis of PTSD is characterized by symptoms of intrusions, avoidance, negative alterations in mood or cognition, and increased arousal following a traumatic experience (DSM-5; American Psychiatric Association, 2013). Although a majority of individuals (more than 70%) will experience a traumatic event in their lifetimes (Benjet et al., 2016); most trauma-exposed individuals do not develop PTSD. Estimated conditional likelihood of PTSD ranges from 8% to 32% (Breslau et al., 1998; Kessler, Sonnega, Bromet, Hughes, & Nelson, 1995) and has been shown to be a function of cumulative trauma exposure, as well as trauma characteristics such as trauma type and timing (Breslau, Chilcoat, Kessler, & Davis, 1999; Brewin, Andrews, & Valentine, 2000; Irish et al., 2008). PTSD has been associated with a number of chronic medical comorbidities (Boscarino, Forsberg, & Goldberg, 2010; Kibler, 2009), as well as higher rates of suicide, disability, and premature mortality (Kessler, 2000). Given the debilitating nature of the disorder, and the fact that only a minority of individuals develop PTSD after a trauma, identifying factors that increase susceptibility has been a hallmark effort of PTSD research. A growing body of research is examining the underlying genetic etiology that contributes to this disorder. In this chapter, we will first briefly summarize the extant knowledge with regard to genetic risk for PTSD, and then discuss the efforts underway by the PTSD working group of the Psychiatric Genomics Consortium (PGC) to identify the genetic architecture underlying susceptibility to PTSD. We will end with a brief review of future directions in the field of PTSD genomics.

Personalized Psychiatry. https://doi.org/10.1016/B978-0-12-813176-3.00023-7

2 Behavioral genetics of PTSD

Behavioral genetic studies of PTSD have aided in quantifying the latent genetic and environmental influences on both trauma exposure and PTSD. Twin studies have suggested that trauma exposure is influenced by genetic factors (referred to as gene-environment correlation, or rGE; True et al., 1993). Specifically, in the Vietnam Era Twin Registry, Lyons et al. (1993) reported heritability estimates ranging from 35% to 47% for combat exposure. This has also been demonstrated in civilians, with estimates for interpersonal traumatic events being modestly heritable (~20%; Stein, Jang, & Livesley, 2002). The heritability of PTSD, after controlling for genetic influences on trauma exposure, has been well-established in the literature (see review by Afifi, Asmundson, Taylor, & Jang, 2010). Heritability estimates for PTSD range from 30% to 72%, with higher heritability estimates found in all-female samples, compared with all-male or combined-sex samples (Lyons et al., 1993; Sartor et al., 2011; Stein et al., 2002). However, due to modest sample sizes, extant studies have been inadequately powered to formally test for quantitative (i.e., differences in magnitude) or qualitative (i.e., differences in the genetic source) sex effects. In addition to estimating heritability, twin studies have also demonstrated that genetic influences on PTSD are shared with other psychiatric disorders (Chantarujikapong et al., 2001; Sartor et al., 2011, 2012). This literature supports the promise of molecular genetic investigations for identifying specific genetic risk.

3 Molecular genetic studies

Developments in the field of molecular genetics, such as the decreasing costs of arrays, greater accessibility of genotyping, and advancing methods for analyzing genetic variants have been instrumental in gene finding efforts for PTSD. To date, molecular studies on PTSD have utilized candidate gene studies, candidate gene by environment (cGxE) studies, and genome-wide association studies (GWAS).

3.1 Candidate gene studies

Candidate gene studies investigate the association of selected genetic variants with a specified phenotype (i.e., PTSD diagnosis). Candidate genes are typically chosen based on prior findings regarding the biology of the phenotype. Given that PTSD has been theorized as a disorder related to an aberrant stress response, many of these variants have focused on genes in the HPA-axis and the locus coeruleus-noradrenergic system. More than 100 candidate gene studies have been conducted to date, examining more than 81 genes (for a review, Sheerin, Lind, Bountress, Nugent, & Amstadter, 2017). Extant studies have identified a wide range of variants, although results across studies are often inconsistent. As the literature has grown, meta-analyses of individual candidate gene studies have increased, and have often resulted in inconsistent findings. A meta-analysis of dopamine candidate gene studies revealed significant effects for *DRD2* and *SLC6A3*, but no overall effect for *COMT* (Li et al., 2016). A meta-analysis of pituitary adenylate cyclase activating polypeptide, specifically *ADCYAP1R1*, underscored the importance of considering sex differences in examining genetic effects associated with PTSD, with significant findings in females, but not males (Lind et al., 2017). In contrast, two meta-analyses of serotonin revealed no overall effect for *5-HTTLPR* (Gressier et al., 2013; Navarro-Mateu, Escamez, Koenen, Alonso, & Sanchez-Meca, 2013). With respect to neurotrophins, meta-analytic findings were conflicting (e.g., *BDNF Val66Met* had a significant effect in Bruenig et al., 2016; but no effect in Wang, 2015).

Beyond inconsistent findings, additional limitations of candidate gene studies are worth noting, including: (1) current limited knowledge related to the biology of PTSD, thereby constraining hypothesized candidate genes; (2) sparse coverage of genes of interest, as most studies examine specific polymorphisms within a gene; (3) small sample sizes, resulting in low power; and (4) heterogeneous quality control approaches (e.g., lack of genomic control for ancestry). These methodologic limitations may account, in part, for the minimal successful replications of candidate gene studies (Tabor, Risch, & Myers, 2002). Finally, candidate gene studies typically do not consider environmental influences on PTSD; characterization of environmental influences may be essential for disentangling the complexities of PTSD.

3.2 cGxE studies

Given the necessary, but not sufficient, role of trauma exposure in the etiology of PTSD, numerous cGxE studies have been conducted as an extension of the candidate gene design. There have been some replications of cGxE associations in the PTSD literature, with researchers identifying the interaction between childhood abuse and *FKBP5* (Binder et al., 2008; Xie et al., 2010), and between *APOE* and combat exposure (Kimbrel et al., 2015; Lyons et al., 2013). These replications underscore the importance of incorporating trauma exposure as part of a PTSD model. Social environment has also been shown to impact the relationship between serotoninergic variants and an individual's increased likelihood of PTSD (Drevo et al., 2016; Koenen et al., 2009). However, a noted challenge of cGxE studies is that incorporating trauma exposure into the model creates a statistical interaction that requires even larger sample sizes for the analyses to be adequately powered (Manuck & McCaffery, 2014; Thomas, 2010). Another challenge of some cGxE analyses is that a dichotomous outcome variable of "case/control" with a logistic regression can lead to increased spurious interaction effects that need to be considered (Eaves, 2006). Future, well-powered cGxE analyses and further replications will continue to contribute to identifying PTSD risk variants, as well as vital environmental influences.

3.3 Genome-wide association studies

In contrast to cGxE analyses and candidate gene association studies, which are based on a priori hypothesizing of the gene's relevance to the phenotype, GWAS use an agnostic hypothesis-generating approach to identify risk variants. To localize risk variants, GWAS compare frequencies of millions of common genetic variants dispersed across the entire genome among cases and controls. To correct for the number of independent association tests performed, the threshold P-value to be reached for significance decreases substantially; standard P-values are on the order of $\sim 10^{-8}$ (Sham & Purcell, 2014). For a polygenic phenotype such as PTSD, where each variant will likely have minimal contribution to disease risk, a very large sample size is needed to reach sufficient power to detect such an association after multiple testing corrections.

Several GWAS of PTSD have been published (Almli et al., 2015; Ashley-Koch et al., 2015; Guffanti et al., 2013; Kilaru et al., 2016; Logue et al., 2013; Nievergelt et al., 2015; Powers et al., 2016; Stein et al., 2016; Wolf et al., 2014; Xie et al., 2013), with some suggestive evidence of replication in independent cohorts, but to date, no genetic variations have been independently replicated across GWAS (Table 1). The paucity of replications across GWAS may be explained by low power, or the complex nature of the PTSD phenotype that includes the heterogeneity in trauma type, ancestry, and sex.

The majority of significant variants identified in these GWAS are found in genes not traditionally included in previous candidate gene approaches, and in fact, none of the candidate genes have been directly supported in the GWAS literature. Instead, novel variants have been identified, allowing for new potential mechanisms to be explored. For example, after the significant genome-wide finding of *RORA* (Logue et al., 2013), subsequent studies were conducted to evaluate the influence of *RORA* on different aspects of PTSD. The *RORA* gene was replicated in an exclusively disaster-exposed sample, underscoring the significance of *RORA* in PTSD while mitigating the concerns of heterogeneous trauma exposure (Amstadter et al., 2013). Further efforts focused the role of *RORA* on neuroprotection after trauma and distress (Miller, Wolf, Logue, & Baldwin, 2013), and aspects of fear and distress more broadly were shown to be related to a variant of the *RORA* gene, providing promising evidence of the functionality of this gene (Miller et al., 2013). The *RORA* literature provides just one example where GWAS findings have led to further discovery and replication, with the hopes of elucidating possible novel biological pathways of PTSD.

4 Psychiatric Genomics Consortium

Although individual GWAS have led to some fruitful mechanistic follow-up studies, the larger GWAS literature has highlighted the critical need for large-scale collaborations to further move the field forward. These collaborations are necessary to generate well-powered analytic samples and yield replicable results (e.g., the inflection point for schizophrenia was reached at approximately 15,000 cases, whereas for depression estimates suggest that it may be reached at 75,000–100,000 cases; Levinson et al., 2014). To address the need for well-powered analytic studies, the Psychiatric Genomic Consortium (PGC) was formed in 2007, and has become the largest psychiatric collaboration to date, including 10 workgroups (e.g., schizophrenia, bipolar disorder, depression) and more than 800 investigators. A key to the overwhelming success of the PGC is the infrastructure it provides to safely and securely receive genome-wide data, to carry out careful and uniform quality control, and to harmonize phenotypic and genotypic data. Moreover, the PGC encourages a culture of collaboration and secondary analyses of the data.

The PGC-PTSD working group was officially established in 2013 as part of a discussion at the "Collaborative Science and PTSD: Maximizing our Understanding of Biomarkers" meeting held at the annual Society of Biological Psychiatry conference, co-sponsored by the National Institute of Mental Health (NIMH) and One Mind. Meeting participants assembled existing GWAS for the first PGC-PTSD analysis, and obtained funding to expand data collection with the goal of continued development in identifying the genetic architecture of PTSD (Logue et al., 2015).

TABLE 1 Heterogeneity of published GWAS studies of PTSD.

	Cohort(s)	N	Phenotype	SNPs[a]-gene
Guffanti et al. (2013)	Detroit Neighborhood Health Study (DNHS)	342 AA 45 EA 26 other	PTSD diagnosis (structured interview)	rs10170218-AC068718.1-lincRNA (entire dataset)
Logue et al. (2013)	Trauma exposed-veterans and their partners	491 EA 84 AA	CAPS-PTSD diagnosis	rs8042149-RORA (EA)
Xie et al. (2013)	Subjects recruited for substance abuse	1578 EA 2766 AA	PTSD diagnosis (DSM-IV)	rs406001-COBL (EA) rs6812849-TLL1 (EA) No GWAS finding (AA)
Wolf et al. (2014)	Trauma exposed-veterans and their partners	484 EA	Clinical symptoms of dissociation	No GWAS finding
Almli et al. (2015)	Systems Biology PTSD Biomarkers Consortium (Military Cohort)	57 Hispanic 45 NHW 35 NHB 6 NHA 4 other	CAPS score-PTSD diagnosis	rs717947-chr4p15
Ashley-Koch et al. (2015)	Iraq/Afghanistan-era veterans and community civilians and veterans from Durham VA and Duke Medical Center	949 NHB 759 NHW	SCID for PTSD diagnosis (DSM-IV) (MIRECC) CAPS for PTSD diagnosis (DSM-IV; cohort of community civilians and veterans)	No GWAS finding
Nievergelt et al. (2015)	Marine Resiliency Study 2376 (MRS-I) 1118 (MRS-II)	2179 EA 640 Hispanic Native American 205 AA 470 other	CAPS-PTSD diagnosis (DSM-IV) for partial and full PTSD.	rs6482463-PRTFDC1 (meta-analyses)
Kilaru et al. (2016)	Grady Trauma Project and Detroit Neighborhood Health Study (DNHS)-(Civilian)	3678 AA	PSS-PTSD diagnosis (DSM-IV criteria)	rs6779753-NLGN1
Powers et al. (2016)	Grady Trauma Project (Civilian)	2600 AA	EDS mPSS-PTSD diagnosis (DSM-IV criteria)	rs6602398-IL2RA (males only) No GWAS finding (females)
Stein et al. (2016)	The New Soldier Study (NSS) 7999 (NSS1) 2835 (NSS2)	5049 EA 1312 AA 1413 Latino	PTSD diagnosis based on PCL and CIDI Screening Scales data	rs159572-ANKRD55 (AA) rs11085374-near ZNF626 (EA possible spurious association) No GWAS finding (Latinos or trans-ancestral meta-analyses)
Duncan et al. (2017)	PGC-PTSD Freeze 1 11 sites 19 data subsets	9954 EA 9691 AA 387 South Africa 698 Latino	PTSD diagnosis (site-specific determination)	No GWAS finding (trans-ethnic or EA meta-analyses) rs139558732-KLHL1 (AA meta-analysis only-probable spurious result)

Abbreviations: AA, African American; CAPS, Clinician Administered PTSD Scale; CIDI, Composite International Diagnostic Interview; EA, European Ancestry; EDS, Emotional Dysregulation; mPSS, Modified PTSD Symptom Scale; NHA, Non-Hispanic Asian; NHB, Non-Hispanic Black; NHW, Non-Hispanic White; PCL, PTSD Checklist; PSS, PTSD Symptom Scale; PTSD, Posttraumatic Stress Disorder; SCID, Structured Clinical Interview for DSM disorders.
[a]Variants reaching GWAS significance.

5 Initial findings from the PGC-PTSD

The first PGC-PTSD data freeze (Freeze 1) included 11 multiethnic studies with 5000 PTSD cases and 15,000 controls, comprising the largest GWAS in PTSD to date (Duncan et al., 2017). To accommodate data sharing restrictions and optimize sample size, GWAS were performed separately for each study, and three main ancestral groups (European, African American, and Latinos), and then combined using meta-analytical approaches (Duncan et al., 2017). This first trans-ethnic GWAS was not yet powered for discovery of specific risk loci, but demonstrated significant genetic influences on PTSD development, with average genetic heritability estimates (h_{SNP}^2) of $\sim$15% in the European subset across both males and females (Duncan et al., 2017). Higher heritability was found in females (h_{SNP}^2 of 29%), similar to overall heritability estimates for schizophrenia and bipolar disorder (Psychiatric GWAS Consortium-Bipolar Disorder Working Group, 2011; Schizophrenia Working Group of the Psychiatric Genomics, 2014), in comparison with heritability estimates for males (which was not distinguishable from zero). Furthermore, previous evidence of overlap of polygenic risk scores between PTSD and other psychiatric disorders (e.g., bipolar disorder; Nievergelt et al., 2015) was replicated.

The PGC-PTSD efforts to increase sample size are currently supported by NIMH, the Department of Defense, the Stanley Center for Psychiatric Genetics, and Cohen Veterans Bioscience. The second data freeze (Freeze 2) occurred at the end of 2017, and includes 32,000 PTSD cases and more than 100,000 trauma-exposed controls from 57 studies worldwide, and is approaching a sample size at which other PGC studies showed first discoveries (i.e., the inflection point noted herein; Gratten, Wray, Keller, & Visscher, 2014).

6 Unique considerations of the PGC-PTSD

Many considerations of managing large consortium data are not unique to the PGC-PTSD working group. Such considerations include: inclusion/exclusion criteria across cohorts; variation in availability of information on comorbidities; and informed consent limitations for data sharing (Bennett et al., 2011). However, due to particular characteristics of the PTSD phenotype, the PGC-PTSD working group is faced with additional methodological considerations including: identifying appropriate PTSD controls (e.g., trauma-exposed controls) and defining a traumatic event; harmonizing across various approaches to measuring PTSD; accounting for the heterogeneity of PTSD and its frequent comorbidity with other disorders; and addressing the critical sex differences and genetic role therein, as well as the diverse ancestry in the PGC-PTSD samples.

Identification of genetic risk for PTSD necessitates use of trauma-exposed controls. In the absence of known trauma exposure, genetic risk in individuals who do not have PTSD is unknown (i.e., they may be at low genetic risk, or may be at high genetic risk that has not been expressed due to lack of trauma exposure); thus, future analyses of the PGC-PTSD will focus on trauma-exposed controls. The workgroup is obtaining item-level data on trauma exposure and PTSD symptoms, which allow for more nuanced examinations to be conducted. Item-level data on trauma exposure measures provide a means of re-categorizing trauma categories, whether by trauma load, number of types of trauma, interpersonal versus non-interpersonal trauma, and combat versus civilian. This will be useful for examining potential cohort effects, particularly those associated with the proportion of males in cohorts often linked to trauma type (e.g., combat trauma exposure studies typically consist of a majority of male cohorts as compared with civilian samples with greater proportions of females and interpersonal traumas). Disentangling this complexity is fundamental to understanding the biological differences of PTSD prevalence rates across sex and trauma type. The PGC-PTSD working group is also making efforts toward harmonization of the PTSD phenotype. The methodology across the 57 study cohorts within the Freeze 2 dataset varied across diagnostic criteria (e.g., DSM-IV, DSM-5, ICD-10), specific measures implemented (e.g., diagnostic interview or self-report measure), and cut-off scores, all of which can add variation in diagnostic determination. The initial two data freezes used data from all study principal investigators related to diagnostic status (lifetime and current, if available), allowing the principal investigators' clinical judgment and methodological discretion to guide coding in their individual cohorts.

Given that PTSD is highly comorbid with other disorders (Pietrzak, Goldstein, Southwick, & Grant, 2011; Young, Lareau, & Pierre, 2014) and has overlapping genetic risk (Chantarujikapong et al., 2001; Sartor et al., 2011, 2012), harmonizing existing data on comorbidities assessed across cohorts within the PGC-PTSD, as well as collaborating across PGC working groups, is crucial. Examining comorbidities across working groups will provide more clarity to our understanding of the genetic underpinnings of PTSD and its comorbidities. For example, polygenic risk scores (further described in the Future Directions section) can be used to examine overlap across disorders.

Another unique challenge faced by the PGC-PTSD working group is how to best account for the diversity in the cohorts. The PGC-PTSD consists of a more diverse sample compared with the other PGC groups, which is a strength of the group. Other GWAS in the PGC to date have relied heavily on studies of individuals of European ancestry (e.g., Major Depressive

Disorder Working Group of the Psychiatric, GWAS Consortium, et al., 2013; Schizophrenia Working Group of the Psychiatric Genomics, 2014) while the PGC-PTSD cases are ~70% European with about 9000 samples of other ancestries. Although the PGC-PTSD contains more diverse samples, racial/ethnic groups are typically analyzed separately, and are then meta-analyzed. As a result, some racial/ethnic groups may still be too low in power to detect effects at present. While diverse samples lay the foundation for detecting genetic associations with PTSD that are relevant to a broader set of individuals (e.g., Nievergelt et al., 2015), an ongoing challenge is that extant methods are not always equipped to appropriately analyze diverse samples, and cannot accommodate analyses of recently admixed populations. Issues regarding sample diversity highlight the need to develop appropriate analytic methods and create a framework that may be used not only in the PGC-PTSD working group, but in other PGC groups as well. More broadly, including a greater diversity of subjects is critical in order to more accurately unpack the genomic architecture of PTSD. For example, controlling for variation in trauma characteristics, the conditional risk of PTSD is 22% higher in African-Americans and 33% lower in Asian/Hawaiian/Pacific Islanders compared with European-Americans (Roberts, Gilman, Breslau, Breslau, & Koenen, 2011). The potential genetic contributions to this racial/ethnic difference in risk clearly remain to be elucidated.

7 Development of the PGC-PTSD working groups

As interest and involvement in the PGC-PTSD working group has expanded, a number of more specific working groups (optimizing expertise of specific PGC-PTSD members) have been developed. These working groups innovatively combine GWAS data with biomarker and other -omics data to answer alternative novel questions regarding the genetic underpinnings of PTSD. To date, there are seven established working groups, five of which will be described in detail here: Gene Expression, Epigenome-Wide Association Studies (EWAS), Neuro-Imaging, Physical Health, and Copy Number Variant (CNV). The Psychophysiology and Systems Informatics working groups are newly formed, and have not set an agenda. These established working groups are at different stages of addressing their specific aims, and as the PGC-PTSD grows, more working groups may be developed.

7.1 Gene expression working group

One of the great strengths of the parent PGC-PTSD working group is that it provides a means by which to integrate multiple different forms of genomic data. For example, in addition to providing genome-wide genotyped data, many cohorts included in the PGC-PTSD working group also collected RNA from peripheral blood or saliva. The Gene Expression working group, founded by Michael A. Hauser, PhD and Allison Ashley-Koch, PhD, coordinates the analysis of these samples using either microarrays or RNA-sequencing analysis (RNA-seq). A total of more than 2000 cases and 2000 controls are currently being analyzed for gene expression signatures associated with PTSD. Chromatin conformation and gene regulatory elements data (i.e., PsychEncode data) generated from brain tissue of PTSD cases and controls will also be analyzed in conjunction with gene expression data from peripheral blood or saliva samples.

Initial planned analyses of the Gene Expression working group include combining microarray data with RNA-seq data. Many methods for analysis of blood expression data adjust for varying cell composition in each sample, yet this approach may not be fully appropriate, as PTSD is expected to induce changes in the immune system. For this reason, the PGC-PTSD Gene Expression working group will use approaches that maintain the key elements of this cell composition, in order to obtain the most detailed view possible of the gene expression changes associated with PTSD. In addition to samples having both GWAS and gene expression data, many have DNA methylation data. This allows multiple cross-platform analyses to be performed.

7.2 EWAS working group

The EWAS working group aims to identify the epigenetic factors associated with change in response to traumatic experiences and PTSD. Epigenetic factors consist of biological mechanisms (e.g., methylation, histone modifications) regulating gene expression without altering the DNA sequence, through environmental influences. Both trauma exposure and PTSD have been shown to alter epigenetic patterns in both animal and human studies (McGowan et al., 2009; Roth, Lubin, Funk, & Sweatt, 2009; Sipahi et al., 2014). Co-founded by Alicia K. Smith, PhD, Monica Uddin, PhD, and Caroline M. Nievergelt, PhD, the EWAS working group has two main goals, which are to (1) develop a common QC protocol and joint analysis plan for epigenome-wide DNA methylation data to facilitate collaboration among working group members (Ratanatharathorn et al., 2017), and (2) identify and evaluate trauma-associated DNA methylation patterns as peripheral biomarkers of PTSD. In pursuit of these goals, the working group has established a shared collection of modular computer programs that are used by participating members to analyze separate cohorts' data. In addition, the working group provides a common space for the storage and sharing of cohort-specific results. EWAS data from almost 1800 cases of current PTSD and trauma-exposed controls have been meta-analyzed to date.

As evidence builds in support of potential biomarkers, the workgroup will be able to draw from a pool of almost 15,000 samples for targeted replication and validation studies. Thus, the EWAS working group is poised to collectively address one of the main challenges of psychiatric genomics, namely the need for large, harmonized samples to adequately power genome-wide analyses. As the EWAS working group moves forward, their goal is to integrate genetic and gene expression data from other workgroups to facilitate a more complete understanding of the molecular architecture and biological underpinnings of PTSD.

7.3 Neuro-imaging working group

The Neuro-Imaging working group focuses on investigating the elevated risk of psychopathology through intermediate phenotypes (or endophenotypes) rather than clinical diagnoses, as endophenotypes may be more likely to aid in understanding mechanisms (Flint, Timpson, & Munafò, 2014). Co-founders Rajendra Morey, MD, Mark Logue, PhD, and Emily Dennis, PhD have built an international collaboration of investigators (Logue et al., 2015) who have linked data from the PGC and the Enhancing Neuroimaging Genetics through Meta-Analysis (ENIGMA) consortium. More than 45 participating sites have formed the ENIGMA-PTSD Consortium Workgroup, and have committed more than 6000 samples; 2000 of which have been aggregated and analyzed. Results from these efforts have shown significantly smaller hippocampal volume, and suggestive evidence of smaller amygdala volume in individuals presenting with PTSD in comparison with trauma-exposed controls (Logue et al., 2017).

The working group is planning to conduct an agnostic gene-by-environment genome-wide interaction study (GEWIS) of relevant structural brain phenotypes where childhood trauma is a major risk factor, with the long-term goal of identifying genetic modulators of brain structure that help early prediction and treatment for a range of psychiatric disorders. Additionally, the most interesting and highly associated loci from the GWAS of neuroimaging phenotypes will be selected for further follow-up (e.g., deep genomic sequencing, subsequent functional analysis) to identify potential causative variants for the associations with imaging phenotypes.

7.4 Physical health working group

The Physical Health working group aims to examine how genetic and epigenetic factors play a role in the comorbidity of PTSD and physical health conditions. Growing research links PTSD to a range of cardiometabolic conditions and diseases of aging, including metabolic syndrome, cardiovascular disease, type 2 diabetes, and dementia (Koenen et al., 2017; Roberts et al., 2015; Sumner et al., 2015; Wolf et al., 2016; Yaffe et al., 2010). However, the extent to which genetic and epigenetic factors are involved in these associations remains unclear. Co-founded by Jennifer A. Sumner, PhD and Erika J. Wolf, PhD, the PGC-PTSD Physical Health working group was developed to address these questions by harnessing the power of collaborative, consortium-based efforts. The goals of this working group include examining the extent to which overlapping genetic factors predispose individuals to both PTSD and physical health conditions, and investigating how genetic factors may contribute to increased risk for (or resilience to) developing these physical health outcomes in individuals with PTSD. In pursuit of these goals, the working group is creating a large dataset of studies with information on PTSD, genetic and epigenetic data, health-related biomarkers (e.g., inflammatory markers, lipids) and intermediates (e.g., body mass index (BMI), blood pressure), cognitive function, and chronic health conditions (e.g., cardiovascular disease, type 2 diabetes, metabolic syndrome).

To date, several cohorts within the PGC-PTSD have genome-wide genotype data and data on physical health-related variables, with a current total of $N = 19,183$ (7115 cases and 12,068 controls). Findings resulting from the working group's initial collaboration suggested that, in European ancestry individuals, there is potential for shared genetic contributions to PTSD and several cardiometabolic traits, including coronary artery disease, BMI, and fasting insulin (Sumner et al., 2017). These findings provide initial support for the idea that the growing number of associations of PTSD and cardiometabolic diseases reported in the literature may be due, in part, to shared genetic contributions, and they will serve as a foundation for the working group's future endeavors.

7.5 CNV working group

The PGC-PTSD CNV working group, co-led by Jonathan Sebat, PhD and Caroline M. Nievergelt, PhD, is focused on identifying rare variants by structural repeats or deletions in stretches of DNA, which have demonstrated larger effect sizes for complex traits and psychiatric disorders as compared with common variants (Cooper, Zerr, Kidd, Eichler, & Nickerson, 2008; Girirajan, Campbell, & Eichler, 2011). The CNV working group is collaborating with the main PGC CNV working group (CNV Working Group of the Psychiatric Genomics Consortium & Schizophrenia Working Groups of the Psychiatric Genomics Consortium, 2016) to establish an important role for rare genetic variants in psychiatric disorders, including schizophrenia and autism (Malhotra & Sebat, 2012). The working group uses the extensive pipelines developed by PGC-CNV to call CNVs from initial intensity files and to perform extensive quality control. Rare CNVs, typically present in less than 1% of subjects, are then extracted in the combined sample. Using a mega-analytical approach across PTSD studies, CNV data analysis includes CNV burden, CNV breakpoint, and gene-based association analyses.

The CNV working group faces unique challenges not faced by other working groups. The characterization of this aspect of the genetic etiology requires not only very large samples, but access to original intensity files. Currently, about 25% of the GWAS data is not shared at the individual level, and additional studies may have difficulties gaining access to the large intensity files from which genotype calls were generated. While GWAS designs are mainly aimed at identifying common variant genetic architecture, characterization of rare CNVs with intermediate to large effects will add a significant contribution to the complete understanding of the genetic etiology of PTSD.

8 Future directions of the PGC-PTSD and PTSD genomics as a whole

The next steps for the PGC-PTSD working group as a whole are to: (1) finalize the GWAS from Freeze 2 data and establish effective strategies to optimize findings given the diverse ancestry of the sample; (2) continue to expand the sample with additional cohorts; (3) conduct GEWIS analyses to investigate environmental influences on the genetic architecture of PTSD; and (4) enhance data harmonization with Cohen Veteran's Bioscience efforts in order to create a deeper phenotype for PTSD. The PGC-PTSD working group continues to rapidly expand; the worldwide collaboration creates a platform for experts to come together with innovative efforts to understand and address the complexities of PTSD. As the collaborative conversations continue, secondary analyses will be performed, and additional working groups will continue to be created to further expand our understanding of the genetic architecture of PTSD. Planned secondary analyses could include, but are not limited to: genome-wide analyses of PTSD symptom severity and cluster severity; examination of more homogeneous trauma populations (e.g., interpersonal trauma, childhood trauma, combat trauma); planned GxE analyses of variants identified by GWAS with critical trauma exposure variables (e.g., childhood trauma), and so forth. In addition, the PGC-PTSD aims to leverage the deep phenotyping available from the current cohorts to dissect PTSD in the broader genetic and environmental context, including gene-environment interplay analyses (e.g., via GEWIS), and GWAS of intermediate phenotypes.

More nuanced and novel questions regarding PTSD can be answered with continued growth in GWAS datasets and ongoing development of novel statistical techniques that can be applied to such data. Techniques such as univariate and bivariate genome-wide complex trait analysis (GCTA) use genome-wide data to determine SNP-level heritability and shared heritability (i.e., informing upon potential shared genetic risk that may explain comorbidities). Cross-trait LD score regression also estimates genetic correlation and can be used to examine comorbidities, using summary statistics from GWAS. Polygenic risk scores (PRS) can help determine whether aggregate genetic risk can predict PTSD, and can also be used in independent replications and across phenotypes (i.e., whether polygenic risk for PTSD is associated with increased risk for another phenotype, such as MDD, providing another avenue of exploring comorbidities). Integrating GWAS analyses with these new thoughtfully organized statistical methods will facilitate the interpretation of the vast amount of existing genomic information, helping to untangle the genetic architecture of PTSD. Moving beyond GWAS, bioinformatics approaches (e.g., gene enrichment and gene pathway analyses), as well as follow-up of genetic findings in laboratory based studies, can help us better understand the mechanisms of risk, and will be critical to advancing our understanding of the etiology of PTSD. Within the PGC-PTSD, proposed work involves approaches such as weighted co-expression network analysis (WGCNA) integrating across methylation, gene expression and SNP data, Mendelian randomization analysis, and machine learning methods.

Though PTSD genomics is in its infancy relative to other psychiatric disorders (e.g., schizophrenia and major depressive disorder), the strides made by the PGC-PTSD working group have resulted in comparable sample sizes, and the working group is quick to make pace with these other disorders. Embodying the proposed goals addressed in the recent overall PGC agenda (Sullivan et al., 2017), the PGC-PTSD working group aims to identify the underlying biology of PTSD, with the hopes of informing the development of innovative therapeutic approaches. As discussed throughout this chapter, PTSD research progresses as the field of genomics continues to advance. The PGC-PTSD working group strives to be at the forefront of that progression to move PTSD research forward.

References

Afifi, T. O., Asmundson, G. J., Taylor, S., & Jang, K. L. (2010). The role of genes and environment on trauma exposure and posttraumatic stress disorder symptoms: A review of twin studies. *Clinical Psychology Review, 30*(1), 101–112. https://doi.org/10.1016/j.cpr.2009.10.002.

Almli, L. M., Stevens, J. S., Smith, A. K., Kilaru, V., Meng, Q., Flory, J., ... Ressler, K. J. (2015). A genome-wide identified risk variant for PTSD is a methylation quantitative trait locus and confers decreased cortical activation to fearful faces. *American Journal of Medical Genetics. Part B, Neuropsychiatric Genetics, 168B*(5), 327–336. https://doi.org/10.1002/ajmg.b.32315.

American Psychiatric Association (2013). *Diagnostic and statistical manual of mental disorders (DSM-5®).* American Psychiatric Pub.

Amstadter, A. B., Sumner, J. A., Acierno, R., Ruggiero, K. J., Koenen, K. C., Kilpatrick, D. G., ... Gelernter, J. (2013). Support for association of RORA variant and post traumatic stress symptoms in a population-based study of hurricane exposed adults. *Molecular Psychiatry, 18*(11), 1148–1149. https://doi.org/10.1038/mp.2012.189.

Ashley-Koch, A. E., Garrett, M. E., Gibson, J., Liu, Y., Dennis, M. F., Kimbrel, N. A., ... Hauser, M. A. (2015). Genome-wide association study of posttraumatic stress disorder in a cohort of Iraq-Afghanistan era veterans. *Journal of Affective Disorders, 184,* 225–234. https://doi.org/10.1016/j.jad.2015.03.049.

Benjet, C., Bromet, E., Karam, E. G., Kessler, R. C., McLaughlin, K. A., Ruscio, A. M., ... Koenen, K. C. (2016). The epidemiology of traumatic event exposure worldwide: Results from the World Mental Health Survey Consortium. *Psychological Medicine, 46*(2), 327–343. https://doi.org/10.1017/S0033291715001981.

Bennett, S. N., Caporaso, N., Fitzpatrick, A. L., Agrawal, A., Barnes, K., Boyd, H. A., ... GENEVA Consortium (2011). Phenotype harmonization and cross-study collaboration in GWAS consortia: The GENEVA experience. *Genetic Epidemiology, 35*(3), 159–173. https://doi.org/10.1002/gepi.20564.

Binder, E. B., Bradley, R. G., Liu, W., Epstein, M. P., Deveau, T. C., Mercer, K. B., ... Ressler, K. J. (2008). Association of FKBP5 polymorphisms and childhood abuse with risk of posttraumatic stress disorder symptoms in adults. *JAMA, 299*(11), 1291–1305. https://doi.org/10.1001/jama.299.11.1291.

Boscarino, J. A., Forsberg, C. W., & Goldberg, J. (2010). A twin study of the association between PTSD symptoms and rheumatoid arthritis. *Psychosomatic Medicine, 72*(5), 481–486. https://doi.org/10.1097/PSY.0b013e3181d9a80c.

Breslau, N., Chilcoat, H. D., Kessler, R. C., & Davis, G. C. (1999). Previous exposure to trauma and PTSD effects of subsequent trauma: Results from the Detroit Area Survey of Trauma. *The American Journal of Psychiatry, 156*(6), 902–907. https://doi.org/10.1176/ajp.156.6.902.

Breslau, N., Kessler, R. C., Chilcoat, H. D., Schultz, L. R., Davis, G. C., & Andreski, P. (1998). Trauma and posttraumatic stress disorder in the community: The 1996 Detroit Area Survey of Trauma. *Archives of General Psychiatry, 55*(7), 626–632.

Brewin, C. R., Andrews, B., & Valentine, J. D. (2000). Meta-analysis of risk factors for posttraumatic stress disorder in trauma-exposed adults. *Journal of Consulting and Clinical Psychology, 68*(5), 748–766.

Bruenig, D., Lurie, J., Morris, C. P., Harvey, W., Lawford, B., Young, R. M., & Voisey, J. (2016). A case-control study and meta-analysis reveal BDNF Val66Met is a possible risk factor for PTSD. *Neural Plasticity, 2016,* 6979435. https://doi.org/10.1155/2016/6979435.

Chantarujikapong, S. I., Scherrer, J. F., Xian, H., Eisen, S. A., Lyons, M. J., Goldberg, J., ... True, W. R. (2001). A twin study of generalized anxiety disorder symptoms, panic disorder symptoms and post-traumatic stress disorder in men. *Psychiatry Research, 103*(2–3), 133–145.

CNV Working Group of the Psychiatric Genomics Consortium, & Schizophrenia Working Groups of the Psychiatric Genomics Consortium. (2016). Contribution of copy number variants to schizophrenia from a genome-wide study of 41,321 subjects. *Nature Genetics, 49*(1), 27–35. https://doi.org/10.1038/ng.3725.

Cooper, G. M., Zerr, T., Kidd, J. M., Eichler, E. E., & Nickerson, D. A. (2008). Systematic assessment of copy number variant detection via genome-wide SNP genotyping. *Nature Genetics, 40*(10), 1199–1203. https://doi.org/10.1038/ng.236.

Drevo, S., Newman, E., Miller, K. E., Davis, J. L., Craig, C., Sheaff, R. J., ... Bell, K. (2016). The role of social environment and gene interactions on development of posttraumatic stress disorder. *Journal of Articles in Support of the Null Hypothesis, 12*(2), 21–30.

Duncan, L. E., Ratanatharathorn, A., Aiello, A. E., Almli, L. M., Amstadter, A. B., Ashley-Koch, A. E., ... Koenen, K. C. (2017). Largest GWAS of PTSD (N = 20 070) yields genetic overlap with schizophrenia and sex differences in heritability. *Molecular Psychiatry.* https://doi.org/10.1038/mp.2017.77.

Eaves, L. J. (2006). Genotype x Environment interaction in psychopathology: Fact or artifact? *Twin Research and Human Genetics, 9*(1), 1–8. https://doi.org/10.1375/183242706776403073.

Flint, J., Timpson, N., & Munafò, M. (2014). Assessing the utility of intermediate phenotypes for genetic mapping of psychiatric disease. *Trends in Neurosciences, 37*(12), 733–741.

Girirajan, S., Campbell, C. D., & Eichler, E. E. (2011). Human copy number variation and complex genetic disease. *Annual Review of Genetics, 45,* 203–226. https://doi.org/10.1146/annurev-genet-102209-163544.

Gratten, J., Wray, N. R., Keller, M. C., & Visscher, P. M. (2014). Large-scale genomics unveils the genetic architecture of psychiatric disorders. *Nature Neuroscience, 17*(6), 782–790. https://doi.org/10.1038/nn.3708.

Gressier, F., Calati, R., Balestri, M., Marsano, A., Alberti, S., Antypa, N., & Serretti, A. (2013). The 5-HTTLPR polymorphism and posttraumatic stress disorder: A meta-analysis. *Journal of Traumatic Stress, 26*(6), 645–653. https://doi.org/10.1002/jts.21855.

Guffanti, G., Galea, S., Yan, L., Roberts, A. L., Solovieff, N., Aiello, A. E., … Koenen, K. C. (2013). Genome-wide association study implicates a novel RNA gene, the lincRNA AC068718.1, as a risk factor for post-traumatic stress disorder in women. *Psychoneuroendocrinology, 38*(12), 3029–3038. https://doi.org/10.1016/j.psyneuen.2013.08.014.

Irish, L., Ostrowski, S. A., Fallon, W., Spoonster, E., Dulmen, M., Sledjeski, E. M., & Delahanty, D. L. (2008). Trauma history characteristics and subsequent PTSD symptoms in motor vehicle accident victims. *Journal of Traumatic Stress, 21*(4), 377–384. https://doi.org/10.1002/jts.20346.

Kessler, R. C. (2000). Posttraumatic stress disorder: The burden to the individual and to society. *The Journal of Clinical Psychiatry, 61*(Suppl. 5), 4–12 discussion 13–14.

Kessler, R. C., Sonnega, A., Bromet, E., Hughes, M., & Nelson, C. B. (1995). Posttraumatic stress disorder in the National Comorbidity Survey. *Archives of General Psychiatry, 52*(12), 1048–1060.

Kibler, J. L. (2009). Posttraumatic stress and cardiovascular disease risk. *Journal of Trauma & Dissociation, 10*(2), 135–150. https://doi.org/10.1080/15299730802624577.

Kilaru, V., Iyer, S. V., Almli, L. M., Stevens, J. S., Lori, A., Jovanovic, T., … Ressler, K. J. (2016). Genome-wide gene-based analysis suggests an association between Neuroligin 1 (NLGN1) and post-traumatic stress disorder. *Translational Psychiatry, 6*, e820. https://doi.org/10.1038/tp.2016.69.

Kimbrel, N. A., Hauser, M. A., Garrett, M., Ashley-Koch, A., Liu, Y., Dennis, M. F., … Beckham, J. C. (2015). Effect of the APOE epsilon4 allele and combat exposure on PTSD among Iraq/Afghanistan-era veterans. *Depression and Anxiety, 32*(5), 307–315. https://doi.org/10.1002/da.22348.

Koenen, K. C., Aiello, A. E., Bakshis, E., Amstadter, A. B., Ruggiero, K. J., Acierno, R., … Galea, S. (2009). Modification of the association between serotonin transporter genotype and risk of posttraumatic stress disorder in adults by county-level social environment. *American Journal of Epidemiology, 169*(6), 704–711. https://doi.org/10.1093/aje/kwn397.

Koenen, K. C., Sumner, J. A., Gilsanz, P., Glymour, M. M., Ratanatharathorn, A., Rimm, E. B., … Kubzansky, L. D. (2017). Post-traumatic stress disorder and cardiometabolic disease: Improving causal inference to inform practice. *Psychological Medicine, 47*(2), 209–225. https://doi.org/10.1017/S0033291716002294.

Levinson, D. F., Mostafavi, S., Milaneschi, Y., Rivera, M., Ripke, S., Wray, N. R., & Sullivan, P. F. (2014). Genetic studies of major depressive disorder: Why are there no genome-wide association study findings and what can we do about it? *Biological Psychiatry, 76*(7), 510–512. https://doi.org/10.1016/j.biopsych.2014.07.029.

Li, L., Bao, Y., He, S., Wang, G., Guan, Y., Ma, D., … Yang, J. (2016). The association between genetic variants in the dopaminergic system and post-traumatic stress disorder: A meta-analysis. *Medicine (Baltimore), 95*(11), e3074. https://doi.org/10.1097/MD.0000000000003074.

Lind, M. J., Marraccini, M. E., Sheerin, C. M., Bountress, K., Bacanu, S. A., Amstadter, A. B., & Nugent, N. R. (2017). Association of posttraumatic stress disorder with rs2267735 in the ADCYAP1R1 gene: A meta-analysis. *Journal of Traumatic Stress, 30*(4), 389–398. https://doi.org/10.1002/jts.22211.

Logue, M. W., Amstadter, A. B., Baker, D. G., Duncan, L., Koenen, K. C., Liberzon, I., … Uddin, M. (2015). The Psychiatric Genomics Consortium posttraumatic stress disorder workgroup: Posttraumatic stress disorder enters the age of large-scale genomic collaboration. *Neuropsychopharmacology, 40*(10), 2287–2297. https://doi.org/10.1038/npp.2015.118.

Logue, M. W., Baldwin, C., Guffanti, G., Melista, E., Wolf, E. J., Reardon, A. F., … Miller, M. W. (2013). A genome-wide association study of posttraumatic stress disorder identifies the retinoid-related orphan receptor alpha (RORA) gene as a significant risk locus. *Molecular Psychiatry, 18*(8), 937–942. https://doi.org/10.1038/mp.2012.113.

Logue, M. W., van Rooij, S. J., Dennis, E. L., Davis, S. L., Hayes, J. P., Stevens, J. S., … Koch, S. B. (2017). Smaller hippocampal volume in posttraumatic stress disorder: A multi-site ENIGMA-PGC study. *Biological Psychiatry, 83*(1), 244–253.

Lyons, M. J., Genderson, M., Grant, M. D., Logue, M., Zink, T., McKenzie, R., … Kremen, W. S. (2013). Gene-environment interaction of ApoE genotype and combat exposure on PTSD. *American Journal of Medical Genetics. Part B, Neuropsychiatric Genetics, 162B*(7), 762–769. https://doi.org/10.1002/ajmg.b.32154.

Lyons, M. J., Goldberg, J., Eisen, S. A., True, W., Tsuang, M. T., Meyer, J. M., & Henderson, W. G. (1993). Do genes influence exposure to trauma? A twin study of combat. *American Journal of Medical Genetics, 48*(1), 22–27. https://doi.org/10.1002/ajmg.1320480107.

Major Depressive Disorder Working Group of the Psychiatric, GWAS Consortium, Ripke, S., Wray, N. R., Lewis, C. M., Hamilton, S. P., Weissman, M. M., … Sullivan, P. F. (2013). A mega-analysis of genome-wide association studies for major depressive disorder. *Molecular Psychiatry, 18*(4), 497–511.

Malhotra, D., & Sebat, J. (2012). CNVs: Harbingers of a rare variant revolution in psychiatric genetics. *Cell, 148*(6), 1223–1241. https://doi.org/10.1016/j.cell.2012.02.039.

Manuck, S. B., & McCaffery, J. M. (2014). Gene-environment interaction. *Annual Review of Psychology, 65*, 41–70. https://doi.org/10.1146/annurev-psych-010213-115100.

McGowan, P. O., Sasaki, A., D'Alessio, A. C., Dymov, S., Labonte, B., Szyf, M., … Meaney, M. J. (2009). Epigenetic regulation of the glucocorticoid receptor in human brain associates with childhood abuse. *Nature Neuroscience, 12*(3), 342–348. https://doi.org/10.1038/nn.2270.

Miller, M. W., Wolf, E. J., Logue, M. W., & Baldwin, C. T. (2013). The retinoid-related orphan receptor alpha (RORA) gene and fear-related psychopathology. *Journal of Affective Disorders, 151*(2), 702–708. https://doi.org/10.1016/j.jad.2013.07.022.

Navarro-Mateu, F., Escamez, T., Koenen, K. C., Alonso, J., & Sanchez-Meca, J. (2013). Meta-analyses of the 5-HTTLPR polymorphisms and posttraumatic stress disorder. *PLoS One, 8*(6), e66227. https://doi.org/10.1371/journal.pone.0066227.

Nievergelt, C. M., Maihofer, A. X., Mustapic, M., Yurgil, K. A., Schork, N. J., Miller, M. W., … Baker, D. G. (2015). Genomic predictors of combat stress vulnerability and resilience in U.S. Marines: A genome-wide association study across multiple ancestries implicates PRTFDC1 as a potential PTSD gene. *Psychoneuroendocrinology, 51*, 459–471. https://doi.org/10.1016/j.psyneuen.2014.10.017.

Pietrzak, R. H., Goldstein, R. B., Southwick, S. M., & Grant, B. F. (2011). Prevalence and Axis I comorbidity of full and partial posttraumatic stress disorder in the United States: Results from Wave 2 of the National Epidemiologic Survey on Alcohol and Related Conditions. *Journal of Anxiety Disorders, 25*(3), 456–465.

Powers, A., Almli, L., Smith, A., Lori, A., Leveille, J., Ressler, K. J., ... Bradley, B. (2016). A genome-wide association study of emotion dysregulation: Evidence for interleukin 2 receptor alpha. *Journal of Psychiatric Research*, *83*, 195–202. https://doi.org/10.1016/j.jpsychires.2016.09.006.

Psychiatric GWAS Consortium-Bipolar Disorder Working Group (2011). Large-scale genome-wide association analysis of bipolar disorder identifies a new susceptibility locus near ODZ4. *Nature Genetics*, *43*(10), 977–983. https://doi.org/10.1038/ng.943.

Ratanatharathorn, A., Boks, M. P., Maihofer, A. X., Aiello, A. E., Amstadter, A. B., Ashley-Koch, A. E., ... Smith, A. K. (2017). Epigenome-wide association of PTSD from heterogeneous cohorts with a common multi-site analysis pipeline. *American Journal of Medical Genetics Part B, Neuropsychiatric Genetics*, *174*(6), 619–630. https://doi.org/10.1002/ajmg.b.32568.

Roberts, A. L., Agnew-Blais, J. C., Spiegelman, D., Kubzansky, L. D., Mason, S. M., Galea, S., ... Koenen, K. C. (2015). Posttraumatic stress disorder and incidence of type 2 diabetes mellitus in a sample of women: A 22-year longitudinal study. *JAMA Psychiatry*, *72*(3), 203–210. https://doi.org/10.1001/jamapsychiatry.2014.2632.

Roberts, A. L., Gilman, S. E., Breslau, J., Breslau, N., & Koenen, K. C. (2011). Race/ethnic differences in exposure to traumatic events, development of post-traumatic stress disorder, and treatment-seeking for post-traumatic stress disorder in the United States. *Psychological Medicine*, *41*(1), 71–83. https://doi.org/10.1017/S0033291710000401.

Roth, T. L., Lubin, F. D., Funk, A. J., & Sweatt, J. D. (2009). Lasting epigenetic influence of early-life adversity on the BDNF gene. *Biological Psychiatry*, *65*(9), 760–769.

Sartor, C. E., Grant, J. D., Lynskey, M. T., McCutcheon, V. V., Waldron, M., Statham, D. J., ... Nelson, E. C. (2012). Common heritable contributions to low-risk trauma, high-risk trauma, posttraumatic stress disorder, and major depression. *Archives of General Psychiatry*, *69*(3), 293–299. https://doi.org/10.1001/archgenpsychiatry.2011.1385.

Sartor, C. E., McCutcheon, V. V., Pommer, N. E., Nelson, E. C., Grant, J. D., Duncan, A. E., ... Heath, A. C. (2011). Common genetic and environmental contributions to post-traumatic stress disorder and alcohol dependence in young women. *Psychological Medicine*, *41*(7), 1497–1505. https://doi.org/10.1017/S0033291710002072.

Schizophrenia Working Group of the Psychiatric Genomics, Consortium. (2014). Biological insights from 108 schizophrenia-associated genetic loci. *Nature*, *511*(7510), 421–427. https://doi.org/10.1038/nature13595.

Sham, P. C., & Purcell, S. M. (2014). Statistical power and significance testing in large-scale genetic studies. *Nature Reviews Genetics*, *15*(5), 335–346. https://doi.org/10.1038/nrg3706.

Sheerin, C. M., Lind, M. J., Bountress, K., Nugent, N. R., & Amstadter, A. B. (2017). The genetics and epigenetics of PTSD: Overview, recent advances, and future directions. *Current Opinion in Psychology*, *14*, 5–11. https://doi.org/10.1016/j.copsyc.2016.09.003.

Sipahi, L., Wildman, D. E., Aiello, A. E., Koenen, K. C., Galea, S., Abbas, A., & Uddin, M. (2014). Longitudinal epigenetic variation of DNA methyltransferase genes is associated with vulnerability to post-traumatic stress disorder. *Psychological Medicine*, *44*(15), 3165–3179. https://doi.org/10.1017/S0033291714000968.

Stein, M. B., Chen, C. Y., Ursano, R. J., Cai, T., Gelernter, J., Heeringa, S. G. ... Army Study to Assess Risk and Resilience in Servicemembers (STARRS) Collaborators. (2016). Genome-wide association studies of posttraumatic stress disorder in 2 cohorts of US Army soldiers. *JAMA Psychiatry*, *73*(7), 695–704. https://doi.org/10.1001/jamapsychiatry.2016.0350.

Stein, M. B., Jang, K. L., & Livesley, W. J. (2002). Heritability of social anxiety-related concerns and personality characteristics: A twin study. *The Journal of Nervous and Mental Disease*, *190*(4), 219–224.

Sullivan, P. F., Agrawal, A., Bulik, C. M., Andreassen, O. A., Borglum, A. D., Breen, G., ... Psychiatric Genomics Consortium (2017). Psychiatric genomics: An update and an agenda. *The American Journal of Psychiatry*. https://doi.org/10.1176/appi.ajp.2017.17030283.

Sumner, J. A., Duncan, L. E., Wolf, E. J., Amstadter, A. B., Baker, D. G., Beckham, J. C., ... Vermetten, E. (2017). Letter to the Editor: Posttraumatic stress disorder has genetic overlap with cardiometabolic traits. *Psychological Medicine*, *47*(11), 2036–2039. https://doi.org/10.1017/S0033291717000733.

Sumner, J. A., Kubzansky, L. D., Elkind, M. S., Roberts, A. L., Agnew-Blais, J., Chen, Q., ... Koenen, K. C. (2015). Trauma exposure and posttraumatic stress disorder symptoms predict onset of cardiovascular events in women. *Circulation*, *132*(4), 251–259. https://doi.org/10.1161/CIRCULATIONAHA.114.014492.

Tabor, H. K., Risch, N. J., & Myers, R. M. (2002). Candidate-gene approaches for studying complex genetic traits: Practical considerations. *Nature Reviews Genetics*, *3*(5), 391–397.

Thomas, D. (2010). Gene–environment-wide association studies: Emerging approaches. *Nature Reviews Genetics*, *11*(4), 259–272.

True, W. R., Rice, J., Eisen, S. A., Heath, A. C., Goldberg, J., Lyons, M. J., & Nowak, J. (1993). A twin study of genetic and environmental contributions to liability for posttraumatic stress symptoms. *Archives of General Psychiatry*, *50*(4), 257–264.

Wang, T. (2015). Does BDNF Val66Met polymorphism confer risk for posttraumatic stress disorder? *Neuropsychobiology*, *71*(3), 149–153. https://doi.org/10.1159/000381352.

Wolf, E. J., Bovin, M. J., Green, J. D., Mitchell, K. S., Stoop, T. B., Barretto, K. M., ... Marx, B. P. (2016). Longitudinal associations between posttraumatic stress disorder and metabolic syndrome severity. *Psychological Medicine*, *46*(10), 2215–2226. https://doi.org/10.1017/S0033291716000817.

Wolf, E. J., Rasmusson, A. M., Mitchell, K. S., Logue, M. W., Baldwin, C. T., & Miller, M. W. (2014). A genome-wide association study of clinical symptoms of dissociation in a trauma-exposed sample. *Depression and Anxiety*, *31*(4), 352–360. https://doi.org/10.1002/da.22260.

Xie, P., Kranzler, H. R., Poling, J., Stein, M. B., Anton, R. F., Farrer, L. A., & Gelernter, J. (2010). Interaction of FKBP5 with childhood adversity on risk for post-traumatic stress disorder. *Neuropsychopharmacology*, *35*(8), 1684–1692. https://doi.org/10.1038/npp.2010.37.

Xie, P., Kranzler, H. R., Yang, C., Zhao, H., Farrer, L. A., & Gelernter, J. (2013). Genome-wide association study identifies new susceptibility loci for posttraumatic stress disorder. *Biological Psychiatry*, *74*(9), 656–663. https://doi.org/10.1016/j.biopsych.2013.04.013.

Yaffe, K., Vittinghoff, E., Lindquist, K., Barnes, D., Covinsky, K. E., Neylan, T., … Marmar, C. (2010). Posttraumatic stress disorder and risk of dementia among US veterans. *Archives of General Psychiatry*, *67*(6), 608–613. https://doi.org/10.1001/archgenpsychiatry.2010.61.

Young, G., Lareau, C., & Pierre, B. (2014). One quintillion ways to have PTSD comorbidity: Recommendations for the disordered DSM-5. *Psychological Injury and Law*, *7*(1), 61–74.

Chapter 24

Genomic contributions to anxiety disorders

Shareefa Dalvie[a,b], Nastassja Koen[a,b] and Dan J. Stein[a,b]

[a]Department of Psychiatry and Mental Health, University of Cape Town, Cape Town, South Africa; [b]South African Medical Research Council (SAMRC) Unit on Risk & Resilience in Mental Disorders, Cape Town, South Africa

1   Introduction

Anxiety disorders (ADs)—characterized by anticipatory anxiety ("anticipation of future threat"), an excess of fear, and related behavioral disturbances (APA, 2013)—are the most commonly-occurring class of mental disorders worldwide, with a lifetime prevalence of 28.8% (Kessler et al., 2005), and may be associated with significantly impaired quality of life (Olatunji, Cisler, & Tolin, 2007). The Diagnostic and Statistical Manual, 5th edition (DSM-5) (APA, 2013) classifies generalized anxiety disorder (GAD), panic disorder (PD), social anxiety disorder (social phobia), and specific phobia as among the key ADs; and nondiagnostic constructs (e.g., anxiety sensitivity, behavioral inhibition) are also relevant when considering a more dimensional approach to these disorders. To date, a significant body of work has investigated genetic and genomic contributions to the etiology of ADs, informed by strong published animal models (e.g., Campos, Fogaça, Aguiar, & Guimarães, 2013; Erhardt & Spoormaker, 2013; Finn, Rutledge-Gorman, & Crabbe, 2003; Jacobson & Cryan, 2010; Lezak, Missig, & Carlezon, 2017; Park et al., 2011). This chapter aims to provide an up-to-date overview of these contributions. We outline evidence from early family and twin studies; basic molecular genetic studies (linkage studies and candidate gene association studies); more comprehensive genome-wide association studies (GWAS); and recent novel approaches including post-GWAS analyses, next-generation sequencing, and systems biology. We also highlight potential future directions for the field, including pharmacogenomics, consortium analyses, and improved diversity of included populations.

2   Family and twin studies

A number of family and twin studies have sought to establish whether anxiety disorders are familial, and to provide estimates of heritability. Metaanalytic evidence of the epidemiology of key anxiety disorders has reported summary odds ratios (ORs) across family studies for illness in first-degree relatives of affected probands, ranging from 4.1 (95% CI: 2.7; 6.1) for four studies of phobias—including specific phobia, social anxiety disorder (SAD), generalized SAD, and agoraphobia; to 6.1 (95% CI: 2.5; 14.9) for two studies of generalized anxiety disorder (GAD). A number of controlled family studies have provided supporting evidence of familial aggregation in panic disorder (PD) (Schumacher et al., 2011); and the summary OR across five studies of PD meeting the metaanalytic inclusion criteria was 5.0 (95% CI: 3.0; 8.2) (Hettema, Neale, & Kendler, 2001; Shimada-Sugimoto, Otowa, & Hettema, 2015).

Metaanalysis of twin studies yielded moderate familial aggregation, with a heritability estimate of 0.48 (95% CI: 0.41; 0.54) for PD (family study data combined with the twin data), with individual (nonshared) environment accounting for the remaining variance in risk (Hettema et al., 2001; Shimada-Sugimoto et al., 2015). Of relevance to the genomics of PD, anxiety sensitivity—the fear of anxiety-related sensations, based on the belief that these sensations are harmful (Reiss & McNally, 1985; Taylor, Jang, Stewart, & Stein, 2008)—has also been found to have a heritable component (Stein, Jang & Livesley, 1999).

Overall, genomic contributions to GAD are considered less substantive than for other anxiety disorders (e.g., Martin, Ressler, Binder, & Nemeroff, 2009)—a heritability estimate of 0.32 (95% CI: 0.24; 0.39), yielded from meta-analysis of twin studies, has been reported for this disorder (Hettema et al., 2001; Shimada-Sugimoto et al., 2015); and one twin study for GAD estimated a heritability of 0.42 (Davies et al., 2015). Although GAD was not included in a recent large

Personalized Psychiatry. https://doi.org/10.1016/B978-0-12-813176-3.00024-9

cross-disorder study from the Psychiatric Genomics Consortium (PGC; Pettersson et al., 2019), given that the internalizing disorders of GAD and major depression have considerable genetic overlap (Amstadter, Maes, Sheerin, Myers, & Kendler, 2016), it is notable that the heritability estimate from sibling data for depression was 0.31. Common (shared) familial environment was found to play a minor role in liability for GAD in women only, with the remaining variance in men and women attributable to individual environment.

SAD and specific phobias were omitted from the metaanalysis by Hettema et al. (2001) due to the relative dearth of large-scale twin studies meeting the authors' inclusion criteria. In more recent metaanalyses across large-scale European cohorts, however, genetic effects have been reported to account for 27% of individual variance, with nonshared environment accounting for most of the remaining variance (Scaini, Belotti, & Ogliari, 2014).

3 Linkage studies

Linkage analysis investigates chromosomal locations to determine whether these regions contain disease genes for a phenotype of interest, usually using a family-based design (Pulst, 1999). Several such studies have been conducted for ADs and these have found a number of chromosomal regions significantly linked to these disorders, including 4q31-q34 (Kaabi et al., 2006), 12q24 (Gragnoli, 2014; Hodgson et al., 2016), 15q (Fyer et al., 2006), and 16q12.1 (Gelernter, Page, Stein, & Woods, 2004). A metaanalysis of genome-wide linkage scans for ADs (two for PD and one for a multivariate anxiety phenotype) identified nominally significant regions on chromosomes 1, 5, 15, and 16 (Webb et al., 2012). Although linkage studies have been shown to have utility in identifying loci with small effect sizes, many of the identified regions for ADs have not been replicated in independent studies.

4 Candidate gene association studies

A good deal of animal and human work has implicated specific candidate genes/polymorphisms in the development of ADs. Genetic animal studies (i.e., gene-finding studies and/or candidate gene manipulation, for example, knockout or transgenic approaches) have provided strong support for the complex, polygenic nature of ADs (Finn et al., 2003). For example, monoamine targets such as monoamine oxidase A (*MAO-A*) and the norepinephrine transporter (*NET*); as well as GABA-related variants and those in the hypothalamic-pituitary-adrenal (HPA) axis, have emerged in rodent models of anxiety-like behavior (Jacobson & Cryan, 2010). However, replication of these findings has been a challenge for the field. To date, no candidate has been definitively established as a disorder-specific risk gene or variant for ADs. For example, while a number of genes—including catechol-*O*-methyl transferase (*COMT*), serotonin 2A receptor (*HTR$_{2A}$*), and *MAO-A*—have been found to be associated with PD in at least two samples (Martin et al., 2009; Smoller, Gardner-Schuster, & Covino, 2008), one recent metaanalysis of 23 widely studied variants reported no strong associations (Howe et al., 2016). Fewer candidate gene association studies of specific phobia have been undertaken, and although *COMT* and *MAO-A* have also been implicated in these disorders (e.g., Smoller et al., 2008), most studies to date have been limited by small sample sizes. Notably, in their systematic review of candidate gene association studies of ADs (or anxiety symptoms), McGrath, Weill, Robinson, Macrae, and Smoller (2012) included only one study of specific phobia that met the inclusion criterion of having a sample size of at least 200 cases (or nuclear families). These authors reported *COMT*, the serotonin transporter gene (*SLC6A4*), and brain-derived neurotrophic factor (*BDNF*) as the three most commonly studied genes across PD, GAD, SAD, and specific phobia.

While few candidate gene studies have been reported specifically for SAD, the phenotypic complexity of this disorder has generated interest in the genomics of anxiety-related constructs, for example, behavioral inhibition (BI)—a heritable temperamental trait characterized by restraint or fearfulness, and withdrawal from unfamiliar stimuli (Clauss & Blackford, 2012; Kagan, Reznick, & Snidman, 1988; Martin et al., 2009; Muris, van Brakel, Arntz, & Schouten, 2011; Stein & Stein, 2008)—with some evidence for an association between a polymorphism in the serotonin transporter gene (*5HTTLPR*) and BI (Arbelle et al., 2003; Battaglia et al., 2005). No candidate genes associated with GAD in single studies have yet been replicated (Martin et al., 2009).

5 Genome-wide association studies

As described in the previous section, linkage and candidate gene studies have not succeeded in identifying definitive causal genes or variants for anxiety and related phenotypes. More recently, genome-wide association studies (GWASs) have been employed to determine the common variants associated with anxiety in a hypothesis free manner. To date, a total of 14 GWASs have been performed for ADs (Table 1). Findings for each of these studies have been varied, with some not

TABLE 1 Genome-wide association studies for anxiety disorders.

PubMed ID	Study	Journal	Study	Disease/trait	Initial sample size	Main findings	P-value significance threshold
19165232	Otowa et al. (2009)	Journal of Human Genetics	Genome-wide association study of panic disorder in the Japanese population	Panic disorder	200 Japanese ancestry cases, 200 Japanese ancestry controls	Seven SNPs associated located in or adjacent to *PKP1, PLEKHG1, TMEM16B, CALCOCO1, SDK2,* and *CLU* (or *APO-J*)	$<1.0 \times 10^{-6}$
20368705	Erhardt et al. (2011)	Molecular Psychiatry	TMEM132D, a new candidate for anxiety phenotypes: evidence from human and mouse studies	Panic disorder	909 European cases and 915 European controls	Two-SNP (located in *TMEM132D*) haplotype associated	$<1.2e^{-7}$
23149450	Otowa et al. (2012)	Translational Psychiatry	Metaanalysis of genome-wide association studies of anxiety disorders for panic disorder in the Japanese population	Panic disorder	718 Japanese cases, 1717 Japanese controls	No genome-wide significant hits	$<5 \times 10^{-8}$
24047446	Schosser et al. (2013)	The World Journal of Biological Psychiatry	Genome-wide association study of co-occurring anxiety in major depression.	Anxiety in major depressive disorder	1080 European ancestry cases, 442 European ancestry controls	No genome-wide significant hits	$<5 \times 10^{-8}$
23565138	Trzaskowski et al. (2013)	PLoS One	First genome-wide association study on anxiety-related behaviors in childhood	Anxiety-related behaviors in childhood	2810 children of European ancestry	No genome-wide significant hits	$<5 \times 10^{-8}$
24278274	Walter et al. (2013)	PLoS One	Performance of Polygenic Scores for Predicting Phobic Anxiety	Phobic anxiety	11,127 participants of European ancestry	No genome-wide significant hits	$<5 \times 10^{-8}$
25390645	Otowa et al. (2014)	PLoS One	Genome-wide and gene-based association studies of anxiety disorders in European and African American samples	Anxiety disorder	324 African American cases, 273 African American controls, 757 European ancestry cases, 940 European ancestry controls	No genome-wide significant hits	$<5 \times 10^{-8}$
26274327	Davies et al. (2015)	PLoS One	Generalized Anxiety Disorder—A Twin Study of Genetic Architecture, Genome-Wide Association	Anxiety sensitivity	730 MZ and DZ female twins	rs13334105 located in *RBFOX1*	$<5 \times 10^{-8}$

Continued

TABLE 1 Genome-wide association studies for anxiety disorders—cont'd

PubMed ID	Study	Journal	Study	Disease/trait	Initial sample size	Main findings	P-value significance threshold
26989097	Coleman et al. (2016)	The British Journal of Psychiatry	Genome-wide association study of response to cognitive-behavioral therapy in children with anxiety disorders and Differential Gene Expression	Response to cognitive-behavioral therapy in anxiety disorder	Up to 902 European ancestry cases, up to 78 cases	No genome-wide significant hits	$<5 \times 10^{-8}$
26754954	Otowa et al. (2016)	Molecular Psychiatry	Metaanalysis of genome-wide association studies of anxiety disorders.	Anxiety disorder	up to 7016 European ancestry cases, up to 14,745 European ancestry controls	rs1709393 located in an uncharacterized noncoding RNA locus and rs1067327 within *CAMKMT*	$<5 \times 10^{-8}$
28167838	Deckert et al. (2017)	Molecular Psychiatry	GLRB allelic variation associated with agoraphobic cognitions, increased startle response and fear network activation: a potential neurogenetic pathway to panic disorder.	Panic disorder/agoraphobia related anxiety phenotype	1370 European ancestry healthy participants	Association found between Agoraphobia Cognition Questionnaire and variants located in *GLRB* (rs7726293, rs191260602)	$<5 \times 10^{-8}$
27159506	Dunn et al. (2017)	American Journal of Medical Genetics. Part B, Neuropsychiatric Genetics	Genome-Wide Association Study (GWAS) of Generalized Anxiety Symptoms in the Hispanic Community Health Study/Study of Latinos	Generalized anxiety disorder	12,282 Hispanic healthy participants	One SNP, rs78602344, located in *THBS2*	$<5 \times 10^{-8}$
28224735	Stein et al. (2017)	American Journal of Medical Genetics. Part B, Neuropsychiatric Genetics	Genetic Risk Variants for Social Anxiety	Social anxiety	11,268 European American, 2622 African American, 3438 Latino American	rs708012 in Europeans and rs78924501 in African Americans	$<5 \times 10^{-8}$
31116379	Meier et al. (2019)	JAMA Psychiatry	Genetic Variants Associated With Anxiety and Stress-Related Disorders A Genome-Wide Association Study and Mouse-Model Study	Anxiety and stress related disorders	12,655 European ancestry cases and 19,225 European ancestry controls	68 variants in *PDE4B*	$<5 \times 10^{-8}$

detecting any GWAS hits (P-value $< 5 \times 10^{-8}$), while others have identified several significant loci for ADs. The largest GWAS to date, consisting of 12,655 cases and 19,225 controls of European ancestry, found the strongest signal for a variant located in *PDE4B*, a gene previously found to be associated with schizophrenia. Notably, this study investigated anxiety- *and* stress-related disorders, which included individuals diagnosed with PTSD (Meier et al., 2019).

The first GWAS for ADs was conducted specifically for PD in a Japanese population group. Even with a relatively modest sample size of 200 patients and the same number of controls, the authors identified seven loci, albeit at a less stringent P-value threshold (P-value $< 1.0 \times 10^{-6}$) than those used in current GWASs (Otowa et al., 2009). However, a follow-up study was unable to replicate these top hits in an independent Japanese case-control group (Otowa et al., 2010). Further, a metaanalysis of two GWASs in Japanese individuals (718 cases and 1717 controls) were not able to detect significant hits for PD (Otowa et al., 2012). Similarly, Erhardt et al. (2011) did not detect any genome-wide significant signals, but found a 2-SNP haplotype (comprising variants in the gene *TMEM132D*) nominally associated with PD in a European sample (Erhardt et al., 2011).

One GWAS has been conducted for GAD. This study, in individuals of Hispanic ancestry ($n = 12,282$), found a significant association with a variant in *THBS2* and GAD, but was not able to replicate this finding in independent datasets. However, the authors did show a significant SNP-based heritability of 7.2% for GAD (Dunn et al., 2017). A GWAS for SAD found significant hits on chromosomes 1 and 6 in African American and European datasets, respectively. This study also found that SAD has a significant SNP-based heritability of 12% in Europeans (Stein et al., 2017). A metaanalysis of seven independent GWASs, comprising 18,000 individuals of European ancestry, found that a variant located in an uncharacterized noncoding RNA locus on chromosome 3q12.3 is associated with ADs when using a case-control study design. The same study identified another SNP within the gene *CAMKMT* located on chromosome 2p21 as being significantly associated with a quantitative factor score for AD (Otowa et al., 2016).

6 Post-GWAS analysis

Most common mental disorders, including ADs, are considered to be polygenic, with contributions from multiple genetic variants—both rare and common—with small effect sizes (Smoller, 2016). Polygenic Risk Score (PRS) analysis utilizes results from a large "discovery" GWAS to calculate a cumulative risk score in a "target" dataset, allowing one to determine whether genetic polymorphisms not achieving genome-wide significance contributes to the disorder of interest (International Schizophrenia Consortium, 2009). Several PRS studies have been conducted for ADs. Otowa et al. (2012) found significant associations between PRS for PD and their target PD dataset at P-value thresholds of less than 0.3 and 0.4, explaining 3.2% and 3.4% of the variance, respectively. Another study did not find an association between PRS generated from anxiety GWAS results and phobic anxiety. In addition, no significant association was found between phobic anxiety and a score based on 31 candidate genes for anxiety (Walter et al., 2013).

Using PRS, it is possible to determine genetic overlap between two different diagnoses or traits. For example, it was found that PRS for depression (using summary statistics from a GWAS on depression) predicts approximately 2.1% of the variance for anxiety in an elderly population group (Demirkan et al., 2011). A PRS for schizophrenia was able to significantly predict anxiety in a group of adolescents, explaining 0.5% of the variance (Jones et al., 2016). Similarly, another study found a significant association between schizophrenia PRS and anxiety in a group of young adults (Sengupta et al., 2017). Thus, it appears that there is genetic overlap between anxiety and other psychiatric diagnoses, such as schizophrenia and depression. Considering the high co-morbidity between these disorders (Buckley, Miller, Lehrer, & Castle, 2009), these findings are not altogether surprising. Another study did not find an association between the PRS for shorter telomere length (a biological indicator of aging) and anxiety (Chang, Prescott, De Vivo, Kraft, & Okereke, 2018).

7 Expression analysis

A limited number of studies have investigated gene expression profiles in ADs. One such study found 631 genes differentially expressed among male cases with anxiety symptoms compared with controls. These genes were enriched for immune-related functions (Wingo & Gibson, 2015). Another study, conducted on discordant monozygotic twins with anxiety sensitivity, found significantly greater expression of *ITM2B*, a gene encoding a transmembrane protein, known to be associated with neurodegenerative disorders (Davies et al., 2015). Interrogation of publicly available transcriptome databases to identify pathways disrupted in GAD revealed dysregulation of carbohydrate metabolism, tight junction pathway, the phosphatidylinositol signaling system, VEGF signaling, long-term potentiation, and the glycolysis pathway (Gormanns et al., 2011). These studies provide interesting avenues for exploring ADs in the future.

8   Epigenetics

Several studies have explored epigenetic modifications in ADs. A total of 230 differentially methylated regions were found in monozygotic twins discordant for anxiety. These regions mapped to genes previously associated with stress-related phenotypes (Alisch et al., 2017). A methylomic study conducted in PD identified 40 CpG sites significantly differentially methylated between cases and controls. Pathway analysis of these sites found an enrichment of genes involved in "positive regulation of lymphocyte activation" (Shimada-Sugimoto et al., 2017). Another methylome study of PD in a European sample found one genome-wide significant signal in the promoter region of the *HECA* gene, specifically in females. This locus was replicated in an independent sample (Iurato et al., 2017). In patients with GAD, severe anxiety was associated with increased methylation at a CpG site located in the gene *ASB1*, a gene involved in cytokine regulation (Emeny et al., 2018). In a study conducted in adolescents, the promoter region of *STK32B*, encoding a signaling protein, had increased methylation in those at high risk for GAD (Ciuculete et al., 2018).

9   Next-generation sequencing

Three studies employing whole-exome sequencing (WES) have been conducted for ADs. One study investigated a Japanese family with PD and identified seven rare protein-altering variants as being potential candidates for this disorder. Further investigation of these genes revealed three novel variants (not present in controls or databases), located in the genes *GANC*, *PLA2G4E*, and *GFRA4*, in a Japanese PD case-control group (Morimoto et al., 2018). Another WES study for PD in a Faroese population group (54 cases and 211 controls) found the gene *DGKH* to have the strongest association with this disorder (Gregersen et al., 2016). *DGKH* has previously been implicated as a candidate gene for bipolar disorder (Baum et al., 2008). In a Caucasian family, with early onset bipolar disorder comorbid with anxiety-spectrum disorders, WES identified rare, likely protein-altering variants present in all three affected members, but not in unaffected family members or background controls. These variants were located in eight genes (*IQUB*, *JMJD1C*, *GADD45A*, *GOLGB1*, *PLSCR5*, *VRK2*, *MESDC2*, and *FGGY*) known to be expressed in the brain (Kerner et al., 2013).

10   Copy number variation

Few studies have considered copy number variations (CNVs) in terms of ADs. One such study did not find a significant increase in rare genome-wide CNVs in Japanese individuals with PD (Kawamura et al., 2011). Similarly, a study conducted on a population sample of children did not find a relationship between burden of CNVs and anxiety (Guyatt et al., 2018). However, 19% of individuals with a 1.6 megabase deletion on chromosome 3q29 were reported to have an AD, and 12.5% with a 22q11.2 deletion to have GAD (Fung et al., 2010; Glassford et al., 2016). Though the role of CNVs in ADs is not currently clear, this area is worth investigating further—particularly considering the small number of published studies in this area and the proposed polygenic architecture of ADs.

11   Pharmacogenomics

Given the variable patient response to first-line anxiolytic pharmacotherapy such as selective serotonin reuptake inhibitors (SSRIs), recent interest has grown in applying pharmacogenomic methods to predict interindividual outcome. While there is an emerging body of pharmacogenomic work focused on the treatment of major depression (e.g., Fabbri, Di Girolamo, & Serretti, 2013; Fabbri & Serretti, 2015; Kato & Serretti, 2010; Ramos et al., 2016), studies of anxiety disorders remain limited. Consistent with work on depression, pharmacodynamic evidence suggests that serotonergic (e.g., *5-HTTLPR*) and neurotrophic (e.g., *BDNF)* candidate variants may moderate individual treatment response (Amitai et al., 2014; Schosser & Kasper, 2009; Tiwari, Souza, & Müller, 2009), but additional work using unbiased methods is needed to confirm these findings. It has also been suggested that functional gene variants in the cytochrome P450 (CYP450) system may have promise in prospectively identifying poor or rapid drug metabolizers, thus informing prospective dose adjustments of psychotropic agents (Amitai et al., 2014; Schosser & Kasper, 2009; Tiwari et al., 2009). Perhaps in the future, such work may identify novel genomic variants to inform more targeted and personalized therapeutic approaches for anxiety disorders, and to minimize adverse effects. However—given the paucity of robust and replicated empirical evidence of the clinical utility and cost-effectiveness of these approaches—their incorporation into standard care is not yet recommended (e.g., see Bousman & Hopwood, 2016; Rosenblat, Lee, & McIntyre, 2017; Zubenko, Sommer, & Cohen, 2018).

12 Future directions

It is increasingly clear that ADs have a complex genetic architecture. While GWASs have begun to delineate this architecture, much further work is needed to fully understand the genetic basis of ADs. Such work will be advanced not only by better-powered GWAS studies, but also by a range of other genomic methods. The dearth of existing literature on, for example, whole-genome epigenetics and gene expression in ADs is notable. Systems biology approaches may ultimately be critical in providing a holistic view of the biology underlying ADs, allowing integration of genomic, epigenomic, and transcriptomic data with clinical features and environmental factors.

The "team science" approach employed by consortia such as the PGC has been crucial to advancing GWAS work (Sullivan, 2018). The anxiety working group of the PGC was first established in 2017. The aims of this group are to conduct GWASs for PD, agoraphobia, GAD, SAD, and specific phobia using (i) categorical case-control samples and (ii) population-based samples with dimensional anxiety measures. This working group aims also to focus on the development of ADs through the lifespan in child and adolescent cohorts (https://www.med.unc.edu/pgc/pgc-workgroups/anxiety-workgroup/). Findings from these well-powered analyses will greatly improve our understanding of the genetics underlying ADs.

Many of the genetic studies that have been conducted for psychiatry in general, including ADs, have utilized samples of European ancestry. There is an underrepresentation of other population groups, particularly those of African ancestry (Dalvie et al., 2015). Future studies will require a more diverse representation of study participants in order to facilitate variant discovery applicable to many more populations.

13 Conclusions

To date, the study of the genomics of ADs in humans has been bolstered by strong (genetic and nongenetic) animal models of anxiety-like behavior. However, while candidate gene association and genome-wide approaches have implicated a number of potential risk variants, replication of findings remains a major challenge. In the future, it is likely that more unbiased, whole-genome methods will be employed, drawing on diverse, large-scale consortia and multisystem biology; and with a view to identifying a range of relevant pathways that may be involved in the pathogenesis of these disorders.

References

Alisch, R. S., Van Hulle, C., Chopra, P., Bhattacharyya, A., Zhang, S. C., Davidson, R. J., … Goldsmith, H. H. (2017). A multi-dimensional characterization of anxiety in monozygotic twin pairs reveals susceptibility loci in humans. *Translational Psychiatry*, *7*(12), 1282. 29225348.

American Psychiatric Association (APA) (2013). *Diagnostic and statistical manual of mental disorders* (5th ed.). Arlington, VA: American Psychiatric Publishing.

Amitai, M., Kronenberg, S., Cohen, T., Frisch, A., Weizman, A., & Apter, A. (2014). Pharmacogenetics and anxiety disorders: Analysis of recent findings. [Article in Hebrew]. *Harefuah*, *153*(3–4) 210–214, 236. 24791568.

Amstadter, A. B., Maes, H. H., Sheerin, C. M., Myers, J. M., & Kendler, K. S. (2016). The relationship between genetic and environmental influences on resilience and on common internalizing and externalizing psychiatric disorders. *Social Psychiatry and Psychiatric Epidemiology*, *51*(5), 669–678. 26687369.

Arbelle, S., Benjamin, J., Golin, M., Kremer, I., Belmaker, R. H., & Ebstein, R. P. (2003). Relation of shyness in grade school children to the genotype for the long form of the serotonin transporter promoter region polymorphism. *The American Journal of Psychiatry*, *160*(4), 671–676. 12668354.

Battaglia, M., Ogliari, A., Zanoni, A., Citterio, A., Pozzoli, U., Giorda, R., … Marino, C. (2005). Influence of the serotonin transporter promoter gene and shyness on children's cerebral responses to facial expressions. *Archives of General Psychiatry*, *62*(1), 85–94. 15630076.

Baum, A. E., Akula, N., Cabanero, M., Cardona, I., Corona, W., Klemens, B., … McMahon, F. J. (2008). A genome-wide association study implicates diacylglycerol kinase eta (DGKH) and several other genes in the etiology of bipolar disorder. *Molecular Psychiatry*, *13*(2), 197–207. 17486107.

Bousman, C. A., & Hopwood, M. (2016). Commercial pharmacogenetic-based decision-support tools in psychiatry. *Lancet Psychiatry*, *3*(6), 585–590. 27133546.

Buckley, P. F., Miller, B. J., Lehrer, D. S., & Castle, D. J. (2009). Psychiatric comorbidities and schizophrenia. *Schizophrenia Bulletin*, *35*(2), 383–402. 19011234.

Campos, A. C., Fogaça, M. V., Aguiar, D. C., & Guimarães, F. S. (2013). Animal models of anxiety disorders and stress. *Revista Brasileira de Psiquiatria*, *35*(Suppl. 2), S101–S111. 24271222.

Chang, S. C., Prescott, J., De Vivo, I., Kraft, P., & Okereke, O. I. (2018). Polygenic risk score of shorter telomere length and risk of depression and anxiety in women. *Journal of Psychiatric Research*, *103*, 182–188. 29883926.

Ciuculete, D. M., Boström, A. E., Tuunainen, A. K., Sohrabi, F., Kular, L., Jagodic, M., ... Schiöth, H. B. (2018). Changes in methylation within the STK32B promoter are associated with an increased risk for generalized anxiety disorder in adolescents. *Journal of Psychiatric Research, 102,* 44–51. 29604450.

Clauss, J. A., & Blackford, J. U. (2012). Behavioral inhibition and risk for developing social anxiety disorder: A meta-analytic study. *Journal of the American Academy of Child and Adolescent Psychiatry, 51*(10), 1066–1075. 23021481.

Coleman, J. R., Lester, K. J., Keers, R., Roberts, S., Curtis, C., Arendt, K., ... Eley, T. C. (2016). Genome-wide association study of response to cognitive-behavioural therapy in children with anxiety disorders. *The British Journal of Psychiatry, 209*(3), 236–243. (2016). 26989097.

Dalvie, S., Koen, N., Duncan, L., Abbo, C., Akena, D., Atwoli, L., ... Koenen, K. C. (2015). Large scale genetic research on neuropsychiatric disorders in African populations is needed. *eBioMedicine, 2*(10), 1259–1261. 26629498.

Davies, M. N., Verdi, S., Burri, A., Trzaskowski, M., Lee, M., Hettema, J. M., ... Spector, T. D. (2015). Generalised anxiety disorder—A twin study of genetic architecture, genome-wide association and differential gene expression. *PLoS One, 10*(8), e0134865. https://doi.org/10.1371/journal.pone.0134865. eCollection. 26274327.

Deckert, J., Weber, H., Villmann, C., Lonsdorf, T. B., Richter, J., Andreatta, M., ... Reif, A. (2017). GLRB allelic variation associated with agoraphobic cognitions, increased startle response and fear network activation: A potential neurogenetic pathway to panic disorder. *Molecular Psychiatry, 22*(10), 1431–1439. 28167838.

Demirkan, A., Penninx, B. W., Hek, K., Wray, N. R., Amin, N., Aulchenko, Y. S., ... Middeldorp, C. M. (2011). Genetic risk profiles for depression and anxiety in adult and elderly cohorts. *Molecular Psychiatry, 16*(7), 773–783. 20567237.

Dunn, E. C., Sofer, T., Gallo, L. C., Gogarten, S. M., Kerr, K. F., Chen, C. Y., ... Smoller, J. W. (2017). Genome-wide association study of generalized anxiety symptoms in the Hispanic Community Health Study/Study of Latinos. *American Journal of Medical Genetics Part B, Neuropsychiatric Genetics, 174*(2), 132–143. 27159506.

Emeny, R. T., Baumert, J., Zannas, A. S., Kunze, S., Wahl, S., Iurato, S., ... Ladwig, K. H. (2018). Anxiety associated increased CpG methylation in the promoter of Asb1: A translational approach evidenced by epidemiological and clinical studies and a murine model. *Neuropsychopharmacology, 43*(2), 342–353. 28540928.

Erhardt, A., Czibere, L., Roeske, D., Lucae, S., Unschuld, P. G., Ripke, S., ... Binder, E. B. (2011). TMEM132D, a new candidate for anxiety phenotypes: Evidence from human and mouse studies. *Molecular Psychiatry, 16*(6), 647–663. 20368705.

Erhardt, A., & Spoormaker, V. I. (2013). Translational approaches to anxiety: Focus on genetics, fear extinction and brain imaging. *Current Psychiatry Reports, 15*(12), 417. https://doi.org/10.1007/s11920-013-0417-9. 24234873.

Fabbri, C., Di Girolamo, G., & Serretti, A. (2013). Pharmacogenetics of antidepressant drugs: An update after almost 20 years of research. *American Journal of Medical Genetics. Part B, Neuropsychiatric Genetics, 162B*(6), 487–520. (2013). 23852853.

Fabbri, C., & Serretti, A. (2015). Pharmacogenetics of major depressive disorder: Top genes and pathways toward clinical applications. *Current Psychiatry Reports, 17*(7), 50. https://doi.org/10.1007/s11920-015-0594-9. 25980509.

Finn, D. A., Rutledge-Gorman, M. T., & Crabbe, J. C. (2003). Genetic animal models of anxiety. *Neurogenetics, 4*(3), 109–135. (2003). 12687420.

Fung, W. L., McEvilly, R., Fong, J., Silversides, C., Chow, E., & Bassett, A. (2010). Elevated prevalence of generalized anxiety disorder in adults with 22q11.2 deletion syndrome. *The American Journal of Psychiatry, 167*(8), 998. 20693476.

Fyer, A. J., Hamilton, S. P., Durner, M., Haghighi, F., Heiman, G. A., Costa, R., ... Knowles, J. A. (2006). A third-pass genome scan in panic disorder: Evidence for multiple susceptibility loci. *Biological Psychiatry, 60*(4), 388–401. 16919526.

Gelernter, J., Page, G. P., Stein, M. B., & Woods, S. W. (2004). Genome-wide linkage scan for loci predisposing to social phobia: Evidence for a chromosome 16 risk locus. *The American Journal of Psychiatry, 161*(1), 59–66. 14702251.

Glassford, M. R., Rosenfeld, J. A., Freedman, A. A., Zwick, M. E., Mulle, J. G., & Unique Rare Chromosome Disorder Support Group. (2016). Novel features of 3q29 deletion syndrome: Results from the 3q29 registry. *American Journal of Medical Genetics. Part A, 170A*(4), 999–1006. 26738761.

Gormanns, P., Mueller, N. S., Ditzen, C., Wolf, S., Holsboer, F., & Turck, C. W. (2011). Phenome-transcriptome correlation unravels anxiety and depression related pathways. *Journal of Psychiatric Research, 45*(7), 973–979. 21255794.

Gragnoli, C. (2014). Proteasome modulator 9 gene SNPs, responsible for anti-depressant response, are in linkage with generalized anxiety disorder. *Journal of Cellular Physiology, 229*(9), 1157–1159. 24648162.

Gregersen, N. O., Lescai, F., Liang, J., Li, Q., Als, T., Buttenschøn, H. N., ... Demontis, D. (2016). Whole-exome sequencing implicates DGKH as a risk gene for panic disorder in the Faroese population. *American Journal of Medical Genetics Part B, Neuropsychiatric Genetics, 171*(8), 1013–1022. 27255576.

Guyatt, A. L., Stergiakouli, E., Martin, J., Walters, J., O'Donovan, M., Owen, M., ... Gaunt, T. R. (2018). Association of copy number variation across the genome with neuropsychiatric traits in the general population. *American Journal of Medical Genetics. Part B, Neuropsychiatric Genetics, 177*(5), 489–502. 29687944.

Hettema, J. M., Neale, M. C., & Kendler, K. S. (2001). A review and meta-analysis of the genetic epidemiology of anxiety disorders. *The American Journal of Psychiatry, 158*(10), 1568–1578. 11578982.

Hodgson, K., Almasy, L., Knowles, E. E., Kent, J. W., Curran, J. E., Dyer, T. D., ... Glahn, D. C. (2016). Genome-wide significant loci for addiction and anxiety. *European Psychiatry, 36,* 47–54. 27318301.

Howe, A. S., Buttenschøn, H. N., Bani-Fatemi, A., Maron, E., Otowa, T., Erhardt, A., ... De Luca, V. (2016). Candidate genes in panic disorder: Meta-analyses of 23 common variants in major anxiogenic pathways. *Molecular Psychiatry, 21*(5), 665–679. 26390831.

International Schizophrenia Consortium, Purcell, S. M., Wray, N. R., Stone, J. L., Visscher, P. M., O'Donovan, M. C., ... Sklar, P. (2009). Common polygenic variation contributes to risk of schizophrenia and bipolar disorder. *Nature, 460*(7256), 748–752. 19571811.

Iurato, S., Carrillo-Roa, T., Arloth, J., Czamara, D., Diener-Hölzl, L., Lange, J., … Erhardt, A. (2017). DNA methylation signatures in panic disorder. *Translational Psychiatry*, *7*(12), 1287. 29249830.

Jacobson, L. H., & Cryan, J. F. (2010). Genetic approaches to modeling anxiety in animals. *Current Topics in Behavioral Neurosciences*, *2*, 161–201. 21309110.

Jones, H. J., Stergiakouli, E., Tansey, K. E., Hubbard, L., Heron, J., Cannon, M., … Zammit, S. (2016). Phenotypic manifestation of genetic risk for schizophrenia during adolescence in the general population. *JAMA Psychiatry*, *73*(3), 221–228. 26818099.

Kaabi, B., Gelernter, J., Woods, S. W., Goddard, A., Page, G. P., & Elston, R. C. (2006). Genome scan for loci predisposing to anxiety disorders using a novel multivariate approach: Strong evidence for a chromosome 4 risk locus. *American Journal of Human Genetics*, *78*(4), 543–553. 16532386.

Kagan, J., Reznick, J. S., & Snidman, N. (1988). Biological bases of childhood shyness. *Science*, *240*(4849), 167–171. 3353713.

Kato, M., & Serretti, A. (2010). Review and meta-analysis of antidepressant pharmacogenetic findings in major depressive disorder. *Molecular Psychiatry*, *15*(5), 473–500. 18982004.

Kawamura, Y., Otowa, T., Koike, A., Sugaya, N., Yoshida, E., Yasuda, S., … Sasaki, T. (2011). A genome-wide CNV association study on panic disorder in a Japanese population. *Journal of Human Genetics*, *56*(12), 852–856. 22011818.

Kerner, B., Rao, A. R., Christensen, B., Dandekar, S., Yourshaw, M., & Nelson, S. F. (2013). Rare genomic variants link bipolar disorder with anxiety disorders to CREB-regulated intracellular signaling pathways. *Frontiers in Psychiatry*, *4*, 154. 24348429.

Kessler, R. C., Berglund, P., Demler, O., Jin, R., Merikangas, K. R., & Walters, E. E. (2005). Lifetime prevalence and age-of-onset distributions of DSM-IV disorders in the National Comorbidity Survey Replication. *Archives of General Psychiatry*, *62*, 593–602.

Lezak, K. R., Missig, G., & Carlezon, W. A. , Jr. (2017). Behavioral methods to study anxiety in rodents. *Dialogues in Clinical Neuroscience*, *19*(2), 181–191. 28867942.

Martin, E. I., Ressler, K. J., Binder, E., & Nemeroff, C. B. (2009). The neurobiology of anxiety disorders: Brain imaging, genetics, and psychoneuroendocrinology. *The Psychiatric Clinics of North America*, *32*(3), 549–575. 19716990.

McGrath, L. M., Weill, S., Robinson, E. B., Macrae, R., & Smoller, J. W. (2012). Bringing a developmental perspective to anxiety genetics. *Development and Psychopathology*, *24*, 1179–1193. 23062290.

Meier, S. M., Trontti, K., Purves, K. L., Als, T. D., Grove, J., Laine, M., … Mors, O. (2019). Genetic Variants Associated With Anxiety and Stress-Related Disorders: A Genome-Wide Association Study and Mouse-Model Study, *JAMA Psychiatry*. https://doi.org/10.1001/jamapsychiatry.2019.1119.

Morimoto, Y., Shimada-Sugimoto, M., Otowa, T., Yoshida, S., Kinoshita, A., Mishima, H., … Ono, S. (2018). Whole-exome sequencing and gene-based rare variant association tests suggest that PLA2G4E might be a risk gene for panic disorder. *Translational Psychiatry*, *8*(1), 41. 29391400.

Muris, P., van Brakel, A. M. L., Arntz, A., & Schouten, E. (2011). Behavioral inhibition as a risk factor for the development of childhood anxiety disorders: A longitudinal study. *Journal of Child and Family Studies*, *20*(2), 157–170. 21475710.

Olatunji, B. O., Cisler, J. M., & Tolin, D. F. (2007). Quality of life in the anxiety disorders: A meta-analytic review. *Clinical Psychology Review*, *27*(5), 5725–5781. 17343963.

Otowa, T., Hek, K., Lee, M., Byrne, E. M., Mirza, S. S., Nivard, M. G., … Hettema, J. M. (2016). Meta-analysis of genome-wide association studies of anxiety disorders. *Molecular Psychiatry*, *21*(10), 1485. 26857599.

Otowa, T., Kawamura, Y., Nishida, N., Sugaya, N., Koike, A., Yoshida, E., … Sasaki, T. (2012). Meta-analysis of genome-wide association studies for panic disorder in the Japanese population. *Translational Psychiatry*, *2*, e186. 23149450.

Otowa, T., Maher, B. S., Aggen, S. H., McClay, J. L., van den Oord, E. J., & Hettema, J. M. (2014). Genome-wide and gene-based association studies of anxiety disorders in European and African American samples. *PLoS One*, *9*(11), e112559. 25390645.

Otowa, T., Tanii, H., Sugaya, N., Yoshida, E., Inoue, K., Yasuda, S., … Sasaki, T. (2010). Replication of a genome-wide association study of panic disorder in a Japanese population. *Journal of Human Genetics*, *55*(2), 91–96. 19960027.

Otowa, T., Yoshida, E., Sugaya, N., Yasuda, S., Nishimura, Y., Inoue, K., … Okazaki, Y. (2009). Genome-wide association study of panic disorder in the Japanese population. *Journal of Human Genetics*, *54*(2), 122–126. 19165232.

Park, C. C., Gale, G. D., de Jong, S., Ghazalpour, A., Bennett, B. J., Farber, C. R., … Smith, D. J. (2011). Gene networks associated with conditional fear in mice identified using a systems genetics approach. *BMC Systems Biology*, *5*, 43. 21410935.

Pettersson, E., Lichtenstein, P., Larsson, H., Song, J., Attention Deficit/Hyperactivity Disorder Working Group of the iPSYCH-Broad-PGC Consortium, Autism Spectrum Disorder Working Group of the iPSYCH-Broad-PGC Consortium, Bipolar Disorder Working Group of the PGC, Eating Disorder Working Group of the PGC, Major Depressive Disorder Working Group of the PGC, Obsessive Compulsive Disorders and Tourette Syndrome Working Group of the PGC, Schizophrenia CLOZUK, Substance Use Disorder Working Group of the PGC, Agrawal, A., … Polderman, T. J. C. (2019). Genetic influences on eight psychiatric disorders based on family data of 4 408 646 full and half-siblings, and genetic data of 333 748 cases and controls. *Psychological Medicine*, *49*(7), 1166–1173. 30221610.

Pulst, S. M. (1999). Genetic linkage analysis. *Archives of Neurology*, *56*(6), 667–672. 10369304.

Ramos, M., Berrogain, C., Concha, J., Lomba, L., García, C. B., & Ribate, M. P. (2016). Pharmacogenetic studies: A tool to improve antidepressant therapy. *Drug Metabolism and Personalized Therapy*, *31*(4), 197–204. 27889704.

Reiss, S., & McNally, R. J. (1985). Expectancy model of fear. In S. Reiss, & R. R. Bootzin (Eds.), *Theoretical issues in behaviour therapy* (pp. 107–121). San Diego, CA: Academic Press.

Rosenblat, J. D., Lee, Y., & McIntyre, R. S. (2017). Does pharmacogenomic testing improve clinical outcomes for major depressive disorder? A systematic review of clinical trials and cost-effectiveness studies. *The Journal of Clinical Psychiatry*, *78*(6), 720–729. 28068459.

Scaini, S., Belotti, R., & Ogliari, A. (2014). Genetic and environmental contributions to social anxiety across different ages: A meta-analytic approach to twin data. *Journal of Anxiety Disorders*, *28*(7), 650–656. 25118017.

Schosser, A., Butler, A. W., Uher, R., Ng, M. Y., Cohen-Woods, S., Craddock, N., ... McGuffin, P. (2013). Genome-wide association study of co-occurring anxiety in major depression. *The World Journal of Biological Psychiatry*, *14*(8), 611–621. 24047446.

Schosser, A., & Kasper, S. (2009). The role of pharmacogenetics in the treatment of depression and anxiety disorders. *International Clinical Psychopharmacology*, *24*(6), 277–288. 19738481.

Schumacher, J., Kristensen, A. S., Wendland, J. R., Nöthen, M. M., Mors, O., & McMahon, F. J. (2011). The genetics of panic disorder. *Journal of Medical Genetics*, *48*(6), 361–368. 21493958.

Sengupta, S. M., MacDonald, K., Fathalli, F., Yim, A., Lepage, M., Iyer, S., ... Joober, R. (2017). Polygenic Risk Score associated with specific symptom dimensions in first-episode psychosis. *Schizophrenia Research*, *184*, 116–121. 27916287.

Shimada-Sugimoto, M., Otowa, T., & Hettema, J. M. (2015). Genetics of anxiety disorders: Genetic epidemiological and molecular studies in humans. *Psychiatry and Clinical Neurosciences*, *69*(7), 388–401. 25762210.

Shimada-Sugimoto, M., Otowa, T., Miyagawa, T., Umekage, T., Kawamura, Y., Bundo, M., ... Sasaki, T. (2017). Epigenome-wide association study of DNA methylation in panic disorder. *Clinical Epigenetics*, *9*, 6. 28149334.

Smoller, J. W. (2016). The genetics of stress-related disorders: PTSD, depression, and anxiety disorders. *Neuropsychopharmacology*, *41*(1), 297–319. (2016). 26321314.

Smoller, J. W., Gardner-Schuster, E., & Covino, J. (2008). The genetic basis of panic and phobic anxiety disorders. *American Journal of Medical Genetics Part C, Seminars in Medical Genetics*, *148C*(2), 118–126. 18412108.

Stein, M. B., Chen, C. Y., Jain, S., Jensen, K. P., He, F., Heeringa, S. G. ... Army STARRS Collaborators. (2017). Genetic risk variants for social anxiety. *American Journal of Medical Genetics. Part B, Neuropsychiatric Genetics*, *174*(2), 120–131. 28224735.

Stein, M. B., Jang, K. L., & Livesley, W. J. (1999). Heritability of anxiety sensitivity: A twin study. *The American Journal of Psychiatry*, *156*(2), 246–251. 9989561.

Stein, M. B., & Stein, D. J. (2008). Social anxiety disorder. *Lancet*, *371*(9618), 1115–1125. (2008). 18374843.

Sullivan, P. F., Agrawal, A., Bulik, C. M., Andreassen, O. A., Børglum, A. D., Breen, G., ... Psychiatric Genomics Consortium. (2018). Psychiatric genomics: An update and an agenda. *The American Journal of Psychiatry*, *175*(1), 15–27. 28969442.

Taylor, S., Jang, K. L., Stewart, S. H., & Stein, M. B. (2008). Etiology of the dimensions of anxiety sensitivity: A behavioral-genetic analysis. *Journal of Anxiety Disorders*, *22*(5), 899–914. 18029140.

Tiwari, A. K., Souza, R. P., & Müller, D. J. (2009). Pharmacogenetics of anxiolytic drugs. *Journal of Neural Transmission (Vienna)*, *116*(6), 667–677. (2009). 19434367.

Trzaskowski, M., Eley, T. C., Davis, O. S., Doherty, S. J., Hanscombe, K. B., Meaburn, E. L., ... Plomin, R. (2013). First genome-wide association study on anxiety-related behaviours in childhood. *PLoS One*, *8*(4), e58676. 23565138.

Walter, S., Glymour, M. M., Koenen, K., Liang, L., Tchetgen Tchetgen, E. J., Cornelis, M., ... Kubzansky, L. D. (2013). Performance of polygenic scores for predicting phobic anxiety. *PLoS One*, *8*(11), e80326. 24278274.

Webb, B. T., Guo, A. Y., Maher, B. S., Zhao, Z., van den Oord, E. J., Kendler, K. S., ... Hettema, J. M. (2012). Meta-analyses of genome-wide linkage scans of anxiety-related phenotypes. *European Journal of Human Genetics*, *20*(10), 1078–1084. 22473089.

Wingo, A. P., & Gibson, G. (2015). Blood gene expression profiles suggest altered immune function associated with symptoms of generalized anxiety disorder. *Brain, Behavior, and Immunity*, *43*, 184–191. (2015). 25300922.

Zubenko, G. S., Sommer, B. R., & Cohen, B. M. (2018). On the marketing and use of pharmacogenetic tests for psychiatric treatment. *JAMA Psychiatry*, *75*(8), 769–770. 29799933.

Proteomics for diagnostic and therapeutic blood biomarker discovery in schizophrenia and other psychotic disorders

David R. Cotter[a], Sophie Sabherwal[a] and Klaus Oliver Schubert[b,c]

[a]Psychiatry, RCSI, Dublin, Ireland, [b]Discipline of Psychiatry, University of Adelaide, Adelaide, SA, Australia, [c]Northern Adelaide Mental Health Services, Northern Adelaide Local Health Network, Lyell McEwin Hospital, Elizabeth Vale, SA, Australia

1 Proteomics for complex psychiatric disorders

Proteomics is defined as the large-scale qualitative and quantitative study of proteomes. Proteomic investigations offer unique biological insights. The expression and function of proteins can be modulated at the DNA transcription stage, or posttranslational-modification stage. In addition, many different proteins can arise as the products of a single gene by alternative splicing and alternative posttranslational modifications of gene transcripts. Neither of these modifications can be accurately predicted by data on nucleic acids, as produced by genomic investigations. In psychiatric disorders such as schizophrenia, where no known Mendelian variants have been identified, but instead variations of many genes confer subtle biological effects in addition to gene-environment interactions and purely environmental factors, the potential value of proteomics is therefore evident. Here, the functional end products of genetic and environmental variation can be directly assessed, and proteomic technologies are now capable of characterizing and identifying posttranslational modifications, protein-protein interactions, and protein turnover.

2 Proteomics and personalized psychiatry

Psychiatric diagnoses are based on clinical symptoms and biological investigations, including neuroimaging investigations and blood tests, which have their main value in terms of out-ruling organic causes for symptoms. These biological tests do not currently contribute in a positive manner to diagnosis within most clinical situations. This is both a clear weakness and an opportunity. As in all medicine, in psychiatry, early identification and treatment are associated with better outcomes. This has become increasingly clear in psychosis research over the past decade, with good evidence that early identification and treatment of subjects with psychiatric disorders, both psychotic and affective, significantly improves their clinical outcome (Larsen et al., 2011). Consequently, over the past decade, there has been a shift in research focus from first onset with psychosis to the so-called "at risk mental state" (ARMS) (Rutigliano et al., 2016) with the aim of identifying vulnerable subjects and offering early treatment to prevent psychosis (Amminger et al., 2010; Clark et al., 2016). However, only 16%–35% UHR subjects go on to convert to psychosis (Cannon et al., 2016; Fusar-Poli et al., 2012), with 50%–65% of these subsequently experiencing nonpsychotic mental disorders, such as depression and anxiety (Kelleher et al., 2012; Rutigliano et al., 2016). Consequently, there is now an increasing focus, not just on the vulnerability to psychotic disorder represented by psychotic experiences (PEs), but on help-seeking, functional impairment (Yung & Lin, 2016), and vulnerability to major psychiatric disorders generally (Rutigliano et al., 2016). The identification of an early biological fingerprint of all of these psychiatric illnesses is theoretically possible, and proteomic methods can contribute to that (Orešič et al., 2011). Specifically, plasma proteomic methods have been applied to age 12 plasma samples and identified a pattern of altered complement proteins among subjects who went on to develop psychotic disorders at age 18 (English et al., 2018).

Biological markers of other clinically important measures should become the focus of the next generation of translational studies. Chief among these, perhaps, is the need to identify biomarkers of clinical outcomes following antipsychotic drug treatment. While few studies have addressed this issue, there is preliminary evidence from a single mass spectrometry-based proteomic investigation for elevated expression of complement pathway proteins among first episode psychosis subjects who had a poor outcomes following amisulpride treatment (Focking et al., 2018). While preliminary, the study demonstrates the power of discovery proteomic methods for personalized psychiatry. Similar methods should now be used to identify proteomic biomarkers of cognitive outcome and treatment response among subjects with, for example, predominantly positive or predominantly negative symptom schizophrenia. In affective disorder, differential-in-gel electrophoresis (DIGE) was used to identify protein changes associated with electro convulsive therapy (ECT), changes that are potentially relevant as biomarkers of prediction of treatment response to ECT and psychotic depression (Glaviano et al., 2014).

Biomarkers identified through proteomic and other methods such as neuroimaging, neuropsychology (Cannon et al., 2016), and more recently, lipidomic (O'Gorman et al., 2017) studies, will likely have improved accuracy, and therefore, clinical value. These biomarkers, when added to other clinical tools reflecting symptom profiles, will hopefully provide a shift toward a precision medicine in psychiatry that has real and tangible impacts on the patients attending new standard clinical services (Fond et al., 2015; Leucht et al., 2015). The following sections provide in depth outlines of the proteomic methods generally used, and a summary of their main findings to date.

3 Mass spectrometry workflows

Mass spectrometry is a high-throughput proteomic technique that works by fragmenting or ionizing peptides, and by sorting these ions by mass-to-charge ratio. Mass spectrometry workflows can be broadly distinguished into *"bottom-up" "top-down,"* and *"discovery,"* or *"targeted"* experiments. "Bottom-up" refers to the reconstruction of protein information from individually identified peptides or fragments, whereas "top-down" approaches are capable of identifying and quantifying intact proteins. "Bottom-up" is currently the more widely used approach for the analysis of multiple proteins in complex samples (for example, blood, postmortem brain, or cerebrospinal fluid), due to its superior sensitivity. Therefore, we will focus on "bottom-up" techniques here. The quantification of proteins in a comparative experiment using a bottom-up approach can be based on "labeled" or "label-free" peptides. Each method has strengths and weaknesses, but most biomarker investigations employ label-free approaches, because they are not limited by the number of labels, and can facilitate a larger number of samples being run in the experiment. This is important, as sample size and statistical power are key factors considered in the design of clinical biomarker investigations.

Discovery proteomic methods are optimized for achieving maximum coverage of the proteome by limiting the number of samples analyzed. In contrast, *targeted* proteomics prioritize sensitivity and throughput by limiting proteome coverage.

4 Mass spectrometry subtypes

Mass spectrometry technology is rapidly advancing in line with the introduction of cutting-edge machinery. There has been increasing variation in the types of acquisition, ionization (for example, electrospray ionization (ESI) and matrix-assisted laser desorption/ionization (MALDI)), mass selection (time-of-flight (TOF) Quadrupole and Orbitrap), and detection used based on our expanding knowledge of the proteome. The introduction of tandem mass spectrometry (MS/MS) that involves multiple stages of mass spectrometry with fragmentation of precursor ions (MS1) to highly specific fragment ions (MS2) has led to significant improvements in the rate and reproducibility of protein identification and quantification. MS/MS can be performed on a specific tandem machine, and also on a single mass analyzer over longer time periods.

5 Acquisition methods for proteomic data

In choosing a proteomic acquisition method, the most important factors include sensitivity, reproducibility, and throughput. Different acquisition methods are capable of discovery and/or targeted analyses in that they are optimal for the analysis of a range of numbers of proteins in complex samples. Currently, data-dependent acquisition (DDA), selected-reaction monitoring (SRM), and data-independent acquisition (DIA) are the most commonly used methods. DDA, which stochastically fragments peptides based on ion abundance, is currently considered the gold standard for discovery experiments due to its ability to quantify a random subset of the entire proteome. Conversely, in SRM, the gold standard for targeted proteomics, only a small subset of the proteome with prespecified targets is selectively quantified, but measurements are highly sensitive and reproducible. Data-independent acquisition (DIA) experiments allow sensitive and reproducible quantitation (almost of SRM quality) on all proteins within a large range of the proteome (specified to incorporate the most proteins possible from the sample).

DIA, first developed about a decade ago, has been gaining much attention in recent years with the newer generation mass spectrometers and the introduction of variations of the technique, such as sequential window acquisition of all

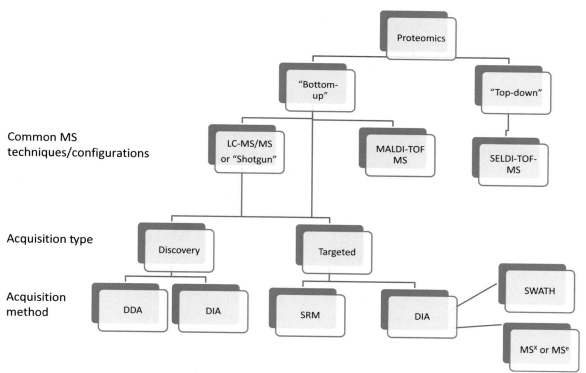

FIG. 1 Understanding mass spectrometry subtypes in relation to one another. *MS*, mass spectrometry; *LC*, liquid chromatography; *MALDI-TOF*, matrix-assisted laser desorption/ionization time of flight; *SELDI-TOF*, surface-enhanced laser desorption/ionization time-of flight; *DDA*, data-dependent acquisition; *DIA*, data-independent acquisition; *SRM*, selected reaction monitoring/parallel reaction monitoring; *SWATH*, sequential window acquisition of all theoretical mass spectra.

theoretical mass spectra (SWATH), multiplexed mass spectrometry (MS^X), and MS^e (Martins-de-Souza, Faça, & Gozzo, 2017). For discovery experiments, DIA has the ability to overcome limitations in reproducibility associated with DDA as a result of its more comprehensive sampling method, and is also more effective in quantifying low abundance proteins due to its nonbiased fragmenting procedure. On the other hand, there are modest sacrifices in sensitivity in DIA experiments compared with DDA due to the increased instrument time required to progress through the isolation windows. To a degree, this shortcoming can be compensated for by the increased signal-to-noise ratio in MS2 spectra, which is the level at which DIA quantifies (Chapman, Goodlett, & Masselon, 2014). As a targeted approach, DIA is inferior to SRM in terms of specificity and sensitivity. However, it overcomes the need for peptide scheduling in advance of running a targeted experiment. Fig. 1 visualizes mass spectrometry subtypes in relation to one another.

In addition to improvements in data acquisition, advanced sample preparation techniques have been developed in tandem with enhancements in mass-spectrometry technology. These include the depletion of high-abundance proteins in order to improve the dynamic range (Echan, Tang, Ali-Khan, Lee, & Speicher, 2005), as well as advanced purification, enrichment, and fractionation techniques, in both brain and blood.

6 Multiplex immunoassays

Antibody-based platforms have been recently developed using microsphere technology to provide researchers with reproducible, quantitative multiplex immunoassay data for hundreds of proteins, from relatively small quantities of blood plasma or serum. Furthermore, the development of biomarker discovery databases has made biomarker discovery, as opposed to purely hypothesis-driven protein quantitation, a feasible objective of immunoassay experiments (Sabherwal, English, Föcking, Cagney, & Cotter, 2016). The issue of antibody availability, and thus the lack of comprehensive proteome coverage, which has long been a significant limitation of any antibody-based biomarker studies, may eventually be overcome with the introduction of antibody-based platforms such as The Human Protein Atlas (www.proteinatlas.org), which contains expression and localization profiles for 86% of the predicted human genome (Sabherwal et al., 2016). Forty-four different human tissues, and annotation data for 83 different cell types are included. The ultimate goal is to continue to extend this analysis to the majority of human proteins (Uhlen, 2005).

In addition to "semidiscovery" experiments, multiplex immunoassays can be used as a validation technique for mass-spectrometry results. For example, MS^e was previously used to identify a serum biomarker panel capable of distinguishing first-onset drug-naive schizophrenia subjects from healthy subjects. This panel was then validated by multiplex immunoassay to create the first commercially available blood-based laboratory test for schizophrenia (Schwarz et al., 2010).

7 A comparison between mass spectrometry and multiplex immunoassay

The two approaches currently driving high-throughput biomarker discovery, multiplex immunoassay and mass spectrometry, often give rise to different findings, due to their unique properties. Here, we will briefly discuss their relative strengths and weaknesses for biomarker discovery.

The analytical specificity of mass spectrometry and its ability to discern one molecule from another is considered as "gold-standard." For most molecules, diagnostic ions that can unequivocally identify the analyte can be isolated. For example, vitamin D is measured on immunoassay platforms using a binding assay that does not discriminate between vitamin D2 (ergocalciferol) and vitamin D3 (cholecalciferol), whereas mass spectrometry can discriminate between the two forms by their different molecular masses (Maunsell, Wright, & Rainbow, 2005).

On the other hand, the reproducibility of mass spectrometry is inferior to that of antibody-based platforms, including multiplex immunoassays. This may reflect both the natural variation in chromatography, and the presence of missing data that arise as a consequence of the DDA sampling procedure in particular. However, it is also a consequence of the true discovery nature of the approach, as opposed to the use of specific antibodies, and also of analytical specificity, which results in the identification of multiple isoforms of the same protein (Sabherwal et al., 2016).

In general, immunoassays allow greater sensitivity of detection than mass spectrometry. On the other hand, SRM type acquisition is now of comparable sensitivity to immunoassays (Doerr, 2010), and immunoassays can be difficult to scale up. Another factor to consider is that immunoassays are limited in terms of antibody availability, and as a result, are also economically dependent on predefined platforms. Overall, immunoassays are less costly than mass spectrometry approaches in relation to instruments and training.

8 Data analysis and bioinformatics

The creation and optimization of data analysis techniques capable of accurately analyzing the volume of data produced by various mass-spectrometry acquisition methods is ongoing. Software packages used in the analysis of DDA proteomics generally use a set of algorithms for peak detection and scoring of peptides, perform mass calibration and database searches for protein identification, quantify identified proteins, and provide summary statistics. Protein quantification can be performed via label-free quantification (LFQ) or intensity-based absolute quantification (iBAQ), and statistical analysis is based on MS1 level intensities (Cox & Mann, 2008).

A commonly used approach for DIA data analysis involves targeted quantification, which employs a spectral library generated by DDA for matching peptides (as chromatogram peaks) and extracting their intensities (as an area under the curve). There is also software available capable of nonspectral library-based quantitation. DIA data can be quantified at the MS1 or MS2 level. DIA derived data, acquired by targeted or nontargeted extraction, poses significant challenges to data analysis. For example, MS2 level DIA data may contain fragments that are shared across multiple co-eluting precursor ions within the same isolation window, creating a difficult problem for quantification. Furthermore, fragment maps will not necessarily be reproducible across multiple runs if the chromatographic elution patterns are distorted by factors such as pressure and temperature changes in the column, or fragment ion interference. Therefore a reliable set of fragments has to be selected carefully before the statistical analysis is performed, which is possible in some bioinformatics software packages that allow the visualization and processing of data before statistical analysis (Egertson, MacLean, Johnson, Xuan, & MacCoss, 2015; Teo et al., 2015).

9 Functional analysis of proteomic data

Pathway analysis software is currently widely used in high-throughput biomarker studies to identify biological pathways of interest from a list of molecular candidates. Examples include ingenuity pathway analysis (IPA), Kyoto encyclopedia of genes and genomes (KEGG), and STRING. Something to consider when carrying out a biomarker study is that the knowledge base of pathway analysis tools is rudimentary. This is particularly true of protein data, as new protein functions and interactions are being discovered at an extremely fast rate. However, as the number and type of functional annotations increase in parallel with technological advances, the utility of pathway analysis and confidence in results is likely to improve.

10 Proteomics for blood biomarker discovery in psychiatry

For the discovery phase of biomarker development, methods such as DDA and SRM are more established (especially in terms of data analysis) and widely accepted within the proteomic community. They do, however, have some significant limitations, some of which can be overcome by DIA approaches (Sajic, Liu, & Aebersold, 2015). There are also currently limitations in validation methods for proteomic experiments; most being comparatively low-throughput, and often costly; for example, enzyme-linked immunosorbent assay (ELISA). DIA approaches have the ability to validate data-dependent discovery proteomic findings in a high-throughput manner, similar to targeted approaches such as SRM (English, Wynne, Cagney, & Cotter, 2014). Therefore, further development of the DIA workflow, for both discovery and targeted experiments, and a surge in its use in biomarker development, is expected.

The increasing use of targeted mass spectrometry instead of antibody-based approaches (for example, SRM, which can detect low abundance compounds with comparable sensitivity) is also foreseen. In this way, the exceptional analytical specificity of mass spectrometry, mentioned previously, could be better utilized. The future of these approaches may also be integrated along with other biomarker discovery approaches in order to provide more accurate and personalized markers of psychotic disorders. In fact, work has already begun on incorporating clinical, socio-environmental, molecular, neuroimaging, and neurophysiological findings for Alzheimer's disease, depressive disorder, and schizophrenia in order to identify a particular multifactorial signature specific to each (Burton, 2011; Kemp, Gordon, John Rush, & Williams, 2008; Kennedy et al., 2012; Shah et al., 2012).

10.1 The search for schizophrenia biomarkers: An example of psychiatric biomarker discovery

There would undoubtedly be value in identifying biological markers of schizophrenia. The clinical outcomes in schizophrenia, a psychiatric disorder affecting up to 1% of the population worldwide, significantly improves in patients who are identified and treated early in their course of illness (Larsen et al., 2011). Symptoms generally emerge in mid-adolescence, but criteria for full disorder are not usually met until late adolescence or adulthood. This progression provides researchers with the opportunity to search for predictive biomarkers in addition to diagnostic markers. Studies involving first-episode and nondrug treated schizophrenia are therefore extremely valuable to biomarker research, in order to allow for more accurate diagnosis, earliest possible intervention and to potentially provide an insight into pathophysiology.

The search for blood biomarkers for brain disease has become increasingly popular in light of recent clinical trials for blood-based biomarkers of the neurodegenerative disorder Alzheimer's disease, and significant studies in Parkinson's disease (Alberio et al., 2012; Hampel et al., 2011). This is of significance to psychiatry research as patient blood is an easily accessible biological sample. In addition, patient blood can be obtained at any stage during the course of illness, unlike cerebrospinal fluid (CSF), for example. Strong evidence now supports an association between systemic abnormalities and schizophrenia pathology, which is also driving the search for peripheral, blood-based biomarkers for the disorder. Such studies have implicated processes such as inflammation, stress response signaling, innate/adaptive immune signaling, and energy metabolism (English et al., 2018; Maes et al., 1997; Marques-Deak, Cizza, & Sternberg, 2005; Upthegrove, Manzanares-Teson, & Barnes, 2014).

In a recent review, drug-free schizophrenia blood biomarker findings found by mass-spectrometry and multiplex immunoassay were collated (Sabherwal et al., 2016) (Table 1). Pathway analysis software was then used to identify top pathways implicated in past studies. These included immune system pathways (communication between innate and adaptive immune cells, hepatic stellate cell activation, atherosclerosis signaling), lipid and glucose metabolism pathways (LXR/RXR activation, FXR/RXR activation and atherosclerosis signaling), blood formation and clotting pathways (hematopoiesis of multi/pluripotent cells and coagulation respectively), and the stress response pathways (glucocorticoid receptor signaling).

A potential limitation of biomarker studies in psychosis is the uncertainty as to whether the findings reflect causality or are epiphenomena. Many studies in schizophrenia attempt to overcome this by identifying and matching cases and controls for potential confounding factors, or adjusting for confounding variables post-hoc. The presence of disorders such as diabetes mellitus, hyperlipidemia, hypertension, cardiovascular or immune diseases, and other neuropsychiatric disorders or substance abuse issues in experimental subjects could significantly influence findings in schizophrenia studies. Of interest, a recent lipidomic study identified significant alteration in lipids, particularly lysophosphatidylcholines (LPCs) in the age 12 blood of subjects who were then well, and later developed psychotic disorder at age 18 (O'Gorman et al., 2017). These findings, along with those of English et al. (2018) who identified complement protein changes in these same subjects at age 12, suggest that a biomarker's fingerprint relevant to psychotic disorder is apparent early, even before formal illness. However, the studies do not, in themselves, inform causality, and whether the lipidomic or the protein changes are causal. Future studies will be able to go back to earlier samples and attempt to determine which changes arise first.

TABLE 1 A total of 47 biomarker candidates were identified in two or more studies, in the same direction, by mass-spectrometry (MS)-based and/or more multiplexed immunoassay (MIA) methods.

Protein name|gene name_accession number (or KEGG identifier)	Method	DOC	Reference
1. Apolipoprotein A1|APOA1_P02647	MIA	↑	Schwarz et al. (2011)
	MIA	↓↓↓	Chan et al. (2015), Domenici et al. (2010), and Schwarz et al. (2012)
	MALDI-TOF-MS	↓	Yang et al. (2006)
	LC-MS[e]	↓	Jaros et al. (2012)
	LC-MS[e]	↓	Levin et al. (2010)
2. Haptoglobin|HP_P00738	LC-MS/MS	↑	Li et al. (2012)
	MALDI-TOF-MS	↑	Yang et al. (2006)
	MIA	↑↑↑	Chan et al. (2015), Ramsey et al. (2013), and Schwarz et al. (2012)
3. Alpha-2 macroglobulin|A2M_P01023	MIA	↑↑↑↑	Chan et al. (2015), Perkins et al. (2015), Ramsey et al. (2013), and Schwarz et al. (2012)
	MIA	↓	Domenici et al. (2010)
4. Carcinoembryonic antigen|CEACAM5_P06731	MIA	↑↑↑↑	Chan et al. (2015), Domenici et al. (2010), Ramsey et al. (2013), and Schwarz et al. (2012)
5. Chromogranin A|CHGA_P10645	MIA	↑↑↑↑	Chan et al. (2015), Guest et al. (2011), Ramsey et al. (2013), and Schwarz et al. (2011)
6. Eotaxin|CCL11_P51671	MIA	↑↑↑↑	Chan et al. (2015), Domenici et al. (2010), Ramsey et al. (2013), and Schwarz et al. (2012)
7. Interleukin-8|IL-8_P10145	MIA	↑↑↑↑	Chan et al. (2015), Domenici et al. (2010), Perkins et al. (2015), and Ramsey et al. (2013)
8. Pancreatic polypeptide|PPY_P01298	MIA	↑↑↑↑	Chan et al. (2015), Guest et al. (2011), Ramsey et al. (2013), and Schwarz et al. (2012)
9. Thyroxine binding globulin|SERPINA7_P05543	MIA	↑↑↑↑	Domenici et al. (2010), Perkins et al. (2015), Ramsey et al. (2013), and Schwarz et al. (2011)
10. Alpha-1 antitrypsin|SERPINA1_P01009	MIA	↑↑↑	Domenici et al. (2010), Ramsey et al. (2013), and Schwarz et al. (2012)
	MALDI-TOF-MS	↑	Yang et al. (2006)
11. Coagulation Factor VII|F7_P08709	MIA	↑↑	Domenici et al. (2010) and Perkins et al. (2015)
		↓↓↓	Schwarz et al. (2012), Ramsey et al. (2013), and Chan et al. (2015)
12. Epidermal growth factor|EGF_P01133	MIA	↑↓↓	Domenici et al. (2010), Ramsey et al. (2013), and Schwarz et al. (2012)
13. Follicle stimulating hormone|FSHR_P23945	MIA	↑↑↑	Chan et al. (2015), Ramsey et al. (2013), and Schwarz et al. (2012)
14. Growth hormone|GH1_P01241	MIA	↑↑	Cheng et al. (2010) and Perkins et al. (2015)
		↓	Domenici et al. (2010)
15. Insulin-like growth factor-binding protein 2|IGFBP2_P18065	MIA	↑↑↑	Chan et al. (2015), Ramsey et al. (2013), and Schwarz et al. (2012)
16. Interleukin-15|IL-15_P40933	MIA	↑↑↑	Domenici et al. (2010), Perkins et al. (2015), and Ramsey et al. (2013)

TABLE 1 A total of 47 biomarker candidates were identified in two or more studies, in the same direction, by mass-spectrometry (MS)-based and/or more multiplexed immunoassay (MIA) methods.—cont'd

Protein name|gene name_accession number (or KEGG identifier)	Method	DOC	Reference
17. Leptin|LEP_P41159	MIA	↑	Domenici et al. (2010)
		↓↓↓	Chan et al. (2015), Cheng et al. (2010), and Ramsey et al. (2013)
18. Macrophage migration inhibitory factor|MIF_P01033	MIA	↑↑↑	Chan et al. (2015), Ramsey et al. (2013), and Schwarz et al. (2012)
19. RANTES|CCL5_P13501	MIA	↑↑↑	Domenici et al. (2010), Ramsey et al. (2013),and Schwarz et al. (2012)
20. Resistin|RETN_Q9HD89	MIA	↓↓↓	Cheng et al. (2010), Harris et al. (2012), and Ramsey et al. (2013)
21. Serum amyloid P|APCS_P02743	MIA	↑↑↑	Domenici et al. (2010), Ramsey et al. (2013), and Schwarz et al. (2011)
22. von Willebrand factor|VWF_P04275	MIA	↑↑↑	Chan et al. (2015), Domenici et al. (2010), and Perkins et al. (2015)
23. Angiotensin converting enzyme|ACE_P12821	MIA	↓↓	Chan et al. (2015) and Ramsey et al. (2013)
24. Apolipoprotein A2|APOA2_P02652	LC-MS	↓	Levin et al. (2010)
	LC-MS[e]	↓	Jaros et al. (2012)
25. Apolipoprotein H/beta-2-glycoprotein|APOH_P02749	MIA	↑↑	Chan et al. (2015) and Domenici et al. (2010)
26. CD40 ligand|CD40LG_P29965	MIA	↓↓	Ramsey et al. (2013) and Schwarz et al. (2012)
27. Complement C3|C3_P01024	MIA	↑↑	Domenici et al. (2010) and Ramsey et al. (2013)
28. Complement factor B|CFB_P00751	MALDI-TOF-MS	↑	Yang et al. (2006)
	LC-MS[e]	↑	Jaros et al. (2012)
29. Connective tissue growth factor|CTGF_P29279	MIA	↑↑	Schwarz et al. (2011) and Schwarz et al. (2012)
30. Fibrinogen|FGA_P02671	MALDI-TOF/TOF-MS	↑	Zhou et al. (2013)
	MIA	↑	Domenici et al. (2010)
31. Glutathione-S-transferase|GST_P09211	MIA	↑↑	Domenici et al. (2010) and Schwarz et al. (2012)
32. Granulocyte macrophage colony stimulating factor|GMCS_P04141	MIA	↓↓	Domenici et al. (2010) and Schwarz et al. (2012)
33. Immunoglobulin A|IgA_P01876	MIA	↓↓	Chan et al. (2015) and Domenici et al. (2010)
34. Insulin|INS_P01308	MIA	↑↑↑	Domenici et al. (2010) and Guest et al. (2011)
35. Interleukin-10|IL-10_P22301	MIA	↑↑	Ramsey et al. (2013) and Chan et al. (2015)
		↓↓	Domenici et al. (2010) and Schwarz et al. (2012)
36. Luteinizing hormone|LH_P01229	MIA	↑↑	Schwarz et al. (2012) and Ramsey et al. (2013)
		↓	Cheng et al. (2010)
37. Macrophage-derived chemokine|MDC_O00626	MIA	↑↑	Domenici et al. (2010) and Ramsey et al. (2013)
38. Prolactin|PRL_P01236	MIA	↑↑	Guest et al. (2011) and Perkins et al. (2015)
39. Prostatic acid phosphatise|PAP_P15309	MIA	↑	Domenici et al. (2010)
		↓↓	Schwarz et al. (2012) and Ramsey et al. (2013)

Continued

TABLE 1 A total of 47 biomarker candidates were identified in two or more studies, in the same direction, by mass-spectrometry (MS)-based and/or more multiplexed immunoassay (MIA) methods.—cont'd

Protein name|gene name_accession number (or KEGG identifier)	Method	DOC	Reference
40. Receptor for advanced glycosylation end products|RAGE_Q15109	MIA	↓↓	Chan et al. (2015) and Ramsey et al. (2013)
41. Serum glutamic oxaloacetic transaminase|SGOT_P00505	MIA	↑↑	Chan et al. (2015) and Schwarz et al. (2012)
		↓	Schwarz et al. (2011)
42. Sortilin|SORT1_Q99523	MIA	↓↓	Ramsey et al. (2013) and Schwarz et al. (2012)
43. Stem cell factor|KITLG_P21583	MIA	↑↓↓	Chan et al. (2015), Domenici et al. (2010), and Schwarz et al. (2012)
44. Tenascin-C|TC_P24821	MIA	↑↑	Chan et al. (2015) and Ramsey et al. (2013)
45. Thrombopoeitin|THPO_P40225	MIA	↑↑	Domenici et al. (2010) and Perkins et al. (2015)
		↓	Schwarz et al. (2012)
46. Tissue inhibitor of metalloproteinases 1|TIMP-1_P01033	MIA	↑↑	Domenici et al. (2010) and Ramsey et al. (2013)
47. Tumor necrosis factor receptor-like 2|TNFR2_Q92956	MIA	↑↑	Domenici et al. (2010) and Ramsey et al. (2013)

All proteins were found to be significantly differentially expressed in the plasma or serum of drug-free schizophrenia (SCZ) subjects. ↑/↓ indicates the direction of change for each compound, per study (arrows are in order of references given). The compounds are arranged in order of number of replications, from highest to lowest. Protein names are listed with corresponding gene name and Uniprot accession number.
Adapted from Sabherwal, S., English, J. A., Föcking, M., Cagney, G., & Cotter, D. R. (2016). Blood biomarker discovery in drug-free schizophrenia: The contribution of proteomics and multiplex immunoassays. *Expert Review of Proteomics*, doi:10.1080/14789450.2016.1252262.

There are relatively few proteomic studies conducted in first-onset drug-naive schizophrenia subjects. This is a result of the low recruitment rates of these patients (approximately 10–30 per year), ultimately leading to longer duration studies and longer storage time for samples that could potentially influence biomarker stability. This issue of recruitment is representative of psychiatry biomarker studies generally. Another factor to consider with regard to recruiting is whether the subjects are representative of the population you wish to study as a whole (for example, that first-onset drug-naive schizophrenia is not representative of the whole population of subjects who go on to develop schizophrenia).

It is important to note that the establishment of biomarkers may not only be of value in the diagnosis and prediction of schizophrenia, but also in the assessment of treatment response, long-term outcome, and clinical phenotypes.

11 Conclusion

In conclusion, despite the fact that high-throughput blood biomarker research in schizophrenia is still in its infancy, studies are beginning to yield consistent and valuable findings. These valuable findings provide an opportunity for the early identification and treatment of these debilitating disorders, and thus lead to better patient outcomes. This is a good example of the evolving role of proteomics in psychiatry. The rapidly advancing mass spectrometry and multiplex immunoassay technologies will no doubt lead to a more accurate, sensitive, reproducible, and comprehensive biological signature across a range of psychiatric disorders. The work will impact, however, not just early diagnosis, but should also be clearly designed to address issues of treatment outcome generally. For example, truly significant clinical impact will be derived from an awareness of protein biomarkers of good response to specific drugs. Further, different clinical phenotypes may respond differentially to treatments. An example of this may be the proposed inflammatory subtype of schizophrenia (Fillman, Sinclair, Fung, Webster, & Shannon Weickert, 2014) and the possibility that an antiinflammatory therapy agent may be more effective among such a subgroup. This is unproven, but is an example of where the next generation of studies may be directed (Fond et al., 2015; Leucht et al., 2015). Finally, it is likely that the best biomarker signatures will be derived not from one modality, but from the use of data from multiple modalities, such that the most discriminatory neuropsychological, neuroimaging, proteomic, and lipidic measures are combined into a single risk measure. This has been done to good effect for risk prediction in the at-risk mental state (Cannon et al., 2016; Clark et al., 2016). Future studies will need to refine this further in terms of transition from an at-risk mental state to psychosis, and to extend to "risk calculators" of treatment response and outcomes of different clinical phenotypes.

References

Alberio, T., Pippione, A. C., Zibetti, M., Olgiati, S., Cecconi, D., Comi, C., … Fasano, M. (2012). Discovery and verification of panels of T-lymphocyte proteins as biomarkers of Parkinson's disease. *Scientific Reports, 2*, 953. https://doi.org/10.1038/srep00953.

Amminger, G. P., Schäfer, M. R., Papageorgiou, K., Klier, C. M., Cotton, S. M., Harrigan, S. M., … Berger, G. E. (2010). Long-chain ω-3 fatty acids for indicated prevention of psychotic disorders. *Archives of General Psychiatry, 67*(2), 146. https://doi.org/10.1001/archgenpsychiatry.2009.192.

Burton, A. (2011). Big science for a big problem: ADNI enters its second phase. *The Lancet Neurology.* https://doi.org/10.1016/S1474-4422(11)70031-X.

Cannon, T. D., Yu, C., Addington, J., Bearden, C. E., Cadenhead, K. S., Cornblatt, B. A., … Kattan, M. W. (2016). An individualized risk calculator for research in prodromal psychosis. *American Journal of Psychiatry, 173*(10), 980–988. https://doi.org/10.1176/appi.ajp.2016.15070890.

Chan, M., Krebs, M.-O., Cox, D., Guest, P., Yolken, R., Rahmoune, H., … Bahn, S. (2015). Development of a blood-based molecular biomarker test for identification of schizophrenia before disease onset. *Translational Psychiatry, 5.* https://doi.org/10.1038/tp.2015.91.

Chapman, J. D., Goodlett, D. R., & Masselon, C. D. (2014). Multiplexed and data-independent tandem mass spectrometry for global proteome profiling. *Mass Spectrometry Reviews, 33*(6), 452–470. https://doi.org/10.1002/mas.21400.

Cheng, T. M. K., Lu, Y.-E., Guest, P. C., Rahmoune, H., Harris, L. W., Wang, L., … Bahn, S. (2010). Identification of targeted analyte clusters for studies of schizophrenia. *Molecular & Cellular Proteomics, 9*(3), 510–522. https://doi.org/10.1074/mcp.M900372-MCP200.

Clark, S. R., Baune, B. T., Schubert, K. O., Lavoie, S., Smesny, S., Rice, S. M., … Amminger, G. P. (2016). Prediction of transition from ultra-high risk to first-episode psychosis using a probabilistic model combining history, clinical assessment and fatty-acid biomarkers. *Translational Psychiatry, 6*(9), e897. https://doi.org/10.1038/tp.2016.170.

Cox, J., & Mann, M. (2008). MaxQuant enables high peptide identification rates, individualized p.p.b.-range mass accuracies and proteome-wide protein quantification. *Nature Biotechnology, 26*(12), 1367–1372. https://doi.org/10.1038/nbt.1511.

Doerr, A. (2010). Targeted proteomics. *Nature Methods, 7*(1), 34. https://doi.org/10.1038/nmeTh.F.284.

Domenici, E., Willé, D. R., Tozzi, F., Prokopenko, I., Miller, S., McKeown, A., … Muglia, P. (2010). Plasma protein biomarkers for depression and schizophrenia by multi analyte profiling of case-control collections. *PLoS One, 5*(2). https://doi.org/10.1371/journal.pone.0009166.

Echan, L. A., Tang, H. Y., Ali-Khan, N., Lee, K., & Speicher, D. W. (2005). Depletion of multiple high-abundance proteins improves protein profiling capacities of human serum and plasma. *Proteomics, 5*(13), 3292–3303. https://doi.org/10.1002/pmic.200401228.

Egertson, J. D., MacLean, B., Johnson, R., Xuan, Y., & MacCoss, M. J. (2015). Multiplexed peptide analysis using data-independent acquisition and skyline. *Nature Protocols, 10*(6), 887–903. https://doi.org/10.1038/nprot.2015-055.

English, J. A., Lopez, L. M., O'Gorman, A., Föcking, M., Hryniewiecka, M., Scaife, C., … Cotter, D. R. (2018). Blood-based protein changes in childhood are associated with increased risk for later psychotic disorder: Evidence from a nested case–control study of the ALSPAC longitudinal birth cohort. *Schizophrenia Bulletin, 44*(2), 297–306. https://doi.org/10.1093/schbul/sbx075.

English, J. A., Wynne, K., Cagney, G., & Cotter, D. R. (2014). Targeted proteomics for validation of biomarkers in early psychosis. *Biological Psychiatry, 76*(6), e7–e9. https://doi.org/10.1016/j.biopsych.2013.11.016.

Fillman, S. G., Sinclair, D., Fung, S. J., Webster, M. J., & Shannon Weickert, C. (2014). Markers of inflammation and stress distinguish subsets of individuals with schizophrenia and bipolar disorder. *Translational Psychiatry, 4*(2). https://doi.org/10.1038/tp.2014.8.

Focking, M., Pollack, T., Dicker, P., Cagney, G., Winter, I., Kahn, R., … Cotter, D. (2018). Proteomic analysis of blood based samples from the OPTiMiSE (optimization of treatment and Managment of schizophrenia in Europe) study point towards complement pathway protein changes. *Schizophrenia Bulletin, 44*(suppl_1), S74–S75.

Fond, G., D'Albis, M. A., Jamain, S., Tamouza, R., Arango, C., Fleischhacker, W. W., … Leboyer, M. (2015). The promise of biological markers for treatment response in first-episode psychosis: A systematic review. *Schizophrenia Bulletin.* https://doi.org/10.1093/schbul/sbv002.

Fusar-Poli, P., Bonoldi, I., Yung, A. R., Borgwardt, S., Kempton, M. J., Valmaggia, L., … McGuire, P. (2012). Predicting psychosis. *Archives of General Psychiatry, 69*(3), 220. https://doi.org/10.1001/archgenpsychiatry.2011.1472.

Glaviano, A., O'Donovan, S. M., Ryan, K., O'Mara, S., Dunn, M. J., & McLoughlin, D. M. (2014). Acute phase plasma proteins are altered by electroconvulsive stimulation. *Journal of Psychopharmacology, 28*(12), 1125–1134. https://doi.org/10.1177/0269881114552742.

Guest, P. C., Schwarz, E., Krishnamurthy, D., Harris, L. W., Leweke, F. M., Rothermundt, M., … Bahn, S. (2011). Altered levels of circulating insulin and other neuroendocrine hormones associated with the onset of schizophrenia. *Psychoneuroendocrinology, 36*(7), 1092–1096. https://doi.org/10.1016/j.psyneuen.2010.12.018.

Hampel, H., Wilcock, G., Andrieu, S., Aisen, P., Blennow, K., Broich, K., … Vellas, B. (2011). Biomarkers for Alzheimer's disease therapeutic trials. *Progress in Neurobiology, 95*(4), 579–593. https://doi.org/10.1016/j.pneurobio.2010.11.005.

Harris, L. W., Pietsch, S., Cheng, T. M. K., Schwarz, E., Guest, P. C., Bahn, S., & Hashimoto, K. (2012). Comparison of peripheral and central schizophrenia biomarker profiles. *PLoS One, 7*(10). https://doi.org/10.1371/journal.pone.0046368.

Jaros, J. A. J., Martins-De-Souza, D., Rahmoune, H., Rothermundt, M., Leweke, F. M., Guest, P. C., & Bahn, S. (2012). Protein phosphorylation patterns in serum from schizophrenia patients and healthy controls. *Journal of Proteomics, 76*, 43–55. https://doi.org/10.1016/j.jprot.2012.05.027.

Kelleher, I., Connor, D., Clarke, M. C., Devlin, N., Harley, M., & Cannon, M. (2012). Prevalence of psychotic symptoms in childhood and adolescence: A systematic review and meta-analysis of population-based studies. *Psychological Medicine, 42*(09), 1857–1863. https://doi.org/10.1017/S0033291711002960.

Kemp, A. H., Gordon, E., John Rush, A., & Williams, L. M. (2008). Improving the prediction of treatment response in depression: Integration of clinical, cognitive, psychophysiological, neuroimaging, and genetic measures. *CNS Spectrums, 13*(12), 1066–1086. quiz 1087-1088. https://doi.org/10.1017/S1092852900017120.

Kennedy, S. H., Downar, J., Evans, K. R., Feilotter, H., Lam, R. W., MacQueen, G. M., … Soares, C. (2012). The Canadian biomarker integration network in depression (CAN-BIND): Advances in response prediction. *Current Pharmaceutical Design, 18*(36), 5976–5989. https://doi.org/10.2174/138161212803523635.

Larsen, T. K., Melle, I., Auestad, B., Haahr, U., Joa, I., Johannessen, J. O., … McGlashan, T. (2011). Early detection of psychosis: Positive effects on 5-year outcome. *Psychological Medicine, 41*(7), 1461–1469. https://doi.org/10.1017/S0033291710002023.

Leucht, S., Winter-Van Rossum, I., Heres, S., Arango, C., Fleischhacker, W. W., Glenthøj, B., … Sommer, I. E. (2015). The optimization of treatment and management of schizophrenia in Europe (OPTiMiSE) trial: Rationale for its methodology and a review of the effectiveness of switching antipsychotics. *Schizophrenia Bulletin, 41*(3), 549–558. https://doi.org/10.1093/schbul/sbv019.

Levin, Y., Wang, L., Schwarz, E., Koethe, D., Leweke, F. M., & Bahn, S. (2010). Global proteomic profiling reveals altered proteomic signature in schizophrenia serum. *Molecular Psychiatry, 15*(11), 1088–1100. https://doi.org/10.1038/mp.2009.54.

Li, Y., Zhou, K., Zhang, Z., Sun, L., Yang, J., Zhang, M., … Wan, C. (2012). Label-free quantitative proteomic analysis reveals dysfunction of complement pathway in peripheral blood of schizophrenia patients: Evidence for the immune hypothesis of schizophrenia. *Molecular BioSystems, 8*(10), 2664–2671. https://doi.org/10.1039/c2mb25158b.

Maes, M., Delange, J., Ranjan, R., Meltzer, H. Y., Desnyder, R., Cooremans, W., & Scharpé, S. (1997). Acute phase proteins in schizophrenia, mania and major depression: Modulation by psychotropic drugs. *Psychiatry Research, 66*(1), 1–11. https://doi.org/10.1016/S0165-1781(96)02915-0.

Marques-Deak, A., Cizza, G., & Sternberg, E. (2005). Brain-immune interactions and disease susceptibility. *Molecular Psychiatry, 10*(3), 239–250. https://doi.org/10.1038/sj.mp.4001732.

Martins-de-Souza, D., Faça, V. M., & Gozzo, F. C. (2017). DIA is not a new mass spectrometry acquisition method. *Proteomics, 17*(7). https://doi.org/10.1002/pmic.201700017.

Maunsell, Z., Wright, D. J., & Rainbow, S. J. (2005). Routine isotope-dilution liquid chromatography-tandem mass spectrometry assay for simultaneous measurement of the 25-hydroxy metabolites of vitamins D2 and D3. *Clinical Chemistry, 51*(9), 1683–1690. https://doi.org/10.1373/clinchem.2005.052936.

O'Gorman, A., Suvitaival, T., Ahonen, L., Cannon, M., Zammit, S., Lewis, G., … Cotter, D. R. (2017). Identification of a plasma signature of psychotic disorder in children and adolescents from the Avon longitudinal study of parents and children (ALSPAC) cohort. *Translational Psychiatry, 7*(9), e1240. https://doi.org/10.1038/tp.2017.211.

Orešič, M., Tang, J., Seppänen-Laakso, T., Mattila, I., Saarni, S. E., Saarni, S. I., … Suvisaari, J. (2011). Metabolome in schizophrenia and other psychotic disorders: A general population-based study. *Genome Medicine, 3*(3). https://doi.org/10.1186/gm233.

Perkins, D. O., Jeffries, C. D., Addington, J., Bearden, C. E., Cadenhead, K. S., Cannon, T. D., … Heinssen, R. (2015). Towards a psychosis risk blood diagnostic for persons experiencing high-risk symptoms: Preliminary results from the NAPLS project. *Schizophrenia Bulletin, 41*(2), 419–428. https://doi.org/10.1093/schbul/sbu099.

Ramsey, J. M., Schwarz, E., Guest, P. C., Van Beveren, N. J. M., Leweke, F. M., Rothermundt, M., … Bahn, S. (2013). Distinct molecular phenotypes in male and female schizophrenia patients. *PLoS One, 8*(11). https://doi.org/10.1371/journal.pone.0078729.

Rutigliano, G., Valmaggia, L., Landi, P., Frascarelli, M., Cappucciati, M., Sear, V., … Fusar-Poli, P. (2016). Persistence or recurrence of non-psychotic comorbid mental disorders associated with 6-year poor functional outcomes in patients at ultra high risk for psychosis. *Journal of Affective Disorders, 203*, 101–110. https://doi.org/10.1016/J.JAD.2016.05.053.

Sabherwal, S., English, J. A., Föcking, M., Cagney, G., & Cotter, D. R. (2016). Blood biomarker discovery in drug-free schizophrenia: The contribution of proteomics and multiplex immunoassays. *Expert Review of Proteomics*. https://doi.org/10.1080/14789450.2016.1252262.

Sajic, T., Liu, Y., & Aebersold, R. (2015). Using data-independent, high-resolution mass spectrometry in protein biomarker research: Perspectives and clinical applications. *Proteomics, Clinical Applications, 9*(3–4), 307–321. https://doi.org/10.1002/prca.201400117.

Schwarz, E., Guest, P. C., Rahmoune, H., Harris, L. W., Wang, L., Leweke, F. M., … Bahn, S. (2012). Identification of a biological signature for schizophrenia in serum. *Molecular Psychiatry, 17*(5), 494–502. https://doi.org/10.1038/mp.2011.42.

Schwarz, E., Guest, P. C., Rahmoune, H., Martins-de-Souza, D., Niebuhr, D. W., Weber, N. S., … Bahn, S. (2011). Identification of a blood-based biological signature in subjects with psychiatric disorders prior to clinical manifestation. *World Journal of Biological Psychiatry, 2975*(June 2011), 1–6. https://doi.org/10.3109/15622975.2011.599861.

Schwarz, E., Izmailov, R., Spain, M., Barnes, A., Mapes, J. P., Guest, P. C., & Bahn, S. (2010). Validation of a blood-based laboratory test to aid in the confirmation of a diagnosis of schizophrenia. *Biomarker Insights, 2010*(5), 39–47. https://doi.org/10.4137/BMI.S4877.

Shah, J., Eack, S. M., Montrose, D. M., Tandon, N., Miewald, J. M., Prasad, K. M., & Keshavan, M. S. (2012). Multivariate prediction of emerging psychosis in adolescents at high risk for schizophrenia. *Schizophrenia Research, 141*(2–3), 189–196. https://doi.org/10.1016/j.schres.2012.08.012.

Teo, G., Kim, S., Tsou, C. C., Collins, B., Gingras, A. C., Nesvizhskii, A. I., & Choi, H. (2015). mapDIA: Preprocessing and statistical analysis of quantitative proteomics data from data independent acquisition mass spectrometry. *Journal of Proteomics, 129*, 108–120. https://doi.org/10.1016/j.jprot.2015.09.013.

Uhlen, M. (2005). A human protein atlas for normal and cancer tissues based on antibody proteomics. *Molecular & Cellular Proteomics, 4*(12), 1920–1932. https://doi.org/10.1074/mcp.M500279-MCP200.

Upthegrove, R., Manzanares-Teson, N., & Barnes, N. M. (2014). Cytokine function in medication-naive first episode psychosis: A systematic review and meta-analysis. *Schizophrenia Research, 155*, 101–108. https://doi.org/10.1016/j.schres.2014.03.005.

Yang, Y., Wan, C., Li, H., Zhu, H., La, Y., Xi, Z., ... He, L. (2006). Altered levels of acute phase proteins in the plasma of patients with schizophrenia. *Analytical Chemistry, 78*(11), 3571–3576. https://doi.org/10.1021/ac051916x.

Yung, A. R., & Lin, A. (2016). Psychotic experiences and their significance. *World Psychiatry: Official Journal of the World Psychiatric Association (WPA), 15*(2), 130–131. https://doi.org/10.1002/wps.20328.

Zhou, N., Wang, J., Yu, Y., Shi, J., Li, X., Xu, B., & Yu, Q. (2013). Mass spectrum analysis of serum biomarker proteins from patients with schizophrenia. *Biomedical Chromatography*, 654–659. https://doi.org/10.1002/bmc.3084.

Chapter 26

Molecular biomarkers in depression: Toward personalized psychiatric treatment

Anand Gururajan[a], John F Cryan[a,b] and Timothy G Dinan[b,c]

[a]*Department of Anatomy & Neuroscience, University College Cork, Cork, Ireland*, [b]*APC Microbiome Ireland, Cork, Ireland*, [c]*Department of Psychiatry & Neurobehavioural Science, University College Cork, Cork, Ireland*

1 Personalized psychiatry—A pipedream or a possibility?

Over the past few decades, rapid advances in high-resolution medical technologies have given clinicians the ability to more accurately predict risk of disease, carry out diagnoses, and implement "tailor-made" therapeutic strategies, a concept known as personalized medicine. This has been particularly successful in several areas of medicine, such as oncology, endocrinology, and neurology. But personalized psychiatry is yet to be a clinical reality and there are several factors for why this is the case. Here we have outlined some of these, and we examine how a focus on biomarkers could improve the status quo and bring personalized psychiatry one step closer to reality.

1.1 Challenges in diagnosis

Depression is currently diagnosed through the use of checklists such as those established by the DSM-V. This checklist contains a list of nine symptoms: depressed mood, loss of interest or pleasure, change in weight or appetite, insomnia or hypersomnia, psychomotor retardation or agitation, fatigue, worthlessness or guilt, impaired concentration, and suicidal ideation or attempts. Patients who present with five or more symptoms within a 2-week period are diagnosed accordingly. But, what about if a patient presents with only three out of the five, and one of these is suicidal thoughts and attempts? According to the current metric, this patient would not be considered depressed, but is nonetheless in need of medical attention. In other words, the DSM does not sufficiently take into account symptom severity. The DSM is also constrained in terms of acknowledging symptoms arising from comorbid conditions and the heterogeneity within subtypes of depression. Overall, utilizing such checklists as a measure of global psychological functioning—how someone feels, what they believe, how they see themselves—is limited, and arguably, takes a rather simplistic view on what it means to be human.

To improve the *status quo*, we need an in-depth understanding of the pathophysiology underscoring the constellation of symptoms associated with depression that manifest to varying degrees. Herein lies the first challenge, one that is currently being addressed by the Research Domain Criteria (RDoC) project initiated by the National Institutes of Mental Health (Insel et al., 2010). This project aims to take a pluralistic, bottom-up, biological approach toward the study of psychiatric disorders such as depression, encouraging a shift away from current black-and-white diagnostic schemes and prevailing modular views of functionally impaired brain regions, aberrant neurotransmission, and their supposed links to specific symptoms of depression. Though the current focus is on its utility as a research tool, we anticipate that the RDoC or a modified version of it, will be used in the context of personalized psychiatry in the near future.

The second challenge that needs to be overcome stems from more than a decade of investigations focused on elucidating the genetic basis for depression. There have been more than a dozen genome-wide association studies (GWAS), but these have yielded little in the way of clinically translatable outputs, or advancing our knowledge on the pathology of this disorder. One of the main reasons for this is the requirement of large sample sizes in order to see any meaningful statistical effect, i.e., a p-value $<5 \times 10^{-8}$; there is ongoing debate as to whether this cut-off score needs to be revised to acknowledge the potential importance of low-frequency variants (Fadista, Manning, Florez, & Groop, 2016). It has been estimated that for depression GWAS, the number of cases and controls required is at least 100,000, roughly 3–5 times more than what was needed for schizophrenia GWAS (Levinson et al., 2014). At first, this might seem to be a significant challenge, but "outsourcing" the tasks of collecting

Personalized Psychiatry. https://doi.org/10.1016/B978-0-12-813176-3.00026-2

samples and clinical data to the private sector is one potential solution. A recent example would be 23andMe, which recently teamed up with Pfizer and the Massachusetts General Hospital to identify 17 SNPs from 75,607 self-reported cases of major depression, and 231,747 controls (Hyde et al., 2016). Notwithstanding the absence of a formal clinical diagnosis, this approach is quick, cost-effective, and provides the future justification for more targeted investigations.

Tangential to the issue of sample size is the need to examine specific patient populations, for example, those suffering from post-partum depression, and this approach may yield more "hits" than "misses." A similar strategy could be employed by focusing on patients that display specific symptoms or endophenotypes that a have putative genetic basis (Goldstein & Klein, 2014). Related to the problem of stratification is the need for more ethnically diverse study populations.

A third factor that is only beginning to be considered in this context of GWAS is the role of the environment and its interactions with the genotype. These studies are known as genome-environment-wide interaction studies, or GEWIS, and there has been one study to date that has used this method to explore the influence of recent stressful life events and the impact of social support; the lack of any significant findings, however, were attributed, again, to the low sample size (Dunn et al., 2016).

It is worth noting that the factors we have discussed here could apply to any study investigating the neurobiology of depression for the purposes of personalizing diagnosis and treatment using other techniques, such as neuroimaging. The underlying theme, however, is the same—the importance of increasing sample size, exploring sample heterogeneity, and the need to acknowledge the influential role of the environment.

1.2 A role for biomarkers?

This leads us to the question, can biomarkers help realize the concept of personalized psychiatry? To answer, we begin by defining biomarkers as objective physiological indicators of normal biological processes, pathogenic processes, or a response to a specific therapeutic intervention (Biomarkers Definitions Working Group, 2001). Rather than linking disease states to pathophysiology that is not fully understood, biomarkers reflect probabilities. This might seem imprecise at first, and no real advancement in terms of improving diagnosis and treatment. But as described herein, the strength of these probabilistic determinations improves when the population initially sampled for their identification is large, and the findings are replicated (validated) across other study centers. Accuracy is further enhanced when biomarkers across multiple modalities (e.g., cytokine levels in the blood, cerebrospinal fluid levels of microRNAs) are used together in the form of a panel, as opposed to relying on a single one. In light of the practical advantages that biomarkers have, especially in terms of being able to assay them quickly and alongside other routine procedures (e.g., blood pressure checks), it has been argued that there is no pressing need to understand their functional relevance on the disorder that is being studied (Paulus, 2015; Pine & Leibenluft, 2015).

Against this background, in this book chapter, we examine evidence on several classes of candidate biomarkers for which there is accumulating evidence that supports their use in clinical settings in the context of depression. We also summarize each section highlighting general limitations of the studies that have been reviewed, and indicate potential strategies to move the field forward.

2 Noncoding RNAs

Over the past few decades, with the advent of high resolution sequencing technologies, it has become possible to study the neuropathology of major depression in the context of transcriptomic dysregulation. In particular, noncoding RNA species, such as microRNAs and long noncoding RNAs, have emerged as key players, particularly in terms of their ability to regulate the expression of a significant proportion of mammalian coding genes (~60% for microRNAs). Furthermore, the high concordance in noncoding RNA expression profiles in the central and peripheral compartments has provided support to their utility as biomarkers of CNS disorders (Liew, Ma, Tang, Zheng, & Dempsey, 2006).

2.1 MicroRNAs

MicroRNAs are short RNAs (18–22 nucleotides), which negatively regulate gene expression by binding to messenger RNAs (mRNAs) and inhibit their translation. Each microRNA can inhibit expression of multiple mRNAs, and each mRNA is the target of multiple microRNAs; as such, microRNAs are critical elements of complex neuronal gene regulatory networks (for review see Ha & Kim, 2014). There have been a number of clinical studies that have reported altered expression of microRNAs in blood samples from patients with depression, as well as from post-mortem analyses of prefrontal cortex and amygdala tissue (see Table 1). However, few have examined (i) whether dysregulation of microRNAs is consistent across

TABLE 1 Noncoding transcriptomic biomarkers.

Study type	Source	Main findings	References
Longitudinal	PMBCs	CON vs MD-UP: miR-589, miR-579, miR-941, miR-133a, miR-494, miR-107, miR-148a, miR-652, miR-425-3p; DOWN: miR-517b, miR-636, miR-1243, miR-381, miR-200c. AD treatment induced changes in miRA expression—UP: miR-20b-3p, miR-433, miR-409-3p, miR-410, miR-485-3p, miR-133a, miR-145; DOWN: miR-331-5p	Belzeaux et al. (2012)
Longitudinal	Whole blood	Escitalopram induced changes in miRNA expression—UP: miR-130b, miR-505, miR-29b-2, miR-26b, miR-22, miR-26a, miR-664, miR-494, let-7d, let-7g, let-7e, let-7f, miR-629, miR-106b, miR-103, miR-191, miR-128, miR-502-3p, miR-374b, miR-132, miR-30d, miR-500, miR-589, miR-183, miR-574-3p, miR-140-3p, miR-335, miR-361-5p; DOWN: miR-34c-5p, miR-770-5p	Bocchio-Chiavetto et al. (2013)
Case-control	Serum	UP: miR-132, miR-182	Li et al. (2013)
Post-mortem, case control	Raphe nuclei, blood	DOWN: miR-135	Issler et al. (2014)
Post-mortem, case-control	Prefrontal cortex, whole blood	DOWN: miR-1202; Citalopram increased miR-1202 expression in blood of responders	Lopez et al. (2014)
Case-control	PMBCs	UP: miR-1972, miR26b, miR-4485, miR-4498, miR-4743	Fan et al. (2014)
Case-control	CSF and whole blood	CSF-DOWN: miR-16; no change in blood	Song et al. (2015)
Case-control	Plasma	UP: miR-320a; DOWN: miR-451a, miR-17-5p, miR-223-3p	Camkurt et al. (2015)
Case-control	Whole blood	DOWN: let-7b, let-7c	Gururajan et al. (2016)
Case-control	Whole blood	DOWN: let-7a, let-7d, let-7f; UP: miR-24, miR-425	Maffioletti et al. (2016)
Case-control	Serum	UP: miR-124	Roy, Dunbar, Shelton, and Dwivedi (2016)
Post-mortem, case-control	Prefrontal cortex, whole blood	UP: miR-146-a-5p, miR-146-b-5p, miR-24-3p, miR-425-3p, miR-3074-5p; Duloxetine decreased miR-146-a-5p, miR-146-b-5p, miR-24-3p, miR-425-3p, miR-3074-5p in blood	Lopez et al. (2017)

central and peripheral compartments and (ii) their functional expression (Gururajan, Clarke, Dinan, & Cryan, 2016). Insights from such approaches will significantly elevate the status of microRNAs as biomarkers for depression, and more importantly, reveal novel drug targets for therapeutic exploitation. Indeed, several clinical studies have shown a putative role for microRNAs in the mechanism of action of commonly prescribed antidepressants, and moreover, baseline expression of some microRNAs is predictive of response (Table 1). The implication here is that baseline microRNA expression assays could be used to ascertain whether a specific antidepressant will be suitable for a patient, and monitoring expression over duration of therapy allows for prognostications. Also, by analyzing the downstream gene expression programs that are inhibited by these microRNAs, novel drug targets may be identified that could lead to pharmacotherapies that are faster-acting and have fewer side effects.

2.2 Long noncoding RNAs

Long noncoding RNAs (lncRNAs, 200 nucleotides in length) are abundantly expressed in the brain with functional roles that are not limited to regulating the expression of protein-coding genes. LncRNAs can also act as precursors to microRNAs, and are involved in chromatin remodeling and intracellular trafficking (Qureshi, Mattick, & Mehler, 2010).

In the context of major depression, one clinical study reported reductions in several peripherally expressed lncRNAs in patients (TCONS_L2_00001212, NONHSAT102891, TCONS_00019174, ENST00000566208, NONHSAG045500, ENST00000517573, NONHSAT034045, NONHSAT142707); the expression of all but one (NONHSAT102891) was normalized following 6 weeks of antidepressant treatment (Cui et al., 2016). Interestingly, hippocampal expression of TCONS_00019174 was found to be reduced in mice that had been exposed to the chronic mild stress (CMS) paradigm (Ni, Liao, Li, Zhang, & Wu, 2017). Furthermore, overexpression of this lncRNA in this brain region rescued CMS-induced anxiety- and depressive-like behaviors, an effect mediated by the Wnt/β-catenin signaling pathway (Ni et al., 2017). More recently, a single nucleotide polymorphism (SNP) for the lncRNAs LINC01108 (rs12526133) and LINC00578 (rs2272260) was found to be associated with major depression; peripheral expression of LINC01108 was higher, whereas LINC00998 was lower in patients relative to controls (Ye et al., 2017). It is worth noting that the clinical studies described here relied on peripheral analytics, and the ethnicity of the population studied was Han. Clearly, there is a gap in the literature with regard to the role of lncRNAs, not only in the pathophysiology of major depression, but also as putative targets for therapeutic exploitation.

2.3 Perspective

Overall, it is unlikely that a single microRNA or lncRNA will be effective as a biomarker for diagnosis of depression. Rather, focus should be given to expression of noncoding RNAs in diseased and healthy states at a network level as regulators of gene expression. Additionally, there is a need to further explore the effects of various therapeutic strategies on the noncoding transcriptome; for example, in the case of microRNAs, the field miRNA-pharmacogenomics has emerged to examine the influence of genetic polymorphisms in the miRNA-pathway on antidepressant response (Rukov, Wilentzik, Jaffe, Vinther, & Shomron, 2014).

3 Coding transcriptomic and proteomic biomarkers

3.1 Neurotransmitters

One of the primary mechanisms by which current antidepressants exert their therapeutic effects is via restoration of an imbalance in expression of neurotransmitters. As such, several studies have reported alterations in the expression of the enzymes involved in their synthesis, their receptors, and their transporters.

3.1.1 Serotonin

Evidence indicates that depression is associated with either hyper- and hyposerotonergic states (Andrews, Bharwani, Lee, Fox, & Thomson, 2015). The peripheral presence of serotonin receptors, their transporter (5HTT), and the critical enzyme involved its synthesis, tryptophan hydroxylase (TPH), confers a distinct advantage in terms of their utility as biomarkers for depression. At the mRNA level, data from post-mortem and clinical studies is equivocal with decreases, increases, and no effects on receptors (5HT1A, 1B, 2B) and 5HTT and TPH expression (Table 2). At a proteomic level, meta-analysis of post-mortem and in vivo imaging studies have revealed decreased 5HTT expression in several brain regions implicated in depression (Kambeitz & Howes, 2015). Importantly, psychotherapy normalized this effect more so in patients with higher baseline symptom severity (Joensuu et al., 2016).

3.1.2 Dopamine

Evidence indicates dysfunctional dopaminergic neurotransmission in the pathology of depression, although currently available antidepressants do not directly affect this system (Dunlop & Nemeroff, 2007). Post-mortem analysis revealed increased expression of the dopamine D4 receptor mRNA in the amygdala of patients with major depression (Xiang et al., 2008); whereas in the periphery, reductions (Rocca et al., 2002), as well and no difference (Iacob et al., 2014) in expression relative to controls have been reported (Iacob et al., 2014; Rocca et al., 2002) (Table 2).

3.1.3 Glutamate

With the increased interest in the potential therapeutic value of ketamine and its analogues for the treatment of depression, the glutamate hypothesis of depression has gained popularity (Dutta, McKie, & Deakin, 2015). Indeed, post-mortem expression of mRNA in the prefrontal cortex for a range of NMDA receptor subunits (NR1, NR2A, N2C, N2D, and

TABLE 2 Neurotransmitter biomarkers.

Biomarker	Study type	Source	Main findings	References
Serotonin receptor/ transporter	Post-mortem	Hippocampus	Decreased expression of 5HT1A mRNA	Lopez, Chalmers, Little, and Watson (1998)
	Case-control	Lymphocytes	No difference in 5HT1A mRNA	Fajardo, Galeno, Urbina, Carreira, and Lima (2003)
	Post-mortem	Hippocampus Prefrontal cortex	Decreased 5HT1A mRNA	López-Figueroa et al. (2004)
	Post-mortem	Raphe nuclei	No difference in TPH levels	Bonkale, Murdock, Janosky, and Austin (2004)
	Post-mortem	Raphe nuclei	Increased TPH expression in patients who had committed suicide	Boldrini, Underwood, Mann, and Arango (2005)
	Case-control	Leukocytes	Increased 5HTT mRNA expression, normalized by 8 weeks of paroxetine treatment	Iga et al. (2005)
	Case-control	Lymphocytes	Reduced 5HTT mRNA expression; more pronounced with the long allelic variant	Lima, Mata, and Urbina (2005)
	Case-control	PMBC	Increased expression of 5HTT mRNA, normalized by 12 weeks of fluoxetine treatment	Tsao, Lin, Chen, Bai, and Wu (2006)
	Case-control	PMBC	Antidepressant-induced increase in 5HTT mRNA	Belzeaux et al. (2010)
	Post-mortem	Raphe nuclei	Increased TPH2 mRNA	Bach-Mizrachi et al. (2005)
Dopamine (D4) receptor	Case-control	Lymphocytes	Decreased D4DR mRNA, normalized with 8 weeks of paroxetine treatment	Rocca et al. (2002)
	Post-mortem	Amygdala	Increased D4DR mRNA	Xiang et al. (2008)
	Case-control	Leukocytes	ECT/isoflurane induced decreased in D4DR mRNA	Iacob et al. (2014)
GABA	Case-control	CSF	Decreased GABA levels	Gold, Bowers Jr., Roth, and Sweeney (1980)
	Case-control	CSF	No difference	Post et al. (1980)
	Case-control	CSF	Decreased GABA levels	Gerner and Hare (1981)
	Case-control	CSF	Decreased GABA levels	Kasa et al. (1982)
	Case-control	CSF	Decreased GABA levels	Gerner et al. (1984)
	Case-control	CSF	Decreased GABA levels	Roy, Dejong, and Ferraro (1991)
	Case-control	Plasma	Decreased GABA levels	Petty and Schlesser (1981)
	Case-control	Plasma	Decreased plasma levels	Petty and Sherman (1984)
	Case-control	Plasma	Decreased plasma levels	Petty, Kramer, Dunnam, and Rush (1990)
	Case-control	Plasma	Decreased plasma levels	Petty, Kramer, Gullion, and John Rush (1992)
	Prospective	Plasma	No change with 4 weeks of desipramine treatment	Petty, Steinberg, Kramer, Fulton, and Moeller (1993)
	Prospective	Plasma	Acute decrease with ECT, responders had higher baseline GABA	Devanand et al. (1995)
	Prospective	Plasma	Increased plasma GABA levels after 10 days of treatment with fluoxetine or s-citalopram	Küçükibrahimoğlu et al. (2009)

N3A) was reportedly increased in patients with depression (Gray, Hyde, Deep-Soboslay, Kleinman, & Sodhi, 2015). However, the absence of receptor expression in the periphery precludes their utility as biomarkers for depression.

3.1.4 GABA

The role of GABA in the pathophysiology of depression and as a target for antidepressant therapies has been well characterized (Luscher, Shen, & Sahir, 2010). Several studies have reported reductions in GABA in the CSF and plasma (Table 2). Plasma levels in particular have been suggested to have a role as biomarkers reflective of treatment response (Küçükibrahimoğlu et al., 2009). Moreover, higher baseline plasma GABA levels were found to be predictive of remission following ECT (Devanand et al., 1995).

3.2 BDNF and growth factors

3.2.1 Brain-derived neurotrophic factor

Brain-derived neurotrophic factor (BDNF) is a neurotrophic with a multitude of critical roles, which include aspects of neurodevelopment, synaptic plasticity, and neurogenesis (Autry & Monteggia, 2012). Peripherally, the expression of BDNF mRNA is lower in patients with depression according to several studies, although one study reported no difference compared with controls (Table 3). Similar equivocal findings have emerged from clinical studies focused on protein levels (Bocchio-Chiavetto et al., 2010; Brunoni, Lopes, & Fregni, 2008; Polyakova et al., 2015). Antidepressants have been reported to have normalized peripheral BDNF expression in most cases, and this usually correlates with symptomatic improvements (Allen et al., 2015; Brunoni et al., 2008; Molendijk et al., 2013; Sen, Duman, & Sanacora, 2008). One study has reported that methylation of the BDNF promoter-associated CpG1 is correlated with clinical diagnosis (Fuchikami et al., 2011), whereas another observed increased methylation of the exon 1 promoter region in patients compared with controls (D'Addario et al., 2013). In the second study, methylation rates were also found to be higher in patients who were on antidepressant therapies. The Val66Met BDNF polymorphism has been described as a risk factor for the onset of psychiatric disorders, including depression, and in this context, recent work has shown that carriers of the Met polymorphism have lower levels of peripheral BDNF, which negatively correlated with both age of onset, and positively correlated with duration of illness, and a previous number of depressive episodes (Colle et al., 2017).

3.2.2 Vascular endothelial growth factor

There is increasing evidence that implicates a role for vascular endothelial growth factor (VEGF) in the neuropathology of depression (Clark-Raymond & Halaris, 2013). Results from clinical studies have reported equivocal results in terms of the expression of VEGF mRNA and protein in patients with depression compared with controls (Clark-Raymond & Halaris, 2013; Iga et al., 2007) (Table 3). Interestingly, preclinical findings have consistently reported reductions in VEGF expression in the brain, but not in the periphery (Bergström, Jayatissa, Mørk, & Wiborg, 2008; Elfving, Plougmann, & Wegener, 2010).

3.2.3 Fibroblast growth factor

There are more than 18 different fibroblast growth factors (FGFs), of which 10 are expressed in and involved in several CNS processes, including neurogenesis and synaptic plasticity (Turner Cortney, Watson Stanley, & Akil, 2012). Postmortem studies have reported dysregulated expression of several FGF transcripts (FGF1, FGFR1, FGF2, FGFR2, FGFR3) in several brain regions implicated in the neuropathology of depression (Table 3). In contrast, the peripheral levels of FGF1 mRNA and FGF2 were not found to be different between patients and controls (Milanesi et al., 2012; Takebayashi, Hashimoto, Hisaoka, Tsuchioka, & Kunugi, 2010).

3.2.4 VGF

The role of VGF in depression is yet to be fully elucidated, but data from preclinical studies have shown that its expression is linked to that of other growth factors, such as BDNF, and that it has antidepressant-like effects in relevant animal models (Malberg & Monteggia, 2008). The peripheral expression of VGF was found to be reduced in patients with depression compared with controls, and the levels were normalized following antidepressant treatment in symptomatic responders (Cattaneo et al., 2012; Cattaneo, Sesta, et al., 2010).

TABLE 3 Transcriptomic and proteomic biomarkers.

Biomarker	Study type	Source	Main findings	References
BDNF	Case-control	Whole blood	No difference in BDNF mRNA between patients and controls	Otsuki et al. (2008)
	Case-control	Lymphocytes	Decreased BDNF mRNA levels in adult and pediatric patients with depression	Pandey et al. (2010)
	Case-control	Leukocytes/serum	Decreased BDNF mRNA normalized by 12 weeks of escitalopram treatment	Cattaneo et al. (2010)
	Case-control	PMBC	Decreased BDNF mRNA	D'Addario et al. (2013)
	Case-control	Lymphocytes	Decreased BDNF mRNA; correlated with symptom severity on the Hamilton Depression Rating Scale	Zhou et al. (2013)
	Case-control	Leukocytes	Decreased BDNF mRNA; more pronounced in patients with treatment-resistant depression	Hong et al. (2014)
	Case-control	Leukocytes	Decreased BDNF levels in Val/Met and Met/Met patients relative to Val/Val patients	Colle et al. (2017)
VEGF	Case-control	Leukocytes	Increased expression of VEGF mRNA levels normalized by 8 weeks of treatment with paroxetine	Iga et al. (2007)
FGF	Post-mortem	Prefrontal, anterior cingulate cortex	Decreased expression of FGF transcripts, but effect less pronounced in patients who were on SSRI treatment	Evans et al. (2004)
	Post-mortem	Locus coeruleus	Decreased expression of mRNA for FGF3 and Trkb in patients	Bernard et al. (2011)
VGF	Case-control	Leukocytes	Decreased expression of VGF mRNA levels normalized by 12 weeks of escitalopram treatment, but only in responders	Cattaneo et al. (2010)
		Leukocytes	Decreased expression of VGF mRNA levels normalized by 8 weeks of escitalopram or nortriptyline treatment, but only in responders	Cattaneo et al. (2013)
HPA axis	Prospective	Plasma	Post-DST nonsuppression was normalized by treatment with tricyclics or cognitive therapy	McKnight, Nelson-Gray, and Barnhill (1992)
	Post-mortem	PVN	Increased expression of CRH mRNA in the PVN of patients	Raadsheer et al. (1995)
	Post-mortem	Hippocampus	Decreased expression of MR mRNA levels in the CA3 region of the hippocampus	Lopez et al. (1998)
	Case-control	Plasma	High CRH levels in all patients compared with controls. Higher in patients with severe depression compared with other groups. No change in ACTH levels	Catalán, Gallart, Castellanos, and Galard (1998)
	Longitudinal	Plasma	Reduced expression of ACTH with 6 weeks of fluoxetine treatment	Inder, Prickett, Mulder, Donald, and Joyce (2001)
	Post-mortem	Hippocampus, prefrontal cortex	Decreased expression of GR mRNA levels in the CA3 region of the hippocampus, layers III–VI of the frontal cortex and layer IV of the temporal cortex	Webster, Knable, O'Grady, Orthmann, and Weickert (2002)
	Case-control	Plasma	High CRH levels in patients. Post-DST nonsuppression in 33% of patients, but not controls	Galard, Catalán, Castellanos, and Gallart (2002)

Continued

TABLE 3 Transcriptomic and proteomic biomarkers.—cont'd

Biomarker	Study type	Source	Main findings	References
	Post-mortem	Locus coeruleus, raphe nuclei	Increased CRH expression in the locus coeruleus and raphe nuclei of patients	Austin, Janosky, and Murphy (2003)
	Case-control/prospective	Plasma	Post-DST nonsuppression in patients associated with symptom severity. Elevated ACTH levels. Normalized following pharmacotherapy and ECT	Kunugi et al. (2005)
	Case-control	Whole blood	Decreased expression of GRα mRNA in depressed patients and patients in remission	Matsubara, Funato, Kobayashi, Nobumoto, and Watanabe (2006)
	Case-control	Serum	BclI polymorphism in the glucocorticoid receptor associated with high baseline ACTH levels and predicted poor responsiveness to SSRI therapy	Brouwer et al. (2006)
	Case-control	Plasma	No difference in plasma ACTH levels	Carroll et al. (2007)
	Case-control/prospective	Plasma	Post-DST nonsuppression in 44% of patients. Normalization in the 72% of patients following pharmacotherapy	Ising et al. (2007)
	Post-mortem	Amygdala, cingulate gyrus	Decreased expression of GRα mRNA in patients compared with controls	Alt et al. (2010)
	Post-mortem	Hippocampus, inferior frontal gyrus	Decreased expression of MR and GR mRNA levels in the hippocampus	Klok et al. (2011)
	Longitudinal	Whole blood	Attenuated increased in mRNA expression of chaperone genes involved in GR response in depressed pregnant women. Decreased GR sensitivity, which positively correlated with symptom severity	Katz et al. (2011)
	Post-mortem	Hippocampus	Decreased MR expression in the anterior, but not posterior hippocampus in patients. No change in GR expression	Medina et al. (2013)
	Case-control	Leukocytes	Decreased expression of GR mRNA levels in patients with depression, which was increased by 12 weeks of antidepressant treatment	Cattaneo et al. (2013)
	Case-control	Leukocytes	No difference in GR mRNA levels between controls and patients with depression and bipolar disorder. Significantly lower levels in patients with depression compared with patients with bipolar disorder	Iacob et al. (2013)
	Case-control	Leukocytes	No difference between controls and patients but ECT/isoflurane treatment decreased GR mRNA levels	Iacob et al. (2014)
Inflammatory biomarkers	Case-control	PMBC	Increased expression of IL-1beta, IL-6, IFN-gamma, and TNFα in patients compared with controls. IFN levels normalized by 12 weeks of fluoxetine treatment	Tsao et al. (2006)
	Case-control	PMBC	Increased expression of IL-10 and no change following 8 weeks of antidepressant treatment	Belzeaux et al. (2010)
	Case-control	Lymphocytes	Decreased expression of ApoER2 in patients compared with controls	Suzuki et al. (2010)

TABLE 3 Transcriptomic and proteomic biomarkers.—cont'd

Biomarker	Study type	Source	Main findings	References
	Case-control	PMBC	Increased expression of mRNAs for PU.1 in unmedicated depressed patients compared with healthy controls and patients with schizophrenia and bipolar disorder	Weigelt et al. (2011)
	Case control	Whole blood	Increased mRNA expression for COX-2, MPO, PLA2G2A, and iNOS in patients compared with controls	Gałecki et al. (2012)
	Case-control	Leukocytes	Increased expression of IL-1beta, IL-6, TNFα, and MIF in patients compared with controls. Responders to antidepressants showed reduced expression of mRNA for IL6	Cattaneo et al. (2013)
	Case-control	Leukocytes	Increased expression of IL-10 and IL-6 in patients compared to controls	Iacob et al. (2013)
	Case-control	Leukocytes	Increased expression of IL-10 in patients was decreased by treatment with ECT/Isoflurane anesthesia	Iacob et al. (2014)
	Longitudinal	Plasma	Higher baseline levels of IL-6, ACT, and CRP were predictive of symptomatology over a 5-year period	Zalli, Jovanova, Hoogendijk, Tiemeier, and Carvalho (2016)
	Case-control	Plasma	Increased peripheral expression of B2RAN2 and endoligin in patients with depression compared with patients with bipolar disorder	Ren et al. (2017)
	Prospective	Serum	BAFF, ICAM-1, eotaxin-1, RAGE, IL-12p40 were associated with response to venlafaxine. ICAM-1, C3, LAP-TGF-b1, eotaxin-2, chemokine CC-4, kallikrein-7, IL-1alpha, IL-15, CD40-ligand, resistin, osteoprogerin, B lymphocyte chemoattractant, were associated with response to imipramine	Chan et al. (2016)
	Case-control	Serum	Decreased expression of MIP-3B, IGFBP-5, and IL-2RA in female patients	Ramsey et al. (2016)
	Case-control	Serum	Increased expression of regulatory T-cells before and after delivery	Krause et al. (2014)
	Case-control	Serum	In dizygotic twin pair, higher differential expression of MPO, IL-6, sTNF RII in brother with depression compared with unaffected brother	Vaccarino et al. (2008)

3.3 Biomarkers associated with the HPA axis

Depression is characterized by aberrant activity of the hypothalamic pituitary adrenal (HPA) axis, although the direction of dysfunction appears to differ between subtypes—hypercortisolism in melancholia, and decreased or normal function in atypical depression (Juruena, Bocharova, Agustini, & Young, 2017). Here we have briefly reviewed findings from studies examining the peripheral components of this axis, in particular, cortisol, corticotropic release factor (CRF), and adreno-corticotrophic hormone (ACTH), and their utility as biomarkers for depression.

3.3.1 Cortisol

Patients with depression have been reported to have abnormal awakening cortisol levels, and disruptions to the diurnal rhythm of cortisol secretion, as well as dysfunctional cortisol response following a stressful task or a pharmacological challenge, such as the dexamethasone suppression test (DST) (Belvederi Murri et al., 2014; Herbert, 2013) (Table 3). The DST is a clinical strategy that is used to assess the status of the HPA axis, particularly in relation to efficacy of negative feedback mechanisms to regulate glucocorticoid release (Dinan, 1994). Briefly, the test involves administering an oral formulation of high or low doses of the glucocorticoid dexamethasone to patients, which is followed by analysis of plasma cortisol the following day, or at later time points (up to 3 days). In the overnight method (dose of 1.5 mg:2300 h, plasma analysis:1600), dexamethasone nonsuppression is defined as baseline plasma cortisol levels in excess of 5 µg/dL or 137 nmol/L. Between 4% and 70% of patients fall under this classification (Dinan, 2001), and patients generally will recover HPA axis function following antidepressant therapy, cognitive behavioral therapies, or the combination of antidepressant and electroconvulsive therapy (Ising et al., 2007; Kunugi et al., 2005; McKnight et al., 1992). Some studies perform a second analysis of the response to a CRF challenge immediately after the collection of the first DST blood sample (Ising et al., 2007; Kunugi et al., 2005; Zobel et al., 2001).

3.3.2 Corticotropic release factor

Corticotropic release factor (CRF) is the hormonal trigger of the HPA axis, and several clinical studies have reported increased expression in the brain, and also in the periphery of patients with depression (Table 3). One study has shown that while at baseline, CRF expression was higher in patients with depression compared with controls; the DST levels were reduced only in patients, but not controls (Galard et al., 2002). These findings are further supported by preclinical investigations with animal models of aspects of depression (Nakase, Kitayama, Soya, Hamanaka, & Nomura, 1998; Plotsky & Meaney, 1993; Zeng, Kitayama, Yoshizato, Zhang, & Okazaki, 2003).

3.3.3 Adrenocorticotrophic hormone

Adrenocorticotrophic hormone (ACTH) is released by the anterior pituitary, and stimulates the release of cortisol into the circulation from the adrenal cortex. While some studies have reported elevated post-DST levels of ACTH in patients with depression (nonsuppressors), others have reported no significant difference compared with controls; ACTH decreased following 6 weeks of treatment with fluoxetine (Table 3) (Carroll et al., 2007; Catalán et al., 1998; Inder et al., 2001; Kunugi et al., 2005). Notably, polymorphisms of the glucocorticoid receptor, as well as the nuclear transporter protein, FKBP5, have been linked to elevated plasma ACTH in patients (Brouwer et al., 2006; Menke et al., 2013). Preclinical studies have consistently reported elevated plasma ACTH in animal models of aspects of depression (Nakase et al., 1998; Zeng et al., 2003).

3.4 Inflammatory biomarkers

The neuropathology of depression is underscored, in part, by a significant inflammatory component, and this has been linked to dysregulation of the HPA-axis (Hodes, Kana, Menard, Merad, & Russo, 2015). Consistent with the complexities of gene by environment interactions in this context, a twin study has also shown larger associations between depression and peripheral markers of inflammation in dizygotic, rather than monozygotic twins (Vaccarino et al., 2008). Case-control studies have consistently reported elevated levels of transcripts for pro-inflammatory cytokines IL-1alpha, IL-1beta, IL-6, IL-8, IL-10, IFN-gamma, MIF, and TNF-alpha (Table 3). In terms of specificity, B2RAN2 and endoligin, both markers of inflammation, were found to be differentially expressed between patients with depression and patients with bipolar disorder (Ren et al., 2017). There have been several studies that have identified markers of inflammation associated with the response to antidepressants such as IL-12 and chemokine CC-4 with venlafaxine and imipramine (Chan et al., 2016), and IL-10 with ECT (Iacob et al., 2014). Similarly, IL-1beta, MIF, and TNF-alpha expression at baseline levels were found to be predictive of nonresponse, whereas IL-6 expression at this time point correlated with antidepressant response (Cattaneo et al., 2013).

A recent study identified several sex-specific serum candidate biomarkers involved in the inflammatory response, which may explain potential differences in prevalence, and suggest underlying variation in pathophysiology; in females these were IGFBP-5, MIP-3B, and IL-2RA (Ramsey et al., 2016). Also in females, one study reported that elevations in pre- and postnatal regulatory T-cells were associated with risk of post-partum depression (PPD) (Krause et al., 2014). Given the high-prevalence of PPD (10%–20%) and the potential long-term consequences to offspring (Perani & Slattery, 2014), early-identification of at-risk women through prenatal screening for biomarkers could ensure they receive adequate treatment.

3.5 Perspective

Based on the clinical evidence that has been presented, there are several candidates for state and trait biomarkers of depression that have shown particular promise. However, as mentioned in the introduction, the lack of consistency among studies in terms of methodology, patient heterogeneity, and insufficient patient metadata makes any further progress difficult. In any case, as described in the previous section on noncoding RNAs, a panel of biomarkers would be a lot more informative than using only one.

4 Genomic biomarkers

4.1 Mitochondrial DNA

Unlike nuclear DNA, mitochondrial DNA (mtDNA) can exist in multiple copies within a single mitochondria. Homoplasmy is defined as when mtDNA sequences are identical in all mitochondria in a given cell. In contrast, heteroplasmy is the state when not all mtDNA are identical, and has been linked to pathological states (Schon, DiMauro, & Hirano, 2012). Recent work has shown peripheral heteroplasmy and higher mtDNA levels in patients with depression, which correlated with previous stress exposure. Similar findings were also observed in a mouse model of stress-induced aspects of depression, but mtDNA levels normalized during a recovery period following the stress. Further research is needed to investigate the mechanistic links between mtDNA and the neuropathology of depression, but the evidence to date from patients with a history of stress and preclinical stress models implicates a role for the HPA-axis (see the previous section).

4.2 Telomeric biomarkers

In the past two decades, there has been increasing interesting in the potential utility of peripheral telomeric DNA sequences as biomarkers of depression (Table 4). Telomeres cap the ends of chromosomes, preventing them from degradation, and fusing with neighboring chromosomes. The length of telomeric sequences, otherwise known as telomere length (TL), is maintained by a negative feedback loop that activates the enzyme telomerase (TEL) when it is either too short, or when it has exceeded target length (Mergny, Riou, Mailliet, Teulade-Fichou, & Gilson, 2002). There are several hypothetical mechanisms that link TL to known pathophysiological processes in depression; in particular, dysregulation of the HPA-axis and inflammation. One mechanism involves the stress-induced oxidative damage of telomere sequences (Wolkowitz et al., 2011), and a second is based on the observation that stress-induced elevations in cortisol reduced expression of the TEL catalytic component, telomerase reverse transcriptase (TERT) (Choi, Fauce, & Effros, 2008). The link between inflammation and TL is unclear, but evidence suggests the involvement of the NF-kB signaling pathway, a pathway that regulates expression of pro-inflammatory cytokines (Wolkowitz et al., 2011).

As shown in Table 4, the evidence in support of telomeres as biomarkers for depression is growing, but there are several critical issues that remain unaddressed; some of these may be applicable to other biomarkers that have been discussed in this chapter. First, TL and TEL expression are non-specific biomarkers in that their dysregulation is also observed in other psychiatric disorders, including schizophrenia (Polho, De-Paula, Cardillo, dos Santos, & Kerr, 2015) and bipolar disorder (Lima et al., 2015). Second, the link between TL and depression symptomatology has yet to be established—how short does TL need to be in order to have a functional effect? Third, it is unclear whether changes in TL or TEL activity seen in depression represent state or trait biomarkers. TL length reduces with age, and such shortening is the default state (Armanios, 2013); but preclinical evidence has shown that antidepressant therapy can restore TEL activity, and subsequently influence TL (Zhou et al., 2011). Last, whether peripheral TL or TEL activity accurately reflects central pathology is unknown due to the paucity of post-mortem data.

5 The gut microbiome

Over the past 2–3 decades, the concept of the gut microbiome as a "second-brain" and its communication with the central nervous system via a so-called bacterial broadband has gained significant attention. The development of sequencing technologies has enabled the identification of bacterial species associated with specific health disorders, including depression (Naseribafrouei et al., 2014). To further support the potentially causal role of the gut microbiome in this context, recent studies have shown that the transplantation of microbiota from patients with depression to germ-free mice resulted in depression-like behaviors (Kelly et al., 2016; Zheng et al., 2016). As such, the prospect of using fecal samples from patients to screen for the presence/abundance of bacterial species known to be associated with depression as biomarkers is worth

TABLE 4 Telomeric biomarkers.

Study type	Source	Main findings	References
Case-control	Leukocytes	Decreased TL in patients with depression. Patient sample included depressed and bipolar disorder patients, but no difference between groups	Simon et al. (2006)
Case-control	Leukocytes	Decreased TL in patients with depression. Sex, age, apolipoprotein E 2, and polymorphisms in MAOA promoter region all have effects on TL	Lung, Chen, and Shu (2007)
Case-control	Leukocytes	TL reduced in patients compared with controls, and was unaffected by ongoing antidepressant therapy	Hartmann, Boehner, Groenen, and Kalb (2010)
Post-mortem	Cerebellum	No difference in TL between patients and controls	Zhang, Cheng, Craig, Redman, and Liu (2010)
Post-mortem	Occipital cortex	No difference in TL between patients and controls	Teyssier, Ragot, Donzel, and Chauvet-Gelinier (2010)
Case-control	Leukocytes	Decreased TL in CVD patients with depression. Study participants were mostly male. Patients diagnosed with MDD had higher anxiety scores and were less physically active	Hoen et al. (2011)
Population	Lymphocytes	No association between TL and depression	Surtees et al. (2011)
Case-control	Leukocytes	No difference between patients and controls. Inverse correlation between telomere length, history of depression, and markers of inflammations	Wolkowitz et al. (2011)
Population	Leukocytes	No association between TL and depressive symptoms	Shaffer et al. (2012)
Case-control	Leukocytes	No association between TL and depression	Teyssier, Chauvet-Gelinier, Ragot, and Bonin (2012)
Case-control	Leukocytes	Short telomere length associated with low basal, low post-DST cortisol levels but high degree of post-DST suppression	Wikgren et al. (2012)
Case-control	Leukocytes	Decreased TL in patients with depression. Patients also showed impaired glucose tolerance	Garcia-Rizo et al. (2013)
Prospective	Leukocytes	No association between TL and depression	Hoen et al. (2012)
Prospective	Leukocytes	No association between depressive symptoms and TL but significant shortening occurred in the youngest age group (37 year olds)	Phillips et al. (2013)
Case-control	Leukocytes	TL reduced in patients compared with controls	Puterman et al. (2013)
Case-control	Leukocytes	Overall reduction in TL in patients with depression, but effect is more pronounced in CD8+ and CD20+ cells than in CD4+ cells	Karabatsiakis, Kolassa, Kolassa, Rudolph, and Dietrich (2014)
Case-control	Leukocytes	TL negatively correlated with symptom severity and illness duration, only in patients with current depression. In both current and remitted patients TL is reduced compared with controls	Verhoeven et al. (2013)
Prospective	Leukocytes	Study was from age 3 to 38 years of age, but first TL measurement was not until age 26. Presence of internalizing disorder between 11 and 38 was predictive of shorter TL in males at 38	Shalev et al. (2014)
Case-control	Leukocytes	No difference in TL between patients and controls; no effect of gender or ethnicity	Needham et al. (2015)
Case-control	Leukocytes	TEL activity was higher in patients and correlated positively with hippocampal volume in patients with depression	Wolkowitz et al. (2015)
Case-control	Leukocytes	Reduced TL in patients with depression compared with controls. Effect was still prevalent in patients who were on antidepressant medication, and after taking into account the effects of early life stress	Cai et al. (2015)

TABLE 4 Telomeric biomarkers.—cont'd

Study type	Source	Main findings	References
Case-control	Leukocytes	Daughters (10–14yrs) of depressed mothers had significantly shorter TL compared with daughters of healthy mothers	Gotlib et al. (2014)
Case-control	Leukocytes	Elderly population sample; no difference in TL between patients and controls; no association between TL and severity, age of onset, cognitive functioning, and antidepressant use	Schaakxs, Verhoeven, Oude Voshaar, Comijs, and Penninx (2015)
Longitudinal	Leukocytes	Exposure to maternal depression as associated with shorter TL in children 4–5 years of age. Shorter TL was also associated with younger paternal age at child's birth	Wojcicki et al. (2015)
Case-control	Leukocytes	No association between TL and depression. In patients, there was a positive correlated between TEL activity and TL	Simon et al. (2015)
Post-mortem	Hippocampus	Decreased TL in patients with depression	Mamdani et al. (2015)
Case-control	Whole blood	TL was reduced in patients with depression who had a history of physical abuse/neglect during childhood, and was associated with reduced TL	Vincent et al. (2017)
Case-control	Leukocytes	TL was reduced in patients with depression, anxiety, and stress-associated disorders. There was no effect of an 8-week mindfulness treatment	Wang et al. (2017)
Case-control/prospective	Leukocytes	TL was reduced comparing incident patients vs healthy controls, but prospective analysis over a mean follow-up 7.6 years revealed no difference	Wium-Andersen, Orsted, Rode, Bojesen, and Nordestgaard (2017)

consideration in future studies. Insights from such research could also lead to the development of psychobiotics, microbes that have a positive effect on mental health (Dinan, Stanton, & Cryan, 2013).

6 Concluding remarks

There are several final points of discussion worth considering in relation to the development of biomarkers for personalized psychiatry. First, as briefly alluded to earlier, validation is critical in order to improve precision, and this could be done sequentially (Gururajan, Clarke, et al., 2016). The functional roles of identified biomarkers are investigated, and this is then followed by the creation of the biomarker panel. This panel is verified and validated in similar, but independent patient groups to determine sensitivity and specificity. Rather than using frequent statistics in clinical studies, which is arguably "black or white," there has been a recent push toward using more Bayesian approaches that examine probabilities based on prior experimental data (Lai, Lavori, Shih, & Sikic, 2012). The biomarkers are then incorporated into clinical practice, and evaluated for whether they have had a positive impact in terms of improved diagnosis, and individualizing therapeutic strategies.

Second, it is important to remember that biomarkers are inherently probabilistic and reductionist, and this is a concept that is often over-looked in media reports. Biomarkers are used to define risk profiles that need to be communicated to patients in a manner that neither raises alarm nor diminish its value. This is especially important when dealing with specific age groups such as children and adolescents; parents need to be given the best possible advice to steer them clear of making hasty decisions that may or may not be in the best interests of their children. Biomarker profiles could also lead to discriminatory practices in specific work-environments, and by insurance providers in the absence of appropriate regulations to protect the privacy and rights of individuals.

Medical diagnostics has the potential to be very lucrative, but any innovation needs to be based on solid clinical evidence, and regulated to prioritize patient health. The case of Theranos exemplifies the absence of the latter two critical factors. What began as an ambitious venture to use smaller volumes of blood in conjunction with cheaper and more accessible diagnostics technologies resulted in repeated failure to comply with industry standards, and resulted in corporate fraud. Against this background, we believe there is a need for companies with an interest in biomarker development to be up front and transparent with their methodologies and data. Similarly, patent law specific to use of biomarker-related technologies should be refined to ensure that intellectual property is protected, while at the same time not placing any

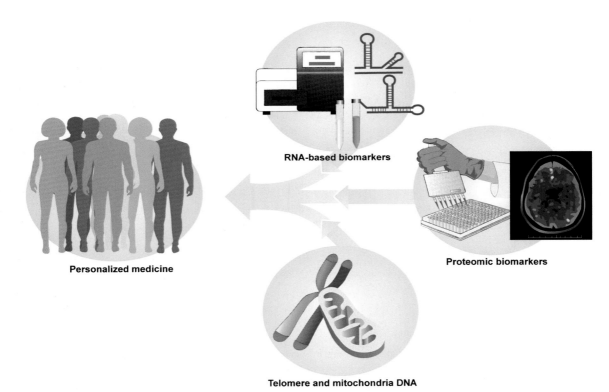

RNA-based biomarkers

Personalized medicine

Proteomic biomarkers

Telomere and mitochondria DNA

FIG. 1 The realization of personalized psychiatry will require synergy of evidence from multiple investigations identifying different types of biomarkers.

further burden on the healthcare sector in terms of cost, access, or ease with which they can be used in the clinical practice (Hopkins & Hogarth, 2012).

We anticipate that personalized biomarkers for depression will be introduced as an objective diagnostic metric within the decade. At a time when drug discovery and development in this area has slowed to a virtual halt, the potential utility of biomarkers as a means to improve diagnostic accuracy, to triage patients based on the probability of response to existing therapies, and serve as candidate targets for new antidepressants will have a definitive positive impact on the cost of care in years to come. As suggested in Fig. 1, this will require a synergy of different approaches that are collaborative and translational, with the goal of bringing personalized psychiatry closer to the clinic.

REFERENCES

Allen, A. P., Naughton, M., Dowling, J., Walsh, A., Ismail, F., Shorten, G., … Clarke, G. (2015). Serum BDNF as a peripheral biomarker of treatment-resistant depression and the rapid antidepressant response: A comparison of ketamine and ECT. *Journal of Affective Disorders, 186,* 306–311.

Alt, S. R., Turner, J. D., Klok, M. D., Meijer, O. C., Lakke, E. A. J. F., DeRijk, R. H., & Muller, C. P. (2010). Differential expression of glucocorticoid receptor transcripts in major depressive disorder is not epigenetically programmed. *Psychoneuroendocrinology, 35,* 544–556.

Andrews, P. W., Bharwani, A., Lee, K. R., Fox, M., & Thomson, J. A. (2015). Is serotonin an upper or a downer? The evolution of the serotonergic system and its role in depression and the antidepressant response. *Neuroscience & Biobehavioral Reviews, 51,* 164–188.

Armanios, M. (2013). Telomeres and age-related disease: How telomere biology informs clinical paradigms. *The Journal of Clinical Investigation, 123,* 996–1002.

Austin, M. C., Janosky, J. E., & Murphy, H. A. (2003). Increased corticotropin-releasing hormone immunoreactivity in monoamine-containing pontine nuclei of depressed suicide men. *Molecular Psychiatry, 8,* 324.

Autry, A. E., & Monteggia, L. M. (2012). Brain-derived neurotrophic factor and neuropsychiatric disorders. *Pharmacological Reviews, 64,* 238–258.

Bach-Mizrachi, H., Underwood, M. D., Kassir, S. A., Bakalian, M. J., Sibille, E., Tamir, H., … Arango, V. (2005). Neuronal tryptophan hydroxylase mRNA expression in the human dorsal and median raphe nuclei: Major depression and suicide. *Neuropsychopharmacology, 31,* 814.

Belvederi Murri, M., Pariante, C., Mondelli, V., Masotti, M., Atti, A. R., Mellacqua, Z., … Amore, M. (2014). HPA axis and aging in depression: Systematic review and meta-analysis. *Psychoneuroendocrinology, 41,* 46–62.

Belzeaux, R., Bergon, A., Jeanjean, V., Loriod, B., Formisano-Tréziny, C., Verrier, L., … Ibrahim, E. C. (2012). Responder and nonresponder patients exhibit different peripheral transcriptional signatures during major depressive episode. *Translational Psychiatry, 2.*

Belzeaux, R., Formisano-Tréziny, C., Loundou, A., Boyer, L., Gabert, J., Samuelian, J.-C., … Ibrahim, E. C. (2010). Clinical variations modulate patterns of gene expression and define blood biomarkers in major depression. *Journal of Psychiatric Research, 44*, 1205–1213.

Bergström, A., Jayatissa, M. N., Mørk, A., & Wiborg, O. (2008). Stress sensitivity and resilience in the chronic mild stress rat model of depression: An in situ hybridization study. *Brain Research, 1196*, 41–52.

Bernard, R., Kerman, I. A., Thompson, R. C., Jones, E. G., Bunney, W. E., Barchas, J. D., … Watson, S. J. (2011). Altered expression of glutamate signaling, growth factor, and glia genes in the locus coeruleus of patients with major depression. *Molecular Psychiatry, 16*, 634–646.

Biomarkers Definitions Working Group (2001). Biomarkers and surrogate endpoints: Preferred definitions and conceptual framework. *Clinical Pharmacology and Therapeutics, 69*, 89–95.

Bocchio-Chiavetto, L., Bagnardi, V., Zanardini, R., Molteni, R., Gabriela Nielsen, M., Placentino, A., … Gennarelli, M. (2010). Serum and plasma BDNF levels in major depression: A replication study and meta-analyses. *The World Journal of Biological Psychiatry, 11*, 763–773.

Bocchio-Chiavetto, L., Maffioletti, E., Bettinsoli, P., Giovannini, C., Bignotti, S., Tardito, D., … Gennarelli, M. (2013). Blood microRNA changes in depressed patients during antidepressant treatment. *European Neuropsychopharmacology: The Journal of the European College of Neuropsychopharmacology, 23*, 602–611.

Boldrini, M., Underwood, M. D., Mann, J. J., & Arango, V. (2005). More tryptophan hydroxylase in the brainstem dorsal raphe nucleus in depressed suicides. *Brain Research, 1041*, 19–28.

Bonkale, W. L., Murdock, S., Janosky, J. E., & Austin, M. C. (2004). Normal levels of tryptophan hydroxylase immunoreactivity in the dorsal raphe of depressed suicide victims. *Journal of Neurochemistry, 88*, 958–964.

Brouwer, J. P., Appelhof, B. C., van Rossum, E. F., Koper, J. W., Fliers, E., Huyser, J., … Hoogendijk, W. J. (2006). Prediction of treatment response by HPA-axis and glucocorticoid receptor polymorphisms in major depression. *Psychoneuroendocrinology, 31*, 1154–1163.

Brunoni, A. R., Lopes, M., & Fregni, F. (2008). A systematic review and meta-analysis of clinical studies on major depression and BDNF levels: Implications for the role of neuroplasticity in depression. *International Journal of Neuropsychopharmacology, 11*, 1169–1180.

Cai, N., Li, Y., Chang, S., Liang, J., Lin, C., Zhang, X., … Flint, J. (2015). Genetic control over mtDNA and its relationship to major depressive disorder. *Current Biology, 25*, 3170–3177.

Camkurt, M. A., Acar, S., Coskun, S., Gunes, M., Gunes, S., Yilmaz, M. F., … Tamer, L. (2015). Comparison of plasma microRNA levels in drug naive, first episode depressed patients and healthy controls. *Journal of Psychiatric Research, 69*, 67–71.

Carroll, B. J., Cassidy, F., Naftolowitz, D., Tatham, N. E., Wilson, W. H., Iranmanesh, A., … Veldhuis, J. D. (2007). Pathophysiology of hypercortisolism in depression. *Acta Psychiatrica Scandinavica, 115*, 90–103.

Catalán, R., Gallart, J. M., Castellanos, J. M., & Galard, R. (1998). Plasma corticotropin-releasing factor in depressive disorders. *Biological Psychiatry, 44*, 15–20.

Cattaneo, A., Bocchio-Chiavetto, L., Zanardini, R., Milanesi, E., Placentino, A., & Gennarelli, M. (2010). Reduced peripheral brain-derived neurotrophic factor mRNA levels are normalized by antidepressant treatment. *International Journal of Neuropsychopharmacology, 13*, 103–108.

Cattaneo, A., Gennarelli, M., Uher, R., Breen, G., Farmer, A., Aitchison, K. J., … Pariante, C. M. (2012). Candidate genes expression profile associated with antidepressants response in the GENDEP study: Differentiating between baseline 'predictors' and longitudinal 'targets'. *Neuropsychopharmacology, 38*, 377.

Cattaneo, A., Gennarelli, M., Uher, R., Breen, G., Farmer, A., Aitchison, K. J., … Pariante, C. M. (2013). Candidate genes expression profile associated with antidepressants response in the GENDEP study: Differentiating between baseline 'predictors' and longitudinal 'targets'. *Neuropsychopharmacology, 38*, 377–385.

Cattaneo, A., Sesta, A., Calabrese, F., Nielsen, G., Riva, M. A., & Gennarelli, M. (2010). The expression of VGF is reduced in leukocytes of depressed patients and it is restored by effective antidepressant treatment. *Neuropsychopharmacology, 35*, 1423.

Chan, M. K., Cooper, J. D., Bot, M., Birkenhager, T. K., Bergink, V., Drexhage, H. A., … Bahn, S. (2016). Blood-based immune-endocrine biomarkers of treatment response in depression. *Journal of Psychiatric Research, 83*, 249–259.

Choi, J., Fauce, S. R., & Effros, R. B. (2008). Reduced telomerase activity in human T lymphocytes exposed to cortisol. *Brain, Behavior, and Immunity, 22*, 600–605.

Clark-Raymond, A., & Halaris, A. (2013). VEGF and depression: A comprehensive assessment of clinical data. *Journal of Psychiatric Research, 47*, 1080–1087.

Colle, R., Trabado, S., David, D. J., Brailly-Tabard, S., Hardy, P., Falissard, B., … Corruble, E. (2017). Plasma BDNF level in major depression: Biomarker of the Val66Met BDNF polymorphism and of the clinical course in met carrier patients. *Neuropsychobiology, 75*, 39–45.

Cui, X., Sun, X., Niu, W., Kong, L., He, M., Zhong, A., … Cheng, Z. (2016). Long non-coding RNA: Potential diagnostic and therapeutic biomarker for major depressive disorder. *Medical Science Monitor, 22*, 5240–5248.

D'Addario, C., Dell'Osso, B., Galimberti, D., Palazzo, M. C., Benatti, B., Di Francesco, A., … Maccarrone, M. (2013). Epigenetic modulation of BDNF gene in patients with major depressive disorder. *Biological Psychiatry, 73*, e6–e7.

Devanand, D. P., Shapira, B., Petty, F., Kramer, G., Fitzsimons, L., Lerer, B., & Sackeim, H. A. (1995). Effects of electroconvulsive therapy on plasma GABA. *Convulsive Therapy, 11*, 3–13.

Dinan, T. G. (1994). Glucocorticoids and the genesis of depressive illness a psychobiological model. *British Journal of Psychiatry, 164*, 365–371.

Dinan, T. G. (2001). Novel approaches to the treatment of depression by modulating the hypothalamic-pituitary-adrenal axis. *Human Psychopharmacology: Clinical and Experimental, 16*, 89–93.

Dinan, T. G., Stanton, C., & Cryan, J. F. (2013). Psychobiotics: A novel class of psychotropic. *Biological Psychiatry, 74*(10), 720–726.

Dunlop, B. W., & Nemeroff, C. B. (2007). The role of dopamine in the pathophysiology of depression. *Archives of General Psychiatry, 64*, 327–337.

Dunn, E. C., Wiste, A., Radmanesh, F., Almli, L. M., Gogarten, S. M., Sofer, T., … Smoller, J. W. (2016). Genome-wide association study (GWAS) and genome-wide environment interaction study (GWEIS) of depressive symptoms in African American and hispanic/latina women. *Depression and Anxiety, 33*, 265–280.

Dutta, A., McKie, S., & Deakin, J. F. W. (2015). Ketamine and other potential glutamate antidepressants. *Psychiatry Research, 225*, 1–13.

Elfving, B., Plougmann, P. H., & Wegener, G. (2010). Differential brain, but not serum VEGF levels in a genetic rat model of depression. *Neuroscience Letters, 474*, 13–16.

Evans, S. J., Choudary, P. V., Neal, C. R., Li, J. Z., Vawter, M. P., Tomita, H., … Akil, H. (2004). Dysregulation of the fibroblast growth factor system in major depression. *Proceedings of the National Academy of Sciences of the United States of America, 101*, 15506.

Fadista, J., Manning, A. K., Florez, J. C., & Groop, L. (2016). The (in)famous GWAS P-value threshold revisited and updated for low-frequency variants. *European Journal of Human Genetics, 24*, 1202–1205.

Fajardo, O., Galeno, J., Urbina, M., Carreira, I., & Lima, L. (2003). Serotonin, serotonin 5-HT1A receptors and dopamine in blood peripheral lymphocytes of major depression patients. *International Immunopharmacology, 3*, 1345–1352.

Fan, H. M., Sun, X. Y., Guo, W., Zhong, A. F., Niu, W., Zhao, L., … Lu, J. (2014). Differential expression of microRNA in peripheral blood mononuclear cells as specific biomarker for major depressive disorder patients. *Journal of Psychiatric Research, 59*, 45–52.

Fuchikami, M., Morinobu, S., Segawa, M., Okamoto, Y., Yamawaki, S., Ozaki, N., … Terao, T. (2011). DNA methylation profiles of the brain-derived neurotrophic factor (BDNF) gene as a potent diagnostic biomarker in major depression. *Plos One, 6*.

Galard, R., Catalán, R., Castellanos, J. M., & Gallart, J. M. (2002). Plasma corticotropin-releasing factor in depressed patients before and after the dexamethasone suppression test. *Biological Psychiatry, 51*, 463–468.

Gałecki, P., Gałecka, E., Maes, M., Chamielec, M., Orzechowska, A., Bobińska, K., … Szemraj, J. (2012). The expression of genes encoding for COX-2, MPO, iNOS, and sPLA2-IIA in patients with recurrent depressive disorder. *Journal of Affective Disorders, 138*, 360–366.

Garcia-Rizo, C., Fernandez-Egea, E., Miller, B. J., Oliveira, C., Justicia, A., Griffith, J. K., … Kirkpatrick, B. (2013). Abnormal glucose tolerance, white blood cell count, and telomere length in newly diagnosed, antidepressant-naïve patients with depression. *Brain, Behavior, and Immunity, 28*, 49–53.

Gerner, R. H., Fairbanks, L., Anderson, G. M., Young, J. G., Scheinin, M., Linnoila, M., … Cohen, D. J. (1984). CSF neurochemistry in depressed, manic, and schizophrenic patients compared with that of normal controls. *The American Journal of Psychiatry, 141*, 1533–1540.

Gerner, R. H., & Hare, T. A. (1981). CSF GABA in normal subjects and patients with depression, schizophrenia, mania, and anorexia nervosa. *The American Journal of Psychiatry, 138*, 1098–1101.

Gold, B. I., Bowers, M. B., Jr., Roth, R. H., & Sweeney, D. W. (1980). GABA levels in CSF of patients with psychiatric disorders. *The American Journal of Psychiatry, 137*, 362–364.

Goldstein, B. L., & Klein, D. N. (2014). A review of selected candidate endophenotypes for depression. *Clinical Psychology Review, 34*, 417–427.

Gotlib, I. H., LeMoult, J., Colich, N. L., Foland-Ross, L. C., Hallmayer, J., Joormann, J., … Wolkowitz, O. M. (2014). Telomere length and cortisol reactivity in children of depressed mothers. *Molecular Psychiatry, 20*, 615.

Gray, A. L., Hyde, T. M., Deep-Soboslay, A., Kleinman, J. E., & Sodhi, M. S. (2015). Sex differences in glutamate receptor gene expression in major depression and suicide. *Molecular Psychiatry, 20*, 1057–1068.

Gururajan, A., Clarke, G., Dinan, T. G., & Cryan, J. F. (2016). Molecular biomarkers of depression. *Neuroscience and Biobehavioral Reviews, 64*, 101–133.

Gururajan, A., Naughton, M. E., Scott, K. A., O'Connor, R. M., Moloney, G., Clarke, G., … Dinan, T. G. (2016). MicroRNAs as biomarkers for major depression: a role for let-7b and let-7c. *Translational Psychiatry, 6*.

Ha, M., & Kim, V. N. (2014). Regulation of microRNA biogenesis. *Nature Reviews Molecular Cell Biology, 15*, 509–524.

Hartmann, N., Boehner, M., Groenen, F., & Kalb, R. (2010). Telomere length of patients with major depression is shortened but independent from therapy and severity of the disease. *Depression and Anxiety, 27*, 1111–1116.

Herbert, J. (2013). Cortisol and depression: Three questions for psychiatry. *Psychological Medicine, 43*, 449–469.

Hodes, G. E., Kana, V., Menard, C., Merad, M., & Russo, S. J. (2015). Neuroimmune mechanisms of depression. *Nature Neuroscience, 18*, 1386–1393.

Hoen, P. W., de Jonge, P., Na, B. Y., Farzaneh-Far, R., Epel, E., Lin, J., … Whooley, M. A. (2011). Depression and leukocyte telomere length in patients with coronary heart disease: Data from the heart and soul study. *Psychosomatic Medicine, 73*, 541–547.

Hoen, P. W., Rosmalen, J. G. M., Schoevers, R. A., Huzen, J., van der Harst, P., & de Jonge, P. (2012). Association between anxiety but not depressive disorders and leukocyte telomere length after 2 years of follow-up in a population-based sample. *Psychological Medicine, 43*, 689–697.

Hong, W., Fan, J., Yuan, C., Zhang, C., Hu, Y., Peng, D., … Fang, Y. (2014). Significantly decreased mRNA levels of BDNF and MEK1 genes in treatment-resistant depression. *NeuroReport, 25*, 753–755.

Hopkins, M. M., & Hogarth, S. (2012). Biomarker patents for diagnostics: problem or solution? *Nature Biotechnology, 30*, 498.

Hyde, C. L., Nagle, M. W., Tian, C., Chen, X., Paciga, S. A., Wendland, J. R., … Winslow, A. R. (2016). Identification of 15 genetic loci associated with risk of major depression in individuals of European descent. *Nature Genetics, 48*, 1031–1036.

Iacob, E., Light, K. C., Tadler, S. C., Weeks, H. R., White, A. T., Hughen, R. W., … Light, A. R. (2013). Dysregulation of leukocyte gene expression in women with medication-refractory depression versus healthy non-depressed controls. *BMC Psychiatry, 13*, 273.

Iacob, E., Tadler, S. C., Light, K. C., Weeks, H. R., Smith, K. W., White, A. T., … Light, A. R. (2014). Leukocyte gene expression in patients with medication refractory depression before and after treatment with ECT or isoflurane anesthesia: A pilot study. *Depression Research and Treatment, 2014*, 12.

Iga, J.-i., Ueno, S.-i., Yamauchi, K., Motoki, I., Tayoshi, S., Ohta, K., … Ohmori, T. (2005). Serotonin transporter mRNA expression in peripheral leukocytes of patients with major depression before and after treatment with paroxetine. *Neuroscience Letters, 389*, 12–16.

Iga, J.-i., Ueno, S.-i., Yamauchi, K., Numata, S., Tayoshi-Shibuya, S., Kinouchi, S., … Ohmori, T. (2007). Gene expression and association analysis of vascular endothelial growth factor in major depressive disorder. *Progress in Neuro-Psychopharmacology and Biological Psychiatry, 31*, 658–663.

Inder, W. J., Prickett, T. C., Mulder, R. T., Donald, R. A., & Joyce, P. R. (2001). Reduction in basal afternoon plasma ACTH during early treatment of depression with fluoxetine. *Psychopharmacology, 156*, 73–78.

Insel, T., Cuthbert, B., Garvey, M., Heinssen, R., Pine, D. S., Quinn, K., ... Wang, P. (2010). Research domain criteria (RDoC): toward a new classification framework for research on mental disorders. *The American Journal of Psychiatry, 167*, 748–751.

Ising, M., Horstmann, S., Kloiber, S., Lucae, S., Binder, E. B., Kern, N., ... Holsboer, F. (2007). Combined dexamethasone/corticotropin releasing hormone test predicts treatment response in major depression—A potential biomarker? *Biological Psychiatry, 62*, 47–54.

Issler, O., Haramati, S., Paul Evan, D., Maeno, H., Navon, I., Zwang, R., ... Chen, A. (2014). MicroRNA 135 is essential for chronic stress resiliency, antidepressant efficacy, and intact serotonergic activity. *Neuron, 83*, 344–360.

Joensuu, M., Ahola, P., Knekt, P., Lindfors, O., Saarinen, P., Tolmunen, T., ... Lehtonen, J. (2016). Baseline symptom severity predicts serotonin transporter change during psychotherapy in patients with major depression. *Psychiatry and Clinical Neurosciences, 70*, 34–41.

Juruena, M. F., Bocharova, M., Agustini, B., & Young, A. H. (2017). Atypical depression and non-atypical depression: Is HPA axis function a biomarker? A systematic review. *Journal of Affective Disorders*, 45–67.

Kambeitz, J. P., & Howes, O. D. (2015). The serotonin transporter in depression: Meta-analysis of in vivo and post mortem findings and implications for understanding and treating depression. *Journal of Affective Disorders, 186*, 358–366.

Karabatsiakis, A., Kolassa, I.-T., Kolassa, S., Rudolph, K. L., & Dietrich, D. E. (2014). Telomere shortening in leukocyte subpopulations in depression. *BMC Psychiatry, 14*, 192.

Kasa, K., Otsuki, S., Yamamoto, M., Sato, M., Kuroda, H., & Ogawa, N. (1982). Cerebrospinal fluid gamma-aminobutyric acid and homovanillic acid in depressive disorders. *Biological Psychiatry, 17*, 877–883.

Katz, E. R., Stowe, Z. N., Newport, D. J., Kelley, M. E., Pace, T. W., Cubells, J. F., & Binder, E. B. (2011). Regulation of mRNA expression encoding chaperone and co-chaperone proteins of the glucocorticoid receptor in peripheral blood: Association with depressive symptoms during pregnancy. *Psychological Medicine, 42*, 943–956.

Kelly, J. R., Borre, Y., O'Brien, C., Patterson, E., El Aidy, S., Deane, J., ... Dinan, T. G. (2016). Transferring the blues: Depression-associated gut microbiota induces neurobehavioural changes in the rat. *Journal of Psychiatric Research, 82*, 109–118.

Klok, M. D., Alt, S. R., Irurzun Lafitte, A. J. M., Turner, J. D., Lakke, E. A. J. F., Huitinga, I., ... DeRijk, R. H. (2011). Decreased expression of mineralocorticoid receptor mRNA and its splice variants in postmortem brain regions of patients with major depressive disorder. *Journal of Psychiatric Research, 45*, 871–878.

Krause, D., Jobst, A., Kirchberg, F., Kieper, S., Hartl, K., Kastner, R., ... Schwarz, M. J. (2014). Prenatal immunologic predictors of postpartum depressive symptoms: A prospective study for potential diagnostic markers. *European Archives of Psychiatry and Clinical Neuroscience, 264*, 615–624.

Küçükibrahimoğlu, E., Saygın, M. Z., Çalışkan, M., Kaplan, O. K., Ünsal, C., & Gören, M. Z. (2009). The change in plasma GABA, glutamine and glutamate levels in fluoxetine- or S-citalopram-treated female patients with major depression. *European Journal of Clinical Pharmacology, 65*, 571–577.

Kunugi, H., Ida, I., Owashi, T., Kimura, M., Inoue, Y., Nakagawa, S., ... Mikuni, M. (2005). Assessment of the dexamethasone/CRH test as a state-dependent marker for hypothalamic-pituitary-adrenal (HPA) axis abnormalities in major depressive episode: A multicenter study. *Neuropsychopharmacology, 31*, 212.

Lai, T. L., Lavori, P. W., Shih, M. C., & Sikic, B. I. (2012). Clinical trial designs for testing biomarker-based personalized therapies. *Clinical Trials (London, England), 9*, 141–154.

Levinson, D. F., Mostafavi, S., Milaneschi, Y., Rivera, M., Ripke, S., Wray, N. R., & Sullivan, P. F. (2014). Genetic studies of major depressive disorder: Why are there no genome-wide association study findings and what can we do about it? *Biological Psychiatry, 76*, 510–512.

Li, Y.-J., Xu, M., Gao, Z.-H., Wang, Y.-Q., Yue, Z., Zhang, Y.-X., ... Wang, P.-Y. (2013). Alterations of serum levels of BDNF-related miRNAs in patients with depression. *Plos One, 8*.

Liew, C. C., Ma, J., Tang, H. C., Zheng, R., & Dempsey, A. A. (2006). The peripheral blood transcriptome dynamically reflects system wide biology: A potential diagnostic tool. *The Journal of Laboratory and Clinical Medicine, 147*, 126–132.

Lima, I. M. M., Barros, A., Rosa, D. V., Albuquerque, M., Malloy-Diniz, L., Neves, F. S., ... de Miranda, D. M. (2015). Analysis of telomere attrition in bipolar disorder. *Journal of Affective Disorders, 172*, 43–47.

Lima, L., Mata, S., & Urbina, M. (2005). Allelic isoforms and decrease in serotonin transporter mRNA in lymphocytes of patients with major depression. *Neuroimmunomodulation, 12*, 299–306.

Lopez, J. F., Chalmers, D. T., Little, K. Y., & Watson, S. J. (1998). A.E. Bennett Research Award. Regulation of serotonin1A, glucocorticoid, and mineralocorticoid receptor in rat and human hippocampus: Implications for the neurobiology of depression. *Biological Psychiatry, 43*, 547–573.

Lopez, J. P., Fiori, L. M., Cruceanu, C., Lin, R., Labonte, B., Cates, H. M., ... Turecki, G. (2017). MicroRNAs 146a/b-5 and 425-3p and 24-3p are markers of antidepressant response and regulate MAPK/Wnt-system genes. *Nature Communications, 8*, 15497.

Lopez, J. P., Lim, R., Cruceanu, C., Crapper, L., Fasano, C., Labonte, B., ... Turecki, G. (2014). miR-1202 is a primate-specific and brain-enriched microRNA involved in major depression and antidepressant treatment. *Nature Medicine, 20*, 764–768.

López-Figueroa, A. L., Norton, C. S., López-Figueroa, M. O., Armellini-Dodel, D., Burke, S., Akil, H., ... Watson, S. J. (2004). Serotonin 5-HT$_{1A}$, 5-HT$_{1B}$, and 5-HT$_{2A}$ receptor mRNA expression in subjects with major depression, bipolar disorder, and schizophrenia. *Biological Psychiatry, 55*, 225–233.

Lung, F.-W., Chen, N. C., & Shu, B.-C. (2007). Genetic pathway of major depressive disorder in shortening telomeric length. *Psychiatric Genetics, 17*, 195–199.

Luscher, B., Shen, Q., & Sahir, N. (2010). The GABAergic deficit hypothesis of major depressive disorder. *Molecular Psychiatry, 16*, 383.

Maffioletti, E., Cattaneo, A., Rosso, G., Maina, G., Maj, C., Gennarelli, M., ... Bocchio-Chiavetto, L. (2016). Peripheral whole blood microRNA alterations in major depression and bipolar disorder. *Journal of Affective Disorders, 200*, 250–258.

Malberg, J. E., & Monteggia, L. M. (2008). VGF, a new player in antidepressant action? *Science Signaling, 1*, pe19.

Mamdani, F., Rollins, B., Morgan, L., Myers, R. M., Barchas, J. D., Schatzberg, A. F., … Sequeira, P. A. (2015). Variable telomere length across post-mortem human brain regions and specific reduction in the hippocampus of major depressive disorder. *Translational Psychiatry, 5*.

Matsubara, T., Funato, H., Kobayashi, A., Nobumoto, M., & Watanabe, Y. (2006). Reduced glucocorticoid receptor alpha expression in mood disorder patients and first-degree relatives. *Biological Psychiatry, 59*, 689–695.

McKnight, D. L., Nelson-Gray, R. O., & Barnhill, J. (1992). Dexamethasone suppression test and response to cognitive therapy and antidepressant medication. *Behavior Therapy, 23*, 99–111.

Medina, A., Seasholtz, A. F., Sharma, V., Burke, S., Bunney, W., Jr., Myers, R. M., … Watson, S. J. (2013). Glucocorticoid and mineralocorticoid receptor expression in the human hippocampus in major depressive disorder. *Journal of Psychiatric Research, 47*, 307–314.

Menke, A., Klengel, T., Rubel, J., Brückl, T., Pfister, H., Lucae, S., … Binder, E. B. (2013). Genetic variation in FKBP5 associated with the extent of stress hormone dysregulation in major depression. *Genes, Brain and Behavior, 12*, 289–296.

Mergny, J.-L., Riou, J.-F., Mailliet, P., Teulade-Fichou, M.-P., & Gilson, E. (2002). Natural and pharmacological regulation of telomerase. *Nucleic Acids Research, 30*, 839–865.

Milanesi, E., Minelli, A., Cattane, N., Cattaneo, A., Mora, C., Barbon, A., … Gennarelli, M. (2012). ErbB3 mRNA leukocyte levels as a biomarker for major depressive disorder. *BMC Psychiatry, 12*, 145.

Molendijk, M. L., Spinhoven, P., Polak, M., Bus, B. A. A., Penninx, B. W. J. H., & Elzinga, B. M. (2013). Serum BDNF concentrations as peripheral manifestations of depression: Evidence from a systematic review and meta-analyses on 179 associations (N = 9484). *Molecular Psychiatry, 19*, 791.

Nakase, S., Kitayama, I., Soya, H., Hamanaka, K., & Nomura, J. (1998). Increased expression of magnocellular arginine vasopressin mRNA in paraventricular nucleus of stress-induced depression-model rats. *Life Sciences, 63*, 23–31.

Naseribafrouei, A., Hestad, K., Avershina, E., Sekelja, M., Linløkken, A., Wilson, R., & Rudi, K. (2014). Correlation between the human fecal microbiota and depression. *Neurogastroenterology & Motility, 26*, 1155–1162.

Needham, B., Mezuk, B., Bareis, N., Lin, J., Blackburn, E., & Epel, E. (2015). Depression, anxiety, and telomere length in young adults: Evidence from the National Health and Nutrition Examination Survey. *Molecular Psychiatry, 20*, 520–528.

Ni, X., Liao, Y., Li, L., Zhang, X., & Wu, Z. (2017). Therapeutic role of long non-coding RNA TCONS_00019174 in depressive disorders is dependent on Wnt/beta-catenin signaling pathway. *Journal of Integrative Neuroscience*.

Otsuki, K., Uchida, S., Watanuki, T., Wakabayashi, Y., Fujimoto, M., Matsubara, T., … Watanabe, Y. (2008). Altered expression of neurotrophic factors in patients with major depression. *Journal of Psychiatric Research, 42*, 1145–1153.

Pandey, G. N., Dwivedi, Y., Rizavi, H. S., Ren, X., Zhang, H., & Pavuluri, M. N. (2010). Brain-derived neurotrophic factor gene and protein expression in pediatric and adult depressed subjects. *Progress in Neuro-Psychopharmacology & Biological Psychiatry, 34*, 645–651.

Paulus, M. P. (2015). Pragmatism instead of mechanism: A call for impactful biological psychiatry. *JAMA Psychiatry, 72*, 631–632.

Perani, C. V., & Slattery, D. A. (2014). Using animal models to study post-partum psychiatric disorders. *British Journal of Pharmacology, 171*, 4539–4555.

Petty, F., Kramer, G. L., Dunnam, D., & Rush, A. J. (1990). Plasma GABA in mood disorders. *Psychopharmacology Bulletin, 26*, 157–161.

Petty, F., Kramer, G. L., Gullion, C. M., & John Rush, A. (1992). Low plasma GABA levels in male patients with depression. *Biological Psychiatry, 32*, 354–363.

Petty, F., & Schlesser, M. A. (1981). Plasma GABA in affective illness. *Journal of Affective Disorders, 3*, 339–343.

Petty, F., & Sherman, A. D. (1984). Plasma GABA levels in psychiatric illness. *Journal of Affective Disorders, 6*, 131–138.

Petty, F., Steinberg, J., Kramer, G. L., Fulton, M., & Moeller, F. G. (1993). Desipramine does not alter plasma GABA in patients with major depression. *Journal of Affective Disorders, 29*, 53–56.

Phillips, A. C., Robertson, T., Carroll, D., Der, G., Shiels, P. G., McGlynn, L., & Benzeval, M. (2013). Do symptoms of depression predict telomere length? Evidence from the West of Scotland Twenty-07 Study. *Psychosomatic Medicine, 75*, 288–296.

Pine, D. S., & Leibenluft, E. (2015). Biomarkers with a mechanistic focus. *JAMA Psychiatry, 72*, 633–634.

Plotsky, P. M., & Meaney, M. J. (1993). Early, postnatal experience alters hypothalamic corticotropin-releasing factor (CRF) mRNA, median eminence CRF content and stress-induced release in adult rats. *Molecular Brain Research, 18*, 195–200.

Polho, G. B., De-Paula, V. J., Cardillo, G., dos Santos, B., & Kerr, D. S. (2015). Leukocyte telomere length in patients with schizophrenia: A meta-analysis. *Schizophrenia Research, 165*, 195–200.

Polyakova, M., Stuke, K., Schuemberg, K., Mueller, K., Schoenknecht, P., & Schroeter, M. L. (2015). BDNF as a biomarker for successful treatment of mood disorders: A systematic and quantitative meta-analysis. *Journal of Affective Disorders, 174*, 432–440.

Post, R. M., Ballenger, J. C., Hare, T. A., Goodwin, F. K., Lake, C. R., Jimerson, D. C., & Bunney, W. E. (1980). Cerebrospinal fluid GABA in normals and patients with affective disorders. *Brain Research Bulletin, 5*, 755–759.

Puterman, E., Epel, E. S., Lin, J., Blackburn, E. H., Gross, J. J., Whooley, M. A., & Cohen, B. E. (2013). Multisystem resiliency moderates the major depression–Telomere length association: Findings from the Heart and Soul Study. *Brain, Behavior, and Immunity, 33*, 65–73.

Qureshi, I. A., Mattick, J. S., & Mehler, M. F. (2010). Long non-coding RNAs in nervous system function and disease. *Brain Research, 1338C*, 20–35.

Raadsheer, F. C., van Heerikhuize, J. J., Lucassen, P. J., Hoogendijk, W. J., Tilders, F. J., & Swaab, D. F. (1995). Corticotropin-releasing hormone mRNA levels in the paraventricular nucleus of patients with Alzheimer's disease and depression. *The American Journal of Psychiatry, 152*, 1372–1376.

Ramsey, J. M., Cooper, J. D., Bot, M., Guest, P. C., Lamers, F., Weickert, C. S., … Bahn, S. (2016). Sex differences in serum markers of major depressive disorder in the Netherlands study of depression and anxiety (NESDA). *PLoS One, 11*.

Ren, J., Zhao, G., Sun, X., Liu, H., Jiang, P., Chen, J., … Zhang, C. (2017). Identification of plasma biomarkers for distinguishing bipolar depression from major depressive disorder by iTRAQ-coupled LC-MS/MS and bioinformatics analysis. *Psychoneuroendocrinology, 86*, 17–24.

Rocca, P., De Leo, C., Eva, C., Marchiaro, L., Milani, A. M., Musso, R., … Bogetto, F. (2002). Decrease of the D4 dopamine receptor messenger RNA expression in lymphocytes from patients with major depression. *Progress in Neuro-Psychopharmacology and Biological Psychiatry, 26,* 1155–1160.

Roy, A., Dejong, J., & Ferraro, T. (1991). CSF GABA in depressed patients and normal controls. *Psychological Medicine, 21,* 613–618.

Roy, B., Dunbar, M., Shelton, R. C., & Dwivedi, Y. (2016). Identification of microRNA-124-3p as a putative epigenetic signature of major depressive disorder. *Neuropsychopharmacology, 42,* 864.

Rukov, J. L., Wilentzik, R., Jaffe, I., Vinther, J., & Shomron, N. (2014). Pharmaco-miR: Linking microRNAs and drug effects. *Briefings in Bioinformatics, 15,* 648–659.

Schaakxs, R., Verhoeven, J. E., Oude Voshaar, R. C., Comijs, H. C., & Penninx, B. W. J. H. (2015). Leukocyte telomere length and late-life depression. *The American Journal of Geriatric Psychiatry, 23,* 423–432.

Schon, E. A., DiMauro, S., & Hirano, M. (2012). Human mitochondrial DNA: Roles of inherited and somatic mutations. *Nature Reviews Genetics, 13,* 878.

Sen, S., Duman, R., & Sanacora, G. (2008). Serum brain-derived neurotrophic factor, depression, and antidepressant medications: Meta-analyses and implications. *Biological Psychiatry, 64,* 527–532.

Shaffer, J. A., Epel, E., Kang, M. S., Ye, S., Schwartz, J. E., Davidson, K. W., … Shimbo, D. (2012). Depressive symptoms are not associated with leukocyte telomere length: Findings from the Nova Scotia Health Survey (NSHS95), a population-based study. *Plos One, 7.*

Shalev, I., Moffitt, T. E., Braithwaite, A. W., Danese, A., Fleming, N. I., Goldman-Mellor, S., … Caspi, A. (2014). Internalizing disorders and leukocyte telomere erosion: A prospective study of depression, generalized anxiety disorder and post-traumatic stress disorder. *Molecular Psychiatry, 19,* 1163.

Simon, N. M., Smoller, J. W., McNamara, K. L., Maser, R. S., Zalta, A. K., Pollack, M. H., … Wong, K.-K. (2006). Telomere shortening and mood disorders: Preliminary support for a chronic stress model of accelerated aging. *Biological Psychiatry, 60,* 432–435.

Simon, N. M., Walton, Z. E., Bui, E., Prescott, J., Hoge, E., Keshaviah, A., … Wong, K.-K. (2015). Telomere length and telomerase in a well-characterized sample of individuals with major depressive disorder compared to controls. *Psychoneuroendocrinology, 58,* 9–22.

Song, M. F., Dong, J. Z., Wang, Y. W., He, J., Ju, X., Zhang, L., … Lv, Y. Y. (2015). CSF miR-16 is decreased in major depression patients and its neutralization in rats induces depression-like behaviors via a serotonin transmitter system. *Journal of Affective Disorders, 178,* 25–31.

Surtees, P. G., Wainwright, N. W. J., Pooley, K. A., Luben, R. N., Khaw, K.-T., Easton, D. F., & Dunning, A. M. (2011). Life stress, emotional health, and mean telomere length in the european prospective investigation into cancer (EPIC)-Norfolk Population Study. *The Journals of Gerontology: Series A, 66A,* 1152–1162.

Suzuki, K., Iwata, Y., Matsuzaki, H., Anitha, A., Suda, S., Iwata, K., … Mori, N. (2010). Reduced expression of apolipoprotein E receptor type 2 in peripheral blood lymphocytes from patients with major depressive disorder. *Progress in Neuro-Psychopharmacology & Biological Psychiatry, 34,* 1007–1010.

Takebayashi, M., Hashimoto, R., Hisaoka, K., Tsuchioka, M., & Kunugi, H. (2010). Plasma levels of vascular endothelial growth factor and fibroblast growth factor 2 in patients with major depressive disorders. *Journal of Neural Transmission, 117,* 1119–1122.

Teyssier, J.-R., Chauvet-Gelinier, J.-C., Ragot, S., & Bonin, B. (2012). Up-regulation of leucocytes genes implicated in telomere dysfunction and cellular senescence correlates with depression and anxiety severity scores. *Plos One, 7.*

Teyssier, J. R., Ragot, S., Donzel, A., & Chauvet-Gelinier, J. C. (2010). Longueur des télomères dans le cortex des patients atteints de troubles dépressifs. *L'Encéphale, 36,* 491–494.

Tsao, C.-W., Lin, Y.-S., Chen, C.-C., Bai, C.-H., & Wu, S.-R. (2006). Cytokines and serotonin transporter in patients with major depression. *Progress in Neuro-Psychopharmacology and Biological Psychiatry, 30,* 899–905.

Turner Cortney, A., Watson Stanley, J., & Akil, H. (2012). The fibroblast growth factor family: Neuromodulation of affective behavior. *Neuron, 76,* 160–174.

Vaccarino, V., Brennan, M. L., Miller, A. H., Bremner, J. D., Ritchie, J. C., Lindau, F., … Hazen, S. L. (2008). Association of major depressive disorder with serum myeloperoxidase and other markers of inflammation: A twin study. *Biological Psychiatry, 64,* 476–483.

Verhoeven, J. E., Révész, D., Epel, E. S., Lin, J., Wolkowitz, O. M., & Penninx, B. W. J. H. (2013). Major depressive disorder and accelerated cellular aging: Results from a large psychiatric cohort study. *Molecular Psychiatry, 19,* 895.

Vincent, J., Hovatta, I., Frissa, S., Goodwin, L., Hotopf, M., Hatch, S. L., … Powell, T. R. (2017). Assessing the contributions of childhood maltreatment subtypes and depression case-control status on telomere length reveals a specific role of physical neglect. *Journal of Affective Disorders, 213,* 16–22.

Wang, X., Sundquist, K., Hedelius, A., Palmer, K., Memon, A. A., & Sundquist, J. (2017). Leukocyte telomere length and depression, anxiety and stress and adjustment disorders in primary health care patients. *BMC Psychiatry, 17,* 148.

Webster, M. J., Knable, M. B., O'Grady, J., Orthmann, J., & Weickert, C. S. (2002). Regional specificity of brain glucocorticoid receptor mRNA alterations in subjects with schizophrenia and mood disorders. *Molecular Psychiatry, 7,* 985.

Weigelt, K., Carvalho, L. A., Drexhage, R. C., Wijkhuijs, A., Wit, H. D., van Beveren, N. J. M., … Drexhage, H. A. (2011). TREM-1 and DAP12 expression in monocytes of patients with severe psychiatric disorders. EGR3, ATF3 and PU.1 as important transcription factors. *Brain, Behavior, and Immunity, 25,* 1162–1169.

Wikgren, M., Maripuu, M., Karlsson, T., Nordfjäll, K., Bergdahl, J., Hultdin, J., … Norrback, K.-F. (2012). Short telomeres in depression and the general population are associated with a hypocortisolemic state. *Biological Psychiatry, 71,* 294–300.

Wium-Andersen, M. K., Orsted, D. D., Rode, L., Bojesen, S. E., & Nordestgaard, B. G. (2017). Telomere length and depression: Prospective cohort study and Mendelian randomisation study in 67 306 individuals. *British Journal of Psychiatry, 210,* 31–38.

Wojcicki, J. M., Heyman, M. B., Elwan, D., Shiboski, S., Lin, J., Blackburn, E., & Epel, E. (2015). Telomere length is associated with oppositional defiant behavior and maternal clinical depression in Latino preschool children. *Translational Psychiatry, 5.*

Wolkowitz, O. M., Mellon, S. H., Epel, E. S., Lin, J., Dhabhar, F. S., Su, Y., … Blackburn, E. H. (2011). Leukocyte telomere length in major depression: Correlations with chronicity, inflammation and oxidative stress-preliminary findings. *Plos One, 6.*

Wolkowitz, O. M., Mellon, S. H., Lindqvist, D., Epel, E. S., Blackburn, E. H., Lin, J., … Mueller, S. (2015). PBMC telomerase activity, but not leukocyte telomere length, correlates with hippocampal volume in major depression. *Psychiatry Research, 232*, 58–64.

Xiang, L., Szebeni, K., Szebeni, A., Klimek, V., Stockmeier, C. A., Karolewicz, B., … Ordway, G. A. (2008). Dopamine receptor gene expression in human amygdaloid nuclei: Elevated D4 receptor mRNA in major depression. *Brain Research, 1207*, 214–224.

Ye, N., Rao, S., Du, T., Hu, H., Liu, Z., Shen, Y., & Xu, Q. (2017). Intergenic variants may predispose to major depression disorder through regulation of long non-coding RNA expression. *Gene, 601*, 21–26.

Zalli, A., Jovanova, O., Hoogendijk, W. J. G., Tiemeier, H., & Carvalho, L. A. (2016). Low-grade inflammation predicts persistence of depressive symptoms. *Psychopharmacology, 233*, 1669–1678.

Zeng, J., Kitayama, I., Yoshizato, H., Zhang, K., & Okazaki, Y. (2003). Increased expression of corticotropin-releasing factor receptor mRNA in the locus coeruleus of stress-induced rat model of depression. *Life Sciences, 73*, 1131–1139.

Zhang, D., Cheng, L., Craig, D. W., Redman, M., & Liu, C. (2010). Cerebellar telomere length and psychiatric disorders. *Behavior Genetics, 40*, 250–254.

Zheng, P., Zeng, B., Zhou, C., Liu, M., Fang, Z., Xu, X., … Xie, P. (2016). Gut microbiome remodeling induces depressive-like behaviors through a pathway mediated by the host's metabolism. *Molecular Psychiatry, 21*, 786.

Zhou, L., Xiong, J., Lim, Y., Ruan, Y., Huang, C., Zhu, Y., … Zhou, X.-F. (2013). Upregulation of blood proBDNF and its receptors in major depression. *Journal of Affective Disorders, 150*, 776–784.

Zhou, Q.-G., Hu, Y., Wu, D.-L., Zhu, L.-J., Chen, C., Jin, X., … Zhu, D.-Y. (2011). Hippocampal telomerase is involved in the modulation of depressive behaviors. *The Journal of Neuroscience, 31*, 12258.

Zobel, A. W., Nickel, T., Sonntag, A., Uhr, M., Holsboer, F., & Ising, M. (2001). Cortisol response in the combined dexamethasone/CRH test as predictor of relapse in patients with remitted depression. A prospective study. *Journal of Psychiatric Research, 35*, 83–94.

FURTHER READING

López, J. F., Chalmers, D. T., Little, K. Y., & Watson, S. J. (1998). Regulation of serotonin$_{1A}$, glucocorticoid, and mineralocorticoid receptor in rat and human hippocampus: implications for the neurobiology of depression. *Biological Psychiatry, 43*, 547–573.

Chapter 27

Neuroimaging biomarkers of late-life major depressive disorder pathophysiology, pathogenesis, and treatment response

Helmet T. Karim[a], Charles F. Reynolds, III[a,b] and Stephen F. Smagula[a,c]

[a]Department of Psychiatry University of Pittsburgh School of Medicine, Pittsburgh, PA, United States; [b]Department of Behavioral and Community Health Sciences, Graduate School of Public Health, University of Pittsburgh, Pittsburgh, PA, United States; [c]Department of Epidemiology, Graduate School of Public Health, University of Pittsburgh, Pittsburgh, PA, United States

Abbreviations

ACC	anterior cingulate
ASL	arterial spin labeling
ASN	anterior salience network
BOLD	blood-oxygen level-dependent response
CBF	cerebral blood flow
dACC	dorsal anterior cingulate
dlPFC	dorsolateral prefrontal cortex
DMN	default mode network
dmPFC	dorsomedial prefrontal cortex
DTI	diffusion tensor imaging
ECN	executive control network
FA	fractional anisotropy
fMRI	functional magnetic resonance imaging
ICA	independent components analysis
IFG	inferior frontal gyrus
IPG	inferior parietal gyrus
LLD	late-life depression
LOOCV	leave-one-out cross-validation
MDD	major depressive disorder
MeFG	medial frontal gyrus
MFG	middle frontal gyrus
MRI	magnetic resonance imaging
MTG	middle temporal gyrus
OFG	orbitofrontal Gyrus
PCC	posterior cingulate
PET	positron emission tomography
PFC	prefrontal cortex
PoCG	post-central gyrus
PreCG	precentral gyrus
rACC	rostral anterior cingulate
SFG	superior frontal gyrus
sgACC	subgenual anterior cingulate
SMA	supplemental motor area

Personalized Psychiatry. https://doi.org/10.1016/B978-0-12-813176-3.00027-4

SMG	supramarginal gyrus
SNRI	serotonin-norepinephrine reuptake inhibitors
SPG	superior parietal gyrus
SSRI	selective serotonin reuptake inhibitors
STG	superior temporal gyrus
SVD	small vessel disease
SVM	support vector machine
vmPFC	ventromedial prefrontal cortex
WMH	white matter hyperintensities

Late-life major depressive disorder (LLD) is an intrinsically debilitating illness that increases health care costs (Katon, Lin, Russo, & Unutzer, 2003; Luppa et al., 2012), amplifies the impact of chronic diseases on disability (Tinetti et al., 2011), and increases risk for chronic diseases (Brown, Stewart, Stump, & Callahan, 2011; Choi, Kim, Marti, & Chen, 2014; Glymour et al., 2012; Jackson & Mishra, 2013; Mezuk, Eaton, Albrecht, & Golden, 2008), suicidality (Conwell, Van Orden, & Caine, 2011), dementia (Cherbuin, Kim, & Anstey, 2015; Diniz, Butters, Albert, Dew, & Reynolds 3rd., 2013; Gatz, Tyas, St John, & Montgomery, 2005), and early death (Cuijpers et al., 2014). With less than 50% of older adults responding to initial first-line pharmacotherapy for LLD (as reviewed in Mulsant, Blumberger, Ismail, Rabheru, & Rapoport, 2014), patient heterogeneity and treatment resistance are the norm, and not the exception. There is a clear need to develop more effective interventions that precisely target LLD's central mechanisms.

Studying biomarkers that indicate LLD's underlying mechanisms can help identify novel treatment targets. Furthermore, establishing objective biomarkers of treatment target engagement could revolutionize treatment allocation strategies and thereby reduce the time between diagnosis and remission. For example, objective biomarkers can be used to indicate which treatment is needed (e.g., cognitive behavioral therapy, selective serotonin reuptake inhibitors, or transcranial magnetic stimulation). Such biomarkers could also be used to assess whether initial doses of a given treatment engage their targets in a way that will lead to remission.

This goal of identifying objective biomarkers of LLD (that map to mechanisms and treatment response) is not new. However, recent research has the benefit of improved neuroimaging contrast, plus greater spatial and temporal resolution (in particular, magnetic resonance imaging or MRI). As a result, there has been increasing evidence for particular objective biomarkers that indicate LLD's underlying mechanisms and treatment responsiveness. However, available evidence does not support a particular method for using MRI biomarkers to personalize LLD treatment.

Patient classification and treatment allocation strategies based on MRI data have not yet been replicated with acceptably high levels of predictive utility (see Section 4). Therefore, while MRI biomarkers hold the promise of personalizing LLD treatment, they are not yet ready for routine clinical applications. Clinical data, such as variability in symptom expression or cognitive function, is more readily available, correlates with LLD's prognosis (Smagula et al., 2015), and indicates whether certain treatments will be efficacious (Kaneriya et al., 2016; Smagula et al., 2016). The best readily available indicator of whether a treatment will lead to remission may be whether symptom change is observed in the first 2 weeks of treatment (Joel et al., 2014). Clinicians personalizing LLD treatments today should systematically treat with available pharmacological and behavioral options, provide close monitoring, and adjust dosing as needed.

Nevertheless, a clear picture is emerging regarding the MRI biomarkers of LLD pathophysiology, pathogenesis, and treatment prognosis. Known biomarkers of LLD's course are found in many, but not all, LLD patients; thus, their presence or absence speaks to the underlying mechanisms of an individual's LLD *that can be the unique targets of personalized treatment*. This literature is the foundation for innovative techniques personalizing medicine based on objective biomarkers of LLD's underlying neurobiological mechanism, and whether treatments engage targets. This chapter provides a narrative review summarizing several decades of neuroimaging biomarker research in LLD. The time is ripe for innovative clinician-scientists to build on this foundation, toward validated strategies for using MRI to individually tailor interventions. We conclude with remarks on the future steps that must be taken to achieve a validated science of personalized medicine for LLD.

1 Neuroimaging biomarkers of LLD pathophysiology

1.1 Section summary

This section reviews neuroimaging biomarkers that differ between patients with prevalent LLD and nondepressed older adults. On average, people with LLD have greater cerebrovascular disease burden, lower white matter integrity, as well as atrophy and reduced metabolism in key cognitive and emotional regulation circuitry. On average, people with LLD show differences in neural engagement during executive and emotional reactivity tasks. In addition, studies comparing patients with LLD and nondepressed controls have found LLD is generally characterized by greater connectivity within the default mode network (DMN), greater connectivity between the DMN and limbic system, and lower connectivity between the DMN and executive control network (ECN). These differences are generally relevant to understanding the neurobiology of prevalent LLD, and therefore, shed light on potential factors mediating LLD or its consequences.

1.2 White matter biomarkers

Early research found white matter hyperintensities (WMH) were more common among older adults with depression (Coffey et al., 1988; Coffey, Figiel, Djang, Saunders, & Weiner, 1989; Dolan, Poynton, Bridges, & Trimble, 1990; Krishnan et al., 1988). WMH are areas of the brain that appear hyperintense (relative to normal appearing white matter) on T2-contrast MRI scans. Post-mortem research indicates that WMHs reflect gliosis and demyelination (Braffman et al., 1988), astrogliosis and oligodendrocytes loss, small vessel loss, myelin damage, and vacuolation (allowing for fluid accumulation) (Murray et al., 2012).

Several studies (Dalby et al., 2010; Greenwald et al., 1996; Kumar, Bilker, Jin, & Udupa, 2000; Sheline et al., 2008; Taylor et al., 2005, 2011) and a recent systematic review have confirmed that, indeed, compared with nondepressed controls, WMHs are more common among patients with LLD (Herrmann, Le Masurier, & Ebmeier, 2008). Patients with LLD have greater WMH burden in the upper cingulum (Taylor et al., 2011), superior longitudinal fasciculus (Dalby et al., 2010; Sheline et al., 2008), as well as fronto-occipital fasciculus, uncinate, and inferior longitudinal fasciculus (Sheline et al., 2008). Other recent research has demonstrated that WMH are related to low positive affect and interpersonal problems in older adults with depression (but not somatic symptoms or depressed affect) (Tully, Debette, Mazoyer, & Tzourio, 2017). These findings support the "vascular depression" hypothesis, which noted that WMH associated with vascular risk factors "predispose, precipitate, or perpetuate" geriatric depressive symptoms (Alexopoulos et al., 1997; Taylor, Aizenstein, & Alexopoulos, 2013).

WMHs disrupt brain connectivity by affecting white matter integrity. Early research showed that depressed patients with clinical vascular disease had reduced brainstem auditory transmission (Kalayam et al., 1997). Diffusion tensor imaging (DTI) research corroborates this finding, showing that WMHs are associated with reduced white matter microstructural integrity (Taylor et al., 2001, 2007). These studies have investigated markers of white matter integrity, namely, fractional anisotropy (FA, measured using DTI; decreased FA is associated to loss of myelin integrity). People with LLD have lower FA on several tracts compared with controls (Alves et al., 2012; Dalby et al., 2010; Sexton et al., 2012; Taylor et al., 2007), particularly in the longitudinal fasciculus (Dalby et al., 2010; Sexton et al., 2012), uncinate fasciculus (Dalby et al., 2010; Sexton et al., 2012; Taylor, Macfall, et al., 2007), posterior cingulum (Alves et al., 2012; Sexton et al., 2012), anterior thalamic radiation (Sexton et al., 2012), and corticospinal tract (Sexton et al., 2012). The longitudinal and uncinate fasciculi connect the executive neural network and fronto-limbic network, respectively, which are thought to mediate cognitive functions and emotion regulation. One study found lower FA in the uncinate fasciculus in early-onset depression compared with late-onset depression (Taylor, Macfall, et al., 2007), which may be related to differences in pathology in early vs late-onset LLD.

A key limitation of some of these studies is that WMH clearly affects FA when present, thus it is not always clear whether the low FA and greater WMH burden observed in LLD are completely independent. In any case, in a meta-analysis of nine studies, the most consistent finding was that people with LLD had lower FA in the dorsolateral prefrontal context and uncinate fasciculus (Wen, Steffens, Chen, & Zainal, 2014); this highlights the important role of disconnection between frontal and limbic structures, and further supports a fronto-limbic disconnection hypothesis.

1.3 Gray matter biomarkers

The pathophysiology of LLD not only involves lower white matter integrity, but differences in the volume and density (volume and thickness of cortex) of key processing hubs in the brain. Several studies have demonstrated that lower gray matter volume and density (especially in fronto-limbic structures) are associated with prevalent depressive symptoms, severity, and age of onset. Lower volumes in LLD (compared with nondepressed controls) have been observed in several regions relevant to emotional and cognitive information processing, including the hippocampus, amygdala, and frontal cortex (specifically orbitofrontal gyrus, OFG, superior frontal gyrus, SFG, middle frontal gyrus, MFG, and anterior cingulate, ACC) (Du et al., 2014).

Lower hippocampal volumes are the most commonly found neuroanatomical correlate in LLD (Andreescu et al., 2008; Ballmaier et al., 2008; Bell-McGinty et al., 2002; Disabato et al., 2014; Egger et al., 2008; Janssen et al., 2004; Lloyd et al., 2004; Sheline, Wang, Gado, Csernansky, & Vannier, 1996; Zhao et al., 2008). The hippocampus is a critical hub for information and memory processing, and is thought to atrophy over prolonged periods of stress and continued depressive symptoms (related to heightened cortisol levels and dysregulation of the hypothalamic-pituitary-adrenal axis in depression). Several studies have investigated volume differences between early-onset and late-onset depression in LLD; half of these studies found no differences (Ballmaier et al., 2008; Burke et al., 2011; Disabato et al., 2014), and the other half have found that either late-onset depression is associated with lower hippocampal volumes (Bell-McGinty et al., 2002; Lloyd et al., 2004) (possibly due to acquired biological factors that influence hippocampal volume), or that the volume was associated with the duration of illness (years lived with depression) (Sheline et al., 1996). This may suggest that low hippocampal volumes are a trait and consequence of depression—however it is possible that low hippocampal volumes could also increase depression risk (as in Olesen et al., 2010).

Another common neuroanatomical correlate is smaller amygdala volumes in LLD compared with nondepressed controls (Andreescu et al., 2008; Burke et al., 2011; Disabato et al., 2014; Egger et al., 2008). The amygdala is involved in processing emotional stimuli, and has long been associated with depression. Frontal regions involved in the regulation of emotions are also smaller in individuals with LLD compared with nondepressed controls. The volumes of the OFG (Andreescu et al., 2008; Lavretsky, Ballmaier, Pham, Toga, & Kumar, 2007; Taylor, Macfall, et al., 2007), ACC (Bell-McGinty et al., 2002; Disabato et al., 2014) (including subgenual aspect, sgACC (Lehmbeck, Brassen, Braus, & Weber-Fahr, 2008)), and superior (Andreescu et al., 2008; Disabato et al., 2014), middle (Bell-McGinty et al., 2002; Disabato et al., 2014), medial frontal gyrii (MeFG) (Andreescu et al., 2008; Egger et al., 2008), as well as the dorsolateral prefrontal cortex (dlPFC) (Chang et al., 2011) have all been associated with LLD (smaller volumes compared with nondepressed controls). Some other studies have also found that the caudate (Krishnan et al., 1993; Smith et al., 2009), putamen (Andreescu et al., 2008; Krishnan et al., 1993), and thalamus (Andreescu et al., 2008; Smith et al., 2009) volumes are smaller in people with LLD compared with healthy controls, which may be related to the processing of rewarding stimuli. These results clearly demonstrate reduced neural resources in circuits related to emotion reactivity, regulation, and executive control in LLD pathophysiology.

1.4 Functional activation biomarkers

Several studies have utilized functional magnetic resonance imaging (fMRI) to understand the association between neural activation and LLD (Aizenstein et al., 2009; Aizenstein et al., 2011; Brassen, Kalisch, Weber-Fahr, Braus, & Buchel, 2008; Wang et al., 2008). FMRI measures the blood-oxygen level-dependent (BOLD) response that follows neural activation due to increased metabolic need in the activated region. Studies have widely used fMRI to examine the regions that activate during particular tasks. These studies replicate an important set of findings from mid-life depression (reviewed in Jaworska, Yang, Knott, & Mac Queen, 2015) that shows greater emotional reactivity in LLD compared with nondepressed controls in limbic regions (e.g., amygdala, sgACC). They have also found a disrupted executive control circuit (e.g., hypoactive dlPFC) in LLD—which has been demonstrated in mid-life.

Disrupted executive function may be associated with vascular burden, as demonstrated in one study using an emotion reactivity task (Aizenstein et al., 2011). In this study, patients with LLD had greater limbic reactivity (in the subgenual anterior cingulate cortex or sgACC) that correlated with greater WMH burden (Aizenstein et al., 2011). During an emotional oddball task, where images are presented continuously and participants respond to a sad/neutral "oddball" image, people with LLD had greater amygdala activity than nondepressed controls (Wang et al., 2008); no behavioral differences in reaction time were observed. Another study using an emotion reactivity task found nondepressed controls exhibited greater OFG and rostral ACC (rACC) activation to negative words (Brassen et al., 2008); LLD patients also had lower activation in the ventromedial PFC (vmPFC) during negative, compared with positive stimuli, which correlated with symptom severity (Brassen et al., 2008). One study found that patients with LLD had lower dlPFC activation during an executive control task (Aizenstein et al., 2009). Another study found that people with LLD had had lower activation while viewing neutral images in frontal executive regions (MFG, supramarginal gyrus or SMG, dACC), as well as in the posterior cingulate (PCC) and angular gyrus (Wang et al., 2008). These studies show a hyperactive limbic system and hypoactive executive control system that may explain poor cognitive control over emotions in people with LLD.

1.5 Functional connectivity biomarkers

An informative and growing literature has utilized resting-state fMRI, where participants are instructed to lie awake in the scanner to observe neural activity during a resting period. The observed patterns of coactivations across regions are replicable, and can be derived using several methods. For example, independent components analysis (ICA) generates stable neural networks such as motor, sensory, anterior salience (ASN), executive control (ECN), and default mode (DMN) networks; or seed-based connectivity analyses that examine the patterns of co-activation with a region chosen to represent one or a part of these networks, e.g., DMN connectivity can be examined using a posterior cingulate cortex or medial frontal gyrus seed, and the ECN can be generated using the left dlPFC or left inferior parietal gyrus (IPG). Several networks have been implicated in LLD pathophysiology (Ajilore et al., 2014; Alexopoulos et al., 2012; Andreescu et al., 2013; Bohr et al., 2012; Cieri et al., 2017; Eyre et al., 2016; Kenny et al., 2010; Liu et al., 2012; Wu et al., 2011), however, possibly due to the differences in the samples recruited (phenotypic heterogeneity), as well as differences in the methods and the measures utilized, there is quite a bit of variability in the resting state connectivity literature.

Of particular interest is the default mode network, which represents the greatest variance of neural activation at rest, and is involved in self-referential thinking. Studies that have either used a core node of the DMN (e.g., PCC) or ICA-based methods have two major findings: (1) greater within-network DMN connectivity (Alexopoulos et al., 2012; Andreescu et al., 2013; Eyre et al., 2016) and DMN-limbic connectivity (Alexopoulos et al., 2012; Wu et al., 2011) in people with LLD compared with nondepressed controls; as well as (2) lower DMN-ECN connectivity (Andreescu et al., 2013; Wu et al., 2011) in people with LLD compared with nondepressed controls (though one study found greater connectivity in LLD (Alexopoulos et al., 2012)).

It is important to note that two studies have found conflicting results regarding ECN connectivity (assessed using either left ICA or seeds [dlPFC] or the dorsal ACC [dACC]). One study found ECN hypoconnectivity (Alexopoulos et al., 2012), while another found ECN hyperconnectivity (Cieri et al., 2017) in people with LLD.

While most research has focused on the DMN and ECN, some has focused on reward networks. These studies highlight the possible role of reward network dysregulation in LLD. For example, one study investigated the connectivity of the caudate, and found that compared with nondepressed controls, people with LLD had hyperconnectivity of the caudate with the ECN, DMN, ASN, and the thalamus (Kenny et al., 2010). One particular study investigated resting state spontaneous local activation and found that people with LLD group had greater spontaneous local activation in the caudate, dACC, dlPFC, angular gyrus, MeFG, and precuneus, but lower activation in the cerebellum, superior temporal gyrus (STG), supplemental motor area (SMA), and postcentral gyrus (PoCG) (Liu, Hu, et al., 2012).

In summary, some literature on functional connectivity in LLD replicates well-known findings in mid-life depression, namely hyperconnectivity of the DMN (thought to be associated with ruminative behavior), hypoconnectivity of the ECN (possibly related to the executive differences or low cognitive control in LLD), and an altered reward network (possibly related to anhedonia).

1.6 Cerebral blood flow biomarkers

Older studies have also attempted to understand the changes in metabolism and cerebral blood flow (CBF) [using positron emission tomography (PET)] that occur in LLD (Awata et al., 1998; Bench, Friston, Brown, Frackowiak, & Dolan, 1993; Curran et al., 1993; Ebmeier, Glabus, Prentice, Ryman, & Goodwin, 1998; Kumar et al., 1993; Lesser et al., 1994; Nobler et al., 2000; Sackeim et al., 1990; Upadhyaya, Abou-Saleh, Wilson, Grime, & Critchley, 1990). These studies indicate, in people with LLD, CBF is lower in a number of limbic and executive regions, which may exacerbate atrophy, as well as disrupt brain activation and connectivity. Using an infusion of a radioligand (which exposes participants to ionizing radiation), PET can be used to measure metabolism or CBF. These studies have found global reductions in metabolism, as well as reductions in prefrontal regions, caudate, and the cingulate in LLD compared with healthy controls. With only one exception (in certain specific regions), almost all of these studies have found that the LLD group exhibited lower metabolism or CBF compared with healthy controls in the ACC (Awata et al., 1998; Bench et al., 1993; Kumar et al., 1993), PCC (Ebmeier et al., 1998; Kumar et al., 1993), frontal cortex (SFG, MFG, inferior frontal gyrus or IFG, dlPFC, OFG) (Awata et al., 1998; Bench et al., 1993; Kumar et al., 1993; Lesser et al., 1994; Nobler et al., 2000; Sackeim et al., 1990; Upadhyaya et al., 1990), motor/sensory cortex (Awata et al., 1998; Kumar et al., 1993), hippocampus(Awata et al., 1998), temporal cortex (Awata et al., 1998; Ebmeier et al., 1998; Kumar et al., 1993; Lesser et al., 1994; Nobler et al., 2000; Sackeim et al., 1990; Upadhyaya et al., 1990), SMG (Awata et al., 1998), parietal cortex (IPG, superior parietal gyrus or SPG) (Kumar et al., 1993; Nobler et al., 2000; Sackeim et al., 1990; Upadhyaya et al., 1990), caudate (Awata et al., 1998; Curran et al., 1993; Kumar et al., 1993), and thalamus (Curran et al., 1993). One study further found that metabolism was lower in late onset compared with early onset in STG (Ebmeier et al., 1998). Due to the low resolution of PET, the spatial specificity of these regions is not well understood, however they do support a hypometabolism of frontal and limbic circuits, as well as global differences. Resting metabolism studies clearly indicate a disrupted metabolic system, which may be related to symptoms and neuropathology of LLD, or possibly even the vascular changes associated with LLD. In recent years, MRI sequences have been developed to measure CBF (arterial spin labeling or ASL), and may help improve our understanding of these changes with greater resolution, and without exposure to ionizing radiation.

2 Neuroimaging biomarkers of LLD pathogenesis
2.1 Section summary

Compared with cross-sectional MRI studies (comparing people with prevalent LLD with nondepressed controls), far less research has investigated how MRI biomarkers differ between people who do or do not develop future depression (i.e., longitudinal studies of depression incidence). Such longitudinal studies are critical to establishing temporality in the relationship between neuro-biomarkers and disease, and thus, understanding true etiological pathways. Biomarker differences between people with LLD and nondepressed controls can be either risk factors for LLD, or nonetiologic pathophysiologic factors (e.g., that emerge as a disease consequence). Preventive interventions must target etiologic factors that precede and increase the risk of disease incidence, whereas targeting disease consequences will not necessarily curtail depression pathogenesis. Most research has focused on markers of cerebrovascular disease, and suggests insults to white matter integrity increases depression risk. Some evidence indicates neurodegeneration may also increase depression risk in older adults.

2.2 White matter integrity and depression development

Longitudinal evidence indicates that WMH burden (Godin et al., 2008; Olesen et al., 2010; Park et al., 2015; Pavlovic et al., 2016) and progression (Firbank et al., 2012; Teodorczuk et al., 2007; van Sloten et al., 2015) predicts incident depression. For example, one study following a cohort of 783 Korean elders found that people with WMHs had an eightfold increased risk of developing MDD over 3 years (Park et al., 2015). But not all studies have found this association (Dotson, Zonderman, Kraut, & Resnick, 2013; Versluis et al., 2006). These mixed findings are potentially due to lack of specificity of the WMH burden measures used, combined with the high prevalence of WMHs in older adults (de Leeuw et al., 2001) and sampling differences. Recently, a large study with a wide set of small vessel disease (SVD) measures found that preexisting infarcts, as well as new infarcts, enlarged perivascular spaces, microbleeds, and WMH progression increased depression risk (van Sloten et al., 2015). Thus, preexisting infarcts and progressive SVD may be drivers of LLD pathogenesis. These associations may be further mediated by reductions in white matter microstructural integrity, which one study found predicted LLD development (Reppermund et al., 2014).

2.3 Regional brain volume and depression development

Available evidence indicates that lower medial temporal lobe volume (Olesen et al., 2010) and decreases in whole brain volume (van Sloten et al., 2015) predict LLD development. These findings suggest avoiding neurodegenerative processes, and preserving the integrity of information processing circuits in particular, are important to preventing LLD.

3 Neuroimaging biomarkers of LLD treatment response
3.1 Section summary

The following section reviews neuroimaging biomarkers comparing patients that remitted or responded to treatment to those that did not. Patients that remit tend to have lower white matter hyperintensity burden (greater white matter integrity) and greater limbic and executive cortical volumes. In general, low executive control during primary executive and emotion reactivity/regulation tasks as well as high emotional neural reactivity predicts worse treatment response. These pathophysiological hallmarks have been shown to improve or normalize following successful treatment. While the literature is mixed on differences in neural networks and metabolism, some have found that the low ECN and high DMN connectivity seems to normalize following successful therapy, and that the CBF in the PFC decreases—although more studies are needed to confirm these findings.

3.2 White matter integrity and treatment response

WMH burden has been associated with nonresponse to pharmacological treatment in a number of studies (Bella et al., 2010; Gunning-Dixon et al., 2010; Hickie et al., 1995; Simpson, Baldwin, Jackson, & Burns, 1998; Sneed et al., 2011), although some studies have failed to find this association (Janssen et al., 2007; Salloway et al., 2002). Positive findings include that total (Bella et al., 2010; Gunning-Dixon et al., 2010; Hickie et al., 1995 ; Sneed et al., 2011), deep/periventricular (Hickie et al., 1995; Sneed et al., 2011) (those located in the deep white matter or near the ventricles), or subcortical (Simpson et al., 1998) WMH burden are lower in treatment responders. Our group recently found that patients who do not remit following pharmacotherapy have an accumulation of WMHs over a 12-week period, whereas remitters do not (Khalaf et al., 2015). For example, levels of the proinflammatory cytokine tumor necrosis factor-alpha correlate with white matter hyperintensity burden in patients with LLD (Smagula et al., 2017). It is possible that, in patients with a high burden of WMHs, traditional pharmacotherapy for LLD does not adequately engage with these underlying mechanisms. For example, certain patients may benefit from separate or adjunctive treatments targeting the drivers of WMH etiology and progression, potentially including tumor necrosis factor-alpha antagonism (with infliximab; which is associated with improvements in depressive symptoms among patients with high levels of inflammation and treatment resistant depression (Raison et al., 2013)).

Fewer studies have investigated the association with FA (using DTI) and treatment outcomes (Alexopoulos et al., 2008; Alexopoulos, Kiosses, Choi, Murphy, & Lim, 2002; Taylor et al., 2008). One study found lower FA (in the rACC, dACC, PCC, precuneus, MFG, hippocampus, parahippocampus, insula, MTG, STG, angular, fusiform, IPG, post-central gyrus, and subthalamic region) was associated with nonresponse to pharmacotherapy (Alexopoulos et al., 2008). This finding has been confirmed in one study that also found lower FA in frontal white matter predicted worse response (Alexopoulos et al., 2002). While another study reported *higher* FA in nonremitters compared with remitters (Taylor et al., 2008), overall, the available evidence suggests that damage to white matter structural integrity decreases the likelihood that traditional pharmacotherapy will lead to remission.

3.3 Regional brain volume and treatment response

Several studies have investigated the association between gray matter volume and treatment nonresponsiveness. These studies have found that lower volumes in the dACC (Gunning et al., 2009), rACC (Gunning et al., 2009), and hippocampus (Hsieh et al., 2002; Zhao et al., 2008) in treatment nonresponders, although one study failed to find any such differences (Janssen et al., 2007). Lower pre-treatment gray matter volume may determine resistance to traditional therapies, in part, because the regions affected in LLD are critical for emotion regulation, executive control, and reward processing. Theoretically, successful treatment results in an eventual rewiring of these critical circuits, and a lower baseline capacity may slow or impair this process. Some studies have investigated whether there were significant changes in volume following treatment. One study reported increased orbital SFG gray matter density that correlated with symptom changes following pharmacologic therapy (Droppa et al., 2017). A study of electroconvulsive therapy response found increases in caudate gray matter density that correlated with improvement in melancholia (Bouckaert et al., 2016), but also found increases in hippocampus, amygdala, insula, and SMG.

3.4 Functional activation and treatment response

Four studies have investigated how pre-treatment neural activation predicts treatment response, while others have investigated the changes that occurred following successful pharmacotherapy (Aizenstein et al., 2009; Brassen et al., 2008; Alexander Khalaf et al., 2016). While the treatments used often theoretically target the limbic system (e.g., to increase serotonin), the circuits affected are often broad, and may be a result of increased limbic reactivity following acute treatment. This bottom-up process may first target limbic structures (due to the high density of serotonin receptors in these regions), but it likely also targets executive controls circuits higher up in the system.

A series of studies have demonstrated differences in activation that normalized following successful treatment. During an executive function task, patients had hypoactive dlPFC activation that normalized in remitters after receiving Paroxetine (a selective serotonin reuptake inhibitor, SSRI) (Aizenstein et al., 2009). During an emotion reactivity task that used negative words, LLD patients exhibited low vmPFC activation that attenuated following successful treatment (using a naturalistic treatment paradigm) (Brassen et al., 2008).

Another study used an emotion regulation task (where participants are asked to actively regulate their emotions to highly negative imagery) to investigate changes in remitters and nonremitters to Venlafaxine (a serotonin-norepinephrine reuptake inhibitor, SNRI) as well as the acute responses that occur (Khalaf et al., 2016). This study reported increased middle temporal gyrus (MTG) activation in remitters, but not in nonremitters, that occurred following a single dose that continued to increase even at the end of the trial (Khalaf et al., 2016). They also reported increased activation in the cuneus, visual cortex, insula, and PCC that was not dependent on the group (Khalaf et al., 2016). Overall, these studies illustrate a normalization of the executive circuits during emotional and executive tasks, and possibly a change in explicit emotion regulation—however, an obvious limitation is the large variability in methods, tasks, and treatments used across these very few studies.

3.5 Functional connectivity and treatment response

Therapies may help reconfigure altered circuits through bottom-up activation of limbic structures, and ultimately the remission may very well depend on a rewiring of higher cortical neural networks. Only a few studies have investigated changes in resting state connectivity following treatment (Alexopoulos et al., 2012; Andreescu et al., 2013; Karim et al., 2016; Wu et al., 2011).

Some studies have shown baseline differences in functional connectivity between remitters and nonremitters, which indicates sources of resistance or response capacity to particular therapies. One study found that baseline DMN connectivity (PCC) was greater in the MeFG in remitters, but lower in the dACC in nonremitters (treated with either SSRIs or SNRIs (Andreescu et al., 2013)). However, another study failed to replicate pretreatment differences in DMN connectivity (Alexopoulos et al., 2012). They did find that the connectivity of the ECN, specifically ECN-dACC, dlPFC, and IPG, were greater in remitters compared with nonremitters treated with escitalopram (SSRI) (Alexopoulos et al., 2012). Another study also reported group differences (that did not change across time), mainly higher ASN (insula)-IFG and MFG connectivity in nonremitters, as well as greater centrality (whole brain connectedness) in the IFG, but lower centrality in the MeFG (Karim et al., 2016) (possibly trait-dependent markers). In general, these reflect greater ASN connectivity with both the ECN and DMN, which may reflect greater anxiety symptoms that need to be individually (or concurrently) targeted.

Additional studies have investigated changes in connectivity following successful therapy. One study investigated DMN connectivity and found that PCC connectivity with both sgACC and dorsomedial prefrontal cortex (dmPFC) connectivity increased in remitters following remission to Paroxetine (SSRI) (Wu et al., 2011). Similarly, another study found that PCC-dlPFC connectivity increased in remitters, while PCC-striatum connectivity increased in nonremitters (Andreescu et al., 2013). This suggests increased DMN to ECN connectivity, as well as increased DMN connectivity with regions involved in emotion monitoring following successful therapy—this may reflect an improvement in executive functioning and monitoring of emotional states.

One study found that in remitters treated with Venlafaxine (SNRI), dlPFC-MTG (between network) connectivity decreased, while dlPFC-precentral gyrus (PreCG, within network) connectivity increased (Karim et al., 2016). They also reported decreased PCC-SMG (between network) and IFG connectivity, as well as increased PCC-MTG and ITG (within network) connectivity in remitters (Karim et al., 2016). Accordingly, the double-dissociation of high ECN and low DMN connectivity at baseline in LLD pathophysiology might be further expanded to include increased within-network connectivity and decreased inter-network connectivity that is altered by successful treatment. Several of these changes in functional connectivity occurred acutely, following a single dose, and continued to change throughout the treatment course. This highlights the potential of using imaging to measure acute neural responses as an indicator of target engagement.

3.6 Cerebral blood flow and treatment response

As the vascular depression hypothesis suggests that LLD is, at least in part, caused by cerebrovascular changes, it is important to understand whether these changes that occur in LLD are reversible. Two studies seem to suggest that, in fact, instead of increasing CBF, successful treatment in remitters (compared with nonremitters) with either electroconvulsive therapy (Nobler et al., 1994) or SSRIs (Nobler et al., 2000) decreased PFC CBF instead. This suggests that either this is either a trait of LLD, that these changes represent distinct neurobiological processes unrelated to treatment response, or even that these changes occur over a longer scale, and thus may eventually normalize over a more extended period.

4 Machine learning to predict LLD and treatment response

While the past few sections have discussed some findings that may help us further understand LLD and treatment response, they lack one specific feature: prediction. The term "predict" is often used synonymously with "associate," however, these are, in fact, quite different. The studies described previously have found associations between neurobiological markers with LLD or treatment response, sometimes explicating the goal of determining if a marker can help a clinician better diagnose or treat a patient. However, it may be necessary to build a statistical or machine learning based model to further understand more complex relationships between the neurobiological factors and LLD or treatment response.

These models can be used in two ways: (1) perform an accurate prediction—then investigate what factors are predictive and use those as markers (this is similar to the previous methods, however, it can be used to understand nonlinear relationships between factors) and (2) perform an accurate prediction, then test on an unseen data set to evaluate validity, and then use the validated model to predict LLD diagnosis or treatment outcome. The first method is fairly straightforward (as it has similar issues to translation, as any biomarker would), however, the second method is complicated by issues regarding the specificity of the model (e.g., the ability to detect true negatives, or how often does the model predict that a nondepressed person has LLD), as well as its sensitivity (e.g., the ability to detect true positives, or how accurately does it predict that a person with LLD has LLD). Such models also raise potential training and ethical issues (e.g., should the prediction be given to the clinician before or after they have made their own judgment? Could this bias clinicians or making them too reliant on a specific technology?). In this section, we briefly review studies that have used neuroimaging to predict diagnosis or treatment response in LLD.

Some studies have used PET, which exposes patients to ionizing radiation, and is therefore limited in its overall use. MRI, however, lends itself to repeated use as it is safe and nonionizing, further, it can be used to measure CBF (many of the studies reviewed on CBF used PET imaging techniques). Thus, we limit our discussion of past studies using MRI instead of PET neuroimaging techniques.

Due to the limited nature of the LLD literature and machine learning, we focus on prediction models in depression generally—but also describe a single finding in LLD (Patel et al., 2015). Several studies have investigated using gray matter density, emotional stimuli, as well as resting state connectivity to predict diagnosis (Costafreda, Chu, Ashburner, & Fu, 2009; Fu et al., 2008; Hahn et al., 2011; Liu et al., 2012; Marquand, Mourao-Miranda, Brammer, Cleare, & Fu, 2008; Mwangi, Matthews, & Steele, 2012; Nouretdinov et al., 2011; Rondina et al., 2014; Zeng et al., 2012) and treatment response (Costafreda et al., 2009; Liu, Guo, et al., 2012; Marquand et al., 2008; Nouretdinov et al., 2011) in mid-life MDD. In general, when predicting diagnosis (MDD vs nondepressed controls), these studies report accuracies ranging from 68% to 94% (mean accuracy 81%), sensitivities (or ability to accurately predict MDD) ranging from 65% to 100% (mean sensitivity 83%), and specificity (or ability to accurately predict nondepressed controls) ranging from 67% to 90% (mean specificity 80%). When predicting remission/response to treatment, these studies reported accuracies ranging from 69% to 90% (mean accuracy 83%), sensitivities ranging from 78% to 89% (mean sensitivity 85%), and specificity ranging from 52% to 90% (mean specificity 80%).

The most commonly utilized algorithms were support vector machines (SVM) for their ability to model nonlinear dynamics in large data sets. A common, but important, set of limitations in each of these studies are their limited sample sizes, and leave-one-out cross-validation (LOOCV) (which is often used in small datasets despite the fact that it is known to be biased toward the selected sample). The best method would be to train the models on one data set, and then test it on an independent data set.

In one study of LLD, a host of factors (neuroimaging and clinical variables) predicted both diagnosis and treatment response (Patel et al., 2015). Using alternating decision trees, Patel et al. found that the mini-mental state examination (a metric of cognitive health), intracranial volume, WMH burden, and age could reliably predict which older adults had MDD and which did not (accuracy 88%, sensitivity 89%, and specificity 86%) (Patel et al., 2015). Further, they found that ASN structural connectivity (using diffusion tensor imaging) as well as dDMN functional connectivity predicted treatment response (accuracy 90%, sensitivity 89%, and specificity 90%). A significant limitation is that they used LOOCV as well as small sample sizes (35 nondepressed controls and 33 LLD where 11 were responders and 13 were nonresponders to treatment). Despite this, these results are promising, and clearly indicate a path forward for machine learning methods.

5   Conclusion and future directions

LLD research has made significant strides that have improved our understanding of the neurobiology of disease and treatment response (summarized in Table 1 and Figs. 1 and 2). However, there has been a significant lag in translating these findings to useful biomarkers. This lag is understandably due to inconsistencies in the literature, which is, in large part, due to the high heterogeneity in depression, relatively small sample sizes, differences in remission criteria, the biomarkers used, how these markers are calculated, and the treatments themselves. Available evidence suggests various neurobiological predictors may be useful, from neurostructural integrity (in both gray and white matter), to neural activation and connectivity. This evidence provides a strong foundation from which to begin developing multivariate biomarker-based prediction algorithms.

As the research of prediction algorithms in mid-life depression is much further along, it is clear from that literature that we are able to both diagnose depression and predict treatment response with a fairly high accuracy, sensitivity, and specificity. The few studies in LLD replicate these findings, and show promise for future work. There are very few studies that have actually used machine-learning approaches to predict treatment response.

In addition, several critical issues may affect the sensitivity of prediction. First, prediction algorithms must be clear on whether they are predicting treatment resistance as a general phenotype (e.g., there are patients that exhibit resistance to many therapies), or predicting treatment promotion as a general phenotype (e.g., there are some that have beneficial characteristics that improve outcome to many therapies). The algorithms that really identify these more general phenotypes can

TABLE 1 Summary of key findings.

Pathophysiology	(1) Strong evidence demonstrates that white matter hyperintensities (WMH), low white matter integrity, and atrophy (especially in the hippocampus, amygdala, and prefrontal cortex) are associated with late-life depression (LLD) (2) Strong evidence shows that people with LLD have altered neural engagement during executive and emotional reactivity tasks (3) Mounting evidence is also defining altered resting-state connectivity associated with LLD, including: greater connectivity within the default mode network (DMN), greater connectivity between the DMN and limbic system, and lower connectivity between the DMN and executive control network (ECN). A smaller literature demonstrates a hyper-connectivity of the reward network (4) There is strong evidence that individuals with LLD have hypo-metabolism of the limbic and prefrontal regions of the brain
Pathogenesis	(1) Markers of cerebrovascular disease affecting white matter have been associated with increased depression risk; however, inconsistencies in the existing literature make it unclear precisely which markers and white matter alterations increase depression risk (2) Based on two studies, neurodegenerative processes may also increase depression risk
Treatment response	(1) Strong evidence shows that WMH are associated with poor treatment response (2) Some studies have shown that low white matter integrity (fractional anisotropy) is associated with poor treatment response (3) Several studies have shown that low hippocampus, anterior cingulate, and insula volumes are predictive of poor response (4) Two studies have demonstrated that hypoactive executive activation normalizes following treatment. Another demonstrated that acute changes in explicit emotion regulation occur following a single dose of medication (5) Some studies have shown greater anterior salience (ASN) connectivity with both the ECN and DMN is related to poor treatment outcome. A couple of studies demonstrated that DMN-ECN connectivity increased following successful treatment. One study demonstrated a double disassociation between ECN and DMN connectivity that occurred following only a single dose of medication (6) Two studies have demonstrated a decrease in CBF in individuals that respond to treatment

be useful, as those that are treatment resistant may not need to go through several antidepressant trials before other treatments are considered, however they may suffer in sensitivity to predicting which particular therapy is best.

By predicting not only remission, but also remission to a particular set of therapies, then the overall specificity of these models can also be further improved. No studies to date (in mid-life or LLD) have investigated the sensitivity of prediction algorithms comparing multiple therapies. While many treatments may result in similar behavioral outcomes (e.g., amelioration of depression symptoms), it is unlikely that the "path to remission" is always identical. Past studies indicate that following treatment, there are changes in functional measures—these may be linked to a neural pathway for remission.

Finally, a few studies have shown that some of these neural changes occur acutely—sometimes only after a single dose/session. By using an approach where functional brain measurements are made before and directly after an acute period of treatment, a neural target can be discovered that, when engaged, may predict future remission. Two studies have already demonstrated that there occur early changes in neural activation and connectivity following a single dose of treatment, and more importantly, that these changes reflect the overall changes that occur at the end of the trial (Karim et al., 2016; Khalaf et al., 2016). This indicates that the brain may be engaged/disengaged (a "switch") early on, and could be used as a predictor of their eventual behavioral response. This not only improves both the sensitivity and specificity, but also reduces wait time—where decisions on particular regimens can be made within as early as a single day, rather than 4 or 5 weeks.

To develop the existing scientific basis into clinically actionable tools, there remains a need for future studies using large samples to definitively characterize known prognostic markers and utilize nonlinear models that allow for more complex interactions between the neural markers. For example, machine-learning models can detect, not only that WMH may reduce treatment responsiveness, but also a compounded effect of low hippocampal volume, which may add nonlinearly to risk (e.g., if having low hippocampal volume and greater WMH burden is not the same as having just one of those risk factors alone). Recent literature has shown that combining multiple sources of clinical information (e.g., cognitive flexibility and symptom expression) improves the ability to predict whether a given treatment will be efficacious for a given patient (Smagula et al., 2016). In the same way, multivariate predictive algorithms using neuroimaging biomarkers can generate more precise predictions; this research has the additional benefit of being able to elucidate the neurobiological mechanisms by which clinical variability associates with LLD's prognosis.

Structural Integrity

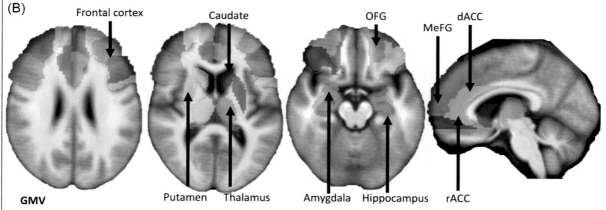

WMH
- Greater WMH predicts development of LLD (Godin et al., 2008)
- Individuals with LLD have greater WMH than non depressed controls (Herrman 2008)
- Greater WMH associated with treatment resistance (Bella et al., 2010)

(A)

SLF

CING

ATR

UNC

ILF

CST

FA
- Individuals with LLD have lower FA than non depressed controls in the superior (SLF) and inferior longitudinal (ILF) fasciculus, uncinate (UNC), cingulum (CING), anterior thalamic radiation (ATR), and corticospinal tract (CST) (Taylor et al., 2001)
- Lower FA at baseline in frontal structural regions connected by the tracts above as well as frontal white matter predict treatment resistance (Alexopoulos et al., 2002)

(B)

Frontal cortex   Caudate   OFG   MeFG   dACC

Putamen   Thalamus   Amygdala   Hippocampus   rACC

GMV
- Lower medial temporal (e.g., hippocampus/parahippocampus) volume predicts development of LLD (Olesen 2010)
- Individuals with LLD have lower gray matter volume/density than non depressed controls in the hippocampus, amygdala, OFG, ACC, SFG, MFG, MeFG, and dlPFC (Du et al., 2014)
- Lower volumes in the dACC, rACC, and hippocampus are associated with treatment resistance (Gunning 2009)

(C)

FIG. 1 Structural markers of LLD with a related reference for each marker (A) WMH on a FLAIR of a single participant are shown. (B, *left*) Each of the tracts implicated in LLD and treatment response as a 3D rendering using the JHU white matter atlas, (B, *three right images*). Each of the implicated tracts overlaid on the T2-image from the JHU atlas. (C) Regions implicated in LLD overlaid on an average structural image of participants with LLD (data from NIMH R01 MH076079).

Near-term studies of treatment response and longer-term studies of depression risk are needed. These studies would help categorize structural and functional markers' changes related to depression and its consequences. Improving methods (through standardization) between processing software used to analyze data can further reduce the variability in the literature—this is especially true in the resting state literature.

Functional

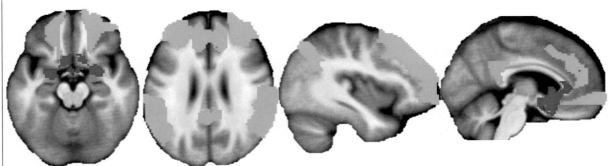

Task based studies

- Compared to non depressed controls, individuals with LLD exhibit hyperactive limbic system (sgACC, Amygdala) while viewing negative images/faces and hypoactive executive system (OFG, rACC, vmPFC, dlPFC, MFG, SMG, dACC) while viewing negative or neutral images/faces as well as during executive control tasks. They also exhibit reduced DMN (PCC, angular) activation while viewing neutral images (Aizenstein et al., 2011; brassen et al., 2008)
- Following treatment, we have observed normalized hypoactive dlPFC during an executive control task and hypoactive vmPFC during an emotion reactivity task. One study observed acute changes in MTG activation during emotion regulation after a single dose of a drug, which continued to increase (Aizenstein et al., 2009)

(A)

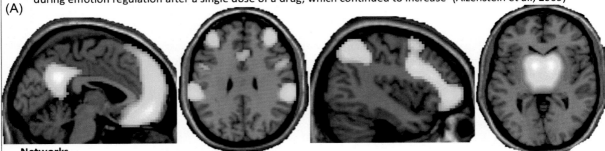

Networks

- Compared to non depressed controls, individuals with LLD exhibit greater DMN, lower DMN-ECN, lower ECN, and greater caudate connectivity (Andreescu et al., 2013)
- Compared to those that are treatment resistant, those that respond to treatment exhibit greater DMN and and ECN, but lower ASN connectivity. Following treatment, DMN connectivity with limbic, striatal, and ECN increase in remitters. Further, remitters also exhibit increased within network and decreased between network connectivity with the DMN and ECN. (Alexopoulos et al., 2012)

(B)

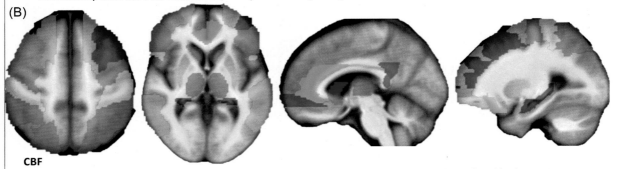

CBF

- Compared to non depressed controls, individuals with LLD exhibit lower CBF in prefrontal and limbic regions (Awata et al., 1998)
- After treatment, remitters further decrease in CBF in prefrontal regions (Nobler et al., 1994)

(C)

FIG. 2 Functional markers of LLD with a related reference for each marker. (A) Regions implicated in LLD as hyperactive (*red*) or hypoactive (*blue*) overlaid on an average structural image of participants with LLD (data from NIMH R01 MH076079). (B) Networks associated with LLD from left to right: DMN, ASN, ECN, and reward (caudate, basal ganglia) overlaid on a template participant (Source: SPM12). (C) Regions with lower CBF in LLD compared with nondepressed controls overlaid on an average structural image of people with LLD of participants with LLD (data from NIMH R01 MH076079).

Understanding which circuits, when affected, lead to specific LLD phenotypes will support rationale intervention approaches targeting the underlying source of dysfunction (e.g., problem-solving training for patients with predisposing executive network dysfunction or mindfulness for patients entering treatment with high salience network dysfunction that has been linked to treatment resistance). In the future, psychiatrists will be given not only a summary of their patient's clinical symptoms, but also neuropsychological assessments and objective neural markers such as total WMH burden, hippocampal, amygdala, and PFC volumes, as well as an assessment of their limbic reactivity and executive connectivity. These multiple sources of information, along with validated multivariate tools, will give a more detailed understanding of the underlying source of the patient's depressive symptoms, and the likelihood a patient will remit to one particular therapy or another. This future will support rationale, targeted prevention approaches, and a dramatic reduction in the number of treatments attempted before a successful regimen is found.

Acknowledgments

This work was supported by T32 MH019986 and K01 MH112683. We would like to thank Dr. Howard Aizenstein for his continued mentorship and guidance on this review.

References

Aizenstein, H. J., Andreescu, C., Edelman, K. L., Cochran, J. L., Price, J., Butters, M. A., … Reynolds, C. F. , 3rd. (2011). fMRI correlates of white matter hyperintensities in late-life depression. *The American Journal of Psychiatry, 168*(10), 1075–1082. https://doi.org/10.1176/appi.ajp.2011.10060853.

Aizenstein, H. J., Butters, M. A., Wu, M., Mazurkewicz, L. M., Stenger, V. A., Gianaros, P. J., … Carter, C. S. (2009). Altered functioning of the executive control circuit in late-life depression: Episodic and persistent phenomena. *The American Journal of Geriatric Psychiatry, 17*(1), 30–42. https://doi.org/10.1097/JGP.0b013e31817b60af.

Ajilore, O., Lamar, M., Leow, A., Zhang, A., Yang, S., & Kumar, A. (2014). Graph theory analysis of cortical-subcortical networks in late-life depression. *The American Journal of Geriatric Psychiatry, 22*(2), 195–206. https://doi.org/10.1016/j.jagp.2013.03.005.

Alexopoulos, G. S., Hoptman, M. J., Kanellopoulos, D., Murphy, C. F., Lim, K. O., & Gunning, F. M. (2012). Functional connectivity in the cognitive control network and the default mode network in late-life depression. *Journal of Affective Disorders, 139*(1), 56–65. https://doi.org/10.1016/j.jad.2011.12.002.

Alexopoulos, G. S., Kiosses, D. N., Choi, S. J., Murphy, C. F., & Lim, K. O. (2002). Frontal white matter microstructure and treatment response of late-life depression: A preliminary study. *The American Journal of Psychiatry, 159*(11), 1929–1932. https://doi.org/10.1176/appi.ajp.159.11.1929.

Alexopoulos, G. S., Meyers, B. S., Young, R. C., Campbell, S., Silbersweig, D., & Charlson, M. (1997). 'Vascular depression' hypothesis. *Archives of General Psychiatry, 54*(10), 915–922. https://doi.org/10.1001/archpsyc.1997.01830220033006.

Alexopoulos, G. S., Murphy, C. F., Gunning-Dixon, F. M., Latoussakis, V., Kanellopoulos, D., Klimstra, S., … Hoptman, M. J. (2008). Microstructural white matter abnormalities and remission of geriatric depression. *The American Journal of Psychiatry, 165*(2), 238–244. https://doi.org/10.1176/appi.ajp.2007.07050744.

Alves, G. S., Karakaya, T., Fusser, F., Kordulla, M., O'Dwyer, L., Christl, J., … Pantel, J. (2012). Association of microstructural white matter abnormalities with cognitive dysfunction in geriatric patients with major depression. *Psychiatry Research, 203*(2–3), 194–200. https://doi.org/10.1016/j.pscychresns.2011.12.006.

Andreescu, C., Butters, M. A., Begley, A., Rajji, T., Wu, M., Meltzer, C. C., … Aizenstein, H. (2008). Gray matter changes in late life depression—A structural MRI analysis. *Neuropsychopharmacology, 33*(11), 2566–2572. https://doi.org/10.1038/sj.npp.1301655.

Andreescu, C., Tudorascu, D. L., Butters, M. A., Tamburo, E., Patel, M., Price, J., … Aizenstein, H. (2013). Resting state functional connectivity and treatment response in late-life depression. *Psychiatry Research, 214*(3), 313–321. https://doi.org/10.1016/j.pscychresns.2013.08.007.

Awata, S., Ito, H., Konno, M., Ono, S., Kawashima, R., Fukuda, H., & Sato, M. (1998). Regional cerebral blood flow abnormalities in late-life depression: Relation to refractoriness and chronification. *Psychiatry and Clinical Neurosciences, 52*(1), 97–105. https://doi.org/10.1111/j.1440-1819.1998.tb00980.x.

Ballmaier, M., Narr, K. L., Toga, A. W., Elderkin-Thompson, V., Thompson, P. M., Hamilton, L., … Kumar, A. (2008). Hippocampal morphology and distinguishing late-onset from early-onset elderly depression. *The American Journal of Psychiatry, 165*(2), 229–237. https://doi.org/10.1176/appi.ajp.2007.07030506.

Bella, R., Pennisi, G., Cantone, M., Palermo, F., Pennisi, M., Lanza, G., … Paolucci, S. (2010). Clinical presentation and outcome of geriatric depression in subcortical ischemic vascular disease. *Gerontology, 56*(3), 298–302. https://doi.org/10.1159/000272003.

Bell-McGinty, S., Butters, M. A., Meltzer, C. C., Greer, P. J., Reynolds, C. F., 3rd, & Becker, J. T. (2002). Brain morphometric abnormalities in geriatric depression: Long-term neurobiological effects of illness duration. *The American Journal of Psychiatry, 159*(8), 1424–1427. https://doi.org/10.1176/appi.ajp.159.8.1424.

Bench, C. J., Friston, K. J., Brown, R. G., Frackowiak, R. S., & Dolan, R. J. (1993). Regional cerebral blood flow in depression measured by positron emission tomography: The relationship with clinical dimensions. *Psychological Medicine, 23*(3), 579–590.

Bohr, I. J., Kenny, E., Blamire, A., O'Brien, J. T., Thomas, A. J., Richardson, J., & Kaiser, M. (2012). Resting-state functional connectivity in late-life depression: Higher global connectivity and more long distance connections. *Frontiers in Psychiatry, 3*, 116. https://doi.org/10.3389/fpsyt.2012.00116.

Bouckaert, F., De Winter, F. L., Emsell, L., Dols, A., Rhebergen, D., Wampers, M., … Vandenbulcke, M. (2016). Grey matter volume increase following electroconvulsive therapy in patients with late life depression: A longitudinal MRI study. *Journal of Psychiatry & Neuroscience, 41*(2), 105–114.

Braffman, B. H., Zimmerman, R. A., Trojanowski, J. Q., Gonatas, N. K., Hickey, W. F., & Schlaepfer, W. W. (1988). Brain MR: Pathologic correlation with gross and histopathology. 2. Hyperintense white-matter foci in the elderly. *AJR American Journal of Roentgenology, 151*(3), 559–566. https://doi.org/10.2214/ajr.151.3.559.

Brassen, S., Kalisch, R., Weber-Fahr, W., Braus, D. F., & Buchel, C. (2008). Ventromedial prefrontal cortex processing during emotional evaluation in late-life depression: A longitudinal functional magnetic resonance imaging study. *Biological Psychiatry, 64*(4), 349–355. https://doi.org/10.1016/j.biopsych.2008.03.022.

Brown, J. M., Stewart, J. C., Stump, T. E., & Callahan, C. M. (2011). Risk of coronary heart disease events over 15 years among older adults with depressive symptoms. *The American Journal of Geriatric Psychiatry, 19*(8), 721–729. https://doi.org/10.1097/JGP.0b013e3181faee19.

Burke, J., McQuoid, D. R., Payne, M. E., Steffens, D. C., Krishnan, R. R., & Taylor, W. D. (2011). Amygdala volume in late-life depression: Relationship with age of onset. *The American Journal of Geriatric Psychiatry, 19*(9), 771–776. https://doi.org/10.1097/JGP.0b013e318211069a.

Chang, C. C., Yu, S. C., McQuoid, D. R., Messer, D. F., Taylor, W. D., Singh, K., … Payne, M. E. (2011). Reduction of dorsolateral prefrontal cortex gray matter in late-life depression. *Psychiatry Research, 193*(1), 1–6. https://doi.org/10.1016/j.pscychresns.2011.01.003.

Cherbuin, N., Kim, S., & Anstey, K. J. (2015). Dementia risk estimates associated with measures of depression: A systematic review and meta-analysis. *BMJ Open. 5*(12)https://doi.org/10.1136/bmjopen-2015-008853.

Choi, N. G., Kim, J., Marti, C. N., & Chen, G. J. (2014). Late-life depression and cardiovascular disease burden: Examination of reciprocal relationship. *The American Journal of Geriatric Psychiatry, 22*(12), 1522–1529. https://doi.org/10.1016/j.jagp.2014.04.004.

Cieri, F., Esposito, R., Cera, N., Pieramico, V., Tartaro, A., & di Giannantonio, M. (2017). Late-life depression: Modifications of brain resting state activity. *Journal of Geriatric Psychiatry and Neurology, 30*(3), 140–150. https://doi.org/10.1177/0891988717700509.

Coffey, C. E., Figiel, G. S., Djang, W. T., Cress, M., Saunders, W. B., & Weiner, R. D. (1988). Leukoencephalopathy in elderly depressed patients referred for ECT. *Biological Psychiatry, 24*(2), 143–161.

Coffey, C. E., Figiel, G. S., Djang, W. T., Saunders, W. B., & Weiner, R. D. (1989). White matter hyperintensity on magnetic resonance imaging: Clinical and neuroanatomic correlates in the depressed elderly. *The Journal of Neuropsychiatry and Clinical Neurosciences, 1*(2), 135–144.

Conwell, Y., Van Orden, K., & Caine, E. D. (2011). Suicide in older adults. *The Psychiatric Clinics of North America, 34*(2), 451–468. https://doi.org/10.1016/j.psc.2011.02.002.

Costafreda, S. G., Chu, C., Ashburner, J., & Fu, C. H. (2009). Prognostic and diagnostic potential of the structural neuroanatomy of depression. *PLoS One. 4*(7)https://doi.org/10.1371/journal.pone.0006353.

Cuijpers, P., Vogelzangs, N., Twisk, J., Kleiboer, A., Li, J., & Penninx, B. W. (2014). Comprehensive meta-analysis of excess mortality in depression in the general community versus patients with specific illnesses. *American Journal of Psychiatry, 171*(4), 453–462. https://doi.org/10.1176/appi.ajp.2013.13030325.

Curran, S. M., Murray, C. M., Van Beck, M., Dougall, N., O'Carroll, R. E., Austin, M. P., … Goodwin, G. M. (1993). A single photon emission computerised tomography study of regional brain function in elderly patients with major depression and with Alzheimer-type dementia. *The British Journal of Psychiatry, 163*, 155–165.

Dalby, R. B., Chakravarty, M. M., Ahdidan, J., Sorensen, L., Frandsen, J., Jonsdottir, K. Y., … Videbech, P. (2010). Localization of white-matter lesions and effect of vascular risk factors in late-onset major depression. *Psychological Medicine, 40*(8), 1389–1399. https://doi.org/10.1017/S0033291709991656.

de Leeuw, F. E., de Groot, J. C., Achten, E., Oudkerk, M., Ramos, L. M., Heijboer, R., … Breteler, M. M. (2001). Prevalence of cerebral white matter lesions in elderly people: A population based magnetic resonance imaging study. The Rotterdam Scan Study. *Journal of Neurology, Neurosurgery, and Psychiatry, 70*(1), 9–14.

Diniz, B. S., Butters, M. A., Albert, S. M., Dew, M. A., & Reynolds, C. F. , 3rd. (2013). Late-life depression and risk of vascular dementia and Alzheimer's disease: Systematic review and meta-analysis of community-based cohort studies. *The British Journal of Psychiatry, 202*(5), 329–335. https://doi.org/10.1192/bjp.bp.112.118307.

Disabato, B. M., Morris, C., Hranilovich, J., D'Angelo, G. M., Zhou, G., Wu, N., … Sheline, Y. I. (2014). Comparison of brain structural variables, neuropsychological factors, and treatment outcome in early-onset versus late-onset late-life depression. *The American Journal of Geriatric Psychiatry, 22* (10), 1039–1046. https://doi.org/10.1016/j.jagp.2013.02.005.

Dolan, R. J., Poynton, A. M., Bridges, P. K., & Trimble, M. R. (1990). Altered magnetic resonance white-matter T1 values in patients with affective disorder. *The British Journal of Psychiatry, 157*, 107–110.

Dotson, V. M., Zonderman, A. B., Kraut, M. A., & Resnick, S. M. (2013). Temporal relationships between depressive symptoms and white matter hyperintensities in older men and women. *International Journal of Geriatric Psychiatry, 28*(1), 66–74. https://doi.org/10.1002/gps.3791.

Droppa, K., Karim, H. T., Tudorascu, D. L., Karp, J. F., Reynolds, C. F., 3rd, Aizenstein, H. J., & Butters, M. A. (2017). Association between change in brain gray matter volume, cognition, and depression severity: Pre- and post-antidepressant pharmacotherapy for late-life depression. *Journal of Psychiatric Research, 95*, 129–134. https://doi.org/10.1016/j.jpsychires.2017.08.002.

Du, M., Liu, J., Chen, Z., Huang, X., Li, J., Kuang, W., … Gong, Q. (2014). Brain grey matter volume alterations in late-life depression. *Journal of Psychiatry & Neuroscience, 39*(6), 397–406.

Ebmeier, K. P., Glabus, M. F., Prentice, N., Ryman, A., & Goodwin, G. M. (1998). A voxel-based analysis of cerebral perfusion in dementia and depression of old age. *NeuroImage, 7*(3), 199–208. https://doi.org/10.1006/nimg.1998.0321.

Egger, K., Schocke, M., Weiss, E., Auffinger, S., Esterhammer, R., Goebel, G., … Marksteiner, J. (2008). Pattern of brain atrophy in elderly patients with depression revealed by voxel-based morphometry. *Psychiatry Research*, *164*(3), 237–244. https://doi.org/10.1016/j.pscychresns.2007.12.018.

Eyre, H. A., Yang, H., Leaver, A. M., Van Dyk, K., Siddarth, P., Cyr, N. S., … Lavretsky, H. (2016). Altered resting-state functional connectivity in late-life depression: A cross-sectional study. *Journal of Affective Disorders*, *189*, 126–133. https://doi.org/10.1016/j.jad.2015.09.011.

Firbank, M. J., Teodorczuk, A., van der Flier, W. M., Gouw, A. A., Wallin, A., Erkinjuntti, T., … O'Brien, J. T. (2012). Relationship between progression of brain white matter changes and late-life depression: 3-year results from the LADIS study. *The British Journal of Psychiatry*, *201*(1), 40–45.

Fu, C. H., Mourao-Miranda, J., Costafreda, S. G., Khanna, A., Marquand, A. F., Williams, S. C., & Brammer, M. J. (2008). Pattern classification of sad facial processing: Toward the development of neurobiological markers in depression. *Biological Psychiatry*, *63*(7), 656–662. https://doi.org/10.1016/j.biopsych.2007.08.020.

Gatz, J. L., Tyas, S. L., St John, P., & Montgomery, P. (2005). Do depressive symptoms predict Alzheimer's disease and dementia? *The Journals of Gerontology Series A, Biological Sciences and Medical Sciences*, *60*(6), 744–747.

Glymour, M. M., Yen, J. J., Kosheleva, A., Moon, J. R., Capistrant, B. D., & Patton, K. K. (2012). Elevated depressive symptoms and incident stroke in Hispanic, African-American, and White older Americans. *Journal of Behavioral Medicine*, *35*(2), 211–220. https://doi.org/10.1007/s10865-011-9356-2.

Godin, O., Dufouil, C., Maillard, P., Delcroix, N., Mazoyer, B., Crivello, F., … Tzourio, C. (2008). White matter lesions as a predictor of depression in the elderly: The 3C-Dijon study. *Biological Psychiatry*, *63*(7), 663–669. https://doi.org/10.1016/j.biopsych.2007.09.006.

Greenwald, B. S., Kramer-Ginsberg, E., Krishnan, R. R., Ashtari, M., Aupperle, P. M., & Patel, M. (1996). MRI signal hyperintensities in geriatric depression. *The American Journal of Psychiatry*, *153*(9), 1212–1215. https://doi.org/10.1176/ajp.153.9.1212.

Gunning, F. M., Cheng, J., Murphy, C. F., Kanellopoulos, D., Acuna, J., Hoptman, M. J., … Alexopoulos, G. S. (2009). Anterior cingulate cortical volumes and treatment remission of geriatric depression. *International Journal of Geriatric Psychiatry*, *24*(8), 829–836. https://doi.org/10.1002/gps.2290.

Gunning-Dixon, F. M., Walton, M., Cheng, J., Acuna, J., Klimstra, S., Zimmerman, M. E., … Alexopoulos, G. S. (2010). MRI signal hyperintensities and treatment remission of geriatric depression. *Journal of Affective Disorders*, *126*(3), 395–401. https://doi.org/10.1016/j.jad.2010.04.004.

Hahn, T., Marquand, A. F., Ehlis, A. C., Dresler, T., Kittel-Schneider, S., Jarczok, T. A., … Fallgatter, A. J. (2011). Integrating neurobiological markers of depression. *Archives of General Psychiatry*, *68*(4), 361–368. https://doi.org/10.1001/archgenpsychiatry.2010.178.

Herrmann, L. L., Le Masurier, M., & Ebmeier, K. P. (2008). White matter hyperintensities in late life depression: A systematic review. *Journal of Neurology, Neurosurgery, and Psychiatry*, *79*(6), 619–624. https://doi.org/10.1136/jnnp.2007.124651.

Hickie, I., Scott, E., Mitchell, P., Wilhelm, K., Austin, M. P., & Bennett, B. (1995). Subcortical hyperintensities on magnetic resonance imaging: Clinical correlates and prognostic significance in patients with severe depression. *Biological Psychiatry*, *37*(3), 151–160. https://doi.org/10.1016/0006-3223(94)00174-2.

Hsieh, M. H., McQuoid, D. R., Levy, R. M., Payne, M. E., Mac Fall, J. R., & Steffens, D. C. (2002). Hippocampal volume and antidepressant response in geriatric depression. *International Journal of Geriatric Psychiatry*, *17*(6), 519–525. https://doi.org/10.1002/gps.611.

Jackson, C. A., & Mishra, G. D. (2013). Depression and risk of stroke in midaged women: A prospective longitudinal study. *Stroke*, *44*(6), 1555–1560. https://doi.org/10.1161/strokeaha.113.001147.

Janssen, J., Hulshoff Pol, H. E., Lampe, I. K., Schnack, H. G., de Leeuw, F. E., Kahn, R. S., & Heeren, T. J. (2004). Hippocampal changes and white matter lesions in early-onset depression. *Biological Psychiatry*, *56*(11), 825–831. https://doi.org/10.1016/j.biopsych.2004.09.011.

Janssen, J., Hulshoff Pol, H. E., Schnack, H. G., Kok, R. M., Lampe, I. K., de Leeuw, F. E., … Heeren, T. J. (2007). Cerebral volume measurements and subcortical white matter lesions and short-term treatment response in late life depression. *International Journal of Geriatric Psychiatry*, *22*(5), 468–474. https://doi.org/10.1002/gps.1790.

Jaworska, N., Yang, X. R., Knott, V., & Mac Queen, G. (2015). A review of fMRI studies during visual emotive processing in major depressive disorder. *The World Journal of Biological Psychiatry*, *16*(7), 448–471. https://doi.org/10.3109/15622975.2014.885659.

Joel, I., Begley, A. E., Mulsant, B. H., Lenze, E. J., Mazumdar, S., Dew, M. A., … Reynolds, C. F., 3rd. (2014). Dynamic prediction of treatment response in late-life depression. *The American Journal of Geriatric Psychiatry*, *22*(2), 167–176. https://doi.org/10.1016/j.jagp.2012.07.002.

Kalayam, B., Alexopoulos, G. S., Musiek, F. E., Kakuma, T., Toro, A., Silbersweig, D., & Young, R. C. (1997). Brainstem evoked response abnormalities in late-life depression with vascular disease. *The American Journal of Psychiatry*, *154*(7), 970–975.

Kaneriya, S. H., Robbins-Welty, G., Smagula, S. F., Karp, J. F., Butters, M. A., Lenze, E. J., … Reynolds, C. F. (2016). Predictors and moderators of remission with aripiprazole augmentation in treatment-resistant late-life depression. *JAMA Psychiatry*, *73*(4), 329–336. https://doi.org/10.1001/jamapsychiatry.2015.3447.

Karim, H. T., Andreescu, C., Tudorascu, D., Smagula, S. F., Butters, M. A., Karp, J. F., … Aizenstein, H. J. (2016). Intrinsic functional connectivity in late-life depression: Trajectories over the course of pharmacotherapy in remitters and non-remitters. *Molecular Psychiatry*. https://doi.org/10.1038/mp.2016.55.

Katon, W. J., Lin, E., Russo, J., & Unutzer, J. (2003). Increased medical costs of a population-based sample of depressed elderly patients. *Archives of General Psychiatry*, *60*(9), 897–903. https://doi.org/10.1001/archpsyc.60.9.897.

Kenny, E. R., O'Brien, J. T., Cousins, D. A., Richardson, J., Thomas, A. J., Firbank, M. J., & Blamire, A. M. (2010). Functional connectivity in late-life depression using resting-state functional magnetic resonance imaging. *The American Journal of Geriatric Psychiatry*, *18*(7), 643–651. https://doi.org/10.1097/JGP.0b013e3181cabd0e.

Khalaf, A., Edelman, K., Tudorascu, D., Andreescu, C., Reynolds, C. F., & Aizenstein, H. (2015). White matter hyperintensity accumulation during treatment of late-life depression. *Neuropsychopharmacology*. https://doi.org/10.1038/npp.2015.158.

Khalaf, A., Karim, H., Berkout, O. V., Andreescu, C., Tudorascu, D., Reynolds, C. F., & Aizenstein, H. (2016). Altered functional magnetic resonance imaging markers of affective processing during treatment of late-life depression. *The American Journal of Geriatric Psychiatry*, *24*(10), 791–801.

Krishnan, K. R., Goli, V., Ellinwood, E. H., France, R. D., Blazer, D. G., & Nemeroff, C. B. (1988). Leukoencephalopathy in patients diagnosed as major depressive. *Biological Psychiatry*, *23*(5), 519–522.

Krishnan, K. R., McDonald, W. M., Doraiswamy, P. M., Tupler, L. A., Husain, M., Boyko, O. B., … Ellinwood, E. H. , Jr. (1993). Neuroanatomical substrates of depression in the elderly. *European Archives of Psychiatry and Clinical Neuroscience*, *243*(1), 41–46.

Kumar, A., Bilker, W., Jin, Z., & Udupa, J. (2000). Atrophy and high intensity lesions: Complementary neurobiological mechanisms in late-life major depression. *Neuropsychopharmacology*, *22*(3), 264–274. https://doi.org/10.1016/S0893-133X(99)00124-4.

Kumar, A., Newberg, A., Alavi, A., Berlin, J., Smith, R., & Reivich, M. (1993). Regional cerebral glucose metabolism in late-life depression and Alzheimer disease: A preliminary positron emission tomography study. *Proceedings of the National Academy of Sciences of the United States of America*, *90*(15), 7019–7023.

Lavretsky, H., Ballmaier, M., Pham, D., Toga, A., & Kumar, A. (2007). Neuroanatomical characteristics of geriatric apathy and depression: A magnetic resonance imaging study. *The American Journal of Geriatric Psychiatry*, *15*(5), 386–394. https://doi.org/10.1097/JGP.0b013e3180325a16.

Lehmbeck, J. T., Brassen, S., Braus, D. F., & Weber-Fahr, W. (2008). Subgenual anterior cingulate cortex alterations in late-onset depression are related to "pessimistic thoughts". *The American Journal of Geriatric Psychiatry*, *16*(3), 248–249. https://doi.org/10.1097/JGP.0b013e318162a0c0.

Lesser, I. M., Mena, I., Boone, K. B., Miller, B. L., Mehringer, C. M., & Wohl, M. (1994). Reduction of cerebral blood flow in older depressed patients. *Archives of General Psychiatry*, *51*(9), 677–686.

Liu, F., Guo, W., Yu, D., Gao, Q., Gao, K., Xue, Z., … Chen, H. (2012). Classification of different therapeutic responses of major depressive disorder with multivariate pattern analysis method based on structural MR scans. *PLoS One*. *7*(7) https://doi.org/10.1371/journal.pone.0040968.

Liu, F., Hu, M., Wang, S., Guo, W., Zhao, J., Li, J., … Chen, H. (2012). Abnormal regional spontaneous neural activity in first-episode, treatment-naive patients with late-life depression: A resting-state fMRI study. *Progress in Neuro-Psychopharmacology & Biological Psychiatry*, *39*(2), 326–331. https://doi.org/10.1016/j.pnpbp.2012.07.004.

Lloyd, A. J., Ferrier, I. N., Barber, R., Gholkar, A., Young, A. H., & O'Brien, J. T. (2004). Hippocampal volume change in depression: Late- and early-onset illness compared. *The British Journal of Psychiatry*, *184*, 488–495.

Luppa, M., Sikorski, C., Motzek, T., Konnopka, A., Konig, H. H., & Riedel-Heller, S. G. (2012). Health service utilization and costs of depressive symptoms in late life—A systematic review. *Current Pharmaceutical Design*, *18*(36), 5936–5957.

Marquand, A. F., Mourao-Miranda, J., Brammer, M. J., Cleare, A. J., & Fu, C. H. (2008). Neuroanatomy of verbal working memory as a diagnostic biomarker for depression. *Neuroreport*, *19*(15), 1507–1511. https://doi.org/10.1097/WNR.0b013e328310425e.

Mezuk, B., Eaton, W. W., Albrecht, S., & Golden, S. H. (2008). Depression and type 2 diabetes over the lifespan: A meta-analysis. *Diabetes Care*, *31*(12), 2383–2390. https://doi.org/10.2337/dc08-0985.

Mulsant, B. H., Blumberger, D. M., Ismail, Z., Rabheru, K., & Rapoport, M. J. (2014). A systematic approach to pharmacotherapy for geriatric major depression. *Clinics in Geriatric Medicine*, *30*(3), 517–534. https://doi.org/10.1016/j.cger.2014.05.002.

Murray, M. E., Vemuri, P., Preboske, G. M., Murphy, M. C., Schweitzer, K. J., Parisi, J. E., … Dickson, D. W. (2012). A quantitative postmortem MRI design sensitive to white matter Hyperintensity differences and their relationship with underlying pathology. *Journal of Neuropathology and Experimental Neurology*, *71*(12), 1113–1122. https://doi.org/10.1097/NEN.0b013e318277387e.

Mwangi, B., Matthews, K., & Steele, J. D. (2012). Prediction of illness severity in patients with major depression using structural MR brain scans. *Journal of Magnetic Resonance Imaging*, *35*(1), 64–71. https://doi.org/10.1002/jmri.22806.

Nobler, M. S., Roose, S. P., Prohovnik, I., Moeller, J. R., Louie, J., Van Heertum, R. L., & Sackeim, H. A. (2000). Regional cerebral blood flow in mood disorders, V.: Effects of antidepressant medication in late-life depression. *The American Journal of Geriatric Psychiatry*, *8*(4), 289–296.

Nobler, M. S., Sackeim, H. A., Prohovnik, I., Moeller, J. R., Mukherjee, S., Schnur, D. B., … Devanand, D. P. (1994). Regional cerebral blood flow in mood disorders, III. Treatment and clinical response. *Archives of General Psychiatry*, *51*(11), 884–897.

Nouretdinov, I., Costafreda, S. G., Gammerman, A., Chervonenkis, A., Vovk, V., Vapnik, V., & Fu, C. H. (2011). Machine learning classification with confidence: Application of transductive conformal predictors to MRI-based diagnostic and prognostic markers in depression. *NeuroImage*, *56*(2), 809–813. https://doi.org/10.1016/j.neuroimage.2010.05.023.

Olesen, P. J., Gustafson, D. R., Simoni, M., Pantoni, L., Östling, S., Guo, X., & Skoog, I. (2010). Temporal lobe atrophy and white matter lesions are related to major depression over 5 years in the elderly. *Neuropsychopharmacology*, *35*(13), 2638–2645. https://doi.org/10.1038/npp.2010.176.

Park, J. H., Lee, S. B., Lee, J. J., Yoon, J. C., Han, J. W., Kim, T. H., … Kim, K. W. (2015). Epidemiology of MRI-defined vascular depression: A longitudinal, community-based study in Korean elders. *Journal of Affective Disorders*, *180*, 200–206. https://doi.org/10.1016/j.jad.2015.04.008.

Patel, M. J., Andreescu, C., Price, J. C., Edelman, K. L., Reynolds, C. F., 3rd, & Aizenstein, H. J. (2015). Machine learning approaches for integrating clinical and imaging features in late-life depression classification and response prediction. *International Journal of Geriatric Psychiatry*, *30*(10), 1056–1067. https://doi.org/10.1002/gps.4262.

Pavlovic, A. M., Pekmezovic, T., Zidverc Trajkovic, J., Svabic Medjedovic, T., Veselinovic, N., Radojicic, A., … Sternic, N. (2016). Baseline characteristic of patients presenting with lacunar stroke and cerebral small vessel disease may predict future development of depression. *International Journal of Geriatric Psychiatry*, *31*(1), 58–65. https://doi.org/10.1002/gps.4289.

Raison, C. L., Rutherford, R. E., Woolwine, B. J., Shuo, C., Schettler, P., Drake, D. F., … Miller, A. H. (2013). A randomized controlled trial of the tumor necrosis factor antagonist infliximab for treatment-resistant depression: The role of baseline inflammatory biomarkers. *JAMA Psychiatry*, *70*(1), 31–41. https://doi.org/10.1001/2013.jamapsychiatry.4.

Reppermund, S., Zhuang, L., Wen, W., Slavin, M. J., Trollor, J. N., Brodaty, H., & Sachdev, P. S. (2014). White matter integrity and late-life depression in community-dwelling individuals: Diffusion tensor imaging study using tract-based spatial statistics. *The British Journal of Psychiatry*, *205*(4), 315–320. https://doi.org/10.1192/bjp.bp.113.142109.

Due to decreasing costs, exome-sequencing and whole genome-sequencing are increasingly popular for the exome/genome-wide identification of deletions and duplications.

3 Copy number variants and study designs

Regardless of the phenotype investigated, scientists can use different study designs for the CNV analysis. The majority of studies apply a case-control design, while some focus on the identification of familial, disease segregating CNVs in multiplex-families. Furthermore, parent-proband trios can be analyzed to identify deletions/duplications that are *de novo* in the patient (not inherited from either parent; Sebat et al., 2009).

Independent of the chosen study design (as outlined herein), three main approaches to identify an association between deletions and/or duplications and the phenotype might be applied: (i) single-marker association analysis; (ii) gene-wise burden analysis; and (iii) genome-wide burden analysis (Sebat et al., 2009). The single-marker association analysis examines and compares the frequency of specific variants between at least two groups (e.g., the patient and the control cohort). The gene-wise burden analysis examines the aggregate frequency of multiple variants within a specific gene or region, and statistically compares this between at least two groups; and the genome-wide burden analysis takes into account the aggregate frequency of all (rare) CNVs detected in the cohorts (e.g., patient and control cohorts; Sebat et al., 2009).

4 Copy number variants and clinical translation efforts

Over the past few years, we have gained unprecedented insight into the biological underpinnings of psychiatric disorders. Particularly, the current understanding of the genetic architecture of schizophrenia has been substantially advanced (CNV and Schizophrenia Working Groups of the Psychiatric Genomics Consortium, 2017; Schizophrenia Working Group of the Psychiatric Genomics Consortium, 2014). The identification of deletions and duplications in specific chromosomal regions with moderate to larger impact on risk (CNV and Schizophrenia Working Groups of the Psychiatric Genomics Consortium, 2017), has increased interest in the clinical translation of these findings (e.g., offering CNV testing to patients with schizophrenia in a diagnostic context). In the schizophrenia CNV analysis conducted by the Psychiatric Genomic Consortium, only a small fraction ($\sim$1%) of patients carried a CNV in one of the eight risk loci that surpassed genome-wide significance (CNV and Schizophrenia Working Groups of the Psychiatric Genomics Consortium, 2017). Even though the risk-associated CNVs are individually rare, they might explain a relatively large proportion of the CNV carrier's liability to his/her psychiatric disorder.

Chromosomal microarray (genome-wide copy-number screening) is a recommended first-tier genetic test for individuals with unexplained developmental delay/intellectual disability, autism spectrum disorders, or multiple congenital anomalies (Miller et al., 2010). Currently, no standardized recommendations exist for the genetic workup of patients with schizophrenia or any other adult onset psychiatric disorder. Several studies have reported a diagnostic yield $\sim$10% when screening patients diagnosed with intellectual disability and a comorbid psychiatric disorder for potentially pathogenic CNVs (Bouwkamp et al., 2017; Lowther et al., 2017; Thygesen et al., 2018; Wolfe et al., 2017). It is of note that these potentially pathogenic CNVs include more CNV loci than the eight loci that surpassed the genome-wide significance threshold in the PGC schizophrenia analysis (CNV and Schizophrenia Working Groups of the Psychiatric Genomics Consortium, 2017). Therefore, it is reasonable to evaluate patients with a psychiatric disorder and comorbid intellectual disability, autism spectrum disorders, or multiple congenital anomalies for an underlying genetic syndrome. Any genetic testing must be carried out within the framework of (local) legal standards, after obtaining written informed consent. If available, it might be useful to offer genetic counseling to the patients and their family members.

Patients with schizophrenia/psychiatric disorder will likely benefit from the identification of a pathogenic, risk-associated CNV. Receiving a molecular diagnosis might provide at least a partial etiological explanation for the symptoms the patient is experiencing. Medical geneticists and genetic counselors are well experienced in communicating uncertainty about penetrance and variable expressivity of the phenotype; genetic counseling supports the patient in understanding the information provided (Costain et al., 2013). Even though the identification of a pathogenic CNV might not have immediate impact on the medical care the patient receives, it might help the patient and family members to understand and accept the psychiatric diagnosis within medical context. Furthermore, it might have reproductive implications for siblings and patients themselves (Costain et al., 2013). However, some healthcare professionals are worried that receiving a genetic test result might also have a negative impact on the patient/family members (e.g., feeling of shame, blame, negative effect on stigma). Therefore, more research is required to evaluate the positive, but also potentially negative implications of receiving a molecular diagnosis.

5 Concluding remarks and future perspectives

With the publication of the genome-wide CNV analysis conducted by the CNV Working Group of the Psychiatric Genomics Consortium, a significant milestone in schizophrenia research was reached (CNV and Schizophrenia Working Groups of the Psychiatric Genomics Consortium, 2017). This study is arguably the most comprehensive work of CNVs relevant to the pathogenesis of schizophrenia (Sullivan, 2017) to date. Likewise, large-scale CNV analyses focusing on other adult onset psychiatric disorders (e.g., affective disorders) are currently being performed within international consortia (e.g., the Psychiatric Genomics Consortium). Their results will provide important insights into the genetic architecture and the underlying biology of these phenotypes. The deletions and duplications identified so far are the "low-hanging fruit." The studies succeeded in their detection due to the CNVs' larger size and their relatively high frequency among the patient cohorts, and their relatively low frequency among the control cohorts (Sullivan, 2017). The established psychiatric risk-associated CNVs are rare. Studies in even larger cohorts will enable both the identification of previously undiscovered risk-associated CNVs, and the analysis of the contribution of common deletions and duplications to the development of schizophrenia and other psychiatric disorders.

Due to technical limitations (e.g., array resolution), CNV studies based on single nucleotide polymorphism arrays may be missing smaller and more complex deletions and duplications (Sullivan, 2017). Whole-genome sequencing is efficient in detecting smaller CNVs (e.g., single exon-affecting deletions and duplications), and might eventually outperform the array-based CNV analyses (Zhou et al., 2018). In the future, the continuously decreasing costs for high-throughput sequencing and cloud computing will allow for whole-genome CNV analyses in larger cohorts.

The identification of risk-associated CNVs promises to propel our mechanistic understanding of neuropsychiatric disorders. Taking advantage of their magnifying effect (larger effect sizes), CNVs are particularly well suited as a model for focused analyses to gain more detailed insight into the disorders underlying gene networks and functional biological pathways (Gur et al., 2017). These focused approaches might include functional genomics, molecular neuroscience, the detailed characterization of patient-derived neuronal models, and animal models (Flaherty & Brennand, 2017; Forsingdal et al., 2018; Sullivan, 2017). Advancing our mechanistic understanding of the risk-associated CNVs might open up new avenues for more tailored treatment approaches, including the development of drugs specifically addressing the underlying biology (Forsingdal et al., 2018).

In the future, diagnostic testing for risk-associated CNVs might be implemented in routine clinical care for patients with adult onset psychiatric disorders. Due to the low frequency of pathogenic CNVs among patients with schizophrenia/adult onset psychiatric disorders, it will be important to establish criteria based on which a decision can be made as to who might be offered a genetic test. Therefore, studies are warranted to identify symptoms that are shared between CNV carriers and non-CNV carriers, and symptoms that are unique to a specific deletion or duplication. Furthermore, it will be important to unravel the mechanisms that influence/determine which phenotype the CNV carrier will develop (from clinically apparently healthy to mildly affected to severe neurodevelopmental disorder). This might have implications for the general medical care of the patients, and might allow for more personalized treatment approaches.

Conflict of interest

The author has no conflicts of interest to declare.

References

Alkan, C., Coe, B. P., & Eichler, E. E. (2011). Genome structural variation discovery and genotyping. *Nature Reviews. Genetics, 12*(5), 363–376. https://doi.org/10.1038/nrg2958.

Bassett, A. S., Chow, E. W., & Weksberg, R. (2000). Chromosomal abnormalities and schizophrenia. *American Journal of Medical Genetics, 97*(1), 45–51.

Bassett, A. S., Marshall, C. R., Lionel, A. C., Chow, E. W., & Scherer, S. W. (2008). Copy number variations and risk for schizophrenia in 22q11.2 deletion syndrome. *Human Molecular Genetics, 17*(24), 4045–4053. https://doi.org/10.1093/hmg/ddn307.

Bassett, A. S., McDonald-McGinn, D. M., Devriendt, K., Digilio, M. C., Goldenberg, P., Habel, A., … International 22q11.2 Deletion Syndrome Consortium. (2011). Practical guidelines for managing patients with 22q11.2 deletion syndrome. *The Journal of Pediatrics, 159*(2), 332–339.e1. https://doi.org/10.1016/j.jpeds.2011.02.039.

Bouwkamp, C. G., Kievit, A. J. A., Markx, S., Friedman, J. I., van Zutven, L., van Minkelen, R., … Kushner, S. A. (2017). Copy number variation in syndromic forms of psychiatric illness: The emerging value of clinical genetic testing in psychiatry. *The American Journal of Psychiatry, 174*(11), 1036–1050. https://doi.org/10.1176/appi.ajp.2017.16080946.

Carson, A. R., Feuk, L., Mohammed, M., & Scherer, S. W. (2006). Strategies for the detection of copy number and other structural variants in the human genome. *Human Genomics, 2*(6), 403–414.

CNV and Schizophrenia Working Groups of the Psychiatric Genomics Consortium. (2017). Contribution of copy number variants to schizophrenia from a genome-wide study of 41,321 subjects. *Nature Genetics, 49*(1), 27–35. https://doi.org/10.1038/ng.3725.

Costain, G., Lionel, A. C., Merico, D., Forsythe, P., Russell, K., Lowther, C., … Bassett, A. S. (2013). Pathogenic rare copy number variants in community-based schizophrenia suggest a potential role for clinical microarrays. *Human Molecular Genetics, 22*(22), 4485–4501. https://doi.org/10.1093/hmg/ddt297.

Degenhardt, F., Priebe, L., Herms, S., Mattheisen, M., Mühleisen, T. W., & Meier, S. (2012). Association between copy number variants in 16p11.2 and major depressive disorder in a German case-control sample. *American Journal of Medical Genetics. Part B, Neuropsychiatric Genetics, 159B*(3), 263–273. https://doi.org/10.1002/ajmg.b.32034.den.

Dunnen, J. T., & White, S. J. (2006). MLPA and MAPH: Sensitive detection of deletions and duplications. *Current Protocols in Human Genetics.* Chapter 7: Unit 7.14. https://doi.org/10.1002/0471142905.hg0714s51.

Feuk, L., Carson, A. R., & Scherer, S. W. (2006). Structural variation in the human genome. *Nature Reviews. Genetics, 7*(2), 85–97. https://doi.org/10.1038/nrg1767.

Flaherty, E. K., & Brennand, K. J. (2017). Using hiPSCs to model neuropsychiatric copy number variations (CNVs) has potential to reveal underlying disease mechanisms. *Brain Research, 1655*, 283–293. https://doi.org/10.1016/j.brainres.2015.11.009.

Forsingdal, A., Jørgensen, T. N., Olsen, L., Werge, T., Didriksen, M., & Nielsen, J. (2018). Can animal models of copy number variants that predispose to schizophrenia elucidate underlying biology? *Biological Psychiatry,* S0006-3223. (18), 31670–31676. https://doi.org/10.1016/j.biopsych.2018.07.004. (Epub ahead of print).

Georgieva, L., Rees, E., Moran, J. L., Chambert, K. D., Milanova, V., Craddock, N., … Kirov, G. (2014). De novo CNVs in bipolar affective disorder and schizophrenia. *Human Molecular Genetics, 23*(24), 6677–6683. https://doi.org/10.1093/hmg/ddu379.

Green, E. K., Rees, E., Walters, J. T., Smith, K. G., Forty, L., Grozeva, D., … Kirov, G. (2016). Copy number variation in bipolar disorder. *Molecular Psychiatry, 21*, 89–93. https://doi.org/10.1038/mp.2014.174.

Grozeva, D., Kirov, G., Ivanov, D., Jones, I. R., Jones, L., Green, E. K., … Wellcome Trust Case Control Consortium (2010). Rare copy number variants: A point of rarity in genetic risk for bipolar disorder and schizophrenia. *Archives of General Psychiatry, 67*(4), 318–327. https://doi.org/10.1001/archgenpsychiatry.2010.25.

Gur, R. E., Bassett, A. S., McDonald-McGinn, D. M., Bearden, C. E., Chow, E., Emanuel, B. S., … Morrow, B. (2017). A neurogenetic model for the study of schizophrenia spectrum disorders: The International 22q11.2 deletion syndrome brain behavior consortium. *Molecular Psychiatry, 22*(12), 1664–1672. https://doi.org/10.1038/mp.2017.161.

Human Genome Structural Variation Working Group, Eichler, E. E., Nickerson, D. A., Altshuler, D., Bowcock, A. M., Brooks, L. D., … Waterston, R. H. (2007). Completing the map of human genetic variation. *Nature, 447*(7141), 161–165. https://doi.org/10.1038/447161a.

Iafrate, A. J., Feuk, L., Rivera, M. N., Listewnik, M. L., Donahoe, P. K., Qi, Y., … Lee, C. (2004). Detection of large-scale variation in the human genome. *Nature Genetics, 36*(9), 949–951. https://doi.org/10.1038/ng1416.

Karayiorgou, M., Morris, M. A., Morrow, B., Shprintzen, R. J., Goldberg, R., Borrow, J., … Lasseter, V. K. (1995). Schizophrenia susceptibility associated with interstitial deletions of chromosome 22q11. *Proceedings of the National Academy of Sciences of the United States of America, 92*(17), 7612–7616.

Lindsay, E. A., Morris, M. A., Gos, A., Nestadt, G., Wolyniec, P. S., Lasseter, V. K., … Pulver, A. E. (1995). Schizophrenia and chromosomal deletions within 22q11.2. *American Journal of Human Genetics, 56*(6), 1502–1503.

Lowther, C., Merico, D., Costain, G., Waserman, J., Boyd, K., Noor, A., … Bassett, A. S. (2017). Impact of IQ on the diagnostic yield of chromosomal microarray in a community sample of adults with schizophrenia. *Genome Medicine, 9*(1), 105. https://doi.org/10.1186/s13073-017-0488-z.

Lupski, J. R. (2007). Genomic rearrangements and sporadic disease. *Nature Genetics, 39*(7 Suppl), S43–S47. https://doi.org/10.1038/ng2084.

Lupski, J. R., & Stankiewicz, P. (2005). Genomic disorders: Molecular mechanisms for rearrangements and conveyed phenotypes. *PLoS Genetics, 1*(6), e49. https://doi.org/10.1371/journal.pgen.0010049.

Martin, C. L., Kirkpatrick, B. E., & Ledbetter, D. H. (2015). Copy number variants, aneuploidies, and human disease. *Clinics in Perinatology, 42*(2), 227–242. vii. https://doi.org/10.1016/j.clp.2015.03.001.

McDonald-McGinn, D. M., Emanuel, B. S., & Zackai, E. H. (1999). 22q11.2 deletion syndrome. In M. P. Adam, H. H. Ardinger, R. A. Pagon, S. E. Wallace, B. LJH, K. Stephens, & A. Amemiya (Eds.), *GeneReviews® [Internet].* Seattle, WA: University of Washington. 1993-2018 [updated 2013 Feb 28].

McDonald-McGinn, D. M., Sullivan, K. E., Marino, B., Philip, N., Swillen, A., Vorstman, J. A., … Bassett, A. S. (2015). 22q11.2 deletion syndrome. *Nature Reviews Disease Primers, 1*, 15071. https://doi.org/10.1038/nrdp.2015.71.

Merikangas, A. K., Corvin, A. P., & Gallagher, L. (2009). Copy-number variants in neurodevelopmental disorders: Promises and challenges. *Trends in Genetics, 25*(12), 536–544. https://doi.org/10.1016/j.tig.2009.10.006.

Michaelson, J. J. (2017). Genetic approaches to understanding psychiatric disease. *Neurotherapeutics, 14*(3), 564–581. https://doi.org/10.1007/s13311-017-0551-x.

Miller, D. T., Adam, M. P., Aradhya, S., Biesecker, L. G., Brothman, A. R., Carter, N. P., … Ledbetter, D. H. (2010). Consensus statement: Chromosomal microarray is a first-tier clinical diagnostic test for individuals with developmental disabilities or congenital anomalies. *American Journal of Human Genetics, 86*(5), 749–764. https://doi.org/10.1016/j.ajhg.2010.04.006.

Mills, R. E., Walter, K., Stewart, C., Handsaker, R. E., Chen, K., Alkan, C., … 1000 Genomes Project. (2011). Mapping copy number variation by population-scale genome sequencing. *Nature, 470*(7332), 59–65. https://doi.org/10.1038/nature09708.

Redon, R., Ishikawa, S., Fitch, K. R., Feuk, L., Perry, G. H., Andrews, T. D., … Hurles, M. E. (2006). Global variation in copy number in the human genome. *Nature, 444*(7118), 444–454. https://doi.org/10.1038/nature05329.

Schizophrenia Working Group of the Psychiatric Genomics Consortium. (2014). Biological insights from 108 schizophrenia-associated genetic loci. *Nature, 511*(7510), 421–427. https://doi.org/10.1038/nature13595.

Sebat, J., Lakshmi, B., Troge, J., Alexander, J., Young, J., Lundin, P., … Wigler, M. (2004). Large-scale copy number polymorphism in the human genome. *Science, 305*(5683), 525–528. https://doi.org/10.1126/science.1098918.

Sebat, J., Levy, D. L., & McCarthy, S. E. (2009). Rare structural variants in schizophrenia: One disorder, multiple mutations; one mutation, multiple disorders. *Trends in Genetics, 25*(12), 528–535. https://doi.org/10.1016/j.tig.2009.10.004.

Shaffer, L. G., & Bejjani, B. A. (2006). Medical applications of array CGH and the transformation of clinical cytogenetics. *Cytogenetic and Genome Research, 115*(3–4), 303–309.

Shaffer, L. G., & Lupski, J. R. (2000). Molecular mechanisms for constitutional chromosomal rearrangements in humans. *Annual Review of Genetics, 34*, 297–329. https://doi.org/10.1146/annurev.genet.34.1.297.

Stankiewicz, P., & Lupski, J. R. (2010). Structural variation in the human genome and its role in disease. *Annual Review of Medicine, 61*, 437–455. https://doi.org/10.1146/annurev-med-100708-204735.

Sugama, S., Namihira, T., Matsuoka, R., Taira, N., Eto, Y., & Maekawa, K. (1999). Psychiatric inpatients and chromosome deletions within 22q11.2. *Journal of Neurology, Neurosurgery, and Psychiatry, 67*(6), 803–806.

Sullivan, P. F. (2010). The psychiatric GWAS consortium: Big science comes to psychiatry. *Neuron, 68*(2), 182–186. https://doi.org/10.1016/j.neuron.2010.10.003.

Sullivan, P. F. (2017). Schizophrenia and the dynamic genome. *Genome Medicine, 9*(1), 22. https://doi.org/10.1186/s13073-017-0416-2.

Thygesen, J. H., Wolfe, K., McQuillin, A., Viñas-Jornet, M., Baena, N., Brison, N., … Vogels, A. (2018). Neurodevelopmental risk copy number variants in adults with intellectual disabilities and comorbid psychiatric disorders. *The British Journal of Psychiatry, 212*(5), 287–294. https://doi.org/10.1192/bjp.2017.65.

Tuzun, E., Sharp, A. J., Bailey, J. A., Kaul, R., Morrison, V. A., Pertz, L. M., … Eichler, E. E. (2005). Fine-scale structural variation of the human genome. *Nature Genetics, 37*(7), 727–732. https://doi.org/10.1038/ng1562.

Wolfe, K., Strydom, A., Morrogh, D., Carter, J., Cutajar, P., Eyeoyibo, M., … Bass, N. (2017). Chromosomal microarray testing in adults with intellectual disability presenting with comorbid psychiatric disorders. *European Journal of Human Genetics, 25*(1), 66–72. https://doi.org/10.1038/ejhg.2016.107.

Zhou, B., Ho, S. S., Zhang, X., Pattni, R., Haraksingh, R. R., & Urban, A. E. (2018). Whole-genome sequencing analysis of CNV using low-coverage and paired-end strategies is efficient and outperforms array-based CNV analysis. *Journal of Medical Genetics*. pii: jmedgenet-2018-105272. https://doi.org/10.1136/jmedgenet-2018-105272 (Epub ahead of print).

Chapter 29

Gene-environment interaction in psychiatry

Hans Jörgen Grabe and Sandra Van der Auwera
Department of Psychiatry and Psychotherapy, University Medicine of Greifswald, Greifswald, Germany

1 Introduction

Adaptation is a driving force of the evolutionary process of the species, and optimal adaptation is the highest individual goal, ensuring optimal individual survival. Thus, the evolutionary process has created biological systems that are able to respond dynamically to environmental challenges. Such challenges are comprised of physical factors such as UV-radiation, caloric restriction, heat, and cold. As the evolutionary process brought complex social systems into being, these social systems became a determinant of survival itself (Brune & Brune-Cohrs, 2006; Emera, Yin, Reilly, Gockley, & Noonan, 2016). Thus, the regulation and the fine-tuning of social interactions became an evolutionary driver of brain systems and, for example, neuroendocrine functions.

All biological systems and biochemical pathways that ensure the adaptability of mammalians and humans are coded by genes that carry common single nucleotides (SNP), rare variants, and other genetic variations such as insertions, deletions, or copy number variations. These genetic variations may contribute to the individual variability of the biological reactivity and response to the environmental challenge, and thus contribute to optimal or impaired individual adaptation.

This individual variability gives rise to susceptibility or resilience when dealing with pathological conditions and disorders. The basic model of gene-environment interaction (GxE) assumes a different reactivity of the biological system under the condition "environmental challenge" (present/absent), dependent on a respective genotype (present/absent). Thus, in humans, the analysis of typical GxEs requires subjects who are exposed/nonexposed to an environmental challenge, and carrying the risk or the nonrisk allele. This is the conceptual point where GxE becomes accessible for future use in personalized psychiatry, as the two relevant factors determining the individual risk are measurable. In Fig. 1, the prototypic concept of GxE is depicted, resulting in four different groups with different risk of disease. Typically, the group of subjects carrying the genetic risk variant and having been exposed to stressful/traumatic events (group d) has the highest risk of disease, whereas the other groups (a–c) have markedly lower risks of disease.

Generally it is assumed, however, that the etiology of mental disorders is multifactorial. Thus, any GxE is embedded within this multifactorial background. Although complex, the concept of genes—posing either risk or resilience toward the environmental exposure—is appealing as it captures the dynamic nature of individual adaptation. The clinical idea behind GxE in psychiatry is that the change of the reactivity of a biological system based on GxE "reaches" the behavioral level, or even the diagnostic level impacting the risk of disorder or disease. However, the differential reactivity of a biological system due to GxE might be better assessed by, for example, physiological, biochemical, or neuropsychological testing than by clinical categories of mental disorders. This concept of "endophenotypes" might constitute a relevant outcome in GxE research (Fig. 1).

It is beyond the possibilities of this chapter to cover extensively the vast literature of GxE in various clinical conditions. However, we will give some insight into the challenges and highlights in GxE in the research of mental disorders.

2 Environmental conditions

The analysis of GxE and its application in personalized psychiatry requires deep insight in the conceptualization and measurement of environmental factors (Grabe, Schulz, et al., 2012). It is evident that many environmental conditions act upon each individual throughout critical phases of mental and biological development in a very individual way. In contrast, the possibilities of assessing and analyzing these conditions are rather simple. We do not yet have any good insight

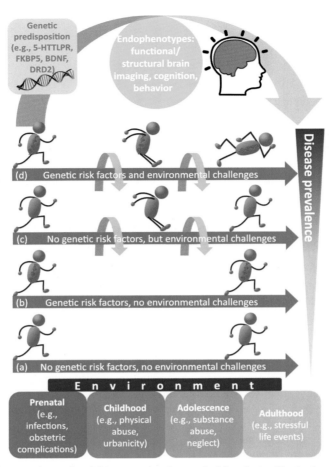

FIG. 1 General model of gene-environment interaction. Subjects carrying the genetic risk variant and having been exposed to stressful/traumatic events (group d) have the highest risk of disease, whereas the other groups (a–c) have markedly lower risks of disease.

into how aversive and supportive influences balance each other and thereby affect possible GxE effects in the statistical models. Multiple environmental domains have been shown to be associated with mental disorders: urban upbringing, migration, stressful life events and early life stress, prenatal infections, obstetric complications, substance abuse or dependence of the parents, as well as the patient himself, and of course more subtle conditions such as parental bonding, quality of care, and learning copings strategies (Grech & van Os, 2017; van Os et al., 2014). Again, this list is not conclusive, but it illustrates the variety of influences that may act upon an individual. From clinical research, we do know that often, multiple negative events in a given period of time are needed to shift the individual from functional to dysfunctional, and from health to illness. Further evidence suggests that "double hits" may play a specific role for disease development—one developmentally early psychological or biological trauma, and a recent event that leads to the actual decompensation of mental functions (Giovanoli et al., 2013; Grabe, Schwahn, Mahler, Schulz, et al., 2012).

3 Candidate genes

The usual approach in GxE research is the analysis of a limited set of genes and polymorphisms that would currently also be the most feasible application for personalized psychiatry. Ideally, the selection of the candidate genes follows a plausible physiological model that connects the biochemical pathway to a broader biological domain, as the serotonin system or the HPA axis that connect biology to distinct human behaviors. In many published studies, several genes and polymorphisms have been analyzed without proper adjustments for multiple testing. Often, results are difficult to compare, as different SNPs from the same gene have been used in different studies. Except from a few candidate genes, it is still largely unclear how the genetic variants might lead to different biological responses in different environmental settings. Complex mechanisms of gene regulation such as DNA methylation, histone acetylation, and further regulatory processing such as noncoding RNAs might be involved (Braff & Tamminga, 2017). Adding these points to the uncertainties in assessment of complex environments and the diagnostic complexity in mental disorders, GxE analyses are definitely challenging. One has to keep these circumstances in mind when interpreting GxE studies.

4 Depressive disorders and anxiety

The amount of GxE papers published in the field of depression and anxiety disorders is too large to give a substantial overview in this chapter. Some recent review papers might add further details (Gottschalk & Domschke, 2017a, 2017b; Sharma, Powers, Bradley, & Ressler, 2016). Here, we present three well studied candidate genes in greater detail, highlighting some of the controversies and related debates.

5 The insertion/deletion polymorphism of the serotonin transporter gene (5-HTTLPR)

One prominent example that has frequently been studied is the insertion/deletion polymorphism of the serotonin transporter gene (5-HTTLPR). The serotonin system per se is involved in the regulation of multiple facets of human behavior such as mood, impulsivity, appetite, sleep, and sexuality. The 5-HTTLPR is thought to impact the transcription rate, thereby influencing the concentration of the serotonin transporter within the presynaptic membrane. In their landmark paper, Caspi et al. found an interaction between the number of life-events and the three genotypes (ss, sl, ll) of the 5-HTTLPR, and the likelihood of increased self-reported depressive symptoms scores or the occurrence of depressive episodes. This interaction was present for adult stressful life-events, as well as for childhood maltreatment (Caspi et al., 2003). Our group could replicate the GxE properties of the 5-HTTLPR in the severity of psychosomatic complaints in the general population (Grabe et al., 2005; Grabe et al., 2011). Further, we could show that the combination of childhood and adult traumatization was important to reveal a risk increasing the interaction effect of the 5-HTTLPR s-allele on current depressive symptoms (Grabe, Schwahn, Mahler, Schulz, et al., 2012). However, in the past 15 years, many studies have been published on the role of 5-HTTLPR in GxE with mixed results. Two metaanalyses revealed overall negative results (Munafo, Durrant, Lewis, & Flint, 2009; Risch et al., 2009). Following the line of methodological criticism, especially on the metaanalysis performed by Risch et al., who only included 14 studies, Karg, Burmeister, Shedden, and Sen (2011)) performed a new metaanalysis including 54 studies. They found strong evidence that the 5-HTTLPR moderates the relationship between stress and depression, with the s-allele being associated with an increased risk of developing depression under stress. Karg et al. further reported that a strong interaction between the s-allele and childhood maltreatment and specific medical conditions was found, but almost never with adult stressful life events. When they restricted their analysis to the studies included in the previous metaanalyses from Munafò and Risch, they found no evidence of interaction. Nevertheless, the difficulties in replicating GxE results with regard to the 5-HTTLPR have stimulated further research activities.

One huge international consortium led by Culverhouse et al. (2017) aimed at determining the magnitude of the GxE interaction of the 5-HTTLPR and the conditions under which it might be observed. A metaanalysis on 31 data sets comprised of 38,802 European ancestry subjects was performed. To reduce heterogeneity, a uniform data analysis script was used for all cohorts. Analyses targeted two stressors (narrow, broad) and two depression outcomes (current, lifetime). The findings did not support a general interaction using a multiplicative statistical model.

From this paper, it follows that if interaction between stressful life-events and the 5-HTTLR do exist, this interaction seems to be present in subgroups, or under certain circumstances only.

Taking these results, the initial expectation that the 5-HTTLPR could serve as a clinically useful biomarker has not proven true. However, from a psychobiological point of view, the 5-HTTLPR is still highly interesting. There are studies that associate the so-call risk allele (s) with behavioral beneficial traits, as with better risk assessment and better social recognition (Homberg & Lesch, 2011). In the study of Reinelt et al. (2015), subjects with low social support carrying at least one short allele reported significantly increased levels of "Sense of Coherence" and resilience, as well as less depressive symptoms than carriers of the l/l genotype. Other studies have found an association of the l/l-genotype with social anxiety disorders (Reinelt et al., 2014), and even against the symptoms of posttraumatic stress disorder (Grabe et al., 2009), concluding that the s-allele might be protective against psychopathological conditions that are frequently associated with depressive disorders. This contributes to a much more complex picture that the multiple environmental risks and protective factors modulate the interaction of the 5-HTTLPR.

6 FKBP5 gene

Another prominent gene in GxE research is the FKBP5 gene. The *FKBP5* gene is located on chromosome 6p21, and codes for FK506 binding protein 51 (FKBP5), a co-chaperone of hsp90, which regulates the glucocorticoid receptor (GR) sensitivity (Binder, 2009). FKBP5 is relevantly expressed in the brain (Gawlik et al., 2006). Functionally, cortisol induces the FKBP5 expression by activation of glucocorticoid-response-elements (Vermeer, Hendriks-Stegeman, van der Burg, van Buul-Offers, & Jansen, 2003). In turn, FKBP5 binding to the GR reduces the GR affinity for cortisol, and diminishes the amount of activated GR translocation to the cell nucleus (Wochnik et al., 2005).

In a recent systematic review, (Wang, Shelton, & Dwivedi, 2018) the effects of the interaction between FKBP5 gene variants and early-life stress, and their associations with stress-related disorders such as major depression and PTSD were examined. Fourteen studies (15,109 participants) were included. Based on the literature, rs1360780, rs3800373, and rs9470080 SNPs were selected within the FKBP5 gene. Their metaanalysis showed that individuals carrying the T allele of rs1360780, C-allele of rs3800373, or T-allele of rs9470080, and having been exposed to early-life trauma, had higher risks for depression or PTSD.

In a recent brain imaging study, we could demonstrate that large clusters of reduced gray matter (GM) volumes (left insula, superior and middle temporal gyrus, putamen, the bilateral hippocampus and olfactory cortex, right putamen, caudate nucleus, pallidum, amygdala) were present in abused TT-carriers (rs1360780) compared with abused non-TT carriers or (Grabe et al., 2016).

Nevertheless, it is still unclear how informative FKBP5 genotypes will be for prevention, clinical use, and individualized treatment. Interestingly, the development of pharmacological compounds blocking FKBP5 action is underway.

7 Brain-derived neurotrophic factor (BDNF)

Another protein that has stimulated neurobiological research is the brain-derived neurotrophic factor (BDNF). The Met allele of the BDNF Val66Met polymorphisms was found to decrease the activity dependent secretion of BDNF by altering the intracellular trafficking. In line with this, the Val66Met was associated with lower hippocampal *N*-acetyl-aspartate levels, and with impaired episodic memory (Egan et al., 2003). Animal research showed that a reduction in BDNF expression in the dentate gyrus reduced neurogenesis, and impacted depression-like behaviors (Taliaz et al., 2011; Taliaz, Stall, Dar, & Zangen, 2010).

Further, Adachi, Barrot, Autry, Theobald, and Monteggia (2008) showed that BDNF-dependent neurogenesis represents an important biological mechanism in response to antidepressant treatment. In support of the relevance of BDNF and the Val66Met polymorphism, evidence from a metaanalysis suggests a better response to antidepressant treatment in subjects carrying the Met allele of the BDNF Val66Met polymorphisms (Kato & Serretti, 2010). There are many lines of experimental evidence that support a tight functional interconnection between the serotonin system and BDNF, especially in neurogenesis and synaptic plasticity (Martinowich & Lu, 2008). Thus, it is reasonable to study gene x gene (GxG), as well as gene x gene x environment interactions (GxGxE) with regard to BDNF, the 5-HTTLPR and stressful life-events. At an early time in the study of GxE, a three-way interaction among BDNFVal66Met, the 5-HTTLPR, and childhood maltreatment was identified, predicting depression (Kaufman et al., 2006). Independent replication studies on this three-way interaction yielded inconsistent results: Kaufman et al. (2006) found significant three-way interactions with the highest scores of depression or distress in the Met allele and s/s genotype, whereas Wichers et al. found the highest distress in the Met allele and s/l genotype group (Wichers et al., 2008). Two larger studies did not replicate any three-way-interaction (Aguilera et al., 2009; Nederhof, Bouma, Oldehinkel, & Ormel, 2010). In more than 2000 individuals, Grabe, Schwahn, Mahler, Appel, et al. (2012) found support for a GxGxE interaction that relevantly impacts the role of the s/s genotype of the 5-HTTLPR in childhood abuse. Depending on the BDNF polymorphism (Val/Val versus Met allele), the s/s genotype showed either protective or risk properties with regard to depressive symptoms.

Given the neurobiological importance of BDNF, numerous brain imaging studies have been performed aiming at elucidating genotype-dependent differences in brain structure. One recent metaanalysis evaluated the impact of rs6265 SNP on hippocampal volumes in neuropsychiatric patients with major depressive disorder, anxiety, bipolar disorder, or schizophrenia (Harrisberger et al., 2015). The metaanalysis was comprised of 18 independent clinical cohorts ($n = 1695$ patients). For each BDNF genotype, the hippocampal volumes were significantly lower in neuropsychiatric patients than in healthy controls, but there was not a genotype-effect on hippocampal volumes. This might indicate that further factors such as GxG or GxGxE interactions are necessary to obtain structural brain differences in BDNF rs6265.

It is noteworthy that BDNF also seems to play important roles in other tissues such as muscle, platelets, and heart muscle.

8   Genome-wide challenges of GxE

Most candidate gene approaches in GxE studies select single variants in specific genes belonging to plausible disease-related pathways. Although this approach seems sensible, there are many challenges. Comparing two studies is often difficult because of differences in sample size, assessments of environmental exposures, phenotype definition, or sample recruitment (Dunn et al., 2015; Mandelli & Serretti, 2013). Moreover, there are no common guidelines on how to set up the mathematical model to perform GxE analyses in MDD. Until now, only three GxE interaction analyses for depression have been performed on a genome-wide level. One was published by Dunn et al. (2016) in a sample of African American and Hispanic women where one genome-wide significant hit was found near *CEP350*, a centrosomal protein that has never been associated with a psychiatric phenotype before. Another study in a Japanese population ($N = 320$) reported the genome-wide hit rs10510057 near RGS10 (Otowa et al., 2016), but given the small sample size, this result should be regarded with caution. In a recent metaanalysis in $N = 3944$ Caucasian subjects, Van der Auwera et al. (2018) did not find any evidence for a genome-wide significant GxE hit, and no supporting evidence for the frequently studied candidate variants.

Thus, it might be prudent to at least partially question some of the former candidate SNP results for MDD and harmonize the different models currently performed in GxE analyses for MDD to perform consistent analyses in samples large enough to identify robust GxE interaction signals.

9   Schizophrenia and bipolar disorder

Recent genome-wide association studies have been very fruitful, especially in highly powered metaanalysis of schizophrenia yielding new genetics signals, but also supporting pathways and mechanisms that have been proposed for decades, such as the dopaminergic system and inflammatory pathways (PGC, 2014). A recent review on GxE studies in schizophrenia and bipolar disorders (BD) identified 11 eligible studies from patients with BD, and 50 studies on schizophrenia spectrum phenotypes. These studies were grouped into five distinct domains in dependence of the environmental factor: (1) gene × cannabis interactions, (2) gene × stress and childhood trauma interactions, (3) gene × season of birth interactions, (4) gene × infectious factors interactions, and (5) gene × obstetric complications interactions (Misiak et al., 2017).

Most of the studies investigating the interactions between cannabis use and genetic factors focused on genetic variants affecting the dopaminergic system. One important dopamine-inactivating enzyme is coded by the catechol-*O*-methyltransferase (COMT) gene. Carriers of the functional 158Val allele with cannabis use in their adolescence were found to have an increased risk of schizophrenia (Caspi et al., 2005). In line with this finding, other studies described higher risk for psychotic symptoms (Ermis et al., 2015; Henquet et al., 2006; Nieman et al., 2016), and earlier age of onset (Estrada et al., 2011). Despite these encouraging studies, other studies have not supported the interaction between the COMT 158Val/Met polymorphism and cannabis use on the risk of psychosis (Kantrowitz et al., 2009; van Winkel, 2011; Zammit et al., 2007), or risk of subclinical psychotic experiences (Zammit, Lewis, Dalman, & Allebeck, 2010) and age of psychosis onset (De Sousa et al., 2013).

The DRD2 rs1076560 T allele had a threefold increase in psychosis risk compared with GG homozygotes in cannabis-using, first-episode psychosis patients compared with controls (Colizzi et al., 2015). Investigating the effects of the rs1360780 polymorphism of the FKBP5 gene in first-episode psychosis patients and healthy controls (Ajnakina et al., 2014) did not find significant interactions between genetic variation in the FKBP5 gene and cannabis use on psychosis. Genetic variations in the gene of the brain-derived neurotrophic factor (BDNF) were associated with earlier age of psychosis onset in female BDNF 66Met allele carriers (Decoster et al., 2011). However, this effect-modification of the BDNF 66Val/Met polymorphism was not seen in males.

Many studies investigating gene × stress interactions in psychotic disorders found positive results. Positive interactions were reported for the BDNF 66 Val/Met polymorphism and parameters of increased risk of psychosis or higher symptom load (Alemany et al., 2016; de Castro-Catala et al., 2016). Additionally, an additive effect of the BDNF 66Met allele and a history of childhood trauma on reduced levels of BDNF mRNA in whole blood was reported. Met carriers who reported high levels of childhood trauma also had significantly reduced hippocampal subfield volumes (Aas et al., 2014).

Four studies have found positive effects for an interaction of genetic polymorphisms within the FKBP5 gene and childhood trauma on psychosis (Ajnakina et al., 2014; Collip et al., 2013; Cristobal-Narvaez et al., 2016; Green et al., 2015). For example, Collip et al. (2013) found a significant interaction between the rs9296158 and rs4713916 polymorphisms of the FKBP5 gene and childhood trauma on psychotic symptoms and cortisol levels in a twin sample that was replicated for the rs4713916 polymorphism in siblings, and for rs9296158 in patients. Ajnakina et al. (2014) reported a significant interaction between the FKBP5 rs1360780 polymorphism and parental separation on psychosis risk. In a newer study, however, an interacting effect of variants within the FKBP5 gene and childhood trauma could be identified in subclinical depression and anxiety, but not for schizotypic or psychotic-like experiences in 808 young adults (de Castro-Catala et al., 2017). One study on the interaction of childhood trauma and the forkhead box protein 2 (FOXP2) gene in predicting a lifetime history of auditory hallucinations found positive results (McCarthy-Jones et al., 2014). Other studies reported on the interactions between recent or daily life stressors and the COMT gene with inconsistent results for either the COMT 158 Met/Met genotype (Peerbooms et al., 2012; van Winkel et al., 2008), or the Val-allele being associated with higher risk of psychotic symptoms (Simons et al., 2009; Stefanis et al., 2013).

As seasonality of birth is associated with the risk of schizophrenia, possibly through infectious/immunological processes, the presence of the HLA-DR1 allele was tested for interaction with seasonality of birth schizophrenia. Narita et al. (2000) reported that the HLA-DR1 allele was associated with increased incidence of winter birth in patients with schizophrenia, which was not replicated in another study (Tochigi et al., 2002). Other studies yielded negative or preliminary positive results for tryptophan hydroxylase 1 (TPH1), the dopamine D4 receptor (DRD4) 7-repeat allele polymorphism, and 5-HTTLPR L/S polymorphism (Chotai, Serretti, Lattuada, Lorenzi, & Lilli, 2003), and for the C677T polymorphism of the methylenetetrahydrofolate reductase (MTHFR) gene, which codes for an essential enzyme in the folate mediated methylation transfer reactions (Muntjewerff, Ophoff, Buizer-Voskamp, Strengman, & den Heijer, 2011). With respect to BD, an association between the HLA-G 14bp ins/del polymorphism and seasonality of birth was found (Debnath et al., 2013).

As infectious diseases may represent a risk factor for the development of schizophrenia, interactions with genetic variants have been tested.

Genetic variants of the glutamate ionotropic receptor NMDA type subunit 2B (GRIN2B) gene and maternal herpes simplex virus type 2 (HSV-2) seropositivity yielded a significant interaction on schizophrenia risk in the offspring (Demontis et al., 2011).

Shirts et al. (2007) found that variations in the MHC Class I Polypeptide-Related Sequence B (MICB) gene may interact with cytomegalovirus (CMV) and herpes simplex virus type 1 (HSV-1) seropositivity, impacting the risk of schizophrenia.

Dickerson et al. (2006) revealed that the COMT 158Val/Val genotype and HSV-1 seropositivity impacted independently on global cognitive performance. Moreover, patients carrying the risk genotype (Val/Val) and being HSV-1 positive had an 85% higher risk of low global cognitive performance. Seropositivity for Toxoplasma gondii revealed and significant interaction with TLR2 gene polymorphism (rs3804099) for the risk of BD (Oliveira et al., 2016). The toll-like receptor (TLR) family plays a fundamental role in pathogen recognition and activation of innate immunity.

Obstetric complications have long been identified as risk-increasing factors for major psychiatric disorders. A few studies so far have investigated putative interactions between genetic factors and obstetric complications for the risk of schizophrenia or BD.

Nicodemus et al. (2008) investigated a set of 13 candidate genes in a family-based study of 116 trios. Twenty-nine probands had at least one serious obstetric complication. Four genes (serine-threonine protein kinase (AKT1), BDNF, dystrobrevin (DTNBP1) and glutamate metabotropic receptor 3 (GRM3) showed significant evidence for gene-by-environment interaction. Ursini et al. (2016) found that methylation at the BDNF rs6265 Val allele in peripheral blood of healthy subjects is associated with hypoxia-related early life events and intermediate phenotypes for schizophrenia. Haukvik et al. (2010) investigated interaction effects of severe hypoxia-related OCs and variation in four hypoxia-regulated schizophrenia susceptibility genes (BDNF, DTNBP1, GRM3, and NRG1) on hippocampal volume in schizophrenia patients and healthy controls. The effects of severe OCs on hippocampal volume were associated with rs13242038 of the GRM3 gene (one out of 32 SNPs studied), but the interaction effect was not specific for schizophrenia.

10 Future relevance of GxE interaction

From the examples described herein, it becomes clear that methodological challenges for the development of diagnostic and therapeutic algorithms for personalized psychiatry are enormous, and the replication of associations is of utmost importance, but often difficult to achieve. Given the heterogeneity of samples, genetic backgrounds, stressors, and phenotypes studied, it is not even clear if nonreplicated GxE effects were truly false positives, or if they indeed occurred in the initial cohort. Nevertheless, independent replications still provide the best empirical support for GxE interactions that generalize to different samples and settings.

GxE interactions for personalized psychiatry in prevention, treatment decision, or risk assessment would require reproducible interaction effects that would allow for sensitivity, specificity, and positive/negative predictive value estimations. In one example for the FKBP5 rs1360780 SNP in the Study of Health in Pomerania (SHIP), we have performed such a promising estimation (Grabe & Schwahn, 2011). Given the large effect sizes demonstrated by Appel et al. (2011), rs1360780 TT-risk allele carriers were included in prediction models for depression in individuals exposed to childhood abuse. At a given prevalence rate of MDD, for example, 30%, the risk prediction (positive predictive value) by genotype and abuse status was about 70%. However, positive and negative predictive values were still far too low to recommend individual risk prediction via this approach. A further refinement of this model was possible via the inclusion of the status "high or low resilience." Considering the inclusion criteria "low resilience," the prevalence rate of MDD increased up to 35% in the selected subjects. Thus, the positive predictive value of the screen-positive subjects (TT-carriers and child abuse) increased to 80%. Although promising, these effects were not replicated in another independent sample.

However, the search for variants with more robust associations is ongoing. Currently, genome-wide approaches are underway to take advantage of the unbiased inclusion of genetic variants for the identification of GxE interactions. Like in a conventional GWAS, this approach gives rise to the expectation to discover new, yet unknown biological pathways that interact with environmental factors. One has to keep in mind that the sample size needed to yield genome-wide significant associations will be enormous, for example, in the range of at least 50–100K subjects for disorders such as major depression.

In psychiatry, the distance between the biological mechanisms and underpinnings and the clinical phenotypes are supposed to be large. This justifies the idea of "endophenotypes" that are much more closely related to biology than the clinical phenotypes themselves. Especially for the field of GxE, this approach could be very informative for personalized psychiatry. Endophenotypes could be derived from, for example, cognitive measures (Burrows & Hannan, 2016), analyses of structural brain imaging data (Grabe et al., 2016), and from functional brain imaging studies (Braff & Tamminga, 2017) (see Fig. 1). Animal studies that allow precise experimental modulation of environmental stressors are of the utmost importance to gain mechanistic insights on how GxE interactions biologically might alter epigenetic regulations, miRNA dynamics, gene expression, and ultimately the structure and function of neural networks (Braun et al., 2017). Also "second-hits" models that capture the temporal dynamics of trauma and biological long-term response will increase our understanding of the nature of mental diseases substantially (Lesse, Rether, Groger, Braun, & Bock, 2017).

Thus, GxE research will most likely not contribute to personalized diagnostics and treatments in the near future, but the mechanistic understanding of the development of mental diseases that emerge from this research approach will significantly influence our conceptual understanding of mental diseases. However, in the longer perspective, findings of GxE research might be implemented in new strategies in personalized psychiatry. For example, it is conceivable that in the future researchers may develop "posttrauma-medications" that will prevent negative biological adaptations after traumatic events (Kao et al., 2016), to develop medications that will stabilize biological processes under conditions of stress, or to develop stress prevention strategies in subjects at higher biological risk of maladaptation.

References

Aas, M., Haukvik, U. K., Djurovic, S., Tesli, M., Athanasiu, L., Bjella, T., et al. (2014). Interplay between childhood trauma and BDNF val66met variants on blood BDNF mRNA levels and on hippocampus subfields volumes in schizophrenia spectrum and bipolar disorders. *Journal of Psychiatric Research*, 59, 14–21. https://doi.org/10.1016/j.jpsychires.2014.08.011.

Adachi, M., Barrot, M., Autry, A. E., Theobald, D., & Monteggia, L. M. (2008). Selective loss of brain-derived neurotrophic factor in the dentate gyrus attenuates antidepressant efficacy. *Biological Psychiatry*, 63(7), 642–649.

Aguilera, M., Arias, B., Wichers, M., Barrantes-Vidal, N., Moya, J., Villa, H., et al. (2009). Early adversity and 5-HTT/BDNF genes: New evidence of gene-environment interactions on depressive symptoms in a general population. *Psychological Medicine*, 39(9), 1425–1432.

Ajnakina, O., Borges, S., Di Forti, M., Patel, Y., Xu, X., Green, P., et al. (2014). Role of environmental confounding in the association between FKBP5 and first-episode psychosis. *Frontiers in Psychiatry*, 5, 84. https://doi.org/10.3389/fpsyt.2014.00084.

Alemany, S., Moya, J., Ibanez, M. I., Villa, H., Mezquita, L., Ortet, G., et al. (2016). Research letter: Childhood trauma and the rs1360780 SNP of FKBP5 gene in psychosis: A replication in two general population samples. *Psychological Medicine*, 46(1), 221–223. https://doi.org/10.1017/S0033291715001695.

Appel, K., Schwahn, C., Mahler, J., Schulz, A., Spitzer, C., Fenske, K., et al. (2011). Moderation of adult depression by a polymorphism in the FKBP5 gene and childhood physical abuse in the general population. *Neuropsychopharmacology*, 36(10), 1982–1991.

Binder, E. B. (2009). The role of FKBP5, a co-chaperone of the glucocorticoid receptor in the pathogenesis and therapy of affective and anxiety disorders. *Psychoneuroendocrinology*, 34(Suppl. 1), S186–S195.

Braff, D. L., & Tamminga, C. A. (2017). Endophenotypes, epigenetics, polygenicity and more: Irv gottesman's dynamic legacy. *Schizophrenia Bulletin*, 43(1), 10–16. https://doi.org/10.1093/schbul/sbw157.

Braun, K., Bock, J., Wainstock, T., Matas, E., Gaisler-Salomon, I., Fegert, J., et al. (2017). Experience-induced transgenerational (re-)programming of neuronal structure and functions: Impact of stress prior and during pregnancy. Review. *Neuroscience and Biobehavioral Reviews*. https://doi.org/10.1016/j.neubiorev.2017.05.021.

Brune, M., & Brune-Cohrs, U. (2006). Theory of mind-evolution, ontogeny, brain mechanisms and psychopathology. Review. *Neuroscience and Biobehavioral Reviews*, 30(4), 437–455. https://doi.org/10.1016/j.neubiorev.2005.08.001.

Burrows, E. L., & Hannan, A. J. (2016). Cognitive endophenotypes, gene-environment interactions and experience-dependent plasticity in animal models of schizophrenia. *Biological Psychology*, *116*, 82–89. https://doi.org/10.1016/j.biopsycho.2015.11.015.

Caspi, A., Moffitt, T. E., Cannon, M., McClay, J., Murray, R., Harrington, H., et al. (2005). Moderation of the effect of adolescent-onset cannabis use on adult psychosis by a functional polymorphism in the catechol-O-methyltransferase gene: Longitudinal evidence of a gene X environment interaction. *Biological Psychiatry*, *57*(10), 1117–1127. https://doi.org/10.1016/j.biopsych.2005.01.026.

Caspi, A., Sugden, K., Moffitt, T. E., Taylor, A., Craig, I. W., Harrington, H., et al. (2003). Influence of life stress on depression: Moderation by a polymorphism in the 5-HTT gene. *Science*, *301*(5631), 386–389.

Chotai, J., Serretti, A., Lattuada, E., Lorenzi, C., & Lilli, R. (2003). Gene-environment interaction in psychiatric disorders as indicated by season of birth variations in tryptophan hydroxylase (TPH), serotonin transporter (5-HTTLPR) and dopamine receptor (DRD4) gene polymorphisms. *Psychiatry Research*, *119*(1–2), 99–111.

Colizzi, M., Iyegbe, C., Powell, J., Ursini, G., Porcelli, A., Bonvino, A., et al. (2015). Interaction between functional genetic variation of DRD2 and cannabis use on risk of psychosis. *Schizophrenia Bulletin*, *41*(5), 1171–1182. https://doi.org/10.1093/schbul/sbv032.

Collip, D., Myin-Germeys, I., Wichers, M., Jacobs, N., Derom, C., Thiery, E., et al. (2013). FKBP5 as a possible moderator of the psychosis-inducing effects of childhood trauma. *The British Journal of Psychiatry*, *202*(4), 261–268. https://doi.org/10.1192/bjp.bp.112.115972.

Cristobal-Narvaez, P., Sheinbaum, T., Rosa, A., Ballespi, S., de Castro-Catala, M., Pena, E., et al. (2016). The interaction between childhood bullying and the FKBP5 gene on psychotic-like experiences and stress reactivity in real life. *PLoS One*, *11*(7). https://doi.org/10.1371/journal.pone.0158809.

Culverhouse, R. C., Saccone, N. L., Horton, A. C., Ma, Y., Anstey, K. J., Banaschewski, T., et al. (2017). Collaborative meta-analysis finds no evidence of a strong interaction between stress and 5-HTTLPR genotype contributing to the development of depression. *Molecular Psychiatry*. https://doi.org/10.1038/mp.2017.44.

de Castro-Catala, M., Pena, E., Kwapil, T. R., Papiol, S., Sheinbaum, T., Cristobal-Narvaez, P., et al. (2017). Interaction between FKBP5 gene and childhood trauma on psychosis, depression and anxiety symptoms in a non-clinical sample. *Psychoneuroendocrinology*, *85*, 200–209. https://doi.org/10.1016/j.psyneuen.2017.08.024.

de Castro-Catala, M., van Nierop, M., Barrantes-Vidal, N., Cristobal-Narvaez, P., Sheinbaum, T., Kwapil, T. R., et al. (2016). Childhood trauma, BDNF Val66Met and subclinical psychotic experiences. Attempt at replication in two independent samples. *Journal of Psychiatric Research*, *83*, 121–129. https://doi.org/10.1016/j.jpsychires.2016.08.014.

De Sousa, K. R., Tiwari, A. K., Giuffra, D. E., Mackenzie, B., Zai, C. C., & Kennedy, J. L. (2013). Age at onset of schizophrenia: Cannabis, COMT gene, and their interactions. *Schizophrenia Research*, *151*(1–3), 289–290. https://doi.org/10.1016/j.schres.2013.10.037.

Debnath, M., Busson, M., Jamain, S., Etain, B., Hamdani, N., Oliveira, J., et al. (2013). The HLA-G low expressor genotype is associated with protection against bipolar disorder. *Human Immunology*, *74*(5), 593–597. https://doi.org/10.1016/j.humimm.2012.11.032.

Decoster, J., van Os, J., Kenis, G., Henquet, C., Peuskens, J., De Hert, M., et al. (2011). Age at onset of psychotic disorder: Cannabis, BDNF Val66Met, and sex-specific models of gene-environment interaction. *American Journal of Medical Genetics. Part B, Neuropsychiatric Genetics*, *156B*(3), 363–369. https://doi.org/10.1002/ajmg.b.31174.

Demontis, D., Nyegaard, M., Buttenschon, H. N., Hedemand, A., Pedersen, C. B., Grove, J., et al. (2011). Association of GRIN1 and GRIN2A-D with schizophrenia and genetic interaction with maternal herpes simplex virus-2 infection affecting disease risk. *American Journal of Medical Genetics. Part B, Neuropsychiatric Genetics*, *156B*(8), 913–922. https://doi.org/10.1002/ajmg.b.31234.

Dickerson, F. B., Boronow, J. J., Stallings, C., Origoni, A. E., Cole, S., Leister, F., et al. (2006). The catechol O-methyltransferase Val158Met polymorphism and herpes simplex virus type 1 infection are risk factors for cognitive impairment in bipolar disorder: Additive gene-environmental effects in a complex human psychiatric disorder. *Bipolar Disorders*, *8*(2), 124–132. https://doi.org/10.1111/j.1399-5618.2006.00288.x.

Dunn, E. C., Brown, R. C., Dai, Y., Rosand, J., Nugent, N. R., Amstadter, A. B., et al. (2015). Genetic determinants of depression: Recent findings and future directions. *Harvard Review of Psychiatry*, *23*(1), 1–18. https://doi.org/10.1097/HRP.0000000000000054.

Dunn, E. C., Wiste, A., Radmanesh, F., Almli, L. M., Gogarten, S. M., Sofer, T., et al. (2016). Genome-Wide Association Study (Gwas) and Genome-Wide by Environment Interaction Study (Gweis) of depressive symptoms in African American and Hispanic/Latina women. *Depression and Anxiety*, *33*(4), 265–280. https://doi.org/10.1002/da.22484.

Egan, M. F., Kojima, M., Callicott, J. H., Goldberg, T. E., Kolachana, B. S., Bertolino, A., et al. (2003). The BDNF val66met polymorphism affects activity-dependent secretion of BDNF and human memory and hippocampal function. *Cell*, *112*(2), 257–269.

Emera, D., Yin, J., Reilly, S. K., Gockley, J., & Noonan, J. P. (2016). Origin and evolution of developmental enhancers in the mammalian neocortex. *Proceedings of the National Academy of Sciences of the United States of America*, *113*(19), E2617–E2626. https://doi.org/10.1073/pnas.1603718113.

Ermis, A., Erkiran, M., Dasdemir, S., Turkcan, A. S., Ceylan, M. E., Bireller, E. S., et al. (2015). The relationship between catechol-O-methyltransferase gene Val158Met (COMT) polymorphism and premorbid cannabis use in Turkish male patients with schizophrenia. *In Vivo*, *29*(1), 129–132.

Estrada, G., Fatjo-Vilas, M., Munoz, M. J., Pulido, G., Minano, M. J., Toledo, E., et al. (2011). Cannabis use and age at onset of psychosis: Further evidence of interaction with COMT Val158Met polymorphism. *Acta Psychiatrica Scandinavica*, *123*(6), 485–492. https://doi.org/10.1111/j.1600-0447.2010.01665.x.

Gawlik, M., Moller-Ehrlich, K., Mende, M., Jovnerovski, M., Jung, S., Jabs, B., et al. (2006). Is FKBP5 a genetic marker of affective psychosis? A case control study and analysis of disease related traits. *BMC Psychiatry*, *6*, 52.

Giovanoli, S., Engler, H., Engler, A., Richetto, J., Voget, M., Willi, R., et al. (2013). Stress in puberty unmasks latent neuropathological consequences of prenatal immune activation in mice. *Science*, *339*(6123), 1095–1099. https://doi.org/10.1126/science.1228261.

Gottschalk, M. G., & Domschke, K. (2017a). Genetics of generalized anxiety disorder and related traits. *Dialogues in Clinical Neuroscience*, *19*(2), 159–168.

Gottschalk, M. G., & Domschke, K. (2017b). Oxytocin and anxiety disorders. *Current Topics in Behavioral Neurosciences*, https://doi.org/10.1007/7854_2017_25.

Grabe, H. J., Lange, M., Wolff, B., Volzke, H., Lucht, M., Freyberger, H. J., et al. (2005). Mental and physical distress is modulated by a polymorphism in the 5-HT transporter gene interacting with social stressors and chronic disease burden. *Molecular Psychiatry, 10*(2), 220–224.

Grabe, H. J., Schulz, A., Schmidt, C. O., Appel, K., Driessen, M., Wingenfeld, K., et al. (2012). A brief instrument for the assessment of childhood abuse and neglect: The childhood trauma screener (CTS). *Psychiatrische Praxis.* https://doi.org/10.1055/s-0031-1298984.

Grabe, H. J., & Schwahn, C. (2011). Interaction between psychosocial environments and genes—What is the clinical relevance? Editorial. *Psychiatrische Praxis, 38*(2), 55–57. https://doi.org/10.1055/s-0030-1265980.

Grabe, H. J., Schwahn, C., Appel, K., Mahler, J., Schulz, A., Spitzer, C., et al. (2011). Update on the 2005 paper: Moderation of mental and physical distress by polymorphisms in the 5-HT transporter gene by interacting with social stressors and chronic disease burden. *Molecular Psychiatry, 16*(4), 354–356.

Grabe, H. J., Schwahn, C., Mahler, J., Appel, K., Schulz, A., Spitzer, C., et al. (2012). Genetic epistasis between the brain-derived neurotrophic factor Val66Met polymorphism and the 5-HTT promoter polymorphism moderates the susceptibility to depressive disorders after childhood abuse. *Progress in Neuro-Psychopharmacology & Biological Psychiatry, 36*(2), 264–270. https://doi.org/10.1016/j.pnpbp.2011.09.010.

Grabe, H. J., Schwahn, C., Mahler, J., Schulz, A., Spitzer, C., Fenske, K., et al. (2012). Moderation of adult depression by the serotonin transporter promoter variant (5-HTTLPR), childhood abuse and adult traumatic events in a general population sample. *American Journal of Medical Genetics. Part B, Neuropsychiatric Genetics, 159B*(3), 298–309. https://doi.org/10.1002/ajmg.b.32027.

Grabe, H. J., Spitzer, C., Schwahn, C., Marcinek, A., Frahnow, A., Barnow, S., et al. (2009). Serotonin transporter gene (SLC6A4) promoter polymorphisms and the susceptibility to posttraumatic stress disorder in the general population. *The American Journal of Psychiatry, 166*(8), 926–933.

Grabe, H. J., Wittfeld, K., Van der Auwera, S., Janowitz, D., Hegenscheid, K., Habes, M., et al. (2016). Effect of the interaction between childhood abuse and rs1360780 of the FKBP5 gene on gray matter volume in a general population sample. *Human Brain Mapping, 37*(4), 1602–1613. https://doi.org/10.1002/hbm.23123.

Grech, A., & van Os, J. (2017). Evidence that the urban environment moderates the level of familial clustering of positive psychotic symptoms. *Schizophrenia Bulletin, 43*(2), 325–331. https://doi.org/10.1093/schbul/sbw186.

Green, M. J., Raudino, A., Cairns, M. J., Wu, J., Tooney, P. A., Scott, R. J., et al. (2015). Do common genotypes of FK506 binding protein 5 (FKBP5) moderate the effects of childhood maltreatment on cognition in schizophrenia and healthy controls? *Journal of Psychiatric Research, 70*, 9–17. https://doi.org/10.1016/j.jpsychires.2015.07.019.

Harrisberger, F., Smieskova, R., Schmidt, A., Lenz, C., Walter, A., Wittfeld, K., et al. (2015). BDNF Val66Met polymorphism and hippocampal volume in neuropsychiatric disorders: A systematic review and meta-analysis. *Neuroscience and Biobehavioral Reviews, 55*, 107–118. https://doi.org/10.1016/j.neubiorev.2015.04.017.

Haukvik, U. K., Saetre, P., McNeil, T., Bjerkan, P. S., Andreassen, O. A., Werge, T., et al. (2010). An exploratory model for G x E interaction on hippocampal volume in schizophrenia; obstetric complications and hypoxia-related genes. *Progress in Neuro-Psychopharmacology & Biological Psychiatry, 34*(7), 1259–1265. https://doi.org/10.1016/j.pnpbp.2010.07.001.

Henquet, C., Rosa, A., Krabbendam, L., Papiol, S., Fananas, L., Drukker, M., et al. (2006). An experimental study of catechol-o-methyltransferase Val158Met moderation of delta-9-tetrahydrocannabinol-induced effects on psychosis and cognition. *Neuropsychopharmacology, 31*(12), 2748–2757. https://doi.org/10.1038/sj.npp.1301197.

Homberg, J. R., & Lesch, K. P. (2011). Looking on the bright side of serotonin transporter gene variation. *Biological Psychiatry, 69*(6), 513–519.

Kantrowitz, J. T., Nolan, K. A., Sen, S., Simen, A. A., Lachman, H. M., & Bowers, M. B., Jr. (2009). Adolescent cannabis use, psychosis and catechol-O-methyltransferase genotype in African Americans and Caucasians. *The Psychiatric Quarterly, 80*(4), 213–218. https://doi.org/10.1007/s11126-009-9108-4.

Kao, C. Y., He, Z., Zannas, A. S., Hahn, O., Kuhne, C., Reichel, J. M., et al. (2016). Fluoxetine treatment prevents the inflammatory response in a mouse model of posttraumatic stress disorder. *Journal of Psychiatric Research, 76*, 74–83. https://doi.org/10.1016/j.jpsychires.2016.02.003.

Karg, K., Burmeister, M., Shedden, K., & Sen, S. (2011). The serotonin transporter promoter variant (5-HTTLPR), stress, and depression meta-analysis revisited: Evidence of genetic moderation. *Archives of General Psychiatry, 68*(5), 444–454.

Kato, M., & Serretti, A. (2010). Review and meta-analysis of antidepressant pharmacogenetic findings in major depressive disorder. *Molecular Psychiatry, 15*(5), 473–500. https://doi.org/10.1038/mp.2008.116.

Kaufman, J., Yang, B. Z., Douglas-Palumberi, H., Grasso, D., Lipschitz, D., et al. (2006). Brain-derived neurotrophic factor-5-HTTLPR gene interactions and environmental modifiers of depression in children. *Biological Psychiatry, 59*(8), 673–680.

Lesse, A., Rether, K., Groger, N., Braun, K., & Bock, J. (2017). Chronic postnatal stress induces depressive-like behavior in male mice and programs second-hit stress-induced gene expression patterns of OxtR and AvpR1a in adulthood. *Molecular Neurobiology, 54*(6), 4813–4819. https://doi.org/10.1007/s12035-016-0043-8.

Mandelli, L., & Serretti, A. (2013). Gene environment interaction studies in depression and suicidal behavior: An update. Review. *Neuroscience and Biobehavioral Reviews, 37*(10 Pt 1), 2375–2397. https://doi.org/10.1016/j.neubiorev.2013.07.011.

Martinowich, K., & Lu, B. (2008). Interaction between BDNF and serotonin: Role in mood disorders. *Neuropsychopharmacology, 33*(1), 73–83.

McCarthy-Jones, S., Green, M. J., Scott, R. J., Tooney, P. A., Cairns, M. J., Wu, J. Q., et al. (2014). Preliminary evidence of an interaction between the FOXP2 gene and childhood emotional abuse predicting likelihood of auditory verbal hallucinations in schizophrenia. *Journal of Psychiatric Research, 50*, 66–72. https://doi.org/10.1016/j.jpsychires.2013.11.012.

Misiak, B., Stramecki, F., Gaweda, L., Prochwicz, K., Sasiadek, M. M., Moustafa, A. A., et al. (2017). Interactions between variation in candidate genes and environmental factors in the etiology of schizophrenia and bipolar disorder: A systematic review. Review. *Molecular Neurobiology.* https://doi.org/10.1007/s12035-017-0708-y.

Munafo, M. R., Durrant, C., Lewis, G., & Flint, J. (2009). Gene x environment interactions at the serotonin transporter locus. *Biological Psychiatry, 65*(3), 211–219.

Muntjewerff, J. W., Ophoff, R. A., Buizer-Voskamp, J. E., Strengman, E., & den Heijer, M. (2011). Effects of season of birth and a common MTHFR gene variant on the risk of schizophrenia. *European Neuropsychopharmacology*, *21*(4), 300–305. https://doi.org/10.1016/j.euroneuro.2010.10.001.

Narita, K., Sasaki, T., Akaho, R., Okazaki, Y., Kusumi, I., Kato, T., et al. (2000). Human leukocyte antigen and season of birth in Japanese patients with schizophrenia. *The American Journal of Psychiatry*, *157*(7), 1173–1175. https://doi.org/10.1176/appi.ajp.157.7.1173.

Nederhof, E., Bouma, E. M., Oldehinkel, A. J., & Ormel, J. (2010). Interaction between childhood adversity, brain-derived neurotrophic factor val/met and serotonin transporter promoter polymorphism on depression: The TRAILS study. *Biological Psychiatry*, *68*(2), 209–212.

Nicodemus, K. K., Marenco, S., Batten, A. J., Vakkalanka, R., Egan, M. F., Straub, R. E., et al. (2008). Serious obstetric complications interact with hypoxia-regulated/vascular-expression genes to influence schizophrenia risk. *Molecular Psychiatry*, *13*(9), 873–877. https://doi.org/10.1038/sj. mp.4002153.

Nieman, D. H., Dragt, S., van Duin, E. D., Denneman, N., Overbeek, J. M., de Haan, L., et al. (2016). COMT Val(158)Met genotype and cannabis use in people with an at risk mental state for psychosis: Exploring gene x environment interactions. *Schizophrenia Research*, *174*(1–3), 24–28. https://doi. org/10.1016/j.schres.2016.03.015.

Oliveira, J., Kazma, R., Le Floch, E., Bennabi, M., Hamdani, N., Bengoufa, D., et al. (2016). Toxoplasma gondii exposure may modulate the influence of TLR2 genetic variation on bipolar disorder: A gene-environment interaction study. *International Journal of Bipolar Disorders*, *4*(1), 11. https://doi. org/10.1186/s40345-016-0052-6.

Otowa, T., Kawamura, Y., Tsutsumi, A., Kawakami, N., Kan, C., Shimada, T., et al. (2016). The first pilot genome-wide gene-environment study of depression in the Japanese population. *PLoS One. 11*(8). https://doi.org/10.1371/journal.pone.0160823.

Peerbooms, O., Rutten, B. P., Collip, D., Lardinois, M., Lataster, T., Thewissen, V., et al. (2012). Evidence that interactive effects of COMT and MTHFR moderate psychotic response to environmental stress. *Acta Psychiatrica Scandinavica*, *125*(3), 247–256. https://doi.org/10.1111/j.1600-0447.2011.01806.x.

PGC. (2014). Biological insights from 108 schizophrenia-associated genetic loci. *Nature*, *511*(7510), 421–427. https://doi.org/10.1038/nature13595.

Reinelt, E., Aldinger, M., Stopsack, M., Schwahn, C., John, U., Baumeister, S. E., et al. (2014). High social support buffers the effects of 5-HTTLPR genotypes within social anxiety disorder. *European Archives of Psychiatry and Clinical Neuroscience*, *264*(5), 433–439. https://doi.org/10.1007/ s00406-013-0481-5.

Reinelt, E., Barnow, S., Stopsack, M., Aldinger, M., Schmidt, C. O., John, U., et al. (2015). Social support and the serotonin transporter genotype (5-HTTLPR) moderate levels of resilience, sense of coherence, and depression. *American Journal of Medical Genetics. Part B, Neuropsychiatric Genetics*, *168B*(5), 383–391. https://doi.org/10.1002/ajmg.b.32322.

Risch, N., Herrell, R., Lehner, T., Liang, K. Y., Eaves, L., Hoh, J., et al. (2009). Interaction between the serotonin transporter gene (5-HTTLPR), stressful life events, and risk of depression: A meta-analysis. *JAMA*, *301*(23), 2462–2471.

Sharma, S., Powers, A., Bradley, B., & Ressler, K. J. (2016). Gene x environment determinants of stress- and anxiety-related disorders. *Annual Review of Psychology*, *67*, 239–261. https://doi.org/10.1146/annurev-psych-122414-033408.

Shirts, B. H., Kim, J. J., Reich, S., Dickerson, F. B., Yolken, R. H., Devlin, B., et al. (2007). Polymorphisms in MICB are associated with human herpes virus seropositivity and schizophrenia risk. *Schizophrenia Research*, *94*(1–3), 342–353. https://doi.org/10.1016/j.schres.2007.04.021.

Simons, C. J., Wichers, M., Derom, C., Thiery, E., Myin-Germeys, I., Krabbendam, L., et al. (2009). Subtle gene-environment interactions driving paranoia in daily life. *Genes, Brain, and Behavior*, *8*(1), 5–12. https://doi.org/10.1111/j.1601-183X.2008.00434.x.

Stefanis, N. C., Dragovic, M., Power, B. D., Jablensky, A., Castle, D., & Morgan, V. A. (2013). Age at initiation of cannabis use predicts age at onset of psychosis: The 7- to 8-year trend. *Schizophrenia Bulletin*, *39*(2), 251–254. https://doi.org/10.1093/schbul/sbs188.

Taliaz, D., Loya, A., Gersner, R., Haramati, S., Chen, A., & Zangen, A. (2011). Resilience to chronic stress is mediated by hippocampal brain-derived neurotrophic factor. *The Journal of Neuroscience*, *31*(12), 4475–4483.

Taliaz, D., Stall, N., Dar, D. E., & Zangen, A. (2010). Knockdown of brain-derived neurotrophic factor in specific brain sites precipitates behaviors associated with depression and reduces neurogenesis. *Molecular Psychiatry*, *15*(1), 80–92.

Tochigi, M., Ohashi, J., Umekage, T., Kohda, K., Hibino, H., Otowa, T., et al. (2002). Human leukocyte antigen-A specificities and its relation with season of birth in Japanese patients with schizophrenia. *Neuroscience Letters*, *329*(2), 201–204.

Ursini, G., Cavalleri, T., Fazio, L., Angrisano, T., Iacovelli, L., Porcelli, A., et al. (2016). BDNF rs6265 methylation and genotype interact on risk for schizophrenia. *Epigenetics*, *11*(1), 11–23. https://doi.org/10.1080/15592294.2015.1117736.

Van der Auwera, S., Peyrot, W. J., Milaneschi, Y., Hertel, J., Baune, B. T., Breen, G., et al. (2018). Genome-wide gene-environment interaction in depression: A systematic evaluation of candidate genes. *American Journal of Medical Genetics. Part B, Neuropsychiatric Genetics*, *177*(1), 40–49. https://doi.org/10.1002/ajmg.b.32593.

van Os, J., Rutten, B. P., Myin-Germeys, I., Delespaul, P., Viechtbauer, W., van Zelst, C., et al. (2014). Identifying gene-environment interactions in schizophrenia: Contemporary challenges for integrated, large-scale investigations. *Schizophrenia Bulletin*, *40*(4), 729–736. https://doi.org/ 10.1093/schbul/sbu069.

van Winkel, R. (2011). Family-based analysis of genetic variation underlying psychosis-inducing effects of cannabis: Sibling analysis and proband follow-up. *Archives of General Psychiatry*, *68*(2), 148–157. https://doi.org/10.1001/archgenpsychiatry.2010.152.

van Winkel, R., Henquet, C., Rosa, A., Papiol, S., Fananas, L., De Hert, M., Peuskens, J., et al. (2008). Evidence that the COMT(Val158Met) polymorphism moderates sensitivity to stress in psychosis: An experience-sampling study. *American Journal of Medical Genetics. Part B, Neuropsychiatric Genetics*, *147B*(1), 10–17. https://doi.org/10.1002/ajmg.b.30559.

Vermeer, H., Hendriks-Stegeman, B. I., van der Burg, B., van Buul-Offers, S. C., & Jansen, M. (2003). Glucocorticoid-induced increase in lymphocytic FKBP51 messenger ribonucleic acid expression: A potential marker for glucocorticoid sensitivity, potency, and bioavailability. *The Journal of Clinical Endocrinology and Metabolism, 88*(1), 277–284.

Wang, Q., Shelton, R. C., & Dwivedi, Y. (2018). Interaction between early-life stress and FKBP5 gene variants in major depressive disorder and post-traumatic stress disorder: A systematic review and meta-analysis. *Journal of Affective Disorders, 225,* 422–428. https://doi.org/10.1016/j.jad.2017.08.066.

Wichers, M., Kenis, G., Jacobs, N., Mengelers, R., Derom, C., Vlietinck, R., et al. (2008). The BDNF Val(66)Met x 5-HTTLPR x child adversity interaction and depressive symptoms: An attempt at replication. *American Journal of Medical Genetics. Part B, Neuropsychiatric Genetics, 147*(1), 120–123.

Wochnik, G. M., Ruegg, J., Abel, G. A., Schmidt, U., Holsboer, F., & Rein, T. (2005). FK506-binding proteins 51 and 52 differentially regulate dynein interaction and nuclear translocation of the glucocorticoid receptor in mammalian cells. *The Journal of Biological Chemistry, 280*(6), 4609–4616.

Zammit, S., Lewis, G., Dalman, C., & Allebeck, P. (2010). Examining interactions between risk factors for psychosis. *The British Journal of Psychiatry, 197*(3), 207–211. https://doi.org/10.1192/bjp.bp.109.070904.

Zammit, S., Spurlock, G., Williams, H., Norton, N., Williams, N., O'Donovan, M. C., et al. (2007). Genotype effects of CHRNA7, CNR1 and COMT in schizophrenia: Interactions with tobacco and cannabis use. *The British Journal of Psychiatry, 191,* 402–407. https://doi.org/10.1192/bjp.bp.107.036129.

Chapter 30

Epigenetics: A new approach to understanding mechanisms in depression and to predict antidepressant treatment response

Helge Frieling, Stefan Bleich and Alexandra Neyazi

Department of Psychiatry, Social Psychiatry and Psychotherapy, Hannover Medical School (MHH), Hannover, Germany

Major depressive disorder (MDD) is one of the most prevalent and leading causes of disability worldwide, affecting more than 300 million people (DALYs GBD, Collaborators H, 2017). MDD places a heavy burden on society and health systems, and also has a profound impact on the quality of life of afflicted persons. It is therefore an important public health priority. However, the efficacy of antidepressant treatment is obviously inefficient, as more than 50% of treated patients fail to reach remission (Lisoway, Zai, Tiwari, & Kennedy, 2018).

Although drug therapy is the treatment most frequently offered to depressed primary care patients, approximately 20% of patients do not fill their prescriptions, and even if they start a course of treatment, they may discontinue early before receiving an adequate dose (Xing, Dipaula, Lee, & Cooke, 2011). Recently, a meta-analysis of randomized controlled trials with second-generation antidepressants, the most commonly prescribed antidepressants, showed response rates defined as at least half of the baseline 17-item Hamilton Rating, below 50% in initial treatment of major depressive disorder (Amick et al., 2015). Moreover, 20% of the patients fail to respond to any intervention (Labermaier, Masana, & Muller, 2013). Although there is no statistically significant difference in effectiveness between first- and second-generation antidepressants in terms of response rates (von Wolff, Holzel, Westphal, Harter, & Kriston, 2013), individuals may vary widely in their response to specific antidepressant treatments. Unfortunately, the selection strategy of antidepressant treatment is mainly guided by trial and error. Thus, the need for markers reliably predicting response to antidepressive treatment and to allow a rational selection of optimal therapy for each patient appears instrumental for a personalized and efficient treatment strategy of depression.

Thus far, few studies have attempted to identify molecular biomarkers that could predict therapeutic response in MDD, and there is no absolute predictor to help guide the selection of antidepressants (Breitenstein, Scheuer, & Holsboer, 2014). The situation is aggravated by the fact that the same antidepressant medications that helped one patient in one depressive episode do not necessarily need to be an effective treatment in the next episode. These dynamic changes in the presentation of depressive symptoms and their response to treatment call for "dynamic" biomarkers. Multiple candidate genes have been identified, and a few genome-wide association studies have been performed, but so far with very limited success. To address the dynamic changes of depressive symptoms and their response to treatment, recent studies have focused on epigenetic mechanisms, as these are modulated by environmental stimuli, and adaptive to different stages of the disorder. Although promising, the field of gene expression and epigenomic biomarkers of antidepressant response is still in its infancy, and needs further development to define useful biomarkers in clinical practice (Belzeaux et al., 2018).

Epigenetics refers to the study of heritable changes in gene expression that occur without a change in DNA sequence. Studies have consistently shown that epigenetic mechanisms provide an "extra" layer of transcriptional control that regulates how genes are expressed. Epigenetic research has gained momentum in depression research after first findings associated early life adversities with long-lasting alterations of DNA methylation of candidate genes for MDD (Murgatroyd et al., 2009; Weaver et al., 2004). Epigenetic patterns are uniquely modified by environmental stimuli, and can biologically integrate nature and nurture influences. Through epigenetic marks, a molecular memory of past gene-environment interactions can be established. There is now evidence that pre-, peri-, and early postnatal adversities, as well as early life stress

Personalized Psychiatry. https://doi.org/10.1016/B978-0-12-813176-3.00030-4

and stressful life events, have an effect on epigenetic patterns, putting them in a unique position as biomarkers for depression to achieve the target of an individualized therapy (Anier et al., 2014; Babenko, Kovalchuk, & Metz, 2015; Devlin, Brain, Austin, & Oberlander, 2010; Fatima, Srivastav, & Mondal, 2017; Unternaehrer et al., 2012). Biomarkers reliably predicting the response to antidepressant therapies would not only significantly reduce costs and identify patients who are eligible, but would also help to develop novel antidepressant treatment procedures for a subgroup of depressed patients resistant to currently approved antidepressant treatments.

1   The main epigenetic processes that may be used as biomarkers for treatment response are DNA modifications, noncoding RNAs, and histone modifications

DNA modifications: DNA methylation is the best-researched epigenetic mechanism, particularly the covalent attachment of a methyl group to the $5'$-carbon of a cytosine residue in cytosine/guanosine-dinucleotides (CpG) of DNA (Chen, Meng, Pei, Zheng, & Leng, 2017; Jurkowska & Jeltsch, 2016; van der Harst, de Windt, & Chambers, 2017). The CpG methylation is facilitated by DNA methyltransferases (DNMTs) utilizing S-adenosylmethionine (SAM) as a substrate for the transmethylation. CpG-rich parts of the DNA, so-called CpG islands, are mainly located in the regulatory parts of the genes, and are usually unmethylated. Methylation of these regions normally represses expression of the gene by two ways: (1). methylated CpGs within transcription binding factor sites of the respective gene promoter can sterically prevent TF binding to the DNA; and (2). methylated DNA attracts a class of DNA binding proteins (MDBs—methylated domain binding proteins) that then associate with further proteins, including histone deacetylases (HDACs), forming a repressor complex at the gene promoter, and leading to a close conformation of the chromatin. DNA methylation is mainly coupled with transcriptional silencing, as it can inhibit the function of DNA binding proteins or transcription factors and "turn off" a gene for a long time period (Smith & Meissner, 2013)—potentially over the boundaries of generations (Rodenhiser & Mann, 2006). However, newer studies suggest that DNA methylation of negative regulatory elements in a gene's promoter can also increase the respective gene's expression. For example, Kang et al. observed a higher DNA methylation status in the promoter region of the gene encoding serotonin transporter (SLC6A4) gene with childhood adversities in MDD patients (Kang et al., 2013). However, they did not find any reliable correlation between SLC6A4 methylation and antidepressant response.

Noncoding RNAs: MicroRNAs (miRNAs) are an expanding class of endogenous 20–25-nucleotide long, noncoding RNA molecules (lncRNAs) that regulate gene expression through post-transcriptional mechanisms, primarily through translational repression of messenger RNA (mRNA) (Narahari, Hussain, & Sreeram, 2017). The resulting effects are highly complex. It is possible that a single mRNA may be regulated by multiple miRNAs; whereas miRNAs can interact with hundreds of mRNAs (Wilczynska & Bushell, 2015). Smalheiser et al. demonstrated that miRNAs are significantly altered in the prefrontal cortex of depressed individuals (Smalheiser et al., 2012). They are also reorganized in a highly coordinated manner. A growing number of publications support that miRNA play key roles in depression, as well as in stress signaling (Condorelli, Latronico, & Cavarretta, 2014; Mendell & Olson, 2012; Nelson, Dimayuga, & Wilfred, 2010; Wang et al., 2017; Wojciechowska, Braniewska, & Kozar-Kaminska, 2017). For example, miR-135a, miR-1202, and miR-124 levels in the blood and brains of depressed human patients are lower compared with controls (Dwivedi, 2017; Issler et al., 2014). Lopez et al. reported miR-1202 as a promising marker of MDD and antidepressant response using small RNA-sequencing in paired samples from MDD patients enrolled in a large, randomized placebo-controlled trial of a serotonin—norepinephrine reuptake inhibitor duloxetine collected before and 8 weeks after treatment (Lopez, Fiori, et al., 2017). Using complementary strategies, they demonstrated that levels of miR146a-5p, miR-146b-5p, miR-24-3p, and miR-425-3p are modified as a function of response to antidepressant treatment, and regulate genes involved in mitogen-activated protein kinase (MAPK) and Wnt signaling pathways that encode a large family of proteins that play key roles as intercellular signaling molecules in development, such as the brain-derived neurotrophic factor (BDNF).

Histone modifications: Epigenetic marking of the genome modulates the binding of transcription activators and repressors or the conformation of chromatin, the DNA packaging around histones. Histone proteins stably associate with DNA to form a repeating structural unit, the chromatin that organizes the genome. Chemical modification of histones modulates the structure of chromatin, and therefore the binding of DNA readers, writers, or erasers. Histone modifications are involved in both silencing and promoting DNA transcription, and include covalent modifications such as methylation, acetylation, or glycosylation (Weaver et al., 2017). Epigenetic modification of histones can be either activating or repressive, depending on the affected amino acid (Cholewa-Waclaw et al., 2016; Quina, Buschbeck, & Di Croce, 2006).

As histone modifications are transient in nature, DNA methylation as a relatively stable modification of the DNA is among the most promising mechanisms to be targeted for biomarker development (Wiedemann, 2011). Almost all of the enzymes involved in these mechanisms have been tested as targets for antidepressant therapy and, in preclinical studies, inhibitors of both DNMTs as well as HDACs have shown antidepressant properties. Some classical antidepressants such as the tricyclic imipramine or the mood stabilizer valproic acid also have HDAC-inhibiting properties. Such "epigenetic" antidepressants are believed to be one of the hottest topics for future therapy development in major depression (for review see Schroeder, Hillemacher, Bleich, & Frieling, 2012).

2 DNA methylation, stress, and depression

DNA methylation mechanisms are required for normal placental and brain development. Alterations in these epigenetic mechanisms are associated with pregnancy complications and low infant birth weight (Filiberto et al., 2014; Kulkarni et al., 2011).

Moreover, stress has a profound impact on DNA methylation. In 2009, Roth and coworkers observed in infant rats that brief and repeated experiences with a stressed dam outside the home cage (maltreatment) alters methylation of DNA within the brain-derived neurotrophic factor BDNF gene in the developing and adult prefrontal cortex (Roth, Lubin, Funk, & Sweatt, 2009). BDNF is a key mediator of activity-dependent processes that have a profound influence on neural development and plasticity. Their results showed long-lasting increase in the BDNF DNA methylation pattern in the adult prefrontal cortex.

Unternaehrer and colleagues examined whether acute psychological stress in humans causes changes in DNA methylation of genes related to brain plasticity and endocrine regulation (Unternaehrer et al., 2012). The study included the results for 76 participants at the age of 61–67 years that underwent the Trier social stress test. Comparisons of quantitative DNA methylation patterns in whole blood before, and 10 and 90 min after the experience of stress revealed increased methylation of the oxytocin receptor (Oxtr) gene pre-stress compared with 10 min poststress, and decreased methylation from 10 min poststress compared with 90 min poststress.

These observations reflect the dynamic nature of DNA methylation regulation in response to psychological stress, an important risk factor for depression (Kendler, Karkowski, & Prescott, 1999).

Furthermore, a genome-wide DNA methylation scan in 39 postmortem frontal cortex MDD samples identified 224 candidate regions with DNA methylation differences >10% (Sabunciyan et al., 2012).

These regions are highly enriched for neuronal growth and development (Sabunciyan et al., 2012). Subsequent DNA methylation studies focused on candidate genes such as brain-derived neurotrophic factor (BDNF) and serotonin transporter (Pishva, Rutten, & van den Hove, 2017).

3 Brain-derived neurotrophic factor and depression

Brain-derived neurotrophic factor (BDNF), a member of the neurotrophic factor family, is a key player in the neurotropic hypothesis of depression and binds to Tropomyosin receptor kinase B (TrkB) receptor (Lopez et al., 2013; Lopez, Pereira, et al., 2017). BDNF has been regarded as an important regulator in the growth, differentiation, maintenance, death/survival, and plasticity of neurons. BDNF protein levels measured in the plasma or serum of patients have been largely tested as potential markers for depression. A number of studies have linked BDNF with both the pathophysiology of depression and the mode of action of antidepressive treatments (Kleimann et al., 2015; Krishnan & Nestler, 2008; Youssef et al., 2018). For example, Fernandes et al. tested the value of BDNF levels as state markers of depression in a recent metaanalysis showing that BDNF levels are consistently reduced during depressive, and also during manic episodes (Fernandes et al., 2011).There is evidence that the reduction of BDNF protein levels in the brain may cause depression.

This "BDNF hypothesis" is supported by a large amount of post-mortem studies showing that depressed patients had lower BDNF levels in brain regions such as the amygdala, anterior cingulate cortex (ACC), caudal brainstem (pons), and hippocampus compared with nondepressed subjects (Banerjee, Ghosh, Ghosh, Bhattacharyya, & Mondal, 2013; Dwivedi et al., 2003; Pandey, 2013; Pandey et al., 2010; Sarchiapone et al., 2008; Tripp et al., 2012; Youssef et al., 2018). Additionally, lower BDNF levels in ACC were found in subjects who had been exposed to early life adversity and/or died by suicide compared with nonsuicide descendants and no reported childhood adversity (Youssef et al., 2018). On the other hand, chronic treatment with antidepressants increases BDNF-mediated signaling (Castren & Kojima, 2017; Krishnan & Nestler, 2008; Monteggia et al., 2004).

Reduced BDNF levels have been also observed in the blood of depressed patients, and these low levels can be reversed following depression treatment (Molendijk et al., 2011; Molendijk et al., 2014; Sen, Duman, & Sanacora, 2008). A nonincrease of serum BDNF seems to be associated with nonresponse to antidepressive treatment (Dreimuller et al., 2012; Tadic et al., 2011). Tadic´ and colleagues have shown that the nonincrease of BDNF in serum or plasma during the first week of antidepressant treatment predicts the final nonresponse and nonremission with high sensitivity, suggesting that early changes in peripheral BDNF may constitute or reflect a necessary prerequisite for final treatment response (Dreimuller et al., 2012; Tadic et al., 2011). It is noteworthy that some studies observed contradicting results in peripheral BDNF levels and depression, for example, in response to electroconvulsive therapy (Gedge et al., 2012; Marano et al., 2007; Rapinesi et al., 2015; Skibinska et al., 2018; Tadic et al., 2014).

4 Methylation patterns as biomarkers for depression

Few studies contributed to the epigenetic profiling of MDD via the analysis of genome-wide DNA methylation patterns. Uddin and co-workers were able to distinguish between healthy controls and patients with major depression using genome-wide methylation profiling (Uddin et al., 2011). Using whole blood-derived genomic DNA from a subset of participants in the Detroit Neighborhood Health Study (DNHS), they applied methylation microarrays to assess genome-wide methylation profiles for more than 14,000 genes in 33 persons who reported a lifetime history of depression and 67 nondepressed adults. Among the genes uniquely methylated or unmethylated in depression were genes from developmental and inflammatory pathways. To our knowledge, these were the first findings of potential DNA methylation biomarkers for depression.

A recent study on methylome-wide associations comprised the biggest sample of MDD patients ($N = 1132$) so far, using a sequencing-based approach that provides near-complete coverage of all 28 million common CpGs in the human genome (Tsankova et al., 2006). Aberg and colleagues could identify a significant overlap between blood and brain tissues ($N = 61$ samples from Brodmann Area 10, BA10).

Interestingly, one of the genes showing an overlap between blood and brain methylation with a distinct methylation pattern in MDD was BDNF.

The complex regulation of BDNF expression underlies epigenetic mechanisms, especially histone modification and DNA methylation (Hing et al., 2012; Kleimann et al., 2015; Tsankova et al., 2006). A number of studies showed that DNA hypermethylation of BDNF not only plays an important role in the pathophysiology of depression, but is also a promising biomarker for diagnosis and prediction of therapeutic response (Fuchikami et al., 2011; Kleimann et al., 2015; Lopez et al., 2013). For example, depression was associated with higher BDNF methylation levels, and BDNF protein levels are reduced in the brain and serum of depressed patients due to DNA methylation of BDNF promoters.

The validity of methylation patterns as biomarkers for depression was further supported by a Japanese study (Fuchikami et al., 2011). Fuchikami et al. investigated DNA methylation of BDNF exons 1 and 4 in the blood of depressed patients and controls, and detected a significant increase in the overall methylation levels of the BDNF exon I promoter in the MDD group (Fuchikami et al., 2011). Moreover, the majority of individual CpG sites (71%) were hypermethylated in the MDD group, confirming a similar pattern. A 2-dimensional hierarchical cluster analysis showed that the observed classification completely matched the clinical diagnosis. The exon 1 region of the BDNF gene was also assessed in a European multicenter study on treatment-resistant depression in 207 MDD patients and 278 controls (Carlberg et al., 2015). Carlberg et al. succeeded in replicating that the CpG site was hypermethylated in patients with MDD.

5 BDNF methylation and antidepressant response

Chen et al. showed in the post-mortem brain of depressed subjects with or without a history of antidepressant treatment compared with controls an increased expression of BDNF-IV (Chen, Ernst, & Turecki, 2011). Responders to treatment with antidepressants displayed a decrease in BDNF exon IV-related trimethylation at histone H3 (Lopez et al., 2013). D'Addario et al. investigated epigenetic mechanisms and the role of BDNF in a group of patients with MDD ($n = 41$) on stable pharmacologic treatment, along with an age-matched group of healthy control subjects ($n = 44$) (D'Addario et al., 2013). The researchers observed a significant reduction of BDNF gene expression in MDD subjects. Additionally, they reported an increase of DNA methylation at BDNF gene promoters in MDD patients compared with control subjects. Data stratified on the basis of pharmacotherapy revealed that patients on antidepressant drugs alone (i.e., selective serotonin and selective norepinephrine reuptake inhibitors) showed higher methylation levels in the promoter region of the BDNF gene compared with patients receiving antidepressant combination therapies with additional mood stabilizers (D'Addario et al., 2013).

In 2014, our working group succeeded in reporting the first preliminary evidence that patients with MDD ($n = 46$) showing hypomethylation of the promoter region of the BDNF gene are unlikely to benefit from antidepressant pharmacotherapy (Tadic et al., 2014). Furthermore, we could show that the methylation status of a single Cytosine-phosphatidyl-Guanine (CpG) site within the promoter region of BDNF exon IV had the same predictive properties when measured before initiation of the antidepressant treatment (Tadic et al., 2014). Patients carrying no methylated copy of the CpG-87 (relative to the first base of exon IV) were unlikely to respond to the treatment during a six-week therapy trial (OR: 9.0). Methylation status of this single CpG also predicted the changes of BDNF plasma levels during the first week of treatment: BDNF plasma levels only increased in those patients showing any methylation at the -87 locus, while in those patients without any methylation, BDNF levels tended to decrease upon antidepressant treatment.

These findings were confirmed in vitro. Employing a reporter-gene assay, we were able to show that 48-h incubation with fluoxetine or venlafaxine only leads to an increase in BDNF expression when the BDNF exon IV promoter fragment investigated was methylated. Antidepressants can also increase BDNF expression via phosphorylation of methyl-CpG-binding protein 2 (MeCP2) (Hutchinson, Deng, Cohen, & West, 2012). Thus, as methylation of the promoter is a prerequisite for specific MeCP2 binding, the results suggest that this mechanism of antidepressant action on BDNF can only be active in carriers of methylation at the relevant CpG site within the promoter. To our knowledge, this is the first report of an epigenetic biomarker of response to antidepressant treatment.

The importance of the epigenetic regulation of BDNF in response to antidepressants is further supported by a prospective study from Lopez and colleagues in 25 treatment-naive MDD patients. The researchers analyzed changes in the methylation of the histone 3 lysine 27 (H3K27me3) associated with BDNF exon IV, where a trimethylation also represses the transcription of the gene (Lopez et al., 2013). Depressed patients responding to an 8-week citalopram treatment exhibited a significant decrease in H3K27 trimethylation of BDNF IV, leading to a significant difference in H3K27me3 expression between responders and nonresponders after 8 weeks of treatment.

Recently, Wang et al. confirmed that the baseline DNA methylation status of BDNF could predict antidepressant treatment response (Wang et al., 2018). The researchers followed 85 depressed Chinese Han patients for 8 weeks after initiating treatment with citalopram. Treatment response was assessed by changes in the Hamilton Depression Rating Scale-17 (HAMD-17) score. Lower mean BDNF DNA methylation was associated with impaired antidepressant response. These findings are harder to interpret, as only mean methylation levels of a small promoter fragment are reported, but show the same tendency as our findings.

Recently, we succeeded with a replication of our initial finding that CpG-87 hypomethylation predicts final nonresponse and nonremission to antidepressant treatment in 126 severely depressed patients taken from the Early Medication Change (EMC) trial conducted by Lieb and Tadic. Patients without a methylated copy of CpG-87 were more than three times less likely to achieve remission upon therapy with escitalopram, venlafaxine, or lithium (Lieb et al. European Psychiatry, accepted for publication). Thus, our BDNF marker is not only the first epigenetic biomarker useful for deciding on the optimal therapeutic strategy—it is also the first epigenetic marker in depression that was replicated in a large prospective trial.

6   BDNF modifications and response to ECT

Furthermore, we conducted a prospective study with patients ($n = 11$) suffering from treatment-resistant major depressive disorder (MDD) investigating whether the promoter methylation of BDNF exons (I, IV and VI) is changed in depression during a series of electroconvulsive therapy (ECT) treatment, and whether there is an association between serum levels of BDNF and the investigated changes in the methylation of the CpG sites (Kleimann et al., 2015). A previous study in a rat model by Tsankova and co-workers has already indicated that chronic upregulation of BDNF transcription after chronic ECT, one of the most effective treatments of major depression, may be mediated through histone-3 acetylation (Tsankova, Kumar, & Nestler, 2004). Their results provided the first in vivo demonstration of the involvement of chromatin remodeling in ECT-induced regulation of gene expression in the brain and treatment of depression. We investigated BDNF expression and promoter methylation of BDNF exon I, IV, and VI in fasting blood samples directly before, 1, and 24 h after ECT sessions in a small pilot study (Kleimann et al., 2015). At the end of the ECT series, four patients were in remission, and six patients had responded to ECT. Remitters had a significantly lower mean methylation rate compared with nonremitters. Additionally, responders had a significantly lower mean promoter methylation rate compared with nonresponders. Analysis of the individual promoter methylation rates revealed that this effect was mainly driven by the promoter methylation of exon I, which was significantly lower in remitters/responders than in nonremitters or nonresponders. We found no significant change in the methylation rates of all investigated BDNF promoters or the BDNF serum levels during the ECT course.

7   Upregulation of BDNF signaling and antidepressant treatment response to ECT

S100A10 (p11) has been implicated in the etiology of depression and in mediating the antidepressant actions of selective serotonin reuptake inhibitors (SSRIs) involving an upregulation BDNF signaling (Svenningsson et al., 2006; Svenningsson, Kim, Warner-Schmidt, Oh, & Greengard, 2013). The levels of p11 mRNA and protein in the brain are down-regulated in depressed humans and in an animal model of depression (Alexander et al., 2010). In a further study, we hypothesized that the promoter methylation of p11 is differently regulated in ECT responders versus nonresponders, and thus could be a putative biomarker of ECT response (Neyazi et al., 2018).

The aim of the study was to identify potential biomarkers for ECT response by developing a valid rat model of pharmacoresistant depression, and then translate the experimental findings to depressive patients in a proof-of-concept (POC) clinical study. In the animal model, we observed that the gene promoter methylation and expression of p11 significantly correlate with the antidepressant effect of ECT. Moreover, we investigated the predictive properties of p11 promoter methylation in two clinical cohorts of patients with pharmacoresistant MDD. In a proof-of-concept clinical trial in 11 patients with refractory MDD, we demonstrated higher p11 promoter methylation in responders to ECT. This finding was replicated in an independent sample of 65 patients with pharmacoresistant MDD. This translational study successfully validated the first molecular biomarker reliably predicting the responsiveness to ECT. Prescreening of this biomarker could help to identify patients eligible for first-line ECT treatment, and also help to develop novel antidepressant treatment procedures for depressed patients resistant to all currently approved antidepressant treatments. For the first time, we were able to show that a translationally and clinically valid biomarker can distinguish between responders and nonresponders of ECT treatment before the first treatment is initiated, which allows for an individualization of treatment strategies in pharmacoresistant depression.

Further studies are needed to elucidate additional markers increasing the predictive power to distinguish between responders and nonresponders to antidepressant treatment, and to identify other treatment options effective for antidepressant-resistant patients.

8 Conclusion and outlook

Epigenetic markers of treatment response are promising biomarkers. In the near future, epigenetic markers might help to biologically define subtypes of depression and guide antidepressant therapies. However, the study of epigenetic markers in psychiatry is still in its infancy, and the current development partially resembles that of genetic testing, with small studies focusing on single candidate genes. The availability of genome-wide screening methods will help to overcome this limitation in due time.

Another obstacle for epigenetic testing in psychiatry is the problem of obtaining measurements from peripheral tissues or fluids such as blood, buccal cells, or saliva, and assuming that similar changes will be detectable in the central nervous system. However, during the past decade, several studies have shown the validity of this approach, making the concept of liquid biopsy also fruitful and promising for study of psychiatric illness.

Rapid development of new sequencing technologies will enable the analysis of genomic and epigenomic patterns associated with the response to certain therapies. Future clinical trials should therefore always include pharmaco-(epi-)genetic testing. Large cohorts with excellent phenotyping and characterization of treatment steps will be needed to enable the use of algorithms detecting patterns specifically associated with treatment response or remission upon a certain treatment.

In conclusion, epigenetic research holds the potential to enable subtyping and response prediction in MDD as well as the facilitation of new drug developments, and thus personalized medicine in psychiatry.

References

Alexander, B., Warner-Schmidt, J., Eriksson, T., Tamminga, C., Arango-Lievano, M., Ghose, S., et al. (2010). Reversal of depressed behaviors in mice by p11 gene therapy in the nucleus accumbens. *Science Translational Medicine, 2*(54). 54ra76.

Amick, H. R., Gartlehner, G., Gaynes, B. N., Forneris, C., Asher, G. N., Morgan, L. C. , et al. (2015). Comparative benefits and harms of second generation antidepressants and cognitive behavioral therapies in initial treatment of major depressive disorder: Systematic review and meta-analysis. *BMJ, 351*, h6019.

Anier, K., Malinovskaja, K., Pruus, K., Aonurm-Helm, A., Zharkovsky, A., & Kalda, A. (2014). Maternal separation is associated with DNA methylation and behavioural changes in adult rats. *European Neuropsychopharmacology: The Journal of the European College of Neuropsychopharmacology, 24* (3), 459–468.

Babenko, O., Kovalchuk, I., & Metz, G. A. (2015). Stress-induced perinatal and transgenerational epigenetic programming of brain development and mental health. *Neuroscience and Biobehavioral Reviews, 48*, 70–91.

Banerjee, R., Ghosh, A. K., Ghosh, B., Bhattacharyya, S., & Mondal, A. C. (2013). Decreased mRNA and protein expression of BDNF, NGF, and their receptors in the hippocampus from suicide: An analysis in human postmortem brain. *Clinical Medicine Insights Pathology, 6*, 1–11.

Belzeaux, R., Lin, R., Ju, C., Chay, M. A., Fiori, L. M., Lutz, P. E. , et al. (2018). Transcriptomic and epigenomic biomarkers of antidepressant response. *Journal of Affective Disorders, 233*, 36–44.

Breitenstein, B., Scheuer, S., & Holsboer, F. (2014). Are there meaningful biomarkers of treatment response for depression? *Drug Discovery Today, 19*(5), 539–561.

Carlberg, L., Schosser, A., Calati, R., Serretti, A., Massat, I., Papageorgiou, K. , et al. (2015). Association study of CREB1 polymorphisms and suicidality in MDD: Results from a European multicenter study on treatment resistant depression. *The International Journal of Neuroscience, 125*(5), 336–343.

Castren, E., & Kojima, M. (2017). Brain-derived neurotrophic factor in mood disorders and antidepressant treatments. *Neurobiology of Disease, 97*(Pt B), 119–126.

Chen, E. S., Ernst, C., & Turecki, G. (2011). The epigenetic effects of antidepressant treatment on human prefrontal cortex BDNF expression. *The International Journal of Neuropsychopharmacology, 14*(3), 427–429.

Chen, D., Meng, L., Pei, F., Zheng, Y., & Leng, J. (2017). A review of DNA methylation in depression. *Journal of Clinical Neuroscience: Official Journal of the Neurosurgical Society of Australasia, 43*, 39–46.

Cholewa-Waclaw, J., Bird, A., von Schimmelmann, M., Schaefer, A., Yu, H., Song, H. , et al. (2016). The role of epigenetic mechanisms in the regulation of gene expression in the nervous system. *The Journal of Neuroscience: The Official Journal of the Society for Neuroscience, 36*(45), 11427–11434.

Condorelli, G., Latronico, M. V., & Cavarretta, E. (2014). microRNAs in cardiovascular diseases: Current knowledge and the road ahead. *Journal of the American College of Cardiology, 63*(21), 2177–2187.

D'Addario, C., Dell'Osso, B., Galimberti, D., Palazzo, M. C., Benatti, B., Di Francesco, A. , et al. (2013). Epigenetic modulation of BDNF gene in patients with major depressive disorder. *Biological Psychiatry, 73*(2), e6–e7.

DALYs GBD, Collaborators H. (2017). Global, regional, and national disability-adjusted life-years (DALYs) for 333 diseases and injuries and healthy life expectancy (HALE) for 195 countries and territories, 1990–2016: A systematic analysis for the Global Burden of Disease Study 2016. *Lancet (London, England), 390*(10100), 1260–1344.

Devlin, A. M., Brain, U., Austin, J., & Oberlander, T. F. (2010). Prenatal exposure to maternal depressed mood and the MTHFR C677T variant affect SLC6A4 methylation in infants at birth. *PLoS One, 5*(8).

Dreimuller, N., Schlicht, K. F., Wagner, S., Peetz, D., Borysenko, L., Hiemke, C. , et al. (2012). Early reactions of brain-derived neurotrophic factor in plasma (pBDNF) and outcome to acute antidepressant treatment in patients with major depression. *Neuropharmacology, 62*(1), 264–269.

Dwivedi, Y. (2017). microRNA-124: A putative therapeutic target and biomarker for major depression. *Expert Opinion on Therapeutic Targets, 21*(7), 653–656.

Dwivedi, Y., Rizavi, H. S., Conley, R. R., Roberts, R. C., Tamminga, C. A., & Pandey, G. N. (2003). Altered gene expression of brain-derived neurotrophic factor and receptor tyrosine kinase B in postmortem brain of suicide subjects. *Archives of General Psychiatry, 60*(8), 804–815.

Fatima, M., Srivastav, S., & Mondal, A. C. (2017). Prenatal stress and depression associated neuronal development in neonates. *International Journal of Developmental Neuroscience, 60*, 1–7.

Fernandes, B. S., Gama, C. S., Cereser, K. M., Yatham, L. N., Fries, G. R., Colpo, G. , et al. (2011). Brain-derived neurotrophic factor as a state-marker of mood episodes in bipolar disorders: A systematic review and meta-regression analysis. *Journal of Psychiatric Research, 45*(8), 995–1004.

Filiberto, A. C., Maccani, M. A., Koestler, D. C., Wilhelm-Benartzi, C., Avissar-Whiting, M., Banister, C. E. , et al. (2014). Birthweight is associated with DNA promoter methylation of the glucocorticoid receptor in human placenta. *Epigenetics, 6*(5), 566–572.

Fuchikami, M., Morinobu, S., Segawa, M., Okamoto, Y., Yamawaki, S., Ozaki, N. , et al. (2011). DNA methylation profiles of the brain-derived neurotrophic factor (BDNF) gene as a potent diagnostic biomarker in major depression. *PLoS One, 6*(8).

Gedge, L., Beaudoin, A., Lazowski, L., du Toit, R., Jokic, R., & Milev, R. (2012). Effects of electroconvulsive therapy and repetitive transcranial magnetic stimulation on serum brain-derived neurotrophic factor levels in patients with depression. *Frontiers in Psychiatry, 3*, 12.

Hing, B., Davidson, S., Lear, M., Breen, G., Quinn, J., McGuffin, P. , et al. (2012). A polymorphism associated with depressive disorders differentially regulates brain derived neurotrophic factor promoter IV activity. *Biological Psychiatry, 71*(7), 618–626.

Hutchinson, A. N., Deng, J. V., Cohen, S., & West, A. E. (2012). Phosphorylation of MeCP2 at Ser421 contributes to chronic antidepressant action. *The Journal of Neuroscience: The Official Journal of the Society for Neuroscience, 32*(41), 14355–14363.

Issler, O., Haramati, S., Paul, E. D., Maeno, H., Navon, I., Zwang, R. , et al. (2014). MicroRNA 135 is essential for chronic stress resiliency, antidepressant efficacy, and intact serotonergic activity. *Neuron, 83*(2), 344–360.

Jurkowska, R. Z., & Jeltsch, A. (2016). Mechanisms and biological roles of DNA methyltransferases and DNA methylation: From past achievements to future challenges. *Advances in Experimental Medicine and Biology, 945*, 1–17.

Kang, H. J., Kim, J. M., Stewart, R., Kim, S. Y., Bae, K. Y., Kim, S. W. , et al. (2013). Association of SLC6A4 methylation with early adversity, characteristics and outcomes in depression. *Progress in Neuro-Psychopharmacology & Biological Psychiatry, 44*, 23–28.

Kendler, K. S., Karkowski, L. M., & Prescott, C. A. (1999). Causal relationship between stressful life events and the onset of major depression. *The American Journal of Psychiatry, 156*(6), 837–841.

Kleimann, A., Kotsiari, A., Sperling, W., Groschl, M., Heberlein, A., Kahl, K. G. , et al. (2015). BDNF serum levels and promoter methylation of BDNF exon I, IV and VI in depressed patients receiving electroconvulsive therapy. *Journal of Neural Transmission (Vienna, Austria: 1996), 122*(6), 925–928.

Krishnan, V., & Nestler, E. J. (2008). The molecular neurobiology of depression. *Nature, 455*, 894.

Kulkarni, A., Dangat, K., Kale, A., Sable, P., Chavan-Gautam, P., & Joshi, S. (2011). Effects of altered maternal folic acid, vitamin B12 and docosahexaenoic acid on placental global DNA methylation patterns in Wistar rats. *PLoS One, 6*(3).

Labermaier, C., Masana, M., & Muller, M. B. (2013). Biomarkers predicting antidepressant treatment response: How can we advance the field? *Disease Markers, 35*(1), 23–31.

Lisoway, A. J., Zai, C. C., Tiwari, A. K., & Kennedy, J. L. (2018). DNA methylation and clinical response to antidepressant medication in major depressive disorder: A review and recommendations. *Neuroscience Letters, 669*, 14–23.

Lopez, J. P., Fiori, L. M., Cruceanu, C., Lin, R., Labonte, B., Cates, H. M. , et al. (2017). MicroRNAs 146a/b-5 and 425-3p and 24-3p are markers of antidepressant response and regulate MAPK/Wnt-system genes. *Nature Communications, 8*, 15497.

Lopez, J. P., Mamdani, F., Labonte, B., Beaulieu, M. M., Yang, J. P., Berlim, M. T. , et al. (2013). Epigenetic regulation of BDNF expression according to antidepressant response. *Molecular Psychiatry*, *18*(4), 398–399.

Lopez, J. P., Pereira, F., Richard-Devantoy, S., Berlim, M., Chachamovich, E., Fiori, L. M. , et al. (2017). Co-variation of peripheral levels of miR-1202 and brain activity and connectivity during antidepressant treatment. *Neuropsychopharmacology: Official Publication of the American College of Neuropsychopharmacology*, *42*(10), 2043–2051.

Marano, C. M., Phatak, P., Vemulapalli, U. R., Sasan, A., Nalbandyan, M. R., Ramanujam, S. , et al. (2007). Increased plasma concentration of brain-derived neurotrophic factor with electroconvulsive therapy: A pilot study in patients with major depression. *The Journal of Clinical Psychiatry*, *68*(4), 512–517.

Mendell, J. T., & Olson, E. N. (2012). MicroRNAs in stress signaling and human disease. *Cell*, *148*(6), 1172–1187.

Molendijk, M. L., Bus, B. A., Spinhoven, P., Penninx, B. W., Kenis, G., Prickaerts, J. , et al. (2011). Serum levels of brain-derived neurotrophic factor in major depressive disorder: State-trait issues, clinical features and pharmacological treatment. *Molecular Psychiatry*, *16*(11), 1088–1095.

Molendijk, M. L., Spinhoven, P., Polak, M., Bus, B. A., Penninx, B. W., & Elzinga, B. M. (2014). Serum BDNF concentrations as peripheral manifestations of depression: Evidence from a systematic review and meta-analyses on 179 associations (N = 9484). *Molecular Psychiatry*, *19*(7), 791–800.

Monteggia, L. M., Barrot, M., Powell, C. M., Berton, O., Galanis, V., Gemelli, T. , et al. (2004). Essential role of brain-derived neurotrophic factor in adult hippocampal function. *Proceedings of the National Academy of Sciences of the United States of America*, *101*(29), 10827–10832.

Murgatroyd, C., Patchev, A. V., Wu, Y., Micale, V., Bockmuhl, Y., Fischer, D. , et al. (2009). Dynamic DNA methylation programs persistent adverse effects of early-life stress. *Nature Neuroscience*, *12*(12), 1559–U108.

Narahari, A., Hussain, M., & Sreeram, V. (2017). MicroRNAs as biomarkers for psychiatric conditions: A review of current research. *Innovations in Clinical Neuroscience*, *14*(1–2), 53–55.

Nelson, P. T., Dimayuga, J., & Wilfred, B. R. (2010). MicroRNA in situ hybridization in the human entorhinal and transentorhinal cortex. *Frontiers in Human Neuroscience*, *4*, 7.

Neyazi, A., Theilmann, W., Brandt, C., Rantamaki, T., Matsui, N., Rhein, M. , et al. (2018). P11 promoter methylation predicts the antidepressant effect of electroconvulsive therapy. *Translational Psychiatry*, *8*(1), 25.

Pandey, G. N. (2013). Biological basis of suicide and suicidal behavior. *Bipolar Disorders*, *15*(5), 524–541.

Pandey, G. N., Dwivedi, Y., Rizavi, H. S., Ren, X., Zhang, H., & Pavuluri, M. N. (2010). Brain-derived neurotrophic factor gene and protein expression in pediatric and adult depressed subjects. *Progress in Neuro-Psychopharmacology & Biological Psychiatry*, *34*(4), 645–651.

Pishva, E., Rutten, B. P. F., & van den Hove, D. (2017). DNA methylation in major depressive disorder. *Advances in Experimental Medicine and Biology*, *978*, 185–196.

Quina, A. S., Buschbeck, M., & Di Croce, L. (2006). Chromatin structure and epigenetics. *Biochemical Pharmacology*, *72*(11), 1563–1569.

Rapinesi, C., Kotzalidis, G. D., Curto, M., Serata, D., Ferri, V. R., Scatena, P. , et al. (2015). Electroconvulsive therapy improves clinical manifestations of treatment-resistant depression without changing serum BDNF levels. *Psychiatry Research*, *227*(2–3), 171–178.

Rodenhiser, D., & Mann, M. (2006). Epigenetics and human disease: Translating basic biology into clinical applications. *CMAJ: Canadian Medical Association Journal = Journal de l'Association Medicale Canadienne*, *174*(3), 341–348.

Roth, T. L., Lubin, F. D., Funk, A. J., & Sweatt, J. D. (2009). Lasting epigenetic influence of early-life adversity on the BDNF gene. *Biological Psychiatry*, *65*(9), 760–769.

Sabunciyan, S., Aryee, M. J., Irizarry, R. A., Rongione, M., Webster, M. J., Kaufman, W. E. , et al. (2012). Genome-wide DNA methylation scan in major depressive disorder. *PLoS One*, *7*(4).

Sarchiapone, M., Carli, V., Roy, A., Iacoviello, L., Cuomo, C., Latella, M. C. , et al. (2008). Association of polymorphism (Val66Met) of brain-derived neurotrophic factor with suicide attempts in depressed patients. *Neuropsychobiology*, *57*(3), 139–145.

Schroeder, M., Hillemacher, T., Bleich, S., & Frieling, H. (2012). The epigenetic code in depression: Implications for treatment. *Clinical Pharmacology and Therapeutics*, *91*(2), 310–314.

Sen, S., Duman, R., & Sanacora, G. (2008). Serum brain-derived neurotrophic factor, depression, and antidepressant medications: Meta-analyses and implications. *Biological Psychiatry*, *64*(6), 527–532.

Skibinska, M., Kapelski, P., Rajewska-Rager, A., Pawlak, J., Szczepankiewicz, A., Narozna, B. , et al. (2018). Brain-derived neurotrophic factor (BDNF) serum level in women with first-episode depression, correlation with clinical and metabolic parameters. *Nordic Journal of Psychiatry*, *72*(3), 191–196.

Smalheiser, N. R., Lugli, G., Rizavi, H. S., Torvik, V. I., Turecki, G., & Dwivedi, Y. (2012). MicroRNA expression is down-regulated and reorganized in prefrontal cortex of depressed suicide subjects. *PLoS One*, *7*(3).

Smith, Z. D., & Meissner, A. (2013). DNA methylation: Roles in mammalian development. *Nature Reviews. Genetics*, *14*(3), 204–220.

Svenningsson, P., Chergui, K., Rachleff, I., Flajolet, M., Zhang, X., El Yacoubi, M. , et al. (2006). Alterations in 5-HT1B receptor function by p11 in depression-like states. *Science*, *311*(5757), 77–80.

Svenningsson, P., Kim, Y., Warner-Schmidt, J., Oh, Y. S., & Greengard, P. (2013). p11 and its role in depression and therapeutic responses to antidepressants. *Nature Reviews. Neuroscience*, *14*(10), 673–680.

Tadic, A., Muller-Engling, L., Schlicht, K. F., Kotsiari, A., Dreimuller, N., Kleimann, A. , et al. (2014). Methylation of the promoter of brain-derived neurotrophic factor exon IV and antidepressant response in major depression. *Molecular Psychiatry*, *19*(3), 281–283.

Tadic, A., Wagner, S., Schlicht, K. F., Peetz, D., Borysenko, L., Dreimuller, N. , et al. (2011). The early non-increase of serum BDNF predicts failure of antidepressant treatment in patients with major depression: A pilot study. *Progress in Neuro-Psychopharmacology & Biological Psychiatry*, *35*(2), 415–420.

Tripp, A., Oh, H., Guilloux, J. P., Martinowich, K., Lewis, D. A., & Sibille, E. (2012). Brain-derived neurotrophic factor signaling and subgenual anterior cingulate cortex dysfunction in major depressive disorder. *The American Journal of Psychiatry*, *169*(11), 1194–1202.

Tsankova, N. M., Berton, O., Renthal, W., Kumar, A., Neve, R. L., & Nestler, E. J. (2006). Sustained hippocampal chromatin regulation in a mouse model of depression and antidepressant action. *Nature Neuroscience*, *9*(4), 519–525.

Tsankova, N. M., Kumar, A., & Nestler, E. J. (2004). Histone modifications at gene promoter regions in rat hippocampus after acute and chronic electroconvulsive seizures. *The Journal of Neuroscience: The Official Journal of the Society for Neuroscience*, *24*(24), 5603–5610.

Uddin, M., Koenen, K. C., Aiello, A. E., Wildman, D. E., de los Santos, R., & Galea, S. (2011). Epigenetic and inflammatory marker profiles associated with depression in a community-based epidemiologic sample. *Psychological Medicine*, *41*(5), 997–1007.

Unternaehrer, E., Luers, P., Mill, J., Dempster, E., Meyer, A. H., Staehli, S. , et al. (2012). Dynamic changes in DNA methylation of stress-associated genes (OXTR, BDNF) after acute psychosocial stress. *Translational Psychiatry*, 2.

van der Harst, P., de Windt, L. J., & Chambers, J. C. (2017). Translational perspective on epigenetics in cardiovascular disease. *Journal of the American College of Cardiology*, *70*(5), 590–606.

von Wolff, A., Holzel, L. P., Westphal, A., Harter, M., & Kriston, L. (2013). Selective serotonin reuptake inhibitors and tricyclic antidepressants in the acute treatment of chronic depression and dysthymia: A systematic review and meta-analysis. *Journal of Affective Disorders*, *144*(1–2), 7–15.

Wang, S. S., Mu, R. H., Li, C. F., Dong, S. Q., Geng, D., Liu, Q. , et al. (2017). microRNA-124 targets glucocorticoid receptor and is involved in depression-like behaviors. *Progress in Neuro-Psychopharmacology & Biological Psychiatry*, *79*(Pt B), 417–425.

Wang, P., Zhang, C., Lv, Q., Bao, C., Sun, H., Ma, G. , et al. (2018). Association of DNA methylation in BDNF with escitalopram treatment response in depressed Chinese Han patients. *European Journal of Clinical Pharmacology*, *74*(8), 1011–1020.

Weaver, I. C. G., Cervoni, N., Champagne, F. A., D'Alessio, A. C., Sharma, S., Seckl, J. , et al. (2004). Epigenetic programming by maternal behavior. *Nature Neuroscience*, *7*(8), 847–854.

Weaver, I. C., Korgan, A. C., Lee, K., Wheeler, R. V., Hundert, A. S., & Goguen, D. (2017). Stress and the emerging roles of chromatin remodeling in signal integration and stable transmission of reversible phenotypes. *Frontiers in Behavioral Neuroscience*, *11*, 41.

Wiedemann, K. (2011). Biomarkers in development of psychotropic drugs. *Dialogues in Clinical Neuroscience*, *13*(2), 225–234.

Wilczynska, A., & Bushell, M. (2015). The complexity of miRNA-mediated repression. *Cell Death and Differentiation*, *22*(1), 22–33.

Wojciechowska, A., Braniewska, A., & Kozar-Kaminska, K. (2017). MicroRNA in cardiovascular biology and disease. *Advances in Clinical and Experimental Medicine*, *26*(5), 865–874.

Xing, S., Dipaula, B. A., Lee, H. Y., & Cooke, C. E. (2011). Failure to fill electronically prescribed antidepressant medications: A retrospective study. *The Primary Care Companion for CNS Disorders*, *13*(1), e1–e7.

Youssef, M. M., Underwood, M. D., Huang, Y. Y., Hsiung, S. C., Liu, Y., Simpson, N. R. , et al. (2018). Association of BDNF Val66Met polymorphism and brain BDNF levels with major depression and suicide. *The International Journal of Neuropsychopharmacology*, *21*(6), 528–538.

Chapter 31

Gene coexpression network and machine learning in personalized psychiatry

Liliana G. Ciobanu, Micah Cearns and Bernhard T. Baune

Discipline of Psychiatry, School of Medicine, The University of Adelaide, Adelaide, SA, Australia

1   Introduction

Despite the high heritability of psychiatric disorders as a class (46.3%) (Polderman et al., 2015), identifying the genes involved in psychiatric illness has remained a formidable challenge. Regardless, attempts to unlock the biological basis of psychiatric disorders in the hope of deriving biologically-based diagnostic tools and new effective drug therapies has continued. To date, genome-wide association studies (GWAS) have identified hundreds of genetic variants of small effect associated with psychiatric disorders (Ikeda et al. 2017; Ripke, Neale, Corvin, & Watters, 2014; Sklar et al., 2011; Wray & Sullivan, 2017), outlining the highly polygenic nature of psychiatric illness. Nevertheless, deriving clinically translatable insights that progress the field beyond symptomatic diagnosis remains a challenge. One suggested explanation for these findings could be due to our limited exploration of the interplay between genes at different levels of biological abstraction. For this reason, evidence suggests that psychiatric illness may be better formulated as a complex system. According to (Bar-Yam, 2002), a complex system is defined as a system whose behavior is inherently challenging to model due to relationships, dependencies, and interactions between their parts and a known system and its environment. Within a systems framework, a psychiatric disorder is most likely not explainable by variation within one gene or biological substrate, but better explained through the interaction of the environment, genes, and their subsequent expression. In this sense, the environment may interact with thousands of genes that are differentially expressed between individuals, resultantly, giving rise to emergent disorders that are greater than the sum of their individual parts. If this is the case, the elucidation of individual candidates will only go so far to explain emergent complex system disorders, an assertion consistent with current findings in psychiatric genetics (McClellan & King, 2010). With this in mind, methodologies are needed that can not only utilize the predictive capacity of individual genes, but subsequent linear, nonlinear and interactive relationships that may give rise to emergent complex system disorders. Classical statistical methods have attempted to achieve this, yet often, their parametric assumptions of independence appear to be directly at odds with the nature of a complex system. Furthermore, classical methods, such as univariate linear regression, struggle to deal with what is known as the "large p small n problem," a problem faced when there are many features (p) in a data set (for example, many genes), yet only a limited number of observations (n) (small sample size) relative to the feature space (West et al., 2001). Given these challenges, alternative methodological frameworks are needed that can both reduce the dimensionality of a genetic feature space, while also allowing for the prediction of clinically relevant outcomes.

Emerging evidence suggests that interactions between both genes and the environment can have a large impact on the phenotypic variability of psychiatric disorders (Kubota, Miyake, & Hirasawa, 2012). Given that gene expression is a product of genetic effects, environmental influences, and epigenetic modifications, studying alterations of mRNA levels in disease represents a promising approach to further our molecular understanding of psychiatric illness. However, a major challenge has remained. How is it that we go about measuring and identifying the expression of which genes are altered in disease specific cells, tissues, and brain regions?

Early studies measuring the differential expression of candidate genes pointed toward many potential targets for complex psychiatric traits. However, the replication of these findings has been limited due to methodological differences, inconsistencies in the diagnostic measures used, and underpowered study designs (Drago, De Ronchi, & Serretti, 2007). Nevertheless, advances in high-throughput technologies have helped facilitate a shift from hypothesis-driven approaches to less biased, data-driven methodologies. A substantial body of research has applied microarray and RNA-seq technologies to investigate disease-associated genome-wide expression alterations in both brain and peripheral tissues. This work has been

Personalized Psychiatry. https://doi.org/10.1016/B978-0-12-813176-3.00031-6

demonstrated in schizophrenia (Cattane et al., 2015; Sanders et al., 2017), bipolar disorder (for review, Seifuddin et al., 2013), and MDD (for review, Ciobanu et al., 2016). While these findings have improved our understanding of the pathophysiological mechanisms in psychiatric disorders, replication of identified candidates has remained a challenge. One proposed reason for the lack of consistent results between studies is attributed to the molecular complexity of psychiatric disorders, which commonly is further compounded by the small sample sizes often employed in this research. The capacity to detect multiple small effects from single molecules at the transcriptome-wide level places large sample size requirements on a study. Given the complex genetic architecture of psychiatric illness and the functional interdependencies between genes, emergent clinical phenotypes are likely to reflect the interactions within a complex network of molecular processes. Therefore, capturing the most important genes that orchestrate the molecular cascades leading to psychiatric pathophysiology is of utmost importance. However, given the methodological constraints discussed herein in traditional analysis, alternative methodologies are needed to further progress the field. A range of these methodologies will be discussed in the following sections.

2 Diagnosis

2.1 Coexpression network analysis to differentiate disease from healthy controls

Gene coexpression network analysis is a technique used to quantify the linear coexpression of multiple genes in relation to disease diagnosis. This methodology allows for effective dimensionality reduction of transcriptome data (decreasing the number of independent variables in a dataset), as well as clustering of interacting genes associated with a phenotype. It can be argued that if two (or more) genes are coexpressed, then the mechanisms regulating their expression must be either the same, or at least similar. Therefore, it is inferred that coexpressed genes are functionally related, and if associated with a disease status, are important contributors to a clinical phenotype.

2.2 Weighted gene coexpression network analysis

Gene clustering is a technique used in coexpression analysis that allows for the identification of subnetworks of convergent biological pathways. The Weighted Gene Coexpression Network Analysis (WGCNA) is one of the most established techniques used in gene clustering. This technique utilizes both microarray and RNA-seq data (Zhang & Horvath, 2005). WGCNA organizes transcriptome data by defining genes as nodes, and the relationships between nodes as edges. Biological networks tend to have a hierarchical structure, such that their nodes can be clustered together into fewer modules of highly interconnected genes. Intermodular connectivity reflects a higher-order structure of biological relationships within a gene network, while intramodular connectivity can identify which genes behave as central hubs in the modules. In coexpression networks, hubs are highly connected genes; therefore, being a hub is an indication of the importance of a gene in a module. Hubs are likely to be key molecular drivers that determine coexpression. Evidence also suggests that they may help to interpret a module, as they likely reflect its predominant biological role.

WGCNA has been successfully applied in many areas of medical research, including psychiatry. For example, WGCNA has been used to differentiate disease cases from healthy controls, and identify brain-based immune system dysregulation in schizophrenia (Mistry, Gillis, & Pavlidis, 2013), postsynaptic density implicated in the pathogenesis of bipolar disorder (Akula, Wendland, Choi, & McMahon, 2015), peripheral inflammation in depression (Malki, Tosto, Jumabhoy, & Lourdusamy, 2013), and transcriptional and splicing dysregulation as underlying mechanisms of neuronal dysfunction in autism spectrum disorder (ASD) (Voineagu et al., 2011). Coexpression network analysis has also been applied to explore overlap and specificity across different disorders compared with healthy controls. For example, comparing gene coexpression patterns between adult and childhood attention-deficit hyperactivity disorder (ADHD), autism spectrum disorder (ASD), major depressive disorder (MDD), and healthy controls, it has been found that immune system dysregulation is involved in both MDD and adult ADHD, and is inversely correlated with a disease status (de Jong et al., 2016). Using WGCNA, neuron differentiation and development pathways in the cerebral cortex have been discovered as potential contributors to the etiologies for both schizophrenia and bipolar disorder (Chen et al., 2013); while distinct molecular signatures have been found for ASD and intellectual disability (ID) (Parikshak et al., 2013). Recent work from Gandal et al. (2016) utilized WGCNA to compare molecular phenotypes across five major psychiatric disorders, including autism (ASD), schizophrenia (SCZ), bipolar disorder (BD), depression (MDD), and alcoholism (AAD). From this comparison, the authors were able to identify a clear pattern of shared and distinct gene-expression perturbations across all conditions. It has been found that neuronal gene coexpression modules were downregulated across ASD, SCZ, and BD, and astrocyte-related modules were predominantly upregulated in ASD and SCZ.

2.3 Differential coexpression network analysis to study differences among gene interconnections

Differential coexpression analysis is a tool that is used to investigate the differences among gene interconnections. This is achieved by calculating the expression correlation change of each gene pair between conditions. Genes that are differentially coexpressed between different conditions are more likely to be regulators, and thus, are likely to explain differences between phenotypes (Amar, Safer, & Shamir, 2013). By comparing the regulatory differences between cases and controls, specific differential networks of genes can be identified in psychiatric disorders. This methodology can be utilized to better understand the dynamic changes in gene regulatory networks in psychiatric illness, as well as comparing network properties across disorders. Thus, Xu et al. (2015) showed that mechanisms underlying MDD and subsyndromal symptomatic depression (SSD) were actually different. The authors found that there was no overlap between the MDD and SSD differentially regulated genes. Furthermore, the authors also found that venlafaxine appeared to have a significant effect on the gene expression profile of MDD patients, but no significant effect on the gene expression profiles of SSD patients. For more information on differential coexpression network analysis, see Hsu, Juan, and Huang (2015).

3 Diagnosis of patient subtypes

3.1 Biclustering for identifying subgroups of patients

Although gene coexpression clustering algorithms have proven useful for studying the molecular complexity of psychiatric disorders, the dominant clustering methodology discussed in the previous section is limited in its ability to detect gene expression patterns that are specific to subgroups of patients. An alternative approach called biclustering may be able to overcome this limitation. Biclustering is an alternative method for detecting differentially coexpressed genes between subgroups of patients, allowing for patient stratification into unique biological subgroups. Biclustering algorithms perform clustering without the need for prior sample group classification, a beneficial characteristic given the uncertainty of diagnostic boundaries across psychiatric disorders. Using this approach, Cha, Hwang, Oh, and Yi (2015) explored the shared molecular basis of five neurodegenerative diseases (Alzheimer's disease, Parkinson's disease, Huntington's disease, multiple sclerosis, amyotrophic lateral sclerosis) and three psychiatric disorders (SCZ, BP, and ASD). The authors found that 4307 genes were coexpressed in multiple brain diseases, while 3409 gene sets were exclusively specified in individual brain diseases. Using the same approach in the field of oncology, Fiannaca, La Rosa, La Paglia, Rizzo, and Urso (2015) were able to identify unique molecular subgroups of breast cancer tumors based on patients' miRNA expression profiles. Interestingly enough, all patients presented with the same clinical profiles. These findings are beneficial for clinical practice, as they may allow for the molecular stratification of both diagnosis and treatment of disease, thus allowing clinicians to tailor treatment strategies to individual patients. At present, biclustering methods show promise to help further elucidate the complexity and biological heterogeneity of complex psychiatric traits. For an extensive overview of the different biclustering algorithms available, refer to (Pontes, Giráldez, & Aguilar-Ruiz, 2015).

4 Personalized treatment

4.1 Gene coexpression networks for understanding treatment response biotypes

Understanding the biological basis of why some patients respond to certain therapies and others do not is essential for advancing personalized care in psychiatry. Due to the ability of gene coexpression analysis to approximate the complex interactions of biological information, it can be argued that this technique may help to inform the isolation of unique treatment response biotypes. Systematic characterization of changes in gene coexpression patterns in responders versus nonresponders may provide biological grounds for developing predictive models that help to minimize therapeutic uncertainty. Furthermore, it may help to reveal novel mechanisms of action that remain unidentified in commonly used psychiatric drugs. Support for this notion has already been demonstrated in the field of oncology to delineate responders and nonresponders to cancer treatment (Hsiao et al., 2016; Yang et al., 2014). In the field of psychiatry, coexpression network analysis of peripheral blood has identified immune-related pathways as important predispositions to antidepressant treatment response in MDD (Belzeaux et al., 2016). Blood-derived gene expression signature was found predictive of clozapine monotherapy in psychosis (Harrison et al., 2016). Furthermore, several gene coexpression modules in patient-derived lymphoblastoid cell lines were discovered as lithium-responsive, indicating widespread effects of lithium on diverse cellular signaling systems, including apoptosis and defense response pathways, protein processing, and response to endoplasmic reticulum stress in bipolar disorder (Breen et al., 2016). As these studies suggest, applying coexpression-based network methods across different disorders and medications can help to yield important insights on the molecular interactions regulating treatment response in complex psychiatric traits.

5 Disease mechanisms

5.1 Gene coexpression network for integrative analyses in psychiatry

Due to the multiple testing burden, large samples sizes are required to detect disrupted genes and pathways when multiple biological processes are analyzed in unison (Chari et al., 2010). Gene coexpression network analysis may help to facilitate this goal by providing an endpoint for the quantification of such processes. For example, Parikshak et al. (2013) found that by intersecting coexpression modules with GWAS loci, they were able to identify ASD genes that tightly coalesced with modules implicated in distinct biological functions during human cortical development, including early transcriptional regulation and synaptic development. Furthermore, when modules of a network are combined with epigenetic information, we can substantially enrich our understanding of the epigenetic interplay between both genes and the environment in psychiatric disorders (Gibney & Nolan, 2010). As such, integrated transcriptome and methylome data derived from peripheral blood was able to identify 43 risk genes that discriminated youth patients and high-risk for bipolar disorder patients from controls (Fries et al., 2017), further demonstrating the utility of integrative analyses in the identification of biomarkers for disease risk.

Combining more than two layers of biological data is a largely unexplored avenue in psychiatric research (Bersanelli et al., 2016). However, Ciuculete et al. (2017) investigated the interplay among 37 psychiatric-related genetic risk variants, as well as shifts in both methylation and mRNA levels in 223 adolescents distinguished as being at risk for the development of psychiatric disorders. Using this approach, the authors were able to detect five SNPs (in *HCRTR1, GAD1, HADC3,* and *FKBP5*), which were associated with eight CpG sites. Three of these CpG sites, cg01089319 (*GAD1*), cg01089249 (*GAD1*), and cg24137543 (*DIAPH1*), manifest in significant gene expression changes and overlap with active regulatory regions in chromatin states of brain tissues. Although these findings are preliminary, further studies employing multistage integrative analysis may help to advance the field and provide novel insights on how genetic variants may modulate risks for the development of specific psychiatric diseases (Ritchie, Holzinger, Li, Pendergrass, & Kim, 2015).

5.2 Shared molecular mechanisms among disorders: Implications for treatment response

There is a growing understanding that many psychiatric diseases share underlying biological mechanisms. The cross-disorder group from the Psychiatric Genomics Consortium (PGC) has provided empirical evidence for the shared genetic etiology of five psychiatric disorders, including schizophrenia, bipolar disorder, major depressive disorder (MDD), autism spectrum disorder (ASD), and attention-deficit/hyperactivity disorder (ADHD) (Lee, Ripke, Neale, & Faraone, 2013). Recently Lotan et al. (2014) curated all GWAS findings for ADHD, anxiety disorders, ASD, bipolar disorder, MDD, and schizophrenia, finding that 22% of identified genes overlapped across two or more disorders. An overlap in underlying biology between different conditions was also observed at the level of gene expression. For example, using microarrays, Iwamoto, Kakiuchi, Bundo, and Ikeda (2004) compared gene expression profiles across different brain regions in bipolar disorder, MDD, and schizophrenia. What they found is that although these mental disorders were molecularly distinctive, there was a relatively large overlap of the gene expression profiles in all three disorders. Using RNA-seq, further support was provided for these findings by Darby, Yolken, and Sabunciyan (2016), finding that hippocampus and orbitofrontal cortex transcriptomes were consistent across diagnostic groups. These fundings may also help to explain why the same medication exerts its effects on patients diagnosed with different disorders.

Although we have outlined many studies that have successfully exploited coexpression network approaches in psychiatric research, this methodology is yet to be fully embraced by the field. New methods and algorithms for gene expression analyses are growing. Including the use of coexpression networks in conjunction with machine learning methodologies (Lareau, White, Oberg, & McKinney, 2015). In the following section, the role of machine learning in both genetics and gene expression for developing predictive models for diagnostic purposes, as well as for treatment response, will be further explored.

6 Future methods: Machine learning

Machine learning (ML) exists at the intersection of computer science, mathematics, and statistics, and is proving to be one such methodology that can handle the modeling of large, complex systems datasets (Iniesta, Stahl, & McGuffin, 2016).

Given the brevity of this chapter, only a very brief explanation will be provided (see Iniesta et al. (2016) for a more extensive review). First, supervised ML algorithms learn from data and improve their performance through experience. This is achieved through learning a set of features (SNPs, gene expression data) that relate to a label. The label can be either binary (response/nonresponse) or multiclass (low-response/moderate-response/high-response). Supervised ML can also be used for regression, where a continuous outcome, such as psychiatric illness severity, can be predicted (Hastie, Tibshirani, & Friedman, 2009). In contrast, unsupervised ML algorithms identify clusters in data derived from a distance metric (such as Euclidean distance) that potentially contains unique characteristics related to an outcome of interest (Ghahramani, 2004). In this section, we will focus on supervised ML models.

6.1 Gene selection and complex trait prediction

There are two main problems that ML may help to resolve in both gene expression and genome-wide association studies. First, given the large p small n problem ($p \gg n$), a methodology is needed that can not only reduce the dimensionality of a genetic feature space, but do so in a way that can lead to a clinically meaningful prediction of a complex trait. For example, using a patient's gene expression data, can we predict whether or not a patient will respond to drug (A) or drug (B)? In differential expression analysis (DEA), univariate statistical tests are often performed to discover quantitative changes in expression levels between case and control phenotypes (Gupta, Dewan, Bharti, & Bhattacharya, 2012). However, according to Okser et al. (2014), regularized ML models have demonstrated an improved ability to select SNPs and differentially expressed genes that are most predictive of complex traits. In ML, a process called regularization penalizes a models' complexity, thus, enabling prediction in individuals outside of a training data set. A beneficial characteristic of this approach is that they search out the most predictive combinations of variants, rather than just individually predictive variants, as in univariate statistical tests. It is therefore not surprising that variants that do not attain genome-wide significance in univariate tests often contribute to the predictive capacity of multilocus ML models (Abraham, Kowalczyk, Zobel, & Inouye, 2013; Evans, Visscher, & Wray, 2009). Furthermore, evidence is beginning to suggest that genetic markers with highly significant and replicated odds ratios derived from genome-wide association studies (GWAS) may actually be poor classifiers of disease (Jakobsdottir, Gorin, Conley, Ferrell, & Weeks, 2009). This is not to suggest that findings from GWAS studies are not useful; they provide valuable information for establishing etiological hypotheses. What it may suggest is that regularized ML models may be more beneficial for the derivation of clinically translatable pathways and variants that evade detection in classical statistical methods when $p \gg n$. Therefore, the use of regularized ML models may lead to greater prediction of disease and overall utility in translational psychiatry.

Support for this hypothesis has been demonstrated by (Wei et al., 2013) in their ML-based prediction of Crohn's disease. Using regularized (lasso) logistic regression, they trained a prediction model while also minimizing a feature space of 10,799 SNPs. Their final model contained 573 SNPs and obtained an area under the curve (AUC) of 0.86. They then compared this model with a traditional log odds model. For dimensionality reduction, they estimated association significance and odds ratios for each variant by using two folds of data. They took all variants where $p < 0.01$ and pruned correlated variants by setting the linkage disequilibrium threshold to $r^2 \frac{1}{4} = 0.8$. The final feature space contained 15,158 SNPs and obtained an AUC of 0.73, a score 13% lower than the penalized logistic regression model, while containing a feature space that was >25 times the size. Thus, achieving both suboptimal predictive performance and decreased computational tractability relative to the regularized ML model.

6.2 Applications for gene expression

The use of ML methodologies for both dimensionality reduction and prediction of complex traits in gene expression studies is less prevalent, however, some support for this methodology has been attained. For example, Tan & Gilbert (2003) trained a decision tree classifier with probes from 6817 genes to predict cancerous colon tumors, attaining an accuracy score of 95% to discern between each type of tumor. Furthermore, using bagged decision trees, they were also able to differentiate between two different types of lung cancer with 93% accuracy. Whether such findings are possible in complex psychiatric traits such as major depression is questionable, yet the methodology is theoretically applicable; and as demonstrated, appears to outperform univariate statistical tests and classical methods for both dimensionality reduction and prediction of complex traits when $p \gg n$. In support of this, preliminary evidence suggests that the use of ML-based methodologies in complex psychiatric trait prediction may be possible, yet much larger studies are still needed. For example, using a prospective design, Guilloux et al. (2014) collected blood samples from a discovery cohort of 34 adult MDD patients with cooccurring anxiety and 33 matched nondepressed controls. Data was collected at baseline, and after 12 weeks of combined citalopram and psychotherapy treatment. Using linear support vector machines trained on gene-expression data from

13 genes, they were able to predict remission/nonremission of MDD with a cross-validated corrected accuracy score of 76.2% (sensitivity = 86.1%, specificity = 59.3%). Much larger studies are still warranted across a range of complex traits and treatments; however, the initial proof of concept is encouraging.

Conclusions

Studying interactions among genes in relation to psychiatric phenotypes using coexpression network analysis is a promising complementary approach to better understand mechanisms of psychiatric disorders, which may lead to developing clinically translatable diagnostic and treatment response biomarkers. However, further research needs to also utilize machine learning algorithms to quantify diagnosis and treatment response prediction. We believe that these methods, combined, can help to advance a rapidly developing field of personalized psychiatry.

References

Abraham, G., Kowalczyk, A., Zobel, J., & Inouye, M. (2013). Performance and robustness of penalized and Unpenalized methods for genetic prediction of complex human disease. *Genetic Epidemiology, 37*(2), 184–195.

Akula, N., Wendland, J. R., Choi, K. H., & McMahon, F. J. (2015). An integrative genomic study implicates the postsynaptic density in the pathogenesis of bipolar disorder. *Neuropsychopharmacology, 41*(3), 886–895.

Amar, D., Safer, H., & Shamir, R. (2013). Dissection of regulatory networks that are altered in disease via differential co-expression. *PLoS Computational Biology, 9*(3).

Bar-Yam, Y. (2002). General features of complex systems. In *Encyclopedia of Life Support Systems.* Oxford: EOLSS UNESCO Publishers.

Belzeaux, R., Lin, C.-W. D., Bergon, Y., Ibrahim, A., EC, T., Tseng, G., & Sibille, E. (2016). Predisposition to treatment response in major depressive episode: A peripheral blood gene coexpression network analysis. *Journal of Psychiatric Research, 81,* 119–126.

Bersanelli, M., Mosca, E., Remondini, D., Giampieri, E., Sala, C., Castellani, G., & Milanesi, L. (2016). Methods for the integration of multi-omics data: Mathematical aspects. *BMC Bioinformatics, 17*(Suppl. 2), 15.

Breen, M. S., White, C. H., Shekhtman, T., Lin, K., Looney, D., Woelk, C. H., & Kelsoe, J. R. (2016). Lithium-responsive genes and gene networks in bipolar disorder patient-derived lymphoblastoid cell lines. *The Pharmacogenomics Journal, 16,* 446–453.

Cattane, N., Minelli, A., Milanesi, E., Maj, C., Bignotti, S., Bortolomasi, M., … Gennarelli, M. (2015). Altered gene expression in schizophrenia: Findings from transcriptional signatures in fibroblasts and blood. *PLoS One, 10*(2), e0116686.

Cha, K., Hwang, T., Oh, K., & Yi, G.-S. (2015). Discovering transnosological molecular basis of human brain diseases using biclustering analysis of integrated gene expression data. *BMC Medical Informatics and Decision Making, 15*(Suppl 1), S7.

Chari, R., Thu, K. L., Wilson, I. M., Lockwood, W. W., Lonergan, K. M., Coe, B. P., … Lam, W. L. (2010). Integrating the multiple dimensions of genomic and epigenomic landscapes of cancer. *Cancer Metastasis Reviews, 29*(1), 73–93.

Chen, C., Cheng, L., Grennan, K., Pibiri, F., Zhang, C., Badner, J. A., … Liu, C. (2013). Two Gene Co-expression Modules Differentiate Psychotics and Controls. *Molecular Psychiatry, 18*(12), 1308–1314.

Ciobanu, L. G., Sachdev, P. S., Trollor, J. N., Reppermund, S., Thalamuthu, A., Mather, K. A., … Baune, B. T. (2016). Differential gene expression in brain and peripheral tissues in depression across the life span: A review of replicated findings. *Neuroscience and Biobehavioral Reviews, 71,* 281–293.

Ciuculete, D. M., Bostrom, A. E., Voisin, S., Philipps, H., Titova, O. E., Bandstein, M., … Schioth, H. B. (2017). A methylome-wide mQTL analysis reveals associations of methylation sites with GAD1 and HDAC3 SNPs and a general psychiatric risk score. *Translational Psychiatry, 7,* e1002.

Darby, M. M., Yolken, R. H., & Sabunciyan, S. (2016). Consistently altered expression of gene sets in postmortem brains of individuals with major psychiatric disorders. *Translational Psychiatry, 6*(9), e890.

de Jong, S., Newhouse, S. J., Patel, H., Lee, S., Dempster, D., Curtis, C., … Breen, G. (2016). Immune signatures and disorder-specific patterns in a cross-disorder gene expression analysis. *The British Journal of Psychiatry, 209*(3), 202–208.

Drago, A., De Ronchi, D., & Serretti, A. (2007). Incomplete coverage of candidate genes: A poorly considered bias. *Current Genomics, 8*(7), 476–483.

Evans, D. M., Visscher, P. M., & Wray, N. R. (2009). Harnessing the information contained within genome-wide association studies to improve individual prediction of complex disease risk. *Human Molecular Genetics, 18*(18), 3525–3531.

Fiannaca, A., La Rosa, M., La Paglia, L., Rizzo, R., & Urso, A. (2015). Analysis of miRNA expression profiles in breast cancer using biclustering. *BMC Bioinformatics, 16*(Suppl. 4), S7.

Fries, G. R., Quevedo, J., Zeni, C. P., Kazimi, I. F., Zunta-Soares, G., Spiker, D. E., … Soares, J. C. (2017). Integrated transcriptome and methylome analysis in youth at high risk for bipolar disorder: A preliminary analysis. *Translational Psychiatry, 7*(3), e1059.

Gandal, M. J., Haney, J., Parikshak, N., Leppa, V., Horvath, S., & Geschwind, D. H. (2018). Shared molecular neuropathology across major psychiatric disorders parallels polygenic overlap. *Science, 359*(6376), 693–697.

Ghahramani, Z. (2004). Unsupervised learning. In O. Bousquet, U. von Luxburg, & G. Rätsch (Eds.), *Advanced Lectures on Machine Learning: ML Summer Schools 2003, Canberra, Australia, February 2–14, 2003, Tübingen, Germany, August 4–16, 2003, Revised Lectures* (pp. 72–112). Heidelberg, Berlin: Springer.

Gibney, E. R., & Nolan, C. M. (2010). Epigenetics and gene expression. *Heredity, 105*(1), 4–13.

Guilloux, J.-P., Bassi, S., Ding, Y., Walsh, C., Turecki, G., Tseng, G., … Sibille, E. (2014). Testing the Predictive Value of Peripheral Gene Expression for Nonremission Following Citalopram Treatment for Major Depression. *Neuropsychopharmacology, 40*, 701–710.

Gupta, R., Dewan, I., Bharti, R., & Bhattacharya, A. (2012). Differential expression analysis for RNA-Seq data. *ISRN Bioinformatics, 2012*.

Harrison, R. N. S., Murray, R. M., Lee, S. H., Paya Cano, J., Dempster, D., Curtis, C. J., … de Jong, S. (2016). Gene-expression analysis of clozapine treatment in whole blood of patients with psychosis. *Psychiatric Genetics, 26*(5), 211–217.

Hastie, T., Tibshirani, R., & Friedman, J. (2009). *The elements of statistical learning: Data mining, inference, and prediction. Springer Series in Statistics.* Springer.

Hsiao, T.-H., Chiu, Y.-C., Hsu, P.-Y., Lu, T.-P., Lai, L.-C., Tsai, M.-H., … Chen, Y. (2016). Differential network analysis reveals the genome-wide landscape of estrogen receptor modulation in hormonal cancers. *Scientific Reports, 6*, 23035.

Hsu, C.-L., Juan, H.-F., & Huang, H.-C. (2015). Functional Analysis and Characterization of Differential Coexpression Networks. *Scientific Reports, 5*, 13295.

Ikeda, M., Takahashi, A., Kamatani, Y., Okahisa, Y., Kunugi, H., Mori, N., … Iwata, N. (2018). A genome-wide association study identifies two novel susceptibility loci and trans population polygenicity associated with bipolar disorder. *Molecular Psychiatry, 23*(3), 639–647.

Iniesta, R., Stahl, D., & McGuffin, P. (2016). Machine learning, statistical learning and the future of biological research in psychiatry. *Psychological Medicine, 46*(12), 2455–2465.

Iwamoto, K., Kakiuchi, C., Bundo, M., Ikeda, K., & Kato, T. (2004). Molecular characterization of bipolar disorder by comparing gene expression profiles of postmortem brains of major mental disorders. *Molecular Psychiatry, 9*(4), 406–416.

Jakobsdottir, J., Gorin, M. B., Conley, Y. P., Ferrell, R. E., & Weeks, D. E. (2009). Interpretation of genetic association studies: Markers with replicated highly significant odds ratios may be poor classifiers. *PLoS Genetics, 5*(2).

Kubota, T., Miyake, K., & Hirasawa, T. (2012). Epigenetic understanding of gene-environment interactions in psychiatric disorders: A new concept of clinical genetics. *Clinical Epigenetics, 4*(1), 1.

Lareau, C. A., White, B. C., Oberg, A. L., & McKinney, B. A. (2015). Differential co-expression network centrality and machine learning feature selection for identifying susceptibility hubs in networks with scale-free structure. *BioData Mining, 8*, 5.

Lee, S. H., Ripke, S., Neale, B. M., & Faraone, S. V. (2013). Genetic relationship between five psychiatric disorders estimated from genome-wide SNPs. *Nature Genetics, 45*(9), 984–994.

Lotan, A., Fenckova, M., Bralten, J., Alttoa, A., Dixson, L., Williams, R. W., & van der Voet, M. (2014). Neuroinformatic analyses of common and distinct genetic components associated with major neuropsychiatric disorders. *Frontiers in Neuroscience, 8*, 331.

Malki, K., Tosto, M. G., Jumabhoy, I., & Lourdusamy, A. (2013). Integrative mouse and human mRNA studies using WGCNA nominates novel candidate genes involved in the pathogenesis of major depressive disorder. *Pharmacogenomics, 14*(16), 1979–1990.

McClellan, J., & King, M. (2010). Genomic analysis of mental illness: A changing landscape. *JAMA, 303*(24), 2523–2524.

Mistry, M., Gillis, J., & Pavlidis, P. (2013). Meta-analysis of gene coexpression networks in the post-mortem prefrontal cortex of patients with schizophrenia and unaffected controls. *BMC Neuroscience, 14*, 105.

Okser, S., Pahikkala, T., Airola, A., Salakoski, T., Ripatti, S., & Aittokallio, T. (2014). Regularized machine learning in the genetic prediction of complex traits. *PLoS Genetics, 10*(11).

Parikshak, N. N., Luo, R., Zhang, A., Won, H., Lowe, J. K., Chandran, V., … Geschwind, D. H. (2013). Integrative functional genomic analyses implicate specific molecular pathways and circuits in autism. *Cell, 155*(5), 1008–1021.

Polderman, T. J. C., Benyamin, B., de Leeuw, C. A., Sullivan, P. F., van Bochoven, A., Visscher, P. M., & Posthuma, D. (2015). Meta-analysis of the heritability of human traits based on fifty years of twin studies. *Nature Genetics, 47*(7), 702–709.

Pontes, B., Giráldez, R., & Aguilar-Ruiz, J. S. (2015). Biclustering on expression data: A review. *Journal of Biomedical Informatics, 57*(Suppl. C), 163–180.

Ripke, S., Neale, B. M., Corvin, A., & Watters, J. T. R. (2014). Biological insights from 108 schizophrenia-associated genetic loci. *Nature, 511*, 421.

Ritchie, M. D., Holzinger, E. R., Li, R., Pendergrass, S. A., & Kim, D. (2015). Methods of integrating data to uncover genotype–phenotype interactions. *Nature Reviews Genetics, 16*, 85.

Sanders, A. R., Drigalenko, E. I., Duan, J., Moy, W., Freda, J., Göring, H. H. H., & Gejman, P. V. (2017). Transcriptome sequencing study implicates immune-related genes differentially expressed in schizophrenia: New data and a meta-analysis. *Translational Psychiatry, 7*(4), e1093.

Seifuddin, F., Pirooznia, M., Judy, J. T., Goes, F. S., Potash, J. B., & Zandi, P. P. (2013). Systematic review of genome-wide gene expression studies of bipolar disorder. *BMC Psychiatry, 13*, 213.

Sklar, P., Ripke, S., Scott, L. J., Andreassen, O. A., Cichon, S., Craddock, N., … Purcell, S. M. (2011). Large-scale genome-wide association analysis of bipolar disorder identifies a new susceptibility locus near ODZ4. *Nature Genetics, 43*(10), 977–983.

Tan, A. C., & Gilbert, D. (2003). Ensemble machine learning on gene expression data for cancer classification. *Applied Bioinformatics, 2*, S75–S83.

Voineagu, I., Wang, X., Johnston, P., Lowe, J. K., Tian, Y., Horvath, S., … Geschwind, D. H. (2011). Transcriptomic analysis of autistic brain reveals convergent molecular pathology. *Nature, 474*(7351), 380–384.

Wei, Z., Wang, W., Bradfield, J., Li, J., Cardinale, C., Frackelton, E., … Hakonarson, H. (2013). Large sample size, wide variant spectrum, and advanced machine-learning technique boost risk prediction for inflammatory bowel disease. *The American Journal of Human Genetics, 92*(6), 1008–1012.

West, M., Blanchette, C., Dressman, H., Huang, E., Ishida, S., Spang, R., … Nevins, J. R. (2001). Predicting the clinical status of human breast cancer by using gene expression profiles. *Proceedings of the National Academy of Sciences, 98*(20), 11462–11467.

Wray, N. R., Ripke, S., Mattheisen, M., Trzaskowski, M., Byrne, E. M., Abdellaoui, A., … et al. (2017). Genome-wide association analyses identify 44 risk variants and refine the genetic architecture of major depression. *Nature Genetics, 50*(5), 668–681.

Xu, F., Yang, J., Chen, J., Wu, Q., Gong, W., Zhang, J., ... Xie, P. (2015). Differential co-expression and regulation analyses reveal different mechanisms underlying major depressive disorder and subsyndromal symptomatic depression. *BMC Bioinformatics*, *16*(1), 112.

Yang, Y., Han, L., Yuan, Y., Li, J., Hei, N., & Liang, H. (2014). Gene co-expression network analysis reveals common system-level properties of prognostic genes across cancer types. *Nature Communications*, *5*.

Zhang, B., & Horvath, S. (2005). A general framework for weighted gene co-expression network analysis. *Statistical Applications in Genetics and Molecular Biology*, *4*.

Chapter 32

Pharmacogenomics of bipolar disorder

Claudia Pisanu[a], Alessio Squassina[a,b], Martin Alda[b] and Giovanni Severino[a]

[a]*Department of Biomedical Sciences, Section of Neuroscience and Clinical Pharmacology, University of Cagliari, Cagliari, Italy,* [b]*Department of Psychiatry, Dalhousie University, Halifax, NS, Canada*

Bipolar disorder (BD) is a common and disabling mental disorder characterized by oscillations of mood and behavior. The lifetime prevalence has been estimated at 1%–2%, but exceeds 2% for subthreshold conditions (Merikangas et al., 2011). Several studies reported decreased life expectancy in BD with 9–14 lost life years compared with the general population (Kessing, Vradi, McIntyre, & Andersen, 2015). The increased mortality is related to suicide, unintentional injuries, and comorbid general medical illnesses, including cardiovascular diseases, diabetes, and chronic obstructive pulmonary disease (Crump, Sundquist, Winkleby, & Sundquist, 2013). Due to the elevated morbidity and mortality, BD has a relevant socioeconomic impact, and is considered to be one of the main causes of disability (Grande, Berk, Birmaher, & Vieta, 2016).

1 Genetic bases of BD

As shown by family and twin studies, BD and other mood disorders aggregate in families and have a high heritability (Schulze, Fangerau, & Propping, 2004). The lifetime morbidity risk for BD in first degree relatives of BD patients has been estimated at 10%–20% (Schulze, 2010). Results of twin and adoption studies show that the heritable component of this mood disorder ranges between 80% and 90%, supporting the hypothesis that genetic factors greatly contribute to the etiology of BD (Craddock & Jones, 1999; Schulze, 2010; Shih, Belmonte, & Zandi, 2004; Smoller & Finn, 2003).

Despite the efforts made to identify the genetic factors involved in the pathogenesis of BD, the genetic underpinnings of this mood disorder are still largely unknown, and appear to be highly heterogeneous (Craddock & Sklar, 2013; Goes, 2016). The most consistent associations were shown for a limited number of candidates, including genes involved in neurotransmission systems or encoding neurotrophins [solute carrier family 6 member 4 (SLC6A4), brain derived neurotrophic factor (BDNF), neuregulin 1 (NRG1), and catechol-*O*-methyltransferase (COMT)], as well as genes involved in neuronal development and implicated in other psychiatric disorders [disrupted in schizophrenia 1 (DISC1), and dystrobrevin binding protein 1 (DTNBP1) (Szczepankiewicz, 2013)]. In the past few years, genome-wide association studies have permitted us to identify a number of top single nucleotide polymorphisms (SNP) associated with BD (for a recent and comprehensive review, see Ikeda, Saito, Kondo, & Iwata, 2017), some of which have also been suggested to play a role in the pathogenesis of other psychiatric phenotypes, such as schizophrenia.

2 Clinical heterogeneity of BD and pharmacological treatments

The broad variability of genes suggested to be implicated in the predisposition to BD is consistent with the high heterogeneity of the clinical presentation of this mood disorder, which represents an important challenge for its management. In this context, an appropriate diagnosis of BD and adequate clinical practice guidelines (CPGs) represent fundamental tools to allow clinicians to prescribe the best treatment (Strakowski, Fleck, & Maj, 2011). While there was some consistency across guidelines on key recommendations, no 'metaconsensus' model for the management of BD exists (Graham, Tavella, & Parker, 2017). Pharmacological treatment, together with psychotherapy and psychoeducation, represents the first line approach in the acute phases of the illness, as well as in the prevention of recurrences. The mainstay of pharmacologic treatment for BD is the class of medications known as mood stabilizers that are comprised of lithium and the anticonvulsants valproate, carbamazepine, and lamotrigine (Tighe, Mahon, & Potash, 2011). As well, atypical antipsychotics are being used increasingly both for acute and long-term treatment of BD (Geddes & Miklowitz, 2013).

Personalized Psychiatry. https://doi.org/10.1016/B978-0-12-813176-3.00032-8

As in the case of other severe psychiatric disorders (depression, anxiety disorders, and schizophrenia), a large proportion of patients do not respond to the prescribed treatment (Cuthbert & Insel, 2013). Part of the reason is due to the fact that major psychiatric diseases are heterogeneous conditions caused by complex interactions among genetic, epigenetic, developmental, and environmental factors (Prendes-Alvarez & Nemeroff, 2018). This helps explain why the effects of the majority of therapeutics used in the management of BD have been discovered by serendipity.

Lithium is considered the mainstay in the treatment of BD, and a substantial body of evidence supports its efficacy (Tighe et al., 2011). Lithium is effective in the acute episodes of both polarities, as well as to prevent recurrences. Moreover, lithium is the only mood stabilizer with proven efficacy in reducing the risk of suicide (Cipriani, Hawton, Stockton, & Geddes, 2013). Although 30% of chronically treated patients show full remission of symptoms, about 70% do not respond sufficiently, and need an addition of or switch to another mood stabilizer (Geddes & Miklowitz, 2013).

Although less effective in the prevention of recurrences of both polarities, the anticonvulsants valproate, lamotrigine, and carbamazepine represent valid alternatives for the treatment of BD. Specifically, lamotrigine is efficacious in the prevention of depressive recurrences, while valproate and carbamazepine are more efficacious in the treatment of mania (Pichler, Hattwich, Grunze, & Muehlbacher, 2015).

While less is known about the other mood stabilizers, in the case of lithium, it has been well documented that clinical response clusters in families (Grof et al., 2002). Lithium responders share distinct clinical characteristics (i.e., episodic course of illness, absence of psychiatric manifestations, and presence of BD in the family history), and have been suggested to represent the core phenotype of BD (Grof et al., 2002; Alda, Grof, Rouleau, Turecki, & Young, 2005). On the basis of these findings, most of the studies investigating the genetic underpinnings of response to mood stabilizers focused on lithium.

3 Pharmacogenetics of lithium: Candidate genes studies

As mentioned in the previous section, lithium—whose efficacy in treating mania stabilizing properties has been first described by John Cade in 1949 (Cade, 1949)—remains the best established long-term treatment for BD (Alda, 2015). A large number of studies investigating genetic variants associated with lithium response focused on the role of genes previously suggested to be implicated in BD or in the mechanism of action of lithium. Among systems explored to identify potential molecular predictors of lithium response, most studies focused on neurotransmission systems, neurothrophins, inositol metabolism, and circadian clock genes. Genes suggested to be associated with lithium response by at least two different studies are reported in Table 1.

3.1 Neurotransmitter systems

Monoamine systems (dopamine, serotonin, and norepinephrine) play a crucial role in the regulation of key functions of the central nervous system (CNS) (Grace, 2016). Disruption of these circuits has been implicated in a number of pathological conditions, including BD (Manji et al., 2003).

The neurotransmitter serotonin plays a crucial role in multiple processes of the CNS, including mood, sleep, cognition, and behavior (Zhou, Engel, & Wang, 2007). The best studied polymorphism in this system is the 5-HTTLPR genetic variant, which is located in the promoter of the serotonin transporter gene (SLC6A4 or 5-HTT). The serotonin transporter is responsible for the reuptake of serotonin from the synaptic cleft into presynaptic neurons (Goldman, Glei, Lin, & Weinstein, 2010). The 5-HTTLPR polymorphism is a 44 bp deletion/insertion that results in two main alleles: a 14-repeat short variant (s), which is associated with reduced transcription and lower serotonin uptake, and a 16-repeat long variant (l), which is associated with higher serotonin uptake (Goldman et al., 2010). The majority of studies showed an association between poor lithium response and the 5-HTTLPR s/s genotype (Rybakowski et al., 2005, 2012; Serretti, Lilli, Mandelli, Lorenzi, & Smeraldi, 2001). Some studies also showed an interaction between 5-HTTLPR s/s or l/s genotypes, and either *BDNF* rs6265 (Serretti et al., 2004) or rs4680 (Rybakowski et al., 2007) in predisposing to poor lithium response. However, contrasting findings have also been reported. Some studies suggested the 5-HTTLPR l/s (Serretti et al., 2004) or s variants (Del Zompo et al., 1999; Tharoor, Kotambail, Jain, Sharma, & Satyamoorthy, 2013) to be associated with good lithium response, while other studies found no significant association (Bremer et al., 2007; Manchia et al., 2009; Michelon et al., 2006). Overall, the results supporting a potential involvement of the serotonin transporter gene in lithium response are not conclusive.

Other important members of the serotonin pathway are comprised of the two isoforms of the tryptophan hydroxylase (TPH) gene—which encodes the rate-limiting enzyme in the synthesis of serotonin—the 5-hydroxytryptamine receptor 1B (HTR1B) and 5-hydroxytryptamine receptor 2A (HTR2A) receptors, as well as the monoamine oxidase A (*MAOA*) gene—which catalyzes the degradation of serotonin, dopamine, and norepinephrine. Results from studies focused on variants located in these genes all provided negative results (Dmitrzak-Weglarz et al., 2005; Manchia et al., 2009;

TABLE 1 Genes suggested to be associated with lithium response by at least two candidate gene studies.

	Gene	Main results	References
Neurotransmission	COMT	Association between rs4680 Met/Met genotype and poor response	Lee and Kim (2010) and Rybakowski et al. (2012)
	DRD1	Poor response and rs4532 G allele or G/G genotype	Rybakowski et al. (2009, 2012)
	SLC6A4	Poor response and 5-HTTLPR s allele or s/s genotype, interaction between 5-HTTLPR genotype and BDNF rs6265 or 4680 genotypes	Serretti et al. (2001, 2002, 2004), and Rybakowski et al. (2005, 2007, 2012)
Neurothrophic factors	BDNF	Good response and rs6265 Val/Met genotype; good response and rs6265/rs988748; response and rs6265 Met/Met genotype in dependency of BD subtype; poor response and rs6265 Val allele; interaction between BDNF rs6265 or rs4680 genotypes and SLC6A45-HTTLPR genotypes	Rybakowski et al. (2005, 2007), Dmitrzak-Weglarz et al. (2008), Wang et al. (2012), and Rybakowski et al. (2012)
	NTRK2	Good response and rs1387923/rs133845 in BD patients with suicidal ideation; good response and rs2769605 A allele	Bremer et al. (2007) and Wang et al. (2013)
Inositol metabolism	INPP1	Good response and rs2067421 after accounting for euphoric/dysphoric mania and history of suicidal ideation; poor response and rs909270-G allele	Bremer et al. (2007) and Mitjans et al. (2015)
Circadian rhythms	GSK3B	Good response and rs334558 C allele; poor response and rs334558 T/T genotype; good response and −50T/C −1727A/T T-A haplotype; poor response and rs11921360-A allele	Benedetti et al. (2005), Lin et al. (2013), Iwahashi et al. (2014), and Mitjans et al. (2015)
	NR1D1 (Rev-Erb-a)	Poor response and rs2314339 T allele good response and NR1D1 rs8192440/CRY1 rs8192440	Campos-de-Sousa et al. (2010) and McCarthy et al. (2011)
	XBP1	Poor response and rs2269577 G/G genotype; good response and rs2269577 C allele	Kakiuchi and Kato (2005) and Masui et al. (2006)

BD, bipolar disorder.

Rybakowski et al., 2012; Serretti, Lorenzi, Lilli, & Smeraldi, 2000; Serretti et al., 2002, 2011; Turecki et al., 1999), except the study conducted by Serretti et al. (1999) which showed an association between the A/A genotype of the *THP1* rs1800532 variant and poor lithium response.

Several studies focused on the role of the dopamine system, which is known to regulate a number of brain functions including mood, cognition, and locomotion (Grace, 2016). Studies focused on this system did not support a major role of genetic variants located in dopaminergic receptors or dopamine metabolizing enzymes in predisposing patients to lithium response (Manchia et al., 2009; Serretti et al., 1999, 2002; Serretti, Lilli, Lorenzi, Franchini, & Smeraldi, 1998; Turecki et al., 1999). Only two studies reported an association between the G allele or the G/G genotype of the dopamine receptor D1 (DRD1) rs4532 variant and poor lithium response (Rybakowski et al., 2012; Rybakowski, Dmitrzak-Weglarz, Suwalska, Leszczynska-Rodziewicz, & Hauser, 2009).

Finally, studies focused on the glutamatergic system mainly provided negative findings (Bremer et al., 2007; Chiesa et al., 2012; Mitjans et al., 2015; Rybakowski et al., 2012; Szczepankiewicz, Skibinska, Suwalska, Hauser, & Rybakowski, 2009), although an association between good lithium response and variants located in the calcium voltage-gated channel auxiliary subunit gamma 2 (CACNG2) gene (which encodes a protein that directly interacts with the AMPA receptor) has been reported in two independent samples (Silberberg et al., 2008).

3.2 Neurothrophic factors

Extensive research has focused on the role of the *BDNF* gene in lithium response, reporting controversial findings. BDNF is the most widely expressed neurotrophin in the mammalian brain, and plays a crucial role in neural development and regulation of synaptic plasticity (Calabrese et al., 2014). On the basis of its physiological role, *BDNF* variants have been extensively investigated as potential biomarkers of response to pharmacological treatments in mood disorders (Polyakova et al., 2015). With regard to lithium, some studies reported an association between poor response and either the Val allele of the functional variant rs6265 (Rybakowski et al., 2012; Serretti et al., 2002), or the Val/Val genotype of the rs6280 variant (Rybakowski et al., 2007). However, other studies reported negative findings (Bremer et al., 2007; Drago et al., 2010; Michelon et al., 2006; Pae et al., 2012). Interestingly, a recent study conducted by Wang et al. (2012) in a Han Chinese sample suggested a differential association between the rs6265 genotypes and lithium response according to BD subtype (Wang et al., 2012). Specifically, the Val/Met and Val/Val genotypes were associated with positive lithium response in patients with BD type 1, and with poor response in BD type 2 patients (Wang et al., 2012).

Contrasting findings have also been reported for the neurotrophic receptor tyrosine kinase 2 (NTRK2) gene, which encodes a membrane receptor that binds neurotrophins (including BDNF) and regulates phosphorylation of members of the MAP kinase (MAPK) signaling pathway. While few studies showed an association between different variants of this gene and good lithium response [e.g. rs1387923/rs133845 in BD patients with suicidal ideation (Bremer et al., 2007)], others provided negative findings (Dmitrzak-Weglarz et al., 2008; Rybakowski et al., 2012).

3.3   Inositol signaling pathway

Inositol depletion was one of the first hypotheses formulated to explain lithium's mechanism of action (Berridge, Downes, & Hanley, 1989). Based on the observation that lithium interferes with inositol metabolism via direct inhibition of key enzymes (inositol monophosphatase and polyphosphatase), a large number of studies investigated the role of genes involved in this system in lithium response. For each target found to be associated with lithium response, negative findings have also been published. For example, the association found between good response and the inositol polyphosphate-1-phosphatase (INPP1) C937A variant in a small sample of 23 BD patients of Norwegian origin (Steen et al., 1998) was not replicated by the same authors in an independent Israeli sample. Similarly, studies investigating the association between lithium response and genetic variants located in the inositol monophosphatase 2 (IMPA2) gene provided contrasting findings (Bremer et al., 2007; Dimitrova et al., 2005; Mitjans et al., 2015; Sjøholt et al., 2004). Turecki et al. (1998) showed an association between BD and a polymorphism located in the phospholipase C, gamma 1 (PLCG1) gene—which encodes an isozyme of phospholipase (PLC). This enzyme catalyzes the formation of inositol 1,4,5-trisphosphate and diacylglycerol from phosphatidylinositol 4,5-bisphosphate, thus playing a crucial role in the inositol metabolism (Turecki et al., 1998). In this study, all BD patients were excellent lithium responders, thus leading to speculation that the investigated dinucleotide (CA)n repeat allele could also be associated with lithium response. A subsequent study only partially confirmed these results, showing that the *PLCG1*-8 allele was more often observed in BD patients responders to lithium, compared with healthy controls (Løvlie, Berle, Stordal, & Steen, 2001). Taken together, available data do not support a major role of genes encoding elements of the inositol signaling pathway in lithium response.

3.4   GSK-3 and circadian clock rhythms

The glycogen synthase kinase 3 (GSK3) enzyme is one of the most investigated targets of lithium. Via inhibition of this enzyme, especially of its beta isoform, lithium is able to interfere with several downstream signaling pathways (Alda, 2015). Some studies showed a significant association between variants located in the *GSK3B* gene and positive response to lithium treatment. An association between the C allele and good response was reported by two different studies (Benedetti et al., 2005; Lin, Huang, & Liu, 2013), while a trend for association between rs6438552 and good response was reported by McCarthy et al. (2011). However, other studies reported negative findings (Bremer et al., 2007; Masui et al., 2006; Michelon et al., 2006; Rybakowski et al., 2012; Szczepankiewicz et al., 2006).

Among the different signaling pathways modulated by the GSK-3 beta enzyme, the circadian clock system is one of the most intriguing, based on findings supporting the involvement of abnormalities in circadian rhythms in the pathogenesis of BD and in lithium's mechanism of action (Alloy, Ng, Titone, & Boland, 2017; Moreira & Geoffroy, 2016). As shown in a recent systematic review, lithium exerts a direct impact on the molecular clock, and its actions are comprised of reduction of amplitude and duration of activity rhythms, as well as delay of the phase of sleep–wakefulness rhythms (Moreira & Geoffroy, 2016).

GSK-3 beta phosphorylates and stabilizes the nuclear receptor Rev-erb alpha, an important negative feedback regulator of the circadian clock (Yin, Wang, Klein, & Lazar, 2006). Lithium is able to exert a direct impact on circadian rhythms via inhibition of GSK-3 beta, which leads to degradation of Rev-erb alpha in the proteasome (Yin et al., 2006). Interestingly, different variants located in the gene encoding Rev-erb alpha (NR1D1) have been associated with lithium response.

Specifically, the rs2314336 T allele has been associated with poor lithium response (Campos-de-Sousa et al., 2010), while the rs819240 variant was found to interact with a polymorphism located in another circadian gene (CRY1) in predisposing patients to lithium response (McCarthy et al., 2011).

Although candidate gene studies identified a relatively high number of putative targets predisposing patients to positive lithium response, the large majority of these results have not been replicated by subsequent investigations, thus limiting the potential future clinical utility of these markers.

4   Pharmacogenetics of lithium: Genome-wide association studies

In the past few years, research on the genetic markers involved in lithium response shifted from candidate gene studies to hypothesis free approaches that permit us to interrogate simultaneously more than 1 million SNPs using whole genome genotyping arrays. The first genome-wide association study (GWAS) of lithium response was conducted by Perlis et al. (2009). In this study, response to lithium was defined as the time to recurrence during lithium treatment. Although no genetic variant met the threshold for genome-wide significance, some of the targets—including one SNP located near the glutamate ionotropic receptor AMPA type subunit 2 (GRIA2) gene—were successfully replicated in an independent sample (Perlis et al., 2009).

Two years later, Squassina et al. (2011) conducted the first GWAS in which lithium response was assessed using the "retrospective criteria of long-term treatment response in research subjects with BD" a validated scale that quantifies the degree of improvement during lithium treatment while accounting for potential confounding factors (Grof et al., 2002). In the GWAS from Squassina and coworkers, no SNP met the genome-wide significant threshold. However, among the top SNPs that were validated in an independent sample, the strongest signal was found for the rs11869731 variant, located in the acid sensing ion channel subunit 2 (ASIC2 or ACCN1) gene (Squassina et al., 2011). This gene encodes a sodium channel that is also permeable to lithium and potassium. Being widely expressed in the CNS, acid-sensing ion channels have been suggested to play a role in the pathophysiology of neurological and psychiatric diseases (Wemmie, Taugher, & Kreple, 2013).

A GWAS conducted in a Han Chinese sample of BD patients found a strong association between the two SNPs rs17026688 and rs17026651, located in the glutamate decarboxylase like 1 (GADL1) gene, and lithium response (Chen et al., 2014). However, to date no study has been able to replicate these findings in samples of patients of either Asian or Caucasian origin (Consortium on Lithium Genetics, 2014). In a recent GWAS conducted by Song et al. (2016) including a very large number of BD patients characterized for lithium response (2698 patients with self-reported and 1176 patients with objectively defined response to the mood stabilizer), the rs116323614 marker located in the spectrin domains 1 (SESTD1) gene, was found to be significant when comparing BD patients responders to lithium with healthy controls (Song et al., 2016). This gene encodes a protein that plays a role in regulation of phospholipids (Miehe et al., 2010).

Finally, the most promising effort to elucidate the complex genetic bases of lithium response is represented by the International Consortium on Lithium Genetics (ConLiGen) (Schulze et al., 2010). ConLiGen was founded in 2008 with the aim of collecting a large sample of BD patients characterized for lithium response using a validated tool (Grof et al., 2002; Manchia et al., 2013). This initiative aims to overcome the limitations affecting the large majority of previous studies, such as limited sample size and different phenotype definitions. Results from the first ConLiGen GWAS study have been published and point to long noncoding RNAs (lncRNA) as potential modulators of lithium response (Hou et al., 2016). In recent years, these functional molecules have gained attention due to their role of regulators of the transcription of the coding portion of the genome, and are increasingly being studied in a number of different pathological conditions. As secondary analyses of the ConLiGen data are currently ongoing, as well as extensions of the sample, this international effort could lead us to undertake important steps toward the clinical implementation of genetic biomarkers of lithium response.

5   Pharmacogenetics of other mood stabilizers

A very limited number of studies investigated potential pharmacogenetic biomarkers of mood stabilizers other than lithium. Among the few studies reporting positive results, the Met/Met genotype of the *COMT* rs4680 SNP was significantly more common in nonresponders compared with responders to mood stabilizers (lithium, valproate or carbamazepine) in manic BD patients (Lee & Kim, 2010).

Another interesting finding reported by a Korean group is the association between the C allele of the functional −116C/ G polymorphism of the X-box binding protein 1 (XBP1) gene and valproate efficacy (Kim, Kim, Lee, & Joo, 2009). *XBP1* encodes for a transcription factor essential for the transcription of major histocompatibility complex (MHC) class II genes

(Ono, Liou, Davidon, Strominger, & Glimcher, 1991). The G allele of the −116C/G variant is associated with reduced expression of the gene. Interestingly, this variant had been previously associated with lithium response, although with an opposite direction of effect (Kakiuchi & Kato, 2005; Masui et al., 2006), suggesting potential differences in the genetic framework predisposing patients to response to different mood stabilizers.

6 Pharmacogenetics of mood stabilizers: Focus on side effects

Pharmacogenetics may also represent a useful tool to identify patients at higher risk of side effects associated with mood stabilizer treatment. However, this field is still in its early phase of development, as it ideally requires well characterized samples of patients for which a long follow-up is available. In a recent study, Tsermpini et al. (2017) investigated the role of putative genetic risk variants in lithium-associated reduction of the estimated GFR (eGFR) in BD patients under long-term lithium therapy. This study suggested that the *ACCN1* rs378448 SNP (which was found to interact with duration of lithium treatment in a model predicting eGFR), could predispose BD patients to renal side effects.

Another promising biomarker of mood stabilizer-induced side effects is represented by the guanine nucleotide-binding protein subunit β-3 (GNB3) gene. Specifically, this gene has been suggested to play a role in metabolic abnormalities associated with treatment with valproate (Chang et al., 2010; Chen et al., 2017). In 2010, the T allele of the rs5443 (C825T) polymorphism was found to be associated with lower body mass index (BMI) and serum leptin levels in BD patients treated with valproate (Chang et al., 2010). This result was recently replicated by a longitudinal study including 100 BD type II patients treated with valproate for 12 weeks (Chen et al., 2017). At the end of the study, carriers of the TT genotype showed lower BMI, smaller waist circumference, and better lipid profiles compared patients with the CT or CC genotypes undergoing the same valproate treatment (Chen et al., 2017).

Carbamazepine causes cutaneous adverse drug reactions (ADRs) in up to 10% of patients (Kaniwa & Saito, 2013), including life-threatening ADRs such as Stevens-Johnson syndrome (SJS) and toxic epidermal necrolysis (TEN). These reactions are significantly more common in patients carrying the human leukocyte antigen (HLA) B*1502 allele, although this association has only been shown in patients of Southeast Asian origin (Kaniwa & Saito, 2013; Yip, Marson, Jorgensen, Pirmohamed, & Alfirevic, 2012). This observation led the Food and Drug Administration (FDA) regulatory agency to recommend genetic screening for patients of Asian ancestry before starting treatment with carbamazepine.

Overall, a scarce number of studies explored how genetic biomarkers could predispose patients to mood stabilizer-induced side effects, and further research is needed to obtain useful biomarkers that could be implemented in the clinical practice.

7 Conclusions and future perspective

Our knowledge of the pathophysiology of BD and pharmacogenomics of lithium salts remains limited. Studies to date report small effects, and most of the findings remain to be replicated. Pharmacogenomics approaches in BD have thus far focused on identifying genetic predictors of treatment response to mood stabilizers, especially lithium. To date, it seems clear that lithium responsive BD patients present common clinical characteristics, and could be characterized by a lower genetic heterogeneity. Although the exact mechanism by which lithium stabilizes mood is still obscure, this drug still remains the gold standard in the management of BD. Thus, the study of the neurobiology and pharmacogenomics of lithium has been very active, and findings look promising for gaining a better understanding and treatment of BD.

Overall, considerable evidence supports a neurodevelopmental contribution to BD (O'Shea & McInnis, 2016). The human brain shows considerable plasticity throughout the lifespan, particularly in the neural stem cell zones of the dentate gyrus and the hippocampal subventricular zone of the lateral ventricles (Corrêa-Velloso et al., 2018). The direct or indirect regulation of cellular plasticity pathway and cell stress protection by lithium can explain its effectiveness in treatment of BD (Machado-Vieira, 2018). Human-induced pluripotent stem cells (iPSC) technology, that is, the ability to reprogram adult somatic tissues to a pluripotent state, will allow us to investigate these and other pathways in viable cell models derived directly from BD patients characterized for lithium response (Di Lullo & Kriegstein, 2017; Pickard, 2017). This approach was recently used by Tobe et al. (2017) to identify a "lithium-response pathway" supporting the involvement of the cytoskeleton regulator phosphorylation of collapsin response mediator protein 2 (CRMP2) in lithium's mechanism of action. The authors showed the ratio between the inactive (phosphorylated) and active (nonphosphorylated) CRMP2 forms to be dysregulated in lithium responsive BD patients, while lithium was shown to be able to normalize this parameter (Tobe et al., 2017).

Stern et al. (2018) studied electrophysiologically hyperexcitable neurons from hippocampal dentate gyrus (DG) differentiated from iPSC-derived fibroblasts from BD patients (Stern et al., 2018). An extensive functional analysis showed large differences in cell parameters between neurons derived from lithium responders compared with lithium nonresponder BD patients. Chronic lithium treatment was also found to reduce the hyperexcitability in lymphoblastoid cell lines derived from lithium responders compared with lithium nonresponders, supporting the validity of these human cell lines to study BD and lithium response.

These innovative approaches will be of great help to advance our understanding of the pathogenesis of BD and of the mechanisms underlying lithium response, and will hopefully allow us to interpret the existing body of literature on lithium pharmacogenetics in light of new knowledge on dynamic mechanisms such as electrophysiological features or cell behaviors in response to different stimuli.

References

Alda, M., Grof, P., Rouleau, G. A., Turecki, G., & Young, L. T. (2005). Investigating responders to lithium prophylaxis as a strategy for mapping susceptibility genes for bipolar disorder. *Progress in Neuro-Psychopharmacology & Biological Psychiatry, 29,* 1038–1045.

Alda, M. (2015). Lithium in the treatment of bipolar disorder: Pharmacology and pharmacogenetics. *Molecular Psychiatry, 20,* 661–670.

Alloy, L. B., Ng, T. H., Titone, M. K., & Boland, E. M. (2017). Circadian rhythm dysregulation in bipolar spectrum disorders. *Current Psychiatry Reports, 19,* 21.

Benedetti, F., Serretti, A., Pontiggia, A., Bernasconi, A., Lorenzi, C., Colombo, C., & Smeraldi, E. (2005). Long-term response to lithium salts in bipolar illness is influenced by the glycogen synthase kinase 3-beta −50 T/C SNP. *Neuroscience Letters, 376,* 51–55.

Berridge, M. J., Downes, C. P., & Hanley, M. R. (1989). Neural and developmental actions of lithium: A unifying hypothesis. *Cell, 59,* 411–419.

Bremer, T., Diamond, C., McKinney, R., Shehktman, T., Barrett, T. B., Herold, C., & Kelsoe, J. R. (2007). The pharmacogenetics of lithium response depends upon clinical co-morbidity. *Molecular Diagnosis & Therapy, 11,* 161–170.

Cade, J. F. (1949). Lithium salts in the treatment of psychotic excitement. *The Medical Journal of Australia, 2,* 349–352.

Calabrese, F., Rossetti, A. C., Racagni, G., Gass, P., Riva, M. A., & Molteni, R. (2014). Brain-derived neurotrophic factor: A bridge between inflammation and neuroplasticity. *Frontiers in Cellular Neuroscience, 8,* 430.

Campos-de-Sousa, S., Guindalini, C., Tondo, L., Munro, J., Osborne, S., Floris, G., … Collier, D. (2010). Nuclear receptor rev-erb-{alpha} circadian gene variants and lithium carbonate prophylaxis in bipolar affective disorder. *Journal of Biological Rhythms, 25,* 132–137.

Chang, H. H., Gean, P. W., Chou, C. H., Yang, Y. K., Tsai, H. C., Lu, R. B., & Chen, P. S. (2010). C825T polymorphism of the GNB3 gene on valproate-related metabolic abnormalities in bipolar disorder patients. *Journal of Clinical Psychopharmacology, 30,* 512–517.

Chen, C. H., Lee, C. S., Lee, M. T., Ouyang, W. C., Chen, C. C., Chong, M. Y., … Consortium, T. B. (2014). Variant GADL1 and response to lithium therapy in bipolar I disorder. *The New England Journal of Medicine, 370,* 119–128.

Chen, P. S., Chang, H. H., Huang, C. C., Lee, C. C., Lee, S. Y., Chen, S. L., … Lu, R. B. (2017). A longitudinal study of the association between the GNB3 C825T polymorphism and metabolic disturbance in bipolar II patients treated with valproate. *The Pharmacogenomics Journal, 17,* 155–161.

Chiesa, A., Crisafulli, C., Porcelli, S., Han, C., Patkar, A. A., Lee, S. J., … Pae, C. U. (2012). Influence of GRIA1, GRIA2 and GRIA4 polymorphisms on diagnosis and response to treatment in patients with major depressive disorder. *European Archives of Psychiatry and Clinical Neuroscience, 262,* 305–311.

Cipriani, A., Hawton, K., Stockton, S., & Geddes, J. R. (2013). Lithium in the prevention of suicide in mood disorders: Updated systematic review and meta-analysis. *BMJ, 346,* f3646.

Consortium on Lithium Genetics, Hou, L., Heilbronner, U., Rietschel, M., Kato, T., Kuo, P. H., … Schulze, T. G. (2014). Variant GADL1 and response to lithium in bipolar I disorder. *The New England Journal of Medicine, 370,* 1857–1859.

Corrêa-Velloso, J. C., Gonçalves, M. C., Naaldijk, Y., Oliveira-Giacomelli, Á., Pillat, M. M., & Ulrich, H. (2018). Pathophysiology in the comorbidity of bipolar disorder and Alzheimer's disease: Pharmacological and stem cell approaches. *Progress in Neuro-Psychopharmacology and Biological Psychiatry, 80,* 34–53.

Craddock, N., & Jones, I. (1999). Genetics of bipolar disorder. *Journal of Medical Genetics, 36,* 585–594.

Craddock, N., & Sklar, P. (2013). Genetics of bipolar disorder. *Lancet, 381,* 1654–1662.

Crump, C., Sundquist, K., Winkleby, M. A., & Sundquist, J. (2013). Comorbidities and mortality in bipolar disorder: A Swedish national cohort study. *JAMA Psychiatry, 70,* 931–939.

Cuthbert, B. N., & Insel, T. R. (2013). Toward the future of psychiatric diagnosis: The seven pillars of RDoC. *BMC Medicine, 11,* 126.

Del Zompo, M., Ardau, R., Palmas, M., Bocchetta, A., Reina, A., & Piccardi, M. P. (1999). Lithium response: Association study with two candidate genes. *Molecular Psychiatry, 4,* S66.

Di Lullo, E., & Kriegstein, A. R. (2017). The use of brain organoids to investigate neural development and disease. *Nature Reviews Neuroscience, 18,* 573–584.

Dimitrova, A., Milanova, V., Krastev, S., Nikolov, I., Toncheva, D., Owen, M. J., & Kirov, G. (2005). Association study of myo-inositol monophosphatase 2 (IMPA2) polymorphisms with bipolar affective disorder and response to lithium treatment. *The Pharmacogenomics Journal, 5,* 35–41.

Dmitrzak-Weglarz, M., Rybakowski, J. K., Suwalska, A., Słopień, A., Czerski, P. M., Leszczyńska-Rodziewicz, A., & Hauser, J. (2005). Association studies of 5-HT2A and 5-HT2C serotonin receptor gene polymorphisms with prophylactic lithium response in bipolar patients. *Pharmacological Reports, 57,* 761–765.

Dmitrzak-Weglarz, M., Rybakowski, J. K., Suwalska, A., Skibinska, M., Leszczynska-Rodziewicz, A., Szczepankiewicz, A., & Hauser, J. (2008). Association studies of the BDNF and the NTRK2 gene polymorphisms with prophylactic lithium response in bipolar patients. *Pharmacogenomics, 9,* 1595–1603.

Drago, A., Serretti, A., Smith, R., Huezo-Diaz, P., Malitas, P., Albani, D., ... Aitchison, K. J. (2010). No association between genetic markers in BDNF gene and lithium prophylaxis in a Greek sample. *International Journal of Psychiatry in Clinical Practice, 14*, 154–157.

Geddes, J. R., & Miklowitz, D. J. (2013). Treatment of bipolar disorder. *Lancet, 381*, 1672–1682.

Goes, F. S. (2016). Genetics of bipolar disorder: Recent update and future directions. *Psychiatric Clinics of North America, 39*, 139–155.

Goldman, N., Glei, D. A., Lin, Y. H., & Weinstein, M. (2010). The serotonin transporter polymorphism (5-HTTLPR): Allelic variation and links with depressive symptoms. *Depression and Anxiety, 27*, 260–269.

Grace, A. A. (2016). Dysregulation of the dopamine system in the pathophysiology of schizophrenia and depression. *Nature Reviews Neuroscience, 17*, 524–532.

Graham, R. K., Tavella, G., & Parker, G. B. (2017). Is there consensus across international evidence-based guidelines for the psychotropic drug management of bipolar disorder during the perinatal period? *Journal of Affective Disorders, 228*, 216–221.

Grande, I., Berk, M., Birmaher, B., & Vieta, E. (2016). Bipolar disorder. *Lancet, 387*, 1561–1572.

Grof, P., Duffy, A., Cavazzoni, P., Grof, E., Garnham, J., MacDougall, M., ... Alda, M. (2002). Is response to prophylactic lithium a familial trait? *The Journal of Clinical Psychiatry, 63*, 942–947.

Hou, L., Heilbronner, U., Degenhardt, F., Adli, M., Akiyama, K., Akula, N., ... Schulze, T. G. (2016). Genetic variants associated with response to lithium treatment in bipolar disorder: A genome-wide association study. *Lancet, 387*, 1085–1093.

Ikeda, M., Saito, T., Kondo, K., & Iwata, N. (2017). Genome-wide association studies of bipolar disorder: A systematic review of recent findings and their clinical implications. *Psychiatry and Clinical Neurosciences*. https://doi.org/10.1111/pcn.12611.

Iwahashi, K., Nishizawa, D., Narita, S., Numajiri, M., Murayama, O., Yoshihara, E., ... Ishigooka, J. (2014). Haplotype analysis of GSK-3β gene polymorphisms in bipolar disorder lithium responders and nonresponders. *Clinical Neuropharmacology, 37*, 108–110.

Kaniwa, N., & Saito, Y. (2013). The risk of cutaneous adverse reactions among patients with the HLA-A* 31:01 allele who are given carbamazepine, oxcarbazepine or eslicarbazepine: A perspective review. *Therapeutic Advances in Drug Safety, 4*, 246–253.

Kakiuchi, C., & Kato, T. (2005). Lithium response and −116C/G polymorphism of XBP1 in Japanese patients with bipolar disorder. *International Journal of Neuropsychopharmacology, 8*, 631–632.

Kessing, L. V., Vradi, E., McIntyre, R. S., & Andersen, P. K. (2015). Causes of decreased life expectancy over the life span in bipolar disorder. *Journal of Affective Disorders, 180*, 142–147.

Kim, B., Kim, C. Y., Lee, M. J., & Joo, Y. H. (2009). Preliminary evidence on the association between XBP1-116C/G polymorphism and response to prophylactic treatment with valproate in bipolar disorders. *Psychiatry Research, 168*, 209–212.

Lee, H. Y., & Kim, Y. K. (2010). Catechol-O-methyltransferase Val158Met polymorphism affects therapeutic response to mood stabilizer in symptomatic manic patients. *Psychiatry Research, 175*, 63–66.

Lin, Y. F., Huang, M. C., & Liu, H. C. (2013). Glycogen synthase kinase 3beta gene polymorphisms may be associated with bipolar I disorder and the therapeutic response to lithium. *Journal of Affective Disorders, 147*, 401–406.

Løvlie, R., Berle, J. O., Stordal, E., & Steen, V. M. (2001). The phospholipase C-gamma1 gene (PLCG1) and lithium-responsive bipolar disorder: Re-examination of an intronic dinucleotide repeat polymorphism. *Psychiatric Genetics, 11*, 41–43.

Machado-Vieira, R. (2018). Lithium, stress, and resilience in bipolar disorder: Deciphering this key homeostatic synaptic plasticity regulator. *Journal of Affective Disorders, 233*, 92–99. https://doi.org/10.1016/j.jad.2017.12.026.

Manchia, M., Congiu, D., Squassina, A., Lampus, S., Ardau, R., Chillotti, C., ... Del Zompo, M. (2009). No association between lithium full responders and the DRD1, DRD2, DRD3, DAT1, 5-HTTLPR and HTR2A genes in a Sardinian sample. *Psychiatry Research, 169*, 164–166.

Manchia, M., Adli, M., Akula, N., Ardau, R., Aubry, J. M., Backlund, L., ... Alda, M. (2013). Assessment of response to lithium maintenance treatment in bipolar disorder: A Consortium on Lithium Genetics (ConLiGen) Report. *PLoS One, 8*.

Manji, H. K., Quiroz, J. A., Payne, J. L., Singh, J., Lopes, B. P., Viegas, J. S., & Zarate, C. A. (2003). The underlying neurobiology of bipolar disorder. *World Psychiatry, 2*, 136–146.

Masui, T., Hashimoto, R., Kusumi, I., Suzuki, K., Tanaka, T., Nakagawa, S., ... Koyama, T. (2006). A possible association between the −116C/G single nucleotide polymorphism of the XBP1 gene and lithium prophylaxis in bipolar disorder. *International Journal of Neuropsychopharmacology, 9*, 83–88.

McCarthy, M. J., Nievergelt, C. M., Shekhtman, T., Kripke, D. F., Welsh, D. K., & Kelsoe, J. R. (2011). Functional genetic variation in the Rev-Erbalpha pathway and lithium response in the treatment of bipolar disorder. *Genes, Brain, and Behavior, 10*, 852–861.

Merikangas, K. R., Jin, R., He, J. P., Kessler, R. C., Lee, S., Sampson, N. A., ... Zarkov, Z. (2011). Prevalence and correlates of bipolar spectrum disorder in the world mental health survey initiative. *Archives of General Psychiatry, 68*, 241–251.

Michelon, L., Meira-Lima, I., Cordeiro, Q., Miguita, K., Breen, G., Collier, D., & Vallada, H. (2006). Association study of the INPP1, 5HTT, BDNF, AP-2beta and GSK-3beta GENE variants and restrospectively scored response to lithium prophylaxis in bipolar disorder. *Neuroscience Letters, 403*, 288–293.

Miehe, S., Bieberstein, A., Arnould, I., Ihdene, O., Rütten, H., & Strübing, C. (2010). The phospholipid-binding protein SESTD1 is a novel regulator of the transient receptor potential channels TRPC4 and TRPC5. *The Journal of Biological Chemistry, 285*, 12426–12434.

Mitjans, M., Arias, B., Jimenez, E., Goikolea, J. M., Saiz, P. A., Garcia-Portilla, M. P., ... Benabarre, A. (2015). Exploring genetic variability at PI, GSK3, HPA, and glutamatergic pathways in lithium response: Association with IMPA2, INPP1, and GSK3B genes. *Journal of Clinical Psychopharmacology, 35*, 600–604.

Moreira, J., & Geoffroy, P. A. (2016). Lithium and bipolar disorder: Impacts from molecular to behavioural circadian rhythms. *Chronobiology International, 33*, 351–373.

O'Shea, K. S., & McInnis, M. G. (2016). Neurodevelopmental origins of bipolar disorder: IPSC models. *Molecular and Cellular Neuroscience, 73*, 63–83.

Ono, S. J., Liou, H. C., Davidon, R., Strominger, J. L., & Glimcher, L. H. (1991). Human X-box-binding protein 1 is required for the transcription of a subset of human class II major histocompatibility genes and forms a heterodimer with c-fos. *Proceedings of the National Academy of Sciences, 88*, 4309–4312.

Pae, C. U., Chiesa, A., Porcelli, S., Han, C., Patkar, A. A., Lee, S. J., … De Ronchi, D. (2012). Influence of BDNF variants on diagnosis and response to treatment in patients with major depression, bipolar disorder and schizophrenia. *Neuropsychobiology, 65*, 1–11.

Perlis, R. H., Smoller, J. W., Ferreira, M. A., McQuillin, A., Bass, N., Lawrence, J., … Purcell, S. (2009). A genomewide association study of response to lithium for prevention of recurrence in bipolar disorder. *The American Journal of Psychiatry, 166*, 718–725.

Pichler, E. M., Hattwich, G., Grunze, H., & Muehlbacher, M. (2015). Safety and tolerability of anticonvulsant medication in bipolar disorder. *Expert Opinion on Drug Safety, 14*, 1703–1724.

Pickard, B. S. (2017). Genomics of lithium action and response. *Neurotherapeutics, 14*, 582–587.

Polyakova, M., Stuke, K., Schuemberg, K., Mueller, K., Schoenknecht, P., & Schroeter, M. L. (2015). BDNF as a biomarker for successful treatment of mood disorders: A systematic & quantitative meta-analysis. *Journal of Affective Disorders, 174*, 432–440.

Prendes-Alvarez, S., & Nemeroff, C. B. (2018). Personalized medicine: Prediction of disease vulnerability in mood disorders. *Neuroscience Letters, 669*, 10–13.

Rybakowski, J. K., Suwalska, A., Czerski, P. M., Dmitrzak-Weglarz, M., Leszczynska-Rodziewicz, A., & Hauser, J. (2005). Prophylactic effect of lithium in bipolar affective illness may be related to serotonin transporter genotype. *Pharmacological Reports, 57*, 124–127.

Rybakowski, J. K., Suwalska, A., Skibinska, M., Dmitrzak-Weglarz, M., Leszczynska-Rodziewicz, A., & Hauser, J. (2007). Response to lithium prophylaxis: Interaction between serotonin transporter and BDNF genes. *American Journal of Medical Genetics, Part B: Neuropsychiatric Genetics, 144B*, 820–823.

Rybakowski, J. K., Dmitrzak-Weglarz, M., Suwalska, A., Leszczynska-Rodziewicz, A., & Hauser, J. (2009). Dopamine D1 receptor gene polymorphism is associated with prophylactic lithium response in bipolar disorder. *Pharmacopsychiatry, 42*, 20–22.

Rybakowski, J. K., Czerski, P., Dmitrzak-Weglarz, M., Kliwicki, S., Leszczynska-Rodziewicz, A., Permoda-Osip, A., … Hauser, J. (2012). Clinical and pathogenic aspects of candidate genes for lithium prophylactic efficacy. *Journal of Psychopharmacology, 26*, 368–373.

Schulze, T. G., Fangerau, H., & Propping, P. (2004). From degeneration to genetic susceptibility, from eugenics to genethics, from Bezugsziffer to LOD score: The history of psychiatric genetics. *International Review of Psychiatry, 16*, 246–259.

Schulze, T. (2010). Genetic research into bipolar disorder: The need for a research framework that integrates sophisticated molecular biology and clinically-informed phenotype characterization. *Psychiatric Clinics of North America, 33*, 67–82.

Schulze, T. G., Alda, M., Adli, M., Akula, N., Ardau, R., Bui, E. T., … McMahon, F. J. (2010). The International Consortium on Lithium Genetics (Con-LiGen): An initiative by the NIMH and IGSLI to study the genetic basis of response to lithium treatment. *Neuropsychobiology, 62*, 72–78.

Serretti, A., Lilli, R., Lorenzi, C., Franchini, L., & Smeraldi, E. (1998). Dopamine receptor D3 gene and response to lithium prophylaxis in mood disorders. *The International Journal of Neuropsychopharmacology, 1*, 125–129.

Serretti, A., Lilli, R., Lorenzi, C., Franchini, L., Di Bella, D., Catalano, M., & Smeraldi, E. (1999). Dopamine receptor D2 and D4 genes, GABA(A) alpha-1 subunit genes and response to lithium prophylaxis in mood disorders. *Psychiatry Research, 87*, 7–19.

Serretti, A., Lilli, R., Mandelli, L., Lorenzi, C., & Smeraldi, E. (2001). Serotonin transporter gene associated with lithium prophylaxis in mood disorders. *The Pharmacogenomics Journal, 1*, 71–77.

Serretti, A., Lorenzi, C., Lilli, R., Mandelli, L., Pirovano, A., & Smeraldi, E. (2002). Pharmacogenetics of lithium prophylaxis in mood disorders: Analysis of COMT, MAO-A, and Gbeta3 variants. *American Journal of Medical Genetics, 114*, 370–379.

Serretti, A., Lorenzi, C., Lilli, R., & Smeraldi, E. (2000). Serotonin receptor 2A, 2C, 1A genes and response to lithium prophylaxis in mood disorders. *Journal of Psychiatric Research, 34*, 89–98.

Serretti, A., Malitas, P. N., Mandelli, L., Lorenzi, C., Ploia, C., Alevizos, B., … Smeraldi, E. (2004). Further evidence for a possible association between serotonin transporter gene and lithium prophylaxis in mood disorders. *The Pharmacogenomics Journal, 4*, 267–273.

Serretti, A., Chiesa, A., Porcelli, S., Han, C., Patkar, A. A., Lee, S. J., … Pae, C. U. (2011). Influence of TPH2 variants on diagnosis and response to treatment in patients with major depression, bipolar disorder and schizophrenia. *Psychiatry Research, 189*, 26–32.

Shih, R. A., Belmonte, P. L., & Zandi, P. P. (2004). A review of the evidence from family, twin and adoption studies for a genetic contribution to adult psychiatric disorders. *International Review of Psychiatry, 16*, 260–283.

Silberberg, G., Levit, A., Collier, D., St Clair, D., Munro, J., Kerwin, R. W., … Navon, R. (2008). Stargazin involvement with bipolar disorder and response to lithium treatment. *Pharmacogenetics and Genomics, 18*, 403–412.

Sjøholt, G., Ebstein, R. P., Lie, R. T., Berle, J. Ø., Mallet, J., Deleuze, J. F., … Steen, V. M. (2004). Examination of IMPA1 and IMPA2 genes in manic-depressive patients: Association between IMPA2 promoter polymorphisms and bipolar disorder. *Molecular Psychiatry, 9*, 621–629.

Smoller, J. W., & Finn, C. T. (2003). Family, twin, and adoption studies of bipolar disorder. *American Journal of Medical Genetics Part C, Seminars in Medical Genetics, 123C*, 48–58.

Song, J., Bergen, S. E., Di Florio, A., Karlsson, R., Charney, A., Ruderfer, D. M., … Belliveau, R. A. (2016). Genome-wide association study identifies SESTD1 as a novel risk gene for lithium-responsive bipolar disorder. *Molecular Psychiatry, 21*, 1290–1297.

Squassina, A., Manchia, M., Borg, J., Congiu, D., Costa, M., Georgitsi, M., … Patrinos, G. P. (2011). Evidence for association of an ACCN1 gene variant with response to lithium treatment in Sardinian patients with bipolar disorder. *Pharmacogenomics, 12*, 1559–1569.

Steen, V. M., Lovlie, R., Osher, Y., Belmaker, R. H., Berle, J. O., & Gulbrandsen, A. K. (1998). The polymorphic inositol polyphosphate 1-phosphatase gene as a candidate for pharmacogenetic prediction of lithium-responsive manic-depressive illness. *Pharmacogenetics, 8*, 259–268.

Stern, S., Santos, R., Marchetto, M. C., Mendes, A. P., Rouleau, G. A., Biesmans, S., … Gage, F. H. (2018). Neurons derived from patients with bipolar disorder divide into intrinsically different sub-populations of neurons, predicting the patients' responsiveness to lithium. *Molecular Psychiatry, 23*(6), 1453–1465. https://doi.org/10.1038/mp.2016.260.

Strakowski, S. M., Fleck, D. E., & Maj, M. (2011). Broadening the diagnosis of bipolar disorder: Benefits vs. risks. *World Psychiatry, 10*, 181–186.

Szczepankiewicz, A., Skibinska, M., Suwalska, A., Hauser, J., & Rybakowski, J. K. (2009). No association of three GRIN2B polymorphisms with lithium response in bipolar patients. *Pharmacological Reports, 61*, 448–452.

Szczepankiewicz, A., Rybakowski, J. K., Suwalska, A., Skibinska, M., Leszczynska-Rodziewicz, A., Dmitrzak-Weglarz, M., … Hauser, J. (2006). Association study of the glycogen synthase kinase-3beta gene polymorphism with prophylactic lithium response in bipolar patients. *The World Journal of Biological Psychiatry, 7*, 158–161.

Szczepankiewicz, A. (2013). Evidence for single nucleotide polymorphisms and their association with bipolar disorder. *Neuropsychiatric Disease and Treatment, 9*, 1573–1582.

Tharoor, H., Kotambail, A., Jain, S., Sharma, P. S., & Satyamoorthy, K. (2013). Study of the association of serotonin transporter triallelic 5-HTTLPR and STin2 VNTR polymorphisms with lithium prophylaxis response in bipolar disorder. *Psychiatric Genetics, 23*, 77–81.

Tighe, S. K., Mahon, P. B., & Potash, J. B. (2011). Predictors of lithium response in bipolar disorder. *Therapeutic Advances in Chronic Disease, 2*, 209–226.

Tobe, B. T. D., Crain, A. M., Winquist, A. M., Calabrese, B., Makihara, H., Zhao, W. N., … Snyder, E. Y. (2017). Probing the lithium-response pathway in hiPSCs implicates the phosphoregulatory set-point for a cytoskeletal modulator in bipolar pathogenesis. *Proceedings of the National Academy of Sciences, 114*, E4462–E4471.

Tsermpini, E., Zhang, Y., Niola, P., Chillotti, C., Ardau, R., Bocchetta, A., … Squassina, A. (2017). Pharmacogenetics of lithium effects on glomerular function in bipolar disorder patients under chronic lithium treatment: A pilot study. *Neuroscience Letters, 638*, 1–4.

Turecki, G., Grof, P., Cavazzoni, P., Duffy, A., Grof, E., Ahrens, B., … Alda, M. (1998). Evidence for a role of phospholipase C-gamma1 in the pathogenesis of bipolar disorder. *Molecular Psychiatry, 3*, 534–538.

Turecki, G., Grof, P., Cavazzoni, P., Duffy, A., Grof, E., Ahrens, B., … Alda, M. (1999). MAOA: Association and linkage studies with lithium responsive bipolar disorder. *Psychiatric Genetics, 9*, 13–16.

Wang, Z., Fan, J., Gao, K., Li, Z., Yi, Z., Wang, L., … Fang, Y. (2013). Neurotrophic tyrosine kinase receptor type 2 (NTRK2) gene associated with treatment response to mood stabilizers in patients with bipolar I disorder. *Journal of Molecular Neuroscience, 50*, 305–310.

Wang, Z., Li, Z., Chen, J., Huang, J., Yuan, C., Hong, W., … Fang, Y. (2012). Association of BDNF gene polymorphism with bipolar disorders in Han Chinese population. *Genes, Brain and Behavior, 11*, 524–528.

Wemmie, J. A., Taugher, R. J., & Kreple, C. J. (2013). Acid-sensing ion channels in pain and disease. *Nature Reviews Neuroscience, 14*, 461–471.

Yin, L., Wang, J., Klein, P. S., & Lazar, M. A. (2006). Nuclear receptor Rev-erbalpha is a critical lithium-sensitive component of the circadian clock. *Science, 311*, 1002–1005.

Yip, V. L., Marson, A. G., Jorgensen, A. L., Pirmohamed, M., & Alfirevic, A. (2012). HLA genotype and carbamazepine-induced cutaneous adverse drug reactions: A systematic review. *Clinical Pharmacology & Therapeutics, 92*, 757–765.

Zhou, M., Engel, K., & Wang, J. (2007). Evidence for significant contribution of a newly identified monoamine transporter (PMAT) to serotonin uptake in the human brain. *Biochemical Pharmacology, 73*, 147–154.

Further Reading

Alural, B., Genc, S., & Haggarty, S. J. (2017). Diagnostic and therapeutic potential of microRNAs in neuropsychiatric disorders: Past, present, and future. *Progress in Neuro-Psychopharmacology & Biological Psychiatry, 73*, 87–103.

da Silva, J., Gonçalves-Pereira, M., Xavier, M., & Mukaetova-Ladinska, E. B. (2013). Affective disorders and risk of developing dementia: Systematic review. *The British Journal of Psychiatry, 202*, 177–186.

Fountoulakis, K. N., Yatham, L., Grunze, H., Vieta, E., Young, A., Blier, P., … Moeller, H. J. (2017). The International College of Neuro-Psychopharmacology (CINP) treatment guidelines for bipolar disorder in adults (CINP-BD-2017), part 2: Review, grading of the evidence, and a precise algorithm. *International Journal of Neuropsychopharmacology, 20*, 121–179.

Hunsberger, J. G., Chibane, F. L., Elkahloun, A. G., Henderson, R., Singh, R., Lawson, J., … Chuang, D. M. (2015). Novel integrative genomic tool for interrogating lithium response in bipolar disorder. *Translational Psychiatry, 5*.

McCarthy, M. J., Nievergelt, C. M., Kelsoe, J. R., & Welsh, D. K. (2012). A survey of genomic studies supports association of circadian clock genes with bipolar disorder spectrum illnesses and lithium response. *PLoS One, 7*.

Takaesu, Y., Inoue, Y., Ono, K., Murakoshi, A., Futenma, K., Komada, Y., & Inoue, T. (2017). Circadian rhythm sleep-wake disorders as predictors for bipolar disorder in patients with remitted mood disorders. *Journal of Affective Disorders, 220*, 57–61.

Chapter 33

Pharmacogenomics of treatment response in major depressive disorder

Joanna M. Biernacka[a,b], Ahmed T. Ahmed[a], Balwinder Singh[a] and Mark A. Frye[a]

[a]Department of Psychiatry and Psychology, Mayo Clinic, Rochester, MN, United States, [b]Department of Health Sciences Research, Mayo Clinic, Rochester, MN, United States

1 Introduction

Major depressive disorder (MDD) is a serious psychiatric illness with a lifetime prevalence of about 12%–17% (Alonso et al., 2004; Andrade et al., 2003; Kessler et al., 2005). According to the World Health Organization (WHO), MDD has become the leading cause of disability worldwide (World Health Organization, 2017). The etiology of MDD is not well understood, but estimates based on twin studies suggest that about 40% of the variation in the liability to MDD is due to additive genetic effects (Sullivan, Neale, & Kendler, 2000). Despite the widespread use of a variety of antidepressant medications to treat MDD, many patients have a suboptimal response to pharmacotherapy, with less than one-third of patients achieving remission after the first adequate antidepressant trial (Trivedi et al., 2006).

Selection of the most appropriate medication for individual patients remains a challenge and, outside of relying on clinical patterns of symptom presentation and comorbid conditions, is usually based on a trial and error approach. Genetic variation contributes to individual differences in response to many medications, including antidepressants (Horstmann & Binder, 2009; Tansey et al., 2013). Lesko and Woodcock defined pharmacogenomics broadly as "the study of interindividual variations in whole-genome or candidate gene single-nucleotide polymorphism (SNP) maps, haplotype markers, and alterations in gene expression or inactivation that might be correlated with pharmacological function and therapeutic response" (Lesko & Woodcock, 2004). Pharmacogenomic studies aim to identify the genetic contributors to drug response, which may shed light on the mechanisms of action of the medications, as well as offer hope for improving clinical care by enabling individualized treatment selection (Mrazek, 2010; Weinshilboum, 2003).

Pharmacogenomic studies of antidepressant treatment response have progressed from candidate gene studies to genome-wide association studies (GWAS). Findings from candidate gene studies point to several genes that may influence antidepressant response, but many candidate gene study results have been inconsistent and difficult to replicate. Although several GWAS have been completed in an attempt to discover new genes contributing to antidepressant treatment outcomes, thus far these studies were generally underpowered, leading to few conclusive findings (Biernacka et al., 2015). Nevertheless, an estimation of the proportion of phenotypic variance explained by SNPs using a mixed linear model analysis of GWAS data demonstrated that common genetic variants may explain ∼40% of the variation in antidepressant response (Tansey et al., 2013), providing compelling motivation for further pharmacogenomic studies.

Recent research has included application of novel approaches, such as polygenic risk score analyses and machine learning methods, as well as incorporation of other biomarker data, to further advance our understanding of genetic influences on antidepressant response. Efforts have also included analyses of more specific intermediate phenotypes related to antidepressant response, which may be under more direct genetic influence. Such phenotypes can include specific symptoms, or components of depression and side effects associated with antidepressant treatment.

While current understanding of genetic contributors to antidepressant response is still very incomplete, several companies now offer pharmacogenomics tests intended to guide antidepressant selection. Such clinical implementation of pharmacogenomics knowledge to improve treatment selection for individual patients is clearly the ultimate goal of pharmacogenomics research. Yet evidence about the effectiveness of the currently available pharmacogenomics tests for guiding antidepressant selection is only beginning to emerge.

In this chapter, we review findings from candidate-gene and genome-wide studies of antidepressant pharmacogenomics, and discuss ongoing research efforts focused on identifying additional genomic variation that impacts

Personalized Psychiatry. https://doi.org/10.1016/B978-0-12-813176-3.00033-X

403

antidepressant response. We then summarize current clinical use of pharmacogenomics testing in treatment of MDD, and research aimed at understanding the clinical effectiveness of such testing. Finally, we discuss future directions in research and clinical implementation of antidepressant pharmacogenomics. As this chapter covers a wide range of topics related to antidepressant pharmacogenomics, specific areas are not covered in great detail. Instead, we offer a broad perspective on the topic, and point readers to other literature with in-depth coverage of specific subtopics.

2 Candidate gene pharmacogenetic studies of antidepressant response

Candidate gene studies of antidepressant response have focused on two types of candidate genes: those that encode proteins involved in drug-specific pharmacokinetics, and those that encode proteins that are known, or hypothesized, to be involved in antidepressant pharmacodynamics. Candidate gene studies have pointed to a number of genes that appear to influence antidepressant treatment outcomes, some of which have been supported by meta-analysis results. Several comprehensive reviews have summarized these studies (Fabbri, Minarini, Niitsu, & Serretti, 2014; Horstmann & Binder, 2009; Licinio & Wong, 2011). Here we present the main highlights of their findings.

2.1 Pharmacokinetic candidate genes

Pharmacokinetics involves processes such as drug absorption, metabolism, distribution, and elimination, which have an impact on the delivery of the drug to the target. Studies of genes involved in antidepressant pharmacokinetics have primarily focused on genes encoding drug metabolizing enzymes, especially certain members of the cytochrome P450 (CYP) gene family, and to a lesser extent, on genes that encode molecules involved in drug transport, particularly certain members of the adenosine triphosphate-binding cassette (ABC) transporter gene family. The CYP genes encode a family of proteins involved in drug metabolism; of the many CYP enzymes, several (i.e., CYP2D6, CYP2C19, CYP2C9, CYP3A4, and CYP1A2) play a particularly prominent role in metabolizing various antidepressants. Many functional variants have been identified in these genes (see https://www.pharmvar.org/ and Gaedigk et al., 2018), which can reduce, sometimes drastically, or increase enzyme levels or activity. Some of these variants have been studied extensively in relation to antidepressant pharmacokinetics.

The *CYP2D6* gene is by far the most studied CYP gene in relation to antidepressants (Kirchheiner & Rodriguez-Antona, 2009; Zanger & Schwab, 2013). Based on the two copies of the *CYP2D6* gene region that an individual carries, they can be classified as poor metabolizers (PMs), intermediate metabolizers (IMs), extensive metabolizers (EMs), and ultrarapid metabolizers (UMs) (Nassan et al., 2016), with some classification schemes using a finer gradation with more levels of metabolism activity. Numerous studies have demonstrated a relationship between *CYP2D6* genetic variation and plasma drug levels for specific medications that are metabolized primarily by CYP2D6. Hicks and colleagues have summarized the findings of many of these studies, and used them to develop selective serotonin reuptake inhibitor (SSRI) and tricyclic antidepressant (TCA) dosing recommendations (Hicks et al., 2015, 2017). Based on such evidence, it has been hypothesized that PMs would have an increased risk of toxic reactions, while UMs might not benefit from treatment at standard doses because they would fail to achieve therapeutic plasma drug levels. For some medications, association of CYP genetic variants (or metabolizer classes defined by genotypes) with adverse side effects has been observed (e.g., QT prolongation, seizures, and death) (Funk & Bostwick, 2013; Nassan et al., 2016), sometimes contributing to drug warning label revisions (Ahmed, Weinshilboum, & Frye, 2018).

While studies looking at functional variants in specific CYP genes have advanced our understanding of antidepressant drug metabolism and its relationship to serum drug levels and side effect burden or specific adverse events, evidence linking these genes to clinical response to many commonly used antidepressants is less compelling. For example, for the commonly used SSRIs, there does not appear to be a strong relationship between CYP-related drug metabolism and clinical response, despite a well-established influence of CYP genetic variation on blood drug levels, suggesting that metabolizer status dependent dose selection might not have a large clinical impact. Nevertheless, a recent large meta-analysis was able to demonstrate a significant association of CYP2C19 activity derived from genotypes with citalopram/escitalopram side effects and percentage symptom improvement and remission (Fabbri et al., 2018).

The ABC transporter p-glycoprotein, encoded by the *ABCB1* gene (also known as *MDR1*), is a member of a superfamily of ABC transporter enzymes that regulate the transport of certain antidepressants across the blood–brain barrier (O'Brien, Dinan, Griffin, & Cryan, 2012). Due to its potential role in modulating the brain concentration of certain antidepressants, *ABCB1* has been studied as a candidate gene in antidepressant response. A series of studies have shown that the central nervous system (CNS) bioavailability of some SSRIs, serotonin norepinephrine reuptake inhibitors (SNRIs), and TCAs appears to be regulated by the *ABCB1*-encoded protein. Results of the studies investigating the effect of *ABCB1* variation on clinical response to these medications have been contradictory, although a meta-analysis provided weak evidence of association of *ABCB1* variants with antidepressant response (Niitsu, Fabbri, Bentini, & Serretti, 2013).

2.2 Pharmacodynamic candidate genes

Current understanding of antidepressant pharmacodynamics, that is, the direct effect of the drug at its target, is far less advanced than our understanding of antidepressant pharmacokinetics. Although the receptor and transporter binding targets are at least partially known for most antidepressants, the complete mechanisms of action of these medications, and how they ultimately impact clinical symptoms, are mostly unknown. Selecting appropriate pharmacodynamic candidate genes thus presents a formidable challenge. Nevertheless, a number of strong candidates have been investigated, mostly because of their role in the monoaminergic system, which has been hypothesized to play a role in depression and antidepressant mechanisms of action (Fabbri, Di Girolamo, & Serretti, 2013).

The serotonin transporter gene, *SLC6A4*, has been the most studied gene in the context of antidepressant response (Licinio & Wong, 2011; Serretti, Kato, De Ronchi, & Kinoshita, 2007). The most widely studied *SLC6A4* polymorphism is a 44 base pair insertion/deletion variant in the promoter region (5-HTTLPR), which results in "short" (S) and "long" (L) alleles; the S allele reduces the transcriptional efficiency of the serotonin transporter gene promoter, resulting in decreased expression of the transporter, and therefore lower serotonin uptake. A number of studies have also investigated antidepressant response in relation to the variable number of 16–17 bp tandem repeats in intron 2 (VNTR) of *SLC6A4*, known as STin2. This variant results in 9-, 10-, and 12-repeat common alleles, and also alters expression of the gene. Meta-analyses led to the conclusion that the L allele is associated with reduced risk of side effects, and predicts better response to SSRIs, particularly in Caucasian populations, while an opposite effect has been observed in Asian populations (Fabbri et al., 2013; Kato & Serretti, 2010; Porcelli, Fabbri, & Serretti, 2012). However, other differences between the studies could have contributed to the different genetic findings across ethnicities, as discussed by Fabbri et al. (2014). Far less is known about the relationship between antidepressant response rates and genotypes at STin2, although some evidence suggests it may also influence antidepressant response (Kato & Serretti, 2010; Niitsu et al., 2013), and the work of Mrazek and colleagues suggested that the haplotype of 5-HTTLPR, STin2 and SNP rs25531 in the promoter region may be associated with remission following citalopram treatment (Mrazek et al., 2009).

In addition to *SLC6A4*, variations in dozens of other pharmacodynamics candidate genes (including *HTR1A*, *HTR2A*, *COMT*, *GNB3*, *CNR1*, *NPY*, *MAOA*, *FKBP5*, and *BDNF*) have also been investigated in relation to antidepressant treatment response (Fabbri et al., 2014; Horstmann & Binder, 2009; Niitsu et al., 2013). For example, it has been reported that MDD patients with the *COMT* 158 val/val genotype treated with antidepressants have a worse response after 4–6 weeks of treatment (Baune et al., 2008). Several papers presented a series of meta-analyses for multiple antidepressant pharmacogenomics candidate genes. Based on meta-analyses at 15 variants in 11 genes (which did not include the 5-HTTLPR), Niitsu et al. concluded that the *BDNF* Val66Met variants was the best single candidate involved in antidepressant response, with an effect on SSRI treatment outcomes (Niitsu et al., 2013). Fabbri and colleagues summarized a broader set of candidate gene studies, including genetic variants with less evidence, and concluded that the involvement of *SLC6A4*, *HTR1A*, *HTR2A*, *BDNF*, and *ABCB1* in SSRI response "seems plausible," and pointing to a few other promising candidates.

The review papers mentioned in this section discussed the contribution of candidate gene studies to our understanding of antidepressant pharmacogenomics, but they have also demonstrated that findings from these studies have been somewhat inconsistent and difficult to interpret. Importantly, the genetic effects observed in these studies explained a small proportion of the variation in clinical drug response. GWAS were thus undertaken in the hopes of identifying additional genetic factors that contribute to antidepressant treatment outcomes.

3 GWAS of antidepressant response

A number of GWAS of antidepressant treatment outcomes were performed beginning in 2009 (Garriock et al., 2010; Ising et al., 2009; Ji et al., 2013; Laje & McMahon, 2011; Uher et al., 2010), focusing on overall treatment outcomes such as "response" (defined as at least a 50% reduction in depression severity score), "remission" (defined as a posttreatment depression score below a prespecified cut-off), or percent reduction in depression severity scores (Table 1). Some of these GWAS datasets have also been used to investigate pharmacogenomics effects on other outcome variables, including blood drug levels (Ji et al., 2014) and side effects (Adkins et al., 2012); such analyses are addressed in the next subsection of this chapter.

Table 1 provides an overview of published GWAS of new cohorts focused on overall clinical outcomes, such as response and remission. The initial GWAS (Garriock et al., 2010; Ising et al., 2009; Ji et al., 2013; Laje & McMahon, 2011;

TABLE 1 Published GWAS studies of primary clinical outcomes of antidepressant treatment

Author/ date	Study name	Analyzed sample size (by ancestry)	Medication(s)	Primary outcome(s)	Top association findings
Ising et al. (2009)	MARS	$N = 339$ (Caucasian)	Various antidepressants	HAMD—early partial response (after 2 weeks), response, remission (at 5 weeks)	rs6989467 5' of CDH17 gene for early partial response ($P = 7.6 \times 10^{-7}$)
Uher et al. (2010)	GENDEP	$N = 706$ (European ancestry)	Nortriptyline ($N = 312$) and escitalopram ($N = 394$)	MADRS—percentage improvement over 12 weeks	rs2500535 (in UST gene) in nortriptyline subset ($P = 3.6 \times 10^{-8}$)
Garriock et al. (2010)	STAR*D	$N = 1491$ (1067 white non-Hispanic, 183 white Hispanic, 241 African American)	Citalopram	QIDS-SR—response and remission	rs6966038 (nearest gene UBE3C) for response ($P = 1.65 \times 10^{-7}$)
Ji et al. (2013)	PGRN-AMPS	$N = 499$ (European ancestry)	Citalopram or escitalopram	QIDS-C16—response and remission at 8 weeks and at last visit	rs11144870 in RFK gene for 8-week response ($P = 1.04 \times 10^{-6}$)
Biernacka et al. (2015)	ISPC	$N = 865$ (567 Asian ancestry, 298 European ancestry)	Citalopram	HAMD—%change and response at week 4	Rs9328202 (near RPS25P7) for % change ($P = 6.15 \times 10^{-7}$).
Tansey et al. (2012)	NEWMEDS (includes GENDEP sample)	$N = 1790$ (European ancestry)	SRIs ($N = 1222$) and NRIs ($N = 568$)	% reduction in depression severity from baseline to end of treatment (based on MADRS, HAMD or BDI)	rs10783282 upstream of ADCY6 ($P = 1.16 \times 10^{-6}$) in analysis of the SRI subset

Studies with no new samples (e.g., meta-analyses of prior publications) and studies of secondary outcomes (e.g., side effects) are not listed. Only genome-wide significant findings, or one most significant finding per study (if there were no genome-wide significant findings) are reported in the last column.
MARS, Munich Antidepressant Response Signature; GENDEP, Genome-Based Therapeutic Drugs for Depression; STAR*D, Sequenced Treatment Alternatives to Relieve Depression; PGRN-AMPS, Pharmacogenomic Research Network—Antidepressant Medication Pharmacogenomic Study; ISPC, International SSRI Pharmacogenomics Consortium; HAMD, Hamilton Depression Rating Scale; MADRS, Montgomery-Åsberg Depression Rating Scale; QIDS-SR, Quick Inventory of Depressive Symptomatology—Self-Reported; QIDS-C16, Quick Inventory of Depressive Symptomatology—Clinician Rated; BDI, Beck Depression Inventory; SRI, serotonin reuptake inhibitors; NRI, noradrenalin reuptake inhibitors; CDH17, Cadherin 17; UST, uronyl 2-sulphotransferase; UBE3C, ubiquitin protein ligase E3C; RFK, riboflavin kinase; RPS25P7, ribosomal protein S25 pseudogene 7; ADCY6, adenylate cyclase 6.

Uher et al., 2010) largely failed to identify genetic variants that were associated with treatment outcomes at a genome-wide statistical significance level, and top findings from these studies have generally failed to replicate in independent samples. The genome-based therapeutic drugs for depression (GENDEP) study was the only one with a finding reaching the standard threshold for genome-wide significance of 5×10^{-8} (rs2500535 in the UST gene, $P = 3.56 \times 10^{-8}$) in an analysis of response in a subset of patients treated with nortriptyline (Uher et al., 2010); however, even this finding does not withstand multiple-testing correction when the number of genome-wide analyses performed in this study is accounted for. Nevertheless, the results of this study led the authors to conclude that efficacy of antidepressants may be predicted by variants in genes that are not traditional candidates, and thus the GWAS approach may be important in uncovering antidepressant pharmacogenetic effects.

It is well known that GWAS of complex psychiatric traits have low power for discovery of relevant variants unless very large samples are analyzed. Several groups have therefore aggregated data across studies of antidepressant pharmacogenomics. Tansey and colleagues described a pharmacogenomic analysis of data from the NEWMEDS ("Novel Methods leading to New Medications in Depression and Schizophrenia") consortium, which included response to serotonergic and noradrenergic antidepressants in more than 2000 European-ancestry individuals with MDD (Tansey et al., 2012). The analyses also included a meta-analysis ($N = 2897$) with data from the sequenced treatment alternatives to relieve depression (STAR*D) study (Garriock et al., 2010), but failed to identify any common genetic variants associated with antidepressant response at a genome-wide significant level. Another meta-analysis combined the results of three prior studies [MARS (Ising et al., 2009), STAR*D (Garriock et al., 2010), and GENDEP (Uher et al., 2010)] to search for genetic variation associated with improvement and remission following treatment with antidepressants (Uher, Investigators, & Investigators, 2013); the authors also performed analyses limited to SSRI (escitalopram or citalopram) response. In the primary analyses including more than 2000 participants, no genome-wide significant associations were detected. A subanalysis of SSRI-treated patients revealed an association between early SSRI response (within 2 weeks of treatment) and a SNP in an intergenic region on chromosome 5. A study reported by the International SSRI Pharmacogenomics Consortium (ISPC) also failed to identify genetic variants with significant evidence of association with SSRI response (Biernacka et al., 2015). In a meta-analysis of the ISPC data with data from two prior GWAS of SSRI response, suggestive evidence of association with response and remission after 4 weeks of treatment was obtained for SNPs in the neuregulin-1 (*NRG1*) gene, which is involved in many aspects of brain development, including neuronal maturation, and contains variants associated with risk for mental disorders such as schizophrenia (Biernacka et al., 2015).

Although the GWAS approach offers the opportunity to discover new contributors to a complex trait, the GWAS completed to date have been underpowered to provide conclusive evidence for the involvement of specific genetic polymorphisms in antidepressant treatment outcomes. One main reason is the relatively small sample size of the current pharmacogenomics studies. Studies of other related psychiatric traits, such as MDD, have demonstrated a highly polygenic genetic architecture with many contributing risk loci with small effects (Wray et al., 2018), and antidepressant response appears to have a similarly complex polygenic architecture (Hodgson, Mufti, Uher, & McGuffin, 2012; Tansey et al., 2013). This polygenic architecture, with many genetic variants of small effects influencing the phenotype, necessitates the analysis of very large samples to detect the relevant loci. However, even with the sample sizes available from completed studies, alternative analytical approaches offer the possibility to gain valuable insights into antidepressant pharmacogenomics.

4   Antidepressant pharmacogenomics study design and definition of treatment outcomes: Impact on research findings and clinical relevance

It is recognized that combining data across many studies to produce very large samples is critical to identifying genetic variants contributing to complex psychiatric traits. However, this approach is particularly challenging in pharmacogenomics studies because of large differences between individual studies. For example, two GWAS of antidepressant response may differ in terms of the patient populations (e.g., differences in illness severity, different clinical subtypes and comorbidities, different ancestral background of study participants), medications (e.g., different classes of medications or different medications within the same class; differences in cotherapies) and/or dosages; or assessments of clinical improvement (e.g., different depression rating scales, different length of treatment or frequency of follow-up) (Ahmed et al., 2018). When data from studies with such differences are combined in a single analysis, the high degree of heterogeneity reduces the power to identify relevant genetic contributors to treatment outcomes. Even in a single-site study of a specific medication, heterogeneity in patient characteristics (e.g., comorbidities, illness severity, symptom profiles, and environmental exposures) can reduce the power to identify genetic variants. However, it should also be noted that although reducing patient and treatment heterogeneity in studies has advantages in terms of statistical power, use of very narrowly defined patient populations in research can lead to findings that lack generalizability to the broad range of patients seen in clinical practice.

Treatment outcome definitions used in research also impact the power of genetic studies and their clinical utility. The majority of pharmacogenomics studies of antidepressant response, particularly the large GWAS, have focused on overall clinical improvement measured by depression rating scales such as the Montgomery-Asperg Depression Rating Scale [MADRS (Montgomery & Asberg, 1979)], the Hamilton Depression Rating Scale [HAM-D (Hamilton, 1960; Rohan et al., 2016)], and the Quick Inventory of Depressive Symptomatology [QIDS (Rush et al., 2003)]. Typically used measures of treatment outcome include "remission" (posttreatment depression score below a prespecified cut-off), and "response" (at least a 50% reduction in depression severity score). Quantitative changes in depression rating scores, such as percent decrease in score, are also often considered. While these are clinically meaningful measures, they encompass a broad spectrum of signs and symptoms of depression, and are thus very heterogeneous with respect to biological underpinnings. Further discoveries of pharmacogenomics effects on antidepressant response can be expected when research focuses on more narrowly defined treatment outcomes, such as changes in specific symptoms or domains of depression, or clinical variables such as cognitive measures. Similarly, investigation of genetic effects on side effects resulting from antidepressant use has shed further light on why some patients do not benefit from treatment.

Studies of genetic effects on more specific outcomes have begun to illuminate antidepressant pharmacogenomics effects. For example, in candidate gene studies, genetic variants of the *SLC6A4* and the serotonin 2A receptor (*HTR2A*) genes were reported to be associated with SSRI side effects (Hicks et al., 2015), whereas a GWAS identified an association of a SNP in the *EMID2* gene ($P = 3.27 \times 10^{-8}$) with the effects of citalopram on vision and hearing (Adkins et al., 2012). Research presented at recent scientific meetings also demonstrates potential gain in power and interpretability of results from considering more specific symptoms or components of depression measured using individual questions on depression rating scales, as opposed to only considering changes in depression severity based on total scores on such scales (Zai et al., 2016).

5 Novel approaches to identifying pharmacogenomic effects and their application in studies of antidepressant response

While traditional candidate gene and GWAS continue to be used in search of specific genetic variants that impact a person's response to antidepressant treatment, researchers have begun using novel analytical approaches to further examine genetic effects on antidepressant response. These approaches include polygenic methods, machine learning, and integration of other types of biomarker data, such as metabolomics, in genetic analyses.

In an early application of the polygenic method in antidepressant pharmacogenomics, a polygenic score derived from a meta-analysis of GENDEP and MARS data accounted for about 1.2% of the variance in certain outcomes in STAR*D, suggesting a weak concordant genetic signal across many polymorphisms (Uher et al., 2013). In a more recently published study, polygenic risk scores for antidepressant treatment response in the STAR*D dataset derived using genetic effect estimates from the GENDEP dataset (and vice versa) failed to predict response (Garcia-Gonzalez et al., 2017; Hodgson et al., 2012). However, another recent analysis suggested that polygenic scores for certain personality traits, such as openness and neuroticism, may contribute to prediction of antidepressant treatment outcomes (Amare et al., 2018). Other researchers have considered machine learning methods to combine effects of variations in multiple genes to predict antidepressant treatment outcomes, but application of such methods is still in the early stages, and requires careful consideration of issues specific to genetic data analysis, such as the potential for confounding by population stratification.

Other novel approaches are focusing on integrating different types of biomarkers in predicting antidepressant response (Lam et al., 2016). Among several types of biomarkers, metabolomics measurements have been investigated in relation to antidepressant response, and the observed associations between treatment outcomes and metabolite levels have been used to guide further searches for genetic variation that contributes to antidepressant response (Gupta et al., 2016). These types of approaches, in combination with newly developed data mining and integrative statistical methods, as well as polygenic analyses, are expected to play a prominent role in discovery of additional genetic factors that influence treatment outcomes, and in the development of tools for individualizing treatment selection.

6 Antidepressant pharmacogenomics resources

As a means to assess and catalog the increasing knowledge regarding pharmacogenomic variation, primarily pharmacokinetic, and its potential clinical impact, the Clinical Pharmacogenomics Implementation Consortium (CPIC) was established as a joint effort between the Pharmacogenomics Knowledgebase (PharmGKB) and the Pharmacogenomics Research Network (PGRN). The goal of CPIC is to develop peer-reviewed guidelines related to clinical use of pharmacogenomic information. CPIC guidelines are currently available for a set of SSRIs and TCAs, though many of the guidelines are based on pharmacokinetic studies with small sample size or studies of healthy controls (Hicks et al., 2015, 2017). Another useful list of clinically relevant pharmacogenetic variations is maintained by the US Food and Drug Administration (FDA; https://www.fda.gov/Drugs/ScienceResearch/ucm572698.htm). Psychotropic drugs have the highest percentage of FDA pharmacogenomics-based drug warning label revisions (Weinshilboum & Wang, 2017). Most commonly, this has focused on the CYP2D6 or CYP2C19 PM phenotype and risk for QTc prolongation with select antidepressants.

7 The emergence of pharmacogenomic tests to individualize treatment of MDD

The potential of pharmacogenomic testing is widely accepted, and has been extensively discussed in psychiatric literature, and despite the somewhat inconclusive results of antidepressant pharmacogenomics research, several genetic tests are now marketed for guiding the selection of various psychotropic medications, including antidepressants. The growing list of available tests includes GeneSight, CNSDose, NeuroIDgenetix, Neuropharmagen, RightMed, and Genecept. These tests are typically based on predictions about potential response to a medication based on a limited number of variants in a handful of genes known to be involved in pharmacokinetics or pharmacodynamics of certain antidepressants (e.g., the CYP genes and the serotonin transporter and receptors). Some of the tests also take potential drug–drug interactions into account (Bradley et al., 2018). Given the high frequency of genetic variants that alter the expression of CYP genes involved in antidepressant metabolism, it is not surprising that attempts to use genetic information to guide antidepressant selection are on the rise.

While these tests are now available for clinical use, evidence for clinical benefits from these tests is only beginning to emerge and is still limited. A few relatively small studies have compared treatment response in groups that received pharmacogenomically-guided treatment versus unguided treatment (i.e., "treatment as usual"). These studies varied in quality of evidence; however, many had serious limitations, such as lack of randomization with an appropriate control, or lack of blinding (Zubenko, Sommer, & Cohen, 2018). Nevertheless, some provided preliminary evidence of improved treatment adherence and treatment outcomes either in overall reduction in depression symptoms or general side effect burden (Winner et al., 2015). Rosenblat and colleagues reviewed these studies, concluding that a limited number of studies have shown promise for pharmacogenomics testing in clinical treatment of MDD (Rosenblat, Lee, & McIntyre, 2017). However, they also noted that the field still lacks replicated evidence of improved health outcomes, or cost-effectiveness, resulting from use of these tests in MDD treatment. It is also worth noting that most of the cited studies were supported by a pharmacogenomic testing company, raising concerns of bias. Large, randomized, blinded, independent trials are now needed to establish the effectiveness of pharmacogenetic testing in psychiatry. It is also important to note that not all available tests are the same, and their effectiveness is not going to be equal for all types of medications. Thus, ongoing evaluation of each newly available or modified test will be needed to ensure that tests used in clinical care are benefiting patients.

Evidence about the potential cost-effectiveness of pharmacogenomic testing is also beginning to emerge. The first two retrospective studies that used relatively large pharmacy claims databases to compare data of patients that had undergone pharmacogenomics testing with data of patients with standard treatment, found that pharmacogenomic testing was associated with increased medication adherence and/or led to significant medication cost reductions (Fagerness et al., 2014; Winner et al., 2015). Subsequent studies supported these conclusions, showing reduced health care utilization and reduced costs associated with pharmacogenomics testing in treatment of depression (Maciel, Cullors, Lukowiak, & Garces, 2018; Perlis, Mehta, Edwards, Tiwari, & Imbens, 2018). In a more general assessment of potential cost-effectiveness of pharmacogenomics testing (not limited to psychiatric medications), Verbelen and colleagues concluded that pharmacogenomics-guided treatment can be a cost-effective and even a cost-saving strategy (Verbelen, Weale, & Lewis, 2017).

Ultimately, clinical implementation of pharmacogenetically-guided treatment selection in psychiatry will not only require development and evaluation of genetic tests, but also use of these tests by clinicians in their practice. A recent survey of primary care clinicians measured their interest in pharmacogenomic clinical decision support tool alerts integrated in electronic medical records (St. Sauver et al., 2016). Results of this study suggested a lack of clinician comfort with integrating pharmacogenomic data into primary care, and emphasized a need for further efforts to educate clinicians about the utility of pharmacogenomic information in clinical practice, as well as refining the clinical decision support tool alerts to make them more user-friendly.

8 Future considerations

Further progress in elucidating pharmacogenomics effects on antidepressant response will require large collaborative research partnerships and careful study design, as well as functional validation of any discovered genetic associations, along with investigation of mechanisms by which particular genes or polymorphisms influence treatment outcomes. There is also a need for evaluating pharmacogenomics effects across different patient populations, including special populations such as adolescent or geriatric patients, and populations of diverse ancestries. Murphy and McMahon noted that the majority of pharmacogenomics studies performed to date have focused on populations of European ancestry, and emphasized that meeting the goals of personalized psychotherapy will require large studies in more diverse populations. Furthermore, because SSRIs are the most commonly used medication class for MDD (Anderson, Pace, Libby, West, & Valuck, 2012; Smith et al., 2008), many of the prior pharmacogenomics studies of antidepressant response, particularly GWAS, have focused on this medication class; ultimately the genetic contribution to the response to other types of antidepressants also needs to be studied comprehensively.

Increased efforts focused on appropriate statistical modeling of expected treatment outcomes are also needed, to optimally combine all factors (genetic and nongenetic) that influence treatment outcomes into clinical decision-support tools. There is also a need for ongoing evaluation of emerging clinical pharmacogenetic tests, along with education of patients and providers about potential benefits, and current limitations, of pharmacogenomics testing. In 2004, Lesko and Woodcock, from the US FDA, noted that "translating pharmacogenomics from bench to bedside (or from discovery to marketability) is a multidisciplinary problem that involves addressing philosophical, societal, cultural, behavioral, and educational differences between the private and public sector, as well as issues unique to drug development, extent of scientific expertise, interdisciplinary communication, and clinical practice." All of these areas require ongoing attention for the successful application of pharmacogenomics testing in treatment of MDD.

9 Conclusion

The past two decades have drastically improved our understanding of antidepressant pharmacogenomics, but also revealed the complexity involved, and demonstrated that prediction of drug response based solely on genetic information still has fairly low accuracy. Nevertheless, given the vast progress being made, there is high hope that pharmacogenomic tests for guiding antidepressant selection will be significantly improved over the next decade, and will begin to have an important positive impact on patient outcomes.

References

Adkins, D. E., Clark, S. L., Aberg, K., Hettema, J. M., Bukszar, J., McClay, J. L., … van den Oord, E. J. (2012). Genome-wide pharmacogenomic study of citalopram-induced side effects in STAR*D. *Translational Psychiatry*, *2*, e129. https://doi.org/10.1038/tp.2012.57.

Ahmed, A. T., Weinshilboum, R., & Frye, M. A. (2018). Benefits of and barriers to pharmacogenomics-guided treatment for major depressive disorder. *Clinical Pharmacology and Therapeutics*, *103*(5), 767–769. https://doi.org/10.1002/cpt.1009.

Alonso, J., Angermeyer, M. C., Bernert, S., Bruffaerts, R., Brugha, T. S., Bryson, H., … Vollebergh, W. A. (2004). Prevalence of mental disorders in Europe: Results from the European Study of the Epidemiology of Mental Disorders (ESEMeD) project. *Acta Psychiatrica Scandinavica. Supplementum*, *109*(S420), 21–27. https://doi.org/10.1111/j.1600-0047.2004.00327.x.

Amare, A. T., Schubert, K. O., Tekola-Ayele, F., Hsu, Y. H., Sangkuhl, K., Jenkins, G., … Baune, B. T. (2018). Association of the polygenic scores for personality traits and response to selective serotonin reuptake inhibitors in patients with major depressive disorder. *Frontiers in Psychiatry*, *9*, 65. https://doi.org/10.3389/fpsyt.2018.00065.

Anderson, H. D., Pace, W. D., Libby, A. M., West, D. R., & Valuck, R. J. (2012). Rates of 5 common antidepressant side effects among new adult and adolescent cases of depression: A retrospective US claims study. *Clinical Therapeutics*, *34*(1), 113–123. https://doi.org/10.1016/j.clinthera.2011.11.024.

Andrade, L., Caraveo-Anduaga, J. J., Berglund, P., Bijl, R. V., De Graaf, R., Vollebergh, W., … Wittchen, H. U. (2003). The epidemiology of major depressive episodes: Results from the International Consortium of Psychiatric Epidemiology (ICPE) Surveys. *International Journal of Methods in Psychiatric Research*, *12*(1), 3–21.

Baune, B. T., Hohoff, C., Berger, K., Neumann, A., Mortensen, S., Roehrs, T., … Domschke, K. (2008). Association of the COMT val158met variant with antidepressant treatment response in major depression. *Neuropsychopharmacology*, *33*(4), 924–932. https://doi.org/10.1038/sj.npp.1301462.

Biernacka, J. M., Sangkuhl, K., Jenkins, G., Whaley, R. M., Barman, P., Batzler, A., … Weinshilboum, R. (2015). The International SSRI Pharmacogenomics Consortium (ISPC): A genome-wide association study of antidepressant treatment response. *Translational Psychiatry*, *5*, e553. https://doi.org/10.1038/tp.2015.47.

Bradley, P., Shiekh, M., Mehra, V., Vrbicky, K., Layle, S., Olson, M. C., … Lukowiak, A. A. (2018). Improved efficacy with targeted pharmacogenetic-guided treatment of patients with depression and anxiety: A randomized clinical trial demonstrating clinical utility. *Journal of Psychiatric Research*, *96*, 100–107. https://doi.org/10.1016/j.jpsychires.2017.09.024.

Fabbri, C., Di Girolamo, G., & Serretti, A. (2013). Pharmacogenetics of antidepressant drugs: An update after almost 20 years of research. *American Journal of Medical Genetics. Part B, Neuropsychiatric Genetics*, *162B*(6), 487–520. https://doi.org/10.1002/ajmg.b.32184.

Fabbri, C., Minarini, A., Niitsu, T., & Serretti, A. (2014). Understanding the pharmacogenetics of selective serotonin reuptake inhibitors. *Expert Opinion on Drug Metabolism & Toxicology*, *10*(8), 1093–1118. https://doi.org/10.1517/17425255.2014.928693.

Fabbri, C., Tansey, K. E., Perlis, R. H., Hauser, J., Henigsberg, N., Maier, W., … Lewis, C. M. (2018). Effect of cytochrome CYP2C19 metabolizing activity on antidepressant response and side effects: Meta-analysis of data from genome-wide association studies. *European Neuropsychopharmacology*, *28*(8), 945–954. https://doi.org/10.1016/j.euroneuro.2018.05.009.

Fagerness, J., Fonseca, E., Hess, G. P., Scott, R., Gardner, K. R., Koffler, M., … Lombard, J. (2014). Pharmacogenetic-guided psychiatric intervention associated with increased adherence and cost savings. *The American Journal of Managed Care*, *20*(5), e146–e156.

Funk, K. A., & Bostwick, J. R. (2013). A comparison of the risk of QT prolongation among SSRIs. *The Annals of Pharmacotherapy*, *47*(10), 1330–1341. https://doi.org/10.1177/1060028013501994.

Gaedigk, A., Ingelman-Sundberg, M., Miller, N. A., Leeder, J. S., Whirl-Carrillo, M., Klein, T. E., & PharmVar Steering, C. (2018). The pharmacogene variation (PharmVar) consortium: Incorporation of the human cytochrome P450 (CYP) allele nomenclature database. *Clinical Pharmacology and Therapeutics*, *103*(3), 399–401. https://doi.org/10.1002/cpt.910.

Garcia-Gonzalez, J., Tansey, K. E., Hauser, J., Henigsberg, N., Maier, W., Mors, O., … Fabbri, C. (2017). Pharmacogenetics of antidepressant response: A polygenic approach. *Progress in Neuro-Psychopharmacology & Biological Psychiatry*, *75*, 128–134. https://doi.org/10.1016/j.pnpbp.2017.01.011.

Garriock, H. A., Kraft, J. B., Shyn, S. I., Peters, E. J., Yokoyama, J. S., Jenkins, G. D., … Hamilton, S. P. (2010). A genomewide association study of citalopram response in major depressive disorder. *Biological Psychiatry, 67*(2), 133–138.

Gupta, M., Neavin, D., Liu, D., Biernacka, J., Hall-Flavin, D., Bobo, W. V., … Weinshilboum, R. M. (2016). TSPAN5, ERICH3 and selective serotonin reuptake inhibitors in major depressive disorder: Pharmacometabolomics-informed pharmacogenomics. *Molecular Psychiatry, 21*(12), 1717–1725. https://doi.org/10.1038/mp.2016.6.

Hamilton, M. (1960). A rating scale for depression. *Journal of Neurology, Neurosurgery, and Psychiatry, 23*, 56–62.

Hicks, J. K., Bishop, J. R., Sangkuhl, K., Muller, D. J., Ji, Y., Leckband, S. G., … Gaedigk, A. (2015). Clinical Pharmacogenetics Implementation Consortium (CPIC) Guideline for CYP2D6 and CYP2C19 genotypes and dosing of selective serotonin reuptake inhibitors. *Clinical Pharmacology and Therapeutics, 98*(2), 127–134. https://doi.org/10.1002/cpt.147.

Hicks, J. K., Sangkuhl, K., Swen, J. J., Ellingrod, V. L., Muller, D. J., Shimoda, K., … Stingl, J. C. (2017). Clinical pharmacogenetics implementation consortium guideline (CPIC) for CYP2D6 and CYP2C19 genotypes and dosing of tricyclic antidepressants: 2016 update. *Clinical Pharmacology and Therapeutics, 102*, 37–44. https://doi.org/10.1002/cpt.597.

Hodgson, K., Mufti, S. J., Uher, R., & McGuffin, P. (2012). Genome-wide approaches to antidepressant treatment: Working towards understanding and predicting response. *Genome Medicine, 4*(6), 52. https://doi.org/10.1186/gm351.

Horstmann, S., & Binder, E. B. (2009). Pharmacogenomics of antidepressant drugs. *Pharmacology & Therapeutics, 124*(1), 57–73. https://doi.org/10.1016/j.pharmthera.2009.06.007.

Ising, M., Lucae, S., Binder, E. B., Bettecken, T., Uhr, M., Ripke, S., … Muller-Myhsok, B. (2009). A genomewide association study points to multiple loci that predict antidepressant drug treatment outcome in depression. *Archives of General Psychiatry, 66*(9), 966–975.

Ji, Y., Biernacka, J. M., Hebbring, S., Chai, Y., Jenkins, G. D., Batzler, A., … Mrazek, D. A. (2013). Pharmacogenomics of selective serotonin reuptake inhibitor treatment for major depressive disorder: Genome-wide associations and functional genomics. *The Pharmacogenomics Journal, 13*(5), 456–463. https://doi.org/10.1038/tpj.2012.32.

Ji, Y., Schaid, D. J., Desta, Z., Kubo, M., Batzler, A. J., Snyder, K., … Weinshilboum, R. M. (2014). Citalopram and escitalopram plasma drug and metabolite concentrations: Genome-wide associations. *British Journal of Clinical Pharmacology, 78*(2), 373–383. https://doi.org/10.1111/bcp.12348.

Kato, M., & Serretti, A. (2010). Review and meta-analysis of antidepressant pharmacogenetic findings in major depressive disorder. *Molecular Psychiatry, 15*(5), 473–500. https://doi.org/10.1038/mp.2008.116.

Kessler, R. C., Berglund, P., Demler, O., Jin, R., Merikangas, K. R., & Walters, E. E. (2005). Lifetime prevalence and age-of-onset distributions of DSM-IV disorders in the National Comorbidity Survey Replication. *Archives of General Psychiatry, 62*(6), 593–602.

Kirchheiner, J., & Rodriguez-Antona, C. (2009). Cytochrome P450 2D6 genotyping: Potential role in improving treatment outcomes in psychiatric disorders. *CNS Drugs, 23*(3), 181–191. https://doi.org/10.2165/00023210-200923030-00001.

Laje, G., & McMahon, F. J. (2011). Genome-wide association studies of antidepressant outcome: A brief review. *Progress in Neuro-Psychopharmacology & Biological Psychiatry,.*

Lam, R. W., Milev, R., Rotzinger, S., Andreazza, A. C., Blier, P., Brenner, C., … Team, C.-B. I. (2016). Discovering biomarkers for antidepressant response: Protocol from the Canadian biomarker integration network in depression (CAN-BIND) and clinical characteristics of the first patient cohort. *BMC Psychiatry, 16*, 105. https://doi.org/10.1186/s12888-016-0785-x.

Lesko, L. J., & Woodcock, J. (2004). Translation of pharmacogenomics and pharmacogenetics: A regulatory perspective. *Nature Reviews. Drug Discovery, 3*(9), 763–769. https://doi.org/10.1038/nrd1499.

Licinio, J., & Wong, M. L. (2011). Pharmacogenomics of antidepressant treatment effects. *Dialogues in Clinical Neuroscience, 13*(1), 63–71.

Maciel, A., Cullors, A., Lukowiak, A. A., & Garces, J. (2018). Estimating cost savings of pharmacogenetic testing for depression in real-world clinical settings. *Neuropsychiatric Disease and Treatment, 14*, 225–230. https://doi.org/10.2147/NDT.S145046.

Montgomery, S. A., & Asberg, M. (1979). A new depression scale designed to be sensitive to change. *The British Journal of Psychiatry, 134*, 382–389.

Mrazek, D. (2010). *Psychiatric pharmacogenomics.* New York, NY: Oxford University Press.

Mrazek, D. A., Rush, A. J., Biernacka, J. M., O'Kane, D. J., Cunningham, J. M., Wieben, E. D., … Weinshilboum, R. M. (2009). SLC6A4 variation and citalopram response. *American Journal of Medical Genetics. Part B, Neuropsychiatric Genetics, 150B*(3), 341–351. https://doi.org/10.1002/ajmg.b.30816.

Nassan, M., Nicholson, W. T., Elliott, M. A., Rohrer Vitek, C. R., Black, J. L., & Frye, M. A. (2016). Pharmacokinetic pharmacogenetic prescribing guidelines for antidepressants: A template for psychiatric precision medicine. *Mayo Clinic Proceedings, 91*(7), 897–907. https://doi.org/10.1016/j.mayocp.2016.02.023.

Niitsu, T., Fabbri, C., Bentini, F., & Serretti, A. (2013). Pharmacogenetics in major depression: A comprehensive meta-analysis. *Progress in Neuro-Psychopharmacology & Biological Psychiatry, 45*, 183–194. https://doi.org/10.1016/j.pnpbp.2013.05.011.

O'Brien, F. E., Dinan, T. G., Griffin, B. T., & Cryan, J. F. (2012). Interactions between antidepressants and P-glycoprotein at the blood-brain barrier: Clinical significance of in vitro and in vivo findings. *British Journal of Pharmacology, 165*(2), 289–312. https://doi.org/10.1111/j.1476-5381.2011.01557.x.

Perlis, R. H., Mehta, R., Edwards, A. M., Tiwari, A., & Imbens, G. W. (2018). Pharmacogenetic testing among patients with mood and anxiety disorders is associated with decreased utilization and cost: A propensity-score matched study. *Depression and Anxiety.* https://doi.org/10.1002/da.22742.

Porcelli, S., Fabbri, C., & Serretti, A. (2012). Meta-analysis of serotonin transporter gene promoter polymorphism (5-HTTLPR) association with antidepressant efficacy. *European Neuropsychopharmacology, 22*(4), 239–258. https://doi.org/10.1016/j.euroneuro.2011.10.003.

Rohan, K. J., Rough, J. N., Evans, M., Ho, S. Y., Meyerhoff, J., Roberts, L. M., & Vacek, P. M. (2016). A protocol for the Hamilton rating scale for depression: Item scoring rules, rater training, and outcome accuracy with data on its application in a clinical trial. *Journal of Affective Disorders, 200*, 111–118. https://doi.org/10.1016/j.jad.2016.01.051.

Rosenblat, J. D., Lee, Y., & McIntyre, R. S. (2017). Does pharmacogenomic testing improve clinical outcomes for major depressive disorder? A Systematic Review of Clinical Trials and Cost-Effectiveness Studies. *The Journal of Clinical Psychiatry*, *78*(6), 720–729. https://doi.org/10.4088/JCP.15r10583.

Rush, A. J., Trivedi, M. H., Ibrahim, H. M., Carmody, T. J., Arnow, B., Klein, D. N., … Keller, M. B. (2003). The 16-item quick inventory of depressive symptomatology (QIDS), clinician rating (QIDS-C), and self-report (QIDS-SR): A psychometric evaluation in patients with chronic major depression. *Biological Psychiatry*, *54*(5), 573–583.

Serretti, A., Kato, M., De Ronchi, D., & Kinoshita, T. (2007). Meta-analysis of serotonin transporter gene promoter polymorphism (5-HTTLPR) association with selective serotonin reuptake inhibitor efficacy in depressed patients. *Molecular Psychiatry*, *12*(3), 247–257. https://doi.org/10.1038/sj.mp.4001926.

Smith, A. J., Sketris, I., Cooke, C., Gardner, D., Kisely, S., & Tett, S. E. (2008). A comparison of antidepressant use in Nova Scotia, Canada and Australia. *Pharmacoepidemiology and Drug Safety*, *17*(7), 697–706. https://doi.org/10.1002/pds.1541.

St. Sauver, J. L., Bielinski, S. J., Olson, J. E., Bell, E. J., Mc Gree, M. E., Jacobson, D. J., … Rohrer Vitek, C. R. (2016). Integrating pharmacogenomics into clinical practice: Promise vs reality. *The American Journal of Medicine*, *129*(10), 1093–1099.e1. https://doi.org/10.1016/j.amjmed.2016.04.009.

Sullivan, P. F., Neale, M. C., & Kendler, K. S. (2000). Genetic epidemiology of major depression: Review and meta-analysis. *The American Journal of Psychiatry*, *157*(10), 1552–1562. https://doi.org/10.1176/appi.ajp.157.10.1552.

Tansey, K. E., Guipponi, M., Hu, X., Domenici, E., Lewis, G., Malafosse, A., … Uher, R. (2013). Contribution of common genetic variants to antidepressant response. *Biological Psychiatry*, *73*(7), 679–682. https://doi.org/10.1016/j.biopsych.2012.10.030.

Tansey, K. E., Guipponi, M., Perroud, N., Bondolfi, G., Domenici, E., Evans, D., … Uher, R. (2012). Genetic predictors of response to serotonergic and noradrenergic antidepressants in major depressive disorder: A genome-wide analysis of individual-level data and a meta-analysis. *PLoS Medicine*. *9*(10). https://doi.org/10.1371/journal.pmed.1001326.

Trivedi, M. H., Rush, A. J., Wisniewski, S. R., Nierenberg, A. A., Warden, D., Ritz, L., … Fava, M. (2006). Evaluation of outcomes with citalopram for depression using measurement-based care in STAR*D: Implications for clinical practice. *The American Journal of Psychiatry*, *163*(1), 28–40. https://doi.org/10.1176/appi.ajp.163.1.28.

Uher, R., Investigators, G., Investigators, M., & Investigators, S. D. (2013). Common genetic variation and antidepressant efficacy in major depressive disorder: A meta-analysis of three genome-wide pharmacogenetic studies. *The American Journal of Psychiatry*, *170*(2), 207–217. https://doi.org/10.1176/appi.ajp.2012.12020237.

Uher, R., Perroud, N., Ng, M. Y., Hauser, J., Henigsberg, N., Maier, W., … McGuffin, P. (2010). Genome-wide pharmacogenetics of antidepressant response in the GENDEP project. *The American Journal of Psychiatry*, *167*(5), 555–564.

Verbelen, M., Weale, M. E., & Lewis, C. M. (2017). Cost-effectiveness of pharmacogenetic-guided treatment: Are we there yet? *The Pharmacogenomics Journal*, *17*(5), 395–402. https://doi.org/10.1038/tpj.2017.21.

Weinshilboum, R. (2003). Inheritance and drug response. *The New England Journal of Medicine*, *348*(6), 529–537.

Weinshilboum, R. M., & Wang, L. (2017). Pharmacogenomics: Precision medicine and drug response. *Mayo Clinic Proceedings*, *92*(11), 1711–1722. https://doi.org/10.1016/j.mayocp.2017.09.001.

Winner, J. G., Carhart, J. M., Altar, C. A., Goldfarb, S., Allen, J. D., Lavezzari, G., … Dechairo, B. M. (2015). Combinatorial pharmacogenomic guidance for psychiatric medications reduces overall pharmacy costs in a 1 year prospective evaluation. *Current Medical Research and Opinion*, *31*(9), 1633–1643. https://doi.org/10.1185/03007995.2015.1063483.

World Health Organization. (2017). "Depression: let's talk" Says WHO, as depression tops list of causes of ill health.

Wray, N. R., Ripke, S., Mattheisen, M., Trzaskowski, M., Byrne, E. M., Abdellaoui, A., … Major Depressive Disorder Working Group of the Psychiatric Genomics (2018). Genome-wide association analyses identify 44 risk variants and refine the genetic architecture of major depression. *Nature Genetics*, *50*(5), 668–681. https://doi.org/10.1038/s41588-018-0090-3.

Zai, G., Alberry, B., Arloth, J., Banlaki, Z., Bares, C., Boot, E., … Kennedy, J. L. (2016). Rapporteur summaries of plenary, symposia, and oral sessions from the XXIIIrd World Congress of Psychiatric Genetics Meeting in Toronto, Canada, 16–20 October 2015. *Psychiatric Genetics*, *26*(6), 229–257. https://doi.org/10.1097/YPG.0000000000000148.

Zanger, U. M., & Schwab, M. (2013). Cytochrome P450 enzymes in drug metabolism: Regulation of gene expression, enzyme activities, and impact of genetic variation. *Pharmacology & Therapeutics*, *138*(1), 103–141. https://doi.org/10.1016/j.pharmthera.2012.12.007.

Zubenko, G. S., Sommer, B. R., & Cohen, B. M. (2018). On the marketing and use of pharmacogenetic tests for psychiatric treatment. *JAMA Psychiatry*, *75*(8), 769–770. https://doi.org/10.1001/jamapsychiatry.2018.0834.

Chapter 34

Genomic treatment response prediction in schizophrenia

Sophie E. Legge, Antonio F. Pardiñas and James T.R. Walters

MRC Centre for Neuropsychiatric Genetics and Genomics, Division of Psychological Medicine and Clinical Neurosciences, Cardiff University, Cardiff, United Kingdom

1 Introduction

Schizophrenia affects approximately 0.7% of the population (McGrath, Saha, Chant, & Welham, 2008). It is characterized by disturbances in cognition, perception, and thought and severely impacts both the quality of life and life expectancy (Owen, Sawa, & Mortensen, 2016). Antipsychotics are the primary treatment for schizophrenia and are effective in the majority of patients, although there is considerable heterogeneity between individuals in symptomatic response and the occurrence of adverse effects while taking these medications. Around 30% of individuals with schizophrenia will remain symptomatic and significantly impaired despite standard antipsychotic treatment, and are considered to have treatment-resistant (or treatment-refractory) schizophrenia (TRS) (Meltzer, 1997; Suzuki et al., 2012). TRS is defined by the National Institute for Health and Care Excellence (NICE) as a lack of adequate response to two sequential trials of antipsychotic treatment of sufficient dose and duration (National Collaborating Centre for Mental Health, 2014). A recent prospective study indicated that there might be two distinct subtypes of TRS: (i) the first group, representing the majority of TRS patients (70%), consists of those that fail to respond to antipsychotics from their onset of illness, and (ii) a smaller subset of patients that show initial response to antipsychotics but then delayed treatment resistance following periods of symptomatic relapse (Lally et al., 2016). It is as yet unclear whether TRS is better conceptualized as a form of illness at the severe end of a psychosis/schizophrenia spectrum or as a more biologically homogeneous subgroup of those with schizophrenia, although recent evidence has supported the latter hypothesis (Barnes & Dursun, 2008; Gillespie, Samanaite, Mill, Egerton, & MacCabe, 2017).

1.1 Burden of TRS

TRS is one of the most disabling forms of mental illness and presents a major clinical management challenge (Conley & Kelly, 2001; Kennedy, Altar, Taylor, Degtiar, & Hornberger, 2014). Individuals with TRS have been shown to be more severely impaired compared to those with treatment-responsive schizophrenia across a range of clinical characteristics and outcomes, including a greater likelihood of a continuous course of illness, poorer cognitive functioning, a higher number of psychiatric inpatient admissions, a greater likelihood of deterioration from their premorbid level of functioning, more frequent detention under the Mental Health Act, and more severe positive and negative symptoms (Kennedy et al., 2014; Legge et al., 2019). Furthermore, patients with TRS have decreased life expectancy due to major physical health problems and increased suicide rates (Conley & Kelly, 2001). It should therefore come as no surprise that the annual direct costs of TRS treatment are higher than for patients with treatment-responsive schizophrenia (estimated at 3- to 11-fold higher), primarily driven by increased hospital admissions (Kennedy et al., 2014).

1.2 Clozapine treatment

The only licensed medication with an indication for TRS is the antipsychotic clozapine. While the mechanism of action of clozapine is still not fully understood, it has been proposed that its efficacy in the context of TRS might be related to the biological underpinnings of treatment resistance, which in turn might implicate different neuronal systems to those affected in the majority of treatment-responsive cases. Clozapine is effective in around 60% of patients with TRS and improves the majority of morbidity and mortality indicators that have been examined (Tiihonen et al., 2017). Nonetheless, clozapine remains widely underprescribed, and for those who do receive clozapine treatment, substantial delays are commonplace (Howes et al., 2012), which is an important shortfall because such delays are associated with poorer long-term outcomes (Harris et al., 2005). When clozapine is not used, TRS patients are often treated with high-dose antipsychotics or polypharmacy. Both approaches have a very limited evidence base in TRS and can result in patients being exposed to substantial risk of adverse effects. Thus, ensuring that suitable patients have timely access to clozapine treatment is widely recognized as an important therapeutic goal in the management of people with schizophrenia (National Collaborating Centre for Mental Health, 2014).

2 Prediction of TRS

Personalized medicine refers to the use of genetic, biological, or wider individual factors to tailor therapeutic interventions, with the aim of achieving "the delivery of the right treatment to the right patient at the right time" (Hall, 1977). The identification of predictors of TRS has the potential to provide insights into the underlying etiology of the condition, and also offers the prospect of enabling clinicians to better identify and appropriately treat those with an increased risk of developing TRS when they first present with psychosis (Carbon & Correll, 2014). The ability to identify those at increased risk of developing TRS could also open the door to novel clinical trial designs to investigate alternative treatment pathways for those at high risk of developing TRS and limit the delays to effective treatments that currently exist.

Research investigating TRS prediction has faced significant challenges, particularly as a result of limitations in suitable sample sizes and the use of diverse outcome measures and definitions, both issues that have historically plagued the field of pharmacogenetics (Gillespie et al., 2017). Characterizations of patient samples in studies of treatment resistance in schizophrenia have varied widely and generally have not conformed with established definitions, either from clinical guidelines (NICE; defined above) (National Collaborating Centre for Mental Health, 2014) or alternative proposed research definitions (Howes et al., 2017). Instead, it has been commonplace for studies to adopt a pragmatic definition for TRS such as "lifetime-ever clozapine treatment" because clozapine is the only licensed treatment for those with TRS in most countries (Nielsen et al., 2016), although such an approach will misclassify those with TRS who have not taken clozapine. An alternative approach has been to attempt to quantitatively assess response to antipsychotics, often using percentage change or threshold scores on symptom measures such as the Positive and Negative Syndrome Scale (PANSS) (Suzuki et al., 2012). Very few studies of TRS prediction have compared patients with strictly defined criteria for treatment resistance and/or treatment responsiveness. Such variations in methodology have limited comparability between studies, presenting a barrier to combining results across studies. That has led to a lack of independent replication of findings of TRS predictors. The agreement of international definitions and guidelines for TRS (Howes et al., 2017) should address some of the previous methodological shortcomings and enable the integration of research results to advance TRS prediction.

Despite the challenges presented above for existing studies of TRS prediction, there have been consistent findings identifying clinical or demographic predictors of TRS. In particular, a younger age at onset of psychosis has been shown to be a significant and important clinical indicator of TRS (Lally, Gaughran, Timms, & Curran, 2016; Legge et al., 2019; Wimberley et al., 2016). Poor premorbid functioning has also been associated with TRS as well as poorer outcomes more widely in schizophrenia (Legge et al., 2019). Other demographic and environmental factors associated with TRS in previous studies include male sex, a longer duration of untreated psychosis, living in less urbanized areas, and comorbid substance-use disorders (Wimberley et al., 2016). The identification of clinical and environmental predictors specific for TRS supports the hypothesis that TRS is at least in part distinct from treatment-responsive schizophrenia.

3 Genomic prediction of TRS

The main focus of this chapter is to provide an overview of the genomic prediction studies of TRS. The identified predictors of TRS will be summarized from studies investigating (i) candidate genes, (ii) common variants (typically over 1% population frequency) via genome-wide association studies (GWAS), (iii) genetic liability for schizophrenia via polygenic risk scores (PRS), and (iv) rare variants (<1% population frequency). We will then discuss the challenges and future directions for the field in genomic prediction of TRS. An overview of different genetic association study designs is given in Fig. 1.

3.1 Heritability of TRS

It has been hypothesized that TRS may be more heritable than treatment-responsive schizophrenia, raising the possibility that there is a greater genetic liability for TRS either due to a larger number of risk alleles (greater polygenicity) or by these

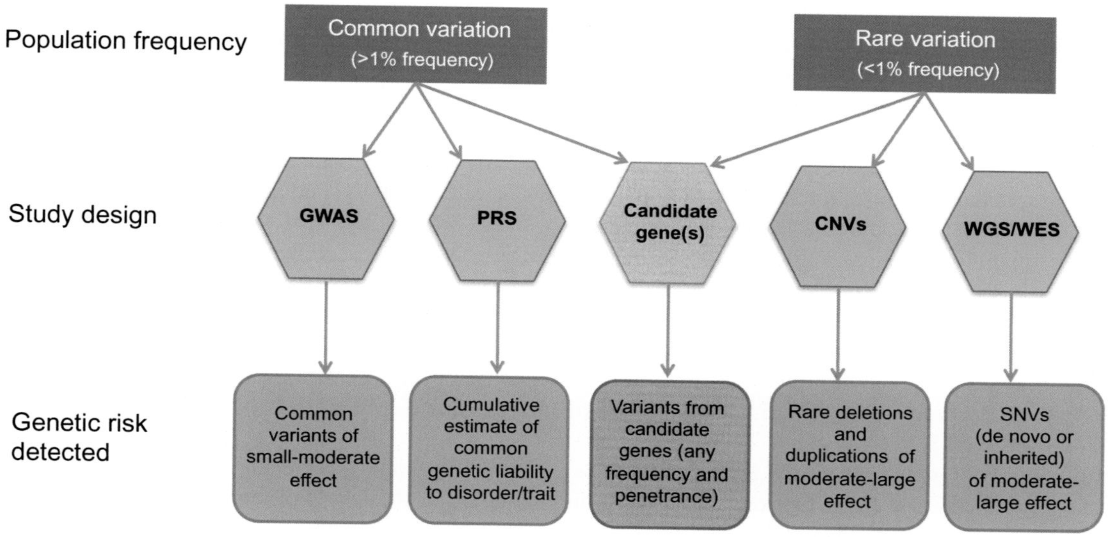

FIG. 1 Summary of genetic association study designs. A summary of commonly used genetic associations study designs. *GWAS*, genome-wide association study; *PRS*, polygenic risk score; *CNVs*, copy number variation; *WGS/WES*, whole genome sequencing/whole exome sequencing.

having stronger effects. Observational family studies have reported higher familial loading scores for schizophrenia in patients with TRS in comparison to healthy controls, whereas equivalent analyses in those with treatment-responsive schizophrenia failed to detect such differences (Joober et al., 2005). In addition, a genetic contribution to TRS is supported by case reports of monozygotic twins showing similar responses to antipsychotic treatment (Mata, Madoz, Arranz, Sham, & Murray, 2001; Wehmeier et al., 2005). Despite these studies being suggestive of TRS having a genetic component, there have been no large-scale replicated evaluations and so no robust evidence of the specific heritability of TRS is available to date. It follows that the heritability of TRS, or indeed antipsychotic response, is yet to be quantified. This is perhaps due to the apparent challenges in conducting suitably powered twin or family studies that would include a requirement for two or more individuals with schizophrenia (within families) and accurate characterization of antipsychotic response or resistance. While recently developed statistical techniques can be used to ascertain heritability from molecular genetic data in case-control studies ("SNP-based heritability" Yang, Zeng, Goddard, Wray, & Visscher, 2017), these require the use of large cohorts of at least several thousand individuals in order to produce reliable estimates, a requirement that would not be fulfilled by any TRS study published to date.

3.2 Candidate studies

Although there have been relatively few pharmacogenetic studies directly investigating TRS, studies of general antipsychotic response should offer insights into the genetic underpinnings of TRS. Previous candidate studies have primarily focused on either the neurotransmitters targeted by antipsychotic medications (chiefly dopamine and to a lesser extent serotonin) or genes encoding the enzymes responsible for drug metabolism (cytochrome P450 family). Among the dopamine-related candidate gene studies, metaanalytic evidence indicates a potential role in antipsychotic response for the Val108Met polymorphism (rs4680) in the Catechol-*O*-Methyltranferase gene *(COMT)*, which plays a key role in dopamine clearance (Huang et al., 2016). There is also some support from a metaanalysis that genetic variation in the dopamine D2 receptor (particularly the −141C Ins/Del polymorphism in *DRD2* Zhang, Lencz, & Malhotra, 2010) may influence antipsychotic treatment response, which seems feasible given that this receptor is the therapeutic target of all currently licensed antipsychotics. Other studies have targeted genetic variation in serotonin system genes with suggestive evidence for the −1438G polymorphism in *HRT2A* (Ellingrod et al., 2003), and 5-HTTLPR, a degenerate repeat polymorphism in *SLC6A4* (Dolzan et al., 2008), playing a putative role in antipsychotic response. *SLC6A4* encodes the serotonin transporter and 5-HTTLPR has been demonstrated to affect the rate of serotonin uptake (Dolzan et al., 2008; Zhang & Malhotra, 2011). Despite these findings, we feel that the evidence can only be considered as preliminary, given that the constituent studies of these metaanalyses employed candidate gene approaches that the field has come to appreciate have important limitations, and none of the reported findings would reach what are now considered robust statistical thresholds for genome-wide significance. It follows that one of the largest systematically conducted studies in the field to date, the Clinical Antipsychotic Trials for

Intervention Effectiveness (CATIE) study ($n = 738$ cases), failed to identify any variants in 74 candidate genes that were significantly associated with discontinuation of olanzapine, quetiapine, risperidone, ziprasidone, or perphenazine (a proxy for nonresponse and/or tolerability) (Need et al., 2009).

More recent candidate studies have directly investigated TRS. A study of 74 candidate genes in 89 patients treated with clozapine (and thus TRS) and 190 schizophrenia patients found that variants within *BDNF*, including the Val66Met (rs6265) polymorphism, were associated with TRS (Zhang et al., 2013). *BDNF*, which interacts with multiple neurotransmitters including dopamine and serotonin, is also implicated in synaptic plasticity, and has been previously associated with therapeutic response in schizophrenia (Krebs et al., 2000). However, the association of variants within *BDNF* was not replicated in subsequent longitudinal studies or a metaanalysis (Cargnin, Massarotti, & Terrazzino, 2016). A significant association between TRS and *DISC1* has been reported (Mouaffak et al., 2011), although this was not replicated in a study of Japanese patients (Hotta et al., 2011). No significant associations have been reported from candidate gene studies specifically relating to serotoninergic genes and TRS (Ji, Takahashi, Branko, et al., 2008; Ji, Takahashi, Saito, et al., 2008), although nominal associations with the *5HT2A* T102C polymorphism have been reported (Anttila et al., 2007; Joober et al., 1999). In 2012, a study of 384 candidate markers from 46 genes failed to identify any variants significantly associated with TRS (Teo et al., 2012).

As with candidate gene approaches in wider neuropsychiatry, candidate pharmacogenetic studies of antipsychotic response have produced few consistent or replicated findings that are at the levels of statistical significance required in modern psychiatric genomics. This is likely due to inadequate statistical power from relatively limited sample sizes, and the considerable variability in experimental design, particularly in definitions of treatment response (van den Oord, Chan, & Aberg, 2018). Advocates of these methodological approaches would argue, perhaps with some justification, that more is known about the metabolism and therapeutic action of antipsychotics than the pathophysiology of neuropsychiatric disorders and hence a more hypothesis-driven (gene-based) rather than a data-driven (genome-wide) approach is appropriate. It has also been argued that sample sizes for pharmacogenetic studies do not need to be as large as case control studies of complex disorders or traits, given the larger predicted effects of pharmacogenetic variants on treatment response (Maranville & Cox, 2015). Whether this observation is generally applicable, the current sample sizes in antipsychotic pharmacogenetic studies are rarely above 1000 cases and have failed to produce genome-wide significant results, which would seem to be a prerequisite to advance the field as far as gene discovery is concerned. Furthermore, it has become clear from the field of schizophrenia genetics that biology-driven candidate gene studies are unlikely to yield robust insights into disease etiology (Farrell et al., 2015) and thus large, well-powered GWAS studies are required to validate and extend the findings that have been reported from candidate gene pharmacogenetic studies of antipsychotic response to date. Indeed, GWAS is being employed productively in other areas of psychiatric pharmacogenetics, such as antidepressant response, yielding promising findings (Jukić Marin, Haslemo, Molden, & Ingelman-Sundberg, 2018) and encouraging initial findings are being reported from such approaches in schizophrenia antipsychotic response (Yu et al., 2018).

3.3 GWAS

A GWAS is a systematic evaluation of polymorphisms throughout the genome (most commonly single nucleotide polymorphisms or SNPs) that aims to evaluate their statistical association to traits or disease. GWAS genotyping arrays rely on an informative backbone of tag SNPs that capture most of the common genetic variation in the genome, and have made large-scale whole-genome genotyping affordable (Visscher, Brown, McCarthy, & Yang, 2012). As with candidate gene studies, there have been few GWAS reports investigating TRS directly.

One of the first GWAS studies for antipsychotic response came from the CATIE study in 738 individuals, which identified an intergenic variant on chromosome 4p15 that predicted the effect of ziprasidone on positive symptoms (McClay et al., 2011), although this did not reach the now accepted genome-wide significance level of $P < 5 \times 10^{-8}$ (see the review of Sham and Purcell (2014) for a discussion on statistical significance in the GWAS framework). Two further GWAS studies for response to iloperidone (Lavedan et al., 2009) and lurasidone (Li, Yoshikawa, Brennan, Ramsey, & Meltzer, 2018) failed to identify any genetic variants at the genome-wide significant level. One of the largest GWAS to date was recently reported from a randomized controlled trial of antipsychotic response to olanzapine, risperidone, quetiapine, aripiprazole, ziprasidone, and haloperidol or perphenazine in a total of 3792 individuals with Han Chinese ancestry (Yu et al., 2018). Yu and colleagues identified five novel genetic associations with general antipsychotic response (*MEGF10*, *SLC1A1*, *PCDH7*, *CNTNAP5*, and *TNIK*) and three associations with drug-specific treatment responses (*CACNA1C* for olanzapine, *SLC1A1* for risperidone, and *CNTN4* for aripiprazole) (Yu et al., 2018). The authors report that these genes are related to synaptic function and neurotransmitter receptors, and support the hypothesis that implicated pharmacogenetic risk variants have a similar effect on response to several different antipsychotics.

There have been three GWAS studies published to date investigating TRS directly, none of which have identified any variants that were significantly associated at the accepted genome-wide significant level, although all sample sizes were small (Koga et al., 2017; Li & Meltzer, 2014; Liou et al., 2012). Adequate sample size is a critical aspect for statistical power in GWAS in order to detect genetic variants that do not typically have large effects because in most polygenic traits, the usual odds-ratios (ORs) of individual SNPs range between 1.05 and 1.3 (Visscher et al., 2017). Association studies in other disorders have demonstrated that as sample sizes increase, so do the number of genetic loci that are detected. For example, the first schizophrenia GWAS from the Psychiatric Genomics Consortium (~50,000 samples) (Ripke et al., 2011) identified seven susceptibility loci compared to 108 loci when the sample size was increased to ~150,000 samples (Schizophrenia Working Group of the Psychiatric Genomics Consortium, 2014). There seems no reason to think that the same shouldn't be true for TRS and antipsychotic treatment response. Thus, it is critical for further research to be undertaken in larger sample sizes in order to reliably detect genetic loci associated with TRS.

3.4 Genetic liability to schizophrenia

Neuropsychiatric traits such as schizophrenia are polygenic in nature, meaning that genetic liability arises through the cumulative effects of many genetic variants, each of a relatively small effect size. The individual SNPs identified by GWAS can be combined into a single continuous variable called a polygenic risk score (PRS), which can be thought of as a measure of the genetic liability (arising from common variants) to that disorder or trait (Lewis & Vassos, 2017). The PRS is calculated using the GWAS SNP results from an independent "training" sample to derive a score for each individual in the target sample, which represents their cumulative genetic risk by summing the number of risk alleles carried by the individual, weighted by the effect size from the training sample (Wray et al., 2014). Given the explanatory power of PRS for susceptibility to schizophrenia, it is reasonable to investigate whether these scores can be informative regarding clinical heterogeneity within the disorder with the aim of stratification.

There have been mixed reports regarding the association of schizophrenia PRS with TRS. A study by Frank and colleagues found that the PRS for schizophrenia was increased in 434 patients treated with clozapine in comparison to 370 patients with no history of clozapine treatment (Frank et al., 2015). A small study investigating response to lurasidone reported a significant association between schizophrenia PRS and improvement in positive symptoms (Li et al., 2018). Zhang and colleagues recently reported the first longitudinal study from a first episode sample (total $n = 510$) and found that patients with a low schizophrenia PRS were more likely to be treatment responders after 12 weeks of antipsychotic treatment than patients with a high PRS.(Zhang et al., 2018) However, three other independent studies, encompassing a similar range of experimental designs and TRS definitions, have found no evidence for an association between schizophrenia PRS and TRS (Legge et al., 2019; Martin & Mowry, 2015; Wimberley et al., 2017).

Whether schizophrenia PRS is increased in individuals with TRS is an important question, given that the results could provide insights into whether TRS should be conceptualized as a form of illness at the severe end of a psychosis/schizophrenia spectrum, or as a more biologically homogeneous subgroup of those with schizophrenia, that is, not influenced by liability to schizophrenia beyond the requirement to have schizophrenia (Barnes & Dursun, 2008; Gillespie et al., 2017). As was the case in pharmacogenetic studies, all PRS-based studies to date have had a total sample size smaller than ~1000 individuals and thus larger samples are needed to provide answers to this question. Nonetheless, even those studies that reported a significant association found that the schizophrenia PRS explained only a small amount of the variance in TRS (maximum value reported $r^2 < 3.7\%$ (Zhang et al., 2018)), suggesting that factors other than broad genetic liability to schizophrenia are likely to provide greater insights into the genetic architecture of TRS.

3.5 Rare variant studies

Compared to healthy controls, individuals with schizophrenia have an increased burden of rare copy number variations (CNVs), which are chromosomal deletions and duplications that range in size from a few kilobases to several megabases of DNA (Rees et al., 2014). Most of the CNVs associated with schizophrenia have been shown to also increase risk for other neurodevelopmental disorders, such as autism spectrum disorders, intellectual deficit, developmental delay, and epilepsy (Kirov et al., 2014). Martin and Mowry reported an increased burden of genome-wide rare copy number duplications in 277 patients with TRS compared to 385 individuals with schizophrenia (Martin & Mowry, 2015). However, this was not replicated by a later study that reported no difference in the burden of rare pathogenic CNVs previously associated with schizophrenia or intellectual disability in individuals with TRS compared to those with treatment-responsive schizophrenia (Legge et al., 2019).

Other studies of rare variations in schizophrenia have focused on single nucleotide variations (SNVs) derived from next-generation sequencing analyses, which may either be inherited or occur as de novo mutations. Ruderfer and colleagues found that patients taking clozapine did not have an enrichment of rare disruptive (protein-altering) variants in genes previously associated with schizophrenia, but did report an excess of rare disruptive variants in gene targets of antipsychotics in those taking clozapine, compared to schizophrenia patients who had not taken clozapine (Ruderfer et al., 2016). In an RCT assessing treatment response to seven antipsychotic medications, Wang and colleagues found that a set of genes associated with reduced N-methyl-D-aspartate (NMDA)—mediated synaptic currents was enriched for rare damaging variants in patients with low treatment response (Wang et al., 2018). This finding was replicated in an independent sample and suggests that glutamatergic signaling at NMDA receptors may play a role in mediating antipsychotic efficacy (Wang et al., 2018).

4 Glutamate hypothesis of TRS

Dysregulation of the neurotransmitters dopamine and glutamate are the two primary neurochemical hypotheses of schizophrenia, and additionally it has been hypothesized that glutamate dysfunction may be particularly relevant for TRS (Moghaddam & Javitt, 2012). Neuroimaging studies have reported evidence that in comparison with schizophrenia patients who are responsive to treatment, those with TRS demonstrate reduced striatal dopamine synthesis (Mouchlianitis, McCutcheon, & Howes, 2016) and elevated anterior cingulate cortex glutamate levels (Demjaha et al., 2014). Reduced striatal dopamine synthesis may explain why TRS patients show at best a partial response to the D2 dopamine receptor blockade that is the target of conventional antipsychotic treatment. Clozapine has been shown to interact with glutamatergic-based signaling mechanisms via actions at the NMDA/glycine receptor (Schwieler, Linderholm, Nilsson-Todd, Erhardt, & Engberg, 2008). Preliminary evidence suggests that high-dose glycine, an NMDA agonist, may reduce negative and cognitive symptoms when used to augment olanzapine or risperidone treatment (Heresco-Levy, Ermilov, Lichtenberg, Bar, & Javitt, 2004). This suggests that other atypical antipsychotics may have different effects on NMDA receptor-mediated neurotransmission compared with clozapine. The development of more comprehensive genetic studies of TRS might help to illuminate this issue, given recent advances in the field of single-cell sequencing, which have already been used to pinpoint specific neuronal subtypes implicated in psychiatric disorders (Skene et al., 2018).

5 Challenges and future directions

There have been relatively few pharmacogenetic studies directly investigating TRS, despite the potential importance of this research. Previous candidate gene studies have failed to produce replicated findings at now established levels of statistical significance. GWAS studies are yet to produce robustly replicated risk loci for TRS, and there has only been one study to date that has been powered enough to identify variants at the genome-wide significant level, although this study targeted general antipsychotic response as an outcome rather than TRS specifically (Yu et al., 2018). A small number of rare variant studies have reported enrichments in gene sets related to antipsychotic targets and NMDA-mediated synaptic currents, but these findings are yet to be replicated. There have been mixed reports regarding the association of schizophrenia PRS with TRS but there is no convincing evidence for a meaningful enrichment of schizophrenia polygenic signal in TRS, suggesting that factors other than genetic susceptibility to broad schizophrenia are likely to provide greater insights into the genetic architecture of TRS. Box 1 gives a summary of the identified risk factors for TRS.

The field of genomic prediction of TRS faces significant challenges related to the likely causal heterogeneity of TRS and the lack of well-characterized samples, problems that have inhibited progress in wider psychiatric pharmacogenomics. The majority of studies to date have been underpowered as a result of small sample sizes and the variability in TRS definition has limited the comparability between studies. There have been relatively few studies comparing patients with strictly defined treatment resistance and treatment responsiveness, a lack of prospective studies, and little independent replication of significant findings. This is partly due to the often prohibitive costs associated with carrying out long-term followup of therapeutic drug trials and the lack of prospective studies with consistent and reliable characterization of treatment response or resistance that also have available genetic data. An encouraging development in this respect involves combining population-based genetic studies with electronic medical records (Wimberley et al., 2016) and targeted recruitment strategies of very large numbers of TRS cases, as exemplified in the CLOZUK study (Pardiñas et al., 2018).

In addition to the limitations of existing research described above, most studies of TRS have been performed in populations of European ancestry, together with the vast majority of complex disorder genomic research. This situation is gradually changing, given the potential scientific advantages of genetic research in non-European populations, but also as a vital

---

BOX 1 Summary of risk factors for TRS

Clinical indicators

- Earlier age of onset of psychosis; this finding has been replicated in several independent studies
- Poor premorbid functioning
- Male sex
- Longer duration of untreated psychosis
- Comorbid substance use disorders

Genetic risk factors

- Val66Met (rs6265) polymorphism in *BDNF;* suggestive evidence from candidate gene studies only and not replicated in metaanalysis.

- Missense variant (rs3738401) in *DISC1;* suggestive evidence from candidate gene studies only and not yet replicated.
- No genome-wide significant variants from three GWAS published to date; all studies were underpowered to detect variants of small effect.
- An increased polygenic risk for schizophrenia; this finding has not been replicated in independent studies
- An increased burden of CNVs; this findings has not been replicated in independent studies
- Enrichment of rare disruptive SNVs in gene targets of antipsychotics; this finding has not yet been replicated

---

step in determining the generalizability or specificity of risk variant findings across populations (Chan, Jin, Loh, & Brunham, 2015).

Box 2 provides an indication of the future direction of TRS studies. Future insights may be gained from investigating the role of genetics of the antipsychotic metabolism (Jukic et al., 2017) and nongenetic factors. For example, the age of onset of psychosis and poor premorbid functioning have been consistently identified as predictors of TRS, suggesting that genetic factors influencing these traits could have relevance to TRS. Also, given its wide use in TRS treatment, improvements in the characterization of the mechanism of action of clozapine might reveal insights into TRS, work that would open up hypotheses for pharmacogenomic and drug-repurposing studies (Gaspar & Breen, 2017).

BOX 2 Future directions for genetic studies of TRS

- Adequately powered studies with large sample sizes
- An agreed definition of TRS that is replicable across study populations
- GWAS and next generation sequencing studies

- Studies in non-European populations
- Prospective, longitudinal cohort studies

---

6 Conclusions

In conclusion, TRS is an important area of need for patients, their relatives, and health services, and it presents the potential for advances in stratification and personalized psychiatry. However, while there is suggestive evidence that TRS may be heritable, this has not been confirmed in large family or molecular genetic studies. Similarly, no robustly replicated risk loci for TRS have been identified to date, largely due to the fact that studies have been underpowered, conducted in relatively small samples, and with limitations in their clinical phenotypic characterization. In order for the field to advance and for further assessment of the clinical utility of genomics in this area of medicine, further well-designed and sufficiently powered research is required before risk variants can be identified with confidence and in order to improve prediction of TRS.

REFERENCES

Anttila, S., Kampman, O., Illi, A., Rontu, R., Lehtimaki, T., & Leinonen, E. (2007). Association between 5-HT2A, TPH1 and GNB3 genotypes and response to typical neuroleptics: A serotonergic approach. *BMC Psychiatry, 7.*

Barnes, T. R., & Dursun, S. (2008). Treatment resistance in schizophrenia. *Psychiatry, 7,* 487–490.

Carbon, M., & Correll, C. U. (2014). Clinical predictors of therapeutic response to antipsychotics in schizophrenia. *Dialogues in Clinical Neuroscience, 16,* 505–524.

Cargnin, S., Massarotti, A., & Terrazzino, S. (2016). BDNF Val66Met and clinical response to antipsychotic drugs: A systematic review and meta-analysis. *European Psychiatry, 33,* 45–53.

Chan, S. L., Jin, S., Loh, M., & Brunham, L. R. (2015). Progress in understanding the genomic basis for adverse drug reactions: A comprehensive review and focus on the role of ethnicity. *Pharmacogenomics, 16,* 1161–1178.

Conley, R. R., & Kelly, D. L. (2001). Management of treatment resistance in schizophrenia. *Biological Psychiatry, 50*, 898–911.

Demjaha, A., Egerton, A., Murray, R. M., Kapur, S., Howes, O. D., Stone, J. M., et al. (2014). Antipsychotic treatment resistance in schizophrenia associated with elevated glutamate levels but normal dopamine function. *Biological Psychiatry, 75*, e11–e13.

Dolzan, V., Serretti, A., Mandelli, L., Koprivsek, J., Kastelic, M., & Plesnicar, B. K. (2008). Acute antipyschotic efficacy and side effects in schizophrenia: Association with serotonin transporter promoter genotypes. *Progress in Neuro-Psychopharmacology & Biological Psychiatry, 32*, 1562–1566.

Ellingrod, V. L., Lund, B. C., Miller, D., Fleming, F., Perry, P., Holman, T. L., et al. (2003). 5-HT2A receptor promoter polymorphism, -1438G/A and negative symptom response to olanzapine in schizophrenia. *Psychopharmacology Bulletin, 37*, 109–112.

Farrell, M. S., Werge, T., Sklar, P., Owen, M. J., Ophoff, R. A., O'Donovan, M. C., et al. (2015). Evaluating historical candidate genes for schizophrenia. *Molecular Psychiatry, 20*, 555–562.

Frank, J., Lang, M., Witt, S. H., Strohmaier, J., Rujescu, D., Cichon, S., et al. (2015). Identification of increased genetic risk scores for schizophrenia in treatment-resistant patients. *Molecular Psychiatry, 20*, 150–151.

Gaspar, H. A., & Breen, G. (2017). Drug enrichment and discovery from schizophrenia genome-wide association results: An analysis and visualisation approach. *Scientific Reports, 7*, 12460.

Gillespie, A. L., Samanaite, R., Mill, J., Egerton, A., & MacCabe, J. H. (2017). Is treatment-resistant schizophrenia categorically distinct from treatment-responsive schizophrenia? a systematic review. *BMC Psychiatry, 17*, 12.

Hall, B.L. Patient placement in a long-term care. California State University (1977).

Harris, M. G., Henry, L. P., Harrigan, S. M., Purcell, R., Schwartz, O. S., Farrelly, S. E., et al. (2005). The relationship between duration of untreated psychosis and outcome: An eight-year prospective study. *Schizophrenia Research, 79*, 85–93.

Heresco-Levy, U., Ermilov, M., Lichtenberg, P., Bar, G., & Javitt, D. C. (2004). High-dose glycine added to olanzapine and risperidone for the treatment of schizophrenia. *Biological Psychiatry, 55*, 165–171.

Hotta, Y., Ohnuma, T., Hanzawa, R., Shibata, N., Maeshima, H., Baba, H., et al. (2011). Association study between disrupted-in-schizophrenia-1 (DISC1) and Japanese patients with treatment-resistant schizophrenia (TRS). *Progress in Neuro-Psychopharmacology & Biological Psychiatry, 35*, 636–639.

Howes, O. D., McCutcheon, R., Agid, O., de Bartolomeis, A., van Beveren, N. J. M., Birnbaum, M. L., et al. (2017). Treatment-resistant schizophrenia: Treatment response and resistance in psychosis (TRRIP) working group consensus guidelines on diagnosis and terminology. *The American Journal of Psychiatry, 174*, 216–229.

Howes, O. D., Vergunst, F., Gee, S., McGuire, P., Kapur, S., & Taylor, D. (2012). Adherence to treatment guidelines in clinical practice: Study of antipsychotic treatment prior to clozapine initiation. *The British Journal of Psychiatry, 201*, 481–485.

Huang, E., Zai, C. C., Lisoway, A., Maciukiewicz, M., Felsky, D., Tiwari, A. K., et al. (2016). Catechol-O-methyltransferase Val158Met polymorphism and clinical response to antipsychotic treatment in schizophrenia and schizo-affective disorder patients: A meta-analysis. *The International Journal of Neuropsychopharmacology, 19*, pyv132.

Ji, X., Takahashi, N., Branko, A., Ishihara, R., Nagai, T., Mouri, A., et al. (2008). An association between serotonin receptor 3B gene (HTR3B) and treatment-resistant schizophrenia (TRS) in a Japanese population. *Nagoya Journal of Medical Science, 70*, 11–17.

Ji, X., Takahashi, N., Saito, S., Ishihara, R., Maeno, N., Inada, T., et al. (2008). Relationship between three serotonin receptor subtypes (HTR3A, HTR2A and HTR4) and treatment-resistant schizophrenia in the Japanese population. *Neuroscience Letters, 435*, 95–98.

Joober, R., Benkelfat, C., Brisebois, K., Toulouse, A., Turecki, G., Lal, S., et al. (1999). T102C polymorphism in the 5HT2A gene and schizophrenia: Relation to phenotype and drug response variability. *Journal of Psychiatry & Neuroscience, 24*, 141–146.

Joober, R., Rouleau, G. A., Lal, S., Bloom, D., Lalonde, P., Labelle, A., et al. (2005). Increased prevalence of schizophrenia spectrum disorders in relatives of neuroleptic-nonresponsive schizophrenic patients. *Schizophrenia Research, 77*, 35–41.

Jukić Marin, M., Haslemo, T., Molden, E., & Ingelman-Sundberg, M. (2018). Impact of CYP2C19 genotype on escitalopram exposure and therapeutic failure: A retrospective study based on 2,087 patients. *The American Journal of Psychiatry, 175*, 463–470.

Jukic, M. M., Opel, N., Strom, J., Carrillo-Roa, T., Miksys, S., Novalen, M., et al. (2017). Elevated CYP2C19 expression is associated with depressive symptoms and hippocampal homeostasis impairment. *Molecular Psychiatry, 22*, 1224.

Kennedy, J. L., Altar, C. A., Taylor, D. L., Degtiar, I., & Hornberger, J. C. (2014). The social and economic burden of treatment-resistant schizophrenia: A systematic literature review. *International Clinical Psychopharmacology, 29*, 63–76.

Kirov, G., Rees, E., Walters, J. T., Escott-Price, V., Georgieva, L., Richards, A. L., et al. (2014). The penetrance of copy number variations for schizophrenia and developmental delay. *Biological Psychiatry, 75*, 378–385.

Koga, A., Bani-Fatemi, A., Hettige, N., Borlido, C., Zai, C., Strauss, J., et al. (2017). GWAS analysis of treatment resistant schizophrenia: Interaction effect of childhood trauma. *Pharmacogenomics, 18*, 663–671.

Krebs, M. O., Guillin, O., Bourdell, M. C., Schwartz, J. C., Olie, J. P., Poirier, M. F., et al. (2000). Brain derived neurotrophic factor (BDNF) gene variants association with age at onset and therapeutic response in schizophrenia. *Molecular Psychiatry, 5*, 558–562.

Lally, J., Ajnakina, O., Di Forti, M., Trotta, A., Demjaha, A., Kolliakou, A., et al. (2016). Two distinct patterns of treatment resistance: Clinical predictors of treatment resistance in first-episode schizophrenia spectrum psychoses. *Psychological Medicine, 46*, 3231–3240.

Lally, J., Gaughran, F., Timms, P., & Curran, S. R. (2016). Treatment-resistant schizophrenia: Current insights on the pharmacogenomics of antipsychotics. *Pharmgenomics Pers Med, 9*, 117–129.

Lavedan, C., Licamele, L., Volpi, S., Hamilton, J., Heaton, C., Mack, K., et al. (2009). Association of the NPAS3 gene and five other loci with response to the antipsychotic iloperidone identified in a whole genome association study. *Molecular Psychiatry, 14*, 804–819.

Legge, S. E., Dennison, C. A., Pardiñas, A. F., Rees, E., Lynham, A. J., Hopkins, L., et al. (2019). Clinical indicators of treatment-resistant psychosis. *The British Journal of Psychiatry.* https://doi.org/10.1192/bjp.2019.120.

Lewis, C. M., & Vassos, E. (2017). Prospects for using risk scores in polygenic medicine. *Genome Medicine, 9*, 96.

Li, J., & Meltzer, H. Y. (2014). A genetic locus in 7p12.2 associated with treatment resistant schizophrenia. *Schizophrenia Research, 159*, 333–339.

Li, J., Yoshikawa, A., Brennan, M. D., Ramsey, T. L., & Meltzer, H. Y. (2018). Genetic predictors of antipsychotic response to lurasidone identified in a genome wide association study and by schizophrenia risk genes. *Schizophrenia Research, 192*, 194–204.

Liou, Y. J., Wang, H. H., Lee, M. T., Wang, S. C., Chiang, H. L., Chen, C. C., et al. (2012). Genome-wide association study of treatment refractory schizophrenia in Han Chinese. *PLoS One, 7*, e33598.

Maranville, J. C., & Cox, N. J. (2015). Pharmacogenomic variants have larger effect sizes than genetic variants associated with other dichotomous complex traits. *The Pharmacogenomics Journal, 16*, 388.

Martin, A. K., & Mowry, B. (2015). Increased rare duplication burden genomewide in patients with treatment-resistant schizophrenia. *Psychological Medicine*, 1–8.

Mata, I., Madoz, V., Arranz, M. J., Sham, P., & Murray, R. M. (2001). Olanzapine: Concordant response in monozygotic twins with schizophrenia. *The British Journal of Psychiatry, 178*, 86.

McClay, J. L., Adkins, D. E., Aberg, K., Stroup, S., Perkins, D. O., Vladimirov, V. I., et al. (2011). Genome-wide pharmacogenomic analysis of response to treatment with antipsychotics. *Molecular Psychiatry, 16*, 76–85.

McGrath, J., Saha, S., Chant, D., & Welham, J. (2008). Schizophrenia: A concise overview of incidence, prevalence, and mortality. *Epidemiologic Reviews, 30*, 67–76.

Meltzer, H. Y. (1997). Treatment-resistant schizophrenia—The role of clozapine. *Current Medical Research and Opinion, 14*, 1–20.

Moghaddam, B., & Javitt, D. (2012). From revolution to evolution: The glutamate hypothesis of schizophrenia and its implication for treatment. *Neuropsychopharmacology, 37*, 4–15.

Mouaffak, F., Kebir, O., Chayet, M., Tordjman, S., Vacheron, M. N., Millet, B., et al. (2011). Association of disrupted in schizophrenia 1 (DISC1) missense variants with ultra-resistant schizophrenia. *The Pharmacogenomics Journal, 11*, 267–273.

Mouchlianitis, E., McCutcheon, R., & Howes, O. D. (2016). Brain-imaging studies of treatment-resistant schizophrenia: A systematic review. *Lancet Psychiatry, 3*, 451–463.

National Collaborating Centre for Mental Health. (2014). *Psychosis and schizophrenia in adults: Treatment and management.* .

Need, A. C., Keefe, R. S., Ge, D., Grossman, I., Dickson, S., McEvoy, J. P., et al. (2009). Pharmacogenetics of antipsychotic response in the CATIE trial: A candidate gene analysis. *European Journal of Human Genetics, 17*, 946–957.

Nielsen, J., Young, C., Ifteni, P., Kishimoto, T., Xiang, Y.-T., Schulte, P. F. J., et al. (2016). Worldwide differences in regulations of clozapine use. *CNS Drugs, 30*, 149–161.

Owen, M. J., Sawa, A., & Mortensen, P. B. (2016). Schizophrenia. *Lancet, 388*, 86–97.

Pardiñas, A. F., Holmans, P., Pocklington, A. J., Escott-Price, V., Ripke, S., Carrera, N., et al. (2018). Common schizophrenia alleles are enriched in mutation-intolerant genes and in regions under strong background selection. *Nature Genetics, 50*, 381–389.

Rees, E., Walters, J. T., Georgieva, L., Isles, A. R., Chambert, K. D., Richards, A. L., et al. (2014). Analysis of copy number variations at 15 schizophrenia-associated loci. *The British Journal of Psychiatry, 204*, 108–114.

Ripke, S., Sanders, A. R., Kendler, K. S., Levinson, D. F., Sklar, P., Holmans, P. A., et al. (2011). Genome-wide association study identifies five new schizophrenia loci. *Nature Genetics, 43*, 969.

Ruderfer, D. M., Charney, A. W., Readhead, B., Kidd, B. A., Kahler, A. K., Kenny, P. J., et al. (2016). Polygenic overlap between schizophrenia risk and antipsychotic response: A genomic medicine approach. *The Lancet Psychiatry, 3*, 350–357.

Schizophrenia Working Group of the Psychiatric Genomics Consortium. (2014). Biological insights from 108 schizophrenia-associated genetic loci. *Nature, 511*, 421–427.

Schwieler, L., Linderholm, K. R., Nilsson-Todd, L. K., Erhardt, S., & Engberg, G. (2008). Clozapine interacts with the glycine site of the NMDA receptor: Electrophysiological studies of dopamine neurons in the rat ventral tegmental area. *Life Sciences, 83*, 170–175.

Sham, P. C., & Purcell, S. M. (2014). Statistical power and significance testing in large-scale genetic studies. *Nature Reviews Genetics, 15*, 335–346.

Skene, N. G., Bryois, J., Bakken, T. E., Breen, G., Crowley, J. J., Gaspar, H. A., et al. (2018). Genetic identification of brain cell types underlying schizophrenia. *Nature Genetics, 50*, 825–833.

Suzuki, T., Remington, G., Mulsant, B. H., Uchida, H., Rajji, T. K., Graff-Guerrero, A., et al. (2012). Defining treatment-resistant schizophrenia and response to antipsychotics: A review and recommendation. *Psychiatry Research, 197*, 1–6.

Teo, C., Zai, C., Borlido, C., Tomasetti, C., Strauss, J., Shinkai, T., et al. (2012). Analysis of treatment-resistant schizophrenia and 384 markers from candidate genes. *Pharmacogenetics and Genomics, 22*, 807–811.

Tiihonen, J., Mittendorfer-Rutz, E., Majak, M., Mehtala, J., Hoti, F., Jedenius, E., et al. (2017). Real-world effectiveness of antipsychotic treatments in a nationwide cohort of 29823 patients with schizophrenia. *JAMA Psychiatry, 74*, 686–693.

van den Oord, E. J. C. G., Chan, R. F., & Aberg, K. A. (2018). Successes and challenges in precision medicine in psychiatry successes and challenges in precision medicine in psychiatry research. *JAMA Psychiatry, 75*, 1269–1270.

Visscher, P. M., Brown, M. A., McCarthy, M. I., & Yang, J. (2012). Five years of GWAS discovery. *American Journal of Human Genetics, 90*, 7–24.

Visscher, P. M., Wray, N. R., Zhang, Q., Sklar, P., McCarthy, M. I., Brown, M. A., et al. (2017). 10 years of GWAS discovery: Biology, function, and translation. *American Journal of Human Genetics, 101*, 5–22.

Wang, Q., Man Wu, H., Yue, W., Yan, H., Zhang, Y., Tan, L., et al. (2018). Effect of damaging rare mutations in synapse-related gene sets on response to short-term antipsychotic medication in Chinese patients with schizophrenia: A randomized clinical trial. *JAMA Psychiatry, 75*, 1261–1269.

Wehmeier, P. M., Gebhardt, S., Schmidtke, J., Remschmidt, H., Hebebrand, J., & Theisen, F. M. (2005). Clozapine: Weight gain in a pair of monozygotic twins concordant for schizophrenia and mild mental retardation. *Psychiatry Research, 133*, 273–276.

Wimberley, T., Gasse, C., Meier, S. M., Agerbo, E., MacCabe, J. H., & Horsdal, H. T. (2017). Polygenic risk score for schizophrenia and treatment-resistant schizophrenia. *Schizophrenia Bulletin, 43*, 1064–1069.

Wimberley, T., Stovring, H., Sorensen, H. J., Horsdal, H. T., MacCabe, J. H., & Gasse, C. (2016). Predictors of treatment resistance in patients with schizophrenia: A population-based cohort study. *Lancet Psychiatry, 3*, 358–366.

Wray, N. R., Lee, S. H., Mehta, D., Vinkhuyzen, A. A., Dudbridge, F., & Middeldorp, C. M. (2014). Research review: Polygenic methods and their application to psychiatric traits. *Journal of Child Psychology and Psychiatry, 55*, 1068–1087.

Yang, J., Zeng, J., Goddard, M. E., Wray, N. R., & Visscher, P. M. (2017). Concepts, estimation and interpretation of SNP-based heritability. *Nature Genetics, 49*, 1304.

Yu, H., Yan, H., Wang, L., Li, J., Tan, L., Deng, W., et al. (2018). Five novel loci associated with antipsychotic treatment response in patients with schizophrenia: A genome-wide association study. *Lancet Psychiatry, 5*, 327–338.

Zhang, J. P., Lencz, T., Geisler, S., DeRosse, P., Bromet, E. J., & Malhotra, A. K. (2013). Genetic variation in BDNF is associated with antipsychotic treatment resistance in patients with schizophrenia. *Schizophrenia Research, 146*, 285–288.

Zhang, J. P., Lencz, T., & Malhotra, A. K. (2010). D2 receptor genetic variation and clinical response to antipsychotic drug treatment: A meta-analysis. *The American Journal of Psychiatry, 167*, 763–772.

Zhang, J. P., & Malhotra, A. K. (2011). Pharmacogenetics and antipsychotics: Therapeutic efficacy and side effects prediction. *Expert Opinion on Drug Metabolism & Toxicology, 7*, 9–37.

Zhang, J. P., Robinson, D., Yu, J., Gallego, J., Fleischhacker, W. W., Kahn, R. S., et al. (2018). Schizophrenia polygenic risk score as a predictor of antipsychotic efficacy in first-episode psychosis. *The American Journal of Psychiatry*. https://doi.org/10.1176/appi.ajp.2018.17121363.

Chapter 35

Personalized treatment in bipolar disorder

Estela Salagre, Eduard Vieta and Iria Grande

Barcelona Bipolar and Depressive Disorders Unit, Institute of Neurosciences, Hospital Clinic, University of Barcelona, IDIBAPS, CIBERSAM, Barcelona, Spain

1 Introduction

Bipolar disorder (BD) is a chronic psychiatric disorder characterized by mood swings between mania and depression (Grande, Berk, Birmaher, & Vieta, 2016). Its lifetime prevalence is about 1%, reaching 2%–4% when subthreshold forms are included (Grande et al., 2016). Although BD is generally characterized by episodic relapses and chronic disability, illness presentation and course can differ widely among individuals (Solomon et al., 2010). Conventional treatment in BD encompasses pharmacotherapy and psychotherapy (Budde, Degner, Brockmoller, & Schulze, 2017). However, about 60% of patients still relapse into depression or mania within two years of treatment initiation (Tohen et al., 2003), suggesting that the actual therapeutic approach may not be optimal yet. Despite the unequivocal usefulness of the implementation of evidence-based medicine guidelines in current daily practice, the therapeutic decision-making process still requires a personalized approach.

The staging approach is currently the most similar model to that of personalized psychiatry (Fernandes & Berk, 2017; Wium-Andersen, Vinberg, Kessing, & McIntyre, 2017). The concept of staging is based on the idea that a certain disease progresses with an identifiable temporal progression: from a prodromal stage to more severe and chronic stages of the illness, and requiring a more complex treatment (Berk et al., 2017). Hence, staging allows us to describe where a patient is within the temporal spectrum of progression of the disorder, and thus refine the diagnosis, adjusting the prognosis, and helping to choose the most appropriate treatment (Berk et al., 2017).

However, an even more personalized way of approaching patient treatment is becoming necessary in BD, as it is happening in other fields of medicine where medicine is moving from a more reactive discipline to a more predictive, personalized, preventive, and participatory medicine ("P4 medicine") (Hood & Friend, 2011). Personalized medicine is often defined as choosing the 'right treatment for the right person at the right time' (Wium-Andersen et al., 2017). In other words, the final goal of personalized treatment in psychiatry is, on one hand, to significantly delay disease onset and, whenever possible, to prevent it (Beckmann & Lew, 2016), and, on the other hand, to provide surveillance measures and therapies precisely tailored to each patient's needs (Alhajji & Nemeroff, 2015; Mirnezami, Nicholson, & Darzi, 2012; Prendes-Alvarez & Nemeroff, 2016). In order to achieve these aims, it is necessary to integrate different levels of information regarding each individual, from patients' clinical and biographical information (i.e., family history and life stressors) (Alhajji & Nemeroff, 2015; Prendes-Alvarez & Nemeroff, 2016), to more biological information, such as molecular and neuroimaging markers, patients' genetic information, and epigenetic modifications. In addition, the coming technological advances will also help in the diagnosis process and in the implementation of personalized treatments.

In this chapter, we will review the current available information on potential tools for a personalized treatment in BD, including clinical factors, molecular and neuroimaging markers, as well as genetic and epigenetic information. We will briefly review the current evidence on the application of technology in diagnosis and treatment of BD, and we will discuss new ethical challenges in personalized psychiatry. Finally, we will summarize main findings and discuss future directions.

2 Tools for personalized treatment in BD

The main goals of personalized psychiatry are to predict disease vulnerability, to improve diagnosis processes, and to optimize treatment selection and monitor response (Alhajji & Nemeroff, 2015). To achieve these goals, it is necessary to integrate a number of clinical and biological variables related to the patient (Fig. 1).

Personalized Psychiatry. https://doi.org/10.1016/B978-0-12-813176-3.00035-3

FIG. 1 Basis and goals of personalized treatment in BD.

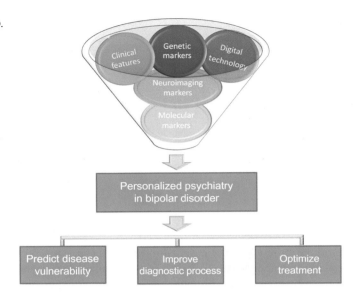

2.1 Clinical and psychobiographical features

Clinical factors are the mainstay of the diagnostic system in psychiatry. Even if pathognomonic features of BD are lacking, evidence suggests that there are a number of clinical factors that have proven to be especially useful to diagnose patients or guide treatment, and thus personalize psychiatry (Wium-Andersen et al., 2017).

Family history of BD is the strongest known risk factor for developing BD (Craddock & Jones, 1999; Salvatore et al., 2014), and it can also influence the course of the bipolar illness (Duffy et al., 2014; Preisig et al., 2016). For example, it has been found that lithium responsiveness in parents is related to a better treatment response and course in the offspring (Duffy et al., 2014). Duffy et al. (2014) also found in exploratory analysis that having a family history of substance use disorders increased the risk of developing a substance use disorder in offspring of parents with BD. Life events can also influence prognosis; for example, lifetime sexual abuse has been related to a worse illness course (Birmaher et al., 2014; Bortolato et al., 2017; Jimenez et al., 2017; Leverich et al., 2002).

Clinical factors might also help to guide treatment selection. More than half of patients with BD have been reported to be more prone to either depressive or manic relapses. This "tendency" has been described as predominant polarity (Colom, Vieta, Daban, Pacchiarotti, & Sanchez-Moreno, 2006; Popovic et al., 2012). Predominant polarity is a valid prognostic parameter with therapeutic implications. Colom et al. (2006) found that depressive polarity was strongly associated with depressive onset of BD, and it was also associated with a higher mean number of suicide attempts. This study also reported a higher prevalence of manic polarity among patients diagnosed with BD I. The identification of a predominant polarity in bipolar patients has led to the development of the Polarity Index, a metric indicating the relative antimanic versus antidepressive preventive efficacy of drugs (Popovic et al., 2012). The Polarity Index may be a valuable tool helping in the choice of maintenance therapy in patients with BD. Likewise, patients with a diagnosis of BD with a seasonal pattern (Goikolea et al., 2007) might benefit from adding light therapy, that is, exposure to artificial light, to standard treatment (Levitt, Lam, & Levitan, 2002), while catatonic features are known to show high response rates to electroconvulsive therapy (Perugi et al., 2017). A personalized evaluation of bipolar patients should also include residual symptoms and cognitive status, as they are related to functionality (Rosa et al., 2009, 2014). Adherence to treatment is another essential issue in a personalized assessment in BD, because patients showing a poor adherence to treatment are perfect candidates for long-acting injectable antipsychotics, especially if they have a predominantly manic polarity course (Vieta et al., 2008).

Similarly, some clinical markers of treatment response to mood stabilizers have been proposed (Squassina, Manchia, & Del Zompo, 2010). Characteristics such an episodic pattern of mania-depression interval, low rates of comorbid conditions, absence of rapid cycling and high age at illness onset may be related to a better response to lithium (Squassina et al., 2010). In turn, responders to valproate have been found to present higher rates of psychosis and mixed states (Calabrese, Fatemi, Kujawa, & Woyshville, 1996; Garnham et al., 2007). Response to lamotrigine seems to be predicted by earlier onset of symptoms, nonepisodic course of illness, comorbidity with panic or substance use disorder, fewer hospitalizations, fewer prior medication trials, and male gender (Obrocea et al., 2002; Passmore et al., 2003). Additionally, lamotrigine seems especially useful in preventing depressive relapse (Carvalho et al., 2014), and it is commonly prescribed in BD patients with depressive polarity (Grande et al., 2012).

2.2 Biological variables

Biological markers are "biological characteristics that can be objectively measured and evaluated as an indicator of normal biological or pathological processes" (Biomarkers Definitions Working Group, 2001). They are a key element of personalized medicine, as they can be used as objective indicators of pathogenic processes or treatment response (Wium-Andersen et al., 2017).

2.2.1 Molecular markers

Over the past years, several molecular markers have been suggested as markers of disease activity, such as neurotrophic factors (de Oliveira et al., 2009; Grande, Fries, Kunz, & Kapczinski, 2010; Walz et al., 2009), pro-inflammatory cytokines (Bai et al., 2015), or compounds such as homocysteine (Salagre et al., 2017) or C-reactive protein (CRP) (Fernandes et al., 2016). The most consistent finding to date has been the decrease in brain-derived neurotrophic factor (BDNF) levels in manic and depressed episodes, which has been confirmed in two independent meta-analyses (Fernandes et al., 2015; Munkholm, Vinberg, & Kessing, 2016). Moreover, it has been suggested that BDNF concentrations could also be used to discriminate unipolar depression from bipolar depression (Fernandes et al., 2009; Li et al., 2014).

In addition, the fact that patients with BD show high levels of inflammatory markers, such as CRP, suggests that BD is associated with low-grade inflammation (Fernandes et al., 2016), therefore raising the question of whether there might be some patients who could benefit from antiinflammatory treatment strategies and from a periodic screening of inflammatory-related conditions such as type 2 diabetes or metabolic syndrome (Teixeira, Salem, Frey, Barbosa, & Machado-Vieira, 2016). Particularly important are molecular markers' predictors of risk of suicide. Levels of the cytokine interleukine-1B (IL-1B) have been found to be increased in subjects with BD and suicide risk compared with subjects with BD without suicide risk and healthy controls (Monfrim et al., 2014). Similarly, low levels of neuropeptide Y in cerebrospinal fluid have been found to be predictive of suicide attempts (Sandberg, Jakobsson, Palsson, Landen, & Mathe, 2014). Although promising, these findings need to be replicated.

2.2.2 Neuroimaging markers

Anatomical and functional neuroimaging markers can help us understand the pathophysiology of the disease, and might ultimately guide diagnosis (Porcu et al., 2016). A recent review on clinical neuroimaging markers in BD reports that the most important findings observed in bipolar patients are the decreased activity of the prefrontal cortex and the increased activity in the amygdala (Porcu et al., 2016). The former might be related to the cognitive and attention deficits observed in these patients, while the second might be associated with mood instability (Phillips & Swartz, 2014; Porcu et al., 2016). Other studies using functional magnetic resonance imaging (fMRI) report that frontal hyperactivation during working memory paradigms may be associated with genetic risk for BD, but results are still preliminary (Drapier et al., 2008; Ladouceur et al., 2013).

In addition to prediction of disease, neuroimaging might be useful to monitor treatment response, as there is some evidence that treatment with mood stabilizers is associated with increased regional gray matter volume, particularly in limbic regions (McDonald, 2015).

2.2.3 Genetic markers

Genetic markers are viewed as fundamental components of personalized medicine, as BD is a highly heritable disease (Prendes-Alvarez & Nemeroff, 2016). However, even if sequencing of the human genome held great promise for personalized medicine, most subsequent linkage studies and candidate gene association studies have not yet found significant associations with BD (Wium-Andersen et al., 2017) (see Box 1 for glossary on genetics).

Candidate gene association studies have focused on multiple candidate genes from the dopaminergic, serotonergic, and noradrenergic systems, but there is no conclusive evidence for association of any of these genes with susceptibility for BD (Prendes-Alvarez & Nemeroff, 2016). Genes from the hypothalamic-pituitary-adrenal (HPA) axis, postulated to play a role in BD, have also been investigated (Alhajji & Nemeroff, 2015); an association between single nucleotide polymorphisms (SNPs) in FK506 binding protein 5 (*FKBP5*), a key component of the glucocorticoid receptor complex, and BD has been identified (Willour et al., 2009), but results have not been replicated. Studies investigating circadian rhythm genes such as

BOX 1 Glossary.

Biobank: A structured resource that holds human biological samples and/or data to facilitate research over time (Wolf et al., 2012).

Bioinformatics: Field of study that uses computation to extract knowledge from biological data. It includes the collection, storage, retrieval, manipulation, and modeling of data for analysis, visualization, or prediction through the development of algorithms and software.

Biological marker: A biological characteristic that is objectively measured and evaluated as an indicator of normal biological or pathological processes, or a response to a therapeutic intervention.

Dynamic treatment regime: Sequence of decision rules, each of which recommends treatment based on features of patient medical history, such as past treatments and outcomes.

Epigenetics: The study of molecular processes that influence the flow of information between a constant DNA sequence and variable gene expression patterns. This includes investigation of nuclear organization, DNA methylation, histone modification, and RNA transcription.

Genetic association study: A genetic association study aims to test whether a given sequence, such as a region of a chromosome, a haplotype, a gene, or an allele, has involvement in controlling the phenotype of a specific trait, metabolic pathway, or disease.

Genetic linkage study: Family-based method used to map a trait to a genomic location by demonstrating co-segregation of the disease with genetic markers of known chromosomal location; locations identified are more likely to contain a causal genetic variant.

Genome-wide association studies (GWAS): Unbiased genome screens of unrelated individuals and appropriately matched controls or parent-affected child trios to establish whether any genetic variant is associated with a trait. These studies typically focus on associations between single-nucleotide polymorphisms (SNPs) and major diseases.

Incidental finding: Unanticipated or unsought for or unrelated findings, secondary variants (Christenhusz, Devriendt, & Dierickx, 2013).

Machine learning: The ability of a machine to improve its performance based on previous results. Machine learning methods enable computers to learn without being explicitly programmed.

Pharmacodynamics: Area of pharmacology concerned with the relationship between a drug's concentration at the site of action and the resulting effect. Factors influencing a drug's pharmacodynamics include the concentration of drug target and the signaling pathways downstream.

Pharmacogenetics: The study of the genetic variation between individuals that affects their response to drugs/pharmaceuticals and other xenobiotics, both therapeutically and in terms of adverse effects.

Pharmacokinetics: Area of pharmacology concerned with the time course of absorption, distribution, metabolism, and excretion of drugs from biological systems in order to understand the effect of the drug, or of the organism, on the drug's impact.

Single nucleotide polymorphism (SNP): A variation at a single position in a DNA sequence among individuals.

Adapted from https://www.nature.com/subjects and https://www.nature.com/scitable/definition.

---

the timeless circadian clock (*TIMELESS*) (Mansour et al., 2006) or the circadian locomotor output cycles kaput (CLOCK) gene (Shi et al., 2008) have not yielded conclusive results, either.

More recently, an increasing number of genome-wide association studies (GWAS) have begun trying to identify the SNPs that are statistically associated with the disease. The collaborative consortia Psychiatric Genomics Consortium (PGC) was founded in 2007 with the aim of conducting high-quality GWAS mega-analyses (Sullivan et al., 2017). These studies have found several common polymorphisms, including variants within the genes calcium channel, voltage-dependent, L type, alpha 1C subunit (*CACNA1C*), Ankyrin-3 (*ANK3*), and *ODZ4* in BD (Ferreira et al., 2008). A key genetic association to gene *SYNE1* for BD has been likewise found (Green et al., 2013; Xu et al., 2014). Nonetheless, further evidence on these genes is needed.

2.2.3.1 Pharmacogenomics

Lack of efficacy of psychopharmacological agents, and side effects, are still a major concern in psychiatry, compromising treatment compliance (Moller & Rujescu, 2010). Nowadays, it is known that the presence of polymorphisms in genes that code for various receptors, drug metabolizing enzymes, and/or transport proteins can influence treatment response and development of adverse events, as they can ultimately affect the pharmacokinetics or pharmacodynamics of drugs (Budde et al., 2017). Consequently, in the past few years, there have been many attempts to identify genetic variants contributing to drug response variability (Budde et al., 2017).

Genes of interest have included the dopamine 2 receptor gene (*DRD2*), dopamine 3 receptor gene (*DRD3*), serotonin 2A and 2C receptor genes (*5HTR2A* and *5HTR2C*), drug metabolizing enzymes, and human leukocyte antigen (HLA) variants (Moore, Hill, & Panguluri, 2014).

Alterations in drug-metabolizing enzymes The hepatic cytochrome P450 enzyme (CYP450) system, particularly the polymorphic CYP450 2D6 isoenzyme, is responsible for the metabolism of most psychotropic drugs (Moore et al., 2014). Patients with polymorphisms of this isoenzyme can be classified as poor, intermediate, extensive, or ultra-rapid metabolizers (Moore et al., 2014). Poor metabolizers have a higher risk of adverse side-effects when treated with psychotropic drugs because of slower metabolism, increased half-life, and consequently, increased bioavailability, which may be especially relevant in drugs with a narrow therapeutic index; on the contrary, ultra-rapid metabolizers are more likely to exhibit subtherapeutic response to normal doses of psychotropic drugs (Zhou, Ingelman-Sundberg, & Lauschke, 2017).

Consequently, multiple pharmacogenomic studies have scrutinized CYP450 2D6 polymorphisms as a predictor of tolerability and side effects of psychiatric treatments (Alhajji & Nemeroff, 2015; Rosenblat, Lee, & McIntyre, 2017). Despite some preliminary data on the potential benefits of using pharmacogenetic tests in patients with mood disorders (Perez et al., 2017; Sanchez-Iglesias et al., 2016), evidence on its clinical utility is still not compelling for recommending routine screening for CYP450 genetic polymorphisms (Rosenblat et al., 2017). However, it might be useful in selected patients with unusual patterns of drug response or unexpected adverse reactions (Foulds, Maggo, & Kennedy, 2016).

Candidate genes related to treatment response Lithium is the first-line treatment for BD, therefore most pharmacogenomic studies on BD have focused on response to lithium prophylaxis (Squassina et al., 2010). Although results arising from linkage studies comparing responders with nonresponders to lithium suggest that chromosome 12q24 and 7q11.2 may be linked to response to lithium in the BD population, further studies have reported no association for specific genes within these regions (Geoffroy, Bellivier, Leboyer, & Etain, 2014).

Candidate-gene approaches have also focused on genes related to biological pathways related to lithium (Squassina et al., 2010), such as the inositol pathway (including the phospholipase C gamma 1 (*PLCG1*) gene (Ftouhi-Paquin et al., 2001; Lovlie, Berle, Stordal, & Steen, 2001), the inositol polyphosphate 1-phosphatase (*INPP1*) gene (Michelon et al., 2006; Steen et al., 1998), the myo-inositol monophosphatase 1 and 2 (*IMPA1* and *IMPA2*) genes (Sjoholt et al., 2004) or the diacylglycerol kinase eta (*DGKH*) gene (Manchia et al., 2009; Squassina et al., 2009)), the circadian signaling system (including the glycogen synthase kinase-3β (*GSK3β*) gene (Michelon et al., 2006; Szczepankiewicz et al., 2006)), the BDNF/Trk signaling pathway (Michelon et al., 2006; Rybakowski et al., 2007; Szczepankiewicz, Skibinska, Suwalska, Hauser, & Rybakowski, 2009) or the cyclic adenosine monophosphate (cAMP) pathway (including the cAMP response element binding protein (*CREB1*, *CREB2* and *CREB3*) genes (Mamdani et al., 2008)). Replications of findings from candidate gene association studies are quite rare, probably due to methodological restraints. Positive findings could be replicated for the *INPP1* gene, the *GSK3β* gene, the nuclear receptor subfamily 1, group D, member 1 (*NR1D1*) gene, the serotonin transporter gene (*SCL6A4*), the catechol-O-methyltransferase (*COMT*) gene, dopamine 1 receptor (*DRD1*) gene, the BDNF gene, and the neurotrophic receptor tyrosine kinase 2 (*NTRK2*) gene.

Regarding GWAS studies, in 2009, Perlis et al. (2009) conducted a GWAS study of response to lithium in a sample of 458 BD patients from the Systematic Treatment Enhancement Program for BD (STEP-BD) cohort, but no significant association was identified. Squassina et al. (2011) performed a GWAS in a sample of 204 Sardinian patients with BD characterized for response to lithium, and found an association of a SNP in intron 1 of the ACCN1 gene with good response to lithium. In a GWAS conducted in a sample of 294 Han Chinese BD patients, the SNPs rs17026688 and rs17026651 of the glutamate decarboxylase like 1 (*GADL1*) gene reached genome-wide significance (Chen et al., 2014). However, this finding could not be replicated, and it should be remarked that these SNPs are rare in non-Asian populations (Budde et al., 2017).

A major limitation of these studies is the relatively small sample size. In order to overcome this limitation, an international consortium was created in 2008 (the International Consortium on Lithium Genetics; ConLiGen) (Schulze et al., 2010). They conducted the largest GWAS study to date on lithium response, which included a sample of 2563 uniformly phenotyped BD patients from more than 20 sites across four continents (Hou et al., 2016). This study reported a genome-wide significant association with a locus of four linked SNPs on chromosome 21; the authors determined that these SNPs might be related to lithium response, although the results need to be replicated. Furthermore, this region contains two genes for long, noncoding RNAs, which seem to play an important role in gene regulation (Hou et al., 2016). Recently, another GWAS study by the ConLiGen consortium demonstrates that BD patients with greater genetic susceptibility for schizophrenia, measured by their schizophrenia polygenic risk score, had a lower probability of showing a favorable long-term response to lithium (Amare et al., 2017). Their results further suggest that inflammatory pathways might be involved in this association. Thus, the authors conclude that the determination of the schizophrenia polygenic risk score prior to initiation of a mood-stabilizer could be useful to guide treatment selection.

Compared with lithium, less evidence is available on the pharmacogenomics of other treatments in BD (Squassina et al., 2010). A candidate gene study examined the role of the Val158Met polymorphism in the COMT gene in response to mood stabilizers including lithium, valproate, or carbamazepine, and found that the Met/Met genotype was more frequent in non-responders to mood stabilizers than in responders (Lee & Kim, 2010). The XBP1-116C/G polymorphism in the X box binding protein-1 (XBP1) gene has also been associated with a better response to maintenance treatment with valproate (Kim, Kim, Lee, & Joo, 2009). Assessing pharmacogenetics of lamotrigine, it was found that polymorphisms in the DRD2 gene, the dopamine β-hydroxylase (*DBH*) gene, the glucocorticoid receptor (*NR3C1*) gene, the histamine H1 receptor (*HRH1*) gene, and the melanocortin 2 receptor (*MCR2*) gene were associated with treatment response (Perlis et al., 2010). Regarding antipsychotics, their pharmacogenetics have been scarcely studied (Davila et al., 2006; Perlis et al., 2010).

Moreover, the inconsistent definition of treatment response is another limitation of current pharmacogenetic studies (Geoffroy et al., 2014). In this regard, in 2002, Grof et al. (2002) developed a rating scale that, on one side, measures the degree of improvement in the course of treatment with lithium and, on the other side, examines clinical factors that might have been mediated in the response to lithium. Through the simultaneous consideration of the degree of response to lithium and possible confounding factors, this scale aims to determine if the overall clinical improvement can be truly attributable to lithium. Also, considering treatment response a quantitative trait instead of a qualitative trait might be more representative of the degree of variation in the treatment response phenotype, and might facilitate the identification of genetic variants involved in intermediate and "extreme" phenotypes (Squassina et al., 2010).

Candidate genes related to adverse events due to treatment Prediction of adverse events is an additional major aim of personalized psychiatry, although evidence regarding pharmacogenomics related to treatment safety in BD is still scarce (Budde et al., 2017).

Results from the STEP-BD genome-wide dataset showed that weight gain caused by mood stabilizers and second-generation antipsychotics was associated with a fat mass and obesity-associated (*FTO*) gene, Tre-2/USP6 gene, BUB2 gene, cdc16 domain family, member 1 (*TBC1D1*) gene, methylene tetrahydrofolate reductase (*MTHFR*) gene, and *HRH1* gene (Creta, Fabbri, & Serretti, 2015). In another study, results suggested that the G-protein beta 3 subunit (*GNB3*) C825T polymorphism might protect against valproate-related metabolic abnormalities (Chang et al., 2010; Chen et al., 2017).

Variants in genes encoding HLA proteins are associated with some idiosyncratic adverse drug reactions, including clozapine-induced agranulocytosis and carbamazepine-induced hypersensitivity reactions (Foulds et al., 2016; Wang et al., 2017). For example, the HLA-B*1502 allele is an important predictor of serious carbamazepine hypersensitivity reactions in Han Chinese populations (Zhang et al., 2011); and as a result, genotyping before starting carbamazepine in this patient group is mandatory according to the United States Federal Drug Administration (Moore et al., 2014).

Regarding genetic risk factors of antidepressant-induced mania in BD, Daray, Thommi, and Ghaemi (2010) reported an association between the serotonin-transporter-linked promoter region (5-HTTLPR) polymorphism and antidepressant induced mania, in a meta-analysis.

2.2.4 Epigenetic markers

Although genetics clearly plays an important role in disease vulnerability, epigenetic modifications might be equally important, as developing BD may be the result of complex gene-environment interactions and, therefore, they should also be considered in personalized psychiatry (Fries et al., 2016; Ludwig & Dwivedi, 2016; Prendes-Alvarez & Nemeroff, 2016). As an example, stressful life events, and particularly childhood trauma, are known to be related to epigenetic modifications, and may increase the risk of developing mood disorders (Brietzke et al., 2012). Methylation of the BDNF gene has also been suggested to play a role in the development of psychiatric disorders (Fries et al., 2016).

Postmortem studies have also confirmed epigenetic modifications in brain samples of psychiatric patients. After analyzing 115 human postmortem brain samples from the frontal lobe, Abdolmaleky et al. (2011) reported significant hypomethylation of the membrane-bound COMT (MB-COMT) promoter in the frontal lobe of brains from individuals with schizophrenia or BD compared with the brains of healthy controls. The same group replicated those findings in peripheral tissues (Nohesara et al., 2011).

2.3 Technology and personalized psychiatry

Technology suitable for the health care sector is rapidly evolving, and these developments are starting to impact medical research in BD and clinical practice. On one hand, modern technology offers large-scale datasets, meaning clinicians are provided billions of data points associated with each individual (Hood & Friend, 2011). This large collection of measurements is known as big data (Librenza-Garcia et al., 2017). Hence, as personalized medicine is envisaged to be intensively data-driven, clinical bioinformatics will be an essential element to support diagnosis, to help finding disease-specific biological markers, and to develop tailored preventive and therapeutic approaches (Beckmann & Lew, 2016). For example, in order to analyze big data, several machine learning methods have been developed during the past few years (Librenza-Garcia et al., 2017). Machine learning methods are ideal for assessing multifactorial disorders, as they facilitate the integration of multiple measurements (e.g., patient clinical records, laboratory testing, brain imaging data, genomic data, proteomics or transcriptomics data, or environmental information) (Librenza-Garcia et al., 2017). They also allow estimating the probability of a particular outcome at an individual level (Librenza-Garcia et al., 2017). For example, recent machine learning studies have aimed to estimate the probability of an individual patient with a mood disorder to attempt suicide (Passos et al., 2016). Other machine-learning approaches seem promising to predict conversion to BD among bipolar offspring (Mourao-Miranda et al., 2012).

These methods also represent a valuable tool for personalizing and improving treatment strategies for BD, as illustrated by another study that aimed to estimate an optimal dynamic treatment regime (DTR) for bipolar depression using Q-learning according to the individual clinical evolution of each patient (Wu, Laber, Lipkovich, & Severus, 2015).

On the other hand, the widespread use of smartphones gives us an opportunity to create mobile phone apps that will finally provide valuable long-term longitudinal data, such as geolocation, voice data, or sleep patterns (Abdullah et al., 2016; Faurholt-Jepsen et al., 2016), that may help in diagnosis and monitoring (Torous, Summergrad, & Nassir Ghaemi, 2016).

Moreover, new devices can help empower patients through mobile device-based noninvasive self-monitoring, and provide them with more tailored and personalized treatment (Beckmann & Lew, 2016; van Os et al., 2017). They can also help clinicians find the optimal dose of psychotropic medication, as well as to identify early signs of response or relapse (van Os et al., 2017). The use of mobile technology to support health care delivery is known as "Mobile Health" ("mHealth") (Menon, Rajan, & Sarkar, 2017). The SIMPLE app (Hidalgo-Mazzei et al., 2016, 2017) is an example of an mHealth smartphone app for BD. It collects information about mood symptoms and offers customized psychoeducation messages in order to empower patients to identify early symptoms and reduce risk of relapse. It proved to be a feasible intervention among BD patients (Hidalgo-Mazzei et al., 2016), with potential positive benefits when applied as an add-on to treatment as usual (Hidalgo-Mazzei et al., 2017). In the same line, a randomized control trial reported that participants with BD assigned to a 10-week mobile phone-based interactive mood monitoring system linking patient-reported mood states with personalized self-management advice showed significantly greater reductions in depressive symptoms compared with those assigned to traditional paper-and-pencil-based mood monitoring. However, these effects were not maintained at 24-week follow-up, once the mobile-phone delivered interactive intervention was concluded (Depp et al., 2015).

3 Ethical issues in personalized psychiatry

Personalized psychiatry in BD involves new ethical challenges, especially concerning predictive genetic testing and big data, and raises a number of ethical, legal, and societal issues (Salari & Larijani, 2017). Regarding genetic testing, the inability to accurately predict a phenotypic outcome due to the complex interactions between genes, along with environmental conditions, limits the application of presymptomatic testing in psychiatry (Demkow & Wolanczyk, 2017).

How to deal with genetic findings or confidentiality should be clearly defined before the implementation of these practices in psychiatry (Salari & Larijani, 2017). Moreover, along with the use of genomic information has come the emergence of biobanks. Ethical concerns about biobanks include sample collection, storage, use, informed consent, identifiability of the samples, sharing samples throughout the world, reidentification, and privacy and confidentiality (Salari & Larijani, 2017). Similar safety and privacy issues arise when dealing with digital data (Armontrout, Torous, Fisher, Drogin, & Gutheil, 2016). Consequently, there is a need to establish a consensus in policies and practices regarding informed consent, return of research results, ownership of specimens and technology developed, utilization, and disposition of remaining specimens after researchers have acquired them (Caulfield & Murdoch, 2017).

4 Summary and future directions

Precision psychiatry in BD aims to merge clinical and environmental factors, molecular and neuroimaging markers, genetics and epigenetics, and technology in an efficient way to guide clinical decisions and predict vulnerability to the disease or treatment response in order to achieve a tailored approach to the patient (Alhajji & Nemeroff, 2015; Prendes-Alvarez & Nemeroff, 2016; Wium-Andersen et al., 2017).

This global approach is indispensable, because biographical factors by themselves, or genetic information by itself, are not sufficient to accurately predict vulnerability to complex diseases such as BD (Fries et al., 2016; Prendes-Alvarez & Nemeroff, 2016). Therefore, personalized psychiatry in BD may especially benefit from information arising from gene by environment studies (Prendes-Alvarez & Nemeroff, 2016; Squassina et al., 2010). Considering the importance of an early intervention in BD in order to change illness prognosis and prevent disability (Vieta et al., 2018), relying on this personalized information will be of the utmost value to future early interventions. Similarly, the observed variability in treatment response to psychotropic drugs is not due to a single genetic variant, but it probably has a complex environmental-gene interaction (Geoffroy et al., 2014; Squassina et al., 2010). Moreover, there is increasing evidence on the influence of the microbiome in human mental health, and its analysis might be a future important factor in personalized medicine, as the microbiome is unique for each individual and dynamic, that is, its composition varies from birth to adulthood, and can also be modified by external factors such as stress, diet, or antibiotic use (Beckmann & Lew, 2016; Salagre, Vieta, & Grande, 2017).

However, a number of obstacles still need to be surmounted before personalized psychiatry can be implemented in clinical practice. Despite the efforts to find biological markers specific to BD, current candidate biological markers are nonspecific, and findings are difficult to replicate (Teixeira et al., 2016). Indeed, current evidence suggests that the identified molecular markers are common across different diseases (Goldsmith, Rapaport, & Miller, 2016). Considering neuroimaging markers, replicability becomes even more difficult, as subtle methodological differences between studies can lead to completely different results (Etkin, 2014). The same is true for genetic data on disease vulnerability and treatment response. Thus, nowadays, the aim of personalized psychiatry is to find a set of molecular, neuroimaging, and genetic markers predictive of BD, rather than finding a single marker causing the disease (Fernandes et al., 2017; Wium-Andersen et al., 2017). The applicability of a panel of biological markers in clinical practice, though, will also depend on cost-effectiveness issues (Alda, 2013; Vieta, 2015).

A more personalized form of psychiatry may need to reconsider current categorical diagnostic systems, and possibly focus on more homogeneous subgroups, which might at the same time help us understand the biological basis of individual symptoms and to increase the probability of identifying genes of interest (Beckmann & Lew, 2016; Foulds et al., 2016; Geoffroy et al., 2014; Mirnezami et al., 2012). In an attempt to translate the heterogeneity of disease presentations to clinical trials, the Research Domain Criteria (RDoC) initiative, driven by The National Institute of Mental Health, seeks to move research toward a multimodal dimensional framework that integrates genetic, molecular, cellular, physiological, motivational, and cognitive data (Insel et al., 2010). Several authors agree on the idea that, for the purpose of personalized medicine in BD, the present diagnostic system faces several limitations (Vieta, 2016; Wium-Andersen et al., 2017). Some authors propose to include complementary dimensional factors, meaning psychopathological constructs that cut across many diagnostic entities, such as impulsivity or suicidality, to the current categorical classification systems as a way to give some nuances to the current categorical model (Vieta & Phillips, 2007). These dimensional factors might be important to predict prognosis or guide treatment, and this approach might be more in line with the personalized approach. For example, predominant polarity might be a critical feature for better characterization of diagnosis and course, given its impact on treatment selection.

Another major challenge of precision psychiatry in BD is handling the huge amount of information provided by electronic health records (Mirnezami et al., 2012). Decision support tools, such as machine learning approaches, present as promising allies, but current evidence on the potential of machine learning in BD is still very preliminary. Small sample sizes and lack of adequate external validation are still major limitations of the available studies (Librenza-Garcia et al., 2017). Nevertheless, in the future, machine learning techniques are expected to help to define clusters of patients who share similar characteristics, and potentially allow us to detect subgroups of patients with similar outcomes (Librenza-Garcia et al., 2017). Another great challenge for the future is to facilitate the implementation of these predictive models into clinical practice, for example, through web-based calculators that could guide decisions on the basis of test results (Librenza-Garcia et al., 2017; Mirnezami et al., 2012).

In conclusion, precision medicine aims to create diagnostic, prognostic, and therapeutic strategies precisely tailored to each patient's requirements, ensuring that patients get the right treatment at the right dose at the right time, with minimum adverse effects and maximum efficacy (Mirnezami et al., 2012). However, available evidence to translate precision psychiatry in BD from a theoretical framework to clinical practice is insufficient, as most of the findings need to be validated, and their clinical usefulness is unclear. Moreover, its implementation will require a cultural shift, with patients viewed as active participants in their own health care (Mirnezami et al., 2012; Vieta, 2015). Also, clinicians will need new technological skills, and both clinicians and patients should be thoroughly educated regarding genotyping tests, as carefully explaining and contextualizing the genetic test results would be essential to avoid misunderstandings (Costa e Silva, 2013; Lebowitz & Ahn, 2017). Furthermore, precision psychiatry may need to solve several new ethical issues before being fully implemented. Despite all these challenges, research in personalized psychiatry is envisaged to keep rapidly progressing so personalized treatment in BD will, in the near future, become a reality.

REFERENCES

Abdolmaleky, H. M., Yaqubi, S., Papageorgis, P., Lambert, A. W., Ozturk, S., Sivaraman, V., & Thiagalingam, S. (2011). Epigenetic dysregulation of HTR2A in the brain of patients with schizophrenia and bipolar disorder. *Schizophrenia Research, 129*(2–3), 183–190. https://doi.org/10.1016/j.schres.2011.04.007.

Abdullah, S., Matthews, M., Frank, E., Doherty, G., Gay, G., & Choudhury, T. (2016). Automatic detection of social rhythms in bipolar disorder. *Journal of the American Medical Informatics Association, 23*(3), 538–543. https://doi.org/10.1093/jamia/ocv200.

Alda, M. (2013). Personalized psychiatry: Many questions, fewer answers. *Journal of Psychiatry & Neuroscience, 38*(6), 363–365. https://doi.org/10.1503/jpn.130221.

Alhajji, L., & Nemeroff, C. B. (2015). Personalized Medicine and Mood Disorders. *The Psychiatric Clinics of North America, 38*(3), 395–403. https://doi.org/10.1016/j.psc.2015.05.003.

Amare, A. T., Schubert, K. O., Hou, L., Clark, S. R., Papiol, S., Heilbronner, U., ... Baune, B. T. (2017). Association of polygenic score for schizophrenia and HLA antigen and inflammation genes with response to lithium in bipolar affective disorder: A genome-wide association study. *JAMA Psychiatry.* https://doi.org/10.1001/jamapsychiatry.2017.3433.

Armontrout, J., Torous, J., Fisher, M., Drogin, E., & Gutheil, T. (2016). Mobile mental health: Navigating new rules and regulations for digital tools. *Current Psychiatry Reports, 18*(10), 91. https://doi.org/10.1007/s11920-016-0726-x.

Bai, Y. M., Su, T. P., Li, C. T., Tsai, S. J., Chen, M. H., Tu, P. C., & Chiou, W. F. (2015). Comparison of pro-inflammatory cytokines among patients with bipolar disorder and unipolar depression and normal controls. *Bipolar Disorders, 17*(3), 269–277. https://doi.org/10.1111/bdi.12259.

Beckmann, J. S., & Lew, D. (2016). Reconciling evidence-based medicine and precision medicine in the era of big data: Challenges and opportunities. *Genome Medicine, 8*(1), 134. https://doi.org/10.1186/s13073-016-0388-7.

Berk, M., Post, R., Ratheesh, A., Gliddon, E., Singh, A., Vieta, E., ... Dodd, S. (2017). Staging in bipolar disorder: From theoretical framework to clinical utility. *World Psychiatry, 16*(3), 236–244. https://doi.org/10.1002/wps.20441.

Biomarkers Definitions Working Group. (2001). Biomarkers and surrogate endpoints: preferred definitions and conceptual framework. *Clinical Pharmacology & Therapeutics, 69*(3), 89–95. https://doi.org/10.1067/mcp.2001.113989.

Birmaher, B., Gill, M. K., Axelson, D. A., Goldstein, B. I., Goldstein, T. R., Yu, H., ... Keller, M. B. (2014). Longitudinal trajectories and associated baseline predictors in youths with bipolar spectrum disorders. *The American Journal of Psychiatry, 171*(9), 990–999. https://doi.org/10.1176/appi.ajp.2014.13121577.

Bortolato, B., Kohler, C. A., Evangelou, E., Leon-Caballero, J., Solmi, M., Stubbs, B., ... Carvalho, A. F. (2017). Systematic assessment of environmental risk factors for bipolar disorder: An umbrella review of systematic reviews and meta-analyses. *Bipolar Disorders, 19*(2), 84–96. https://doi.org/10.1111/bdi.12490.

Brietzke, E., Kauer Sant'anna, M., Jackowski, A., Grassi-Oliveira, R., Bucker, J., Zugman, A., ... Bressan, R. A. (2012). Impact of childhood stress on psychopathology. *Revista Brasileira de Psiquiatria, 34*(4), 480–488.

Budde, M., Degner, D., Brockmoller, J., & Schulze, T. G. (2017). Pharmacogenomic aspects of bipolar disorder: An update. *European Neuropsychopharmacology, 27*(6), 599–609. https://doi.org/10.1016/j.euroneuro.2017.02.001.

Calabrese, J. R., Fatemi, S. H., Kujawa, M., & Woyshville, M. J. (1996). Predictors of response to mood stabilizers. *Journal of Clinical Psychopharmacology, 16*(2 (Suppl. 1)), 24s–31s.

Carvalho, A. F., Quevedo, J., McIntyre, R. S., Soeiro-de-Souza, M. G., Fountoulakis, K. N., Berk, M., ... Vieta, E. (2014). Treatment implications of predominant polarity and the polarity index: A comprehensive review. *International Journal of Neuropsychopharmacology. 18*(2). https://doi.org/10.1093/ijnp/pyu079.

Caulfield, T., & Murdoch, B. (2017). Genes, cells, and biobanks: Yes, there's still a consent problem. *PLoS Biology. 15*(7). https://doi.org/10.1371/journal.pbio.2002654.

Chang, H. H., Gean, P. W., Chou, C. H., Yang, Y. K., Tsai, H. C., Lu, R. B., & Chen, P. S. (2010). C825T polymorphism of the GNB3 gene on valproate-related metabolic abnormalities in bipolar disorder patients. *Journal of Clinical Psychopharmacology, 30*(5), 512–517. https://doi.org/10.1097/JCP.0b013e3181f03f50.

Chen, P. S., Chang, H. H., Huang, C. C., Lee, C. C., Lee, S. Y., Chen, S. L., ... Lu, R. B. (2017). A longitudinal study of the association between the GNB3 C825T polymorphism and metabolic disturbance in bipolar II patients treated with valproate. *The Pharmacogenomics Journal, 17*(2), 155–161. https://doi.org/10.1038/tpj.2015.96.

Chen, C. H., Lee, C. S., Lee, M. T., Ouyang, W. C., Chen, C. C., Chong, M. Y., ... Cheng, A. T. (2014). Variant GADL1 and response to lithium therapy in bipolar I disorder. *The New England Journal of Medicine, 370*(2), 119–128. https://doi.org/10.1056/NEJMoa1212444.

Christenhusz, G. M., Devriendt, K., & Dierickx, K. (2013). Secondary variants—In defense of a more fitting term in the incidental findings debate. *European Journal of Human Genetics, 21*(12), 1331–1334. https://doi.org/10.1038/ejhg.2013.89.

Colom, F., Vieta, E., Daban, C., Pacchiarotti, I., & Sanchez-Moreno, J. (2006). Clinical and therapeutic implications of predominant polarity in bipolar disorder. *Journal of Affective Disorders, 93*(1–3), 13–17. https://doi.org/10.1016/j.jad.2006.01.032.

Costa e Silva, J. A. (2013). Personalized medicine in psychiatry: New technologies and approaches. *Metabolism, 62*(Suppl. 1), S40–S44. https://doi.org/10.1016/j.metabol.2012.08.017.

Craddock, N., & Jones, I. (1999). Genetics of bipolar disorder. *Journal of Medical Genetics, 36*(8), 585–594.

Creta, E., Fabbri, C., & Serretti, A. (2015). Genetics of second-generation antipsychotic and mood stabilizer-induced weight gain in bipolar disorder: Common and specific effects of key regulators of fat-mass homoeostasis genes. *Pharmacogenetics and Genomics, 25*(7), 354–362. https://doi.org/10.1097/fpc.0000000000000144.

Daray, F. M., Thommi, S. B., & Ghaemi, S. N. (2010). The pharmacogenetics of antidepressant-induced mania: A systematic review and meta-analysis. *Bipolar Disorders, 12*(7), 702–706. https://doi.org/10.1111/j.1399-5618.2010.00864.x.

Davila, R., Zumarraga, M., Basterreche, N., Arrue, A., Zamalloa, M. I., & Anguiano, J. B. (2006). Influence of the catechol-O-methyltransferase Val108/158Met polymorphism on the plasma concentration of catecholamine metabolites and on clinical features in type I bipolar disorder—A preliminary report. *Journal of Affective Disorders, 92*(2–3), 277–281. https://doi.org/10.1016/j.jad.2006.02.009.

de Oliveira, G. S., Cereser, K. M., Fernandes, B. S., Kauer-Sant'Anna, M., Fries, G. R., Stertz, L., … Kapczinski, F. (2009). Decreased brain-derived neurotrophic factor in medicated and drug-free bipolar patients. *Journal of Psychiatric Research, 43*(14), 1171–1174. https://doi.org/10.1016/j.jpsychires.2009.04.002.

Demkow, U., & Wolanczyk, T. (2017). Genetic tests in major psychiatric disorders-integrating molecular medicine with clinical psychiatry—Why is it so difficult? *Translational Psychiatry. 7*(6). https://doi.org/10.1038/tp.2017.106.

Depp, C. A., Ceglowski, J., Wang, V. C., Yaghouti, F., Mausbach, B. T., Thompson, W. K., & Granholm, E. L. (2015). Augmenting psychoeducation with a mobile intervention for bipolar disorder: A randomized controlled trial. *Journal of Affective Disorders, 174*, 23–30. https://doi.org/10.1016/j.jad.2014.10.053.

Drapier, D., Surguladze, S., Marshall, N., Schulze, K., Fern, A., Hall, M. H., … McDonald, C. (2008). Genetic liability for bipolar disorder is characterized by excess frontal activation in response to a working memory task. *Biological Psychiatry, 64*(6), 513–520. https://doi.org/10.1016/j.biopsych.2008.04.038.

Duffy, A., Horrocks, J., Doucette, S., Keown-Stoneman, C., McCloskey, S., & Grof, P. (2014). The developmental trajectory of bipolar disorder. *The British Journal of Psychiatry, 204*(2), 122–128. https://doi.org/10.1192/bjp.bp.113.126706.

Etkin, A. (2014). Neuroimaging and the future of personalized treatment in psychiatry. *Depression and Anxiety, 31*(11), 899–901. https://doi.org/10.1002/da.22325.

Faurholt-Jepsen, M., Busk, J., Frost, M., Vinberg, M., Christensen, E. M., Winther, O., … Kessing, L. V. (2016). Voice analysis as an objective state marker in bipolar disorder. *Translational Psychiatry. 6*, https://doi.org/10.1038/tp.2016.123.

Fernandes, B. S., & Berk, M. (2017). Staging in bipolar disorder: One step closer to precision psychiatry. *Revista Brasileira de Psiquiatria, 39*(2), 88–89. https://doi.org/10.1590/1516-4446-2017-3902.

Fernandes, B. S., Gama, C. S., Kauer-Sant'Anna, M., Lobato, M. I., Belmonte-de-Abreu, P., & Kapczinski, F. (2009). Serum brain-derived neurotrophic factor in bipolar and unipolar depression: A potential adjunctive tool for differential diagnosis. *Journal of Psychiatric Research, 43*(15), 1200–1204. https://doi.org/10.1016/j.jpsychires.2009.04.010.

Fernandes, B. S., Molendijk, M. L., Kohler, C. A., Soares, J. C., Leite, C. M., Machado-Vieira, R., … Carvalho, A. F. (2015). Peripheral brain-derived neurotrophic factor (BDNF) as a biomarker in bipolar disorder: A meta-analysis of 52 studies. *BMC Medicine, 13*, 289. https://doi.org/10.1186/s12916-015-0529-7.

Fernandes, B. S., Steiner, J., Molendijk, M. L., Dodd, S., Nardin, P., Goncalves, C. A., … Berk, M. (2016). C-reactive protein concentrations across the mood spectrum in bipolar disorder: A systematic review and meta-analysis. *Lancet Psychiatry, 3*(12), 1147–1156. https://doi.org/10.1016/s2215-0366(16)30370-4.

Fernandes, B. S., Williams, L. M., Steiner, J., Leboyer, M., Carvalho, A. F., & Berk, M. (2017). The new field of 'precision psychiatry'. *BMC Medicine, 15*(1), 80. https://doi.org/10.1186/s12916-017-0849-x.

Ferreira, M. A., O'Donovan, M. C., Meng, Y. A., Jones, I. R., Ruderfer, D. M., Jones, L., … Craddock, N. (2008). Collaborative genome-wide association analysis supports a role for ANK3 and CACNA1C in bipolar disorder. *Nature Genetics, 40*(9), 1056–1058. https://doi.org/10.1038/ng.209.

Foulds, J. A., Maggo, S. D., & Kennedy, M. A. (2016). Personalised prescribing in psychiatry: Has pharmacogenomics delivered on its promise? *The Australian and New Zealand Journal of Psychiatry, 50*(6), 509–510. https://doi.org/10.1177/0004867416640099.

Fries, G. R., Li, Q., McAlpin, B., Rein, T., Walss-Bass, C., Soares, J. C., & Quevedo, J. (2016). The role of DNA methylation in the pathophysiology and treatment of bipolar disorder. *Neuroscience and Biobehavioral Reviews, 68*, 474–488. https://doi.org/10.1016/j.neubiorev.2016.06.010.

Ftouhi-Paquin, N., Alda, M., Grof, P., Chretien, N., Rouleau, G., & Turecki, G. (2001). Identification of three polymorphisms in the translated region of PLC-gamma1 and their investigation in lithium responsive bipolar disorder. *American Journal of Medical Genetics, 105*(3), 301–305.

Garnham, J., Munro, A., Slaney, C., Macdougall, M., Passmore, M., Duffy, A., … Alda, M. (2007). Prophylactic treatment response in bipolar disorder: Results of a naturalistic observation study. *Journal of Affective Disorders, 104*(1–3), 185–190. https://doi.org/10.1016/j.jad.2007.03.003.

Geoffroy, P. A., Bellivier, F., Leboyer, M., & Etain, B. (2014). Can the response to mood stabilizers be predicted in bipolar disorder? *Frontiers in Bioscience (Elite Edition), 6*, 120–138.

Goikolea, J. M., Colom, F., Martinez-Aran, A., Sanchez-Moreno, J., Giordano, A., Bulbena, A., & Vieta, E. (2007). Clinical and prognostic implications of seasonal pattern in bipolar disorder: A 10-year follow-up of 302 patients. *Psychological Medicine, 37*(11), 1595–1599. https://doi.org/10.1017/s0033291707000864.

Goldsmith, D. R., Rapaport, M. H., & Miller, B. J. (2016). A meta-analysis of blood cytokine network alterations in psychiatric patients: Comparisons between schizophrenia, bipolar disorder and depression. *Molecular Psychiatry, 21*(12), 1696–1709. https://doi.org/10.1038/mp.2016.3.

Grande, I., Balanza-Martinez, V., Jimenez-Arriero, M., Iglesias Lorenzo, F. G., Franch Valverde, J. I., de Arce, R., … Vieta, E. (2012). Clinical factors leading to lamotrigine prescription in bipolar outpatients: Subanalysis of the SIN-DEPRES study. *Journal of Affective Disorders, 143*(1–3), 102–108. https://doi.org/10.1016/j.jad.2012.05.035.

Grande, I., Berk, M., Birmaher, B., & Vieta, E. (2016). Bipolar disorder. *Lancet, 387*(10027), 1561–1572. https://doi.org/10.1016/s0140-6736(15)00241-x.

Grande, I., Fries, G. R., Kunz, M., & Kapczinski, F. (2010). The role of BDNF as a mediator of neuroplasticity in bipolar disorder. *Psychiatry Investigation, 7*(4), 243–250. https://doi.org/10.4306/pi.2010.7.4.243.

Green, E. K., Grozeva, D., Forty, L., Gordon-Smith, K., Russell, E., Farmer, A., … Craddock, N. (2013). Association at SYNE1 in both bipolar disorder and recurrent major depression. *Molecular Psychiatry, 18*(5), 614–617. https://doi.org/10.1038/mp.2012.48.

Grof, P., Duffy, A., Cavazzoni, P., Grof, E., Garnham, J., MacDougall, M., … Alda, M. (2002). Is response to prophylactic lithium a familial trait? *Journal of Clinical Psychiatry, 63*(10), 942–947.

Hidalgo-Mazzei, D., Mateu, A., Reinares, M., Murru, A., Del Mar Bonnin, C., Varo, C., … Colom, F. (2016). Psychoeducation in bipolar disorder with a SIMPLe smartphone application: Feasibility, acceptability and satisfaction. *Journal of Affective Disorders, 200*, 58–66. https://doi.org/10.1016/j.jad.2016.04.042.

Hidalgo-Mazzei, D., Reinares, M., Mateu, A., Juruena, M. F., Young, A. H., Perez-Sola, V., … Colom, F. (2017). Is a SIMPLe smartphone application capable of improving biological rhythms in bipolar disorder? *Journal of Affective Disorders, 223*, 10–16. https://doi.org/10.1016/j.jad.2017.07.028.

Hood, L., & Friend, S. H. (2011). Predictive, personalized, preventive, participatory (P4) cancer medicine. *Nature Reviews Clinical Oncology, 8*(3), 184–187. https://doi.org/10.1038/nrclinonc.2010.227.

Hou, L., Heilbronner, U., Degenhardt, F., Adli, M., Akiyama, K., Akula, N., … Schulze, T. G. (2016). Genetic variants associated with response to lithium treatment in bipolar disorder: A genome-wide association study. *Lancet, 387*(10023), 1085–1093. https://doi.org/10.1016/s0140-6736(16)00143-4.

Insel, T., Cuthbert, B., Garvey, M., Heinssen, R., Pine, D. S., Quinn, K., … Wang, P. (2010). Research domain criteria (RDoC): Toward a new classification framework for research on mental disorders. *The American Journal of Psychiatry, 167*(7), 748–751. https://doi.org/10.1176/appi.ajp.2010.09091379.

Jimenez, E., Sole, B., Arias, B., Mitjans, M., Varo, C., Reinares, M., … Benabarre, A. (2017). Impact of childhood trauma on cognitive profile in bipolar disorder. *Bipolar Disorders*. https://doi.org/10.1111/bdi.12514.

Kim, B., Kim, C. Y., Lee, M. J., & Joo, Y. H. (2009). Preliminary evidence on the association between XBP1-116C/G polymorphism and response to prophylactic treatment with valproate in bipolar disorders. *Psychiatry Research, 168*(3), 209–212. https://doi.org/10.1016/j.psychres.2008.05.010.

Ladouceur, C. D., Diwadkar, V. A., White, R., Bass, J., Birmaher, B., Axelson, D. A., & Phillips, M. L. (2013). Fronto-limbic function in unaffected offspring at familial risk for bipolar disorder during an emotional working memory paradigm. *Developmental Cognitive Neuroscience, 5*, 185–196. https://doi.org/10.1016/j.dcn.2013.03.004.

Lebowitz, M. S., & Ahn, W. K. (2017). Blue genes? Understanding and mitigating negative consequences of personalized information about genetic risk for depression. *Journal of Genetic Counseling*. https://doi.org/10.1007/s10897-017-0140-5.

Lee, H. Y., & Kim, Y. K. (2010). Catechol-O-methyltransferase Val158Met polymorphism affects therapeutic response to mood stabilizer in symptomatic manic patients. *Psychiatry Research, 175*(1–2), 63–66. https://doi.org/10.1016/j.psychres.2008.09.011.

Leverich, G. S., McElroy, S. L., Suppes, T., Keck, P. E., Jr., Denicoff, K. D., Nolen, W. A., … Post, R. M. (2002). Early physical and sexual abuse associated with an adverse course of bipolar illness. *Biological Psychiatry, 51*(4), 288–297.

Levitt, A. J., Lam, R. W., & Levitan, R. (2002). A comparison of open treatment of seasonal major and minor depression with light therapy. *Journal of Affective Disorders, 71*(1–3), 243–248.

Li, Z., Zhang, C., Fan, J., Yuan, C., Huang, J., Chen, J., … Fang, Y. (2014). Brain-derived neurotrophic factor levels and bipolar disorder in patients in their first depressive episode: 3-Year prospective longitudinal study. *The British Journal of Psychiatry, 205*(1), 29–35. https://doi.org/10.1192/bjp.bp.113.134064.

Librenza-Garcia, D., Kotzian, B. J., Yang, J., Mwangi, B., Cao, B., Pereira Lima, L. N., … Passos, I. C. (2017). The impact of machine learning techniques in the study of bipolar disorder: A systematic review. *Neuroscience and Biobehavioral Reviews, 80*, 538–554. https://doi.org/10.1016/j.neubiorev.2017.07.004.

Lovlie, R., Berle, J. O., Stordal, E., & Steen, V. M. (2001). The phospholipase C-gamma1 gene (PLCG1) and lithium-responsive bipolar disorder: Re-examination of an intronic dinucleotide repeat polymorphism. *Psychiatric Genetics, 11*(1), 41–43.

Ludwig, B., & Dwivedi, Y. (2016). Dissecting bipolar disorder complexity through epigenomic approach. *Molecular Psychiatry, 21*(11), 1490–1498. https://doi.org/10.1038/mp.2016.123.

Mamdani, F., Alda, M., Grof, P., Young, L. T., Rouleau, G., & Turecki, G. (2008). Lithium response and genetic variation in the CREB family of genes. *American Journal of Medical Genetics Part B: Neuropsychiatric Genetics, 147b*(4), 500–504. https://doi.org/10.1002/ajmg.b.30617.

Manchia, M., Squassina, A., Congiu, D., Chillotti, C., Ardau, R., Severino, G., & Del Zompo, M. (2009). Interacting genes in lithium prophylaxis: Preliminary results of an exploratory analysis on the role of DGKH and NR1D1 gene polymorphisms in 199 Sardinian bipolar patients. *Neuroscience Letters, 467*(2), 67–71. https://doi.org/10.1016/j.neulet.2009.10.003.

Mansour, H. A., Wood, J., Logue, T., Chowdari, K. V., Dayal, M., Kupfer, D. J., … Nimgaonkar, V. L. (2006). Association study of eight circadian genes with bipolar I disorder, schizoaffective disorder and schizophrenia. *Genes, Brain, and Behavior, 5*(2), 150–157. https://doi.org/10.1111/j.1601-183X.2005.00147.x.

McDonald, C. (2015). Brain structural effects of psychopharmacological treatment in bipolar disorder. *Current Neuropharmacology, 13*(4), 445–457.

Menon, V., Rajan, T. M., & Sarkar, S. (2017). Psychotherapeutic applications of mobile phone-based technologies: A systematic review of current research and trends. *Indian Journal of Psychological Medicine, 39*(1), 4–11. https://doi.org/10.4103/0253-7176.198956.

Michelon, L., Meira-Lima, I., Cordeiro, Q., Miguita, K., Breen, G., Collier, D., & Vallada, H. (2006). Association study of the INPP1, 5HTT, BDNF, AP-2beta and GSK-3beta GENE variants and restrospectively scored response to lithium prophylaxis in bipolar disorder. *Neuroscience Letters, 403*(3), 288–293. https://doi.org/10.1016/j.neulet.2006.05.001.

Mirnezami, R., Nicholson, J., & Darzi, A. (2012). Preparing for precision medicine. *The New England Journal of Medicine, 366*(6), 489–491. https://doi.org/10.1056/NEJMp1114866.

Moller, H. J., & Rujescu, D. (2010). Pharmacogenetics–genomics and personalized psychiatry. *European Psychiatry, 25*(5), 291–293. https://doi.org/10.1016/j.eurpsy.2009.12.015.

Monfrim, X., Gazal, M., De Leon, P. B., Quevedo, L., Souza, L. D., Jansen, K., … Kaster, M. P. (2014). Immune dysfunction in bipolar disorder and suicide risk: Is there an association between peripheral corticotropin-releasing hormone and interleukin-1beta? *Bipolar Disorders, 16*(7), 741–747. https://doi.org/10.1111/bdi.12214.

Moore, T. R., Hill, A. M., & Panguluri, S. K. (2014). Pharmacogenomics in psychiatry: Implications for practice. *Recent Patents on Biotechnology*, *8*(2), 152–159.

Mourao-Miranda, J., Oliveira, L., Ladouceur, C. D., Marquand, A., Brammer, M., Birmaher, B., … Phillips, M. L. (2012). Pattern recognition and functional neuroimaging help to discriminate healthy adolescents at risk for mood disorders from low risk adolescents. *PLoS ONE*. *7*(2)https://doi.org/10.1371/journal.pone.0029482.

Munkholm, K., Vinberg, M., & Kessing, L. V. (2016). Peripheral blood brain-derived neurotrophic factor in bipolar disorder: A comprehensive systematic review and meta-analysis. *Molecular Psychiatry*, *21*(2), 216–228. https://doi.org/10.1038/mp.2015.54.

Nohesara, S., Ghadirivasfi, M., Mostafavi, S., Eskandari, M. R., Ahmadkhaniha, H., Thiagalingam, S., & Abdolmaleky, H. M. (2011). DNA hypomethylation of MB-COMT promoter in the DNA derived from saliva in schizophrenia and bipolar disorder. *Journal of Psychiatric Research*, *45*(11), 1432–1438. https://doi.org/10.1016/j.jpsychires.2011.06.013.

Obrocea, G. V., Dunn, R. M., Frye, M. A., Ketter, T. A., Luckenbaugh, D. A., Leverich, G. S., … Post, R. M. (2002). Clinical predictors of response to lamotrigine and gabapentin monotherapy in refractory affective disorders. *Biological Psychiatry*, *51*(3), 253–260.

Passmore, M. J., Garnham, J., Duffy, A., MacDougall, M., Munro, A., Slaney, C., … Alda, M. (2003). Phenotypic spectra of bipolar disorder in responders to lithium versus lamotrigine. *Bipolar Disorders*, *5*(2), 110–114.

Passos, I. C., Mwangi, B., Cao, B., Hamilton, J. E., Wu, M. J., Zhang, X. Y., … Soares, J. C. (2016). Identifying a clinical signature of suicidality among patients with mood disorders: A pilot study using a machine learning approach. *Journal of Affective Disorders*, *193*, 109–116. https://doi.org/10.1016/j.jad.2015.12.066.

Perez, V., Salavert, A., Espadaler, J., Tuson, M., Saiz-Ruiz, J., Saez-Navarro, C., … Menchon, J. M. (2017). Efficacy of prospective pharmacogenetic testing in the treatment of major depressive disorder: Results of a randomized, double-blind clinical trial. *BMC Psychiatry*, *17*(1), 250. https://doi.org/10.1186/s12888-017-1412-1.

Perlis, R. H., Adams, D. H., Fijal, B., Sutton, V. K., Farmen, M., Breier, A., & Houston, J. P. (2010). Genetic association study of treatment response with olanzapine/fluoxetine combination or lamotrigine in bipolar I depression. *Journal of Clinical Psychiatry*, *71*(5), 599–605. https://doi.org/10.4088/JCP.08m04632gre.

Perlis, R. H., Smoller, J. W., Ferreira, M. A., McQuillin, A., Bass, N., Lawrence, J., … Purcell, S. (2009). A genomewide association study of response to lithium for prevention of recurrence in bipolar disorder. *The American Journal of Psychiatry*, *166*(6), 718–725. https://doi.org/10.1176/appi.ajp.2009.08111633.

Perugi, G., Medda, P., Toni, C., Mariani, M. G., Socci, C., & Mauri, M. (2017). The role of electroconvulsive therapy (ECT) in bipolar disorder: Effectiveness in 522 patients with bipolar depression, mixed-state, mania and catatonic features. *Current Neuropharmacology*, *15*(3), 359–371. https://doi.org/10.2174/1570159x14666161017233642.

Phillips, M. L., & Swartz, H. A. (2014). A critical appraisal of neuroimaging studies of bipolar disorder: Toward a new conceptualization of underlying neural circuitry and a road map for future research. *The American Journal of Psychiatry*, *171*(8), 829–843. https://doi.org/10.1176/appi.ajp.2014.13081008.

Popovic, D., Reinares, M., Goikolea, J. M., Bonnin, C. M., Gonzalez-Pinto, A., & Vieta, E. (2012). Polarity index of pharmacological agents used for maintenance treatment of bipolar disorder. *European Neuropsychopharmacology*, *22*(5), 339–346. https://doi.org/10.1016/j.euroneuro.2011.09.008.

Porcu, M., Balestrieri, A., Siotto, P., Lucatelli, P., Anzidei, M., Suri, J. S., … Saba, L. (2016). Clinical neuroimaging markers of response to treatment in mood disorders. *Neuroscience Letters*. https://doi.org/10.1016/j.neulet.2016.10.013.

Preisig, M., Strippoli, M. P., Castelao, E., Merikangas, K. R., Gholam-Rezaee, M., Marquet, P., … Vandeleur, C. L. (2016). The specificity of the familial aggregation of early-onset bipolar disorder: A controlled 10-year follow-up study of offspring of parents with mood disorders. *Journal of Affective Disorders*, *190*, 26–33. https://doi.org/10.1016/j.jad.2015.10.005.

Prendes-Alvarez, S., & Nemeroff, C. B. (2016). Personalized medicine: Prediction of disease vulnerability in mood disorders. *Neuroscience Letters*. https://doi.org/10.1016/j.neulet.2016.09.049.

Rosa, A. R., Magalhaes, P. V., Czepielewski, L., Sulzbach, M. V., Goi, P. D., Vieta, E., … Kapczinski, F. (2014). Clinical staging in bipolar disorder: Focus on cognition and functioning. *The Journal of Clinical Psychiatry*, *75*(5), e450–e456. https://doi.org/10.4088/JCP.13m08625.

Rosa, A. R., Reinares, M., Franco, C., Comes, M., Torrent, C., Sanchez-Moreno, J., … Vieta, E. (2009). Clinical predictors of functional outcome of bipolar patients in remission. *Bipolar Disorders*, *11*(4), 401–409. https://doi.org/10.1111/j.1399-5618.2009.00698.x.

Rosenblat, J. D., Lee, Y., & McIntyre, R. S. (2017). Does pharmacogenomic testing improve clinical outcomes for major depressive disorder? A systematic review of clinical trials and cost-effectiveness studies. *The Journal of Clinical Psychiatry*, *78*(6), 720–729. https://doi.org/10.4088/JCP.15r10583.

Rybakowski, J. K., Suwalska, A., Skibinska, M., Dmitrzak-Weglarz, M., Leszczynska-Rodziewicz, A., & Hauser, J. (2007). Response to lithium prophylaxis: Interaction between serotonin transporter and BDNF genes. *American Journal of Medical Genetics Part B: Neuropsychiatric Genetics*, *144b*(6), 820–823. https://doi.org/10.1002/ajmg.b.30420.

Salagre, E., Vieta, E., & Grande, I. (2017). The visceral brain: Bipolar disorder and microbiota. *Revista de Psiquiatría y Salud Mental*, *10*(2), 67–69. https://doi.org/10.1016/j.rpsm.2017.02.001.

Salagre, E., Vizuete, A. F., Leite, M., Brownstein, D. J., McGuinness, A., Jacka, F., … Fernandes, B. S. (2017). Homocysteine as a peripheral biomarker in bipolar disorder: A meta-analysis. *European Psychiatry*, *43*, 81–91. https://doi.org/10.1016/j.eurpsy.2017.02.482.

Salari, P., & Larijani, B. (2017). Ethical issues surrounding personalized medicine: A literature review. *Acta Medica Iranica*, *55*(3), 209–217.

Salvatore, P., Baldessarini, R. J., Khalsa, H. M., Vazquez, G., Perez, J., Faedda, G. L., … Tohen, M. (2014). Antecedents of manic versus other first psychotic episodes in 263 bipolar I disorder patients. *Acta Psychiatrica Scandinavica*, *129*(4), 275–285. https://doi.org/10.1111/acps.12170.

Chapter 36

Genetic testing in psychiatry: State of the evidence

Chad A. Bousman[a,b,c,d,e,h], Lisa C. Brown[f], Ajeet B. Singh[g], Harris A. Eyre[g,h,i,j,k] and Daniel J. Müller[l,m]

[a]Department of Medical Genetics, University of Calgary, Calgary, AB, Canada, [b]Department of Psychiatry, University of Calgary, Calgary, AB, Canada, [c]Department of Physiology & Pharmacology, University of Calgary, Calgary, AB, Canada, [d]Alberta Children's Hospital Research Institute, Calgary, AB, Canada, [e]Hotchkiss Brain Institute, Calgary, AB, Canada, [f]Myriad Neuroscience, Mason, OH, United States, [g]IMPACT SRC, School of Medicine, Deakin University, Geelong, VIC, Australia, [h]Innovation Institute, Texas Medical Center, Houston, TX, United States, [i]Department of Psychiatry, University of Melbourne, Melbourne, VIC, Australia, [j]Discipline of Psychiatry, University of Adelaide, Adelaide, SA, Australia, [k]Brainstorm Lab for Mental Health Innovation, Stanford Medical School, Stanford University, Palo Alto, CA, United States, [l]Centre for Addiction and Mental Health, Campbell Family Research Institute, Toronto, ON, Canada, [m]Department of Psychiatry, University of Toronto, Toronto, ON, Canada

1 Introduction

Since the completion of the Human Genome Project in 2003, the field of psychiatric genetics has grown exponentially, leading to increases in both the depth and breadth of knowledge about the genetic factors related to the development and treatment of psychiatric disorders. Although the true underlying genetic factors involved in most common psychiatric disorders (i.e., major depression, anxiety, bipolar, and schizophrenia) and their effective treatment remain to be fully elucidated, efforts to translate what has been discovered into clinical practice are underway. These translational efforts align with the precision medicine approach, which aims to customize healthcare based on an individual's genetic, environmental, and lifestyle factors (National Research Council (U.S.). Committee on A Framework for Developing a New Taxonomy of Disease, 2011), and holds promise for better patient outcomes and wellness through precision prevention, diagnosis, and treatment of psychiatric disorders.

Genetic testing offers one avenue from which more precise prevention, diagnosis, and treatment of psychiatric disorders could emerge. Given the now relatively low, and decreasing cost of genetic testing, cost-effectiveness is becoming more justifiable, and is likely to accelerate clinical use. However, for precision psychiatry to become a reality, a solid evidence base from which this "precision" can be derived is required.

In this chapter, we provide a concise overview of the evidence for genetic testing in psychiatry, with a specific focus on the most common psychiatric disorders, rather than disorders that already have established genetic testing in place, such as neurodevelopmental (e.g., Fragile X syndrome, phenylketonuria) or neurodegenerative (e.g., Huntington's disease) disorders that are often seen by psychiatrists. We will also identify the current gaps in knowledge related to genetic testing, and offer ways forward for research that, in part, will inform translation and adoption of genetic testing in the psychiatric clinical setting.

2 Types of genetic tests

From an implementation perspective, genetic testing can be broadly classified as either clinical or direct-to-consumer. Both types of testing can be of high quality (i.e., reliable and valid) and provide useful information that, when used correctly, can affect patient outcomes (i.e., utility). However, only clinical genetic testing requires initiation and/or mediation of testing by a healthcare provider. Herein, we focus exclusively on clinical genetic testing, as the evidence base for direct-to-consumer testing (e.g., 23andMe) in psychiatry is severely limited.

Within the clinical genetic testing classification, tests can be further classified by their function. A number of classifications have been defined (NIH, 2017), but the most relevant to psychiatry relate to the identification of "at-risk" individuals (predictive), assisting with diagnosis, or guiding optimal treatment. In the subsequent sections, we present the evidence base for each of these genetic testing types as they relate to psychiatry.

Personalized Psychiatry. https://doi.org/10.1016/B978-0-12-813176-3.00036-5

3 Current evidence base

Quality evidence is vital to the future of genetic testing in psychiatry. Although a number of frameworks for evaluating the evidence base for genetic tests exist, consensus on the evidentiary requirements are lacking (Morrison & Boudreau, 2012). However, common to all frameworks is the need to establish analytical validity, clinical validity, and clinical utility. A valid test must provide an accurate result, and both analytical and clinical validity are measures of this accuracy. Analytical validity refers to how well the tests can detect the genetic variant(s) it intends to detect, and clinical validity refers to how well the variant(s) are related to the presence, absence, or risk of a particular outcome. Whereas, clinical utility is a measure of the applicability (e.g., efficacy, effectiveness) and feasibility (e.g., acceptability, efficiency, affordability) of the test for improving patient outcomes. In the following sections, particular attention is given to evidence related to the clinical validity and clinical utility of genetic tests, as data related to analytical validity is typically not published. However, genetic testing laboratories in the U.S., for example, must demonstrate analytical validity to receive certification under the Clinical Laboratory Improvement Amendments, and similar international certification standards, and as such, analytical validity is typically implied.

3.1 Genetic tests for identifying "at risk" individuals

The ability to identify or screen for individuals "at risk" for psychiatric illness has obvious clinical and public health implications. The majority of research in this area has focused on singular or combinatorial nongenetic factors (e.g., neuroimaging, cognitive, clinical), and has recently led to the development of several online risk calculators for the psychotic spectrum (Cannon et al., 2016; Fusar-Poli et al., 2017) and bipolar (Hafeman et al., 2017) disorders. However, to our knowledge, no risk calculator or screening instrument includes genetic information other than family history of a psychiatric disorder, which is an indirect measure of genetic risk.

One example of genetic research related to the identification of "at risk" individuals has focused on transition to psychosis among individuals already deemed to be at ultra-high risk due to family history, subthreshold symptomatology, and/or declines in functioning. Among these ultra-high risk individuals, genetic variation in catechol-o-methyltransferase (*COMT*) (McIntosh et al., 2007), neuregulin 1 (*NRG1*) (Bousman et al., 2013; Hall et al., 2006; Kéri, Kiss, & Kelemen, 2009), D-amino acid oxidase activator (*DAOA*) (Bousman et al., 2013; Mössner et al., 2010), and interleukin 1 beta (*IL1B*) (Bousman, Lee, et al., 2017) have been associated with risk for transition to psychosis. However, the positive predictive values (i.e., probability that transition occurs when the allele is present) of the variants identified to date are either suboptimal, or have not been replicated, hindering their clear utility in clinical practice. Thus, the current evidence base for genetic testing for the purpose of identifying individuals "at risk" for a psychiatric disorder is extremely limited, and currently has limited clinical utility.

3.2 Genetic tests for assisting with diagnosis

Classification of major psychiatric disorders depends on a set of criteria used to assign individuals to a diagnostic category based on relevant clinical similarities. The two major psychiatric disorder classification systems (DSM and ICD) are polythetic, in that two individuals may have the same phenotypic classification, but exhibit or report variations in characteristic symptoms in contrast to monothetic, in which the members would be identical in all characteristics (Millon, 1991). As such, a genetic test to assist with the process of assigning and/or ruling out a diagnosis could be of clinical value.

The challenge, however, is that the common psychiatric disorders are genetically complex, following a polygenic inheritance pattern (i.e., caused by multiple genes) in which single variants associated with a disorder have poor penetrance and lack diagnostic specificity. In fact, the Psychiatric Genomics Consortium (PGC) has shown that common genetic variation explains, at best, about 7% of the variation in diagnosis liability (Schizophrenia Working Group of the Psychiatric Genomics Consortium, 2014), and the Cross-Disorder Group of PGC has shown high to moderate genetic correlation among most of the common psychiatric disorders (Lee et al., 2013). Thus, it may not be surprising that genetic-based risk classifiers for autism (Pramparo et al., 2015; Skafidas et al., 2014), schizophrenia (Liu et al., 2017), bipolar (Chuang & Kuo, 2017), and depression (Wong, Dong, Andreev, Arcos-Burgos, & Licinio, 2012) have not demonstrated strong evidence for translation into clinical settings, and as such, the prospects for common genetic variation to be useful in aiding diagnosis in the near future are slim.

An argument could be made for the use of rare copy number variants (CNVs) in assisting with diagnosis. CNVs result in chromosomal microdeletions or microduplications, and can substantially increase risk for a psychiatric disorder (Malhotra & Sebat, 2012). However, similar to common genetic variants, they lack a strong degree of diagnostic specificity, and as such are likely to have limited utility. However, the Genetic Testing Working Group of the International Society of

TABLE 1 Gene-drug interaction evidence, dosing guideline, and drug label resources.—cont'd

Resource	Website	Evidence levels	Level description
FDA EMA PMDA HCSC	fda.gov ema.europa.eu/ema/ pmda.go.jp/english/ canada.ca/en/health-canada/services/drugs-health-products.html	Testing required	The label states or implies that some sort of gene, protein or chromosomal testing, including genetic testing, functional protein assays, cytogenetic studies, etc., should be conducted before using this drug. This requirement may only be for a particular subset of patients. Labels that state the variant is an indication for the drug, as implying a test requirement. If the label states a test "should be" performed, this is also interpreted as a requirement.
		Testing recommended	The label states or implies that some sort of gene, protein or chromosomal testing, including genetic testing, functional protein assays, cytogenetic studies, etc., is recommended before using this drug. This recommendation may only be for a particular subset of patients. Labels that say testing "should be considered" to be recommending testing.
		Actionable	The label does not discuss genetic or other testing for gene/protein/chromosomal variants, but does contain information about changes in efficacy, dosage or toxicity due to such variants. The label may mention contraindication of the drug in a particular subset of patients but does not require or recommend gene, protein or chromosomal testing.
		Informative	The label mentions a gene or protein is involved in the metabolism or pharmacodynamics of the drug, but there is no information to suggest that variation in these genes/proteins leads to different response.

CPIC, Clinical Pharmacogenetics Implementation Consortium; *CPNDS*, Canadian Pharmacogenomics Network for Drug Safety; *DPWG*, Dutch Pharmacogenetics Working Group; *EMA*, European Medicines Agency; *FDA*, US Food and Drug Administration; *HCSC*, Health Canada (Santé Canada); *PGRN*, Pharmacogenomics Research Network; *PGx*, Pharmacogenetics; *PharmGKB*, Pharmacogenomics Knowledgebase; *PMDA*, Pharmaceuticals and Medical Devices Agency, Japan.

are classified as extensive or normal metabolizers, while those carrying two alleles that code for inactive or absent enzyme function are defined as poor metabolizers. Furthermore, individuals who carry one normal and one inactive/absent allele are classified as intermediate metabolizers, and those with a gain of function allele, gene duplications and/or multiplications of active alleles are defined as "rapid" or "ultrarapid" metabolizers. However, it is worth noting that genotype-predicted "normal" or "intermediate" metabolism can be converted to phenotypically "poor," "rapid," or "ultrarapid" metabolism by factors other than genotype (e.g., co-medications, co-morbidities, diet, smoking), an often underappreciated phenomenon known as phenoconversion (Shah & Smith, 2015).

The bulk of the evidence between these two P-450 genes and antidepressant dosing is linked to selective serotonin reuptake inhibitors (SSRIs) and tricyclic antidepressants (TCAs), although evidence is emerging for serotonin-norepinephrine reuptake inhibitors (SNRIs). This evidence has prompted the development of dosing guidelines for TCAs and SSRIs by the FDA, Health Canada, CPIC (Hicks et al., 2015; Hicks et al., 2016), and the Royal Dutch Association for the Advancement of Pharmacy—Pharmacogenetics Working Group (DPWG) (Swen et al., 2011), the latter guidelines including the SNRI, venlafaxine. However, it should be noted that the level of evidence according to the PharmGKB (Table 1) for venlafaxine is Level 2B, whereas all other antidepressants with guidelines have Level 1 evidence. Thus, one should not assume the evidence base for any two gene-drug interactions are equivalent based solely on the existence of dosing guidelines. Likewise, it should not be assumed that gene-drug pairs covered by two or more guidelines will provide the same guidance, particularly for pairs involving *CYP2D6*, where consensus is lacking across different consortia/expert groups. For example, a clinician intending to prescribe amitriptyline to an individual genotyped as *CYP2D6 *10/*10* would be guided to give a normal starting dose based on the CPIC guidelines, but if consulting the DPWG guidelines, a 40% dose reduction would be recommended (Hicks, Swen, & Gaedigk, 2014). On the other hand, if an individual had a *CYP2C19* genotype of **1/*17*, CPIC would recommend an alternative therapy; whereas DPWG guidelines would advise the normal starting dose for amitriptyline (Bank et al., 2017).

3.3.3 CYP2D6—Antipsychotics

Relative to antidepressants, the pharmacogenetics of antipsychotic medications are less pronounced. However, the FDA, Health Canada, European Medicines Agency, and DPWG have deemed the level of evidence among *CYP2D6* and several antipsychotics (aripiprazole, haloperidol, risperidone, and zuclopenthixol) sufficient to develop dosing guidelines (Swen et al., 2011). Other guideline groups (i.e., CPIC, CPNDS) have not followed suit, although it is expected that CPIC will be drafting similar guidelines in the near future. Interestingly, the PharmGKB level of evidence for interactions among *CYP2D6* and these four antipsychotics are low (level 3), with the exception of risperidone (level 2A). This is not to say that existing guidelines are not justified, but it does suggest that different evidence thresholds are likely used to develop guidelines across groups/consortia, and as such, guidelines published by these groups may not be equivalent or interchangeable.

3.3.4 Promising gene-drug interactions

There are a number of promising gene-drug pairs relevant to psychiatry that do not currently carry testing recommendations and/or appear in guidelines, but have an evidence base that suggests they may in the future. Table 2 provides a list of these gene-drug pairs, along with a summary of the current evidence. Of note, all of the gene-drug pairs mentioned in Table 2 were selected based on the presence of a moderate (level 2B) or higher level of evidence according to PharmGKB, and several of these pairs (denoted with "*" in Table 2) are currently being considered for guideline development by CPIC.

3.3.5 Commercial pharmacogenetic testing

Numerous commercial pharmacogenetic-based decision support tools have emerged based on the pharmacogenetic evidence highlighted herein, and the well documented challenges clinicians face in medication selection and dosing (Bousman & Hopwood, 2016). These tools vary in the gene-drug pairs that are tested, but all include *CYP2D6* and *CY2C19*; aligned with FDA, Health Canada, CPIC, and DPWG guidelines associated with these genes, and antidepressant/antipsychotic selection and dosing. However, it should be noted that the *CYP2D6* and *CY2C19* alleles tested by these commercial panels varies substantially, and as such, are not necessarily equivalent (Bousman, Jaksa, & Pantelis, 2017). In addition, only a minority (<20%) of commercial gene panels in psychiatry test for HLA-B*15:02 and/or HLA-A*31:01 alleles (Bousman & Hopwood, 2016), despite requirements/recommendations for testing, and the availability of international guidelines. Further, more than half (54%) of genes included on many of these testing panels have limited or preliminary evidence for their clinical utility in psychiatry (Fig. 1), and the medication recommendation algorithms that genotyping results are subjected to differ among companies. That said, results from four published small randomized controlled trials (RCTs) suggest the use of this testing to guide medication decisions in practice can improve patient outcomes (Bradley et al., 2017; Pérez et al., 2017; Singh, 2015; Winner, Carhart, Altar, Allen, & Dechairo, 2013), particularly among individuals with moderate-severe major depressive disorder. Moreover, a number of open-label and retrospective studies have echoed the results from the RCTs (for review see: Peterson et al., 2017), and there is preliminary evidence suggesting these commercial tools could be cost-effective (Berm et al., 2016; Brown, Lorenz, Li, & Dechairo, 2017; Rosenblat et al., 2017). However, the evidence base for these tools is still emerging, and independent (noncommercial) validation of study findings has yet to be established. In addition, there are no professional or clinical guidelines to assist clinicians who are contemplating the use of these tools in their practice, and the availability of such testing may be limited depending on the region of the world in which a clinician's practice operates.

4 Strategies for strengthening the evidence base

The current evidence base supporting genetic testing in psychiatry suggests testing is not ready for "prime-time," with the exception of pharmacogenetic testing in specific clinical situations. As such, in the following section, we highlight a few strategies that will assist the field in generating the evidence required for genetic testing to move toward wide clinical adoption. We pay particular attention to pharmacogenetic testing given the availability and growing use of this testing in psychiatry.

4.1 Phenotype definition and measurement

The Achille's heel in psychiatric research is often the adequate definition and measurement of the phenotype of interest. Whether the phenotype of interest is "high-risk status," diagnosis, or treatment response, there are considerable variations in how investigators can define and measure these phenotypes, which can result in heterogeneity within a sample,

TABLE 2 Promising gene-drug interactions.

Gene-drug pair	Clinical Association type	Evidence level		Drug labels			
		PharmaGKB	CPIC	FDA	EMA	PMDA	HCSC
CYP2D6— Atomoxetine[a]	Efficacy/toxicity	2A	B	Actionable	–	Actionable	Actionable
CYP2B6— Methadone[a]	Dosage	2A	B	–	–	–	–
CYP2D6— Risperidone[a]	Pharmacokinetic	2A	B	Informative	–	–	Informative
DRD2— Risperidone	Efficacy	2A	C	–	–	–	–
CYP2D6— Venlafaxine	Efficacy/toxicity	2A	B	Informative	–	–	–
SLC6A4— Citalopram/ escitalopram	Efficacy	2A	B/C	–	–	–	–
HTR2A— Antidepressants	Efficacy	2B	D	–	–	–	–
HTR2C— Antipsychotics	Toxicity	2B	D	–	–	–	–
FKBP5— Antidepressants	Efficacy	2B	D	–	–	–	–
DRD2/ANKK1— Antipsychotics	Toxicity	2B	D	–	–	–	–
MC4R— Antipsychotics	Toxicity	2B	C	–	–	–	–
GRIK4— Antidepressants	Efficacy	2B	D	–	–	–	–
ABCB1— Methadone	Dosage	2B	C/D	–	–	–	–
SCN1A— Carbamazepine	Dosage	2B	B	–	–	–	–
EPHX1— Carbamazepine	Dosage	2B	D	–	–	–	–
COMT/TXNRD2— SSRIs	Efficacy	2B	C	–	–	–	–

CPIC, Clinical Pharmacogenetics Implementation Consortium; *EMA*, European Medicines Agency; *FDA*, US Food and Drug Administration; *HCSC*, Health Canada (Santé Canada); *PGRN*, Pharmacogenomics Research Network; *PGx*, Pharmacogenetics; *PharmGKB*, Pharmacogenomics Knowledgebase; *PMDA*, Pharmaceuticals and Medical Devices Agency, Japan.
[a]*CPIC guidelines expected in the future.*

potentially concealing underlying biological mechanisms and making comparisons between studies difficult. Although it would be naive to believe that all investigators will agree to use the same definitions and measures for a particular phenotype, we feel it is reasonable for a consensus to be developed on the minimum "caseness" criteria, and measurements required for a particular phenotype, without necessarily advocating for the exclusive use of any one measure or measurement procedure. Consensus efforts such as remission criteria in schizophrenia (Andreasen et al., 2005), antidepressant response and remission in depression (Riedel et al., 2010), and more recently, the diagnosis of treatment-resistance

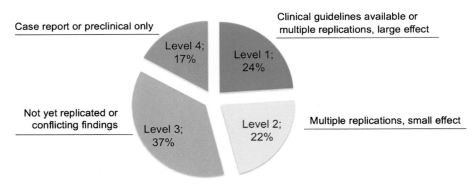

FIG. 1 Proportion of genes included on commercial pharmacogenetic testing panels that meet each of the four PharmGKB levels of evidence criteria. *Data derived from Bousman, C. A., & Hopwood, M. (2016). Commercial pharmacogenetic-based decision-support tools in psychiatry. Lancet Psychiatry, 3(6), 585–590. https://doi.org/10.1016/S2215-0366(16)00017-1.*

schizophrenia (Howes et al., 2017) are good examples. However, the degree to which these consensus criteria are adopted by investigators varies, and as a result, can hinder cross-comparison of research studies, and ultimately the strength of the evidence base.

4.2 Pharmacogenetic testing panel and reporting standardization

A specific concern for current pharmacogenetic testing is the lack of standardization from the perspective of both gene panel content and results reporting. Panels range in size from one to 30 genes, and even when the same gene is included between two or more panels, the specific genetic variant tested within that gene may differ (Bousman & Hopwood, 2016; Bousman, Jaksa, et al., 2017). Due to these gene panel differences, the probability of inter-test agreement for a particular patient is modest, and as such, these tests are not interchangeable. Thus, the selection of which pharmacogenetic test to use is not trivial. Fortunately, recommendations for pharmacogenetic testing results reporting have recently been developed (Caudle et al., 2017; Kalman et al., 2016), but similar recommendations for gene panel membership do not exist. Based on the current evidence, it would seem reasonable to recommend the inclusion of *CYP2D6*, *CYP2C19*, *HLA-A*, and *HLA-B* on all testing panels. However, recommending which alleles (and allele phenotype calls) should be included for each of these genes, particularly *CYP2D6* and *CYP2C19*, will require the convening of an expert panel to arrive at consensus.

4.3 Innovative clinical trials

While implementation of pharmacogenetic testing in clinical practice has been demonstrated to be feasible at large-scale (Herbert et al., 2018; Müller, Kekin, Kao, & Brandl, 2013), there is debate on how best to assess the clinical validity and utility of pharmacogenetic-based decision support tools. One side has argued that rigorous double-blind RCTs are required (Janssens & Deverka, 2014), while an emerging alternative side has argued that despite RCTs being viewed as the "gold-standard," they are not particularly appropriate, because internal validity (i.e., experimental rigor) is maximized at the expense of external validity (i.e., clinical generalizability) (Dhanda, Guzauskas, Carlson, Basu, & Veenstra, 2017; Frieden, 2017; Gillis & Innocenti, 2014; Ratain & Johnson, 2014). Although there is certainly a need for RCTs to demonstrate the clinical validity of pharmacogenetic testing, they are not ideal for establishing clinical utility or usefulness. To determine clinical utility, trial designs must evaluate these tests in a manner that reflects how they will be used in the typical clinical setting (Miller, 2006). Good clinical practice would recommend patients are not blinded, and that treatment decisions are discussed between providers and patients as part of a shared decision making process (Arandjelovic et al., 2017). Thus, there is a need for pragmatic clinical trials with sufficient follow-up (i.e., 6 months or more) that can maximize external validity while maintaining sufficient internal rigor; approximating the "real world" implementation of pharmacogenetic testing (Frieden, 2017).

Related to trial design, it is also evident that previous and ongoing trials of pharmacogenetic testing have a number of biases that need to be addressed in the future. First, all trials have been conducted by the manufacturers of the tests, and as such, independent evaluation is desperately needed. Second, the trials to date have included primarily middle-aged, depressed females of European descent (Peterson et al., 2017). Inclusion of men and ethnic minorities are needed in future trials to determine generalizability of testing in these populations. This is particularly important for pharmacogenetic testing because the allelic frequencies of most pharmacokinetic and pharmacodynamic genes vary widely by ethnicity (Gunes & Dahl, 2008; Ozawa et al., 2004; Van Booven et al., 2010); reducing or increasing the relevance of some alleles included on current testing panels. Third, there is currently no "gold standard" pharmacogenetic test in psychiatry. Head-to-head comparative trials will be required to determine superiority of a particular tool. However, given the large number of tools available, and the lack of any previous or ongoing head-to-head trials, it is unlikely that a gold standard will be determined in the short-term. Finally, there is a need for quality cost-effectivene analyses to be imbedded into future trials of pharmacogenetic testing. If clinical utility is favorable, but cost-effectiveness cannot be established, payers are less likely to approve the use of pharmacogenetic testing. Initial economic studies have been promising, but the quality of these studies has been questioned (Rosenblat et al., 2017). Encouragingly, there are at least six trials (NCT02466477, NCT02286440, NCT02189057, NCT02770339, NCT03113890, NCT02634177) underway that will, in part, address the preceding issues. These trials involve independent investigators, pragmatic design elements, diverse ethnic populations, multiple sites, and/or economic outcome measures.

5   Conclusion

There are formidable challenges that must be overcome prior to the global use of genetic testing in psychiatry. The identification of individuals "at-risk" for, and the diagnosis of, psychiatric disorders are not likely to benefit from predictive or diagnostic genetic testing in the immediate future. However, the use of pharmacogenetic tests to optimize pharmacotherapy shows promise. The challenges of implementing these tests in clinical practice are many (see Chapter 37; Liu et al., for detailed discussion), but the success of implementation efforts will ultimately depend on the strength of the underlying evidence base.

References

Abbasi, J. (2016). Getting pharmacogenomics into the clinic. *JAMA, 316*(15), 1533–1535. https://doi.org/10.1001/jama.2016.12103.

Amstutz, U., Shear, N. H., Rieder, M. J., Hwang, S., Fung, V., Nakamura, H., ... CPNDS Clinical Recommendation Group (2014). Recommendations for HLA-B*15:02 and HLA-A*31:01 genetic testing to reduce the risk of carbamazepine-induced hypersensitivity reactions. *Epilepsia, 55*(4), 496–506. https://doi.org/10.1111/epi.12564.

Andreasen, N. C., Carpenter, W. T., Kane, J. M., Lasser, R. A., Marder, S. R., & Weinberger, D. R. (2005). Remission in schizophrenia: Proposed criteria and rationale for consensus. *The American Journal of Psychiatry, 162*(3), 441–449. https://doi.org/10.1176/appi.ajp.162.3.441.

Arandjelovic, K., Eyre, H. A., Lenze, E., Singh, A. B., Berk, M., & Bousman, C. (2017). The role of depression pharmacogenetic decision support tools in shared decision making. *Journal of Neural Transmission (Vienna).* https://doi.org/10.1007/s00702-017-1806-8.

Bank, P. C., Caudle, K. E., Swen, J. J., Gammal, R. S., Whirl-Carrillo, M., Klein, T. E., ... Guchelaar, H. J. (2017). Comparison of the guidelines of the clinical pharmacogenetics implementation consortium and the Dutch pharmacogenetics working group. *Clinical Pharmacology and Therapeutics.* https://doi.org/10.1002/cpt.762.

Berm, E. J., Looff, M., Wilffert, B., Boersma, C., Annemans, L., Vegter, S., ... Postma, M. J. (2016). Economic evaluations of pharmacogenetic and pharmacogenomic screening tests: A systematic review. Second update of the literature. *PLoS ONE. 11*(1). https://doi.org/10.1371/journal.pone.0146262.

Bousman, C. A., & Hopwood, M. (2016). Commercial pharmacogenetic-based decision-support tools in psychiatry. *Lancet Psychiatry, 3*(6), 585–590. https://doi.org/10.1016/S2215-0366(16)00017-1.

Bousman, C. A., Jaksa, P., & Pantelis, C. (2017). Systematic evaluation of commercial pharmacogenetic testing in psychiatry: A focus on CYP2D6 and CYP2C19 allele coverage and results reporting. *Pharmacogenetics and Genomics, 27*(11), 387–393. https://doi.org/10.1097/FPC.0000000000000303.

Bousman, C. A., Lee, T. Y., Kim, M., Lee, J., Mostaid, M. S., Bang, M., ... Kwon, J. S. (2017). Genetic variation in cytokine genes and risk for transition to psychosis among individuals at ultra-high risk. *Schizophrenia Research.* https://doi.org/10.1016/j.schres.2017.08.040.

Bousman, C. A., Yung, A. R., Pantelis, C., Ellis, J. A., Chavez, R. A., Nelson, B., ... Foley, D. L. (2013). Effects of NRG1 and DAOA genetic variation on transition to psychosis in individuals at ultra-high risk for psychosis. *Translational Psychiatry, 3*, e251. https://doi.org/10.1038/tp.2013.23.

Bradley, P., Shiekh, M., Mehra, V., Vrbicky, K., Layle, S., Olson, M. C., ... Lukowiak, A. A. (2017). Improved efficacy with targeted pharmacogenetic-guided treatment of patients with depression and anxiety: A randomized clinical trial demonstrating clinical utility. *Journal of Psychiatric Research, 96*, 100–107. https://doi.org/10.1016/j.jpsychires.2017.09.024.

Brown, L. C., Lorenz, R. A., Li, J., & Dechairo, B. M. (2017). Economic utility: Combinatorial pharmacogenomics and medication cost savings for mental health care in a primary care setting. *Clinical Therapeutics, 39*(3), 592–602. e591. https://doi.org/10.1016/j.clinthera.2017.01.022.

Cannon, T. D., Yu, C., Addington, J., Bearden, C. E., Cadenhead, K. S., Cornblatt, B. A., ... Kattan, M. W. (2016). An individualized risk calculator for research in prodromal psychosis. *The American Journal of Psychiatry, 173*(10), 980–988. https://doi.org/10.1176/appi.ajp.2016.15070890.

Caudle, K. E., Dunnenberger, H. M., Freimuth, R. R., Peterson, J. F., Burlison, J. D., Whirl-Carrillo, M., … Hoffman, J. M. (2017). Standardizing terms for clinical pharmacogenetic test results: Consensus terms from the Clinical Pharmacogenetics Implementation Consortium (CPIC). *Genetics in Medicine*, *19*(2), 215–223. https://doi.org/10.1038/gim.2016.87.

Chuang, L. C., & Kuo, P. H. (2017). Building a genetic risk model for bipolar disorder from genome-wide association data with random forest algorithm. *Scientific Reports*, *7*. https://doi.org/10.1038/srep39943.

Collins, F. S., & McKusick, V. A. (2001). Implications of the Human Genome Project for medical science. *JAMA*, *285*(5), 540–544.

de Leon, J., & Spina, E. (2016). What is needed to incorporate clinical pharmacogenetic tests into the practice of psychopharmacotherapy? *Expert Review of Clinical Pharmacology*, *9*(3), 351–354. https://doi.org/10.1586/17512433.2016.1112737.

Dhanda, D. S., Guzauskas, G. F., Carlson, J. J., Basu, A., & Veenstra, D. L. (2017). Are evidence standards different for genomic- vs. clinical-based precision medicine? A quantitative analysis of individualized warfarin therapy. *Clinical Pharmacology and Therapeutics*, *102*(5), 805–814. https://doi.org/10.1002/cpt.663.

Drew, L. (2016). Pharmacogenetics: The right drug for you. *Nature*, *537*(7619), S60–S62. https://doi.org/10.1038/537S60a.

Drozda, K., Müller, D. J., & Bishop, J. R. (2014). Pharmacogenomic testing for neuropsychiatric drugs: Current status of drug labeling, guidelines for using genetic information, and test options. *Pharmacotherapy*, *34*(2), 166–184. https://doi.org/10.1002/phar.1398.

Dubovsky, S. L. (2016). The limitations of genetic testing in psychiatry. *Psychotherapy and Psychosomatics*, *85*(3), 129–135. https://doi.org/10.1159/000443512.

Ferrell, P. B., Jr., & McLeod, H. L. (2008). Carbamazepine, HLA-B*1502 and risk of Stevens-Johnson syndrome and toxic epidermal necrolysis: US FDA recommendations. *Pharmacogenomics*, *9*(10), 1543–1546. https://doi.org/10.2217/14622416.9.10.1543.

Frieden, T. R. (2017). Evidence for health decision making—Beyond randomized, controlled trials. *The New England Journal of Medicine*, *377*(5), 465–475. https://doi.org/10.1056/NEJMra1614394.

Fusar-Poli, P., Rutigliano, G., Stahl, D., Davies, C., Bonoldi, I., Reilly, T., & McGuire, P. (2017). Development and validation of a clinically based risk calculator for the transdiagnostic prediction of psychosis. *JAMA Psychiatry*, *74*(5), 493–500. https://doi.org/10.1001/jamapsychiatry.2017.0284.

García-González, J., Tansey, K. E., Hauser, J., Henigsberg, N., Maier, W., Mors, O., … Major Depressive Disorder Working Group of the Psychiatric Genomic Consortium. (2017). Pharmacogenetics of antidepressant response: A polygenic approach. *Progress in Neuro-Psychopharmacology & Biological Psychiatry*, *75*, 128–134. https://doi.org/10.1016/j.pnpbp.2017.01.011.

Gillis, N. K., & Innocenti, F. (2014). Evidence required to demonstrate clinical utility of pharmacogenetic testing: The debate continues. *Clinical Pharmacology and Therapeutics*, *96*(6), 655–657. https://doi.org/10.1038/clpt.2014.185.

Green, E. D., & Guyer, M. S. (2011). Charting a course for genomic medicine from base pairs to bedside. *Nature*, *470*(7333), 204–213. https://doi.org/10.1038/nature09764.

Gunes, A., & Dahl, M. L. (2008). Variation in CYP1A2 activity and its clinical implications: Influence of environmental factors and genetic polymorphisms. *Pharmacogenomics*, *9*(5), 625–637. https://doi.org/10.2217/14622416.9.5.625.

Hafeman, D. M., Merranko, J., Goldstein, T. R., Axelson, D., Goldstein, B. I., Monk, K., … Birmaher, B. (2017). Assessment of a person-level risk calculator to predict new-onset bipolar spectrum disorder in youth at familial risk. *JAMA Psychiatry*, *74*(8), 841–847. https://doi.org/10.1001/jamapsychiatry.2017.1763.

Hall, J., Whalley, H. C., Job, D. E., Baig, B. J., McIntosh, A. M., Evans, K. L., … Lawrie, S. M. (2006). A neuregulin 1 variant associated with abnormal cortical function and psychotic symptoms. *Nature Neuroscience*, *9*(12), 1477–1478. https://doi.org/10.1038/nn1795.

Herbert, D., Neves-Pereira, M., Baidya, R., Cheema, S., Groleau, S., Shahmirian, A., … Kennedy, J. (2018). Genetic testing as a supporting tool in prescribing psychiatric medication—Design and protocol of the IMPACT study. *Journal of Psychiatric Research*, *96*, 265–272 [in press].

Hicks, J. K., Bishop, J. R., Sangkuhl, K., Müller, D. J., Ji, Y., Leckband, S. G., … Clinical Pharmacogenetics Implementation Consortium (2015). Clinical pharmacogenetics implementation consortium (CPIC) guideline for CYP2D6 and CYP2C19 genotypes and dosing of selective serotonin reuptake inhibitors. *Clinical Pharmacology and Therapeutics*, *98*(2), 127–134. https://doi.org/10.1002/cpt.147.

Hicks, J. K., Sangkuhl, K., Swen, J. J., Ellingrod, V. L., Müller, D. J., Shimoda, K., … Stingl, J. C. (2016). Clinical pharmacogenetics implementation consortium guideline (CPIC) for CYP2D6 and CYP2C19 genotypes and dosing of tricyclic antidepressants: 2016 update. *Clinical Pharmacology and Therapeutics*. https://doi.org/10.1002/cpt.597.

Hicks, J. K., Swen, J. J., & Gaedigk, A. (2014). Challenges in CYP2D6 phenotype assignment from genotype data: A critical assessment and call for standardization. *Current Drug Metabolism*, *15*(2), 218–232.

Howes, O. D., McCutcheon, R., Agid, O., de Bartolomeis, A., van Beveren, N. J., Birnbaum, M. L., … Correll, C. U. (2017). Treatment-resistant schizophrenia: Treatment response and resistance in psychosis (TRRIP) working group consensus guidelines on diagnosis and terminology. *The American Journal of Psychiatry*, *174*(3), 216–229. https://doi.org/10.1176/appi.ajp.2016.16050503.

ISPG, G. T. W. G. (2017). *Genetic testing statement. Retrieved from. https://ispg.net/genetic-testing-statement/*.

Janssens, A. C., & Deverka, P. A. (2014). Useless until proven effective: The clinical utility of preemptive pharmacogenetic testing. *Clinical Pharmacology and Therapeutics*, *96*(6), 652–654. https://doi.org/10.1038/clpt.2014.186.

Kalman, L. V., Agúndez, J., Appell, M. L., Black, J. L., Bell, G. C., Boukouvala, S., … Zanger, U. M. (2016). Pharmacogenetic allele nomenclature: International workgroup recommendations for test result reporting. *Clinical Pharmacology and Therapeutics*, *99*(2), 172–185. https://doi.org/10.1002/cpt.280.

Kéri, S., Kiss, I., & Kelemen, O. (2009). Effects of a neuregulin 1 variant on conversion to schizophrenia and schizophreniform disorder in people at high risk for psychosis. *Molecular Psychiatry*, *14*(2), 118–119. https://doi.org/10.1038/mp.2008.1.

Leckband, S. G., Kelsoe, J. R., Dunnenberger, H. M., George, A. L., Tran, E., Berger, R., ... Clinical Pharmacogenetics Implementation Consortium (2013). Clinical Pharmacogenetics Implementation Consortium guidelines for HLA-B genotype and carbamazepine dosing. *Clinical Pharmacology and Therapeutics*, 94(3), 324–328. https://doi.org/10.1038/clpt.2013.103.

Lee, S. H., Ripke, S., Neale, B. M., Faraone, S. V., Purcell, S. M., Perlis, R. H., ... IIBDGC (2013). Genetic relationship between five psychiatric disorders estimated from genome-wide SNPs. *Nature Genetics*, 45(9), 984–994. https://doi.org/10.1038/ng.2711.

Liu, C., Bousman, C. A., Pantelis, C., Skafidas, E., Zhang, D., Yue, W., & Everall, I. P. (2017). Pathway-wide association study identifies five shared pathways associated with schizophrenia in three ancestral distinct populations. *Translational Psychiatry*, 7(2), e1037. https://doi.org/10.1038/tp.2017.8.

Malhotra, D., & Sebat, J. (2012). CNVs: Harbingers of a rare variant revolution in psychiatric genetics. *Cell*, 148(6), 1223–1241. https://doi.org/10.1016/j.cell.2012.02.039.

McIntosh, A. M., Baig, B. J., Hall, J., Job, D., Whalley, H. C., Lymer, G. K., ... Johnstone, E. C. (2007). Relationship of catechol-O-methyltransferase variants to brain structure and function in a population at high risk of psychosis. *Biological Psychiatry*, 61(10), 1127–1134. https://doi.org/10.1016/j.biopsych.2006.05.020.

Miller, M. (2006). The seductiveness of evidence. *Journal of Substance Abuse Treatment*, 30(2), 91–92. https://doi.org/10.1016/j.jsat.2005.11.001.

Millon, T. (1991). Classification in psychopathology: Rationale, alternatives, and standards. *Journal of Abnormal Psychology*, 100(3), 245–261.

Morrison, A., & Boudreau, R. (2012). *Evaluation frameworks for genetic tests [Environmental Scan issue 37]*. Ottawa: Canadian Agency for Drugs and Technologies in Health.

Mössner, R., Schuhmacher, A., Wagner, M., Quednow, B. B., Frommann, I., Kühn, K. U., ... Maier, W. (2010). DAOA/G72 predicts the progression of prodromal syndromes to first episode psychosis. *European Archives of Psychiatry and Clinical Neuroscience*, 260(3), 209–215. https://doi.org/10.1007/s00406-009-0044-y.

Müller, D. J., Kekin, I., Kao, A. C., & Brandl, E. J. (2013). Towards the implementation of CYP2D6 and CYP2C19 genotypes in clinical practice: Update and report from a pharmacogenetic service clinic. *International Review of Psychiatry*, 25(5), 554–571. https://doi.org/10.3109/09540261.2013.838944.

National Research Council (U.S.). Committee on A Framework for Developing a New Taxonomy of Disease (2011). *Toward precision medicine: Building a knowledge network for biomedical research and a new taxonomy of disease*. Washington, DC: National Academies Press.

NIH, G. H. R. (2017). *Help me understand genetics: Genetic testing, October 24, 2017*. Retrieved from. ghr.nlm.nih.gov.

Ozawa, S., Soyama, A., Saeki, M., Fukushima-Uesaka, H., Itoda, M., Koyano, S., ... Sawada, J. (2004). Ethnic differences in genetic polymorphisms of CYP2D6, CYP2C19, CYP3As and MDR1/ABCB1. *Drug Metabolism and Pharmacokinetics*, 19(2), 83–95.

Padmanabhan, S. (2014). *Handbook of pharmacogenomics and stratified medicines*. .

Pérez, V., Salavert, A., Espadaler, J., Tuson, M., Saiz-Ruiz, J.,Sáez-Navarro, C. ... AB-GEN Collaborative Group. (2017). Efficacy of prospective pharmacogenetic testing in the treatment of major depressive disorder: Results of a randomized, double-blind clinical trial. *BMC Psychiatry*, 17(1), 250. https://doi.org/10.1186/s12888-017-1412-1.

Peterson, K., Dieperink, E., Anderson, J., Boundy, E., Ferguson, L., & Helfand, M. (2017). Rapid evidence review of the comparative effectiveness, harms, and cost-effectiveness of pharmacogenomics-guided antidepressant treatment versus usual care for major depressive disorder. *Psychopharmacology*, 234(11), 1649–1661. https://doi.org/10.1007/s00213-017-4622-9.

Pramparo, T., Pierce, K., Lombardo, M. V., Carter Barnes, C., Marinero, S., Ahrens-Barbeau, C., ... Courchesne, E. (2015). Prediction of autism by translation and immune/inflammation coexpressed genes in toddlers from pediatric community practices. *JAMA Psychiatry*, 72(4), 386–394. https://doi.org/10.1001/jamapsychiatry.2014.3008.

Ratain, M. J., & Johnson, J. A. (2014). Meaningful use of pharmacogenetics. *Clinical Pharmacology and Therapeutics*, 96(6), 650–652. https://doi.org/10.1038/clpt.2014.188.

Riedel, M., Möller, H. J., Obermeier, M., Schennach-Wolff, R., Bauer, M., Adli, M., ... Seemüller, F. (2010). Response and remission criteria in major depression—A validation of current practice. *Journal of Psychiatric Research*, 44(15), 1063–1068. https://doi.org/10.1016/j.jpsychires.2010.03.006.

Rosenblat, J. D., Lee, Y., & McIntyre, R. S. (2017). Does pharmacogenomic testing improve clinical outcomes for major depressive disorder? A systematic review of clinical trials and cost-effectiveness studies. *Journal of Clinical Psychiatry*, 78(6), 720–729. https://doi.org/10.4088/JCP.15r10583.

Schizophrenia Working Group of the Psychiatric Genomics Consortium (2014). Biological insights from 108 schizophrenia-associated genetic loci. *Nature*, 511(7510), 421–427. https://doi.org/10.1038/nature13595.

Shah, R. R., & Smith, R. L. (2015). Addressing phenoconversion: The Achilles' heel of personalized medicine. *British Journal of Clinical Pharmacology*, 79(2), 222–240. https://doi.org/10.1111/bcp.12441.

Singh, A. B. (2015). Improved antidepressant remission in major depression via a pharmacokinetic pathway polygene pharmacogenetic report. *Clinical Psychopharmacology and Neuroscience*, 13(2), 150–156. https://doi.org/10.9758/cpn.2015.13.2.150.

Skafidas, E., Testa, R., Zantomio, D., Chana, G., Everall, I. P., & Pantelis, C. (2014). Predicting the diagnosis of autism spectrum disorder using gene pathway analysis. *Molecular Psychiatry*, 19(4), 504–510. https://doi.org/10.1038/mp.2012.126.

Swen, J. J., Nijenhuis, M., de Boer, A., Grandia, L., Maitland-van der Zee, A. H., Mulder, H., ... Guchelaar, H. J. (2011). Pharmacogenetics: From bench to byte—An update of guidelines. *Clinical Pharmacology and Therapeutics*, 89(5), 662–673. https://doi.org/10.1038/clpt.2011.34.

Van Booven, D., Marsh, S., McLeod, H., Carrillo, M. W., Sangkuhl, K., Klein, T. E., & Altman, R. B. (2010). Cytochrome P450 2C9-CYP2C9. *Pharmacogenetics and Genomics*, 20(4), 277–281. https://doi.org/10.1097/FPC.0b013e3283349e84.

Winner, J. G., Carhart, J. M., Altar, C. A., Allen, J. D., & Dechairo, B. M. (2013). A prospective, randomized, double-blind study assessing the clinical impact of integrated pharmacogenomic testing for major depressive disorder. *Discovery Medicine*, 16(89), 219–227.

Wong, M. L., Dong, C., Andreev, V., Arcos-Burgos, M., & Licinio, J. (2012). Prediction of susceptibility to major depression by a model of interactions of multiple functional genetic variants and environmental factors. *Molecular Psychiatry*, *17*(6), 624–633. https://doi.org/10.1038/mp.2012.13.

Yip, V. L., Marson, A. G., Jorgensen, A. L., Pirmohamed, M., & Alfirevic, A. (2012). HLA genotype and carbamazepine-induced cutaneous adverse drug reactions: A systematic review. *Clinical Pharmacology and Therapeutics*, *92*(6), 757–765. https://doi.org/10.1038/clpt.2012.189.

Yip, V. L., & Pirmohamed, M. (2017). The HLA-A*31:01 allele: Influence on carbamazepine treatment. *Pharmacogenomics and Personalized Medicine*, *10*, 29–38. https://doi.org/10.2147/PGPM.S108598.

Chapter 37

Opportunities and challenges of implementation models of pharmacogenomics in clinical practice

Jonathan C.W. Liu[a,b,a], Ilona Gorbovskaya[a,a], Chad Bousman[c], Lisa C. Brown[d] and Daniel J. Müller[a,b,e]

[a]Pharmacogenetics Research Clinic, Centre for Addiction and Mental Health, Toronto, Canada, [b]Department of Pharmacology & Toxicology, University of Toronto, Toronto, ON, Canada, [c]Departments of Medical Genetics, Psychiatry, and Physiology & Pharmacology, University of Calgary, Calgary, AB, Canada, [d]Myriad Neuroscience, Mason, OH, United States, [e]Department of Psychiatry, University of Toronto, Toronto, ON, Canada

1 Introduction

Pharmacogenetics is the study of pharmacotherapy and the genetic variants that influence a patient's response and adverse reactions to medications. The conceptual framework of pharmacogenetics was developed in the 1950's, with one example being the observed variable response of debrisoquine oxidation, which was attributed to genetic and inheritable patterns. It was later identified that variants in the gene coding for the cytochrome P450 (CYP) 2D6 enzyme were associated with the variable response of debrisoquine oxidation (Gonzalez et al., 1988). Pharmacogenetics has since expanded throughout many clinical disciplines, including psychiatry, where pharmacotherapy has remained particularly challenging (Rush et al., 2006). With the completion of the Human Genome Project in 2003, it was anticipated that this would accelerate precision medicine and provide a new avenue for improving psychiatric diagnosis and the trial and error pharmacological treatment approach. Psychiatric disorders are genetically complex, involving rare and common polymorphisms of various genes, and follow polygenic inheritance. Consequently, the true underlying genetic factors for many psychiatric disorders are yet to be elucidated. As described in *the previous chapter*, the current evidence base and clinical validity for genetic tests to diagnose and identify individuals at risk for psychiatric conditions is promising, but has remained limited. In contrast, the biological mechanisms and genes involved with psychopharmacotherapy outcomes are better understood, particularly pharmacokinetic mechanisms (e.g., metabolism). Gene variants in 'core' phase I drug metabolizing enzymes (e.g., CYP2D6, CYP2C19) are the most well characterized among the numerous potential genes and variants thought to be involved in the pharmacodynamics and pharmacokinetics of psychiatric medications. There are numerous studies linking pharmacokinetic and pharmacodynamic genes to therapy outcomes, highlighted in *previous chapters*, some of which have shown consistently high levels of evidence. However, for various reasons, pharmacogenetic testing is still not a standard practice in psychiatry. In this chapter, we present the challenges of implementing pharmacogenetic testing in a psychiatric clinical setting. In addition, we introduce a successful pharmacogenetic implementation model developed at the Centre for Addiction and Mental Health (CAMH), which has already enrolled and genotyped more than 10,000 individuals as of winter 2017/18.

2 Barriers to implementing pharmacogenetic tests

2.1 Access to laboratory and analytical validity

In order to implement pharmacogenetic testing, an institution would need to ensure that its lab has the accreditation to provide clinical information (e.g., Clinical Laboratory Improvement Amendments (CLIA) and College of American Pathologists (CAP) in the U.S.). However, if the institution does not have access to an accredited lab internally, using an offsite laboratory can potentially discourage the use of pharmacogenetics, as outside laboratories might be hard to find,

---

a. These authors contributed equally to this work.

Personalized Psychiatry. https://doi.org/10.1016/B978-0-12-813176-3.00037-7

449

and might prolong delivery time. Smaller clinics or office practices without the strong infrastructure of institutions, especially in rural areas, might be particularly disadvantaged.

With the rapid advances in this field, commercial companies have developed pharmacogenetic testing services for clinics and their patients. However, these commercial tests vary substantially in terms of their gene and variant content, algorithms used to translate test results into clinical recommendations, and cost (Bousman & Hopwood, 2016). This variation in testing can, in part, be attributed to the absence of testing and interpretation standards, as well as minimal independent regulatory oversight. Only one commercial pharmacogenomic test has published on its analytical validity, or accuracy of genotyping the genes and alleles included in the test panel (Jablonski et al., 2018). Most medications are metabolized by more than one pharmacokinetic enzyme, and have promiscuous targets of action, therefore some tests utilize combinatorial algorithms. These algorithms combine the effects of multiple pharmacokinetic and/or pharmacodynamic genetic variants to provide recommendations for a single medication. This is in contrast to a panel of many genes, where medication recommendations are made based on the effect of one gene at a time (Altar et al., 2015; Smith, 2017).

Pharmacogenetic tests are considered special medical devices, and therefore fall under the U.S. Food and Drug Administration's (FDA) authority. The extent to which a test is regulated is determined by how it comes to market. Manufacturers must receive FDA approval if the tests are marketed as a commercial kit and sold to multiple labs (e.g., AmpliChip) (NHGRI, 2018). However, genetic tests developed by and used exclusively within clinical laboratories, known as laboratory-developed tests (LDT), are not regulated by the FDA, and are only accredited by local/national bodies (e.g., CLIA, CAP). The FDA chooses not to regulate such tests, and therefore these LDTs are being used in the clinic without having their analytical and clinical validity assessed. The importance of regulation ensures that tests have analytical validity along with clinical utility. Regulation creates a standard that, as of now, does not exist. However, the FDA has recently been made aware of the growing concerns of unregulated tests on public health, and is modifying its approach when it comes to genetic testing (NHGRI, 2018). The interesting conversation is "how" the FDA would regulate the tests, as not all psychiatric pharmacogenetic tests are alike; therefore it is important to evaluate them on their individual merits. However, there are concerns over the use of these tests, because it is unclear if the FDA will continue to allow the use of laboratory-developed tests (Arwood, Chumnumwat, Cavallari, Nutescu, & Duarte, 2016).

2.2 Establishing genotyping

In addition to laboratory access, the genes and gene variants used to test patient samples need to be considered. Establishing which genes and variants should be included on a test panel is not always straightforward, and can be challenging. For example, CYP2D6 is involved in the metabolism of most antidepressants and some antipsychotic medications (Pouget, Shams, Tiwari, & Mueller, 2014), but there are more than one hundred known variants in CYP2D6 that affect metabolism of medication, and each lab or company analyzes different genetic variants (Bousman, Jaksa, & Pantelis, 2017) and interpret the corresponding phenotypes differently (Bousman et al., 2017; Eum, Lee, & Bishop, 2016; Gaedigk et al., 2018). Metabolic phenotypes are typically characterized as normal/extensive, intermediate, poor, or (ultra)rapid metabolizers, based on the combined allele haplotype. However, the various gene pairs used to predict these metabolic phenotypes (i.e., 2D6 *1/*4 is intermediate metabolizer) have varying degrees of evidence, thus assigning a metabolizer status is still based on probabilities, and remains a subject of intense debate (Eum et al., 2016). Therefore, while CYP2D6 may be chosen for its known relevance, the variants chosen to be tested within genes, and the interpretation of those variants, still differ from one laboratory's test to another (Bousman et al., 2017). Newly discovered allelic (or haplotypic) variations are still being reported, and need to be validated for their clinical relevance and usefulness in clinical practice. It is important to note that very few pharmacodynamic genes have shown clinical validity and utility, so it is important for clinicians to be educated on the evidence supporting the utilization of specific genes to guide treatment decisions.

Standardization of the alleles being tested can allow interchangeability of data, and help promote psychiatric pharmacogenetic testing. While standardization may be beneficial for offsite communication, geographical population should also be considered when deciding on the alleles and variants to be tested (Gaedigk, Sangkuhl, Whirl-Carrillo, Klein, & Leeder, 2017). It has been shown that the frequencies of certain genetic polymorphisms involved with pharmacokinetics and pharmacodynamics can vary by ethnicity. For example, the reduced function allele, 2D6*17, is found at a higher frequency in Africans, but is almost completely absent in Asian populations. However, the current literature and evidence reviews are primarily based on the "Caucasian" group, and less on African-Americans or Asians (Gaedigk et al., 2017). This begs the questions of "how" these variants are interpreted and applied, and "who" makes those recommendations.

Furthermore, the information and/or interpretation of test results provided by the commercial tests vary (Bousman et al., 2017). Guidelines for test results reporting have been published (Caudle et al., 2017; Kalman et al., 2016) along with international dosing guidelines for some gene-drug pairs (e.g., CYP2D6-antidepressants) (CPIC Guidelines, n.d.). However, some guidelines may not correspond with the alleles commercially tested (Lu, Lewis, & Traylor, 2017), and many gene-drug pairs do not yet have published dosing guidelines. A consensus has yet to be reached on which genetic variants should be tested in commercial platforms (Bousman et al., 2017), and efforts by various entities is warranted. Recently, the Association for Molecular Pathology published recommendations for clinical CYP2C19 genotyping allele selection (Pratt et al., 2018). The variants tested are likely to have varying degrees of sensitivity due to allele frequency differences across populations (Gaedigk et al., 2017; Lu et al., 2017). Nevertheless, commercial pharmacogenetic testing is evolving and has, in part, facilitated the integration of pharmacogenetic testing into the clinical setting.

2.3 Cost effectiveness

When adopting pharmacogenetic testing into the clinic, cost effectiveness of the tests needs to be considered. If the tests are clinically valuable, but the cost of the test is greater than the benefit, institutions would be hesitant to adopt any innovative product. Fortunately, rapid advances in genetics allow lower costs of genotyping, paving the way for pharmacogenetic tests to be a part of standard healthcare. Many economic evaluation studies have shown that pharmacogenetic testing can have a positive impact on both healthcare quality and cost (Verbelen, Weale, & Lewis, 2017). Due to increased accuracy of treatment, patients would experience fewer adverse effects, and healthcare utilization such as hospital visits, decreased medication costs, and improvement in work productivity would yield increased savings for patients, employers, and the healthcare system (Winner et al., 2015). This is especially relevant in psychiatry, where the time required to identify the ideal medication at the ideal dose can take several weeks to months. The potential high direct and indirect costs of prescribing multiple medications that might only have low therapeutic benefits can, to some extent, be avoided through the guidance of pharmacogenetics (Trautmann, Rehm, & Wittchen, 2016). For example, it has been modeled that using pharmacogenetic-guided medication management for depression could potentially save USD$3962 annually per patient (includes costs of medical services and prescriptions, and missed productivity costs) (Maciel, Cullors, Lukowiak, & Garces, 2018).

The attitudes and interest in pharmacogenetic testing will likely increase with lower costs attached to testing. However, while the actual cost of providing genotypic information has been steadily decreasing due to technological advances, and reports suggesting that pharmacogenetics can be a cost-effective and cost-saving strategy (Verbelen et al., 2017), most insurance companies are not covering these costs. Insurance coverage of pharmacogenetic tests remains an important deciding factor in whether testing will become a part of standard clinical practice. A systematic review showed that the policies of the five largest U.S. private payers lack coverage of multigene tests (Phillips et al., 2017). Three-hundred and thirteen genetic tests were included in the 55 policies of the five payers. Among these policies, 22% covered all genetic tests mentioned in the policy, 51% covered none, and 27% covered some, but not all (Phillips et al., 2017). Two of the 55 policies mention pharmacogenetic testing, specifically, genotyping for CYP450 polymorphisms. According to both policies, identifying genetic variation in CYPs that may be linked to the response to antipsychotics and selective serotonin reuptake inhibitors are deemed experimental and investigational, and are therefore not covered by the policies (Phillips et al., 2017). In the U.S., only one psychiatric pharmacogenetic test has a local coverage decision from an important payer (Centers for Medicare and Medicaid Services, 2018). With additional research showing the evidence for and healthcare savings of pharmacogenetics-guided psychiatric medication management, it is anticipated that insurance coverage of these tests will increase (Verbelen et al., 2017). Even with coverage for some genes, the benefit of commercial tests is the presentation and interpretation of this complex information and how it relates to medication and treatment decisions.

2.4 Ethical considerations

Implementing widespread psychiatric testing can present with certain concerns from an ethical perspective. It is widely believed that these issues are of less concern in pharmacogenetics compared with disease risk genetics because actionable pharmacogenetic variants are not typically associated with disease risk. Nevertheless, ethical issues still exist, as most genes have pleiotropic effects. Given that most insurers do not cover testing, there will be disparities in who will be able to access and afford testing. While some companies may have plans to help cover for the test, the majority do not. Another concern regarding implementation of pharmacogenetic testing is the legal implication of delivering or not delivering pharmacogenetic-guided prescriptions. Additionally, if the clinician prescribes a drug that is not recommended, but an adverse drug reaction occurs, would the clinician be liable? This may draw providers away from implementing pharmacogenetic testing in their clinics (Marchant & Lindor, 2013). Yet another concern might be related to over-interpretation of genetic tests: despite the fact that measuring the metabolic profile can guide medication selection, treatment success is not guaranteed. In the case where a patient may continue to experience pharmacotherapy failure despite having completed genetic testing, the patient may begin to experience a nocebo effect - undesirable symptoms induced by the patient's expectation of detriment or harm (Eknoyan, Hurley, & Taber, 2013).

The use of pharmacogenetic testing has also raised questions regarding incidental findings. Genotyping a patient's DNA sample will produce results outside of the intended use, in this case, the metabolic profile. The decision to convey such results has become an important ethical issue in genetics research. Such findings can potentially identify risks of varying degrees for a number of medical conditions. There is no clear consensus among patients, providers, and researchers under which circumstances should incidental findings be discussed (Appelbaum et al., 2016; Appelbaum & Benston, 2017; Henrikson, Burke, & Veenstra, 2008). Some groups argue that information should only be disclosed if the results could significantly affect the individual (Henrikson et al., 2008). Problems may occur due to the differences in what is perceived as significant by the physician, the patient, and their relatives and partners. Because the preference to return secondary findings varies, most commentators suggest that the patients and research participants should mostly decide themselves (Appelbaum et al., 2016). Such a decision would require some sort of consent form to be used, but the content of the form has raised many concerns (Appelbaum & Benston, 2017).

2.5 Providers, payers, and industry perspectives

A few studies have evaluated the challenges for widespread implementation of pharmacogenomic testing (Ahmed, Weinshilboum, & Frye, 2018; Ciarleglio & Ma, 2017; Keeling et al., 2017). A study in Canada found that there were four major barriers to testing, including the availability of published clinical evidence supporting the use of testing, regulatory approval, guidelines, and consensus from experts (Amara, Blouin-Bougie, Bouthillier, & Simard, 2017). Cost is also an important consideration for both patients and providers when considering a pharmacogenomic test. A patient must weigh the cost of the test and the potential benefit from guided treatment. While some economic cost savings data exists, testing is not widely covered by payers.

Payers share similar concerns as providers, including the evidence from RCTs and clinical utility studies' inclusion in guidelines, and downstream economic cost versus savings (Keeling et al., 2017). Regulatory approval and guidelines are two issues that are especially difficult to address for pharmacogenomic testing in psychiatry. In the U.S., the FDA does not regulate or approve pharmacogenomic testing in psychiatry, making it difficult to gain endorsement from a regulatory agency. When it comes to guidelines, The American Psychiatric Association guidelines are considered the gold standard for treatment of psychiatric illness, but have not been updated for major depressive disorder (the diagnosis in which testing has clinical evidence) since 2010 (APA 3rd Edition, Work Group on Major Depressive Disorder, 2010). Fortunately, the Canadian Network for Mood and Anxiety Treatments (CANMAT) updates major depression treatment guidelines more frequently, offering the potential to evaluate and include pharmacogenomic testing for specific patients in the future (Lam et al., 2016).

While pharmacogenomics is represented in some guidelines, there is still no recommendation regarding which patients to test and what genes/genetic variants to test for. The Dutch Pharmacogenetics Working Group (DPWG) has announced that the group will produce robust guidelines specific to pharmacogenomic testing over the next 5 years (Swen et al., 2018). The barriers for providers and payers are also shared by the commercial testing companies, and reimbursement and inclusion in guidelines is crucial for adoption of testing. Education surrounding pharmacogenomic testing is also a challenge for commercial companies. The study by Amara et al. (2017) found that a majority of the surveyed providers did not receive pharmacogenetic training in medical school or during postgraduate study (77.5% and 86.3%, respectively). Often, most available education on pharmacogenomic testing in psychiatry is provided by the commercial entities themselves, which is subject to marketing bias. In order for pharmacogenomic testing to become the standard of care, providers, payers, researchers, and commercial entities need to address and overcome these barriers.

3 Education and opportunities to implement pharmacogenetic tests

Healthcare professionals interact with patients daily, therefore they are in the advantageous position of being able to help educate patients about pharmacogenetic testing. Before educating patients about pharmacogenetic testing, healthcare members themselves must be knowledgeable on the subject. Given that clinical implementation of pharmacogenetic testing is still in its early stages, many healthcare professionals may not be fully aware of what pharmacogenetics services are offered in their healthcare setting, or how to use the results if they are available. Education plays a key role in the testing process to ensure proper implementation of the information. Both healthcare professionals and patients should be well informed about the pharmacogenetic process and its limitations, especially with respect to evaluating the evidence supporting the genes, specific tests, how to interpret the test, and how to integrate into practice in conjunction with clinical expertise.

3.1 Education—Patients

Because pharmacogenetic testing is often seen as a preventative measure with limited risks, the amount of information provided during the consent process might be very limited (Mills, Voora, Peyser, & Haga, 2013). A standardized consent form can help ensure that essential pieces of information are included to properly educate the patient about the test, allowing them to make an informed decision. It was suggested that at least five main topics should be discussed: (1) the purpose of testing and the role of genes in drug response; (2) test risks, benefits, limitations, and alternatives; (3) an emphasis on the genetic basis of the test; (4) future benefits of testing; and (5) opting out/consequences (Mills et al., 2013; Roche & Berg, 2015). Standardization of terms, with the objective to characterize allele function and inferred phenotypes, should help with education, reporting, and sharing results throughout clinics (Caudle et al., 2017). Development of education and counseling measures for patients have received positive responses (Mills, Ensinger, Callanan, & Haga, 2017). Through education, patients can learn what genetic tests are, and understand what information they yield.

3.2 Education—Physicians

The use of pharmacogenetic tests relies heavily on the attitudes of physicians who are the intersection among patients, pharmacists, and geneticists. The potential of being able to tailor medication to a patient's genetic profile provides many benefits; however, the acceptance of pharmacogenetic data varies among physicians. Some agree with the notion that it can be beneficial for specific patients, but many physicians still express lower levels of confidence and knowledge of the process (Walden et al., 2015). At the University of Illinois at Chicago (UIC), an established pharmacogenetic program, the genotype-guided dose recommendations were not always followed by physicians for two reasons: (1) they did not see a recommendation in the electronic health record, or (2) they disagreed with the recommendation (Nutescu et al., 2013). Thus, implementation of pharmacogenetic testing might remain challenging if physicians have limited knowledge, and if their attitudes would be affected negatively in the process. The findings of Cogent Research (2011) poll of 1000 U.S. adults revealed that 90% of participants lacked confidence in their physician's ability to understand and use the genomic information. Additionally, another study found that providers' recommendations after pharmacogenetic testing consisted of not using the test results at all (60%), and saying it should be kept for the future (40%) (Vassy et al., 2018). Ongoing education of clinicians should be provided and advertised to help improve responses toward pharmacogenetic implementation in the clinic, and clinicians have a duty to take advantage of these educational opportunities. Additionally, the knowledge gained by the physicians can help reassure patients by addressing their concerns regarding pharmacogenetic testing.

3.3 Education—Nurses

Nurses are in the front lines of interacting with patients. By providing care to patients and treating adverse effects, nurses should also be knowledgeable about the purpose, possibilities, and limitations of pharmacogenetics (Burke, Love, Jones, & Fife, 2016). In situations where the testing is performed in a hospital, the nurse may be responsible for discussing the results of the tests with their patients. Proper training and education can allow nurses to effectively and efficiently fulfill this role, and provide high-quality patient care (Haga & Mills, 2015).

3.4 Education—Pharmacists

With the intention of implementing pharmacogenetic testing as a common practice, pharmacists are placed in a position where they would be well poised to provide pharmacogenetic-related information and counseling regarding a patient's medication. Implementation of pharmacogenetic practices was found to be as low as 4% among pharmacists in North America, the UK, and Asia (Yau, Binti, Aziz, & Haque, 2015). However, the majority of pharmacists have a high interest in pharmacogenetic testing, and are more willing to discuss it with patients when they are provided with training and education on the topic (Yau et al., 2015). Some pharmacogenetic test administrators are already offering testing via pharmacists through pharmacy systems.

4 Pharmacogenetic testing's role in collaborative care and shared decision making

The traditional laboratory-developed tests (LDT) provide information about a patient's genotype and predicted metabolic phenotype, but do not provide any clinical interpretation. It is unrealistic to expect healthcare members to remember what specific genotypes mean (e.g., haplotype combinations of $CYP2D6$). Therefore, commercial pharmacogenetic-based decision support tools have been designed to offer user-friendly genetic test reports that suggest drug selection and dosage, and possibly flag potential drug-drug or drug-gene-drug interactions (Bousman & Hopwood, 2016). Institutions that have already established pharmacogenetic practices have highlighted the benefits of clinical decision support tools (Arandjelovic et al., 2017; Johnson, 2013). However, the evidence base for these tools is still limited, and requires further independent validation. In other words: the analytical validity, clinical validity, and clinical usefulness of these tools are not yet fully elucidated (Bousman & Hopwood, 2016). Despite this, pharmacogenetic decision support tools are becoming widely adopted in psychiatry in the absence of any other predictive tools to optimize medication treatment.

5 The "IMPACT" study at the Centre for Addiction and Mental Health in Toronto

The Individualized Medicine: Pharmacogenetics Assessment and Clinical Treatment (IMPACT) study has been implementing pharmacogenetic testing within a psychiatric clinic since 2011 (Herbert et al., 2018), following positive results of a pilot study (Müller, Kekin, Kao, & Brandl, 2013). The Ontario-wide research study was developed at The Centre for Addiction and Mental Health (CAMH). IMPACT introduces pharmacogenetic biological rationale and supportive clinical testing into Canadian healthcare settings by incorporating known pharmacogenetic data with psychiatric care. To be compatible with the dynamics of both psychiatric clinics in hospitals and primary healthcare systems, the study recruited both inpatients and outpatients with a broad spectrum of mental health issues from a variety of clinical settings, mainly in Ontario. IMPACT uses a self-designed Clinical Information Management System (CIMS) as a database that can also serve to schedule appointments, send reports to providers, and record follow-ups.

The initial phase of IMPACT consisted of a cross-sectional assessment and a prospective eight-week study. A panel of cytochrome P450 genes was used to assess participants' suitability for a particular medication. For full details on the specific gene variants tested, please refer to Herbert et al. (2018). Pharmacotherapies were classified into three categories: (green) use as directed, (yellow) use with caution, and (red) use with caution and frequent monitoring. At the baseline visit, DNA genotyping samples were obtained via buccal swabs, the Beck Depression Inventory (BDI), and a revised version of the Udvalg for Kliniske Undersøgelser (UKU) side effects scale were administered to patients. Various other tests were also administered, depending on the patient's mental health status. IMPACT included secondary outcome measures, such as healthcare resource utilization, productivity losses, healthcare costs, and cost-effectiveness of pharmacogenetic testing. The first and second follow-ups occur at 4 and 8 weeks, in which the participants complete similar tests to those conducted during the baseline visit. After completion of the study, data was analyzed to compare individuals who were receiving drug treatment in correlation with their genetic data, versus those who were receiving treatment not correlated with their genetic profile.

In the second phase of the IMPACT study, a pharmacogenetic decision support tool was developed that reported test results and provided clinical interpretation. Reports are sent to participating physicians to help guide the physician's prescribing patterns. After receiving the reports, any changes in prescribing patterns are noted, such as change in dose or pharmacotherapy. In addition, symptom changes associated with medication modifications based on genetic data are assessed. Healthcare utilization is assessed by linking pharmacogenetic data to variables indicated by the Institute for Clinical Evaluative Sciences (ICES), which evaluates healthcare delivery and outcomes. Some variables used by ICES include rates of office visits, emergency room visits, and hospitalizations. Participating physicians also complete the Pharmacogenetics in Psychiatry Follow-up Questionnaire (PIP-FQ) to assess their experience with the pharmacogenetic report (Walden et al., 2015; Walden et al., 2018).

The pilot analysis ($n = 177$) showed an increase in overall patient functioning and decrease in the number of reported symptoms and side effects following pharmacogenetic-guided treatment ($P = .04$). Additionally, PIP-FQ surveys indicated 80% of participating physicians were at least satisfied with the genetic information provided, and 76% of physicians had at least a satisfactory understanding of the pharmacogenetic report, while only 6% of participating physicians did not find the report useful for guiding treatment for their patients. When asked about the clinical outcome of their patients, 23% of physicians indicated improvement in their patients, and 41% indicated no change in patients following pharmacogenetics testing. Notably, not a single physician reported deterioration in patient condition following pharmacogenetic testing (Walden et al., 2018). This would not only suggest that testing has been beneficial in a substantial patient population, but that not a single individual deteriorated as a result of recommendations based on genetic testing.

6 Conclusion

Implementation of pharmacogenetics in the clinic is challenging but feasible, as indicated by numerous examples, including the IMPACT study. As discussed in other chapters of this book, clinical relevance is highest for CYP2D6 and CYP2C19 genes with respect to most psychiatric medications. Thus, all pharmacogenetics labs and commercial tests have included these genes in their testing kits, although gene variants that are analyzed, and related interpretations still vary. Other genes and gene variants are being used despite their lower level of evidence, as there are no common standards yet. Efforts should be focused on expanding the current evidence base for selected genes and gene variants to improve the clinical validity and utility of pharmacogenetic tests. Gene variants might also be combined in additive models (or algorithms) to create more elaborate decision support tools. However, such tools would need to be evaluated further for their clinical validity and utility. Costs might differ substantially among companies, and reimbursement by insurers is not standardized. However, as genomics becomes more integrated into the healthcare infrastructure, pharmacogenetic testing will become a core clinical service, and coverage of cost could increase. Although attitudes toward clinical pharmacogenetic tests are generally positive, there is a need to boost educational efforts for all members of the healthcare team, including providers, nurses, pharmacists, and patients. Incorporating homogenous training programs into the healthcare infrastructure will help educate healthcare members on the use, limitations, and interpretation of pharmacogenetic tests. Ethical concerns need to be addressed, although attitudes are mostly favorable for pharmacogenetic testing. Further efforts and additional research is therefore needed in order to achieve clinical implementation of pharmacogenetic testing as a standard of care in the treatment of psychiatric disorders.

References

Ahmed, A. T., Weinshilboum, R., & Frye, M. A. (2018). Benefits of and barriers to pharmacogenomics-guided treatment for major depressive disorder. *Clinical Pharmacology and Therapeutics*. https://doi.org/10.1002/cpt.1009.

Altar, C. A., Carhart, J. M., Allen, J. D., Hall-Flavin, D. K., Dechairo, B. M., & Winner, J. G. (2015). Clinical validity: Combinatorial pharmacogenomics predicts antidepressant responses and healthcare utilizations better than single gene phenotypes. *The Pharmacogenomics Journal*, 15, 443–451. https://doi.org/10.1038/tpj.2014.85.

Amara, N., Blouin-Bougie, J., Bouthillier, D., & Simard, J. (2017). On the readiness of physicians for pharmacogenomics testing: An empirical assessment. *The Pharmacogenomics Journal*. https://doi.org/10.1038/tpj.2017.22.

Appelbaum, P. S., & Benston, S. (2017). Anticipating the ethical challenges of psychiatric genetic testing. *Current Psychiatry Reports*, 19(39). https://doi.org/10.1007/s11920-017-0790.

Appelbaum, P. S., Fyer, A., Klitzman, R. L., Martinez, J., Parens, E., Zhang, Y., & Chung, W. K. (2016). Researchers' views on informed consent for return of secondary results in genomic research. *Genetics in Medicine*, 17(8), 644–650. https://doi.org/10.1038/gim.2014.163.

Arandjelovic, K., Eyre, H. A., Lenze, E., Singh, A. B., Berk, M., & Bousman, C. (2017). The role of depression pharmacogenetic decision support tools in shared decision making. *Journal of Neural Transmission*. https://doi.org/10.1007/s00702-017-1806-8.

Arwood, M., Chumnumwat, S., Cavallari, L., Nutescu, E., & Duarte, J. (2016). Implementing pharmacogenomics at your institution: Establishment and overcoming implementation challenges. *Clinical and Translational Science*, 9(5), 233–245. https://doi.org/10.1111/cts.12404.

Bousman, C. A., & Hopwood, M. (2016). Commercial pharmacogenetic-based decision support tools in psychiatry. *Lancet Psychiatry*, 3(6), 585–590. https://doi.org/10.1016/S2215-0366(16)00017-1.

Bousman, C. A., Jaksa, P., & Pantelis, C. (2017). Systematic evaluation of commercial pharmacogenetic testing in psychiatry: A focus on CYP2D6 and CYP2C19 allele coverage and results reporting. *Pharmacogenetic and Genomics*, 27(11), 387–393. https://doi.org/10.1097/FPC.0000000000000303.

Burke, E., Love, R., Jones, P., & Fife, T. (2016). Pharmacogenetic testing: Application in mental health prescribing. *Journal of the American Psychiatric Nurses Association*, 22(3), 185–191. https://doi.org/10.1177/1078390316641488.

Caudle, K. E., Dunnenberger, H. M., Freimuth, R. R., Peterson, J. F., Burlison, J. D., Whirl-Carrillo, M., … Hoffman, J. M. (2017). Standardizing terms for clinical pharmacogenetic test results: Consensus terms from the Clinical Pharmacogenetics Implementation Consortium (CPIC). *Genetics in Medicine*, 19(2), 215–223. https://doi.org/10.1038/gim.2016.87.

Centers for Medicare and Medicaid Services (2018). Pharmacogenomic testing for warfarin response. Retrieved from: https://www.cms.gov/Medicare/Coverage/Coverage-with-Evidence-Development/Pharmacogenomic-Testing-for-Warfarin-Response.html.

Ciarleglio, A. E., & Ma, C. (2017). The Daniel K. Inouye college of pharmacy scripts: Precision medicine through the use of pharmacogenomics: Current status and barriers to implementation. *Hawai'i Journal of Medicine and Public Health*, 76(9), 265–269.

Clinical Pharmacogenetics Implementation Consortium. *CPIC Guidelines. Available from:* https://cpicpgx.org/guidelines. *Accessed 30 November 2017.*

Cogent Research. (2011). *Americans skeptical of physicians' knowledge of genomics.* .

Eknoyan, D., Hurley, R. A., & Taber, K. H. (2013). The neurobiology of placebo and nocebo: How expectations influence treatment outcomes. *Journal of Neuropsychiatry*, 25(4), 250–254. https://doi.org/10.1176/appi.neuropsych.13090207.

Eum, S., Lee, A. M., & Bishop, J. R. (2016). Pharmacogenetic tests for antipsychotic medications: Clinical implication and considerations. *Dialogues in Clinical Neuroscience*, 18(3), 323–337.

Gaedigk, A., Ingelman-Sundberg, M., Miller, N. A., Leeder, J. S., Whirl-Carrillo, M., Klein, T. E., & the PharmVar Steering Committee. (2018). The pharmacogene variation (PharmVar) consortium: Incorporation of the human cytochrome P450 (CYP) allele nomenclature database. *Clinical Pharmacology & Therapeutics*, 103(3), 299–401. https://doi.org/10.1002/cpt.910.

Gaedigk, A., Sangkuhl, K., Whirl-Carrillo, M., Klein, T., & Leeder, J. S. (2017). Prediction of CYP2D6 phenotype from genotype across world populations. *Genetics in Medicine*, *19*, 69–76. https://doi.org/10.1038/gim.2016.80.

Gonzalez, F. J., Skoda, R. C., Kimura, S., Umeno, M., Zanger, U. M., Nerbert, D. W., ... Meyer, U. A. (1988). Characterization of the common genetic defect in humans deficient in debrisoquine metabolism. *Nature*, *331*(4), 442–446.

Haga, S. B., & Mills, R. (2015). Nurses' communication of pharmacogenetic test results as part of discharge care. *Pharmacogenomics*, *16*(3), 251–256.

Henrikson, N. B., Burke, W., & Veenstra, D. L. (2008). Ancillary risk information and pharmacogenetic tests: Social and policy implications. *The Pharmacogenomics Journal*, *8*, 85–89. https://doi.org/10.1038/sj.tpj.6500457.

Herbert, D., Neves-Pereira, M., Baidya, R., Cheema, S., Groleau, S., Shahmirian, A., ... Kennedy, J. L. (2018). Genetic testing as a supporting tool in prescribing psychiatric medication: Design and protocol of the IMPACT study. *Journal of Psychiatric Research*, *96*, 265–272. https://doi.org/10.1016/j.jpsychires.2017.09.002.

Jablonski, M. R., King, N., Wang, Y., Winner, J. G., Watterson, L. R., Gunselman, S., & Dechiro, B. M. (2018). Analytical validation of a psychiatric pharmacogenetic test. *Personalized Medicine*. https://doi.org/10.2217/pme-2017-0094.

Johnson, J. A. (2013). Pharmacogenetics in clinical practice: How far have we come and where are we going? *Pharmacogenomics*, *14*(7), 835–843. https://doi.org/10.2217/pgs.13.52.

Kalman, L. V., Agundez, J., Appel, M. L., Black, J. L., Bell, G. C., Boukouvala, S., et al. (2016). Pharmacogenetic allele nomenclature: International workgroup recommendations for test result reporting. *Clinical Pharmacology and Therapeutics*, *99*(2), 172–185. https://doi.org/10.100/cpt.280.

Keeling, N. J., Rosenthal, M. M., West-Strum, D., Patel, A. S., Haidar, C. E., & Hoffman, J. M. (2017). Preemptive pharmacogenetic testing: Exploring the knowledge and perspectives of US payers. *Genetics in Medicine*. https://doi.org/10.1038/gim.2017.181.

Lam, R. W., McIntosh, D., Wang, J., Enns, M. W., Kolivakis, T., Michalak, E. E., et al. (2016). Canadian network for mood and anxiety treatments (CANMAT) 2016 clinical guidelines for the management of adults with major depressive disorder: Section 1. Disease burden and principles of care. *Canadian Journal of Psychiatry*, *61*(9), 510–523. https://doi.org/10.1177/0706743716659416.

Lu, M., Lewis, C. M., & Traylor, M. (2017). Pharmacogenetic testing through the direct-to-consumer genetic testing company 23andMe. *BMC Medical Genomics*, *10*(47), 1–8. https://doi.org/10.1186/s12920-017-0283-0.

Maciel, A., Cullors, A., Lukowiak, A. A., & Garces, J. (2018). Estimating cost savings of pharmacogenetic testing for depression in real-world clinical settings. *Neuropsychiatric Disease and Treatment*, *14*, 225–230. https://doi.org/10.2147/NDT.S145046.

Marchant, G. E., & Lindor, R. A. (2013). Personalized medicine and genetic malpractice. *Genetics in Medicine*, *15*(12), 921–922.

Mills, R., Ensinger, M., Callanan, N., & Haga, S. B. (2017). Development and initial assessment of a patient education video about pharmacogenetics. *Journal of Personalized Medicine*, *7*, 4. https://doi.org/10.3390/jpm7020004.

Mills, R., Voora, D., Peyser, B., & Haga, S. B. (2013). Delivering pharmacogenetic testing in a primary care setting. *Pharmacogenomics and Personalized Medicine*, *6*, 105–112. https://doi.org/10.2147/PGPM.S50598.

Müller, D. J., Kekin, I., Kao, A. C. C., & Brandl, E. J. (2013). Towards the implementation of CYP2D6 and CYP 2C19 genotypes in clinical practice: Update and report from a pharmacogenetic service clinic. *International Review of Psychiatry*, *25*(5), 554–571. https://doi.org/10.3109/09540261.2013.838944.

National Human Genome Research Institute (NHGRI). (2018). Regulation of genetic tests. Retrieved from: https://www.genome.gov/10002335/regulation-of-genetic-tests/.

Nutescu, E. A., Drozda, K., Bress, A. P., Galanter, W. L., Stevenson, J., Stamos, T. D., ... Cavallari, L. H. (2013). Feasibility of implementing a comprehensive warfarin pharmacogenetics service. *Pharmacotherapy*, *33*(11), 11156–11164. https://doi.org/10.1002/phar.1329.

Phillips, K. A., Deverka, P. A., Trosman, J. R., Douglas, M. P., Chambers, J. D., Weldon, C. B., & Dervan, A. P. (2017). Payer coverage policies for multigene tests. *Nature Biotechnology*, *35*(7), 614–617. https://doi.org/10.1038/nbt.3912.

Pouget, J. G., Shams, T. A., Tiwari, A. K., & Mueller, D. J. (2014). Pharmacogenetics and outcome with antipsychotic drugs. *Dialogues in Clinical Neuroscience*, 555–566.

Pratt, V. M., Del Tredici, A. L., Hachad, H., Ji, Y., Kalman, L. V., Scott, S. A., & Weck, K. E. (2018). Recommendations for clinical CYP2C19 genotyping allele selection: A report of the association for molecular pathology. *The Journal of Molecular Diagnostics*. https://doi.org/10.1016/j.jmoldx.2018.01.011.

Roche, M. I., & Berg, J. S. (2015). Incidental findings with genomic testing: Implications for genetic counseling practice. *Current Genetic Medicine Reports*, *3*(4), 166–176. https://doi.org/10.1007/s40142-015-0075-9.

Rush, A. J., Trivedi, M. H., Wisniewski, S. R., Nierenberg, A. A., Stewart, J. W., Warden, D., ... Fava, M. (2006). Acute and longer-term outcomes in depressed outpatients requiring one or several treatment steps: A STAR*D report. *American Journal of Psychiatry*, *163*, 913–918.

Smith, R. M. (2017). Advancing psychiatric pharmacogenomics using drug development paradigms. *Pharmacogenomics*, *18*(15), 1459–1467. https://doi.org/10.2217/pgs-2017-0104.

Swen, J. J., Nijenhuis, M., van Rhenen, M., de Boer-Veger, N. J., Buunk, A., Houwink, E. J., ... Guchelaar, H. (2018). Pharmacogenetic information in clinical guidelines—The European perspective. *Clinical Pharmacology and Therapeutics*. https://doi.org/10.1002/cpt.1049.

Trautmann, S., Rehm, J., & Wittchen, H. (2016). The economic costs of mental disorders: Do our societies react appropriately to the burden of mental disorders? *EMBO Reports*, *17*(9), 1245–1249. https://doi.org/10.15252/embr.201642951.

Vassy, J. L., Davis, J. K., Kirby, C., Richardson, I. J., Green, R. C., McGuire, A. L., & UbeL, P. A. (2018). How primary care providers talk to patients about genome sequencing results: Risk, rationale, and recommendation. *Journal of General Internal Medicine*. https://doi.org/10.1007/s11606-017-4295-4.

Verbelen, M., Weale, M. E., & Lewis, C. M. (2017). Cost-effectiveness of pharmacogenetic-guided treatment: Are we there yet? *The Pharmacogenomics Journal, 17*(5), 395–402. https://doi.org/10.1038/tpj.2017.21.

Walden, L. M., Brandl, E. J., Changasi, A., Sturgess, J. E., Soibel, A., Notario, J., … Müller, D. J. (2015). Physicians' opinions following pharmacogenetic testing for psychotropic medication. *Psychiatry Research, 229*(3), 913–918. https://doi.org/10.1016/j.psychres.2015.07.032.

Walden, L. M., Brandl, E. J., Tiwari, A. K., Cheema, S., Freeman, N., Braganza, N., … Müller, D. J. (2018). Genetic testing for CYP2D6 and CYP2C19 suggests improved outcome for antidepressant and antipsychotic medication. *Psychiatry Research*, [in press].

Winner, J. G., Carhart, J. M., Altar, C. A., Goldfarb, S., Allen, J. D., Lavezzari, G., … Dechairo, B. M. (2015). Combinatorial pharmacogenomic guidance for psychiatric medications reduces overall pharmacy costs in a 1 year prospective evaluation. *Current Medical Research and Opinion, 31*(9), 1633–1643. https://doi.org/10.1185/03007995.2015.1063483.

Work Group on Major Depressive Disorder. (2010). *Practice guideline for the treatment of patients with major depressive disorder* (3rd ed.). Washington, DC: American Psychiatric Association.

Yau, A., Binti, A., Aziz, A., & Haque, M. (2015). Knowledge, attitude and practice concerning pharmacogenomics among pharmacists: A systematic review. *Journal of Young Pharmacists, 7*(3). https://doi.org/10.5530/jyp.2015.3.3.

Further reading

Christenhusz, G. M., Devriendt, K., & Dierickx, K. (2013). To tell or not to tell? A systematic review of ethical reflections on incidental findings arising in genetics contexts. *European Journal of Human Genetics, 21*, 248–255. https://doi.org/10.1038/ejhg.2012.130.

Chapter 38

Metabolomics in psychiatry

Renee-Marie Ragguett[a] and Roger S. McIntyre[a,b,c,d]

[a]Mood Disorders Psychopharmacology Unit, University Health Network, Toronto, ON, Canada, [b]Department of Psychiatry, University of Toronto, Toronto, ON, Canada, [c]Department of Pharmacology, University of Toronto, Toronto, ON, Canada, [d]Institute of Medical Science, University of Toronto, Toronto, ON, Canada

The diagnostic process within psychiatry has experienced only marginal changes during the 21st century, wherein psychopathology has been the driving factor behind diagnoses. While psychiatric disorders are largely recognized as complex given their heterogeneous symptomatology presentations, it would not be suspect that the traditional nosological systems, such as the Diagnostic and Statistic Manual 5 (DSM-5), would not capture the idiosyncrasies of these disorders (Hyman, 2010). Toward developing a more comprehensive way of studying psychiatric disorders, the National Institute of Mental Health (NIMH) has developed the Research Domain Criteria (RDoC), which aims to classify psychiatric disorders using both behavioral and neurobiological measures (National Institute of Mental Health, 2015). Furthermore, the increased application of objective measurements in the psychiatric disorder classification could prove to be beneficial toward steering away from subjective psychopathological diagnosis methods, and inform personalized medicine in psychiatry.

The application of "omics" has been suggested as a potential contributor toward both advancing medicine and informing personalized medicine, due to its unique and specialized approaches (Chen & Snyder, 2013). Metabolomics is a subcategory of the "omics" category, which is involved with profiling the metabolites present in various biological fluids and/or tissues. Specific metabolites can be defined as small molecules (<1500 Da), wherein they can be endogenous metabolites (i.e., produced by the human body) or exogenous metabolites (i.e., introduced to the human body, e.g., through pharmaceutical agents). Specifically, metabolomics has demonstrated potential applications in many areas in psychiatry, including diagnosis and treatment outcomes. Moreover, if psychiatric disorders could be categorized with the use of metabolomic biomarkers, this objective measure has the potential to be used throughout the occurrence of the disorders, providing many opportunities for not only monitoring disease prognosis, but also drug response.

1 Why metabolomics in psychiatry?

The study of metabolomics, and thus, the metabolome (i.e., all metabolites found within a given sample) is unique from other "omes," such as the proteome and the genome, as the metabolome is dynamic due to the impact of exogenous activity on the metabolic processes. The metabolome is therefore a reflection of the genome, proteome, and environment. As such, temporal and mealtime variations, for example, have shown to have an effect on the metabolome (Kim et al., 2014). It has been further suggested that the presence of specific metabolic traits could serve as an intermediate between genes and disease pathophysiology, or possibly act as a biomarker for these traits (Suhre & Gieger, 2012). Biomarkers have been described as "a characteristic that is objectively measured and evaluated as an indicator of normal biological processes, pathogenic processes, or pharmacologic responses to a therapeutic intervention" by the National Institute of Health Biomarkers Definitions Working Group (Biomarkers Definitions Working Group, 2001). Biomarkers are widely used throughout other medical fields, but their use in psychiatry, however, has been limited by various aspects of the disorders. For example, the complexity of the biological bases of psychiatric disorders, which has yet to be elucidated; the diagnosis of psychiatric disorders, which rely heavily on observed symptomatology as opposed to physical measurable evidence; and inaccessibility of brain tissue, that is, the tissue most associated with the phenotypic presentation of the psychiatric disorder (Venkatasubramanian & Keshavan, 2016).

Conveniently, metabolomics offers a way in which to examine the status of the brain, and, correspondingly, the whole body in a relatively noninvasive fashion. However, the application of these processes can only be accepted under the idea that a psychiatric illness can have an affect on the whole body. This concept has been supported by the exploration and study of biomarkers in the blood that have been shown to reflect the presence and status of certain psychiatric diseases (Guest,

Personalized Psychiatry. https://doi.org/10.1016/B978-0-12-813176-3.00038-9

TABLE 1 Metabolomic strategies and their potential applications in psychiatry

Metabolomic strategy	Potential applications in psychiatry
Metabolomic profiling	- Disease diagnosis and classification - Disease monitoring - Drug discovery and development - Personalized drug treatment - Treatment response
Preclinical metabolomic profiling	- Drug discovery and development
Metabolomics + other "omics"	- Disease diagnosis and classification

2016; Niculescu et al., 2015). While blood is a common media for performing metabolomic assays, other suitable biological samples include urine and tissue extracts (Beckonert et al., 2007). The application of metabolic biomarkers in psychiatry can not only serve a diagnostic purpose, they could also serve as therapeutic targets, and given the dynamic nature of the metabolome, be both prognostic and sensitive to treatment outcomes (Table 1).

2 Application of metabolomics in psychiatry

The use of metabolomics in psychiatry is an emerging technique and field of study that aims to inform personalized medicine. While many advances are still required within this field to reach the level of sophistication needed for personalized medicine, there have been many promising advances over recent years, a number of which are outlined as follows.

2.1 Disease diagnosis

In studying the metabolome, we hope to identify a unique set of disease biomarkers that could aid in both the diagnosis and classification of diseases. There is an advantage for the use of metabolomics during the disease diagnosis and identification stage of a psychiatric illness, as the population is often drug naive. Being drug naive is beneficial, given the influential nature of the environment on the metabolome; specifically, any medication being taken can modify the metabolome, and as such, can present confounding results when attempting to detect the presence of an illness. The importance of using pharmaceutical treatment-naive participants for disease diagnosis was outlined by Cai et al. (2012), who explored the levels of neurotransmitters and their metabolites using plasma and urine from 11 first-episode neuroleptic-naive schizophrenia patients receiving risperidone monotherapy over 6 weeks, and 11 matched controls. Treatment with risperidone monotherapy resulted in various metabolic changes including plasma glycine, plasma progesterone, and blood ketone bodies demonstrating the changes in the metabolome occurring alongside drug treatment. Furthermore, the schizophrenia patient neurotransmitter profiles, which are known to be dysregulated alongside the psychiatric disorder, showed similarities to controls during risperidone monotherapy treatment (Cai et al., 2012). It is therefore evident that there is an importance in using treatment-naive patients in metabolomic studies related to disease diagnosis and identification, as pharmaceutical treatment has demonstrated changes in the metabolic profile, wherein these changes could be minimal, such as a modification of metabolite ratios when compared with controls, or major, such that the metabolomic profile of those with a psychiatric illness resembles healthy controls, which would ultimately result in a false negative diagnosis when relying on these markers.

In addition to the longitudinal approach taken by Cai et al. (2012), wherein their treatment-naive patients were then treated with a pharmaceutical agent, another suggested approach to remove the potential limitation of only examining those with the disorder who are undergoing treatment is to look at those who are treatment-naive (usually first-episode patients), and those with recurrent symptoms who are undergoing treatment, and controls. This approach has been explored by Yang et al. (2013), and in this case, they stated that the metabolic profiles of first-episode schizophrenia patients were similar to those of recurrent schizophrenia patients, thus allowing their metabolic markers identified to be applicable for diagnosis. Specifically, they identified serum metabolite markers of glycerate, eicosenoic acid, β-hydroxybutyrate, pyruvate, and cystine, showing a unique profile in those with schizophrenia in relation to healthy controls (Yang et al., 2013).

2.2 Treatment response

2.2.1 Pharmacometabolomics

Pharmacometabolomics is a subcategory of metabolomics that is particularly interested in (a) predicting the individual's response to a specific pharmacological agent, and/or (b) monitoring the response to a pharmacological agent. The applications of pharmacometabolomics could eventually result in individualized drug therapy, wherein there is a predicted most efficient pharmaceutical agent that would be prescribed based on an individual's metabolome. This concept relies on the premise that drug response is reflected in the metabolome.

Pharmacometabolomics can be applied when monitoring pharmaceutical treatment responses, as the administration of many pharmaceutical agents results in metabolic changes (Kaddurah-Daouk, Weinshilboum, & Pharmacometabolomics Research Network, 2015). It is possible to typify the response phenotype (e.g., poor response and good response) by its associated metabolite profile. Villasenor et al. (2014) explored the metabolic profile of 22 bipolar disorder patients treated with ketamine, and found differences in the metabolic profiles between responders and nonresponders. Specifically, the differences were found in an area that is not inherently targeted by the administration of ketamine, that is, mitochondrial beta-oxidation of fatty acids. These results suggest that those with a specific metabolic profile would have a preferred response to ketamine (Villasenor et al., 2014). Similarly, Rotroff et al. have demonstrated increases in phosphoethanolamines postketamine treatment in those with major depressive disorder who have shown a reduction in depressive symptomatology, as opposed to those with major depressive disorder postketamine treatment who have demonstrated a lesser response (i.e., less or no reduction in depressive symptomatology) (Rotroff et al., 2016). It is therefore noted that there are favorable metabolic profiles for specific pharmaceutical treatments, and treatment response can also be demonstrated on a metabolic level. In addition, treatment response by metabotype need not only be completed for psychiatric symptomatology, it could also be applied for comorbid effects, such as weight gain, often seen with the use of atypical antipsychotics (Kaddurah-Daouk et al., 2007), ultimately allowing for a full profile of treatment response covering both primary effects (i.e., those associated with the symptoms you wish to treat) and secondary effects (i.e., comorbid effects alongside pharmaceutical use). Interestingly, when identifying response biomarker candidates, human studies are not required. Weckmann et al. conducted a metabolic profiling of ketamine in murine models, and identified various metabolites that met the quality for biomarkers of ketamine response (Weckmann, Labermaier, Asara, Muller, & Turck, 2014).

Another approach by which to monitor treatment response would likely be through pharmacometabolomics-informed pharmacogenomics. The importance of including pharmacometabolomics in pharmacogenomics stems from the in-depth personalized information generated from the metabolomic profile (Neavin, Kaddurah-Daouk, & Weinshilboum, 2016). Pharmacometabolomics-informed pharmacogenomics studies within psychiatry are emerging, and have demonstrated promising results in advancing this field. For example, Gupta et al. (2016) explored this strategy in 306 major depressive disorder patients treated with citalopram or escitalopram, in order to identify genes that have an association with metabolites that are associated with the response to selective serotonin reuptake inhibitors by combining both metabolic assays and genome-wide association studies (GWAS). They have identified the expression of SNPs *TSPAN5* and *ERICH3*, which can alter the concentration of plasma serotonin. Gupta et al. further noted that there are very few studies, and very little is known about *ERICH3* (Gupta et al., 2016). This finding ultimately demonstrates the usefulness of metabolomic techniques and its combination with other fields, particularly "omic" fields, such as genomics, in treatment response.

2.2.2 Drug discovery and development

Through pharmacometabolomics, there is an ongoing effort to map drug pathways. This is beneficial, as the full metabolic pathways for many pharmaceutical agents currently prescribed for psychiatric disorders are not all known (e.g., aripiprazole). In mapping pharmaceutical pathways and discovering biomarkers for disease presence, we are able to create pharmaceutical agents that can ideally target the appropriate areas. For example, the metabolomic profile for ketamine and esketamine drug action was recently mapped in those with major depressive disorder, and in doing so, Rotroff et al. found novel metabolomic changes further informing the way in which these two agents have antidepressant properties (Rotroff et al., 2016).

2.3 Disease monitoring

Metabolites identified for use in disease diagnosis and pharmacometabolomics can be used for disease monitoring. The concept would be as follows: metabolites identified to be associated with disease symptomatology would be altered during the course of the disease, representing the phenotypic state at the time of the metabolic profiling (e.g., disease severity, remission, or relapse). For example, Kaddurah-Daouk et al. (2012) contrasted the metabolic profile of those with major depressive disorder ($n = 14$), remitted major depressive disorder (rMDDe) ($n = 14$), and healthy controls ($n = 18$) in order to determine if each group would have a unique metabolic state characteristic of their psychiatric phenotype. Specifically, the group compared the levels and ratio of metabolites within the tryptophan, tyrosine, purine, and methionine pathways across the three groups. It was demonstrated across many metabolites that there were intragroup metabolic profile variations. For example, there were various unique metabolite ratios per group, and levels of specific metabolites were associated with disease severity. While there were differences observed in these groups, they were sometimes demonstrated in opposing fashions to current literature which, as suggested by the authors, could represent their small sample size, and large heterogeneity in the major depressive disorder group (Kaddurah-Daouk et al., 2012). While there were some limitations in this study, it does provide a promising starting point for examining differences in metabolic profiles during various points in a psychiatric disorder.

3 Limitations of metabolomics in psychiatry

3.1 Environmental variations

Given that the metabolome can and does vary with the environment, it is critical when completing metabolite analyses, particularly in comparison studies, to take samples under the same conditions each time a sample is taken from an individual. There are many environmental components that can be controlled for, including the time when the sample is taken, and if the sample is taken during fasting conditions. However, given the complexity of the metabolome, and its dynamics occurring alongside exogenous factors, there can exist complex relationships with elements of the environment that are less controllable overall. For example, specific ingredients in personal care products can have an affect on the metabolome. This has been demonstrated in rats using levels of these products that would mimic the exposure that humans would have. The study further demonstrated that the effect on the metabolome associated with the exposure to these ingredients varied based on developmental age (Houten et al., 2016). As it is well understood that the metabolome varies through development, it would therefore be required to further control for age differences in comparison studies, or in detecting for disorder presence. Ultimately, many samples representative of a wide range of ages would be required.

3.2 Unknown metabolites

As metabolomics is a growing field, there are near-constant discoveries of new unknown metabolites that become apparent in nontargeted metabolomic studies. Nontargeted metabolomic studies aim to explore any apparent metabolite in the given sample, therefore there may be both unknown and known metabolites, in comparison with a targeted metabolite study that aims to monitor the status of a particular set of known metabolites. Unfortunately, these unknown metabolites appear often during nontargeted exploration of metabolic profiles due to the novel nature of the technique. For example, metabolite 18,225 was increased when participants with MDD were administered ketamine (Rotroff et al., 2016). As the structure of these metabolites are unknown, there is a gap in understanding the complete picture of discovered metabolic profiles. This is evident in the case of metabolite 18,255, as its increase alongside ketamine administration has been observed, however, not much else is known about it. Identifying unknown metabolites is costly and time consuming, usually requiring the application of highly technical methodologies, and further, there is a prioritization for the identification of unknown metabolites based on importance (Bowen & Northen, 2010). There are currently mass spectrometry databases that can aid in the identification of unknown metabolites; however, there are inconsistencies in these databases due to different instruments used to generate the data. Furthermore, many of the unknown metabolites are not present in these databases (Bowen & Northen, 2010). Improving the process of identifying and categorizing unknown metabolites is an important future direction, as they may provide not only the complete metabolic profile, but also provide novel treatment targets.

3.3 Availability and application of metabolomic assays in clinic

Single metabolite assays are completed routinely in clinical practice; however, the power of metabolomics lies in its ability to examine many metabolites. There are often limitations in the computational tools required to process large amounts of data, such as those required for interpreting large amount of metabolomic samples (Goodacre, Vaidyanathan, Dunn, Harrigan, & Kell, 2004). As such, several advances have been made in software packages that allow for an in-depth analysis of the metabolites (Kosmides, Kamisoglu, Calvano, Corbett, & Androulakis, 2013). Furthermore, in clinical practice, the downfall of this highly detailed data would be the time required to process the data and receive results. This may prove to be problematic if instantaneous results are needed in practice (metabolomics for the masses: the future of metabolomics in a personalized world, Trivedi, Hollywood, & Goodacre, 2017). Toward exploring this avenue, a clinical metabolic phenome center has been developed in London, England, with a goal of providing dynamic patient-centered care through metabolomics (Lindon & Nicholson, 2014).

4 Future vistas

4.1 The human metabolome project

There are various metabolome databases available; however, the largest is the Human Metabolome Database (HMDB) (www.hmdb.ca), which is a comprehensive online database of the human metabolome. The number of metabolite entries are continuously growing, and, as of 2017, contains ~114,100 metabolite entries. The HMDB works to combine various types of data, including protein sequences, clinical data, and disease links to compounds. As more metabolites are discovered and various information surrounding them continues to grow, the HMDB grows as well. The HMDB has been used for facilitating research for many studies, and will likely continue to be the main database for storing and organizing the metabolome for some time (Wishart et al., 2013).

4.2 Integration of other "omics"

The integration of many omics techniques has been suggested in order to typify the complexities of psychiatric disorders. For example, many psychiatric disorders have also been described using an epigenetic approach, that is, psychopathology results from a combination of genes and environment, therefore the metabolome alone would not be able to capture this complexity. However, it is possible that there may be methylome-metabolome-type associations as well, as it has been demonstrated that DNA methylation can have an impact on metabolites (Petersen et al., 2014; Ptak & Petronis, 2010). As such, while the power of the metabolome is in the quantity and specificity of data it produces, such as big data, we can further say that the metabolome also has deep data power, as it can be combined with other omics, which can also be represented as big data.

4.3 Personalized medicine

Developments in technology throughout the 21st century, and their applications to psychiatry, have resulted in major changes in the way in which we treat and monitor prognosis. For example, the application of telemedicine, allowing for distance health care, and digitalized screening tools, for example, the THINC-it tool which allows for rapid and accurate screening of a patient's cognitive state in major depressive disorder. These advances are patient centered, allowing streamlined processes inside and outside of the clinic (Herendeen & Deshpande, 2014; McIntyre et al., 2017). As metabolomics continues to develop, the goal would be for metabolomics to be applied in patient centered care, particularly to inform personalized medicine in psychiatry. Metabolomics, along with other omics, has the possibility to integrate an objective diagnosis and monitoring alongside various psychiatric disorders, tremendously advancing these processes in the 21st century.

References

Beckonert, O., Keun, H. C., Ebbels, T. M., Bundy, J., Holmes, E., Lindon, J. C., & Nicholson, J. K. (2007). Metabolic profiling, metabolomic and metabolomic procedures for NMR spectroscopy of urine, plasma, serum and tissue extracts. *Nature Protocols, 2*(11), 2692–2703. https://doi.org/10.1038/nprot.2007.376.

Biomarkers Definitions Working Group. (2001). Biomarkers and surrogate endpoints: Preferred definitions and conceptual framework. *Clinical Pharmacology and Therapeutics, 69*(3), 89–95. https://doi.org/10.1067/mcp.2001.113989.

Bowen, B. P., & Northen, T. R. (2010). Dealing with the unknown: Metabolomics and metabolite atlases. *Journal of the American Society for Mass Spectrometry, 21*(9), 1471–1476. https://doi.org/10.1016/j.jasms.2010.04.003.

Cai, H. L., Li, H. D., Yan, X. Z., Sun, B., Zhang, Q., Yan, M., ... Ye, H. S. (2012). Metabolomic analysis of biochemical changes in the plasma and urine of first-episode neuroleptic-naive schizophrenia patients after treatment with risperidone. *Journal of Proteome Research, 11*(8), 4338–4350. https://doi.org/10.1021/pr300459d.

Chen, R., & Snyder, M. (2013). Promise of personalized omics to precision medicine. *Wiley Interdisciplinary Reviews. Systems Biology and Medicine, 5*(1), 73–82. https://doi.org/10.1002/wsbm.1198.

Goodacre, R., Vaidyanathan, S., Dunn, W. B., Harrigan, G. G., & Kell, D. B. (2004). Metabolomics by numbers: Acquiring and understanding global metabolite data. *Trends in Biotechnology, 22*(5), 245–252. https://doi.org/10.1016/j.tibtech.2004.03.007.

Guest, P. C. (2016). *Biomarkers and mental illness: It's not all in the mind.* Springer.

Gupta, M., Neavin, D., Liu, D., Biernacka, J., Hall-Flavin, D., Bobo, W. V., ... Weinshilboum, R. M. (2016). TSPAN5, ERICH3 and selective serotonin reuptake inhibitors in major depressive disorder: Pharmacometabolomics-informed pharmacogenomics. *Molecular Psychiatry, 21*(12), 1717–1725. https://doi.org/10.1038/mp.2016.6.

Herendeen, N., & Deshpande, P. (2014). Telemedicine and the patient-centered medical home. *Pediatric Annals, 43*(2), e28–e32. https://doi.org/10.3928/00904481-20140127-07.

Houten, S. M., Chen, J., Belpoggi, F., Manservisi, F., Sanchez-Guijo, A., Wudy, S. A., & Teitelbaum, S. L. (2016). Changes in the metabolome in response to low-dose exposure to environmental chemicals used in personal care products during different windows of susceptibility. *PLoS ONE, 11*(7). https://doi.org/10.1371/journal.pone.0159919.

Hyman, S. E. (2010). The diagnosis of mental disorders: The problem of reification. *Annual Review of Clinical Psychology, 6*, 155–179. https://doi.org/10.1146/annurev.clinpsy.3.022806.091532.

Kaddurah-Daouk, R., McEvoy, J., Baillie, R. A., Lee, D., Yao, J. K., Doraiswamy, P. M., & Krishnan, K. R. (2007). Metabolomic mapping of atypical antipsychotic effects in schizophrenia. *Molecular Psychiatry, 12*(10), 934–945. https://doi.org/10.1038/sj.mp.4002000.

Kaddurah-Daouk, R., Weinshilboum, R., & Pharmacometabolomics Research Network. (2015). Metabolomic signatures for drug response phenotypes: Pharmacometabolomics enables precision medicine. *Clinical Pharmacology and Therapeutics, 98*(1), 71–75. https://doi.org/10.1002/cpt.134.

Kaddurah-Daouk, R., Yuan, P., Boyle, S. H., Matson, W., Wang, Z., Zeng, Z. B., ... Drevets, W. (2012). Cerebrospinal fluid metabolome in mood disorders-remission state has a unique metabolic profile. *Scientific Reports, 2*, 667. https://doi.org/10.1038/srep00667.

Kim, K., Mall, C., Taylor, S. L., Hitchcock, S., Zhang, C., Wettersten, H. I., ... Weiss, R. H. (2014). Mealtime, temporal, and daily variability of the human urinary and plasma metabolomes in a tightly controlled environment. *PLoS ONE, 9*(1). https://doi.org/10.1371/journal.pone.0086223.

Kosmides, A. K., Kamisoglu, K., Calvano, S. E., Corbett, S. A., & Androulakis, I. P. (2013). Metabolomic fingerprinting: Challenges and opportunities. *Critical Reviews in Biomedical Engineering, 41*(3), 205–221.

Lindon, J. C., & Nicholson, J. K. (2014). The emergent role of metabolic phenotyping in dynamic patient stratification. *Expert Opinion on Drug Metabolism & Toxicology, 10*(7), 915–919. https://doi.org/10.1517/17425255.2014.922954.

McIntyre, R. S., Best, M. W., Bowie, C. R., Carmona, N. E., Cha, D. S., Lee, Y., ... Harrison, J. (2017). The THINC-integrated tool (THINC-it) screening assessment for cognitive dysfunction: Validation in patients with major depressive disorder. *Journal of Clinical Psychiatry, 78*(7), 873–881. https://doi.org/10.4088/JCP.16m11329.

National Institute of Mental Health. (2015). *Strategic plan for research.* Retrieved 1 August 2017.

Neavin, D., Kaddurah-Daouk, R., & Weinshilboum, R. (2016). Pharmacometabolomics informs pharmacogenomics. *Metabolomics, 12*(7). https://doi.org/10.1007/s11306-016-1066-x.

Niculescu, A. B., Levey, D., Le-Niculescu, H., Niculescu, E., Kurian, S. M., & Salomon, D. (2015). Psychiatric blood biomarkers: Avoiding jumping to premature negative or positive conclusions. *Molecular Psychiatry, 20*(3), 286–288. https://doi.org/10.1038/mp.2014.180.

Petersen, A. K., Zeilinger, S., Kastenmuller, G., Romisch-Margl, W., Brugger, M., Peters, A., ... Suhre, K. (2014). Epigenetics meets metabolomics: An epigenome-wide association study with blood serum metabolic traits. *Human Molecular Genetics, 23*(2), 534–545. https://doi.org/10.1093/hmg/ddt430.

Ptak, C., & Petronis, A. (2010). Epigenetic approaches to psychiatric disorders. *Dialogues in Clinical Neuroscience, 12*(1), 25–35.

Rotroff, D. M., Corum, D. G., Motsinger-Reif, A., Fiehn, O., Bottrel, N., Drevets, W. C., ... Kaddurah-Daouk, R. (2016). Metabolomic signatures of drug response phenotypes for ketamine and esketamine in subjects with refractory major depressive disorder: New mechanistic insights for rapid acting antidepressants. *Translational Psychiatry, 6*(9), e894. https://doi.org/10.1038/tp.2016.145.

Suhre, K., & Gieger, C. (2012). Genetic variation in metabolic phenotypes: Study designs and applications. *Nature Reviews. Genetics, 13*(11), 759–769. https://doi.org/10.1038/nrg3314.

Trivedi, D. K., Hollywood, K. A., & Goodacre, R. (2017). Metabolomics for the masses: The future of metabolomics in a personalized world. *New Horizons in Translational Medicine, 3*(6), 294–305. https://doi.org/10.1016/j.nhtm.2017.06.001.

Venkatasubramanian, G., & Keshavan, M. S. (2016). Biomarkers in psychiatry—A critique. *Annals of Neurosciences, 23*(1), 3–5. https://doi.org/10.1159/000443549.

Villasenor, A., Ramamoorthy, A., Silva dos Santos, M., Lorenzo, M. P., Laje, G., Zarate, C., Jr., ... Wainer, I. W. (2014). A pilot study of plasma metabolomic patterns from patients treated with ketamine for bipolar depression: Evidence for a response-related difference in mitochondrial networks. *British Journal of Pharmacology, 171*(8), 2230–2242. https://doi.org/10.1111/bph.12494.

Weckmann, K., Labermaier, C., Asara, J. M., Muller, M. B., & Turck, C. W. (2014). Time-dependent metabolomic profiling of ketamine drug action reveals hippocampal pathway alterations and biomarker candidates. *Translational Psychiatry, 4*, e481. https://doi.org/10.1038/tp.2014.119.

Wishart, D. S., Jewison, T., Guo, A. C., Wilson, M., Knox, C., Liu, Y., ... Scalbert, A. (2013). HMDB 3.0—The human metabolome database in 2013. *Nucleic Acids Research, 41*(Database issue), D801–D807. https://doi.org/10.1093/nar/gks1065.

Yang, J., Chen, T., Sun, L., Zhao, Z., Qi, X., Zhou, K., ... Wan, C. (2013). Potential metabolite markers of schizophrenia. *Molecular Psychiatry, 18*(1), 67–78. https://doi.org/10.1038/mp.2011.131.

Chapter 39

Real-time fMRI brain-computer interface: A tool for personalized psychiatry?

David E.J. Linden

Faculty of Health, Medicine and Life Sciences, School of Mental Health and Neuroscience, Maastricht University, Maastricht, The Netherlands; Division of Psychological Medicine and Clinical Neurosciences, School of Medicine, Cardiff University, Cardiff, United Kingdom

1 Introduction

Since its invention 20 years ago, functional magnetic resonance imaging (fMRI) has become one of the most widely used, and probably the most publicly visible noninvasive technique to measure brain activation. fMRI has played a central role in the development of cognitive neuroscience, and several new fields, including social neuroscience, neuroeconomics, and genetic imaging, which may not have developed had it not been for the unique opportunities afforded by fMRI. The particular strengths of this technique are in its spatial resolution and fidelity, ability to reach deep subcortical structures, and whole-brain coverage, enabling the mapping of functionally connected networks, and the extraction of information from activation patterns that are distributed across different brain areas. In the psychiatric domain, fMRI has made major contributions to the understanding of psychopathology, and the effects of risk genes on cognitive and affective networks (Linden, 2012). In neurology, fMRI has become a central technique for mapping neuroplasticity; for example, in recovery from stroke (Seitz, 2010), and for presurgical mapping. However, fMRI has not yet fulfilled its translational potential, and there is, as of today, no established diagnostic, prognostic, or therapeutic use of this technique for any of the mental disorders. Factors that may have held back the clinical development of fMRI include its lack of molecular resolution (where radionuclide imaging has an advantage), semi-quantitative nature, dependence on control/baseline conditions (with the exception of resting-state fMRI), and difficulty controlling for physiological confounds (cardiovascular, respiratory, and eye and head movements).

fMRI-based neurofeedback (fMRI-NF) has the potential to open up radically new paths to translation. During fMRI-NF training, participants receive feedback on their brain activity in real time, and are instructed to change this activation. This change is normally a simple up- or down-regulation, but could also entail changing the activation difference between two areas, their correlation, or the output of a multivariate pattern classification algorithm. The "hemodynamic" delay between neural activity and the vascular response that contributes to the fMRI signal, which is approximately 5 s, does not pose an obstacle when participants are informed of this delay (Weiskopf et al., 2004). The technical requirements include direct network access to the scanner hardware, software that processes the incoming data in real-time and computes incremental online statistics, and software for the conversion of the real-time activation data into sensory stimulation (e.g., visual, tactile) for the participant. The relevant workflows for such a "brain computer interface" (BCI) have been established on systems from all major MRI scanner manufacturers, and several custom-made and commercial software packages have been developed for online preprocessing and statistical analysis of fMRI data. Over the past 10 years, fMRI-NF has been used to train healthy participants in the self-regulation of motor, sensory, language, and emotion areas (Weiskopf, 2011). In analogy to the effects of other noninvasive or invasive brain stimulation techniques, one should expect that such self-regulation also has consequences at the behavioral, cognitive, and affective level. However, clinical applications have been sparse, and had initially focused on chronic pain (Decharms et al., 2005). Compared with other neurofeedback techniques (with EEG or MEG), and with noninvasive physical stimulation techniques (tDCS and TMS), fMRI-NF has the advantage of higher spatial resolution and localization accuracy, and better access to deep limbic and subcortical structures. Compared with deep-brain stimulation, fMRI-NF has the advantage of noninvasiveness and spatial flexibility (although it is not intended to replace established DBS indications, e.g., in Parkinson's disease [PD]). Finally, compared with all external stimulation techniques, neurofeedback has the attractive characteristic of enabling the patients themselves to control their brain activity, and thus contributing to their experience of self-efficacy, which is an important therapeutic factor in many neuropsychiatric disorders (Bandura, 1997).

Personalized Psychiatry. https://doi.org/10.1016/B978-0-12-813176-3.00039-0

fMRI-NF has been boosted by developments in MR physics, allowing for fast acquisition of high quality data sets, and statistics, allowing for online calculation of univariate and multivariate statistics. It is now a mature experimental technique, and the exciting question for the next decade will be whether it can make a true translational contribution in medicine. The central challenges are identifying the symptoms and disorders that will respond to fMRI-NF, adapting the treatment protocols to the neural networks involved in each of them, evaluating the underlying neuroplastic mechanisms, and devising training strategies that enable sustainable long-term effects.

2 Principles and methods

Attempts to train humans and animals to regulate their own brain activity, feeding back signals from noninvasive (electroencephalography [EEG]) or invasive recordings, go back to the 1960s (reviewed by Birbaumer & Cohen, 2007). The theoretical principles were largely derived from operant conditioning, whereby the participant learns the optimal strategy through the contingencies between their actions and a reward. In an animal, the reward would have to be a primary reward. For example, the desired neural response would be reinforced by the delivery of food or drink. In humans, the information about the achievement of a particular neural target (e.g., increasing the ratio between theta and alpha power of the EEG by 20%) could itself be the reward. Whether this reward reinforces a particular mental strategy that leads to the desired physiological response, or the physiological response itself is a futile question because the two cannot be separated in any meaningful way under the assumptions of psychophysical unity. It has been reported that humans can achieve reliable self-control over parameters of their EEG, for example, the topography of slow cortical potentials, the alpha-theta ratio, or the ratio between sensorimotor rhythm and theta activity, through such operant conditioning.

FMRI-based neurofeedback (fMRI-NF) can be conducted in a similar fashion, where participants are blind to the functional relevance of the targeted activation, and essentially aim to achieve self-regulation by trial and error (Weiskopf et al., 2004). However, fMRI-NF can also harness the considerable knowledge about the neural basis of particular mental and cognitive processes that the past 20 years of functional brain mapping have achieved, and introduce a cognitive component into the learning strategy. For example, in our study on upregulation of the supplementary motor area in Parkinson's disease (Subramanian et al., 2011), patients were told about the motor planning functions of the target area, and informed that motor imagery might be one viable strategy to up-regulate it. Giving patients initial hints about potential strategies for the up-regulation training, which they can then refine through an operant conditioning protocol with fMRI-NF, can considerably reduce scanning time, and provide patients with tangible results within one scanning session (Subramanian et al., 2011).

3 Clinical applications of fMRI-NF

Clinical studies of EEG neurofeedback (EEF-NF) training have been conducted in epilepsy, attention deficit/hyperactivity disorder (ADHD), depression and anxiety, as well as other neurological and psychiatric disorders (Birbaumer & Cohen, 2007; Hammond, 2005). Although some of the initial results were promising, the only field where EEG-NF has entered clinical practice is ADHD (using a variety of target parameters, e.g., theta/beta ratio or slow cortical potentials). For ADHD, several published trials found positive effects on symptoms as measured by parent or teacher questionnaires or clinical assessments (Lofthouse et al., 2011). One limitation of most trials has been the lack of blinding of participants to treatment condition, and the general difficulties of setting up of double-blind controlled trials of NF training (Lansbergen et al., 2011), and a recent properly blinded trial did not find any superiority for the active EEG-neurofeedback intervention (Schönenberg et al., 2017). Although these difficulties will apply similarly in fMRI-NF, there are good reasons to explore its use in clinical conditions, including and beyond those that may respond to EEG-NF. Because of the strong cognitive component, fMRI-NF may achieve its neurophysiological targets faster than EEG-NF, and thus enhance patient motivation and compliance. Furthermore, fMRI-NF can directly target specific brain areas or networks that have been implicated in neuropsychiatric disorders, either because they show a dysfunction that might cause the disorder, or because they may compensate for a primary dysfunction. In the first case, the aim of neurofeedback would be to restore the function to normal levels, and in the second, to promote the recruitment of the compensatory network. Because the plastic effects of neurofeedback training are not confined to the area or network whose activation is explicitly used as the target for training, both processes can very well happen in tandem, making it a versatile tool for redressing imbalances in and between brain networks.

Up to now, the clinical potential of fMRI-NF to improve symptoms or change behavior has only been explored in a small number of published studies, all with small patient numbers. FMRI-NF, targeting the anterior cingulate cortex, has shown effects on chronic pain in patients with fibromyalgia (Decharms et al., 2005). Two out of six patients with chronic tinnitus improved after training down-regulation of auditory cortex activity (Haller, Birbaumer, & Veit, 2010). In our study of Parkinson's disease (PD), we trained patients in the early stages of the disease to up-regulate their supplementary motor area (SMA) (Subramanian et al., 2011). The choice of target area was based on pathophysiological models implicating underactivity of the SMA in PD (Nachev, Kennard, & Husain, 2008), and based on our observation that reliable up-regulation can be achieved through motor imagery. The five patients who received fMRI-NF, but not five control patients who engaged in motor imagery without feedback, achieved reliable SMA up-regulation, and improved motor fluency and clinical symptoms (Subramanian et al., 2011). In a randomized, controlled trial with the same fMRI-NF design, but addition of actual exercise components for both the NF and a control group, we found a similar size of clinical effects in the active group, but no statistical superiority over the control intervention (Subramanian et al., 2016).

In the first psychiatric application, we trained patients with depression to up-regulate brain networks responsive to positive affective stimuli. This paradigm was modeled on our previous work with healthy participants, which had shown that the neurofeedback component is required for reliable control over emotion networks (Johnston et al., 2011). The eight patients who underwent this fMRI-NF protocol for four sessions improved significantly on the 17-item Hamilton Rating Scale for Depression, and this clinical improvement was not observed in eight control patients who engaged in a protocol of positive emotional imagery (matched to the fMRI-NF protocol for intervention and assessment times and affective stimuli) outside the scanner (Linden et al., 2012). A recent randomized controlled trial pitting up-regulation training of the amygdala against up-regulation of a control area (the intraparietal sulcus region) found clinical improvement in patients with depression that was significantly stronger in the active (amygdala upregulation) than the control intervention (Young et al., 2017). We have recently replicated the clinical improvement after up-regulation training of areas responsive to positive emotions seen in our earlier work (Linden et al., 2012) although similar improvement was seen in a control neurofeedback intervention targeting the parahippocampal place area (Mehler et al., 2018). The feasibility of fMRI-NF has also been demonstrated for schizophrenia (up-regulation of anterior insula) (Ruiz et al., 2011) and stroke (upregulation of ventral premotor cortex) (Sitaram et al., 2012), but data on clinical improvement are not available. FMRI-NF has also been piloted for several substance use disorders (Fovet, Jardri, & Linden, 2015; Hartwell et al., 2013; Karch et al., 2015; Kirsch et al., 2016), and formal clinical trials are under way (Cox et al., 2016; Gerchen et al., 2018).

4 A personalized approach

Neurofeedback, like most psychotherapeutic approaches, entails mobilization of personal resources and strategies. For example, each individual will have different strategies for increasing or decreasing activation in a particular brain region. Although the set of potential strategies may be more limited for some areas (e.g., there is only a limited range of familiar upper limb movements that people are likely to try imagining when asked to up-regulate higher motor areas), there would be a very wide range of imageries and thoughts that could be evoked to self-regulate limbic activation, and this is inextricably linked to each individual's biography and disposition. In this sense, neurofeedback is a highly personalized tool, even if, for its clinical evaluation, standardized criteria (such as attaining a certain level of self-regulation) have to be defined.

Beyond this general feature of personalization, neurofeedback protocols can also explicitly incorporate personalized elements at several levels:

- Personalized neural targets
- Personalized stimulus material
- Personalized homework or other adjunct elements

4.1 Personalized neural targets

Defining personalized neural targets comes closest to the "personalized medicine" approach discussed elsewhere in this book. It is predicated on the assumption that pathological activation levels or patterns (too much, too little, or wrongly coupled) can be identified in individual patients (or subgroups of patients), which would be similar to finding a specific molecular marker or highly penetrant genetic risk variant. Identification of such altered brain activation patterns through a diagnostic fMRI scan could then be used to stratify patients for different neurofeedback protocols, targeting the respective dysfunctional network. This approach would require identification of such diagnostic patterns of brain activation in patients (and ideally also the "healthy" target pattern to aim for). An increasing body of work is looking at the reliability of such network markers based on resting-state fMRI, and at some point, it might become possible to identify patients with, say, depression, reliably based on their resting-state fMRI signal, or even subtype them into diagnostic subgroups. However, the relevant features may not be ideal targets for fMRI-NF because of their slow evolution over time. Resting-state fMRI patterns are driven by low-frequency oscillations of the BOLD signal ($<0.1\,Hz$), and thus it would take several tens of seconds for a particular mental strategy to be reflected in a change of resting pattern. However, delayed feedback protocols have been developed, and for some tasks, may actually provide better outcomes than immediate feedback, so the problem of target signals that are slow to change is surmountable.

Most fMRI-NF studies do not define their activation targets through resting-state patterns, though, but use a functional localizer to identify a functional cluster or network that then becomes the target for up- or down-regulation in subsequent neurofeedback training runs. This is another way of personalizing the neural target, although it is not necessarily demonstrated that the patient's activation would be different to that of a control group. Here the idea is that a functional network needs to be identified at an individual level that should be targeted—because it may play a role in a dysfunction, or may be involved in compensatory processes—although sometimes more standardized constraints are applied. For example, when identifying individual networks responsive to alcohol cues, which can vary greatly from participant to participant (and even within participants from scanning session to scanning session), one could implement anatomical constraints, for example, by only searching within a mask of the basal ganglia, limbic, and paralimbic areas.

4.2 Personalized stimuli and other elements of the neurofeedback therapy

Such functional localizers commonly use material that is relevant to the disease process or the putative functional deficit. Examples include positive affective pictures for localization of emotionally responsive regions in depression (Linden et al., 2012), contamination scenes in contamination anxiety (Scheinost et al., 2013), pictures of spiders in arachnophobia (Zilverstand et al., 2015), pictures of food to train regulation of food craving (Ihssen et al., 2016), or pictures of addictive cues for use in substance (and potentially also behavioral) addictions (Sokunbi et al., 2014). Although the desired activation patterns (and accompanying physiological responses, e.g., arousal) can often be induced with generic stimuli for many of these protocols, a personalized approach is preferable. If the aim of the neurofeedback process is directly to target dysfunctional activation patterns that arise during (and presumably contribute to) clinical symptoms, it will be necessary to induce these symptoms, or at least subclinical surrogates or certain key component processes, during the scanning, of course with the appropriate consent, permissions, and clinical cover in place. Symptom induction is generally most effective with personalized stimuli. For example, patients with obsessive compulsive disorder (OCD) may be invited to take pictures of the scenarios that trigger their symptoms, which can then be used for symptom induction (De Putter, Van Yper, & Koster, 2017). Exposure to triggering events—generally through imagination, but increasingly also through use of virtual reality—is also a key component of psychological interventions for PTSD, and personalized scenarios or scripts can be used to localize disease-relevant areas. Another related application is their presentation during the neurofeedback runs: here, patients will be instructed to train a neural target, for example, down-regulation of amygdala activity during exposure to personalized triggers (trauma words, Nicholson et al., 2017) with the overall therapeutic aim of decreasing their salience. FMRI-NF would thus be incorporated in a wider, personalized extinction therapy.

Related concepts have been implemented in fMRI-NF protocols for substance use disorders. In alcohol dependence, for example, both patients' preferred drinks and their individual motivation goal identified during therapy vary considerably among individuals, and both robust neural activations and meaningful psychological engagement likely depends on the individually tailored selection of pictures, which may require prescanning picture rating sessions (Cox et al., 2016). If these personalized pictures are then also presented during the fMRI-NF training, and indicate training success by varying their size (Sokunbi et al., 2014), the neurofeedback intervention assumes elements of a personalized craving reduction program.

5 Concluding remarks

These examples may show how personalized elements that are already routinely used in psychological therapies—in sensu, in vivo, or in virtuo exposure for anxiety disorders and OCD; identification of individualized motivational goals for addiction—can be incorporated in neurofeedback training protocols. Combined with the individual definition of target brain areas/networks and the individualized strategies that patients will use to achieve their targets, the use of customized stimulus makes neurofeedback a highly personalized intervention. This may be attractive for patients who are interested in incorporating neural elements in their symptom reduction training, and more generally want to explore the relationship between their brain activation and mental life, but it also poses challenges for the accurate description of component processes and evaluation of their implementation, for example, through LOGIC models (Quinn et al., 2016). Even with these personalized features, neurofeedback can still be standardized and manualized in a manner sufficient to allow formal testing in randomized controlled trials.

Acknowledgment

Some of this material has previously been published in: Goebel R., & Linden D. (2014). Neurofeedback with real-time functional MRI. In C. Mulert & M. Shenton (Eds.), *MRI in psychiatry*. Springer, Berlin: Heidelberg.

References

Bandura, A. (1997). *Self-efficacy: The exercise of control.* New York: W.H. Freeman. ix, 604p. ISBN 0716726262 (cased) 0716728508 (pbk).

Birbaumer, N., & Cohen, L. (2007). Brain-computer interfaces: Communication and restoration of movement in paralysis. *The Journal of Physiology, 579* (Pt. 3), 621–636. ISSN 0022-3751. Available from: http://www.ncbi.nlm.nih.gov/entrez/query.fcgi?cmd=Retrieve&db=PubMed&dopt=Citation&list_uids=17234696.

Cox, W. M. , et al. (2016). Neurofeedback training for alcohol dependence versus treatment as usual: Study protocol for a randomized controlled trial. *Trials, 17*(1), 480. ISSN 1745-6215. Available from: https://www.ncbi.nlm.nih.gov/pubmed/27716290.

De Putter, L. M., Van Yper, L., & Koster, E. H. (2017). Obsessions and compulsions in the lab: A meta-analysis of procedures to induce symptoms of obsessive-compulsive disorder. *Clinical Psychology Review, 52,* 137–147. ISSN 1873-7811. Available from: https://www.ncbi.nlm.nih.gov/pubmed/28119197.

Decharms, R. , et al. (2005). Control over brain activation and pain learned by using real-time functional MRI. *Proceedings of the National Academy of Sciences of the United States of America, 102*(51), 18626–18631. ISSN 0027-8424. Available from: http://www.ncbi.nlm.nih.gov/entrez/query.fcgi?cmd=Retrieve&db=PubMed&dopt=Citation&list_uids=16352728.

Fovet, T., Jardri, R., & Linden, D. (2015). Current issues in the use of fMRI-based neurofeedback to relieve psychiatric symptoms. *Current Pharmaceutical Design, 21*(23), 3384–3394. ISSN 1873-4286. Available from: http://www.ncbi.nlm.nih.gov/pubmed/26088117.

Gerchen, M. F. , et al. (2018). The SyBil-AA real-time fMRI neurofeedback study: Protocol of a single-blind randomized controlled trial in alcohol use disorder. *BMC Psychiatry, 18*(1), 12. ISSN 1471-244X. Available from: https://www.ncbi.nlm.nih.gov/pubmed/29343230.

Haller, S., Birbaumer, N., & Veit, R. (2010). Real-time fMRI feedback training may improve chronic tinnitus. *European Radiology, 20*(3), 696–703. ISSN 1432-1084. Available from: http://www.ncbi.nlm.nih.gov/pubmed/19760238.

Hammond, D. (2005). Neurofeedback with anxiety and affective disorders. *Child and Adolescent Psychiatric Clinics of North America, 14*(1), 105–123. vii. ISSN 1056-4993. Available from: http://www.ncbi.nlm.nih.gov/entrez/query.fcgi?cmd=Retrieve&db=PubMed&dopt=Citation&list_uids=15564054.

Hartwell, K. J. , et al. (2013). Real-time fMRI in the treatment of nicotine dependence: A conceptual review and pilot studies. *Psychology of Addictive Behaviors, 27*(2), 501–509. ISSN 1939-1501. Available from: https://www.ncbi.nlm.nih.gov/pubmed/22564200.

Ihssen, N. , et al. (2016). Neurofeedback of visual food cue reactivity: A potential avenue to alter incentive sensitization and craving. *Brain Imaging and Behavior*, ISSN 1931-7565. Available from: http://www.ncbi.nlm.nih.gov/pubmed/27233784.

Johnston, S. , et al. (2011). Upregulation of emotion areas through neurofeedback with a focus on positive mood. *Cognitive, Affective, & Behavioral Neuroscience, 11*(1), 44–51. ISSN 1531-135X. Available from: http://www.ncbi.nlm.nih.gov/pubmed/21264651.

Karch, S. , et al. (2015). Modulation of craving related brain responses using real-time fMRI in patients with alcohol use disorder. *PLoS ONE, 10*(7) ISSN 1932-6203. Available from: http://www.ncbi.nlm.nih.gov/pubmed/26204262.

Kirsch, M. , et al. (2016). Real-time functional magnetic resonance imaging neurofeedback can reduce striatal cue-reactivity to alcohol stimuli. *Addiction Biology, 21*(4), 982–992. ISSN 1369-1600. Available from: https://www.ncbi.nlm.nih.gov/pubmed/26096546.

Lansbergen, M. M. , et al. (2011). ADHD and EEG-neurofeedback: A double-blind randomized placebo-controlled feasibility study. *Journal of Neural Transmission, 118*(2), 275–284. ISSN 1435-1463. Available from: http://www.ncbi.nlm.nih.gov/pubmed/21165661.

Linden, D. E. (2012). The challenges and promise of neuroimaging in psychiatry. *Neuron, 73*(1), 8–22. ISSN 1097-4199. Available from: http://www.ncbi.nlm.nih.gov/pubmed/22243743.

Linden, D. E. J. , et al. (2012). Real-time self-regulation of emotion networks in patients with depression. *PLoS ONE, 7,* e38115.

Lofthouse, N. , et al. (2011). A review of neurofeedback treatment for pediatric ADHD. *Journal of Attention Disorders*, ISSN 1557-1246. Available from: http://www.ncbi.nlm.nih.gov/pubmed/22090396.

Mehler, D. M. A., Sokunbi, M. O., Habes, I., Barawi, K., Subramanian, L., Range, M., ... Linden, D. E. J. (2018). Targeting the affective brain-a randomized controlled trial of real-time fMRI neurofeedback in patients with depression. *Neuropsychopharmacology, 43*(13), 2578–2585.

Nachev, P., Kennard, C., & Husain, M. (2008). Functional role of the supplementary and pre-supplementary motor areas. *Nature Reviews Neuroscience, 9* (11), 856–869. ISSN 1471-0048. Available from: http://www.ncbi.nlm.nih.gov/entrez/query.fcgi?cmd=Retrieve&db=PubMed&dopt=Citation&list_uids=18843271.

Nicholson, A. A. , et al. (2017). The neurobiology of emotion regulation in posttraumatic stress disorder: Amygdala downregulation via real-time fMRI neurofeedback. *Human Brain Mapping, 38*(1), 541–560. ISSN 1097-0193. Available from: https://www.ncbi.nlm.nih.gov/pubmed/27647695.

Quinn, L. , et al. (2016). Development and delivery of a physical activity intervention for people with Huntington disease: Facilitating translation to clinical practice. *Journal of Neurologic Physical Therapy, 40*(2), 71–80. ISSN 1557-0584. Available from: https://www.ncbi.nlm.nih.gov/pubmed/26863152.

Ruiz, S. , et al. (2011). Acquired self-control of insula cortex modulates emotion recognition and brain network connectivity in schizophrenia. *Human Brain Mapping*, ISSN 1097-0193. Available from: http://www.ncbi.nlm.nih.gov/pubmed/22021045.

Scheinost, D. , et al. (2013). Orbitofrontal cortex neurofeedback produces lasting changes in contamination anxiety and resting-state connectivity. *Translational Psychiatry, 3,* e250. ISSN 2158-3188. Available from: http://www.ncbi.nlm.nih.gov/pubmed/23632454.

Schönenberg, M. , et al. (2017). Neurofeedback, sham neurofeedback, and cognitive-behavioural group therapy in adults with attention-deficit hyperactivity disorder: A triple-blind, randomised, controlled trial. *Lancet Psychiatry, 4*(9), 673–684. ISSN 2215-0374. Available from: https://www.ncbi.nlm.nih.gov/pubmed/28803030.

Seitz, R. J. (2010). How imaging will guide rehabilitation. *Current Opinion in Neurology, 23*(1), 79–86. ISSN 1473-6551. Available from: http://www.ncbi.nlm.nih.gov/pubmed/19926990.

Sitaram, R. , et al. (2012). Acquired control of ventral premotor cortex activity by feedback training: An exploratory real-time FMRI and TMS study. *Neurorehabilitation and Neural Repair, 26*(3), 256–265. ISSN 1552-6844. Available from: http://www.ncbi.nlm.nih.gov/pubmed/21903976.

Sokunbi, M. O. , et al. (2014). Real-time fMRI brain-computer interface: Development of a "motivational feedback" subsystem for the regulation of visual cue reactivity. *Frontiers in Behavioral Neuroscience, 8*, 392. ISSN 1662-5153. Available from: http://www.ncbi.nlm.nih.gov/pubmed/25505392.

Subramanian, L. , et al. (2011). Real-time functional magnetic resonance imaging neurofeedback for treatment of Parkinson's disease. *The Journal of Neuroscience, 31*(45), 16309–16317. ISSN 1529-2401. Available from: http://www.ncbi.nlm.nih.gov/pubmed/22072682.

Subramanian, L. , et al. (2016). Functional magnetic resonance imaging neurofeedback-guided motor imagery training and motor training for Parkinson's disease: Randomized trial. *Frontiers in Behavioral Neuroscience, 10*, 111. Available from: https://www.ncbi.nlm.nih.gov/pubmed/27375451.

Weiskopf, N. (2011). Real-time fMRI and its application to neurofeedback. *NeuroImage*, ISSN 1095-9572. Available from: http://www.ncbi.nlm.nih.gov/pubmed/22019880.

Weiskopf, N. , et al. (2004). Self-regulation of local brain activity using real-time functional magnetic resonance imaging (fMRI). *Journal of Physiology, Paris, 98*(4–6), 357–373. ISSN 0928-4257. Available from: http://www.ncbi.nlm.nih.gov/entrez/query.fcgi?cmd=Retrieve&db=PubMed&dopt=Citation&list_uids=16289548.

Young, K. D. , et al. (2017). Randomized clinical trial of real-time fMRI amygdala neurofeedback for major depressive disorder: Effects on symptoms and autobiographical memory recall. *The American Journal of Psychiatry*, ISSN 1535-7228. Available from: https://www.ncbi.nlm.nih.gov/pubmed/28407727.

Zilverstand, A. , et al. (2015). fMRI neurofeedback facilitates anxiety regulation in females with spider phobia. *Frontiers in Behavioral Neuroscience, 9*, 148. ISSN 1662-5153. Available from: http://www.ncbi.nlm.nih.gov/pubmed/26106309.

Chapter 40

How functional neuroimaging can be used for prediction and evaluation in psychiatry

Beata R. Godlewska and Catherine J. Harmer

Oxford Health NHS Foundation Trust, Warneford Hospital, Oxford, United Kingdom

Abbreviations

5HT1A	serotonin receptor 1A
ACC	anterior cingulate cortex
APD	antipsychotic drug
ARMS	at-risk mental states
BD	bipolar disorder
BD I	bipolar disorder type I
BD II	bipolar disorder type II
CBT	cognitive-behavioral treatment
CHR	clinical high risk
dlPFC	dorsolateral prefrontal cortex
DMN	default mode network
EEG	electroencephalography
FC	functional connectivity
fMRI	functional magnetic resonance imaging
GAD	generalized anxiety disorder
IFG	inferior frontal gyrus (IFG)
iSPOT-D	International Study to Predict Optimized Treatment in Depression
MDD	major depressive disorder
MEG	magnetoencephalography
mOFC	medial orbitofrontal cortex
mPFC	medial prefrontal cortex
OCD	obsessive-compulsive disorder
OFC	orbitofrontal cortex
PET	positron emission tomography
PFC	prefrontal cortex
pgACC	pregenual anterior cingulate cortex
PTSD	posttraumatic stress disorder
rACC	rostral anterior cingulate cortex
RDoC	Research Domain Criteria
ROI	region of interest
rsFC	resting state functional connectivity
SAD	social anxiety disorder
SCH	schizophrenia
SMA	supplementary motor area
SNRI	serotonin norepinephrine reuptake inhibitor
SPECT	single-photon emission computed tomography
SSRI	selective serotonin reuptake inhibitor
STG	superior temporal gyrus

Personalized Psychiatry. https://doi.org/10.1016/B978-0-12-813176-3.00040-7

SVC	super vector clustering
UHR	ultra-high risk
vlPFC	ventrolateral prefrontal cortex
vmPFC	ventromedial prefrontal cortex

Psychiatry is in great need of biomarkers, measures defined as "characteristics that are objectively measured and evaluated as an indicator of normal biologic processes, pathogenic processes, or pharmacologic responses to a therapeutic intervention" (Biomarkers Definitions Working Group, 2001). Such markers are still developing in psychiatry, and unlike many other health problems, such as diabetes or hypertension, mental health conditions are diagnosed mainly on the basis of observation, and accounts provided by the patients and people who know them well. This process, by definition, depends on good communication, symptoms awareness, and good memory of the facts, as well as strong expertise and skills of the clinician. These are all prone to uncertainties and errors, hence, there is a strong need for objective biological measures, such as functional neuroimaging.

Functional neuroimaging refers to a number of techniques that allow for imaging of the brain function in vivo, and form useful tools for exploring the relationship between activity of specific brain areas and clinically relevant issues. The most commonly used approaches include functional magnetic resonance imaging (fMRI), positron emission tomography (PET), electroencephalography (EEG), magnetoencephalography (MEG), and single-photon emission computed tomography (SPECT). Functional studies have employed a number of paradigms, allowing access to different types of information on the function of the brain. Specific functions can be assessed with the use of appropriate tasks, while resting state measurements can provide information on unconstrained network function, and have the benefit of being cognitively less demanding. The focus may be on preselected networks or structures, so-called regions of interest (ROIs), chosen because of the known or suspected pathology, or on the whole brain in a more exploratory way, with data-driven approaches creating an opportunity for new insights not restricted by predefined hypotheses. Currently, psychiatric disorders are perceived as malfunctions of widely distributed networks, which is a shift from an initial focus on single structures.

To be clinically useful, data from brain maps, representing functional patterns of activity linked to a function or state of interest, need to be translated back to an individual. In other words, to be useful, a clinician should be able to answer a relevant clinical question with a simple brain scan that would show the pattern of brain activity typical for, for example, a specific diagnosis or a treatment most likely to work for this given person. This translation has not been achieved yet, but remains a core aim for ongoing research programs (e.g., Trivedi et al., 2016).

While the holy grail of individualized diagnosis and treatment plans in psychiatry is still elusive, there has been an increasing use of functional neuroimaging to understand the neural underpinnings of psychiatric disorders and their treatment. It is hoped that this mechanistic approach can serve as a framework for treatment prediction and outcome research.

1 Understanding the neural basis of mental health conditions

Studies dating back to the early 1990s (Mayberg, 1997) allowed the conceptualization of MDD according to different neurobiological models (Graham et al., 2013). The classic fronto-limbic model (Drevets, 2001; Mayberg, 2003) focuses on the inability of frontal cortices to control overactive limbic structures responsible for the quick automatic processing of salient emotional stimuli, with the rostral and subgenual anterior cingulate cortex (ACC) being the key mediating structures. This overactive state has been linked to depressed mood and—characteristic for MDD—bias toward negative interpretation of emotionally salient stimuli (Miskowiak et al., 2015). The cortico-striatal-thalamic loop model focuses on the functions of these structures in reward guided learning, decision making, and emotional regulation and appraisal, which are especially relevant in the context of anhedonia (Fettes, Schulze, & Downar, 2017; Peters, Dunlop, & Downar, 2016). The third model focuses on dysfunction in the default mode network (DMN), involved in internally oriented attention and self-referential thinking. Hyperactivity and hyperconnectivity within DMN, and between DMN and the frontoparietal attention network, have been observed in MDD (Whitfield-Gabrieli & Ford, 2015). Some studies suggest that patients with MDD have a decreased ability to deactivate DMN during tasks, which is thought to reflect increased self-focus and rumination; this is, however, still a subject of debate (Wang, Öngür, Auerbach, & Yao, 2016). Although there has been some variability among studies, a meta-analysis confirmed that all three models are well supported by the weight of evidence across experimental studies (Graham et al., 2013). Contemporary views increasingly emphasize the need to shift research focus from clinical diagnoses to dimensional approaches, in which units to be investigated (so called dimensions, domains, or constructs) are defined not by clinical diagnosis, but by current understanding of biological and behavioral underpinnings of emotion, cognition, motivation, and social behavior (e.g., Research Domain Criteria) (Insel, 2014). This shift is accompanied by a move from investigating single structures to looking at large-scale networks as neural substrates for domains and individual symptoms, and has been observed in MDD (Waters & Mayberg, 2017; Williams, 2016) and other psychiatric conditions.

Anxiety disorders such as generalized anxiety disorder (GAD), social anxiety disorder (SAD), and specific phobias share many similarities with MDD, such as inadequate control of overactive limbic/paralimbic regions (in particular the amygdala, insula, and hippocampus) by PFC (Bandelow et al., 2016). In regard to similarities with MDD, it is interesting that SSRIs are effective across conditions. Obsessive-compulsive disorder (OCD), despite a possibly different neurocircuitry, with the most prominent dysfunction affecting fronto-striato-thalamic circuit (Harrison et al., 2013), is also treated with SSRIs.

Functional findings in bipolar disorder (BD) provide neural substrates for its hallmark clinical symptoms: abnormalities in emotional regulation, cognition, and reward processing (Pan, Keener, Hassel, & Phillips, 2009; Phillips & Swartz, 2014). Reduced ventrolateral prefrontal cortex (vlPFC) activity, and its functional connectivity (FC) with the amygdala, is believed to result in inadequate top-down control, leading to abnormal processing and regulation of all emotions across mood states. Another group of findings, which may explain increased attention to positive stimuli in mania, is decreased positive effective connectivity (EC) between the orbitofrontal cortex (OFC) and amygdala, accompanied by increased amygdala, striatum, and medial PFC (mPFC) function, while processing happy facial expressions (Almeida et al., 2009; Phillips & Swartz, 2014; Surguladze et al., 2010). Performance of nonemotional cognitive tasks was linked to elevated function of the amygdala, OFC, and temporal cortex, suggesting increased attention to emotionally salient information, even in nonemotional contexts. Reward processing dysfunctions—another key BD characteristic—were associated with increased ventral striatum, vlPFC, and OFC function during reward processing across mood states (Phillips & Swartz, 2014). Manic and depressive states may also feature other abnormalities specific to those states (Brooks III & Vizueta, 2014; Strakowski et al., 2011).

Schizophrenia (SCH) was acknowledged early on as a disorder of network dysfunctions, with this view fitting well with the diversity of its symptomatic presentations. Disruptions in connectivity were observed within and between distributed and regional circuits, affecting most pathways in the brain, suggesting diffuse pathology. These abnormalities were shown across stages of the disorder, from the asymptomatic "at risk" period to chronic schizophrenia (Ruiz, Birbaumer, & Sitaram, 2013). Two phenomena have been consistently replicated: prevalence of structural and functional hypoconnectivity over hyperconnectivity, and the central role of PFC (Fornito, Zalesky, Pantelis, & Bullmore, 2012; Pettersson-Yeo, Allen, Benetti, McGuire, & Mechelli, 2011; Zhou, Fan, Qiu, & Jiang, 2015). Hypofrontality has long been hypothesized as a neural substrate of cognitive deficits and negative symptoms. Research confirmed frontal lobe dysfunction, especially during task performance, with patients failing to activate PFC to the level necessary to perform the task as well as healthy individuals. Altered PFC FC was related to symptom dimensions: frontotemporal hypoconnectivity to positive symptoms, in particular, auditory hallucinations (Hoffman & Hampson, 2011; Lawrie et al., 2002); frontostriatal hypoconnectivity, important in light of PFC control over dopamine-regulating regions (Howes & Kapur, 2009), to positive symptoms and cognitive deficits (Quide, Morris, Shepherd, Rowland, & Green, 2013; Yoon, Minzenberg, Raouf, D'Esposito, & Carter, 2013); PFC-striatum-thalamus loop disruption to negative symptoms, and difficulties in initiating mental activity (Semkovska, Bedard, & Stip, 2001); and disruption of the executive frontoparietal network to cognitive control, executive function, and memory deficits (Zhou et al., 2015). Hyperconnectivity and failure to deactivate DMN have been linked to impairments in self-referential processing, reality monitoring, deficits in theory of mind, and thought disturbances, as well as positive symptoms' severity and working memory impairments (Whitfield-Gabrieli et al., 2009).

2   Identification of individuals at high risk of developing a psychiatric condition

Identification of individuals at an increased risk of developing a mental illness later in life is of great importance, as it would allow selective introduction of therapeutic interventions before conversion to a fully symptomatic state, or even early prevention before any prodromal symptoms manifest, for example, by targeted psychological, lifestyle, or pharmacological interventions. At-risk status is defined by a history of mental health illness in the family, especially in first-degree relatives, due to sharing the same genes, or on the basis of the presence of early nonspecific symptoms, which often precede the onset of a fully symptomatic condition. For example, having a parent with depression increases the likelihood of depression by 40%, while subthreshold psychotic symptoms confer an increased risk of conversion to full psychosis (Kaymaz et al., 2012).

Some of the findings in those so-called ultra-high-risk (UHR), or at-risk mental states (ARMS), individuals, identified by functional neuroimaging, are similar to those present in diagnosable conditions. Changes in fMRI response, although possibly less pronounced, are often observed within the same structures, yet UHR individuals also differ in many aspects, including directions of functional changes and/or differences in structures involved. A hypothesis to explain those differences argues that asymptomatic UHR activates compensatory mechanisms that prevent the development of symptoms, and allows the maintenance of performance until the system stops coping with increased demands. Interestingly, a similar use of compensatory mechanisms has been suggested for remitted individuals at risk of a relapse (e.g., Norbury, Godlewska, & Cowen, 2014).

In the area of psychosis, the Personal Assessment and Crisis Evaluation (PACE) (Yung et al., 2007) criteria are a frequently used screening tool based on the presence of prodromal symptoms, such as attenuated or brief intermittent psychotic symptoms, and/or significant recent deterioration in global functioning in the context of high genetic risk. A meta-analysis of working memory n-back studies concluded that psychosis-UHR showed widespread dysfunctions, suggestive of generalized neural vulnerability, with four key dysfunctional cortical regions: the right inferior parietal lobule, the left medial frontal gyrus, the left superior temporal gyrus (STG), and the frontopolar cortex in the right superior frontal gyrus (Dutt et al., 2015). Other studies suggested a compensatory increase in the function of the ACC, occipito-parietal, inferior frontal, and parahippocampal regions (Fusar-Poli, Broome, Matthiasson, et al., 2011, Fusar-Poli, Broome, Woolley, et al., 2011).

Individuals at risk of bipolar disorder (BD-UHR) were characterized by functional abnormalities in the structures shown to be affected in fully symptomatic BD, including PFC, limbic regions, basal ganglia, and inferior frontal gyrus (IFG) (Arts, Jabben, Krabbendam, & van Os, 2008). These differences often mirrored those seen in BD, such as reduced prefrontal control of the amygdala during emotional processing, or impaired rsFC between PFC, limbic structures, and basal ganglia. Differences were also observed, likely representing compensatory mechanisms. For example, during cognitive tasks, BD-UHR showed hyperactivation of the vlPFC (Drapier et al., 2008).

Risk of MDD (MDD-UHR) has also been associated with changes in fMRI responses compared with low-risk participants. The findings include, for example, dysfunctions of DMN (Chai et al., 2016; Norbury, Mannie, & Cowen, 2011), diminished response to emotional faces expressing fear in dlPFC, with maintained normal amygdala response (Mannie, Taylor, Harmer, Cowen, & Norbury, 2011), no response to the emotional Stroop in pgACC (Mannie et al., 2008), or an increased activation in the hippocampus-related neural networks during a memory task (Mannie, Harmer, Cowen, & Norbury, 2014). As seen across disorders, compensatory processes also seem to occur in those at risk with maintained performance (Mannie, Harmer, Cowen, & Norbury, 2010).

Together, these results suggest it may be possible to use fMRI to explore mechanisms underlying risk and resilience to psychiatric disorders. However, whether they would show sufficient sensitivity and specificity to predict individual trajectory, above other methods such as subsymdromal symptoms, remains to be tested.

3 Differential diagnosis

Assistance in the diagnostic process would undoubtedly be one of the most important applications of biomarkers in everyday clinical practice. As psychiatric diagnoses are made practically solely on the basis of clinical assessments, they are prone to errors (Birur, Kraguljac, Shelton, & Lahti, 2017). Diagnostic mistakes can be made, for example, when two disorders have similar clinical presentations (e.g., bipolar and unipolar depression), or when symptoms pointing at a particular diagnosis have not manifested yet (e.g., BD before the onset of the first hypomanic or manic episode). Indeed, misdiagnosing bipolar depression as unipolar depression (MDD) is not an uncommon clinical scenario, especially in cases where depressive episodes are present ahead of the first manic or hypomanic episode, or hypomanic symptoms are missed or not reported (Perlis et al., 2004). About 60% of bipolar depressed patients are initially diagnosed as MDD, and the diagnosis of BD is, on average, delayed by 10 years (Hirschfeld, Lewis, & Vornik, 2003). This often leads to applying treatments that, at best, are ineffective, and at worst, have iatrogenic effects, such as precipitation of manic or hypomanic states by antidepressants. The role for functional neuroimaging in differentiating between unipolar and bipolar depression has been highlighted, along with an increase in the number of studies directly comparing these depressive subtypes. However, this number is still limited, and emerging patterns need to be interpreted carefully due to a number of inconsistencies. Some more consistent findings suggest that compared with MDD, bipolar I depressed patients may show elevated amygdala response to happy and ambiguous sad facial expressions, and reduced bottom-up amygdala-vmPFC effective connectivity in response to overt happy facial expressions. MDD patients, as compared with bipolar I depressed individuals, may exhibit elevated amygdala response to negative emotional faces, and decreased response to positive ones (Brooks & Vizueta, 2014; Grotegerd et al., 2014). Dysfunctional response of vlPFC circuitry to negative stimuli, and activation of attention network in the presence of ambiguous, neutral faces, was only shown in MDD. Sensitivity of response to types of emotions, their ambiguity, and presentation as supra- or subliminal, suggests that differences in paradigms may be at least partly responsible for inconsistencies. The two types of depression were also differentiated by large-scale network alterations, with depressed bipolar individuals exhibiting increased FC in the frontoparietal network (externally-oriented executive network), while MDD was characterized by an increased FC in introspective and self-referential DMN and reduced FC to DMN from cingulo-opercular networks, detecting the need to switch between internal and external demands (Goya-Maldonado et al., 2015). A failure to deactivate DMN during tasks was greater in BD (Rodríguez-Cano et al., 2017).

There is a paucity of, and a great need for, studies including depressed bipolar II individuals as a separate group (Marchand, Lee, Johnson, Gale, & Thatcher, 2013; Rastelli, Cheng, Weingarden, Frank, & Swartz, 2013); it would not be correct to assume, without further investigations, that findings in BDI can be extrapolated to BDII, while the diagnostic process can be extremely challenging in this very group, and could strongly benefit from objective biomarkers.

Functional neuroimaging can also be helpful in exploring a much debated concept of the continuum of psychoses, suggested by overlapping presentations, genetic findings, and effective treatments for different disorders with psychotic features (DeRosse & Karlsgodt, 2015; Lawrie, Hall, McIntosh, Owens, & Johnstone, 2010). Task-based fMRI studies seem to suggest a gradual impairment in function of overlapping brain networks, from healthy individuals through BD to SCH in cognitive tasks, and from HC, through SCH to BD in emotional tasks (Birur et al., 2017). MPFC, a part of DMN, was proposed as the key dysfunctional hub shared by acutely ill patients with SCH and BD (Kuhn & Gallinat, 2013; Ongur et al., 2010). Global brain connectivity measures (connectivity strength measured for every region of the brain with every other region of the brain) showed the connectivity strength to be the lowest in SCH, intermediate in BD, and the strongest in HC (Argyelan et al., 2014). The ongoing debate over whether schizoaffective disorder should be considered as a separate diagnostic entity, or as SCH with clear mood symptoms, fits this discussion (Evans et al., 1999). A systematic review concluded that schizoaffective disorder resembles SCH more closely than BD, and suggested it to be either a subtype or SCH, or lie on the continuum spectrum, closer to SCH than BD (Madre et al., 2016). However, schizoaffective patients were rarely studied as a separate group, and this question is still open.

A promising new approach is machine learning, in which algorithms are trained on large pools of participants to recognize patterns of neural activity corresponding to particular clinical events (Orrù, Pettersson-Yeo, Marquand, Sartori, & Mechelli, 2012). Such patterns can serve as templates that can be applied to an individual patient's data in order to guide diagnosis or treatment. SVC algorithms trained on functional neuroimaging data were shown to differentiate between patients with clear clinical pictures of bipolar depression and MDD, and among SCH, BD, and HC with accuracy exceeding 90% (Fu & Costafreda, 2013). In the case of less clear clinical presentations, where such tools would be especially valuable, the classification was less accurate, for example, 79% in cases of euthymic BD patients (Fu & Costafreda, 2013). This shows the potential of machine learning algorithms, but also the need for further work, including identification of the functional paradigms and ways of training algorithms to create a useful tool for ambiguous clinical scenarios.

4 Treatment choice and mechanisms of drug action

Functional neuroimaging is an excellent tool for exploration of mechanisms of action of medications and psychological therapies, allowing observation of neural changes over treatment. Despite the wide use of treatments, little is known about mechanisms behind their efficacy. Such knowledge would increase understanding of mechanisms underlying symptoms formation, and inform drug development. Functional neuroimaging can also help in identification of treatment response biomarkers. Currently, treatments are applied on a "trial and error" basis, and often, many attempts are needed to find a successful one, while some patients do not respond at all (Warden, Rush, Trivedi, & Fava, 2007). Delays in response translate into an increased burden of the disorder at an individual and societal level. The ultimate goal in this line of research is identification of biomarkers that will clearly indicate which treatment should be used in an individual patient to achieve symptomatic response in the shortest possible time. Such predictive biomarkers may represent general response measures, or markers of response to individual treatments. Although putative biomarkers are not yet ready to be applied in clinical settings, functional neuroimaging has already made initial progress in this area.

The largest number of studies was conducted in MDD. It was suggested that, in general terms, CBT works mainly by increasing activity in executive control and dorsal attention networks, while pharmacotherapies, and CBT to a lesser extent, work by reducing overreactivity of ventral cortico-limbic structures, a mechanism shared by MDD and anxiety disorders (DeRubeis, Siegle, & Hollon, 2008). Pregenual ACC (pgACC), which has been proposed as the functional hub between those dorsal and ventral networks, has been identified as a general predictor of remission to both pharmacological and psychological treatments. Increased pretreatment pgACC activity was linked to better treatment response across treatments, as well as imaging and analysis methods (Fu, Steiner, & Costafreda, 2013; Pizzagalli, 2011). The other key structure, the amygdala, has been consistently implicated in mechanisms of treatment efficacy, while the findings on its role in response prediction were inconsistent (Fu et al., 2013). PgACC pretreatment function seems to be a good general predictor of response to all antidepressant treatments, regardless of their nature.

Clinically, however, more specific markers of response are needed. This topic is being increasingly researched, with some promising findings. Baseline right anterior insula metabolism was shown to discriminate between responders to CBT and the selective serotonin reuptake inhibitor (SSRI) escitalopram (McGrath et al., 2013). Dunlop, Rajendra, Craighead, & Mayberg, 2017 showed a role for rsFC between sgACC and the dorsal midbrain, frontal operculum, and vmPFC in differentiating between responders to medication and responders to CBT. General classification of patients into potential pharmacotherapy or CBT responders would already be clinically useful. Current research, however, also aims to predict response to different antidepressant drugs. A large study compared SSRIs, escitalopram and sertraline, and an SNRI, venlafaxine, in patients not distinguishable on the basis of their clinical features. It showed that pretreatment amygdala hyporeactivity to subliminal happiness and threat was a general predictor of good response (75% accuracy), while pretreatment hyperreactivity to subliminal sadness was a predictor of poor response to venlafaxine (81% accuracy) (Williams et al., 2015). In the same sample, pretreatment dlPFC activation in a cognitive task was a general predictor of remission, while pretreatment inferior parietal cortex activation allowed differentiating between SSRI and SNRI responders (Gyurak et al., 2016). Although these findings need to be replicated, they clearly show the potential of functional neuroimaging in clarifying mechanisms of action of different therapies, and in identifying biomarkers of response. Another study shows that additional factors may play an important role: higher pretreatment dlPFC activity was linked to clinical improvement, but only in people with no history of childhood abuse, while such history seemed to abolish this effect (Miller et al., 2015).

The most commonly applied paradigm consisted of testing brain function before and after a few weeks of treatment. Although this approach allows exploring mechanisms of change, it doesn't explain whether neural changes preceded mood improvement, or vice versa. A recent neurocognitive model suggests that the crucial neural shift and bias normalization happen early on in treatment, and mood improvement follows (Pringle, Browning, Cowen, & Harmer, 2011). This hypothesis was supported by studies showing that in subsequent responders to escitalopram, a decrease in neural response to fearful versus happy facial expressions across the network of structures relevant to MDD, including the amygdala, occurred already after 7 days of treatment, in the absence of mood improvement (Godlewska, Browning, Norbury, Cowen, & Harmer, 2016; Godlewska, Norbury, Cowen, & Harmer, 2012).

In anxiety disorders, in line with the mechanisms of drug efficacy described in the preceding paragraphs, ACC-amygdala function was shown to be a treatment target, and a predictor of treatment outcome across anxiety diagnoses (Bandelow et al., 2016), with reduced pretreatment amygdala function being predictive of good response to CBT in PTSD, and venlafaxine in GAD, and increased mPFC function being associated with good response to BT/CBT in OCD and PTSD, and venlafaxine in GAD. Reduced pretreatment activity in OFC, and greater activity in the caudate and posterior cingulate cortex was associated with better response to SSRIs, and neurosurgery in PTSD and GAD.

Conditions other than MDD have been much less researched. In the absence of replicated findings, up to date research is more a "proof-of-concept" of functional neuroimaging feasibility in exploration of treatment-related issues in these conditions.

In BD, the majority of studies focused on children and adolescents. While this is an important population to study, it remains to be seen if the findings are generalizable to populations with later onset. The findings include an increase in insula and sgACC response to emotional stimuli after, respectively, risperidone and divalproex treatment (Pavuluri, Passarotti, Fitzgerald, Wegbreit, & Sweeney, 2012), a negative correlation between pretreatment function of BA 47 with improvement in manic symptoms during ziprasidone treatment (Schneider et al., 2012), or a negative correlation between change in amygdala activation to negative emotional stimuli and improvement in depressive scores over lamotrigine treatment (Chang, Wagner, Garrett, Howe, & Reiss, 2008). In a pediatric sample it was shown that normalization of increased dlPFC function in response to an emotional words task was already present after 16 weeks of pharmacotherapy, while in amygdala, ACC, vlPFC and ventral striatum, this effect was seen only after 3 years (Yang et al., 2013). This observation suggests that some findings can be missed due to an insufficient length of follow-up. Studies in adults showed a potential of psychological treatments—mindfulness-based cognitive therapy (Ives-Deliperi, Howells, Stein, Meintjes, & Horn, 2013) and psychoeducation (Favre et al., 2013)—to increase PFC activation in mildly depressed or remitted bipolar patients. Recent studies showed that better response to lithium therapy was predicted by higher baseline FC between the temporal pole and amygdala, and by lower baseline binding potential of 5HT1A in the raphe (Spuhler et al., 2017), and was linked to subcortical and amygdala function (Strakowski et al., 2011).

Despite a large diversity of findings on neural effects of antipsychotic drugs (APD) in SCH, two consistent themes emerge: the key role of frontal regions for treatment effect, and an association of symptoms reduction with normalization, or an increase in PFC function (Kani, Shinn, Lewandowski, & Ongür, 2017). This was observed across a number of PFC areas, including dlPFC, vlPFC, and ACC, and mirrored by FC studies, which showed APD-induced changes in FC between frontal regions and other areas in the brain. APD were also shown to influence FC from subcortical structures to other brain regions (Kani et al., 2017).

In summary, functional neuroimaging has shown to be a valuable tool to explore neurocognitive mechanisms underpinning different mental health conditions. The challenge remains to translate this approach to the clinic. This may involve support for diagnosis between similar clinical presentations, the prediction of treatment response and prognosis, as well as for the development and screening of novel treatments. To support the application of mechanisms in general to an individual requires large scale, and replication, as well as increased focus on the pathophysiology of core processes across disorders.

5 General remarks

Functional neuroimaging, allowing observation of brain processes in vivo, has greatly advanced the understanding of neural mechanisms underlying diverse phenomena related to psychiatric disorders. The groundwork for exploring how this knowledge can be translated into clinical applications has been laid; however, no functional neuroimaging biomarkers are yet ready to be used in everyday psychiatric practice. Before this happens, the findings need to be consistently replicated, including multiple population samples representing "real life" patients from different socio-economic backgrounds, and affected by common comorbidities (Pepe, Feng, Janes, Bossuyt, & Potter, 2008). Only then can standardized clinical assays be developed. Most putative biomarkers have not achieved high enough accuracy and consistency to warrant their reliable practical use.

Difficulties identifying robust biomarkers are linked to the complex nature of psychiatric disorders. An important issue is heterogeneity of the studied groups. Classification of mental health disorders is still based on the clinical picture, and traditional diagnostic labels are widely used, even if it is increasingly accepted that traditional diagnoses label individuals whose symptoms result from different underlying pathologies. This may, in turn, lead to blurring of the findings. It has been increasingly acknowledged that a dimensional approach should be used instead (McArthur, 2017), with a few studies already exploring domains of symptoms rather than traditional diagnoses. A large study (1188 functional datasets), using machine learning algorithms, classified depressed patients into four subtypes on the basis of dysfunctions in FC between frontostriatal and limbic networks (Drysdale et al., 2017). These subtypes were associated with different symptomatic profiles and response to transcranial magnetic stimulation. To tackle this issue, alternative sets of criteria have been proposed. Among the most popular are Research Domain Criteria (RDoC) (Insel, 2014), which aim at integrating multiple—genetic, molecular, cellular, and neural—measures to create a more comprehensive understanding of human functioning in health and illness. The focus has been shifted from a diagnostic label to one of the five main domains that span across diagnostic entities, and can be assessed with laboratory, imaging, and behavioral tests.

Homogeneity of the samples in terms of underlying pathologies is crucial for machine learning approaches. The algorithms, to be successfully applied, first need to learn from samples in which abnormalities are clear, and correspond to clinical scenarios of interest, so that they can be used as a template against which events in question are checked and correctly classified. An erroneous inclusion of subjects will disturb the learning process, and result in less accurate predictions (Fu & Costafreda, 2013).

Another aspect to consider is differences in study designs. Paradigms vary widely across the studies, while it was shown that the type of the task used and the way it is presented may affect the outcome. For example, some tasks will better activate certain brain structures than others, so the lack of response may not represent a real phenomenon, but simply reflect the use of the task not fit for the purpose. Faces expressing one emotion may better activate, for example, amygdala, than those expressing other emotions, and the level of this activation—or its lack—may vary depending on whether emotions are presented sub- or supraliminally. Studies also vary in their technical aspects, including the scanner and imaging sequences used, and the way data is analyzed. Although robust differences should survive such differences, changes in mental health disorders are often quite subtle, hence such variability may lead to the lack of consistency. For these reasons, interpretation of pooled studies using different paradigms needs to be done with caution. Modest numbers of participants in single studies; at best, tens of patients, mean that studies are often underpowered, and pose a question about generalizability of findings to wider populations.

An important question relates to the nature of abnormalities, that is, whether they are expressions of a state and will change alongside symptomatic changes, or whether they represent a constant feature, that is, a trait. Longitudinal studies following individuals from the UHR stage through symptomatic states to remission could address this question, unfortunately, they are extremely difficult to conduct, and hence, rarely performed, and usually underpowered. Caution is needed when interpreting findings of studies including individuals in different "states," for example, manic and depressed BD patients, or medicated and unmedicated subjects. The influence of medication is another important issue as pharmacotherapy may—and was shown to—affect brain function. At the same time, it has been suggested that those patients with severe conditions, such as schizophrenia or BD who remain well without treatment, are not likely to be representative of the wider population of patients (Brooks & Vizueta, 2014).

Future research has the potential to tackle these issues and lead to a step change in diagnosis, treatment, and treatment development. Although it is not yet clear which measures, or combinations, are the most reliable tools, the speed of advances in technology and fast accumulating knowledge hold a strong promise of clinically applicable findings in the not-too-distant-future.

References

Almeida, J. R., Versace, A., Mechelli, A., Hassel, S., Quevedo, K., Kupfer, D. J., & Phillips, M. L. (2009). Abnormal amygdala-prefrontal effective connectivity to happy faces differentiates bipolar from major depression. *Biological Psychiatry*, *66*, 451–459.

Argyelan, M. , et al. (2014). Resting-state fMRI connectivity impairment in schizophrenia and bipolar disorder. *Schizophrenia Bulletin*, *40*, 100–110.

Arts, B., Jabben, N., Krabbendam, L., & van Os, J. (2008). Meta-analyses of cognitive functioning in euthymic bipolar patients and their first-degree relatives. *Psychological Medicine*, *38*, 771–785.

Bandelow, B., Baldwin, D., Abelli, M., Altamura, C., Dell'Osso, B., Domschke, K., … Riederer, P. (2016). Biological markers for anxiety disorders, OCD and PTSD—A consensus statement. Part I: Neuroimaging and genetics. *The World Journal of Biological Psychiatry*, *17*, 321–365.

Biomarkers Definitions Working Group. (2001). Biomarkers and surrogate endpoints: Preferred definitions and conceptual framework. *Clinical Pharmacology and Therapeutics*, *69*, 89–95.

Birur, B., Kraguljac, N. V., Shelton, R. C., & Lahti, A. C. (2017). Brain structure, function, and neurochemistry in schizophrenia and bipolar disorder—A systematic review of the magnetic resonance neuroimaging literature. *NPJ Schizophrenia*, *3*, 15.

Brooks, J. O., III, & Vizueta, N. (2014). Diagnostic and clinical implications of functional neuroimaging in bipolar disorder. *Journal of Psychiatric Research*, *57*, 12–25.

Chai, X. J., Hirshfeld-Becker, D., Biederman, J., Uchida, M., Doehrmann, O., Leonard, J. A., … Whithfield-Gabrieli, S. (2016). Altered intrinsic brain architecture in children at familial risk of major depression. *Biological Psychiatry*, *80*, 849–858.

Chang, K. D., Wagner, C., Garrett, A., Howe, M., & Reiss, A. (2008). A preliminary functional magnetic resonance imaging study of prefrontal-amygdalar activation changes in adolescents with bipolar depression treated with lamotrigine. *Bipolar Disorders*, *10*, 426–431.

DeRosse, P., & Karlsgodt, K. H. (2015). Examining the psychosis continuum. *Current Behavioral Neuroscience Reports*, *2*, 80–89.

DeRubeis, R. J., Siegle, G. J., & Hollon, S. D. (2008). Cognitive therapy versus medication for depression: Treatment outcomes and neural mechanisms. *Nature Reviews. Neuroscience*, *9*, 788–796.

Drapier, D., Surguladze, S., Marshall, N., Schulze, K., Fern, A., Hall, M. H., … McDonald, C. (2008). Genetic liability for bipolar disorder is characterized by excess frontal activation in response to a working memory task. *Biological Psychiatry*, *64*, 513–520.

Drevets, W. C. (2001). Neuroimaging and neuropathological studies of depression: Implications for the cognitive–emotional features of mood disorders. *Current Opinion in Neurobiology*, *11*, 240–249.

Drysdale, A. T., Grosenick, L., Downar, J., Dunlop, K., Mansouri, F., Meng, Y., … Liston, C. (2017). Resting-state connectivity biomarkers define neurophysiological subtypes of depression. *Nature Medicine*, *23*, 28–38.

Dunlop, B. W., Rajendra, J. K., Craighead, W. E., & Mayberg, H. S. (2017). Functional connectivity of the subcallosal cingulate cortex and differential outcomes to treatment with cognitive-behavioral therapy or antidepressant medication for major depressive disorder. *The American Journal of Psychiatry*, *174*, 533–545.

Dutt, A., Tseng, H. H., Fonville, L., Drakesmith, M., Su, L., Evans, J., … David, A. S. (2015). Exploring neural dysfunction in 'clinical high risk' for psychosis: A quantitative review of fMRI studies. *Journal of Psychiatric Research*, *61*, 122–134.

Evans, J. D., Heaton, R. K., Paulsen, J. S., McAdams, L. A., Heaton, S. C., & Jeste, D. V. (1999). Schizoaffective disorder: A form of schizophrenia or affective disorder? *Journal of Clinical Psychiatry*, *60*, 874–882.

Favre, P., Baciu, M., Pichat, C., De Pourtales, M. A., Fredembach, B., Garcon, S., … Polosan, M. (2013). Modulation of fronto-limbic activity by the psychoeducation in euthymic bipolar patients. A functional MRI study. *Psychiatry Research*, *214*, 285–295.

Fettes, P., Schulze, L., & Downar, J. (2017). Cortico-striatal-thalamic loop circuits of the orbitofrontal cortex: Promising therapeutic targets in psychiatric illness. *Frontiers in Systems Neuroscience*, *27*, 25.

Fornito, A., Zalesky, A., Pantelis, C., & Bullmore, E. T. (2012). Schizophrenia, neuroimaging and connectomics. *NeuroImage*, *62*, 2296–2314.

Fu, C. H. Y., & Costafreda, S. G. (2013). Neuroimaging-based biomarkers in psychiatry: Clinical opportunities of a paradigm shift. *Canadian Journal of Psychiatry*, *58*, 499–508.

Fu, C. H., Steiner, H., & Costafreda, S. G. (2013). Predictive neural biomarkers of clinical response in depression: A meta-analysis of functional and structural neuroimaging studies of pharmacological and psychological therapies. *Neurobiology of Disease*, *52*, 75–83.

Fusar-Poli, P., Broome, M. R., Matthiasson, P., Woolley, J. B., Mechelli, A., Johns, L. C., … McGuire, P. (2011). Prefrontal function at presentation directly related to clinical outcome in people at ultrahigh risk of psychosis. *Schizophrenia Bulletin*, *37*, 189–198.

Fusar-Poli, P., Broome, M. R., Woolley, J. B., Johns, L. C., Tabraham, P., Bramon, E., … McGuire, P. (2011). Altered brain function directly related to structural abnormalities in people at ultra high risk of psychosis: Longitudinal VBM-fMRI study. *Journal of Psychiatric Research*, *45*, 190–198.

Godlewska, B. R., Browning, M., Norbury, R., Cowen, P. J., & Harmer, C. J. (2016). Early changes in emotional processing as a marker of clinical response to SSRI treatment in depression. *Translational Psychiatry*, *6*, e957.

Godlewska, B. R., Norbury, R., Cowen, P. J., & Harmer, C. J. (2012). Short-term SSRI treatment normalizes amygdala hyperactivity in depressed patients. *Psychological Medicine*, *41*, 2609–2617.

Goya-Maldonado, R., Brodmann, K., Keil, M., Trost, S., Dechent, P., & Gruber, O. (2015). Differentiating unipolar and bipolar depression by alterations in large-scale brain networks. *Human Brain Mapping*, *37*, 808–818.

Graham, J., Salimi-Khorshidi, G., Hagan, C., Walsh, N., Goodyer, I., Lennox, B., & Suckling, J. (2013). Meta-analytic evidence for neuroimaging models of depression: State or trait? *Journal of Affective Disorders*, *151*, 423–431.

Grotegerd, D., Stuhrmann, A., Kugel, H., Schmidt, S., Redlich, R., Zwanzger, P., … Dannlowski, U. (2014). Amygdala excitability to subliminally presented emotional faces distinguishes unipolar and bipolar depression: An fMRI and pattern classification study. *Human Brain Mapping*, *35*, 2995–3007.

Gyurak, A., Patenaude, B., Korgaonkar, M. S., Grieve, S. M., Williams, L. M., & Etkin, A. (2016). Frontoparietal activation during response inhibition predicts remission to antidepressants in patients with major depression. *Biological Psychiatry, 79*, 274–281.

Harrison, B. J., Pujol, J., Cardoner, N., Deus, J., Alonso, P., Lopez-Sola, M., Contreras-Rodriguez, O., Real, E., Segalas, C., Blanco-Hinojo, L. , et al. (2013). Brain corticostriatal systems and the major clinical symptom dimensions of obsessive-compulsive disorder. *Biological Psychiatry, 73*, 321–328.

Hirschfeld, R. M. A., Lewis, L., & Vornik, L. A. (2003). Perceptions and impact of bipolar disorder: How far have we really come? Results of the national depressive and manic depressive association: 2000 survey of individuals with bipolar disorder. *The Journal of Clinical Psychiatry, 64*, 161–174.

Hoffman, R. E., & Hampson, M. (2011). Functional connectivity studies of patients with auditory verbal hallucinations. *Frontiers in Human Neuroscience, 6*, 6.

Howes, O. D., & Kapur, S. (2009). The dopamine hypothesis of schizophrenia: Version III—The final common pathway. *Schizophrenia Bulletin, 35*, 549–562.

Insel, T. R. (2014). The NIMH research domain criteria (RDoC) project: Precision medicine for psychiatry. *The American Journal of Psychiatry, 171*, 395–397.

Ives-Deliperi, V. L., Howells, F., Stein, D. J., Meintjes, E. M., & Horn, N. (2013). The effects of mindfulness-based cognitive therapy in patients with bipolar disorder: A controlled functional MRI investigation. *Journal of Affective Disorders, 150*, 1152–1157.

Kani, A. S., Shinn, A. K., Lewandowski, K. E., & Ongür, D. (2017). Converging effects of diverse treatment modalities on frontal cortex in schizophrenia: A review of longitudinal functional magnetic resonance imaging studies. *Journal of Psychiatric Research, 84*, 256–276.

Kaymaz, N., Drukker, M., Lieb, R., Wittchen, H. U., Werbeloff, N., Weiser, M., … van Os, J. (2012). Do subthreshold psychotic experiences predict clinical outcomes in unselected non-help-seeking population-based samples? A systematic review and meta-analysis, enriched with new results. *Psychological Medicine, 42*, 2239–2253.

Kuhn, S., & Gallinat, J. (2013). Resting-state brain activity in schizophrenia and major depression: A quantitative meta-analysis. *Schizophrenia Bulletin, 39*, 358–365.

Lawrie, S. M., Buechel, C., Whalley, H. C., Frith, C. D., Friston, K. J., & Johnstone, E. C. (2002). Reduced frontotemporal functional connectivity in schizophrenia associated with auditory hallucinations. *Biological Psychiatry, 51*, 1008–1011.

Lawrie, S., Hall, J., McIntosh, A. M., Owens, D. G. C., & Johnstone, E. C. (2010). The 'continuum of psychosis': Scientifically unproven and clinically impractical. *British Journal of Psychiatry, 197*, 423–425.

Madre, M., Canales-Rodriguez, E. J., Ortiz-Gil, J., Murru, A., Torrent, C., Bramon, E., … Amann, B. L. (2016). Neuropsychological and neuroimaging underpinnings of schizoaffective disorder: A systematic review. *Acta Psychiatrica Scandinavica, 134*, 16–30.

Mannie, Z., Harmer, C. J., Cowen, P. J., & Norbury, R. (2010). A functional magnetic resonance imaging study of verbal working memory in young people at increased familial risk of depression. *Biological Psychiatry, 67*, 471–477.

Mannie, Z., Harmer, C. J., Cowen, P. J., & Norbury, R. (2014). Structural and functional imaging of the hippocampus in young people at familial risk of depression. *Psychological Medicine, 44*, 2939–2948.

Mannie, Z., Norbury, R., Murphy, S. E., Inkster, B., Harmer, C. J., & Cowen, P. J. (2008). Affective modulation of anterior cingulate cortex in young people at increased familial risk of depression. *The British Journal of Psychiatry, 192*, 356–361.

Mannie, Z., Taylor, M. J., Harmer, C. J., Cowen, P. J., & Norbury, R. (2011). Frontolimbic responses to emotional faces in young people at familial risk of depression. *Journal of Affective Disorders, 130*, 127–132.

Marchand, W. R., Lee, J. N., Johnson, S., Gale, P., & Thatcher, J. (2013). Differences in functional connectivity in major depression versus bipolar II depression. *Journal of Affective Disorders, 150*, 527–532.

Mayberg, H. S. (1997). Limbic-cortical dysregulation: A proposed model of depression. *Journal of Neuropsychiatry & Clinical Neurosciences, 9*, 471–481.

Mayberg, H. S. (2003). Modulating dysfunctional limbic-cortical circuits in depression: Towards development of brain-based algorithms for diagnosis and optimised treatment. *British Medical Bulletin, 65*, 193–207.

McArthur, R. A. (2017). Aligning physiology with psychology: Translational neuroscience in neuropsychiatric drug discovery. *Neuroscience & Biobehavioral Reviews, 76*, 4–21.

McGrath, C. L., Kelley, M. E., Holtzheimer, P. E., III, Dunlop, B. W., Craighead, W. D., Franco, A. R., … Mayberg, H. (2013). Toward a neuroimaging treatment selection biomarker for major depressive disorder. *JAMA Psychiatry, 70*, 821–829.

Miller, S., McTeague, L. M., Gyurak, A., Patenaude, B., Williams, L. M., Grieve, S. M., … Etkin, A. (2015). Cognition-childhood maltreatment interactions in the prediction of antidepressant outcomes in major depressive disorder patients: Results from the iSPOT-D trial. *Depression and Anxiety, 32*, 594–604.

Miskowiak, K. W., Glerup, L., Vestbo, C., Harmer, C. J., Reinecke, A., Macoveanu, J., … Vinberg, M. (2015). Different neural and cognitive response to emotional faces in healthy monozygotic twins at risk of depression. *Psychological Medicine, 45*, 1447–1458.

Norbury, R., Godlewska, B., & Cowen, P. J. (2014). When less is more: A functional magnetic resonance imaging study of verbal working memory in remitted depressed patients. *Psychological Medicine, 44*, 1197–1203.

Norbury, R., Mannie, Z., & Cowen, P. J. (2011). Imaging vulnerability for depression. *Molecular Psychiatry, 16*, 1067–1068.

Ongur, D., Lundy, M., Greenhouse, I., Shinn, A. K., Menon, V., Cohen, B. M., & Renshaw, P. F. (2010). Default mode network abnormalities in bipolar disorder and schizophrenia. *Psychiatry Research, 183*, 59–68.

Orrù, G., Pettersson-Yeo, W., Marquand, A. F., Sartori, G., & Mechelli, A. (2012). Using support vector machine to identify imaging biomarkers of neurological and psychiatric disease: A critical review. *Neuroscience and Biobehavioral Reviews, 36*, 1140–1152.

Pan, L., Keener, M. T., Hassel, S., & Phillips, M. L. (2009). Functional neuroimaging studies of bipolar disorder: Examining the wide clinical spectrum in the search for disease endophenotypes. *International Review of Psychiatry, 21*, 368–379.

Pavuluri, M. N., Passarotti, A. M., Fitzgerald, J. M., Wegbreit, E., & Sweeney, J. A. (2012). Risperidone and divalproex differentially engage the fronto-striato-temporal circuitry in pediatric mania: A pharmacological functional magnetic resonance imaging study. *Journal of the American Academy of Child and Adolescent Psychiatry, 51*, 157–170.

Pepe, M. S., Feng, Z., Janes, H., Bossuyt, P. M., & Potter, J. D. (2008). Pivotal evaluation of the accuracy of a biomarker used for classification or prediction: Standards for study design. *Journal of the National Cancer Institute, 100*, 1432–1438.

Perlis, R. H., Miyahara, S., Marangell, L. B., Wisniewski, S. R., Ostacher, M., DelBello, M. P., … Nierenberg, A. (2004). Long-term implications of early onset in bipolar disorder: Data from the first 1000 participants in the systematic treatment enhancement program for bipolar disorder (STEP-BD). *Biological Psychiatry, 55*, 875–881.

Peters, S. K., Dunlop, K., & Downar, J. (2016). Cortico-striatal-thalamic loop circuits of the salience network: A central pathway in psychiatric disease and treatment. *Frontiers in Systems Neuroscience, 27*, 104.

Pettersson-Yeo, W., Allen, P., Benetti, S., McGuire, P., & Mechelli, A. (2011). Dysconnectivity in schizophrenia: Where are we now? *Neuroscience & Biobehavioral Reviews, 35*, 1110–1124.

Phillips, M. L., & Swartz, H. A. (2014). A critical appraisal of neuroimaging studies of bipolar disorder: Toward a new conceptualization of underlying neural circuitry and a road map for future research. *The American Journal of Psychiatry, 171*, 829–843.

Pizzagalli, D. A. (2011). Frontocingulate dysfunction in depression: Toward biomarkers of treatment response. *Neuropsychopharmacology, 36*, 183–206.

Pringle, A., Browning, M., Cowen, P. J., & Harmer, C. J. (2011). A cognitive neuropsychological model of antidepressant drug action. *Progress in Neuro-Psychopharmacology & Biological Psychiatry, 35*, 1586–1592.

Quide, Y., Morris, R. W., Shepherd, A. M., Rowland, J. E., & Green, M. J. (2013). Task-related fronto-striatal functional connectivity during working memory performance in schizophrenia. *Schizophrenia Research, 150*, 468–475.

Rastelli, C. P., Cheng, Y., Weingarden, J., Frank, E., & Swartz, H. A. (2013). Differences between unipolar depression and bipolar II depression in women. *Journal of Affective Disorders, 150*, 1120–1124.

Rodríguez-Cano, E., Alonso-Lana, S., Sarró, S., Fernández-Corcuera, P., Goikolea, J. M., Vieta, E., … Pomarol-Clotet, E. (2017). Differential failure to deactivate the default mode network in unipolar and bipolar depression. *Bipolar Disorders, 19*, 386–395.

Ruiz, S., Birbaumer, N., & Sitaram, R. (2013). Abnormal neural connectivity in schizophrenia and fMRI-brain-computer interface as a potential therapeutic approach. *Frontiers in Psychiatry, 4*, 17.

Schneider, M. R., Adler, C. M., Whitsel, R., Weber, W., Mills, N. P., Bitter, S. M., … DelBello, M. P. (2012). The effects of ziprasidone on prefrontal and amygdalar activation in manic youth with bipolar disorder. *Israel Journal of Psychiatry and Related Sciences, 49*, 112–120.

Semkovska, M., Bedard, M. A., & Stip, E. (2001). Hypofrontality and negative symptoms in schizophrenia: Synthesis of anatomic and neuropsychological knowledge and ecological perspectives. *Encephale, 27*, 405–415.

Spuhler, K., Huang, C., Ananth, M., Bartlett, E., Ding, J., He, X., … Parsey, R. (2017). Using PET and MRI to assess pretreatment markers of lithium treatment responsiveness in bipolar depression. *Journal of Nuclear Medicine, 58*, 138.

Strakowski, S. M., Eliassen, J. C., Lamy, M., Cerullo, M. A., Allendorfer, J. B., Madore, M., … Adler, C. M. (2011). Functional magnetic resonance imaging brain activation in bipolar mania: Evidence for disruption of the ventrolateral prefrontal-amygdala emotional pathway. *Biological Psychiatry, 69*, 381–388.

Surguladze, S. A., Marshall, N., Schulze, K., Hall, M. H., Walshe, M., Bramon, E., … McDonald, C. (2010). Exaggerated neural response to emotional faces in patients with bipolar disorder and their first-degree relatives. *NeuroImage, 53*, 58–64.

Trivedi, M. H., McGrath, P. J., Fava, M., Parsey, R. V., Kurian, B. T., Phillips, M. L., … Weissman, M. M. (2016). Establishing moderators and bio-signatures of antidepressant response in clinical care (EMBARC): Rationale and design. *Journal of Psychiatric Research, 78*, 11–23.

Wang, X., Öngür, D., Auerbach, R. P., & Yao, S. (2016). Cognitive vulnerability to major depression: View from the intrinsic network and cross-network interactions. *Harvard Review of Psychiatry, 24*, 188–201.

Warden, D., Rush, A. J., Trivedi, M. H., Fava, M., & Wisniewski, S. R. (2007). The STAR*D Project results: A comprehensive review of findings. *Current Psychiatry Reports, 9*, 449–459.

Waters, A. C., & Mayberg, H. S. (2017). Brain-based biomarkers for the treatment of depression: Evolution of an idea. *Journal of the International Neuropsychological Society, 23*, 870–880.

Whitfield-Gabrieli, S., & Ford, J. M. (2015). Default mode network activity and connectivity in psychopathology. *Annual Review of Clinical Psychology, 8*, 49–76.

Whitfield-Gabrieli, S., Thermenos, H. W., Milanovic, S., Tsuang, M. T., Faraone, S. V., McCarley, R. W., … Seidman, L. J. (2009). Hyperactivity and hyperconnectivity of the default network in schizophrenia and in first-degree relatives of persons with schizophrenia. *Proceedings of the National Academy of Sciences of the United States of America, 106*, 1279–1284.

Williams, L. M. (2016). Defining biotypes for depression and anxiety based on large-scale circuit dysfunction: A theoretical review of the evidence and future directions for clinical translation. *Depression and Anxiety, 34*, 9–24.

Williams, L. M., Korgaonkar, M. S., Song, Y. C., Paton, R., Eagles, S., Goldstein-Piekarski, A., … Etkin, A. (2015). Amygdala reactivity to emotional faces in the prediction of general and medication-specific responses to antidepressant treatment in the randomized iSPOT-D trial. *Neuropsychopharmacology, 40*, 2398–2408.

Yang, H., Lu, L. H., Wu, M., Stevens, M., Wegbreit, E., Fitzgerald, J., … Pavuluri, M. N. (2013). Time course of recovery showing initial prefrontal cortex changes at 16 weeks, extending to subcortical changes by 3 years in pediatric bipolar disorder. *Journal of Affective Disorders, 150*, 571–577.

Yoon, J. H., Minzenberg, M. J., Raouf, S., D'Esposito, M., & Carter, C. S. (2013). Impaired prefrontal-basal ganglia functional connectivity and substantia nigra hyperactivity in schizophrenia. *Biological Psychiatry, 74*, 122–129.

Yung, A. R., McGorry, P. D., Francey, S. M., Nelson, B., Baker, K., Phillips, L. J., … Amminger, G. P. (2007). PACE: A specialized service for young people at risk of psychotic disorders. *Medical Journal of Australia, 18*, S43.

Zhou, Y., Fan, L., Qiu, C., & Jiang, T. (2015). Prefrontal cortex and the dysconnectivity hypothesis of schizophrenia. *Neuroscience Bulletin, 31*, 207–219.

Further reading

Costafreda, S. G., Fu, C. H. Y., Picchioni, M., Toulopoulou, T., McDonald, M., Kravariti, E., … McGuire, P. K. (2011). Pattern of neural responses to verbal fluency shows diagnostic specificity for schizophrenia and bipolar disorder. *BMC Psychiatry, 11*, 18.

Salvador, R., Martinez, A., Pomarol-Clotet, E., Sarro, S., Suckling, J., & Bullmore, E. (2007). Frequency based mutual information measures between clusters of brain regions in functional magnetic resonance imaging. *NeuroImage, 35*, 83–88.

Schneider, M. R., Klein, C. C., Weber, W., Bitter, S. M., Elliott, K. B., Strakowski, S. M., … DelBello, M. P. (2014). The effects of carbamazepine on prefrontal activation in manic youth with bipolar disorder. *Psychiatry Research, 223*, 268–270.

Wisniewski, S. R. (2007). The STAR*D project results: A comprehensive review of findings. *Current Psychiatry Reports, 9*, 449–459.

Chapter 41

Neuroimaging, genetics, and personalized psychiatry: Developments and opportunities from the ENIGMA consortium

Lianne Schmaal[a,b], Christopher R.K. Ching[c], Agnes B. McMahon[c], Neda Jahanshad[c,d,h] and Paul M. Thompson[c,d,e,f,g,h]

[a]Orygen, The National Centre of Excellence in Youth Mental Health, Parkville, VIC, Australia, [b]Centre for Youth Mental Health, The University of Melbourne, Parkville, VIC, Australia, [c]Imaging Genetics Center, Mark & Mary Stevens Neuroimaging and Informatics Institute, Keck School of Medicine of USC, Los Angeles, CA, United States, [d]Department of Neurology, University of Southern California, Los Angeles, CA, United States, [e]Department of Psychiatry, University of Southern California, Los Angeles, CA, United States, [f]Department of Radiology, University of Southern California, Los Angeles, CA, United States, [g]Department of Pediatrics, University of Southern California, Los Angeles, CA, United States, [h]Department of Biomedical Engineering, University of Southern California, Los Angeles, CA, United States

1 Introduction

Current intervention strategies for treating mental illnesses are not particularly effective. For example, cognitive behavioral therapy and pharmacotherapy—the two most common treatments for depression, which is one of the most common and prevalent psychiatric illnesses worldwide—are both only moderately effective. Only around one-third of patients with depression benefit from the first antidepressant they are prescribed, and more than 40% of patients do not achieve remission, even after two optimally delivered trials of antidepressant medications (Nierenberg et al., 2006). In addition, only around 50% of people with major depressive disorder (MDD) respond to psychotherapy (Cuijpers et al., 2014).

The overall limited treatment response may, in part, be explained by the fact that mental illnesses are complex heterogeneous disorders with differing underlying biological mechanisms that require treatments that can target these differing mechanisms. Unfortunately, these targeted treatments are currently not available. Over the past few decades, there have been increased efforts to find markers that can assist in optimizing individualized care for people with mental illness. This "precision medicine" approach aims to personalize prevention and intervention strategies to individual people by taking into account an individual's characteristics such as their age, sex, genes, personality, blood markers, lifestyle, and environment. Biomarkers have been of particular interest in this respect, as measurable indicators of pathological biological processes associated with mental disorders could help to establish biologically based diagnoses. Biomarkers may also support the detection and development of novel treatments with innovative mechanisms of action, monitor drug effects, and predict who might benefit most from treatments targeting specific biological processes.

Biomarkers are defined as objective biological measures that can be *diagnostic*, that is, index a biological process that discriminates between health and disease, or between different diseases. They may also be *prognostic* or *predictive*, that is, reflect a biological process associated with progression of a disease or treatment response (Atkinson et al., 2001). *Diagnostic* biomarkers may assist in (differential) diagnosis of mental illnesses and provide biological targets for development of new treatments. For example, differentiation between MDD and bipolar disorder (BD) in an early phase is important, as misdiagnosis may result in inadequate pharmacological therapy. However, current diagnostic tools poorly distinguish between the depressed episode of MDD and BD due to comparable symptom profiles (Goodwin et al., 2008). Diagnostic biomarkers may improve psychiatric diagnoses by underpinning them with pathophysiological evidence, thereby aligning psychiatric classification with classification systems used in other areas of medicine (Moffitt et al., 2008). *Prognostic* biomarkers may be used to predict the onset of a mental disorder, which is of particular importance in child and adolescent psychiatry, as well as the course of mental illness, while *predictive* biomarkers can be used to predict treatment response.

Personalized Psychiatry. https://doi.org/10.1016/B978-0-12-813176-3.00041-9

The majority of studies aimed at identifying clinically useful biomarkers have focused on genetic variants and alterations in brain structure and function associated with mental disorders. Technological advances in genetic sequencing and neuroimaging over the past few decades have fueled the search for biomarkers. In the early 2000s, hope was expressed that these technological advances would help to identify new targets for treatment and lead to a biologically supported psychiatric classification system. More than a decade later, we have to conclude that genetics and neuroimaging have not yet fulfilled their promise in identifying clinically relevant biomarkers for mental disorders (e.g., Fond et al., 2015; Gadad et al., 2018), although much progress has been made in identifying genetic or imaging markers that differentiate patients at the group level. Despite intensive genetic and neuroimaging research in the last decades, we still have a limited understanding of the exact genetic and neurobiological underpinnings of mental disorders. Key issues that have hampered progress include, but are not limited to (e.g., see Kapur, Phillips, & Insel, 2012 for a discussion of other key issues), underpowered studies and a lack of reproducible findings.

2   Underpowered studies and the crisis of reproducibility

Historically, most genetic and neuroimaging studies in psychiatry have been conducted with small sample sizes. As a consequence, many studies are underpowered—resulting in a lower probability of finding "true" effects (low-powered studies produce more false negatives than high-powered studies), and an exaggerated estimate of the effect size when a true effect is discovered (Button et al., 2013). If, for example, the true effect has a small effect size—which is often the case for differences between patients and controls in genetic risk or brain measures—only studies with small sample sizes that, by chance, overestimate the magnitude of the effect will pass the threshold for discovery (Ioannidis, 2008). Underpowered studies hamper replication of significant findings, because if the effect sizes of the original study are inflated, power calculations to determine the sample size for a replication study will be too optimistic, and replication studies will tend to show smaller effect sizes that may not reach the threshold for statistical significance (Miller, 2009). Hence, underpowered studies have led to inconsistent and poorly replicated genetic and neuroimaging findings in psychiatry (e.g., Müller et al., 2017; Sullivan, 2007). Use of larger sample sizes or performing meta-analyses are considered to be important ways to counteract these issues. However, recruiting a large sample is often difficult, because of limited access to patient populations or scanning facilities, and the high costs associated with acquiring large sets of genetic or neuroimaging data. Problems with retrospective meta-analyses are the potential overrepresentation of positive findings in the published literature (the "file drawer" problem), and a lack of harmonization of data processing and statistical analysis methods across the different studies included in the meta-analysis.

Worldwide pooling of existing genetic and/or neuroimaging datasets represents a highly effective alternative to tackle issues associated with underpowered studies and poor reproducibility, because it (i) makes optimal use of valuable and costly existing datasets, (ii) collates large datasets relatively cheaply, (iii) controls publication bias, (iv) allows standardization of protocols for data processing and analysis, and (v) combines the expertise of hundreds of professionals in the fields of neuroimaging, psychiatry, and mathematics. The Psychiatric Genomics Consortium (PGC) provides one good example of how worldwide sharing of datasets can accelerate the discovery of the biological bases that contribute to mental illnesses, and can serve as potential new drug targets. Although most complex disorders, including mental illnesses, are moderately to highly heritable (ranging from 30% to 80%; Geschwind & Flint, 2015), effects of individual genetic variants are small, and individually, tend to explain less than 1% of the trait variance, or the risk for mental disorders (e.g., Farrell et al., 2015; Johnson et al., 2017). Traditionally, genetic studies focused on investigating associations between a priori hypothesized single genes and mental disorders—so-called *candidate gene* studies. However, candidate gene studies have produced very extensive and conflicting results on gene associations for many mental disorders, with high rates of false positive findings and low rates of replication. Therefore, in the past decade, the focus in psychiatric genetic research has shifted to genome-wide association (GWAS) studies, which are hypothesis-free data-driven studies that scan millions of common variants across the whole genome to identify significant associations between individual genetic variations and mental disorders. Very large sample sizes—in the order of tens of thousands of patients and controls—are typically needed for GWAS discovery and replication due to the small effect sizes of individual genetic polymorphisms, and due to the large number of statistical comparisons across genetic markers (usually 500K to 1M) that require correction for multiple comparisons. To address this, consortia such as the Psychiatric Genomics Consortium (PGC) have emerged, that involve data sharing across many genotyped samples worldwide. These efforts have led to significant breakthroughs in the identification of numerous novel genetic risk loci for mental disorders. For example, in 2014, 128 independent genome-wide significant single nucleotide polymorphisms (SNPs) across 108 genomic loci were identified in patients with schizophrenia in a GWAS meta-analysis of data from more than 36,000 patients and 113,000 controls by the PGC (Ripke et al., 2014). Similar progress in other psychiatric disorders has been published (Hou et al., 2016; Wray, Sullivan, & Wray, 2018) or is underway (Sullivan et al., 2018).

In this chapter, we present and discuss another worldwide data sharing initiative: the Enhancing NeuroImaging Genetics through Meta-Analysis (ENIGMA) consortium. ENIGMA was initiated to address issues related to small sample sizes in imaging genetics and neuroimaging research. One goal of ENIGMA is to identify genetic influences on brain structure and function, as intermediate phenotypes for mental and neurological disorders. A second goal is to study patterns of brain abnormalities in mental and neurological disorders that are robust and replicable across many different samples worldwide. First, we provide an overview of the history, aims, and organizational structure of ENIGMA, followed by findings on genetic variants of brain measures and structural brain alterations associated with mental disorders. Next, we will discuss progress to date, as well as the potential of future ENIGMA work with regard to identifying clinically useful biomarkers. Finally, we discuss some of the main challenges faced by worldwide data sharing initiatives such as ENIGMA, which involve sociological, ethical, and technical considerations.

3 The enhancing neuroimaging genetics through meta-analysis consortium

The ENIGMA Consortium was founded in 2009 with the aim of boosting statistical power to detect genetic influences on brain measures, and identify disease effects on the brain that are replicable across many samples worldwide. As of April 2018, ENIGMA has brought together scientists and datasets from 39 countries working collaboratively to study factors that influence brain structure and function in health and disease, using magnetic resonance imaging (MRI) and other neuroimaging modalities (Fig. 1).

ENIGMA has published the largest genetic studies of the brain, in partnership with other consortia (Adams et al., 2016; Hibar et al., 2017; Satizabal et al., 2017), mapping genome-wide effects of more than a million genetic loci in more than 35,000 brain MRI scans. In addition, ENIGMA has published neuroimaging studies that contain two hundred times as many samples as studies that were considered large and well-powered just 5 years ago ($N \sim 50$), and is now publishing studies with more than 10,000 individuals. The current organizational structure of ENIGMA is shown in Fig. 2.

ENIGMA's initial aim was to perform genome-wide analyses to identify common genetic variants that affect brain structure. Because many successful GWAS studies of complex traits, including psychiatric disorders, required samples of more than 75,000 people, many researchers hoped that brain measures might offer a more efficient way to discover genes involved in psychiatric disorders. Psychiatric disorders are highly heterogeneous, with clinical factors and biological mechanisms likely to differ substantially, even among people diagnosed with the same disorder. Most major mental illnesses are

Locations of ENIGMA members across the world

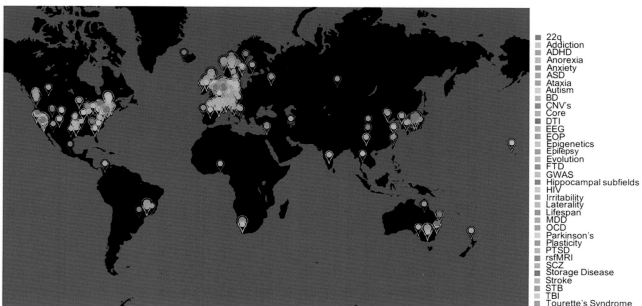

FIG. 1 Map of ENIGMA members across the world, by working group; an interactive version is available online at http://enigma.usc.edu/about-2/map/.

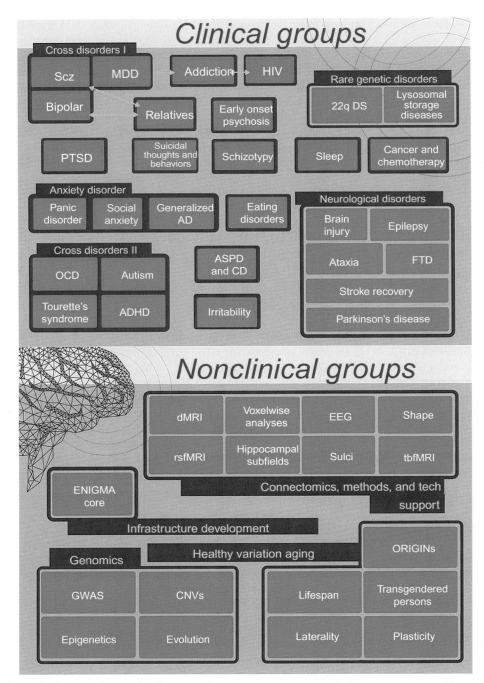

FIG. 2 ENIGMA is organized as a set of working groups that perform or support international studies on specific topics. The Clinical Working groups study specific diseases of the brain, including psychiatric and neurological disorders, ranging from affective disorders to substance use disorders, anxiety, PTSD, and even infectious diseases such as HIV/AIDS, and monogenetic disorders such as 22q Deletion Syndrome. In parallel, Technical working groups develop harmonized analytic approaches for genomic data (e.g., GWAS, copy number variants, and epigenetic analyses) and various kinds of neuroimaging data (including MRI, diffusion-weighted imaging, resting state functional MRI, and EEG).

thought to have a complex genetic architecture with polygenic influences and gene-by-environment interaction effects. These underlying complexities make it very challenging to identify genetic variants that are robustly associated with the disease, as their effects may also depend on complex interactions between multiple genes of mostly small to modest effect. In line with previous observations in other medical conditions such as heart disease (Cohen, Boerwinkle, Mosley, & Hobbs, 2006) and diabetes (Walters et al., 2010), it was assumed that genetic association is stronger at the level of biological substrates related to the psychiatric disorder than a psychiatric diagnosis itself. Therefore, genetic risk factors for psychiatric disorders may be more easily detected by examining intermediate phenotypes such as brain measures obtained from MRI (Gottesman & Gould, 2003). Intermediate phenotypes, or endophenotypes, may have a simpler genetic architecture than a clinical diagnosis (Geschwind & Flint, 2015), and brain measures may also offer a more precise, objective, and reproducible phenotype than a clinical diagnostic scale (Potkin et al., 2009). Thus, genome-wide screening of brain measures was considered to be a relatively efficient approach to identify otherwise weak or unobservable genetic effects on complex phenotypes of interest (e.g., psychiatric disorder). As collecting a sufficient amount of brain imaging data for GWAS analysis may be difficult or too costly for any one individual site, the need for a global community effort in imaging genetics became clear, and the ENIGMA consortium was initiated to combine existing genomic and imaging data around the globe.

Building on ENIGMA's initial successes in imaging genetics, which are further described below, disease "working groups" were initiated to study patterns of brain abnormalities in major psychiatric, neurodevelopmental, neurological, and neurogenetic disorders. In this chapter, we specifically focus on the psychiatric working groups. Since the initiation of the first ENIGMA psychiatric working groups in 2012 that focused on MDD, bipolar disorder, schizophrenia, and attention deficit hyperactivity disorder (ADHD), more than 20 additional psychiatric working groups were formed. For a complete overview of the different psychiatric working groups, see Table 1. These working groups typically analyze imaging data from 5000 to 10,000 people; eight of the mental disorder working groups have recently published the largest imaging studies to date of the disorders they study (schizophrenia, bipolar disorder, MDD, post-traumatic stress disorder (PTSD), obsessive compulsive disorder (OCD), ADHD, autism spectrum disorders, substance use disorder, see Table 1). Projects conducted within and across the ENIGMA psychiatric working groups are designed to: (1) identify imaging markers of specific mental disorders that are consistent across many samples worldwide; (2) investigate how factors including age, sex, and disease characteristics (e.g., stage, severity and duration of mental illness, age of onset, and medication use) moderate these brain alterations; and (3) identify common and unique patterns of brain alterations across different mental disorders.

The ENIGMA psychiatric working groups are supported by ENIGMA Methods working groups. These working groups are dedicated to developing and large-scale testing of imaging and genomics methods. For example, some of these ENIGMA Methods working groups develop standardized protocols to harmonize processing and quality assurance of imaging data, including structural MRI, diffusion-weighted MRI, resting state functional MRI and electroencephalography (EEG), across all samples in order to reduce statistical heterogeneity and researcher degrees of freedom, and to ensure or evaluate reproducibility. Brain measures derived from the ENIGMA protocols have shown good reliability and heritability (Acheson et al., 2017; Adhikari, Jahanshad, Reynolds, Cox, & Nichols, 2018; Jahanshad et al., 2013; Kochunov et al., 2015). ENIGMA is dedicated to "open science" and, therefore, ENIGMA protocols are publicly available on the ENIGMA website, and have been widely used in hundreds of ongoing and published projects.

ENIGMA has successfully harmonized imaging measures across existing datasets from around the world to achieve unprecedented power in detecting brain alterations associated with several mental disorders. These efforts have also driven the largest, and most successful, replicated findings of specific genetic variants that associate with brain structure. The findings of these studies will be discussed next.

4 ENIGMA imaging genomics and genome-wide association studies

By pooling brain MRI scans and genome-wide genetic data from thousands of individuals, ENIGMA performed a coordinated analysis with the CHARGE Consortium to discover the first genetic loci associated with the volume of the hippocampus—the brain's key center for learning and memory (Stein et al., 2012). Over recent years, ENIGMA and CHARGE have worked to expand this analysis—the total sample size now includes more than 25,000 individual samples. By adding more data, the consortia have replicated earlier findings and identified new genetic risk factors for smaller hippocampal volumes (Hibar et al., 2015; Hibar et al., 2017). These genomic risk factors may provide some insight into the biological underpinnings and mechanisms underlying diseases that affect the hippocampus, and the structure and function of the brain, as evidenced by a growing number of collaborative ENIGMA GWAS studies, including of all subcortical volumes (Hibar et al., 2015; Satizabal et al., 2017), cortical surface area and thickness (Grasby et al., 2018); and global brain function as determined by EEG power (Smit et al., 2018).

Beyond GWAS, the ENIGMA-CNV working group also began to perform the largest imaging studies of rare genetic variants (Sonderby et al., 2018). Although CNVs are implicated in autism, epilepsy, and other disorders, their rareness makes it hard to perform a concerted study of their effects. In one study, the 16p11.2 distal CNV affected intracranial volume, and the putamen and pallidum specifically, a pattern replicated in data from the DeCODE Genetics consortium.

TABLE 1 Overview ENIGMA GWAS and psychiatric disorder working groups, including current sample sizes (July 2018), number of sites, and publications per working group

Group	Project	N of samples	N of sites	Reference	Journal
Genome-wide association	Cortical	51,238	58	Grasby et al. (2018)	BioRxiv
	Subcortical	30,717	50	Hibar et al. (2015)	Nature
Major depressive disorder	Cortical	10,105	20	Schmaal et al. (2017)	Molecular Psychiatry
	Subcortical	8927	15	Schmaal et al. (2016)	Molecular Psychiatry
	DTI	2907	18	van Velzen et al. (2019)	Molecular Psychiatry
	Childhood adversity	3106	12	Frodl et al. (2017)	Journal of Psychiatric Research
	Suicidality	3097	20	Renteria et al. (2017)	Translational Psychiatry
Obsessive compulsive disorder	Cortical	3665	27	Boedhoe et al. (2018)	American Journal of Psychiatry
	Subcortical	3589	35	Boedhoe et al. (2017)	American Journal of Psychiatry
Schizophrenia	Cortical	9572	39	van Erp et al. (2018)	Biological Psychiatry
	Subcortical	4568	15	van Erp et al. (2016)	Molecular Psychiatry
	DTI	4322	29	Kelly et al. (2018)	Molecular Psychiatry
	Positive symptoms	1987	17	Walton et al. (2017)	Acta Psychiatry
	Negative symptoms	1985	17	Walton et al. (2017)	Psychological Medicine
Autism spectrum disorder	Cortical and subcortical	3222	49	Van Rooij et al. (2018)	American Journal of Psychiatry
Attention deficit hyperactivity disorder	Cortical	4200	36	Hoogman et al. (2019)	American Journal of Psychiatry
	Subcortical	3242	23	Hoogman et al. (2017)	The Lancet Psychiatry
Bipolar disorder	Cortical	6503	28	Hibar et al. (2018)	Molecular Psychiatry
	Subcortical	4304	20	Hibar et al. (2016)	Molecular Psychiatry
Posttraumatic stress disorder	Subcortical	1868	16	Logue et al. (2018)	Biological Psychiatry
Substance use disorders	Cortical and subcortical	3420	23	Mackey et al. (2018)	American Journal of Psychiatry
Anxiety disorders	Cortical and subcortical	4210	69		

Additional, more recent discoveries were made for the effect of the 1p21.1 distal CNV in intracranial volume, cortical surface area, and caudate volume—deletion ($n = 16$), and duplication carriers ($n = 9$) of the 1q21.1 CNV were compared with noncarriers ($n = 19,678$) from the same scanner sites. This CNV is associated with delayed development, schizophrenia, congenital heart defects, and obesity, and the brain phenotypes may at least partially explain the various neurodevelopmental abnormalities observed in 1q21.1 distal carriers.

Although ENIGMA's recent GWAS have discovered more than 100 genome-wide significant variants associated with brain structural measures derived from MRI, the fact that tens of thousands of scans were needed to discover these variants suggests that initial hopes were unfounded that brain imaging GWAS would be much more efficient than psychiatric GWAS in discovering genetic markers associated with traits of interest. Recently, however, Holland and colleagues (Holland et al., 2018) performed a comprehensive analysis of many neuroimaging and psychiatric traits, modeling the sample sizes needed to discover common variants that account for various fractions of the genetic variance for each of the traits assessed. They concluded that some traits, such as HDL cholesterol levels in the blood, do indeed have a simpler genetic architecture than traits such as schizophrenia or MDD. As a result, smaller samples are sufficient to discover genetic loci that account for a given fraction of the genetic variance. Initial evidence suggests that some neuroimaging measures may indeed have a simpler genetic architecture than cognitive metrics such as intellectual attainment, with more of the genetic variance accounted for by a smaller number of genetic loci. Of all the quantitative traits evaluated, educational attainment had the highest polygenicity—slightly larger than that for schizophrenia; the SNP "discoverability" was lower for MDD than for bipolar disorder or schizophrenia, lending insight into the vast sample sizes needed in GWAS for some psychiatric disorders.

5   ENIGMA psychiatric neuroimaging studies

With the largest combined imaging samples of individuals with mental disorders, ENIGMA's psychiatric working groups are well positioned to identify patterns of brain alterations associated with mental disorders, and to test their replicability and reliability across many different samples worldwide. The first ENIGMA studies of mental disorders were conducted using a meta-analysis approach. Following this approach, investigators participating in ENIGMA psychiatric working groups ran harmonized ENIGMA processing, quality assurance (QA), and statistical analysis protocols on their data locally, and sent back the QA output and analysis results to a central analysis site where the data was combined with data from other sites for a meta-analysis. In this way, large sample sizes can be acquired without the sometimes burdensome requirements of large-scale data transfers or issues related to sharing of individual raw imaging data. More recently, a mega-analysis approach that aggregates de-identified, individual-level data, has also been adopted by many ENIGMA psychiatric working groups.

ENIGMA meta- and mega-analyses of subcortical volume alterations have shown robust evidence for a smaller hippocampus across various mental disorders, including schizophrenia (SZ; van Erp et al., 2016), bipolar disorder (BD; Hibar et al., 2016), MDD (Schmaal et al., 2016), OCD (Boedhoe et al., 2017), PTSD (Logue et al., 2018), substance and alcohol use disorders (Mackey et al., 2018), and ADHD (Hoogman et al., 2017), but not autism spectrum disorder (ASD; Van Rooij et al., 2018). The hippocampus plays a key role in the formation of new memories, as well as learning and emotion processing (Desmedt, Marighetto, Richter-Levin, & Calandreau, 2015; Frodl et al., 2006; Hickie et al., 2005). The hippocampus is particularly prone to effects of chronic stress associated with a range of mental disorders. Chronic stress induces elevated glucocorticoid levels due to chronic hyperactivity of the hypothalamic-pituitary-adrenal (HPA) axis, and the hippocampus has a particularly high expression of glucocorticoid receptors (Sapolsky, Krey, & McEwen, 1984). Chronic elevated levels of glucocorticoids may promote atrophy of the hippocampus via remodeling and downregulation of growth factors, including brain-derived neurotrophic factor (Campbell & MacQueen, 2003). Additional volume reductions were less consistent across mental disorders, with smaller volumes of the amygdala and nucleus accumbens observed in SZ, ASD, ADHD, and substance use disorders; of the thalamus in SZ, BD, and alcohol use disorder; of the putamen in ASD, ADHD, and alcohol use disorder; and smaller volumes of the pallidum in ASD and alcohol use disorder, but larger pallidum volumes in SZ and adults with OCD (Boedhoe et al., 2017; Hibar et al., 2016; Hoogman et al., 2017; Mackey et al., 2018; van Erp et al., 2016; Van Rooij et al., 2018).

In addition, the structure of frontal, temporal, and parietal brain regions seems to be affected in multiple mental disorders, including regions such as the dorsolateral prefrontal cortex, orbitofrontal cortex, cingulate cortex, insula, inferior parietal cortex and temporal gyri, as shown by studies produced by the ENIGMA SZ, BD, MDD, Addiction, and OCD working groups (Boedhoe et al., 2018; Hibar et al., 2018; Mackey et al., 2018; Schmaal et al., 2017; van Erp et al., 2018). These regions have been implicated in a wide range of cognitive functions, including emotion processing and regulation (dorsolateral prefrontal cortex, cingulate cortex, temporal gyri; Kret & Ploeger, 2015), social cognition (dorsolateral prefrontal cortex, temporal gyri; Fernández, Mollinedo-Gajate, & Peñagarikano, 2018), motivation and reward (orbitofrontal cortex, anterior cingulate cortex; Haber, 2016), interoception (insula, anterior cingulate cortex; Seth & Friston, 2016), and decision-making (dorsolateral prefrontal cortex, cingulate cortex, orbitofrontal cortex, inferior parietal cortex; Starcke & Brand, 2012), that may contribute to the broad spectrum of emotional, cognitive, and behavioral disturbances observed in these mental disorders. Interestingly, ASD was associated with lower cortical gray matter thickness in temporal brain regions, in line with findings in SZ, BD, MDD, and substance use disorders, but with greater cortical thickness in frontal regions (Van Rooij et al., 2018). In general, the largest and most widespread effects of structural brain alterations were found in SZ (maximum Cohen's *d* effect size 0.53), followed by BD and OCD (maximum Cohen's *d* effect size 0.30), ASD and ADHD (maximum Cohen's *d* effect size 0.20), and MDD, substance use disorders, and PTSD (maximum Cohen's *d* effect size 0.14–0.17) compared with healthy controls. Some of the published findings of ENIGMA psychiatric working groups are summarized in Fig. 3.

The ENIGMA psychiatric working groups have recently started to analyze measures from other neuroimaging modalities than structural MRI, for example, diffusion MRI, which allows investigation of structural white matter connections between different brain regions, and measures of white matter microstructure. The first results from the ENIGMA SZ working group show that schizophrenia is not only associated with widespread alterations in the structure of brain regions, but also with widespread alterations in structural connections between these brain regions (Kelly et al., 2018). Other working groups are finalizing their diffusion MRI analyses at the time of writing this chapter. Structural imaging protocols and tools have thus far been the mainstay of ENIGMA, driving more than 20 research papers to date. Although our previous findings have provided important insights into structural brain abnormalities in mental disorders, it remains unclear how alterations observed in individual regions interact and contribute to disturbances in brain *functioning*. Therefore, an

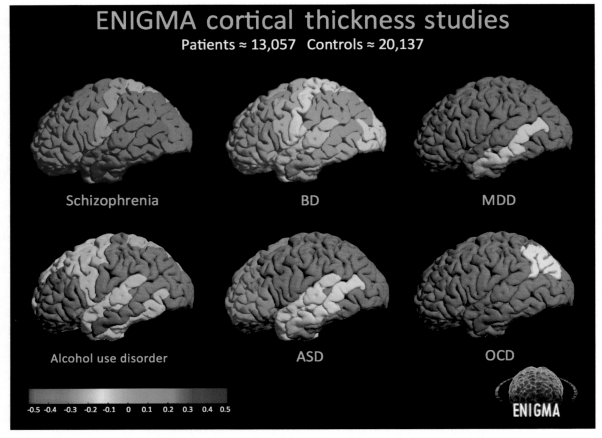

FIG. 3 Global studies of brain disease. ENIGMA recently published the largest neuroimaging studies of six psychiatric disorders, including schizophrenia, bipolar disorder (BD), major depressive disorder (MDD), alcohol use disorder, autism spectrum disorder, and obsessive compulsive disorder, investigating the differences in cortical thickness in individuals with psychiatric disorders compared with healthy controls. The color bar represents Cohen's *d* effect sizes.

important next step is to extend our previous work by investigating the impact of various mental disorders on functional brain circuitries.

To date, ENIGMA's work has provided more definitive answers to questions of the extent of structural brain abnormalities in mental disorders by addressing issues of poor replication, unreliable results, and an overestimation of effect sizes in previously underpowered studies. These findings of structural brain alterations associated with mental disorders from the ENIGMA consortium can (i) help to prioritize brain measures for future analyses aimed at unraveling underlying cellular, molecular, and genetic mechanisms of brain abnormalities in mental disorders, (ii) identify novel treatment targets, and (iii) generate new hypotheses about the impact of mental disorders on the integrity of the brain (or vice versa) that can be tested in future, more targeted studies (e.g., using a longitudinal design).

6 Toward personalized psychiatry

The work conducted by the ENIGMA psychiatric working groups described herein were derived from comparisons between people with and without a mental illness, as determined by current symptom-based classification schemes, such as the Diagnostic and Statistical Manual of Mental Disorders (DSM) or ICD. However, these DSM or ICD-based categories of mental disorders have been criticized in recent years, as they have shown low diagnostic validity and lack of specificity (e.g., Kapur et al., 2012; Kotov et al., 2017; McGorry, Nelson, Goldstone, & Yung, 2010; Stein, Lund, & Nesse, 2013). The main criticism is that current classification schemes assume that mental disorders are discrete entities. However, many mental disorders are instead complex, highly heterogeneous disorders, and distinct pathophysiological mechanisms may cause a similar presentation of symptoms in different people, which would require different treatments. Thus, if people are merely classified on the basis of the presence of an DSM or ICD diagnosis, distinct pathophysiological mechanisms underlying different subgroups of patients may remain undetected. Consequently, different groups of people with the same diagnosis of a mental disorder, but with distinct underlying mechanisms, cannot be distinguished, and, hence, cannot be stratified to different interventions. Not surprisingly, reducing the diagnostic heterogeneity of mental disorders is widely recognized as the next frontier in progressing research into underlying pathophysiological mechanisms and treatment allocation for people suffering from mental disorders.

To address the issue of the heterogeneity of mental disorders, the ENIGMA studies have investigated whether structural brain alterations are specifically linked to distinct subgroups of individuals characterized by different stages of brain development and different clinical characteristics, such as age of onset or disease stage. For example, distinct patterns of cortical structural brain alterations were observed in adolescents compared with adults with MDD. Gray matter thickness in the orbitofrontal cortex, anterior and posterior cingulate, insula, and temporal lobes was lower in adult MDD patients than controls. In contrast, adolescents with MDD showed no detectable abnormalities in cortical thickness, at least at the group level, but did show global reductions in cortical surface area (Schmaal et al., 2017). Similarly, children with OCD had a larger thalamus and thinner inferior and superior parietal cortices, whereas adults with OCD were characterized by a smaller hippocampus, a larger pallidum, lower surface area of the transverse temporal cortex, and a thinner inferior parietal cortex (Boedhoe et al., 2017, 2018). Distinct patterns, depending on age, were also observed in people with ADHD, with a smaller accumbens, amygdala, caudate, hippocampus, putamen, and intracranial volume in children, a smaller hippocampus in adolescents, and no subcortical structural brain abnormalities in adults (Hoogman et al., 2017). In ASD, the largest differences in cortical thickness occurred around adolescence (Van Rooij et al., 2018), whereas in BD and SZ, the largest differences in cortical thickness were found with increasing age (Hibar et al., 2018; van Erp et al., 2018). With regard to sex, males and females with MDD, SZ, ADHD, OCD, and ASD did not show different patterns of structural brain alterations (Boedhoe et al., 2017, 2018; Hoogman et al., 2017; Schmaal et al., 2016, 2017; van Erp et al., 2016, 2018; Van Rooij et al., 2018). However, a larger thalamus was a specific characteristic of adult females with BD (Hibar et al., 2016), and less thinning of frontal and temporal cortices of adolescent females with BD (Hibar et al., 2018). In addition, hippocampal volume reductions were more pronounced in females compared with males with PTSD (Logue et al., 2018). The heterogeneity observed in structural brain alterations associated with mental disorders can further be explained by differences in disease characteristics. More pronounced structural brain abnormalities were observed in subgroups of people with more than one episode, and an earlier age of onset of depression (Schmaal et al., 2016), a longer duration of illness in schizophrenia and bipolar disorder (Hibar et al., 2018; van Erp et al., 2016, 2018), and with higher symptom severity in ASD (Van Rooij et al., 2018) and PTSD (Logue et al., 2018).

Other sources of heterogeneity may be less disorder-specific, and affect the extent of brain alterations in a similar way across various mental disorders. For example, childhood adversity is a risk factor for various mental disorders (Kessler et al., 2010). Although the effects of childhood adversity on brain structure have been examined separately in people with depression (Frodl et al., 2017) and in PTSD (Logue et al., 2018) within the ENIGMA MDD and PTSD working groups, efforts are currently underway to examine the transdiagnostic effects of childhood adversity. In addition, suicidal thoughts and behaviors represent another transdiagnostic disorder that may be associated with a specific, but similar, pattern of brain alterations across disorders (Renteria et al., 2017). The ENIGMA Suicidal Thoughts and Behaviors working group—the first transdiagnostic working group within ENIGMA—was recently established to investigate shared and unique brain suicidality related brain abnormalities across different mental disorders (http://enigma.ini.usc.edu/ongoing/enigma-stb/). Suicidal behaviors are also highly etiologically heterogeneous and multifactorial in nature. Therefore, the ENIGMA Suicidal Thoughts and Behaviors working group aims to identify biopsychosocial subtypes of suicidal ideation and suicide attempt, thereby identifying distinct trajectories of suicidal behaviors in groups of people that differ in, for example, age, sex, and a constellation of various other neuroanatomical, psychosocial, cognitive, and clinical characteristics.

The preceding findings from the ENIGMA psychiatric working groups indicate that structural alterations in mental disorders depend on specific characteristics, such as stage of brain development, childhood trauma, and clinical characteristics, such as stage of the disorder. These findings provide valuable insights into the biological heterogeneity of mental disorders.

Machine learning methods are becoming increasingly more popular for evaluating the diagnostic or predictive value of biomarkers in previously unseen individuals (e.g., "new" patients). ENIGMA offers a unique framework to evaluate brain imaging measures as potential diagnostic biomarkers for various mental disorders, because the inclusion of many samples worldwide reflects the broad range of clinical heterogeneity in mental disorders, which is essential for creating more generalizable clinical decision support tools. In addition, the inclusion of many datasets allows the evaluation of how well a predictive model developed in a subset of datasets generalizes to new patients (other datasets). ENIGMA psychiatric working groups have evaluated patterns of structural brain alterations as a potential diagnostic biomarker of MDD (Zhu et al., 2017) or bipolar disorder (Nunes et al., 2018) in individual patients. Identifying a diagnostic biomarker for depression could be particularly relevant, as there is only minimal agreement among psychiatrists on who does and does not have major depressive disorder. For example, the field trials for the DSM-5 demonstrated an intraclass kappa of 0.28, which means that highly trained specialist psychiatrists under study conditions were only able to agree that a patient has depression between 4% and 15% of the time (Regier et al., 2013). The findings of the ENIGMA MDD and ENIGMA BD working groups showed that diagnostic status (having a mental disorder vs. not having a mental disorder) could be predicted in individual patients with accuracies ranging between 60% and 70% (Nunes et al., 2018; Zhu et al., 2017), which is not yet sufficient for these models to become clinically useful. More work is needed to improve the diagnostic classification accuracies, perhaps by including additional (biological) predictors to the model, including data from other imaging modalities, as well as genetic or plasma markers, or even mobile sensor data.

7 Challenges of large-scale data-sharing

Worldwide data-sharing initiatives such as ENIGMA are not without their challenges. Some of these challenges encompass ethical and computational issues with regard to data sharing, as well as science and data sharing policies that may vary from one research institute to another, from country to country, or even from continent to continent, and may change over time (e.g., the recent implementation of the General Data Protection Regulation [GDPR] in the European Union). This may restrict some researchers from sharing raw neuroimaging data, although sharing de-identified, individual-level data may still be feasible.

Another challenge that ENIGMA is facing is the lack of harmonization in data collection. When combining already collected data across worldwide samples, data collection protocols are not prospectively harmonized. Clinical assessments, therefore, differ across studies, which limits the analysis of sources of heterogeneity, and perhaps subsequently, the identification of biomarkers. In addition, neuroimaging data were collected using different MRI scanners and different sequences, which may introduce noise and further complicate the search for robust biomarkers.

By pooling data across many samples worldwide, ENIGMA psychiatric working groups include combined samples that are highly heterogeneous. In the case of MDD, for example, the working group performs studies encompassing a range of depressive phenotypes—from very mild to severe, and a broad range of previous treatments received. We argue that it is precisely for this reason that this work furnishes an important transition toward real-world clinical populations, including patients recruited from community and primary care settings, where most people with mental illness reside. By combining a large number of datasets that were collected across the world, we have had the opportunity to better estimate the effect sizes of structural brain alterations in people with mental disorders. As it turns out, many of these structural brain alterations show lower effect sizes than previously assumed based on previously published studies. However, as discussed earlier in this chapter, many of the high effect sizes observed in prior studies may have been driven by small sample sizes and the "file drawer" problem. Therefore, our ENIGMA findings more likely reflect the true effect sizes of structural brain alterations in a representative population of people with mental disorders with a broad range of illness severity. Nonetheless, it has become clear that because of these small effect sizes, we are actively searching for novel imaging measures, including fMRI measures, or a combination of measures that outperform the current structural MRI measures, because effect sizes between a Cohen's d of 1.5 and 3 are likely to be required for a biomarker to be clinically useful, depending on the nature of the application (Castellanos, Di Martino, Craddock, Mehta, & Milham, 2013). Future ENIGMA studies should further investigate whether combining these neuroimaging modalities with other neuroimaging and/or clinical, psychosocial, and other biological data modalities could result in clinically useful diagnostic or predictive tools.

8 Conclusions

8.1 Future directions

Findings from the ENIGMA psychiatric working groups indicate that mental disorders are associated with structural alterations in a highly dynamic way, depending on specific characteristics, such as stage of brain development and stage of the disorder. One way to better characterize the impact of different stages of brain development on brain alterations in mental disorders is by defining "normal" variability in brain structure in healthy populations. This can be achieved by mapping the normative association between age and brain structure across the entire lifespan in people without mental or neurological illnesses. These normative charts, or brain charts, may then be used to help in identifying individuals with mental disorders that may deviate from this normative pattern. This can be compared with the use of growth charts to map child development in terms of height and weight as a function of age, where deviations from a normal growth trajectory manifest as outliers within the normative range at each age. The ENIGMA Lifespan working group is currently developing these normative charts using cross-sectional data from healthy volunteers from many samples worldwide. These models could be further improved by including data on longitudinal brain changes in healthy individuals from samples included in the ENIGMA Plasticity working group. The normative charts can subsequently be used by the ENIGMA psychiatric working groups to help in identifying patients who deviate from appropriate normative charts.

Importantly, the measures derived from the normative charts are modeled as deviations in individual patients (i.e., individual differences instead of group averages), which is a critical step toward personalized psychiatry. Much current neuroimaging research still focuses on characterizing group differences using the mean and standard deviation of derived measures. Crucially, this approach assumes homogeneity among members of (psychiatric) groups, with the different members resembling the "average." Group-based findings obtained using this approach are not sufficiently informative about the individual standing in front of a clinician. To identify diagnostic, prognostic, or predictive biomarkers that can be translated into clinical practice, it is necessary to shift the design of experiments from testing for clinical differences among groups to individual-level prediction of relevant outcomes, such as diagnostic status or treatment response.

In addition, the ENIGMA studies investigating the value of brain measures as diagnostic biomarkers for single disorders are an important first step, but perhaps a clinically more useful question is whether these brain measures can be used as biomarkers for differential diagnosis, or for predicting future outcomes, such as treatment response or course trajectory. Although ENIGMA may not provide the ideal platform to test the latter, because of a lack of longitudinal data and a lack in harmonization of treatment protocols in the few treatment samples included in ENIGMA, machine learning efforts are underway to evaluate structural brain measures as biomarkers for differentiating among different disorders such as depression, bipolar disorder, and schizophrenia. Still, it may prove too difficult to identify diagnostic biomarkers for broad categories such as DSM or ICD diagnoses that likely include subgroups of people with different patterns of underlying pathology. One way to address this is to identify biologically more meaningful subgroups based on different patterns of brain alterations. However, a key challenge remains as to how to detect relevant biomarkers that underlie biologically meaningful classifications of mental illness without the use of a priori defined labels such as DSM diagnoses. A potential solution for identifying neurobiologically relevant subtypes of mental disorders without using a priori defined clinical labels is to apply data-driven methods to brain measures. Such data-driven methods, also referred to as unsupervised machine learning techniques, have been successfully used to stratify mental disorders based on brain imaging measures. For example, using a hierarchical clustering approach, a recent study identified neurobiological subtypes of adult patients with a MDD diagnosis that were defined by distinct patterns of dysfunctional brain connectivity (Drysdale et al., 2016). Each neurobiological subtype was associated with a distinct clinical symptom profile. Importantly, these neurobiological subtypes predicted treatment response to repetitive transcranial magnetic stimulation more effectively than symptoms alone (Drysdale et al., 2016). Various efforts are underway within the ENIGMA psychiatric working groups to identify neurobiologically relevant subtypes of mental disorders based on structural or diffusion-weighted MRI measures.

8.2 Conclusions

Current intervention strategies for treating mental illness are not always effective, which can partly be explained by the fact that mental illnesses are complex, pathophysiologically heterogeneous disorders. Identifying diagnostic, predictive, and prognostic biomarkers of mental disorders is widely recognized as the next frontier in progressing research into underlying pathophysiological mechanisms and treatment allocation of people suffering from mental illness. Group-level analyses in mental disorders, especially those in genetics and neuroimaging, have encountered issues in statistical power and reproducibility. In order to overcome these issues, the ENIGMA consortium aims to consolidate efforts across many researchers worldwide, and increase sample sizes. Over the past 10 years since its initiation, ENIGMA has grown rapidly, and ENIGMA working groups have published the largest samples sizes in neuroimaging studies for nine major brain disorders (Table 1), which provided more definitive answers to questions of the extent of structural brain abnormalities in mental disorders. Currently, various efforts are underway within ENIGMA to address the heterogeneity in mental disorders, by identifying more pathophysiologically homogeneous subgroups of patients. Moreover, because of the inclusion of many samples worldwide, reflecting the broad range of clinical heterogeneity in mental disorders, ENIGMA offers a unique framework to evaluate brain imaging measures as potential diagnostic biomarkers for various mental disorders. Some of the current ENIGMA work focuses on individual-level—as opposed to group-level—prediction of relevant outcomes, which is a critical step toward personalized psychiatry.

Acknowledgment

The ENIGMA studies reported here were supported by a range of public and private agencies worldwide, and by a grant from the NIH (U54 EB020403).

References

Acheson, A., Wijtenburg, S. A., Rowland, L. M., Winkler, A., Mathias, C. W., Hong, L. E., … Dougherty, D. M. (2017). Reproducibility of tract-based white matter microstructural measures using the ENIGMA-DTI protocol. *Brain and Behavior*, *7*(2), 1–10.

Adams, H. H. H., Hibar, D. P., Chouraki, V., Stein, J. L., Nyquist, P. A., Rentería, M. E., … Thompson, P. M. (2016). Novel genetic loci underlying human intracranial volume identified through genome-wide association. *Nature Neuroscience*, *19*(12), 1569–1582.

Adhikari, B. M., Jahanshad, N., Reynolds, R. C., Cox, R. W., & Nichols, T. E. (2018). Heritability estimates on resting state fMRI data using ENIGMA analysis pipeline. *Pacific Symposium on Biocomputing*, *23*, 307–318.

Atkinson, A. J., Colburn, W. A., DeGruttola, V. G., DeMets, D. L., Downing, G. J., Hoth, D. F., … Zeger, S. L. (2001). Biomarkers and surrogate end-points: Preferred definitions and conceptual framework. *Clinical Pharmacology and Therapeutics*, *69*(3), 89–95.

Boedhoe, P. S. W., Schmaal, L., Abe, Y., Alonso, P., Ameis, S. H., Anticevic, A., … Van Den Heuvel, O. A. (2018). Cortical abnormalities associated with pediatric and adult obsessive-compulsive disorder: Findings from the enigma obsessive-compulsive disorder working group. *American Journal of Psychiatry*, *175*(5), 453–462.

Boedhoe, P. S. W., Schmaal, L., Abe, Y., Ameis, S. H., Arnold, P. D., Batistuzzo, M. C., … Van Den Heuvel, O. A. (2017). Distinct subcortical volume alterations in pediatric and adult OCD: A worldwide meta- and mega-analysis. *American Journal of Psychiatry*, *174*(1), 60–69.

Button, K. S., Ioannidis, J. P. A., Mokrysz, C., Nosek, B. A., Flint, J., Robinson, E. S. J., & Munafò, M. R. (2013). Power failure: Why small sample size undermines the reliability of neuroscience. *Nature Reviews Neuroscience*, *14*(5), 365–376.

Campbell, S., & MacQueen, G. (2003). The role of the hippocampus in the pathophysiology of major depression. *Journal of Psychiatry & Neuroscience*, *29*(6), 417–426.

Castellanos, F. X., Di Martino, A., Craddock, R. C., Mehta, A. D., & Milham, M. P. (2013). Clinical applications of the functional connectome. *NeuroImage*, *80*, 527–540.

Cohen, J. C., Boerwinkle, E., Mosley, T. H., & Hobbs, H. H. (2006). Sequence variations in PCSK9, low LDL, and protection against coronary heart disease. *Heart Disease*, *354*, 1264–1272.

Cuijpers, P., Karyotaki, E., Weitz, E., Andersson, G., Hollon, S. D., & Van Straten, A. (2014). The effects of psychotherapies for major depression in adults on remission, recovery and improvement: A meta-analysis. *Journal of Affective Disorders*, *159*(2014), 118–126.

Desmedt, A., Marighetto, A., Richter-Levin, G., & Calandreau, L. (2015). Adaptive emotional memory: The key hippocampal-amygdalar interaction. *Stress*, *18*(3), 297–308.

Drysdale, A. T., Grosenick, L., Downar, J., Dunlop, K., Mansouri, F., Meng, Y., … Liston, C. (2016). Resting-state connectivity biomarkers define neurophysiological subtypes of depression. *Nature Medicine*, *23*(1), 28–38.

Farrell, M. S., Werge, T., Sklar, P., Owen, M. J., Ophoff, R. A., O'donovan, M. C., … Sullivan, P. F. (2015). Evaluating historical candidate genes for schizophrenia. *Molecular Psychiatry*, *20*(5), 555–562.

Fernández, M., Mollinedo-Gajate, I., & Peñagarikano, O. (2018). Neural circuits for social cognition: Implications for autism. *Neuroscience*, *370*, 148–162.

Fond, G., D'Albis, M. A., Jamain, S., Tamouza, R., Arango, C., Fleischhacker, W. W., … Leboyer, M. (2015). The promise of biological markers for treatment response in first-episode psychosis: A systematic review. *Schizophrenia Bulletin, 41*(3), 559–573.

Frodl, T., Janowitz, D., Schmaal, L., Tozzi, L., Dobrowolny, H., Stein, D. J., … Grabe, H. J. (2017). Childhood adversity impacts on brain subcortical structures relevant to depression. *Journal of Psychiatric Research, 86*, 58–65.

Frodl, T., Schaub, A., Banac, S., Charypar, M., Jäger, M., Kümmler, P., … Meisenzahl, E. M. (2006). Reduced hippocampal volume correlates with executive dysfunctioning in major depression. *Journal of Psychiatry & Neuroscience, 31*(5), 316–323.

Gadad, B. S., Jha, M. K., Czysz, A., Furman, J. L., Mayes, T. L., Emslie, M. P., & Trivedi, M. H. (2018). Peripheral biomarkers of major depression and antidepressant treatment response: Current knowledge and future outlooks. *Journal of Affective Disorders, 233*, 3–14.

Geschwind, D. H., & Flint, J. (2015). Genetics and genomics of psychiatric disease. *Science, 349*(6255), 719–724.

Goodwin, G. M., Anderson, I., Arango, C., Bowden, C. L., Henry, C., Mitchell, P. B., … Wittchen, H. U. (2008). ECNP consensus meeting. Bipolar depression. Nice, March 2007. *European Neuropsychopharmacology, 18*(7), 535–549.

Gottesman, I. I., & Gould, T. D. (2003). The endophenotype concept in psychiatry: Etymology and strategic intentions. *American Journal of Psychiatry, 160*(4), 636–645.

Grasby, K. L., Jahanshad, N., Painter, J. N., Colodro-Conde, L., Bralten, J., Hibar, D. P., … Medland, S. E. (2018). *The genetic architecture of the human cerebral cortex.* bioRxiv. Retrieved from *https://www.biorxiv.org/content/early/2018/09/09/399402.*

Haber, S. N. (2016). Corticostriatal circuitry. *Dialogues in Clinical Neuroscience, 18*(1), 7–21.

Hibar, D. P., Adams, H. H. H., Jahanshad, N., Chauhan, G., Stein, J. L., Hofer, E., … Ikram, M. A. (2017). Novel genetic loci associated with hippocampal volume. *Nature Communications, 8*, 13624.

Hibar, D. P., Stein, J. L., Renteria, M. E., Arias-Vasquez, A., Desrivières, S., Jahanshad, N., … Medland, S. E. (2015). Common genetic variants influence human subcortical brain structures. *Nature, 520*(7546), 224–229.

Hibar, D. P., Westlye, L. T., Doan, N. T., Jahanshad, N., Cheung, J. W., Ching, C. R. K., … Andreassen, O. A. (2018). Cortical abnormalities in bipolar disorder: An MRI analysis of 6503 individuals from the ENIGMA bipolar disorder working group. *Molecular Psychiatry, 2*(4), 932–942.

Hibar, D. P., Westlye, L. T., Van Erp, T. G. M., Rasmussen, J., Leonardo, C. D., Faskowitz, J., … Andreassen, O. A. (2016). Subcortical volumetric abnormalities in bipolar disorder. *Molecular Psychiatry, 21*(12), 1710–1716.

Hickie, I., Naismith, S., Ward, P. B., Turner, K., Scott, E., Mitchell, P., … Parker, G. (2005). Reduced hippocampal volumes and memory loss in patients with early- and late-onset depression. *The British Journal of Psychiatry, 186*, 197–202.

Holland, D., Frei, O., Fan, C.-C., Shadrin, A. A., Smeland, O. B., Sundar, V. S., … Dale, A. M. (2018). *Beyond SNP heritability: Polygenicity and discoverability estimated for multiple phenotypes with a univariate Gaussian mixture model.* bioRxiv. Retrieved from *http://biorxiv.org/content/early/2018/06/07/133132.abstract.*

Hoogman, M., Bralten, J., Hibar, D. P., Mennes, M., Zwiers, M. P., Schweren, L. S. J., … Franke, B. (2017). Subcortical brain volume differences in participants with attention deficit hyperactivity disorder in children and adults: A cross-sectional mega-analysis. *The Lancet Psychiatry, 4*(4), 310–319.

Hoogman, M., Muetzel, R., Guimaraes, J. P., Shumskaya, E., Mennes, M., … Zwiers, M. P. Franke (2019). Brain imaging of the cortex in ADHD: A coordinated analysis of large-scale clinical and population-based samples. *American Journal of Psychiatry, 176*(7), 531–542.

Hou, L., Bergen, S. E., Akula, N., Song, J., Hultman, C. M., Landén, M., … McMahon, F. J. (2016). Genome-wide association study of 40,000 individuals identifies two novel loci associated with bipolar disorder. *Human Molecular Genetics, 25*(15), 3383–3394.

Ioannidis, J. P. A. (2008). Why most discovered true associations are inflated. *Epidemiology, 19*(5), 640–648.

Jahanshad, N., Kochunov, P. V., Sprooten, E., Mandl, R. C., Nichols, T. E., Almasy, L., … Glahn, D. C. (2013). Multi-site genetic analysis of diffusion images and voxelwise heritability analysis: A pilot project of the ENIGMA-DTI working group. *NeuroImage, 81*, 455–469.

Johnson, E. C., Border, R., Melroy-Greif, W. E., de Leeuw, C. A., Ehringer, M. A., & Keller, M. C. (2017). No evidence that schizophrenia candidate genes are more associated with schizophrenia than noncandidate genes. *Biological Psychiatry, 82*(10), 702–708.

Kapur, S., Phillips, A. G., & Insel, T. R. (2012). Why has it taken so long for biological psychiatry to develop clinical tests and what to do about it. *Molecular Psychiatry, 17*(12), 1174–1179.

Kelly, S., Jahanshad, N., Zalesky, A., Kochunov, P., Agartz, I., Alloza, C., … Donohoe, G. (2018). Widespread white matter microstructural differences in schizophrenia across 4322 individuals: Results from the ENIGMA schizophrenia DTI working group. *Molecular Psychiatry, 23*(5), 1261–1269.

Kessler, R. C., McLaughlin, K. A., Green, J. G., Gruber, M. J., Sampson, N. A., Zaslavsky, A. M., … Williams, D. R. (2010). Childhood adversities and adult psychopathology in the WHO world mental health surveys. *British Journal of Psychiatry, 197*(5), 378–385.

Kochunov, P., Jahanshad, N., Marcus, D., Winkler, A., Sprooten, E., Nichols, T. E., … Van Essen, D. C. (2015). Heritability of fractional anisotropy in human white matter: A comparison of human connectome project and ENIGMA-DTI data. *NeuroImage, 111*, 300–301.

Kotov, R., Waszczuk, M. A., Krueger, R. F., Forbes, M. K., Watson, D., Clark, L. A., … Zimmerman, M. (2017). The hierarchical taxonomy of psychopathology (HiTOP): A dimensional alternative to traditional nosologies. *Journal of Abnormal Psychology, 126*(4), 454–477.

Kret, M. E., & Ploeger, A. (2015). Emotion processing deficits: A liability spectrum providing insight into comorbidity of mental disorders. *Neuroscience and Biobehavioral Reviews, 52*, 153–171.

Logue, M. W., van Rooij, S. J. H., Dennis, E. L., Davis, S. L., Hayes, J. P., Stevens, J. S., … Morey, R. A. (2018). Smaller hippocampal volume in posttraumatic stress disorder: A multisite ENIGMA-PGC study: Subcortical volumetry results from posttraumatic stress disorder consortia. *Biological Psychiatry, 83*(3), 244–253.

Mackey, S., Allgeier, N., Chaarani, B., Spechler, P., Orr, C., Bunn, J., … Garavan, H. (2019). Mega-analysis of grey matter volume in substance dependence: General and substance-specific regional effects. *American Journal of Psychiatry, 176*(2), 119–128.

McGorry, P. D., Nelson, B., Goldstone, S., & Yung, A. R. (2010). Clinical staging: A heuristic and practical strategy for new research and better health and social outcomes for psychotic and relate mood disorders. *Canadian Journal of Psychiatry, 55*(8), 486–497.

Miller, J. (2009). What is the probability of replicating a statistically significant effect? *Psychonomic Bulletin and Review, 16*(4), 617–640.

Moffitt, T. E., Arseneault, L., Jaffee, S. R., Kim-Cohen, J., Koenen, K. C., Odgers, C. L., … Viding, E. (2008). Research review: DSM-V conduct disorder: Research needs for an evidence base. *Journal of Child Psychology and Psychiatry and Allied Disciplines, 49*(1), 3–33.

Müller, V. I., Cieslik, E. C., Serbanescu, I., Laird, A. R., Fox, P. T., & Eickhoff, S. B. (2017). Altered brain activity in unipolar depression revisited: Meta-analyses of neuroimaging studies. *JAMA Psychiatry, 74*(1), 47–55.

Nierenberg, A. A., Fava, M., Trivedi, M. H., Wisniewski, S. R., Thase, M. E., McGrath, P. J., … Rush, A. J. (2006). A comparison of lithium and T3 augmentation following two failed medication treatments for depression: A STAR*D report. *American Journal of Psychiatry, 163*(9), 1519–1530.

Nunes, A., Schnack, H. G., Ching, C. R. K., Agartz, I., Akudjedu, T., Alda, M., … Hajek, T. (2018). Using structural MRI to identify bipolar disorders—13 site machine learning study in 3020 individuals from the ENIGMA bipolar disorders working group. *Molecular Psychiatry.* https://doi.org/10.1038/s41380-018-0228-9 epub ahead of print.

Potkin, S. G., Guffanti, G., Lakatos, A., Turner, J. A., Kruggel, F., Fallon, J. H., … Macciardi, F. (2009). Hippocampal atrophy as a quantitative trait in a genome-wide association study identifying novel susceptibility genes for Alzheimer's disease. *PLoS ONE, 4*(8).

Regier, D. A., Narrow, W. E., Clarke, D. E., Kraemer, H. C., Kuramoto, S. J., Kuhl, E. A., & Kupfer, D. J. (2013). DSM-5 field trials in the United States and Canada, part II: Test-retest reliability of selected categorical diagnoses. *American Journal of Psychiatry, 170*(1), 59–70.

Renteria, M. E., Schmaal, L., Hibar, D. P., Couvy-Duchesne, B., Strike, L. T., Mills, N. T., … Hickie, I. B. (2017). Subcortical brain structure and suicidal behaviour in major depressive disorder: A meta-analysis from the ENIGMA-MDD working group. *Molecular Psychiatry, 7.*

Ripke, S., Neale, B. M., Corvin, A., Walters, J. T. R., Farh, K. H., Holmans, P. A., … O'Donovan, M. C. (2014). Biological insights from 108 schizophrenia-associated genetic loci. *Nature, 511*(7510), 421–427.

Sapolsky, R. M., Krey, L. C., & McEwen, B. S. (1984). Glucocorticoid-sensitive hippocampal neurons are involved in terminating the adrenocortical stress response. *Proceedings of the National Academy of Sciences of the United States of America, 81*(19), 6174–6177.

Satizabal, C. L., Adams, H. H. H., Hibar, D. P., White, C. C., Stein, J. L., Scholz, M., … Ikram, M. A. (2017). *Genetic architecture of subcortical brain structures in over 40,000 individuals worldwide.* bioRxiv. Retrieved from *http://biorxiv.org/content/early/2017/08/28/173831.abstract.*

Schmaal, L., Hibar, D. P., Saemann, P. G., Hall, G. B., Baune, B. T., & Jahanshad, N. (2017). Cortical abnormalities in adults and adolescents with major depression based on brain scans from 20 cohorts worldwide in the ENIGMA major depressive disorder working group. *Molecular Psychiatry, 22*(6), 900–909.

Schmaal, L., Veltman, D. J., van Erp, T. G. M., Sämann, P. G., Frodl, T., Jahanshad, N., … Hibar, D. P. (2016). Subcortical brain alterations in major depressive disorder: Findings from the ENIGMA major depressive disorder working group. *Molecular Psychiatry, 21*(6), 806–812.

Seth, A. K., & Friston, K. J. (2016). Active interoceptive inference and the emotional brain. *Philosophical Transactions of the Royal Society, B: Biological Sciences, 371*(1708), 20160007.

Smit, D. J. A., Wright, M. J., Meyers, J. L., Martin, N. G., Ho, Y. Y. W., Malone, S. M., … Boomsma, D. I. (2018). Genome-wide association analysis links multiple psychiatric liability genes to oscillatory brain activity. *Human Brain Mapping.* https://doi.org/10.1002/hbm.24238 epub ahead of print.

Sonderby, I. E., Gústafsson, O., Doan, N. T., Hibar, D. P., Martin-Brevet, S., … Andreassen, O. A. (2018). Dose response of the 16p11.2 distal copy number variant on intracranial volume and basal ganglia. *Molecular Psychiatry,* [in press], https://doi.org/10.1038/s41380-018-0118-1.

Starcke, K., & Brand, M. (2012). Decision making under stress: A selective review. *Neuroscience and Biobehavioral Reviews, 36*(4), 1228–1248.

Stein, D. J., Lund, C., & Nesse, R. M. (2013). Classification systems in psychiatry: Diagnosis and global mental health in the era of DSM-5 and ICD-11. *Current Opinion in Psychiatry, 26*(5), 493–497.

Stein, J. L., Medland, S. E., Vasquez, A. A., Hibar, D. P., Senstad, R. E., Winkler, A. M., … Thompson, P. M. (2012). Identification of common variants associated with human hippocampal and intracranial volumes. *Nature Genetics, 44*(5), 552–561.

Sullivan, P. (2007). Spurious genetic associations. *Biological Psychiatry, 61*(10), 1121–1126.

Sullivan, P. F., Agrawal, A., Bulik, C. M., Andreassen, O. A., Børglum, A. D., Breen, G., … Sklar, P. (2018). Psychiatric genomics: An update and an agenda. *American Journal of Psychiatry, 175*(1), 15–27.

van Erp, T. G. M., Hibar, D. P., Rasmussen, J. M., Glahn, D. C., Pearlson, G. D., Andreassen, O. A., … Turner, J. A. (2016). Subcortical brain volume abnormalities in 2028 individuals with schizophrenia and 2540 healthy controls via the ENIGMA consortium. *Molecular Psychiatry, 21*(4), 547–553.

van Erp, T. G. M., Walton, E., Hibar, D. P., Schmaal, L., Jiang, W., Glahn, D. C., … Turner, J. A. (2018). Cortical brain abnormalities in 4474 individuals with schizophrenia and 5098 control subjects via the enhancing neuro imaging genetics through meta analysis (ENIGMA) consortium. *Biological Psychiatry.* https://doi.org/10.1016/j.biopsych.2018.04.023 epub ahead of print.

Van Rooij, D., Anagnostou, E., Arango, C., Auzias, G., Behrmann, M., Busatto, G. F., … Buitelaar, J. K. (2018). Cortical and subcortical brain morphometry differences between patients with autism spectrum disorder and healthy individuals across the lifespan: Results from the ENIGMA ASD working group. *American Journal of Psychiatry, 175*(4), 359–369.

van Velzen, L. S., Kelly, S., Isaev, D., Aleman, A., Aftanas, L. I., … Bauer, J. Schmaal (2019). White matter disturbances in major depressive disorder: A coordinated analysis across 20 international cohorts in the ENIGMA MDD working group. *Molecular Psychiatry.* https://doi.org/10.1038/s41380-019-0477-2 [epub ahead of print].

Walters, R. G., Jacquemont, S., Valsesia, A., De Smith, A. J., Martinet, D., Andersson, J., … Beckmann, J. S. (2010). A new highly penetrant form of obesity due to deletions on chromosome 16p11.2. *Nature, 463*(7281), 671–675.

Walton, E., Hibar, D. P., van Erp, T. G. M., Potkin, S. G., Roiz-Santiañez, R., Crespo-Facorro, B., … Ehrlich, S. (2017). Positive symptoms associate with cortical thinning in the superior temporal gyrus via the ENIGMA schizophrenia consortium. *Acta Psychiatrica Scandinavica, 135*(5), 439–447.

Wray, N., Sullivan, P. F., & The Major Depressive Disorder Working Group of the Psychiatric Genomics Consortium. (2018). Genome-wide association analyses identify 44 risk variants and refine the genetic architecture of major depression. *Nature Genetics, 50*(5), 668–681.

Zhu, D., Li, Q., Riedel, B. C., Jahanshad, N., Hibar, D. P., Veer, I. M., ... Thompson, P. M. (2017). Large-scale classification of major depressive disorder via distributed Lasso. *Proceedings of SPIE*, 10160. https://doi.org/10.1117/12.2256935.

Further reading

Walton, E., Hibar, D. P., van Erp, T. G. M., Potkin, S. G., Roiz-Santiañez, R., Crespo-Facorro, B., ... Ehlrich, S. (2018). Prefrontal cortical thinning links to negative symptoms in schizophrenia via the ENIGMA consortium. *Psychological Medicine, 48*(1), 82–94.

Chapter 42

Applying a neural circuit taxonomy in depression and anxiety for personalized psychiatry

Leanne M. Williams[a,b] and Andrea N. Goldstein-Piekarski[a,b]

[a]Department of Psychiatry and Behavioral Sciences, Stanford University, Stanford, CA, United States; [b]Sierra-Pacific Mental Illness Research, Education, and Clinical Center (MIRECC), Veterans Affairs Palo Alto Health Care System, Palo Alto, CA, United States

1 Getting a handle on our terminology for precision psychiatry

The concept of "precision psychiatry" can be considered aligned with the overall move toward measurement-based care and precision medicine. Three aspects of precision medicine have been described by at least three other interchangeably used terms: *stratified medicine, personalized medicine,* and *precision health,* respectively (Coalition PM, 2014; Fernandes, Williams, Steiner, et al., 2017; Sciences AoM, 2015). Stratified medicine focuses on identifying subgroups of patients who will benefit from treatments as a step toward a fully personalized approach that tailors treatments to individual people. Personalized medicine has focused on harnessing new advances in genomics to select treatment options with the greatest likelihood of success. Precision health can be thought of as a major new frontier, expanding our breakthroughs in precision medicine to a wider concept of health and prevention that goes beyond a focus on disease.

In 2015, the Obama Administration launched a major research effort aimed at improving health and changing the way we treat disease (Secretary OotP, 2015a). In this initiative, precision medicine is described as "an innovative approach that takes into account individual differences in people's genes, environments, and lifestyles," thereby allowing doctors to tailor treatment to individual patients (Secretary OotP, 2015a). When launching the Precision Medicine Initiative, President Obama said: "Precision medicine—in some cases, people call it personalized medicine – gives us one of the greatest opportunities for new medical breakthroughs that we have ever seen" (Secretary OotP, 2015b).

This federal Precision Medicine Initiative is paralleled by two federal research efforts focused specifically on psychiatry and neurosciences. First, the "BRAIN Initiative" is aimed at developing neuro-technologies for demystifying brain disorders, including psychiatric disorders (Markoff, 2013). This depth of understanding of the brain will only strengthen our ability to precisely identify dysfunction at the level of a single patient. Second, the U.S. National Institute of Mental Health (NIMH) is pioneering the Research Domain Criteria (RDoC) project (Insel, 2010), which has initiated a research approach to generating a neurobiologically valid framework for classifying psychiatric disorders, and for generating novel interventions related to neurobiological underpinnings. Together, the Precision Medicine, BRAIN, and RDoC initiatives will support and promote great advances in precision psychiatry.

A major challenge for precision medicine in psychiatry is that psychiatry does not yet use measurement to track the equivalent of vitals (Harding, Rush, Arbuckle, Trivedi, & Pincus, 2011), or image the organ of interest (the brain). Although modern neuroimaging techniques have generated many insights about types of brain circuit dysfunction that underlie psychiatric disorders, these insights have not been systematically linked to prediction of clinical outcomes, and have not been delivered into the hands of clinicians as an actionable system for improving people's lives. That precision psychiatry is new to the game in precision medicine is not surprising given the complexity of the organ and behavior of interest. However, psychiatry can benefit tremendously by including advanced diagnostic and therapeutic technologies that form an integral part of other clinical specialties.

Personalized Psychiatry. https://doi.org/10.1016/B978-0-12-813176-3.00042-0

2 Foundations for a precision psychiatry taxonomy for depression and anxiety based on neural circuits

With advances in brain imaging techniques with sufficient spatial and temporal resolution to quantify neural connections in vivo, we have the foundations for formulating an understanding of mental illness as disorders of brain functioning (Williams, 2016, 2017). Here, the focus is on depression and anxiety as an illustration of such a formulation.

.The term "neural circuit" has typically referred to how one neuron communicates with another through synaptic connections and transmission (Yuste, 2015). Here, the term "large-scale neural circuit" is used to refer to the macroscale neural organization. In brain imaging studies, these macroscale circuits have commonly been referred to as "networks" (e.g., the "default mode network") (Greicius, Krasnow, Reiss, & Menon, 2003; Raichle, 2015).

Neural circuits can be considered a pertinent scale of measurement from which to delineate a neurobiology of human mental disorder. Circuits integrate across different levels and measures of brain function, but still reflect the complexity of the brain. Circuits are engaged by specific human cognitive, emotional, and self-reflective functions, and offer promise for defining appropriate animal homologues. It is likely that most of the human brain involves multiple parallel circuits that are interdigitated such that each cortical lobe contains components of multiple circuits (Felleman & Van Essen, 1991; Mesulam, 1998). This organization may have occurred with the expansion of the association cortex in humans relative to non-human primates (Mesulam, 1998). Mood and anxiety disorders may be possible maladaptive consequences of this expansion.

Researchers have identified intrinsic neural circuits that support domain-general processes of self-reflection, salience perception, and alertness (Fig. 1; Buckner, Krienen, & Yeo, 2013), as well as conflict monitoring, attention, sensori-motor, visual, and auditory processes (Fox, Snyder, Vincent, et al., 2005; Gordon, Laumann, Adeyemo, et al., 2014; Lindquist & Barrett, 2012; Oosterwijk, Lindquist, Anderson, et al., 2012; Power, Cohen, Nelson, et al., 2011; Seeley, Menon, Schatzberg, et al., 2007; Sheline, Price, Yan, & Mintun, 2010; Spreng, Sepulcre, Turner, Stevens, & Schacter, 2013; Yeo, Krienen, Sepulcre, et al., 2011). The intrinsic architecture has been demonstrated using large-scale functional connectivity analysis of hundreds of brain regions that have been identified using parcelation and meta-analysis, and that define major brain systems at rest and across many task-evoked states (e.g., Cole, Bassett, Power, Braver, & Petersen, 2014). These circuits may be observed in the task-free state and during task-evoked conditions. During rest, the default mode circuit tends to upregulated, and other circuits, downregulated (Cole et al., 2014; Fox et al., 2005). Specific task states (such as those designed to probe reactivity to potential threat or reward) engage more specialized functional components of these circuits (e.g., Castelli, Happe, Frith, & Frith, 2000; Haber & Knutson, 2010; Rushworth, Mars, & Sallet, 2013; Touroutoglou, Andreano, Barrett, & Dickerson, 2015; White, Coniston, Rogers, & Frith, 2011; Williams, Das, Liddell, et al., 2006) (Fig. 1).

In the illustrative formulation, dysfunctions are considered in relation to six circuits: default mode, salience, negative affect, positive affect, attention, and cognitive control.

Research to date has understandably focused on case:control comparisons of diagnostic groups of mood and anxiety disorder defined by traditional symptom criteria. These previous studies have also focused on activation within specific brain regions of interest, and typically on one imaging modality at a time. While the emphasis has been on regional activation, there has been a noticeable expansion of functional connectivity studies of depression. Findings from case:control studies tend to be inconsistent. This inconsistency is not surprising, given the heterogeneity of depression and anxiety. It is feasible that the heterogeneity reflects the contributions, depending on each sample, of different profiles of neural hyper-reactivity and hyper-reactivity, along with hypoconnectivity and hyper-connectivity, compared with healthy peers.

3 "Default Mode" circuit

The default mode circuit (usually referred to as the "default mode network") encompasses the anterior medial prefrontal cortex (amPFC), posterior cingulate cortex (PCC), and angular gyrus (AG) (Greicius et al., 2003; Greicius, Supekar, Menon, & Dougherty, 2009) (Fig. 1; Table 1). This circuit is probed under task-free conditions, and typically when participants are instructed to reflect on their freely generated thoughts (Seeley et al., 2007; Shulman et al., 1997). Anterior and posterior regions may represent sub-networks of the default mode circuit (for review, Mulders, van Eijndhoven, Schene, Beckmann, & Tendolkar, 2015). The default mode circuit is engaged, even during "rest" periods that occur between task stimuli (Korgaonkar, Ram, Williams, Gatt, & Grieve, 2014). By extent of default, more network functioning may also be quantified as a summary composite, and used to locate an individual, relative to a normative distribution (Ball, Goldstein-Piekarski, Gatt, & Williams, 2017).

TABLE 1 A summary of the current state of knowledge for macroscale circuits involved in core human functions, the functional role of these circuits, the nature of their functional disruptions in mood and anxiety disorders, corresponding structural alterations and the contribution of these circuits to understanding treatment outcomes

Circuit	Role	Functional alterations in depression and anxiety	Structural alterations in depression and anxiety	Treatment outcomes
Default mode: Anterior middle frontal cortex (amPFC), posterior cingulate cortex (PCC), and angular gyrus (AG) (Greicius et al., 2003, 2009)	Self-referential thinking at rest (Greicius, Flores, Menon, et al., 2007; Shulman, Fiez, Corbetta, et al., 1997)	*Hypoconnectivity:* Posterior hypoconnectivity correlated with over-general memory (Zhu, Wang, Xiao, et al., 2012) and treatment sensitivity in MDD (for review, Dichter, Gibbs, & Smoski, 2014). mPFC-AG hypo-connectivity in SAD (Qiu, Liao, Ding, et al., 2011) *Hyper-connectivity:* Anterior medial hyper-connectivity in MDD (Greicius et al., 2007; Kaiser, Andrews-Hanna, Wager, & Pizzagalli, 2015; Sheline et al., 2010), correlated with rumination in MDD (Hamilton, Furman, Chang, et al., 2011; Zhu et al., 2012) and treatment resistance in MDD (de Kwaasteniet, Rive, Ruhe, et al., 2015; Dichter et al., 2014; Wu, Andreescu, Butters, et al., 2011) Hyper-connectivity of the default mode with the attention circuit in MDD for meta-analysis (Kaiser et al., 2015)	*Gray matter:* Reductions in MDD (Grieve, Korgaonkar, Koslow, Gordon, & Williams, 2013; Singh, Kesler, Hadi Hosseini, et al., 2013) *White matter:* Hypoconnectivity in MDD (Korgaonkar, Fornito, Williams, & Grieve, 2014; Qin, Wei, Liu, et al., 2014)	*Hypoconnectivity:* Hypoconnectivity predicts nonremission to common antidepressants in older adults with depression (Andreescu, Tudorascu, Butters, et al., 2013) *Intact connectivity:* Patients with depression correlated with remittance with antidepressants (Liston, Chen, Zebley, et al., 2014) Functional connectivity between subgenual ACC and default mode posterior cingulate predictive of response to TMS (Philip, Barredo, van't Wout-Frank, et al., 2018)
Salience: Dorsal anterior cingulate cortex (dACC), anterior insula (aI), Temporal Pole (TP), and sublenticular extended amygdala (SLEA) (Seeley et al., 2007; Oosterwijk et al., 2012)	Detecting salient changes (Seeley et al., 2007)	*Hypoconnectivity:* Amygdala-insula hypoconnectivity in MDD (Veer, Beckmann, van Tol, et al., 2010) Insula hypoconnectivity in MDD correlated with overall symptom severity (Manoliu, Meng, Brandl, et al., 2014) Amygdala—ACC hypoconnectivity in SAD (Arnold Anteraper, Triantafyllou, Sawyer, et al., 2014) Amygdala hypoconnectivity correlated with avoidance symptoms (Liao, Qiu, Gentili, et al., 2010) *Hyper-connectivity:* Hyper-connectivity of the salience with the default mode circuit in MDD (Manoliu et al., 2014), correlated with severity of depressive rumination (Hamilton et al., 2011)		*Right anterior insula metabolism:* Greater resting state metabolism associated with better response to escitalopram, poor response to CBT (Dunlop, Kelley, McGrath, Craighead, & Mayberg, 2015; McGrath, Kelley, Dunlop, et al., 2014) *Insula hyporeactivity:* Hyporeactivity to IAPS stimuli associated with response to fluoxetine/olanzapine (Rizvi, Salomons, Konarski, et al., 2013) *Hyper-connectivity:* Rostral ACC hyper-connectivity predicts response to placebo (Sikora, Heffernan, Avery, et al., 2016) *Intrinsic functional connectivity:* Insula and dorsal ACC connectivity changes correlated with response to rTMS (Philip et al., 2018)

TABLE 1 A summary of the current state of knowledge for macroscale circuits involved in core human functions, the functional role of these circuits, the nature of their functional disruptions in mood and anxiety disorders, corresponding structural alterations and the contribution of these circuits to understanding treatment outcomes—cont'd

Circuit	Role	Functional alterations in depression and anxiety	Structural alterations in depression and anxiety	Treatment outcomes
Threat: Amygdala, dorsal, rostral, and subgenual ACC, mPFC and insula (Seeley et al., 2007; Oosterwijk et al., 2012)	Threat reactivity and regulation (Williams et al., 2006)	*Altered activation for threat:* Amygdala hyper-activation for threat faces in MDD, GAD, SAD (Blair, Shaywitz, Smith, et al., 2008; Prater, Hosanagar, Klumpp, Angstadt, & Phan, 2013) and anxiety traits (Clauss, Avery, VanDerKlok, et al., 2014; Etkin, Klemenhagen, Dudman, et al., 2004) ACC hypoactivation to threat faces in GAD and SAD (Blair, Geraci, Smith, et al., 2012; Etkin, Prater, Hoeft, Menon, & Schatzberg, 2010; Palm, Elliott, McKie, Deakin, & Anderson, 2011) and amygdala hypoactivation to threat faces in MDD (Lawrence, Williams, Surguladze, et al., 2004; Matthews, Strigo, Simmons, Yang, & Paulus, 2008; Thomas, Drevets, Dahl, et al., 2001; Williams, Korgaonkar, Song, et al., 2015) *Hypoconnectivity for threat:* Amygdala-ACC hypoconnectivity to fear in MDD (Matthews et al., 2008; Musgrove, Eberly, Klimes-Dougan, et al., 2015), SAD (Prater et al., 2013), and GAD (Etkin et al., 2010; Etkin, Prater, Schatzberg, Menon, & Greicius, 2009) Amygdala hypoactivation to fear/anger is a general predictor of response to antidepressants, and amygdala hyper-activation to sad is a differential predictor of nonresponse to SNRI antidepressants (Williams et al., 2015)	*Gray matter:* Reduced hippocampal gray matter in MDD and GAD (Bossini, Tavanti, Calossi, et al., 2008; Kempton, Salvador, Munafo, et al., 2011; Moon, Kim, & Jeong, 2014; Mueller, Aouidad, Gorodetsky, et al., 2013; Zhao, Du, Huang, et al., 2014) *White matter connectivity:* Reduced uncinate fasciculus white matter connections in MDD (Steffens, Taylor, Denny, Bergman, & Wang, 2011)	*Hyporeactivity:* Amygdala hypo-reactivity and smaller amygdala volume associated with response to escitalopram, sertraline, and paroxetine (Li, Lin, Chou, et al., 2010; Ruhe, Booij, Veltman, Michel, & Schene, 2012; Williams et al., 2015) *Hyper-connectivity:* Hyper-connectivity between amygdala and ACC distinguishes nonremitters to SSRI escitalopram (Fu, Williams, Cleare, et al., 2004) *Functional connectivity:* Subgenual ACC functional connectivity attenuated by rTMS (Taylor, Ho, Abagis, et al., 2018)

Reward: Ventral striatum, orbitofrontal cortex (OFC), dACC, and mPFC regions (Berridge & Kringelbach, 2008; Haber & Knutson, 2010)	Sensitivity to and anticipation of reward stimuli	*Striatal hypoactivation:* Anhedonic MDD: Striatal hypo-activation for happy faces (Keedwell, Andrew, Williams, Brammer, & Phillips, 2005a) (for review, Treadway & Zald, 2011; for meta-analysis, Zhang, Chang, Guo, Zhang, & Wang, 2013) and monetary tasks (Treadway & Zald, 2011) (for meta-analysis, Hamilton, Etkin, Furman, et al., 2012); mPFC hypoactivation for positively valenced stimuli (Mitterschiffthaler, Kumari, Malhi, et al., 2003) *Altered frontal activation: ACC/MPFC/OFC/* hyper-activation for happy faces (Keedwell et al., 2005a; Keedwell, Andrew, Williams, Brammer, & Phillips, 2005b; Mitterschiffthaler et al., 2003; Zhang et al., 2013), reward anticipation and reward outcomes (Dichter, Kozink, McClernon, & Smoski, 2012) in MDD	*Gray matter:* Reduced striatal volume in MDD (Kim, Hamilton, & Gotlib, 2008; Pizzagalli, Holmes, Dillon, et al., 2009) *White matter:* Reduced white matter connectivity in MDD (Sacchet, Prasad, Foland-Ross, et al., 2014)	*Increased activation:* Higher activation of ACC associated with greater symptom reduction with behavioral activation therapy for depression (Carl, Walsh, Eisenlohr-Moul, et al., 2016) Increased nucleus accumbens activation associated with response to venlafaxine and fluoxetine (Heller, Johnstone, Light, et al., 2013) *Hyper-connectivity:* Hyper-connectivity between the striatum and the ventral medial region of the orbitofrontal cortex associated with response to rTMS (Downar, Geraci, Salomons, et al., 2014)
Attention: Medial superior frontal cortices (msPFC), aI, anterior inferior parietal lobule (aIPL), and precuneus (PCu) (Gordon et al., 2014)	Alertness and sustained attention (Fornito, Harrison, Zalesky, & Simons, 2012)	*Hypoconnectivity:* Hypoconnectivity in MDD (Liao et al., 2010; Qiu et al., 2011; Veer et al., 2010) *Hyper-connectivity:* Frontoparietal hyper-connectivity with the striatal node of the reward circuit in SAD (Arnold Anteraper et al., 2014)	*White matter:* Reduced frontoparietal diffusion centrality in MDD (Qin et al., 2014)	*Intrinsic functional connectivity:* Connectivity between intraparietal sulcus and orbital frontal cortex predicts somatic symptom improvement with behavioral activation therapy for depression (Crowther, Smoski, Minkel, et al., 2015) Increased perfusion and connectivity between attention circuit and amygdala associated with response to CBT (Shou, Yang, Satterthwaite, et al., 2017; Sosic-Vasic, Abler, Gron, Plener, & Straub, 2017)

Continued

TABLE 1 A summary of the current state of knowledge for macroscale circuits involved in core human functions, the functional role of these circuits, the nature of their functional disruptions in mood and anxiety disorders, corresponding structural alterations and the contribution of these circuits to understanding treatment outcomes—cont'd

Circuit	Role	Functional alterations in depression and anxiety	Structural alterations in depression and anxiety	Treatment outcomes
Cognitive control: Dorsolateral prefrontal cortex (DLPFC), ACC, precentral gyrus (PCG), dorsal parietal cortex (DPC) (Niendam, Laird, Ray, et al., 2012)	Working memory and selective attention (Chen, Chang, Oathes, et al., 2013; Niendam et al., 2012)	*Hypoactivation:* DLPFC/ACC hypoactivation in MDD (Elliott, Baker, Rogers, et al., 1997; Elliott, Sahakian, Herrod, Robbins, & Paykel, 1997; Holmes & Pizzagalli, 2008; Korgaonkar, Grieve, Etkin, Koslow, & Williams, 2013; Siegle, Thompson, Carter, Steinhauer, & Thase, 2007; Vasic, Walter, Sambataro, & Wolf, 2009; Vilgis, Chen, Silk, Cunnington, & Vance, 2014) and in social anxiety (Koric, Volle, Seassau, et al., 2012) and induced anxious mood (Fales, Barch, Burgess, et al., 2008) *Hypoconnectivity:* DLPFC-ACC hypoconnectivity in MDD (Holmes & Pizzagalli, 2008; Vasic et al., 2009) *Hyper-activation:* Hyper-activation in MDD, suggesting compensation to achieve normal cognitive performance (Fitzgerald, Srithiran, Benitez, et al., 2008; Harvey, Fossati, Pochon, et al., 2005; Matsuo, Glahn, Peluso, et al., 2007; Walter, Wolf, Spitzer, & Vasic, 2007)	*Gray matter:* Reduced DLPFC and ACC gray matter in adult MDD (Grieve et al., 2013) and late-life MDD (Chang, Yu, McQuoid, et al., 2011)	*Activation:* Go-NoGo activation of middle and inferior gyrus within the DLPFC associated with depression symptom improvement with SSRIs (Langenecker, Kennedy, Guidotti, et al., 2007). For the DLFPC, greater Go-NoGo inhibition predicts worse outcomes for escitalopram and sertraline (Gyurak, Patenaude, Korgaonkar, et al., 2016), and less inhibition predicts remission on the SNRI venlafaxine-XR (Gyurak et al., 2016) GoNoGo-elicited dACC hyper-activation associated with poorer response to escitalopram and duloxetine (Crane, Jenkins, Bhaumik, et al., 2017) *Intrinsic functional connectivity:* Increased DLPFC-amygdala connectivity associated with group CBT treatment of depression in adolescents (Straub, Metzger, Plener, et al., 2017)

3.1 Default mode circuit disruptions in depression and anxiety

Most commonly, depression has been associated with over-activation and hyper-connectivity relative to controls within the default mode network (Greicius et al., 2007; Kaiser et al., 2015; Sheline et al., 2010). Hyper-connectivity of the default mode circuit with subgenual ACC in MDD has been associated with higher levels of maladaptive rumination about depressive thoughts (Hamilton et al., 2011; Zhu et al., 2012). There is also evidence for *hypo*connectivity of the default mode circuit, particularly in posterior medial cortex regions, in MDD relative to controls that are correlated with clinical indicators of over-general autobiographical memory (Zhu et al., 2012). Within those with depression, a data-driven approach using community detection algorithms has identified two unique subtypes of depression, differentiated by the presence or absence of PCC-anterior cingulate cortex (ACC)/mPFC connectivity (Price, Gates, Kraynak, Thase, & Siegle, 2017).

Anatomical abnormalities might contribute to default mode circuit hyper- and hypofunction. Structurally, MDD has been associated with decreased regional gray matter connectivity (Singh et al., 2013) and loss of white matter connectivity (Korgaonkar, Fornito, et al., 2014) within the default mode circuit, particularly within the posterior sub-network. Wide-spread reductions in gray matter have also been observed across regions of the default mode circuit and in nodes within interacting circuits (Grieve et al., 2013). Specifically, MDD patients show reduced gray matter volume in ACC and anterior medial regions of the prefrontal cortex, and in parietooccipital regions, consistent with components of the default mode circuit, as well as in striatal and limbic components of the affective circuits (Grieve et al., 2013).

3.2 Default mode circuit and treatment implications

The observations of multiple default mode network connectivity profiles in depression suggest distinct implications for treatment.

One promising line of treatment research indicates that knowing about the type of default mode circuit dysfunction at a pretreatment baseline is important for predicting which patients are likely or not to be responsive to a specific type of treatment. This knowledge is important for ultimately adding objective measures to the armamentarium of clinicians, and getting people to the right treatment early. Default mode circuit hypoconnectivity between anterior and posterior nodes relative to controls at baseline appears to characterize a form of depression that is resistant to typical first-line medications (Goldstein-Piekarski, Staveland, Ball, et al., 2018). Pretreatment hypoconnectivity of the default mode also predicts non-remission to commonly prescribed antidepressants in older adults with depression (Andreescu et al., 2013). By contrast, relatively intact default mode connectivity, indistinguishable from healthy controls, characterizes patients with depression who go on to remit to antidepressant treatment. These findings suggest that a certain degree of intact connectivity may be necessary for antidepressant action. Complementing these findings, distinct profiles of functional connectivity between the subgenual ACC and default mode posterior cingulate may be predictive of treatment response to the newly FDA approved technique, transcranial magnetic stimulation (TMS) (Liston et al., 2014; Philip et al., 2018). TMS may act to improve depression symptomology in this subtype by normalizing hyper-connectivity within the default mode network (Liston et al., 2014). Together, these findings lend support to a circuit-guided approach to helping tailor choice of intervention.

4 "Salience" circuit

The "salience" circuit is defined by core nodes in the ACC, anterior insula (aI), and sublenticular extended amygdala (Oosterwijk et al., 2012; Seeley et al., 2007) (Fig. 1). This circuit is implicated in the detection of salient interoceptive and exteroceptive stimuli (Seeley et al., 2007).

4.1 Salience circuit disruptions in depression and anxiety

Within the salience circuit, insula hyper-reactivity has been observed in MDD when evoked by sadness and disgust (Sprengelmeyer, Steele, Mwangi, et al., 2011; Suslow, Konrad, Kugel, et al., 2010) (for review, Stuhrmann, Suslow, & Dannlowski, 2011) (Table 1). Heightened insula reactivity is positively correlated with severity of depressive symptoms (Lee, Seong Whi, Hyung Soo, et al., 2007), suggesting a bias toward salient and mood-congruent stimuli. Individuals with generalized social anxiety disorder also show exaggerated insula reactivity when attending to salient emotional faces (Klumpp, Post, Angstadt, Fitzgerald, & Phan, 2013). These functional activation differences might be due, in part, to structural deficits. For example, MDD patients show a loss of insula gray matter, which is negatively correlated with symptom severity (Sprengelmeyer et al., 2011).

In regard to functional connectivity, profiles of both hyper- and hypoconnectivity have been observed in depression and anxiety. Insula hypoconnectivity within the salience circuit has been observed in depression, social anxiety disorder, and in panic disorder (for review, Mulders et al., 2015; Peterson, Thome, Frewen, & Lanius, 2014). Insula hypoconnectivity has been inversely associated with symptom severity (Mulders et al., 2015). In generalized anxiety, a weakening of the normal connectivity between the insula and the ACC has been observed, specifically when the patient is required to focus attention on salient emotional faces presented among neutral stimuli (such as shapes) (Klumpp et al., 2013).

Hypoconnectivity between the insula and amygdala has also been reported in MDD (Veer et al., 2010) and correlated with overall symptom severity (Manoliu et al., 2014) (for review, Mulders et al., 2015). Amygdala hypoconnectivity has been more specifically correlated with avoidance symptoms in social anxiety disorder (Liao et al., 2010). Correspondingly, hypoconnectivity between the amygdala and ACC has also been observed in social anxiety disorder (Arnold Anteraper et al., 2014).

Hyper-connectivity between the insula and anterior nodes of the default mode circuit has been positively correlated with symptoms of nervous apprehension, and reported in both depression (for review, Mulders et al., 2015) and social anxiety disorder (for review, Peterson et al., 2014).

Dorsal nodes of the salience circuit show both hyper- and hypoconnectivity with the posterior precuneus node of the attention circuit (for meta-analysis, Kaiser et al., 2015). The direction of altered connectivity between salience and other circuits may fluctuate with the nature of interoceptive or exteroceptive stimuli.

4.2 Salience circuit and treatment implications

Relatively few studies have focused on the salience circuit with respect to treatment outcome. Nonetheless, the small number of studies already indicates promise in utilizing profiles of salience circuit dysfunction to guide differential treatment choices to a broad array of depression treatment modalities, including psychotherapy, SSRIs, and TMS. For example, greater right anterior insula resting state metabolism as measured by positron emotion tomography (PET) imaging has been associated with a better response to escitalopram, but a poor response to CBT (Dunlop et al., 2015; McGrath et al., 2014; McGrath, Kelley, Holtzheimer, et al., 2013). Response to fluoxetine/olanzapine has also been associated with a pre-treatment hyporeactivity of the insula elicited by IAPS stimuli (Rizvi et al., 2013). Salience circuit function may also help differentiate patients for whom harnessing the value of placebo conditions can be clinically beneficial. Hyper-connectivity of the salience circuit, specifically with the rostral ACC (rACC), has been found to predict symptom alleviation in response to a placebo condition (Sikora et al., 2016).

Salience circuit connectivity may be changed by interventions designed to noninvasively target neural modulation. Changes in intrinsic functional connectivity in components of the salience network, including the insula and dorsal ACC (dACC), identified by a whole-brain multivoxel pattern activation (MVPA) analysis were correlated with response to rTMS (Philip et al., 2018). These findings involve changes in connectivity between the salience circuit and other circuits, consistent with the role of the salience circuit role in orienting attention toward meaningful stimuli, and in attentional biases in depression. As an example, rTMS has been found to decrease functional connectivity between the salience circuit and the hippocampus, and this decrease is correlated with depressive symptom improvement (Philip et al., 2018). Relatively greater resting state functional connectivity between the insula and middle temporal gyrus has also been found to predict response to psychotherapy when regions of interest are defined by those that differ in connectivity between depressed patients and healthy controls (Crowther et al., 2015).

5 Affective circuits

Affective circuits are commonly probed by biologically relevant stimuli such as facial expressions of potential threat (fear, anger) and social reward (happy).

6 Negative Affect

The circuit engaged by negative affect comprises subcortical nodes in the amygdala, brainstem regions, hippocampus, and insula, and both dorsal and ventral prefrontal nodes—dorsal medial prefrontal cortex (dmPFC) and dACC connections as well as ventral mPFC (vMPFC) and ventral (subgenual and pregenual)-rACC connections (Kober, Barrett, Joseph, et al., 2008; Robinson, Krimsky, Lieberman, et al., 2014; Fig. 1; Table 1). In light of their commonly observed co-activation (Kober et al., 2008), the negative affect circuit might subserve the perception of negative emotion cues and the salience circuit, the arousal aspects of feeling these emotions.

In the task-free state, hyper-connectivity between the anterior (subgenual ACC and dMPFC) nodes of the negative affect circuit and the default mode has been observed in depression (Hamilton, Farmer, Fogelman, & Gotlib, 2015; Sheline et al., 2010). This state of intrinsic hyper-connectivity is thought to drive rumination and the negative attributions that underlie negative biases and negative mood.

Threat-evoked components of the negative affective circuits comprise the amygdala, hippocampus, insula, and both dorsal and ventral portions of the prefrontal cortex, including the dorsal medial prefrontal cortex (dmPFC) and its dorsal ACC connections, and the ventral mPFC (vMPFC) and its ventral (subgenual and pregenual)-rostral ACC connections (Kober et al., 2008; Robinson et al., 2014; Fig. 1; Table 1). The dorsal prefrontal sub-circuit has been preferentially implicated in appraisal and expression of emotion, and may be considered an "aversive amplification" sub-network (Robinson et al., 2014) that serves to boost the processing of signals of potential threat (Robinson et al., 2014). Complementing this function, the ventral sub-circuit has been implicated in automatic regulation of negative emotion (Etkin et al., 2010; Kober et al., 2008) (for review, Phillips, Drevets, Rauch, & Lane, 2003; for meta-analysis, Kober et al., 2008). These components overlap with components of the salience circuit, and they may both be engaged by the arousal and valence properties of threat stimuli, respectively. These sub-networks may be engaged, even in the absence of conscious sensory awareness (Williams et al., 2006) (for meta-analysis, Kober et al., 2008). In light of their commonly observed co-activation (Kober et al., 2008), the negative affect circuit might subserve the perception of negative emotion cues and the salience circuit, the arousal aspects of feeling these emotions.

6.1 Negative affective circuit disruptions in depression and anxiety

Altered threat processing, involving amygdala-ACC activation and connectivity, have been observed across multiple diagnostic categories (for review, Kim, Loucks, Palmer, et al., 2011; Phillips et al., 2003; Price & Drevets, 2010). Amygdala over-reactivity elicited by implicit or non-conscious processing of threat-related stimuli has been reported in current depressive disorder (Peluso, Glahn, Matsuo, et al., 2009; Sheline, Barch, Donnelly, et al., 2001; Surguladze, Brammer, Keedwell, et al., 2005; Williams et al., 2015; Yang, Simmons, Matthews, et al., 2010) (for review, Jaworska, Yang, Knott, & Macqueen, 2014), generalized anxiety disorder (Fonzo, Ramsawh, Flagan, et al., 2015), generalized social phobia/anxiety disorder (Fonzo et al., 2015; Phan, Fitzgerald, Nathan, & Tancer, 2006; Stein, Goldin, Sareen, Zorrilla, & Brown, 2002), specific phobia (Killgore, Britton, Schwab, et al., 2014), and panic disorder (Fonzo et al., 2015; Killgore et al., 2014). Excessive amygdala activity elicited by masked threat faces has also been associated with a dimension of trait anxiety, and with neuroticism in otherwise healthy people (Etkin et al., 2004), consistent with a trait-like phenotype of hyper-reactivity to sources of threat. Hypo-activation of the ACC has been observed in generalized anxiety disorder (Blair et al., 2012; Etkin et al., 2010; Palm et al., 2011) and generalized social anxiety (Blair et al., 2012). The amygdala is also more generally engaged by biologically significant emotion. In addition to the findings for threat, MDD has also been associated with mood-congruent hyper-reactivity of the amygdala evoked by sad faces (Arnone, McKie, Elliott, et al., 2012; Fu et al., 2004; Victor, Furey, Fromm, Ohman, & Drevets, 2010).

Alterations in amydala activation may also reflect a reduction in connectivity between the amygdala and subgenual/ ventral ACC, observed during implicit processing of threat-related faces in unmedicated MDD (Matthews et al., 2008; Musgrove et al., 2015), generalized social anxiety disorder (Prater et al., 2013) and generalized anxiety disorder (Etkin et al., 2010). A lack of connectivity elicited during the conscious evaluation of threat has also been observed between the amygdala and prefrontal regions, including the ACC (Clauss et al., 2014), OFC (Sladky, Hoflich, Kublbock, et al., 2013), mPFC (Hahn, Stein, Windischberger, et al., 2011; Prater et al., 2013) for social anxiety disorder. Disruptions in amygdala-ACC functional connectivity might also have a basis in disruptions to white matter connectivity. For example, MDD has been associated with reduction in the uncinate fasciculus white matter connections that support functional communication between the amygdala and ACC (Steffens et al., 2011).

6.2 Negative affective circuit and treatment implications

Pretreatment hyporeactivity of the amygdala during implicit processing of threat-related faces has been found to predict subsequent response to escitalopram and sertraline, with around 75% cross-validated accuracy (Williams, Korgaonkar, Song, et al., 2015b). Amygdala hypoactivation to negative emotional faces has also been associated with treatment response to another SSRI, paroxetine (Ruhe et al., 2012). By contrast, nonresponders had relative hyper-reactivity of the amygdala at the pretreatment baseline (Williams et al., 2015). Complementing these findings, smaller amygdala volume at pretreatment baseline has also been associated with remission of depressive symptoms following pharmacotherapy (Li et al., 2010). Response to the SSRI escitalopram after 6 weeks has also been associated with decreases in amygdala activation early in treatment (7 days) (Godlewska, Browning, Norbury, Cowen, & Harmer, 2016). Relatedly, functional hyper- (rather than hypo-) connectivity between the amygdala and ACC during the processing of fearful faces has also been found to distinguish non-remitters to the SSRI escitalopram (Matthews et al., 2008), whereas remitters have not been found to differ from healthy controls on amygdala connectivity (Vai, Bulgarelli, Godlewska, et al., 2016).

Supplementing these findings for prediction based on baseline affective circuit function, a third line of mechanistically oriented research has demonstrated changes in circuit function post-pharmacotherapy. These findings suggest that circuit function is a viable target endpoint for intervention studies. For studies showing hypoactivation or hypoconnectivity at baseline, both SSRIs (escitalopram and sertraline), and the SNRI venlafaxine-xr have been found to increase amygdala activation to implicit threat stimuli (Williams et al., 2015). Moreover, the SSRI fluoxetine increased amygdala-ACC connectivity elicited by such stimuli (Chen, Suckling, Ooi, et al., 2008). Other studies report that, following treatment with SSRIs (sertraline, escitalopram, citalopram, fluoxetine, or paroxetine), there is attenuation of baseline amygdala hyperactivation elicited by negative valence emotional faces (Arnone et al., 2012; Fu et al., 2004; Victor et al., 2010), particularly for responders (Godlewska, Norbury, Selvaraj, Cowen, & Harmer, 2012; Ruhe et al., 2012). Together, both studies of both prediction and pre-post treatment change suggest that the direction of disruption (hypo- or hyper-) may depend on the task used to probe the affective circuit, such that both directions of disruption are relevant for future study. In one illustrative rTMS study to date, stimulation (relative to sham) in patients with major depression was found to attenuate subgenual ACC (sgACC) resting state functional connectivity with the negative affective network, most apparent in responders to stimulation (Taylor et al., 2018).

7 Positive affect circuit: "Reward"

Reward processing components of the affective circuits are defined by the striatal nucleus accumbens, and ventral tegmental areas (collectively referred to as "the striatum") and their projections to the orbitofrontal cortex (OFC) and mPFC (Berridge & Kringelbach, 2008). These regional components are preferentially engaged by different types of reward processing, including sensitivity to the presence of salient reward stimuli and the anticipation of these stimuli. There are also connections between the striatum and the amygdala, consistent with interactions between the processing of threat and reward and of significant stimuli that encompass multiple valences (Haber & Knutson, 2010).

7.1 Reward circuit disruptions in depression and anxiety

Across studies, hypoactivation of the striatum has been identified as a robust characteristic of at least some patients with depression, especially those who report experiences of anhedonia (for meta-analysis, Hamilton et al., 2012; for review, Treadway & Zald, 2011). Such hypoactivation in depression is elicited not only by primary signals of social reward (such as happy faces), but also by tasks that rely on reward-motivated decision-making (Treadway & Zald, 2011). Striatal hypoactivation also characterizes adolescents at risk of depression (Gotlib, Hamilton, Cooney, et al., 2010), suggesting that a trait-like disruption to reward circuits may contribute to the development of mood disorder. Consistent with the possibility of a trait-like biotype for altered reward circuitry and anhedonia, depression has also been associated with gray matter loss in the striatum (Kim et al., 2008; Pizzagalli et al., 2009). In addition, depression has been associated with increased white matter connectivity in bilateral corticospinal tracts, a structural alteration that might underlie some aspects of the striatal and motor functional disruptions in this disorder (Sacchet et al., 2014).

For socially rewarding facial expressions of happiness, hypoactivation of the amygdala has also been observed in unmedicated MDD (Williams et al., 2015), generalized anxiety disorder (Blair et al., 2008), panic disorder (Ottaviani, Cevolani, Nucifora, et al., 2012), and obsessive compulsive disorder (OCD) (Cannistraro, Wright, Wedig, et al., 2004), and may reflect a further neural characteristic of transdiagnostic anhedonia. Frontally, in remitted depression, *hyper*-activation of the OFC, medial prefrontal/midfrontal regions and ACC has also been observed in response to happy faces (Keedwell et al., 2005a; Mitterschiffthaler et al., 2003), reward outcomes (Dichter et al., 2012), and reward anticipation (for meta-analysis) (Zhang et al., 2013). Frontal hyper-activation might reflect an adaption accompanying striatal hypoactivation. However, the opposing finding of medial frontal hypoactivation for positive valence processing in anhedonic female patients has also been observed (Keedwell et al., 2005a; Mitterschiffthaler et al., 2003).

7.2 Reward circuit and treatment implications

Arguably, because reward circuit dysfunction does not appear to be modulated by typical SSRIs (Dunlop & Nemeroff, 2007), to our knowledge there have been no studies to date assessing whether baseline reward circuit dysfunction in

MDD predicts antidepressant outcomes. There have been studies of behavioral therapy and of rTMS. When probed at baseline by a reward task, lower relative to higher sustained activation of the ACC has been associated with greater symptom reduction in response to behavioral activation therapy for depression (Carl et al., 2016). For rTMS, responders to stimulation are characterized by relative hyper-connectivity between the striatum and the ventral medial region of the broader orbitofrontal cortex, as well as other regions related to reward function (Downar et al., 2014). Other mechanistically oriented studies have investigated antidepressants in regard to their impact on modulating reward circuits. The antidepressants venlafaxine and fluoxetine have been associated with an increase in nucleus accumbens activation that also track increases in experienced positive emotion (Heller et al., 2013). Corresponding increases in accumbens-middle prefrontal connectivity have similarly been associated with increases in self-reported positive emotion (Heller et al., 2013).

Notably, treatment-resistant patients experiencing characteristics of anhedonia do not tend to benefit from antidepressants that act on serotonin (Dunlop & Nemeroff, 2007). In cases in which there is an underlying disruption to reward circuits, a rationale alternative to consider would be antidepressants that facilitate plasticity in striatal dopamine pathways. In animal models, PET shows that pramipexole binds to extrastriatal dopamine receptors, and modulates striatal function when probed by a reward task (Dunlop & Nemeroff, 2007). Bupropion is also thought to act on dopamine and modulate striatal function (Dunlop & Nemeroff, 2007). Positive affect (reward) circuit function is a promising probe for target-driven, mechanistically focused, intervention studies that test whether there is a relationship among positive affect circuit dysfunction, dopamine-related plasticity, and phenotypes characterized by anhedonia features.

8 Attention circuit

The frontoparietal "attention" circuit is defined by the medial superior frontal cortices, lateral prefrontal cortex, anterior inferior parietal lobule, and precuneus (Fornito et al., 2012; Gordon et al., 2014). The anterior insula is also involved in the integration of dorsal and ventral components of frontoparietal attention circuits. This circuit is implicated in alertness, sustained attention, and the support of recollection (Fornito et al., 2012; Gordon et al., 2014). It interacts closely with the default mode circuit to configure the switching from resting to task-context processing (Fornito et al., 2012; Vincent, Kahn, Snyder, Raichle, & Buckner, 2008).

8.1 Attention circuit disruptions in depression and anxiety

There has also been relatively little work done on disruptions to the frontoparietal attention circuit in depression and anxiety. However, several studies have observed hypoconnectivity within the attention circuit in MDD and in social anxiety (Liao et al., 2010; Qiu et al., 2011; Veer et al., 2010). Such hypoconnectivity within the attention circuit has been correlated with a specific behavioral profile of false alarm errors (e.g., responding to "no go" stimuli as if they are "go" stimuli) in anxiety disorder, suggesting a biotype of inattention and impairments in vigilance.

8.2 Attention circuit and treatment implications

Response to psychotherapy may be linked to attention network function. Intrinsic functional connectivity between the left intraparietal sulcus and orbital frontal cortex has been found to predict somatic symptom improvement following behavioral activation treatment for depression (Crowther et al., 2015). It is possible that psychotherapy action may be, in part, mediated by changes in attention circuit function. Increases in perfusion (as measured by arterial spin labeling MRI), as well as connectivity between the attention circuit and the amygdala, have been observed in response to CBT (Shou et al., 2017; Sosic-Vasic et al., 2017).

9 Cognitive control circuit

The "cognitive control" circuit comprises the DLPFC, ACC, dorsal parietal cortex (DPC), and precentral gyrus (Table 1). Together, these regions and their interconnectivity are implicated in the support of higher cognitive functions such as working memory and selective attention (for meta-analysis, Niendam et al., 2012; evidence from convergent neuroimaging methods, Cole & Schneider, 2007). Under task-specific demands, the cognitive control circuit is implicated in cognitive flexibility (Roalf, Ruparel, Gur, et al., 2014).

9.1 Cognitive control circuit disruptions in depression and anxiety

Dysfunction of the cognitive control circuit may be elicited by tasks that require effortful selective processing of relevant stimuli and inhibition of irrelevant stimuli, such as in a working memory task. Hypo-activation of the DLPFC and dorsal anterior cingulate cortex (dACC) during cognitive tasks, and in stress-induced situations, has been found in depressed patients and in social anxiety (Elliott, Baker, et al., 1997; Elliott, Sahakian, et al., 1997; Korgaonkar et al., 2013; Siegle et al., 2007). Induced anxiety has also been related to persistent DLPFC hypo-activation during working memory performance (Fales et al., 2008). Hypoactivity in defining nodes of the cognitive control circuit has been observed in adolescents with depression, and found to persist after recovery in adult and later-life depression (Aizenstein, Butters, Wu, et al., 2009; Elliott, Baker, et al., 1997; Halari, Simic, Pariante, et al., 2009; Harvey et al., 2005), suggesting that this type of dysfunction may have a trait-like status. This trait-like status is also suggested by the presence of reductions in gray matter volume of the same DLPFC and ACC regions in younger and older adults with MDD (Chang et al., 2011; Grieve et al., 2013).

Cognitive control problems in depression may also involve problems suppressing default circuit function, reflected in positive correlations (rather than anticorrelation) between DLFPC cognitive control regions and posterior cingulate default mode regions (Bartova, Meyer, Diers, et al., 2015; Sheline et al., 2010). Suggesting a second type of cognitive control circuit dysfunction, some depressed patients show *hyper-* (rather than hypo-) activation of the DLPFC during working memory and executive function tasks. DLPFC hyper-activation has been observed in depression during tasks with an increasing cognitive demand, but in the absence of behavioral deficits in performing the task (Fitzgerald et al., 2008; Harvey et al., 2005; Holmes, MacDonald 3rd, Carter, et al., 2005; Hugdahl, Rund, Lund, et al., 2004; Matsuo et al., 2007; Wagner, Sinsel, Sobanski, et al., 2006; Walter et al., 2007). In this context, hyper-activation may reflect an attempt at compensation to retain normal cognitive behavior (Fitzgerald et al., 2008; Harvey et al., 2005). Over-activity in both the rostral and dorsal portions of the ACC (Harvey et al., 2005; Rose, Simonotto, & Ebmeier, 2006; Wagner et al., 2006), as well as DLPFC-ACC hyper-connectivity, has also been observed in MDD when performance is similar to that of controls. Hyper-activation in regions of the cognitive control circuitry has been observed in adolescents with depression (Harvey et al., 2005) and in medicated (Harvey et al., 2005; Rose et al., 2006; Walter et al., 2007) and unmedicated (Matsuo et al., 2007) individuals with MDD, and it persists in the ACC after remission (Schoning, Zwitserlood, Engelien, et al., 2009). *Hyper-*connectivity of the DLPFC and cingulate has also been observed in MDD during working memory tasks (Holmes & Pizzagalli, 2008; Vasic et al., 2009).

9.2 Cognitive control circuit and treatment implications

Two studies have found that activation of attention circuit regions elicited by a Go-NoGo task is associated with pharmacotherapy treatment response. Depression symptom improvement following an SSRI was associated with pretreatment activation of the bilateral insula, middle frontal gyrus, and inferior frontal gyrus to successful rejections (Langenecker et al., 2007). Go-NoGo activation in attention regions may also help distinguish those who would respond to different classes of pharmacotherapy. Remission on SSRIs has been associated with relatively greater inhibitory response in the inferior parietal cortex, and SNRI remission with relatively lower inhibitory response (Gyurak et al., 2016).

Comparative hypoactivation of the DLPFC region of the cognitive control circuit, elicited during the inhibition condition of a Go-NoGo task, has been found to predict worse outcomes for SSRIs (escitalopram and sertraline) (Gyurak et al., 2016), as well as for behavioral therapy for PTSD (Falconer, Allen, Felmingham, Williams, & Bryant, 2013). There is also evidence that DLPFC hypoactivation pretreatment predicts poorer outcomes for the SNRI venlafaxine-XR in treating depression (Gyurak et al., 2016), although there is also evidence that patients with poor cognitive control function benefit from venlafaxine-XR (Tian, Du, Spagna, et al., 2016). In the task-free state, relatively lower functional connectivity within the cognitive control circuit has been found to predict poor response to the SSRI escitalopram in an older adult population (Alexopoulos, Hoptman, Kanellopoulos, et al., 2012). Additionally, depression symptom improvement following an SSRI has been associated with pretreatment activation of the bilateral insula, middle frontal gyrus, and inferior frontal gyrus to successful rejections during a Go-NoGo task (Langenecker et al., 2007). Greater dACC activation to commission errors during a parametric version of the Go-NoGo task also was associated with poorer treatment outcome to escitalopram and duloxetine (Crane et al., 2017). Greater engagement of the DLPFC during an emotional face task, both before and after paroxetine, was also associated with better treatment response (Ruhe et al., 2012). Go-NoGo activation in overlapping attention regions may also help distinguish those who would respond to different classes of pharmacotherapy. Remission on SSRIs has been associated with relatively greater inhibitory response in the inferior parietal cortex, and SNRI remission with relatively lower inhibitory response (Gyurak et al., 2016). Structurally, nonremission has been associated with reduced volume in the DLPFC compared with controls (Li et al., 2010).

Psychotherapy and pharmacotherapy both have been shown to alter these cognitive control profiles, and in some cases, the degree of change has been associated with improved outcome. For example, increased intrinsic DLPFC—amygdala functional connectivity (Straub et al., 2017) was observed following group CBT treatment in depressed adolescents. Moreover, the degree of change in DLPFC—amygdala connectivity was positively correlated with the degree of depression symptom improvement, possibly suggesting that these changes in connectivity may be mediating treatment effects. CBT has also been shown to increase DLPFC perfusion as measured by continuous arterial spin labeling independent of symptom improvement (Sosic-Vasic et al., 2017). With respect to pharmacotherapy, treatment with SSRIs, including escitalopram and paroxetine, has been found to increase DLPFC activation during a cognitive (Fales, Barch, Rundle, et al., 2009) and emotional task (Ruhe et al., 2012). However, the extent of the degree of change in DLPFC engagement has not been associated with the extent of symptom improvement.

10 Future directions to close the clinical translational gap

To accelerate progress toward a clinically applicable precision psychiatry model of depression and anxiety, that builds on circuit findings to date, there is a need for standardized protocols, normative data, multimodal data integration, new computational models, and expansion to prospective intervention studies.

10.1 Standardized protocols

Currently, our understanding of the role that brain circuits and their activation play in clinical dysfunction is limited, in part, by inconsistent findings stemming from a lack of standardization across research protocols. To advance the field of precision psychiatry, it will be necessary to undertake larger, multi-site investigations made possible by the use of standardized protocols, integrative analytic models, and databases (Korgaonkar et al., 2013; Siegle, 2011). This approach has been implemented with success in several imaging studies to date (Korgaonkar et al., 2013; Trivedi, McGrath, Fava, et al., 2016). While novel imaging protocols are important for new scientific discoveries, standardized protocols will be essential for the future viability of routine scans for mental health assessment.

10.2 Normative data

In order to incorporate neural circuit taxonomies into practice, we need to establish thresholds for normative vs abnormal function that apply to an individual patient. It will be important to define the normative distribution of neural circuit function in healthy individuals (e.g., Ball et al., 2017), and identify thresholds for overt disorder and failures of function. Methods for establishing the reproducibility of imaging data across people, sites, and time will also be needed (Biswal, Mennes, Zuo, et al., 2010).

10.3 Integration across modalities within the same patients

To further refine circuit taxonomies, it will be essential to integrate imaging data with other modalities. We could consider imaging as just one starting point for directing such effort. To refine classifications based on neural circuits, there is a need to consider the relations among activation, connectivity, and structure within a circuit, as well as the more nuanced interactions between circuits. In parallel, there is a need to dive in depth into understanding the precise ways that brain circuits relate to behavior and symptoms: which specific symptoms reflect the activation of particular circuits, are there symptoms that reflect a "final common pathway" as the outcome of multiple different types of circuit disruptions, does a change in a brain circuits predict a change in symptoms, and so on. An integrated understanding of how neural circuit dysfunctions are modulated by more distal factors, such as genetic variation, genetic expression, life events, and their interaction will also be paramount. These multi-modal efforts will necessarily accumulate increasingly "big data." In turn, big data requires computational innovation.

10.4 Computational innovation

Integrating imaging datasets with other rich data in a manner that translates insights into clinically meaningful outcomes requires special attention to computational approaches. Broadly speaking, there are two complementary approaches: data-driven and theory-driven (for review and synthesis see Huys, Maia, & Frank, 2016; Redish & Gordon, 2016). Regardless of the exact approach, it is increasingly recognized (as it is in other areas of medicine), that demonstration of the reproducibility of findings is key to ultimately having sufficient confidence to warrant clinical translational.

10.5 Prospective, circuit-guided interventions

Short-term progress can be made using precision psychiatry to better match patients to existing treatments and/or prevention strategies. Prospective trials, guided by circuit targets, will be important for accelerating progress. Longer-term progress in precision psychiatry will come not only from prospectively improving existing interventions, but also generating new ones. As neurobiologically based subtypes of psychiatric disorders are refined, new medications can be developed to target specific deficits. As one illustration of circuit-based approaches, neurostimulation technologies, such as repetitive transcranial magnetic stimulation (rTMS), allow for intervention strategies that can capitalize on emerging circuit models. rTMS, a non-invasive high-frequency stimulation of the cortex, was initially conceptualized as an intervention for treatment-resistant depression due to the low response and remission rates for rTMS in early trials. However, applications of rTMS are rapidly expanding to allow precise tailoring based on individual patients' circuit-level dysfunction, with corresponding increases in response and remission rates (e.g., Downar & Daskalakis, 2013).

11 Conclusions

Advances in human neuroimaging of circuits involved in self-reflective, affective, and cognitive functions provide the foundations for formulating a neural circuit understanding of depression and anxiety. Such an approach offers a tangible means to advance precision tools and taxonomies for classification of subtypes, and for tailoring inventions, according to underlying neurobiological dysfunctions. Circuit subtypes also provide objective targets from which to guide prospective, personalized intervention studies, and to inform an understanding of underlying mechanisms. Ultimately, these approaches will accelerate the translation of knowledge into the routine practice of neural circuit-informed precision psychiatry. To close the translational gap between current knowledge and translation into the clinic, important future directions for our field are a focus on building circuit-based norms and standardized protocols, accelerating computational approaches that can fuse imaging and phenotype data in a manner that is clinically interpretable, and establishing collaborative efforts to launch prospective circuit-guided trials that move us into the realm of actionable outcomes.

Acknowledgments

We thank David Choi for contributions to the referencing for the chapter and Melissa Shiner for contributions to the future for the chapter.

Psychiatry has not yet been in the game for precision medicine, in contrast to cardiovascular disease and other chronic diseases. Cardiovascular care is further ahead in linking precise insights about the organs of interest to treatment indications. For example, the EKG can be used to identify types of arrhythmia (too fast, too slow, irregular) and to indicate specific treatments (pacemaker), and the angiogram to confirm the presence of blockage (emboli, stroke, myocardial infarction) and to indicate other treatments (lifestyle changes, medications, surgery).

This situation is now changing rapidly as we witness the emergence of precision psychiatry informed by neuroscience. With advances in technologies for imaging the human brain in action, and for integrating other biomarkers, we are now identifying types of brain circuit dysfunction that underlie depression and anxiety, analogous to types of heart dysfunction that underlie phenotypes of cardiovascular disease. An important next step is to accelerate translational research addressed at the question of how we apply these new insights within an actionable system for improving people's lives.

The goals of precision psychiatry are three-fold: precise classification (i.e., a specific understanding of the pathophysiology of each individual patient—what has gone wrong), precise treatment planning (i.e., tailoring treatment plans in a personalized manner—how can we fix what has gone wrong), and precise prevention (i.e., targeted and tailored prevention strategies—how can we keep things from going wrong).

The first of these three goals, precise classification, hinges on the identification of subtype profiles (or "biotypes") that coherently map neurobiological disruptions onto symptoms and behaviors, take into account life experience and context, and are relevant to guiding treatment choices (Williams, 2016, 2017). These profiles and biotypes may, in some instances, align with our rich symptom classifications currently defined by the Diagnostic and Statistical Manual of Mental Disorders, 5th edition (Publishing D-APAJAAP, 2013), and in other cases, may cut across diagnostic classifications and, in others, may reflect unforeseen, novel subtypes. Precision psychiatry, a rapidly emerging field, encompasses the discovery of these types, as well as their application to treatment and prevention. Although the scientific insights and treatment models on which precision psychiatry is based are well-established, the integration across science and practice, and the dissemination of scientific insights into clinical practice, is at the forefront of our field.

Here, the primary focus is on the organ of dysfunction in psychiatry: the brain. In particular, the focus is on large-scale neural circuits probed by functional neuroimaging. Necessarily, there is also a need to incorporate other biomarker domains, such as genetics, life history, and cognition. Genetic variants and life experience have an important role in modulating neural circuit function. In turn, neural circuit functions are expressed in

Normative function within the circuits of interest

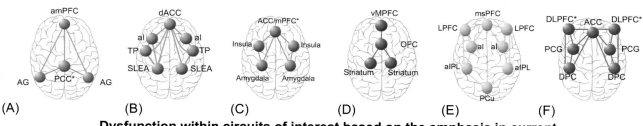

Dysfunction within circuits of interest based on the emphasis in current

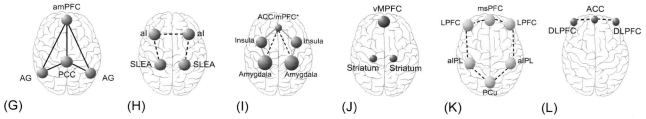

FIG. 1 An illustration of our circuits and circuit dysfunction of interest. Panels A–F show normative function within the default mode node network (A) defined by the anterior medial prefrontal cortex (amPFC), the posterior cingulate cortex (PCC), and the angular gyrus (AG), salience circuit (B) defined by the anterior insula (aI), dorsal ACC (dACC), sublenticular extended amygdala (SLEA) and temporal parietal junction (TP), the negative affect circuit (C) defined by the amygdala, insula and ACC/medial prefrontal cortex* (encompassing the dACC, subgenual ACC, and medial PFC), the positive affect circuit (D) comprising the ventral and dorsal striatum, orbitofrontal cortex (OFC) and ventromedial PFC (vmPFC), the attention circuit (E) defined by the middle superior. PFC (msPFC), lateral prefrontal cortex (LPFC), anterior inferior parietal lobule (aIPL) and precuneus (PCu) and cognitive control circuit (F), defined by the DLPFC* (encompassing the dorsolateral prefrontal cortex, anterior prefrontal cortex, and inferior frontal cortex), ACC, precentral gyrus (PCG), and dorsal parietal cortex (DPC). Panels G through L illustrate dysfunctions of these circuits observed in mood disorders, focusing on the findings emphasized by our current accumulated knowledge. Not shown is the distinct dysfunction of the negative affect circuit associated with hyporeactivity to sad-related negative stimuli. The size of the regions illustrated here represents the direction of dysfunction in activation (smaller = hypoactivation, larger = hyper-activation). The style of the connections between regions represents the direction of dysfunction in functional connectivity (thick lines = hyper-connectivity, dashed lines = hypoconnectivity).

how individuals perform on tests of general and emotional cognition. The synthesis of evidence presented here focuses on depression and anxiety. A unifying premise is that we have sufficient evidence to develop a new taxonomy for depression and anxiety that considers these disorders as types of disruptions in underlying neural circuits (Fig. 1). It is proposed that, through quantification of neural circuits, we can start laying the foundations of a neural-circuit guided precision psychiatry that informs classification and treatment planning.

References

Aizenstein, H. J., Butters, M. A., Wu, M., et al. (2009). Altered functioning of the executive control circuit in late-life depression: Episodic and persistent phenomena. *The American Journal of Geriatric Psychiatry, 17*, 30–42.

Alexopoulos, G. S., Hoptman, M. J., Kanellopoulos, D., et al. (2012). Functional connectivity in the cognitive control network and the default mode network in late-life depression. *Journal of Affective Disorders, 139*, 56–65.

Andreescu, C., Tudorascu, D. L., Butters, M. A., et al. (2013). Resting state functional connectivity and treatment response in late-life depression. *Psychiatry Research, 214*, 313–321.

Arnold Anteraper, S., Triantafyllou, C., Sawyer, A. T., et al. (2014). Hyper-connectivity of subcortical resting-state networks in social anxiety disorder. *Brain Connectivity, 4*, 81–90.

Arnone, D., McKie, S., Elliott, R., et al. (2012). Increased amygdala responses to sad but not fearful faces in major depression: Relation to mood state and pharmacological treatment. *The American Journal of Psychiatry, 169*, 841–850.

Ball, T. M., Goldstein-Piekarski, A. N., Gatt, J. M., & Williams, L. M. (2017). Quantifying person-level brain network functioning to facilitate clinical translation. *Translational Psychiatry, 7*, e1248.

Bartova, L., Meyer, B. M., Diers, K., et al. (2015). Reduced default mode network suppression during a working memory task in remitted major depression. *Journal of Psychiatric Research, 64*, 9–18.

Berridge, K. C., & Kringelbach, M. L. (2008). Affective neuroscience of pleasure: Reward in humans and animals. *Psychopharmacology, 199*, 457–480.

Biswal, B. B., Mennes, M., Zuo, X. N., et al. (2010). Toward discovery science of human brain function. *Proceedings of the National Academy of Sciences of the United States of America, 107*, 4734–4739.

Blair, K. S., Geraci, M., Smith, B. W., et al. (2012). Reduced dorsal anterior cingulate cortical activity during emotional regulation and top-down attentional control in generalized social phobia, generalized anxiety disorder, and comorbid generalized social phobia/generalized anxiety disorder. *Biological Psychiatry*, *72*, 476–482.

Blair, K., Shaywitz, J., Smith, B. W., et al. (2008). Response to emotional expressions in generalized social phobia and generalized anxiety disorder: Evidence for separate disorders. *The American Journal of Psychiatry*, *165*, 1193–1202.

Bossini, L., Tavanti, M., Calossi, S., et al. (2008). Magnetic resonance imaging volumes of the hippocampus in drug-naive patients with post-traumatic stress disorder without comorbidity conditions. *Journal of Psychiatric Research*, *42*, 752–762.

Buckner, R. L., Krienen, F. M., & Yeo, B. T. (2013). Opportunities and limitations of intrinsic functional connectivity MRI. *Nature Neuroscience*, *16*, 832–837.

Cannistraro, P. A., Wright, C. I., Wedig, M. M., et al. (2004). Amygdala responses to human faces in obsessive-compulsive disorder. *Biological Psychiatry*, *56*, 916–920.

Carl, H., Walsh, E., Eisenlohr-Moul, T., et al. (2016). Sustained anterior cingulate cortex activation during reward processing predicts response to psychotherapy in major depressive disorder. *Journal of Affective Disorders*, *203*, 204–212.

Castelli, F., Happe, F., Frith, U., & Frith, C. (2000). Movement and mind: A functional imaging study of perception and interpretation of complex intentional movement patterns. *NeuroImage*, *12*, 314–325.

Chang, C. C., Yu, S. C., McQuoid, D. R., et al. (2011). Reduction of dorsolateral prefrontal cortex gray matter in late-life depression. *Psychiatry Research*, *193*, 1–6.

Chen, A. C., Oathes, D. J., Chang, C., et al. (2013). Causal interactions between fronto-parietal central executive and default-mode networks in humans. *Proceedings of the National Academy of Sciences of the United States of America*, *110*, 19944–19949.

Chen, C. H., Suckling, J., Ooi, C., et al. (2008). Functional coupling of the amygdala in depressed patients treated with antidepressant medication. *Neuropsychopharmacology*, *33*, 1909–1918.

Clauss, J. A., Avery, S. N., VanDerKlok, R. M., et al. (2014). Neurocircuitry underlying risk and resilience to social anxiety disorder. *Depression and Anxiety*, *31*, 822–833.

Coalition PM (2014). *The case for personalized medicine.*

Cole, M. W., Bassett, D. S., Power, J. D., Braver, T. S., & Petersen, S. E. (2014). Intrinsic and task-evoked network architectures of the human brain. *Neuron*, *83*, 238–251.

Cole, M. W., & Schneider, W. (2007). The cognitive control network: Integrated cortical regions with dissociable functions. *NeuroImage*, *37*, 343–360.

Crane, N. A., Jenkins, L. M., Bhaumik, R., et al. (2017). Multidimensional prediction of treatment response to antidepressants with cognitive control and functional MRI. *Brain*, *140*, 472–486.

Crowther, A., Smoski, M. J., Minkel, J., et al. (2015). Resting-state connectivity predictors of response to psychotherapy in major depressive disorder. *Neuropsychopharmacology*, *40*, 1659–1673.

de Kwaasteniet, B. P., Rive, M. M., Ruhe, H. G., et al. (2015). Decreased resting-state connectivity between neurocognitive networks in treatment resistant depression. *Frontiers in Psychiatry*, *6*, 28.

Dichter, G. S., Gibbs, D., & Smoski, M. J. (2014). A systematic review of relations between resting-state functional-MRI and treatment response in major depressive disorder. *Journal of Affective Disorders*, *172c*, 8–17.

Dichter, G. S., Kozink, R. V., McClernon, F. J., & Smoski, M. J. (2012). Remitted major depression is characterized by reward network hyperactivation during reward anticipation and hypoactivation during reward outcomes. *Journal of Affective Disorders*, *136*, 1126–1134.

Downar, J., & Daskalakis, Z. J. (2013). New targets for rTMS in depression: A review of convergent evidence. *Brain Stimulation*, *6*, 231–240.

Downar, J., Geraci, J., Salomons, T. V., et al. (2014). Anhedonia and reward-circuit connectivity distinguish nonresponders from responders to dorsomedial prefrontal repetitive transcranial magnetic stimulation in major depression. *Biological Psychiatry*, *76*, 176–185.

Dunlop, B. W., Kelley, M. E., McGrath, C. L., Craighead, W. E., & Mayberg, H. S. (2015). Preliminary findings supporting insula metabolic activity as a predictor of outcome to psychotherapy and medication treatments for depression. *The Journal of Neuropsychiatry and Clinical Neurosciences*, *27*, 237–239.

Dunlop, B. W., & Nemeroff, C. B. (2007). The role of dopamine in the pathophysiology of depression. *Archives of General Psychiatry*, *64*, 327–337.

Elliott, R., Baker, S. C., Rogers, R. D., et al. (1997). Prefrontal dysfunction in depressed patients performing a complex planning task: A study using positron emission tomography. *Psychological Medicine*, *27*, 931–942.

Elliott, R., Sahakian, B. J., Herrod, J. J., Robbins, T. W., & Paykel, E. S. (1997). Abnormal response to negative feedback in unipolar depression: Evidence for a diagnosis specific impairment. *Journal of Neurology, Neurosurgery, and Psychiatry*, *63*, 74–82.

Etkin, A., Klemenhagen, K. C., Dudman, J. T., et al. (2004). Individual differences in trait anxiety predict the response of the basolateral amygdala to unconsciously processed fearful faces. *Neuron*, *44*, 1043–1055.

Etkin, A., Prater, K. E., Hoeft, F., Menon, V., & Schatzberg, A. F. (2010). Failure of anterior cingulate activation and connectivity with the amygdala during implicit regulation of emotional processing in generalized anxiety disorder. *The American Journal of Psychiatry*, *167*, 545–554.

Etkin, A., Prater, K. E., Schatzberg, A. F., Menon, V., & Greicius, M. D. (2009). Disrupted amygdalar subregion functional connectivity and evidence of a compensatory network in generalized anxiety disorder. *Archives of General Psychiatry*, *66*, 1361–1372.

Falconer, E., Allen, A., Felmingham, K. L., Williams, L. M., & Bryant, R. A. (2013). Inhibitory neural activity predicts response to cognitive-behavioral therapy for posttraumatic stress disorder. *The Journal of Clinical Psychiatry*, *74*, 895–901.

Fales, C. L., Barch, D. M., Burgess, G. C., et al. (2008). Anxiety and cognitive efficiency: Differential modulation of transient and sustained neural activity during a working memory task. *Cognitive, Affective, & Behavioral Neuroscience*, *8*, 239–253.

Fales, C. L., Barch, D. M., Rundle, M. M., et al. (2009). Antidepressant treatment normalizes hypoactivity in dorsolateral prefrontal cortex during emotional interference processing in major depression. *Journal of Affective Disorders, 112*, 206–211.

Felleman, D. J., & Van Essen, D. C. (1991). Distributed hierarchical processing in the primate cerebral cortex. *Cerebral Cortex, 1*, 1–47.

Fernandes, B. S., Williams, L. M., Steiner, J., et al. (2017). The new field of 'precision psychiatry'. *BMC Medicine, 15*, 80.

Fitzgerald, P. B., Srithiran, A., Benitez, J., et al. (2008). An fMRI study of prefrontal brain activation during multiple tasks in patients with major depressive disorder. *Human Brain Mapping, 29*, 490–501.

Fonzo, G. A., Ramsawh, H. J., Flagan, T. M., et al. (2015). Common and disorder-specific neural responses to emotional faces in generalised anxiety, social anxiety and panic disorders. *The British Journal of Psychiatry, 206*(3), 206–215.

Fornito, A., Harrison, B. J., Zalesky, A., & Simons, J. S. (2012). Competitive and cooperative dynamics of large-scale brain functional networks supporting recollection. *Proceedings of the National Academy of Sciences of the United Sates of America, 109*, 12788–12793.

Fox, M. D., Snyder, A. Z., Vincent, J. L., et al. (2005). The human brain is intrinsically organized into dynamic anticorrelated functional networks. *Proceedings of the National Academy of Sciences of the United States of America, 102*, 9673–9678.

Fu, C. H., Williams, S. C., Cleare, A. J., et al. (2004). Attenuation of the neural response to sad faces in major depression by antidepressant treatment: A prospective, event-related functional magnetic resonance imaging study. *Archives of General Psychiatry, 61*, 877–889.

Godlewska, B. R., Browning, M., Norbury, R., Cowen, P. J., & Harmer, C. J. (2016). Early changes in emotional processing as a marker of clinical response to SSRI treatment in depression. *Translational Psychiatry, 6*, e957.

Godlewska, B. R., Norbury, R., Selvaraj, S., Cowen, P. J., & Harmer, C. J. (2012). Short-term SSRI treatment normalises amygdala hyperactivity in depressed patients. *Psychological Medicine, 42*, 2609–2617.

Goldstein-Piekarski, A. N., Staveland, B. R., Ball, T. M., et al. (2018). Intrinsic functional connectivity predicts remission on antidepressants: A randomized controlled trial to identify clinically applicable imaging biomarkers. *Translational Psychiatry, 8*, 57.

Gordon, E. M., Laumann, T. O., Adeyemo, B., et al. (2014). Generation and evaluation of a cortical area parcellation from resting-state correlations. *Cerebral Cortex, 26*, 288–303.

Gotlib, I. H., Hamilton, J. P., Cooney, R. E., et al. (2010). Neural processing of reward and loss in girls at risk for major depression. *Archives of General Psychiatry, 67*, 380–387.

Greicius, M. D., Flores, B. H., Menon, V., et al. (2007). Resting-state functional connectivity in major depression: Abnormally increased contributions from subgenual cingulate cortex and thalamus. *Biological Psychiatry, 62*, 429–437.

Greicius, M. D., Krasnow, B., Reiss, A. L., & Menon, V. (2003). Functional connectivity in the resting brain: A network analysis of the default mode hypothesis. *Proceedings of the National Academy of Sciences of the United States of America, 100*, 253–258.

Greicius, M. D., Supekar, K., Menon, V., & Dougherty, R. F. (2009). Resting-state functional connectivity reflects structural connectivity in the default mode network. *Cerebral Cortex, 19*, 72–78.

Grieve, S. M., Korgaonkar, M. S., Koslow, S. H., Gordon, E., & Williams, L. M. (2013). Widespread reductions in gray matter volume in depression. *Neuroimage Clinical, 3*, 332–339.

Gyurak, A., Patenaude, B., Korgaonkar, M. S., et al. (2016). Frontoparietal activation during response inhibition predicts remission to antidepressants in patients with major depression. *Biological Psychiatry, 79*, 274–281.

Haber, S. N., & Knutson, B. (2010). The reward circuit: Linking primate anatomy and human imaging. *Neuropsychopharmacology, 35*, 4–26.

Hahn, A., Stein, P., Windischberger, C., et al. (2011). Reduced resting-state functional connectivity between amygdala and orbitofrontal cortex in social anxiety disorder. *NeuroImage, 56*, 881–889.

Halari, R., Simic, M., Pariante, C. M., et al. (2009). Reduced activation in lateral prefrontal cortex and anterior cingulate during attention and cognitive control functions in medication-naive adolescents with depression compared to controls. *Journal of Child Psychology and Psychiatry, 50*, 307–316.

Hamilton, J. P., Etkin, A., Furman, D. J., et al. (2012). Functional neuroimaging of major depressive disorder: A meta-analysis and new integration of base line activation and neural response data. *The American Journal of Psychiatry, 169*, 693–703.

Hamilton, J. P., Farmer, M., Fogelman, P., & Gotlib, I. H. (2015). Depressive rumination, the default-mode network, and the dark matter of clinical neuroscience. *Biological Psychiatry, 78*, 224–230.

Hamilton, J. P., Furman, D. J., Chang, C., et al. (2011). Default-mode and task-positive network activity in major depressive disorder: Implications for adaptive and maladaptive rumination. *Biological Psychiatry, 70*, 327–333.

Harding, K. J. K., Rush, J., Arbuckle, M., Trivedi, M. H., & Pincus, H. A. (2011). Measurement-based care in psychiatric practice: A policy framework for implementation. *Journal of Clinical Psychiatry, 72*, 1136–1143.

Harvey, P. O., Fossati, P., Pochon, J. B., et al. (2005). Cognitive control and brain resources in major depression: An fMRI study using the n-back task. *NeuroImage, 26*, 860–869.

Heller, A. S., Johnstone, T., Light, S. N., et al. (2013). Relationships between changes in sustained fronto-striatal connectivity and positive affect in major depression resulting from antidepressant treatment. *The American Journal of Psychiatry, 170*, 197–206.

Holmes, A. J., MacDonald, A., 3rd, Carter, C. S., et al. (2005). Prefrontal functioning during context processing in schizophrenia and major depression: An event-related fMRI study. *Schizophrenia Research, 76*, 199–206.

Holmes, A. J., & Pizzagalli, D. A. (2008). Response conflict and frontocingulate dysfunction in unmedicated participants with major depression. *Neuropsychologia, 46*, 2904–2913.

Hugdahl, K., Rund, B. R., Lund, A., et al. (2004). Brain activation measured with fMRI during a mental arithmetic task in schizophrenia and major depression. *The American Journal of Psychiatry, 161*, 286–293.

Huys, Q. J. M., Maia, T. V., & Frank, M. J. (2016). Computational psychiatry as a bridge from neuroscience to clinical applications. *Nature Neuroscience, 19*, 404–413.

Insel, T. (2010). The data of diagnosis: New approaches to psychiatric classification. *Psychiatry: Interpersonal & Biological Processes, 73,* 4p.

Jaworska, N., Yang, X. R., Knott, V., & Macqueen, G. (2014). A review of fMRI studies during visual emotive processing in major depressive disorder. *The World Journal of Biological Psychiatry, 16*(7), 448–471.

Kaiser, R. H., Andrews-Hanna, J. R., Wager, T. D., & Pizzagalli, D. A. (2015). Large-scale network dysfunction in major depressive disorder: A meta-analysis of resting-state functional connectivity. *JAMA Psychiatry, 72,* 603–611.

Keedwell, P. A., Andrew, C., Williams, S. C., Brammer, M. J., & Phillips, M. L. (2005a). The neural correlates of anhedonia in major depressive disorder. *Biological Psychiatry, 58,* 843–853.

Keedwell, P. A., Andrew, C., Williams, S. C., Brammer, M. J., & Phillips, M. L. (2005b). A double dissociation of ventromedial prefrontal cortical responses to sad and happy stimuli in depressed and healthy individuals. *Biological Psychiatry, 58,* 495–503.

Kempton, M. J., Salvador, Z., Munafo, M. R., et al. (2011). Structural neuroimaging studies in major depressive disorder. Meta-analysis and comparison with bipolar disorder. *Archives of General Psychiatry, 68,* 675–690.

Killgore, W. D., Britton, J. C., Schwab, Z. J., et al. (2014). Cortico-limbic responses to masked affective faces across ptsd, panic disorder, and specific phobia. *Depression and Anxiety, 31,* 150–159.

Kim, M. J., Hamilton, J. P., & Gotlib, I. H. (2008). Reduced caudate gray matter volume in women with major depressive disorder. *Psychiatry Research, 164,* 114–122.

Kim, M. J., Loucks, R. A., Palmer, A. L., et al. (2011). The structural and functional connectivity of the amygdala: From normal emotion to pathological anxiety. *Behavioural Brain Research, 223,* 403–410.

Klumpp, H., Post, D., Angstadt, M., Fitzgerald, D. A., & Phan, K. L. (2013). Anterior cingulate cortex and insula response during indirect and direct processing of emotional faces in generalized social anxiety disorder. *The Biology of Mood and Anxiety Disorders, 3,* 7.

Kober, H., Barrett, L. F., Joseph, J., et al. (2008). Functional grouping and cortical-subcortical interactions in emotion: A meta-analysis of neuroimaging studies. *NeuroImage, 42,* 998–1031.

Korgaonkar, M. S., Fornito, A., Williams, L. M., & Grieve, S. M. (2014). Abnormal structural networks characterize major depressive disorder: A connectome analysis. *Biological Psychiatry, 76,* 567–574.

Korgaonkar, M. S., Grieve, S. M., Etkin, A., Koslow, S. H., & Williams, L. M. (2013). Using standardized fMRI protocols to identify patterns of prefrontal circuit dysregulation that are common and specific to cognitive and emotional tasks in major depressive disorder: First wave results from the iSPOT-D study. *Neuropsychopharmacology, 38,* 863–871.

Korgaonkar, M. S., Ram, K., Williams, L. M., Gatt, J. M., & Grieve, S. M. (2014). Establishing the resting state default mode network derived from functional magnetic resonance imaging tasks as an endophenotype: A twins study. *Human Brain Mapping, 35,* 3893–3902.

Koric, L., Volle, E., Seassau, M., et al. (2012). How cognitive performance-induced stress can influence right VLPFC activation: An fMRI study in healthy subjects and in patients with social phobia. *Human Brain Mapping, 33,* 1973–1986.

Langenecker, S. A., Kennedy, S. E., Guidotti, L. M., et al. (2007). Frontal and limbic activation during inhibitory control predicts treatment response in major depressive disorder. *Biological Psychiatry, 62,* 1272–1280.

Lawrence, N. S., Williams, A. M., Surguladze, S., et al. (2004). Subcortical and ventral prefrontal cortical neural responses to facial expressions distinguish patients with bipolar disorder and major depression. *Biological Psychiatry, 55,* 578–587.

Lee, B. T., Seong Whi, C., Hyung Soo, K., et al. (2007). The neural substrates of affective processing toward positive and negative affective pictures in patients with major depressive disorder. *Progress in Neuro-Psychopharmacology & Biological Psychiatry, 31,* 1487–1492.

Li, C. T., Lin, C. P., Chou, K. H., et al. (2010). Structural and cognitive deficits in remitting and non-remitting recurrent depression: A voxel-based morphometric study. *NeuroImage, 50,* 347–356.

Liao, W., Qiu, C., Gentili, C., et al. (2010). Altered effective connectivity network of the amygdala in social anxiety disorder: A resting-state FMRI study. *PLoS One, 5,* .

Lindquist, K. A., & Barrett, L. F. (2012). A functional architecture of the human brain: Emerging insights from the science of emotion. *Trends in Cognitive Sciences, 16,* 533–540.

Liston, C., Chen, A. C., Zebley, B. D., et al. (2014). Default mode network mechanisms of transcranial magnetic stimulation in depression. *Biological Psychiatry, 76,* 517–526.

Manoliu, A., Meng, C., Brandl, F., et al. (2014). Insular dysfunction within the salience network is associated with severity of symptoms and aberrant inter-network connectivity in major depressive disorder. *Frontiers in Human Neuroscience, 7,* 930.

Markoff, J. (2013). Obama seeking to boost study of human brain. *The New York Times, 17,* 2013.

Matsuo, K., Glahn, D. C., Peluso, M. A., et al. (2007). Prefrontal hyperactivation during working memory task in untreated individuals with major depressive disorder. *Molecular Psychiatry, 12,* 158–166.

Matthews, S. C., Strigo, I. A., Simmons, A. N., Yang, T. T., & Paulus, M. P. (2008). Decreased functional coupling of the amygdala and supragenual cingulate is related to increased depression in unmedicated individuals with current major depressive disorder. *Journal of Affective Disorders, 111,* 13–20.

McGrath, C. L., Kelley, M. E., Dunlop, B. W., et al. (2014). Pretreatment brain states identify likely nonresponse to standard treatments for depression. *Biological Psychiatry, 76,* 527–535.

McGrath, C. L., Kelley, M. E., Holtzheimer, P. E., et al. (2013). Toward a neuroimaging treatment selection biomarker for major depressive disorder. *JAMA Psychiatry, 70,* 821–829.

Mesulam, M. M. (1998). From sensation to cognition. *Brain, 121*(Pt 6), 1013–1052.

Mitterschiffthaler, M. T., Kumari, V., Malhi, G. S., et al. (2003). Neural response to pleasant stimuli in anhedonia: An fMRI study. *Neuroreport, 14,* 177–182.

Moon, C. M., Kim, G. W., & Jeong, G. W. (2014). Whole-brain gray matter volume abnormalities in patients with generalized anxiety disorder: Voxel-based morphometry. *Neuroreport, 25,* 184–189.

Mueller, S. C., Aouidad, A., Gorodetsky, E., et al. (2013). Gray matter volume in adolescent anxiety: An impact of the brain-derived neurotrophic factor Val(66)Met polymorphism? *Journal of the American Academy of Child and Adolescent Psychiatry, 52,* 184–195.

Mulders, P. C., van Eijndhoven, P. F., Schene, A. H., Beckmann, C. F., & Tendolkar, I. (2015). Resting-state functional connectivity in major depressive disorder: A review. *Neuroscience and Biobehavioral Reviews, 56,* 330–344.

Musgrove, D., Eberly, L., Klimes-Dougan, B., et al. (2015). Impaired bottom-up effective connectivity between amygdala and subgenual anterior cingulate cortex in unmedicated adolescents with major depression: Results from a dynamic causal modeling analysis. *Brain Connectivity, 5*(10), 608–619.

Niendam, T. A., Laird, A. R., Ray, K. L., et al. (2012). Meta-analytic evidence for a superordinate cognitive control network subserving diverse executive functions. *Cognitive, Affective, & Behavioral Neuroscience, 12,* 241–268.

Oosterwijk, S., Lindquist, K. A., Anderson, E., et al. (2012). States of mind: emotions, body feelings, and thoughts share distributed neural networks. *NeuroImage, 62,* 2110–2128.

Ottaviani, C., Cevolani, D., Nucifora, V., et al. (2012). Amygdala responses to masked and low spatial frequency fearful faces: A preliminary fMRI study in panic disorder. *Psychiatry Research, 203,* 159–165.

Palm, M. E., Elliott, R., McKie, S., Deakin, J. F., & Anderson, I. M. (2011). Attenuated responses to emotional expressions in women with generalized anxiety disorder. *Psychological Medicine, 41,* 1009–1018.

Peluso, M. A., Glahn, D. C., Matsuo, K., et al. (2009). Amygdala hyperactivation in untreated depressed individuals. *Psychiatry Research, 173,* 158–161.

Peterson, A., Thome, J., Frewen, P., & Lanius, R. A. (2014). Resting-state neuroimaging studies: A new way of identifying differences and similarities among the anxiety disorders? *Canadian Journal of Psychiatry, 59,* 294–300.

Phan, K. L., Fitzgerald, D. A., Nathan, P. J., & Tancer, M. E. (2006). Association between amygdala hyperactivity to harsh faces and severity of social anxiety in generalized social phobia. *Biological Psychiatry, 59,* 424–429.

Philip, N. S., Barredo, J., van't Wout-Frank, M., et al. (2018). Network mechanisms of clinical response to transcranial magnetic stimulation in posttraumatic stress disorder and major depressive disorder. *Biological Psychiatry, 83,* 263–272.

Phillips, M. L., Drevets, W. C., Rauch, S. L., & Lane, R. (2003). Neurobiology of emotion perception II: Implications for major psychiatric disorders. *Biological Psychiatry, 54,* 515–528.

Pizzagalli, D. A., Holmes, A. J., Dillon, D. G., et al. (2009). Reduced caudate and nucleus accumbens response to rewards in unmedicated individuals with major depressive disorder. *The American Journal of Psychiatry, 166,* 702–710.

Power, J. D., Cohen, A. L., Nelson, S. M., et al. (2011). Functional network organization of the human brain. *Neuron, 72,* 665–678.

Prater, K. E., Hosanagar, A., Klumpp, H., Angstadt, M., & Phan, K. L. (2013). Aberrant amygdala-frontal cortex connectivity during perception of fearful faces and at rest in generalized social anxiety disorder. *Depression and Anxiety, 30,* 234–241.

Price, J. L., & Drevets, W. C. (2010). Neurocircuitry of mood disorders. *Neuropsychopharmacology, 35,* 192–216.

Price, R. B., Gates, K., Kraynak, T. E., Thase, M. E., & Siegle, G. J. (2017). Data-driven subgroups in depression derived from directed functional connectivity paths at rest. *Neuropsychopharmacology, 42*(13), 2623–2632.

Publishing D-APAJAAP (2013). *Diagnostic and statistical manual of mental disorders.* .

Qin, J., Wei, M., Liu, H., et al. (2014). Abnormal brain anatomical topological organization of the cognitive-emotional and the frontoparietal circuitry in major depressive disorder. *Magnetic Resonance in Medicine, 72,* 1397–1407.

Qiu, C., Liao, W., Ding, J., et al. (2011). Regional homogeneity changes in social anxiety disorder: A resting-state fMRI study. *Psychiatry Research, 194,* 47–53.

Raichle, M. E. (2015). The brain's default mode network. *Annual Review of Neuroscience, 38,* .

Redish, A. D., & Gordon, J. A. (2016). *Computational psychiatry: New perspectives on mental illness.* Cambridge, Massachusetts: The MIT Press.

Rizvi, S. J., Salomons, T. V., Konarski, J. Z., et al. (2013). Neural response to emotional stimuli associated with successful antidepressant treatment and behavioral activation. *Journal of Affective Disorders, 151,* 573–581.

Roalf, D. R., Ruparel, K., Gur, R. E., et al. (2014). Neuroimaging predictors of cognitive performance across a standardized neurocognitive battery. *Neuropsychology, 28,* 161–176.

Robinson, O. J., Krimsky, M., Lieberman, L., et al. (2014). The dorsal medial prefrontal (anterior cingulate) cortex–amygdala aversive amplification circuit in unmedicated generalised and social anxiety disorders: An observational study. *The Lancet Psychiatry, 1,* 294–302.

Rose, E. J., Simonotto, E., & Ebmeier, K. P. (2006). Limbic over-activity in depression during preserved performance on the n-back task. *NeuroImage, 29,* 203–215.

Ruhe, H. G., Booij, J., Veltman, D. J., Michel, M. C., & Schene, A. H. (2012). Successful pharmacologic treatment of major depressive disorder attenuates amygdala activation to negative facial expressions: A functional magnetic resonance imaging study. *The Journal of Clinical Psychiatry, 73,* 451–459.

Rushworth, M. F. S., Mars, R. B., & Sallet, J. (2013). Are there specialized circuits for social cognition and are they unique to humans? *Current Opinion in Neurobiology, 23,* 436–442.

Sacchet, M. D., Prasad, G., Foland-Ross, L. C., et al. (2014). Characterizing white matter connectivity in major depressive disorder: Automated fiber quantification and maximum density paths. *Proceedings of IEEE International Symposium on Biomedical Imaging, 11,* 592–595.

Schoning, S., Zwitserlood, P., Engelien, A., et al. (2009). Working-memory fMRI reveals cingulate hyperactivation in euthymic major depression. *Human Brain Mapping, 30,* 2746–2756.

Sciences AoM (2015). *Stratified, personalised or P4 medicine: A new direction for placing the patient at the centre of healthcare and health education.* .

Secretary OotP (2015a). *Fact sheet: President Obama's precision medicine initiative.* .

Secretary OotP (2015b). *Remarks by the President on precision medicine.* .

Seeley, W. W., Menon, V., Schatzberg, A. F., et al. (2007). Dissociable intrinsic connectivity networks for salience processing and executive control. *The Journal of Neuroscience, 27,* 2349–2356.

Sheline, Y. I., Barch, D. M., Donnelly, J. M., et al. (2001). Increased amygdala response to masked emotional faces in depressed subjects resolves with antidepressant treatment: An fMRI study. *Biological Psychiatry, 50,* 651–658.

Sheline, Y. I., Price, J. L., Yan, Z., & Mintun, M. A. (2010). Resting-state functional MRI in depression unmasks increased connectivity between networks via the dorsal nexus. *Proceedings of the National Academy of Sciences of the United States of America, 107,* 11020–11025.

Shou, H., Yang, Z., Satterthwaite, T. D., et al. (2017). Cognitive behavioral therapy increases amygdala connectivity with the cognitive control network in both MDD and PTSD. *Neuroimage Clinical, 14,* 464–470.

Shulman, G. L., Fiez, J. A., Corbetta, M., et al. (1997). Common blood flow changes across visual tasks: II. Decreases in cerebral cortex. *Journal of Cognitive Neuroscience, 9,* 648–663.

Siegle, G. J. (2011). Beyond depression commentary: Wherefore art thou, depression clinic of tomorrow? *Clinical Psychology: Science and Practice, 18,* 305–310.

Siegle, G. J., Thompson, W., Carter, C. S., Steinhauer, S. R., & Thase, M. E. (2007). Increased amygdala and decreased dorsolateral prefrontal BOLD responses in unipolar depression: Related and independent features. *Biological Psychiatry, 61,* 198–209.

Sikora, M., Heffernan, J., Avery, E. T., et al. (2016). Salience network functional connectivity predicts placebo effects in major depression. *Biological Psychiatry: Cognitive Neuroscience and Neuroimaging, 1,* 68–76.

Singh, M. K., Kesler, S. R., Hadi Hosseini, S. M., et al. (2013). Anomalous gray matter structural networks in major depressive disorder. *Biological Psychiatry, 74,* 777–785.

Sladky, R., Hoflich, A., Kublbock, M., et al. (2013). Disrupted effective connectivity between the amygdala and orbitofrontal cortex in social anxiety disorder during emotion discrimination revealed by dynamic causal modeling for fMRI. *Cerebral Cortex, 25*(4), 895–903.

Sosic-Vasic, Z., Abler, B., Gron, G., Plener, P., & Straub, J. (2017). Effects of a brief cognitive behavioural therapy group intervention on baseline brain perfusion in adolescents with major depressive disorder. *Neuroreport, 28,* 348–353.

Spreng, R. N., Sepulcre, J., Turner, G. R., Stevens, W. D., & Schacter, D. L. (2013). Intrinsic architecture underlying the relations among the default, dorsal attention, and frontoparietal control networks of the human brain. *Journal of Cognitive Neuroscience, 25,* 74–86.

Sprengelmeyer, R., Steele, J. D., Mwangi, B., et al. (2011). The insular cortex and the neuroanatomy of major depression. *Journal of Affective Disorders, 133,* 120–127.

Steffens, D. C., Taylor, W. D., Denny, K. L., Bergman, S. R., & Wang, L. (2011). Structural integrity of the uncinate fasciculus and resting state functional connectivity of the ventral prefrontal cortex in late life depression. *PLoS One, 6,* .

Stein, M. B., Goldin, P. R., Sareen, J., Zorrilla, L. T., & Brown, G. G. (2002). Increased amygdala activation to angry and contemptuous faces in generalized social phobia. *Archives of General Psychiatry, 59,* 1027–1034.

Straub, J., Metzger, C. D., Plener, P. L., et al. (2017). Successful group psychotherapy of depression in adolescents alters fronto-limbic resting-state connectivity. *Journal of Affective Disorders, 209,* 135–139.

Stuhrmann, A., Suslow, T., & Dannlowski, U. (2011). Facial emotion processing in major depression: A systematic review of neuroimaging findings. *The Biology of Mood and Anxiety Disorders, 1,* 10.

Surguladze, S., Brammer, M. J., Keedwell, P., et al. (2005). A differential pattern of neural response toward sad versus happy facial expressions in major depressive disorder. *Biological Psychiatry, 57,* 201–209.

Suslow, T., Konrad, C., Kugel, H., et al. (2010). Automatic mood-congruent amygdala responses to masked facial expressions in major depression. *Biological Psychiatry, 67,* 155–160.

Taylor, S. F., Ho, S. S., Abagis, T., et al. (2018). Changes in brain connectivity during a sham-controlled, transcranial magnetic stimulation trial for depression. *Journal of Affective Disorders, 232,* 143–151.

Thomas, K. M., Drevets, W. C., Dahl, R. E., et al. (2001). Amygdala response to fearful faces in anxious and depressed children. *Archives of General Psychiatry, 58,* 1057–1063.

Tian, Y. H., Du, J., Spagna, A., et al. (2016). Venlafaxine treatment reduces the deficit of executive control of attention in patients with major depressive disorder. *Scientific Reports, 6,* 28028.

Touroutoglou, A., Andreano, J. M., Barrett, L. F., & Dickerson, B. C. (2015). Brain network connectivity-behavioral relationships exhibit trait-like properties: Evidence from hippocampal connectivity and memory. *Hippocampus, 25*(12), 1591–1598.

Treadway, M. T., & Zald, D. H. (2011). Reconsidering anhedonia in depression: Lessons from translational neuroscience. *Neuroscience and Biobehavioral Reviews, 35,* 537–555.

Trivedi, M. H., McGrath, P. J., Fava, M., et al. (2016). Establishing moderators and biosignatures of antidepressant response in clinical care (EMBARC): Rationale and design. *Journal of Psychiatric Research, 78,* 11–23.

Vai, B., Bulgarelli, C., Godlewska, B. R., et al. (2016). Fronto-limbic effective connectivity as possible predictor of antidepressant response to SSRI administration. *The Journal of the European College of Neuropsychopharmacology, 26,* 2000–2010.

Vasic, N., Walter, H., Sambataro, F., & Wolf, R. C. (2009). Aberrant functional connectivity of dorsolateral prefrontal and cingulate networks in patients with major depression during working memory processing. *Psychological Medicine, 39,* 977–987.

Veer, I. M., Beckmann, C. F., van Tol, M. J., et al. (2010). Whole brain resting-state analysis reveals decreased functional connectivity in major depression. *Frontiers in Systems Neuroscience, 4*, 41.

Victor, T. A., Furey, M. L., Fromm, S. J., Ohman, A., & Drevets, W. C. (2010). Relationship between amygdala responses to masked faces and mood state and treatment in major depressive disorder. *Archives of General Psychiatry, 67*, 1128–1138.

Vilgis, V., Chen, J., Silk, T. J., Cunnington, R., & Vance, A. (2014). Frontoparietal function in young people with dysthymic disorder (DSM-5: Persistent depressive disorder) during spatial working memory. *Journal of Affective Disorders, 160*, 34–42.

Vincent, J. L., Kahn, I., Snyder, A. Z., Raichle, M. E., & Buckner, R. L. (2008). Evidence for a frontoparietal control system revealed by intrinsic functional connectivity. *Journal of Neurophysiology, 100*, 3328–3342.

Wagner, G., Sinsel, E., Sobanski, T., et al. (2006). Cortical inefficiency in patients with unipolar depression: An event-related FMRI study with the Stroop task. *Biological Psychiatry, 59*, 958–965.

Walter, H., Wolf, R. C., Spitzer, M., & Vasic, N. (2007). Increased left prefrontal activation in patients with unipolar depression: An event-related, parametric, performance-controlled fMRI study. *Journal of Affective Disorders, 101*, 175–185.

White, S. J., Coniston, D., Rogers, R., & Frith, U. (2011). Developing the Frith-Happé animations: A quick and objective test of Theory of Mind for adults with autism. *Autism Research, 4*, 149–154.

Williams, L. M. (2016). Precision psychiatry: A neural circuit taxonomy for depression and anxiety. *Lancet Psychiatry, 3*, 472–480.

Williams, L. M. (2017). Defining biotypes for depression and anxiety based on large-scale circuit dysfunction: A theoretical review of the evidence and future directions for clinical translation. *Depression and Anxiety, 34*, 9–24.

Williams, L. M., Das, P., Liddell, B. J., et al. (2006). Mode of functional connectivity in amygdala pathways dissociates level of awareness for signals of fear. *The Journal of Neuroscience, 26*, 9264–9271.

Williams, L. M., Korgaonkar, M. S., Song, Y. C., et al. (2015a). Amygdala reactivity to emotional faces in the prediction of general and medication-specific responses to antidepressant treatment in the randomized iSPOT-D trial. *Neuropsychopharmacology: Nature Publishing Group, 40*(10), 2398–2408.

Wu, M., Andreescu, C., Butters, M. A., et al. (2011). Default-mode network connectivity and white matter burden in late-life depression. *Psychiatry Research, 194*, 39–46.

Yang, T. T., Simmons, A. N., Matthews, S. C., et al. (2010). Adolescents with major depression demonstrate increased amygdala activation. *Journal of the American Academy of Child and Adolescent Psychiatry, 49*, 42–51.

Yeo, B. T., Krienen, F. M., Sepulcre, J., et al. (2011). The organization of the human cerebral cortex estimated by intrinsic functional connectivity. *Journal of Neurophysiology, 106*, 1125–1165.

Yuste, R. (2015). From the neuron doctrine to neural networks. *Nature Reviews Neuroscience, 16*, 487–497.

Zhang, W. N., Chang, S. H., Guo, L. Y., Zhang, K. L., & Wang, J. (2013). The neural correlates of reward-related processing in major depressive disorder: A meta-analysis of functional magnetic resonance imaging studies. *Journal of Affective Disorders, 151*, 531–539.

Zhao, Y. J., Du, M. Y., Huang, X. Q., et al. (2014). Brain grey matter abnormalities in medication-free patients with major depressive disorder: A meta-analysis. *Psychological Medicine, 44*, 2927–2937.

Zhu, X., Wang, X., Xiao, J., et al. (2012). Evidence of a dissociation pattern in resting-state default mode network connectivity in first-episode, treatment-naive major depression patients. *Biological Psychiatry, 71*, 611–617.

Chapter 43

Multimodal modeling for personalized psychiatry

Scott R. Clark[a], Micah Cearns[a], Klaus Oliver Schubert[a,b] and Bernhard T. Baune[c,d,e]

[a]Discipline of Psychiatry, University of Adelaide, Adelaide, SA, Australia, [b]Northern Adelaide Local Health Network, Mental Health Service, Adelaide, SA, Australia, [c]Department of Psychiatry and Psychotherapy, University of Münster, Münster, Germany, [d]Department of Psychiatry, Melbourne Medical School, The University of Melbourne, Melbourne, Australia, [e]The Florey Institute of Neuroscience and Mental Health, The University of Melbourne, Parkville, VIC, Australia

1 Introduction

Personalized medicine can be defined as the use of an individual's unique characteristics, including environmental exposures, symptoms, and biological markers to more accurately predict vulnerability to disease, response to specific therapies, and long term outcomes (Ozomaro et al., 2013). In psychiatry, this task is complicated by the syndromic nature of diagnosis, as represented in the Diagnostic and Statistical Manual of Mental Disorders (DSM) (Scarr et al., 2015). These diagnostic categories have been developed without a detailed understanding of underlying biological mechanisms (Faucher & Goyer, 2015). While they have provided the basis for the development of a science of psychiatry, they are not sufficient or reliable descriptions of individual presentations, medication response, illness, or functional trajectories to progress personalized treatment (Goldstein-Piekarski, Williams, & Humphreys, 2016; Ogasawara, Nakamura, Kimura, Aleksic, & Ozaki, 2018; Regier, Narrow, Clarke, et al., 2013; Wardenaar & de Jonge, 2013). For subpopulations within heterogeneous diagnostic categories, significant treatment effects or relationships between predictors and outcomes may either be diluted (Mulder, Singh, Hamilton, et al., 2018) or nonspecific, mapping across sub-syndromes (Fusar-Poli, Bonoldi, Yung, et al., 2012; Hartmann, Nelson, Spooner, et al., 2017). The resultant uncertainty in psychiatric diagnosis and treatment effectiveness can then lead to either a wait and watch approach, where syndromic thresholds are not met, but intervention may be beneficial, or multiple trials of different medications with unnecessary side effects and illness burden due to poor efficacy (Clark, Schubert, & Baune, 2015; Norbury & Seymour, 2018; Schubert, Clark, & Baune, 2015).

To truly progress personalized psychiatry, the sensitivity and specificity of illness and outcome stratification relies on identifying better predictors of clinically relevant outcomes. However, studies have consistently shown that individual predictors only explain small amounts of variance (Schubert et al., 2015). For example, individual demographic and clinical variables commonly measured in standardized scales have limited accuracy for the prediction of the onset of psychosis (Fusar-Poli et al., 2012; Mechelli, Lin, Wood, et al., 2017), function after first episode of psychosis (Santesteban-Echarri, Paino, Rice, et al., 2017), relapse after antipsychotic cessation (Bowtell, Ratheesh, McGorry, Killackey, & O'Donoghue, 2017a), return to work after depression (Ervasti, Joensuu, Pentti, et al., 2017), response to antipsychotics in schizophrenia (Carbon & Correll, 2014), electroconvulsive therapy response (Haq, Sitzmann, Goldman, Maixner, & Mickey, 2015), and lithium response in mood disorders (Kleindienst, Engel, & Greil, 2005).

Multivariate models that combine clinical predictors can explain greater variance in mental health outcomes. Particularly, machine learning (ML) shows promise to outperform clinician decision-making (Dwyer, Falkai, & Koutsouleris, 2018). For example, Chekroud, Zotti, Shehzad, et al. (2016) achieved an accuracy of ~65% for the prediction of response to antidepressants in major depressive disorder using only questionnaire data, compared with a sample of clinicians who predicted response with only 49.3% accuracy. The addition of biological markers to clinical data, from a growing range of "modes," including blood based (genomics, proteomics, metabolomics, and lipidomics), imaging (e.g., structural and functional magnetic resonance imaging (MRI)), and electrophysiology may offer promise to improve this accuracy. However, complex relationships exist between predictors and outcomes, leading to higher order and nonlinear emergent properties between biology and behavior (Dwyer et al., 2018; Kendler, 2008). Complexity originates in the multiple levels of scale (molecular, structural, physiological, psychological, behavioral), redundancy present in biological

Personalized Psychiatry. https://doi.org/10.1016/B978-0-12-813176-3.00043-2

systems (Bedse, Hartley, Neale, et al., 2017; Leistritz, Weiss, Bar, et al., 2013; Lytton, Arle, Bobashev, et al., 2017), and the significant potential for seemingly unrelated diseases or outcomes to be associated with any one marker (Wang, Yang, Gelernter, & Zhao, 2015). For example, in psychosis, biological markers may be differentially related to specific phenotypes within broader syndromal categories, and these categories may map to similar or divergent long-term outcomes (see Fig.1). Differentiation should also be made between markers of "state" that identify an emerging change in risk, such as an acute episode, and markers of "trait" (e.g., marker 3 in Fig. 1). For example, studies of peripheral inflammation show that cytokines IL-1β, IL-6, and TGF-β are elevated in first-episode psychosis, and relapse and normalize with treatment, suggesting that they are state markers. In comparison IL-12, IFNG, TNF, and sIL-2R are elevated during acute exacerbation, but remain elevated after antipsychotic treatment, suggesting they are trait markers (Miller, Buckley, Seabolt, Mellor, & Kirkpatrick, 2011).

Multimodal, multivariate modeling techniques that can quantify both linear and nonlinear relationships among biomarkers, standardized clinical data, and outcomes of interest may be better suited to describe such complexity. Emerging evidence shows that the addition of biological variables can improve the accuracy of outcome prediction across different disorders (Clark et al., 2015; Koutsouleris, Kambeitz-Ilankovic, Ruhrmann, et al., 2018; Lee, Hermens, Naismith, et al., 2015; Moser, Doucet, Lee, et al., 2018; Schubert et al., 2015). For example, Koutsouleris et al. (2018) recently used ML to analyze clinical, general function, and imaging data from patients at clinical high risk (CHR) for psychosis, and others with recent-onset major depression, and found that adding neuroimaging to clinical data produced a 1.9-fold and 10.5-fold increase accuracy of prediction of social function at 1 year, respectively.

This chapter will review the nature and collection of different modes of predictor data, and then discuss the construction of multimodal prediction models in psychiatry, providing an overview on how models could be used to improve clinical practice.

2 Modes of data in psychiatry

2.1 Clinical data

The starting point for all psychiatric assessments is the review of collateral and medical record information and subsequent patient interview. Historical and clinical data obtained though these processes provides important phenotypic and risk information for the prediction of psychiatric outcomes (Alda, 2013). Biomarkers are only meaningful when interpreted in the context of the patient's history and presentation, therefore, clinical information should form the basis of multimodal prediction models (Sox, Blatt, Higgins, & Marton, 2013). Subtypes of clinical data include demographics, items from personal and family history, presenting symptoms, and cognition and function. Familial history of illness and gender are proxies for genetic and environmental risk. For example, first-degree family history of schizophrenia carries significant risk for psychotic, mood, and substance disorders (DeVylder & Lukens, 2013). Age is also a risk factor due to

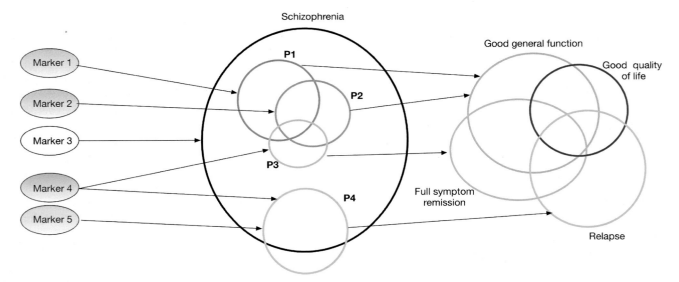

FIG. 1 Map of potential associations between hypothetical biomarkers and disease phenotypes.

developmental trajectories, and later, higher risk of neurodegenerative processes. For example, excessive programmed synaptic pruning is thought to occur in adolescents that develop psychosis (Sekar, Bialas, de Rivera, et al., 2016), and early age of onset is associated with more severe forms of mental illness (Immonen, Jaaskelainen, Korpela, & Miettunen, 2017). Environmental exposures, such as inter-uterine infections, inflammation, birth trauma, later substance use, and psychological trauma may be either predisposing or precipitating factors that impact outcomes through epigenetic and direct mechanisms (Depino, 2018; Hamlyn, Duhig, McGrath, & Scott, 2013; Heim & Binder, 2012; Messias, Chen, & Eaton, 2007; Palmisano & Pandey, 2017). Environmental and genetic interactions are also reflected in measures of personality and behavioral adjustment. These factors may determine the level of resilience to life stressors and symptoms of mental illness (Bowtell, Eaton, Thien, et al., 2017b; Carbon & Correll, 2014; Sanchez-Roige, Gray, Mac Killop, Chen, & Palmer, 2018). A number of important predictors of patient outcomes are collected in the psychiatric history including comorbid diagnoses, number and length of hospitalizations, treatment response, and risk behaviors such as self-harm, violence, and substance abuse (De Carlo, Calati, & Serretti, 2016; Hachtel, Harries, Luebbers, & Ogloff, 2018; Kautzky, Dold, Bartova, et al., 2018; Large, Mullin, Gupta, Harris, & Nielssen, 2014; Liu, Sareen, Bolton, & Wang, 2016; Messer, Lammers, Muller-Siecheneder, Schmidt, & Latifi, 2017). Comorbidity, need for hospitalization and poor treatment response are all proxies for more severe and enduring illness, while substance use, violence, and self-harm are indicators of individual risk for poor outcome. Clinical data also includes assessments of current mental state and function. Salient psychiatric phenomenology is recorded in the mental state examination. Information on impairments in day to day function is documented in the history of the patient's presenting complaint. Validated symptom and functional scales, which are discussed as follows, offer more reliable and comprehensive sources of this data.

2.1.1 Collection and quantification of clinical data in psychiatry

Demographic and historical data can be collected by participant self-report, clinician administered questionnaires, or from existing data sources via medical record review or database linkage. Questionnaires can range in complexity from simple data on specific illness events, such as the age of first diagnosis, to integrated scales for course of illness, medication response over time, as well as complex scales that assess details of past events such as trauma, suicidal ideation, and self-harm (Grof, Duffy, Cavazzoni, et al., 2002; Posner, Brown, Stanley, et al., 2011).

Symptom scales are designed for either categorical diagnosis, or to quantify continuous illness severity. The reliability and validity of these measures is improved by using either structured or semistructured interviews that consist of sets of questions to elicit specific symptomatology in a standardized manner. Responses are then scored in terms of diagnostic criteria or symptom severity, depending on the focus of the measure. The Mini-International Neuropsychiatric Interview (MINI) (Sheehan, Lecrubier, Sheehan, et al., 1998) and the Structured Clinical Interview for DSM (SCID) (First, Williams, Karg, & Spitzer, 2015) are examples of widely used diagnostic interviews. In contrast, the Structured Clinical Interview for Positive and Negative Syndrome Scale (SCI-PANSS) is the gold standard assessment for severity of psychotic symptoms (Opler, Yavorsky, & Daniel, 2017), while the Structured Interview Guide for the Hamilton Depression Rating Scale (SIGH-D) is a common measure of depressive symptoms (Williams, 1988). The SCI-PANSS and SIGH-D also have validated subscales that are useful in defining phenotypes and predicting outcomes (Kay, 1990; Williams, 1988). For example, the negative symptoms subscale from the PANSS is predictive of poor long-term function in schizophrenia (Harvey, Khan, & Keefe, 2017).

2.2 Cognitive and general function

The assessment of *cognitive function* involves the use of standardized tests to analyze performance across different domains of day-to-day thinking. Both paper-based and computer-based cognitive batteries are widely used in psychiatric research (Bakkour, Samp, Akhras, et al., 2014). Scores can be categorized into a number of subdomains, such as processing speed, working and long-term memory, language, constructional ability, and executive functions, including sequencing, planning, and decision making (Bakkour et al., 2014). Both trait and state impairments of cognitive function are predictive of general function in mental illness (Baune, Li, & Beblo, 2013; Baune & Malhi, 2015). The pattern of performance in subdomains and severity of impairment varies among individuals, depending on baseline intelligence, schooling, and age. Patients with schizophrenia are the most severely affected, followed by schizoaffective, bipolar disorder, and then major depression (Barch, 2009; Bora & Pantelis, 2015). Specific cognitive phenotypes may be trans-diagnostic predictors of long-term functional outcomes (Lee et al., 2015).

Measures of baseline *quality of life and general function* are also predictive of subsequent functional outcomes (Alvarez-Jimenez, Gleeson, Henry, et al., 2012; Boyer et al., 2013; Santesteban-Echarri et al., 2017). Self-report and clinician rated quality of life and functional scales range from simple global assessments, such as the Global Assessment of Functioning Scale (GAF) (American Psychiatric Association, 1994) to more detailed multidomain scales, such as the Functioning Assessment Short Test (Gonzalez-Ortega, Rosa, Alberich, et al., 2010) or the Short Form 36 quality of life scale (McHorney, Ware Jr., & Raczek, 1993). Objective measures of function carry less reporting bias, but are costly and time consuming to implement. Validation of self-report is important, as patients may under or overestimate symptoms and performance due to lack of insight (Brown & Velligan, 2016; Harvey, Velligan, & Bellack, 2007). More robust, composite self-report functional measures may be obtained by combining scales using clustering algorithms (Lin, Wood, Nelson, et al., 2011; Olagunju, Clark, & Baune, 2018).

2.3 Peripheral biomarkers

Primary pathology in psychiatric illness occurs within the central nervous system, out of reach of noninvasive biological sampling techniques. However, valid links have been established between brain and peripheral markers obtained from blood or saliva (Harris et al., 2012; Sekar et al., 2016; Walton, 2018). These peripheral samples provide rich epigenomic (Hoffmann, Sportelli, Ziller, & Spengler, 2017), genomic (Smoller et al., 2018; Sullivan, Agrawal, Bulik, et al., 2018), transcriptomic (Forstner, Hofmann, Maaser, et al., 2015; Reinbold, Forstner, Hecker, et al., 2018), proteomic (Cooper, Ozcan, Gardner, et al., 2017; Schubert, Stacey, Arentz, et al., 2018), metabolomic, and lipidomic data (Davison, O'Gorman, Brennan, & Cotter, 2017) for the development of specific markers, marker panels (Chan, Gottschalk, Haenisch, et al., 2014), or risk scores (e.g., polygenetic risk (International Consortium on Lithium Genetics (ConLi +Gen) et al., 2018; Bogdan, Baranger, & Agrawal, 2018; Mistry, Harrison, Smith, Escott-Price, & Zammit, 2018)) for mental health outcomes. Interestingly, some of the markers described seem to be transdiagnostic, suggesting common pathways for different diagnostic outcomes (Pinto, Moulin, & Amaral, 2017). At molecular levels of biological reduction, associations with emergent behavior and symptoms are small (Sullivan et al., 2018), and may be contingent on new techniques in systems biology, such as expression quantitative trait loci (eQTLs), canonical pathway, and network analysis (see Fig. 2). Such techniques can integrate genomic and proteomic data to identify the key genomic regulators of biological pathways associated with mental illness. Critical pathways, rather than individual markers of small association, may be more robust markers of risk as a pathway subsumes multiple points and types of genomic risk into a single measure (Chan, Cooper, Heilmann-Heimbach, et al., 2017; Ciobanu, Sachdev, Trollor, et al., 2016; Lee, Marchionni, Andrews, et al., 2017; Schubert et al., 2018; Stacey, Schubert, Clark, et al., 2018). Promising blood-based protein panels have been

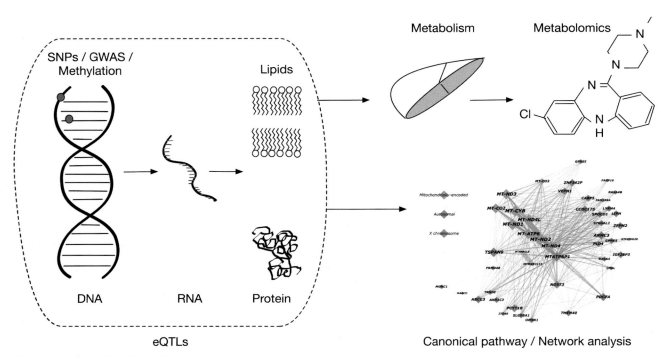

FIG. 2 Overview of blood-based biomarkers.

developed to aid the prediction of a first psychotic episode (Perkins, Jeffries, Addington, et al., 2015), diagnosis in schizophrenia (Schwarz, Guest, Rahmoune, et al., 2012a) and bipolar disorder (Haenisch, Cooper, Reif, et al., 2016), and medication response and relapse in schizophrenia (Schwarz, Guest, Steiner, Bogerts, & Bahn, 2012b). The clinical utility and reliability of these panels needs replication in larger studies.

Genome-wide association studies (GWAS) using single nucleotide polymorphism (SNP) analysis, RNA sequencing, and mass spectrometry of protein expression, allow for simultaneous analysis of thousands of markers. However, the technology for these omics approaches remains complex, expensive, and lacks reliability to scale to daily clinical practice (Blum, Mousavi, & Emili, 2018; Breen, Stein, & Baldwin, 2016; Craft, Chen, & Nairn, 2013; Dalvie, Koen, McGregor, et al., 2016; Sullivan et al., 2018). Replication of findings has not been consistent for many markers (Scarr et al., 2015), and due to the small effect size, larger samples are required to attain adequate statistical power. Currently, these techniques have been used to develop targeted assays for key subsets of markers that could, with further validation, be more reliably and cost-effectively implemented into practice (Schubert et al., 2018).

2.4 Brain imaging and electrophysiology

Abnormalities in both structural and functional brain imaging are associated with psychiatric diagnosis, medication response, and functional outcomes (Dazzan, Arango, Fleischacker, et al., 2015; Dietsche, Kircher, & Falkenberg, 2017; Fu & Costafreda, 2013; Wylie, Smucny, Legget, & Tregellas, 2016). Structural MRI can define and compare various brain parameters, including regional thickness, volume, curvature, surface area, density, and integrity in gray and white matter. For example, across longitudinal studies in clinical high risk of psychosis and schizophrenia, there appears to be cortical gray matter loss in temporal and frontal regions in the prodromal period, followed by frontal and thalamic gray matter loss and frontal cortical thinning following first episode psychosis, worsening into chronic schizophrenia, particularly in those with poor outcomes (Dietsche et al., 2017; Kambeitz-Ilankovic, Meisenzahl, Cabral, et al., 2016). In comparison, MRI diffusion tensor imaging shows specific patterns of disruption in the integrity of axon fiber tracts associated with diagnosis and outcomes in major depressive disorder (Kambeitz, Cabral, Sacchet, et al., 2017; Tae, Ham, Pyun, Kang, & Kim, 2018).

Functional brain imaging can quantify changes in the activation of brain regions during specific cognitive tasks or in resting state. Specific techniques include functional MRI (fMRI) (Wylie et al., 2016), positron emission tomography (PET) (Piel, Vernaleken, & Rosch, 2014), near infra-red spectroscopy NIRS (Kumar et al., 2017), electroencephalography (EEG), and magnetoencephalography (MEG) (Alamian, Hincapie, Pascarella, et al., 2017). Specific EEG measures include event-related potentials (ERPs) and quantitative electroencephalography (qEEG) (Buoli et al., 2016; Duncan, Barry, Connolly, et al., 2009). All of these methods have returned findings related to prediction of psychiatric diagnosis, medication response, and functional outcomes. Useful indices of brain connectivity can be extracted from structural (anatomical) and functional MRI (brain region co-activation), EEG, and MEG (geographic synchronization of brain wave oscillations) (Du, Fu, & Calhoun, 2018; Houck, Cetin, Mayer, et al., 2017; Vecchio, Miraglia, & Maria, 2017). Interestingly, resting state functional MRI and diffusion tensor imaging show higher accuracy for the diagnosis of major depressive disorder, in comparison with structural MRI measures (Kambeitz et al., 2017). In general, MRI-based measures have high spatial, but poorer temporal resolution, while EEG and MEG have high temporal, and poorer spatial resolution. The combination of structural and functional measures has the potential to identify integrated patterns or signatures of change that explain more variance in illness and functional outcomes (Schultz, Fusar-Poli, Wagner, et al., 2012). For example, Cabral, Kambeitz-Ilankovic, Kambeitz, et al. (2016) found that the combination of MRI structural and resting state functional connectivity improved the categorization of patients with schizophrenia from controls.

3 Multivariate modeling

Given the vast range of data types and statistical methods, the construction of a multimodal model is a complex exercise. The process can be broken down into four broad steps: (1) study formulation and design; (2) variable selection and coding; (3) model estimation/parameter optimization, and (4) model validation (Steyerberg & Vergouwe, 2014). Steps 2–4 can be broadly grouped into model training and testing.

3.1 Study formulation and design

The construction of a multivariate prognostic model requires data from at least two timepoints, including a measurement of two or more predictors at baseline, and an outcome measure at some later time of follow up. The task of these models is either classification into categories (e.g., differentiating patients with depression from control or medication responders from nonresponders using MRI data), or prediction of continuous outcomes (regression, e.g., predicting depressive symptoms from MRI data) (Mwangi, Matthews, & Steele, 2012; Mwangi, Tian, & Soares, 2014; Patel, Khalaf, & Aizenstein, 2016). In terms of overall design, prospective longitudinal studies are better able to quantify the temporal relationship between events of interest and important confounding factors (Mann, 2003; Moons, Royston, Vergouwe, Grobbee, & Altman, 2009). More complex longitudinal designs with multiple sampling points can provide detailed descriptions of evolving illness trajectories (Schubert, Clark, Van, Collinson, & Baune, 2017). In contrast, retrospective case–control studies using available medical records are easier to implement at lower cost. However, confounding is more difficult to measure and missing data is common due to variation in day to day practice. There are also limitations on the interpretation of items collected for clinical purposes other than the modeling in question (Weiskopf, Hripcsak, Swaminathan, & Weng, 2013; Weiskopf & Weng, 2013).

Outcome measures can be categorical (e.g., diagnosis, remission, or death) or continuous (e.g., symptom or functional scores). When selecting a specific outcome measure, attention should be given to reliability, clinical validity, and sensitivity to capture relevant changes (Coster, 2013). For example, the vast majority of clinical trials in psychiatry have focused on symptom reduction or remission as the primary endpoint. However, function is not necessarily linked to symptom burden, and measures of function that include assessments of quality of life along with day to day performance may more accurately describe real world illness impact and outcomes (Alvarez-Jimenez et al., 2012; Brown & Velligan, 2016; Lin et al., 2011; Olagunju et al., 2018). Ultimately, practical limitations may determine the final choice of outcome measure and the conclusions that can be drawn from the modeling exercise.

Predictors can also be categorical (e.g., diagnosis, gender) or continuous (e.g., brain MRI gray matter volume, or serum protein level). The selection of predictor variables can be performed a priori based on existing literature (Waljee, Higgins, & Singal, 2013), or by using ML techniques to extract the optimal subset from all data available (Dwyer et al., 2018).

3.2 Model training and testing

Generalizability is key to the utility of any prediction model—the model must be accurate beyond the sample in which it is derived (Dwyer et al., 2018). In the absence of external replication in a different population (Collins, de Groot, Dutton, et al., 2014), cross-validation reduces the risk of model overfitting to one data set by ensuring the model is developed and tested in different sub-samples of available data (Moons, Kengne, Woodward, et al., 2012; Steyerberg et al., 2001; Steyerberg & Vergouwe, 2014). One simplistic approach is to manually divide the available data into training and testing samples. A more robust approach is to use k-fold cross-validation, dividing the sample into k-folds, training on some, and testing on the others. In addition, repeated runs of *k*-fold cross-validation are recommended to avoid opportune splits in the data that may inflate accuracy estimates (Kohavi, 1995). Advances in ML techniques and statistical programming have set new benchmarks for cross-validation. All analysis can now be conducted within a modeling "pipeline" that embeds all steps required for prediction within cross-validation folds to avoid overly optimistic estimates from data leakage (Dwyer et al., 2018). Data leakage occurs when there is unintentional transference of information about the test data into the training data that invalidates generalizability claims outside the study sample (Dwyer et al., 2018). Unfortunately, the use of cross-validation has not been systematic in medical literature. Moving forward, standards have been developed to critically evaluate predictive modeling literature (Collins et al., 2014; Collins, Reitsma, Altman, & Moons, 2015; Studerus, Ramyead, & Riecher-Rossler, 2017).

3.2.1 Variable selection and coding

The selection of independent variables and any reclassification or recoding needs to be carefully considered before commencing multivariate analysis. The inclusion of too many predictors may tune model parameters too closely to noise in the sample (overfitting, see http://scikit-learn.org/stable/model_selection.html) (Babyak, 2004). Overfit models do not generalize well outside the derivation sample. For example, a rule of thumb for regression models is 1 predictor for each 10 events of interest, although the optimum number may depend on specific study characteristics (Ogundimu, Altman, & Collins, 2016). Thus, more complex models with great numbers of predictors require larger sample sizes. Some evidence suggests that the optimal feature space (number of predictors) should depend on the biases and heuristics of the algorithm used for prediction (Tang & Alelyani, 2014). To avoid overfitting, multiple factors can be combined into single variables based on underlying hypotheses; for example, using total symptom scores rather than sub-scores, or items from structured assessments, or collapsing categories with infrequent occurrence into a single score (e.g., different comorbidities into a comorbidity index) (Babyak, 2004; Royston, Moons, Altman, & Vergouwe, 2009). There are numerous automated-feature selection algorithms that allow for the exclusion of highly correlated variables, low variance variables, and variables that do not contribute strongly to predictive accuracy. Many of these methods allow variables to be ranked in terms of their importance to the overall model prediction (Huys, Maia, & Frank, 2016; Iniesta, Stahl, & McGuffin, 2016; Mwangi et al., 2014; Spratt, Ju, & Brasier, 2013). Penalized regression methods such as lasso, ridge, and the elastic net automatically weight the contributions of predictors by adjusting coefficients toward zero, effectively reducing the influence of those that do not independently contribute to model accuracy, and in the process, shrinking the total number of variables in the model (Huys et al., 2016; Okser et al., 2014; Pavlou, Ambler, Seaman, et al., 2015). Penalized regression is particularly useful for analysis of data sets with large numbers of predictors; for example, the use of GWAS data with 10,000 SNP variables to predict a mental health outcome in 1000 people (Wu, Chen, Hastie, Sobel, & Lange, 2009; Zhou, Sehl, Sinsheimer, & Lange, 2010; Zou & Hastie, 2005).

Missing data in predictors can be dealt with via listwise or pairwise deletion of cases, particularly if the percentage of cases with missing data is small (i.e., 5% or less (Azur, Stuart, Frangakis, & Leaf, 2011)). However, deletion results in reduced sample size, study power, and can also introduce selection bias when the score on a variable is related to non-random missingness (e.g., participants scoring high on psychotic symptoms are more likely to miss assessments) (Enders, 2006). Missing data can be imputed (estimated) based on sample characteristics. Traditional data imputation techniques, such as mean and mode imputation, also carry risk of introducing bias (Enders, 2006). Regression-based imputation, such as multivariate imputation by chained equations (MICE), is more robust and flexible (Azur et al., 2011). Learning-based approaches such as k nearest neighbors and bagged decision trees are gaining popularity, and appear to outperform mean and mode imputation also (Valdiviezo & Van Aelst, 2015).

3.2.2 Model selection (estimation)

There are a vast array of predictive algorithms available for multimodal modeling, and the specific technique chosen to operationalize a modeling task should reflect the data available, and the underlying research question (Iniesta et al., 2016). Techniques vary in ability to manage missing data, reduce feature numbers, prevent overfitting, in the type of decision boundary, and the clarity or ease at which the role/strength of predictors in the model can be accessed (see Table 43.1). Some models are more resilient to missing data. For example, naive Bayesian models can skip over missing attributes when calculating distributions, and are potentially the least likely to be affected (Kalousis & Hilario, 2000). Tree-based algorithms such as random forests deal specifically with missing data as part of their implementation (Loh, 2014). The field is rapidly evolving, and new implementations of existing algorithms (e.g., support vector machines (SVMs)) have been designed specifically to manage large data sets with a significant percentage of missing values (Razzaghi, Roderick, Safro, & Marko, 2016). Generally, for small sample sizes, low complexity models with a small number of parameters, such as naive Bayes or penalized regression (ridge, lasso, and the elastic net) will provide a better fit to the data. As sample size increases, higher complexity models can be considered (random forests, gradient boosted machines, neural networks) (Raudys & Jain, 1991). For classification problems, the decision boundary that separates patients into groups can be linear, or more complex and nonlinear (Yan & Xu, 2009). General linear model (GLM) based regression models and linear SVMs are not able to accurately represent nonlinear boundaries. In these cases, random forest, simple Bayes' or the use of a radial basis function (RBF) kernel for SVM is appropriate. The RBF is to used transform nonlinear relationships into linear ones represented in higher dimensional space (Scholkopf, Sung, Burges, et al., 1997). In complex health workflows, understanding the value of each variable in a multimodal model is important for decisions on the ordering of testing. Complex models such as neural networks and nonlinear SVMs do not provide information on the contribution of individual predictors (Altmann, Tolosi, Sander, & Lengauer, 2010). In contrast, beta weights in GLM and Gini coefficients in random forest provide some indication of the GLM—General linear regression model effect size for each variable. Using Bayesian expected value of information techniques, the utility of collecting each mode can be specifically determined (Yokota & Thompson, 2004).

Ultimately, it is difficult to select one specific model that will best fit the data without testing a number of suitable candidates (known as the "No Free Lunch" theorem (Wolpert, 1996)). Therefore, when using ML approaches, multiple models should be trained and tested. Permutation testing to assess the probability that the observed model accuracy occurred purely by chance should also be performed (Ojala & Garriga, 2010). As previously discussed, all predictive modeling applications should be implemented using pipeline architecture to avoid the leakage of data that may be predictive of the dependent variable. This is possible through the use of pipelines in Scikit-learn http://scikit-learn.org/stable/ and the Caret package in R https://topepo.github.io/caret/data-splitting.html).

TABLE 43.1 Comparison of common prediction algorithms.

Algorithm	Description	Handles missing data	Handles overfitting/feature reduction	Handles nonlinear decision boundaries	Contribution of predictors easily interpreted	Use case	Example
GLM (nonregularized)	Includes logistic and multiple regression	No	Yes—through stepwise procedures	No	Accessible via regression coefficients	Where simple linear models including covariates are sufficient	Widely used and established efficacy in neuroimaging (Chen, Adleman, Saad, Leibenluft, & Cox, 2014)
Penalized regression models (lasso, ridge, and the elastic net)	Extension of GLM; includes regularization parameters to protect against overfitting. Lasso tends to zero coefficients for variables that are highly correlated. Ridge will shrink coefficients, but not to zero. The elastic net combines both procedures for balance	No	Yes—adjusts for overfitting through penalization	No	Accessible via regression coefficients	When the number of independent variables is high, particularly if greater than the number of observations. When coefficient values are needed but not *P*-values	The use of rich clinical and/or genetic data to predict antidepressant response in major depression (Chekroud et al., 2016; Maciukiewicz, Marshe, Hauschild, et al., 2018)
Naïve Bayes	Uses Bayes Theorem to combine predictors to determine outcome probability, assumes variable independence	Yes	Overfitting is unlikely unless sample size is extremely small	Yes	Yes—with simple techniques such as the odds ratio form of Bayes' Rule that utilize likelihood ratios (Clark, Baune, Schubert, et al., 2016)	Assumes independent variables are noncorrelated, but still functions well when this assumption is violated. Useful when only a small amount of training data is available. Good first pass model as trains quickly	Prediction of outcomes in psychotic illness (Clark et al., 2015, 2016; Schubert et al., 2017)

Random forest	Multiple tree models built with random subsamples and different initial variables	Yes	More decision trees may provide protection against overfitting. Uses Gini coefficients and mean decrease in impurity to reduce features	Yes	Yes—Gini coefficients and mean decrease in impurity (Altmann et al., 2010)	When a nonlinear decision boundary provides a better fit to the data. When the importance of individual variables needs to be known	Prediction of severity of ADHD severity based on stress and genetic data (van der Meer, van Hoekstra, van Donkelaar, et al., 2017)
Support vector machines	Hyperplanes constructed in multidimensional space that separate cases of different outcomes	No	Linear kernel—No. Radial basis Function kernel—Yes	Linear kernel—No. Radial Basis Function kernel—Yes	Linear kernel—Yes. Radial basis function kernel—No (Altmann et al., 2010)	Linear kernel—When a simple model is needed for a small dataset. Radial basis function kernel—When a nonlinear decision boundary is needed on a large sample	Widely used and established efficacy in neuroimaging (Orrù, Pettersson-Yeo, Marquand, Sartori, & Mechelli, 2012)
Artificial neural networks	Nodes (neurons) are connected in multiple layers and weighted during training	No	Yes—but overfitting may arise from too many hidden inner model layers (Srivastava, Hinton, Krizhevsky, Sutskever, & Salakhutdinov, 2014)	Yes	No	When large amounts of training data are available and individual contributions from variables do not need to be known	Combining structural and functional neuroimaging and cognitive data for diagnosis and outcome in mild cognitive impairment, Alzheimer's disease, psychosis, ADHD, and temporal lobe epilepsy (Vieira, Pinaya, & Mechelli, 2017)

3.2.3 Model assessment

Model performance can be described in terms of variance explained (R^2 (Scarr et al., 2015)) or root mean squared error (RMSE) in regression, while for classification, discriminative ability (sensitivity and specificity), positive and negative predicted value, area under the receiver operating curve (ROC-AUC), F1, and goodness-of-fit statistics for calibration (among others) can be used (Steyerberg, Vickers, Cook, et al., 2010). ROC-AUC plots provide the ability to assess the tradeoff between true and false positive rates when establishing the predicted probability threshold for models (the cut-off between a positive and a negative prediction of outcome). The Youden index represents the best balance of sensitivity and specificity within the ROC-AUC, and is one option for use as a model decision threshold (Youden, 1950).

3.3 Limitations

The accuracy of predictive models is restrained by current syndromal diagnosis and limited understanding of the biological basis of psychopathology. External replication of models has been limited, and there is high variability in the rigor of published studies (Studerus et al., 2017). Additionally, outcome measures need to be well defined and relevant to individuals' day to day quality of life and function, rather than just focusing on symptomatology (Harvey, 2013). ML techniques have revolutionized the analysis of complex data sets, providing rigor through complex cross-validation and permutation testing, and the reduction of data leakage via pipeline architecture. However, one major problem for models developed within restrictive pipelines is flexibility for individual decision making in complex clinical environments. Clinical assessment occurs as a sequential process of data collection, followed by decisions to investigate further or treat (Collins et al., 2015; Sox et al., 2013). Specifically, psychiatric assessment begins with a clinical interview, including historical and mental state assessment, and progresses to specific investigations. Complete data for all predictors is only likely to be collected for more complex cases where the outcomes are more uncertain (i.e., there is a longer series of tests/assessments if diagnosis or prognosis remains unclear). Additionally, the availability of advanced biomarker technologies, such as genomic and imaging analysis, will vary greatly across clinical sites, producing variation in the types of data available. Multimodal models then need to be flexible enough to combine and manage available data effectively at a series of clinically relevant decision points, informing the decision to test or treat. Some subgroups of patients are likely to have different assessment requirements and unique diagnostic workflows depending on presenting features. The optimal set of modes and markers for a given sub-phenotype, such as schizophrenia with prominent negative symptoms or treatment resistant major depression, may be dependent on underlying mechanisms, severity, and the stage of illness.

3.4 Conclusions and future directions

The field of multimodal multivariate modeling in psychiatry is rapidly evolving with -omic, imaging, and ML techniques. Large (10,000–100,000s) samples of well phenotyped patients across diagnoses are required for subgroup analysis to better understand the relationship between biology and outcome. Emerging evidence suggests that some genomic and physiological predictors may be transdiagnostic, indicating that the inclusion of environmental and resilience factors may be important for the prediction of final patient phenotype. Most analyses compare cross-sectional predictors and outcomes, rather longitudinal change in markers, symptoms, and function. Illness and functional trajectories may have more utility in accurately predicting long term mental health outcomes (Schubert et al., 2015). For example, the evolution of depressive symptoms across adolescence can be described by common mood trajectory classes mediated by biological and environmental risk factors (Schubert et al., 2017).

On a technical level, the importance of pipeline methodologies and feature reduction techniques to limit overfitting of data is now apparent, and more standardized use may help the field move forward from the current "crisis" of failures in external replication (Dwyer et al., 2018). At a practical level, the flexibility of a modeling method has implications for its utility in clinical practice. Simpler probabilistic models, such as naive Bayes', may have the advantage of modular implementation. The odds-ratio form of the Bayes rule (Gale, Glue, & Gallagher, 2013; McGee, 2002) offers a relatively simple method to combine multivariate data in a probabilistic fashion that approximates the stepwise accumulation of data collected in the diagnostic process (Clark, 2009; Clark et al., 2015, 2016; Schubert et al., 2015; Sox et al., 2013). Using this type of modeling, we demonstrated that predictive accuracy at first presentation with clinical high risk of psychosis or first episode psychosis could be improved by combining several modes of assessment (e.g., cognitive, neuroimaging, electrophysiology, or fatty acid biomarkers). In these studies, participants could only be effectively categorized into low-, medium-, and high-risk groups when multiple modes of investigation were added to clinical data. Potentially, probabilistic decision curve analysis can be used to derive more complex decision thresholds based on the relative weight of harms and benefits of true and false positive and negative classifications (Hatcher, 2005; Steyerberg & Vergouwe, 2014; Vickers & Elkin, 2006). If the effect size or change in outcome probability beyond clinical assessment for each biomarker is known, then investigations can be ordered based on the relative value of new information that they provide. Following, the optimum sequence of investigations can be derived and implemented (Clark, 2009; Clark et al., 2015, 2016; Schubert et al., 2015; Sox et al., 2013; Wardenaar & de Jonge, 2013). Future research will need to combine such approaches with rigorous cross-validation procedures. One solution has been to develop models for each combination of assessment modes factorial (e.g., clinical plus cognitive, clinical plus MRI, clinical plus cognitive plus MRI) and select the best model for a given scenario based on performance measures (Koutsouleris et al., 2018). Potentially, the core workflows for the assessment of a particular presentation could be covered by individual models, but the practicality of implementation in complex real-world clinical scenarios remains to be seen.

In conclusion, the emerging field of multimodal multivariate modeling, combining clinical and biological variables, offers promise for the development of personalized psychiatric treatment. Given the variability, biases, and error in clinical decision-making practices, there is potential for models that are relatively simple or of modest accuracy to generate real improvements in individual outcomes. As technologies develop, the next stage in this journey has already commenced, with the implementation of cloud-based decision support platforms for personalized psychiatric treatment decisions (Spring Health, n.d.). Larger datasets, new biological insights, and rigorous cross-validation processes to improve validity will likely drive the field forward.

References

Alamian, G., Hincapie, A. S., Pascarella, A., et al. (2017). Measuring alterations in oscillatory brain networks in schizophrenia with resting-state MEG: State-of-the-art and methodological challenges. *Clinical Neurophysiology, 128*, 1719–1736.

Alda, M. (2013). Personalized psychiatry: Many questions, fewer answers. *Journal of Psychiatry & Neuroscience, 38*, 363–365.

Altmann, A., Tolosi, L., Sander, O., & Lengauer, T. (2010). Permutation importance: A corrected feature importance measure. *Bioinformatics, 26*, 1340–1347.

Alvarez-Jimenez, M., Gleeson, J. F., Henry, L. P., et al. (2012). Road to full recovery: Longitudinal relationship between symptomatic remission and psychosocial recovery in first-episode psychosis over 7.5 years. *Psychological Medicine, 42*, 595–606.

American Psychiatric Association. (1994). *Diagnostic and Statistical Manual of Mental Disorders* (4th ed.). Washington, D.C.: American Psychiatric Association.

Azur, M. J., Stuart, E. A., Frangakis, C., & Leaf, P. J. (2011). Multiple imputation by chained equations: What is it and how does it work? *International Journal of Methods in Psychiatric Research, 20*, 40–49.

Babyak, M. A. (2004). What you see may not be what you get: A brief, nontechnical introduction to overfitting in regression-type models. *Psychosomatic Medicine, 66*, 411–421.

Bakkour, N., Samp, J., Akhras, K., et al. (2014). Systematic review of appropriate cognitive assessment instruments used in clinical trials of schizophrenia, major depressive disorder and bipolar disorder. *Psychiatry Research, 216*, 291–302.

Barch, D. M. (2009). Neuropsychological abnormalities in schizophrenia and major mood disorders: Similarities and differences. *Current Psychiatry Reports, 11*, 313–319.

Baune, B. T., & Malhi, G. S. (2015). A review on the impact of cognitive dysfunction on social, occupational, and general functional outcomes in bipolar disorder. *Bipolar Disorders, 17*(Suppl 2), 41–55.

Baune, B. T., Li, X., & Beblo, T. (2013). Short- and long-term relationships between neurocognitive performance and general function in bipolar disorder. *Journal of Clinical and Experimental Neuropsychology, 35*, 759–774.

Bedse, G., Hartley, N. D., Neale, E., et al. (2017). Functional redundancy between canonical endocannabinoid signaling Systems in the Modulation of anxiety. *Biological Psychiatry, 82*, 488–499.

Blum, B. C., Mousavi, F., & Emili, A. (2018). Single-platform 'multi-omic' profiling: Unified mass spectrometry and computational workflows for integrative proteomics-metabolomics analysis. *Molecular Omics, 14*, 307–319.

Bogdan, R., Baranger, D. A. A., & Agrawal, A. (2018). Polygenic risk scores in clinical psychology: Bridging genomic risk to individual differences. *Annual Review of Clinical Psychology, 14*, 119–157.

Bora, E., & Pantelis, C. (2015). Meta-analysis of cognitive impairment in first-episode bipolar disorder: Comparison with first-episode schizophrenia and healthy controls. *Schizophrenia Bulletin, 41*, 1095–1104.

Bowtell, M., Ratheesh, A., McGorry, P., Killackey, E., & O'Donoghue, B. (2017a). Clinical and demographic predictors of continuing remission or relapse following discontinuation of antipsychotic medication after a first episode of psychosis. A systematic review. *Schizophrenia Research, 197*, 9–18.

Bowtell, M., Eaton, S., Thien, K., et al. (2017b). Rates and predictors of relapse following discontinuation of antipsychotic medication after a first episode of psychosis. *Schizophrenia Research, 195*, 231–236.

Boyer, L., Millier, A., Perthame, E., Aballea, S., Auquier, P., & Toumi, M. (2013). Quality of life is predictive of relapse in schizophrenia. *BMC Psychiatry, 13*, 15.

Breen, M. S., Stein, D. J., & Baldwin, D. S. (2016). Systematic review of blood transcriptome profiling in neuropsychiatric disorders: Guidelines for biomarker discovery. *Human Psychopharmacology, 31*, 373–381.

Brown, M. A., & Velligan, D. I. (2016). Issues and developments related to assessing function in serious mental illness. *Dialogues in Clinical Neuroscience, 18*, 135–144.

Buoli, M., Caldiroli, A., Cumerlato Melter, C., Serati, M., de Nijs, J., & Altamura, A. C. (2016). Biological aspects and candidate biomarkers for psychotic bipolar disorder: A systematic review. *Psychiatry and Clinical Neurosciences, 70*, 227–244.

Cabral, C., Kambeitz-Ilankovic, L., Kambeitz, J., et al. (2016). Classifying schizophrenia using multimodal multivariate pattern recognition analysis: Evaluating the impact of individual clinical profiles on the neurodiagnostic performance. *Schizophrenia Bulletin, 42*(Suppl 1), S110–S117.

Carbon, M., & Correll, C. U. (2014). Clinical predictors of therapeutic response to antipsychotics in schizophrenia. *Dialogues in Clinical Neuroscience, 16*, 505–524.

Chan, M. K., Gottschalk, M. G., Haenisch, F., et al. (2014). Applications of blood-based protein biomarker strategies in the study of psychiatric disorders. *Progress in Neurobiology, 122*, 45–72.

Chan, M. K., Cooper, J. D., Heilmann-Heimbach, S., et al. (2017). Associations between SNPs and immune-related circulating proteins in schizophrenia. *Scientific Reports, 7*, 12586.

Chekroud, A. M., Zotti, R. J., Shehzad, Z., et al. (2016). Cross-trial prediction of treatment outcome in depression: A machine learning approach. *Lancet Psychiatry, 3*, 243–250.

Chen, G., Adleman, N. E., Saad, Z. S., Leibenluft, E., & Cox, R. W. (2014). Applications of multivariate modeling to neuroimaging group analysis: A comprehensive alternative to univariate general linear model. *NeuroImage, 99*, 571–588.

Ciobanu, L. G., Sachdev, P. S., Trollor, J. N., et al. (2016). Differential gene expression in brain and peripheral tissues in depression across the life span: A review of replicated findings. *Neuroscience and Biobehavioral Reviews, 71*, 281–293.

Clark, S. R. (2009). *Decision support for the treatment of community-acquired pneumonia.* Adelaide: University of Adelaide, Department of Medicine, Health Informatics Unit, University of Adelaide.

Clark, S. R., Schubert, K. O., & Baune, B. T. (2015). Towards indicated prevention of psychosis: Using probabilistic assessments of transition risk in psychosis prodrome. *Journal of Neural Transmission, 122*, 155–169.

Clark, S. R., Baune, B. T., Schubert, K. O., et al. (2016). Prediction of transition from ultra-high risk to first-episode psychosis using a probabilistic model combining history, clinical assessment and fatty-acid biomarkers. *Translational Psychiatry, 6*.

Collins, G. S., de Groot, J. A., Dutton, S., et al. (2014). External validation of multivariable prediction models: A systematic review of methodological conduct and reporting. *BMC Medical Research Methodology, 14*, 40.

Collins, G. S., Reitsma, J. B., Altman, D. G., & Moons, K. G. (2015). Transparent reporting of a multivariable prediction model for individual prognosis or diagnosis (TRIPOD). *Annals of Internal Medicine, 162*, 735–736.

Cooper, J. D., Ozcan, S., Gardner, R. M., et al. (2017). Schizophrenia-risk and urban birth are associated with proteomic changes in neonatal dried blood spots. *Translational Psychiatry, 7*, 1290.

Coster, W. J. (2013). Making the best match: Selecting outcome measures for clinical trials and outcome studies. *The American Journal of Occupational Therapy, 67*, 162–170.

Craft, G. E., Chen, A., & Nairn, A. C. (2013). Recent advances in quantitative neuroproteomics. *Methods, 61*, 186–218.

Dalvie, S., Koen, N., McGregor, N., et al. (2016). Toward a global roadmap for precision medicine in psychiatry: Challenges and opportunities. *OMICS, 20*, 557–564.

Davison, J., O'Gorman, A., Brennan, L., & Cotter, D. R. (2017). A systematic review of metabolite biomarkers of schizophrenia. *Schizophrenia Research,*.

Dazzan, P., Arango, C., Fleischacker, W., et al. (2015). Magnetic resonance imaging and the prediction of outcome in first-episode schizophrenia: A review of current evidence and directions for future research. *Schizophrenia Bulletin, 41*, 574–583.

De Carlo, V., Calati, R., & Serretti, A. (2016). Socio-demographic and clinical predictors of non-response/non-remission in treatment resistant depressed patients: A systematic review. *Psychiatry Research, 240*, 421–430.

Depino, A. M. (2018). Perinatal inflammation and adult psychopathology: From preclinical models to humans. *Seminars in Cell & Developmental Biology, 77*, 104–114.

DeVylder, J. E., & Lukens, E. P. (2013). Family history of schizophrenia as a risk factor for axis I psychiatric conditions. *Journal of Psychiatric Research, 47*, 181–187.

Dietsche, B., Kircher, T., & Falkenberg, I. (2017). Structural brain changes in schizophrenia at different stages of the illness: A selective review of longitudinal magnetic resonance imaging studies. *The Australian and New Zealand Journal of Psychiatry, 51*, 500–508.

Du, Y., Fu, Z., & Calhoun, V. D. (2018). Classification and prediction of brain disorders using functional connectivity: Promising but challenging. *Frontiers in Neuroscience, 12*, 525.

Duncan, C. C., Barry, R. J., Connolly, J. F., et al. (2009). Event-related potentials in clinical research: Guidelines for eliciting, recording, and quantifying mismatch negativity, P 300, and N400. *Clinical Neurophysiology, 120*, 1883–1908.

Dwyer, D. B., Falkai, P., & Koutsouleris, N. (2018). Machine learning approaches for clinical psychology and psychiatry. *Annual Review of Clinical Psychology, 14*, 91–118.

Enders, C. K. (2006). A primer on the use of modern missing-data methods in psychosomatic medicine research. *Psychosomatic Medicine, 68*, 427–436.

Ervasti, J., Joensuu, M., Pentti, J., et al. (2017). Prognostic factors for return to work after depression-related work disability: A systematic review and meta-analysis. *Journal of Psychiatric Research, 95*, 28–36.

Faucher, L., & Goyer, S. (2015). RDoC: Thinking outside the DSM box without falling into a reductionist trap. In S. Demazeux, & P. Singy (Eds.), *Vol. 10. The DSM-5 in perspective.* Dordrecht: Springer.

First, M. B., Williams, J. B., Karg, R. S., & Spitzer, R. L. (2015). *Structured Clinical Interview for DSM-5 Disorders, Clinician Version (SCID-5-CV).* Arlington, VA: American Psychiatric Association.

Forstner, A. J., Hofmann, A., Maaser, A., et al. (2015). Genome-wide analysis implicates microRNAs and their target genes in the development of bipolar disorder. *Translational Psychiatry, 5.*

Fu, C. H., & Costafreda, S. G. (2013). Neuroimaging-based biomarkers in psychiatry: Clinical opportunities of a paradigm shift. *Canadian Journal of Psychiatry, 58,* 499–508.

Fusar-Poli, P., Bonoldi, I., Yung, A. R., et al. (2012). Predicting psychosis: Meta-analysis of transition outcomes in individuals at high clinical risk. *Archives of General Psychiatry, 69,* 220–229.

Gale, C., Glue, P., & Gallagher, S. (2013). Bayesian analysis of posttest predictive value of screening instruments for the psychosis high-risk state. *JAMA Psychiatry, 70,* 880–881.

Goldstein-Piekarski, A. N., Williams, L. M., & Humphreys, K. (2016). A trans-diagnostic review of anxiety disorder comorbidity and the impact of multiple exclusion criteria on studying clinical outcomes in anxiety disorders. *Translational Psychiatry, 6.*

Gonzalez-Ortega, I., Rosa, A., Alberich, S., et al. (2010). Validation and use of the functioning assessment short test in first psychotic episodes. *The Journal of Nervous and Mental Disease, 198,* 836–840.

Grof, P., Duffy, A., Cavazzoni, P., et al. (2002). Is response to prophylactic lithium a familial trait? *The Journal of Clinical Psychiatry, 63,* 942–947.

Hachtel, H., Harries, C., Luebbers, S., & Ogloff, J. R. (2018). Violent offending in schizophrenia spectrum disorders preceding and following diagnosis. *The Australian and New Zealand Journal of Psychiatry, 52*(8), 782–792.

Haenisch, F., Cooper, J. D., Reif, A., et al. (2016). Towards a blood-based diagnostic panel for bipolar disorder. *Brain, Behavior, and Immunity, 52,* 49–57.

Hamlyn, J., Duhig, M., McGrath, J., & Scott, J. (2013). Modifiable risk factors for schizophrenia and autism–shared risk factors impacting on brain development. *Neurobiology of Disease, 53,* 3–9.

Haq, A. U., Sitzmann, A. F., Goldman, M. L., Maixner, D. F., & Mickey, B. J. (2015). Response of depression to electroconvulsive therapy: A meta-analysis of clinical predictors. *The Journal of Clinical Psychiatry, 76,* 1374–1384.

Harris, L. W., Pietsch, S., Cheng, T. M., Schwarz, E., Guest, P. C., & Bahn, S. (2012). Comparison of peripheral and central schizophrenia biomarker profiles. *PLoS One, 7.*

Hartmann, J. A., Nelson, B., Spooner, R., et al. (2017). *Broad clinical high-risk mental state (CHARMS): Methodology of a cohort study validating criteria for pluripotent risk.* Early Intervention Psychiatry.

Harvey, P. D. (2013). Assessment of everyday functioning in schizophrenia: Implications for treatments aimed at negative symptoms. *Schizophrenia Research, 150,* 353–355.

Harvey, P. D., Velligan, D. I., & Bellack, A. S. (2007). Performance-based measures of functional skills: Usefulness in clinical treatment studies. *Schizophrenia Bulletin, 33,* 1138–1148.

Harvey, P. D., Khan, A., & Keefe, R. S. E. (2017). Using the positive and negative syndrome scale (PANSS) to define different domains of negative symptoms: Prediction of everyday functioning by impairments in emotional expression and emotional experience. *Innovations in Clinical Neuroscience, 14,* 18–22.

Hatcher, S. (2005). Decision analysis in psychiatry. *British Journal of Psychiatry, 166,* 184–190.

Heim, C., & Binder, E. B. (2012). Current research trends in early life stress and depression: Review of human studies on sensitive periods, gene-environment interactions, and epigenetics. *Experimental Neurology, 233,* 102–111.

Hoffmann, A., Sportelli, V., Ziller, M., & Spengler, D. (2017). Epigenomics of major depressive disorders and schizophrenia: Early life decides. *International Journal of Molecular Sciences, 18*(8), E1711.

Houck, J. M., Cetin, M. S., Mayer, A. R., et al. (2017). Magnetoencephalographic and functional MRI connectomics in schizophrenia via intra- and inter-network connectivity. *NeuroImage, 145,* 96–106.

Huys, Q. J., Maia, T. V., & Frank, M. J. (2016). Computational psychiatry as a bridge from neuroscience to clinical applications. *Nature Neuroscience, 19,* 404–413.

Immonen, J., Jaaskelainen, E., Korpela, H., & Miettunen, J. (2017). Age at onset and the outcomes of schizophrenia: A systematic review and meta-analysis. *Early Intervention in Psychiatry, 11,* 453–460.

Iniesta, R., Stahl, D., & McGuffin, P. (2016). Machine learning, statistical learning and the future of biological research in psychiatry. *Psychological Medicine, 46,* 2455–2465.

International Consortium on Lithium Genetics (ConLi+Gen), Amare, A. T., Schubert, K. O., et al. (2018). Association of polygenic score for schizophrenia and HLA antigen and inflammation genes with response to lithium in bipolar affective disorder: A genome-wide association study. *JAMA Psychiatry, 75,* 65–74.

Kalousis, A., & Hilario, M. (2000). Supervised knowledge discovery from incomplete data. In N. F. F. Ebecken (Ed.), *Vol. 15, Data Mining II* (pp. 269–278). Seville, Spain: WIT Press.

Kambeitz, J., Cabral, C., Sacchet, M. D., et al. (2017). Detecting neuroimaging biomarkers for depression: A meta-analysis of multivariate pattern recognition studies. *Biological Psychiatry, 82,* 330–338.

Kambeitz-Ilankovic, L., Meisenzahl, E. M., Cabral, C., et al. (2016). Prediction of outcome in the psychosis prodrome using neuroanatomical pattern classification. *Schizophrenia Research, 173,* 159–165.

Kautzky, A., Dold, M., Bartova, L., et al. (2018). Refining prediction in treatment-resistant depression: Results of machine learning analyses in the TRD III sample. *The Journal of Clinical Psychiatry, 79.*

Kay, S. R. (1990). Positive-negative symptom assessment in schizophrenia: Psychometric issues and scale comparison. *The Psychiatric Quarterly, 61,* 163–178.

Kendler, K. S. (2008). Explanatory models for psychiatric illness. *The American Journal of Psychiatry, 165,* 695–702.

Kleindienst, N., Engel, R., & Greil, W. (2005). Which clinical factors predict response to prophylactic lithium? A systematic review for bipolar disorders. *Bipolar Disorders, 7,* 404–417.

Kohavi, R. (1995). A study of cross-validation and bootstrap for accuracy estimation and model selection. *International Joint Conference on Artificial Intelligence, 14*, 1137–1145.

Koutsouleris, N., Kambeitz-Ilankovic, L., Ruhrmann, S., et al. (2018). Prediction models of functional outcomes for individuals in the clinical high-risk state for psychosis or with recent-onset depression: A multimodal, multisite machine learning analysis. *JAMA Psychiatry, 75*(11), 1156–1172.

Kumar, V., Shivakumar, V., Chhabra, H., Bose, A., Venkatasubramanian, G., & Gangadhar, B. N. (2017). Functional near infra-red spectroscopy (fNIRS) in schizophrenia: A review. *Asian Journal of Psychiatry, 27*, 18–31.

Large, M., Mullin, K., Gupta, P., Harris, A., & Nielssen, O. (2014). Systematic meta-analysis of outcomes associated with psychosis and co-morbid substance use. *The Australian and New Zealand Journal of Psychiatry, 48*, 418–432.

Lee, R. S., Hermens, D. F., Naismith, S. L., et al. (2015). Neuropsychological and functional outcomes in recent-onset major depression, bipolar disorder and schizophrenia-spectrum disorders: A longitudinal cohort study. *Translational Psychiatry, 5*.

Lee, B. J., Marchionni, L., Andrews, C. E., et al. (2017). Analysis of differential gene expression mediated by clozapine in human postmortem brains. *Schizophrenia Research, 185*, 58–66.

Leistritz, L., Weiss, T., Bar, K. J., et al. (2013). Network redundancy analysis of effective brain networks: A comparison of healthy controls and patients with major depression. *PLoS One, 8*.

Lin, A., Wood, S. J., Nelson, B., et al. (2011). Neurocognitive predictors of functional outcome two to 13 years after identification as ultra-high risk for psychosis. *Schizophrenia Research, 132*, 1–7.

Liu, Y., Sareen, J., Bolton, J. M., & Wang, J. L. (2016). Development and validation of a risk prediction algorithm for the recurrence of suicidal ideation among general population with low mood. *Journal of Affective Disorders, 193*, 11–17.

Loh, W. (2014). Fifty years of classification and regression trees. *International Statistical Review, 82*, 329–348.

Lytton, W. W., Arle, J., Bobashev, G., et al. (2017). Multiscale modeling in the clinic: Diseases of the brain and nervous system. *Brain Informatics, 4*, 219–230.

Maciukiewicz, M., Marshe, V. S., Hauschild, A. C., et al. (2018). GWAS-based machine learning approach to predict duloxetine response in major depressive disorder. *Journal of Psychiatric Research, 99*, 62–68.

Mann, C. J. (2003). Observational research methods. Research design II: Cohort, cross sectional, and case-control studies. *Emergency Medicine Journal, 20*, 54–60.

McGee, S. (2002). Simplifying likelihood ratios. *Journal of General Internal Medicine, 17*, 646–649.

McHorney, C. A., Ware, J. E., Jr., & Raczek, A. E. (1993). The MOS 36-item short-form Health survey (SF-36): II. Psychometric and clinical tests of validity in measuring physical and mental health constructs. *Medical Care, 31*, 247–263.

Mechelli, A., Lin, A., Wood, S., et al. (2017). Using clinical information to make individualized prognostic predictions in people at ultra high risk for psychosis. *Schizophrenia Research, 184*, 32–38.

van der Meer, D., Hoekstra, P. J., van Donkelaar, M., et al. (2017). Predicting attention-deficit/hyperactivity disorder severity from psychosocial stress and stress-response genes: A random forest regression approach. *Translational Psychiatry, 7*.

Messer, T., Lammers, G., Muller-Siecheneder, F., Schmidt, R. F., & Latifi, S. (2017). Substance abuse in patients with bipolar disorder: A systematic review and meta-analysis. *Psychiatry Research, 253*, 338–350.

Messias, E. L., Chen, C. Y., & Eaton, W. W. (2007). Epidemiology of schizophrenia: Review of findings and myths. *The Psychiatric Clinics of North America, 30*, 323–338.

Miller, B. J., Buckley, P., Seabolt, W., Mellor, A., & Kirkpatrick, B. (2011). Meta-analysis of cytokine alterations in schizophrenia: Clinical status and antipsychotic effects. *Biological Psychiatry, 70*, 663–671.

Mistry, S., Harrison, J. R., Smith, D. J., Escott-Price, V., & Zammit, S. (2018). The use of polygenic risk scores to identify phenotypes associated with genetic risk of bipolar disorder and depression: A systematic review. *Journal of Affective Disorders, 234*, 148–155.

Moons, K. G., Royston, P., Vergouwe, Y., Grobbee, D. E., & Altman, D. G. (2009). Prognosis and prognostic research: What, why, and how? *BMJ, 338*, b375.

Moons, K. G., Kengne, A. P., Woodward, M., et al. (2012). Risk prediction models: I. development, internal validation, and assessing the incremental value of a new (bio)marker. *Heart, 98*, 683–690.

Moser, D. A., Doucet, G. E., Lee, W. H., et al. (2018). Multivariate associations among behavioral, clinical, and multimodal imaging phenotypes in patients with psychosis. *JAMA Psychiatry, 75*(4), 386–395.

Mulder, R., Singh, A. B., Hamilton, A., et al. (2018). The limitations of using randomised controlled trials as a basis for developing treatment guidelines. *Evidence-Based Mental Health, 21*, 4–6.

Mwangi, B., Matthews, K., & Steele, J. D. (2012). Prediction of illness severity in patients with major depression using structural MR brain scans. *Journal of Magnetic Resonance Imaging, 35*, 64–71.

Mwangi, B., Tian, T. S., & Soares, J. C. (2014). A review of feature reduction techniques in neuroimaging. *Neuroinformatics, 12*, 229–244.

Norbury, A., & Seymour, B. (2018). Response heterogeneity: Challenges for personalised medicine and big data approaches in psychiatry and chronic pain. *F1000Res, 7*, 55.

Ogasawara, K., Nakamura, Y., Kimura, H., Aleksic, B., & Ozaki, N. (2018). Issues on the diagnosis and etiopathogenesis of mood disorders: Reconsidering DSM-5. *Journal of Neural Transmission (Vienna), 125*, 211–222.

Ogundimu, E. O., Altman, D. G., & Collins, G. S. (2016). Adequate sample size for developing prediction models is not simply related to events per variable. *Journal of Clinical Epidemiology, 76*, 175–182.

Ojala, M., & Garriga, G. C. (2010). Permutation tests for studying classifier performance. *Journal of Machine Learning Research, 11*, 1833–1863.

Okser, S., Pahikkala, T., Airola, A., Salakoski, T., Ripatti, S., & Aittokallio, T. (2014). Regularized machine learning in the genetic prediction of complex traits. *PLoS Genetics, 10*.

Olagunju, A. T., Clark, S. R., & Baune, B. T. (2018). Clozapine and psychosocial function in schizophrenia: A systematic review and meta-analysis. *CNS Drugs, 32*(11), 1011–1023.

Opler, M. G. A., Yavorsky, C., & Daniel, D. G. (2017). Positive and negative syndrome scale (PANSS) training: Challenges, solutions, and future directions. *Innovations in Clinical Neuroscience, 14*, 77–81.

Orru, G., Pettersson-Yeo, W., Marquand, A. F., Sartori, G., & Mechelli, A. (2012). Using support vector machine to identify imaging biomarkers of neurological and psychiatric disease: A critical review. *Neuroscience and Biobehavioral Reviews, 36*, 1140–1152.

Ozomaro, U., Wahlestedt, C., & Nemeroff, C. B. (2013). Personalized medicine in psychiatry: Problems and promises. *BMC Medicine, 11*, 132.

Palmisano, M., & Pandey, S. C. (2017). Epigenetic mechanisms of alcoholism and stress-related disorders. *Alcohol, 60*, 7–18.

Patel, M. J., Khalaf, A., & Aizenstein, H. J. (2016). Studying depression using imaging and machine learning methods. *NeuroImage: Clinical, 10*, 115–123.

Pavlou, M., Ambler, G., Seaman, S. R., et al. (2015). How to develop a more accurate risk prediction model when there are few events. *BMJ, 351*.

Perkins, D. O., Jeffries, C. D., Addington, J., et al. (2015). Towards a psychosis risk blood diagnostic for persons experiencing high-risk symptoms: Preliminary results from the NAPLS project. *Schizophrenia Bulletin, 41*, 419–428.

Piel, M., Vernaleken, I., & Rosch, F. (2014). Positron emission tomography in CNS drug discovery and drug monitoring. *Journal of Medicinal Chemistry, 57*, 9232–9258.

Pinto, J. V., Moulin, T. C., & Amaral, O. B. (2017). On the transdiagnostic nature of peripheral biomarkers in major psychiatric disorders: A systematic review. *Neuroscience and Biobehavioral Reviews, 83*, 97–108.

Posner, K., Brown, G. K., Stanley, B., et al. (2011). The Columbia-suicide severity rating scale: Initial validity and internal consistency findings from three multisite studies with adolescents and adults. *The American Journal of Psychiatry, 168*, 1266–1277.

Raudys, S. J., & Jain, A. K. (1991). Small sample size effects in statistical pattern recognition: Recommendations for practitioners. *IEEE Transactions on Pattern Analysis and Machine Intelligence, 13*, 252–264.

Razzaghi, T., Roderick, O., Safro, I., & Marko, N. (2016). Multilevel weighted support vector machine for classification on healthcare data with missing values. *PLoS One, 11*.

Regier, D. A., Narrow, W. E., Clarke, D. E., et al. (2013). DSM-5 field trials in the United States and Canada, part II: Test-retest reliability of selected categorical diagnoses. *The American Journal of Psychiatry, 170*, 59–70.

Reinbold, C. S., Forstner, A. J., Hecker, J., et al. (2018). Analysis of the influence of microRNAs in Lithium response in bipolar disorder. *Frontiers in Psychiatry, 9*, 207.

Royston, P., Moons, K. G., Altman, D. G., & Vergouwe, Y. (2009). Prognosis and prognostic research: Developing a prognostic model. *BMJ, 338*, b604.

Sanchez-Roige, S., Gray, J. C., Mac Killop, J., Chen, C. H., & Palmer, A. A. (2018). The genetics of human personality. *Genes, Brain, and Behavior, 17*.

Santesteban-Echarri, O., Paino, M., Rice, S., et al. (2017). Predictors of functional recovery in first-episode psychosis: A systematic review and meta-analysis of longitudinal studies. *Clinical Psychology Review, 58*, 59–75.

Scarr, E., Millan, M. J., Bahn, S., et al. (2015). Biomarkers for psychiatry: The journey from fantasy to fact, a report of the 2013 CINP think tank. *The International Journal of Neuropsychopharmacology, 18*.

Scholkopf, B., Sung, K., Burges, C. J. C., et al. (1997). Comparing support vector machines with Gaussian kernels to radial basis function classifiers. *IEEE Transactions on Signal Processing, 45*, 2758–2765.

Schubert, K. O., Clark, S. R., & Baune, B. T. (2015). The use of clinical and biological characteristics to predict outcome following First episode psychosis. *The Australian and New Zealand Journal of Psychiatry, 49*, 24–35.

Schubert, K. O., Clark, S. R., Van, L. K., Collinson, J. L., & Baune, B. T. (2017). Depressive symptom trajectories in late adolescence and early adulthood: A systematic review. *The Australian and New Zealand Journal of Psychiatry, 51*, 477–499.

Schubert, K. O., Stacey, D., Arentz, G., et al. (2018). Targeted proteomic analysis of cognitive dysfunction in remitted major depressive disorder: Opportunities of multi-omics approaches towards predictive, preventive, and personalized psychiatry. *Journal of Proteomics, 188*, 63–70.

Schultz, C. C., Fusar-Poli, P., Wagner, G., et al. (2012). Multimodal functional and structural imaging investigations in psychosis research. *European Archives of Psychiatry and Clinical Neuroscience, 262*(Suppl 2), S97–106.

Schwarz, E., Guest, P. C., Rahmoune, H., et al. (2012a). Identification of a biological signature for schizophrenia in serum. *Molecular Psychiatry, 17*, 494–502.

Schwarz, E., Guest, P. C., Steiner, J., Bogerts, B., & Bahn, S. (2012b). Identification of blood-based molecular signatures for prediction of response and relapse in schizophrenia patients. *Translational Psychiatry, 2*.

Sekar, A., Bialas, A. R., de Rivera, H., et al. (2016). Schizophrenia risk from complex variation of complement component 4. *Nature, 530*, 177–183.

Sheehan, D. V., Lecrubier, Y., Sheehan, K. H., et al. (1998). The Mini-International Neuropsychiatric Interview (M.I.N.I.): The development and validation of a structured diagnostic psychiatric interview for DSM-IV and ICD-10. *Journal of Neural Transmission, 59*(Suppl 20), 22–33 (quiz 34-57).

Smoller, J. W., Andreassen, O. A., Edenberg, H. J., Faraone, S. V., Glatt, S. J., & Kendler, K. S. (2018). Psychiatric genetics and the structure of psychopathology. *Molecular Psychiatry, 24*(3), 409–420.

Sox, H. C., Blatt, M. A., Higgins, M. C., & Marton, K. I. (2013). *Medical decision making* (2nd ed.). West Sussex: John Wiley & Sons.

Spratt, H., Ju, H., & Brasier, A. R. (2013). A structured approach to predictive modeling of a two-class problem using multidimensional data sets. *Methods, 61*, 73–85.

Spring Health. n.d. Retrieved from https://www.springhealth.com; (Accessed 11 November 2018).

Srivastava, N., Hinton, G., Krizhevsky, A., Sutskever, I., & Salakhutdinov, R. (2014). Dropout: A simple way to prevent neural networks from overfitting. *Journal of Machine Learning Research, 15*, 1929–1958.

Stacey, D., Schubert, K. O., Clark, S. R., et al. (2018). A gene co-expression module implicating the mitochondrial electron transport chain is associated with long-term response to lithium treatment in bipolar affective disorder. *Translational Psychiatry, 8*, 183.

Steyerberg, E. W., & Vergouwe, Y. (2014). Towards better clinical prediction models: Seven steps for development and an ABCD for validation. *European Heart Journal, 35*, 1925–1931.

Steyerberg, E. W., Harrell, F. E., Jr., Borsboom, G. J., Eijkemans, M. J., Vergouwe, Y., & Habbema, J. D. (2001). Internal validation of predictive models: Efficiency of some procedures for logistic regression analysis. *Journal of Clinical Epidemiology, 54*, 774–781.

Steyerberg, E. W., Vickers, A. J., Cook, N. R. et al. (2010). Assessing the performance of prediction models: A framework for traditional and novel measures. *Epidemiology, 21*, 128–138.

Studerus, E., Ramyead, A., & Riecher-Rossler, A. (2017). Prediction of transition to psychosis in patients with a clinical high risk for psychosis: A systematic review of methodology and reporting. *Psychological Medicine, 47*, 1163–1178.

Sullivan, P. F., Agrawal, A., Bulik, C. M., et al. (2018). Psychiatric genomics: An update and an agenda. *The American Journal of Psychiatry, 175*, 15–27.

Tae, W. S., Ham, B. J., Pyun, S. B., Kang, S. H., & Kim, B. J. (2018). Current clinical applications of diffusion-tensor imaging in neurological disorders. *Journal of Clinical Neurology, 14*, 129–140.

Tang, J., & Alelyani, S. H. L. (2014). Feature selection for classification: A review. In *Data classification: Algorithms and applications* (pp. 37–58). New York: Chapman & Hall/CRC.

Valdiviezo, H., & Van Aelst, S. (2015). Tree-based prediction on incomplete data using imputation or surrogate decisions. *Information Sciences, 311*, 163–181.

Vecchio, F., Miraglia, F., & Maria, R. P. (2017). Connectome: Graph theory application in functional brain network architecture. *Clinical Neurophysiology Practice, 2*, 206–213.

Vickers, A. J., & Elkin, E. B. (2006). Decision curve analysis: A novel method for evaluating prediction models. *Medical Decision Making: An International Journal of the Society for Medical Decision Making, 26*, 565–574.

Vieira, S., Pinaya, W. H., & Mechelli, A. (2017). Using deep learning to investigate the neuroimaging correlates of psychiatric and neurological disorders: Methods and applications. *Neuroscience and Biobehavioral Reviews, 74*, 58–75.

Waljee, A., Higgins, P., & Singal, A. (2013). A primer on predictive models. *Clinical and Translational Gastroenterology, 4*.

Walton, E. L. (2018). Saliva biomarkers in neurological disorders: A "spitting image" of brain health? *Biomedical Journal, 41*, 59–62.

Wang, Q., Yang, C., Gelernter, J., & Zhao, H. (2015). Pervasive pleiotropy between psychiatric disorders and immune disorders revealed by integrative analysis of multiple GWAS. *Human Genetics, 134*, 1195–1209.

Wardenaar, K. J., & de Jonge, P. (2013). Diagnostic heterogeneity in psychiatry: Towards an empirical solution. *BMC Medicine, 11*, 201.

Weiskopf, N. G., & Weng, C. (2013). Methods and dimensions of electronic health record data quality assessment: Enabling reuse for clinical research. *Journal of the American Medical Informatics Association, 20*, 144–151.

Weiskopf, N. G., Hripcsak, G., Swaminathan, S., & Weng, C. (2013). Defining and measuring completeness of electronic health records for secondary use. *Journal of Biomedical Informatics, 46*, 830–836.

Williams, J. B. (1988). A structured interview guide for the Hamilton depression rating scale. *Archives of General Psychiatry, 45*, 742–747.

Wolpert, D. (1996). The lack of a priori distinctions between learning algorithms. *Neural Computation, 8*(7), 1341–1390.

Wu, T. T., Chen, Y. F., Hastie, T., Sobel, E., & Lange, K. (2009). Genome-wide association analysis by lasso penalized logistic regression. *Bioinformatics, 25*, 714–721.

Wylie, K. P., Smucny, J., Legget, K. T., & Tregellas, J. R. (2016). Targeting functional biomarkers in schizophrenia with neuroimaging. *Current Pharmaceutical Design, 22*, 2117–2123.

Yan, Z., & Xu, C. (2009). Comparing support vector machines with Gaussian kernels to radial basis function classifiers. In *8th IEEE international conference on cognitive informatics* (p. 2009). Kowloon, Hong Kong, China: IEEE.

Yokota, F., & Thompson, K. M. (2004). Value of information literature analysis: A review of applications in health risk management. *Medical Decision Making, 24*, 287–298.

Youden, W. J. (1950). Index for rating diagnostic tests. *Cancer, 3*, 32–35.

Zhou, H., Sehl, M. E., Sinsheimer, J. S., & Lange, K. (2010). Association screening of common and rare genetic variants by penalized regression. *Bioinformatics, 26*, 2375–2382.

Zou, H., & Hastie, T. (2005). Regularization and variable selection via the elastic net. *Journal of the Royal Statistical Society: Series B, 67*, 301–320.

Chapter 44

Standardized biomarker and biobanking requirements for personalized psychiatry

Catherine Toben[a], Victoria K. Arnet[a], Anita Lo[a], Pamela H. Saunders[b] and Bernhard T. Baune[c,d,e]

[a]Discipline of Psychiatry, University of Adelaide, Adelaide, SA, Australia, [b]SABR Manager, SAHMRI, Adelaide, SA, Australia, [c]Department of Psychiatry and Psychotherapy, University of Münster, Münster, Germany, [d]Department of Psychiatry, Melbourne Medical School, The University of Melbourne, Melbourne, VIC, Australia, [e]The Florey Institute of Neuroscience and Mental Health, The University of Melbourne, Parkville, VIC, Australia

1 Personalized psychiatry—from diagnosis to treatment response

Worldwide, the increasing phenomenon of mental health disorders has devastating effects on individuals, families, and communities, with depression being projected to be the leading cause of global disability by 2030. As neuropsychiatric disorders are complex and heterogeneous in nature, prognosis, as well as optimum treatment plans, are often ineffective. Currently, patients rely solely on detailed clinical assessments that take time and effort, and that can be incomplete with regard to subtypes (Remick, Sadovnick, Gimbarzevsky, Lam, & Zis, 1993).

The emerging field of personalized psychiatry aims to circumvent this by ultimately integrating genetic and blood-based biomarker panels to assist in diagnosis and design of individualized therapeutic treatment regimes. Blood-based biomarkers require standardized specimen collection and storage protocols. To this end, a biorepository, or biobank, is critical for the accrual of large-scale multimodal data sets containing human biological material and associated clinical data capturing a broad range of commonly encountered clinical psychopathologies. Furthermore, recent innovative methods and technologies have enabled detection and characterization of the underlying molecular and cellular pathophysiology of mental health disorders. In this chapter, we will focus on the large-scale storage of peripheral biomarker material.

2 Fundamentals of biobanking relevant to mental health disorders

2.1 Implementation of a business plan to ensure sustainability

Biobanking is a relatively new field within psychiatry. The establishment of a biobank requires implementation of a business plan, and an organizational structure tailored to the jurisdiction of the facility in which it will reside. For example, a privately governed facility will have different governing requirements, compared with a publicly supported and governed institute. Importantly, sustainability of a biobank will be ensured by minimizing risk factors pertaining to negative economic influences that could lead to its demise. To this end, the potential areas for revenue should be identified and actioned. The business plan for a biobank should define the following structures and committees.

2.1.1 Financial provisions

Consideration should be given to implementing an itemized financial plan that will account for costs, revenue, and funds received (Gonzalez-Sanchez, Lopez-Valeiras, & García-Montero, 2014; McDonald, Velasco, & Ilasi, 2010). Financial support and investment can be achieved via grants from various sources, including philanthropic, government, and institutional. To further support the ongoing maintenance of the biobank, travel grants for further education, as well as equipment grants, are important. In order for the psychiatric biobank to be viable, it is paramount to ensure that any services, including storage/handling and bioanalytical analyses provided to users, are cost recovered. Researchers should include costs for high-quality and standardized sample collection/transport and storage, as well as bioanalytical analyses within their grant applications (Gonzalez-Sanchez et al., 2014).

2.1.2   Organizational structure and governance model

The development of an organizational hierarchy and governance committee for the psychiatric biobank will ensure the aims and services are conducted efficiently. This may include health and medical researchers/professionals, as well as a legal representation (Chalmers et al., 2016). To assure accountability is maintained with stakeholders (researchers, collaborators, general public), processes to allow for transparency and traceability of biospecimen management are required. These include quality control, a defined management plan for access and custodianship of biospecimens, as well as an ethical framework (McDonald et al., 2010). An example includes the access committee that would assess requests for sample usage. Ideally, this committee would include a lay member committed to reflecting the general public perception on sample and associated data usage.

2.1.3   Biobank succession and disassembly

Funding cuts or natural or man-made disasters pose a considerable risk to the maintenance and longevity of the psychiatric biobank. In order to prepare for these factors, it is important to construct a detailed action plan for forced biobank closure, and potential loss of irreplaceable biospecimens and data (Stephens & Dimond, 2015). It is also important to define an ownership succession plan. For example, upon closure of a biobank, the samples/data collected by a biobank should be allocated to a nominated custodian within the institution. However, samples/data managed by the biobank would be returned to the owner of the samples.

2.1.4   Reporting strategies

As part of quality control management, Human Research Ethics Committees (HREC) are an integral part of a sustainable psychiatric biobank. Ongoing consultation and transparent communication of procedures and protocols will help to ensure the viability of the biobank. Furthermore, a comprehensive report may encourage funding from additional stakeholders. As part of the community consultation process, consideration to publicly release the annual report could also be made.

2.1.5   Consumer engagement

The discussion in relation to consumer engagement is ongoing. Consumers as "Citizen Scientists" is a growing concept, enabling members of the general public to be involved in psychiatric research. Individuals need no longer be passive human "subjects," but can be engaged in innovative ways, including digital media platforms over time, and recognized as active, interested, and valued research participants in the field of psychiatry (Kaye et al., 2015).

2.2   Regulatory governance

2.2.1   Ethical considerations

Research involving human participants requires some form of ethical review by an independent committee. Informed consent, confidentiality, secondary use of samples, data over time, and return of results are processes that require standardized ethical governance. In the act of ethically acquiring participants, a consent form and information sheet is essential (Participant Information and Consent Forms (PCIF)). The PCIF should be given appropriate thought, as reconsenting is a very disruptive, difficult, costly, and time-consuming process. For example, complications may arise through outright refusal of an updated consent form by the participant, or loss of the participant at follow up (e.g. death). Both broad and dynamic consent forms are flexible, and include unspecified future use of biospecimens and associated data subject to ethical approval (Kaye et al., 2015). This provides researchers with room for future analytical access, while the dynamic consent provides more interaction with the participant, usually via web-based platforms (Chalmers et al., 2016; Graham et al., 2014).

2.2.2   Generation of global participant data networks within ethical and legislative frameworks

Participant clinical data is maintained by ethical guidelines and legal frameworks for the individual's safety and protection. In an era in which biobanks operate within global networks, it is important for researchers and biobanks to abide by the varied legal frameworks within multiple jurisdictions across countries (Chalmers et al., 2016).

Added to this is the dynamic nature of legal frameworks that may be reformed within different countries regarding biospecimens and associated data collection, storage, and sharing (Chalmers et al., 2016; Shenkin et al., 2017). The ownership and access of participant data and biospecimens is complicated by disparate digital information dynamic legislations. This has the potential to lead to unforeseen risks regarding data access and disclosure of data to third parties (Dove, 2015). Importantly for international collaborations, forethought is required from the beginning to ensure inclusion of adequate information on PCIF, as well as strategies for maintaining data security, participant privacy, and confidentiality on different servers, and meeting data protection law requirements (Bauchner, Golub, & Fontanarosa, 2016; International Compilation of Human Research Standards 2018 Edition, 2018; National Health and Medical Research Council (NHMRC), ARC, & AVCC, 2007). For example, sharing genetic data, as generated from genome-wide association studies (GWAS) (Eiseman, Bloom, Brower, Clancy, & Olmsted, 2003; Zhao & Castellanos, 2016), among international consortia requires comprehensive ethical review. A balance is required among participant privacy, confidentiality, and informed consent with regard to sharing data, especially from publicly funded projects. This is important to maintain research integrity and transparency while maximizing the research value from collected data. This is pertinent to the psychiatric field where a large proportion of collected data is of a highly sensitive nature. Furthermore, participant vulnerability must be taken into account, for example, due to cognitive impairments or research within pediatric cohorts (DuBois, Bante, & Hadley, 2011; Zhao & Castellanos, 2016). Implementation of the preceding safeguards for the potential use of data, including biospecimens, will uphold the participant's privacy within local and global governance frameworks, regardless of location (Dove, 2015).

Striving toward international harmonization requires abiding by minimum ethical and legislative standards for biospecimen collection, data management, and collection to ensure sample quality and strengthen study power across multiple consortia. Minimum normative subject data sets should include standard metadata (Fig. 1), which will facilitate high quality research across consortia within personalized psychiatric medicine.

2.2.3 Participant consent form requirements in psychiatry

A consent form and information sheet should include:

- Permission for future use and application of the samples, and all associated data (not sample type or study specific)
- Assurance in the case of participant withdrawal that medical care will not be compromised
- Option for participant to withdraw at any time
- Permission to contact participant in the future
- State clearly who owns the data (usually the institution)

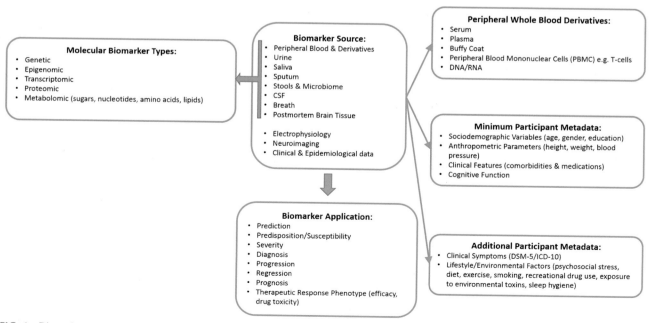

FIG. 1 Biomarker types, sources, and application in personalized psychiatry.

- Statements to clarify consequences of potential commercialization from the use of samples and/or associated data. For example, study participant or their families will not be the recipient of any monies in relation to commercial discoveries
- Address strategy about incidental findings

2.2.4   Incidental findings

One of the most discussed issues relating to biobanking is the return of incidental findings to the participant. According to NHMRC guidelines, it is a best practice for the biobank to develop a defensible plan for the case in which a researcher may discover a significant genetic finding (National Health and Medical Research Council (NHMRC) et al., 2007 (updated 2015)). The biobanks ethics and governance should determine whether or not the biobank will inform the participant of any incidental findings. A comprehensive procedure involving how the finding will be retested and communicated should be developed if the participant is to be informed.

2.3   Sample management and infrastructure for high quality storage of biospecimens

In establishing a biobank with large-scale human biological material and associated data, it is essential that analytical processes and procedures are standardized (Fig. 2). This will ensure sample integrity is maintained for any future analyses, as well as allowing for replication of experiments, and hence, validation of results.

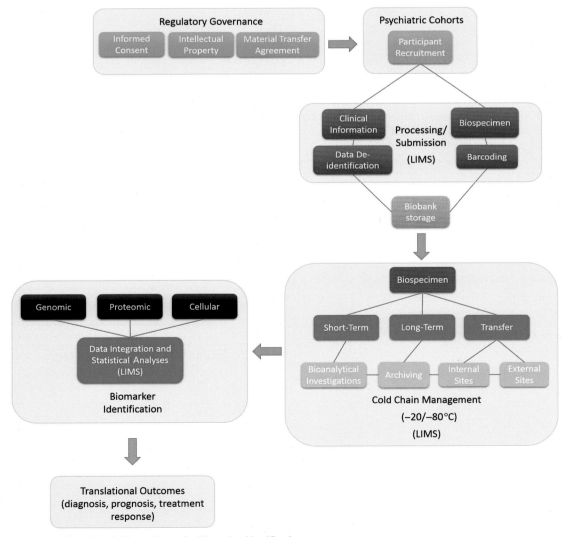

FIG. 2   Management of psychiatric biospecimens for biomarker identification.

2.3.1 Importance of quality control and standardized protocols for a biobank

The development of standard operating procedures (SOPs) needs to include standardization of all aspects, from sample collection to biobanking, with internal, and potentially external, auditing of procedures being important for quality control. Information on method of collection, processing, and accessioning should be included and recorded to account for variations within a cohort. Upholding sample integrity will assist with valid and reproducible biomarker results.

2.3.2 Methods for deidentification of participant data

In conjunction with an ethics committee, the type of participant identification will be determined; individually identifiable, reidentifiable, or nonidentifiable. The type of participant sample identification chosen determines the method of designing and assigning the new identifier. Usually, a participant will be assigned a globally unique Personal Participant Identifier (PPID), independent of factors that may distinguish the participant. This process can be enhanced by using a barcode with a human readable format.

2.3.3 Biobanking data management and integration strategies

A biobank needs to be able to track, generate, and update data related to the biospecimen efficiently. Many Laboratory Information Management System (LIMS) software options are available to biobanks that allow improved productivity in workflow and data management. Due to funding issues, Excel spreadsheets are commonly used to manage biobanking data. As the number of samples collected grows exponentially, the amount of associated data does also, and comprehensive reporting is required. Furthermore, data security is important to protect participant privacy and information (International Society for Biological and Environmental Repositories (ISBER), 2011). Therefore, spreadsheets would need to be replaced with an appropriate software system. Factors in assessing which LIMS is suitable include:

- Available budget (short- and long-term)
- Projected size of the biobank (sample quantity, type, and retention period)
- Envisaged staff usage over the life of the biobank (how many user access licenses are required and how many are available with the selected software)
- Opportunity for ongoing IT support access
- Compatibility between software systems to be used by the biorepository and collaborators
- Accessibility options and security rules around data access
- Where the data is allowed to be stored; local server, local cloud, or international owned cloud servers

2.3.4 Psychiatric biospecimen storage and handling

The choice of sample storage is defined by cost effective usage of available facilities. Factors to consider include:

- Infrastructure running and maintenance costs
- Capacity (volumes of sample × sample quantity)
- Period of sample storage (short-, medium-, and long-term)

Temperature controlled environments, while handling samples outside of storage, should also be incorporated into SOPs, such as dry ice and cryocarts. High infrastructure costs associated with biobanking not only come from the purchase of the specialized equipment, but also from the high ongoing operational costs, including scheduled maintenance, uninterrupted power supply, generator back-up power supply, and monitoring systems. Accessibility to the storage facility within the biobank should be limited and applied through physical security systems. Ideally, access to individual freezers should be restricted to "owner" users and centrally monitored.

Other aspects of sample management that need to be considered and standardized in both procedure and compliance within the facility are oxygen sensors with alarms and associated rescue equipment, and also ambient air temperature monitors with alarms to reduce equipment compressors' stress. For audit purposes, a backup monitoring system for individual equipment needs to be a system independent of the manufacturer's self-monitoring system. Backup storage availability needs to be predetermined in the event of equipment failure with a detailed SOP for monitoring sample movement.

Variations in temperature need to be recorded in cold chain management (from the time the sample is first stabilized to the time the sample is used and depleted, as well as time spent in transportation) (Moore et al., 2011). Regular internal and external testing and auditing can be carried out to assess degradation of samples over specific time frames. Other measures to ensure quality of biospecimens include reducing freeze-thaw cycles by using minimal volume aliquots for 1–3 freeze-thaw cycles. Details of freeze-thaw cycles, including frequency and methods, are to be recorded. Quality control of biospecimens can also be audited independently by groups such as the Integrated Biobank of Luxembourg (IBBL).

2.3.5 Recording of biospecimen information

To enable future replication of results, the recording of all aspects of biobanking requires strict documentation of all data generated through the total management of the samples. This commences at the time of collection of the parent sample, transportation, processing, initial stabilization, long-term cold chain management, shipping, and distribution. When retrieving samples for shipping, depletion of an individual's original stock of samples is not recommended. To assist in standardizing annotation of biospecimen types and collection methods, the Standard Pre-analytical Coding for Biospecimens (SPREC) codes can be used (Lehmann et al., 2012).

2.3.6 Sharing biospecimens

An inherent aspect of biobanking is the sharing of stored samples. In fact "specimen locators," otherwise known as "specimen catalogues," are now commonly available on biobank websites. These assist researchers in locating specific specimens related to their field of research. Consequently, sharing of samples and/or associated clinical data includes both data recently collected with collaborators, as well as retrospectively collected samples. Samples that have been permitted for sharing will need to follow a process whereby a letter of intent (LOI) would be submitted to the biobank access committee. If approved for sharing, the next step would include ratification of a material transfer agreement (MTA).

2.3.7 Transfer of biospecimens to internal and external sites

Standardized transportation protocols are to be derived from local regulatory bodies, such as the International Air Transport Association (IATA), which is, for example, Australia's aviation transport authority. These protocols will stipulate that distribution of biospecimens and associated clinical data locally, nationally, or internationally will require current import/export permit requirements specific to the biospecimen. To further ensure standardization in transport, it is recommended that biobanking staff be trained in accredited transport of specimen courses such as those offered by the Civil Aviation Safety Authority (CASA).

The chosen courier company is required to provide total quality of control for biospecimens in transit from time of collection to receipt of samples. In the case of frozen samples, a best practice is the use of a dry shipper with a temperature probe, a dry ice monitoring and retrieval transportation audit log (Moore et al., 2011), as well as real time sample location acknowledgement via email. Costs of biospecimen transportation to/from the biobank should be recovered from the researcher (McDonald et al., 2010). Any events that may compromise sample integrity, and thereby future outcome of analyses, need to be recorded and communicated with all stakeholders.

2.3.8 Associated clinical data

To allow for comprehensive clinical data collection, the PCIF annotation needs to include the different sources of data collection, including the health department, data registries, and any specific health professionals. The methodology for collection of clinical data may be through paper records that require digitalization, manual data entry, or via data downloads from health department records, pathology departments, data registries, and other authorized sources. Patient data collection from patient medical records has, in the past, been a time-consuming process, however with more modern and streamlined patient electronic health records, platforms such as South Australia's Enterprise Patient Administration System (EPAS) clinical data collection will become less time consuming and complicated, and more cost effective. Access to clinical data can be provided in association with biospecimen assay results, or independently of biospecimen distribution. In the latter case, the biobank can consider charging for a separate service.

3 Reproducible identification of biomarker signatures within mental health disorders

Biomarkers are objectively measured characteristics providing an indication of a biological process (normal or pathological) or response to an intervention. In the field of psychiatry, investigation into biomarkers has focused on biological specimens including peripheral blood, saliva, urine, and hair analyses. In this chapter, we will focus on the identification of those biomarkers from peripheral blood that are associated with inflammation, oxidative stress, and blood brain barrier integrity. For biomarkers to ultimately be applied in clinical practice, certain conditions need to be fulfilled. This includes the use of high sensitivity and specific biomarker assays, as well as output of accurate and reproducible results.

The importance of a multicombination biomarker panel or signature will improve the specificity and sensitivity required for stratification of heterogeneous psychiatric disorders.

Discovery of novel psychiatric biomarkers requires an integrated approach among the clinic, biobank, and researchers. As the realm of psychiatry presents complexity in terms of a heterogeneous phenotype, it is important to standardize collection of biospecimens and biomarker analyses. As in any experimental design, there are a number of factors that must be addressed in order to establish a robust plan of research. All stages of the biomarker discovery workflow are to be considered, especially from initial concept, to sample collection, and biomarker validation (Fig. 3). The scope of this chapter will focus on describing key aspects within the first three stages of the biomarker discovery workflow, and noting factors pertinent to psychiatry.

Technological progress has increased the variety of biomarker types and sources that may be assessed and analyzed. This includes molecular (DNA, RNA, and protein) and cellular, as well as other metabolites. Each type of biomarker has its own strengths and limitations, with many methodologies still under development. As proteins appear to be easily detectable in point-of-care devices, proteomic studies remain one of the most promising sources of biomarker discovery within psychiatry.

It is important to note that automation within the discovery process should be included where possible and practical to:

- Minimize introduction of errors, including type I and II
- Increase efficiency of handling large numbers of samples
- Minimize missing and/or misplacement of samples

3.1 Biomarker discovery in psychiatry—considerations, stage I

Aim: Determine type of biomarker(s) to be identified, as well as the application (e.g. diagnostic, prognostic, predictive of therapeutic outcome, etc.) and disorder of interest.

Some of the factors necessary to identify within the first stage, experimental design, include; clinical case, potential biomarker application, overall biomarker identification approach (global or hypothesis driven), details about the type of biomarker to be identified, and the number of participants required for statistical power.

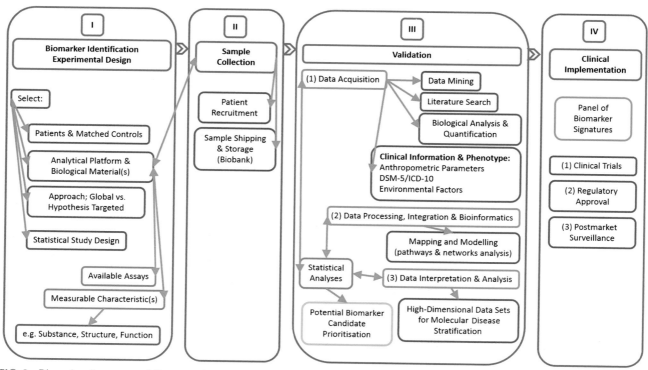

FIG. 3 Biomarker discovery workflow toward precision psychiatry (stages I to IV).

Other considerations include:

- Have the samples already been collected?
- Is collected sample integrity appropriate for the type of biomarkers to be investigated?
- Is there sufficient amount of sample for the biomarker assessment techniques available?
- Are the available biomarker assessment techniques specific and sensitive?
- If following a hypothesis driven approach, is there enough evidence to suggest the biomarker candidate could sufficiently stratify participants?
- Are there strategies to minimize bias, such as blindness and randomization?

Securing these details will help to direct, define, streamline, and standardize downstream protocols and procedures.

3.2 Biomarker discovery in psychiatry—considerations stage II

Aim: (a) Appropriate participant characterization (b) sample collection, processing, and storage maintains integrity of biomarkers of interest while ensuring sample and participant details are recorded appropriately.

3.2.1 Participant recruitment

In the field of psychiatry, despite an increasing prevalence in mental illness, there are inherent challenges associated with securing participants for a study. For example, participants with psychoses can demonstrate characteristics of unreliability as part of their presentation, and are therefore less likely to attend follow-up time points. Establishing bidirectional lines of communication with the researchers and the whole clinical team, including receptionists, is a useful strategy to gain and maintain access to more eligible participants, and mitigate effects of participant dropout. In addition, media and digital platforms may be used within an ethical framework to also raise awareness and widen exposure to increase recruitment opportunities.

Appropriate subject groups, including healthy controls and matched controls, as well as calculated required sample sizes defined in the study design of Stage I, should be standardized. Strategies should be included to address the potential of failing to meet predetermined sample size.

3.2.2 Biospecimen collection and processing

Safety concerns should be addressed with standard operating procedures (SOPs), including appropriate personal protective equipment (PPE), such as mandatory eye protection, laboratory gowns, gloves, and use of biohazard cabinets (e.g. infectious diseases may be transmitted via particular biological samples, such as peripheral blood). Appropriate action plans should also be developed in conjunction with health, safety, and wellbeing (HSW) committees within facilities where samples will be collected, processed, and analyzed.

Other factors relating to sample collection and processing to define and standardize include:

- Appropriate time point(s) of collection are established for the study
- Ensure any participant requirements prior to sample collection have been met (examples include fasting status, taking medication, etc.)
- Appropriate consumables chosen; collection materials, preservatives, and buffers to enhance biomarker measurement while minimizing potential artefacts and errors downstream
- Certain factors may need to be controlled for, as the stability of some biomarkers are low. This includes time and other environmental factors, such as temperature.
- Additionally, there may be factors that may affect biomarker detection, and should be controlled for at sample collection, processing, and storage stages. Examples include participant fasting status, deproteination of sample (e.g. when observing levels of oxidative stress marker), quality of collected biospecimen (hemolysis of blood may interfere with certain biomarker detection), tube design, and sterility may compromise sample integrity or ability for accurate detection, and so forth.

3.2.3 Biospecimen shipping and storage for preservation of biomarker integrity

Important factors of shipping and storage requirements of samples should be defined and standardized. Most factors relate to maintaining biomarker integrity, and therefore sample integrity. Storage requirements are sample specific, and depend on how long the samples may need to be stored before use. These items include:

- Control critical environmental factors, such as temperature, pH, use of RNase/DNase free tubes
- Distance and time are important factors when samples must be transferred to multiple sites (collection and/or processing and/or biomarker detection). Specialized couriers may be required, especially to deliver biological and

temperature sensitive samples in a safe and timely manner (e.g. when dry ice is required to maintain −80°C frozen state of samples).

- To minimize biomarker degradation within liquid biopsy samples induced by multiple freeze-thaw cycles, it is recommended to divide the collected and processed sample into aliquots at the point of processing. The volume chosen is based on types and number of biomarker detection technique(s) planned to be used.
- Strategies to maintain security and stability of sample until use in biomarker validation (Stage III). This includes back-up freezers, back-up energy generation, and secure long-term and short-term storage security (e.g. card access).

3.3 Biomarker discovery in psychiatry—considerations, stage III

Aim: (a) Confidently measure, analyze, and identify biomarker(s) of interest within collected samples, and (b) integrate observed biological results with associated participant data to form clinically relevant inferences (specified at Stage I).

(1) Data Acquisition

A variety of data and information may be acquired through data mining and literature searches. However, biological analysis and quantification of biomarkers based on collected and processed samples is essential in identifying reproducible and clinically relevant biomarkers.

When investigating molecular biomarkers, appropriate bioanalytical infrastructure access and maintenance is required. This includes equipment such as calibrated pipettes, centrifuge, microplate readers, and temperature-controlled storage facilities, and so forth. If multiple sites of analysis are involved, it is important to conserve critical equipment parameters and standards, such as regularly serviced pipettes, types of pipettes used, and centrifuge spinning variables. Collaborations with equipment providers can provide valuable advice and access to additional resources.

Biological assays are a popular method of molecular biomarker detection and quantification. Depending on the biomarker of interest, bioanalytical assays may be sourced in validated kit formats. Regardless of format, quality assurance is extremely important, and in addition to the studies' experimental controls (i.e. healthy controls), other appropriate internal quality control(s) (QCs) within the assay must be selected (i.e. sample spiked with positive and negative QC). Addition of QCs also offers a means to assess intra- and inter-assay reproducibility. Specificity and sensitivity (minimizes high background noise within a cohort) are additional performance characteristics that must be considered when deciding which assay to select. Appropriate replication of samples must also be confirmed. For example, triplicates are a common minimum in immunoassays. However, if samples are extremely limited, duplicates are adequate. In other techniques, such as genomic sequencing, replicates are not usually required. Further validation for the use of a manufactured bioanalytical kit requires measurements for dilution linearity and recovery. Further parameters, including limit of quantification, limit of detection, specificity, and selectivity are generally assessed by the manufacturer (Andreasson et al., 2015). Error and variation thresholds need to be determined in the context of study design and biomarker detection method. The co-efficient of variation (CV) under 20% is adequate for immunoassays in the proteomic biomarker validation phase (Guidance for Industry: Bioanalytical Method Validation, 2001). However, biomarker detection methods require a tighter CV limit for clinical implementation. Removal of bias at the assay level is another item to consider and standardize. In order for this to be achieved, digital randomization of sample orders for biomarker analysis is a recommended strategy. Statistical methods must also be evaluated in reference to the selected assay before establishment of the study protocol. Application for the appropriate regression model for an assay's standard curve can be further confirmed by a curve fit of $R^2 > 0.95$.

Establishing standardized formatting and location of associated sample data generated by the assay, as well as potentially useful sample information, saves time in data collation, analysis, result reporting, and future reference. Potential sample information of note could include sample integrity and unusual clinical information.

(2) Data processing, Integration, and Bioinformatics and (3) Data interpretation and Analysis

Data processing, integration, and interpretation are the final stages of initial biomarker discovery validation. It is an area where there are challenges and demands with increasingly high-dimensional, complex, and large data sets. However, coding experts, software packages (including statistical and network modeling packages), and emerging trends toward machine learning ("method of data analysis that automates analytical model building") are helping to efficiently and effectively identify and perform the required correlative and mapping models to aid in the identification of clinically relevant biomarker signatures (Alawieh et al., 2012).

Some variables affecting biomarker research are impractical to standardize at the earlier recruitment stages. These include confounding variables, such as medication and comorbidities that can, however, be controlled for using statistical techniques. Sound and robust inferential statistical analyses are required within biomarker research. Methodologies are relatively standardized in the field, with traditional null hypothesis significance testing most commonly used. Specific tests employed are based on the question at hand, and the types of variables at play. Sample size is also important, however, the statistical power required to ensure that the clinical use of the biomarkers is powerful enough to influence decision-making processes is yet to be agreed upon within the medical community. Interestingly, other methods, such as Bayesian statistics, may more easily communicate confidence and percentages of likelihood than traditional *P*-values. Additionally, the size of effect is slowly gaining use in the field of psychiatry (McDermott et al., 2013; Robotti, Manfredi, & Marengo, 2014). Candidate biomarkers require further validation before commercialization and clinical use (Fig. 3, stage IV). These processes progressing toward clinical implementation are highly regulated and entail a series of preclinical phases. They include functional *in vitro* assays and external and reproducible verification (DeSilva et al., 2003; Guidance for Industry: Bioanalytical Method Validation, 2001; Marini et al., 2014).

3.3.1 Reporting and sharing results

Standardized reporting is beneficial as it ensures consistency and reproducibility, and all required details are included. Internal reporting procedures for future reference and audits are useful for both laboratory and biobank records, as well as updating collaborators. Sharing results externally, for example, by publication and/or potential commercialization, requires consideration throughout the biomarker discovery process (Bauchner et al., 2016). Increasingly, fund providers will request analyzed data to be published via open source, and prior to subsequent funding (Shenkin et al., 2017). There are individual requirements specific to each avenue, beyond the scope of this chapter.

3.4 Future vision for biomarker discovery and validation in psychiatry

A multimodal biomarker assessment-based approach will enable clinicians to better stratify the heterogeneous course of psychiatric disorders into clinical subgroups, and help advance the development of personalized psychiatric diagnostic and treatment strategies. Research in the field of clinically relevant psychiatric biomarkers is growing. However, standardization in this field is important to reduce bias and errors, as well as increase sensitivity and specificity, which consequently removes barriers toward biomarker signatures gaining clinical implementation.

Approval of an appropriate panel of biomarkers is still in its infancy due to the relatively long lag time between translational validation and commercialization. For this to be realized, integration among biobanking staff, clinicians, basic science researchers, and bioinformaticians is required. Psychiatric-focused biobanks are the cornerstone to supporting significant neuropsychiatric translational research, and moving toward precision-based medical care.

References

Alawieh, A., Zaraket, F., Li, J.-L., Nokkari, A., Razafsha, M., Fadlallah, B., … Kobeissy, F. (2012). Systems biology, bioinformatics, and biomarkers in neuropsychiatry. *Frontiers in Neuroscience*, 6(187). https://doi.org/10.3389/fnins.2012.00187.

Andreasson, U., Perret-Liaudet, A., van Waalwijk van Doorn, L. J. C., Blennow, K., Chiasserini, D., Engelborghs, S., … Teunissen, C. E. (2015). A practical guide to immunoassay method validation. *Frontiers in Neurology*, 6(179). https://doi.org/10.3389/fneur.2015.00179.

Bauchner, H., Golub, R. M., & Fontanarosa, P. B. (2016). Data sharing: An ethical and scientific imperative. *JAMA*, *315*(12), 1238–1240. https://doi.org/10.1001/jama.2016.2420.

Chalmers, D., Nicol, D., Kaye, J., Bell, J., Campbell, A. V., Ho, C. W. L., … Whitton, T. (2016). Has the biobank bubble burst? Withstanding the challenges for sustainable biobanking in the digital era. *BMC Medical Ethics*, *17*(1), 39. https://doi.org/10.1186/s12910-016-0124-2.

DeSilva, B., Smith, W., Weiner, R., Kelley, M., Smolec, J., Lee, B., … Celniker, A. (2003). Recommendations for the bioanalytical method validation of ligand-binding assays to support pharmacokinetic assessments of macromolecules. *Pharmaceutical Research*, *20*(11), 1885–1900.

Dove, E. S. (2015). Biobanks, data sharing, and the drive for a global privacy governance framework. *The Journal of Law, Medicine & Ethics*, *43*(4), 675–689. https://doi.org/10.1111/jlme.12311.

DuBois, J., Bante, H., & Hadley, W. B. (2011). Ethics in psychiatric research: A review of 25 years of NIH-funded empirical research projects. *AJOB Primary Research*, *2*(4), 5–17. https://doi.org/10.1080/21507716.2011.631514.

Eiseman, E., Bloom, G., Brower, J., Clancy, N., & Olmsted, S. (2003). *Case studies of existing human tissue repositories: "Best Practices" for a biospecimen resource for the genomic and proteomic era.* Santa Monica, CA: RAND Corporation.

Gonzalez-Sanchez, M. B., Lopez-Valeiras, E., & García-Montero, A. C. (2014). Implementation of a cost-accounting model in a biobank: Practical implications. *Pathobiology*, *81*(5-6), 286–297.

Graham, C. E., Molster, C., Baynam, G. S., Bushby, K., Hansson, M., Mascalzoni, D., … Dawkins, H. J. (2014). Current trends in biobanking for rare diseases: A review. *Journal of Biorepository Science for Applied Medicine*, 2, 49–61. https://doi.org/10.2147/BSAM.S46707.

Guidance for Industry: Bioanalytical Method Validation. (2001). Rockville, MD. Retrieved from: //www.fda.gov/downloads/Drugs/Guidance/ucm070107.pdf.

International Compilation of Human Research Standards, 2018 Edition. (2018). Office for Human Research Protections, US Department of Health and Human Services, USA.

International Society for Biological and Environmental Repositories (ISBER). (2011). *Best practices for repositories: Collection, storage, retrieval and distribution of biological materials for research* (3rd ed.). Vancouver, Canada: International Society for Biological and Environmental Repositories (ISBER).

Kaye, J., Whitley, E. A., Lund, D., Morrison, M., Teare, H., & Melham, K. (2015). Dynamic consent: A patient interface for twenty-first century research networks. *European Journal of Human Genetics, 23*(2), 141–146. https://doi.org/10.1038/ejhg.2014.71.

Lehmann, S., Guadagni, F., Moore, H., Ashton, G., Barnes, M., Benson, E., … Environmental Repositories Working Group on Biospecimen Science. (2012). Standard preanalytical coding for biospecimens: Review and implementation of the Sample PREanalytical Code (SPREC). *Biopreservation and Biobanking, 10*(4), 366–374. https://doi.org/10.1089/bio.2012.0012.

Marini, J. C., Anderson, M., Cai, X. Y., Chappell, J., Coffey, T., Gouty, D., … Bowsher, R. R. (2014). Systematic verification of bioanalytical similarity between a biosimilar and a reference biotherapeutic: Committee recommendations for the development and validation of a single ligand-binding assay to support pharmacokinetic assessments. *The AAPS Journal, 16*(6), 1149–1158. https://doi.org/10.1208/s12248-014-9669-5.

McDermott, J. E., Wang, J., Mitchell, H., Webb-Robertson, B.-J., Hafen, R., Ramey, J., & Rodland, K. D. (2013). Challenges in biomarker discovery: Combining expert insights with statistical analysis of complex omics data. *Expert Opinion on Medical Diagnostics, 7*(1), 37–51. https://doi.org/10.1517/17530059.2012.718329.

McDonald, S. A., Velasco, E., & Ilasi, N. T. (2010). Business process flow diagrams in tissue bank informatics system design, and identification and communication of best practices: The pharmaceutical industry experience. *Biopreservation and Biobanking, 8*(4), 203–209. https://doi.org/10.1089/bio.2010.0020.

Moore, H. M., Kelly, A. B., Jewell, S. D., McShane, L. M., Clark, D. P., Greenspan, R., … Vaught, J. (2011). Biospecimen reporting for improved study quality (BRISQ). *Cancer Cytopathology, 119*(2), 92–102. https://doi.org/10.1002/cncy.20147.

National Health and Medical Research Council (NHMRC), ARC, & AVCC. (2007) National Statement on Ethical Conduct in Human Research 2007 (Updated May 2015). Canberra: The National Health and Medical Research Council Retrieved from: //www.nhmrc.gov.au/guidelines-publications/e72.

Remick, R., Sadovnick, A. D., Gimbarzevsky, B., Lam, R., Zis, P., & Huggins, M. (1993). Obtaining a family psychiatric history: Is it worth the effort? *Canadian Journal of Psychiatry, 38*, 590–594. https://doi.org/10.1177/070674379303800904.

Robotti, E., Manfredi, M., & Marengo, E. (2014). Biomarkers discovery through multivariate statistical methods: A review of recently developed methods and applications in proteomics. *Journal of Proteomics & Bioinformatics, S3*, 003. https://doi.org/10.4172/jpb.S3-003.

Shenkin, S. D., Pernet, C., Nichols, T. E., Poline, J.-B., Matthews, P. M., van der Lugt, A., … Wardlaw, J. M. (2017). Improving data availability for brain image biobanking in healthy subjects: Practice-based suggestions from an international multidisciplinary working group. *Neuro Image, 153*(Supplement C), 399–409. https://doi.org/10.1016/j.neuroimage.2017.02.030.

Stephens, N., & Dimond, R. (2015). Closure of a human tissue biobank: Individual, institutional, and field expectations during cycles of promise and disappointment. *New Genetics and Society, 34*(4), 417–436. https://doi.org/10.1080/14636778.2015.1107469.

Zhao, Y., & Castellanos, F. X. (2016). Annual research review: Discovery science strategies in studies of the pathophysiology of child and adolescent psychiatric disorders—Promises and limitations. *Journal of Child Psychology and Psychiatry, 57*(3), 421–439. https://doi.org/10.1111/jcpp.12503.

Chapter 45

Ethical, policy, and research considerations for personalized psychiatry

Ryan Abbott[a,b], Donald D. Chang[c] and Harris A. Eyre[c,d,e,f,g,h]

[a]School of Law, University of Surrey, Guildford, United Kingdom, [b]David Geffen School of Medicine at UCLA, Los Angeles, CA, United States, [c]The University of Queensland School of Medicine, Ochsner Clinical School, New Orleans, LA, United States, [d]Texas Medical Center, Innovation Institute, Yarraville, VIC, Australia, [e]Department of Psychiatry, University of Melbourne, Parkville, VIC, Australia, [f]Disciplines of Psychiatry, University of Adelaide, Adelaide, SA, Australia, [g]IMPACT SRC, School of Medicine, Deakin University, Geelong, VIC, Australia, [h]Brainstorm Lab for Mental Health Innovation, Department of Psychiatry and Behavioral Sciences, Stanford University, Palo Alto, CA, United States

1 Introduction

Personalized medicine focuses on an individual's unique attributes to improve disease risk assessment, diagnostic accuracy, and treatment response (Ozomaro et al., 2013). It is a broad, multidisciplinary field that focuses on a patient's genetic makeup, lifestyle, and environment. Personalized psychiatry research has been spurred by a major National Institute of Mental Health (NIMH) project called the Research Domain Criteria (RDoC). This project is an effort to develop a new diagnostic approach for psychiatry that combines biological, behavioral, and social factors. It has highlighted the need for diverse data and dimensional approaches for better diagnosis and treatment of mental disorders.

There are some instructive examples of clinical advances in the field of personalized psychiatry, particularly when it comes to leveraging digital phenotyping, proteomics, and genomics. Digital phenotyping involves collecting sensor, keyboard, as well as voice and speech data from smartphones to measure behavior, cognition, and mood (Insel, 2017). One study found certain speech features predicted development of psychosis with 100% accuracy, outperforming classification from clinical interviews (Bedi et al., 2015). It also found that speech features were significantly correlated with prodromal symptoms. The clinical tools utilizing digital phenotyping have been proliferating rapidly given the increasing use of smartphones.

Using proteomics, the Major Depressive Disorder Score (MDDScore) test uses a probabilistic statistical approach to differentiate depressed and nondepressed individuals (Bilello et al., 2015). It has demonstrated a sensitivity and specificity exceeding 90%, and it is one of the few biomarker-based assays for depression whose effectiveness has been validated in a replication study. The nine biomarkers it measures are associated with changes observed during major depression that take place in the neurotrophic, metabolic, inflammatory, and hypothalamic–pituitary–adrenal axis pathways. An algorithm calculates a sex-adjusted score on a scale of 1–9 indicating the probability that a patient has depression, based on the pattern of these biomarkers' concentrations, as well as body-mass index. The test is now undergoing further independent validation.

Finally, the entire human genome can now be sequenced for less than $1000, and that cost continues to decline. Greater affordability of genetic testing has coincided with government initiatives in the United States and the United Kingdom to further develop the evidence base for genetically guided prescribing (Collins & Varmus, 2015). Increased access to genetic testing, coupled with a growing evidence base for prescribing based on an individual's genetic profile, will likely impact standards of care, and may result in a new clinical paradigm powered by pharmacogenetics (Abbott et al., 2017).

This chapter draws insights from established bodies of literature in medical ethics, policy, and research to propose guidelines for the development of personalized psychiatry. Personalized psychiatry also has its own lessons for more established fields.

2 Ethics

Medical ethics is the branch of knowledge that deals with moral principles governing the practice of medicine and professional conduct (Kerridge, Loew, & Stewart, 2013). It guides the development of regulations and policies, and suggests

Personalized Psychiatry. https://doi.org/10.1016/B978-0-12-813176-3.00045-6

appropriate behaviors and attitudes toward patient care. Core tenets of medical ethics include the principles of autonomy (patient's right to self-govern), beneficence (do good), nonmaleficence (do no harm), and justice (fair distribution) (Beauchamp and Childress, 2015). The consideration and application of medical ethics in all clinical settings is an integral part of a healthcare provider's duty to patients, the profession, and the public.

Many of the ethical issues in personalized psychiatry are encountered in psychiatry generally. For example, whether and to what extent mental illness impacts a patient's ability to make a decision (Robertson, 2012). However, personalized psychiatry alters some existing dynamics, and also poses some distinct ethical challenges. For example, technological advances, combined with decreasing costs and greater access to existing technologies, have resulted in a deluge of patient digital, proteomic, and genomic data. This presents new considerations for clinicians: Should data be obtained and shared with patients? How accurate is this data, and is it predictive of clinical outcomes? Does biological and digital information need to be shared with a patient's family members? Will providing information cause distress or embarrassment? Could it be used to discriminate? These considerations will be discussed in the context of big data and health disparities.

2.1 Big data

23AndMe is a genetic testing company whose business model provides direct-to-consumer (DTC) genetic testing. In April 2017, the company's Personal Genome Service Genetic Health Risk (GHR) test became the first FDA approved DTC genetic test for 10 diseases, including Parkinson's and late-onset Alzheimer's (U.S. Food and Drug Administration, 2017b). The company's tests are developed, in part, from its extensive and proprietary database, which contains genetic information from the more than 2 million consumers who have used its services. Consumers who elected to see "what your DNA says about you" (www.23andme.com) may receive an overwhelming amount of genetic information, sometimes without the context of professional genetic counseling and interpretation. Consumers are also providing their data to 23AndMe, potentially with limited restrictions on its use. The success of 23AndMe and the FDA's approval highlights the shift toward increased patient autonomy, and away from medical paternalism.

Such uses of big data make individual genomic information more accessible at the consumer level, and may encourage dialogue and shared decision making with physicians. Shared decision making has been shown to improve patient outcomes (Shay & Lafata, 2015). Physicians are able to order similar tests on patients, for example, using pharmacogenetic Decision Support Tools (DSTs) that may help physicians to select appropriate pharmaceutical interventions. Whether ordered by patients or physicians, knowledge of genetic susceptibilities and predispositions to pharmaceutical efficacy and adverse events may improve outcomes. All of this is more likely given the growing evidence base for the use of DSTs in psychiatry (Abbott et al., 2017). Beyond improving individual outcomes, big data can contribute to our understanding of intervention safety and efficacy to improve prescribing patterns, drug development, and pharmacovigiliance (Abbott, 2013).

There are some important considerations to weigh against these potential benefits. Critics have argued that allowing individuals to review their genetic information without consulting a physician or genetic counselor can be harmful (Annas & Elias, 2014). While some genetic associations with disease are widely accepted, these associations only inform on risk of disease development—they are not diagnostic of the disease itself, an important distinction that can result in misinterpretation or misguided treatment (McPherson, 2006). This reinforces the importance of patient education and knowledgeable providers. However, patient and provider education is not a panacea. The rapid development of technology has outpaced our ability to fully interpret available information (Marx, 2013).

Genetics tests have also been marketed while concerns persist about the accuracy of such tests. Indeed, in 2014, the FDA issued a warning letter to 23AndMe to halt its marketing until further clinical validation had occurred. As clinical practice transitions to a greater utilization of pharmacogenetics, it is important to determine the appropriate evidentiary standard for widespread adoption of genetic testing. On one hand, an overly conservative review of novel technologies can impede translational benefits. On the other, innovation without adequate oversight can lead to commercialization of ineffective, costly, and unsafe products, which can result in waste and patient harm (Abbott & Stevens, 2014).

As more precision medicine tests enter the market and have a greater impact on patient care and financing, such tests are likely to be subject to stricter regulation. There will be fewer commercial opportunities for unproven tests. Indeed, the FDA is already considering stronger regulations for so-called Lab Developed Tests (LDTs) in the wake of products that have lacked adequate evidence and controls, and for which companies have falsified data (U.S. Food and Drug Administration, 2017c). Such tests are not harmless; they may cause people to initiate unnecessary treatment, or to delay or forego necessary treatment.

Privacy is also a core concern associated with big data and personalized psychiatry. Should consumers share their results with providers, families, and employers? Are healthcare providers able to take action based on the results? Should the use of such data be permitted for research, and by private companies such as 23AndMe? These issues remain controversial.

2.2 Health disparities

The novel technologies utilized in personalized psychiatry risk widening health disparities as access to these services may be limited by socioeconomic status. For example, critics have argued that pharmacogenomics will further widen health disparities on a global scale by advancing the "boutique-style" of personalized healthcare in developed countries, but not in less developed countries that have limited access (Chadwick, 2014; Daar & Singer, 2005). On the other hand, pharmacogenomics may potentially decrease health disparities as patients will be treated based on their own genetic profiles, instead of relying on clinical trial data. Clinical trials, particularly Phase III clinical trials, have historically been disproportionately composed of male, Caucasian patients (Fisher & Kalbaugh, 2011). It is difficult to assess the full impact of personalized psychiatry, given that the field is still young.

Who should carry the fiscal responsibility of genetic testing? Until the technologies being used for personalized psychiatry have sufficient clinical validation, consumers are accessing services that may not be safe and effective. As such, while this may be tolerated in the name of patient autonomy and self-determination, it may not be something payors should be required to fund. Concerns about direct-to-consumer advertising and patient demand for unproven technologies draw similarities to those raised with off-label drug use: how can access to potential treatments be balanced with the evidence-based regulation approval process in government (Abbott & Ayres, 2014)? Similar to off-label drug use, the principle of maleficence (do no harm) would argue that these technologies undergo a rigorous and exhaustive process in order to gain government approval. Yet, the time required for this approval process may come at a cost of delayed access and treatment for patients in need. Once a treatment is evidence based though, payor coverage of treatments should be provided for patients, along with any other medically necessary treatments (Abbott & Stevens, 2014).

Precision medicine has some special considerations for informed consent and privacy. Patients need to understand that testing provides a probabilistic outcome as to the relative risk of a favorable or undesirable outcome of medication use. This information has to be married with the other considerations that shape the risk–benefit analysis in prescribing decisions. Patients may be concerned about psychological distress as a result of learning about their own genetic profile, particularly if testing goes beyond evaluation for medication response. Patients may also be concerned about loss of confidentiality and discrimination. Providers should be aware that in the U.S., the Genetic Information Nondiscrimination Act of 2008 (GINA) largely prohibits the use of genetic information in health insurance and employment settings. In any event, as with any protected health information, clinicians should keep genetic information confidential for both ethical considerations and compliance with the Health Insurance Portability and Accountability Act (HIPAA).

Providers may need to discuss the importance of telling family members about the results of genetic testing with patients. Knowledge of familial medication response variability is generally less clinically important than knowledge of familial predisposition to disease, but it may still influence treatment decisions. That is particularly the case with genetic vulnerability to a serious adverse drug reaction, such as Stevens-Johnson Syndrome (SJS). The relevance of test results for family members should be flagged with patients prior to testing. Prescribers should strongly encourage patients to disclose relevant information such as heightened risk of SJS to family members. The World Psychiatric Association Ethics Committee recommends that, "psychiatrists involved in genetic research or counseling shall be mindful of the fact that the implications of genetic information are not limited to the individual from whom it was obtained, and that its disclosure can have negative and disruptive effects on the families and communities of the individuals concerned" (Robertson, 2012, p. 59). However, in the absence of mandatory notification legislation, the decision of whether to share information is generally for the patient to make.

3 Health policy

Appropriate public policies are critical to promoting industry development and to fully realizing the vision of personalized psychiatry. A few key policy topics are considered here.

3.1 Biobanking

A biobank is an organization that acquires and stores biological samples. These samples can be used for clinical care, but are most often stored for research purposes. Biobanks are a valuable resource for personalized psychiatry because the samples and associated proteomic, genomic, and metabolomic data represent large amounts of information for analysis.

However, there are ethical concerns associated with biobanking related to privacy, autonomy, informed consent, and benefit sharing. Individual genetic information may reveal potentially sensitive information about a person's health. Patients have historically had biological samples used without their consent, and there are also concerns about benefit sharing from the commercialization and use of biological materials, as well as misuse of biological knowledge (Abbott, 2014).

The Electronic Medical Records and Genomics Network (eMERGE) is a consortium of biobanks that are on the cutting edge of genomic research that include the Marshfield, Mayo, Vanderbilt, Northwestern University, and Group Health biobanks (Gottesman et al., 2013). The goal of this network is to link multiple sources of genetic and electronic data. This consortium is instructive regarding policy development for biobanking, given that they are developing best practices for informatics, electronic phenotyping, genomic medicine implementation, and ethical and regulatory research issues. For further details, see Gottesman et al., 2013.

3.2 High quality clinical trials

The robust development of personalized psychiatry requires clear policy guidance on the evidentiary standard needed to adopt novel technologies. Three essential criteria to consider are analytical validity, clinical validity, and clinical utility (Haddow & Palomaki, 2004).

Analytical validity is a solution's ability to accurately detect and measure a biomarker of interest. In psychiatry, this may include genomic, digital, or other biomarkers. Key concerns are around test result repeatability under identical and different conditions, and validity in real-world settings. Clinical validity involves considerations around sensitivity, specificity, accuracy, positive, and negative predictive values. Clinical utility involves considerations around patient outcomes, support of diagnosis and treatment monitoring for physicians, decision-making impact and guidance, as well as familial and social impacts (i.e., does the test identify at-risk family members, high-risk ethnicities, and does it have impacts on health systems/budgets?).

Evidence-based medicine is an approach to medical practice intended to optimize decision making by emphasizing the use of evidence from well-designed and well-conducted research. The development of evidence-based approaches in personalized psychiatry is important to ensure the conscientious, explicit, judicious, and reasonable use of personalized psychiatry (Howick et al., 2009). An example of a useful framework is the Oxford Evidence-based Medicine Framework, a hierarchical framework for the best evidence in a particular field.

3.3 Reimbursement

Patient access to, and further development of, personalized psychiatry will depend in large part on payor reimbursement policies. Adequate reimbursement will ensure private sector investments into R&D, and ensure the translation of technologies into global markets. This is important given the noted inequality in mental health outcomes between low and high socioeconomic strata (World Health Organization and Calouste Gulbenkian Foundation, 2014).

Transparent reimbursement criteria are vital to development efforts. Developers of novel solutions need to understand the evidentiary burden required to qualify for reimbursement. Given the rapidity of personalized psychiatry innovations, it has been difficult for payors to keep up with robust and clear criteria for reimbursement. A key component of reimbursement requires cost-effectiveness data justifying the economic impacts of novel solutions (Abbott & Stevens, 2014). It is important that industry engage with the move toward value-based health care, and that it can demonstrate sufficient clinical and economic value to payors (Porter, 2009).

3.4 Regulation

Regulatory agencies globally have a role to play in ensuring the appropriate development and clinical use of technologies that are safe and effective for their respective communities. Balance must be struck between over-regulation, which will stifle innovation by raising costs 'to market,' and under-regulation, which insufficiently scrutinizes new technologies (Epstein & Abbott, 2014).

Regulatory agencies have struggled in recent times with negotiating the rapid growth of personalized medicine and personalized psychiatry. The FDA, for example, has recently announced the creation of a Digital Health Unit, to support in assessing what clinical validation should be for medical software (U.S. Food and Drug Administration, 2017a). This is particularly relevant to personalized psychiatry with the growth of digital phenotyping companies using digital biomarkers such as voice and facial analytics. The FDA is also involved in the potential increased regulation of the lab developed test sector (U.S. Food and Drug Administration, 2017c). This is currently under the purview of the Clinical Laboratory Improvement Amendments (CLIA), and involves psychiatric pharmacogenetics. Clearly, regulatory agencies must consider and stratify the analysis of these technologies on a risk/benefit basis, and the recent guidelines for lab developed tests and software-as-a-medical-device appear to explore this in more depth.

3.5 Information system upgrades

Health information system upgrades are an important pillar of the development of personalized psychiatry. Sufficient infrastructure is needed to collect increasing amounts of patient data, transmit it appropriately between providers and health care actors, and use such data in an ethically appropriate manner for research. The challenge physicians have in integrating and interpreting the complex and large amounts of data is fierce. Hence, interoperability of data management systems, standardized approaches to data analysis, and machine learning techniques will be essential to improving personalized psychiatry (Abbott, 2017). Organizations such as the Global Alliance for Genomics in Health are working on nonelectronic health records focused on interoperable genomics data exchange. Interoperability of electronic health records is similarly important to ensuring the effective and efficient transfer of health information for patients (Khoury, Iademarco, & Riley, 2016).

4 Medical research

Conventional clinical research models are designed around a representative population, with the results being extrapolated to the general patient population. However, this approach has deficiencies. It is often not possible to conduct rigorous trials that evaluate all possible subgroups (e.g., age, gender, sex, race, ethnicity, underlying genetics). It is also difficult to extrapolate from a group response to individual patients due to genetic, environmental, and lifestyle variability.

Here, we discuss a few medical research considerations that are relevant to personalized psychiatry.

4.1 Research considerations for personalized psychiatry

Big data: Discussed previously in the context of ethical considerations, big data also has significant research implications. Rather than designing trials based on a representative population, analysis of data collected from various health care information bases can be used to determine whether patients who share similar characteristics are more or less likely than the general population to respond (Abbott, 2013). The benefit of this approach is that, due to its algorithmic and statistical base, big data approaches may allow for more flexibility and comparators than conventional research models—a significant advantage, given the complexity of personalized medicine. This has been driving interest in this area, as underscored by the 200 million USD National Precision Medicine Initiative announced by President Obama in 2015, which includes development of a national gene database (Reardon, 2015).

N-of-1 trials: N-of-1 trials involve frequent monitoring of an individual's response to single or multiple treatment regimens (Schork, 2015). The classic example of an N-of-1 trial is when a physician therapeutically trials different treatments on an individual patient. The strength of an N-of-1 trial lies in its ability to longitudinally follow a single patient's response to different treatments over time. This approach thus controls, to an extent, for genetic, environmental, and lifestyle variables. One of the most difficult aspects of psychiatry is interpreting how the complex interplay of variables in a patient's profile contribute to their clinical presentation. N-of-1 trials may mitigate some of these difficulties by accounting for these factors as the patient is standardized to him or herself. Importantly, N-of-1 trials have the potential to be applied for personalized psychiatry research, as it allows for a study of each individual response to treatment. Indeed, a number of N-of-1 trials have been conducted in neuropsychiatric disorders (Gabler, Duan, Vohra, & Kravitz, 2011). Moreover, even though the focus is on an individual patient, broader conclusions can be drawn when there are enough aggregates of N-of-1 trials, as patients can be stratified based on commonalities of their medical history (Gabler et al., 2011; Schork, 2015). The success of N-of-1 trials has been demonstrated in an Australian study on treatment modalities for osteoarthritis and chronic pain (Scuffham et al., 2010). The study showed that the N-of-1 trial approach ultimately proved to be effective at determining effective treatments, despite relatively high up-front costs.

Biobanking: Biobanks facilitate research on biological and phenotypic ties, which may allow for improved personalized medicine outcomes (Kinkorová, 2015; Liu & Pollard, 2015). However, there are a number of challenges associated with research, some of which, in particular, pertain to personalized psychiatry. First, all research conducted using biobanks faces the challenge of interpreting and correlating phenotypic data with its biological root. In psychiatry, this can be particularly difficult, as it is well-recognized that diagnosis of psychiatric disorders can be assessed in a variety of different ways, with variability and relatively poor interrater reliability (Clark, Watson, & Reynolds, 1995; Kendell & Jablensky, 2003).

Obtaining informed consent from psychiatric patients for collecting biological samples may also be a challenge, in that they may be more likely to lack capacity to consent or be vulnerable to exploitation.

Biobanks are well-positioned to address a significant deficit in psychiatric management: biomarkers (Fernandes et al., 2017; Ozomaro et al., 2013). Historically, psychiatric diseases have been diagnosed based on behavioral symptoms. While numerous biomarkers have been proposed to aid in diagnosis, there is a still a paucity of robust biomarkers. Biobanks may help to address this unmet need by allowing the ability to access, compare, and validate large pools of candidate biomarkers. The notion that it is likely no single biomarker will definitively be diagnostic for a psychiatric condition further underscores the value of biobanks. Moreover, it is becoming widely accepted that research strategies employing "-omic" techniques (e.g. genomics, proteomics, etc.) can lead to an improved understanding of the pathophysiology of psychiatric disorders—more so than traditional laboratory bench approaches (Fernandes et al., 2017). Biobanks provide the platform for these "-omic" techniques, and can thus help drive personalized psychiatry research.

Multi-disciplinary Collaboration: There is disparity between the rapid pace of genomic research and lack of entry of personalized medicine into clinical environments (Abbasi, 2016). To address this, some centers have adopted a specific multidisciplinary task force dedicated to the strategic implementation and effective use of personalized medicine. One such example is the personalized medicine task team set up at the Radboud University Nijmegen Medical Centre in the Netherlands (Evers et al., 2012). The team included specialists in a variety of areas including urology, neurology, and psychology (Evers et al., 2012). Specific aims for the team included developing strategies and databases for integrating personalized medicine at the center, evaluating approaches for patient involvement, and conducting personalized medicine research (Evers et al., 2012). The assembly of such a team underscores that the full realization of the therapeutic potential for personalized medicine may require changes in how we approach translational research. Historically, translational research focused on a specific biological pathway or elucidating a mechanism at one point in the pathogenesis of a disease. However, personalized medicine illustrates how complex the interplay between genes, proteins, and the environment can be (Haiech & Kilhoffer, 2012). Thus, in addition to pathophysiology, accounting for a patient's environment, lifestyle, and mental health will be necessary to advance translational research. Indeed, it has been noted that one of the most promising ways forward toward personalized medicine is through multidisciplinary and collaborative research (Qattan, Demonacos, & Krstic-Demonacos, 2012).

5 Conclusion

It has historically been challenging to connect phenotypic and biological data in psychiatry. The emergence of personalized psychiatry offers a promising platform and resource for gaining deeper insights into mental illnesses, with the possibility of improving patient outcomes. Yet this new field raises new questions and concerns, not only about scientific research, but also about ethics and policy.

As novel ethical challenges arise related to how much information to provide to patients, use of patient data, consent from patients with diminished capacity, work with vulnerable populations, lack of therapeutic evidence, implications for family members, and access to medicine, physicians should be guided by the fundamental principles of medical ethics. Considering dilemmas from the perspective of patient autonomy, beneficence, nonmaleficence, and justice will provide a useful framework for resolving conflicts. In situations in which these values are in conflict, physicians can also look to examples of how analogous problems have been dealt with in general psychiatric practice—for instance, whether to provide medications off-label without evidentiary support, and how much information to provide family members about mental health diagnoses and risks. Providers should also strive to always treat patients and other caregivers with dignity and respect, and to provide care with honesty. There is a further need for medical ethicists to consider the challenges of personalized psychiatry, and to develop consensus guidelines on how best novel technologies can be developed and implemented in ways that maximize social value and promote optimal patient care.

Developing personalized psychiatry-friendly policies will require input from clinicians, researchers, health system administrators, reimbursement agencies, governments, research funding agencies, and the private sector. The goal is for the development and clinical implementation of high-value solutions that reduce the cost of care and improve outcomes. There is also a need for ongoing physician education. Particularly in the absence of a strong regulatory environment, physicians need to understand the evidence supporting the use of precision medicine, and the need for clinical utility in the absence of otherwise compelling indications.

In this chapter, we have outlined some of the key ethical, policy, and research considerations in personalized psychiatry. As the field continues to progress, developing appropriate strategies to address these issues will be of critical importance so that physicians, policy makers, and patients can make well-informed decisions about medical care.

References

Abbasi, J. (2016). Getting pharmacogenomics into the clinic. *JAMA*, 1–3. https://doi.org/10.1001/jama.2016.12103.

Abbott, R. (2013). Big data and pharmacovigilance: Using health information exchanges to revolutionize drug safety. *Iowa Law Review*, *99*(1), 225–292.

Abbott, R. (2014). *Documenting traditional medical knowledge*. World Intellectual Property Organization.

Abbott, R. (2017). The sentinel initiative as a cultural commons. *Governing the Medical Commons*, *823*(110), 1–39. https://doi.org/10.2139/ssrn.2795860.

Abbott, R., & Ayres, I. (2014). Evidence and extrapolation: Mechanisms for regulating off-label uses of drugs and devices. *Duke Law Journal*, *64*(3), 377–435.

Abbott, R., & Stevens, C. (2014). Redefining medical necessity: A consumer-driven solution to the U.S. Health Care Crisis. *Loyola of Los Angeles Law Review*, 47(943), 1–24.

Abbott, R., Chang, D. D., Eyre, H. A., Bousman, C. A., Merrill, D., & Lavretsky, H. (2017). Pharmacogenetic decision support tools: A new paradigm for late-life depression? *The American Journal of Geriatric Psychiatry*, *26*(2), 125–133. https://doi.org/10.1016/j.jagp.2017.05.012.

Annas, G. J., & Elias, S. (2014). 23andMe and the FDA. *New England Journal of Medicine*, *370*(11), 985–988. https://doi.org/10.1056/NEJMp1316367.

Beauchamp, T. L., & Childress, J. F. (2015). Principles of biomedical ethics, 7th edition. *Occupational Medicine*, *65*(1), 88–89. https://doi.org/10.1093/occmed/kqu158.

Bedi, G., Carrillo, F., Cecchi, G. A., Slezak, D. F., Sigman, M., Mota, N. B., … Corcoran, C. M. (2015). Automated analysis of free speech predicts psychosis onset in high-risk youths. *NPJ Schizophrenia*. *1*, https://doi.org/10.1038/npjschz.2015.30.

Bilello, J. A., Thurmond, L. M., Smith, K. M., Pi, B., Rubin, R., Wright, S. M., … Papakostas, G. I. (2015). MDDScore: Confirmation of a blood test to aid in the diagnosis of major depressive disorder. *The Journal of Clinical Psychiatry*, *76*(2), e199–e206. https://doi.org/10.4088/JCP.14m09029.

Chadwick, R. (2014). The ethics of personalized medicine: A philosopher's perspective. *Personalized Medicine*, *11*(1).

Clark, L. A., Watson, D., & Reynolds, S. (1995). Diagnosis and classification of psychopathology: Challenges to the current system and future directions. *Annual Review of Psychology*, *46*, 121–153. https://doi.org/10.1146/annurev.ps.46.020195.001005.

Collins, F. S., & Varmus, H. (2015). A new initiative on precision medicine. *New England Journal of Medicine*, *372*(9), 793–795. https://doi.org/10.1056/NEJMp1500523.

Daar, A. S., & Singer, P. A. (2005). Pharmacogenetics and geographical ancestry: Implications for drug development and global health. *Nature Reviews. Genetics*, *6*(3), 241–246. https://doi.org/10.1038/nrg1559.

Epstein, R., & Abbott, R. (2014). FDA involvement in off-label use: Debate between Richard Epstein and Ryan Abbott. *Sw. L. Rev. 44*, 1.

Evers, A. W. M., Rovers, M. M., Kremer, J. A. M., Veltman, J. A., Schalken, J. A., Bloem, B. R., & van Gool, A. J. (2012). An integrated framework of personalized medicine: From individual genomes to participatory health care. *Croatian Medical Journal*, *53*(4), 301–303. https://doi.org/10.3325/cmj.2012.53.301.

Fernandes, B. S., Williams, L. M., Steiner, J., Leboyer, M., Carvalho, A. F., & Berk, M. (2017). The new field of "precision psychiatry". *BMC Medicine*, *15*(1), 80. https://doi.org/10.1186/s12916-017-0849-x.

Fisher, J. A., & Kalbaugh, C. A. (2011). Challenging assumptions about minority participation in US clinical research. *American Journal of Public Health*, *101*(12), 2217–2222. https://doi.org/10.2105/AJPH.2011.300279.

Gabler, N. B., Duan, N., Vohra, S., & Kravitz, R. L. (2011). N-of-1 trials in the medical literature: A systematic review. *Medical Care*, *49*(8), 761–768. https://doi.org/10.1097/MLR.0b013e318215d90d.

Gottesman, O., Kuivaniemi, H., Tromp, G., Faucett, W. A., Li, R., Manolio, T. A., … eMERGE Network. (2013). The electronic medical records and genomics (eMERGE) Network: Past, present, and future. *Genetics in Medicine: Official Journal of the American College of Medical Genetics*, *15*(10), 761–771. https://doi.org/10.1038/gim.2013.72.

Haddow, J. E., & Palomaki, G. E. (2004). ACCE: A model process for evaluating data on emerging genetic tests. In M. J. Khoury, J. Little, & W. Burke (Eds.), *Human genome epidemiology: A scientific foundation for using genetic information to improve health and prevent disease* (pp. 217–233). Oxford University Press.

Haiech, J., & Kilhoffer, M.-C. (2012). Personalized medicine and education: The challenge. *Croatian Medical Journal*, *53*, 298–300. https://doi.org/10.3325/cmj.2012.53.298.

Howick, J., Phillips, B., Ball, C., Sackett, D., Badenoch, D., Straus, S., … Dawes, M. (2009). *Levels of evidence and grades of recommendation*. Oxford Center for evidence-based Medicine.

Insel, T. R. (2017). Digital phenotyping: Technology for a new Science of behavior. *JAMA*. *94301*, https://doi.org/10.1001/JAMA.2017.11295.

Kendell, R., & Jablensky, A. (2003). Distinguishing between the validity and utility of psychiatric diagnoses. *The American Journal of Psychiatry*, *160*(1), 4–12. https://doi.org/10.1176/appi.ajp.160.1.4.

Kerridge, I., Loew, M., & Stewart, C. (2013). Ethics and law for the health professions. In *Federation press* (4th ed., pp. 124–133).

Khoury, M. J., Iademarco, M. F., & Riley, W. T. (2016). Precision public health for the era of precision medicine. *American Journal of Preventive Medicine*, *50*(3), 398–401. https://doi.org/10.1016/j.amepre.2015.08.031.

Kinkorová, J. (2015). Biobanks in the era of personalized medicine: Objectives, challenges, and innovation: Overview. *The EPMA Journal*, *7*, 4. https://doi.org/10.1186/s13167-016-0053-7.

Liu, A., & Pollard, K. (2015). Biobanking for. *Personalized Medicine*, *7*, 55–68. https://doi.org/10.1007/978-3-319-20579-3_5.

Marx, V. (2013). Biology: The big challenges of big data. *Nature*, *498*(7453), 255–260. https://doi.org/10.1038/498255a.

McPherson, E. (2006). Genetic diagnosis and testing in clinical practice. *Clinical Medicine & Research*, *4*(2), 123–129. https://doi.org/10.3121/cmr.4.2.123.

Ozomaro, U., Wahlestedt, C., Nemeroff, C. B., Myers, A., Nemeroff, C., Mehta, R., … Sheridan, J. (2013). Personalized medicine in psychiatry: Problems and promises. *BMC Medicine*, *11*(1), 132. https://doi.org/10.1186/1741-7015-11-132.

Porter, M. E. (2009). A strategy for health care reform–toward a value-based system. *The New England Journal of Medicine*, *361*(2), 109–112. https://doi.org/10.1056/NEJMp0904131.

Qattan, M., Demonacos, C., & Krstic-Demonacos, M. (2012). Roadmap to personalized medicine. *Croatian Medical Journal*, *53*(4), 294–297.

Reardon, S. (2015). Obama to seek $215 million for precision-medicine plan. *Nature*. https://doi.org/10.1038/nature.2015.16824.

Robertson, M. (2012). An overview of psychiatric ethics. *Health Education and Training Institute*, 1–66. https://doi.org/10.1080/09540269874592.

Schork, N. J. (2015). Personalized medicine: Time for one-person trials. *Nature*, *520*(7549), 609–611. https://doi.org/10.1038/520609a.

Scuffham, P. A., Nikles, J., Mitchell, G. K., Yelland, M. J., Vine, N., Poulos, C. J., … Glasziou, P. (2010). Using N-of-1 trials to improve patient management and save costs. *Journal of General Internal Medicine*, *25*(9), 906–913. https://doi.org/10.1007/s11606-010-1352-7.

Shay, L. A., & Lafata, J. E. (2015). Where is the evidence? A systematic review of shared decision making and patient outcomes. *Medical Decision Making: An International Journal of the Society for Medical Decision Making*, *35*(1), 114–131. https://doi.org/10.1177/0272989X14551638.

U.S. Food and Drug Administration. (2017a). *Digital health*. U.S. Food and Drug Administration.

U.S. Food and Drug Administration. (2017b). *FDA allows marketing of first direct-to-consumer tests that provide genetic risk information for certain conditions*. U.S. Food and Drug Administration.

U.S. Food and Drug Administration. (2017c). *Laboratory developed tests*. U.S. Food and Drug Administration.

World Health Organization and Calouste Gulbenkian Foundation. (2014). *Social determinants of mental health*: (pp. 1–54). World Health Organization. https://doi.org/10.3109/09540261.2014.928270.

Chapter 46

The future of personalized psychiatry

Bernhard T. Baune

Department of Psychiatry, University of Münster, Münster, Germany; Department of Psychiatry, Melbourne Medical School, The University of Melbourne, Melbourne, VIC, Australia; The Florey Institute of Neuroscience and Mental Health, The University of Melbourne, Parkville, VIC, Australia

With current progress in biotechnology, molecular neurobiology, systems biology, mental health neuroscience, bioinformatics, prediction modeling, measurement-based psychiatry, and digitalization, further developments toward understanding the complexity of mental illness and enhanced translation into clinical practice can be expected in the coming decade. In this context, it is crucial to avoid misconceptions about personalized psychiatry by restricting it to pharmacogenetics/pharmacogenomics. Rather, a wider range of promising technologies such as pharmacotranscriptomics, pharmacoproteomics, pharmacometabolomics, neuroimaging, and EEG as well as objective measures of deep phenotyping for short-term treatment response and longer-term course of illness prediction should be considered. Going beyond a pharmacological approach, an extension of personalized psychiatry into the therapeutic fields of neurostimulation (e.g., ECT) and psychotherapy should be considered.

A restriction to the term of *"precision psychiatry"* should be avoided since personalized psychiatry is conceptually based on the complex underpinnings of mental illness between the biology of brain diseases interacting with a complex psychosocial environment, the exposome, patient perspectives, and participation in a changing health care system environment, as opposed to precision psychiatry that primarily focuses on the biological (e.g., genomic) components of mental illness.

Conceptualized as a bio-psychosocial model of mental illness, personalized psychiatry has the potential to become a day-to-day aid for the physician to make data-supported informed decisions about treatment indication, treatment selection, and treatment response evaluation. Personalized psychiatry may also extend into newer fields such as presymptomatic (preventive) interventions embedded into an integrated health care system that combines traditional medicine and preventive approaches to mental health.

Future key features of personalized psychiatry will likely include the development of automated systems and decision-making abilities for rational therapies while the physician continues to make final decisions about diagnosis and treatments, having such technological aids at hand. The prospect of a successful implementation of treatment response prediction modeling will likely increase when using multiple modes of objective and subjective medical and patient-oriented data. Only through the availability of different types of health-related information about an individual patient can a truly personalized approach to mental health be realized. Notably, none of these developments will replace the ever-important patient-physician relationship. In fact, the need for excellent professional, interpersonal, and therapeutic skills, including empathy, will need to be at the highest level when realizing personalized psychiatry in practice. New lines of professional training needs and job opportunities will likely arise for the mental health workforce as new professional standards underpin the translation of personalized psychiatry into clinical practice. It is important to note that there is no need to wait for another 15–20 years to realize the potential of personalized psychiatry. On the contrary, as new technologies, interventions, and developments become available, these should be implemented where deemed appropriate. As personalized psychiatry approaches continue to translate into clinical practice over time, their use should depend on the personal judgments and decisions of the treating physician in each case.

In conclusion, progress in personalized psychiatry and related technologies justifies an optimistic view. There will be significantly more activity related to personalized psychiatry in the clinical, psychological, social, biopharmaceutical, and technology sectors. Research funding agencies in, for example, the European Union and the United States, have realized this potential and have started putting out larger-scale calls for the area of personalized psychiatry. Importantly, several of these calls emphasize the translational aspects of personalized psychiatry, giving this young field the hope to turn the appealing idea of personalized psychiatry into a long-needed changing endeavor of mental health practice.

Personalized Psychiatry. https://doi.org/10.1016/B978-0-12-813176-3.00046-8

Index

Note: Page numbers followed by *f* indicate figures, *t* indicate tables, and *b* indicate boxes.